5TH EDITION

Biological Safety

PRINCIPLES AND PRACTICES

5TH EDITION

Biological Safety

PRINCIPLES AND PRACTICES

EDITED BY

DAWN P. WOOLEY
Wright State University, Dayton, Ohio

KAREN B. BYERS
Dana Farber Cancer Institute, Boston, Massachusetts

Washington, DC

Library of Congress Cataloging-in-Publication Data

Names: Wooley, Dawn P., editor. | Byers, Karen B., editor.
Title: Biological safety : principles and practices / edited by Dawn P. Wooley, Department of Neuroscience, Cell Biology and Physiology, Wright State University, Dayton, OH, Karen B. Byers, Dana Farber Cancer Institute, Boston, MA.
Description: 5th edition. | Washington, DC : ASM Press, [2017] | Includes index.
Identifiers: LCCN 2017000395 (print) | LCCN 2017004110 (ebook) | ISBN 9781555816209 (print) | ISBN 9781555819637 (ebook)
Subjects: LCSH: Microbiological laboratories—Safety measures. | Biological laboratories—Safety measures.
Classification: LCC QR64.7 .L33 2017 (print) | LCC QR64.7 (ebook) | DDC 570.289—dc23
LC record available at https://lccn.loc.gov/2017000395

doi:10.1128/9781555819637

Printed in Canada

10 9 8 7 6 5 4 3 2

Address editorial correspondence to: ASM Press, 1752 N St., N.W., Washington, DC 20036-2904, USA.
Send orders to: ASM Press, P.O. Box 605, Herndon, VA 20172, USA.
Phone: 800-546-2416; 703-661-1593. Fax: 703-661-1501.
E-mail: books@asmusa.org
Online: http://www.asmscience.org

Contents

INTRODUCTION

SECTION IV. ADMINISTRATIVE CONTROL

SECTION V. SPECIAL ENVIRONMENTS

Contributors

Dann Adair
Conviron, Pembina, North Dakota

Deborah M. Anderson
Laboratory for Infectious Disease Research and Department of Veterinary Pathobiology, University of Missouri, Columbia, Missouri

Paul E. Anderson
Laboratory for Infectious Disease Research and Department of Veterinary Pathobiology, University of Missouri, Columbia, Missouri

Matthew J. Arduino
Division of Healthcare Quality Promotion, Centers for Disease Control and Prevention, Atlanta, Georgia

John T. Balog
U.S. Food and Drug Administration, Office of Operations, Employee Safety and Environmental Management, Silver Spring, Maryland

Timothy Baszler
Washington State University, Paul G. Allen School for Global Animal Health, Pullman, Washington

Allan Bennett
Public Health England, Biosafety, Porton, Salisbury, Wiltshire, United Kingdom

David S. Bressler
Centers for Global Health, Centers for Disease Control and Prevention, Atlanta, Georgia

LouAnn C. Burnett
International Biological and Chemical Threat Reduction, Sandia National Laboratories, Albuquerque, New Mexico

Karen Brandt Byers
Dana Farber Cancer Institute, Boston, Massachusetts

Charles H. Calisher
Arthropod-borne and Infectious Diseases Laboratory, Department of Microbiology, Immunology and Pathology, College of Veterinary Medicine and Biomedical Sciences, Colorado State University, Fort Collins, Colorado

Darin S. Carroll
Poxvirus and Rabies Branch, Division of High Consequence Pathogens and Pathology, Centers for Disease Control and Prevention, Atlanta, Georgia

Mary L. Cipriano
Abbott Laboratories, North Chicago, Illinois (retired)

J. Patrick Condreay
pc Biosafety Consulting Services, LLC, Carrboro, North Carolina

Jonathan T. Crane
HDR, Inc., Atlanta, Georgia

Marian Downing
Abbott Laboratories, North Chicago, Illinois (retired)

Elizabeth Gilman Duane
Environmental Health and Engineering Inc., Needham, Massachusetts

David C. Eagleson
The Baker Company, Inc., Sanford, Maine

Richard C. Fink
Environmental Health and Engineering Inc., Needham, Massachusetts, and Pfizer (retired)

Diane O. Fleming
Biological Safety Professional (retired), Mitchellville, Maryland

Lance Gaudette
The Baker Company, Inc., Sanford, Maine

Tanya Graham
Biosafety Consulting for Veterinary Medicine, LLC, Esteline, South Dakota

Paul A. Granato
Department of Pathology, SUNY Upstate Medical University, Syracuse, New York, and Laboratory Alliance of Central New York, LLC, Liverpool, New York

A. Lynn Harding
Biosafety Consultant, Chattanooga, Tennessee

J. Ross Hawkins
Division of Advanced Therapies, National Institute for Biological Standards and Control a centre of the Medicines and Healthcare Regulatory Agency, South Mimms, Herts, United Kingdom

Robert J. Hawley
Consultant, Biological Safety and Security, Frederick, Maryland

Robert A. Heckert
Robert Heckert Consulting, Palm Desert, California

Kara F. Held
The Baker Company, Inc., Sanford, Maine

Barbara L. Herwaldt
Centers for Disease Control and Prevention, Parasitic Diseases Branch, Atlanta, Georgia

Stephen Higgs
Biosecurity Research Institute, Kansas State University, Manhattan, Kansas

Yan-Jang S. Huang
Department of Diagnostic Medicine/Pathobiology, College of Veterinary Medicine, Kansas State University, Manhattan, Kansas

Debra L. Hunt
Duke University, Durham, North Carolina

Ruth Irwin
Information Systems for Biotechnology, Virginia Polytechnic Institute & State University, Blacksburg, Virginia

Sean G. Kaufman
Behavioral-Based Improvement Solutions, Woodstock, Georgia

Lon V. Kendall
Department of Microbiology, Immunology and Pathology and Laboratory Animal Resources, Colorado State University, Fort Collins, Colorado

Thomas A. Kost
GlaxoSmithKline Research and Development, Research Triangle Park, North Carolina (retired)

Joseph P. Kozlovac
USDA ARS Office of National Programs, Animal Production & Protection, Beltsville, Maryland

Patricia Lambrecht
Department of Plant Pathology, University of Nebraska-Lincoln, Lincoln, Nebraska

James Lawler
Navy Medical Research Center, Clinical Research, Fort Detrick, Maryland

Travis R. McCarthy
Laboratory for Infectious Disease Research, University of Missouri, Columbia, Missouri

Nicole Vars McCullough
3M, Personal Safety Division, Saint Paul, Minnesota

Claudia A. Mickelson
EHS Office, Massachusetts Institute of Technology, Cambridge, Massachusetts (retired)

Melissa A. Morland
University of Maryland, Baltimore, Baltimore, Maryland

Jason Paragas
Lawrence Livermore National Laboratory, Global Security, Livermore, California

Simon Parks
Biosafety, Air and Water Microbiology Group, Public Health England, Porton Down, Wiltshire, United Kingdom

Ami A. Patel
Laboratory for Infectious Disease Research, University of Missouri, Columbia, Missouri

Michael A. Pentella
Massachusetts Department of Public Health, State Public Health Laboratory, Jamaica Plain, Massachusetts

Janet S. Peterson
Biosafety Consultant, Ellicott City, Maryland

Brian R. Petuch
Global Safety & Environment, Merck, West Point, Pennsylvania

Wanda Phipatanakul
Boston Children's Hospital, Harvard Medical School, Boston, Massachusetts

Charles W. Quint, Jr.
The Baker Company, Inc., PO Sanford, Maine

Ryan F. Relich
Division of Clinical Microbiology, Indiana University Health Pathology Laboratory, and
Department of Pathology and Laboratory Medicine, Indiana University School of Medicine, Indianapolis, Indiana

Jonathan Y. Richmond
Bsafe.us, Southport, North Carolina

Michelle Rozo
Navy Medical Research Center, Clinical Research, Fort Detrick, Maryland

Wiley A. Schell
Department of Medicine, Division of Infectious Diseases and International Health, Duke University, Durham, North Carolina

James M. Schmitt
Occupational Medical Service, National Institutes of Health, Bethesda, Maryland

Clare Shieber
Public Health England, Biosafety, Air and Water Microbiology Group, Porton, Salisbury, Wiltshire, United Kingdom

James W. Snyder
Department of Pathology and Laboratory Medicine, University of Louisville, Louisville, Kentucky

Glyn N. Stacey
Division of Advanced Therapies, National Institute for Biological Standards and Control a centre of the Medicines and Healthcare Regulatory Agency, Blanche Lane, South Mimms, Herts, United Kingdom

David G. Stuart
The John M. Eagleson, Jr. Institute, Kennebunk, Maine

Danielle Tack
Poxvirus and Rabies Branch, Centers for Disease Control and Prevention, Atlanta, Georgia

Sue A. Tolin
Department of Plant Pathology, Physiology and Weed Science, Virginia Polytechnic Institute & State University, Blacksburg, Virginia

Theresa D. Bell Toms
Leidos Biomedical Research Inc., National Cancer Institute at Frederick, Frederick, Maryland

Dana L. Vanlandingham
Department of Diagnostic Medicine/Pathobiology, College of Veterinary Medicine, Kansas State University, Manhattan, Kansas

Anne K. Vidaver
Department of Plant Pathology, University of Nebraska-Lincoln, Lincoln, Nebraska

Robert A. Wood
Department of Pediatrics, Division of Allergy and Immunology, Johns Hopkins University, Baltimore, Maryland

Dawn P. Wooley
Department of Neuroscience, Cell Biology and Physiology, Wright State University, Dayton, Ohio

Abbey K. Woolverton
Department of Epidemiology and Biostatistics, Milken Institute School of Public Health, George Washington University, Washington, DC

Christopher J. Woolverton
Department of Biostatistics, Environmental Health Science and Epidemiology, College of Public Health, Kent State University, Kent, Ohio

Foreword

On October 29, 1997, a non-human primate research worker was transferring macaques from a transport cage to a squeeze cage preceding a routine annual physical. One of the macaques became agitated, and as he jumped, his tail flicked material from the bottom of the cage into the face and eye of the researcher. On December 10, 1997, that vivacious and talented 22-year-old worker, Elizabeth "Beth" Griffin, died as a result of that innocuous event.

Beth's death was initiated by an ocular exposure to the Herpes simian B virus (*Macacine herpesvirus 1*). Her case was the first known exposure to be the result of something other than a bite or a scratch. An Agnes Scott College graduate, Beth—a dancer—died from an encephalitic disease that first paralyzed her from the neck down before finally causing her death.

Beth's death gained national attention in the U.S. media. It was a featured story on a network newsmagazine. The incident gained international attention in the world of research. The world—especially the research world—wanted to know how such a thing could ever happen and what could be done to ensure it never happened again.

A number of things could have been done that would have meant this story would never be read. There were systematic failures in the occupational health response to her exposure. There were failures in the health care system. There were things Beth could have done, such as wear goggles while handling the monkeys or use the nearby eyewash stations within 5 minutes of her exposure. An emergency response measure could have provided a simple postexposure prophylactic prescription taken shortly after her incident. These actions and others as elements of an institutional culture of safety—Prevention, Detection, and Response—could have changed everything.

Two years after her death, Beth's family established a nonprofit foundation to increase safety and occupational health awareness for people who worked with non-human primates. With the collaborative assistance of organizations such as the Association of Primate Veterinarians (APV), the American Association for Laboratory Animal Science (AALAS), and the American College of Laboratory Animal Medicine (ACLAM), many changes were made in processes and responses to exposures. Many people working in non-human primate research environments began carrying cards, quickly tagged as "Beth Cards," that informed medical personnel to take specific measures to rule out B virus exposure first—not last—if the person was exhibiting certain viral symptoms.

In 2003, the world became gripped in an outbreak of a disease called SARS (severe acute respiratory syndrome). The outbreak began in China, but because of mobility the disease soon began popping up elsewhere. As Beth's death had been a tipping point for safety awareness in working with non-human primates, the SARS outbreak and the global response of expanding laboratory capacity to detect and identify emerging infectious diseases became a massive springboard for biosafety.

The "Amerithrax" incident of 2001 had already sparked international attention to practices used in working with certain biological agents. The concepts of biosafety and biosecurity preceded all of these incidents by decades, but never had there been such total community attention to the potential risks of biological exposures.

At the encouragement of those groups with whom we had already collaborated, the Elizabeth Griffin Research Foundation reached out with our "no more Beth Griffin tragedies" message to the American Biological Safety Association to assist in highlighting awareness of and response to the exposure risks that those who work with biological agents face on a rather routine basis. With their assistance—and that of a growing number of similar professional organizations around the world—biosafety is a front-burner issue in conducting safe and responsible science. Much has been done to increase the awareness, research, and application of sound protocols that both reduce the risk of exposure and improve the quality of response to an exposure should one occur. The very truth that you are reading this book on biosafety and biosecurity is proof enough of how far this has come.

Good science is safe science. If the science isn't safe, it isn't good. Nothing can be more damaging to the reputation of a research institution or to the public view of the value of science than a bungled exposure issue or the appearance of cutting corners on safety in order to accomplish something. Biological risks are very different from many others in that they are most often not immediately evident, due to incubation periods. There are no immediate detection capabilities as with chemical or radiation risks, since biological manifestation may easily be delayed and often misdiagnosed. Compound those issues with the fact that many biological agents have highly contagious, often lethal capabilities, and we quickly see it's not just the laboratory worker at risk.

Watchfulness, attention, caution, and prudence are all required whenever someone does anything that places individuals beyond themselves at risk. To engage in biological research requires that you exercise caution and follow protocols, not only for your safety but also for the safety of the community and world that surrounds you. It is not an option or a luxury. It is a necessity. Every risk, no matter how small it may seem, must be considered, assessed, and properly mitigated. The techniques of safety and security are every bit as important as the techniques used in your research.

Before getting into the technical nuts and bolts of biosafety and biosecurity, please keep these basics in mind.

1. Everyone who works with biological agents in any capacity should discuss their work with their personal physician. You are quite possibly the zebra among a stable of horses.
2. Remember that most people drown in shallow water. While much attention is required to higher-risk agents, most laboratory-acquired infections (LAIs) occur when working with what are thought to be lower-risk agents. Most LAI deaths are attributed to Level 2 agents, not Level 3 or 4.
3. Learn from near-misses. Encourage nonpunitive conversations about things that "almost happened." The "almost happened" events are likely to recur, so learn from them.
4. Compliance is a by-product of safe research. It is not the purpose of safe research.
5. Be a role model of biosafety and biosecurity. Create atmospheres where being safe appears the most natural thing to do.
6. Link up with the biosafety personnel at your institution. Learn from them.
7. If you think there's a safer way, don't just think it. Prove it by research, demonstrate it, and share what you learned with the biosafety community.
8. Commit to never letting a Beth Griffin tragedy happen wherever you may be.

We adhere to the words spoken by Thomas Huxley at the opening of The Johns Hopkins University in Baltimore, Maryland. In his remarks, Huxley noted that "the end of life is not knowledge, but action." On behalf of the Elizabeth R. Griffin Research Foundation and our collaborative partners worldwide, we encourage that you not just learn the material in this book but act upon, promote, and add to this body of knowledge throughout your scientific career.

Caryl P. Griffin, MDiv, *President and Founder*
James Welch, *Executive Director*
Elizabeth R. Griffin Foundation
www.ergriffinresearch.org

Preface

It is with a great sense of honor and reverence that we take over the reins of editing this book from our esteemed colleagues, Diane O. Fleming and Debra L. Hunt. It is our hope that this 5th edition of *Biological Safety: Principles and Practices* remains the main text in the field of biosafety. We are indebted to the many authors who have contributed to this edition. This book serves as a valuable resource not only for biosafety professionals, but also for students, staff, faculty, and clinicians who are working with or around potentially biohazardous materials in research laboratories, medical settings, and industrial environments. Those who supervise biosafety or laboratory staff members will also benefit from this book.

We decided to keep the overall structure similar to the previous edition, with five major sections. Eight new chapters were added on the following topics: molecular agents, arthropod vector biocontainment, aerobiology, training programs, veterinary and greenhouse biosafety, field studies, and clinical laboratories. Biosafety Practices is not a separate chapter in this edition; the concepts have been incorporated into relevant chapters. Similarly, the information on prions was incorporated into the new chapter on molecular agents. The title of the last section was changed from "Special Considerations" to "Special Environments" and some chapters were moved out of this section to keep the focus on unique settings encountered in biosafety practice. Since regulatory guidelines are always changing, we have directed our readers to online sources for the most up-to-date information. Chapters have been made to be more fluid and stand-alone by minimizing references to other chapters. We are fortunate to have color in this new edition.

Both of this edition's editors are Certified Biosafety Professionals, but we came to the field of biosafety through different avenues, giving us complementary perspectives on the topic. Dawn Wooley became intensely interested in biosafety during her graduate days at Harvard while researching the newly discovered AIDS viruses. These were the days before there were important administrative controls such as the Bloodborne Pathogen Standard. In trying to protect herself and others around her from these newly emerging pathogens, Dawn developed a love for the field of biosafety that has persisted until today. Karen Byers developed a keen interest in biosafety while working with measles in Harvard research laboratories. An appointment to the Institutional Biosafety Committee inspired her to become a biosafety professional. She is very grateful for Lynn Harding's mentorship and the opportunities for professional development and leadership provided by colleagues in the American Biological Safety International (ABSA).

Professional organizations such as ABSA, the American Society of Microbiology (ASM), the American Public Health Association (APHL), the Clinical and Laboratory

Standards Institute (CLSI), and the American Association for Laboratory Animal Science (AALAS) have played a key role in fostering the development and implementation of evidence-based biosafety practice. The Foreword to this edition reminds us of the importance of this endeavor.

Gregory W. Payne, Senior Editor, ASM Press, was instrumental in pushing for the update of this book, and he provided much-needed guidance and inspiration. We thank Ellie Tupper and Lauren Luethy for their expert assistance with the production of this book.

We hope that our readers enjoy the book as much as we have appreciated the opportunity to work on it for you and the rest of the biosafety community. Be safe!

Dawn P. Wooley
Karen B. Byers

SECTION

I

Hazard Identification

The Microbiota of Humans and Microbial Virulence Factors

1

PAUL A. GRANATO

"The 1990s have been marked by a renewed recognition that our human species is still locked in a Darwinian struggle with our microbial and viral predators." Although this unreferenced quotation was made by Nobel Laureate Joshua Lederberg, as he was discussing the acquired immunodeficiency syndrome (AIDS) and multidrug-resistant *Mycobacterium tuberculosis* epidemics that emerged in the early 1990s, his comment could also apply to almost any infectious disease process that has occurred since the recognition of the germ theory of disease in the late 1880s. For as we journey through the 21st century, and despite the advances of modern medicine and the continual development of new vaccines and anti-infective therapeutic agents, the human species continues to battle microbial predators in this Darwinian struggle for survival.

MICROBIOTA AND THE HUMAN GENOME PROJECT

The human normal flora consists of an ecological community of commensal, symbiotic, and pathogenic microorganisms in dynamic balance that literally share and inhabit our body spaces throughout life. In 2001, Lederberg (1) coined the term "microbiota" to describe these microbial communities that were characterized by using cultural methods. Subsequently, in 2008, the Human Microbiome Project (HMP) was funded by the National Institutes of Health to use noncultural methods to study how changes in the human microbiome are associated with health and disease (2). The HMP used genetic-based, molecular methods, such as metagenomics and genome sequencing, to characterize all microbes present in a body site, even those that could not be cultured. As such, by using metagenomics (which provides a broad genetic perspective on a single microbial community) and extensive whole-genome sequencing (which provides a genetic perspective on individual microorganisms in a given microbial community), the HMP provided a more comprehensive understanding of the microorganisms that inhabit a particular body site through genetic analysis.

The HMP studies (3) have shown that even healthy individuals differ remarkably in the microbes that occupy body sites such as the skin, mouth, intestine, and vagina.

Much of this diversity remains unexplained, although diet, environment, host genetics, and early microbial exposure have all been implicated. These studies have also led some investigators to conclude that the human microbiome may play a role in autoimmune diseases like diabetes, rheumatoid arthritis, muscular dystrophy, multiple sclerosis, fibromyalgia, and perhaps some cancers (4). Others have proposed that a particular mix of microbes in the intestine may contribute to common obesity (5–7). It has also been shown that some of the microbes in the human body can modify the production of neurotransmitters in the brain that may possibly modify schizophrenia, depression, bipolar disorder, and other neurochemical imbalances (8).

DYNAMICS OF THE HOST–PARASITE RELATIONSHIP

The dynamics of this host–parasite relationship for survival are in a continual state of change. In health, a balance exists between the host and the microbe that allows for the mutual survival and coexistence of both. This balance is best maintained when humans have operative host defense mechanisms and are not exposed to any particular infectious microbial agent. The three major host defense mechanisms that must be operative to maintain this balance and the health of the human host are (i) intact skin and mucous membranes, (ii) a functional group of phagocytic cells consisting principally of the reticuloendothelial system (RES), and (iii) the ability to produce a humoral immune response. Defects in any one or combination or all of these host defense mechanisms will shift the balance in favor of the microbe and predispose the host to the risk of developing an infectious disease process. For example, breaks in skin or mucous membranes due to accidents, trauma, surgery, or thermal injury may serve as a portal of entry for microorganisms to produce infection. In addition, the inability to phagocytize microorganisms effectively by the RES due to lymphoma or leukemia and the inability to produce functional humoral antibodies due to defects in plasma cells or exposure to immunosuppressive agents (i.e., drugs, irradiation, etc.) may also predispose to the development of infection. This balance in favor of the microbe may be shifted back toward the host through the use of antimicrobial agents and/or the administration of vaccines for the treatment and prevention of disease. Unfortunately, as these agents or selective pressures may adversely affect the survival of the microbe, these developments are often followed by a shift in balance back in favor of the ever-adaptable microbe by, perhaps, acquiring new mechanisms for producing human disease or resisting the action of an antimicrobial agent.

The microbial world consists of bacteria, fungi, viruses, and protozoa that represent over several hundred thousand known species. The great majority of these, however, are not involved in any dynamic relationship with the human host because they are incapable of surviving or causing disease in humans. By comparison, those microorganisms that are involved in the dynamic relationship with the host are limited in number, consisting of fewer than 1,000 known microbial species. It is this limited group of microorganisms that is the focus of discussion in this chapter.

The relationships that exist between the human host and the microbial world are varied and complex. When a microorganism that is capable of causing disease becomes established in the body, this process is called an infection, and an infection that produces symptoms in a human is called an infectious disease. By contrast, persistence of microorganisms in a particular body site (such as the normal microbial flora, as is discussed in a subsequent section of this chapter) is often referred to as colonization rather than infection. Importantly, infection or colonization does not necessarily lead to the development of an infectious disease. If host defenses are adequate, a person may be infected by a disease-causing microorganism for an indefinite period without any signs or symptoms of disease. Such individuals are referred to as asymptomatic carriers or simply carriers who have asymptomatic or subclinical infection. These asymptomatic carriers serve as important reservoirs for transmission of the infecting organisms to susceptible hosts who may subsequently develop symptomatic disease.

The ability of certain microorganisms to infect or cause disease depends on the susceptibility of the host, and there are notable species differences in host susceptibility for many infections. For instance, dogs do not get measles and humans do not get distemper. Thus, the term pathogenicity, which is defined as the ability of a microorganism to cause disease, must be qualified according to the host species involved. Microorganisms that do not normally produce disease in the healthy human host are often called saprophytes, commensals, or nonpathogens.

In recent years, increasing numbers of infectious diseases have been caused by microorganisms that were previously considered nonpathogenic. These infectious diseases often develop in patients whose surface/barrier, cellular, or immunologic defenses are compromised by such things as trauma, genetic defects, underlying disease, or immunosuppressive therapy. Microorganisms that are frequent causes of disease only in the immunocompromised host or when skin or mucosal surfaces or barriers are breached are called opportunistic pathogens. Opportunistic pathogens are often saprophytes that rarely cause disease in individuals with functional host defense mechanisms.

Pathogenicity refers to the ability of a microorganism to cause disease, and virulence provides a quantitative measure of this property. Virulence factors refer to the properties that enable a microorganism to establish itself on or within a host and enhance the organism's ability to produce disease. Virulence is not generally attributable to a single discrete factor but depends on several parameters related to the organism, the host, and their interaction. Virulence encompasses two general features of a pathogenic microorganism: (i) invasiveness, or the ability to attach, multiply, and spread in tissues, and (ii) toxigenicity, the ability to produce substances that are injurious to human cells. Highly virulent, moderately virulent, and avirulent strains may occur within a single species of organisms.

The microorganisms that cause human infectious diseases are acquired from two major sources or reservoirs: those acquired from outside the body, called exogenous reservoirs, and those infectious diseases that result from microorganisms that inhabit certain body sites, called endogenous reservoirs. Most exogenous infections are acquired from other individuals by direct contact, by aerosol transmission of infectious respiratory secretions, by ingestion of contaminated food or drink, or indirectly through contact with contaminated inanimate objects (often called fomites). Some exogenous infections may also be acquired by puncture of the skin during an insect or animal bite and, perhaps, by occupational exposure from sharps. Endogenous infections occur more commonly than exogenous infections and are acquired from microorganisms that reside normally on various body sites (called normal commensal flora) gaining access to anatomic sites that are normally sterile in health.

NORMAL MICROBIAL FLORA

The terms "normal microbial flora," "normal commensal flora," "indigenous flora," and "microbiota" are often used synonymously to describe microorganisms that are frequently found in particular anatomic sites in healthy individuals, whereas the term "microbiome" refers to their genomes. This microbial flora is associated with the skin and mucous membranes of every human from shortly after birth until death and represents an extremely large and diverse population of microorganisms. The healthy adult consists of about 10 trillion cells and routinely harbors at least 100 trillion microbes (9). The entire microbiome accounts for about 1% to 3% of the total human body mass (10) with some weight estimates ranging as high as 3 pounds or 1,400 grams. The constituents and numbers of the flora vary in different anatomic sites and sometimes at different ages. They comprise microorganisms whose morphologic, physiologic, and genetic properties allow them to colonize and multiply under the conditions that exist in a particular body site, to coexist with other colonizing organisms, and to inhibit competing intruders. Thus, each anatomic site that harbors a normal microbial flora presents a particular environmental niche for the development of a unique microbial ecosystem.

Local physiologic and environmental conditions at various body sites determine the nature and composition of the normal flora that exists there. These conditions are sometimes highly complex, differing from site to site, and sometimes vary with age. Some of these local anatomic conditions include the amounts and types of nutrients available for microbial growth, pH, oxidation reduction potentials, and resistance to local antibacterial substances, such as bile, lysozyme, or short-chain fatty acids. In addition, many bacteria have a remarkable affinity for specific types of epithelial cells to which they adhere and on which they multiply. This adherence, which is mediated by the presence of bacterial pili/fimbriae or other microbial surface components, allows the microbe to attach to specific receptor sites found on the surface of certain epithelial cells. Through this mechanism of adherence, microorganisms are permitted to grow and multiply while avoiding removal by the flushing effects of surface fluids and peristalsis. Various microbial interactions also determine their relative prevalence in the flora. Some of these interactions include competition for nutrients and inhibition of growth by the metabolic products produced by other microorganisms in the ecosystem (for example, the production of hydrogen peroxide, antibiotics, and/or bacteriocins).

The normal microbial flora plays an important role in health and disease. In health, for example, the normal microbial flora of the intestine participates in human nutrition and metabolism. Certain intestinal bacteria synthesize and secrete vitamin K, which can then be absorbed by the bowel for use in the human. In addition, the metabolism of several key compounds involves excretion from the liver into the intestine and their return from there to the liver. This enterohepatic circulatory loop is particularly important for the metabolism of steroids and bile salts. These substances are excreted through the bile in conjugated form as glucuronides or sulfates but cannot be reabsorbed in this form. Certain members of the bacterial intestinal flora make glucuronidases and sulfatases that can deconjugate these compounds, thereby allowing their reabsorption and use by the human host (11, 12). Another beneficial role of the normal microbial flora is the antigenic stimulation of the host's immune system. Although the various classes of the immunoglobulins produced from this antigenic exposure are usually present in low concentrations, their presence plays an important role in host defense. In particular, various classes of the immunoglobulin A (IgA) group of antibodies produced in response to this antigenic stimulation are secreted through mucous membranes. The role of these immunoglobulins is not well understood, but they may contribute to host defense by interfering with the colonization of deeper tissues by certain normal flora organisms.

Perhaps one of the most important roles of the normal microbial flora is to help prevent infectious disease following exposure to potential microbial pathogens. The normal commensal flora has the physical advantage of previous occupancy on skin and mucous membranes. Many of these commensal microorganisms adhere to epithelial binding sites, thereby preventing attachment to that receptor site by a potential microbial pathogen. As is discussed later in this chapter, certain pathogens that are incapable of adhering to their specific epithelial receptors are incapable of causing human disease. In addition, some commensal microorganisms are capable of producing antibiotics, bacteriocins, or other products that may be inhibitory or lethal to pathogenic microorganisms. The collective effect of the normal flora's ability to adhere to epithelial receptor sites and to produce antimicrobial substances plays an important role in maintaining the health of the host following exposure to a potential microbial pathogen.

The normal microbial flora, although important for the maintenance of human health, is a critical factor in human infectious disease. Because the human body is colonized with diverse and large populations of microorganisms as part of one's normal flora, the three major host defense mechanisms (intact mechanical surfaces, RES, and immune system) must be continually operative and functional for the maintenance of human health in this continually dynamic relationship between the host and parasite. On occasion, normal flora organisms may gain entry into normally sterile body sites, or defects in one or more of the host's defense mechanisms may result in the development of symptomatic infection from one or more of these organisms.

These endogenous human infections occur more frequently than those that are acquired from an exogenous source. In general, physicians see more patients with infectious diseases acquired from one's normal microbial flora than those infectious disease processes that are acquired from outside the body (13). It is for these reasons that clinicians and clinical microbiologists must be knowledgeable as to the various microbes that reside as the normal flora in different anatomic sites.

In medicine, it is often said, "Common things occur commonly." Knowing the normal microbial flora at a particular anatomic site is often useful in predicting the likely etiologic agents of infection when a neighboring tissue becomes infected from an endogenous source. Therefore, the normal microbial flora for various anatomic sites is reviewed in the following section. Because the residents of the normal microbial flora may vary with the age of the host, this discussion also addresses the normal flora typically found in both healthy newborns and adults when differences in microbial ecosystems may exist.

Skin

Human skin is a complex microbial ecosystem. The healthy fetus is sterile *in utero* until the birth membranes rupture. During and after birth, the infant's skin is exposed to the mother's genital tract flora, to skin flora from the mother and other individuals who handle the baby, and to a variety of microorganisms acquired by direct contact of the baby with the environment. During the infant's first few days of life, the nature of its microbial skin flora often reflects chance exposure to microorganisms that can grow on particular sites in the absence of microbial competitors. Subsequently, as the infant is exposed to a full range of human environmental organisms, those best adapted to survive on particular skin sites predominate and establish themselves as part of the resident skin flora. Thereafter, the normal microbial flora resembles that of adult individuals.

The pH of the skin is usually about 5.6. This factor alone may be responsible for inhibiting the establishment of many microbial species. Despite this, skin provides excellent examples of various microenvironments. Some areas are moist, such as the toe webs and perineum, whereas some areas are relatively dry, such as the forearm. Sebaceous glands found on the face, scalp, and upper chest and back produce an abundance of lipids on the skin, whereas other areas, such as the axillae, produce specialized secretions from apocrine glands. Eccrine glands, also called merocrine glands or simply sweat glands, are found in the skin of virtually all anatomic sites of the body. These glands produce a clear, odorless secretion consisting primarily of water and saline that is induced following exposure to high temperature or exercise. As a result of these differences in microenvironments, quantitative differences in microbial flora occur in each of the three major regions of skin: (i) axilla, perineum, and toe webs; (ii) hands, face, and trunk; and (iii) arms and legs (14). These quantitative differences are the result of differences in skin surface temperature and moisture content as well as the presence of different concentrations of skin surface lipids that may be inhibitory or lethal to various groups of microorganisms at each of these skin sites (15).

The major groups of microorganisms that are normal residents of skin, even though their numbers may vary as influenced by the microenvironment, include various genera of bacteria and the lipophilic yeasts of the genus *Malassezia*. Nonlipophilic yeasts, such as *Candida* species, are also inhabitants of the skin (14). Other bacterial species may be found less commonly on the skin, and some of these include hemolytic streptococci (especially in children), atypical mycobacteria, and *Bacillus* species.

The predominant bacterial inhabitants of the skin are the coagulase-negative staphylococci, micrococci, saprophytic *Corynebacterium* species (diphtheroids), and *Propionibacterium* species. Among this group, *Propionibacterium acnes* is the best studied because of its association with acne vulgaris. *P. acnes* is found briefly on the skin of neonates, but true colonization begins during the 1 to 3 years prior to sexual maturity, when numbers rise from less than 10 CFU/cm^2 to about 10^6 CFU/cm^2, chiefly on the face and upper thorax (16). Various species of coagulase-negative staphylococci are found as normal inhabitants of skin, and some of these include *Staphylococcus epidermidis*, *S. capitis*, *S. warneri*, *S. hominis*, *S. haemolyticus*, *S. lugdunensis*, and *S. auricularis* (17–20). Some of these staphylococci demonstrate ecological niche preferences at certain anatomic sites. For example, *S. capitis* and *S. auricularis* show an anatomic preference for the head and the external auditory meatus, respectively, whereas *S. hominis* and *S. haemolyticus* are found principally in areas where there are numerous apocrine glands, such as the axillae and pubic areas (17). *Staphylococcus aureus* regularly inhabits the external nares of about 30% of healthy individuals and the perineum, axillae, and toe webs of about 15%, 5%, and 2%, respectively, of healthy people (14). *Micrococcus* spp., particularly *Micrococcus luteus*, are also found on the skin, especially in women and children, where they may be present in large numbers. *Acinetobacter* spp. are found on the skin of about 25% of the population in the axillae, toe webs, groin, and antecubital fossae. Other Gram-negative bacilli are found more rarely on the skin, and these include *Proteus* and *Pseudomonas* in the toe webs and *Enterobacter* and *Klebsiella* on the hands. Saprophytic mycobacteria may occasionally be found on the skin of the external auditory canal and of the genital and axillary regions, whereas hemolytic streptococci tend to colonize the skin of children but not adults (14).

The principal fungal flora is *Malassezia,* a yeast. Dermatophytic fungi may also be recovered from the skin in the absence of disease, but it is unclear whether they represent the normal flora or transient colonizers. Carriage of *Malassezia* spp. probably reaches 100% in adults, but proper determination of carriage rates is obscured by the difficulty of growing some species of these lipophilic yeasts in the laboratory (14).

Members of the skin microflora live both on the skin surface in the form of microcolonies and in the ducts of hair follicles and sebaceous glands (14). Wolff et al. (21) proposed that *Malassezia* species live near the opening of the duct, the staphylococci further down, and the propionibacteria near the sebaceous glands. A more recent study (22), however, suggests that all three microbial groups are more evenly distributed throughout the follicles. In any event, organisms in the follicles are secreted onto the skin surface along with the sebum, but staphylococci, at least, also exist in microcolonies on the surface. These microcolonies may be of various sizes and are larger (10^3 to 10^4 cells per microcolony) on areas such as the face than on the arms (10^1 to 10^2 cells per microcolony) (14).

Washing may decrease microbial skin counts by 90%, but normal numbers are reestablished within 8 h (23). Abstinence from washing does not lead to an increase in numbers of bacteria on the skin. Normally, 10^3 to 10^4 organisms are found per square centimeter. However, counts may increase to 10^6/cm^2 in more humid areas, such as the groin and axilla. Small numbers of bacteria are dispersed from the skin to the environment, but certain individuals may shed up to 10^6 organisms in 30 min of exercise. Many of the fatty acids found on the skin may be bacterial products that inhibit colonization by other species. The flora of hair is similar to that of the skin (24).

Eye

The normal microbial flora of the eye contains many of the bacteria found on the skin. However, the mechanical action of eyelids and the washing effect of the eye secretions that contain the bacteriolytic enzyme lysozyme serve to limit the populations of microorganisms normally found on the eye. The predominant normal microbial flora of the eye consists of coagulase-negative staphylococci, diphtheroids, and, less commonly, saprophytic *Neisseria* species and viridans group streptococci.

Ear

The microbiota of the external ear is similar to that of skin, with coagulase-negative staphylococci and *Corynebacterium* species predominating. Less frequently found are *Bacillus*, *Micrococcus*, and saprophytic species of *Neisseria* and mycobacteria. Normal flora fungi include *Aspergillus*, *Alternaria*, *Penicillium*, and *Candida*.

Respiratory Tract

Nares

In the course of normal breathing, many kinds of microbes are inhaled through the nares to reach the upper respiratory tract. Among these are aerosolized normal soil inhabitants as well as pathogenic and potentially pathogenic bacteria, fungi, and viruses. Some of these microorganisms are filtered out by the hairs in the nose, whereas others may land on moist surfaces of the nasal passages, where they may be subsequently expelled by sneezing or blowing one's nose. Generally, in health these airborne microorganisms are transient colonizers of the nose and do not establish themselves as part of the resident commensal flora.

The external 1 cm of the external nares is lined with squamous epithelium and has a flora similar to that found on the skin, except that *S. aureus* is commonly carried as the principal part of the normal flora in some individuals. Approximately 25% to 30% of healthy adults in the community harbor this organism in their anterior nares at any given time, 15% permanently and the remaining 15% transiently (25).

Nasopharynx

Colonization of the nasopharynx occurs soon after birth following aerosol exposure of microorganisms from the respiratory tract from those individuals who are in close contact with the infant (i.e., the mother, other family members, etc.). The normal microbial flora of the infant establishes itself within several months and generally remains unchanged throughout life. The nasopharynx has a flora similar to that of the mouth (see below) and is the site of carriage of potentially pathogenic bacteria such as *Neisseria meningitidis*, *Branhamella catarrhalis*, *Streptococcus pneumoniae*, *S. aureus*, and *Haemophilus influenzae* (25).

The respiratory tract below the level of the larynx is protected in health by the actions of the epiglottis and the peristaltic movement of the ciliary blanket of the columnar epithelium. Thus, only transiently inhaled organisms are encountered in the trachea and larger bronchi. The accessory sinuses are normally sterile and are protected in a similar fashion, as is the middle ear, by the epithelium of the eustachian tubes.

Gastrointestinal Tract

Mouth

Colonization of the mouth begins immediately following birth when the infant is exposed to the microorganisms in the environment, and the numbers present increase rapidly in the first 6 to 10 h after birth (26). During the first few days, several species appear sporadically as transients, many of them not being suitable for the oral environment. During this period, the oral mucosa becomes colonized by its first permanent residents; these are derived mainly from the mouth of the mother and other persons in contact with the infant (26, 27). The child is continuously exposed to transmission of oral bacteria from family members by direct and indirect contact (the latter, for example, via spoons and feeding bottles), as well as by airborne transmission. The various members of the resident microflora become established gradually during the first years of life as growth conditions become suitable for them. This microbial succession is caused by environmental changes related to the host, such as tooth eruption or dietary changes, as well as to microbial interrelations due to, for example, the initial colonizers reducing tissue redox potentials or supplying growth factors.

During the first months of life, the oral microflora mainly inhabits the tongue and is dominated by streptococci, with small numbers of other genera such as *Neisseria*, *Veillonella*, *Lactobacillus*, and *Candida*. *Streptococcus salivarius* is regularly isolated from the baby's mouth starting from the first day of life, and often the bacteriocin types are identical to those of the mother (28). *Streptococcus sanguinis* colonizes the teeth soon after eruption (29), whereas *Streptococcus mutans* colonizes much more slowly over several years, starting in pits and fissures and spreading to proximal and other surfaces of the teeth (30). Colonization with *S. mutans* and lactobacilli is correlated with dental caries (29, 31), and, in fact, their establishment can be inhibited or delayed by caries-preventive measures in the infants' mothers (32). Dental caries result from the ability of these bacteria to produce biofilms that adhere to the tooth surface. Biofilms and their relationship to microbial virulence will be discussed later in the Virulence Factors and Mechanisms section of this chapter.

As dental plaque forms on the erupting teeth, the oral microflora becomes more complex and predominately anaerobic. Studies of 4- to 7-year-olds have shown the plaque microflora in the gingival area to be similar to that in adults, with motile rods and spirochetes observed by direct microscopy, and the same species of *Actinomyces*, *Bacteroides*, *Capnocytophaga*, *Eikenella*, etc., recovered by cultural techniques (33–36). In studies of 7- to 19-year-olds, the prevalence of some organisms and the proportions they constitute of the flora seem, however, to differ with age and hormonal status. Thus, *Prevotella* species and spirochetes increase around puberty, while *Actinomyces naeslundii* and *Capnocytophaga* spp. tend to decrease with increasing age of the children.

In healthy adults, the resident oral microflora consists of more than 200 Gram-positive and Gram-negative bacterial species as well as several different species of mycoplasmas, yeasts, and protozoa. Only about 100 oral species of bacteria have known genus species names based upon biochemical and physiologic characteristics (37). With the eruption of teeth and the development of gingival crevices, anaerobic bacteria emerge as the principal flora of the mouth. Concentrations of bacteria vary from approximately 10^8 CFU/ml in the saliva to 10^{12} CFU/ml in the gingival crevices around teeth, with the anaerobic bacteria outnumbering the aerobic bacteria by a ratio of a least 100:1.

The mouth has several different habitats where microorganisms can grow. Each habitat has its own unique environment and is populated by a characteristic community of microorganisms consisting of different populations of various species in each ecosystem. Each species performs a certain functional role as part of the microbial community. Some of the major ecosystems may be found on mucosal surfaces of the palate, gingiva, lips, cheeks, and floor of the mouth, the papillary surface of the tongue, and tooth surfaces, with their associated dental plaque, gingival pockets, etc. To remain in the mouth, the microorganisms must adhere to the oral surfaces, resist being eliminated with the stream of saliva swallowed, and grow under the different conditions prevailing at each site. Such sites can harbor extremely numerous and complex microbial communities. For detailed and comprehensive information, the reader is referred to the review by Theilade (37).

In general, streptococcal species constitute 30% to 60% of the bacterial flora of the surfaces within the mouth. These are primarily viridans group streptococci: *S. salivarius*, *S. mutans*, *S. sanguinis*, and *S. mitis*, found on the teeth and in dental plaque. Specific binding to mucosal cells or to tooth enamel has been demonstrated with these organisms. Bacterial plaque developing on the teeth may contain as many as 10^{11} streptococci per gram in addition to actinomycetes and *Veillonella* and *Bacteroides* species. Anaerobic organisms, such as *Prevotella melaninogenica*, treponemes, fusobacteria, clostridia, propionibacteria, and peptostreptococci, are present in gingival crevices, where the oxygen concentration is less than 0.5%. Many of these organisms are obligate anaerobes and do not survive in higher oxygen concentrations. The natural habitat of the pathogenic species *Actinomyces israelii* is the gingival crevice. Among the fungi, species of *Candida* and *Geotrichum* are found in 10% to 15% of individuals (37).

Esophagus

Little attention has been given to characterizing the normal microflora of the esophagus. Essentially, the esophagus is a transit route for food passing from the mouth to the stomach, with approximately 1.5 liters of saliva swallowed per day (38, 39). Although much of this stimulated saliva is swallowed with food, there is a resting rate of saliva secretion estimated to be about 20 ml/h (38), and this saliva is swallowed as fluid. In addition, nasal secretions containing the microbial flora of that site may also be swallowed, introducing salt-tolerant organisms, such as staphylococci, from the anterior and posterior nares. Consequently, normal flora mouth and nasal microorganisms will be recovered from the esophagus, but it is uncertain whether these organisms represent transient colonization or an established microflora.

Stomach

As for the esophagus, oral and nasal normal flora microorganisms, as well as microorganisms ingested in food and drink, are swallowed into the stomach. However, the vast majority is destroyed following exposure to the gastric acid (pH 1.8 to 2.5) (40). Concentrations of bacteria in the healthy stomach are generally low, less than 10^3 CFU/ml, and are composed primarily of relatively acid-resistant species, such as gastric helicobacters, streptococci, staphylococci, lactobacilli, fungi, and even smaller numbers of peptostreptococci, fusobacteria, and *Bacteroides* species (41–43). Gram-positive organisms predominate in the stomach, with a striking absence of Enterobacteriaceae as well as *Bacteroides* and *Clostridium* species.

The gastric flora can become more complex when the ability to achieve an acid pH is altered by the buffering action of food, by hypochlorhydria due to an intrinsic pathogenic process or surgery (40), or by the medicinal use of proton pump inhibitors, such as omeprazole. In the newborn, the stomach secretes very little gastric acid and does not achieve optimal acid secretion rates until 15 to 20 days after birth (41). Consequently, during the first few days of life, the stomach does not constitute a microbicidal barrier to gut colonization.

Intestine

A fecal flora is acquired soon after birth (44). The composition of the early flora depends on a number of factors, including the method of delivery, the gestational age of the newborn infant, and whether the infant is breast- or bottle-fed.

After vaginal delivery, the newborn gut is first colonized by facultative organisms acquired from the mother's vaginal flora, mainly *Escherichia coli* and streptococci (44). The guts of infants delivered by cesarean section are usually colonized by Enterobacteriaceae other than *E. coli* with a composition resembling the environmental flora of the delivery room (45). Anaerobes appear within the first week or two of life and are acquired more uniformly and more rapidly in bottle-fed than in breast-fed babies. Virtually 100% of full-term, bottle-fed, vaginally delivered infants have an anaerobic flora within the first week of life, with *Bacteroides fragilis* predominating, whereas only 59% of similarly delivered but breast-fed infants have anaerobes at this time, and less than 10% harbor *B. fragilis* (46). Breast-fed infants have a marked predominance of *Bifidobacterium* spp. in their colons that exceed the number of Enterobacteriaceae 100- to 1,000-fold (47).

The nature of the gut flora may be influenced by the nutrient content of breast or cow's milk, compared to that of infant formulas that are fortified with nutrients such as iron. The presence of iron seems to stimulate a complex flora composed of Enterobacteriaceae, *Clostridium* species, and *Bacteroides* species. The low-iron breast or cow's milk diet selects for a simple flora composed predominately of *Bifidobacterium* species and *Lactobacillus* species (48, 49). In breast-fed infants, the *Bifidobacterium* population increases in the first few weeks of life to become the stable and dominant component of the fecal flora until the weaning period (50, 51). The properties of breast milk that promote the dominance of Gram-positive bacilli in the feces are not known with certainty but no doubt involve both nutritional and immunologic factors.

Weaning produces significant changes in the composition of the gut flora resulting in increased numbers of *E. coli*, *Streptococcus*, *Clostridium*, *Bacteroides*, and *Peptostreptococcus* species. After weaning, a more stable adult-type flora occurs, in which the number of *Bacteroides* organisms equals or exceeds the number of *Bifidobacterium* organisms, with *E. coli* and *Clostridium* counts decreasing (16).

In adults, the composition of the fecal flora appears to vary more from individual to individual than it does in particular subjects studied over time (41, 43, 52). Bacteria make up most of the flora in the colon and account for up to 60% of the dry mass of feces (53). From 300 (54) to 1,000 (55) different bacterial species reside in the gut, with most estimates at about 500 (56–58). However, it is likely that 99% of the intestinal bacteria are represented by 30 to 40 species (59). Fungi and protozoa also make up part of the gut flora, but little is known about their activities.

The numbers and types of bacteria found in the small intestine depend on the flow rate of intestinal contents. When stasis occurs, the small intestine may contain an extensive, complex microbial flora. Normally, flow is brisk enough to wash the microbial flora through to the distal ileum and colon before the microorganisms multiply. Consequently, the types and numbers of microbes encountered in the duodenum, the jejunum, and the initial portions of ileum are similar to those found in the stomach and on average comprise 10^3 CFU/ml (60–63). Anaerobes only slightly outnumber facultative organisms, with streptococci, lactobacilli, yeasts, and staphylococci also found.

As the ileocecal valve is approached, the number and variety of Gram-negative bacteria begin to increase (34, 42, 64). Coliforms are found consistently, and the numbers of both Gram-positive and Gram-negative anaerobic organisms (such as *Bifidobacterium*, *Clostridium*, *Bacteroides*, and *Fusobacterium*) rise sharply to 10^5 to 10^6 CFU/ml on average. In the adult colon, another dramatic increase in the microbial flora occurs as soon as the ileocecal valve is crossed. Here, the number of microorganisms present approaches the theoretical limits of packing cells in space. Nearly one-third of the dry weight of feces consists of bacteria, with each gram of stool containing up to 10^{11} to 10^{12} organisms (65). This microbial number is about 1 log greater than the total number of cells in the entire human body (66, 67).

Over 98% of the organisms found in the colon are strict anaerobes, with the anaerobes outnumbering aerobes 1,000- to 10,000-fold. The distribution of the major genera of organisms found in the colon per gram of feces is as follows: *Bacteroides*, 10^{10} to 10^{11}; *Bifidobacterium*, 10^{10} to 10^{11}; *Eubacterium*, 10^{10}; *Lactobacillus*, 10^7 to 10^8; coliforms, 10^6 to 10^8; aerobic and anaerobic streptococci, 10^7 to 10^8; *Clostridium*, 10^6; and yeasts at variable numbers (24). Thus, more than 90% of the fecal flora consists of *Bacteroides* and *Bifidobacterium*. Intensive studies of the colonic microbial flora have shown that the average healthy adult harbors well over 200 given species of bacteria alone.

Benefits of intestinal flora

The intestinal microbiota performs many important functions for the host to maintain health and life. Without gut flora, the human body would not be able to utilize some of the undigested carbohydrates consumed because some gut flora possess enzymes that human cells lack for hydrolyzing certain polysaccharides (55). In addition, bacteria can ferment carbohydrates to produce acetic acid, propionic acid, and butyric acid that can be used by host cells to provide a major source of useful energy and nutrients (58, 59). Intestinal bacteria can also assist in absorbing dietary minerals such as calcium, magnesium, and iron (54). Gut bacteria can enhance the absorption and storage of lipids (55) and produce essential vitamins, such as vitamin K, that are subsequently absorbed by the intestine for use by the human host.

The normal gut microbiota plays a role in defense against infection by preventing harmful bacterial species from colonizing the gut through competitive exclusion, an activity often referred to as the "barrier effect." Harmful bacterial species, such as *Clostridium difficile*, the overgrowth of which can cause pseudomembranous colitis, are unable to grow excessively due to competition from helpful gut flora. These microorganisms adhere to the mucosal lining of the intestine, thereby preventing the attachment and potential overgrowth of potentially pathogenic species (54). Gut flora also play important roles in establishing the host's systemic immunity (54, 56, 57), preventing allergies (68), and preventing inflammatory bowel disease, such as Crohn's disease (69).

Genitourinary Tract

Urethra

The only portion of the urinary tract in both males and females that harbors a normal microbial flora is the distal 1 to 2 cm of the urethra. The remainder of the urinary tract is sterile in health. The microbial flora of the distal portion of the urethra consists of various members of the Enterobacteriaceae, with *E. coli* predominating. Lactobacilli, diphtheroids, alpha-hemolytic and nonhemolytic streptococci, enterococci, coagulase-negative staphylococci, *Peptostreptococcus* species, and *Bacteroides* species are also found. In addition, *Mycoplasma hominis, Ureaplasma urealyticum, Mycobacterium smegmatis,* and *Candida* species may also be recovered from this anatomic site in health (25).

Vagina

The normal microbial flora of the vagina varies according to hormonal influences at different ages (70). At birth, the vulva of a newborn is sterile, but after the first 24 h of life, it gradually acquires a rich and varied flora of saprophytic organisms, such as diphtheroids, micrococci, and nonhemolytic streptococci. After 2 to 3 days, estrogen from the maternal circulation induces the deposition of glycogen in the vaginal epithelium, which favors the growth of lactobacilli. The lactobacilli produce acid from glycogen that lowers the pH of the vagina, and a resultant microbial flora develops that resembles that in a pubertous female.

The low pH created by the lactic acid produced by lactobacilli serves as an important host defense mechanism in puberty by preventing the growth of potential vaginal pathogens such as *Gardnerella vaginalis, Mobiluncus* spp., *Neisseria gonorrhoeae,* and *S. aureus* (71–74). In addition, lactobacilli help to prevent colonization of potentially pathogenic microorganisms by avidly adhering to receptor sites on the vaginal epithelium, thereby preventing attachment of pathogenic microorganisms and reducing the possibility of infection (75). In addition, up to 98% of lactobacilli may also produce hydrogen peroxide, which has been shown to inactivate human immunodeficiency virus type 1 (HIV-1), herpes simplex virus 2, *Trichomonas vaginalis, G. vaginalis,* and *E. coli* (76, 77). Collectively, the production of lactic acid and hydrogen peroxide by lactobacilli serves as important host defense mechanisms in preventing many vaginal infections.

After the passively transferred estrogen is excreted, the glycogen disappears, with the resultant loss of lactobacilli as the predominant vaginal flora and the increase of pH to a physiologic or slightly alkaline level. At this time, the normal microbial flora is mixed, nonspecific, and relatively scanty and contains organisms derived from the floras of the skin and colon. At puberty, the glycogen reappears in the vaginal epithelium and the adult microbial flora is established. The predominant flora of the vagina in puberty consists of anaerobic bacteria in concentrations of 10^7 to 10^9 CFU/ml of vaginal secretion; these outnumber the aerobic bacteria 100-fold. The major groups of microorganisms represented include lactobacilli, diphtheroids, micrococci, coagulase-negative staphylococci, *Enterococcus faecalis,* microaerophilic and anaerobic streptococci, mycoplasmas, ureaplasmas, and yeasts. During pregnancy, the anaerobic microflora decreases significantly, whereas the numbers of aerobic lactobacilli increase 10-fold (78, 79).

The vaginal flora in postmenopausal women is poorly studied. Specimens are often difficult to obtain from healthy women in this category because they seldom present to a physician unless with some gynecological problem and because the amount of vaginal secretion produced and available for sampling is greatly reduced. However, at least one report (80) documents a significant decrease in lactobacilli in the vaginal flora in postmenopausal women due to the lack of circulating estrogen and the resultant decrease in glycogen in the vaginal mucosa.

VIRULENCE FACTORS AND MECHANISMS

The factors that determine the initiation, development, and outcome of an infection involve a series of complex and shifting interactions between the host and the parasite, which can vary with different infecting microorganisms. In general, humans are able to resist infection by having functional host defense mechanisms. On occasion, defects in host defense mechanisms or exposure to a particularly virulent microbial agent may predispose to the development of an infectious disease. The microbial factors that contribute to the virulence of a microorganism can be divided into three major categories: (i) those that promote colonization of host surfaces, (ii) those that evade the host's immune system and promote tissue invasion, and (iii) those that produce toxins that result in tissue damage in the human host. Pathogenic microorganisms may have any, or all, of these factors.

Colonization Factors

Adherence

Most infections are initiated by the attachment or adherence of the microbe to host tissue, followed by microbial replication to establish colonization. This attachment can be relatively nonspecific or can require the interaction between structures on the microbial surfaces and specific receptors on host cells. This adherence phenomenon is particularly important in the mouth, small intestine, and urinary bladder, where mucosal surfaces are washed continually by fluids. In these areas, only microorganisms that can adhere to the mucosal surface can colonize that site.

Bacteria adhere to tissues by having pili and/or adhesins. Pili or fimbriae are rod-shaped structures that consist primarily of an ordered array of a single protein subunit called pilin. The tip of the pilus mediates adherence of bacteria by attaching to a receptor molecule on the host cell surface that is composed of carbohydrate residues of either glycoproteins or glycolipids. The binding of the pilus to its host target cell can be quite specific and accounts for the tissue tropism associated with certain bacterial infections. Bacterial pili are easily broken and lost and have to be continually regenerated by the bacterium. An important function of pilus replacement, at least for some bacteria, is that it provides a way for the bacterium to evade the host's immune response. Host antibodies that bind to the tips of pili physically block the pili from binding to their host cell targets. Some bacteria can evade this immune defense by growing pili of different antigenic types, thereby rendering the host's immune response ineffective. For example, *N. gonorrhoeae* can produce over 50 pilin types that make it virtually impossible for the host to mount an antibody response that prevents colonization (81).

Bacterial adherence can also be accomplished by a process involving bacterial cell-surface structures known as adhesins and complementary receptors on the surface of host cells. These adhesins, also known as afimbrial adhesins, are proteins that promote the tighter binding of bacteria to host cells following initial binding by pili. The mechanisms used by a microorganism to adhere to a host cell dictate its ability to enter the cell and set in motion a number of physiologic events. An elegant example of microbial attachment followed by a sequence of pathological effects is that of enteropathogenic *E. coli*. Following initial adhesion, intracellular calcium levels increase, activating actin-severing enzymes and protein kinases, which then lead to vesiculation and disruption of the microvilli. The bacteria are then able to attach to the epithelium in a more intimate fashion, allowing maximal activation of protein kinases. This results in major changes to the cytoskeleton and alterations in the permeability of the membrane to ions. Changes in ion permeation result in ion secretion and reduction in absorption, resulting in the secretory diarrhea that is the hallmark of this disease. It has been found that a majority of enteropathogenic *E. coli* isolates contain a large plasmid that codes for its adhesive properties (82).

Biofilms

Microbial biofilms develop when microorganisms adhere irreversibly to a submerged surface and produce extracellular polymers that facilitate adhesion and provide a structural matrix. The surface may be living tissue, such as teeth or mucosal cells, or inert, nonliving material, such as indwelling medical devices that have been inserted into the body. Most biofilms are caused by bacteria, but they can also be caused by fungi, particularly yeast. These biofilms are complex aggregates of extracellular polymers produced by the microorganism growing on a solid animate or inanimate surface that are characterized by a chemical heterogenicity and structural diversity. On human tissue, the first or basal layer of bacteria or yeast attaches directly to the surface of the host cells and other layers of the microorganism are attached to the basal layer by a polysaccharide matrix. Biofilms have been detected in the vagina, mouth, and intestine, and, in fact, the resident microfloras of these sites may largely be organized into biofilms. These dense mats of organisms may help explain the barrier function of these sites in protection of the host. However, the formation of biofilms may also be the prelude to disease. For example, dental plaque is a biofilm that is known to cause disease, such as caries and gingivitis, and *Pseudomonas aeruginosa* has been shown to establish pathogenic biofilms in the lungs of cystic fibrosis patients.

Biofilms may also form on foreign objects that have been implanted in the human host or come in repeated contact with human tissue. Biofilms can develop on virtually any indwelling medical device, such as central venous catheters and needleless connectors, endotracheal tubes, intrauterine devices, mechanical heart valves, pacemakers, prosthetic joints, and urinary catheters. Indeed, hospital-acquired infections in patients with such indwelling medical devices are generally preceded by the formation of a biofilm on the surface of the foreign object. Microorganisms within biofilms are imbedded within the extracellular polymer matrix, which makes them highly resistant to antibiotic treatment. For this reason, individuals with such infections invariably require surgical replacement of the prosthesis or removal of the catheter or central line because these infections are refractory to antimicrobial therapy. Biofilm formation on embedded plastic and stainless steel devices provides yet another example of well-intentioned iatrogenic activities that continue to create new niches for microorganisms to exploit as causes of human infection.

Iron acquisition mechanisms

Once a microorganism adheres to a body site, it has an obligate requirement for iron for its subsequent growth and multiplication. Although the human body contains a plentiful supply of iron, the majority is not easily accessible to microorganisms. The concentration of usable iron is particularly low because lactoferrin, transferrin, ferritin, and hemin bind most of the available iron, and the free iron remaining is far below the level required to support microbial growth (81). Thus, microorganisms have evolved a number of mechanisms for the acquisition of iron from their environments (83). Microorganisms produce siderophores that chelate iron with a very high affinity and that compete effectively with transferrin and lactoferrin to mobilize iron for microbial use. In addition, some microbial species can utilize host iron complexes directly without the production of siderophores. For example, *Neisseria* species possess specific receptors for transferrin and can remove iron from transferrin at the cell surface; *Yersinia pestis* can use heme as a sole source of iron; *Vibrio vulnificus* can utilize iron from the hemoglobin-haptoglobin complex; and *H. influenzae* can use hemoglobin, hemoglobin-haptoglobin, heme-hemopexin, and heme-albumin complexes as iron sources. Another mechanism for iron acquisition is the production of hemolysins, which act to release iron complexed to intracellular heme and hemoglobin.

Motility

Some mucosal surfaces, such as the mouth, stomach, and small intestine, are protected from microbial colonization because they are constantly being washed with fluids. Other mucosal surfaces, such as the colon or vagina, are relatively stagnant areas. In either case, microorganisms that can move directionally toward a mucosal surface will have a better chance of contacting host surfaces than nonmotile organisms. Although motility due to flagella and that due to chemotaxis are appealing candidates as virulence factors, in only a few cases (e.g., *Helicobacter pylori* and *Vibrio cholerae*) has motility been proven to be an important factor for virulence (81).

Evading the Host's Immune System

Capsules

A capsule is a loose, relatively unstructured network of polymers that covers the surface of a microorganism. Most of the well-studied capsules are composed of polysaccharides, but capsules can also be made of proteins or protein–carbohydrate mixtures. The role of capsules in microbial virulence is to protect the organism from complement activation and phagocyte-mediated destruction. Although the host will normally make antibodies directed against the bacterial capsule, some bacteria are able to subvert this response by having capsules that resemble host polysaccharides.

Cryptococcus neoformans is an encapsulated pathogenic fungus. The mechanism by which the capsule of *C. neoformans* enables the organism to evade host defenses is the presentation of a surface not recognized by phagocytes. Although the capsule of *C. neoformans* is a potent activator of the alternative complement pathway, in cryptococcal sepsis, massive activation of complement by capsular polysaccharides can lead to marked depletion of serum complement components and the subsequent loss of serum opsonic capacity. Other immunosuppressive effects that have been attributed to the presence of capsules include downregulation of cytokine secretion, inhibition of leukocyte accumulation, induction of suppressor T cells and suppressor factors, inhibition of antigen presentation, and inhibition of lymphoproliferation.

IgA proteases

Microorganisms that reach mucosal surfaces may often encounter secretory IgA antibody, which can inhibit their adherence and growth on the epithelium. Certain bacteria that reside and/or cause disease on these mucosal surfaces are able to evade the action of secretory antibody by producing IgA proteases that inactivate IgA antibody. The actual role of IgA proteases in virulence is not well understood, and there is some controversy about their importance; however, the unusual specificity of these enzymes suggests that they must play some role in colonization of mucosal surfaces (81). Examples of pathogenic bacteria capable of producing IgA proteases include *H. influenzae, S. pneumoniae, N. meningitidis,* and *N. gonorrhoeae.*

Intracellular residence

Invasive organisms penetrate anatomic barriers and either enter cells or pass through them to disseminate within the body. To survive under these conditions, some organisms have developed special virulence factors that enable them to avoid or disarm host phagocytes. One such antiphagocytic strategy prevents the migration of phagocytes to the site where organisms are growing or limits their effectiveness once there. Some microbes are capable of producing toxic proteins that kill phagocytes once they have arrived, whereas others have developed the ability to survive after phagocytosis by polymorphonuclear cells, monocytes, or macrophages. Strategies for surviving phagocytosis include escaping from the phagosome before it merges with the lysosome, preventing phagosome–lysosome fusion from occurring, or, after fusion, enzymatically dissolving the phagolysosome membrane and escaping. *Toxoplasma gondii* is a classic example of an organism that is a successful intracellular parasite. After entry, *T. gondii* resides within a phagosome vacuole that is permanently made incapable of infusion with other intracellular organelles, including lysosomes. The parasite's survival within this vacuole depends on maintaining the appropriate pH, excluding lysosomal contents, and activating specific mechanisms necessary for nutrient acquisition while contained inside the vacuole (84).

Serum resistance

Resistance to the lytic effects of complement is almost a universal requirement for pathogens that traverse mucosal or skin barriers but remain in the extracellular environment. The lytic effect of serum on Gram-negative organisms is complement mediated and can be initiated by the classical or alternative pathway. One of the principal targets of complement is the lipopolysaccharide (LPS) layer of Gram-negative bacteria. Some pathogens are called "serum resistant" and have evolved defense mechanisms that include (i) failure to bind and activate complement, (ii) shedding of surface molecules that activate the complement system, (iii) interruption of the complement cascade before the formation of the C5b-C9 conplex, and (iv) enhancement in the formation of nonlytic complexes. Many of the microbes that are able to cause systemic infections, such as certain strains of *Salmonella* and *E. coli*, are serum resistant, emphasizing the importance of this trait.

Toxins

Toxins produced by certain microorganisms during growth may alter the normal metabolism of human cells with damaging and sometimes deleterious effects on the host. Toxins are traditionally associated with bacterial diseases, but may also play important roles in diseases caused by fungi, protozoa, and helminths. Two major types of bacterial toxins exist—exotoxins and endotoxins. Exotoxins are proteins that are usually heat labile and are generally secreted into the surrounding medium or tissue. However, some exotoxins are bound to the bacterial surface and are released upon cell death and lysis. In contrast, endotoxins are LPSs of the outer membrane of Gram-negative bacteria.

Exotoxins

Exotoxins are produced by a variety of organisms, including Gram-positive and Gram-negative bacteria, and can cause disease through several mechanisms. First, exotoxins may be produced in and consumed along with food. Disease produced by these exotoxins is generally self-limiting because the bacteria do not remain in the body, thus eliminating the toxin source. Second, bacteria growing in a wound or tissue may produce exotoxins that cause damage to the surrounding tissues of the host, contributing to the spread of infection. Third, bacteria may colonize a wound or mucosal surface and produce exotoxins that enter the bloodstream and affect distant organs and tissues. Toxins that attack a variety of different cell types are called cytotoxins, whereas those that attack specific cell types are designated by the cell type or organ affected, such as a neurotoxin, leukotoxin, or hepatotoxin. Exotoxins can also be named for the species of bacteria that produce them or for the disease with which they are associated, such as cholera toxin, Shiga toxin, diphtheria toxin, and tetanus toxin. Toxins are also named on the basis of their activities, for example, adenylate cyclase and lecithinase, whereas others are simply given letter designations, such as *P. aeruginosa* exotoxin A.

Five major groups of bacterial exotoxins are known, and they are reviewed in detail elsewhere (85, 86). These exotoxins are typically categorized on the basis of their mechanisms of action: they damage cell membranes, inhibit protein synthesis, activate second-messenger pathways, inhibit the release of neurotransmitters, or activate the host immune response. Some of the exotoxins are also known as A-B toxins because the portion of the toxin that binds to a host cell receptor (portion B, or binding portion) is separate from the portion that mediates the enzyme activity responsible for its toxicity (portion A, or active portion). Two structural types of A-B toxins exist. The simplest kind is synthesized as a single protein with a disulfide bond. A more complex type of A-B toxin has a binding portion that is composed of multiple subunits but is still attached to the A portion by the disulfide bond. The disulfide bonds are broken when the B portion binds to a specific host cell-surface molecule and the A portion is transported into the host cell. Thus, the B portion of the molecule determines the host cell specificity of the toxin. For example, if the B portion binds specifically to the cell receptors found only on the surface of neurons, the toxin will be a specific neurotoxin. Generally speaking, without cell receptor specificity, the A portion of these toxins could kill many cell types if it were to gain entry into the cells. Once having entered the host cell, the A portion becomes enzymatically active and exerts its toxic effect. The A portion of most exotoxins affects the cyclic adenosine monophosphate (cAMP) levels in the host cell by ribosylating the protein that controls cAMP. This causes the loss of control of ion flow, which results in the loss of water from the host tissue into the lumen of the intestine, causing diarrhea. Other toxins have A portions that cleave host cell ribosomal RNA (rRNA), thereby shutting down protein synthesis, as occurs with diphtheria toxin (85, 86).

Another type of exotoxin, called membrane-disrupting toxin, lyses host cells by disrupting the integrity of their plasma membranes. There are two types of membrane-disrupting toxins. One is a protein that inserts itself into the host cell membrane by using cholesterol as a receptor and forms channels or pores, allowing cytoplasmic contents to leak out and water to enter. The second type of membrane-disrupting exotoxin consists of phospholipases. These enzymes remove the charged head group from the phospholipids of the cell membrane, which destabilizes the membrane and causes cell lysis. These enzymes are appropriately referred to as cytotoxins.

Some bacterial exotoxins serve as superantigens by acting directly on T cells and antigen-presenting cells (APCs) of the immune system. Impairment of the immunologic functions of these cells by toxin can lead to serious human disease. One large family of toxins in this category is the pyrogenic toxin superantigens, whose important biological activities include potent stimulation of the immune cell system, pyrogenicity, and enhancement of endotoxin shock. Examples of bacterial exotoxins that function as superantigens include the staphylococcal and streptococcal exotoxins that are discussed in detail elsewhere (85–87).

In general, these bacterial superantigens exert their effect by forming a bridge between major histocompatibility complex (MHC) class II of macrophages or other APCs and receptors or T cells that interact with the class II MHC. Normally, APCs process protein antigens by cleaving them into peptides and displaying one of the resulting peptides in a complex with MHC class II on the APC surface. Only a few helper T cells will have receptors that recognize this particular MHC–peptide complex, so only a few T cells will be stimulated. This T-cell simulation causes them to produce cytokines, such as interleukin-2 (IL-2), that stimulate T-cell proliferation and T-cell interaction with B cells, resulting in antibody production by B cells. Superantigens are not processed by proteolytic digestion inside APCs but bind directly to MHC class II on the APC surface. Because superantigens do this indiscriminately, many APCs will have superantigen molecules bound to their surfaces. The superantigen also binds T cells indiscriminately and thus forms many more APC–T helper cell pairs than would normally be found. Thus, instead of APCs stimulating 1 in 10,000 T cells (the normal response to an antigen), as many as 1 in 5 T cells can be stimulated by the bridging action of the superantigens. The superantigen's action causes the release of excessively high levels of IL-2 that produce symptoms of nausea, vomiting, fever, and malaise. Excessive IL-2 production also results in the excess production of other cytokines that can lead to shock (85).

Much more is known about the biology and pathophysiology of exotoxins. Readers who wish additional and more detailed information are referred to the following excellent article and book citations (86, 88–90).

Endotoxin

Endotoxin is the LPS component of the outer membrane of Gram-negative bacteria. Its toxic lipid portion (lipid A) is embedded in the outer membrane, with its core antigen extending outward from the bacterial surface. Endotoxins are heat stable, destroyed by formaldehyde, and relatively less toxic than many exotoxins. Lipid A exerts its effects when bacteria lyse by binding to plasma proteins and then interacting with receptors on monocytes, macrophages, and other host cells, thereby forcing the production of cytokines and the activation of the complement and coagulation cascades. The result of these events is an increase in host body temperature, a decrease in blood pressure, damage to vessel walls, disseminated intravascular coagulation, and a decrease in blood flow to essential organs such as the lung, kidney, and brain, leading to organ failure. Activation of the coagulation cascade leads to insufficiency of clotting components, resulting in hemorrhage and further organ damage. Superantigens can also greatly enhance the host's susceptibility to endotoxic shock by acting synergistically with endotoxin to further augment the release of inflammatory cytokines that are lethal to cells of the immune system (87).

Hydrolytic enzymes

Many pathogenic organisms produce extracellular enzymes such as hyaluronidase, proteases, DNases, collagenase, elastinase, and phospholipases that are capable of hydrolyzing host tissues and disrupting cellular structure. Although not normally considered classic exotoxins, these enzymes can destroy host cells as effectively as exotoxins and are frequently sufficient to initiate clinical disease. For example, *Aspergillus* species secrete a variety of proteases that function as virulence factors by degrading the structural barriers of the host, thereby facilitating the invasion of tissues (91). Other examples are the hyaluronidase and gelatinase enzymes that have been long associated with virulent enterococci. Hyaluronidase-producing enterococci have been implicated as the cause of periodontal disease due to their disruption of the intercellular cementing substances of the epithelium (92). Reports of hyaluronidase in other microorganisms describe it as a spreading factor in *Ancylostoma duodenale* cutaneous larva migrans (93) and as an important factor in the dissemination of *Treponema pallidum* (94).

CONCLUSION

The dynamics of the host–parasite relationship are in a constant state of change throughout life as the balance shifts between states of health and disease. Accordingly, Joshua Lederberg's words continue to resonate today with as much relevance as they did in the 1990s because our human species is still locked in this Darwinian struggle for survival with our microbial and viral predators.

References

1. **Lederberg J, McCray AT.** 2001. 'One sweet 'omics'—a genealogical treasury of words. *Scientist* **15:**8.
2. **The NIH NMP Working Group; Peterson J et al.** 2009. The NIH human microbiome project. *Genome Res* **19:**2317–2323.
3. **NIH/National Human Genome Research Institute.** 2012. Human microbiome project: diversity of human microbes greater than previously predicted. *Science Daily* https://www.sciencedaily.com/releases/2010/05/100520141214.htm.
4. **Wu S, Rhee K-J, Albesiano E, Rabizadeh S, Wu X, Yen H-R, Huso DL, Brancati FL, Wick E, McAllister F, Housseau F, Pardoll DM, Sears CL.** 2009. A human colonic commensal promotes colon tumorigenesis via activation of T helper type 17 T cell responses. *Nat Med* **15:**1016–1022.
5. **Ridaura VK, Faith JJ, Rey FE, Cheng J, Duncan AE, Kau AL, Griffin NW, Lombard V, Henrissat B, Bain JR, Muehlbauer MJ, Ilkayeva O, Semenkovich CF, Funai K, Hayashi DK, Lyle BJ, Martini MC, Ursell LK, Clemente JC, Van Treuren W, Walters WA, Knight R, Newgard CB, Heath AC, Gordon JI.** 2013. Gut microbiota from twins discordant for obesity modulate metabolism in mice. *Science* **341:**1241214.
6. **Turnbaugh PJ, Ley RE, Mahowald MA, Magrini V, Mardis ER, Gordon JI.** 2006. An obesity-associated gut microbiome with increased capacity for energy harvest. *Nature* **444:**1027–1031.

7. **Turnbaugh PJ, Hamady M, Yatsunenko T, Cantarel BL, Duncan A, Ley RE, Sogin ML, Jones WJ, Roe BA, Affourtit JP, Egholm M, Henrissat B, Heath AC, Knight R, Gordon JI.** 2009. A core gut microbiome in obese and lean twins. *Nature* **457:**480–484.
8. **Bravo JA, Forsythe P, Chew MV, Escaravage E, Savignac HM, Dinan TG, Bienenstock J, Cryan JF.** 2011. Ingestion of *Lactobacillus* strain regulates emotional behavior and central GABA receptor expression in a mouse via the vagus nerve. *Proc Natl Acad Sci USA* **108:**16050–16055.
9. **Davis CP.** 1996. Normal flora, p 113–119. *In* Baron S (ed), *Medical Microbiology*, 4th ed. The University of Texas Medical Branch at Galveston, Galveston, TX.
10. **National Human Genome Research Institute (NHGRI).** 2012. NIH human microbiome project defines normal bacterial makeup of the body. National Institutes of Health. https://www.nih.gov/news-events/news-releases/nih-human-microbiome-project-defines-normal-bacterial-makeup-body.
11. **Bokkenheuser VD, Winter J.** 1983. Biotransformation of steroids, p 215. *In* Hentges DJ (ed), *Human Intestinal Microflora in Health and Disease*. Academic Press, New York.
12. **Wilson KH.** 1999. The gastrointestinal biota, p 629. *In* Yamada T, Alpers DH, Laine L, Owyang C, Powell DW (ed), *Textbook of Gastroenterology*, 3rd ed. Lippincott Williams & Wilkins, Baltimore, MD.
13. **Eisenstein BI, Schaechter M.** 1993. Normal microbial flora, p 212. *In* Schaechter M, Medoff G, Eisenstein BI (ed), *Mechanisms of Microbial Disease*, 2nd ed. Williams and Wilkins, Baltimore, MD.
14. **Noble WC.** 1990. Factors controlling the microflora of the skin, p 131–153. *In* Hill MJ, Marsh PD (ed), *Human Microbial Ecology*. CRC Press, Inc, Boca Raton, FL.
15. **McGinley KJ, Webster GF, Ruggieri MR, Leyden JJ.** 1980. Regional variations in density of cutaneous propionibacteria: correlation of Propionibacterium acnes populations with sebaceous secretion. *J Clin Microbiol* **12:**672–675.
16. **Mevissen-Verhage EA, Marcelis JH, de Vos MN, Harmsen-van Amerongen WC, Verhoef J.** 1987. *Bifidobacterium, Bacteroides*, and *Clostridium* spp. in fecal samples from breast-fed and bottle-fed infants with and without iron supplement. *J Clin Microbiol* **25:**285–289.
17. **Kloos WE.** 1986. Ecology of human skin, p 37–50. *In* Maardh PA, Schleifer KH (ed), *Coagulase-Negative Staphylococci*. Almyqvist & Wiksell International, Stockholm, Sweden.
18. **Kloos WE.** 1997. Taxonomy and systematics of staphylococci indigenous to humans, p 113–137. *In* Crossley KB, Archer GL (ed), *The Staphylococci in Human Disease*. Churchill Livingstone, New York.
19. **Kloos WE.** 1998. Staphylococcus, p 577–632. *In* Collier L, Balows A, Sussman M (ed), *Topley & Wilson's Microbiology and Microbial Infections*, 9th ed, vol 2. Edward Arnold, London.
20. **Kloos WE, Schleifer KH, Gotz F.** 1991. The genus *Staphylococcus*, p 1369–1420. *In* Balows A, Truper HG, Dworkin M, Harder W, Schleifer KH (ed), *The Prokaryotes*, 2nd ed. Springer-Verlag, New York.
21. **Wolff HH, Plewig G, Januschke E.** 1976. Ultrastruktur der Mikroflora in Follikeln und Komedonen. [Ultrastructure and microflora in follicles and comedones.] *Hautarzt* **27:**432–440.
22. **Leeming JP, Holland KT, Cunliffe WJ.** 1984. The microbial ecology of pilosebaceous units isolated from human skin. *J Gen Microbiol* **130:**803–807.
23. **Evans CA.** 1976. The microbial ecology of human skin, p 121–128. *In* Stiles HM, Loesche WJ, O'Brien TC (ed), *Microbial Aspects of Dental Caries,* vol. 1 (special supplement to *Microbiology Abstracts—Bacteriology)*. Information Retrievable, Inc., New York.
24. **Gallis HA.** 1988. Normal flora and opportunistic infections, p 339. *In* Joklik WK, Willett HP, Amos DB, Wilfert CM (ed), *Zinsser Microbiology*, 19th ed. Appleton & Lange, Norwalk, CT.
25. **Sherris JC.** 1984. Normal microbial flora, p 50–58. *In* Sherris JC, Ryan KJ, Ray CG, Plorde JJ, Corey L, Spizizen J (ed), *Medical Microbiology: an Introduction to Infectious Diseases*. Elsevier Science Publishing, New York.
26. **Socransky SS, Manganiello SD.** 1971. The oral microbiota of man from birth to senility. *J Periodontol* **42:**485–496.
27. **Tannock GW, Fuller R, Smith SL, Hall MA.** 1990. Plasmid profiling of members of the family *Enterobacteriaceae*, lactobacilli, and bifidobacteria to study the transmission of bacteria from mother to infant. *J Clin Microbiol* **28:**1225–1228.
28. **Tagg JR, Pybus V, Phillips LV, Fiddes TM.** 1983. Application of inhibitor typing in a study of the transmission and retention in the human mouth of the bacterium *Streptococcus salivarius*. *Arch Oral Biol* **28:**911–915.
29. **Carlsson J, Grahnén H, Jonsson G.** 1975. Lactobacilli and streptococci in the mouth of children. *Caries Res* **9:**333–339.
30. **Ikeda T, Sandham HJ.** 1971. Prevalence of *Streptococcus mutans* on various tooth surfaces in Negro children. *Arch Oral Biol* **16:**1237–1240.
31. **Ikeda T, Sandham HJ, Bradley EL Jr.** 1973. Changes in *Streptococcus mutans* and lactobacilli in plaque in relation to the initiation of dental caries in Negro children. *Arch Oral Biol* **18:**555–566.
32. **Köhler B, Andréen I, Jonsson B.** 1984. The effect of caries-preventive measures in mothers on dental caries and the oral presence of the bacteria *Streptococcus mutans* and lactobacilli in their children. *Arch Oral Biol* **29:**879–883.
33. **Delaney JE, Ratzan SK, Kornman KS.** 1986. Subgingival microbiota associated with puberty: studies of pre-, circum-, and postpubertal human females. *Pediatr Dent* **8:**268–275.
34. **Frisken KW, Tagg JR, Laws AJ, Orr MB.** 1987. Suspected periodontopathic microorganisms and their oral habitats in young children. *Oral Microbiol Immunol* **2:**60–64.
35. **Moore LVH, Moore WE, Cato EP, Smibert RM, Burmeister JA, Best AM, Ranney RR.** 1987. Bacteriology of human gingivitis. *J Dent Res* **66:**989–995.
36. **Wojcicki CJ, Harper DS, Robinson PJ.** 1987. Differences in periodontal disease-associated microorganisms of subgingival plaque in prepubertal, pubertal and postpubertal children. *J Periodontol* **58:**219–223.
37. **Theilade E.** 1990. Factors controlling the microflora of the healthy mouth, p 1–54. *In* Hill MJ, Marsh PD (ed), *Human Microbial Ecology*. CRC Press, Inc, Boca Raton, FL.
38. **Bartholomew B, Hill MJ.** 1984. The pharmacology of dietary nitrate and the origin of urinary nitrate. *Food Chem Toxicol* **22:**789–795.
39. **Parsons DS.** 1971. Salt transport. *J Clin Pathol* **24**(Suppl. 5)**:**90–98.
40. **Drasar BS, Shiner M, McLeod GM.** 1969. Studies on the intestinal flora. I. The bacterial flora of the gastrointestinal tract in healthy and achlorhydric persons. *Gastroenterology* **56:**71–79.
41. **Gorbach SL.** 1971. Intestinal microflora. *Gastroenterology* **60:** 1110–1129.
42. **Gorbach SL, Nahas L, Lerner PI, Weinstein L.** 1967a. Studies of intestinal microflora. I. Effects of diet, age, and periodic sampling on numbers of fecal microorganisms in man. *Gastroenterology* **53:**845–855.
43. **Gorbach SL, Plaut AG, Nahas L, Weinstein L, Spanknebel G, Levitan R.** 1967b. Studies of intestinal microflora. II. Microorganisms of the small intestine and their relations to oral and fecal flora. *Gastroenterology* **53:**856–867.
44. **Roberts AK.** 1988. The development of the infant faecal flora. Ph.D. thesis. Council for National Academic Awards.
45. **Neut C, Bezirtzoglou E, Romand C, Beeren H, Delcroix M, Noel AM.** 1987. Bacterial colonization of the large intestine in newborns delivered by caesarian section. *Zentralbl Bakteriol. Hyg A* **266:**330–337.
46. **Keusch GT, Gorbach SL.** 1995. Enteric microbial ecology and infection, p 1115–1130. *In* Haubrich WS, Schaffner F, Berk JE (ed), *Gastroenterology*, 5th ed. W. B. Saunders Co, Philadelphia.

47. **Benno Y, Sawada K, Mitsuoka T.** 1984. The intestinal microflora of infants: composition of fecal flora in breast-fed and bottle-fed infants. *Microbiol Immunol* **28:**975–986.
48. **Hall MA, Cole CB, Smith SL, Fuller R, Rolles CJ.** 1990. Factors influencing the presence of faecal lactobacilli in early infancy. *Arch Dis Child* **65:**185–188.
49. **Smith HW, Crabb WE.** 1961. The faecal bacterial flora of animals and man: its development in the young. *J Pathol Bacteriol* **82:**53–66.
50. **Mata LJ, Urrutia JJ.** 1971. Intestinal colonization of breast-fed children in a rural area of low socioeconomic level. *Ann NY Acad Sci* **176:**93–109.
51. **Mitsuoka T, Kaneuchi C.** 1977. Ecology of the bifidobacteria. *Am J Clin Nutr* **30:**1799–1810.
52. **Donaldson RM Jr.** 1964. Normal bacterial populations of the intestine and their relationship to intestinal function. *N Engl J Med* **270:**938–945, 994–1001, 1050–1056.
53. **Stephen AM, Cummings JH.** 1980. The microbial contribution to human faecal mass. *J Med Microbiol* **13:**45–56.
54. **Guarner F, Malagelada JR.** 2003. Gut flora in health and disease. *Lancet* **361:**512–519.
55. **Sears CL.** 2005. A dynamic partnership: celebrating our gut flora. *Anaerobe* **11:**247–251.
56. **Steinhoff U.** 2005. Who controls the crowd? New findings and old questions about the intestinal microflora. *Immunol Lett* **99:**12–16.
57. **O'Hara AM, Shanahan F.** 2006. The gut flora as a forgotten organ. *EMBO Rep* **7:**688–693.
58. **Gibson GR.** 2004. Fibre and effects on probiotics (the prebiotic concept). *Clin Nutr Suppl* **1:**25–31.
59. **Beaugerie L, Petit J-C.** 2004. Antibiotic-associated diarrhoea. *Best Pract Res Clin Gastroenterol* **18:**337–352.
60. **Cregan J, Hayward NJ.** 1953. The bacterial content of the healthy small intestine. *BMJ* **1:**1356–1359.
61. **Finegold SM, Sutter VL, Mathison GE.** 1983. Normal indigenous intestinal flora, p 3–31. *In* Hentges DJ (ed), *Human Intestinal Microflora in Health and Disease.* Academic Press, New York.
62. **Justesen T, Nielsen OH, Jacobsen IE, Lave J, Rasmussen SN.** 1984. The normal cultivable microflora in upper jejunal fluid in healthy adults. *Scand J Gastroenterol* **19:**279–282.
63. **Plaut AG, Gorbach SL, Nahas L, Weinstein L, Spanknebel G, Levitan R.** 1967. Studies of intestinal microflora. 3. The microbial flora of human small intestinal mucosa and fluids. *Gastroenterology* **53:**868–873.
64. **Simon GL, Gorbach SL.** 1984. Intestinal flora in health and disease. *Gastroenterology* **86:**174–193.
65. **MacNeal WJ, Latzer LL, Kerr JE.** 1909. The fecal bacteria of healthy men. I. Introduction and direct quantitative observations. *J Infect Dis* **6:**123–169.
66. **Birkbeck J.** 1999. Colon cancer: the potential involvement of the normal flora, p 262–294. *In* Tannock GW (ed), *Medical Importance of the Normal Flora.* Kluwer Academic Publishers, London.
67. **Shanahan F.** 2002. The host-microbe interface within the gut. *Best Pract Res Clin Gastroenterol* **16:**915–931.
68. **Björkstén B, Sepp E, Julge K, Voor T, Mikelsaar M.** 2001. Allergy development and the intestinal microflora during the first year of life. *J Allergy Clin Immunol* **108:**516–520.
69. **Guarner F, Malagelada J-R.** 2003. Role of bacteria in experimental colitis. *Best Pract Res Clin Gastroenterol* **17:**793–804.
70. **Ison CA.** 1990. Factors affecting the microflora of the lower genital tract of healthy women, p 111–130. *In* Hill MJ, Marsh PD (ed), *Human Microbial Ecology.* CRC Press, Inc, Boca Raton, FL.
71. **Graver MA, Wade JJ.** 2011. The role of acidification in the inhibition of *Neisseria gonorrhoeae* by vaginal lactobacilli during anaerobic growth. *Ann Clin Microbiol Antimicrob* **10:**8.
72. **Matu MN, Orinda GO, Njagi ENM, Cohen CR, Bukusi EA.** 2010. In vitro inhibitory activity of human vaginal lactobacilli against pathogenic bacteria associated with bacterial vaginosis in Kenyan women. *Anaerobe* **16:**210–215.
73. **Skarin A, Sylwan J.** 1986. Vaginal lactobacilli inhibiting growth of *Gardnerella vaginalis, Mobiluncus* and other bacterial species cultured from vaginal content of women with bacterial vaginosis. *Acta Pathol Microbiol Immunol Scand [B]* **948:**399–403.
74. **Strus M, Malinowska M, Heczko PB.** 2002. In vitro antagonistic effect of *Lactobacillus* on organisms associated with bacterial vaginosis. *J Reprod Med* **47:**41–46.
75. **Boris S, Barbés C.** 2000. Role played by lactobacilli in controlling the population of vaginal pathogens. *Microbes Infect* **2:**543–546.
76. **O'Hanlon DE, Moench TR, Cone RA.** 2011. In vaginal fluid, bacteria associated with bacterial vaginosis can be suppressed with lactic acid but not hydrogen peroxide. *BMC Infect Dis* **11:**200.
77. **Baeten JM, Hassan WM, Chohan V, Richardson BA, Mandaliya K, Ndinya-Achola JO, Jaoko W, McClelland RS.** 2009. Prospective study of correlates of vaginal *Lactobacillus* colonisation among high-risk HIV-1 seronegative women. *Sex Transm Infect* **85:**348–353.
78. **Goplerud CP, Ohm MJ, Galask RP.** 1976. Aerobic and anaerobic flora of the cervix during pregnancy and the puerperium. *Am J Obstet Gynecol* **126:**858–868.
79. **Lindner JGEM, Plantema FHF, Hoogkamp-Korstanje JAA.** 1978. Quantitative studies of the vaginal flora of healthy women and of obstetric and gynaecological patients. *J Med Microbiol* **11:**233–241.
80. **Cruikshank R, Sharman A.** 1934. The biology of the vagina in the human subject. II. The bacterial flora and secretion of the vagina at various age periods and their relation to glycogen in the vaginal epithelium. *J Obstet Gynaecol Br Emp* **32:**208–226.
81. **Salyers AA, Whitt DD.** 1994. Virulence factors that promote colonization, p 30–46. *In* Salyers AA, Whitt DD (ed), *Bacterial Pathogenesis: a Molecular Approach.* ASM Press, Washington, DC.
82. **Baldini MM, Kaper JB, Levine MM, Candy DCA, Moon HW.** 1983. Plasmid-mediated adhesion in enteropathogenic Escherichia coli. *J Pediatr Gastroenterol Nutr* **2:**534–538.
83. **Litwin CM, Calderwood SB.** 1993. Role of iron in regulation of virulence genes. *Clin Microbiol Rev* **6:**137–149.
84. **Schaechter M, Eisenstein BI.** 1993. Genetics of bacteria, p 57–76. *In* Schaechter M, Medoff G, Eisenstein BI (ed), *Mechanisms of Microbial Disease,* 2nd ed. Williams & Wilkins, Baltimore, MD.
85. **Salyers AA, Whitt DD.** 1994. Virulence factors that damage the host, p 47–60. *In* Salyers AA, Whitt DD (ed), *Bacterial Pathogenesis: a Molecular Approach.* ASM Press, Washington, DC.
86. **Schmitt CK, Meysick KC, O'Brien AD.** 1999. Bacterial toxins: friends or foes? *Emerg Infect Dis* **5:**224–234.
87. **Kotb M.** 1995. Bacterial pyrogenic exotoxins as superantigens. *Clin Microbiol Rev* **8:**411–426.
88. **Brogden KA, Roth JA, Stanton TB, Bolin CA, Minion FC, Wannemuehler MJ (ed).** 2000. *Virulence Mechanisms of Bacterial Pathogens.* ASM Press, Washington, DC.
89. **Cossart P, Boquet P, Normark S, Rappuoli R (ed).** 2000. *Cellular Microbiology.* ASM Press, Washington, DC.
90. **Salyers AA, Whitt DD (ed).** 1994. *Bacterial Pathogenesis: a Molecular Approach.* ASM Press, Washington, DC.
91. **Kothary MH, Chase T Jr, Macmillan JD.** 1984. Correlation of elastase production by some strains of *Aspergillus fumigatus* with ability to cause pulmonary invasive aspergillosis in mice. *Infect Immun* **43:**320–325.
92. **Rosan B, Williams NB.** 1964. Hyaluronidase production by oral enterococci. *Arch Oral Biol* **9:**291–298.
93. **Hotez PJ, Narasimhan S, Haggerty J, Milstone L, Bhopale V, Schad GA, Richards FF.** 1992. Hyaluronidase from infective *Ancylostoma* hookworm larvae and its possible function as a virulence factor in tissue invasion and in cutaneous larva migrans. *Infect Immun* **60:**1018–1023.
94. **Fitzgerald TJ, Repesh LA.** 1987. The hyaluronidase associated with *Treponema pallidum* facilitates treponemal dissemination. *Infect Immun* **55:**1023–1028.

2

Indigenous Zoonotic Agents of Research Animals

LON V. KENDALL

Laboratory animals have played a major role in advancing biomedical research and will continue to be important for identifying fundamental mechanisms of disease and exploring the efficacy and safety of novel therapies. The health status of these animals can have a direct impact on the validity and value of the research results, as well as on the health and safety of those who work with them. Husbandry practices are established in reputable laboratories to protect both the animals and the personnel that work with them.

WORKING WITH ZOONOTIC AGENTS

Biosecurity and Biocontainment

"Biosecurity" is the term commonly used when referring to maintaining the health status of a research animal and is a conscious effort to detect, prevent, contain, and eradicate adventitious agents in laboratory animals (1). This differs from the definition used when referring to biosecurity in the context of Select Agents, in which biosecurity represents the security measures designed to prevent the loss, theft, misuse, diversion, or intentional release of pathogens and toxins (2). Biocontainment is the conscious effort to contain and prevent exposures of personnel and the environment to biohazardous agents. Containment is typically maintained through four primary control points—engineering controls, personal protective equipment (PPE), standard operating procedures (SOPs), and administrative controls (2). In this context, the biohazardous agents are often a known entity, such as a culture flask of *Yersinia pestis* or an animal model infected with *Burkholderia pseudomallei*. Another common term in animal facilities is "barrier." Barriers are typically designed to prevent the unwanted entry of agents into the animal facility and maintain animals in a pathogen-free status (3). Many of the practices used for biosecurity, biocontainment, or barrier operations are synonymous with a different purpose. For example, personnel working with animals maintained in a barrier facility typically wear personal protective clothing, such as laboratory coat, gloves, hair net, respiratory mask, and shoe covers, to minimize the fomite potential of the person working with the animals. The same personal protective clothing used in biocontainment has the primary function of preventing exposure from the animals to the personnel. As a general rule, laboratory animal facilities operate at an animal biosafety level 2 (BSL2) as outlined in the *Biosafety in Microbiological and Biomedical Laboratories,* 5th ed. (BMBL), which includes use of SOPs, appropriate training of individuals, use of PPE, minimization of aerosol productions through the use of cage change stations, and hygiene and sanitation practices (2) to maintain biosecurity, reduce personnel exposure to allergens, and maintain the health of the research animals. Because the practices of operating to maintain biosecurity are similar to biocontainment, they also protect personnel from the potential zoonotic agents of laboratory animals.

Zoonotic Potential

Zoonoses are diseases that are naturally transmitted between vertebrate animals and humans. Working with laboratory animals has inherent risks, because zoonotic agents can be transmitted via skin contact, inhalation, ingestion, and ocular exposure (4, 5). Although the most common occupational hazard when working with laboratory animals is allergen exposure, and developing a significant allergic response to animal dander, hair, saliva, urine, serum, bedding, or other allergens affects up to 44% of animal workers (6), other hazards, such as potential exposure to zoonotic diseases, are possible. Over the past several decades, institutions have taken several steps to reduce the potential for zoonotic disease. Improvements in laboratory animal husbandry and science have reduced the number of pathogens in or on animals in the laboratory setting. Most research animal facilities in the United States categorize their animals into a hierarchy of potential hazard on the basis of the animal species, source, and health quality.

The highest index of suspicion as a source for zoonotic hazards is in first-generation wild-caught animals. Additional weight is given to non-human primate (NHP) species due to their close phylogenetic relationship to humans and the diseases they have in common (7). In this hierarchy of potential hazard, the wild-caught animals are followed by random-source animals or known-source animals that have not been raised in a controlled, disease-limited environment under an adequate program of veterinary care. The source of the animals and factors contributing to the likelihood of their exposure to zoonotic pathogens through contact with other animal populations or conspecifics (same species) with endemic infections must be taken into account with both wild-caught and random-source animals. Research programs involved with the maintenance of these animals require effective programs of disease detection, diagnosis, treatment, control, and prevention. There is a special obligation to investigate the particular zoonotic hazards that might be associated with the use of wild mammals or birds, or their fresh carcasses, before initiating full-scale research or teaching efforts involving these species.

The majority of rodents used in biomedical research and teaching are acquired from vendors that maintain biosecurity to produce pathogen-free animal models. Potential zoonotic hazards are associated with many laboratory animals, but the actual transmission of zoonotic disease has become uncommon due to the increased use of animals specifically bred for research over many generations. Such animals generally represent a reduced hazard, with a few exceptions. The majority of small laboratory animals, such as mice, rats, and rabbits, used in research in the United States have been produced commercially in highly controlled environments under the oversight of a rigorous veterinary care program. Due to extended disease surveillance and eradication efforts in these settings, these animal species now have very few or none of the zoonotic diseases associated with their wild counterparts.

These animals are regularly screened for the presence of adventitious agents, including potential zoonotic agents (Table 1). Many of the large species, such as cats, dogs, and ferrets, are acquired from purpose-bred facilities that maintain strict biosecurity to minimize disease in their colonies. However, there are still some species, particularly the larger species, such as NHPs, wild-caught animals, and acquired livestock, that pose the greatest zoonotic threat (8, 9). Many institutions prescreen animals when zoonotic diseases are a concern. For example, those that use sheep and goat models will frequently prescreen them for *Coxiella burnetii* prior to shipment to the facility and will not accept known positive animals. With enhanced biosecurity practices and the use of pathogen-free and prescreened animals prior to arrival, the risk of acquiring a naturally occurring zoonotic disease while working with laboratory animals is very low. Nonetheless, natural pathogens can still enter the animal facility by way of research personnel, husbandry staff, newly arrived animals, insects, and vermin.

Infection indicates the presence of microbes that may be pathogens, opportunists, or commensals. However, infection is not synonymous with disease. The inapparent but significant effects of microorganisms in animals that appear to be normal and healthy may actually render them unsuitable as research subjects (10). A list of the more common adventitious agents infecting laboratory rodents is presented in Table 1 (11). These agents typically result in subclinical disease, but they can have a significant, negative impact on research results. None of the agents listed in Table 1 are considered zoonotic diseases, and the reader is referred to more authoritative text on diseases of laboratory animals for additional information (10, 12).

TABLE 1.

Natural pathogens of common laboratory animals[a]

Species	Agent
Cats	*Chlamydia felis*
	Dermatophytes[b]
	Intestinal nematodes
	Fleas
	Feline calicivirus
	Feline coronavirus
	Feline herpesvirus
	Feline immunodeficiency virus
	Feline leukemia virus
	Feline parvovirus
Mice and rats	Cilia-associated respiratory bacillus
	Citrobacter rodentium
	Clostridium piliforme
	Corynebacterium kutscheri
	Corynebacterium bovis
	Helicobacter spp.
	Klebsiella pneumoniae
	Mycoplasma pulmonis
	Pasteurella pneumotropica
	Pseudomonas aeruginosa
	Salmonella enterica[b]
	Staphylococcus aureus
	Streptococcus pneumoniae
	Pneumocystis carinii
	Acariasis
	Encephalitozoon cuniculi
	Intestinal protozoa
	Oxyurids, pinworms
	Adenovirus
	Cytomegalovirus
	Ectromelia virus
	Parvovirus (MVM, MPV, RPV, H-1, KRV)
	Lymphocytic choriomeningitis virus[b]
	Coronavirus (MHV, RCV)
	Rotavirus
	Reovirus 3
	Sendai virus
	Theiler's murine encephalomyelitis virus
Swine	*Actinobacillus pleuropneumoniae*
	Bordetella bronchiseptica
	Clostridium perfringens type C
	Erysipelothrix rhusiopathiae
	Haemophilus parasuis
	Lawsonia intracellularis
	Leptospira spp.
	Mycoplasma hyopneumoniae
	Pasteurella multocida[b]
	Streptococcus suis[b]
	Intestinal helminths
	Intestinal protozoa
	Encephalomyocarditis virus
	Hemagglutinating encephalomyelitis virus
	Porcine circovirus
	Porcine enteroviruses
	Porcine parvovirus
	Porcine rotavirus
Swine (cont.)	Swine herpesvirus, pseudorabies
	Swine influenza virus
	Transmissible gastroenteritis virus
	Porcine respiratory virus
Dogs	*Bordetella bronchiseptica*
	Brucella canis[b]
	Campylobacter jejuni[b]
	Dermatophytes[b]
	Malassezia pachydermatis
	Cryptosporidium parvum[b]
	Intestinal nematodes
	Dirofilaria immitis
	Intestinal protozoa
	Canine adenovirus
	Canine coronavirus
	Canine distemper virus
	Parainfluenza virus 2
	Canine parvovirus
Non-human primates	*Campylobacter* spp.[b]
	Shigella flexneri[b]
	Streptococcus pneumoniae
	Balantidium coli[b]
	Entamoeba histolytica[b]
	Intestinal nematodes
	Herpesvirus (B virus)[b]
	Respiratory syncytial virus[b]
	Rotavirus
	Simian hemorrhagic fever virus
	Simian immunodeficiency viruses
	Simian retrovirus type D
	Simian T-cell leukemia virus
Rabbits	*Bordetella bronchiseptica*
	Cilia-associated respiratory bacillus
	Clostridium piliforme
	Clostridium spiroforme
	Francisella tularensis
	Listeria monocytogenes
	Pasteurella multocida
	Staphylococcus aureus
	Treponema paraluis-cuniculi
	Dermatophytes
	Cheyletiella parasitivorax
	Cryptosporidium parvum
	Encephalitozoon cuniculi
	Hepatic coccidiosis
	Intestinal coccidiosis
	Intestinal helminths
	Adenovirus
	Cottontail rabbit papillomavirus
	Lapine parvovirus
	Myxoma virus
	Rabbit enteric coronavirus
	Rabbit hemorrhagic disease virus
	Rabbit oral papillomavirus
	Rotavirus

[a]Adapted from references 10 and 12.
[b]Potential zoonotic.

TABLE 2.

Centers for Disease Control and Prevention notifiable diseases in the United States that are zoonotic (of the 62 notifiable infectious conditions, 11 are zoonotic)[a]

Anthrax
Brucellosis
Cryptosporidiosis
Giardiasis
Hantavirus pulmonary syndrome
Psittacosis
Q fever
Rabies
Salmonellosis
Shigellosis
Tuberculosis

[a]Adapted from CDC (13).

Occupational Acquired Infections

Many of the zoonotic agents encountered in laboratory animals are not reportable, which makes assessing the exposure risk difficult. The Centers for Disease Control and Prevention (CDC) lists 62 nationally notifiable infectious conditions, and only 11 of those are identified as conditions that may be acquired from laboratory animals (Table 2) (13, 14). Successful transmission of a zoonotic pathogen requires three key elements: a source of the agent, a susceptible host, and a means to transmit disease (15). Aside from an infected host, zoonotic agents may be present in the environment where infected animals are housed. These potential sources of infection may originate from contaminated walls, floors, cages, bedding, equipment, supplies, feed, and water. Careful consideration should be aimed at minimizing the risk of exposure to these areas as well as to the animals.

Host susceptibility can be influenced by a number of variables, including vaccination status, underlying illness, immunosuppression, and pregnancy. None of these conditions is solely adequate to preclude individuals from working with animals, because adequate control measures are usually available.

Transmission of these zoonotic agents may occur by three possible means. The first is contact transmission, which occurs through ingestion, cutaneous, percutaneous, or mucous membrane exposure. The second is aerosol transmission in which the agent is transferred through the air and deposited on the mucous membranes or in the respiratory tract. The third is vector-borne transmission, which is unlikely to occur in a laboratory setting where natural vectors of diseases are typically not present.

The prevalence of zoonotic disease transmission in the laboratory animal and veterinary settings is difficult to determine, as many are not reportable. However, there have been some surveys conducted in an attempt to gauge the prevalence of such infections (4, 5). The most recent national survey of laboratory animal workers regarding zoonotic disease was performed in 2004 (4). Of the 1,367 responses evaluated, 23 people reported 28 cases of infection with a zoonotic disease within the past 5 years. On the basis of statistical analysis, this equated to approximately 45 cases per 10,000 worker-years at risk. The most frequent zoonotic exposures were dermatophytes (ringworm, 9 out of 28 reported cases). The following agents were identified with no more than two cases of each reported: *Coxiella burnetii, Giardia* spp., *Pasteurella* spp., *Mycobacterium* spp., *Clostridium difficile,* cat scratch disease, ectoparasites, influenza virus, rhinovirus, simian foamy virus, and herpes B virus. Laboratory rodents were the most common source of zoonotic disease reported (17%), followed by dogs, cats, and NHPs (14% each). The primary routes of infection in this report were skin contact (39%), animal bites (18%), inhalation (14%), and other routes, such as splashes to the mucous membranes (7%), needle stick (4%), and unspecified (11%).

The incidence of nonfatal occupational injuries for veterinary assistants and laboratory animal care workers in 2011 was 366/10,000, with 113/10,000 related to incidents involving animals or insects; however, zoonotic diseases are not specifically identified (16). A 2012 survey of veterinarians in Oregon reported several zoonoses, including dermatophytes (54%), *Giardia* spp. (13%), cat scratch disease (15%), cryptosporidiosis (7%), and the following at less than 5%—sarcoptic mange, roundworm infestation, campylobacteriosis, listeriosis, salmonellosis, leptospirosis, pasteurellosis, tularemia, psittacosis, brucellosis, coxiellosis (Q fever), tuberculosis, toxoplasmosis, and histoplasmosis. Cats accounted for over 55% of the exposures, followed by cattle (13%) and dogs (11%) and to a lesser extent birds, horses, small ruminants, and other animal species (5). These were reported exposures that occurred during the veterinarian's career. The incidence of zoonotic disease identified from laboratory animals appears to be low, similar to the incidence of laboratory-acquired infections in health care professionals. A 1986 survey of hospital workers suggested 3.5/1,000 full-time equivalents workers infections occurred, and in a 1995 survey of hospitals in the United Kingdom, the rate was 18/100,000 worker years (17, 18). Although the incidence of zoonotic diseases acquired in the laboratory setting is rare, good practices should be implored to reduce the risk further, particularly when working with larger species. The National Association of State Public Health Veterinarians published a compendium of veterinary standard precautions for preventing zoonotic diseases (15), which highlights infection control practices to minimize exposure to infectious materials. Many of these practices can be readily translated to the laboratory environment, and are consistent with animal biosafety level 2 (ABSL2) practices outlined in BMBL (2).

Reducing Zoonotic Risks

The standard precautions to minimize exposure are based on personal hygiene, protective clothing, and preventing animal-related injuries (15). The majority of laboratory animal facilities work with animals at an ABSL2 level as outlined in BMBL (2). These precautions involve the use of dedicated PPE, which typically includes a laboratory coat and gloves as a minimum standard. Gloves provide an additional layer of protection to the skin and should be worn when coming into contact with animals, their bodily fluids, and caging. They should be changed between groups of animals and between clean and dirty procedures. Once gloves are removed, hands should be thoroughly washed with antimicrobial soap and water to remove agents mechanically and reduce their ability to replicate. This could be followed by a 60–95% alcohol-based hand sanitizer. This should always be performed after working with animals and removing gloves. Sleeved garments are recommended to be used with gloves, such as a laboratory coat or long-sleeved scrubs dedicated to animal care. This will minimize the chances for the skin on the forearms to become exposed to bites or scratches. When using long-sleeved garments, it is best to have a cuffed sleeve to keep the sleeve in position and to prevent smaller rodents from escaping up the arm.

Facial protection should be used whenever potential for splashes or sprays may occur to prevent the exposure of mucous membranes of the eyes, nose, and mouth. Surgical masks may provide some protection against mucous membrane exposure, but they do not provide adequate respiratory protection from particulate antigens such as viruses and bacteria. Respirators certified by the National Institute for Occupational Safety and Health (NIOSH) are typically worn in laboratory animal facilities to minimize the exposure to respiratory pathogens. Personnel wearing approved N95 respirators in the animal facility should be medically cleared and properly fitted for optimal effectiveness. Other items, such as footwear or foot protection and head covers, may be used by research facilities as a standard of practice when working with animals for biosecurity or biocontainment purposes to further reduce the risk of disease transmission (15).

Precautions should also be made to prevent animal-related injuries (15). Personnel should use appropriate practices to handle animals to minimize their stress and hence minimize their desire to bite or scratch. Physical restraints, bite-resistant gloves, acclimation, and training are all methods to minimize animal-related injuries. One common practice in laboratory animal facilities for handling animals infected with an ABSL3 agent is to use tongs or forceps to transfer mice during a cage change. The forceps or tongs are dipped in disinfectant, and the mice are grasped by the base of the tail and placed in a new cage. This virtually eliminates the risk of the animal handler getting bitten by the mouse during the cage change. For larger species, methods could be used to train the animals for handling or gentle restraint. This is particularly useful when working with NHPs, dogs, cats, and other larger species.

Environmental infection control is another key aspect to consider to minimize the exposure to zoonotic agents. Routine cleaning and disinfection are common practices in the laboratory animal facility for both biosecurity and biocontainment practices. The disinfectant used should be specific for the pathogen and used according to the label with proper contact time. The most common disinfectants used in the laboratory animal facility are quaternary ammonium compounds, hydrogen peroxide-based compounds, and chlorine-based compounds. Animal holding facilities are designed with materials that are nonporous and easily cleaned to help facilitate disinfection. In addition, they are designed to replace the air very frequently, having 10–15 air changes per hour. This reduces the contaminants (allergens, odors, agents) in the room. Laboratory animal facilities provide a dedicated break room for eating and storing human food. Food or drink for personnel is never permitted in a laboratory animal facility.

A properly managed laboratory animal facility has a robust and highly functional occupational health program. These programs identify the risks for personnel working with animals and the means to mitigate the risk. When working with a known infectious agent, particularly one that is zoonotic, the risk assessment should consider a number of variables. The characteristics of the agents should be evaluated, such as the dose-response relationship, virulence, communicability, prevalence, route of exposure, and shedding, as well as the stability of the agent in the environment and its susceptibility to disinfection. Additional consideration should be given on the basis of the availability of prophylaxis or therapy to the agent (7, 9). A medical questionnaire and physician consult are a key component of the program. Depending on the risk assessment, vaccinations may be required or recommended. Many facilities manage and document exposure as required by Occupational Safety and Health Administration (OSHA). This provides management with an excellent tool to help identify areas that are in need of improvement to prevent future exposures. Critical to an infection control program are staff training and education. The staff working with animals needs to be properly educated and trained on how to handle animals on a daily basis, understanding the risks associated with working with animals and steps to mitigate those risks to protect themselves. In addition to environmental and administrative controls, the exposure to zoonotic agents in the laboratory environment can be reduced with proper personal protective clothing and equipment, animal handling technique, and hand washing.

Despite the rarity of encountering zoonotic agents in contemporary research animal colonies, several specific agents have persisted, and there are instances where deficiencies in organizational procedures, facility design, and equipment have contributed to the reintroduction of zoonotic pathogens into the laboratory environment. Specific examples of agents that have continued to persist due to ineffective elimination efforts are *Macacine herpesvirus* (herpes B virus of macaque monkeys), *C. burnetii* (Q fever) among small ruminants, and cat scratch disease fever in domestic felines. Lymphocytic choriomeningitis is a viral zoonosis that has been introduced into healthy animal colonies through contaminated tumors and cell lines and through entry of feral animals (9). In a contemporary research animal environment dominated by expensive, sensitive, and often irreplaceable, genetically manipulated mutant mice, the impact of colony contamination of this type can be devastating. Even with the best veterinary support, the technical difficulties and expenses related to the detection of chronic subclinical or latent infections can prevent progress in the elimination of zoonotic infections in research animals. Institutions need a team approach involving laboratory safety, occupational health, research, veterinary, and facility management personnel to achieve an optimal program for reduction of risks associated with zoonoses in laboratory animal facilities.

The greater potential for acquiring a zoonotic disease when working with laboratory animals occurs while working with animals experimentally infected with the agent. These are often infectious models investigating the pathogenesis of disease and/or to discover novel therapeutics or prevention measures. Under these circumstances, the infected animals are handled following biocontainment practices outlined in the BMBL, which reduces the transmission potential.

ZOONOTIC DISEASES ASSOCIATED WITH COMMONLY USED LABORATORY ANIMALS

This chapter focuses on the zoonotic diseases caused by select indigenous agents of common laboratory animals that may pose an occupational hazard to animal handlers. Table 3 provides a more comprehensive list of zoonotic agents and their animal hosts. The intent is to inform those working in animal facilities, including clinical and other research scientists and biological safety personnel, about potential zoonotic pathogens associated with animals used in laboratory research. The information presented is primarily obtained from these excellent resources on zoonotic diseases (2, 8, 14, 19–21) with additional references provided at the beginning of each section. This information is organized by agent category (protozoan, bacterial, viral, fungal, and parasitic) and includes reservoir, incidence (if available), mode of transmission, clinical signs, diagnosis, and prevention. An understanding of how the more common zoonotic agents present in the animal hosts and how they are transmitted to personnel will allow biosafety professionals to perform risk assessment to mitigate against exposure.

Protozoan Diseases

Amebiasis

Amebiasis (14, 19, 20) is an intestinal disease in humans caused by a protozoan parasite, *Entamoeba histolytica.* NHPs, such as macaques, baboons, and squirrel monkeys, are the primary laboratory animals that may become infected with *E. histolytica*. This protozoan can be found in clinically normal NHPs, or it may cause severe hemorrhagic or catarrhal diarrhea. The incidence of *E. histolytica* in NHPs may be up to 30%. Transmission is primarily through the ingestion of amebic cysts found in the feces of infected animals. Cysts are shed periodically in the feces of infected animals and may persist in the environment for up to 12 days or up to 30 days in the water. The protozoan cysts can be identified by microscopic examination of fecal wet mounts or fecal flotation with zinc sulfate. Proper identification of the cysts is necessary so that the cysts are not confused with nonpathogenic amoebas. Infected animals can be treated with antiprotozoal agents such as metronidazole. Water can be treated with 10 ppm of chlorine to kill amoebic cysts.

Human beings are a natural host of *E. histolytica* and are likely a source of infection in NHPs. The natural disease in humans is primarily of concern in the tropical and subtropical regions of the world, with an estimated 10% of the world population infected with *Entamoeba* spp. Infection occurs 2–6 weeks after exposure and may be subclinical or result in diarrhea and abdominal pain. Humans with amoebic dysentery may recover spontaneously or develop a relapsing diarrhea. Hepatitis with abscess formation due to *E. histolytica* may develop, and rupture of the abscess may result in seeding of the amoeba to the lung, skin, or peritoneal cavity. The diagnosis of amoebiasis can be made by fecal examination and can be treated with metronidazole, paromomycin, or diloxanide furoate. Proper sanitation and personal hygiene in a laboratory animal environment and ABSL2 practices will prevent transmission.

Balantidiasis

Balantidiasis (14, 19, 20, 22) is caused by the large, ciliated protozoan *Balantidium coli.* It is most common in swine, with an incidence of 40–80%, and may also be found in NHPs with an incidence up to 60%. Transmission is primarily through the ingestion of trophozoites or cysts. Infected swine are typically subclinical, whereas other

TABLE 3.

Zoonotic diseases and their reservoir hosts[a]

Agent	Host
Viruses	
Eastern equine encephalitis	V, B, H
Western equine encephalitis	V, H
Venezuelan equine encephalitis	V, H, R
Chikungunya	V, N, B
Japanese encephalitis virus	V, B, P, H
Dengue virus	V, N
Yellow fever virus	V, N
West Nile virus	V, B, H
St. Louis encephalitis virus	V, H
La Crosse virus	V, Rb
Rift Valley fever virus	V, R., S, G, C
Hantavirus	R
Lymphocytic choriomeningitis virus	R, Gp
Lassa virus	R
Marburg	N
Ebola	N
Rabies	All mammals
Vesicular stomatitis virus	V, H, C, P
Newcastle disease virus	B
Hendra virus	H, Bat
Nipah virus	P, Bat
Influenza	P, B, Ft
Swine vesicular disease	P
Herpesvirus B	N
Monkeypox	N, Pd
Vaccina virus	C, S, G
Orf	S, G
Prions	C, S, G, Cd
Protozoa	
Entamoeba histolytica	N, D, R
Balantidium coli	P, N, Rb, R
Cryptosporidium spp.	C, S, G, R, Gp
Giardia spp.	C, S, G, D, R, F
Leishmania spp.	V, R, D
Enterocytozoon bieneusi	P, N
Encephalitozoon spp.	P, C, D, B, Rb, R
Pleistophora spp.	Fish
Trypanosoma brucei	V, P, C, D, S, N
Toxoplasma gondii	F, P, Rb
Ascarids	
Cheyletiella spp.	D, F, Rb
Pneumonyssus spp.	N
Sarcoptes spp.	D, F, H, C, P
Bacteria	
Bacillus anthracis	C, S, G, H, P, D
Bartonella spp.	D, F
Brucella spp.	C, S, G, P, D
Agent	**Virus**
Bacteria	
Campylobacter spp.	C, S, D, F, B, R, N
Capnocytophaga spp.	D, F, Rb
Chlamydia spp.	B
Clostridium spp.	C, S, G, P, H, D, F, N
Escherichia coli	C, S, G, P, D, F, N
Erysipelothrix rhusiopathiae	P, Turkey, Fish
Burkholderia spp.	C, H, S, G, R, F
Leptospira spp.	R, P, D, C
Listeria spp.	C, S, G
Mycobacterium tuberculosis complex	N, C, P
Nontuberculous *Mycobacterium* spp.	Fish, Amphibian
Pasteurella spp.	B, F, D, Rb, R, S, C, Gp
Yersinia pestis	V, R, F
Streptobacillus moniliformis	Rats, Gp
Coxiella burnetii	S, G, C, F
Salmonella spp.	C, P, N, R, Rpt, Poultry
Streptococcus suis	P
Francisella tularensis	Rb, F
Yersinia spp.	P, D, F, N
Helicobacter pylori	P, C, D, F
Fungi	
Microsporidium sp.	P, F, D, Poultry, H, Rb, Gp, R
Trichophyton spp.	R, Gp, Rb, D, F, H
Pneumocystis carinii	R, N, P, S, G, D
Sporothrix schenckii	F, D
Parasites	
Clonorchis sinensis	F, D, Rat
Dicrocoelium dendriticum	V, Rb, C
Fasciola hepatica	V, C, S
Fasciolopsis buski	V, P
Opisthorchis felineus	V, F
Paragonimus westermani	V, D, F, P
Schistosoma mansoni	V, N, R, P
Echinococcus multilocularis	D, F, R
Echinococcus granulosus	D, S, P
Rodentolepis nana	R
Taenia saginata	C
Taenia solium	P
Capillaria aerophila	D, C
Filaria spp.	V, N
Dirofilaria immitis	V, D
Gnathostoma spp.	D, F
Toxocara canis	D, F
Baylisascaris procyonis	D, F
Oesophagostomum spp.	N
Strongyloides spp.	D, F, N

[a]B, birds; C, cattle; D, dogs; F, cats; Ft, ferrets; G, goats; Gp, guinea pigs; H, horses; N, non-human primates; P, pigs; Rb, rabbits; S, sheep; V, insect vectors. Adapted from references 14, 19, and 82.

species, such as NHPs, may be subclinical or present with diarrhea or dysentery. Infected animals can be diagnosed with *B. coli* by identifying the trophozoites on fecal flotation or wet-mount preparations. It can be readily identified by its characteristic oval shape covered with cilia and two distinct nuclei. Water sources may also be contaminated; however, typical chlorination does not appear to be effective at killing the cysts.

Human infections of *Balantidium* occur worldwide, but are more prevalent in temperate or warm climates and are endemic in several Asian countries and South America. Most humans appear to be resistant to infection and often are subclinical depending on other host risk factors. Acute disease may result in ulcerative colitis with diarrhea, abdominal pain, tenesmus, nausea, and vomiting. The diagnosis can be made by fecal examination. Treatment of asymptomatic carriers and those who present with clinical disease can be accomplished with tetracycline, metronidazole, or iodoquinol. Proper sanitation, personal hygiene in a laboratory environment, and ABSL2 practices will prevent transmission.

Cryptosporidiosis

Cryptosporidiosis (14, 19, 20, 23–25) can be caused by a number of species of the coccidian protozoan *Cryptosporidium*. There are over 13 named species, as well as a cryptosporidia parvum-like parasite identified in over 150 different species of animals. The most significant human pathogen and most common zoonotic species is *Cryptosporidium parvum*. Other species of cryptosporidia that have been reported to infect humans include *C. baileyi* (from birds), *C. felis* (cats), *C. meleagridis* (turkeys), and an unidentified species from dogs. Other hosts of *Cryptosporidium* spp. include wild mammals, ferrets, guinea pigs, NHPs, reptiles, and fish. Whereas there is some host specificity, cross-infection is possible. *C. parvum* primarily infects cattle, but also pigs, sheep, and goats. It is estimated that this protozoan is present in 90% of U.S. dairies, and up to 50% of the dairy calves will shed oocysts.

Transmission to people is by the fecal-oral route. Cryptosporidiosis causes clinical disease in neonatal animals, resulting in a watery diarrhea. *C. parvum* typically infects calves in dairies at 8–15 days of age, and therefore young calves should be considered infected with the protozoa. Adult animals are typically asymptomatic. Oocysts can survive for up to 140 days in the environment and are resistant to many common disinfectants, including chlorine, iodophores, and formaldehyde; however, hydrogen peroxide and ammonium hydroxide can reduce infectivity 1,000-fold. Cryptosporidia can be diagnosed using fecal wet mounts to identify the oocysts. Commercially available assays are available to detect *C. parvum* antigen, and it can be identified by histologic evaluation of the small intestine or by molecular diagnostics. The disease is usually self-limiting, and supportive care is indicated. Animals could be treated with metronidazole, paromomycin, or azithromycin.

The occurrence of cryptosporidia in humans is worldwide and typically associated with outbreaks from contaminated feed and water. The 50% infectious dose in humans is 50–100 oocysts with an incubation period ranging from 4 to 28 days; many individuals that work with young cattle have become infected with clinical disease. Infections from NHPs and cats are also reported. In the research animal environment, this likely occurs during the cage cleaning process when fecal matter is aerosolized during the washing process and ingested inadvertently. Individuals 1–9 years old and 30–39 are more susceptible to infection and developing clinical disease. In the United States, cryptosporidial infection is more common in the northern states (Wisconsin, Minnesota, North Dakota, South Dakota), with an increased incidence from June through October. *C. parvum* infects the small intestine and causes a watery diarrhea, with abdominal pain, nausea, anorexia, fever, and weight loss. Without intervention, the disease is self-limiting in 1–2 weeks. Immunocompromised individuals are more severely affected. The diagnosis of human cryptosporidiosis can be made similarly by fecal examination, antigen detection, histology, and molecular diagnostics. Supportive therapy is indicated in immunocompetent individuals, supplemented with antimicrobial therapy in immunocompromised individuals. Proper sanitation and personal hygiene in a laboratory environment are important to prevent infections. ABSL2 practices with the addition of face shield or mask to minimize mucous membrane exposure will prevent transmission.

Toxoplasmosis

Toxoplasmosis (14, 19, 20, 26) is caused by an intracellular coccidian protozoan that has a worldwide distribution and has been found in most warm-blooded animals. The definite host is cats, the major reservoir in research animals, although other animals such as mice, rats, dogs, sheep, pigs, and chickens can serve as an intermediate host. Cats acquire infection from other cats or ingestion of an intermediate host. An estimated 30–80% of cats are seropositive to *Toxoplasma gondii*, suggesting exposure to the protozoa. Cats support the life cycle of the protozoa in their intestinal tract where infectious oocytes are produced and are usually asymptomatic. Intermediate hosts develop tissue cysts that are infective when consumed. Rare clinical signs include uveitis, neurologic signs, vomiting, diarrhea, dyspnea, and anorexia. Infected cats shed oocysts in their feces for 1–2 weeks after infection. Oocyst shedding can be intermittent. It takes 1–5 days for oocytes to become sporulated and infectious to people. Experimentally infected animals may develop infectious bradyzoites encysted in the tissues, which may serve as a source of exposure to laboratory personnel. Oocysts are very stable in the environment and may remain infective for months to years. Hence prompt removal of cat feces greatly reduces the risk of *T. gondii* exposure. Toxoplasmosis is diagnosed by identification of oocysts during a fecal examination, isolation of tachyzoites from aqueous humor, or serologic testing [enzyme-linked immunosorbent assay (ELISA), immunofluorescence assay (IFA)]. Infected cats can be treated with pyrimethamine and sulfonamide therapy.

T. gondii exposure in people is very common, with approximately one-third of the world population having antibody titers. The seroprevalence in the United States is up to 35%, and it is up to 75% in Africa and South America. The primary means of transmission in people is through cyst-containing meat, which is unlikely to occur in the laboratory. Fecal-oral transmission through exposure to cat feces is likely. There is a 10- to 25-day incubation period in people, and 80–90% of infected individuals will be asymptomatic. Symptoms in humans can include lymphadenopathy, fever, headache, malaise, and, rarely, encephalopathy, chorioretinitis, pneumonia, and myocarditis. Most symptoms resolve in a few weeks in immunocompetent individuals. Immunocompromised individuals may develop more severe disease, including encephalitis and chorioretinitis. Infections in pregnant women can have severe consequences, including abortion, stillbirth, and preterm delivery. Congenital infections may manifest with microcephalus, hydrocephalus, chorioretinitis, and hepatosplenomegaly. *T. gondii* is primarily diagnosed by serology. People can be treated with pyrimethamine sulfadiazine. All cats should be considered carriers of *T. gondii*, and daily removal of cat feces prior to sporulation of infected oocysts minimizes transmission in the laboratory. Proper sanitation, personal hygiene in a laboratory environment, daily removal of cat feces from the litter box, and ABSL2 practices will prevent transmission.

Giardiasis

Giardiasis (14, 19, 20, 26) is caused by the flagellated protozoan *Giardia lamblia,* also known as *G. intestinalis* or *G. duodenalis*. Many domestic and laboratory animals may serve as a reservoir host for giardia, including cats, dogs, NHPs, pigs, and small ruminants. The incidence of giardia is 2–15% in dogs and cats, and 9–39% in swine and ruminants. The incidence can be higher in a kennel environment. There is some controversy whether or not giardia is a zoonotic pathogen. Historically *Giardia* spp. were classified on the basis of their host origin, but more recent studies have demonstrated a lack of host specificity for some species of giardia. Different genotypes have been identified that infect multiple hosts. For example, genotype A is able to infect people, livestock, cats, dogs, beavers, guinea pigs, and lemurs, whereas genotype B infects people, lemurs, chinchillas, dogs, beavers, and rats. Infected animals may be asymptomatic or develop diarrhea, bloating, abdominal cramping, anorexia, nausea, and weight loss. Transmission occurs through the ingestion of infectious cysts shed in the feces. The diagnosis of giardia can be made by identification of trophozoites or cysts on fecal wet-mount preparations or antigen detection assays. Because cysts are shed intermittently, fecal samples from at least 3 consecutive days should be evaluated. Infected animals can be treated with metronidazole.

There were 19,000 cases of human giardia from 2006 to 2008, more commonly in children less than 10 and adults from 35 to 44 years of age. The primary source of infection in people is from contaminated water. The incubation period of giardia is 7–10 days following exposure, and the disease presents with similar clinical signs in people as it does in animals. Cysts may persist for months in the environment, but are susceptible to 20% commercial bleach disinfection and quaternary ammonium disinfectants. In addition, cysts are susceptible to desiccation. Proper sanitation, personal hygiene in a laboratory environment, and ABSL2 practices will prevent transmission.

Bacterial Diseases

Brucellosis

Brucellosis (14, 19, 20, 27–31) is caused by a small, Gram-negative coccobacillus. There are four species of *Brucella* that may be associated with zoonotic disease in a research setting: *B. abortus*, *B. canis*, *B. suis*, and *B. melitensis*. The reservoir host is specific for each species. Dogs are the host for *B. canis*, cattle are the host for *B. abortus*, pigs are the host for *B. suis*, and sheep and goats are the host for *B. melitensis*. Due to the Cooperative State–Federal Brucellosis Eradication Program, the incidence of brucellosis in agricultural cattle and swine in the United States is very low, and the United States is considered brucellosis free except in the Greater Yellowstone Area (for *B. abortus*) and Texas (for *B. suis*). However, wild ruminants and feral swine continue to serve as a reservoir for *Brucella* spp., which can infect domestic livestock. *B. melitensis* is considered a foreign animal disease and is not present in the United States except at the Mexico-Texas border. *B. canis* is the most significant *Brucella* species to which personnel working with laboratory animals may be exposed. Dogs can also become infected with *B. abortus* and *B. suis* following consumption or exposure to infected livestock placenta or fetuses. Canine brucellosis has an incidence of 1–8% in dog production colonies. Overt clinical signs in infected animals are primarily associated with the reproductive tract, including infertility, abortion, and vaginal discharge. Males may present with orchitis, prostatitis, and epididymitis. Other systemic signs of infection in dogs may also be present, such as lymphadenopathy, arthritis, uveitis, and discospondylitis. The disease is transmitted in animals through breeding or oronasal contact with vaginal discharges or infected placenta or fetuses, which contain high numbers of organisms (10^{10} organisms/ml). Milk and urine from infected animals may also serve as a mode of transmission. Animals become infected 1–4 weeks postexposure, and can remain infected for several years. A more common source of infection in laboratory animals is experimentally infected rodent models. Brucellosis can be diagnosed by culture of infected material on selective media, molecular assays, or serologic assays with the rapid agglutination test the most sensitive. Infected animals can be treated with tetracycline, fluorquinolone, and aminoglycoside antibiotic therapy. A U.S. Department of Agriculture (USDA)-approved vaccine is available for cattle, but there is no vaccine available for dogs.

Brucellosis in people is a worldwide disease and is closely linked to the disease in animals, with more than 500,000 cases reported annually. There were just over 100 reported cases in the United States in 2009. *Brucella* is one of the more common laboratory-acquired infections, albeit rarely associated with transmission from infected animals. Research personnel can become infected following direct contact with infected placenta or fetuses through the mucous membranes, skin lesions, or aerosolization of infected materials. The infective dose is 10–100 organisms. The incubation period in people may range from 1 week to 3 months. The clinical presentation can be similar regardless of the species, with more severe disease occurring with the more virulent *B. melitensis*, followed by *B. suis*, *B. abortus*, and *B. canis* the least virulent. Infected humans may present with headache, backache, depression, lymphadenopathy, hepatosplenomegaly, nausea, vomiting, joint pain, pneumonia, uveitis, and urogenital infections. Definitive diagnosis is by culture of infected tissues or blood. Serologic assays are available for diagnosis, which are not species specific and cross-react with a number of bacterial agents. Treatment can be provided with doxycycline and streptomycin for 6 weeks. Controlling the source of animals to brucellosis-free herds and colonies and ABSL2 practices will prevent transmission for routine practices. Known infected animals should be handled with ABSL3 precautions. ABSL3 containment, practices, and facilities for all manipulations of cultures, experimental animal studies, or known infected animals are required.

Campylobacteriosis

Campylobacter sp. is a microaerophilic, Gram-negative rod, and is one of the more common causes of diarrhea in animals (14, 19, 20, 32, 33). The most frequent causes of zoonotic disease are *C. jejuni* and *C. coli*, which have been isolated from dogs, cats, ferrets, guinea pigs, hamsters, NHPs, livestock, and several species of birds. *C. upsaliensis* may also be carried by young dogs and cats and causes disease in people. Young animals are more susceptible to infection and readily shed the organism, thus becoming a source of infection to people. Infected animals are typically asymptomatic carriers. Naïve animals may develop clinical disease consisting of a watery or mucohemorrhagic diarrhea, vomiting, and inappetence. *Campylobacter* spp. are transmitted by the fecal-oral route through contaminated feed or water, or direct contact with infected animals. Fecal cultures on selective media and microaerophilic conditions provide a definitive diagnosis. Presumptive diagnosis can be made using dark-field microscopy of fresh fecal samples. Antimicrobial therapy with erythromycin may curtail clinical signs, but animals may remain carriers.

Campylobacter diarrhea is a leading cause of diarrhea in people worldwide, with an estimated 45–870 cases/100,000 people. Although all people are susceptible to disease, it is more prevalent in children less than 5 with a secondary peak in young adults. Those that work with reservoir host animals have increased risk. Infection in people occurs via the fecal-oral route through contaminated meat or drinking water. The infectious dose is less than 500 organisms. In the laboratory environment, exposure occurs from contaminated excretions. Individuals who are immunosuppressed, male, and young are more susceptible. The incubation time is 3–5 days, and infected people develop an acute enteritis with fever, headache, nausea, and chills. Diarrhea is characterized as watery with a putrid odor, containing bile, blood, mucus, and/or pus. Systemic infection may result in 0.1% of cases and carries a poor prognosis. Systemic disease is most commonly associated with *C. fetus* infections. Definitive diagnosis is through fecal cultures on selective media at 42°C. Molecular and serological methods are also available. Treatment consists of fluid and electrolyte replacement and erythromycin, doxycycline, and ciprofloxacin antimicrobial therapy. Good personal hygiene, frequent hand washing, good sanitation, and ABSL2 practices will prevent transmission.

Capnocytophagosis

Capnocytophagosis (14, 19, 20, 34, 35) is caused by *Capnocytophaga canimorsus* from dogs and *C. cynodegmi* from cats. There are eight other species of *Capnocytophaga*, but only *C. canimorsus* causes severe disease in humans. It is a commensal Gram-negative bacillus found in the oral cavity of dogs and cats with a prevalence of 16–18%. There are no clinical signs associated with the bacteria in dogs and cats, unless they become infected from another animal's bite. Diagnosis is difficult and slow, requiring 2–7 days for visible colonies to appear on blood agar. Enriched media may enhance growth and identification. Molecular assays are available. The disease is transmitted to people primarily by dog bites and to a lesser extent cat bites and animal licks of an open wound. The incubation period is 1–8 days following a dog bite. People will initially develop local inflammation, followed by cellulitis, pain, purulent discharge, and lymphadenopathy. Septicemia may present as fever, myalgia, nausea, and headache. The mortality rate can be 30%. Humans can be treated with amoxicillin-clavulanate antibiotics. Animal bites are very common and represent 1% of all emergency room visits. Proper restraint methods and recognition of dog and cat aggression can minimize the chance of being bitten. Good personal hygiene and frequent hand washing, good sanitation, and ABSL2 practices will prevent transmission.

Cat scratch disease

Bartonella spp. are small, curved Gram-negative rods. There are nine species or subspecies of *Bartonella* that are considered zoonotic. *Bartonella henselae* is the most important and is considered the causative agent of cat scratch disease (14, 19, 20, 36, 37). Other species of *Bartonella* are known to infect rodents, but unlikely to occur in research facilities, given the pathogen-free status of many rodent colonies. *Bartonella* spp. are transmitted among cats by arthropod vectors, including most commonly cat fleas and less commonly ticks. The incidence of bacteremia in cats is 5–40% in the United States, with similar findings in other countries. Bacteremia can be present for up to 2 years in cats, and infected cats may remain asymptomatic. If clinical signs are present, they may be transient and unapparent to the research personnel. These include small abscesses at site of inoculation, transient fever, lethargy, lymphadenomegaly, myalgia, and, less commonly, reproductive failure and endocarditis. Because of the waxing and waning bacteremia, diagnosis can be challenging. Definitive diagnosis is on the basis of culture or molecular diagnostics of a blood sample. Long-term antibiotic therapy with enrofloxacin, doxycycline, rifampin, or azithromycin may eliminate the bacteria from the host. Prevention in cats is aimed at controlling the flea vector.

There are approximately 25,000 cases of human bartonellosis annually in the United States. In more than 90% of the cases, a history of cat scratch or bite wound can be identified. It is believed that dried, infected flea feces are transmitted to people by the claws during a scratch. The initial site of inoculation may develop a small erythematous papule. Within 1–2 weeks, a regional lymphadenopathy develops, which regresses in 6 weeks. Individuals that develop systemic disease present with fever and malaise. Occasionally, individuals may develop generalized lymphadenopathy, encephalitis, osteomyelitis, arthritis, hepatitis, pneumonia, and blindness. Immunodeficient individuals are at risk for developing bacillary angiomatosis and bacillary peliosis, which is a more serious presentation of the disease. The diagnosis is on the basis of history, clinical signs, and culture or molecular identification of the organism. People can be treated by antibiotic therapy with azithromycin, doxycycline, or ciprofloxacin. Eliminating the vector from research facilities and proper restraint techniques to minimize scratches greatly reduce the risk of disease in research personnel. Good personal hygiene, frequent hand washing, good sanitation, and ABSL2 practices will prevent transmission.

Chlamydiosis

Chlamydial bacteria are small, obligate intracellular, Gram-negative coccobacilli. The family is divided into two genera, *Chlamydia* and *Chlamydophila*. Three *Chlamydophila* species—*C. psittaci* (avian), *C. abortus* (ruminants), and *C. felis* (cats)—are the most common *Chlamydophila* species that are zoonotic (14, 19, 20, 38, 39). *C. psittaci* has a worldwide distribution and a high infection rate in psittacine birds and pigeons. Infected birds may have latent infections with periodic shedding that is typically related to a stressful event. The most frequent clinical signs that develop are anorexia, diarrhea, respiratory distress, sinusitis, conjunctivitis, and yellow-colored droppings. *C. abortus* causes enzootic abortion in sheep, and an estimated 8% of flocks are infected. *C. felis* causes conjunctivitis. Diagnosis can be made with serologic assays, molecular diagnostics, and histologic examination of tissues. Culture of *Chlamydia* requires cell culture media or embryonated eggs. Tetracycline is the drug of choice for treating infected animals. *Chlamydia* is susceptible to common disinfectants, including quaternary ammonia compounds and bleach.

C. psittaci is the most important animal chlamydia transmitted to people. The incidence of disease in people is likely underestimated due to the difficulty in diagnosis; nonetheless from 1996 through 2001, there were 165 reported cases in the United States and 1,620 cases in the United Kingdom. The incubation period is 7–21 days, but may extend to 3 months. People typically become infected by inhaling dried feces or respiratory secretions from infected birds. Human symptoms are variable, ranging from subclinical infections to mild respiratory infections to severe atypical pneumonias with high fever, headache, and multiorgan failure. Less severe disease may last 7–10 days, and severe disease may last 3–7 weeks. Diagnosis is demonstrated by an increase in antibody responses as tested by serologic assays (complement fixation, ELISA) or molecular diagnostics. Doxycycline or tetracycline is the therapy of choice. Acquiring birds from chlamydia-free flocks can reduce the risks. Good personal hygiene, frequent hand washing, good sanitation, and ABSL2 practices are generally adequate to prevent transmission. However, ABSL3 procedures are warranted with activities with a high potential for aerosol production.

Erysipeloid

Erysipelothrix rhusiopathiae is a Gram-positive, nonmotile, non-spore-forming bacterium (14, 19, 20, 40, 41). The primary hosts are pigs, but other animals, including sheep, rabbits, poultry, and fish, have been identified as carriers. It is estimated that 30–50% of healthy swine carry *E. rhusiopathiae* in their tonsils and lymphoid tissues. Erysipelas in pigs presents as an acute, subacute, or chronic infection. Acute disease may lead to septicemia and sudden death. Subacute disease is characterized by diamond-skin lesions, which may resolve with no other sequelae or may progress to necrosis of the skin. The chronic form is characterized by arthritis and endocarditis. Bacteria can be shed in the feces, urine, saliva, and nasal secretions, which can contaminate bedding and feed. The bacterium is viable in the environment for 1–5 months. *E. rhusiopathiae* can be definitively diagnosed by culture. Molecular diagnostics and serologic testing are also available. Antimicrobial therapy with penicillin can readily treat the bacteria. A commercial vaccine is available for use in swine herds and turkey flocks.

Erysipelas infection in people is not common. Exposure occurs from contaminated materials or animals, and the bacteria penetrate the skin through small breaks. The incubation period is 2–7 days with two clinical presentations. The localized cutaneous form is the most common and causes a bright red-purple skin discoloration that later develops into blistering and sloughing. The diffuse form is characterized by widespread erythema, lymphadenitis, arthralgia, and myalgia. Endocarditis may also occur. Diagnosis is on the basis of history and culture, and the infection is readily responsive to penicillin. Vaccinations are not rewarding in preventing the disease in people. Good personal hygiene, frequent hand washing, good sanitation, and ABSL2 practices will prevent transmission.

Leptospirosis

Leptospira spp. are Gram-negative spiral bacteria with hook-shaped ends. Disease is caused by *L. interrogans* and divided into serovars on the basis of epidemiological criteria (14, 19, 20, 42–46). The primary serovars and their reservoir hosts are *L. icterohaemorrhagia* (rats and dogs), *L. pomona* and *L. bratislava* (swine), *L. hardjo* (cattle), *L. canicola* (dogs), and *L. ballum* (rats and mice). Within laboratory rodents, the incidence of leptospiral infections is very rare given the pathogen-free status of rodents that are acquired for research purposes. The last reported case in a research colony occurred in the 1980s. The incidence in wild rodents, which may occasionally be used as laboratory animals, is significant. A survey of wild rats in Detroit demonstrated a 90% incidence. Leptospirosis is relatively uncommon in modern swine facilities with appropriate sanitation practices. The incidence in dairy cattle herds has been reported to be up to 59%. The incidence in dogs is relatively low, with about 230 cases described from 2000 through 2006. The disease in animals is variable depending on the serovar and the hosts. In general mice and rats are asymptomatic whereas dogs may develop a fever, hematuria, and hepatic and renal disease; livestock develop reproductive failure. Spirochetes are carried in the renal tubules and shed in the urine of reservoir animals.

The diagnosis can be made with serologic testing demonstrating an increase in antibody responses, examination of urine by dark-field microscopy demonstrating spirochetes, or molecular assays. Culture requires a selective semisolid medium from multiple urine samples due to intermittent shedding samples, and can take up to 7 days to identify colonies. Vaccination programs in herd and kennel settings help control the disease. Antimicrobial therapy with penicillin, tetracycline, and azithromycin eliminates leptospiremia in dogs.

Leptospirosis is probably the most common and widespread zoonotic disease in the world, with an incidence up to 100 per 100,000 in Asia, India, and Latin America. The incidence is increasing in North America. Transmission occurs primarily through skin abrasions and mucous membranes via direct contact with infected urine or tissues. The incubation period is 5–14 days and follows a diphasic course. The first phase occurs 4–7 days postinoculation and presents with a sudden fever, headache, malaise, conjunctival hyperemia, cough, and hemorrhage. Hepatosplenomegaly, abdominal pain, jaundice, nausea, and diarrhea may also occur. The early antibody response resolves symptoms in 3–5 days. The second phase occurs in 50% of individuals in which spirochetes are excreted in the urine, and new symptoms develop, including jaundice, renal failure, anemia, hemorrhage, pneumonia, myocarditis, meningitis, iridocyclitis, and hepatic failure. The mortality rate is 5–40% among severe cases. The diagnosis is on the basis of history and identification of the organism by microscopy, molecular assays, or culture from blood, cerebrospinal fluid (CSF), urine, or biopsy tissues. Penicillin and doxycycline are the antibiotics of choice for treating leptospiral infections. Leptospirosis was reinstated as a nationally reportable disease in 2013. Good personal hygiene, frequent hand washing, good sanitation, and ABSL2 practices will prevent transmission.

Mycobacterial diseases

Mycobacterial infections (14, 19, 20, 47–53) are caused by bacilli of the genus *Mycobacterium*, which are nonmotile, nonsporulating, and acid fast and possess thick, lipid-rich cell walls. Species are identified on the basis of culture, biochemical, and molecular characteristics. There are several species of *Mycobacterium* that can infect animals. The *Mycobacterium tuberculosis* complex includes *M. tuberculosis, M. bovis, M. africanum, M. microti, M. caprae,* and *M. pinnipedii*. Reservoir hosts for these bacteria include NHPs, cattle, voles, swine, and seals. The incidence of naturally occurring mycobacterial disease in imported NHPs was found to be about 1% from 1990 through 1993. Similarly, the percent of cattle identified as positive for mycobacterial disease at slaughter and the prevalence of disease in wildlife populations is less than 1%. Other causes of mycobacterial disease include *M. avium-intracellulare* from birds and atypical mycobacterial species, including *M. marinum, M. fortuitum,* and *M. chelonae,* from amphibians, fish, and reptiles. The mycobacterial disease can be acquired by animals from contaminated soil or water, or from direct contact with infected animals. NHPs acquire infections from contaminated soil and water or other infected animals. The clinical signs associated with mycobacteriosis can vary from asymptomatic carriers to significant disease, depending on the species of mycobacteria and the immune status of the hosts. In severe cases, animals may have pulmonary disease, anorexia, chronic weight loss, or lymphadenopathy. Infection in specific organs such as the gastrointestinal tract or central nervous system can also result in clinical signs such as diarrhea or neurologic symptoms. Cutaneous lesions are common with atypical mycobacterial infections. The presence of *Mycobacterium* is most commonly diagnosed by using the intradermal tuberculin skin test in asymptomatic animals. Reactors to the skin test may require further evaluation with thoracic radiographs. Serologic assays are available for use in NHPs such as the interferon-γ (PRIMAGAN) test. Identification of acid-fast organisms on cytology samples is diagnostic. Molecular diagnostics and histologic evaluation of tissues can confirm diagnosis. Mycobacteria grow slowly and form colonies after 3 weeks of culture on solid media. In the laboratory setting, *Mycobacterium*-positive animals are typically culled from the colony.

Tuberculosis occurs worldwide, with an estimated incidence of one in three individuals. Cattle and NHPs represent principal sources of the zoonotic disease from *M. bovis* and *M. tuberculosis*, respectively. Infections with other mycobacteria are usually through contaminated soil and water. Because of the difficulty in identifying positive animals, laboratory-acquired infections occur occasionally. Mycobacteria primarily infect people by aerosol exposure and have a 4- to 6-week incubation period. Similar to animals, tuberculosis may be asymptomatic and only detectable by a positive skin test. Pulmonary disease may develop several years after infection and produce a bloody cough. Extrapulmonary disease may be seen in virtually any organ, including the lymph nodes, pleura, genitourinary tract, skeletal system, meninges, and central nervous system. Progressive pulmonary disease can be fatal within 5 years in 50% of untreated cases. Although there are no known reports of direct transmission of mycobacteria from amphibians or fish to people, water contaminated by infected animals likely occurs as the bacteria come in contact with skin wounds or abrasions. Cutaneous lesions are common when exposed to the atypical mycobacteria such as *M. marinum*, particularly from water sources. Infected abrasions in the skin can lead to granulomas or ulcers on the skin that typically resolve. The diagnosis can be made on the basis of clinical symptoms, identification of acid-fast bacilli on sputum samples or biopsy samples, skin test, culture, and serologic and molecular assays. Infected individuals with tuberculosis complex may be treated with long-term (6–12 months) antibiotic therapy, including rifampin, isoniazid, and pyrazinamide. The cell wall component makes *Mycobacterium* spp. resistant to alkali and detergents, but they are susceptible to phenolic compounds and hydrogen peroxide disinfectants. Prevention includes education of personnel, regular testing of animal care personnel and colony animals, particularly NHPs, and isolation and/or elimination of positive animals. As a general rule, ABSL2 precautions and practices will minimize exposure. Known positive NHPs should be housed at ABSL3 level with appropriate respiratory protection to prevent aerosol exposure.

Pasteurellosis

Pasteurella multocida is a Gram-negative rod and is the *Pasteurella* species that primarily causes zoonotic disease in people (14, 19, 20). The reservoir hosts are primarily dogs, cats, and rabbits, although birds and rodents may also be carriers. Approximately 75% of cats and 55% of dogs harbor the organism in the nasopharynx. Up to 90% of healthy rabbits may also be carriers. *P. multocida* is considered to be an opportunistic pathogen of laboratory mice and rats, and many vendors have eliminated it from their colonies; nonetheless, they may still be carriers. The disease is typically asymptomatic. Severe disease may be characterized by respiratory signs, rhinitis, otitis, subcutaneous and visceral abscesses, and genital tract infections. The disease in rabbits is very significant, and many commercial rabbit suppliers produce *Pasteurella*-free rabbits. *Pasteurella* causes atrophic rhinitis in pigs. Definitive diagnosis can be made by culture of lesions on blood agar. Antimicrobial therapy may provide remission of disease and alleviation of clinical signs, but they are likely to recur, particularly in rabbits.

Transmission of *P. multocida* to people is primarily from bite wounds. The frequency of *P. multocida* isolated from dog and cat bite wounds is 20–50% and 75%, respectively. The incubation period is 2–14 days. Local inflammation and cellulitis may develop in minor cases, whereas severe cases may develop lymphadenopathy, osteomyelitis, myocarditis, and pneumonia. Routine culture methods readily identify the bacterium, and molecular diagnostics are also available. Penicillin, doxycycline, or amoxicillin-clavulanic acid are the antimicrobials of choice to treat infections. Proper restraint techniques to minimize bites will greatly reduce the risk of disease in research personnel. Good personal hygiene, frequent hand washing, good sanitation, and ABSL2 practices will prevent transmission.

Rat-bite fever

Streptobacillus moniliformis and *Spirillum minus* are the two bacteria responsible for rat-bite fever (14, 19, 20, 54, 55). *S. moniliformis* is a Gram-negative rod, and *S. minus* is a Gram-negative spiral bacterium. Both organisms are found in the upper respiratory tract and oral cavity of rodents, primarily rats. *S. minus* has also been isolated from ferrets, pigs, dogs, cats, and NHPs. Surveys from wild rodent populations suggest a prevalence of *S. moniliformis* to be 50–100% and *S. minus* to be 0–25%. It is typically asymptomatic, but disease has been associated with infection in guinea pigs, causing lymphadenitis, and endocarditis and arthritis have been associated with disease in NHPs. *S. minus* does not grow *in vitro* and requires molecular diagnostics for confirmation. *S. moniliformis* is a facultative anaerobe and grows slowly on enriched artificial media. Because it is asymptomatic in rodents, there is no need for treatment. Although commercially available rodents are typically pathogen free, research colonies could become infected from wild rodents that enter the facility or from contaminated feed or bedding.

The incidence of rat-bite fever is unknown because it is not a reportable disease, but rats represent 1% of all animal bites treated, and a report from 1980 suggested there were 40,000 rat bites with 2% leading to infection. There continue to be reports of laboratory-acquired infections and infections from rodents acquired from pet stores. People become primarily infected through the bite of a rat. Infections with *S. minus* have an incubation period of 2–3 weeks, and *S. moniliformis* has an incubation period of 3–5 days. Both present with fever, chills, myalgia, and skin rash at the site of the bite, which resolves within a few days. Arthritis can occur in 50% of the cases of *S. moniliformis*. More severe cases present with lymphadenitis, sore throat, endocarditis, pneumonia, hepatitis, and meningitis, which can be fatal in 10% of the cases if left untreated. Both agents are readily susceptible to penicillin and tetracycline. Proper restraint techniques to minimize bites will greatly reduce the risk of disease in research personnel. Good personal hygiene, frequent hand washing, good sanitation, and ABSL2 practices will prevent transmission.

Q fever

Query fever (Q fever) is caused by *Coxiella burnetii,* which is a ubiquitous, pleomorphic, obligate intracellular bacterium (14, 19, 20, 56–63). It has a worldwide distribution and infects a broad range of animals. Two cycles of infection can be identified, one in domestic animals and the other natural infection in ticks. The reservoir includes many wild and domestic animals; however, domestic ruminants represent the most frequent source of human infections. Pet dogs and cats may represent potential sources in urban populations. Serologic surveys demonstrated a seroprevalence of 41% in goats, 16% in sheep, 4% in cattle, and up to 20% in dogs and cats. *C. burnetii* causes abortion and stillbirths in infected animals, but is otherwise asymptomatic with persistent shedding. Animals shed the bacteria in milk, urine, and feces, and in very high concentration in birth by-products (up to 10^9 organisms/g). Animals become infected from tick bite inoculation or from ingestion or inhalation of contaminated milk, urine, or placenta. The diagnosis is on the basis of serologic evaluation of the phase I and phase II antigens to differentiate chronic and acute disease, respectively. The bacteria can also be identified by cell culture inoculation, molecular assays, or identification of the bacteria from impression smears with a Gimenez stain. Tetracycline is the antibiotic of choice in treating animals. The organism is very stable in the environment and can survive up to a year; it is resistant to routine chemical disinfection, including bleach, phenolic compounds, formalin, and quaternary ammonium compounds.

Q fever has a worldwide distribution. In the United States, there were 110–170 cases reported from 2005 through 2010 with a seroprevalence of 3%. Transmission to people usually occurs by inhalation of bacteria from dried contaminated birth by-products. As few as 10 organisms can cause infection in people. Individuals working on farms are at a greater risk, although there have been reports of Q fever acquired in the laboratory setting. It is the second most commonly reported laboratory-acquired infection. There are three forms of infections in people. The first is subclinical with seroconversion. Acute Q fever is characterized by febrile illness with hepatitis and pneumonia, which occurs 2–3 weeks after exposure and resolves in 1–2 weeks. Infection in pregnant women may lead to premature births, stillbirths, or abortions. Chronic Q fever appears mainly as endocarditis, which occurs in individuals with preexisting valvular disease. The diagnosis can be made by serology with a convalescent and acute titer demonstrating a 4-fold increase in phase II antigen. It can be cultured from the buffy coat of infected individuals. It is challenging to acquire Q fever-free ruminants for research because the incidence of seroprevalence among herds is high. Prescreening of source animals prior to arrival and minimizing the use of gravid females can significantly reduce exposure. However, care should always be exercised when working with ruminants. Good personal hygiene, frequent hand washing, good sanitation, and ABSL2 practices will prevent transmission. Respiratory protection may be considered as additional PPE to minimize aerosolization exposure. Experimentally infected animals should be maintained under ABSL3 conditions. A Q fever vaccine is licensed in Australia, and a vaccine is available from the specialized immunization program from U.S. Army Medical Research Institute of Infectious Diseases (USAMRIID) for personnel at high risk.

Salmonellosis

Salmonella species are flagellated, non-spore-forming, Gram-negative bacilli with over 2,400 serovars (14, 19, 20, 64–69). The most common serovars associated with zoonotic disease are *S. enterica* serovar Enteriditis (*S.* Enteriditis) and *S. enterica* serovar Typhimurium (*S.* Typhimurium). *Salmonella* has a wide host range. Most of the rodent species used in research are pathogen free and do not pose a significant risk to research personnel; however, they may become infected with contaminated feed or other environmental sources. *Salmonella* is uncommonly found in the intestinal tract of healthy dogs and cats, with an incidence of 2–6%. There is an increased risk of *Salmonella* when working with NHPs because it is one of the more common causes of diarrhea in NHPs. The incidence of *Salmonella* in swine is 3%, but could be up to 33%. Incidence in cattle production facilities is less than 1%. The percentage of *Salmonella*-positive birds ranges from 5% to 100%, and 94% of reptiles are carriers of *Salmonella*. The disease can present as a self-limiting diarrhea, severe hemorrhagic gastroenteritis, vomiting, and lethargy. Septicemia may occur, leading to hepatic and splenic abscesses. Animal hosts can also be asymptomatic carriers. Transmission is by the fecal-oral route, and the diagnosis of salmonellosis can be made by fecal or blood culture using a selective enriched medium or molecular diagnostics. Animals with clinical disease are usually treated with supportive therapy. Antibiotic use is somewhat controversial and may lead to prolonged periods of communicability. Vaccination programs can be used in poultry and livestock operations to manage the disease.

Salmonella occurs worldwide and is typically acquired from ingesting contaminated food or water; however, direct contact with infected animals may also occur. *S.* Typhimurium is the most common cause of salmonellosis in people in the United States, followed by *S.* Enteriditis. Most human infections present following a short incubation period of 12–48 hours, but it may take up to 5 days. A relatively large infectious dose (10^3–10^5 organisms/ml) is required. Acute disease is characterized by a sudden onset of nausea, vomiting, and watery diarrhea, which resolves in 2–5 days. Complicated systemic infection may occur, leading to septicemia, arthritis, osteomyelitis, meningitis, and endocarditis. Diagnosis is on the basis of culture and is most effective in the first hours after onset. Therapy for infected people is typically supportive care. Severe cases may require antibiotic therapy including ciprofloxacin and amoxicillin. Prevention in the laboratory environment depends on rapid identification of infected animals and their treatment or removal. Good personal hygiene, frequent hand washing, good sanitation, and ABSL2 practices will prevent transmission.

Shigellosis

Shigella spp. are Gram-negative, facultative anaerobic, non-spore-forming rod-shaped bacteria (14, 19, 20, 70). People serve as the primary reservoir host for *Shigella* and can infect NHPs, which can in turn infect other NHPs and humans. NHPs can harbor *S. dysenteriae*, *S. flexneri*, and *S. sonnei*. They may be asymptomatic or present with a hemorrhagic diarrhea, gingivitis, abortion, and air sac infection. It is one of the most common causes of diarrhea in NHPs. Transmission occurs from direct or indirect fecal-oral routes, including flies, which may be a vector. Diagnosis is on the basis of culture using a selective enriched medium, similar to that used for diagnosing *Salmonella*. Treatment often includes supportive therapy and antibiotics. The carrier state in NHPs has been reported to be eliminated through specific antibiotic therapy reducing the zoonotic potential.

Shigella is a common cause of diarrhea in people with an incidence of 30 cases/100,000 population, with children being more susceptible. As few as 10–100 organisms can cause disease in people, which is characterized by an acute onset of diarrhea with fever, nausea, and abdominal cramping. Illness lasts 4–5 days and can be treated with trimethoprim sulfa antibiotics. Good personal hygiene, frequent hand washing, good sanitation, and ABSL2 practices will prevent transmission.

Viral Diseases

Herpesviruses

There are several herpesviruses identified in NHPs and other research animals; however, the majority of them are limited to their host species and are not reported to infect humans. For example, NHPs may be infected with alphaherpesviruses such as simian varicella virus, beta-herpesviruses such as *Rhesus cytomegalovirus*, and gammaherpesviruses such as Rhesus macaque rhadinovirus. However, *Macacine herpesvirus* can cause fatalities in humans.

Macacine herpesvirus *(B Virus)*

Macacine herpesvirus (formerly known as *Cercopithecine herpesvirus* 1, herpes B, monkey B virus, *Herpesvirus simiae*, and herpesvirus B) is an alphaherpesvirus commonly identified in NHPs of the genus *Macaca* (14, 19, 20, 71–74). The incidence in captive-bred and wild macaques can reach a seroprevalence of 100%, with approximately 2% shedding virus at any given point. In its natural host, *Macacine herpesvirus* is generally asymptomatic or may result in mild clinical disease characterized by lingual and/or labial vesicles or ulcers that resolve in 1–2 weeks. The virus persists in the sensory ganglion and can be reactivated with stressors, such as the onset of the breeding season, relocation, or immune suppression. Transmission among macaque species is through primary contact with conjunctival, genital, and mucous membrane. Diagnosis can be made through viral isolation, serological testing, or molecular diagnostics. Treatment of infected macaques is not necessary because lesions resolve in 1–2 weeks.

B virus transmission to people occurs through direct contact with saliva, ocular fluid, genital secretions, or scratches from infected monkeys. Indirect contact can come through contaminated needles, cell cultures, and possibly contaminated cages. Early local signs appear 48 hours after exposure, with erythema, ulcers, and pain at the site of infection and lymphadenopathy. Systemic signs appear in 1–3 weeks characterized by paresthesia, muscle weakness, conjunctivitis, and dysphagia. The signs can progress to encephalitis, with fever causing an ascending paralysis and subsequent respiratory paralysis. Encephalitis develops in 90% of the cases of people with B virus infection and 75% are fatal when appropriate care is not received. There has been one case of human-to-human transmission in which a spouse was treating her husband's B virus-induced lesions with hydrocortisone cream. Since the virus was first identified in 1933, there have been 40–50 documented cases. The transmission of B virus to research personnel in recent years seems to be an infrequent event likely due to the enhanced use of personal protective clothing at most facilities housing macaques.

B virus diagnosis can be made by viral isolation, serology, or molecular diagnosis. Following exposure, the area should be immediately cleaned with an appropriate disinfectant, such as chlorohexidine or povidone iodine, for a minimum of 15 minutes. Mucous membrane exposure should be flushed with sterile saline or water for 15 minutes. Antivirals, such as acyclovir or ganciclovir, are recommended following exposure until a confirmatory test of the animal can be completed. The research community has been developing herpes B virus-free macaque colonies with varying success, so herpes B virus animals could be acquired. However, given the difficulty of diagnosing B virus, care should be taken when working with any macaque species. ABSL2 practices with additional protection of the mucous membranes through the use of eye protection or an appropriate face shield will minimize exposure. If working with macaque species, facilities should have a postexposure plan to mitigate exposures. ABSL3 capabilities are required when working with material from which B virus is being cultured, and ABSL4 capabilities are required for experimental infections of macaques or other animal models.

Hantaviruses

Hantaviruses belong to the family *Bunyaviridae* and have worldwide distribution (14, 19, 20, 75–77). While most *Bunyaviridae* are transmitted through obligate intermediate arbovirus vectors, there are a number of hantaviruses that are carried by rodents that may infect humans, including the Old World hantaviruses (Hantaan virus, Seoul virus, Puumala virus, Dobrava virus) resulting in a hemorrhagic fever with renal syndrome (HFRS), and New World hantaviruses (Sin Nombre virus, Andes virus, New York virus) resulting in hantavirus pulmonary syndrome (HPS). The hantaviruses are transmitted directly from their rodent host. The primary reservoir is wild rodents from the genera *Apodemus, Clethrionomys, Mus, Rattus, Pitymys, Microtus, Peromyscus, Sigmodon*, and *Oryzomys*. The seroprevalence of wild rodents is up to 45%, and it is estimated that 25% of the wild rodent population may be infected with a hantavirus. In the reservoir host, the disease is subclinical. Serologic assays and molecular diagnostics can be used for diagnosis.

There are about 200,000 annual cases of HFRS, and there have been about 300 cases of HPS since its discovery in 1993. The incidence of laboratory-acquired infections is higher in Asia. There are no reports of acquiring the disease in the United States from laboratory rats, other than from experimentally infected rats. The virus is transmitted through the inhalation of infectious aerosols and is excreted in large amounts in the urine, saliva, feces, and respiratory secretions. The incubation period is 1–4 weeks with early clinical signs characterized by sudden fever, malaise, myalgia, nausea, vomiting, and headaches. HFRS can lead to thrombocytopenia with hemorrhagic conjunctivitis, intestinal bleeding, and hematuria. HPS can lead to respiratory failure due to capillary leakage into the lungs. The fatality rate of HFRS is up to 15% and HPS is greater than 40%. The diagnosis is on the basis of clinical signs, serologic testing, and molecular diagnostics. Early treatment with antivirals (ribavirin) can hasten the course of infection. Symptomatic treatment to maintain renal and respiratory function is necessary. In the laboratory setting, controlling wild rodent populations is critical to minimize the exposure to pathogen-free research colonies. Considering most laboratory populations are not infected with hantaviruses, ABSL2 precautions are sufficient. In experimentally infected rodents, ABSL2 precautions are sufficient when experimentally infected rodents are known not to excrete the virus. However, ABSL3 precautions should be taken when working with samples from potentially infected rodents (such as in a field study), and ABSL4 precautions are necessary when working with infected rodents that are permissive for chronic infections.

Lymphocytic choriomeningitis

Lymphocytic choriomeningitis virus (LCMV) is a member of the *Arenaviridae* with a predilection for rodent reservoirs (14, 19, 20). The house mouse is the natural reservoir for the virus, whereas laboratory mice and hamsters are the most significant source of LCMV in a research setting. The virus is transmitted *in utero* and results in a subclinical infection with chronic viremia and viruria. Exposure of laboratory animals to the virus occurs primarily through the introduction of contaminated tumor cell lines. The diagnosis is made by virus isolation, molecular diagnostics, and serology.

LCMV has a worldwide distribution in people, with a seroprevalence of 2–10%. It is transmitted to people by parenteral inoculation, inhalation, or contamination of mucous membranes. People develop a flu-like syndrome 1–3 weeks after exposure. The disease can become more severe in some individuals and is characterized by maculopapular rash, lymphadenopathy, and meningitis. It is rarely fatal. LCMV infections during pregnancy may result in ocular abnormalities, macrocephaly, and microcephaly. The antiviral ribavirin has been beneficial in the treatment of other arenaviral infections but has limited evaluation in LCMV infections. Supportive care is indicated. Wild rodent-control programs are critical to minimize contamination of research rodent colonies, and screening of cell lines used for *in vivo* studies should be strongly considered. ABSL2 precautions and practices will minimize exposure. Given the potential for aerosol transmission, ABSL3 practices should be used when the virus is known to be present and there is a potential for producing aerosols.

Orf (contagious ecthyma)

Orf virus is a *Parapoxvirus* found in sheep, goats, and wild ungulates (14, 19, 20, 78, 79). It is endemic in many sheep and goat herds in the United States and worldwide. The disease affects all ages of animals but is most frequent and severe in younger animals. The disease is characterized by proliferative pustule encrustation on the lips, nostrils, mucous membranes, and urogenital orifices. Infection does not reliably confer protection against reinfection. Transmission is by direct contact and fomites. The virus can persist in the environment for up to 12 years.

Contact with diseased sheep or goats leads to infection through skin breaks in people and exposure to characteristic lesions, which are high in viral particles. The incubation period in people is 3–7 days, and lesions develop typically on the hands, arms, or face. The lesions are similar to those in sheep, forming a maculopapular rash or pustule progressing to a proliferative nodule with central umbilication. Lesions persist for 3–6 weeks and regress spontaneously with minimal scarring. More severe cases may rarely develop lymphadenopathy or generalized lesions. Diagnosis is on the basis of clinical history and characteristic lesions. Molecular diagnostics may be used to aid in diagnosis. ABSL2 precautions when working with sheep and goats, including the use of gloves, are sufficient to minimize infection.

Poxviruses

There are two poxviruses of NHPs that may cause disease in people, albeit these are rarely acquired in modern laboratory settings (14, 19, 20, 79–81).

Monkeypox virus is an *Orthopoxvirus* that can infect NHPs. The virus is isolated primarily from animals originating from Africa with squirrels and other rodents serving as the primary reservoir. There have been rare reports of monkeypox virus infecting laboratory primates, most occurring prior to 1990. The disease is transmitted by cutaneous inoculation or inhalation and has a 12-day incubation period. NHPs develop a fever followed by cutaneous eruptions 4–5 days later. These typically occur on the limbs, truck, face, lips, and buccal cavity. Within infected colonies, the disease can spread rapidly, with high morbidity. People develop a fever, malaise, and prostration, followed by a maculopustular rash and lymphadenopathy. The disease is usually self-limiting and lesions resolve. Supportive therapy may include antivirals. There have been no reported cases of monkeypox being transmitted from laboratory NHPs to research personnel.

The first reported monkeypox outbreak in the United States occurred in 2003 and was traced back to the importation of an infected Gambian giant pouched rat that infected a prairie dog pet supplier. Subsequently there were 71 individuals exposed to infected prairie dogs, and 26% of the patients were hospitalized. The diagnosis can be made on the basis of clinical history, characteristic lesions, and histologic assessment of infected skin. Serology and viral isolation can also be performed. This disease resembles smallpox. Smallpox vaccination provides protective immunity in people and has been used to treat NHPs; however, the eradication of smallpox has reduced the number of vaccinated individuals over the years.

Tanapox virus is a *Yatapoxvirus* that primarily infects NHPs of the genus *Presbytis* in African and captive macaques in the United States. In both NHPs and people, the disease is characterized by a circumscribed, elevated red lesion on the eyelids, face, body, or genitalia. People may also experience fever, headache, and malaise. The infection can spread rapidly by direct contact through a NHP colony, and people are exposed from viral exposure through skin abrasions. The lesions spontaneously regress in 4–5 weeks. Poxvirus diseases are uncommon in research colonies of NHPs, and exposure can be minimized using ABSL2 precautions.

Rabies

Rabies is caused by a rhabdovirus and has a broad reservoir host range of small mammals, including dogs, cats, skunks, raccoons, and bats (14, 19, 20). Wild-caught rodents and rabbits that may be brought into the laboratory setting (or studied in the wild) may carry rabies and present a risk to personnel. Wildlife account for over 90% of the rabid animals in the United States. Viral transmission occurs primarily from the bite of a rabid animal, but infected saliva that enters through skin abrasions or mucous membranes may also transmit the disease. Infected animals display a change in behavior and aggression, with death 4–10 days after the onset of clinical signs due to paralysis. Infected animals should be euthanized because there is no treatment for the disease.

The incidence of rabies in people in the United States is about 100 cases per year, but is more than 30,000 per year in Asia and Africa. The fatality rate has declined to one to two cases per year in the United States. One in five individuals develops rabies after exposure if not protected by vaccination or postexposure therapy. The incubation period varies from 10 days to 3 months, with initial signs occurring 2–4 days after infection and consisting of nausea, vomiting, headache, paresthesia, and pain at the site of the bite. The disease can progress to restlessness, ptyalism, dysphagia, rage, insomnia, convulsions, spasms, muscle twitching, and hydrophobia. Flaccid paralysis of the head and limb muscles may occur. Death is usually the result of respiratory and cardiac failure. Clinical cases of rabies are almost always fatal.

The diagnosis is on the basis of clinical signs and history of exposure. Definitive diagnosis is on the basis of evaluation of histologic sections of brain with demonstrative Negri bodies. These can be identified antemortem in people in corneal imprints and biopsy specimens. Viral culture and molecular diagnostics are also available. Prevention with prophylactic rabies vaccine in individuals at high risk and use of vaccinated or pathogen-free animals will reduce the potential exposure in the research setting. Laboratory dogs and cats are typically vaccinated for rabies or acquired from closed colonies that minimize their potential exposure to rabies. Animals suspected of having rabies should be quarantined and euthanized if they display clinical signs. ABSL2 precautions and practices will minimize exposure

Fungal Infections

Dermatomycosis

There are a number of dermatophytes that can result in ringworm in people and animals (14, 19, 20). These include *Microsporum canis*, *Trichophyton mentagrophytes*, and *Trichophyton verrucosum*. *M. canis* is prevalent in dogs, cats, and NHPs, *T. mentagrophytes* is prevalent in rodents and rabbits, and *T. verrucosum* is prevalent in livestock. The carrier incidence of *M. canis* in dogs and cats is 2–15%, depending on the geographic location of the animals. A 1981 survey of laboratory rodents found a subclinical incidence of 1–5% in mice, rats, guinea pigs, and rabbits. Dermatophytes are transmitted by direct contact with infected lesions or fomites. The incubation period is several days to weeks, with the most common clinical presentation being hair loss, scaling, and crusting with pruritus. People can develop ringworm-like lesions on the head or skin. The diagnosis can be made by performing a Wood's lamp examination of the affected areas, and potassium hydroxide (KOH) preparations on the hair samples can demonstrate fungal elements. Dermatophyte test media can be used, and they require a 5- to 21-day incubation period. The disease is typically self-limiting, and infected animals and people can be treated with topical and systemic antifungals, such as ketoconazole or griseofulvin. ABSL2 precautions and practices will minimize exposure.

Parasitic Infections

There are many helminth parasite infections that animals can harbor that can result in zoonotic disease (14, 19, 20). However, the chances of this occurring in a laboratory environment are very unlikely. This is due to the health status of the animals, the routine preventative care of the animals, and the frequency of sanitation that minimizes the potential for the helminth ova to embryonate. Two helminth parasites are notable to mention. *Rodentolepis nana* (formerly known as *Hymenolepis nana*) is a cestode parasite of rodents with a direct life cycle. The cestode resides in the small intestine of rodents and is often subclinical but may lead to impactions. The direct life cycle has a 14- to 16-day prepatent period, and auto-reinfection is common. People are infected via ingestion of the ova. *R. nana* is the only human tapeworm that does not require an intermediate host. Severe infection in people can result in abdominal pain and diarrhea. Diagnosis is on the basis of fecal flotation and demonstration of ova. Praziquantel can be used to treat infections.

Strongyloides is a nematode parasite capable of direct human infection that commonly infects dogs, cats, and NHPs. These infections can be subclinical or lead to enterocolitis in animals. Infection in people occurs from percutaneous exposure of the filariform larvae. Light infections are typically subclinical, whereas more severe cases may present with abdominal pain, diarrhea, urticaria, and weight loss. Diagnosis is on the basis of fecal flotation using the Baermann technique. Treatment with antiparasitics, such as benzimidazoles, thiabendazole, albendazole, and ivermectin, can be used.

Animals can harbor a number of ectoparasites with zoonotic potential. Similar to the helminth parasites, animal health, preventative health programs, rigorous sanitation practices, and vermin control programs minimize the potential for these to be of concern in the laboratory setting. Diagnosis can be obtained by direct or microscopic examination of the skin or hair samples, or from bedding. ABSL-2 standard practices and precautions are adequate to prevent enteric exposure to endoparasites and minimize exposures to ectoparasites.

CONCLUSION

The occurrence of zoonotic diseases acquired from laboratory animals is a very uncommon event in modern laboratory animal facilities. The emphasis on the use of personal protective clothing to provide both biosecurity and biosafety, rigorous sanitation practices, and the acquisition of animals of known health status significantly reduce personnel exposure to zoonotic diseases. This chapter provides a brief overview of the potential zoonotic diseases that may be encountered when working with common laboratory animals. However, there is a greater risk when working with animal models experimentally infected with zoonotic agents. An understanding of the risks, the life cycle, disease transmission, and appropriate biosafety level procedures will further reduce the risk to personnel.

References

1. **Lipman NS.** 2009. Rodent facilities and caging systems, p 265–288. *In* Hessler JR, Lehner NDM (ed), *Planning and Designing Research Animal Facilities*. Academic Press, Burlington, MA.
2. **Center for Disease Control and Prevention and National Institutes of Health.** 2007. *Biosafety in Microbiological and Biomedical Laboratories*, 5th ed. HHS Publication no. (CDC) 21-112. http://www.cdc.gov/biosafety/publications/bmbl5/BMBL.pdf
3. **Hessler JR.** Barrier housing for rodents, p 335–346. *In* Hessler JR, Lehner NDM (ed), *Planning and Designing Research Animal Facilities*. Academic Press, Burlington, MA.
4. **Weigler BJ, Di Giacomo RF, Alexander S.** 2005. A national survey of laboratory animal workers concerning occupational risks for zoonotic diseases. *Comp Med* **55:**183–191.
5. **Jackson J, Villarroel A.** 2012. A survey of the risk of zoonoses for veterinarians. *Zoonoses Public Health* **59:**193–201.
6. **Jeal H, Jones M.** 2010. Allergy to rodents: an update. *Clin Exp Allergy* **40:**1593–1601.
7. **National Research Council.** 2003. *Occupational Health and Safety in the Care and Use of Nonhuman Primates*. National Academy Press, Washington, DC.
8. **Fox JG, Newcomer CE, Rozmiarek H.** 2002. Selected zoonoses, p 1059–1106. *In* Fox JG, Anderson LC, Loew FM, Quimby FW (ed), *Laboratory Animal Medicine*, 2nd ed. Academic Press, Burlington, MA.
9. **National Research Council.** 1997. Zoonoses, p 106–122. *In Occupational Health and Safety in the Care and Use of Research Animals*. National Academy Press, Washington, DC.
10. **Baker DG.** 2003. *Natural Pathogens of Laboratory Animals: Their Effects on Research*. ASM Press, Washington, DC.
11. **Carty AJ.** 2008. Opportunistic infections of mice and rats: jacoby and Lindsey revisited. *ILAR J* **49:**272–276.
12. **Percy DH, Barthold SW.** 2007. *Pathology of Laboratory Rodents and Rabbits*, 3rd ed. Blackwell Publishing, Ames, IA.
13. **Center for Disease Control and Prevention.** 2012. 2012 Nationally notifiable diseases and conditions and current case definitions. Atlanta, GA. Available at http://www.cdc.gov/mmwr/preview/mmwrhtml/mm6153a1.htm. Accessed September 30, 2015.
14. **Hankenson FC, Johnston NA, Weigler BJ, Di Giacomo RF.** 2003. Zoonoses of occupational health importance in contemporary laboratory animal research. *Comp Med* **53:**579–601.
15. **Scheftel JM, Elchos BL, Cherry B, DeBess EE, Hopkins SG, Levine JF, Williams CJ, Bell MR, Dvorak GD, Funk RH, Just SD, Samples OM, Schaefer EC, Silvia CA, National Association of State Public Health Veterinarians (NASPVH).** 2010. Compendium of veterinary standard precautions for zoonotic disease prevention in veterinary personnel: National Association of State Public Health Veterinarians Veterinary Infection Control Committee 2010. *J Am Vet Med Assoc* **237:**1403–1422.
16. **Bureau of Labor Statistics. United States Department of Labor.** 2013. Nonfatal cases involving days away from work: selected characteristics. Available at: www.bls.gov. Accessed July 3, 2013.
17. **Miller JM, Astles R, Baszler T, Chapin K, Carey R, Garcia L, Gray L, Larone D, Pentella M, Pollock A, Shapiro DS, Weirich E, Wiedbrauk D, Biosafety Blue Ribbon Panel, Centers for Disease Control and Prevention (CDC).** 2012. Guidelines for safe work practices in human and animal medical diagnostic laboratories. *MMWR Suppl* **61**(Suppl)**:**1–102.
18. **Walker D, Campbell D.** 1999. A survey of infections in United Kingdom laboratories, 1994–1995. *J Clin Pathol* **52:**415–418.
19. **Krauss H, Weber A, Appel M, Enders B, Isenberg HD, Schiefer HG, Slenczka W, Von Graevenitz A, Zahner H.** 2003. *Zoonoses Infectious Diseases Transmissible from Animals to Humans*, 3rd ed. ASM Press, Washington, DC.
20. **Newcomer CE.** 2000. Zoonoses, p 121–150. *In* Fleming DO, Hung DL (ed), *Biological Safety Principles and Practice*, 3rd ed. ASM Press, Washington, DC.
21. **Hawker J, Begg N, Blair I, Reintjes R, Weinberg J, Ekdahl K.** 2012. *Communicable Disease Control and Health Protection Handbook*, 3rd ed. Wiley-Blackwell, West Sussex, UK.
22. **Schuster FL, Ramirez-Avila L.** 2008. Current world status of Balantidium coli. *Clin Microbiol Rev* **21:**626–638.
23. **Hlavsa MC, Watson JC, Beach MJ.** 2005. Cryptosporidiosis surveillance—United States 1999–2002. *MMWR Surveill Summ* **54**(SS01)**:**1–8.
24. **Ramirez NE, Ward LA, Sreevatsan S.** 2004. A review of the biology and epidemiology of cryptosporidiosis in humans and animals. *Microbes Infect* **6:**773–785.
25. **Weir SC, Pokorny NJ, Carreno RA, Trevors JT, Lee H.** 2002. Efficacy of common laboratory disinfectants on the infectivity of Cryptosporidium parvum oocysts in cell culture. *Appl Environ Microbiol* **68:**2576–2579.
26. **Esch KJ, Petersen CA.** 2013. Transmission and epidemiology of zoonotic protozoal diseases of companion animals. *Clin Microbiol Rev* **26:**58–85.
27. **Atluri VL, Xavier MN, de Jong MF, den Hartigh AB, Tsolis RM.** 2011. Interactions of the human pathogenic Brucella species with their hosts. *Annu Rev Microbiol* **65:**523–541.
28. **Greene CE, Carmichael LE.** 2006. Canine brucellosis, p 369–381. *In* Greene CE (ed), *Infectious Diseases of the Dog and Cat*, 3rd ed. Elsevier, St. Louis, MO.
29. **Seleem MN, Boyle SM, Sriranganathan N.** 2010. Brucellosis: a re-emerging zoonosis. *Vet Microbiol* **140:**392–398.
30. **Traxler RM, Lehman MW, Bosserman EA, Guerra MA, Smith TL.** 2013. A literature review of laboratory-acquired brucellosis. *J Clin Microbiol* **51:**3055–3062.
31. **United States Department of Agriculture.** 2011. National Brucellosis Slaughter Surveillance Plan. June 30, 2011. http://www.aphis.usda.gov/animal_health/animal_diseases/brucellosis/downloads/nat_bruc_slaughter_surv_plan.pdf. Accessed July 29, 2013.
32. **Fox JG.** 2006. Enteric bacterial infections, p 339–343. *In* Greene CE (ed), *Infectious Diseases of the Dog and Cat*, 3rd ed. Elsevier, St. Louis, MO.
33. **Moore JE, Corcoran D, Dooley JSG, Fanning S, Lucey B, Matsuda M, McDowell DA, Mégraud F, Millar BC, O'Mahony R, O'Riordan L, O'Rourke M, Rao JR, Rooney PJ, Sails A, Whyte P.** 2005. Campylobacter. *Vet Res* **36:**351–382.

34. **Gaastra W, Lipman LJA.** 2010. Capnocytophaga canimorsus. *Vet Microbiol* **140:**339–346.
35. **Oehler RL, Velez AP, Mizrachi M, Lamarche J, Gompf S.** 2009. Bite-related and septic syndromes caused by cats and dogs. *Lancet Infect Dis* **9:**439–447.
36. **Chomel BB, Kasten RW.** 2010. Bartonellosis, an increasingly recognized zoonosis. *J Appl Microbiol* **109:**743–750.
37. **Guptill-Yoran L.** 2006. Feline bartonellosis, p 511–518. *In* Greene CE (ed), *Infectious Diseases of the Dog and Cat,* 3rd ed. Elsevier, St. Louis, MO.
38. **Longbottom D, Coulter LJ.** 2003. Animal chlamydioses and zoonotic implications. *J Comp Pathol* **128:**217–244.
39. **Rohde G, Straube E, Essig A, Reinhold P, Sachse K.** 2010. Chlamydial zoonoses. *Dtsch Arztebl Int* **107:**174–180.
40. **Veraldi S, Girgenti V, Dassoni F, Gianotti R.** 2009. Erysipeloid: a review. *Clin Exp Dermatol* **34:**859–862.
41. **Wang Q, Chang BJ, Riley TV.** 2010. Erysipelothrix rhusiopathiae. *Vet Microbiol* **140:**405–417.
42. **Evangelista KV, Coburn J.** 2010. Leptospira as an emerging pathogen: a review of its biology, pathogenesis and host immune responses. *Future Microbiol* **5:**1413–1425.
43. **Goldstein RE.** 2010. Canine leptospirosis. *Vet Clin North Am Small Anim Pract* **40:**1091–1101.
44. **Greene CE, Sykes JE, Brown CA, Hartmann K.** 2006. Leptospirosis, p 402–417. *In* Greene CE (ed), *Infectious Diseases of the Dog and Cat,* 3rd ed. Elsevier, St. Louis, MO.
45. **Guerra MA.** 2009. Leptospirosis. *J Am Vet Med Assoc* **234:**472–478, 430.
46. **Hartskeerl RA, Collares-Pereira M, Ellis WA.** 2011. Emergence, control and re-emerging leptospirosis: dynamics of infection in the changing world. *Clin Microbiol Infect* **17:**494–501.
47. **Centers for Disease Control and Prevention (CDC).** 1993. Tuberculosis in imported nonhuman primates—United States, June 1990-May 1993. *MMWR Morb Mortal Wkly Rep* **42:**572–576.
48. **Garcia MA, Yee J, Bouley DM, Moorhead R, Lerche NW.** 2004. Diagnosis of tuberculosis in macaques, using whole-blood in vitro interferon-gamma (PRIMAGAM) testing. *Comp Med* **54:**86–92.
49. **Nahid P, Menzies D.** 2012. Update in tuberculosis and nontuberculous mycobacterial disease 2011. *Am J Respir Crit Care Med* **185:**1266–1270.
50. **O'Brien DJ, Schmitt SM, Berry DE, Fitzgerald SD, Vanneste JR, Lyon TJ, Magsig D, Fierke JS, Cooley TM, Zwick LS, Thomsen BV.** 2004. Estimating the true prevalence of Mycobacterium bovis in hunter-harvested white-tailed deer in Michigan. *J Wildl Dis* **40:**42–52.
51. **Saggesse MD (ed.).** 2012. Mycobacteriosis. Vet Clin North Am 42:1–131.
52. **Sakamoto K.** 2012. The pathology of Mycobacterium tuberculosis infection. *Vet Pathol* **49:**423–439.
53. **United States Department of Agriculture.** 2009. Bovine tuberculosis, infected cattle detected at slaughter and number of affected cattle herds, United States, 2003–2009. Available at http://www.aphis.usda.gov/animal_health/animal_diseases/tuberculosis/ downloads/tb_erad.pdf. Accessed August 7, 2013.
54. **Elliott SP.** 2007. Rat bite fever and Streptobacillus moniliformis. *Clin Microbiol Rev* **20:**13–22.
55. **Gaastra W, Boot R, Ho HTK, Lipman LJA.** 2009. Rat bite fever. *Vet Microbiol* **133:**211–228.
56. **Angelakis E, Raoult D.** 2010. Q fever. *Vet Microbiol* **140:**297–309.
57. **Anderson A, Bijlmer H, Fournier PE, Graves S, Hartzell J, Kersh GJ, Limonard G, Marrie TJ, Massung RF, McQuiston JH, Nicholson WL, Paddock CD, Sexton DJ, Center for Disease Control and Prevention.** 2013b. Diagnosis and management of Q fever—United States, Recommendations from CDC and the Q fever working group. *MMWR Recomm Rep* **62**(RR-03)**:**1–30.
58. **Delsing CE, Warris A, Bleeker-Rovers CP.** 2012. Q fever: still more queries than answers. *Adv Exp Med Biol* **719:**133–143.
59. **Fournier PE, Marrie TJ, Raoult D.** 1998. Diagnosis of Q fever. *J Clin Microbiol* **36:**1823–1834.
60. **Georgiev M, Afonso A, Neubauer H, Needham H, Thiery R, Rodolakis A, Roest H, Stark K, Stegeman J, Vellema P, van der Hoek W, More S.** 2013. Q fever in humans and farm animals in four European countries, 1982 to 2010. *Euro Surveill* **18:**20407. http://www.eurosurveillance.org/ViewArticle.Aspx?ArticleID=20407. Accessed August 2, 2013.
61. **Greene CE, Breitschwerdt EB.** 2006. Rocky Mountain spotted fever, murine typhuslike disease, rickettsialpox, typhus, and Q fever, p 243–245. *In* Greene CE (ed), *Infectious Diseases of the Dog and Cat,* 3rd ed. Elsevier, St. Louis, MO.
62. **Maurin M, Raoult D.** 1999. Q fever. *Clin Microbiol Rev* **12:** 518–553.
63. **McQuiston JH, Childs JE.** 2002. Q fever in humans and animals in the United States. *Vector Borne Zoonotic Dis* **2:**179–191.
64. **Foley, S.L., A.M. Lynne, R. Nayak.** 2008. Salmonella challenges: prevalence in swine and poultry and potential pathogenicity of such isolates. J Anim Sci **86**(E Suppl)**:**E149–E162.
65. **Marks SL, Rankin SC, Byrne BA, Weese JS.** 2011. Enteropathogenic bacteria in dogs and cats: diagnosis, epidemiology, treatment, and control. *J Vet Intern Med* **25:**1195–1208.
66. **Rhoades JR, Duffy G, Koutsoumanis K.** 2009. Prevalence and concentration of verocytotoxigenic *Escherichia coli, Salmonella enterica* and *Listeria monocytogenes* in the beef production chain: a review. *Food Microbiol* **26:**357–376.
67. Salmonellosis. http://www.compliance.iastate.edu/ibc/guide/zoonoticfactsheets/ Salmonellosis.pdf. Accessed September 9, 2013.
68. **Stevens MP, Humphrey TJ, Maskell DJ.** 2009. Molecular insights into farm animal and zoonotic Salmonella infections. *Philos Trans R Soc Lond B Biol Sci* **364:**2709–2723.
69. **Weese JS.** 2011. Bacterial enteritis in dogs and cats: diagnosis, therapy, and zoonotic potential. *Vet Clin North Am Small Anim Pract* **41:**287–309.
70. **Burgos-Rodriguez AG.** 2011. Zoonotic diseases of primates. *Vet Clin North Am Exot Anim Pract* **14:**557–575, viii.
71. **Centers for Disease Control and Prevention (CDC).** 1998. Fatal Cercopithecine herpesvirus 1 (B virus) infection following a mucocutaneous exposure and interim recommendations for worker protection. *MMWR Morb Mortal Wkly Rep* **47:**1073–1076, 1083.
72. **Cohen JI, Davenport DS, Stewart JA, Deitchman S, Hilliard JK, Chapman LE, B Virus Working Group.** 2002. Recommendations for prevention of and therapy for exposure to B virus (cercopithecine herpesvirus 1). *Clin Infect Dis* **35:**1191–1203.
73. **Elmore D, Eberle R.** 2008. Monkey B virus (Cercopithecine herpesvirus 1). *Comp Med* **58:**11–21.
74. **Estep, R.D., I. Messaoudi, S.W. Wong.** 2010. Simian herpesviruses and their risk to humans. Vaccine **28S2:**B78–B84.
75. **Krüger DH, Schönrich G, Klempa B.** 2011. Human pathogenic hantaviruses and prevention of infection. *Hum Vaccin* **7:** 685–693.
76. **Mir MA.** 2010. Hantaviruses. *Clin Lab Med* **30:**67–91.
77. **Simmons JH, Riley LK.** 2002. Hantaviruses: an overview. *Comp Med* **52:**97–110.
78. **Haig DM, Mercer AA.** 1998. Ovine diseases. Orf. *Vet Res* **29:**311–326.
79. **Lewis-Jones S.** 2004. Zoonotic poxvirus infections in humans. *Curr Opin Infect Dis* **17:**81–89.
80. **Centers for Disease Control and Prevention (CDC).** 2003. Multistate outbreak of monkeypox—Illinois, Indiana, and Wisconsin, 2003. *MMWR Morb Mortal Wkly Rep* **52:**537–540.
81. **Essbauer S, Pfeffer M, Meyer H.** 2010. Zoonotic poxviruses. *Vet Microbiol* **140:**229–236.
82. **Center for Disease Control and Prevention.** 2013a. http://www.cdc.gov/. Accessed July 30, 2013.

Biological Safety Considerations for Plant Pathogens and Plant-Associated Microorganisms of Significance to Human Health

3

ANNE K. VIDAVER, SUE A. TOLIN, AND PATRICIA LAMBRECHT

Cross-kingdom or interkingdom pathogenic microorganisms are now recognized more commonly (1, 2). There are an increasing number of organisms, and occasionally even the same strains of an organism, that can colonize and/or infect both plants and humans. Pathogenic microorganisms demonstrating cross-kingdom host ranges may have been overlooked for several reasons. Taxonomic name changes may have resulted in misidentification. In addition, a different name may have been used to describe the same microorganism (2, 3). A primary reason for this failure to recognize nontraditional pathogenic microorganisms may be due to little earlier knowledge that some plant pathogens can grow at temperatures of body extremities and external parts of the body, which may be as low as 33.2°C (4). The thermal growth tolerance for most fungi is significant at 37°C or below (5). Fungi associated with human disease are speculated to have originated from asymptomatic or diseased plants (6). However, mammalian defense systems of innate temperature are a potent nonspecific defense against most fungi (5) and bacteria. As to classification of cross-kingdom microorganisms, Hubalek (7) suggested that emerging human infectious diseases could be grouped into those transmissible between humans (anthroponoses), those transmissible from animals to humans (zoonoses), and those transmissible to humans from the environment as sapronoses. However, the term "sapronoses" was presented as those diseases having an environmental reservoir (organic matter, soil, and plants). Other treatises define sapronoses as diseases whose source is only an abiotic substrate (nonliving environment). Hubalek (7) did not address diseases of humans transmissible from plants, but the term "phytoses" has been used in presentations by the Centers for Disease Control and Prevention (CDC) (R. V. Tauxe, personal communication), although "phytonoses" would be consistent with Hubalek's classification system (7).

The many texts and manuals that deal with methods of working with plant pathogens and associated microbes do not provide any cautions or statements with respect to potential risk to human health (8–21), except for a one-page statement in a handbook (22). In medical texts, there are no cautionary statements about the potential for occupational exposure to cross-kingdom plant pathogens or plant-associated organisms, other than as allergens (23, 24). For access to the literature, as well as to journals covering plant diseases or associations, the reader is referred to common texts in plant pathology. Our goal in this chapter is to capture some of the published information that links organisms in plant disease reports with those in clinical and other medical reports and to show that phytonoses are not limited to familiar human pathogens contaminating foods.

Commonalities in gene sequences and function among pathogens of plants, animals, and humans are no longer surprising (1, 2, 25–31). In addition, some plant-associated microorganisms that may prevent plant disease nevertheless can cause human allergies or disease. Such biocontrol agents are regulated commercially in the U.S. by the Environmental Protection Agency (EPA) for environmental risks but not for risks to humans. Hence, it is prudent to assess safety issues with respect to human exposure to plant-associated microorganisms, including viruses, for laboratory and confined uses. Keeping with the purpose of this book, our major focus is on risk characterization and mitigation of worker exposure during specimen examination, culturing, inoculation of plants, and diagnosis of plant pathogens known to affect human health.

A modest number of effectors produced by both bacteria and fungi are recognized pathogenicity factors (1, 2). The extent to which the same genes operate in both plants and humans is not clear. For bacteria, commonalities among plant, animal, or human pathogens are most evident in type III secretion pathways (1, 25, 32, 33) and virulence factors in pseudomonads (29, 30, 31). In most cases, virulence or pathogenicity factors in common still await discovery. In fungi, commonalities are at the structural, morphological, biochemical, and genetic levels (2, 24, 34, 35).

Gene sequences of some plant viruses are sufficiently homologous to other viruses to classify them into higher taxa whose members are predominantly human or animal viruses, namely *Picornavirales* and *Mononegavirales*, including *Rhabdoviridae*, *Bunyaviridae*, and *Reoviridae* (36). The close relationship of these and other viruses is not surprising given the small size of viral genomes, similarity in functionalities, and uniqueness of viral hallmark genes (37, 38).

The emergence or reemergence of human diseases caused by microorganisms is due to many factors (39, 40). In humans, for example, both *Burkholderia cepacia* and *Pseudomonas aeruginosa* have become important pathogens in cystic fibrosis; both are also an infrequent cause of infection in non-cystic fibrosis patients (41, 42). The emergence of newly identified fungal pathogens and the reemergence of previously uncommon fungal diseases of humans (mycoses) are attributed to an increase in the number of susceptible individuals, such as bone marrow and organ transplant recipients, cancer patients being treated with chemotherapy, critically ill persons, very-low-birth-weight infants, and persons with certain other infections, notably human immunodeficiency virus (HIV) (43). Clinically relevant mycoses may occur in healthy, immunocompetent individuals as well (44).

PLANT AND HUMAN CROSS-KINGDOM PATHOGENS

There are increasing numbers of plant-pathogenic microbial organisms associated with human diseases or maladies (2). Of the more than 500 species of bacteria isolated from human infections (45), about 5% are also known as plant pathogens or as biocontrol agents associated with plants. Reports of human infections for 28 bacterial species are listed in Table 1. Of these species, seven are Gram positive, with three being *Bacillus* spp. The remaining 21 are Gram negative and include multiple species in several genera and single species in a few other genera. Homologues of bacterial genes coding for virulence or pathogenicity factors in common between crossover pathogens are becoming better known (1, 2, 31, 46). Only a few direct references combine plant and human microbial associations (1, 2). However, Kirzinger et al. (1) speculated that some plant pathogens, such as *Pantoea stewartii*, may be found in a cross-kingdom disease. This prevalent bacterial pathogen of corn is not known to present a human malady.

Of the more than 1,400 recognized pathogens of humans, 300 are species of fungi reported to have been isolated from humans with infectious systemic diseases (45, 47), with about 12 associated with serious diseases. At least 54 are also known as plant pathogens (Table 2). Additional plant hosts of plant-pathogenic fungi can be found at http://nt.ars-grin.gov/fungaldatabases/index.cfm (48). Of these, most are in the phylum *Ascomycota*, popularly known as ascomycetes (49). Multiple species in the genera *Alternaria*, *Aspergillus*, *Bipolaris*, *Colletotrichum*, *Curvularia*, and *Fusarium* have been implicated in human disease. *Fusarium moniliforme*, *F. oxysporum*, *F. solani*, and *F. verticillioides* are the most common in human infections (50). Twelve additional genera are represented by one species each. Many of these plant pathogens are among the fungal genera included in a review of fungi that are linked to allergic asthma and other diseases in humans (24). This suggests specificity in evolution of pathogenicity as well as susceptibility in humans, which may simply be the ability of humans to meet the nutritional and asexual reproductive capacity or other characteristics of these fungi and to respond to allergens in spores or mycelia. Many plant-pathogenic fungi are associated with mycotoxicoses, but these are not included in Table 2 because entry is via consumption of contaminated

TABLE 1.

Taxa of bacterial pathogens and saprophytes of plants associated with human disease or maladies[a]

Taxon	Plant disease/association	Human disease/association
Agrobacterium radiobacter (syn. *Rhizobium radiobacter*)	Plant-associated bacteria, rhizosphere; registered biocontrol agent for crown gall, strain K84 (Galltrol A, AgBioChem Inc., Orinda, CA) in fruit, nut, and ornamental nursery stock and strain K1026 (Nogall, Bio-Care Technology Pty Ltd., Somersby, New South Wales, Australia) for control of crown gall on fruit and nut trees, caneberries, roses, and other ornamentals	Opportunist pathogen (100) Bacterial endophthalmitis (101); bacteremia (102), endocarditis (103), peritonitis (104), and urinary tract infections (105, 106)
Agrobacterium tumefaciens (syn. *Rhizobium tumefaciens*)	Agent of crown gall with wide host range (107, 108)	Peritonitis (109, 110), bacteremias (102), and urinary tract infection (111)
Bacillus megaterium	White blotch of wheat and bacterial wetwood of poplar and elm (112)	Oral mucosal inflammation (113)
Bacillus circulans	Date palm disease (114)	Peritonitis (115), serious nongastrointestinal infections in animals, and diarrheal enterotoxin production in human cells (116, 117)
Bacillus pumilus	Bacterial blotch of immature Balady peach (118); registered biocontrol agent, strain GB34 (Yield Shield, Gustafson, Plano, TX) for control of soilborne fungal pathogens causing root disease in soybean	Oral mucosal inflammation (113, 119)
Burkholderia cepacia	Sour skin of onion (120, 121) and cavity disease of *Agaricus bitorquis* (122, 123); phytoremediation (124) and endophyte (125)	Bacteremia (126); pulmonary complex (127, 128); serious respiratory pathogen in cystic fibrosis patients (28, 79); bacteremia, cardiac cirrhosis and cellulitis (129), and endophthalmitis (130)
Burkholderia cenocepacia	Rot of fruit (131)	Septicemia (1)
Burkholderia gladioli	Slippery skin of onion (132); decay of *Gladiolus* spp., *Iris* spp., and rice; leaf spot and blight of *Asplenium nidus* (133); bacterial disease of *Dendrobium* spp. orchid (134)	Bacteremia (135), pneumonia (136), and cervical adenitis (137)
Burkholderia glumae	Rice rot (138)	Chromic granulomatous disease (1)
Burkholderia plantarii	Seedling blight of rice, gladiolus, iris (138)	Meliodosis (1)
Burkholderia pseudomallei	Tomato rot (139)	Meliodosis, glanders (1)
Clostridium butyricum	Wetwood of poplar (140) and disease of hornbeam (141)	Necrotizing enterocolitis in babies (142)
Clostridium histolyticum	Plant associated (143)	Gas gangrene (myonecrosis) and necrotic lesions (144)
Curtobacterium flaccumfaciens pv. *flaccumfaciens*	Bean wilt and blight;saprophyte (145)	Septic arthritis (146)
Enterobacter cloacae	Wetwood on elm, internal decay of onion (147), and rhizome rot of edible ginger (148); grey kernel of macadamia (149); biocontrol agent (150–152)	Septicemia and respiratory tract infections (153) and gas gangrene (154)
Erwinia persicina (syn. *Erwinia nulandii*)	Necrosis in fruits, vegetables (155), necrosis of bean pods and seeds (156)	Urinary tract infection (157)
Klebsiella pneumoniae	Endophyte; many plant hosts, including maize (158)	Pneumonia (159), bacteremia (160), and meningitis (161)
Klebsiella variicola	Plant-associated on banana, rice, sugarcane, and maize (162)	Bacteremia and urinary tract infection (162)
Stenotrophomonas maltophilia	Plant-associated biocontrol bacterium and plant pathogen (163)	Bacteremia and respiratory tract infections (164, 165)
Pantoea agglomerans (syn. *Enterobacter agglomerans*, *Erwinia herbicola*)	Pathogen of *Wisteria* and onion; wetwood of elm; black flesh of pineapple and grapefruit; spot disease and frost damage on corn, soy, and clover; and disease of millet (166); saprophyte	Nosocomial/opportunistic infections and septic arthritis (167–170)
Pantoea ananatis	Eucalyptus, maize, rice pathogen; various symptoms; epiphyte, endophyte (1, 171)	Septicemia (see 1, 172)

(*continued*)

TABLE 1.

Taxa of bacterial pathogens and saprophytes of plants associated with human disease or maladies[a] *(Continued)*

Taxon	Plant disease/association	Human disease/association
Pseudomonas aeruginosa	Onion rot (173) and *Arabidopsis* rot	Burn wound infections and pneumonia (30, 42, 174) and meningitis, bacteremia, and sepsis (175)
Pseudomonas fluorescens	Registered biocontrol for *Erwinia amylovora* (176) on apple, cherry, almond, peach, and pear (Blight Ban A506, Frost Technology Corporation, Burr Ridge, IL); frost protection on fruit crops, almond, tomato, and potato to reduce frost-forming bacteria on leaves and blossoms (Frostban, Frost Technology Corporation)	Bacteremia (177)
Pseudomonas putida	Plant saprophyte with potential for application in biological control of plant pathogens, bioremediation, and production of bioplastics (178)	Nosocomial infections (179); meningitis (180); and bacteremia, pneumonia, and sepsis (175)
Rathayibacter toxicus (syn. *Clavibacter toxicus*)	Gummosis of cereals (181)	*Rathayibacter* poisoning; death of livestock associated with consumption of *Rathayibacter*-infected annual ryegrass (181); human disease speculative (182)
Serratia ficaria	Plant associated in fig trees (biological cycle of fig wasp) (183)	Organ infections in fig tree zones (184) and endophthalmitis, gallbladder empyema, and septicemia (185)
Serratia marcescens	Alfalfa crown and root rot (186), cucurbit yellow vine disease (187), and endophytic colonization of rice (188)	Respiratory tract infections, urinary tract infections, and bacteremia (189) and conjunctivitis, endocarditis, meningitis, and wound infections (190)
Xanthomonas campestris pv. campestris	Black rot of crucifers and wilt and blight stump rot of broccoli, cabbage, cauliflower, brussel sprouts, kale, mustard, radish, rutabaga, sunflower, stock, and turnip (8, 109)	Bacteremia (191)

[a]Unidentified or inadequately identified species of *Microbacterium* and *Streptomyces* have been reported for clinical infections (192, 193), and both identified and inadequately identified species have been associated with plants (109, 194, 195).

foods and unlikely to occur in a laboratory setting (51). A few human pathogens have been shown to infect the model plant *Arabidopsis thaliana* under artificial conditions (1, 2, 52), but these are not listed. Some presumed plant pathogens listed by Kirzinger et al. (1) are not listed here because we were unable to find the plant reference.

The question of whether plant viruses can cross the kingdom barrier and be pathogenic to humans has been recently posed by Balique et al. (53) and Mandal and Jain (54). There is renewed interest in this area, but the current consensus is that they only infect plants and therefore do not cause disease in humans (53). Metagenomics has greatly enhanced exploration of viruses associated with human disease (55, 56), and the concept of a complex human virome has emerged (57, 58). The shift in diagnostics toward molecular assays in clinical laboratories and in environmental samples has enabled discovery of increasing complex "viromes" in ecosystems. The relevance of a viral discovery by metagenomics to human health is discussed by Lipkin and Anthony (59) and is assigned a hierarchy of confidence levels, in which possible, probable, and confirmed causal relationships are concluded depending on additional biological data, including fulfillment of Koch's postulates.

Pathways for exposure of humans to phytoviruses are abundant, because they are highly prevalent in the environment, particularly in vegetables and fruits and in waters and soils (53). The highest reported incidence, and perhaps posing potential risks to humans, are those plant viruses that have no known vector and are extremely stable in soil, water, and nonliving plant sources. The first virus isolated (and crystallized), tobacco mosaic virus (TMV), has these properties, as do other members of the genus *Tobamovirus*, family *Virgaviridae*. TMV was associated with pulmonary disease in smokers in 1967 (60) (Table 3).

Recent studies confirmed that the TMV genome gained access to the human body through smoking, as TMV RNA and infectious virus were detected in saliva (61). The presence of serum anti-TMV antibodies has been verified and shown to be higher in smokers than in non-smokers (62). Human cells into which TMV RNA or cowpea mosaic virus was introduced directly have shown cytopathic effects, but no evidence of productive infections (Table 3).

TABLE 2.

Taxa of fungal pathogens and saprophytes of plants associated with human disease or maladies[a]

Taxon	Plant disease/association	Human disease/association
Phylum *Ascomycota*		
Alternaria alternata	Wide host range; causes leaf spots, blights, damping off, and stem and fruit rots (108). Black mold of tomato, leaf spot and ear and root of pea, seedling foliage blight of rot of maize, blight of foliage and pod of pea, seedling foliage blight of sugarcane (196), black point on wheat, dark flecks on *Dendrobium*, fruit spot on papaya, and leaf spot on garden bean.	Phaeohyphomycosis (197), mycotic keratitis, cutaneous (198) and visceral infections, and osteomyelitis; palatel ulcers (see 199)
Alternaria tenuissima	Strawberry fruit rot (200), leaf spot of broad bean (201), leaf spot of high bush blueberry (202), leaf spot of *Amaranthus hybridus* (203), and leaf spot of papaya	Phaeohyphomycosis (204, 205), sinusitis, and ulcerated cutaneous and visceral infections (206)
Aspergillus candidus	Decay of apple fruits (207), root rots	Cerebral aspergillosis (208), cutaneous aspergillosis, endocarditis, endophthalmitis, hepatosplenic aspergillosis, meningitis, myocarditis, onychomycosis, osteomyelitis, otomycosis, pulmonary aspergillosis (209), and sinusitis
Aspergillus flavus	Pathogen and saprophyte; has many hosts and causes such diseases as ear and kernel rot of maize (210, 211), yellow mold of peanut (212), and boll rot on cotton (213) Biopesticide: *A. flavus* strain AF36 (Arizona Cotton Research and Protection Council, Phoenix, AZ), a nontoxin-producing strain registered (EPA) on cotton fields in Texas and Arizona for control of strains of *A. flavus* which produce aflatoxin; *A. flavus* strain NRRL 21882, registered for use in peanut crops to control aflatoxin-producing strains of *A. flavus* (Circle One Global, Inc., Shellman, GA)	Systemic aspergillosis (214), endocarditis (215), and keratitis (216)
Aspergillus glaucus	Corn ear and kernel rot (217)	Cerebral, cutaneous, hepatosplenic, and pulmonary aspergillosis; endocarditis; endophthalmitis; meningitis; myocarditis; onychomycosis; osteomyelitis; otomycosis; and sinusitis (218)
Aspergillus niger	Black mold of peanut (217) and onion (219) and maize ear rot (217)	Aspergilloma (fruiting body in tissue), otomycoses (220), and pulmonary aspergillosis (221)
Aspergillus oryzae	Saprophyte and mycotoxin producer (222)	Necrotizing scleritis (223) and bronchopulmonary aspergillosis (224)
Aureobasidium pullulans	Russet of apple fruit (225) and d'Anjou pear (226) and stem break and browning of flax	Various opportunistic mycoses, pulmonary mycoses, scleritis (227), and phaeohyphomycosis (228)
Bipolaris australiensis	Leaf spot and crown and root rot of turfgrass, (www.apsnet.org/publications/commonnames/Pages/Turfgrasses.aspx)	Phaeohyphomycosis, allergic and chronic sinusitis, keratitis, endophthalmitis (229), endocarditis, osteomyelitis, meningitis, encephalitis, peritonitis, and pulmonary infection (230)
Bipolaris hawaiiensis	Leaf and culm lesions on Callides Rhodesgrass (231) and Bermuda grass disease (232)	Endophthalmitis, phaeohyphomycotic orbitopathy, sinusitis, and granulomatous encephalitis (233)
Bipolaris spicifera	Leaf spot of cotton (www.apsnet.org/publications/commonnames/Pages/cotton.aspx)	Phaeohyphomycosis (234), fungal endarteritis (235), meningitis (236), and peritonitis (237)
Chaetomium globosum	Disease in tomato (238) and infection of barley roots	Cerebral phaeohyphomycosis (239), onychomycosis (240), and pneumonia (241)
Cladosporium oxysporum	Leaf spots and blights of many plants and leaf spot of pepper (242)	Phaeohyphomycosis (243)
Colletotrichum coccodes	Black dot of tomato (244) and potato (245)	Phaeohyphomycosis (246)
Colletotrichum crassipes	Rots of tomato, other hosts (48)	Azole resistance; cutaneous phaeohyphomycosis (99)

(continued)

TABLE 2.

Taxa of fungal pathogens and saprophytes of plants associated with human disease or maladies[a] *(Continued)*

Taxon	Plant disease/association	Human disease/association
Colletotrichum dematium	Anthracnose of grains (48)	Azole resistance; keratitis (99)
Colletotrichum gloeosporioides	Anthracnose on many fruits and plantation crops (109), including anthracnose of papaya leaves (247) and avocado, poplar, aspen, and cottonwood shoot blight, fruit rot on apple and berries of coffee, and dieback of citrus	Keratitis (248) and phaeohyphomycosis (246); azoleresistance; cutaneous phaeohyphomysis (99) and keratitis (96)
Colletotrichum graminicola	Anthracnose leaf blight and stem rot of maize and sorghum (249)	Azole resistance; corneal pathogen (99)
Coniothyrium fuckelii	Stem blight, dieback, and canker of *Rosa* spp. and strawberry (*Fragaria*); black root rot and cane blight of *Rubus* spp. (250)	Mycotic keratitis (251) and liver infection (252)
Curvularia bracyspora	Leaf spot disease of *Rosa* spp. (253)	Necrotizing cutaneous infection (254) and mycotic keratitis (255)
Curvularia clavata	Leaf spot of maize (256)	Invasive sinusitis and cerebritis (257) and human skin infection (258)
Curvularia geniculata	Banana leaf spot (259) and melting out of turfgrasses (108)	Mycotic keratitis (260) and maduromycotic mycetomas in animals (261)
Curvularia lunata	Leaf rot of rice (262), leaf spot of bentgrass (263), melting out of turfgrasses (108), leaf spot of maize (264), and leaf spot of cotton (265)	Cerebral phaeohyphomycosis (266), systemic cutaneous infection (267), and allergic fungal rhinosinusitis (268)
Curvularia pallescens	Leaf spot of sugarcane (269), leaf spot and ear rot of maize (270), brown spot of asparagus (271), and leaf spot of rubber (272)	Phaeohyphomycosis (273)
Curvularia senegalensis	Seedling foliage blight on sugarcane (196) and leaf spot of maize and other hosts (274)	Mycotic keratitis (275)
Cylindrocarpon lichenicola	Postharvest fruit invasion and corm rot of *Colocasia esculenta* (taro) (276)	Disseminated infection (277) and keratomycosis (278)
Diaporthe phaseolorum (*Phomopsis phaseoli*)	Soybean and other diseases (279)	Eumycetoma (280)
Drechslera biseptata	*Drechslera* leaf spot of turfgrasses and black point in wheat grains (281), mycotoxin producer (282)	Brain abscess (283)
Fusarium anthophilum	Eucalyptus decline, sunflower wilt (284)	Isolated from several local, invasive disseminated infections (50)
Fusarium clamydosporum	Root rot and wilt of *Coleus forskohlii* (285) and blight of kangaroo paw (*Anigozanthos* spp.) (286)	Invasive infection (287)
Fusarium dimerum	One of several agents of fig endosepsis (288)	Disseminated infection (289), endocarditis (290), and eye infection (291)
Fusarium fujikuroi	Bakanae disease (foolish seedling disease in rice) (292)	Keratitis (293)
Fusarium incarnatum	One of several agents of fig endosepsis (288); walnut canker (294)	Disseminated infection (289), endocarditis (290), and eye infection (291)
Fusarium musae	Banana fruit pathogen (295)	Superficial and opportunistic, disseminated infections (296)
Fusarium moniliforme	Ear, root, and stalk rot and seedling blight of maize (108), sugarcane wilt complex; and pseudostem heart rot of banana (297), wide host range	Human fusariosis, local and systemic (298)
Fusarium napiforme	Sorghum rot (48)	Superficial, invasive and disseminated diseases (50)
Fusarium nygamai	Fruit rots (48)	Superficial, invasive and disseminated diseases (50)
Fusarium oxysporum	Wilts and blights on a wide range of vegetable and plantation crops, ornamentals, small grains (299), and turfgrasses, including potato, sugarcane, bean, cowpea, and *Musa* spp. (300), and corm and root rots (301)	Disseminated fusariosis (302), skin and nail infection (303), pneumonia (304), onychomycosis (305), and keratitis (293)

TABLE 2.
(Continued)

Taxon	Plant disease/association	Human disease/association
Fusarium proliferatum	Leaf, sheath, stem spots, damping off, and flower spots on *Dendrobium* and *Cattleya* orchid; head blight in wheat and other small-grain cereals (299); wilt and dieback of date palm (306); ear rot of maize (307)	Disseminated infection in immunosuppressed individuals (308), suppurative thrombophlebitis (309), and esophageal cancer (310)
Fusarium sacchari	Stalk rot, corn, rice (48)	Superficial, invasive and disseminated diseases (50)
Fusarium solani	Yellows, fruit rots, seedling rots, root rots, and damping off on a wide range of hosts; fungal root rot of banana (311); stem canker of sweet potato, black walnut, and poinsettia (108)	Invasive furiosis (312, 313), onychomycosis (305), and keratitis (96)
Fusarium vasinfectum	Pepper, cotton, bean diseases (314)	Superficial, invasive and disseminated disease (50)
Fusarium verticillioides	Ear rot of maize (307); fruit rot, sorghum, millet (48)	Superficial, invasive and disseminated diseases (50); esophageal cancer (310)
Lasiodiplodia theobromae	Fruit and stem rot of papaya (315); canker of dogwood (316); kumquat dieback (317); black kernel rot of maize; crown, finger, stalk, and peduncle rot of banana (318); and collar rot of peanut (319)	Subcutaneous abscess (320), ophthalmic mycoses (321), onychomycosis, and phaeohyphomycosis (322)
Lecythophora hoffmannii	Soft rots and decay of the surface layers of natural and preservative-treated timber (323)	Chronic sinusitis (324)
Paecilomyces variotii	Dieback and canker of pistachio (325)	Pneumonia (326), central nervous system infection (327), and peritonitis (328)
Phaeoacremonium inflatipes	Woody plants, wilt and decline (329)	Phaeohyphomycosis (329)
Phaeoacremonium parasiticum	Woody plants, wilt and decline (329)	Phaeohyphomycosis (subcutaneous infections to disseminated disease) (329)
Phaeoacremonium rubrigenum	Woody plants, wilt and decline (329)	Phaeohyphomycosis (329)
Phoma eupyrena	Blight of fir and pine species (330)	Cutaneous lesions (331)
Phylum *Zygomycota*		
Mucor circinelloides	Fruit rot of *Luffa acutangula* (332) and mucor rot of mango	Zygomycosis (333) and gangrenous mucormycosis (334)
Rhizopus oryzae	Fruit rots of pineapple, mango, and carrot (www.ismpminet.org/resources/ common/names)	Pulmonary zygomycosis (335)
Rhizopus stolonifer	Pre- and postharvest soft rots of many fruits, vegetables, and crops; sunflower head rot (336); seedling blight on lupine (108)	Zygomycosis (337)

[a] Not included are zoonotic fungi, those from animals that can cause infections in people, fungal biocontrol agents, and commercial fungi used in brewing or baking. Members of these classes can cause human clinical disease. Toxigenic fungi and mycotoxins are not addressed.

To determine baseline levels of RNA viruses associated with human gastroenteritis, such as noroviruses, shotgun-cloned cDNA libraries from fecal samples were sequenced from two healthy individuals (63). Surprisingly, 42 viral species were detected, 35 of which were viruses known to cause diseases of consumable fruits, vegetables, and cereals, suggesting food consumption contributed to the RNA virome of human feces. The most abundant virus was pepper mild mottle virus (PMMoV; genus *Tobamovirus*) (64). The four next most abundant viruses were cereal or grass viruses of three *Tombusviridae* genera—*Avenavirus* (oat chlorotic stunt virus), *Maculavirus* (maize chlorotic mottle virus), *Panicovirus* (panicum mosaic virus), and one *Tymoviridae* genus *Marifivirus* (oat blue dwarf virus). Additional sampling detected PMMoV in over 75% of all human fecal samples taken in southern California and in Singapore. PMMoV particles that were infectious to peppers were recovered from fecal extracts and from processed hot pepper sauces, demonstrating the high stability of this virus and a potential dietary source (63). However, the authors concluded that evidence for replication of PMMoV was lacking, as negative-strand RNA was not detected by strand-specific PCR (63). PMMoV and another tobamovirus, cucurbit-infecting kyuri green mottle mosaic virus, were detected in human fecal samples in Japan (65). When these plant viruses were present, the norovirus concentration was higher, leading to speculation of an interaction between the viruses (65); however, the investigators stopped short of suggesting that plant viruses replicated in human gut epithelial cells.

TABLE 3.

Selected taxa of plant viruses associated with human diseases and cellular responses

Taxon	Plant disease/association	Human disease/association
Family *Virgaviridae*, Genus *Tobamovirus*		
Tobacco mosaic virus (TMV)	Mosaic disease of tobacco, tomato, pepper and other hosts; historical model virus (338, 339)	Detected in sputum and thoracentesis fluids of smokers with pulmonary disease (59) Induces endoplasmic reticulum stress-related autophagy in human epithelial carcinoma (HeLa) cells (340) Anti-TMV antibodies in smokers (61)
Pepper mild mottle virus	Mottling of pepper leaves and severe distortion of fruits (63, 341, 342); abundant in processed hot pepper sauces (65, 66) and in water (85, 343, 344)	Abundant in feces of patients with fever, abdominal pains, and pruritus immune responses (65); associated with human viruses in river water and in human fecal samples (345)
Order *Picornavirales*, Family *Secoviridae*, Subfamily *Comovirinae*, Genus *Comovirus*		
Cowpea mosaic virus	Mosaic disease of cowpea and other legumes (346) modified for use for vaccine delivery (347)	Coat protein targets surface vimentin of human umbilical vein endothelial cells (HUVEC), HeLa cells, and KB cells (348, 349)
Family *Phycodnaviridae*		
Acanthocystis turfacea chlorella virus 1 (ATCV-1)	Infects *Chlorella heliozoae*, a blue-green alga, an endosymbiont of the heliozoon *A. turfacea* (71)	Human oropharyngeal virome, associated with impaired cognitive functions in humans and mice (70, 72) Persists in macrophages and induces inflammatory factors (350)
Phaeodactylum tricornutum virus (PtV)	Infects the aquatic diatom *P. tricornutum*	Consistently detected in cervico-vaginal secretions from women with colpitis, uterus fibroids, and uterine cervix erosion, who had been exposed to Black Sea water during prior beach visits (74)

A direct or indirect pathogenic role for plant viruses in humans was suggested by Colson et al. (66) after finding abundant sequences of the PMMoV in human fecal samples (Table 3). PMMoV RNA sequences were also reported from over 50% of processed hot pepper sauce or powder samples (66). It is speculative whether the symptoms, particularly from dermatology patients, are caused by the virus or are associated with the consumption of spicy foods. It is not clear whether PMMoV replicated in the human intestine, even though it was found in high concentrations and predicted by Colson et al. (66). This report led to examination of pepper sauces from many sites, including China, where PPMoV was detected by RT-PCR in 42 pepper sauces from the 10 main manufacturing provinces, and all remained infectious to tobacco (67).

In nature, plant-to-plant dissemination of most plant viruses is by biological vectors. Some vector-borne viruses accumulate and circulate through their vectors, including virus transmitted by aphids (*Luteoviridae*, *Nanovirus*), beetles (*Comovirus, Sobemovirus*), leafhoppers (*Secoviridae*), and whiteflies (*Geminiviridae*), but are not considered to replicate in insect cells (68). Many of these viruses can be present in the bat guano virome, reflecting the insect diet of bats (69). Certain other plant viruses are considered cross-kingdom pathogens because they are known to replicate in specific arthropod vectors. These viruses are members of one genus of *Bunyaviridae* (*Tospovirus*) in thrips, two genera of *Rhabdoviridae* in aphids or leafhoppers (*Cytorhabdovirus, Nucleorhabdovirus*), and three genera of *Reoviridae* in leafhoppers (Subfamilies *Sedoreovirinae*, including *Phytoreovirus*, and *Spinareovirinae*, including *Fijivirus* and *Oryzavirus*) (36, 70). Although these viruses are the most likely to cross kingdoms and affect humans, there are no published reports of this occurring. Other genera in these three families, as well as viruses in the families *Asfarviridae*, *Flaviviridae*, and *Togaviridae* that do not contain plant viruses, cause human encephalitis and hemorrhagic fevers and have mosquito or tick vectors (36).

A virus in the family *Phycodnaviridae*, Acanthocystis turfacea chlorella virus 1 (ATCV-1), which infects certain eukaryotic green algal species (Table 3), has been detected as part of the human oropharyngeal virome, and is associated with certain mental disorders of humans, including impairment of cognitive functions and major depression (71). Viral sequences were detected in humans by throat swabs, and the presence of ATCV-1 was associated with decreased performance on cognitive assessments of visual processing and visual motor speed. In mouse models, inoculation in the intestinal tract resulted in detectable deficits in memory and attention while navigating mazes. These large viruses are among the largest known with a double-stranded DNA (dsDNA) genome of about 370 kb and replicate in unicellular, chlorella-like green algae (72). The association of viruses with schizophrenia was also shown to differ with the oropharyngeal phageome (73).

Another megavirus has also been detected in human clinical material (Table 3). A *Phycodnavirus* that grew in the diatom *Phaeodactylum tricornutum* was detected from 41 samples, each consisting of clinical material pooled from two to five women having the same diagnosis of either colpitis, uterine fibroids, or erosion of uterus cervix. The common factor among the women was that all 182 were on Black Sea beaches 2 to 5 months before disease onset, suggesting viral transfer from an aquatic ecosystem to a probable role in human gynecological diseases (74).

RISK ASSESSMENT AND BIOSAFETY LEVELS

As the *NIH Guidelines* note (20), risk assessment is ultimately a subjective process. By the standard that agents are not associated with disease in healthy adult humans, almost all the plant pathogens and plant-associated microorganisms are in NIH Risk Group 1, for which biosafety level 1 (BSL1) is usually recommended. However, as noted in Tables 1–3, a few plant pathogens should be viewed as more problematic in that some strains of some species may infect both immunocompromised hosts and, more rarely, immunocompetent hosts. Sources of pathogens for microbial infection and contamination of plants include infected or infested seed, wind-driven inoculum, contaminated harvest machinery and containers, irrigation water, and postharvest handling (75). Humans can be exposed to microorganisms via any of these environmental sources. Exposure of humans is likely to increase during laboratory procedures of pathogen isolation and culturing, common diagnostic procedures conducted in plant disease clinics and in classrooms. The airborne spores produced by fungi can initiate mycoses if inhaled by humans. By-products of plant pathogens in food, such as mycotoxins, can also cause illness (51). Allergic reactions to plant pathogens and products such as toxins, while known to us, have not been well documented (76). Allergens from some plant-pathogenic fungi, primarily *Alternaria* and *Fusarium* species, were included in reviews by Horner et al. (77) and Simon-Nobbe et al. (78). In addition, the pathogen *Burkholderia pseudomallei*, a pathogen of tomato under artificial conditions (79), in humans is considered a Select Agent by the U.S. government and requires stringent operating conditions (see www.Selectagents.gov).

With the prevalent paradigm of specificity among plant and human pathogens, it is not surprising that the literature is sparse on comparative connections among taxonomic groups of pathogens of different hosts (29, 30, 32). Few studies have examined the capacity of microorganisms to cause disease in both plants and animals (1, 2, 29, 30, 31, 53, 80). It has long been recognized that certain plant-infecting viruses are able to propagate in insect vectors and that viruses in the same family (e.g., *Bunyaviridae*, *Reoviridae*, and *Rhabdoviridae*) are pathogens of vertebrates (68, 70). Recent evidence from viral, bacterial, and fungal evolution and diversity research provides avenues for further speculation on origins of host specificity and of cross-kingdom viruses (38, 81–83) and other microbes (84).

The rise of synthetic biology raises the possibilities of accidental and deliberate construction of known and new pathogens, including those with cross-kingdom potential. At present, we are not aware of such constructs. How novel taxa would be dealt with is also not clear.

Reports of bona fide human pathogens as endophytes (organisms able to colonize, live, and often grow without plant symptoms) or contaminants, especially in produce and other dietary sources (65, 66, 85, 86), emphasize the need for prudent use of precautionary steps in handling diseased plant specimens to minimize infection with agents of concern to humans. The bacterial species *Campylobacter*, *Listeria monocytogenes*, *Shigella*, *Salmonella enterica*, and *Escherichia coli* O157:H7 and several enteric viruses are recent pathogens of concern associated with edible plants (e.g., references 85, 87). With the infectious dose of some pathogens as low as 10 cells, e.g., *E. coli* O157:H7, even fresh produce with a pristine appearance can be suspect.

Risk Potential for Laboratory Personnel

As mentioned previously, healthy adults are not normally at risk of being infected by plant-associated microorganisms, but infections can occur (Tables 1–3), as well as allergic reactions. Good laboratory practices and some specific suggestions for dealing with plant-associated microorganisms are found in the methods manuals previously cited, the *NIH Guidelines*, and in plant diagnostic clinics associated with the U.S. National Plant Diagnostic Network (www.npdn.org). Immunocompromised adults, e.g., transplant recipients and those with immunodeficiencies (genetic or microbial), and persons with allergic sensitivities should take particular care in handling plant pathogens and microorganisms associated with plants. Large-scale cultures, aerosol-generating procedures, the use of needles and syringes, and direct contact with skin wounds are examples of activities that may increase the risk of exposure and infection.

The risk of plant viruses to humans is not considered in biosafety manuals. Plant viruses may pose a risk in the laboratory in the event that there is an open wound or percutaneous injury through which a virus may enter. Theoretically, a needlestick from a hollow-bore needle attached to a syringe containing a plant virus could provide a mode of entry into the human host, but not into receptive cells. Animals such as rabbits have been used for decades to produce high-titered antibodies specific against the intravenously or intramuscularly injected plant viruses, with no evidence of replication. The immune response is to the foreign antigenic protein, not the result of an infection (88). Anecdotally, plant virologists have been known to possess antibodies to viruses with which they have worked and have shown allergic sensitivities following the inhalation of viral aerosols. It is unlikely that plant viruses could infect and become viremic in humans. If a plant virus were able to replicate in the wounded cell, the end result is likely to be the creation of a new virus that may or may not cause disease in the infected host, because plant viruses lack known cellular receptor-binding proteins that may be acquired from a viral sequence preexisting in the cell. There has been one suggestion that a plant nanovirus had recombined with picorna-like viral RNA to form circoviruses capable of infecting vertebrates (89). The host switch recombination event is postulated to have occurred in a vertebrate when it was exposed to sap from a nanovirus-infected plant. Dimitrov (90) and Baranowski et al. (91) reviewed virus entry and the evolution of cell recognition by viruses and suggested that minimal changes in viral genomes may trigger changes in receptor usage for virus entry, but neither review considered plant viruses. Positive-strand RNA plant viruses are known to employ the surface of various subcellar membranes in the process of replication and virion synthesis (92). Studies of replication of brome mosaic virus in yeast have also demonstrated that host membranes are modified during a productive infection (93).

A review by Ritzenthaler (94) considered cell-to-cell movement parallels between human retroviruses, herpesviruses, and plant viruses, explaining the specific interaction between virus-encoded proteins and plasmodesmata. Interestingly, viruses replicating in both plant and insect cells use the same mechanism of forming tubules between adjacent cells through which virus particles move.

Containment

Samples of plant material obtained for isolation of pathogens or biocontrol agents are protected from contamination by using aseptic techniques, including surface sterilization of seeds, leaves, stems, and roots. Given the increased risk to people, it would be prudent to use disposable gloves and/or alcohol-based hand sanitizers in handling specimens. In isolation procedures, materials may be ground or sliced to obtain the putative organism directly or after concentration in centrifuge tubes; buffers are usually added to such materials to obtain suitable suspensions and to provide optimal pH and ionic composition for stabilizing the structure of the pathogen. Some pathogens are obligate parasites or simply cannot be cultured; such organisms are not currently known to present an infectious risk to humans, but they may be allergenic. After microorganisms are isolated, experimental procedures frequently involve the generation of aerosols, e.g., by flaming, low- or high-pressure spraying, or inoculating plants with various mechanical devices. Plants may need to be wounded prior to or during inoculation with such materials as silicon carbide or Carborundum, which itself is a corneal and respiratory irritant.

Air-purifying particulate filter respirators that are effective against intake of particles 2 μm or larger from the surrounding air should be considered for all personnel and recommended for at-risk groups who may be exposed to potentially infectious or allergenic aerosols. The organisms known to have been associated with a disease condition in humans, and thus of particular concern, are listed in Tables 1–3. The N95 National Institute for Occupational Safety and Health (NIOSH) series of filters are recommended. They are easy to wear, disposable, and of modest cost. In scientific supply catalogs, these filters have NIOSH approval numbers with the prefix TC-21C. Surgical masks are not suitable because they fail performance criteria for protection against airborne contaminants. Further information on respiratory protection may be found in other sources.

The recommended containment level for all plant-associated microorganisms is BSL1 or BSL2 conditions in the laboratory or growth chamber, and BL1-P or BL2-P plant containment conditions in the greenhouse, as delineated in the *NIH Guidelines* (20). These principles of containment are applicable to both wild-type and recombinant organisms and are design based to protect laboratory workers. When BL3-P conditions are required by regulatory agencies, the reason is for minimizing escape of the pathogen and for protection of the environment, not people.

Guidelines for using microorganisms in fieldwork or natural ecosystems are provided by the U.S. Department of Agriculture (USDA) (21) and can be found in the primary literature on plant pathology. The texts and methods manuals mentioned in the introduction may be consulted for organismal isolation, survival, growth, decontamination, and plant inoculation. Each organism–plant interaction is unique and requires special conditions of plant susceptibility, such as plant age, tissue specificity, temperature, humidity, and photoperiod, for achieving an infection that mimics natural conditions. Specific insect vectors may be necessary to inoculate plants with some viruses that cannot be mechanically transmitted. For several purposes, tobacco (*Nicotiana tabacum* or *N. benthamiana*) and the weed *A. thaliana* are considered model plants for testing putative plant pathogens, akin to mouse models for human pathogens. There is only one documented case of endangerment of field populations of plants known to have occurred as a result of the use of plant pathogens in contained facilities (95). Nevertheless, the USDA's Plant Protection Act regulations (Section 7 C.F.R. part 330) require that unless a plant pathogen has been isolated locally, permits for its use must be obtained and containment conditions specified therein must be followed. These conditions are aimed at preventing damage to plants through environmental dissemination, rather than at protecting the human worker.

For medical personnel, the *NIH Guidelines* or national guidelines of other countries take precedence in situations in which human diseases or maladies are caused by taxa that also are known as plant pathogens.

Disposal

In the laboratory or greenhouse, autoclaving cultures and pathogen-infected material, or otherwise rendering them biologically inactive, is routine. In gardens and experimental fields, as well as commercial areas, timely chemical treatments for a few bacteria, many fungi, and some insect vectors or wild hosts decrease the inoculum. Chemicals are not available or cost effective for controlling many plant pathogens, and biocontrol agents are few. Other management practices to decrease the inoculum, and thus decrease exposure, are crop rotation, planting of resistant varieties (where available), adjusting planting dates, and plowing under infected or infested plant material. These practices decrease inoculum by the process of competition with other microorganisms in the soil, where many plant pathogens are poor survivors. Composting can also be effective.

Movement of Plant Pathogens

If the isolated microorganism is known to be a plant pathogen, irrespective of risk (no official risk groups are delineated for risk of a pathogen to plants in the United States, or for risk to humans), a permit is required from the USDA Animal and Plant Health Inspection Service (APHIS) to move the agent from one location, state, or country to another. Packaging, storage, and transportation of plant pathogens are also under the aegis of APHIS rules and regulations (http://www.aphis.usda.gov/plant_health), as well as the U.S. Department of Commerce. However, suspected or unknown pathogens can be sent by ordinary mail to laboratories for diagnostic purposes, as is done for human clinical specimens.

CONCLUSION

More cross-kingdom microorganisms are likely to become known because of greater human exposure to potential cross-kingdom pathogens. As shown here, such microorganisms are increasingly more likely to be recognized by the medical profession. Humans, both healthy and immunocompromised, experience changes in diets, food and nonfood plant production, changes in food preparation, new nonfood plant introductions (flowers, shrubs, trees), and increased human travel. Climate change may also play a role in novel cross-kingdom exposures. Furthermore, the production and use of synthetic biology may also expose people to novel cross-kingdom pathogens.

Greater caution should be exercised by those working with plant pathogens and biocontrol agents than has routinely been taken. Specifically, aerosol generation in plant and culture experimentation should be minimized or carefully controlled. Eye protection is recommended in view of the commonality of eye infections, especially by fungi (96). In addition to concern about the etiologic agent, multiple antibiotic resistance is exhibited by some of these microorganisms, such as *B. cepacia* (80) and *Stenotrophomonas maltophilia* (97). For fungi, mortality from invasive infections is much greater than for bacterial pathogens (98) due to the limited treatment options available (99). Immunocompromised adults and persons with allergic sensitivities or open wounds should take particular care in handling plant pathogens and microorganisms associated with plants. Development of rapid and reliable methods for identifying viral, bacterial, and fungal pathogens will make diagnosis easier. New therapeutic approaches for treating invasive bacterial and fungal infections of humans, as well as virus-associated allergies, are continuing medical challenges.

We acknowledge advice and suggestions from several colleagues, including Drs. S. Everhhart, H. Hallen-Adams, and G. Yuen. A.K.V. and S.A.T. strongly express our appreciation for the contributions of P. Lambrecht to the earlier edition and the update of this chapter and for her assistance in its preparation prior to her death.

References

1. **Kirzinger MWB, Nadarasah G, Stavrinides J.** 2011. Insights into cross-kingdom plant pathogenic bacteria. *Genes (Basel)* **2:** 980–997.
2. **van Baarlen P, van Belkum A, Summerbell RC, Crous PW, Thomma BPHJ.** 2007. Molecular mechanisms of pathogenicity: how do pathogenic microorganisms develop cross-kingdom host jumps? *FEMS Microbiol Rev* **31:**239–277.
3. **Baddley JW, Mostert L, Summerbell RC, Moser SA.** 2006. *Phaeoacremonium parasiticum* infections confirmed by beta-tubulin sequence analysis of case isolates. *J Clin Microbiol* **44:** 2207–2211.
4. **Sund-Levander M, Forsberg C, Wahren LK.** 2002. Normal oral, rectal, tympanic and axillary body temperature in adult men and women: a systematic literature review. *Scand J Caring Sci* **16:** 122–128.

5. **Robert VA, Casadevall A.** 2009. Vertebrate endothermy restricts most fungi as potential pathogens. *J Infect Dis* **200:**1623–1626.
6. **Revankar SG, Sutton DA.** 2010. Melanized fungi in human disease. *Clin Microbiol Rev* **23:**884–928.
7. **Hubálek Z.** 2003. Emerging human infectious diseases: anthroponoses, zoonoses, and sapronoses. *Emerg Infect Dis* **9:**403–404.
8. **Agrios GN.** 2005. *Plant Pathology*, 5th ed. Elsevier Academic Press, Burlington, MA.
9. **Dhingra OD, Sinclair JB.** 1995. *Basic Plant Pathology Methods*, 2nd ed. CRC Press, Boca Raton, FL.
10. **Fahy PC, Persley GJ.** 1983. *Plant Bacterial Diseases. A Diagnostic Guide.* Academic Press, New York.
11. **Kahn RP, Mathur SB (ed).** 1999. *Containment Facilities and Safeguards for Exotic Plant Pathogens and Pests.* APS Press, St. Paul, MN.
12. **Klement Z, Rudolph K, Sands DC (ed).** 1990. *Methods in Phytobacteriology.* Akademiai Kiado, Budapest, Hungary.
13. **Rechcigl NA, Rechcigl JE.** 1997. *Environmentally Safe Approaches to Crop Disease Control.* CRC Press, Boca Raton, FL.
14. **Saettler AL, Schaad NW, Roth DA (ed).** 1989. *Detection of Bacteria in Seed and Other Planting Material.* APS Press, St. Paul, MN.
15. **Schaad NW (ed).** 2001. *Laboratory Guide for Identification of Plant Pathogenic Bacteria*, 3rd ed. APS Press, St. Paul, MN.
16. **Schumann GL, D'Arcy CJ.** 2010. *Essential plant pathology*, 2nd ed. APS Press, St. Paul, MN.
17. **Singleton LL, Mihail JD, Rush CM (ed).** 1992. *Methods for Research on Soilborne Phytopathogenic Fungi.* APS Press, St. Paul, MN.
18. **Tuite J.** 1990. *Teachers supplement: Laboratory Exercises of Methods in Plant Pathology: Fungi and Bacteria.* Purdue University, West Lafayette, IN.
19. **Centers for Disease Control and Prevention and National Institutes of Health.** 1999. *Biosafety in Microbiological and Biomedical Laboratories*, 4th ed. U.S. Government Printing Office, Washington, DC.
20. **NIH (National Institutes of Health).** 2002. *Guidelines for Research Involving Recombinant DNA Molecules.* NIH Guidelines 59 CFR 34472 (July 5, 1994), as amended. http://osp.od.nih.gov/office-biotechnology-activities/biosafety/nih-guidelines.
21. **U.S. Department of Agriculture, Office of Agricultural Biotechnology.** 1992. Supplement to Minutes. Guidelines recommended to USDA by the Agricultural Biotechnology Research Advisory Committee, December 3–4, 1991 Guidelines for Research Involving Planned Introduction into the Environment of Genetically Modified Organisms. Document 91-04. http://www.aphis.usda.gov/brs/pdf/abrac%201991.
22. **Ritchie BJ.** 2002. Biosafety in the laboratory, p 379–383. *In* Waller JM, Lenné JM, Waller SJ (ed), *Plant Pathologist's Pocketbook*, 3rd ed. CABI Bioscience, Egham, UK.
23. **Horner WE, Helbling A, Salvaggio JE, Lehrer SB.** 1995. Fungal allergens. *Clin Microbiol Rev* **8:**161–179.
24. **Simon-Nobbe B, Denk U, Pöll V, Rid R, Breitenbach M.** 2008. The spectrum of fungal allergy. *Int Arch Allergy Immunol* **145:** 58–86.
25. **Alfano JR, Collmer A.** 2004. Type III secretion system effector proteins: double agents in bacterial disease and plant defense. *Annu Rev Phytopathol* **42:**385–414.
26. **Cao H, Baldini RL, Rahme LG.** 2001. Common mechanisms for pathogens of plants and animals. *Annu Rev Phytopathol* **39:** 259–284.
27. **Gorbalenya AE, Donchenko AP, Blinov VM, Koonin EV.** 1989. Cysteine proteases of positive strand RNA viruses and chymotrypsin-like serine proteases. A distinct protein superfamily with a common structural fold. *FEBS Lett* **243:**103–114.
28. **Govan JR, Hughes JE, Vandamme P.** 1996. *Burkholderia cepacia*: medical, taxonomic and ecological issues. *J Med Microbiol* **45:**395–407.
29. **Govan JR, Vandamme P.** 1998. Agricultural and medical microbiology: a time for bridging gaps. *Microbiology* **144:**2373–2375.
30. **Rahme LG, Stevens EJ, Wolfort SF, Shao J, Tompkins RG, Ausubel FM.** 1995. Common virulence factors for bacterial pathogenicity in plants and animals. *Science* **268:**1899–1902.
31. **Tan M-W.** 2002. Cross-species infections and their analysis. *Annu Rev Microbiol* **56:**539–565.
32. **Hueck CJ.** 1998. Type III protein secretion systems in bacterial pathogens of animals and plants. *Microbiol Mol Biol Rev* **62:** 379–433.
33. **Preston GM, Studholme DJ, Caldelari I.** 2005. Profiling the secretomes of plant pathogenic Proteobacteria. *FEMS Microbiol Rev* **29:**331–360.
34. **Hall N, Keon JPR, Hargreaves JA.** 1999. A homologue of a gene implicated in the virulence of human fungal diseases is present in a plant fungal pathogen and is expressed during infection. *Physiol Mol Plant Pathol* **55:**69–73.
35. **Procop GW, Roberts GD.** 1998. Laboratory methods in basic mycology, p 871–961. *In* Forbes BA, Sahm DF, Weissfeld AS (ed), *Bailey and Scott's Diagnostic Microbiology*, 10th ed. Mosby, St. Louis, MO.
36. **King AMQ, Adams MJ, Carstens EB, Lefkowitz EJ (ed).** 2012. *Virus taxonomy: classification and nomenclature of viruses: Ninth Report of the International Committee on Taxonomy of Viruses.* Elsevier Academic Press, San Diego, CA.
37. **Dolja VV, Koonin EV.** 2011. Common origins and host-dependent diversity of plant and animal viromes. *Curr Opin Virol* **1:** 322–331.
38. **Koonin EV, Dolja VV, Krupovic M.** 2015. Origins and evolution of viruses of eukaryotes: the ultimate modularity. *Virology* **479–480:** 2–25.
39. **Vidaver AK.** 1996. Emerging and re-emerging infectious diseases. *ASM News* **62:**583–585.
40. **IOM (Institute of Medicine).** 2011. *Fungal Diseases: An Emerging Threat to Human, Animal and Plant Health.* The National Academies Press, Washington, DC.
41. **Holmes A, Nolan R, Taylor R, Finley R, Riley M, Jiang RZ, Steinbach S, Goldstein R.** 1999. An epidemic of *Burkholderia cepacia* transmitted between patients with and without cystic fibrosis. *J Infect Dis* **179:**1197–1205.
42. **Vikram HR, Shore ET, Venkatesh PR.** 1999. Community acquired *Pseudomonas aeruginosa* pneumonia. *Conn Med* **63:**271–273.
43. **Dixon DM, McNeil MM, Cohen ML, Gellin BG, La Montagne JR.** 1996. Fungal infections: a growing threat. *Public Health Rep* **111:**226–235.
44. **Pontón J, Rüchel R, Clemons KV, Coleman DC, Grillot R, Guarro J, Aldebert D, Ambroise-Thomas P, Cano J, Carrillo-Muñoz AJ, Gené J, Pinel C, Stevens DA, Sullivan DJ.** 2000. Emerging pathogens. *Med Mycol* **38**(Suppl 1)**:**225–236.
45. **Taylor LH, Latham SM, Woolhouse ME.** 2001. Risk factors for human disease emergence. *Philos Trans R Soc Lond B Biol Sci* **356:**983–989.
46. **Alfano JR, Collmer A.** 2001. Mechanisms of bacterial pathogenesis in plants: familiar foes in a foreign kingdom, p 179–226. *In* Groisman EA (ed), *Principles of Bacterial Pathogenesis.* Academic Press, San Diego, CA.
47. **Pfaller MA, Diekema DJ.** 2010. Epidemiology of invasive mycoses in North America. *Crit Rev Microbiol* **36:**1–53.
48. **Farr DF, Rossman AY.** Fungal Databases. Systematic Mycology and Microbiology Laboratory, ARS, USDA. https://nt.ars-grin.gov/fungaldatabases.
49. **Berbee ML.** 2001. The phylogeny of plant and animal pathogens in the Ascomycota. *Physiol Mol Plant Pathol* **59:**165–187.
50. **Nucci M, Anaissie E.** 2007. *Fusarium* infections in immunocompromised patients. *Clin Microbiol Rev* **20:**695–704.
51. **Bennett JW, Klich M.** 2003. Mycotoxins. *Clin Microbiol Rev* **16:**497–516.

52. **van Baarlen P, van Belkum A, Thomma BPHJ.** 2007. Disease induction by human microbial pathogens in plant-model systems: potential, problems and prospects. *Drug Discov Today* **12:**167–173.
53. **Balique F, Lecoq H, Raoult D, Colson P.** 2015. Can plant viruses cross the kingdom border and be pathogenic to humans? *Viruses* **7:**2074–2098.
54. **Mandal B, Jain RK.** 2010. Can plant virus infect human being? *Indian J Virol* **21:**92–93.
55. **Rosario K, Breitbart M.** 2011. Exploring the viral world through metagenomics. *Curr Opin Virol* **1:**289–297.
56. **Mokili JL, Rohwer F, Dutilh BE.** 2012. Metagenomics and future perspectives in virus discovery. *Curr Opin Virol* **2:**63–77.
57. **Popgeorgiev N, Temmam S, Raoult D, Desnues C.** 2013. Describing the silent human virome with an emphasis on giant viruses. *Intervirology* **56:**395–412.
58. **Abeles SR, Pride DT.** 2014. Molecular bases and role of viruses in the human microbiome. *J Mol Biol* **426:**3892–3906.
59. **Lipkin WI, Anthony SJ.** 2015. Virus hunting. *Virology* **479–480:** 194–199.
60. **LeClair RA.** 1967. Recovery of culturable tobacco mosaic virus from sputum and thoracentesis fluids obtained from cigarette smokers with a history of pulmonary disease. *Am Rev Respir Dis* **95:**510–511.
61. **Balique F, Colson P, Barry AO, Nappez C, Ferretti A, Moussawi KA, Ngounga T, Lepidi H, Ghigo E, Mege JL, Lecoq H, Raoult D.** 2013. *Tobacco mosaic virus* in the lungs of mice following intra-tracheal inoculation. *PLoS One* **8:**e54993.
62. **Liu R, Vaishnav RA, Roberts AM, Friedland RP.** 2013. Humans have antibodies against a plant virus: evidence from tobacco mosaic virus. *PLoS One* **8:**e60621.
63. **Zhang T, Breitbart M, Lee WH, Run J-Q, Wei CL, Soh SWL, Hibberd ML, Liu ET, Rohwer F, Ruan Y.** 2006. RNA viral community in human feces: prevalence of plant pathogenic viruses. *PLoS Biol* **4:**e3.
64. **Wetter C, Conti M, Altschuh D, Tabillion R, van Regenmortel MHV.** 1984. Pepper mild mottle virus, a tobamovirus infecting pepper in Sicily. *Phytopathology* **74:**405–410.
65. **Nakamura S, Yang C-S, Sakon N, Ueda M, Tougan T, Yamashita A, Goto N, Takahashi K, Yasunaga T, Ikuta K, Mizutani T, Okamoto Y, Tagami M, Morita R, Maeda N, Kawai J, Hayashizaki Y, Nagai Y, Horii T, Iida T, Nakaya T.** 2009. Direct metagenomic detection of viral pathogens in nasal and fecal specimens using an unbiased high-throughput sequencing approach. *PLoS One* **4:**e4219.
66. **Colson P, Richet H, Desnues C, Balique F, Moal V, Grob J-J, Berbis P, Lecoq H, Harlé J-R, Berland Y, Raoult D.** 2010. Pepper mild mottle virus, a plant virus associated with specific immune responses, Fever, abdominal pains, and pruritus in humans. *PLoS One* **5:**e10041.
67. **Peng J, Shi B, Zheng H, Lu Y, Lin L, Jiang T, Chen J, Yan F.** 2015. Detection of pepper mild mottle virus in pepper sauce in China. *Arch Virol* **160:**2079–2082.
68. **Blanc S.** 2007. Virus transmission—getting out and in, p 1–28. *In* Waigmann E, Heinlein M (ed), *Virus Transport in Plants.* Springer-Verlag, Berlin and Heidelberg.
69. **Li L, Victoria JG, Wang C, Jones M, Fellers GM, Kunz TH, Delwart E.** 2010. Bat guano virome: predominance of dietary viruses from insects and plants plus novel mammalian viruses. *J Virol* **84:**6955–6965.
70. **Nault LR.** 1997. Arthropod transmission of plant viruses: a new synthesis. *Ann Entomol Soc Am* **90:**521–541.
71. **Yolken RH, Jones-Brando L, Dunigan DD, Kannan G, Dickerson F, Severance E, Sabunciyan S, Talbot CC Jr, Prandovszky E, Gurnon JR, Agarkova IV, Leister F, Gressitt KL, Chen O, Deuber B, Ma F, Pletnikov MV, Van Etten JL.** 2014. Chlorovirus ATCV-1 is part of the human oropharyngeal virome and is associated with changes in cognitive functions in humans and mice. *Proc Natl Acad Sci USA* **111:**16106–16111.
72. **Van Etten JL, Dunigan DD.** 2012. Chloroviruses: not your everyday plant virus. *Trends Plant Sci* **17:**1–8.
73. **Yolken RH, Severance EG, Sabunciyan S, Gressitt KL, Chen O, Stallings C, Origoni A, Katsafanas E, Schweinfurth LAB, Savage CLG, Banis M, Khushalani S, Dickerson FB.** 2015. Metagenomic sequencing indicates that the oropharyngeal phageome of individuals with schizophrenia differs from that of controls. *Schizophr Bull* **41:**1153–1161.
74. **Stepanova OA, Solovyova YV, Solovyov AV.** 2011. Results of algae viruses search in human clinical material. *Ukrainica Bioorganica Acta* **9:**53–56.
75. **Scholthof KB.** 2003. One foot in the furrow: linkages between agriculture, plant pathology, and public health. *Annu Rev Public Health* **24:**153–174.
76. **Hall N, Keon JPR, Hargreaves JA.** 1999. A homologue of a gene implicated in the virulence of human fungal diseases is present in a plant fungal pathogen and is expressed during infection. *Physiol Mol Plant Pathol* **55:**69–73.
77. **Horner WE, Helbling A, Salvaggio JE, Lehrer SB.** 1995. Fungal allergens. *Clin Microbiol Rev* **8:**161–179.
78. **Simon-Nobbe B, Denk U, Pöll V, Rid R, Breitenbach M.** 2008. The spectrum of fungal allergy. *Int Arch Allergy Immunol* **145:** 58–86.
79. **Lee YH, Chen Y, Ouyang X, Gan YH.** 2010. Identification of tomato plant as a novel host model for *Burkholderia pseudomallei*. *BMC Microbiol* **10:**28.
80. **Wigley P, Burton NF.** 1999. Genotypic and phenotypic relationships in *Burkholderia cepacia* isolated from cystic fibrosis patients and the environment. *J Appl Microbiol* **86:**460–468.
81. **Dolja VV, Koonin EV.** 2011. Common origins and host-dependent diversity of plant and animal viromes. *Curr Opin Virol* **1:**322–331.
82. **Gibbs AJ, Fargette D, García-Arenal F, Gibbs MJ.** 2010. Time—the emerging dimension of plant virus studies. *J Gen Virol* **91:**13–22.
83. **Roossinck MJ.** 2011. The big unknown: plant virus biodiversity. *Curr Opin Virol* **1:**63–67.
84. **Morris CE, Bardin M, Kinkel LL, Moury B, Nicot PC, Sands DC.** 2009. Expanding the paradigms of plant pathogen life history and evolution of parasitic fitness beyond agricultural boundaries. *PLoS Pathog* **5:**e1000693.
85. **Brandl MT.** 2006. Fitness of human enteric pathogens on plants and implications for food safety. *Annu Rev Phytopathol* **44:** 367–392.
86. **Haramoto E, Kitajima M, Kishida N, Konno Y, Katayama H, Asami M, Akiba M.** 2013. Occurrence of pepper mild mottle virus in drinking water sources in Japan. *Appl Environ Microbiol* **79:**7413–7418.
87. **Cheong S, Lee C, Song SW, Choi WC, Lee CH, Kim S-J.** 2009. Enteric viruses in raw vegetables and groundwater used for irrigation in South Korea. *Appl Environ Microbiol* **75:**7745–7751.
88. **Van Regenmortel MHV.** 1982. *Serology and Immunochemistry of Plant Viruses.* Academic Press, New York.
89. **Gibbs MJ, Weiller GF.** 1999. Evidence that a plant virus switched hosts to infect a vertebrate and then recombined with a vertebrate-infecting virus. *Proc Natl Acad Sci USA* **96:**8022–8027.
90. **Dimitrov DS.** 2004. Virus entry: molecular mechanisms and biomedical applications. *Nat Rev Microbiol* **2:**109–122.
91. **Baranowski E, Ruiz-Jarabo CM, Domingo E.** 2001. Evolution of cell recognition by viruses. *Science* **292:**1102–1105.
92. **Xu K, Nagy PD.** 2014. Expanding use of multi-origin subcellular membranes by positive-strand RNA viruses during replication. *Curr Opin Virol* **9:**119–126.
93. **Diaz A, Wang X.** 2014. Bromovirus-induced remodeling of host membranes during viral RNA replication. *Curr Opin Virol* **9:** 104–110.
94. **Ritzenthaler C.** 2011. Parallels and distinctions in the direct cell-to-cell spread of the plant and animal viruses. *Curr Opin Virol* **1:** 403–409.

95. **McKeen WE.** 1989. *Blue Mold of Tobacco*. APS Press, St. Paul, MN.
96. **Thomas PA, Kaliamurthy J.** 2013. Mycotic keratitis: epidemiology, diagnosis and management. *Clin Microbiol Infect* **19:**210–220.
97. **Denton M, Kerr KG.** 1998. Microbiological and clinical aspects of infection associated with *Stenotrophomonas maltophilia*. *Clin Microbiol Rev* **11:**57–80.
98. **Engelhard D.** 1998. Bacterial and fungal infections in children undergoing bone marrow transplantation. *Bone Marrow Transplant* **21**(Suppl 2)**:**S78–S80.
99. **Serfling A, Wohlrab J, Deising HB.** 2007. Treatment of a clinically relevant plant-pathogenic fungus with an agricultural azole causes cross-resistance to medical azoles and potentiates caspofungin efficacy. *Antimicrob Agents Chemother* **51:**3672–3676.
100. **Edmond MB, Riddler SA, Baxter CM, Wicklund BM, Pasculle AW.** 1993. *Agrobacterium radiobacter*: a recently recognized opportunistic pathogen. *Clin Infect Dis* **16:**388–391.
101. **Miller JM, Novy C, Hiott M.** 1996. Case of bacterial endophthalmitis caused by an *Agrobacterium radiobacter*-like organism. *J Clin Microbiol* **34:**3212–3213.
102. **Southern PM Jr.** 1996. Bacteremia due to *Agrobacterium tumefaciens* (*radiobacter*). Report of infection in a pregnant woman and her stillborn fetus. *Diagn Microbiol Infect Dis* **24:**43–45.
103. **Plotkin GR.** 1980. *Agrobacterium radiobacter* prosthetic valve endocarditis. *Ann Intern Med* **93:**839–840.
104. **Melgosa Hijosa M, Ramos Lopez MC, Ruiz Almagro P, Fernandez Escribano A, Luque de Pablos A.** 1997. *Agrobacterium radiobacter* peritonitis in a Down's syndrome child maintained on peritoneal dialysis. *Perit Dial Int* **17:**515.
105. **Namdari H, Hamzavi S, Peairs RR.** 2003. *Rhizobium* (*Agrobacterium*) *radiobacter* identified as a cause of chronic endophthalmitis subsequent to cataract extraction. *J Clin Microbiol* **41:**3998–4000.
106. **Dunne WM Jr, Tillman J, Murray JC.** 1993. Recovery of a strain of *Agrobacterium radiobacter* with a mucoid phenotype from an immunocompromised child with bacteremia. *J Clin Microbiol* **31:**2541–2543.
107. **Moore LW, Warren G.** 1979. *Agrobacterium radiobacter* strain 84 and biological control of crown gall. *Annu Rev Phytopathol* **17:**163–179.
108. **Westcott C.** 2001. *Westcott's Plant Disease Handbook*, 6th ed. Revised by R. Kenneth Horst. Kluwer Academic Publishers, Boston, MA.
109. **Ramirez FC, Saeed ZA, Darouiche RO, Shawar RM, Yoffe B.** 1992. *Agrobacterium tumefaciens* peritonitis mimicking tuberculosis. *Clin Infect Dis* **15:**938–940.
110. **Alnor D, Frimodt-Meller N, Espersen F, Frederiksen W.** 1994. Infections with the unusual human pathogens *Agrobacterium species* and *Ochrobactrum anthropi*. *Clin Infect Dis* **18:**914–920.
111. **Hulse M, Johnson S, Ferrieri P.** 1993. *Agrobacterium* infections in humans: experience at one hospital and review. *Clin Infect Dis* **16:**112–117.
112. **Murdoch CW, Campana RJ.** 1983. Bacterial species associated with wetwood of elm. *Phytopathology* **73:**1270–1273.
113. **Rubinstein I, Pedersen GW.** 2002. Bacillus species are present in chewing tobacco sold in the United States and evoke plasma exudation from the oral mucosa. *Clin Diagn Lab Immunol* **9:**1057–1060.
114. **Leary JV, Chun WWW.** 1989. Pathogenicity of *Bacillus circulans* to seedlings of date palm (*Phoenix dactylifera*). *Plant Dis* **73:**353–354.
115. **Berry N, Hassan I, Majumdar S, Vardhan A, McEwen A, Gokal R.** 2004. *Bacillus circulans* peritonitis in a patient treated with CAPD. *Perit Dial Int* **24:**488–489.
116. **Rowan NJ, Caldow G, Gemmell CG, Hunter IS.** 2003. Production of diarrheal enterotoxins and other potential virulence factors by veterinary isolates of *Bacillus* species associated with nongastrointestinal infections. *Appl Environ Microbiol* **69:**2372–2376.
117. **Deva AK, Narayan KG.** 1989. Enterotoxigenicity of *Bacillus circulans*, *B. coagulans* and *B. stearothermophilus*. *Indian J Comp Microbiol Immunol Infect Dis* **10:**80–87.
118. **Saleh OI, Huang PY, Huang JS.** 1997. *Bacillus pumilus*, the cause of bacterial blotch of immature Balady peach in Egypt. *J Phytopathol Berl* **145:**447–453.
119. **Suominen I, Andersson M, Hallaksela A-M, Salkinoja-Salonen MS.** 1999. Identifying toxic Bacillus pumilus from industrial contaminants, food, paperboard and live trees, p 100–105. *In* Tuijtelaars ACJ, Samson RA, Rombouts FM, Notermans S (ed), *Food Microbiology and Food Safety into the Next Millennium*. Proceedings of the 17th International Conference of the International Committee on Food Microbiology and Hygiene, Veldoven, The Netherlands, September 13–17, 1999.
120. **Burkholder WH.** 1950. Sour skin, a bacterial rot of onion bulbs. *Phytopathology* **40:**115–117.
121. **Yohalem DS, Lorbeer JW.** 1997. Distribution of *Burkholderia cepacia* phenotypes by niche, method of isolation and pathogenicity to onion. *Ann Appl Biol* **130:**467–479.
122. **Gill WM, Cole ALJ.** 1992. Cavity disease of *Agaricus bitorquis* caused by *Pseudomonas cepacia*. *Can J Microbiol* **38:**394–397.
123. **Alameda M, Mignucci JS.** 1998. *Burkholderia cepacia*, causal agent of bacterial blotch of oyster mushroom. *J Agric Univ P R* **82:**109–110.
124. **Glick BR.** 2004. Teamwork in phytoremediation. *Nat Biotechnol* **22:**526–527.
125. **Hinton DM, Bacon CW.** 1995. *Enterobacter cloacae* is an endophytic symbiont of corn. *Mycopathologia* **129:**117–125.
126. **Woods CW, Bressler AM, LiPuma JJ, Alexander BD, Clements DA, Weber DJ, Moore CM, Reller LB, Kaye KS.** 2004. Virulence associated with outbreak-related strains of Burkholderia cepacia complex among a cohort of patients with bacteremia. *Clin Infect Dis* **38:**1243–1250.
127. **De Boeck K, Malfroot A, Van Schil L, Lebecque P, Knoop C, Govan JR, Doherty C, Laevens S, Vandamme P, Belgian Burkholderia cepacia Study Group.** 2004. Epidemiology of Burkholderia cepacia complex colonisation in cystic fibrosis patients. *Eur Respir J* **23:**851–856.
128. **Courtney JM, Dunbar KE, McDowell A, Moore JE, Warke TJ, Stevenson M, Elborn JSM.** 2004. Clinical outcome of *Burkholderia cepacia* complex infection in cystic fibrosis adults. *J Cyst Fibros* **3:**93–98.
129. **Lau SM, Yu WL, Wang JH.** 1999. Cardiac cirrhosis with cellulitis caused by *Burkholderia cepacia* bacteremia. *Clin Infect Dis* **29:**447–448.
130. **Pathengay A, Raju B, Sharma S, Das T, Endophthalmitis Research Group.** 2005. Recurrent endophthalmitis caused by *Burkholderia cepacia*. *Eye (Lond)* **19:**358–359.
131. **Lee Y-A, Chan C-W.** 2007. Molecular typing and presence of genetic markers among strains of banana finger-tip rot pathogen, *Burkholderia cenocepacia*, in Taiwan. *Phytopathology* **97:**195–201.
132. **Kishun R, Swarup J.** 1981. Growth studies on *Pseudomonas gladioli* pv. *allicola* pathogenic to onion. *Indian J Mycol Plant Pathol* **11:**247–250.
133. **Chase AR, Miller JW, Jones JB.** 1984. Leaf spot and blight of *Asplenium nidus* caused by *Pseudomonas gladioli*. *Plant Dis* **68:**344–347.
134. **Chuenchitt S, Dhirabhava W, Karnjanarat S, Buangsuwon D, Uematsu T.** 1983. A new bacterial disease on orchids *Dendrobium* sp. caused by *Pseudomonas gladioli*. *Kasetsart J Nat Sci* **17:**26–36.
135. **Shin JH, Kim SH, Shin MG, Suh SP, Ryang DW, Jeong MH.** 1997. Bacteremia due to *Burkholderia gladioli*: case report. *Clin Infect Dis* **25:**1264–1265.
136. **Ross JP, Holland SM, Gill VJ, DeCarlo ES, Gallin JI.** 1995. Severe *Burkholderia* (*Pseudomonas*) *gladioli* infection in chronic granulomatous disease: report of two successfully treated cases. *Clin Infect Dis* **21:**1291–1293.

137. **Graves M, Robin T, Chipman AM, Wong J, Khashe S, Janda JM.** 1997. Four additional cases of *Burkholderia gladioli* infection with microbiological correlates and review. *Clin Infect Dis* **25:** 838–842.
138. **Maeda Y, Shinohara H, Kiba A, Ohnishi K, Furuya N, Kawamura Y, Ezaki T, Vandamme P, Tsushima S, Hikichi Y.** 2006. Phylogenetic study and multiplex PCR-based detection of *Burkholderia plantarii*, *Burkholderia glumae* and *Burkholderia gladioli* using gyrB and rpoD sequences. *Int J Syst Evol Microbiol* **56:**1031–1038.
139. **Lee YH, Chen Y, Ouyang X, Gan YH.** 2010. Identification of tomato plant as a novel host model for *Burkholderia pseudomallei*. *BMC Microbiol* **10:**28.
140. **Schink B, Ward JC, Zeikus JG.** 1981. Microbiology of wetwood: importance of pectin degradation and *clostridium* species in living trees. *Appl Environ Microbiol* **42:**526–532.
141. **Gvozdiak RI, Khodos SF, Lipshits VV.** 1976. Biolohichni vlastyvosti Clostridium butyricum V. phytopathogenicum var. nova—zbudnyka zakhvoriuvannia hraba (in Ukrainian). [Biological properties of *Clostridium butyricum* V. *phytopathogenicum* var. nova, the causative agent of hornbeam disease]. *Mikrobiol Zh* **38:**288–292. (In Russian.)
142. **Howard FM, Bradley JM, Flynn DM, Noone P, Szawatkowski M.** 1977. Outbreak of necrotising enterocolitis caused by *Clostridium butyricum*. *Lancet* **310:**1099–1102.
143. **Wells JM, Butterfield JE, Revear LG.** 1993. Identification of bacteria associated with postharvest diseases of fruits and vegetables by cellular fatty acid composition: an expert system for personal computers. *Phytopathology* **83:**445–455.
144. **Brazier JS, Gal M, Hall V, Morris TE.** 2004. Outbreak of Clostridium histolyticum infections in injecting drug users in England and Scotland. Eurosurveillance Mon 9:15–16.. http://www.eurosurveillance.org/em/v09n09/0909–221.asp.
145. **Collins MD, Jones D.** 1983. Reclassification of *Corynebacterium flaccumfaciens, Corynebacterium betae, Corynebacterium oortii* and *Corynebacterium poinsettiae* in the genus *Curtobacterium*, as *Curtobacterium flaccumfaciens* comb. nov. *J Gen Microbiol* **129:**3545–3548.
146. **Francis MJ, Doherty RR, Patel M, Hamblin JF, Ojaimi S, Korman TM.** 2011. *Curtobacterium flaccumfaciens* septic arthritis following puncture with a Coxspur Hawthorn thorn. *J Clin Microbiol* **49:**2759–2760.
147. **Bishop AL, Davis RM.** 1990. Internal decay of onions caused by *Enterobacter cloacae*. *Plant Dis* **74:**692–694.
148. **Nishijima KA, Alvarez AM, Hepperly PR, Shintaku MH, Keith LM, Sato DM, Bushe BC, Armstrong JW, Zee FT.** 2004. Association of *Enterobacter cloacae* with rhizome rot of edible ginger in Hawaii. *Plant Dis* **88:**1318–1327.
149. **Nishijima KA, Wall MM, Siderhurst MS.** 2007. Demonstrating pathogenicity of *Enterobacter cloacae* on Macadamia and identifying associated volatiles of gray kernel of macadamia in Hawaii. *Plant Dis* **91:**1221–1228.
150. **Punja ZK.** 1997. Comparative efficacy of bacteria, fungi and yeasts as biological control agents for diseases of vegetable crops. *Can J Plant Pathol* **19:**315–323.
151. **Wilson CL, Franklin JD, Pusey PL.** 1987. Biological control of *Rhizopus* rot of peach with *Enterobacter cloacae*. *Phytopathology* **77:**303–305.
152. **Watanabe K, Abe K, Sato M.** 2000. Biological control of an insect pest by gut-colonizing *Enterobacter cloacae* transformed with ice nucleation gene. *J Appl Microbiol* **88:**90–97.
153. **Jochimsen EM, Frenette C, Delorme M, Arduino M, Aguero S, Carson L, Ismaïl J, Lapierre S, Czyziw E, Tokars JI, Jarvis WR.** 1998. A cluster of bloodstream infections and pyrogenic reactions among hemodialysis patients traced to dialysis machine waste-handling option units. *Am J Nephrol* **18:**485–489.
154. **Fata F, Chittivelu S, Tessler S, Kupfer Y.** 1996. Gas gangrene of the arm due to *Enterobacter cloacae* in a neutropenic patient. *South Med J* **89:**1095–1096.
155. **Hao MV, Brenner DJ, Steigerwalt AG, Kosako Y, Komagata K.** 1990. *Erwinia persicinus*, a new species isolated from plants. *Int J Syst Bacteriol* **40:**379–383.
156. **Brenner DJ, Rodrigues Neto JR, Steigerwalt AG, Robbs CF.** 1994. *"Erwinia nulandii"* is a subjective synonym of *Erwinia persicinus*. *Int J Syst Bacteriol* **44:**282–284.
157. **O'Hara CM, Steigerwalt AG, Hill BC, Miller JM, Brenner DJ.** 1998. First report of a human isolate of *Erwinia persicinus*. *J Clin Microbiol* **36:**248–250.
158. **Dong Y, Chelius MK, Brisse S, Kozyrovska N, Podschun R, Triplett EW.** 2003. Comparisons between two *Klebsiella*: the plant endophyte *K. pneumoniae* 342 and a clinical isolate, *K. pneumoniae* MGH78578. *Symbiosis* **35:**247–259.
159. **Prince SE, Dominger KA, Cunha BA, Klein NC.** 1997. *Klebsiella pneumoniae* pneumonia. *Heart Lung* **26:**413–417.
160. **Kang CI, Kim SH, Kim DM, Park WB, Lee KD, Kim HB, Oh MD, Kim EC, Choe KW.** 2004. Risk factors for and clinical outcomes of bloodstream infections caused by extended-spectrum beta-lactamase-producing *Klebsiella pneumoniae*. *Infect Control Hosp Epidemiol* **25:**860–867.
161. **Tang LM, Chen ST.** 1994. *Klebsiella* pneumoniae meningitis: prognostic factors. *Scand J Infect Dis* **26:**95–102.
162. **Rosenblueth M, Martínez L, Silva J, Martínez-Romero E.** 2004. *Klebsiella variicola*, a novel species with clinical and plant-associated isolates. *Syst Appl Microbiol* **27:**27–35.
163. **Suckstorff I, Berg G.** 2003. Evidence for dose-dependent effects on plant growth by Stenotrophomonas strains from different origins. *J Appl Microbiol* **95:**656–663.
164. **Denton M, Kerr KG.** 1998. Microbiological and clinical aspects of infection associated with *Stenotrophomonas maltophilia*. *Clin Microbiol Rev* **11:**57–80.
165. **Goss CH, Mayer-Hamblett N, Aitken ML, Rubenfeld GD, Ramsey BW.** 2004. Association between *Stenotrophomonas maltophilia* and lung function in cystic fibrosis. *Thorax* **59:** 955–959.
166. **Frederickson DE, Monyo ES, King SB, Odvody GN.** 1997. A disease of pearl millet in Zimbabwe caused by *Pantoea agglomerans*. *Plant Dis* **81:**959.
167. **Bennett SN, McNeil MM, Bland LA, Arduino MJ, Villarino ME, Perrotta DM, Burwen DR, Welbel SF, Pegues DA, Stroud L, Zeitz PS, Jarvis WR.** 1995. Postoperative infections traced to contamination of an intravenous anesthetic, propofol. *N Engl J Med* **333:**147–154.
168. **Bicudo EL, Macedo VO, Carrara MA, Castro FFS, Rage RI.** 2007. Nosocomial outbreak of *Pantoea agglomerans* in a pediatric urgent care center. *Braz J Infect Dis* **11:**281–284.
169. **Cruz AT, Cazacu AC, Allen CH.** 2007. *Pantoea agglomerans*, a plant pathogen causing human disease. *J Clin Microbiol* **45:** 1989–1992.
170. **Kratz A, Greenberg D, Barki Y, Cohen E, Lifshitz M.** 2003. *Pantoea agglomerans* as a cause of septic arthritis after palm tree thorn injury; case report and literature review. *Arch Dis Child* **88:**542–544.
171. **Coutinho TA, Venter SN.** 2009. *Pantoea ananatis*: an unconventional plant pathogen. *Mol Plant Pathol* **10:**325–335.
172. **De Baere T, Verhelst R, Labit C, Verschraegen G, Wauters G, Claeys G, Vaneechoutte M.** 2004. Bacteremic infection with *Pantoea ananatis*. *J Clin Microbiol* **42:**4393–4395.
173. **Cother EJ, Darbyshire B, Brewer J.** 1976. *Pseudomonas aeruginosa*: cause of internal brown rot of onion. *Phytopathology* **66:** 828–834.
174. **Johansen HK, Kovesi TA, Koch C, Corey M, Høiby N, Levison H.** 1998. *Pseudomonas aeruginosa* and *Burkholderia cepacia* infection in cystic fibrosis patients treated in Toronto and Copenhagen. *Pediatr Pulmonol* **26:**89–96.
175. **Torii K, Noda Y, Miyazaki Y, Ohta M.** 2003. An unusual outbreak of infusion-related bacteremia in a gastrointestinal disease ward. *Jpn J Infect Dis* **56:**177–178.

176. **Johnson KB, Stockwell VO.** 1998. Management of fire blight: a case study in microbial ecology. *Annu Rev Phytopathol* **36:** 227–248.
177. **Hsueh PR, Teng LJ, Pan HJ, Chen YC, Sun CC, Ho SW, Luh KT.** 1998. Outbreak of *Pseudomonas fluorescens* bacteremia among oncology patients. *J Clin Microbiol* **36:**2914–2917.
178. **Nelson KE, Weinel C, Paulsen IT, Dodson RJ, Hilbert H, Martins dos Santos VAP, Fouts DE, Gill SR, Pop M, Holmes M, Brinkac L, Beanan M, DeBoy RT, Daugherty S, Kolonay J, Madupu R, Nelson W, White O, Peterson J, Khouri H, Hance I, Chris Lee P, Holtzapple E, Scanlan D, Tran K, Moazzez A, Utterback T, Rizzo M, Lee K, Kosack D, Moestl D, Wedler H, Lauber J, Stjepandic D, Hoheisel J, Straetz M, Heim S, Kiewitz C, Eisen JA, Timmis KN, Düsterhöft A, Tümmler B, Fraser CM.** 2002. Complete genome sequence and comparative analysis of the metabolically versatile *Pseudomonas putida* KT2440. *Environ Microbiol* **4:**799–808.
179. **Lombardi G, Luzzaro F, Docquier JD, Riccio ML, Perilli M, Colì A, Amicosante G, Rossolini GM, Toniolo A.** 2002. Nosocomial infections caused by multidrug-resistant isolates of *pseudomonas putida* producing VIM-1 metallo-β-lactamase. *J Clin Microbiol* **40:**4051–4055.
180. **Ghosh K, Daar S, Hiwase D, Nusrat N.** 2000. Primary *Pseudomonas* meningitis in an adult, splenectomized, multitransfused thalassaemia major patient. *Haematologia (Budap)* **30:**69–72.
181. **Riley IT, Ophel KM.** 1992. *Clavibacter toxicus* sp. nov., the bacterium responsible for annual ryegrass toxicity in Australia. *Int J Syst Bacteriol* **42:**64–68.
182. **Edgar J.** 2004. Future impact of food safety issues on animal production and trade: implications for research. *Aust J Exp Agric* **44:**1073–1076.
183. **Grimont PAD, Grimont F, Starr MP.** 1979. *Serratia ficaria* sp. nov., a bacterial species associated with Smyrna figs and the fig wasp *Blastophaga psenes*. *Curr Microbiol* **2:**277–282.
184. **Anahory T, Darbas H, Ongaro O, Jean-Pierre H, Mion P.** 1998. *Serratia ficaria*: a misidentified or unidentified rare cause of human infections in fig tree culture zones. *J Clin Microbiol* **36:**3266–3272.
185. **Badenoch PR, Thom AL, Coster DJ.** 2002. *Serratia ficaria* endophthalmitis. *J Clin Microbiol* **40:**1563–1564.
186. **Lukezic L, Hildebrand DC, Schroth M, Shinde A.** 1982. Association of *Serratia marcescens* with crown rot of alfalfa in Pennsylvania. *Phytopathology* **72:**714–718.
187. **Bruton BD, Mitchell F, Fletcher J, Pair SD, Wayadande A, Melcher U, Brady J, Bextine B, Popham TW.** 2003. *Serratia marcescens*, a phloem-colonizing, squash bug-transmitted bacterium: causal agent of cucurbit yellow vine disease. *Plant Dis* **87:** 937–944.
188. **Gyaneshwar P, James EK, Mathan N, Reddy PM, Reinhold-Hurek B, Ladha JK.** 2001. Endophytic colonization of rice by a diazotrophic strain of *Serratia marcescens*. *J Bacteriol* **183:**2634–2645.
189. **Ostrowsky BE, Whitener C, Bredenberg HK, Carson LA, Holt S, Hutwagner L, Arduino MJ, Jarvis WR.** 2002. *Serratia marcescens* bacteremia traced to an infused narcotic. *N Engl J Med* **346:**1529–1537.
190. **Su LH, Ou JT, Leu HS, Chiang PC, Chiu YP, Chia JH, Kuo AJ, Chiu CH, Chu C, Wu TL, Sun CF, Riley TV, Chang BJ, Infection Control Group.** 2003. Extended epidemic of nosocomial urinary tract infections caused by *Serratia marcescens*. *J Clin Microbiol* **41:**4726–4732.
191. **Li ZX, Bian ZS, Zheng HP, Yue YS, Yao JY, Gong YP, Cai MY, Dong XZ.** 1990. First isolation of *Xanthomonas campestris* from the blood of a Chinese woman. *Chin Med J (Engl)* **103:** 435–439.
192. **Funke G, Haase G, Schnitzler N, Schrage N, Reinert RR.** 1997. Endophthalmitis due to *Microbacterium* species: case report and review of *microbacterium* infections. *Clin Infect Dis* **24:**713–716.
193. **Carey J, Motyl M, Perlman DC.** 2001. Catheter-related bacteremia due to *Streptomyces* in a patient receiving holistic infusions. *Emerg Infect Dis* **7:**1043–1045.
194. **Kaku H.** 2004. Histopathology of red stripe of rice. *Plant Dis* **88:** 1304–1309.
195. **Zinniel DK, Lambrecht P, Harris NB, Feng Z, Kuczmarski D, Higley P, Ishimaru CA, Arunakumari A, Barletta RG, Vidaver AK.** 2002. Isolation and characterization of endophytic colonizing bacteria from agronomic crops and prairie plants. *Appl Environ Microbiol* **68:**2198–2208.
196. **Rott P, Comstock JC.** 2000. Seedling foliage blights, p 190–206. *In* Rott P, Bailey R, Comstock JC, Croft B, Saumtally S (ed), *A Guide to Sugarcane Diseases*. CIRAD Publication Service, Montpellier, France.
197. **Duffill MB, Coley KE.** 1993. Cutaneous phaeohyphomycosis due to *Alternaria alternata* responding to itraconazole. *Clin Exp Dermatol* **18:**156–158.
198. **Ono M, Nishigori C, Tanaka C, Tanaka S, Tsuda M, Miyachi Y.** 2004. Cutaneous alternariosis in an immunocompetent patient: analysis of the internal transcribed spacer region of rDNA and Brm2 of isolated *Alternaria alternata*. *Br J Dermatol* **150:**773–775.
199. **Revankar SG, Sutton DA.** 2010. Melanized fungi in human disease. *Clin Microbiol Rev* **23:**884–928.
200. **Howard CM, Albregts EE.** 1973. A strawberry fruit rot caused by *Alternaria tenuissima*. *Phytopathology* **63:**938–939.
201. **Honda Y, Rahman MZ, Islam SZ, Muroguchi N.** 2001. Leaf spot disease of broad bean caused by *Alternaria tenuissima* in Japan. *Plant Dis* **85:**95.
202. **Milholland RD.** 1995. *Alternaria* leaf spot and fruit rot, p 18. *In* Caruso FL, Ramsdell DC (ed), *Compendium of Blueberry and Cranberry Diseases*. APS Press, St. Paul, MN.
203. **Blodgett JT, Swart WJ.** 2002. Infection, colonization and disease of *Amaranthus hybridus* leaves by the *Alternaria tenuissima* group. *Plant Dis* **86:**1199–1205.
204. **Romano C, Fimiani M, Pellegrino M, Valenti L, Casini L, Miracco C, Faggi E.** 1996. Cutaneous phaeohyphomycosis due to *Alternaria tenuissima*. *Mycoses* **39:**211–215.
205. **Romano C, Valenti L, Miracco C, Alessandrini C, Paccagnini E, Faggi E, Difonzo EM.** 1997. Two cases of cutaneous phaeohyphomycosis by *Alternaria alternata* and *Alternaria tenuissima*. *Mycopathologia* **137:**65–74.
206. **Rossmann SN, Cernoch PL, Davis JR.** 1996. Dematiaceous fungi are an increasing cause of human disease. *Clin Infect Dis* **22:** 73–80.
207. **Thind TS, Saksena SB, Agrawal SC.** 1976. Post harvest decay of apple fruits incited by *Aspergillus candidus* in Madhya Pradesh. *Indian Phytopathol* **29:**318.
208. **Linares G, McGarry PA, Baker RD.** 1971. Solid solitary aspergillotic granuloma of the brain. Report of a case due to *Aspergillus candidus* and review of the literature. *Neurology* **21:**177–184.
209. **Krysinska-Traczyk E, Dutkiewicz J.** 2000. *Aspergillus candidus*: a respiratory hazard associated with grain dust. *Ann Agric Environ Med* **7:**101–109.
210. **St. Leger RJ, Screen SE, Shams-Pirzadeh B.** 2000. Lack of host specialization in Aspergillus flavus. *Appl Environ Microbiol* **66:**320–324.
211. **Smart MG, Wicklow DT, Caldwell RW.** 1990. Pathogenesis in Aspergillus ear rot of maize: light microscopy of fungal spread from wounds. *Phytopathology* **80:**1287–1294.
212. **Pitt JI, Dyer SK, McCammon S.** 1991. Systemic invasion of developing peanut plants by *Aspergillus flavus*. *Lett Appl Microbiol* **13:**16–20.
213. **Brown RL, Cleveland TE, Cotty PJ, Mellon JE.** 1992. Spread of *Aspergillus flavus* in cotton bolls, decay of intercarpellary membranes and production of fungal pectinases. *Phytopathology* **82:** 462–467.
214. **Yamada K, Mori T, Irie S, Matsumura M, Nakayama M, Hirano T, Suda K, Oshimi K.** 1998. [Systemic aspergillosis caused by

an aflatoxin-producing strain of *Aspergillus* in a post-bone marrow transplant patient with acute myeloid leukemia]. *Rinsho Ketsueki* **39:**1103–1108.

215. **Rao K, Saha V.** 2000. Medical management of *Aspergillus flavus* endocarditis. *Pediatr Hematol Oncol* **17:**425–427.
216. **Thomas PA, Kaliamurthy J.** 2013. Mycotic keratitis: epidemiology, diagnosis and management. *Clin Microbiol Infect* **19:** 210–220.
217. **Nyvall RF.** 1979. *Field Crop Diseases Handbook.* AVI Publishing Company, Inc, Westport, CT.
218. **O'Shaughnessy EM, Forrest GN, Walsh TJ.** 2004. Invasive aspergillosis in patients with hematologic malignancies: recent advances and new challenges. *Curr Treat Options Infect Dis* **5:** 507–515.
219. **Tanaka K, Nonaka F.** 1977. Studies on the rot of onion bulbs by Aspergillus niger. *Proc Assoc Plant Prot Kyushu* **23:**36–39.
220. **Mishra GS, Mehta N, Pal M.** 2004. Chronic bilateral otomycosis caused by *Aspergillus niger. Mycoses* **47:**82–84.
221. **Yamaguchi M, Nishiya H, Mano K, Kunii O, Miyashita H.** 1992. Chronic necrotising pulmonary aspergillosis caused by *Aspergillus niger* in a mildly immunocompromised host. *Thorax* **47:**570–571.
222. **Geiser DM, Dorner JW, Horn BW, Taylor JW.** 2000. The phylogenetics of mycotoxin and sclerotium production in *Aspergillus flavus* and *Aspergillus oryzae. Fungal Genet Biol* **31:**169–179.
223. **Stenson S, Brookner A, Rosenthal S.** 1982. Bilateral endogenous necrotizing scleritis due to *Aspergillus oryzae. Ann Ophthalmol* **14:**67–72.
224. **Akiyama K, Takizawa H, Suzuki M, Miyachi S, Ichinohe M, Yanagihara Y.** 1987. Allergic bronchopulmonary aspergillosis due to *Aspergillus oryzae. Chest* **91:**285–286.
225. **Matteson Heidenreich MC, Corral-Garcia MR, Momol EA, Burr TJ.** 1997. Russet of apple fruit caused by *Aureobasidium pullulans* and *Rhodotorula glutinis. Plant Dis* **81:**337–342.
226. **Spotts RA, Cervantes LA.** 2002. Involvement of *Aureobasidium pullulans* and *Rhodotorula glutinis* in russet of d'Anjou pear fruit. *Plant Dis* **86:**625–628.
227. **Gupta V, Chawla R, Sen S.** 2001. *Aureobasidium pullulans* scleritis following keratoplasty: a case report. *Ophthalmic Surg Lasers* **32:**481–482.
228. **Kaczmarski EB, Liu Yin JA, Tooth JA, Love EM, Delamore IW.** 1986. Systemic infection with *Aureobasidium pullulans* in a leukaemic patient. *J Infect* **13:**289–291.
229. **Chalet M, Howard DH, McGinnis MR, Zapatero I.** 1986. Isolation of *Bipolaris australiensis* from a lesion of viral vesicular dermatitis on the scalp. *J Med Vet Mycol* **24:**461–465.
230. **Flanagan KL, Bryceson AD.** 1997. Disseminated infection due to *Bipolaris australiensis* in a young immunocompetent man: case report and review. *Clin Infect Dis* **25:**311–313.
231. **Sonoda RM.** 1991. *Exserohilum rostratum* and *Bipolaris hawaiiensis* causing leaf and culm lesions on Callide Rhodesgrass. *Proc Soil Crop Sci Soc Fla* **50:**28–30.
232. **Pratt RG.** 2001. Occurrence and virulence of *Bipolaris hawaiiensis* on bermudagrass (*Cynodon dactylon*) on poultry waste application sites in Mississippi. *Plant Dis* **85:**1206.
233. **Morton SJ, Midthun K, Merz WG.** 1986. Granulomatous encephalitis caused by *Bipolaris hawaiiensis. Arch Pathol Lab Med* **110:**1183–1185.
234. **McGinnis MR, Campbell G, Gourley WK, Lucia HL.** 1992. Phaeohyphomycosis caused by *Bipolaris spicifera*: an informative case. *Eur J Epidemiol* **8:**383–386.
235. **Ogden PE, Hurley DL, Cain PT.** 1992. Fatal fungal endarteritis caused by *Bipolaris spicifera* following replacement of the aortic valve. *Clin Infect Dis* **14:**596–598.
236. **Latham RH.** 2000. Bipolaris spicifera meningitis complicating a neurosurgical procedure. *Scand J Infect Dis* **32:**102–103.
237. **Bava AJ, Fayad A, Céspedes C, Sandoval M.** 2003. Fungal peritonitis caused by *Bipolaris spicifera. Med Mycol* **41:**529–531.
238. **Geraldi MAP, Ito MF, Ricci A Jr, Paradela FO, Nagal H.** 1980. *Chaetomium globosum* Kunze, causal agent of a new tomato disease. *Summa Phytopathol* **6:**79–84.
239. **Anandi V, John TJ, Walter A, Shastry JC, Lalitha MK, Padhye AA, Ajello L, Chandler FW.** 1989. Cerebral phaeohyphomycosis caused by *Chaetomium globosum* in a renal transplant recipient. *J Clin Microbiol* **27:**2226–2229.
240. **Hattori N, Adachi M, Kaneko T, Shimozuma M, Ichinohe M, Iozumi K.** 2000. Case report. Onychomycosis due to *Chaetomium globosum* successfully treated with itraconazole. *Mycoses* **43:**89–92.
241. **Yeghen T, Fenelon L, Campbell CK, Warnock DW, Hoffbrand AV, Prentice HG, Kibbler CC.** 1996. *Chaetomium pneumonia* in patient with acute myeloid leukaemia. *J Clin Pathol* **49:**184–186.
242. **Hammouda AM.** 1992. A new leaf spot of pepper caused by *Cladosporium oxysporum. Plant Dis* **76:**536–537.
243. **Romano C, Bilenchi R, Alessandrini C, Miracco C.** 1999. Case report. Cutaneous phaeohyphomycosis caused by *Cladosporium oxysporum. Mycoses* **42:**111–115.
244. **Dillard HR, Cobb AC.** 1997. Disease progress of black dot on tomato roots and reduction in incidence with foliar applied fungicides. *Plant Dis* **81:**1439–1442.
245. **Andrivon D, Lucas JM, Guérin C, Jouan B.** 1998. Colonization of roots, stolons, tubers and stems of various potato (*Solanum tuberosum*) cultivars by the black dot fungus *Colletotrichum coccodes. Plant Pathol* **47:**440–445.
246. **O'Quinn RP, Hoffmann JL, Boyd AS.** 2001. *Colletotrichum* species as emerging opportunistic fungal pathogens: a report of 3 cases of phaeohyphomycosis and review. *J Am Acad Dermatol* **45:**56–61.
247. **Dickman MD, Alvarez AM.** 1983. Latent infection of papaya caused by *Colletotrichum gloeosporioides. Plant Dis* **67:**748–750.
248. **Yamamoto N, Matsumoto T, Ishibashi Y.** 2001. Fungal keratitis caused by *Colletotrichum gloeosporioides. Cornea* **20:**902–903.
249. **Bergstrom GC, Nicholson RL.** 1999. The biology of corn anthracnose. *Plant Dis* **83:**596–608.
250. **Heimann MF, Boone DM.** 1983. Raspberry (*Rubus*) disorders: cane blight and spur blight [*Leptosphaeria coniothyrium, Coniothyrium fuckelii,* Wisconsin]. Cooperative Extension Programs, University of Wisconsin–Extension, Madison, WI.
251. **Laverde S, Moncada LH, Restrepo A, Vera CL.** 1973. Mycotic keratitis; 5 cases caused by unusual fungi. *Sabouraudia* **11:** 119–123.
252. **Kiehn TE, Polsky B, Punithalingam E, Edwards FF, Brown AE, Armstrong D.** 1987. Liver infection caused by *Coniothyrium fuckelii* in a patient with acute myelogenous leukemia. *J Clin Microbiol* **25:**2410–2412.
253. **Kore SS, Bhide VP.** 1976. A first report of *Curvularia brachyspora* Boedijn inciting leaf-spot disease of rose. *Curr Sci* **45:**74.
254. **Torda AJ, Jones PD.** 1997. Necrotizing cutaneous infection caused by *Curvularia brachyspora* in an immunocompetent host. *Australas J Dermatol* **38:**85–87.
255. **Marcus L, Vismer HF, van der Hoven HJ, Gove E, Meewes P.** 1992. Mycotic keratitis caused by *Curvularia brachyspora* (Boedjin). A report of the first case. *Mycopathologia* **119:**29–33.
256. **Mandokhot AM, Basu Chaudhary KCB.** 1972. A new leaf spot of maize incited by *Curvularia clavata. Neth J Plant Pathol* **78:**65–68.
257. **Ebright JR, Chandrasekar PH, Marks S, Fairfax MR, Aneziokoro A, McGinnis MR.** 1999. Invasive sinusitis and cerebritis due to *Curvularia clavata* in an immunocompetent adult. *Clin Infect Dis* **28:**687–689.
258. **Gugnani HC, Okeke CN, Sivanesan A.** 1990. *Curvularia clavata* as an etiologic agent of human skin infection. *Lett Appl Microbiol* **10:**47–49.
259. **Meredith DS.** 1963. Some graminicolous fungi associated with spotting of banana leaves in Jamaica. *Ann Appl Biol* **51:**371–378.
260. **Georg LK.** 1964. *Curvularia geniculata,* a cause of mycotic keratitis. *J Med Assoc State Ala* **33:**234–236.

261. **Bridges CH.** 1957. Maduromycotic mycetomas in animals; Curvularia geniculata as an etiologic agent. *Am J Pathol* **33:**411–427.
262. **Lakshmanan P.** 1992. Sheath rot of rice incited by *Curvularia lunata* in Tamal Nadu, India. *Trop Pest Manage* **38:**107.
263. **Muchovej JJ, Couch HB.** 1987. Colonization of bentgrass turf by *Curvularia lunata* after leaf clipping and heat stress. *Plant Dis* **71:**873–875.
264. **Ito MF, Paradela F, Soave J, Sugimori MH.** 1979. Leaf spot caused in maize (*Zea mays* L.) by *Curvularia lunata. Summa Phytopathol* **5:**181–184.
265. **Gour HN, Dube HC.** 1975. Production of pectic enzymes by *Curvularia lunata* causing leaf spot of cotton. *Proc Indian Natl Sci Acad Part B.* **41:**480–485.
266. **Carter E, Boudreaux C.** 2004. Fatal cerebral phaeohyphomycosis due to *Curvularia lunata* in an immunocompetent patient. *J Clin Microbiol* **42:**5419–5423.
267. **Tessari G, Forni A, Ferretto R, Solbiati M, Faggian G, Mazzucco A, Barba A.** 2003. Lethal systemic dissemination from a cutaneous infection due to *Curvularia lunata* in a heart transplant recipient. *J Eur Acad Dermatol Venereol* **17:**440–442.
268. **Taj-Aldeen SJ, Hilal AA, Schell WA.** 2004. Allergic fungal rhinosinusitis: a report of 8 cases. *Am J Otolaryngol* **25:**213–218.
269. **Rao GP, Singh SP, Singh M.** 1992. Two alternative hosts of *Curvularia pallescens*, the leaf spot causing fungus of sugarcane. *Trop Pest Manage* **38:**218.
270. **Lal S, Tripathi HS.** 1977. Host range of *Curvularia pallescens,* the incitant of leaf spot of maize. *Indian J Mycol Plant Pathol* **7:** 92–93.
271. **Salleh R, Safinat A, Julia L, Teo CH.** 1996. Brown spot caused by *Curvularia* spp., a new disease of asparagus. *Biotrophia* **9:**26–37.
272. **Rajalakshmy VK.** 1976. Leaf spot of rubber caused by *Curvularia pallescens* Boedijn. *Curr Sci* **24:**530.
273. **Agrawal A, Singh SM.** 1995. Two cases of cutaneous phaeohyphomycosis caused by *Curvularia pallescens. Mycoses* **38:** 301–303.
274. **Yang S.** 1973. Isolation and effect of temperature on spore germination, radial growth and pathogenicity of *Curvularia senegalensis. Phytopathology* **63:**1540–1541.
275. **Guarro J, Akiti T, Horta RA, Morizot Leite-Filho LA, Gené J, Ferreira-Gomes S, Aguilar C, Ortoneda M.** 1999. Mycotic keratitis due to *Curvularia senegalensis* and *in vitro* antifungal susceptibilities of *Curvularia* spp. *J Clin Microbiol* **37:**4170–4173.
276. **Usharani P, Ramarao P.** 1981. Corm rot of *Colocasia esculenta* caused by *Cylindrocarpon lichenicola. Indian Phytopathol* **34:** 381–382.
277. **Rodríguez-Villalobos H, Georgala A, Beguin H, Heymans C, Pye G, Crokaert F, Aoun M.** 2003. Disseminated infection due to *Cylindrocarpon* (*Fusarium*) *lichenicola* in a neutropenic patient with acute leukaemia: report of a case and review of the literature. *Eur J Clin Microbiol Infect Dis* **22:**62–65.
278. **Mangiaterra M, Giusiano G, Smilasky G, Zamar L, Amado G, Vicentín C.** 2001. Keratomycosis caused by *Cylindrocarpon lichenicola. Med Mycol* **39:**143–145.
279. **Pioli RN, Morandi EN, Martínez MC, Lucca F, Tozzini A, Bisaro V, Hopp HE.** 2003. Morphologic, molecular, and pathogenic characterization of *Diaporthe phaseolorum* variability in the core soybean-producing area of Argentina. *Phytopathology* **93:**136–146.
280. **Iriart X, Binois R, Fior A, Blanchet D, Berry A, Cassaing S, Amazan E, Papot E, Carme B, Aznar C, Couppié P.** 2011. Eumycetoma caused by *Diaporthe* phaseolorum (*Phomopsis phaseoli*): a case report and a mini-review of *Diaporthe*/*Phomopsis* spp invasive infections in humans. *Clin Microbiol Infect* **17:**1492–1494.
281. **Fischl G, Szunics L, Bakonyi J.** 1993. Black points in wheat grains. *Novenytermeles* **42:**419–429.
282. **Leach CM, Tulloch M.** 1972. World-wide occurrence of the suspected mycotoxin producing fungus *Drechslera biseptata* with grass seed. *Mycologia* **64:**1357–1359.
283. University of Adelaide, Adelaide, Australia. Mycology online. http://www.mycology.adelaide.edu.au/Fungal_Descriptions/Hyphomycetes_(dematiaceous)/Drechslera/index.html.
284. **Sharfun-Nahar MM.** 2006. Pathogenicity and transmission studies of seed-borne *Fusarium* species (Sec. *Liseola* and *Sporotrichiella*) in sunflower. *Pak J Bot* **38:**487–492.
285. **Boby VU, Bagyaraj DJ.** 2003. Biological control of root-rot of *Coleus forskohlii* Briq. using microbial inoculants. *World J Microbiol* **19:**175–180.
286. **Satou M, Ichinoe M, Fukumoto F, Tezuka N, Horiuchi S.** 2001. Fusarium blight of kangaroo paw (*Anigozanthos* spp.) caused by *Fusarium chlamydosporum* and *Fusarium semitectum. J Phytopathol* **149:**203–206.
287. **Segal BH, Walsh TJ, Liu JM, Wilson JD, Kwon-Chung KJ.** 1998. Invasive infection with *Fusarium chlamydosporum* in a patient with aplastic anemia. *J Clin Microbiol* **36:**1772–1776.
288. **Michailides TJ, Morgan DP, Subbarao KV.** 1996. Fig endosepsis: an old disease still a dilemma for California growers. *Plant Dis* **80:**828–841.
289. **Austen B, McCarthy H, Wilkins B, Smith A, Duncombe A.** 2001. Fatal disseminated *fusarium* infection in acute lymphoblastic leukaemia in complete remission. *J Clin Pathol* **54:**488–490.
290. **Camin AM, Michelet C, Langanay T, de Place C, Chevrier S, Guého E, Guiguen C.** 1999. Endocarditis due to *Fusarium dimerum* four years after coronary artery bypass grafting. *Clin Infect Dis* **28:**150.
291. **Vismer HF, Marasas WF, Rheeder JP, Joubert JJ.** 2002. *Fusarium dimerum* as a cause of human eye infections. *Med Mycol* **40:**399–406.
292. **Hwang IS, Kang WR, Hwang DJ, Bae SC, Yun SH, Ahn IP.** 2013. Evaluation of bakanae disease progression caused by *Fusarium fujikuroi* in *Oryza sativa* L. *J Microbiol* **51:**858–865.
293. **Chang DC, Grant GB, O'Donnell K, Wannemuehler KA, Noble-Wang J, Rao CY, Jacobson LM, Crowell CS, Sneed RS, Lewis FMT, Schaffzin JK, Kainer MA, Genese CA, Alfonso EC, Jones DB, Srinivasan A, Fridkin SK, Park BJ, Fusarium Keratitis Investigation Team, Fusarium Keratitis Investigation Team.** 2006. Multistate outbreak of Fusarium keratitis associated with use of a contact lens solution. *JAMA* **296:**953–963.
294. **Seta S, Gonzalez M, Lori G.** 2004. First report of walnut canker caused by *Fusarium incarnatum* in Argentina. *Plant Pathol* **53:** 248.
295. **Van Hove F, Waalwijk C, Logrieco A, Munaut F, Moretti A.** 2011. *Gibberella musae* (*Fusarium musae*) sp. nov., a recently discovered species from banana is sister to *F. verticillioides. Mycologia* **103:**570–585.
296. **Triest D, Stubbe D, De Cremer K, Piérard D, Detandt M, Hendrickx M.** 2015. Banana infecting fungus, *Fusarium musae,* is also an opportunistic human pathogen: are bananas potential carriers and source of fusariosis? *Mycologia* **107:**46–53.
297. **Jones DR, Lomeiro EO.** 2000. Pseudostem heart rot, p 166–167. *In* Jones DR (ed), *Diseases of Banana, Abaca and Enset.* CABI Publishing, Wallingford, UK.
298. **Dignani MC, Anaissie E.** 2004. Human fusariosis. *Clin Microbiol Infect* **10**(Suppl 1):67–75.
299. **Bottalico A, Perrone G.** 2002. Toxigenic *Fusarium* species and mycotoxins associated with head blight in small-grain cereals in Europe. *Eur J Plant Pathol* **108:**611–624.
300. **Raabe RD, Conners LL, Martinez AP.** 1981. *Checklist of plant disease in Hawaii: including records of microorganisms, principally fungi, found in the state. Information text series 022.* Hawaii College of Tropical Agriculture and Human Resources, University of Hawaii at Manoa.
301. **Lucas JA, Dickinson CH.** 1998. *Plant Pathology and Plant Pathogens,* 3rd ed. Blackwell Science, Oxford, UK.
302. **Sander A, Beyer U, Amberg R.** 1998. Systemic *Fusarium oxysporum* infection in an immunocompetent patient with an adult

respiratory distress syndrome (ARDS) and extracorporal membrane oxygenation (ECMO). *Mycoses* **41:**109–111.

303. **Romano C, Miracco C, Difonzo EM.** 1998. Skin and nail infections due to *Fusarium oxysporum* in Tuscany, Italy. *Mycoses* **41:** 433–437.
304. **Rodriguez-Villalobos H, Aoun M, Heymans C, De Bruyne J, Duchateau V, Verdebout JM, Crokaert F.** 2002. Cross reaction between a pan-Candida genus probe and *Fusarium* spp. in a fatal case of *Fusarium oxysporum* pneumonia. *Eur J Clin Microbiol Infect Dis* **21:**149–152.
305. **Godoy P, Nunes E, Silva V, Tomimori-Yamashita J, Zaror L, Fischman O.** 2004. Onychomycosis caused by *Fusarium solani* and *Fusarium oxysporum* in São Paulo, Brazil. *Mycopathologia* **157:**287–290.
306. **Abdalla MY, Al-Rokibah A, Moretti A, Mulè G.** 2000. Pathogenicity of toxigenic *Fusarium proliferatum* from date palm in Saudi Arabia. *Plant Dis* **84:**321–324.
307. **Bush BJ, Carson ML, Cubeta MA, Hagler WM, Payne GA.** 2004. Infection and fumonisin production by *Fusarium verticillioides* in developing maize kernels. *Phytopathology* **94:**88–93.
308. **Summerbell RC, Richardson SE, Kane J.** 1988. *Fusarium proliferatum* as an agent of disseminated infection in an immunosuppressed patient. *J Clin Microbiol* **26:**82–87.
309. **Murray CK, Beckius ML, McAllister K.** 2003. *Fusarium proliferatum* superficial suppurative thrombophlebitis. *Mil Med* **168:** 426–427.
310. **Picot A, Barreau C, Pinson-Gadais L, Caron D, Lannou C, Richard-Forget F.** 2010. Factors of the *Fusarium verticillioides*-maize environment modulating fumonisin production. *Crit Rev Microbiol* **36:**221–231.
311. **Jones DR, Stover RH.** 2000. Fungal root rot, p 162–163. *In* Jones DR (ed), *Diseases of Banana, Abaca and Enset.* CABI Publishing, Wallingford, UK.
312. **Repiso T, García-Patos V, Martin N, Creus M, Bastida P, Castells A.** 1996. Disseminated fusariosis. *Pediatr Dermatol* **13:**118–121.
313. **Bushelman SJ, Callen JP, Roth DN, Cohen LM.** 1995. Disseminated *Fusarium solani* infection. *J Am Acad Dermatol* **32:** 346–351.
314. **Davis RM, Colyer PD, Rothrock CS, Kochman JK.** 2006. Fusarium wilt of cotton: population diversity and implications for management. *Plant Dis* **90:**692–703.
315. **Dantas SAF, Oliveira SMA, Michereff SJ, Nascimento LC, Gurgel LMS, Pessoa WRLS.** 2003. Post harvest fungal diseases in papaya and orange marketed in the distribution centre of Recife. *Fitopatol Bras* **28:**528–533.
316. **Mullen JM, Gilliam CH, Hagan AK, Morgan Jones G.** 1991. Canker of dogwood caused by *Lasiodiplodia theobromae*, a disease influenced by drought stress or cultivar selection. *Plant Dis* **75:**886–889.
317. **Ko WH, Wang IT, Ann PJ.** 2004. *Lasiodiplodia theobromae* as a causal agent of kumquat dieback in Taiwan. *Plant Dis* **88:**1383.
318. **Anthony S, Abeywickrama K, Dayananda R, Wijeratnam SW, Arambewela L.** 2004. Fungal pathogens associated with banana fruit in Sri Lanka, and their treatment with essential oils. *Mycopathologia* **157:**91–97.
319. **Phipps PM, Porter DM.** 1998. Collar rot of peanut caused by *Lasiodiplodia theobromae. Plant Dis* **82:**1205–1209.
320. **Maslen MM, Collis T, Stuart R.** 1996. *Lasiodiplodia theobromae* isolated from a subcutaneous abscess in a Cambodian immigrant to Australia. *J Med Vet Mycol* **34:**279–283.
321. **Thomas PA.** 2003. Current perspectives on ophthalmic mycoses. *Clin Microbiol Rev* **16:**730–797.
322. **Kindo AJ, Pramod C, Anita S, Mohanty S.** 2010. Maxillary sinusitis caused by *Lasiodiplodia theobromae. Indian J Med Microbiol* **28:**167–169.
323. **Bugos RC, Sutherland JB, Adler JH.** 1988. Phenolic compound utilization by the soft rot fungus *Lecythophora hoffmannii. Appl Environ Microbiol* **54:**1882–1885.
324. **Marriott DJ, Wong KH, Aznar E, Harkness JL, Cooper DA, Muir D.** 1997. *Scytalidium dimidiatum* and *Lecythophora hoffmannii*: unusual causes of fungal infections in a patient with AIDS. *J Clin Microbiol* **35:**2949–2952.
325. **Ashkan SM, Abusaidi D, Ershad D.** 1997. Etiological study of dieback and canker of pistachio nut tree in Rafsangan. *Iranian J Plant Pathol* **33:**15–26.
326. **Byrd RP Jr, Roy TM, Fields CL, Lynch JA.** 1992. *Paecilomyces varioti* pneumonia in a patient with diabetes mellitus. *J Diabetes Complications* **6:**150–153.
327. **Kantarcioğlu AS, Hatemi G, Yücel A, De Hoog GS, Mandel NM.** 2003. *Paecilomyces variotii* central nervous system infection in a patient with cancer. *Mycoses* **46:**45–50.
328. **Wright K, Popli S, Gandhi VC, Lentino JR, Reyes CV, Leehey DJ.** 2003. *Paecilomyces peritonitis*: case report and review of the literature. *Clin Nephrol* **59:**305–310.
329. **Mostert L, Groenewald JZ, Summerbell RC, Robert V, Sutton DA, Padhye AA, Crous PW.** 2005. Species of *Phaeoacremonium* associated with infections in humans and environmental reservoirs in infected woody plants. *J Clin Microbiol* **43:**1752–1767.
330. **Kliejunas JT, Allison JR, McCain AH, Smith RS Jr.** 1985. *Phoma* blight of fir and Douglas fir seedlings in a California nursery. *Plant Dis* **69:**773–775.
331. **Bakerspigel A, Lowe D, Rostas A.** 1981. The isolation of *Phoma eupyrena* from a human lesion. *Arch Dermatol* **117:**362–363.
332. **Singh H, Singh RS, Chohan JS.** 1974. Fruit rot of *Luffa acutangula* by *Mucor circinelloides. Indian J Mycol Plant Pathol* **4:** 99–100.
333. **Chandra S, Woodgyer A.** 2002. Primary cutaneous zygomycosis due to *Mucor circinelloides. Australas J Dermatol* **43:**39–42.
334. **Boyd AS, Wiser B, Sams HH, King LE.** 2003. Gangrenous cutaneous mucormycosis in a child with a solid organ transplant: a case report and review of the literature. *Pediatr Dermatol* **20:** 411–415.
335. **Eisen DP, Robson J.** 2004. Complete resolution of pulmonary *Rhizopus oryzae* infection with itraconazole treatment: more evidence of the utility of azoles for zygomycosis. *Mycoses* **47:** 159–162.
336. **Yang SM, Morris JB, Unger PW, Thompson TE.** 1979. Rhizopus head rot of cultivated sunflower, *Helianthus annuus*, in Texas USA. *Plant Dis Rep* **63:**833–835.
337. **González A, del Palacio A, Cuétara MS, Gómez C, Carabias E, Malo Q.** 1996. Zigomicosis: revisión de 16 casos. [Zygomycosis: review of 16 cases]. *Enferm Infecc Microbiol Clin* **14:** 233–239.
338. **Harrison BD, Wilson TMA.** 1999. Milestones in the research on tobacco mosaic virus. *Philos Trans R Soc Lond B Biol Sci* **354:** 521–529.
339. **Zaitlin M.** 2000. Tobacco mosaic virus. Descriptions of plant viruses No. 370. Association of Applied Biologists. http://www.dpvweb.net/dpv/showdpv.php?dpvno=370.
340. **Li L, Wang L, Xiao R, Zhu G, Li Y, Liu C, Yang R, Tang Z, Li J, Huang W, Chen L, Zheng X, He Y, Tan J.** 2012. The invasion of tobacco mosaic virus RNA induces endoplasmic reticulum stress-related autophagy in HeLa cells. *Biosci Rep* **32:**171–184.
341. **Alonso E, García Luque I, Avila-Rincón MJ, Wicke B, Serra MT, Díaz-Ruiz JR.** 1989. A tobamovirus causing heavy losses in protected pepper crops in Spain. *J Phytopathol* **125:**67–76.
342. **Jarret RL, Gillaspie AG, Barkley NA, Pinnow DL.** 2008. The occurrence and control of *Pepper mild mottle virus* in the USDA/ARS Capsicum germplasm collection. *Seed Tech* **30:**26–36.
343. **Rosario K, Symonds EM, Sinigalliano C, Stewart J, Breitbart M.** 2009. *Pepper mild mottle virus* as an indicator of fecal pollution. *Appl Environ Microbiol* **75:**7261–7267.
344. **Han T-H, Kim S-C, Kim S-T, Chung C-H, Chung J-Y.** 2014. Detection of norovirus genogroup IV, klassevirus, and pepper mild mottle virus in sewage samples in South Korea. *Arch Virol* **159:** 457–463.

345. **Hamza IA, Jurzik L, Überla K, Wilhelm M.** 2011. Evaluation of pepper mild mottle virus, human picobirnavirus and Torque teno virus as indicators of fecal contamination in river water. *Water Res* **45:**1358–1368.
346. **Van Kammen A, van Lent J, Wellink H.** 2001. Cowpea mosaic virus. Descriptions of Plant Viruses No. 378. Association of Applied Biologists. http://www.dpvweb.net/dpv/showdpv.php?dpvno=378.
347. **Brennan FR, Jones TD, Hamilton WD.** 2001. Cowpea mosaic virus as a vaccine carrier of heterologous antigens. *Mol Biotechnol* **17:**15–26.
348. **Koudelka KJ, Destito G, Plummer EM, Trauger SA, Siuzdak G, Manchester M.** 2009. Endothelial targeting of cowpea mosaic virus (CPMV) via surface vimentin. *PLoS Pathog* **5:** e1000417.
349. **Plummer EM, Manchester M.** 2013. Endocytic uptake pathways utilized by CPMV nanoparticles. *Mol Pharm* **10:**26–32.
350. **Petro TM, Agarkova IV, Zhou Y, Yolken RH, Van Etten JL, Dunigan DD.** 2015. Response of mammalian macrophages to challenge with the chlorovirus *Acanthocystis turfacea* chlorella virus 1. *J Virol* **89:**12096–12107.

4

Laboratory-Associated Infections

KAREN BRANDT BYERS AND A. LYNN HARDING

Publications on laboratory-associated infections (LAIs) provide critical information for prevention strategies. The review of actual case studies illustrates the importance of adhering to biosafety protocols and may trigger changes in laboratory procedures. Singh has stated that it is time for a centralized system for reporting, analyzing, and communicating "lessons learned" about LAIs to be developed (1). Surveys on some subsets of laboratory workers, and case reports on individual LAIs, have been published; however, without a centralized system, it is impossible to assess the true incidence of LAIs. In addition, the underreporting of such infections is widely acknowledged due to fear of reprisal and the stigma associated with such events (2).

To address the need for collecting additional LAI data, experts have recommended that a "nonpunitive surveillance and reporting system with the potential for anonymity" be implemented in the United States for diagnostic laboratories (3) and high-containment research laboratories (4). This has not been implemented for all LAIs; however, mandatory reporting for infections with Select Agents, a class of infectious agents that pose a severe threat to human and animal health, has been instituted in the United States. Data are available on the LAIs that occurred with Select Agents in the United States for the years 2004–2012. Between 2004 and 2010, approximately 10,000 researchers had access to Select Agents; eight primary LAIs occurred, no secondary LAIs, and no fatalities. In addition, three clinical laboratories reported LAI with *Brucella*, a Select Agent (5).

Systematic reporting of LAI would support the assessment of containment practices and the development of evidence-based biosafety measures (6). Until more comprehensive data are available, a literature review such as this one provides updated information about the types of microorganisms and exposures responsible for LAIs and generates awareness that laboratory workers continue to be at risk of infection.

EPIDEMIOLOGIC STUDIES OF LAI

In this chapter, we review the LAIs reported in the literature since 1979, compare them to the data from the 1930–1978 Pike and Sulkin surveys, and present a summary of the agents, routes of exposure, and types of activities, as well as the host and environmental factors available on LAIs. Incidence data, when available, are included. This summary was compiled to support the continued development of biosafety programs to minimize the occupational risk of infection from diagnostic, research, teaching, and production activities. Information on hazard assessment and prevention strategies is available in other chapters.

Epidemiology is defined as the study of the distribution and determinants of diseases and injuries in human populations. That is, epidemiology is concerned with the extent and types of illnesses and injuries in groups of people and with the factors that influence their distribution. In the context of this chapter, the illnesses are LAIs, and the factors analyzed include the infectious agent, the activities conducted when the exposure occurred (clinical, research, field, teaching, animal work), and the routes of transmission. Inherent in the definition of epidemiology is the necessity for measuring the amount of disease in a population or occupation by relating the number of cases to a population base. Cases or events that fit the case definition are identified and counted; the number of cases in a potentially exposed population (attack rates, or incidence) can then be calculated and compared to the rates of occurrence in other populations. Comparisons of the data from clinical, research, and production laboratories from published case studies may be misleading, because attack rates for a given population are rarely available.

Incidence

Estimates from the Pike data suggested that the risk for researchers was six to seven times greater than that for hospital and public health laboratory workers; the calculated attack rate for researchers was 4.1 per 1,000 (7). Reid, in 1957, reported that the incidence of tuberculosis among laboratory personnel working with *Mycobacterium tuberculosis* was three times higher than among those not working with the agent (8). Philips, in 1965, estimated that the frequency of LAIs (using available U.S. and European data) resulted in the expected number of one to five infections per million working hours (9). A 1971 survey of laboratory-acquired cases of tuberculosis, shigellosis, brucellosis, and hepatitis in England and Wales reported an annual incidence rate of 43 infections per 1,000 medical laboratory employees (10).

Since these early studies, only a few surveys, primarily in clinical laboratories, have provided incidence data. Grist (11, 12) reported that the incidence for clinical microbiologists was 9.4 infections per 1,000 employees. Jacobson et al. surveyed supervisors of 1,191 clinical laboratory workers in Utah to determine the LAIs that occurred between 1978 and 1982. The annual incidence of LAIs in that population was 3 per 1,000 employees (13). Further analysis of the data indicated that the annual incidence in small laboratories (less than 25 employees) was greater than that in large laboratories (5.0 versus 1.5 per 1,000) and that approximately 1% of all microbiologists reported an LAI. Vesley and Hartmann surveyed LAIs among 4,202 public health and 2,290 hospital clinical laboratory employees (14). The annual incidence rate for all full-time employees was calculated at 1.4 infections per 1,000 employees for public health laboratories and 3.5 infections per 1,000 employees in hospital laboratories. Incidence in microbiology laboratories was higher, at 2.7 per 1,000 staff in public health laboratories and 4.0 per 1,000 employees in hospital microbiology labs. The relatively low incidence in the Vesley and Hartman study was attributed to safety awareness and improvements in safety devices. A Japanese survey of clinical laboratory workers in 306 hospitals cited an annual incidence rate of 2.0 infections per 1,000 persons (15).

Baron and Miller conducted a voluntary online survey of clinical microbiology laboratory directors between 2002 and 2004 (16). Forty-five LAIs were reported by directors of 88 U.S. hospital laboratories of various sizes and three reference laboratories. The calculated incidence of LAIs was compared with incidence of infection in the general population aged 30–59. Clinical microbiologists were at high risk for *Brucella* infections, with calculated incidences of 641 cases per 100,000 laboratory technologists, compared to 0.08 cases per 100,000 in the general population. The incidence of *Neisseria meningitidis* was 25.1 per 100,000 for microbiologists compared to 0.6 per 100,000 for the general population; for *Escherichia coli* O157H7, it was 83 versus 0.96. The risk of *Shigella* infection was the same in both populations (6 per 100,000) and nearly the same for *Coccidioides* (13.6 per 100,000 microbiologists versus 12 per 100,000 general population). The incidence of *Salmonella* and *Clostridium difficile* infections was lower in microbiologists than in the general population. For *Salmonella*, the incidence for clinical laboratory staff was 1.5, compared to the incidence in the general population of 17.9 per 100,000; for *C. difficile* the incidence was 0.2 compared to 8 per 100,000 in the general population (16).

Interventions Based on LAIs

Many reports also document responsible internal review of the factors contributing to the LAI. An example would be the report on a LAI that occurred when a technician was subculturing a collection of *N. meningitidis* isolates. In compliance with the existing laboratory protocols, the infected technician conducted the work on the open bench and used glass Pasteur pipettes to remove colonies from the frozen agar surface. Analysis of the laboratory procedures after the LAI was confirmed resulted in protocol revisions to prevent recurrence. After the LAI, laboratory procedures were updated. Subculturing the *N. meningitidis* isolates is conducted in the biosafety cabinet, soft cotton swabs have replaced the glass Pasteur pipettes, and the vaccine against *N. meningitidis* is offered to staff (17). In another incident, a glass lyophilization vial punctured a researcher's gloved hand, resulting in a buffalopox LAI. To prevent recurrence, the freezing temperature for the vials prior to lyophilization was reduced to −60°C and a better-quality glass vial is now used (18). The "lessons learned" from these incidents were published to prevent additional LAIs in staff conducting similar procedures.

In 2011, the Centers for Disease Control and Prevention (CDC) reported on 109 cases linked to clinical and teaching microbiology laboratories. The cases were identified through PulseNet, the electronic monitoring system for foodborne outbreaks in the United States. Twelve percent of the cases required hospitalization, and there was one fatality. Fifty-four cases responded to CDC surveys, and 65% reported association with a university/college/community college teaching laboratory (19). An investigation revealed that the facilities and safety policies in all of the teaching laboratories were essentially equivalent. However, students in laboratories without LAIs were more familiar with biosafety training and the symptoms of infection associated with the agents studied. This led to the design of a poster for the student population illustrating that objects such pens, lab notebooks, and cell phones or personal music devices used in the lab can be a source of infection. The poster also promoted effective preventive measures such as hand washing to prevent transmission of LAI. In addition to advice for students, the CDC *Salmonella* outbreak website has a message for teachers and clinical supervisors on prevention of LAI. Another outbreak linked to clinical and teaching microbiology laboratories was reported in 2014, with 41 cases, 36% hospitalization, and no fatalities (20). The poster has provided supervisors with an excellent training tool; however, it is an ongoing challenge to reinforce appropriate work practices with each new class of inexperienced students.

LAIs: THE CONTINUUM

LAIs were not a new phenomenon in the 20th century, as historical accounts of typhoid, *Brucella*, and tetanus were recorded as early as 1885, 1887, and 1893, respectively (21, 22). Reports from the 1930s and 1940s demonstrate that microbial agents were potentially hazardous to individuals within the laboratory and posed some risk to those working in close proximity to the laboratories (23, 24).

LAIs are defined as all infections acquired through laboratory or laboratory-related activities regardless of whether they are symptomatic (overt) or asymptomatic (subclinical) in nature. In 1950, Sulkin and Pike circulated a questionnaire to 5,000 laboratories in the United States, including those associated with state and local health departments, accredited hospitals, private schools of medicine and veterinary science, undergraduate teaching institutions, manufacturers of biologic products, and various government agencies. The questionnaire solicited information on unreported infections resulting from laboratory work and slightly more than half of those surveyed responded (23, 24). During the period 1930–1975, these authors published cumulative data describing 3,921 LAIs in the United States and other countries (25, 26). In 1978, Pike added 158 new infections, bringing the total to 4,079 documented LAIs, 168 of which were fatal (27). As an apparent reflection of the work being performed in the responding laboratories at that time, bacteria accounted for 1,704 of the infections, viruses for 1,179, rickettsia for 598, fungi for 354, chlamydia for 128, and parasites for 116. Bacteria or viruses were associated with more than two-thirds of the lethal and nonlethal infections. *Brucella*, *Coxiella burnetii*, hepatitis B virus, *Salmonella enterica* serovar Typhi, *Francisella tularensis*, and *M. tuberculosis*, as shown in Table 1, were the infections reported most frequently (28). While the risk of infection with these agents remained, Pike noted that most (96%) *Brucella* and typhoid fever cases and 60% of hepatitis cases were reported before 1955. Unfortunately, between 1979 and 2015, *Brucella* spp., *M. tuberculosis*, *C. burnetii*, *Salmonella* spp., and hepatitis B virus have remained among the top 10 causes of reported LAIs (see Table 1).

In an attempt to extend the Sulkin and Pike LAI data, 475 references published between 1979 and 2015 were reviewed to determine the microorganisms associated with laboratory infections, the primary function of the facilities in which the infections occurred, and the type of work activity associated with the event. To be included in this survey, an infection had to result from laboratory work, and the infected individual had to be a laboratory worker or another person who inadvertently was exposed (by being in the area) as a result of work with the infectious agents or infected animals. Secondary infections were also noted in this literature survey and are defined here as LAIs transmitted by a laboratory worker to a person not associated with, or in the vicinity of, the laboratory, such as a family member or health care provider. These secondary infections were not included in the primary LAI count unless they were responsible for a fatality. A tertiary infection results from transmission from a secondary infection. During the period 1979–2015,

TABLE 1.

Comparison of 10 most commonly reported LAIs

1930–1978[a]				1979–2015			
Rank	Agent[b]	No. LAIs	No. deaths	Rank	Agent[b]	No. LAIs	No. deaths
1	*Brucella* spp.	426	5	1	*Brucella* spp.	378	4[c]
2	*Coxiella burnetii*	280	1	2	*Mycobacterium tuberculosis*	255	0
3	Hepatitis B	268	3	3	Arboviruses[d]	222	3
4	*Salmonella enterica* serovar Typhi	258	20	4	*Salmonella* spp.	212	2[e]
5	*Francisella tularensis*	225	2	5	*Coxiella burnetii*	205	3
6	*Mycobacterium tuberculosis*	194	4	6	Hantavirus	189	1
7	*Blastomyces dermatitidis*	162	0	7	Hepatitis B virus	113	1
8	Venezuelan equine encephalitis virus	146	1	8	*Shigella* spp.	88	0
9	*Chlamydia psittaci*	116	9	9	Human immunodeficiency virus	48	Not known
10	*Coccicioides immitis*	93	10	10	*Neisseria meningitidis*	43	13
		2,168	48			1,753	24

[a]Adapted from reference 27.
[b]Not included are 113 cases of hemorrhagic fever contracted from wild rodents in one laboratory in Russia in 1962 (486).
[c]All deaths are aborted fetuses.
[d]Typical arboviruses and orbiviruses, rhabdoviruses, and arenaviruses that are associated with arthropods or have zoonotic cycles (233), with additional arboviral reports added.
[e]One death was a secondary exposure case (47).

publications described 2,376–2,392 symptomatic infections with 42 deaths, 19 secondary infections, and 8 tertiary infections. The overt infections were caused by bacteria (51%), viruses (32%), rickettsia (9%), parasites (7%), and fungi (<1%). This information is summarized in Table 2 by category of agent and available published information on asymptomatic, secondary, and tertiary LAIs is also included. If the rickettsial infections are included with bacteria, as is the case today, 60% of the LAIs are bacterial in origin.

Laboratory Function

The distribution of the symptomatic infections according to type of work performed in a facility is shown in Table 3. Clinical (diagnostic) and research laboratories account for (17% and 59%, respectively) of the symptomatic infections reported in the earlier Sulkin and Pike surveys and (42% and 36%, respectively) in the 1979–2015 survey. It appears that more LAIs from clinical laboratories are being reported in recent years. The increases in reported clinical infections may be due in part to a more active employee health program. Another explanation is that, during the early stages of culture identification, personnel are working with unknowns and may not be using adequate containment procedures. Clinical microbiology staff rely on physician notification that a sample is suspected of containing a pathogen transmitted by the aerosol route to avoid working on the open bench; unfortunately, the presence of a Risk Group 3 pathogen is not always suspected or the notification does not always occur. For example, despite a hospital policy requiring

TABLE 2.

Total LAIs 1979–2015

Catgory of agent	Symptomatic	Asymptomatic	Total primary LAIs	Deaths	Secondary infections	Tertiary infections
Bacteria	1,212–1,226	142	1,354–1,368	21	12	3
Rickettsiae	205	269	474	1	0	0
Viruses	764–766	439	1,203–1,205	19	7	5
Parasites	170	4	174	0	0	0
Fungi	25–26	0	25	0	0	0
Total	2,376–2,392	854	3,230–3,246	41	19	8

TABLE 3.

Number of LAIs associated with indicated primary work purpose

	Clinical		Research		Production		Teaching		Site not listed		Field	Total		
	1930–1975[a]	1979–2015	1930–1975	1979–2015	1930–1975	1979–2015	1930–1975	1979–2015	1930–1975	1979–2015	1979–2015	1930–1975	1979–2015	1930–2015
Bacteria	396	783	914	122	40	81	69	181	378	45–59	1	1,797	1,212–1,226	3,009–3,023
Rickettsiae	27	1	455	204	18	0	0	0	73	0		573	205	778
Viruses	173	215	706	497	73	9	15	13	82	9–10	16	1,049	760–761	1,809–1,810
Parasites	18	5	70	77	0	0	4	81	23	6	1	115	170	285
Fungi	43	4	155	16	2	0	18	1	135	4–5	0	353	25–26	378–379
Unspecified	20	—	7	0	1	0		0	6			34	—	34
Total	677	1,008	2,307	916	134	90	106	276	697	58–74	18	3,921	2,372–2,388	6,293–6,309

[a]Adapted from reference 26.

notification, no precautionary statements accompanied the samples submitted from a suspected *F. tularensis* infection. As a result of the missing notification, 12 microbiology staff and two autopsy staff received antibiotic postexposure prophylaxis (PEP) (29). In addition, because notifications advising the need for biosafety level 3 (BSL3) practices are infrequent, clinical laboratory personnel may be unaware of the procedural changes required to safely handle atypical specimens (30).

Underreporting

Table 3 provides data on the accumulated reports of LAIs over the past 85 years and the setting in which the LAIs occurred. It is widely accepted (24, 31) that the numbers represent a substantial underestimation of the extent of LAIs. Many scientists (32) and safety professionals can recount numerous unrecorded cases. With improvements in containment practices and equipment, and occupational health programs providing immunization and post-prophylaxis therapy, one would expect that the number of infections due to bacterial, rickettsial, and fungal agents would be decreasing. This appears to have occurred in the clinical laboratories of the United Kingdom, where the incidence of LAI was 62.7 cases per 100,000 person-years in 1988–1989 surveys (11) compared to 16.2 cases per 100,000 person-years reported in the 1994–1995 surveys (33).

RECENT INFORMATION ON WORKPLACE EXPOSURES

Bacterial LAIs

Table 4 summarizes information on the bacterial infections published in the literature during the period 1979–2015. During the past 36 years, 1,212 symptomatic LAIs, 142 asymptomatic infections, 12 secondary infections, and 3 tertiary infections due to bacteria were reported. The most frequently reported bacterial infections were *Brucella* spp. (389–393 LAIs), *M. tuberculosis* (243–246 LAIs), *Salmonella* (133–137 LAIs), *Shigella* (90 LAIs), *N. meningitidis* (43 LAIs), and *Chlamydia* (20 LAIs). A range is reported for a few species to include the data consolidated from two surveys conducted in Belgium for the period 2007–2012 (34).

Twenty-two fatalities due to bacterial LAIs occurred; 13 fatal LAIs were due to *N. meningitidis* (33, 35–41); 4 involved pregnancies that resulted in aborted fetuses as a consequence of LAIs with *Brucella melitensis* (42–45); 3 were due to *Salmonella* spp., one of which was a secondary infection (19, 46, 47); and one each for the attenuated (48) and wild-type strain of *Yersinia pestis* (49).

Secondary infections, the transmission of a LAI to another person outside the work environment, are rare. Between 1979 and 2015, 12 secondary bacterial infections and three tertiary bacterial infections occurred. A primary LAI with *Shigella sonnei* in a clinical laboratory resulted in the secondary transmission to a grandchild, with tertiary infections in three additional relatives (50). Four secondary infections in children under age 4 occurred with *Salmonella* Typhimurium in a 2011 outbreak associated with teaching or clinical laboratories (19). Two separate incidents of secondary *Brucella* infections were attributed to sexual transmission (51, 52). A microbiologist prepared dinner and transmitted *Salmonella* to his wife and son (47). A lactating mother with an LAI transmitted *Leptospira interrogans* to her infant through breast milk (53). Two secondary transmissions of *Bordetella pertussis* occurred (54).

TABLE 4.

Bacterial LAI references 1979–2015

Agent	Number of LAI		References
	Overt	Subclinical	
Bacteria			
Bacillus anthracis	1	1	88, 155
Bacillus Calmette-Guérin	2	0	34, 147
Bacillus cereus	1	0	325
Bacteroides asaccharolyticus	1		326
Bartonella henselae	3	0	34, 158
Bordetella pertussis	12	0	54, 327
Brucella spp.	389–393	24	5, 6, 16, 30, 34, 42–45, 51, 52, 56–62, 69, 75, 76, 93, 94, 96–124, 130, 328–341; J. Suen, 32nd Biol. Safety Conf.,[a] 1989; D.T. Brayman, 32nd Biol. Safety Conf.,[a] 1989
Burkholderia pseudomallei and *B. mallei*	3	3	78, 318, 342
Campylobacter spp.	5–6	0	12, 15, 34, 343, 344
Chlamydia spp.	20	20	93, 142–145, 345–347; K. Peterson, 25th Biol. Safety Conf.,[a] 1982
Clostridium difficile	3	0	16, 89
Corynebacterium diphtheriae	2	0	67, 68
Corynebacterium equi	1	0	348
Enterobacter aerogenes	1	0	349
Escherichia coli O157 and SP88, *Klebsiella*	22	0	16, 33, 64–66, 148, 151, 152, 332, 350–356
Francisella tularensis	11	0	5, 86, 87, 357–361
Gastrospirillum hominis	1	0	90
Haemophilus ducreyi	2	0	13, 362
Helicobacter pylori	4	0	363–365
Leptospira interrogans	8	0	53, 93, 366–368
Listeria monocytogenes	2	0	6, 34
Mycobacterium bovis	1	0	369
Mycobacterium kansasii	1	0	370
Mycobacterium leprae	1	0	156
Mycobacterium tuberculosis	255–259	96	11–13, 15, 33, 34, 93, 125, 126, 129–137, 139, 371–378; D. Robbins, 40th Biol. Safety Conf.,[a] 1997; D. Vesley, 42nd Biol. Safety Conf.,[a] 1999
Mycoplasma pneumoniae	4	0	15, 34
Neisseria gonorrhoeae	7	0	379–383; R. Hackney, 28th Biol. Safety Conf.,[a] 1985
Neisseria meningitidis	43	1	16, 17, 35–41, 63, 80–84, 351, 384–388
Pasteurella multocida	2	0	146
Salmonella spp.	212	0	11, 12, 15–17, 19, 34, 46, 47, 70–73, 77, 93, 131, 132, 149, 332, 389–395
Shigella spp.	88	0	11–13, 16, 33, 34, 50, 130–132, 150, 248, 332, 351, 396–399; D. Vesley, 30th Biol. Safety Conf.,[a] 1987; H. Mathews, 42nd Biol. Safety Conf.,[a] 1999
Staphylococcus spp.	18	0	13, 16, 33, 55, 332, 400, 401; D. Vesley, 30th Biol. Safety Conf.,[a] 1987
Streptobacillus moniliformis	2	0	74, 402
Streptococcus spp.	12	0	12, 13, 131, 403–406
Vibrio cholerae and *V. parahaemolyticus*	4	0	91, 131, 407, 408
Yersinia pestis	2	0	49, 85
Total bacteria	**1,212–1,226**	**142**	

(continued)

TABLE 4.
(Continued)

Agent	Number of LAI: Overt	Subclinical	References
Rickettsia			
Coxiella burnetii	195	267	121, 159, 160, 409–416
Rickettsia typhi and other typhus groups	10	2	161–165, 417
Total *Rickettsia*	205	269	

[a]References for Biol. Safety Conf. are abstracts of meetings sponsored by the American Biological Safety Association International, Mundelein, IL.

The asymptomatic infections included serological evidence without clinical symptoms for organisms such as *Brucella* and *Chlamydia* as well as six incidences of nasal carriage of methicillin-resistant *Staphylococcus aureus* (MRSA) without clinical symptoms. In one case, the same strain of MRSA worked with in the laboratory was isolated from the nasal passages of both a laboratory worker and the worker's cat (55).

Infections in clinical laboratories

A majority of the bacterial LAIs occurred in clinical laboratories, and four of these occurred in veterinary diagnostic laboratories. A survey of 88 hospital microbiology laboratories and three national reference laboratories was conducted for the years 2002–2004 (16). In that 2-year period, the bacterial LAIs reported were *Shigella* (15 LAIs), *Brucella* (7 LAIs), *Salmonella* spp. (7 LAIs), *S. aureus* (6 LAIs, with 5 of these being methicillin-resistant), *N. meningitidis* (4 LAIs), *E. coli* O157:H7 (2 LAIs), and *C. difficile* (1 LAI).

It should be noted that a single unexpected isolation of *Brucella* can result in multiple exposures. Analysis of two cases of *Brucella* in clinical microbiologists revealed that 146 staff were potentially exposed to the same samples; antibiotic PEP was offered to those at high risk of exposure, and serology indicated that no one other than the two previously identified LAIs had been infected (56). Sufficient data have been collected from *Brucella* LAIs to define criteria for staff at high risk of infection and a strategy for offering antibiotics for PEP (56). This was important for the proper follow-up to the clinical proficiency test of 2007 that involved sending an "unknown" sample to 1,317 laboratories in the United States and Canada. The laboratories were informed that the samples should be handled in a Class II biosafety cabinet using BSL3 practices to identify the sample. Unfortunately, 916 laboratory workers were exposed to the sample, identified as *Brucella* RB51, including 679 staff (74%) with high-risk exposures and 237 (26%) with low-risk exposures. PEP was offered, and no cases of brucellosis were reported (30). Other case reports document the variable and sometimes long incubation period prior to the development of clinical symptoms of brucellosis resulting from exposures in clinical laboratories (43, 57–61). Recurrent infections are also an issue with *Brucella*; an example is a recurrent case of acute hepatitis in a clinical microbiologist (62). The U.S. Select Agent program documented three *Brucella* infections between 2004 and 2010 in clinical laboratories (5).

There are 43 published LAIs of *N. meningitidis* infection; 13 were fatal. Thirty-seven LAIs from *N. meningitidis* occurred in clinical laboratories; 11 were fatal. Between 1985 and 1999, five microbiologists in England and Wales were infected (35). Sejvar reviewed 16 cases of *N. meningitidis* LAIs between 1985 and 2000; 8 were fatal (41). This prompted the CDC to recommend that samples drawn from sterile sites (blood, cerebrospinal fluid, inner ear fluid) be handled only in the biosafety cabinet, even for the initial plating. The Sejvar article was submitted in a successful appeal to the New Zealand Department of Labor to evaluate whether a serious case of *N. meningitidis* was occupationally acquired; genetic mapping indicated that the infection was an LAI (63).

E. coli O157:H7 and *E. coli* VTEC O117:K1:H7 have a low infectious dose, and infections occur even when no laboratory errors or incidents are identified (64, 65). However, one report on four LAIs with O157 underscores the importance of adhering to strict BSL2 practices and established laboratory policies (66). A *Corynebacterium diphtheriae* LAI occurred in an experienced lab technician who took an advanced training course and, without incident, performed a Gram stain and toxicity tests (67); another *C. diphtheriae* LAI occurred during a proficiency testing exercise (68).

Infections in teaching laboratories

Teaching laboratories were the setting for a number of bacterial infections. Twenty-seven students and their teacher were infected with *Brucella* in a veterinary teaching laboratory in China (69). Twenty-four students in college laboratories studying clinical microbiology in the United States were infected with *S.* Typhimurium (19); some of the students were also working in a clinical laboratory (Gaines, personal communication). Two of the LAIs in this cluster occurred at a university that, 3 years earlier, had removed S. Typhimurium from the curriculum due to safety concerns. During the outbreak investigation, it was determined that the stock culture of *Citrobacter freundii* used for student identification of "unknown specimens" was actually a mislabeled S. Typhimurium culture (70). Two additional cases were reported in 2013 (71). Another student contracted S. Typhimurium enteritis associated with erythema nodosum and reactive arthritis caused by the strain used in her microbiology class (72). Typhoid fever with serious complications developed 3 weeks after a student participated in a classroom identification exercise involving *S. enterica* serovar Typhi (73). A single case of *Streptobacillus moniliformis* occurred when a psychology undergraduate student was bitten by a rat (74).

Infections in production laboratories

A total of 73 LAIs were associated with production of vaccines against *B. pertussis* (R. McKinney et al., 28th Biol. Safety Conf., 1985) (54), *Brucella* (60, 75, 76), and *Salmonella* (77). The materials from a spill cleanup in a *Salmonella* poultry vaccine plant infected 21 staff (77). Fifteen workers at a *Brucella* S19 plant had active brucellosis, and 6 had asymptomatic infections (76). Twenty-two workers at a *Brucella* rev-1 vaccine plant developed brucellosis, and 6 had asymptomatic infections (60). One *Burkholderia pseudomallei* infection occurred when a culture mistakenly believed to be *Pseudomonas cepacia* was sonicated on the open bench for enzyme preparation (78).

An outbreak occurred from the wind-borne spread of anthrax from a military microbiology plant in Russia. The 77 individuals, 68 of whom died as a result of exposure to *Bacilllus anthracis*, were not included in this survey (79).

Infections in research laboratories

A total of 116 bacterial infections were reported from research facilities. Six cases of *N. meningitidis* infection occurred in research laboratories (39, 80–84). Fatalities resulted from an LAI with *N. meningitidis* serotype B (39) and from an LAI with an attenuated strain of *Y. pestis* (85). The U.S. Select Agent program reported three *B. melitensis* LAIs and four *F. tularensis* LAIs. Three researchers in the same laboratory were infected by handling *F. tularensis* incorrectly assumed to be the avirulent strain (86). In another case, video surveillance tapes of work in the biosafety cabinet were analyzed after diagnosis of an LAI with *F. tularensis*; the unvaccinated researcher wore insufficient respiratory protection and disposed of contaminated waste materials outside the biosafety cabinet (87). A researcher did not notice that a paper towel was contaminated with spores before disposing of it outside the biosafety cabinet; the follow-up nasal culture was positive for *B. anthracis* (88).

Case studies of graduate students infected while conducting research activities include a Ph.D. student infected with *C. difficile* (89). An infection with *Gastrospirillum hominis* was attributed to ingestion (90). The infected researcher did not wear gloves during the dissection of a cat stomach and also was splattered on the face and glasses with material from the tissue bath (90). Another student was infected during the supervised cleanup of a spilled shaker flask of *Vibrio cholerae*; this was the first case of indigenous cholera reported in Austria in 50 years (91).

Means of exposure for bacterial LAIs

Sniffing plates for identification purposes is frequently cited as a route of aerosol exposure. An experimental evaluation of this risk was conducted using overnight cultures and an air sampler (92). The highest number of organisms from a 4-minute air sample was from *S. aureus* (12.5 CFU/ml). *Bacillus* spp., *B. pseudomallei*, *Pseudomonas aeruginosa*, and nonpathogenic *E. coli* produced 6.25 CFU/ml. The authors conclude that the risk is low, because a sniff is estimated to inhale 50 to 200 ml of air; however, the recommendation is avoidance if a pathogen transmitted by the airborne route is suspected (3). Baron and Miller (16) state that the attribution of sniffing plates as a cause of LAI may be historical; subculturing, preparing smears, and performing catalase assays are more likely sources of aerosol transmission.

Brucella infections

Aerosol exposures to *Brucella* occurred in clinical, production, and research laboratories, with infections documented in staff that were not working directly with the organism. Although mucous membrane exposures from splashed cultures (42, 44, 45) and parenteral exposures (42, 93) were reported, these cause less than 20% of the exposure incidents (94).

Brucella *infections in clinical laboratories.* In addition to *Brucella* exposure incidents described previously in this chapter (56, 95), 31% of the staff in a large U.S. clinical laboratory were infected when one slant was subcultured on the open bench (61). *Brucella* is an endemic zoonotic in many countries, which increases the risk of infection in the general population and in clinical laboratory staff. For example, 2.5% of the positive aerobic blood culture results from 2002–2009 in an area of southern Israel were positive for *B. melitensis*. Because a significant amount of handling occurs prior to isolate identification, it is recommended that, in countries where *Brucella* is endemic, all positive blood cultures be manipulated in the biosafety cabinet to prevent exposures (96). Routine identification procedures conducted on the open bench resulted in 7 LAIs in Saudia Arabia (97), 12 LAIs in Turkey (98), 7 LAIs in Israel (94), and 38 LAIs in Iran (99). An LAI in Australia resulted from a *Brucella suis* isolation; the samples were from a hunter of feral pigs (100). In Beijing, a nonendemic area of China, a laboratory technician was infected by an isolate from an acute case of brucellosis; the index case was a fur-maker from Inner Mongolia (101). Travel-associated infections in tourists were the source of seven LAIs in clinical laboratory staff in Germany (102). LAIs with *B. melitensis* in clinical laboratories in France, Turkey, Canada, and Saudi Arabia (98, 103–109) were attributed to sniffing plates.

There are several reports of misidentification of *Brucella* from commercial bacterial identification kits; this resulted in exposures when antibiotic sensitivity testing procedures were conducted on the open bench (57, 109–112). According to two reports, the clinical laboratory worker was infected while culturing the blood of a lab worker with a LAI (110, 113). A similar event occurred when, following misidentification, index patient cultures were forwarded to reference laboratories where two staff members were exposed and developed brucellosis (114, 115). Bouza et al., in a retrospective study of *Brucella* infections in clinical laboratory staff in Spain, analyzed 75 LAIs; 62 occurred between 1980 and 1999 and were included in this survey (116). These investigators found that the number of LAIs correlated with the number of *Brucella* isolations in the laboratory. Attack rates were: 6.4% in labs with less than five isolations of *Brucella* per year, 13.9% for labs with 5–10 isolates per year, 21.4% for labs with 11–20 isolates per year, and 46% for labs with more than 20 isolations of *Brucella* spp. per year. The route of exposure was airborne in 68 cases, contact with skin in 1 case, and unknown in 6 cases (116). A veterinarian became infected while isolating *Brucella* from dromedary camel milk (117). One of the nine technicians isolating *Brucella* from goats in a veterinary microbiology laboratory in Malaysia was infected (118).

A literature review and analysis of English-language reports of *Brucella* LAIs provided information on exposure risk and PEP (119). Routine identification activities resulted in 88% of the infections. Staff identified as having had high-risk exposures were 9.3 times more likely to develop an LAI as staff in the low-risk exposure group. The CDC has revised the risk classifications for *Brucella* exposures as follows: "High—All persons manipulating a *Brucella* isolation in a class II biosafety cabinet without using biosafety level 3 precautions or on an open bench and any person present within a 5-ft. radius of these activities; all persons present in a laboratory room during widespread aerosol-generating procedures. Low—All persons present in a laboratory room at a distance greater than 5 ft. from manipulation of a *Brucella* isolate but without high-risk exposures as defined above. None—if all handling and testing of a *Brucella* isolate was done in a Class II biosafety cabinet using biosafety level 3 precautions" (120).

Brucella *Infections in Production Laboratories.* Analysis of 22 symptomatic and 6 asymptomatic LAIs from the production of live *B. melitensis* Rev-1 veterinary vaccine revealed a 17.1% attack rate for all staff and a 39.5% attack rate for staff working in areas with open windows above the exhaust from a production area (60). Fifteen symptomatic and 5 asymptomatic infections resulted from working in a *Brucella* S19 plant; aerosol is assumed to be the major route of exposure because only 5 individuals recalled exposure incidents (76).

***Brucella* infections in research laboratories**

When one polystyrene centrifuge tube containing *Brucella abortus* shattered during transport, 11 researchers and 1 administrative staff member were infected (58). In the United States, research with *Brucella* is conducted under strict regulations because it is a potential agent for bioterrorism; three of the reported six infections with *Brucella* occurred in research laboratories. One LAI resulted from conjunctival exposure while decontaminating an aerosol chamber used for rodent experiments with aerosolized *Brucella* (121). *Brucella* isolated from a marine mammal caused an LAI despite appropriate use of BSL3 procedures (122). Antigen production from the M-strain of *Brucella canis*, which is avirulent in dogs, also caused an LAI (123). A postgraduate student became infected from cultures collected for study of bovine miscarriage (124).

Mycobacterium tuberculosis infections

Mycobacterium tuberculosis *infections in clinical settings.* Twenty-three percent of the bacterial infections were due to *M. tuberculosis*. Workers were exposed to infectious aerosols from defective or improperly certified biosafety cabinets (125–127), the absence of biosafety cabinets (128), a defective ventilation system (11), autopsies, and preparing tissue sections (11, 12, 129–135). Analysis of 28 LAIs with *M. tuberculosis* in Japan indicated that 25 occurred in laboratories that did not have a biosafety cabinet (128). One laboratory reported that locating commonly used laboratory equipment in a mycobacteriology laboratory also resulted in exposure of personnel not working with *M. tuberculosis* (136); this was also a factor in some clinical laboratories in Japan (128). The aerosols generated in postmortem analysis of a dog (134) and by the use of pressurized refrigerant for cryosectioning human specimens (129) resulted in LAIs for staff involved in those procedures, but not for staff involved in clinical care of the unsuspected canine or human case of tuberculosis. One LAI was traced to the inadequacy of the heat-inactivation step (20 minutes at 80°C) for the large inoculum required for phenol-chloroform extraction used for IS*6110* restriction fragment length polymorphism (RFLP) typing analysis of samples (137). Three workers in a medical waste processing facility were infected with *M. tuberculosis*; in one case, the strain was matched to a patient in a hospital that sent waste to the facility (138).

A study of seroconversion rates in 17 Canadian hospitals reported an overall annual risk of tuberculin conversion at 1% per year for laboratory technicians. Included in this study were measurements of room air-exchange rates for the microbiology and pathology laboratories, as well the length of time required for diagnosis of tuberculosis. Menzies found that the risk of seroconversion was higher for staff working in laboratories with low air-exchange rates and in hospitals with delayed diagnosis of tuberculosis in admitted patients (133). Parenteral exposures to *M. tuberculosis* have also been documented (136, 139). While transferring a sample of *M. tuberculosis* for drug susceptibility testing, a microbiology laboratory technician sustained a needlestick and developed a cutaneous infection at the site (140).

Mycobacterium tuberculosis *infections in research.* A leaking pressure valve on an aerosol exposure chamber for infecting rodents with *M. tuberculosis* resulted in three subclinical infections (141). In another incident, two of the three researchers working with drug-resistant *M. tuberculosis* became infected; the third became tuberculin purified protein derivative (PPD) positive. This was considered an aerosol exposure because respirators were removed within the animal room after cage changing (D. Robbins, 40th Biol. Safety Conf., 1997).

Neisseria meningitidis infections

Neisseria meningitidis *infections in clinical laboratories.* Sejvar et al. reviewed 16 infections with *N. meningitidis* that occurred in clinical microbiology laboratories; 9 were serogroup B, and 7 were serogroup C (41). Fourteen of the staff had made suspensions of the organisms on the open bench; 2 had done the procedure behind a splash shield. In contrast to most of the *Brucella* exposures, only the staff member who worked with the specimen became infected. The microbiologists who became infected with *N. meningitidis* had all conducted routine procedures, such as making a suspension or doing a catalase test. This indicates transmission by droplet, not aerosol. Unfortunately, eight cases, or 50%, were fatal (41). One technician in a bacteriology laboratory had *N. meningitidis* group C cultured from her right elbow and *Salmonella enterica* serovar Enteritidis cultured from her right knee (17).

Neisseria meningitidis *in research laboratories.* Work on the open bench resulted in a fatality to a researcher working with serogroup B (39). An unvaccinated undergraduate student conducting a summer research project handled cultures on the open bench and became infected (81). Treatment of an LAI required that a researcher undergo amputation of her legs, left arm, and the digits of her right hand. Initially, the infection was considered to be community acquired by the Department of Labor, but subsequent analysis confirmed the workplace origin of the infection (63, 82). The Sejvar et al. article (41) was the basis for the successful appeal to the Department of Labor (63). A researcher became infected with Z5463, *N. meningitidis* group A strain, as a result of work in a defective biosafety cabinet (83).

Aerosol exposures to Chlamydia

Sonication of *Chlamydia trachomatis* cultures on the open bench caused seven LAIs (142; K. Peterson, 25th Biol. Safety Conf., 1982). Aerosolized *C. trachomatis* L/34/bu serovar caused atypical pneumonia in two members of the same laboratory; one handled the organism directly, but the other did not (143). In a teaching laboratory, the aerosols from the contaminated plumage of a flying pigeon infected the instructor with *C. psittaci* (144). In preparation for an avian influenza study in 3-week-old turkeys, 1-day-old turkeys were placed in a negative-pressure chamber and cared for by a veterinary scientist. At 2 weeks of age, the turkeys developed a respiratory infection, and, simultaneously, the researcher was infected with *C. psittaci* genotypes D, F and E/B (145).

Other routes of transmission for bacterial LAIs

Parenteral exposures. A needle used to aliquot *Pasteurella multocida* isolated from a fowl cholera epidemic caused a severe inflammation infection of hand and arm; a needlestick from the Clemson strain of *P. multocida* produced only a very mild local infection (146). A puncture from a Pasteur pipette contaminated with bacillus Calmette-Guérin (BCG) resulted in a carpal tunnel syndrome in a laboratory technician (147).

Ingestion. Contamination of the hands resulting in subsequent ingestion is the probable mode of transmission for enteric pathogens. LAIs associated with this route of transmission were *Salmonella* (124 LAIs), *Shigella* (85 LAIs), pathogenic *E. coli* (18 LAIs), *Vibrio* spp. (6 LAIs), *C. difficile* (3 LAIs), and *Listeria monocytogenes* (1 LAI). *Shigella* was the most commonly reported LAI in surveys of clinical laboratories in the United States and the United Kingdom (13, 16, 33). Many enteric pathogens have a low infectious dose, and hand-washing procedures may not remove all pathogens. In one incident, a child visiting the laboratory touched a culture of *E. coli* O157. The child's hands were immediately washed by her parent; however, a serious infection occurred (148). A food-borne outbreak can result in a large number of cultures submitted to a laboratory, which increases the risk of infection for staff. This was a factor in an LAI with *E. coli* O157 (66) and three *Shigella* infections (33). Strict compliance with laboratory policies for wearing and removing gloves, as well as hand washing after glove removal, must be required of all staff to prevent contamination of phones, computer keyboards, etc. (66).

The distribution of *S.* Typhi as a proficiency testing exercise resulted in a number of *S.* Typhi LAIs associated with poor work practices, such as mouth pipetting, smoking, and eating in the lab (46, 149). The importance of automatic faucets or foot pedals is illustrated by the 19 *S.* Typhi LAIs that occurred when a student in a clinical laboratory contaminated the hand-washing sink faucets (150) and by an outbreak of three simultaneous *S. sonnei* infections when only one staff member had handled the culture (16). Cleaning a biosafety cabinet was the only potential exposure to *S. sonnei* for one LAI (16). Twelve infections occurred when a disgruntled employee contaminated pastries in the staff break room using a stock strain of *Shigella* (16). *E. coli* O157 exposures caused 17 LAIs; 13 were in clinical laboratories settings. Five *E. coli* O157 LAIs occurred in research settings (151–153). Needlesticks resulted in cutaneous infections with *M. tuberculosis* (140) and *Neisseria gonorrhea* (154). A cutaneous case of *B. anthracis* resulted from transport of contaminated vials without wearing gloves (155). A cutaneous infection with *Bacillus cereus* (325) and a possible *Mycobacterium leprae* cutaneous infection were also reported (156).

Rickettsial LAIs

For the sake of consistency with the Sulkin and Pike LAI surveys, rickettsia is considered as a separate category of agents rather than being included with bacteria as is current practice. Between 1979 and 2015, there were literature reports of 205 symptomatic rickettsial infections with one death. During this period, *C. burnetii*, the etiologic agent of Q fever, was the fifth most common cause of all LAIs and accounted for 195 symptomatic rickettsial infections and one fatality (see Table 4 for references). When asymptomatic infections are added, the total becomes 405. Eleven rickettsial infections were identified as belonging to the Typhus Group—*Rickettsia typhi* (8 LAIs), *Rickettsia conorii* (2 LAIs), and *Rickettsia tsutsugamushi* (1 LAI). No secondary infections were noted from rickettsial infections.

Research involving the use of sheep in hospitals and medical school laboratories continues to expose laboratory and nonlaboratory personnel to *C. burnetii*. Antibody titers against *C. burnetii* in three research staff were reported in 2009 and are recorded in Table 4; however, the interpretation of the titers is in dispute (121).

Means of exposure for rickettsial LAIs

When publications identify the mode of transmission, Q fever infections were attributed to inhalation. A total of 189 *C. burnetii* LAIs were associated with zoonotic transmission from naturally infected asymptomatic sheep. The infected personnel either worked with the sheep or were in some proximity to sheep during their workday. Sheep may carry the organism in their blood, urine, feces, tissue, and milk. It has been estimated that the placenta of infected sheep may contain 10^9 organisms per gram of tissue and 10^5 organisms per gram of milk (157). Wedum et al. (22) noted that the infectious dose for 25–50% of human volunteers for *C. burnetii* by inhalation is only 10 organisms (22). This knowledge may have positively impacted husbandry practices for sheep used in research in the United States. A survey of U.S. members of the Association for Assessment and Accreditation of Laboratory Animal Care, International (AAALAC) who worked with research animals between 1999 and 2004 indicated only one confirmed case of Q fever (158). One case of Q fever resulted from necropsy of aborted fetuses of cattle, sheep, goats, pigs, and horses (T. Graham, Am. Biol. Safety Conf., 2014). The two nonzoonotic *C. burnetii* infections were attributed to exposure to a human placenta (159) and a leaking biosafety cabinet filter (160). The remaining rickettsial infections were associated with parenteral (161, 162), mucous membrane (163), and inhalation or unknown transmissions. The known sources of the exposures were an eye and lip splash from opening a microcentrifuge tube (163), sonication of infected cells on the open bench (164), and needlesticks. Other sources were not identified except that the agent was being worked with (165).

Viral LAIs

A total of 764 overt viral infections with 19 fatalities were reported between 1979 and 2015; references are listed in Table 5. The fatalities resulted from arboviruses (3 LAIs), hantavirus (2 LAIs), filovirus (1 LAI), *Macacine herpesvirus 1* (formerly called cercopithecine herpesvirus, CHV-1, or herpes B virus) (5 LAIs), hepatitis B virus (1 LAI), hepatitis C virus (1 LAI), Ebola virus (2 LAIs), Marburg virus (2 LAIs), severe acute respiratory syndrome coronavirus (SARS-CoV) (1 LAI), and one fetal abortion caused by a maternal parvovirus infection. The fatal hantavirus infections resulted from field studies with bank voles in Finland (166) and with rodents in West Virginia (167).

A groundbreaking genomic mapping study of Ebola virus contains a statement honoring the contribution of five authors who succumbed to the disease prior to the publication (168). In addition to sample collection, these authors were actively involved in Ebola patient care, and the fifth was also caring for an infected family member (169). Due to the difficulty of assessing exposure during the challenging field conditions of the 2014 Ebola epidemic, these infections are noted here but not included in the final total of viral LAI.

The seven secondary infections were due to a novel adenovirus titi monkey adenovirus (TMAdV) (1 LAI), *Macacine herpesvirus 1* (1 LAI), Marburg virus (1 LAI), a vaccine strain of poliovirus (1 LAI), SARS (2 LAIs), and Zika virus (1 LAI). Two of the 19 LAI fatalities listed were secondary infections resulting from the autopsy of a Marburg virus LAI and a mother providing care for a SARS LAI. The secondary polio infection occurred in the immunized child of a worker accidentally exposed to Mahoney prototype vaccine of poliovirus in a vaccine production facility. The stool isolate from the infected child demonstrated complete nucleotide sequence identity with the virus strain used for vaccine production (170). The Zika virus secondary infection occurred in the wife of a field virologist who collected mosquitoes in Senegal and developed symptoms after return to the United States. This case, and the infections of the two field virologists on this study, were the first identification of Zika virus in the Western world (171). An outbreak of a novel adenovirus (TMadV) in a titi monkey colony caused a respiratory illness in a researcher and a secondary infection in a household member (172). It is notable that one primary SARS LAI led to two secondary infections (mother and nurse); this led to five tertiary infections that resulted from contact with a nurse who became a secondary case while caring for a primary LAI (173).

Of the 759 viral LAIs, 497 (65%) occurred in research laboratories, 215 (28%) in clinical laboratories, 16 (2%) in field work, and 9 (1%) in production laboratories. A total of 460 asymptomatic infections were also reported between 1979 and 2015. Refer to Table 5 for citations for these viral infections. It should be noted that "research" activities include laboratory studies with animal models and field studies.

Retroviruses

LAIs with retroviruses were first described in 1988. Retroviral infections associated with human immunodeficiency virus (HIV), simian immunodeficiency virus (SIV), simian foamy virus (SFV), and simian D retrovirus (SDR) have been reported since then. In the United States, between 1985 and 2015, there were 17 confirmed occupationally acquired cases of HIV in clinical laboratory technicians and 21 possible cases; in addition, 4 researchers were infected handling HIV cultures (174). The Ippolito review provides details on occupationally acquired cases through 1997 (175). An HIV LAI is listed in the Belgian survey (34), and a seroconversion to HIV in a clinical laboratory worker was also reported in the literature (176).

Two technicians seroconverted to SIV while handling samples from nonhuman primates (NHPs), and one of the SIV-infected individuals may have been persistently infected (177). Seroconversions to spumavirus have been documented in staff occupationally exposed to NHPs. Details are provided in the discussion of zoonotic infections associated with NHP studies.

Poxviruses

Twenty-seven LAIs resulted from research activities with poxviruses between 1986 and 2015, and 23 were due to vaccinia virus. Recombinant viruses constructed from the Western Reserve strain of vaccinia virus caused nine LAIs (178–182); one LAI was caused by recombinant New York City Board of Health (NYCBOH) strain (183) and one by recombinant racoonpox virus (184). Immune responses to the insert in the vaccinia virus construct were demonstrated in three of the LAIs (180, 184, 185). Although thymidine kinase deletion mutants are less pathogenic in mice than the parent vaccinia virus strain, nine LAIs were caused by these deletion mutants (180–183, 186, 187).

Sixteen exposures to vaccinia virus were reported to the CDC Poxvirus Team in a 3-year period (183). Five of the exposures reported to the CDC were from eye splash, seven were from needlesticks, two occurred in an animal care facility, one occurred as a result of tube leakage, and one was unknown. Ten exposures did not result in infection; however, 5 LAIs occurred and 4 of these required hospitalization. All of the infected staff members had not complied with the requirement of smallpox vaccination within 10 years (183). A case report that provides more detail on one of the needlestick exposures listed in the CDC report is available (188). Two individuals were working on immunizing mice in the small space of a 1.2-meter biosafety cabinet. After placing an immunized mouse in its cage, the hand of one individual was scratched by the needle being held by the other individual. The scrape went through the glove and skin, and, although the plunger of the syringe was not depressed, the individual developed an LAI that required hospitalization. Procedures were revised, and now, when sharps are used, only one person may work in a biosafety cabinet. Double gloves are also required, and Occupational Health requires vaccination or a signed declination form (188). In 2015, a needlestick with wild-type Western Reserve strain resulted in an LAI in a recently immunized individual (179). In addition, two LAIs from cowpox virus (189, 190) and one each from raccoonpox virus (184) and buffalopox virus (18) were reported.

Zoonotic viral infections

Analysis of the viral LAIs associated with animal activities demonstrates how critical it is for laboratory staff to understand the potential for zoonotic infections in animal models. Between 1979 and 2015, there were 219 overt infections, 2 fatalities, and 180 seroconversions associated with zoonotic viral infections that were not experimentally introduced to the research model. The overt zoonotic LAIs were caused by hantavirus (188 LAIs), *Macacine*

TABLE 5.

Viral LAI references 1979–2015

Microorganism	Number of LAIs		References[a]
	Overt	Subclinical	
Adenovirus, novel TMAdV	2	2	172
Adenovirus, novel BaAdV-1	0	5	214
BaAdV-2	0	6	214
Adenovirus type 5 and adeno-associated virus	1	0	351
African horsesickness virus	4	5	232
Arboviruses and other viruses in SALS survey[b]	192	122	418
Bovine papular stomatitis virus	5	0	419
Buffalopox virus	1	0	18
Calcivirus	2	0	220
Chikungunya virus[c]	3	2	160, 234
Coxsackie type A24 virus	2	0	259, 420
Cowpox virus	2	0	189, 190
Creutzfeldt-Jakob virus	3	0	371, 421, 422
Dengue virus	7	0	226, 234, 247, 248, 423, 424
Dhori virus	5	0	236
Dugbe virus[c]	1	0	234
Ebola virus	9	0	216, 217, 425
Ebola-related virus	0	42	206–208
Echo virus	3	0	426–428
Ganjam virus[c]	5	0	429, 430
Hantavirus[c]	189	74	166, 167, 192–203, 237, 431–433
Hepatitis A virus	5	0	15
Hepatitis B virus	113–114	147	12, 13, 15, 34, 130, 131, 222, 225, 434–436
Hepatitis C virus (formerly non-A, non-B)	34	0	13, 15, 33, 130, 132, 225, 255, 267, 434–436; D. Vesley, 30th Biol. Safety Conf., 1987
Herpesvirus including zoster	6	0	12, 34, 130; D. Vesley, 30th Biol. Safety Conf., 1987
Human immunodeficiency virus	48	0	34, 174–176, 257
Influenza A virus	6	0	34, 261
Influenza B virus	1	0	260
Junin virus[c]	1	0	437
Kyanasur Forest disease virus	1	0	236
Lymphochoriomeningitis virus[c]	6	0	204, 256; A. Braun, 47th Biol. Safety Conf., 2004
Macacine herpesvirus 1 (CHV-1, B virus)	11	0	205, 240–246, 438
Machupo virus[c]	1	0	238
Marburg virus	2	0	215, 439, 440
Mayaro virus	1	0	441
Mimivirus	1	0	231
Newcastle disease virus	1	0	442
Norwalk virus	1	0	443
Orf virus	2	0	249
Orungo virus	0	3	234
Parvovirus	10–11	1	34, 444, 445
Poliovirus	1	0	170

(continued)

TABLE 5.

Viral LAI references 1979–2015 *(Continued)*

Microorganism	Number of LAIs: Overt	Number of LAIs: Subclinical	References[a]
Rabies virus	1	0	34
Rabbitpox virus	1	0	183
Raccoon pox virus (recombinant)	1	0	184
Rift Valley fever virus[c]	0	2	234
Rocio virus[a]	1	0	R. Gershon, 27th Biol. Safety Conf., 1984
Rubella virus	6	0	15
Sabia virus	2	0	235, 446
SARS-CoV	6	0	173, 227–229
Semliki Forest virus[c]	1	0	447
Simian foamy virus	2	20	210–213, 448–450
Simian immunodeficiency virus	0	4	177, 251, 258, 451
Simian type D retrovirus	0	2	209
SPH114202	1	0	452
Swine influenza virus	2	0	239
Tacaribe virus	1	0	453
Tick-borne meningoencephalitis virus[c]	1	0	454
Vaccinia virus	23	2	160, 178, 183, 185–188, 254, 262–264, 455
Varicella virus	1	0	15
Venezuelan equine encephalitis virus[c]	4	0	160, 234, 456
Vesicular stomatitis virus	1	0	457
Vesivirus	2	0	219
Wesselsbron virus[c]	3	0	234
West Nile virus[c]	6	0	191, 218, 253, 458
Zika virus	2	0	171
Total viral LAIs	759–760	460	

[a]References for Biological Safety Conference are meetings sponsored by the American Biological Safety Association, Mundelien, IL.
[b]Typical arboviruses, orbiviruses, rhabdoviruses, and arenaviruses associated with arthropods or that have zoonotic cycles.
[c]Additional infections with this virus listed in the SALS report (233).

herpesvirus 1 (10 LAIs), lymphochoriomeningitis virus (LCMV) (6 LAIs), influenza A virus (5 LAIs), West Nile virus (5 LAIs), orf virus (2 LAIs), Ebola virus (1 LAI), and a novel adenovirus in titi monkeys, TMAdV (1 LAI). The fatalities were caused by hantavirus in a graduate student doing field research in West Virginia (167) and West Nile virus in a veterinary student doing a postmortem on a horse (191).

Zoonotic infections from rodents in animal colonies

The 189 hantavirus transmissions and 36 subclinical infections occurred among researchers who thought they were working with uninfected rodents. Rodent colonies may be infected by feral animals, and this may explain hantavirus infections in Argentina (192), Belgium (193), China (194–196), France (197), Japan (198, 199), Korea (200), the United Kingdom (201, 202), and Singapore (203). This association was confirmed when Seoul virus was identified in the wild rat population of Yunnan, China, and identified as the cause of a hemorrhagic fever with renal syndrome (HFRS) outbreak in students using rats for research (195). One lot of rats produced by a commercial vendor and supplied to three colleges in Yunnan, China, infected a researcher with a reassortant of Hantaan virus that had not been previously described; subclinical infections with the same reassortant were also detected in 5 students who had been bitten and 11 animal care staff (196). Another example of zoonotic infection involved eight animal handlers and junior scientists who were exposed to LCMV while working with nude mice (204). In this incident, the mice were inadvertently infected by an LCMV-contaminated tumor cell line. The serological monitoring program for sentinel animals in the facility had lapsed for 6 months. An animal care technician was diagnosed with LCMV 3 months after he became ill, when the sentinel mice in the animal room seroconverted. The source of infection was traced to an LCMV-contaminated cell line (A. Braun, 47th Biol. Safety Conf., 2004).

Zoonotic infections associated with non-human primate studies

Macacine herpesvirus 1, formerly CHV-1 or herpes B virus, was transmitted from NHPs to 11 caretakers or researchers, resulting in a fatality and a secondary infection (see Table 5 for references). One fatality resulted from an eyesplash exposure to urine and feces from a caged NHP; eye protection was not worn and rinsing the eye was not attempted for 45 minutes because observation was considered a low-risk activity (205). A researcher dissecting optic nerve from a sample of NHP tissue was hospitalized to receive treatment with intravenous, high-dose ganciclovir and then discharged with a life-long prescription for oral valacyclovir to keep the latent virus in check (K. Johnson and T. Winters, Harvard School of Public Health Grand Rounds, 2012).

Asymptomatic infections with an Ebola-related filovirus were reported in 42 animal handlers (206–208). Two seroconversions to the zoonotic SDV have also been reported in animal handlers; one of the individuals was also infected with spumavirus from a bite in a separate incident (209). Seroconversions to the zoonotic spumavirus, or SFV, were documented in 20 animal handlers or persons working with NHPs (210). The significance of these seroconversions to simian retroviruses is not well understood. The virus causes a latent infection, and seropositivity has been documented 10 years after a bite from a mandrill and 22 years after a macaque bite (211). In one case, SFV was isolated from a culture of peripheral blood monocytes obtained from a healthy animal caretaker who seroconverted to SFV 20 years prior to the virus isolation (212). Switzer et al. commented "although SFV is nonpathogenic in naturally infected NHPs, the significance of SFV infection in humans is poorly defined. The introduction of SFV infections is of concern because changes in the pathogenicity of simian retroviruses following cross-species infection are well documented, since both HIV-1 and HIV-2 emerged from benign SIV infections in the natural primate hosts. To date information on this subject is inadequate to come to any conclusions; however, the importance of long term follow-up on these exposures has been recognized and has been initiated by the Centers for Disease Control and Prevention" (213). Novel zoonotic adenoviruses have also been identified in titi monkey (TMAdV) and baboon colonies (BaAdV-1 and BaAdV-2). TMAdV caused a primary infection and a secondary in a household member (172). Eleven seroconversions to BaAdV-1 and BaAdV-2 occurred in staff who worked with a baboon colony (214).

Experimentally infected animals

In comparison, there were reports of only 20 symptomatic infections, and no asymptomatic infections, from work with experimentally infected animals. The LAIs from experimentally infected animals were caused by cowpox virus (1 LAI), Ebola virus (1 LAI), LCMV (3 LAIs), Marburg virus (1 LAI), swine influenza (2 LAIs), vaccinia virus (7 LAIs), Venezuelan equine encephalitis virus (3 LAIs), and West Nile virus (2 LAIs). The inoculation of guinea pigs with Marburg virus and Ebola virus each resulted in a fatal LAI (215, 216).

Viral LAIs in field work

One overt infection with a new strain of Ebola virus was reported in a research worker who autopsied a wild chimpanzee to determine the cause of death (217). Fortunately, that infection did not result in the fatal hemorrhagic disease associated with other filovirus infections (Marburg virus and Ebola virus) in Europe and Africa. Transmission of West Nile virus occurred during field collection of blue jays (218) and horse autopsy (191); five investigators were infected with influenza A virus during an investigation of seal deaths (219). A field researcher examining seals (220) and one examining sea lions were painfully infected with marine calciviruses that cause lesions (220). Two cases of hantaviral infection occurred in field studies. A graduate student evaluating the impact of forestry practices on small mammals was fatally infected in West Virginia; a technician working on a similar study in California was also infected (221). In effort to evaluate the occupational risk for field studies, 995 mammology conference attendees who had exposure to rodents in North America provided samples for a survey of antibody levels. Antibodies against Sin Nombre virus were found in four persons, and two had antibodies to Arroyo or Guanarito virus (221). None of the seropositive individuals had worn any personal protective equipment prior to the U.S. hantavirus outbreak.

Viral LAIs in research and clinical activities

Sixty-seven percent of the viral LAIs occurred in research facilities. Arboviruses and other vector-borne viruses in both research and field settings accounted for 223 of the LAIs with three fatalities (see Table 5 for arbovirus references). A total of 215 viral infections occurred in clinical laboratories between 1979 and 2015. The 114 hepatitis B virus LAIs reported in the scientific literature are undoubtedly the tip of the iceberg, because one study calculated the attack rates for clinical laboratory technicians at 70% prior to the systematic introduction of hepatitis B virus vaccine (222). Following the implementation of the Occupational Safety and Health Administration (OSHA) Bloodborne Pathogen Standard in the United States (223), significant reductions in workplace transmission of hepatitis B virus have resulted from the availability of hepatitis B virus immunization, the use of "universal precautions," or consistent BSL2 containment practices, needles with safety devices, and improved sharps disposal (224). This experience is mirrored in the report from the Wroclaw region of Poland, where 323 health care workers were infected with hepatitis B virus between 1990 and 2002; 30 of the 323 cases were clinical laboratory staff. Since 2002, the number of cases reported annually in that region decreased due to the increasing number of vaccinated staff. However, the number of hepatitis C virus infections was increasing (225). In the absence of a hepatitis C virus vaccine or a recommended PEP regimen, strict adherence to safety practices is the most effective defense against infection with hepatitis C virus.

Means of exposure for viral LAIs

It is often not possible to determine the exact cause of a LAI; only that it has occurred and the information available is that the individual "worked with" the agent. One example would be a researcher in a nonendemic area who performed the initial infection of a mosquito colony by feeding dengue-infected blood under an artificial membrane. The researcher was bitten by an escaped, but unengorged, mosquito. The virus later isolated from the infected researcher was 98.9% homologous with the laboratory strain. The authors of the case study did not rule out percutaneous exposure from the mouthparts of a mosquito; however, mucocutaneous infection from infectious blood droplets was also considered a possible route of exposure (226). In another example, the San Miguel sea lion virus serotype 5 (SMSV-5) infected a laboratory worker who collected oropharyngeal secretions from the mouths and teeth of diseased seals, isolated the virus from cell cultures, and did gradient purifications to concentrate the virus (220). The activity that resulted in exposure and clinical infection with SMSV-5 is not known; however, this LAI is considered the first documented case of a new human disease.

Aerosol exposures

Many of the viral LAIs resulted from inhalation of infectious virus. SARS was transmitted to four researchers in China when incompletely inactivated infectious materials were removed from the BSL3 laboratory for further analysis at BSL2 (173, 227). In Singapore, another SARS infection occurred due to cross-contamination of a culture of West Nile virus with SARS, and subsequent handling of the SARS-infected West Nile preparation at BSL2 (228). One SARS LAI occurred in Taiwan when liquid waste leaking from a biohazard bag in a BSL4 lab was cleaned up; the researcher wore inadequate personal protective equipment and did not use a disinfectant effective against the virus (229). Exposure to aerosolized Mayaro virus occurred during a sucrose-acetone antigen extraction procedure that involved dehydration of the product with a vacuum pump (230). A technician performing Western blots with patient samples from a pneumonia outbreak unfortunately became the first documented infection with *Mimivirus*, which is a virus isolated from amoebae in a cooling tower (231).

Four serious LAIs and five seroconversions to African horsesickness virus were the first indication that this virus could infect humans. In this case, individuals were exposed to a aerosols from a dried powder vaccine that were released when vials broke as they were being filled (232). Historically, about 20% of arbovirus infections are attributed to inhalation exposures (233). Some of the inhalation transmissions from arboviruses included infections due to the spread of Wesselbron virus from work in another room, to opening a blender containing Dugbe virus-infected mouse brains without precautions (234), and to a centrifuge bottle containing Sabia virus cracking during a run (235). Five infections with Dhori virus resulted from inhalation of aerosols generated when opening flasks (236). Enteropathica endemica was transmitted, presumably by aerosol, from close contact with bank voles (237). The hantavirus LAIs were attributed to aerosol exposure from handling mice and preparing tumor samples from mice that may have been infected by feral animals. The LCMV infections occurred when personnel handled mice that had been inadvertently infected by a contaminated cell line (204). A clinical laboratory worker was exposed to Machupo virus aerosols when a blood tube broke in the centrifuge (238); that technician developed Bolivian hemorrhagic fever. The wearing of dust masks instead of respirators resulted in two LAIs with swine influenza in staff collecting nasal cultures from infected pigs (239).

Parenteral exposures

Laboratory and wild animals were the source of many of the viral LAIs, with parenteral exposures due to bites, scratches, and accidents with sharps resulting in infections. Monkey bites and/or scratches transmitted *Spumavirus* (SFV) to 23 NHP handlers; that route of exposure also transmitted *Macacine herpesvirus 1,* formerly CHV-1 or herpes B virus, to 10 animal handlers and researchers resulting in 4 deaths (205, 240–246, R. Rebar, personal communication). The bite of infected mosquitoes transmitted dengue virus (247, 248), Zika virus (171), and chikungunya virus (234). Another case of chikungunya virus was transmitted by needlestick (160), and biting shrews transmitted Mokola virus (234). During a procedure to insert gavage tubes, two researchers were bitten by sheep and infected with orf virus (249). Six NHP animal handlers developed antibodies to filovirus antigens; four had evidence of recent infection, and one of these sustained a scalpel cut during the autopsy of an infected NHP (208).

In addition to phlebotomy, parenteral exposures to HIV in clinical settings were associated with handling broken blood tubes and a broken capillary tube (175). A parenteral exposure in a laboratory producing large quantities of concentrated HIV occurred when a blunt cannula was used to clean a centrifuge rotor (250). An HIV researcher was infected with the virus by needlestick (174). The first report of SIV infection in a human being resulted from a deep puncture wound (251). West Nile virus LAIs were caused by needlestick (252); parenteral exposures were received during necropsy of an infected bird and preparation of an infected mouse brain (253). A needlestick during a viral purification procedure (254) or animal inoculation (182, 188) each caused a vaccinia LAI; a needlestick also caused the raccoonpox virus infection (184). A puncture from a glass lyophilization vial transmitted buffalopox virus (18). Hepatitis C virus was transmitted when a sample tube containing infected blood broke and cut two fingers (255). The bite of a rat infected with cowpox virus transmitted the virus (189). LCMV infection was confirmed by serology on blood drawn 5–7 years after exposure to Armstrong clone 53b (1 case) and clone 13 (3 cases) (256).

Mucocutaneous exposures

Mucocutaneous exposures to HIV in clinical settings were the result of splatter from an apheresis machine, sink disposal of blood samples (175), a blood analyzer apparatus (257), and opening a Vacutainer tube (176). In production laboratories, mucocutanous exposures to HIV occurred when concentrated virus splashed a worker in the face or seeped through gloves in contact with leakage from a centrifuge (250). A nonintact skin exposure to SIV was also reported (258). One worker contracted conjunctivitis due to an eye splash while pipetting dilutions of coxsackie virus (259). Despite immediate lavage, conjunctivitis occurred in a researcher who sustained splatter in the eye while injecting mice with influenza B virus (260). A conjunctival infection with influenza A virus occurred when a seal sneezed in the face of a researcher; four other field workers were also infected with influenza A virus while doing autopsies to investigate seal deaths (261). A veterinary student who removed the brain and spinal cord of a pony infected with West Nile virus neuroinvasive strain 2 became infected; mucous membrane infection by droplet is suspected (191). The secondary case of *Macacine herpesvirus 1* occurred when the wife of an infected worker applied his contaminated cortisone cream to her nonintact skin (244). Three of the vaccinia LAIs were the result of failure to wear gloves (160, 180, 262). An investigator was infected with a strain of vaccinia virus that was a contaminant of the viral stock he was working with (178). Two vaccinia virus LAIs were caused by inadvertent contact with contaminated surfaces (181, 263). The exact cause of an eye infection could not be determined; however, plates were opened for analysis on the bench, eye protection was not worn, and glove use may not have been consistent (264). Similarly, an infection with cowpox is attributed to handling contaminated reagents or touching contaminated surfaces; cowpox virus DNA was found on many lab surfaces and as a contaminant in other viral stocks. The infected person was not working directly with the virus and, for this reason, had not chosen to become vaccinated (190). In four instances, staff members seeking medical attention for infections did not initially disclose the fact that their research may have exposed them to with vaccinia (186, 187, 262, 263). Fortunately, there were no nosocomial transmissions.

Parasitic LAIs

A total of 170 symptomatic, 4 asymptomatic, and 2 secondary parasitic infections representing 6 different genera and 20 species were reported during this period. The activities resulting in LAI were research (76 LAIs) veterinary teaching (81 LAIs), diagnostic in a clinical laboratory (3 LAIs), field studies (1 LAI), and not specified (9 LAIs). The agents responsible for the infections were *Cryptosporidium* (96 LAIs), *Leishmania* (15 LAIs), *Trypanosoma* (26 LAIs), *Toxoplasma* (15 LAIs), *Plasmodium* (17 LAIs), and *Schistosoma* (1 LAI) (see Table 6 for specific references). In addition, Brener reported personal knowledge of 50 cases of laboratory-associated *Trypanosoma cruzi*, with one fatality; however, details on dates and types of exposure were not available so these were not included in this survey (32).

TABLE 6.

Parasitic LAI references 1979–2015

Agent	LAI overt	LAI subclinical	References
Cryptosporidium spp.	96	2	265–269, 287, 459–465
Leishmania spp.	15	0	270, 276, 284–286, 289, 292, 294, 466, 467
Plasmodium cynomolgi	4	0	280, 468
Plasmodium falciparum	9	0	270–273, 288, 293, 469, 470
Plasmodium vivax	4	0	272, 280
Schistosoma mansoni	1	0	277
Toxoplasma gondii	15	2	34, 279–283, 291, 471–475
Trypanosoma spp.	26	0	34, 270, 275, 280, 290, 476–479
Total	170	4	

Means of exposure to parasitic infections

The most common means of acquiring the reported parasitic LAIs were ingestion and parenteral exposure, generally needlesticks associated with animal inoculation. With one exception, an airborne infection (265), all of the *Cryptosporidium* infections were associated with ingestion of the infectious microorganisms. The *Cryptosporidium* infections occurred in veterinary students and were zoonotic transmissions through contact with infected calves. Two secondary infections occurred in spouses who washed the veterinary student's contaminated clothing (266, 267). Two separate outbreaks were from veterinary laboratory exercises involving the removal of calves from artificial wombs (268, 269). The veterinary medical schools where the outbreaks occurred adopted preventive measures such as requiring testing of calves used for teaching, on-site doffing of personal protective equipment, and providing facilities for hand washing and changing clothing (268).

Until the LAIs were diagnosed, researchers had assumed that *Plasmodium cynomolgi* only infected monkeys; however, mosquito bites in the infected monkey colony transmitted the parasite to humans (270). Mosquito bites in insectories also transmitted *Plasmodium falciparum* and *Plasmodium vivax* (271–274).

The most common source of the exposure to parasitic agents was working with infected animals, insects, or ectoparasites, and accidents related to sharps and spills or splashes accounted for the remainder. Some of the activities associated with the transmission of parasitic infection include working without gloves (275–278) or eye protection (279–283), injecting animals (284, 285), recapping needles (286), smelling or being sprayed with stomach contents (265, 287), being bitten by infected mosquitoes (271–274, 288; H. Mathews, 42nd Biol. Safety Conf., 1999), and numerous needlesticks (283, 289, 290). Unique exposures were associated with assuming that the strain being handled was avirulent (291), being on immunosuppressive therapy while working with infectious materials (292), and puncturing a thumb while pressing a glass hematocrit tube into clay sealant (293). One infection was contracted during an unrelated field study of birds (294).

Fungal LAIs

Only 25 fungal LAIs were found in the literature review; the references are listed in Table 7. There were seven cases of *Trichophyton mentagrophytes*, and four to five additional dermatophyte infections, including *Trichophyton verrucosum* and *Microsporum canis*, were reported (34). Three cases each of *Blastomyces dermatiditis* (295, 296) and *Coccidioides immitis* (5, 16) (H. Mathews, 42nd Biol. Safety Conf., 1999) were reported. *Sporothrix schenckii* caused two LAIs (297, 298); one LAI each were caused by *Arthroderma benhamiae* (299), *Encephalitozoon cuniculi* (300), *Penicillium marneffei* (301), and *Trichophyton simii* (302). Three infections occurred in clinical or public health laboratories (16, 295), while 18 occurred during research activities and 4 to 5 cases in an unspecified location.

Means of exposure to fungal infections

Fungal LAIs resulted from cutaneous, parenteral, inhalation, and mucous membrane exposures. Handling lab rats

TABLE 7.

Fungal LAI references 1979–2015

Fungus	Overt	Subclinical	References
Arthroderma benhamiae	1	0	299
Blastomyces dermatiditis	3	0	295, 296
Coccidioides immitis	3	0	5, 16 H. Mathews, 42nd Biol. Safety Conf.,[a] 1999
Dermatophytes, including *Trichophyton verrucosum, Microsporum canis*	4–5	0	34
Encephalitozoon cuniculi	1	0	300
Histoplasma capsulatum	1	0	480
Paracoccidioides brasiliensis	1	0	481
Penicillium marneffei	1	0	301
Sporothrix schenckii	2	0	297, 298
Trichophyton mentagrophytes	7	0	303, 304
Trichophyton simii	1	0	302
Total	25–26	0	

[a]References for Biol. Safety Conf. are abstracts of meetings sponsored by the American Biological Safety Association International, Mundelein, IL.

or guinea pigs that were not experimentally infected resulted in seven zoonotic LAIs (303, 304). Cutaneous infections occurred when two drops of liquid culture spilled on the bandage over a cut (298) and when one drop of culture fell on the skin while filtering a fungal mat (302). Two LAIs resulted from cuts acquired during tissue sectioning in pathology (295). An immunosuppressed student visited a mycology laboratory where the agent was handled (301) and became infected with *P. marneffei*. Failure to wear gloves or use correct work practices (297; H. Mathews, 42nd Biol. Safety Conf., 1999) resulted in LAI. Contaminated hands rubbing the eye probably inoculated the mucous membrane of the lower eyelid (299). Culture supernatant containing spores of *E. cuniculi* splashed into the eyes of a laboratory worker; this incident resulted in a severe eye disease with one cornea still clouded a year later (300).

Allergic reactions should also be an occupational concern for those working with fungal agents. Aerosol exposure during release of a pressurized canister used in the isolation of lysosomal enzymes from slime mold resulted in rhinoconjunctivitis and asthma in a research microbiologist (487). Ten staff required medical attention for allergic responses to *Penicillium citrinum* after working for 1 day on production of adenosine triphosphate (ATP) (305).

Role of Infectious Aerosols in LAIs

Laboratory studies on potential sources of infection have focused on hazards produced from routine microbiological techniques. Table 8 lists data from several studies on the number of viable particles recovered within 2 feet of a work area, on the basis of an extensive series of air sampling determinations. Aerosols present two means of worker exposure—through minute respirable airborne particles and by the disposition of larger heavy droplets onto surfaces, equipment, and personnel. The data in Table 9 indicate that standard laboratory procedures can generate aerosolized particles that are respirable and, therefore, potentially hazardous to the laboratory workers and to others in the vicinity. However, the mere presence of organisms in the air is insufficient to cause disease. For infection to be initiated, the infectious dose, a means of exposure, and a susceptible host are all required. The FDA cautions that published infectious doses do not take into account the wide variability in the virulence of the strain and host susceptibility. For example, the published infectious dose for *E. coli* O157 is as low as 10 organisms (306); however, other strains of *E. coli* require an infectious dose of 10^8 (22) to initiate infection.

Occupational Health Programs

The continued development of occupational health programs will result in a minimization or reduction of LAIs through increased access to preemployment counseling on host factors, immunization, and timely PEP. In the United States, occupational health programs provide vaccination against hepatitis B virus and PEP for hepatitis B virus and human immunodeficiency virus (HIV); this is required for compliance with the OSHA Bloodborne Pathogen Standard.

TABLE 8.

Concentration and particle size of aerosols created during representative laboratory techniques[a]

Operation	No. of viable colonies[b]	Particle size[c] (μm)
Mixing culture with:		
Pipette	6.6	2.3 ± 1.0
Vortex mixer used with 5 ml of culture in capped tube for 15 seconds	0.0	0.0
Vortex mixer used with 10 ml of culture in capped tube until culture overflow hit rotating head	9.4	4.8 ± 1.9
Use of blender:		
Top on	119.6	1.9 ± 0.7
Top off	1,500.0	1.7 ± 0.5
Use of a sonicator:	6.3	4.8 ± 1.6
Lyophilized cultures:		
Opened carefully	134.0	10.0 ± 4.3
Dropped and broken	4,838.0	10.0 ± 4.8

[a]Adapted from reference 482.
[b]Mean number of viable colonies per cubic foot of air sampled.
[c]Count median diameter of particle.

TABLE 9.

Infectious dose for humans[a]

Disease or Agent	Dose[b]	Route of Inoculation
Coxsackie A21 virus	≤18[c]	Inhalation
Escherichia coli	10^8	Ingestion
Francisella tularensis	10	Inhalation
Giardia lamblia	10–100 cysts[d]	Ingestion
Influenza A2 virus	≤790[c]	Inhalation
Malaria	10	Intravenous
Measles	0.2[c,e]	Inhalation
Mycobacterium tuberculosis	<10[f]	Inhalation
Poliovirus 1	2[c,e,g]	Ingestion
Q fever	10	Inhalation
Salmonella Typhi	10^5	Ingestion
Scrub typhus	3	Intradermal
Shigella flexneri	180	Ingestion
Shigellosis	10^9	Ingestion
Treponema pallidum	57	Intradermal
Venezuelan encephalitis virus	1[c,h]	Subcutaneous
Vibrio cholerae	10^8	Ingestion

[a]Adapted from reference 22.
[b]Dose in number of organisms unless otherwise indicated.
[c]Median infectious tissue culture dose.
[d]Adapted from reference 483.
[e]In children.
[f]Adapted from references 484 and 485.
[g]Plaque-forming units.
[h]Guinea pig infective unit.

Vaccination programs

When available for the agents studied, vaccination programs are very effective in preventing or minimizing the severity of LAIs. An analysis of 16 vaccinia virus exposure incidents reported to the CDC between 2005 and 2008 revealed that only four individuals had received the recommended smallpox vaccination within the previous 10 years (183). Prior to an LAI with vaccinia virus, the occupational health service at an academic institution offered counseling only to individuals seeking vaccination. After an LAI occurred, the policy was amended. Now all staff working with vaccinia virus are required to receive vaccination counseling, the vaccine is offered to all staff without contraindications, and declinations must be documented (178).

Vaccination is also recommended for microbiologists at risk of exposure to *N. meningitidis* (307). Previously, vaccinations against serogroup B were not available in the United States. However, as of August 15, 2015, the CDC recommends an additional vaccination against serogroup B for microbiologists handling *N. meningitidis* cultures (see http://www.cdc.gov/vaccines/hcp/vis/vis-statements/mening-serogroup.html). To provide appropriate immunizations, staff with potential laboratory exposure must be referred to an occupational health service. The immunization status of an undergraduate summer student was not confirmed prior to work with *N. meningitidis*; the student assured the principal investigator, and admitting hospital physicians when he became ill, that he had been immunized. However, the student was misinformed; he had not been vaccinated and PCR results on cerebrospinal fluid were positive for *N. meningitidis* serogroup A (81). Vaccination is also recommended for laboratory staff routinely exposed to *S.* Typhi.

Vaccination is an important tool in the prevention of infectious agents with a low infectious dose by the aerosol route. A historical review of LAI at the United States Army Medical Research Institute of Infectious Diseases (USAMRIID) between 1943 and 1969 indicates that the introduction of biosafety cabinets reduced the risk of infection with anthrax, glanders, and plague. However, even after the introduction of biosafety cabinets, infections with *F. tularensis* continued at the average rate of 15 per year, for Venezuelan equine encephalitis at 1.9 per year, and Q fever at 3.4 per year. In contrast, between 1989 and 2002, only 5 LAIs resulted from 289 reported exposures. The LAIs were glanders, Q fever, vaccinia, chikungunya, and Venezuelan equine encephalitis. The Venezuelan equine encephalitis, vaccinia, and Q fever LAIs were instances where infection occurred in vaccinated staff; however, the Q fever symptoms were mild. Also, vaccination status was considered in the evaluation for PEP for low-risk exposures (159).

Specialized programs are required for high-containment laboratories and include physical and mental fitness for duty, as well as the capacity to quarantine exposed staff. Guidance is available on management of exposures to some Ebola virus (308).

Postexposure prophylaxis

Guidelines for PEP after occupational exposures to hepatitis B and C viruses and HIV are available and emphasize that such exposures should be considered medical emergencies requiring prompt evaluation and response (309, 310). Several publications reported incidents where PEP probably minimized or eliminated acute laboratory infections with *B. abortus* (58, 75), *B. melitensis* (109, 311), *B. pseudomallei* (312, 313), *F. tularensis* (29), and *M. tuberculosis* (488). However, prevention of exposures should be emphasized; side effects that inhibit completion of the prophylactic antibiotic treatment for *B. melitensis* (311, 314) and for *B. pseudomallei* (313) have been reported. In 2013, CDC published revised guidelines for PEP after *Brucella* exposures (120). All potentially exposed staff should receive serological monitoring, symptom surveillance, and daily temperature self-checks; staff with high-risk exposures should receive PEP (120). The CDC reviewed 153 incidents that resulted in 1,724 exposures to *Brucella* between 2008 and 2011; 839 were high-risk exposures (see risk classification in section on *Brucella* in Clinical Laboratories). Only five LAIs occurred; four of the infected microbiologists had not taken PEP and the fifth worker started the antibiotic 21 days after the exposure incident (120).

Rusnak et al. describe a proactive occupational health response to an incident that was considered a low risk of exposure to *B. anthracis* spores in immunized staff (88). A staff member removed a rack of anthrax cultures from the incubator and transported it on a cart to a coworker seated at the biosafety cabinet. The paper towel covering the top of flasks was contaminated with spores; this was noted only after the towel had been discarded outside of the biosafety cabinet. The occupational health service obtained nose and throat swabs from both potentially exposed staff members and initiated a short course of antibiotic therapy. The next day, 6 CFU of *B. anthracis* grew from the nasal passage sample from one researcher, so both were prescribed the full 1-month course of antibiotic therapy for PEP (88).

Guidance on PEP for many agents is available on the CDC website (www.cdc.gov). References on PEP for high-consequence exposures include advice for *B. pseudomallei* (160, 308, 315, 316).

Host factors

Another critical function of occupational health programs involves advising staff of preexisting medical conditions that may put them at greater risk for serious consequences from exposure to an infectious agent. The risk and severity of infection may be influenced by concurrent diseases, medical conditions, drugs that alter the host defense, allergic hypersensitivity, inability to receive a certain vaccine, and reproductive issues. These risk factors need to be recognized and addressed before initiating work with infectious agents (317).

Some workers may face increased risks for certain infections that alter or impair normal host defense mechanisms. For example, host defenses provided by healthy intact skin can be disrupted by diseases such as chronic dermatitis, eczema, and psoriasis, thus providing a portal of entry in the absence of personal protective clothing. Achlorhydric individuals are more susceptible to *Vibrio* infections (306), and individuals with heart valve problems should not be exposed to *C. burnetii*. Antibiotic therapy may suppress gastrointestinal flora, increasing the possibility of colonization by a foreign or resistant population of microorganisms. Deficiencies in the immune system function can place workers at a higher risk of occupational infection. Immunodeficiency may result from certain connective tissue diseases, cancer chemotherapy, or HIV infection (317). Other causes of immunodeficiency include steroid treatment for medical conditions such as asthma, inflammatory bowel disease, and acute viral infection. Pregnancy brings the potential for mild immunodeficiency, especially for the developing fetus. The risk of infection is also increased with diabetes; this was considered a factor in the first glanders infection in the United States since 1945, an LAI with *B. mallei* (318). Diabetes may also have been a factor in the unexpected fatality due to an attenuated, pigment-negative strain of *Y. pestis*, KIMD27, as well as undiagnosed hereditary hemochromatosis. The attenuation of KIMD27 is based on the strain's inability to acquire iron; it is thought that the iron overload in the scientist's blood may have provided the strain with the sufficient iron to recover virulence (85). It is known that individuals with hemochromatosis are more susceptible to infection with at least 32 organisms, including Gram-negative organisms, Gram-positive, or acid-fast bacteria, fungi, and parasites (319). Occupational risks associated with the reproductive system may involve exposures during pregnancy that result in adverse outcomes such as spontaneous abortion and birth defects. Infertility can occur in either sex. Male exposures can cause damage to sperm, transmission of toxic agents in seminal fluid, or infection of the pregnant woman from her partner's contaminated clothing. Breast-feeding may also be a source of infection. More commonly, concerns are directed to the potential congenital infection of a fetus, *in utero* or during delivery, as a result of a pregnant woman acquiring a work-related infection. Exposure to microorganisms known to cause congenital or neonatal infections, such as *Brucella*, *Cytomegalovirus*, hepatitis B virus, herpes simplex virus, HIV, LCMV, parvovirus, *L. monocytogenes*, rubella virus, *Treponema pallidum*, and toxoplasmas, is a distinct possibility in laboratory work (320). There is also a link between Zika virus and microcephaly (http://wwwnc.cdc.gov/travel/notices/alert/zika-virus-central-america).

In microbiological and biomedical laboratories, workers can also develop allergies to proteins (biological products derived from raw materials, fermentation products, or enzymes), chemicals, and the dander or aerosolized urine products of animals (321, 322).

Behavioral factors

Regarding occupational exposures to pathogenic microorganisms, the worker is key in controlling the safe outcome of any operation. He or she handles the agent, performs experiments, operates equipment, handles animals, disposes of infectious waste, and, when necessary, cleans up spills. The worker must come to the workplace prepared to function successfully. This means having adequate education, technical experience, and safety training to understand a task or project and perform it safely; being able to focus on the work so that inattention or random distractions do not lead to accidents; and being motivated to adhere to safe work practices. Philipps (323) and Martin (324) both discussed behavioral factors associated with laboratory safety. In a study conducted at Fort Detrick in a large microbiological research laboratory, Phillips described various characteristics associated with accident-prone and accident-free individuals. The study found that individuals in the 20- to 29-year age group had an abnormally high accident rate and that women had slightly fewer accidents than men. Unfortunately, the biomedical workforce is usually young and innovative and falls into the higher accident group. Sixty-five percent of all accidents in Phillips' study were due to human error, and 20% were due to equipment problems. The remaining 15% were ultimately attributed to "unsafe acts," which could also be considered human error. Although not always acknowledged as having a role in LAIs, behavioral factors need to be taken into consideration.

CONCLUSION

This review was written to support biosafety programs by heightening awareness that LAIs continue to occur and by providing data for reinforcement of biosafety practices in daily operations. The authors of publications cited in this chapter have made an important contribution to the objective risk assessment and the development of improved biosafety practices. The safety culture in a workplace must encourage the reporting of exposure incidents to determine whether steps can be taken to prevent recurrence. The risk of LAIs can be minimized if laboratory staff are aware of the potential for exposure to infectious agents and work continuously with biosafety and occupational health professionals to accomplish this goal.

References

1. **Singh K.** 2011. It's time for a centralized registry of laboratory-acquired infections. *Nat Med* **17:**919.
2. **Sewell DL.** 2000. Laboratory-acquired Infections. *Clin Microbiol Newsl* **22:**73–77.
3. **Miller JM, Astles R, Baszler T, Chapin K, Carey R, Garcia L, Gray L, Larone D, Pentella M, Pollock A, Shapiro DS, Weirich E, Wiedbrauk D, Biosafety Blue Ribbon Panel, Centers for Disease Control and Prevention (CDC).** 2012. Guidelines for safe work practices in human and animal medical diagnostic laboratories. Recommendations of a CDC-convened, Biosafety Blue Ribbon Panel. *MMWR Suppl* **61**(Suppl)**:**1–102.
4. **Trans-Federal Task Force on Optimizing Biosafety and Biocontainment Oversight.** 2009. Final report. http://www.ars.usda.gov/is/br/bbotaskforce/biosafety-FINAL-REPORT-092009.pdf.
5. **Henckel RD, Miller T, Weyant RS.** 2012. Monitoring Select Agent theft, loss, and release reports in the United States—2004–2010. *Appl Biosaf* **17:**171–180.
6. **Kimman TG, Smit E, Klein MR.** 2008. Evidence-based biosafety: a review of the principles and effectiveness of microbiological containment measures. *Clin Microbiol Rev* **21:**403–425.
7. **Sulkin SE, Pike RM.** 1951. Laboratory-acquired infections. *J Am Med Assoc* **147:**1740–1745.
8. **Reid DD.** 1957. Incidence of tuberculosis among workers in medical laboratories. *BMJ* **2:**10–14.
9. **Phillips GB.** 1965a. Microbiological hazards in the laboratory, Part 1. Control. *J Chem Educ* **42:**A43–A48.
10. **Harrington JM, Shannon HS.** 1976. Incidence of tuberculosis, hepatitis, brucellosis, and shigellosis in British medical laboratory workers. *BMJ* **1:**759–762.
11. **Grist NR.** 1981. Infection hazards in clinical laboratories. *Scott Med J* **26:**197–198.
12. **Grist NR.** 1983. Infections in British clinical laboratories 1980–81. *J Clin Pathol* **36:**121–126.
13. **Jacobson JT, Orlob RB, Clayton JL.** 1985. Infections acquired in clinical laboratories in Utah. *J Clin Microbiol* **21:**486–489.
14. **Vesley D, Hartmann HM.** 1988. Laboratory-acquired infections and injuries in clinical laboratories: a 1986 survey. *Am J Public Health* **78:**1213–1215.
15. **Masuda T, Isokawa T.** 1991. [Biohazard in clinical laboratories in Japan]. *Kansenshogaku Zasshi* **65:**209–215.
16. **Baron EJ, Miller JM.** 2008. Bacterial and fungal infections among diagnostic laboratory workers: evaluating the risks. *Diagn Microbiol Infect Dis* **60:**241–246.
17. **Athlin S, Vikerfors T, Fredlund H, Olcén P.** 2007. Atypical clinical presentation of laboratory-acquired meningococcal disease. *Scand J Infect Dis* **39:**911–913.
18. **Riyesh T, Karuppusamy S, Bera BC, Barua S, Virmani N, Yadav S, Vaid RK, Anand T, Bansal M, Malik P, Pahuja I, Singh RK.** 2014. Laboratory-acquired buffalopox virus infection, India. *Emerg Infect Dis* **20:**324–326.
19. **Centers for Disease Control and Prevention.** 2011. Multistate Outbreak of *Salmonella typhimurium* infections associated with exposure to clinical and teaching microbiology laboratories. http://www.cdc.gov/salmonella/typhimurium-laboratory/index.html.
20. **Centers for Disease Control and Prevention.** 2014. Human *Salmonella* Typhimurium infections linked to exposure to clinical and teaching microbiology laboratories. http://www.cdc.gov/salmonella/typhimurium-labs-06-14/index.html.
21. **Anonymous.** 1988. Microbiological safety cabinets and laboratory acquired infection. *Lancet* **2:**844–845.
22. **Wedum AG, Barkley WE, Hellman A.** 1972. Handling of infectious agents. *J Am Vet Med Assoc* **161:**1557–1567.
23. **Sulkin SE, Pike RM.** 1951. Survey of laboratory-acquired infections. *Am J Public Health Nations Health* **41:**769–781.
24. **Pike RM, Sulkin SE.** 1952. Occupational hazards in microbiology. *Sci Mon* **75:**222–227.
25. **Pike RM, Sulkin SE, Schulze ML.** 1965. Continuing importance of laboratory-acquired infections. *Am J Public Health Nations Health* **55:**190–199.
26. **Pike RM.** 1976. Laboratory-associated infections: summary and analysis of 3921 cases. *Health Lab Sci* **13:**105–114.
27. **Pike RM.** 1978. Past and present hazards of working with infectious agents. *Arch Pathol Lab Med* **102:**333–336.
28. **Pike RM.** 1979. Laboratory-associated infections: incidence, fatalities, causes, and prevention. *Annu Rev Microbiol* **33:**41–66.
29. **Shapiro DS, Schwartz DR.** 2002. Exposure of laboratory workers to *Francisella tularensis* despite a bioterrorism procedure. *J Clin Microbiol* **40:**2278–2281.
30. **Centers for Disease Control and Prevention (CDC).** 2008. Update: potential exposures to attenuated vaccine strain *Brucella abortus* RB51 during a laboratory proficiency test—United States and Canada, 2007. *MMWR Morb Mortal Wkly Rep* **57:**36–39.
31. **Collins CH, Kennedy DA.** 1999. *Laboratory-Acquired Infections: History, Incidence, Causes and Preventions*, 4th ed. Butterworth Heinemann, Oxford.

32. **Brener Z.** 1987. Laboratory-acquired Chagas disease: comment. *Trans R Soc Trop Med Hyg* **81:**527.
33. **Walker D, Campbell D.** 1999. A survey of infections in United Kingdom laboratories, 1994–1995. *J Clin Pathol* **52:**415–418.
34. **Willemark N, Van Vaerenbergh B, Descamps E, Brosius B, Dai Do Thi C, Leunda A, Baldo A, Herman P.** 2015. *Laboratory-Acquired Infections in Belgium (2007–2012).* Institut Scientifique de Santé Publique, Brussels, Belgium. http://www.biosafety.be/CU/PDF/2015_Willemarck_LAI%20report%20Belgium_2007_2012_Final.pdf.
35. **Boutet R, Stuart JM, Kaczmarski EB, Gray SJ, Jones DM, Andrews N.** 2001. Risk of laboratory-acquired meningococcal disease. *J Hosp Infect* **49:**282–284.
36. **Bremner DA.** 1992. Laboratory acquired meningococcal septicaemia. *Aust Microbiol* **13:**A106.
37. **Centers for Disease Control (CDC).** 1991. Laboratory-acquired meningococcemia—California and Massachusetts. *MMWR Morb Mortal Wkly Rep* **40:**46–47, 55.
38. **Centers for Disease Control and Prevention (CDC).** 2002. Laboratory-acquired meningococcal disease—United States, 2000. *MMWR Morb Mortal Wkly Rep* **51:**141–144.
39. **Sheets CD, Harriman K, Zipprich J, Louie JK, Probert WS, Horowitz M, Prudhomme JC, Gold D, Mayer L.** 2014. Fatal meningococcal disease in a laboratory worker—California, 2012. *MMWR Morb Mortal Wkly Rep* **63:**770–772.
40. **Paradis JF, Grimard D.** 1994. Laboratory-acquired invasive meningococcus—Quebec. *Can Commun Dis Rep* **20:**12–14.
41. **Sejvar JJ, Johnson D, Popovic T, Miller JM, Downes F, Somsel P, Weyant R, Stephens DS, Perkins BA, Rosenstein NE.** 2005. Assessing the risk of laboratory-acquired meningococcal disease. *J Clin Microbiol* **43:**4811–4814.
42. **Al-Aska AK, Chagla AH.** 1989. Laboratory-acquired brucellosis. *J Hosp Infect* **14:**69–71.
43. **Georghiou PR, Young EJ.** 1991. Prolonged incubation in brucellosis. *Lancet* **337:**1543.
44. **Young EJ.** 1983. Human brucellosis. *Rev Infect Dis* **5:**821–842.
45. **Young EJ.** 1991. Serologic diagnosis of human brucellosis: analysis of 214 cases by agglutination tests and review of the literature. *Rev Infect Dis* **13:**359–372.
46. **Blaser MJ, Hickman FW, Farmer JJ III, Brenner DJ, Balows A, Feldman RA.** 1980. *Salmonella typhi*: the laboratory as a reservoir of infection. *J Infect Dis* **142:**934–938.
47. **Blaser MJ, Lofgren JP.** 1981. Fatal salmonellosis originating in a clinical microbiology laboratory. *J Clin Microbiol* **13:**855–858.
48. **Centers for Disease Control and Prevention (CDC).** 2011. Fatal laboratory-acquired infection with an attenuated *Yersinia pestis* strain—Chicago, Illinois, 2009. *MMWR Morb Mortal Wkly Rep* **60:**201–205.
49. **Wong D, Wild MA, Walburger MA, Higgins CL, Callahan M, Czarnecki LA, Lawaczeck EW, Levy CE, Patterson JG, Sunenshine R, Adem P, Paddock CD, Zaki SR, Petersen JM, Schriefer ME, Eisen RJ, Gage KL, Griffith KS, Weber IB, Spraker TR, Mead PS.** 2009. Primary pneumonic plague contracted from a mountain lion carcass. *Clin Infect Dis* **49:**e33–e38.
50. **De Schrijver KAL, Bertrand S, Collard JM, Eilers K, De Schrijver K, Lemmens A, Bertrand S, Collard JM, Eilers K.** 2007. [Een laboratoriuminfectie met *Shigella sonnei* gevolgd door een cluster secundaire infectiessecundaire infecties] Abstract in English: Outbreak of *Shigella sonnei* in a clinical microbiology laboratory with secondary infections in the community). *Tijdschr Geneeskd* **63:**686–690.
51. **Ruben B, Band JD, Wong P, Colville J.** 1991. Person-to-person transmission of *Brucella melitensis*. *Lancet* **337:**14–15.
52. **Goossens H, Marcelis L, Dekeyser P, Butzler JP.** 1983. *Brucella melitensis*: person-to-person transmission? *Lancet* **321:**773.
53. **Bolin CA, Koellner P.** 1988. Human-to-human transmission of *Leptospira interrogans* by milk. *J Infect Dis* **158:**246–247.
54. **U.S. Department of Health and Human Services Public Health Service.** 1999. *Biosafety in Microbiological and Biomedical Laboratories. Centers for Disease Control and Prevention and National Institutes of Health te.* U.S. Government Printiing Office, Washington, D.C.
55. **Jager MM, Murk JL, Pique R, Wulf MW, Leenders AC, Buiting AG, Bogaards JA, Kluytmans JA, Vandenbroucke-Grauls CM.** 2010. Prevalence of carriage of meticillin-susceptible and meticillin-resistant *Staphylococcus aureus* in employees of five microbiology laboratories in The Netherlands. *J Hosp Infect* **74:**292–294.
56. **Centers for Disease Control and Prevention (CDC).** 2008. Laboratory-acquired brucellosis—Indiana and Minnesota, 2006. *MMWR Morb Mortal Wkly Rep* **57:**39–42.
57. **Batchelor BI, Brindle RJ, Gilks GF, Selkon JB.** 1992. Biochemical mis-identification of *Brucella melitensis* and subsequent laboratory-acquired infections. *J Hosp Infect* **22:**159–162.
58. **Fiori PL, Mastrandrea S, Rappelli P, Cappuccinelli P.** 2000. *Brucella abortus* infection acquired in microbiology laboratories. *J Clin Microbiol* **38:**2005–2006.
59. **Gerberding JL, Romero JM, Ferraro MJ.** 2008. Case 34-2008. A 58-year-old woman with neck pain and fever. *N Engl J Med* **359:**1942–1949.
60. **Ollé-Goig JE, Canela-Soler J.** 1987. An outbreak of *Brucella melitensis* infection by airborne transmission among laboratory workers. *Am J Public Health* **77:**335–338.
61. **Staszkiewicz J, Lewis CM, Colville J, Zervos M, Band J.** 1991. Outbreak of *Brucella melitensis* among microbiology laboratory workers in a community hospital. *J Clin Microbiol* **29:** 287–290.
62. **Ozaras R, Celik AD, Demirel A.** 2004. Acute hepatitis due to brucellosis in a laboratory technician. *Eur J Intern Med* **15:**264.
63. **Emrys G.** 2008. Report into the investigation of ESR meningitis infection case of Dr. Jeannette Adu-Bobie. http://www.dol.govt.nz/news/media/2008/adu-bobie-report.asp.
64. **Burnens AP, Zbinden R, Kaempf L, Heinzer I, Nicolet J.** 1993. A case of laboratory acquired infection with *Escherichia coli O157:H7*. *Zentralbl Bakteriol* **279:**512–517.
65. **Olesen B, Jensen C, Olsen K, Fussing V, Gerner-Smidt P, Scheutz F.** 2005. *VTEC O117:K1:H7*. A new clonal group of E. coli associated with persistent diarrhoea in Danish travellers. *Scand J Infect Dis* **37:**288–294.
66. **Spina N, Zansky S, Dumas N, Kondracki S.** 2005. Four laboratory-associated cases of infection with *Escherichia coli O157:H7*. *J Clin Microbiol* **43:**2938–2939.
67. **Thilo W, Kiehl W, Geiss HK.** 1997. A case report of laboratory-acquired diphtheria. *Euro Surveill* **2:**67–68.
68. **Laboratory PHS.** 1998. Throat infection with toxigenic *Corynebacterium diptheriae*. *Commun Dis Wkly Rep* **8:**60–61.
69. **Yanyu L.** 2011. Dean, secretary deposed after group infection. http://www.chinadaily.com.cn/china/2011-09/04/content_13614791.htm
70. **M.A. Said SS. J. Wright-Andoh, R. Myers, J. Razeq, D. Blythe.** 2011. *Salmonella enterica serotype Typhimurium* gastrointestinal illness associated with a university microbiology course—Maryland, abstr. 62nd Epidemic Intelligence Service (EIS) Conference. **62**: 03.
71. **Centers for Disease Control and Prevention (CDC).** 2013. Salm*onella typhimurium* infections associated with a community college microbiology laboratory—Maine, 2013. *MMWR Morb Mortal Wkly Rep* **62:**863.
72. **Steckelberg JM, Terrell CL, Edson RS.** 1988. Laboratory-acquired *Salmonella typhimurium* enteritis: association with erythema nodosum and reactive arthritis. *Am J Med* **85:**705–707.
73. **Hoerl D, Rostkowski C, Ross SL, Walsh TJ.** 1988. Typhoid fever acquired in a medical teaching laboratory. *Lab Med* **19:**166–168.
74. **Centers for Disease Control (CDC).** 1984. Rat-bite fever in a college student—California. *MMWR Morb Mortal Wkly Rep* **33:**318–320.

75. **Montes J, Rodriguez MA, Martin T, Martin F.** 1986. Laboratory-acquired meningitis caused by *Brucella abortus* strain 19. *J Infect Dis* **154:**915–916.
76. **Wallach JC, Ferrero MC, Victoria Delpino M, Fossati CA, Baldi PC.** 2008. Occupational infection due to *Brucella abortus S19* among workers involved in vaccine production in Argentina. *Clin Microbiol Infect* **14:**805–807.
77. **Centers for Disease Control and Prevention (CDC).** 2007. *Salmonella serotype enteritidis* infections among workers producing poultry vaccine—Maine, November-December 2006. *MMWR Morb Mortal Wkly Rep* **56:**877–879.
78. **Schlech WF III, Turchik JB, Westlake RE Jr, Klein GC, Band JD, Weaver RE.** 1981. Laboratory-acquired infection with *Pseudomonas pseudomallei (melioidosis)*. *N Engl J Med* **305:**1133–1135.
79. **Meselson M, Guillemin J, Hugh-Jones M, Langmuir A, Popova I, Shelokov A, Yampolskaya O.** 1994. The Sverdlovsk anthrax outbreak of 1979. *Science* **266:**1202–1208.
80. **Bhatti AR, DiNinno VL, Ashton FE, White LA.** 1982. A laboratory-acquired infection with *Neisseria meningitidis*. *J Infect* **4:** 247–252.
81. **Kessler AT, Stephens DS, Somani J.** 2007. Laboratory-acquired serogroup A meningococcal meningitis. *J Occup Health* **49:**399–401.
82. **New Zealand Herald.** 2005. Scientist loses limbs to meningococcal disease. http://www.nzherald.co.nz/nz/news/article.cfm?c_id=1&objectid=10120376.
83. **Omer H, Rose G, Jolley KA, Frapy E, Zahar JR, Maiden MC, Bentley SD, Tinsley CR, Nassif X, Bille E.** 2011. Genotypic and phenotypic modifications of *Neisseria meningitidis* after an accidental human passage. *PLoS One* **6:**e17145.
84. **ProMED-mail.** 2009. Meningitis, meningococcal, USA (Massachusetts). http://www.promedmail.org/post/20091112.3924.
85. **Centers for Disease Control and Prevention (CDC).** 2011. Fatal laboratory-acquired infection with an attenuated Yersinia pestis Strain—Chicago, Illinois, 2009. *MMWR Morb Mortal Wkly Rep* **60:**201–205.
86. **Barry M.** 2005. Report of pneumonic tularemia in three Boston University researchers—November 2004–March 2005. Boston Public Health Commission, Boston, MA. http://cbc.arizona.edu/sites/default/files/Boston_Univerity_Tularemia_report_2005.pdf
87. **Eckstein M.** 2010. Army-broken procedures led to lab infection. http://www.fredericknewspost.com/archive/article_5e4539ea-7902-5b13-8d37-8c7b6ade7538.html.
88. **Rusnak J, Boudreau E, Bozue J, Petitt P, Ranadive M, Kortepeter M.** 2004. An unusual inhalational exposure to *Bacillus anthracis* in a research laboratory. *J Occup Environ Med* **46:**313–314.
89. **Bouza E, Martin A, Van den Berg RJ, Kuijper EJ.** 2008. Laboratory-acquired *clostridium difficile* polymerase chain reaction ribotype 027: a new risk for laboratory workers? *Clin Infect Dis* **47:**1493–1494.
90. **Lavelle JP, Landas S., Mitros F, Conklin JL.** 1994. Acute gastritis associated with spiral organisms from cats. Dig Dis Sci 39: 744–750.
91. **Huhulescu S, Leitner E, Feierl G, Allerberger F.** 2010. Laboratory-acquired *Vibrio cholerae O1* infection in Austria, 2008. *Clin Microbiol Infect* **16:**1303–1304.
92. **Barkham T, Taylor MB.** 2002. Sniffing bacterial cultures on agar plates: a useful tool or a safety hazard? *J Clin Microbiol* **40:**3877.
93. **Miller CD, Songer JR, Sullivan JF.** 1987. A twenty-five year review of laboratory-acquired human infections at the National Animal Disease Center. *Am Ind Hyg Assoc J* **48:**271–275.
94. **Yagupsky P, Peled N, Riesenberg K, Banai M.** 2000. Exposure of hospital personnel to *Brucella melitensis* and occurrence of laboratory-acquired disease in an endemic area. *Scand J Infect Dis* **32:**31–35.
95. **Centers for Disease Control and Prevention (CDC).** 2008. Laboratory-acquired brucellosis—Indiana and Minnesota, 2006. *MMWR Morb Mortal Wkly Rep* **57:**39–42.
96. **Shemesh AA, Yagupsky P.** 2012. Isolation rates of *Brucella melitensis* in an endemic area and implications for laboratory safety. *Eur J Clin Microbiol Infect Dis* **31:**441–443.
97. **Kiel FW, Khan MY.** 1993. Brucellosis among hospital employees in Saudi Arabia. *Infect Control Hosp Epidemiol* **14:**268–272.
98. **Ergönül O, Celikbaş A, Tezeren D, Güvener E, Dokuzoğuz B.** 2004. Analysis of risk factors for laboratory-acquired brucella infections. *J Hosp Infect* **56:**223–227.
99. **Hasanjani Roushan MR, Mohrez M, Smailnejad Gangi SM, Soleimani Amiri MJ, Hajiahmadi M.** 2004. Epidemiological features and clinical manifestations in 469 adult patients with brucellosis in Babol, Northern Iran. *Epidemiol Infect* **132:**1109–1114.
100. **Eales KM, Norton RE, Ketheesan N.** 2010. Brucellosis in northern Australia. *Am J Trop Med Hyg* **83:**876–878.
101. **Jiang H, Fan M, Chen J, Mi J, Yu R, Zhao H, Piao D, Ke C, Deng X, Tian G, Cui B.** 2011. MLVA genotyping of Chinese human *Brucella melitensis* biovar 1, 2 and 3 isolates. *BMC Microbiol* **11:**256.
102. **Al Dahouk S, Neubauer H, Hensel A, Schöneberg I, Nöckler K, Alpers K, Merzenich H, Stark K, Jansen A.** 2007. Changing epidemiology of human brucellosis, Germany, 1962–2005. *Emerg Infect Dis* **13:**1895–1900.
103. **Grammont-Cupillard M, Berthet-Badetti L, Dellamonica P.** 1996. Brucellosis from sniffing bacteriological cultures. *Lancet* **348:**1733–1734.
104. **Demirdal T, Demirturk N.** 2008. Laboratory-acquired brucellosis. *Ann Acad Med Singapore* **37:**86–87.
105. **Memish ZA, Alazzawi M, Bannatyne R.** 2001. Unusual complication of breast implants: brucella infection. *Infection* **29:**291–292.
106. **Memish ZA, Venkatesh S.** 2001. Brucellar epididymo-orchitis in Saudi Arabia: a retrospective study of 26 cases and review of the literature. *BJU Int* **88:**72–76.
107. **Memish ZA, Mah MW.** 2001. Brucellosis in laboratory workers at a Saudi Arabian hospital. *Am J Infect Control* **29:**48–52.
108. **Aloufi AD, Memish ZA, Assiri AM, McNabb SJN.** 2016. Trends of reported human cases of brucellosis, Kingdom of Saudi Arabia, 2004–2012. *J Epidemiol Glob Health* **6:**11–18.
109. **Robichaud S, Libman M, Behr M, Rubin E.** 2004. Prevention of laboratory-acquired brucellosis. *Clin Infect Dis* **38:**e119–e122.
110. **Chusid MJ, Russler SK, Mohr BA, Margolis DA, Hillery CA, Kehl KC.** 1993. Unsuspected brucellosis diagnosed in a child as a result of an outbreak of laboratory-acquired brucellosis. *Pediatr Infect Dis J* **12:**1031–1033.
111. **Peiris V, Fraser S, Fairhurst M, Weston D, Kaczmarski E.** 1992. Laboratory diagnosis of brucella infection: some pitfalls. *Lancet* **339:**1415–1416.
112. **Public Health Service Laboratory.** 1991. Microbiological test strip (API20NE) identifies *Brucella melitensis* as *Moraxella phenylpyruvica*. *CDR (Lond Engl Wkly)* **1:**165.
113. **Noviello S, Gallo R, Kelly M, Limberger RJ, DeAngelis K, Cain L, Wallace B, Dumas N.** 2004. Laboratory-acquired brucellosis. *Emerg Infect Dis* **10:**1848–1850.
114. **Gruner E, Bernasconi E, Galeazzi RL, Buhl D, Heinzle R, Nadal D.** 1994. Brucellosis: an occupational hazard for medical laboratory personnel. Report of five cases. *Infection* **22:**33–36.
115. **Luzzi GA, Brindle R, Sockett PN, Solera J, Klenerman P, Warrell DA.** 1993. Brucellosis: imported and laboratory-acquired cases, and an overview of treatment trials. *Trans R Soc Trop Med Hyg* **87:**138–141.
116. **Bouza E, Sánchez-Carrillo C, Hernangómez S, González MJ, The Spanish Co-operative Group for the Study of Laboratory-acquired Brucellosis.** 2005. Laboratory-acquired brucellosis: a Spanish national survey. *J Hosp Infect* **61:**80–83.
117. **Schulze zur Wiesch J, Wichmann D, Sobottka I, Rohde H, Schmoock G, Wernery R, Schmiedel S, Dieter Burchard G, Melzer F.** 2010. Genomic tandem repeat analysis proves laboratory-acquired brucellosis in veterinary (camel) diagnostic laboratory in the United Arab Emirates. *Zoonoses Public Health* **57:**315–317.

118. **Hartady T, Saad MZ, Bejo SK, Salisi MS.** 2014. Clinical human brucellosis in Malaysia: a case report. *Asian Pac J Trop Dis* **4:**150–153.
119. **Traxler RM, Lehman MW, Bosserman EA, Guerra MA, Smith TL.** 2013. A literature review of laboratory-acquired brucellosis. *J Clin Microbiol* **51:**3055–3062.
120. **Traxler RM, Guerra MA, Morrow MG, Haupt T, Morrison J, Saah JR, Smith CG, Williams C, Fleischauer AT, Lee PA, Stanek D, Trevino-Garrison I, Franklin P, Oakes P, Hand S, Shadomy SV, Blaney DD, Lehman MW, Benoit TJ, Stoddard RA, Tiller RV, De BK, Bower W, Smith TL.** 2013. Review of brucellosis cases from laboratory exposures in the United States in 2008 to 2011 and improved strategies for disease prevention. *J Clin Microbiol* **51:**3132–3136.
121. **United States Government Accountability Office.** 2009. High-containment laboratories national strategy for oversight is needed. GAO-09-1045T. http://www.gao.gov/new.items/d09574.pdf.
122. **Brew SD, Perrett LL, Stack JA, MacMillan AP, Staunton NJ.** 1999. Human exposure to Brucella recovered from a sea mammal. *Vet Rec* **144:**483.
123. **Wallach JC, Giambartolomei GH, Baldi PC, Fossati CA.** 2004. Human infection with M-strain of *Brucella canis*. *Emerg Infect Dis* **10:**146–148.
124. **Arlett PR.** 1996. A case of laboratory acquired brucellosis. *BMJ* **313:**1130–1132.
125. **Shireman PK.** 1992. Endometrial tuberculosis acquired by a health care worker in a clinical laboratory. *Arch Pathol Lab Med* **116:**521–523.
126. **Müller HE.** 1988. Laboratory-acquired mycobacterial infection. *Lancet* **332:**331.
127. **Clark RP, Rueda-Pedraza ME, Teel LD, Salkin IF, Mahoney W.** 1988. Microbiological safety cabinets and laboratory acquired infection. *Lancet* **332:**844–845.
128. **Goto M, Yamashita T, Misawa S, Komori T, Okuzumi K, Takahashi T.** 2007. [Current biosafety in clinical laboratories in Japan: report of questionnaires' data obtained from clinical laboratory personnel in Japan]. *Kansenshogaku Zasshi* **81:**39–44.
129. **Centers for Disease Control (CDC).** 1981. Tuberculous infection associated with tissue processing—California. *MMWR Morb Mortal Wkly Rep* **30:**73–74.
130. **Grist NR, Emslie J.** 1985. Infections in British clinical laboratories, 1982–3. *J Clin Pathol* **38:**721–725.
131. **Grist NR, Emslie JA.** 1987. Infections in British clinical laboratories, 1984–5. *J Clin Pathol* **40:**826–829.
132. **Grist NR, Emslie JA.** 1989. Infections in British clinical laboratories, 1986–87. *J Clin Pathol* **42:**677–681.
133. **Menzies D, Fanning A, Yuan L, FitzGerald JM, Canadian Collaborative Group in Nosocomial Transmission of Tuberculosis.** 2003. Factors associated with tuberculin conversion in Canadian microbiology and pathology workers. *Am J Respir Crit Care Med* **167:**599–602.
134. **Posthaus H, Bodmer T, Alves L, Oevermann A, Schiller I, Rhodes SG, Zimmerli S.** 2011. Accidental infection of veterinary personnel with *Mycobacterium tuberculosis* at necropsy: a case study. *Vet Microbiol* **149:**374–380.
135. **Templeton GL, Illing LA, Young L, Cave D, Stead WW, Bates JH.** 1995. The risk for transmission of *Mycobacterium tuberculosis* at the bedside and during autopsy. *Ann Intern Med* **122:** 922–925.
136. **Peerbooms PGH, van Doornum GJJ, van Deutekom H, Coutinho RA, van Soolingen D.** 1995. Laboratory-acquired tuberculosis. *Lancet* **345:**1311–1312.
137. **Bemer-Melchior P, Drugeon HB.** 1999. Inactivation of *Mycobacterium tuberculosis* for DNA typing analysis. *J Clin Microbiol* **37:**2350–2351.
138. **Angela M, Weber YB, Mortimer VD.** 2000. A tuberculosis outbreak among medical waste workers. *J Am Biol Saf Assoc* **5:**70–80.
139. **Kao AS, Ashford DA, McNeil MM, Warren NG, Good RC.** 1997. Descriptive profile of tuberculin skin testing programs and laboratory-acquired tuberculosis infections in public health laboratories. *J Clin Microbiol* **35:**1847–1851.
140. **Belchior I, Seabra B, Duarte R.** 2011. Primary inoculation skin tuberculosis by accidental needle stick. *BMJ Case Rep* **2011:** bcr1120103496.
141. **Washington State Department of Labor and Industries Region 2—Seattle Office.** 2004. Inspection report on laboratory associated infections due to *Mycobacterium tuberculosis*. Inspection307855056. http://www.sunshine-project.org/idriuwmadchamber.pdf.
142. **Bernstein DI, Hubbard T, Wenman WM, Johnson BL Jr, Holmes KK, Liebhaber H, Schachter J, Barnes R, Lovett MA.** 1984. Mediastinal and supraclavicular lymphadenitis and pneumonitis due to *Chlamydia trachomatis* serovars L1 and L2. *N Engl J Med* **311:**1543–1546.
143. **Paran H, Heimer D, Sarov I.** 1986. Serological, clinical and radiological findings in adults with bronchopulmonary infections caused by *Chlamydia trachomatis*. *Isr J Med Sci* **22:**823–827.
144. **Marr JJ.** 1983. The professor and the pigeon. Psittacosis in the groves of academe. *Mo Med* **80:**135–136.
145. **Van Droogenbroeck C, Beeckman DS, Verminnen K, Marien M, Nauwynck H, Boesinghe LT, Vanrompay D.** 2009. Simultaneous zoonotic transmission of *Chlamydophila psittaci* genotypes D, F and E/B to a veterinary scientist. *Vet Microbiol* **135:**78–81.
146. **Olson LD.** 1980. Accidental penetration of hands with virulent and avirulent *Pasteurella multocida* of turkey origin. *Avian Dis* **24:**1064–1066.
147. **Janier M, Gheorghiu M, Cohen P, Mazas F, Duroux P.** 1982. [Carpal tunnel syndrome due to mycobacterium bovis BCG (author's transl)]. *Sem Hop* **58:**977–979.
148. **Salerno AE, Meyers KE, McGowan KL, Kaplan BS.** 2004. Hemolytic uremic syndrome in a child with laboratory-acquired *Escherichia coli O157:H7*. *J Pediatr* **145:**412–414.
149. **Blaser MJ, Feldman RA.** 1980. Acquisition of typhoid fever from proficiency-testing specimens. *N Engl J Med* **303:**1481.
150. **Mermel LA, Josephson SL, Dempsey J, Parenteau S, Perry C, Magill N.** 1997. Outbreak of *Shigella sonnei* in a clinical microbiology laboratory. *J Clin Microbiol* **35:**3163–3165.
151. **Bavoil PM.** 2005. Federal indifference to laboratory-acquired infections. *ASM News* **71:**1. (Letter.)
152. **Rangel JM, Sparling PH, Crowe C, Griffin PM, Swerdlow DL.** 2005. Epidemiology of *Escherichia coli O157:H7* outbreaks, United States, 1982–2002. *Emerg Infect Dis* **11:**603–609.
153. **Kozlovac J, Gurtler J.** 2015. E. coli 0157:H7 case study. Abstr. USDA ARS 3rd International Biosafety & Biocontainment Symposium: Biorisk Management in a One-Health World, Baltimore, MD.
154. **Vraneš J, Lukšić I, Knežević J, Ljubin-Sternak S** 2015. Health care associated cutaneous abscess—a rare form of primary gonococcal infection. *Am J Med Case Rep* **3:**88–90.
155. **Centers for Disease Control and Prevention (CDC).** 2002. Update: cutaneous anthrax in a laboratory worker—Texas, 2002. *MMWR Morb Mortal Wkly Rep* **51:**482.
156. **Bhatia VN.** 1990. Possible multiplication of *M. leprae* (?) on skin and nail bed of a laboratory worker. *Indian J Lepr* **62:**226–227.
157. **Welsh HH, Lennette EH, Abinanti FR, Winn JF.** 1951. Q fever in California. IV. Occurrence of *Coxiella burnetii* in the placenta of naturally infected sheep. *Public Health Rep* **66:**1473–1477.
158. **Weigler BJ, Di Giacomo RF, Alexander S.** 2005. A national survey of laboratory animal workers concerning occupational risks for zoonotic diseases. *Comp Med* **55:**183–191.
159. **Ossewaarde JM, Hekker AC.** 1984. [Q fever infection probably caused by a human placenta]. *Ned Tijdschr Geneeskd* **128:**2258–2260.
160. **Rusnak JM, Kortepeter MG, Aldis J, Boudreau E.** 2004. Experience in the medical management of potential laboratory exposures to agents of bioterrorism on the basis of risk assessment

at the United States Army Medical Research Institute of Infectious Diseases (USAMRIID). *J Occup Environ Med* **46:**801–811.

161. **Halle S, Dasch GA.** 1980. Use of a sensitive microplate enzyme-linked immunosorbent assay in a retrospective serological analysis of a laboratory population at risk to infection with typhus group rickettsiae. *J Clin Microbiol* **12:**343–350.
162. **Perna A, Di Rosa S, Intonazzo V, Sferlazzo A, Tringali G, La Rosa G.** 1990. Epidemiology of boutonneuse fever in western Sicily: accidental laboratory infection with a rickettsial agent isolated from a tick. *Microbiologica* **13:**253–256.
163. **Norazah A, Mazlah A, Cheong YM, Kamel AG.** 1995. Laboratory acquired murine typhus—a case report. *Med J Malaysia* **50:**177–179.
164. **Oh M, Kim N, Huh M, Choi C, Lee E, Kim I, Choe K.** 2001. Scrub typhus pneumonitis acquired through the respiratory tract in a laboratory worker. *Infection* **29:**54–56.
165. **Woo JH, Cho JY, Kim YS, Choi DH, Lee NM, Choe KW, Chang WH.** 1990. A case of laboratory-acquired murine typhus. *Korean J Intern Med* **5:**118–122.
166. **Israeli E.** 2014. [A hantavirus killed an Israeli researcher: hazards while working with wild animals]. *Harefuah* **153:**443–444, 499.
167. **Sinclair JR, Carroll DS, Montgomery JM, Pavlin B, McCombs K, Mills JN, Comer JA, Ksiazek TG, Rollin PE, Nichol ST, Sanchez AJ, Hutson CL, Bell M, Rooney JA.** 2007. Two cases of hantavirus pulmonary syndrome in Randolph County, West Virginia: a coincidence of time and place? *Am J Trop Med Hyg* **76:**438–442.
168. **Gire SK, et al.** 2014. Genomic surveillance elucidates Ebola virus origin and transmission during the 2014 outbreak. *Science* **345:**1369–1372.
169. **Koebler J.** 2016. Five scientists died of Ebola while working on a single study of the virus., Update 9/3/14. Motherboard. http://motherboard.vice.com/read/five-scientists-died-of-ebola-while-working-on-a-single-study-on-the-virus.
170. **Mulders MN, Reimerink JH, Koopmans MP, van Loon AM, van der Avoort HG.** 1997. Genetic analysis of wild-type poliovirus importation into The Netherlands (1979–1995). *J Infect Dis* **176:**617–624.
171. **Foy BD, Kobylinski KC, Chilson Foy JL, Blitvich BJ, Travassos da Rosa A, Haddow AD, Lanciotti RS, Tesh RB.** 2011. Probable non-vector-borne transmission of Zika virus, Colorado, USA. *Emerg Infect Dis* **17:**880–882.
172. **Chen EC, Yagi S, Kelly KR, Mendoza SP, Maninger N, Rosenthal A, Spinner A, Bales KL, Schnurr DP, Lerche NW, Chiu CY.** 2011. Cross-species transmission of a novel adenovirus associated with a fulminant pneumonia outbreak in a new world monkey colony. *PLoS Pathog* **7:**e1002155.
173. **Heymann DL, Aylward RB, Wolff C.** 2004. Dangerous pathogens in the laboratory: from smallpox to today's SARS setbacks and tomorrow's polio-free world. *Lancet* **363:**1566–1568.
174. **Joyce MP, Kuhar D, Brooks JT.** 2015. Notes from the field: occupationally acquired HIV infection among health care workers—United States, 1985–2013. *MMWR Morb Mortal Wkly Rep* **63:**1245–1246.
175. **Ippolito G, Puro V, Heptonstall J, Jagger J, De Carli G, Petrosillo N.** 1999. Occupational human immunodeficiency virus infection in health care workers: worldwide cases through September 1997. *Clin Infect Dis* **28:**365–383.
176. **Eberle J, Habermann J, Gürtler LG.** 2000. HIV-1 infection transmitted by serum droplets into the eye: a case report. *AIDS* **14:**206–207.
177. **Centers for Disease Control (CDC).** 1992. Seroconversion to simian immunodeficiency virus in two laboratory workers. *MMWR Morb Mortal Wkly Rep* **41:**678–681.
178. **Centers for Disease Control and Prevention (CDC).** 2008. Laboratory-acquired vaccinia exposures and infections—United States, 2005–2007. *MMWR Morb Mortal Wkly Rep* **57:**401–404.
179. **Hsu CH, Farland J, Winters T, Gunn J, Caron D, Evans J, Osadebe L, Bethune L, McCollum AM, Patel N, Wilkins K, Davidson W, Petersen B, Barry MA, Centers for Disease Control and Prevention (CDC).** 2015. Laboratory-acquired vaccinia virus infection in a recently immunized person—Massachusetts, 2013. *MMWR Morb Mortal Wkly Rep* **64:**435–438.
180. **Jones L, Ristow S, Yilma T, Moss B.** 1986. Accidental human vaccination with vaccinia virus expressing nucleoprotein gene. *Nature* **319:**543.
181. **Mempel M, Isa G, Klugbauer N, Meyer H, Wildi G, Ring J, Hofmann F, Hofmann H.** 2003. Laboratory acquired infection with recombinant vaccinia virus containing an immunomodulating construct. *J Invest Dermatol* **120:**356–358.
182. **Openshaw PJ, Alwan WH, Cherrie AH, Record FM.** 1991. Accidental infection of laboratory worker with recombinant vaccinia virus. *Lancet* **338:**459.
183. **MacNeil A, Reynolds MG, Damon IK.** 2009. Risks associated with vaccinia virus in the laboratory. *Virology* **385:**1–4.
184. **Rocke TE, Dein FJ, Fuchsberger M, Fox BC, Stinchcomb DT, Osorio JE.** 2004. Limited infection upon human exposure to a recombinant raccoon pox vaccine vector. *Vaccine* **22:**2757–2760.
185. **Eisenbach C, Neumann-Haefelin C, Freyse A, Korsukéwitz T, Hoyler B, Stremmel W, Thimme R, Encke J.** 2007. Immune responses against HCV-NS3 after accidental infection with HCV-NS3 recombinant vaccinia virus. *J Viral Hepat* **14:**817–819.
186. **Centers for Disease Control and Prevention (CDC).** 2009. Laboratory-acquired vaccinia virus infection—Virginia, 2008. *MMWR Morb Mortal Wkly Rep* **58:**797–800.
187. **Health and Safety Executive.**2003. Incidents—Lessons to be learnt. Accidental infection with vaccinia virus. http://www.hse.gov.uk/biosafety/gmo/acgm/acgm32/paper8.htm.
188. **Korioth-Schmitz B, Affeln D, Simon SL, Decaneas WM, Schweon GB, Wong M, Gardner A.** 2015. Vaccinia virus—laboratory tool with a risk of laboratory-acquired infection. *Appl Biosaf* **20:**6–11.
189. **Marennnikova SS, Zhukova OA, Manenkova GM, Ianova NN.** 1988. [Laboratory-confirmed case of human infection with ratpox (cowpox)]. *Zh Mikrobiol Epidemiol Immunobiol* **6:**30–32.
190. **McCollum AM, Austin C, Nawrocki J, Howland J, Pryde J, Vaid A, Holmes D, Weil MR, Li Y, Wilkins K, Zhao H, Smith SK, Karem K, Reynolds MG, Damon IK.** 2012. Investigation of the first laboratory-acquired human cowpox virus infection in the United States. *J Infect Dis* **206:**63–68.
191. **Venter M, Steyl J, Human S, Weyer J, Zaayman D, Blumberg L, Leman PA, Paweska J, Swanepoel R.** 2010. Transmission of West Nile virus during horse autopsy. *Emerg Infect Dis* **16:**573–575.
192. **Weissenbacher MC, Cura E, Segura EL, Hortal M, Baek LJ, Chu YK, Lee HW.** 1996. Serological evidence of human hantavirus infection in Argentina, Bolivia and Uruguay. *Medicina (B Aires)* **56:**17–22.
193. **Desmyter J, Johnson KM, Deckers C, Leduc JW, Brasseur F, Van Ypersele De Strihou C. .** 1983. Laboratory rat associated outbreak of haemorrhagic fever with renal syndrome due to Hantaan-like virus in Belgium. *Lancet* 33**2:**1445–1448.
194. **Wang GD.** 1985. [Outbreak of hemorrhagic fever with renal syndrome caused by a laboratory animal (white rat) infection]. *Zhonghua Liu Xing Bing Xue Za Zhi* **6:**233–235.
195. **Zhang YZ, Dong X, Li X, Ma C, Xiong HP, Yan GJ, Gao N, Jiang DM, Li MH, Li LP, Zou Y, Plyusnin A.** 2009. Seoul virus and hantavirus disease, Shenyang, People's Republic of China. *Emerg Infect Dis* **15:**200–206.
196. **Zhang Y, Zhang H, Dong X, Yuan J, Zhang H, Yang X, Zhou P, Ge X, Li Y, Wang LF, Shi Z.** 2010. Hantavirus outbreak associated with laboratory rats in Yunnan, China. *Infect Genet Evol* **10:**638–644.

197. **Douron E, Moriniere B, Matheron S, Girard PM, Gonzalez J-P, Hirsch F, McCormick JB.** 1984. HFRS after a wild rodent bite in the Haute-Savoie—and risk of exposure to Hantaan-like virus in a Paris laboratory. *Lancet* **323:**676–677.
198. **Kawamata J, Yamanouchi T, Dohmae K, Miyamoto H, Takahaski M, Yamanishi K, Kurata T, Lee HW.** 1987. Control of laboratory acquired hemorrhagic fever with renal syndrome (HFRS) in Japan. *Lab Anim Sci* **37:**431–436.
199. **Umenai T, Woo Lee P, Toyoda T, Yoshinaga K, Horiuchi T, Wang Lee H, Saito T, Hongo M, Ishida N.** 1979. Korean haemorrhagic fever in staff in an animal laboratory. *Lancet* **313:**1314–1316.
200. **Cho SH, Yun YS, Kang D, Kim S, Kim IS, Hong ST.** 1999. Laboratory-acquired infections with hantavirus at a research unit of medical school in Seoul, 1996. *Korean J Prev Med* **32:**269–275.
201. **Lloyd G, Bowen ET, Jones N, Pendry A.** 1984. HFRS outbreak associated with laboratory rats in UK. *Lancet* **323:**1175–1176.
202. **Lloyd G, Jones N.** 1986. Infection of laboratory workers with hantavirus acquired from immunocytomas propagated in laboratory rats. *J Infect* **12:**117–125.
203. **Wong TW, Chan YC, Yap EH, Joo YG, Lee HW, Lee PW, Yanagihara R, Gibbs CJ Jr, Gajdusek DC.** 1988. Serological evidence of hantavirus infection in laboratory rats and personnel. *Int J Epidemiol* **17:**887–890.
204. **Dykewicz CA, Dato VM, Fisher-Hoch SP, Howarth MV, Perez-Oronoz GI, Ostroff SM, Gary H Jr, Schonberger LB, McCormick JB.** 1992. Lymphocytic choriomeningitis outbreak associated with nude mice in a research institute. *JAMA* **267:** 1349–1353.
205. **Centers for Disease Control and Prevention (CDC).** 1998. Fatal Cercopithecine herpesvirus 1 (B virus) infection following a mucocutaneous exposure and interim recommendations for worker protection. *MMWR Morb Mortal Wkly Rep* **47:**1073–1076, 1083.
206. **Centers for Disease Control (CDC).** 1990. Update: filovirus infection in animal handlers. *MMWR Morb Mortal Wkly Rep* **39:** 221.
207. **Centers for Disease Control (CDC).** 1990. Update: ebola-related filovirus infection in nonhuman primates and interim guidelines for handling nonhuman primates during transit and quarantine. *MMWR Morb Mortal Wkly Rep* **39:**22–24, 29–30.
208. **Centers for Disease Control (CDC).** 1990. Update: filovirus infection associated with contact with nonhuman primates or their tissues. *MMWR Morb Mortal Wkly Rep* **39:**404–405.
209. **Lerche NW, Switzer WM, Yee JL, Shanmugam V, Rosenthal AN, Chapman LE, Folks TM, Heneine W.** 2001. Evidence of infection with simian type D retrovirus in persons occupationally exposed to nonhuman primates. *J Virol* **75:**1783–1789.
210. **Centers for Disease Control and Prevention (CDC).** 1997. Nonhuman primate spumavirus infections among persons with occupational exposure—United States, 1996. *MMWR Morb Mortal Wkly Rep* **46:**129–131.
211. **Mouinga-Ondémé A, Betsem E, Caron M, Makuwa M, Sallé B, Renault N, Saib A, Telfer P, Marx P, Gessain A, Kazanji M.** 2010. Two distinct variants of simian foamy virus in naturally infected mandrills (Mandrillus sphinx) and cross-species transmission to humans. *Retrovirology* **7:**105.
212. **Schweizer M, Falcone V, Gänge J, Turek R, Neumann-Haefelin D.** 1997. Simian foamy virus isolated from an accidentally infected human individual. *J Virol* **71:**4821–4824.
213. **Switzer WM, Bhullar V, Shanmugam V, Cong ME, Parekh B, Lerche NW, Yee JL, Ely JJ, Boneva R, Chapman LE, Folks TM, Heneine W.** 2004. Frequent simian foamy virus infection in persons occupationally exposed to nonhuman primates. *J Virol* **78:** 2780–2789.
214. **Chiu CY, Yagi S, Lu X, Yu G, Chen EC, Liu M, Dick EJ Jr, Carey KD, Erdman DD, Leland MM, Patterson JL.** 2013. A novel adenovirus species associated with an acute respiratory outbreak in a baboon colony and evidence of coincident human infection. *MBio* **4:**e00084-13.
215. **Alibek K, Handelman S.** 1999. *Biohazard.* Dell Publishing of Random House, New York.
216. **International Society for Infectious Diseases.** 2004. Ebola lab accident death—Russia (Siberia). Archive number 20040522.1377, http://www.promedmail.org.
217. **Le Guenno B, Formenty P, Wyers M, Gounon P, Walker F, Boesch C.** 1995. Isolation and partial characterisation of a new strain of Ebola virus. *Lancet* **345:**1271–1274.
218. **Fonseca K, Prince GD, Bratvold J, Fox JD, Pybus M, Preksaitis JK, Tilley P.** 2005. West Nile virus infection and conjunctival exposure. *Emerg Infect Dis* **11:**1648–1649.
219. **Smith AW, Iversen PL, Skilling DE, Stein DA, Bok K, Matson DO.** 2006. Vesivirus viremia and seroprevalence in humans. *J Med Virol* **78:**693–701.
220. **Smith AW, Berry ES, Skilling DE, Barlough JE, Poet SE, Berke T, Mead J, Matson DO.** 1998. In vitro isolation and characterization of a calicivirus causing a vesicular disease of the hands and feet. *Clin Infect Dis* **26:**434–439.
221. **Fulhorst CF, Milazzo ML, Armstrong LR, Childs JE, Rollin PE, Khabbaz R, Peters CJ, Ksiazek TG.** 2007. Hantavirus and arenavirus antibodies in persons with occupational rodent exposure. *Emerg Infect Dis* **13:**532–538.
222. **Skinhøj P, Søeby M.** 1981. Viral hepatitis in Danish health care personnel, 1974–78. *J Clin Pathol* **34:**408–411.
223. **United States Department of Labor.** 1991. Occupational exposure to bloodborne pathogens. Fed Reg **56:**64175–64182.
224. **Mahoney FJ, Stewart K, Hu H, Coleman P, Alter MJ.** 1997. Progress toward the elimination of hepatitis B virus transmission among health care workers in the United States. *Arch Intern Med* **157:**2601–2605.
225. **Wacławik J, Gasiorowski J, Inglot M, Andrzejak R, Gładysz A.** 2003. [Epidemiology of occupational infectious diseases in health care workers]. *Med Pr* **54:**535–541.
226. **Britton S, van den Hurk AF, Simmons RJ, Pyke AT, Northill JA, McCarthy J, McCormack J.** 2011. Laboratory-acquired dengue virus infection—a case report. *PLoS Negl Trop Dis* **5:**e1324.
227. **World Health Organization Western Pacific Region.** 2004. *Summary of China's Investigation into the April outbreak.* World Health Organization Western Pacific Region, Manila, Philippines.
228. **Lim PL, Kurup A, Gopalakrishna G, Chan KP, Wong CW, Ng LC, Se-Thoe SY, Oon L, Bai X, Stanton LW, Ruan Y, Miller LD, Vega VB, James L, Ooi PL, Kai CS, Olsen SJ, Ang B, Leo YS.** 2004. Laboratory-acquired severe acute respiratory syndrome. *N Engl J Med* **350:**1740–1745.
229. **World Health Organization Western Pacific Region.** 2003. Severe acute respinratory syndrome (SARS) in Taiwan, China 17 December 2003. http://www.who.int/csr/don/2003_12_17/en/.
230. **Junt T, Heraud JM, Lelarge J, Labeau B, Talarmin A.** 1999. Determination of natural versus laboratory human infection with Mayaro virus by molecular analysis. *Epidemiol Infect* **123:** 511–513.
231. **Raoult D, Renesto P, Brouqui P.** 2006. Laboratory infection of a technician by mimivirus. *Ann Intern Med* **144:**702–703.
232. **van der Meyden CH, Erasmus BJ, Swanepoel R, Prozesky OW.** 1992. Encephalitis and chorioretinitis associated with neurotropic African horsesickness virus infection in laboratory workers. Part I. Clinical and neurological observations. *S Afr Med J* **81:** 451–454.
233. **Scherer W.** 1980. Laboratory safety for arboviruses and certain other viruses of vertebrates. *Am J Trop Med Hyg* **29:**1359–1381.
234. **Tomori O, Monath TP, O'Connor EH, Lee VH, Cropp CB.** 1981. Arbovirus infections among laboratory personnel in Ibadan, Nigeria. *Am J Trop Med Hyg* **30:**855–861.
235. **Barry M, Russi M, Armstrong L, Geller D, Tesh R, Dembry L, Gonzalez JP, Khan AS, Peters CJ.** 1995. Brief report: treatment

of a laboratory-acquired Sabiá virus infection. *N Engl J Med* **333:** 294–296.

236. **Gaidomovich YSA, Burenko M, Leschinskaya HV.** 2000. Human laboratory acquired arbo-, arena-, and hantavirus. *Appl Biosaf* **5:**5–11.
237. **Brummer-Korvenkontio M, Vaheri A, Hovi T, von Bonsdorff CH, Vuorimies J, Manni T, Penttinen K, Oker-Blom N, Lähdevirta J.** 1980. *Nephropathia epidemica:* detection of antigen in bank voles and serologic diagnosis of human infection. *J Infect Dis* **141:**131–134.
238. **Centers for Disease Control and Prevention (CDC).** 1994. Bolivian hemorrhagic fever—El Beni Department, Bolivia, 1994. *MMWR Morb Mortal Wkly Rep* **43:**943–946.
239. **Wentworth DE, McGregor MW, Macklin MD, Neumann V, Hinshaw VS.** 1997. Transmission of swine influenza virus to humans after exposure to experimentally infected pigs. *J Infect Dis* **175:**7–15.
240. **Artenstein AW, Hicks CB, Goodwin BS Jr, Hilliard JK.** 1991. Human infection with B virus following a needlestick injury. *Rev Infect Dis* **13:**288–291.
241. **Davenport DS, Johnson DR, Holmes GP, Jewett DA, Ross SC, Hilliard JK.** 1994. Diagnosis and management of human B virus (Herpesvirus simiae) infections in Michigan. *Clin Infect Dis* **19:**33–41.
242. **Freifeld AG, Hilliard J, Southers J, Murray M, Savarese B, Schmitt JM, Straus SE.** 1995. A controlled seroprevalence survey of primate handlers for evidence of asymptomatic herpes B virus infection. *J Infect Dis* **171:**1031–1034.
243. **Holmes GP, et al.** 1990. B virus (Herpesvirus simiae) infection in humans: epidemiologic investigation of a cluster. *Ann Intern Med* **112:**833–839.
244. **Centers for Disease Control and Prevention.** 1987. B-virus infection in humans—Pensacola, Florida. MMWR Morb Mortal Wkly Rep 36:289–290, 295–286.
245. **Centers for Disease Control (CDC).** 1989. B virus infections in humans—Michigan. *MMWR Morb Mortal Wkly Rep* **38:** 453–454.
246. **Scinicariello F, English WJ, Hilliard J.** 1993. Identification by PCR of meningitis caused by herpes B virus. *Lancet* **341:**1660–1661.
247. **Ilkal MA, Dhanda V, Rodrigues JJ, Mohan Rao CV, Mourya DT.** 1984. Xenodiagnosis of laboratory acquired dengue infection by mosquito inoculation & immunofluorescence. *Indian J Med Res* **79:**587–590.
248. **Wu H-S, Wu W-C, Kuo H-S.** 2009. A three-year experience to implement laboratory biosafety regulations in Taiwan. *Appl Biosaf* **14:**33–36.
249. **Moore DM, MacKenzie WF, Doepel F, Hansen TN.** 1983. Contagious ecthyma in lambs and laboratory personnel. *Lab Anim Sci* **33:**473–475.
250. **Ippolito G, Puro V, Petrosillo N, De Carli G.** 1999. Surveillance of occupational exposure to bloodborne pathogens in health care workers: the Italian national programme. *Euro Surveill* **4:** 33–36.
251. **Khabbaz RF, Rowe T, Heneine WM, Kaplan JE, Folks TM, Schable CA, George JR, Pau C, Parekh BS, Curran JW, Schochetman G, Lairmore MD, Murphey-Corb M.** 1992. Simian immunodeficiency virus needlestick accident in a laboratory worker. *Lancet* **340:**271–273.
252. **Venter M, Burt FJ, Blumberg L, Fickl H, Paweska J, Swanepoel R.** 2009. Cytokine induction after laboratory-acquired West Nile virus infection. *N Engl J Med* **360:**1260–1262.
253. **Centers for Disease Control and Prevention (CDC).** 2002. Laboratory-acquired West Nile virus infections—United States, 2002. *MMWR Morb Mortal Wkly Rep* **51:**1133–1135.
254. **Moussatché N, Tuyama M, Kato SE, Castro AP, Njaine B, Peralta RH, Peralta JM, Damaso CR, Barroso PF.** 2003. Accidental infection of laboratory worker with vaccinia virus. *Emerg Infect Dis* **9:**724–726.
255. **Ertem GT, Tulek N, Oral B, Kinikli S.** 2005. Therapy of acute hepatitis C with interferon-alpha2b plus ribavirin in a health care worker. *Acta Gastroenterol Belg* **68:**104–106.
256. **Kotturi MF, Swann JA, Peters B, Arlehamn CL, Sidney J, Kolla RV, James EA, Akondy RS, Ahmed R, Kwok WW, Buchmeier MJ, Sette A.** 2011. Human CD8+ and CD4+ T cell memory to lymphocytic choriomeningitis virus infection. *J Virol* **85:**11770–11780.
257. **DeCarli G, Perry J, Jagger J.** 2004. Occupational co-infection with HIV and HCV. *Adv Expo Prev* **7:**13–18.
258. **Khabbaz RF, Heneine W, George JR, Parekh B, Rowe T, Woods T, Switzer WM, McClure HM, Murphey-Corb M, Folks TM.** 1994. Brief report: infection of a laboratory worker with simian immunodeficiency virus. *N Engl J Med* **330:**172–177.
259. **Langford MP, Stanton GJ, Barber JC, Baron S.** 1979. Early-appearing antiviral activity in human tears during a case of picornavirus epidemic conjunctivitis. *J Infect Dis* **139:**653–658.
260. **Ando Y, Iwasaki T, Terao K, Nishimura H, Tamura S.** 2001. Conjunctivitis following accidental exposure to influenza B virus/Shangdong/07/97. *J Infect* **42:**223–224.
261. **Webster RG, Geraci J, Petursson G, Skirnisson K.** 1981. Conjunctivitis in human beings caused by influenza A virus of seals. *N Engl J Med* **304:**911.
262. **Loeb M, Zando I, Orvidas MC, Bialachowski A, Groves D, Mahoney J.** 2003. Laboratory-acquired vaccinia infection. *Can Commun Dis Rep* **29:**134–136.
263. **Wlodaver CG, Palumbo GJ, Waner JL.** 2004. Laboratory-acquired vaccinia infection. *J Clin Virol* **29:**167–170.
264. **Lewis FM, Chernak E, Goldman E, Li Y, Karem K, Damon IK, Henkel R, Newbern EC, Ross P, Johnson CC.** 2006. Ocular vaccinia infection in laboratory worker, Philadelphia, 2004. *Emerg Infect Dis* **12:**134–137.
265. **Højlyng N, Holten-Andersen W, Jepsen S.** 1987. Cryptosporidiosis: a case of airborne transmission. *Lancet* **330:**271–272.
266. **Pohjola S, Oksanen H, Jokipii L, Jokipii AM.** 1986. Outbreak of cryptosporidiosis among veterinary students. *Scand J Infect Dis* **18:**173–178.
267. **Reif JS, Wimmer L, Smith JA, Dargatz DA, Cheney JM.** 1989. Human cryptosporidiosis associated with an epizootic in calves. *Am J Public Health* **79:**1528–1530.
268. **Philpott MS, Fautin CH, Bird KE, O'Reilly KL.** 2015. A laboratory-associated outbreak of cryptosporidiosis. *Appl Biosaf* **20:**130–136.
269. **Drinkard LN, Halbritter A, Nguyen GT, Sertich PL, King M, Bowman S, Huxta R, Guagenti M.** 2015. Notes from the field: outbreak of cryptosporidiosis among veterinary medicine students—Philadelphia, Pennsylvania, February 2015. *MMWR Morb Mortal Wkly Rep* **64:**773.
270. **Herwaldt BL, Juranek DD.** 1993. Laboratory-acquired malaria, leishmaniasis, trypanosomiasis, and toxoplasmosis. *Am J Trop Med Hyg* **48:**313–323.
271. **Centers for Disease Control and Prevention.** 1984. Malaria surveillance annual summary, 1982. *MMWR Morb Mortal Wkly Rep*
272. **Mali S, Steele S, Slutsker L, Arguin PM, Centers for Disease Control and Prevention (CDC).** 2008. Malaria surveillance—United States, 2006. *MMWR Surveill Summ* **57:**24–39.
273. **Cullen KA, Arguin PM, Division of Parasitic Diseases and Malaria, Center for Global Health, Centers for Disease Control and Prevention (CDC).** 2013. Malaria surveillance—United States, 2011. *MMWR Surveill Summ* **62:**1–17.
274. **Cullen KA, Arguin PM, Centers for Disease Control and Prevention (CDC).** 2014. Malaria surveillance—United States, 2012. *MMWR Surveill Summ* **63:**1–22.
275. **Centers for Disease Control and Prevention.** 1980. Chagas disease—Michigan. *Morb Mortal Wkly Rep* **29:**147–148.
276. **Sampaio RN, de Lima LM, Vexenat A, Cuba CC, Barreto AC, Marsden PD.** 1983. A laboratory infection with *Leishmania braziliensis braziliensis*. *Trans R Soc Trop Med Hyg* **77:**274.

277. **Van Gompel A, Van den Enden E, Van den Ende J, Geerts S.** 1993. Laboratory infection with *Schistosoma mansoni*. *Trans R Soc Trop Med Hyg* **87:**554.
278. **Partanen P, Turunen HJ, Paasivuo RT, Leinikki PO.** 1984. Immunoblot analysis of *Toxoplasma gondii* antigens by human immunoglobulins G, M, and A antibodies at different stages of infection. *J Clin Microbiol* **20:**133–135.
279. **Hermentin K, Hassl A, Picher O, Aspöck H.** 1989. Comparison of different serotests for specific Toxoplasma IgM-antibodies (ISAGA, SPIHA, IFAT) and detection of circulating antigen in two cases of laboratory acquired Toxoplasma infection. *Zentralbl Bakteriol Mikrobiol Hyg [A]* **270:**534–541.
280. **Herwaldt BL.** 2001. Laboratory-acquired parasitic infections from accidental exposures. *Clin Microbiol Rev* **14:**659–688.
281. **Johnson M, Broady K, Angelici MC, Johnson A.** 2003. The relationship between nucleoside triphosphate hydrolase (NTPase) isoform and Toxoplasma strain virulence in rat and human toxoplasmosis. *Microbes Infect* **5:**797–806.
282. **Parker SL, Holliman RE.** 1992. Toxoplasmosis and laboratory workers: a case-control assessment of risk. *Med Lab Sci* **49:** 103–106.
283. **Villavedra M, Battistoni J, Nieto A.** 1999. IgG recognizing 21-24 kDa and 30-33 kDa tachyzoite antigens show maximum avidity maturation during natural and accidental human toxoplasmosis. *Rev Inst Med Trop Sao Paulo* **41:**297–303.
284. **Delgado O, Guevara P, Silva S, Belfort E, Ramirez JL.** 1996. Follow-up of a human accidental infection by *Leishmania (Viannia) braziliensis* using conventional immunologic techniques and polymerase chain reaction. *Am J Trop Med Hyg* **55:**267–272.
285. **Sadick MD, Locksley RM, Raff HV.** 1984. Development of cellular immunity in cutaneous leishmaniasis due to *Leishmania tropica*. *J Infect Dis* **150:**135–138.
286. **Evans TG, Pearson RD.** 1988. Clinical and immunological responses following accidental inoculation of *Leishmania donovani*. *Trans R Soc Trop Med Hyg* **82:**854–856.
287. **Blagburn BL, Current WL.** 1983. Accidental infection of a researcher with human Cryptosporidium. *J Infect Dis* **148:**772–773.
288. **Williams JL, Innis BT, Burkot TR, Hayes DE, Schneider I.** 1983. Falciparum malaria: accidental transmission to man by mosquitoes after infection with culture-derived gametocytes. *Am J Trop Med Hyg* **32:**657–659.
289. **Freedman DO, MacLean JD, Viloria JB.** 1987. A case of laboratory acquired Leishmania donovani infection; evidence for primary lymphatic dissemination. *Trans R Soc Trop Med Hyg* **81:**118–119.
290. **Hofflin JM, Sadler RH, Araujo FG, Page WE, Remington JS.** 1987. Laboratory-acquired Chagas disease. *Trans R Soc Trop Med Hyg* **81:**437–440.
291. **Baker CC, Farthing CP, Ratnesar P.** 1984. Toxoplasmosis, an innocuous disease? *J Infect* **8:**67–69.
292. **Knobloch J, Demar M.** 1997. Accidental *Leishmania mexicana* infection in an immunosuppressed laboratory technician. *Trop Med Int Health* **2:**1152–1155.
293. **Jensen JB, Capps TC, Carlin JM.** 1981. Clinical drug-resistant falciparum malaria acquired from cultured parasites. *Am J Trop Med Hyg* **30:**523–525.
294. **Felinto de Brito ME, Andrade MS, de Almeida ÉL, Medeiros ÂCR, Werkhäuser RP, Araújo AIF, Brandão-Filho SP, Paiva de Almeida AM, Gomes Rodrigues EH.** 2012. Occupationally acquired american cutaneous leishmaniasis. *Case Rep Dermatol Med* **2012:**279517.
295. **Larson DM, Eckman MR, Alber RL, Goldschmidt VG.** 1983. Primary cutaneous (inoculation) blastomycosis: an occupational hazard to pathologists. *Am J Clin Pathol* **79:**253–255.
296. **Tenenbaum MJ, Greenspan J, Kerkering TM, Utz JP.** 1982. Blastomycosis. *Crit Rev Microbiol* **9:**139–163.
297. **Cooper CR, Dixon DM, Salkin IF.** 1992. Laboratory-acquired sporotrichosis. *J Med Vet Mycol* **30:**169–171.
298. **Ishizaki H, Ikeda M, Kurata Y.** 1979. Lymphocutaneous sporotrichosis caused by accidental inoculation. *J Dermatol* **6:** 321–323.
299. **Mochizuki T, Watanabe S, Kawasaki M, Tanabe H, Ishizaki H.** 2002. A Japanese case of tinea corporis caused by *Arthroderma benhamiae*. *J Dermatol* **29:**221–225.
300. **van Gool T, Biderre C, Delbac F, Wentink-Bonnema E, Peek R, Vivarès CP.** 2004. Serodiagnostic studies in an immunocompetent individual infected with *Encephalitozoon cuniculi*. *J Infect Dis* **189:**2243–2249.
301. **Hilmarsdottir I, Coutellier A, Elbaz J, Klein JM, Datry A, Guého E, Herson S.** 1994. A French case of laboratory-acquired disseminated *Penicillium marneffei* infection in a patient with AIDS. *Clin Infect Dis* **19:**357–358.
302. **Kamalam A, Thambiah AS.** 1979. *Trichophyton simii* infection due to laboratory accident. *Dermatologica* **159:**180–181.
303. **Contreras-Barrera ME, Moreno-Coutiño G, Torres-Guerrero DE, Aguilar-Donis A, Arenas R.** 2009. Eritema multiforme secundario a infección por Trichophyton mentagrophytes. [Erythema multiforme secondary to cutaneous Trichophyton mentagrophytes infection]. *Rev Iberoam Micol* **26:**149–151.
304. **Hironaga M, Fujigaki T, Watanabe S.** 1981. *Trichophyton mentagrophytes* skin infections in laboratory animals as a cause of zoonosis. *Mycopathologia* **73:**101–104.
305. **Shi ZC, Lei PC.** 1986. Occupational mycoses. *Br J Ind Med* **43:**500–501.
306. **United States Food and Drug Agency.**2009. *Bad Bug Book*. http://www.fda.gov/Food/FoodSafety/FoodborneIllness/FoodborneIllnessFoodbornePathogens Natural Toxins/ BadBugBook /ucm071372.
307. **Cohn AC, MacNeil JR, Clark TA, Ortega-Sanchez IR, Briere EZ, Meissner HC, Baker CJ, Messonnier NE, Centers for Disease Control and Prevention (CDC).** 2013. Prevention and control of meningococcal disease: recommendations of the Advisory Committee on Immunization Practices (ACIP). *MMWR Recomm Rep* **62**(RR-2)**:**1–28.
308. **Kortepeter MG, Martin JW, Rusnak JM, Cieslak TJ, Warfield KL, Anderson EL, Ranadive MV.** 2008. Managing potential laboratory exposure to ebola virus by using a patient biocontainment care unit. *Emerg Infect Dis* **14:**881–887.
309. **Kuhar DT, Henderson DK, Struble KA, Heneine W, Thomas V, Cheever LW, Gomaa A, Panlilio AL, US Public Health Service Working Group.** 2013. Updated US Public Health Service guidelines for the management of occupational exposures to human immunodeficiency virus and recommendations for postexposure prophylaxis. *Infect Control Hosp Epidemiol* **34:**875–892. [corrected in Infect Control Hosp Epidemiol 2013;Nov;34:1238 (Note: Dosage error in article text.)]
310. **U.S. Public Health Service.** 2001. Updated U.S. Public Health Service Guidelines for the Management of Occupational Exposures to HBV, HCV, and HIV and Recommendations for Postexposure Prophylaxis. *MMWR Recomm Rep* **50**(RR-11)**:**1–52.
311. **Gannon CK.** 2003. Anatomy of an exposure: a hospital lab's recovery of *Brucella melitensis*. *MLO Med Lab Obs* **35:**22–25.
312. **Centers for Disease Control and Prevention (CDC).** 2004. Laboratory exposure to *Burkholderia pseudomallei*—Los Angeles, California, 2003. *MMWR Morb Mortal Wkly Rep* **53:**988–990.
313. **Cahn A, Koslowsky B, Nir-Paz R, Temper V, Hiller N, Karlinsky A, Gur I, Hidalgo-Grass C, Heyman SN, Moses AE, Block C.** 2009. Imported melioidosis, Israel, 2008. *Emerg Infect Dis* **15:** 1809–1811.
314. **Maley MW, Kociuba K, Chan RC.** 2006. Prevention of laboratory-acquired brucellosis: significant side effects of prophylaxis. *Clin Infect Dis* **42:**433–434.
315. **Rusnak JM, Kortepeter MG, Hawley RJ, Boudreau E, Aldis J, Pittman PR.** 2004. Management guidelines for laboratory exposures to agents of bioterrorism. *J Occup Environ Med* **46:** 791–800.

316. **Benoit TJ, Blaney DD, Gee JE, Elrod MG, Hoffmaster AR, Doker TJ, Bower WA, Walke HT, Centers for Disease Control and Prevention (CDC).** 2015. Melioidosis cases and selected reports of occupational exposures to *Burkholderia pseudomallei*—United States, 2008–2013. *MMWR Surveill Summ* **64:**1–9.
317. **Goldman R (ed).** 1995. *Medical Surveillance Program.* Marcel Dekker, Inc, New York..
318. **Centers for Disease Control and Prevention (CDC).** 2000. Laboratory-acquired human glanders—Maryland, May 2000. *MMWR Morb Mortal Wkly Rep* **49:**532–535.
319. **Weinberg ED.** 1999. Iron loading and disease surveillance. *Emerg Infect Dis* **5:**346–352.
320. **Bolyard EA, Tablan OC, Williams WW, Pearson ML, Shapiro CN, Deitchmann SD, Hospital Infection Control Practices Advisory Committee.** 1998. Guideline for infection control in healthcare personnel, 1998. *Infect Control Hosp Epidemiol* **19:**407–463.
321. **Agrup G, Belin L, Sjöstedt L, Skerfving S.** 1986. Allergy to laboratory animals in laboratory technicians and animal keepers. *Br J Ind Med* **43:**192–198.
322. **Committee AIHAB.** 1995. Biogenic allergens, p 44–48, *Biosafety Reference Manual,* 2nd ed. American Industrial Hygiene Association, Fairfax, VA.
323. **Phillips GB.** 1965. *Causal Factors in Microbiology Laboratory Accidents and Infections.* National Technical Information Service, Fort Detrick, MD.
324. **Martin JC.** 1980. Behavior factors in laboratory safety: personnel characteristics and modification of unsafe acts, p 321–342. *In* Fuscaldo AA, Erlick BJ, Hindman B (ed), *Laboratory Safety: Theory and Practice.* Academic Press, New York.
325. **Kaiser J.** 2011. Updated: University of Chicago Microbiologist Infected from Possible Lab Accident. *Science,* AAAS online. http://www.sciencemag.org/news/2011/09/updated-university-chicago-microbiologist-infected-possible-lab-accident.
326. **Mansheim BJ, Kasper DL.** 1979. Detection of anticapsular antibodies to *Bacteroides asaccharolyticus* in serum from rabbits and humans by use of an enzyme-linked immunosorbent assay. *J Infect Dis* **140:**945–951.
327. **Burstyn DG, Baraff LJ, Peppler MS, Leake RD, St Geme J Jr, Manclark CR.** 1983. Serological response to filamentous hemagglutinin and lymphocytosis-promoting toxin of *Bordetella pertussis. Infect Immun* **41:**1150–1156.
328. **Al Dahouk S, Nöckler K, Hensel A, Tomaso H, Scholz HC, Hagen RM, Neubauer H.** 2005. Human brucellosis in a nonendemic country: a report from Germany, 2002 and 2003. *Eur J Clin Microbiol Infect Dis* **24:**450–456.
329. **Breton I, Burucoa C, Grignon B, Fauchere JL, Becq Giraudon B.** 1995. Brucellose acquise au laboratoire. [Laboratory-acquired brucellosis.] *Med Mal Infect* **25:**549–551.
330. **Elidan J, Michel J, Gay I, Springer H.** 1985. Ear involvement in human brucellosis. *J Laryngol Otol* **99:**289–291.
331. **Fabiansen C, Knudsen JD, Lebech AM.** 2008. [Laboratory-acquired brucellosis]. *Ugeskr Laeger* **170:**2161.
332. **Grist NR, Emslie JA.** 1991. Infections in British clinical laboratories, 1988-1989. *J Clin Pathol* **44:**667–669.
333. **Marianelli C, Petrucca A, Pasquali P, Ciuchini F, Papadopoulou S, Cipriani P.** 2008. Use of MLVA-16 typing to trace the source of a laboratory-acquired Brucella infection. *J Hosp Infect* **68:**274–276.
334. **Martin-Mazuelos E, Nogales MC, Florez C, Gómez-Mateos JM, Lozano F, Sanchez A.** 1994. Outbreak of *Brucella melitensis* among microbiology laboratory workers. *J Clin Microbiol* **32:** 2035–2036.
335. **Podolak E.** 2010. Researcher suspended for authorized experiments. *Biotechniques.* http://www.biotechniques.com/news/Researcher-suspended-for-unauthorized-experiments/biotechniques-296880.html
336. **Rees RK, Graves M, Caton N, Ely JM, Probert WS.** 2009. Single tube identification and strain typing of *Brucella melitensis* by multiplex PCR. *J Microbiol Methods* **78:**66–70.
337. **Rodrigues AL, Silva SK, Pinto BL, Silva JB, Tupinambás U.** 2013. Outbreak of laboratory-acquired *Brucella abortus* in Brazil: a case report. *Rev Soc Bras Med Trop* **46:**791–794.
338. **Sayin-Kutlu S, Kutlu M, Ergonul O, Akalin S, Guven T, Demiroglu YZ, Acicbe O, Akova M, Occupational Infectious Diseases Study Group.** 2012. Laboratory-acquired brucellosis in Turkey. *J Hosp Infect* **80:**326–330.
339. **Smith JA, Skidmore AG, Andersen RG.** 1980. Brucellosis in a laboratory technologist. *Can Med Assoc J* **122:**1231–1232.
340. **Wansbrough L.** 2010. Brucella in Hospital Laboratory Workers: Epi Update, a publication of the Bureau of Immunology, June 2010. Florida Department of Health. http://floridahealth.gov/diseases-and-conditions/disease-reporting-and management/florida-epidemic-intelligence-service/_documents/2009-2011/documents/june2010epiupdate.pdf.
341. **Wünschel M, Olszowski AM, Weissgerber P, Wülker N, Kluba T.** 2011. [Chronic brucellosis: a rare cause of septic loosening of arthroplasties with high risk of laboratory-acquired infections]. *Z Orthop Unfall* **149:**33–36.
342. **Ashdown LR.** 1992. Melioidosis and safety in the clinical laboratory. *J Hosp Infect* **21:**301–306.
343. **Oates JD, Hodgin UG Jr.** 1981. Laboratory-acquired *Campylobacter enteritis. South Med J* **74:**83.
344. **Penner JL, Hennessy JN, Mills SD, Bradbury WC.** 1983. Application of serotyping and chromosomal restriction endonuclease digest analysis in investigating a laboratory-acquired case of *Campylobacter jejuni* enteritis. *J Clin Microbiol* **18:**1427–1428.
345. **Hyman CL, Augenbraun MH, Roblin PM, Schachter J, Hammerschlag MR.** 1991. Asymptomatic respiratory tract infection with *Chlamydia pneumoniae TWAR. J Clin Microbiol* **29:**2082–2083.
346. **Surcel HM, Syrjälä H, Leinonen M, Saikku P, Herva E.** 1993. Cell-mediated immunity to *Chlamydia pneumoniae* measured as lymphocyte blast transformation in vitro. *Infect Immun* **61:** 2196–2199.
347. **Tuuminen T, Salo K, Surcel HM.** 2002. A casuistic immunologic response in primary and repeated *Chlamydophila pneumoniae* infections in an immunocompetent individual. *J Infect* **45:**202–206.
348. **Egawa T, Hara H, Kawase I, Masuno T, Asari S, Sakurai M, Kishimoto S.** 1990. Human pulmonary infection with *Corynebacterium equi. Eur Respir J* **3:**240–242.
349. **Johanson RE.** 2004. *Enterobacter aerogenes* needlestick leads to improved biological management system. *Appl Biosaf* **9:** 65–67.
350. **Booth L, Rowe B.** 1993. Possible occupational acquisition of *Escherichia coli O157* infection. *Lancet* **342:**1298–1299.
351. **Campbell MJ.** 2015. Characterizing accidents, exposures, and laboratory-acquired infections reprted to the National Institutes of Health Office of Biotechnology Activities (NIH/OBA) Division Under the NIH Guidelines for Work with Recombinant DNA materials from 1976–2010. *Appl Biosaf* **20:**12–26.
352. **Ostroff SM, Kobayashi JM, Lewis JH.** 1989. Infections with *Escherichia coli O157:H7* in Washington State. The first year of statewide disease surveillance. *JAMA* **262:**355–359.
353. **Parry SH, Abraham SN, Feavers IM, Lee M, Jones MR, Bint AJ, Sussman M.** 1981. Urinary tract infection due to laboratory-acquired *Escherichia coli:* relation to virulence. *Br Med J (Clin Res Ed)* **282:**949–950.
354. **Public Health Service Laboratory.** 1996. Escherichia coli O 157 infection acquired in the laboratory. *Commun Dis Rep CDR Wkly* **6:**239.
355. **Gilbert GL.** 2015. Laboratory testing in management of patients with suspected Ebolavirus disease: infection control and safety. *Pathology* **47:**400–402.

356. **Rao GG, Saunders BP, Masterton RG.** 1996. Laboratory acquired verotoxin producing *Escherichia coli (VTEC)* infection. *J Hosp Infect* **33:**228–230.
357. **Donnelly TM, Behr M.** 2000. Laboratory-acquired lymphadenopathy in a veterinary pathologist. *Lab Anim (NY)* **29:**23–25.
358. **Hornick R.** 2001. Tularemia revisited. *N Engl J Med* **345:**1637–1639.
359. **Janovská S, Pávková I, Reichelová M, Hubálek M, Stulík J, Macela A.** 2007. Proteomic analysis of antibody response in a case of laboratory-acquired infection with *Francisella tularensis subsp. tularensis. Folia Microbiol (Praha)* **52:**194–198.
360. **Lam ST, Sammons-Jackson W, Sherwood J, Ressner R.** 2012. Laboratory-acquired tularemia successfully treated with ciprofloxacin. *Infect Dis Clin Pract* **20:**204–207.
361. **Mailles A, V. Vaillant, V. Bilan.** 2013. Bilan de 10 annees de surbeillance de la tularemie chez l'homme en France. Institue de Veile Sanitaire, Legal depot September 2013.
362. **Trees DL, Arko RJ, Hill GD, Morse SA.** 1992. Laboratory-acquired infection with *Haemophilus ducreyi* type strain CIP 542. *Med Microl Lett* **1:**330–337.
363. **Matysiak-Budnik T, Briet F, Heyman M, Mégraud F.** 1995. Laboratory-acquired *Helicobacter pylori* infection. *Lancet* **346:** 1489–1490.
364. **Raymond J, Bingen E, Brahimi N, Bergeret M, Kalach N.** 1996. Randomly amplified polymorphic DNA analysis in suspected laboratory *Helicobacter pylori* infection. *Lancet* **347:**975.
365. **Takata T, Shirotani T, Okada M, Kanda M, Fujimoto S, Ono J.** 1998. Acute hemorrhagic gastropathy with multiple shallow ulcers and duodenitis caused by a laboratory infection of *Helicobacter pylori. Gastrointest Endosc* **47:**291–294.
366. **Broughton ES, Flack LE.** 1986. The susceptibility of a strain of *Leptospira interrogans* serogroup icterohaemorrhagiae to amoxycillin, erythromycin, lincomycin, tetracycline, oxytetracycline and minocycline. *Zentralbl Bakteriol Mikrobiol Hyg [A]* **261:**425–431.
367. **Gilks CF, Lambert HP, Broughton ES, Baker CC.** 1988. Failure of penicillin prophylaxis in laboratory acquired leptospirosis. *Postgrad Med J* **64:**236–238.
368. **Sugunan AP, Natarajaseenivasan K, Vijayachari P, Sehgal SC.** 2004. Percutaneous exposure resulting in laboratory-acquired leptospirosis—a case report. *J Med Microbiol* **53:**1259–1262.
369. **Cooke MM, Gear AJ, Naidoo A, Collins DM.** 2002. Accidental *Mycobacterium bovis* infection in a veterinarian. *N Z Vet J* **50:** 36–38.
370. **Brutus JP, Lamraski G, Zirak C, Hauzeur JP, Thys JP, Schuind F.** 2005. Septic monoarthritis of the first carpo-metacarpal joint caused by *Mycobacterium kansasii. Chir Main* **24:**52–54.
371. **Weber T, Tumani H, Holdorff B, Collinge J, Palmer M, Kretzschmar HA, Felgenhauer K.** 1993. Transmission of Creutzfeldt-Jakob disease by handling of dura mater. *Lancet* **341:** 123–124.
372. **Washington State Department of Labor and Industries.** 2004. Region 2- Seattle Office. Inspection report on laboratory associated infections due to *Mycobacterium tuberculosis*. Inspection 307855056. http://www.sunshine-project.org/idriuwmadchamber .pdf.
373. **Duray PH, Flannery B, Brown S.** 1981. Tuberculosis infection from preparation of frozen sections. *N Engl J Med* **305:**167.
374. **Leyten EM, Mulder B, Prins C, Weldingh K, Andersen P, Ottenhoff TH, van Dissel JT, Arend SM.** 2006. Use of enzyme-linked immunospot assay with *Mycobacterium tuberculosis*-specific peptides for diagnosis of recent infection with *M. tuberculosis* after accidental laboratory exposure. *J Clin Microbiol* **44:**1197–1201.
375. **Mazurek GH, Cave MD, Eisenach KD, Wallace RJ Jr, Bates JH, Crawford JT.** 1991. Chromosomal DNA fingerprint patterns produced with IS6110 as strain-specific markers for epidemiologic study of tuberculosis. *J Clin Microbiol* **29:**2030–2033.
376. **Sharma VK, Kumar B, Radotra BD, Kaur S.** 1990. Cutaneous inoculation tuberculosis in laboratory personnel. *Int J Dermatol* **29:**293–294.
377. **Sugita M, Tsutsumi Y, Suchi M, Kasuga H.** 1989. High incidence of pulmonary tuberculosis in pathologists at Tokai University Hospital: an epidemiological study. *Tokai J Exp Clin Med* **14:**55–59.
378. **Alonso-Echanove J, Granich RM, Laszlo A, Chu G, Borja N, Blas R, Olortegui A, Binkin NJ, Jarvis WR.** 2001. Occupational transmission of *Mycobacterium tuberculosis* to health care workers in a university hospital in Lima, Peru. *Clin Infect Dis* **33:**589–596.
379. **Bruins SC, Tight RR.** 1979. Laboratory-acquired gonococcal conjunctivitis. *JAMA* **241:**274.
380. **Centers for Disease Control (CDC).** 1981. Gonococcal eye infections in adults—California, Texas, Germany. *MMWR Morb Mortal Wkly Rep* **30:**341–343.
381. **Malhotra R, Karim QN, Acheson JF.** 1998. Hospital-acquired adult gonococcal conjunctivitis. *J Infect* **37:**305.
382. **Podgore JK, Holmes KK.** 1981. Ocular gonococcal infection with minimal or no inflammatory response. *JAMA* **246:**242–243.
383. **Zajdowicz TR, Kerbs SB, Berg SW, Harrison WO.** 1984. Laboratory-acquired gonococcal conjunctivitis: successful treatment with single-dose ceftriaxone. *Sex Transm Dis* **11:**28–29.
384. **Christen G, Tagan D.** 2004. Infection à Neisseria meningitidis acquise en laboratoire. [Laboratory-acquired Neisseria meningitidis infection]. *Med Mal Infect* **34:**137–138.
385. **Guibourdenche M, Darchis JP, Boisivon A, Collatz E, Riou JY.** 1994. Enzyme electrophoresis, sero- and subtyping, and outer membrane protein characterization of two *Neisseria meningitidis* strains involved in laboratory-acquired infections. *J Clin Microbiol* **32:**701–704.
386. **Petty BG, Sowa DT, Charache P.** 1983. Polymicrobial polyarticular septic arthritis. *JAMA* **249:**2069–2072.
387. **Public Health Service Laboratory.** 1992. Laboratory-acquired meningococcal infection. *Commun Dis Rep CDR Wkly* **2:**39.
388. **Woods JP, Cannon JG.** 1990. Variation in expression of class 1 and class 5 outer membrane proteins during nasopharyngeal carriage of *Neisseria meningitidis. Infect Immun* **58:**569–572.
389. **Ashdown L, Cassidy J.** 1991. Successive *Salmonella Give* and *Salmonella Typhi* infections, laboratory-acquired. *Pathology* **23:** 233–234.
390. **Barker A, Duster M, Van Hoof S, Safdar N.** 2015. Nontyphoidal *Salmonella*: an occupational hazard for clinical laboratory workers. *Appl Biosaf* **20:**72–74.
391. **Holmes MB, Johnson DL, Fiumara NJ, McCormack WM.** 1980. Acquisition of typhoid fever from proficiency-testing specimens. *N Engl J Med* **303:**519–521.
392. **Koay AS, Jegathesan M, Rohani MY, Cheong YM.** 1997. Pulsed-field gel electrophoresis as an epidemiologic tool in the investigation of laboratory acquired *Salmonella typhi* infection. *Southeast Asian J Trop Med Public Health* **28:**82–84.
393. **Lester A, Mygind O, Jensen KT, Jarløv JO, Schønheyder HC.** 1994. [Typhoid and paratyphoid fever in Denmark 1986–1990. Epidemiologic aspects and the extent of bacteriological follow-up of patients]. *Ugeskr Laeger* **156:**3770–3775.
394. **Mermin JH, Townes JM, Gerber M, Dolan N, Mintz ED, Tauxe RV.** 1998. Typhoid fever in the United States, 1985–1994: changing risks of international travel and increasing antimicrobial resistance. *Arch Intern Med* **158:**633–638.
395. **Thong K-L, Cheong Y-M, Pang T.** 1996. A probable case of laboratory-acquired infection with salmonella typhi: evidence from phage typing, antibiograms, and analysis by pulsed-field gel electrophoresis. *Int J Infect Dis* **1:**95–97.
396. **Dadswell JV.** 1983. Laboratory acquired shigellosis. *Br Med J (Clin Res Ed)* **286:**58.
397. **Aleksić S, Bockemühl J, Degner I.** 1981. Imported shigellosis: aerogenic *Shigella boydii 74* (Sachs A 12) in a traveller followed

by two cases of laboratory-associated infections. *Tropenmed Parasitol* **32:**61–64.

398. **Kolavic SA, Kimura A, Simons SL, Slutsker L, Barth S, Haley CE.** 1997. An outbreak of *Shigella dysenteriae* type 2 among laboratory workers due to intentional food contamination. *JAMA* **278:**396–398.
399. **Van Bohemen CG, Nabbe AJ, Zanen HC.** 1985. IgA response during accidental infection with *Shigella flexneri*. *Lancet* **326:**673.
400. **Gosbell IB, Mercer JL, Neville SA.** 2003. Laboratory-acquired *EMRSA-15* infection. *J Hosp Infect* **54:**324–325.
401. **Wagenvoort JH, De Brauwer EI, Gronenschild JM, Toenbreker HM, Bonnemayers GP, Bilkert-Mooiman MA.** 2006. Laboratory-acquired meticillin-resistant *Staphylococcus aureus* (MRSA) in two microbiology laboratory technicians. *Eur J Clin Microbiol Infect Dis* **25:**470–472.
402. **Anderson LC, Leary SL, Manning PJ.** 1983. Rat-bite fever in animal research laboratory personnel. *Lab Anim Sci* **33:**292–294.
403. **Hawkey PM, Pedler SJ, Southall PJ.** 1980. *Streptococcus pyogenes:* a forgotten occupational hazard in the mortuary. *BMJ* **281:**1058.
404. **Kurl DN.** 1981. Laboratory-acquired human infection with group A type 50 streptococci. *Lancet* **318:**752.
405. **Little JS, O'Reilly MJ, Higbee JW, Camp RA.** 1984. Suppurative flexor tenosynovitis after accidental self-inoculation with *Streptococcus pneumoniae type I*. *JAMA* **252:**3003–3004.
406. **Raviglione MC, Tierno PM, Ottuso P, Klemes AB, Davidson M.** 1990. Group G streptococcal meningitis and sepsis in a patient with AIDS. A method to biotype group G streptococcus. *Diagn Microbiol Infect Dis* **13:**261–264.
407. **Anonymous.** 1991. Sveskt kolerafall trolig smitta i laboratorium. [A Swedish case of cholera of probable laboratory origin.] *Lakartidningen* **89:**3668.
408. **Lee KK, Liu PC, Huang CY.** 2003. *Vibrio parahaemolyticus* infectious for both humans and edible mollusk abalone. *Microbes Infect* **5:**481–485.
409. **Graham T.** 2013. I am the laboratory-acquired infection. Abstr., Biological Safety Conference, American Biological Safety Association, Mundelein, IL.
410. **Graham CJ, Yamauchi T, Rountree P.** 1989. Q fever in animal laboratory workers: an outbreak and its investigation. *Am J Infect Control* **17:**345–348.
411. **Hall CJ, Richmond SJ, Caul EO, Pearce NH, Silver IA.** 1982. Laboratory outbreak of Q fever acquired from sheep. *Lancet* **319:**1004–1006.
412. **Hamadeh GN, Turner BW, Trible W Jr, Hoffmann BJ, Anderson RM.** 1992. Laboratory outbreak of Q fever. *J Fam Pract* **35:**683–685.
413. **Henning K, Hotzel H, Peters M, Welge P, Popps W, Theegarten D.** 2009. [Unanticipated outbreak of Q fever during a study using sheep, and its significance for further projects]. *Berl Munch Tierarztl Wochenschr* **122:**13–19.
414. **Meiklejohn G, Reimer LG, Graves PS, Helmick C.** 1981. Cryptic epidemic of Q fever in a medical school. *J Infect Dis* **144:**107–113.
415. **Simor AE, Brunton JL, Salit IE, Vellend H, Ford-Jones L, Spence LP.** 1984. Q fever: hazard from sheep used in research. *Can Med Assoc J* **130:**1013–1016.
416. **Whitney EA, Massung RF, Kersh GJ, Fitzpatrick KA, Mook DM, Taylor DK, Huerkamp MJ, Vakili JC, Sullivan PJ, Berkelman RL.** 2013. Survey of laboratory animal technicians in the United States for *Coxiella burnetii* antibodies and exploration of risk factors for exposure. *J Am Assoc Lab Anim Sci* **52:**725–731.
417. **Herrero JI, Ruiz R, Walker DH.** 1993. La técnica de western immunoblotting en situaciones atípicas de infección por *Rickettsia conorii*. Presentación de 2 casos. [The western immunoblotting technique in atypical situations of *Rickettsia conorii* infection. Presentation of 2 cases.] *Enferm Infecc Microbiol Clin* **11:**139–142.
418. **The Subcommittee on Arbovirus Laboratory Safety of the American Committee on Arthropod-Borne Viruses.** 1980. Laboratory safety for arboviruses and certain other viruses of vertebrates. *Am J Trop Med Hyg* **29:**1359–1381.
419. **Schnurrenberger PR, Swango LJ, Bowman GM, Luttgen PJ.** 1980. Bovine papular stomatitis incidence in veterinary students. *Can J Comp Med* **44:**239–243.
420. **Langford MP, Anders EA, Burch MA.** 2015. Acute hemorrhagic conjunctivitis: anti-coxsackievirus A24 variant secretory immunoglobulin A in acute and convalescent tear. *Clin Ophthalmol* **9:**1665–1673.
421. **Sitwell L, Lach B, Atack E, Atack D, Izukawa D.** 1988. Creutzfeldt-Jakob disease in histopathology technicians. *N Engl J Med* **318:**853–854.
422. **Miller DC.** 1988. Creutzfeldt-Jakob disease in histopathology technicians. *N Engl J Med* **318:**853–854.
423. **Chen LH, Wilson ME.** 2004. Transmission of dengue virus without a mosquito vector: nosocomial mucocutaneous transmission and other routes of transmission. *Clin Infect Dis* **39:**e56–e60.
424. **Okuno Y, Fukunaga T, Tadano M, Fukai K.** 1982. Serological studies on a case of laboratory dengue infection. *Biken J* **25:**163–170.
425. **Le Guenno B.** 1995. Emerging viruses. *Sci Am* **273:**56–64.
426. **Mertens T, Hager H, Eggers HJ.** 1982. Epidemiology of an outbreak in a maternity unit of infections with an antigenic variant of Echovirus 11. *J Med Virol* **9:**81–91.
427. **Hager H, Mertens T, Eggers HJ.** 1980. [An epidemic in a maternity unit caused by an echo virus 11 variant (author's transl)]. *MMW Munch Med Wochenschr* **122:**619–622.
428. **Spalton DJ, Palmer S, Logan LC.** 1980. Echo 11 conjunctivitis. *Br J Ophthalmol* **64:**487–488.
429. **Sudeep AB, Jadi RS, Mishra AC.** 2009. Ganjam virus. *Indian J Med Res* **130:**514–519.
430. **Rao CV, Dandawate CN, Rodrigues JJ, Rao GL, Mandke VB, Ghalsasi GR, Pinto BD.** 1981. Laboratory infections with Ganjam virus. *Indian J Med Res* **74:**319–324.
431. **Centers for Disease Control and Prevention.** 1994. Laboratory management of agents associated with hantavirus pulmonary syndrome: interim biosafety guidelines. *MMWR Recomm Rep* **43**(RR-7)**:**1–7.
432. **Lee HW, Johnson KM.** 1982. Laboratory-acquired infections with Hantaan virus, the etiologic agent of Korean hemorrhagic fever. *J Infect Dis* **146:**645–651.
433. **Weissenbacher MC, Merani MS, Hodara VL, de Villafañe G, Gajdusek DC, Chu YK, Lee HW.** 1990. Hantavirus infection in laboratory and wild rodents in Argentina. *Medicina (B Aires)* **50:**43–46.
434. **Anderson RA, Woodfield DG.** 1982. Hepatitis B virus infections in laboratory staff. *N Z Med J* **95:**69–71.
435. **Sampliner R, Bozzo PD, Murphy BL.** 1984. Frequency of antibody to hepatitis B in a community hospital laboratory. *Lab Med* **15:**256–257.
436. **Takahashi K, Miyakawa Y, Gotanda T, Mishiro S, Imai M, Mayumi M.** 1979. Shift from free "small" hepatitis B e antigen to IgG-bound "large" form in the circulation of human beings and a chimpanzee acutely infected with hepatitis B virus. *Gastroenterology* **77:**1193–1199.
437. **Weissenbacher MC, Edelmuth E, Frigerio MJ, Coto CE, de Guerrero LB.** 1980. Serological survey to detect subclinical Junín virus infection in laboratory personnel. *J Med Virol* **6:**223–226.
438. **Johnson K, Winters T.** 2012. Herpes B virus: Implications in lab workers, travelers, and pet owners. Harvard School of Public Health Grand Rounds. http://www.hsph.harvard.edu/oemr/files/2012/08/Grand-Rounds-Herpes-B-Final-Johnson-Winters.pdf.
439. **Beer B, Kurth R, Bukreyev A.** 1999. Characteristics of Filoviridae: marburg and Ebola viruses. *Naturwissenschaften* **86:**8–17.

440. **Nikiforov VV, Turovskiĭ Iul, Kalinin PP, Akinfeeva LA, Katkova LR, Barmin VS, Riabchikova EI, Popkova NI, Shestopalov AM, Nazarov VP, et al.** 1994. [A case of a laboratory infection with Marburg fever]. *Zh Mikrobiol Epidemiol Immunobiol* **May–June:**104–106.
441. **Pedrosa PB, Cardoso TA.** 2011. Viral infections in workers in hospital and research laboratory settings: a comparative review of infection modes and respective biosafety aspects. *Int J Infect Dis* **15:**e366–e376.
442. **Morgan C.** 1987. Import of animal viruses opposed after accident at laboratory. *Nature* **328:**8.
443. **Erdman DD, Gary GW, Anderson LJ.** 1989. Serum immunoglobulin A response to Norwalk virus infection. *J Clin Microbiol* **27:**1417–1418.
444. **Cohen BJ, Brown KE.** 1992. Laboratory infection with human parvovirus B19. *J Infect* **24:**113–114.
445. **Shiraishi H, Sasaki T, Nakamura M, Yaegashi N, Sugamura K.** 1991. Laboratory infection with human parvovirus B19. *J Infect* **22:**308–310.
446. **Coimbra TL, Nassar ES, de Souza LT, Ferreira IB, Rocco IM, Burattini MN, da Rosa AT, Vasconcelos PF, Pinheiro FP, LeDuc JW, Rico-Hesse R.** 1994. New arenavirus isolated in Brazil. *Lancet* **343:**391–392.
447. **Willems WR, Kaluza G, Boschek CB, Bauer H, Hager H, Schütz HJ, Feistner H.** 1979. Semliki forest virus: cause of a fatal case of human encephalitis. *Science* **203:**1127–1129.
448. **Heneine W, Switzer WM, Sandstrom P, Brown J, Vedapuri S, Schable CA, Khan AS, Lerche NW, Schweizer M, Neumann-Haefelin D, Chapman LE, Folks TM.** 1998. Identification of a human population infected with simian foamy viruses. *Nat Med* **4:**403–407.
449. **Schweizer M, Turek R, Hahn H, Schliephake A, Netzer KO, Eder G, Reinhardt M, Rethwilm A, Neumann-Haefelin D.** 1995. Markers of foamy virus infections in monkeys, apes, and accidentally infected humans: appropriate testing fails to confirm suspected foamy virus prevalence in humans. *AIDS Res Hum Retroviruses* **11:**161–170.
450. **von Laer D, Neumann-Haefelin D, Heeney JL, Schweizer M.** 1996. Lymphocytes are the major reservoir for foamy viruses in peripheral blood. *Virology* **221:**240–244.
451. **Centers for Disease Control (CDC).** 1992. Anonymous survey for simian immunodeficiency virus (SIV) seropositivity in SIV-laboratory researchers—United States, 1992. *MMWR Morb Mortal Wkly Rep* **41:**814–815.
452. **Vasconcelos PF, Travassos da Rosa AP, Rodrigues SG, Tesh R, Travassos da Rosa JF, Travassos da Rosa ES.** 1993. Infecção humana adquirida em laboratório causada pelo virus SP H 114202 (Arenavirus: família Arenaviridae): aspectos clínicos e laboratoriais. [Laboratory-acquired human infection with SP H 114202 virus (Arenavirus: Arenaviridae family): clinical and laboratory aspects]. *Rev Inst Med Trop Sao Paulo* **35:**521–525.
453. **Flanagan ML, Oldenburg J, Reignier T, Holt N, Hamilton GA, Martin VK, Cannon PM.** 2008. New world clade B arenaviruses can use transferrin receptor 1 (TfR1)-dependent and -independent entry pathways, and glycoproteins from human pathogenic strains are associated with the use of TfR1. *J Virol* **82:**938–948.
454. **Avšič-Županc T, Poljak M, Matičič M, Radšel-Medvešček A, LeDuc JW, Stiasny K, Kunz C, Heinz FX.** 1995. Laboratory acquired tick-borne meningoencephalitis: characterisation of virus strains. *Clin Diagn Virol* **4:**51–59.
455. **Costa GB, Moreno EC, de Souza Trindade G, Studies Group in Bovine Vaccinia.** 2013. Neutralizing antibodies associated with exposure factors to Orthopoxvirus in laboratory workers. *Vaccine* **31:**4706–4709.
456. **Fillis CA, Calisher CH.** 1979. Neutralizing antibody responses of humans and mice to vaccination with Venezuelan encephalitis (TC-83) virus. *J Clin Microbiol* **10:**544–549.
457. **Reif JS, Webb PA, Monath TP, Emerson JK, Poland JD, Kemp GE, Cholas G.** 1987. Epizootic vesicular stomatitis in Colorado, 1982: infection in occupational risk groups. *Am J Trop Med Hyg* **36:**177–182.
458. **New York State Department of Health.** 2001. West Nile Virus Update- January 1, 2001–December 31, 2001.
459. **Anderson BC, Donndelinger T, Wilkins RM, Smith J.** 1982. Cryptosporidiosis in a veterinary student. *J Am Vet Med Assoc* **180:**408–409.
460. **Centers for Disease Control (CDC).** 1982. Human cryptosporidiosis—Alabama. *MMWR Morb Mortal Wkly Rep* **31:** 252–254.
461. **Current WL, Reese NC, Ernst JV, Bailey WS, Heyman MB, Weinstein WM.** 1983. Human cryptosporidiosis in immunocompetent and immunodeficient persons. Studies of an outbreak and experimental transmission. *N Engl J Med* **308:**1252–1257.
462. **Gait R, Soutar RH, Hanson M, Fraser C, Chalmers R.** 2008. Outbreak of cryptosporidiosis among veterinary students. *Vet Rec* **162:**843–845.
463. **Levine JF, Levy MG, Walker RL, Crittenden S.** 1988. Cryptosporidiosis in veterinary students. *J Am Vet Med Assoc* **193:** 1413–1414.
464. **Preiser G, Preiser L, Madeo L.** 2003. An outbreak of cryptosporidiosis among veterinary science students who work with calves. *J Am Coll Health* **51:**213–215.
465. **Reese NC, Current WL, Ernst JV, Bailey WS.** 1982. Cryptosporidiosis of man and calf: a case report and results of experimental infections in mice and rats. *Am J Trop Med Hyg* **31:** 226–229.
466. **Dillon NL, Stolf HO, Yoshida EL, Marques MEA.** 1993. Leishmaniose cutânea acidental. [Accidental cutaneous leishmaniasis]. *Rev Inst Med Trop Sao Paulo* **35:**385–387.
467. **Herwaldt BL.** 2006. Protozoa and helminths. *In* Fleming DO, Hunt DL (ed), *Biological Safety: Principles and Practices*, 4th ed. ASM Press, Washington, DC.
468. **Druilhe P, Trape JF, Leroy JP, Godard C, Gentilini M.** 1980. [Two accidental human infections by Plasmodium cynomolgi bastianellii. A clinical and serological study]. *Ann Soc Belg Med Trop* **60:**349–354.
469. **Bending MR, Maurice PD.** 1980. Malaria: a laboratory risk. *Postgrad Med J* **56:**344–345.
470. **Grist NR.** 1981. Hepatitis and other infections in clinical laboratory staff, 1979. *J Clin Pathol* **34:**655–658.
471. **Hajeer AH, Balfour AH, Mostratos A, Crosse B.** 1994. *Toxoplasma gondii:* detection of antibodies in human saliva and serum. *Parasite Immunol* **16:**43–50.
472. **Hermentin K, Picher O, Aspöck H, Auer H, Hassl A.** 1983. A solid-phase indirect haemadsorption assay (SPIHA) for detection of immunoglobulin M antibodies to *Toxoplasma gondii:* application to diagnosis of acute acquired toxoplasmosis. *Zentralbl Bakteriol Mikrobiol Hyg [A]* **255:**380–391.
473. **Payne RA, Joynson DH, Balfour AH, Harford JP, Fleck DG, Mythen M, Saunders RJ.** 1987. Public Health Laboratory Service enzyme linked immunosorbent assay for detecting Toxoplasma specific IgM antibody. *J Clin Pathol* **40:**276–281.
474. **Peters SE, Gourlay Y, Seaton A.** 2002. *Listeria meningitis* as a complication of chemoprophylaxis against laboratory acquired toxoplasma infection: a case report. *J Infect* **44:**126.
475. **Woodison G, Balfour AH, Smith JE.** 1993. Sequential reactivity of serum against cyst antigens in *Toxoplasma* infection. *J Clin Pathol* **46:**548–550.
476. **Añez N, Carrasco H, Parada H, Crisante G, Rojas A, Gonzalez N, Ramirez JL, Guevara P, Rivero C, Borges R, Scorza JV.** 1999. Acute Chagas' disease in western Venezuela: a clinical, seroparasitologic, and epidemiologic study. *Am J Trop Med Hyg* **60:**215–222.
477. **Emeribe AO.** 1988. *Gambiense trypanosomiasis* acquired from needle scratch. *Lancet* **331:**470–471.

478. **Herbert WJ, Parratt D, Van Meirvenne N, Lennox B.** 1980. An accidental laboratory infection with trypanosomes of a defined stock. II. Studies on the serological response of the patient and the identity of the infecting organism. *J Infect* **2:**113–124.
479. **Receveur MC, LeBras M, Vincendeau P.** 1993. Laboratory-acquired Gambian trypanosomiasis. *N Engl J Med* **329:**209–210.
480. **Buitrago MJ, Gonzalo-Jimenez N, Navarro M, Rodriguez-Tudela JL, Cuenca-Estrella M.** 2011. A case of primary cutaneous histoplasmosis acquired in the laboratory. *Mycoses* **54:** e859–e861.
481. **Loth EA, Dos Santos JH, De Oliveira CS, Uyeda H, Simão RD, Gandra RF.** 2015. Infection caused by the yeast form of *Paracoccidioides brasiliensis. JMM Case Rep* **2.** doi:10.1099/jmmcr.0.000016
482. **Kenny MT, Sabel FL.** 1968. Particle size distribution of Serratia marcescens aerosols created during common laboratory procedures and simulated laboratory accidents. *Appl Microbiol* **16:** 1146–1150.
483. **Blacklow NR, Dolin R, Fedson DS, Dupont H, Northrup RS, Hornock RB, Chanock RM.** 1972. Acute infectious nonbacterial gastroenteritis: etiology and pathogenesis. *Ann Intern Med* **76:**993–1008.
484. **Riley RL.** 1957. Aerial dissemination of pulmonary tuberculosis. *Am Rev Tuberc* **76:**931–941.
485. **Riley RL.** 1961. Airborne pulmonary tuberculosis. *Bacteriol Rev* **25:**243–248.
486. **Kulagin SM, Fedorova NI, Ketiladze ES.** 1962. Laboratory outbreak of hemorrhagic fever with renal syndrome: clinoco-epidemiological characteristics. *Zh Mikrobiol Epidemiol Immunobiol* **33:**121–126.
487. **Gottlieb SJ, Garibaldi E, Hutcheson PS, Slavin RG.** 1993. Occupational asthma to the slime mold Dictyostelium discoideum. J Occup Med 35:1231–1235.
488. **Shireman PK.** 1992. Endometrial tuberculosis acquired by a health care worker in a clinical laboratory. Arch Pathol Lab Med 116:521–523.

SECTION II

Hazard Assessment

5

Risk Assessment of Biological Hazards

DAWN P. WOOLEY AND DIANE O. FLEMING

INTRODUCTION

Biological risk assessment is a challenging process because the variables cannot always be measured quantitatively and subjective judgments must often be made. There is a complex interaction between agents, activities, and people in a constantly changing environment. Work with biohazardous agents, or materials suspected of containing such agents, needs to be assessed for the risk they pose to the individual, the community, and the environment. A biohazardous agent is an infectious agent or other substance produced by a living organism that causes disease in another living organism. Whether the work is performed at a research, clinical, teaching, or large-scale production facility, a risk assessment should be performed to provide the information needed to eliminate risk or reduce it to an acceptable level. The assessment of risk needs to be carried out by knowledgeable people using professional judgment and common sense. By using valid information about the specific agent and taking into account any additional risks posed by the specific procedures and equipment, the evaluator should be able to identify the most appropriate work practices, personal protective equipment, and facilities to protect people and the environment. A risk assessment should be done before work begins and should be repeated when changes are to be made in agents, practices, employees, or facilities. The risk assessment for work with biohazardous agents must take into account not only the agent but also the host and environment. This chapter focuses on agent- and activity-based risk assessments for general work not involving Select Agents. Host factors are addressed briefly, but they are more appropriately covered by occupational medicine.

RISK AND RISK ASSESSMENT

In the context of biosafety, risk may be defined as the probability of a negative health consequence as the result of exposure to a biohazardous agent. Risk assessment refers to the process of identifying hazards, evaluating what might happen, estimating the likelihood of it happening, and considering the consequences of it happening. Risk management is the process of implementing steps to mitigate risk. In the context of biomedical laboratories, risk focuses primarily on preventing laboratory-acquired infections. Whenever possible, quantitative measurements, such as infectious dose, should be used. Qualitative data, such as the phenotype of a genetically modified organism, must also be considered. In any given accident, rarely is there a single cause. Generally, multiple factors combine in such a way as to cause an accident. Therefore, eliminating just one factor may have prevented an infection from occurring, like breaking a link in the chain of infection.

A good strategy to use in assessing and managing risk is the Search-Evaluate-Execute approach, which can be easily remembered by using the acronym SEE (adapted from reference 1). The first two steps are mental processing steps that are part of the assessment process, while the third step is an action step that is part of the management process. *Search* means to identify the biological hazards that exist in the environment in which the proposed work will be conducted. Typical environments include research or clinical laboratories, medical or field settings, vivaria, and production facilities. *Evaluate* means to think about how the various factors could interact to cause an accident. This step requires some level of expertise because a novice worker might not be able to anticipate all of the potential dangers. *Execute* means to implement a safety plan. Make certain that the appropriate equipment, facilities, and personal protective gear are available and that the best practices are employed, taking into account practical considerations. For example, implementing what seemingly appears to be the safest practice may inadvertently cause the work practice to become cumbersome, thus creating a new hazard. Other safety measures that may be executed as part of a safety plan include occupational health screening, preexposure vaccination, and postexposure prophylaxis. Using a strategy such as SEE cannot eliminate all risk, but it could help to reduce the chance for an accident.

THE EVALUATOR

Assessing the risks associated with biohazardous agents, or obtaining such an assessment, is the responsibility of the supervisor or his or her designee, as determined by the institution. An appropriately trained professional (microbiologist, biosafety officer, sanitarian, industrial hygienist, infection control practitioner, veterinarian, etc.) is needed to assess the risk associated with the agent. The American Biological Safety Association (ABSA) International, Mundelein, IL, maintains a list of registered and certified biological safety professionals who may be able to provide advice to those who lack such expertise. Guidance may be obtained online from the Centers for Disease Control and Prevention (CDC) or the National Institutes of Health (NIH) Office of Science Policy (OSP). Specific advice on or interpretations of the *NIH Guidelines for Research Involving Recombinant or Synthetic Nucleic Acid Molecules*, also known as the *NIH Guidelines* (2), may be obtained from staff at NIH/OSP, who may, in turn, seek expertise from its Recombinant DNA Advisory Committee (RAC), which includes a Biosafety Working Group. The risks associated with specialized protocols may be addressed best by the workers most familiar with the equipment and procedures to be used. This is especially true as new technologies create unexpected problems that could lead to unanticipated novel risks.

Risk-based control of biohazardous agents should be made as flexible as possible to allow a knowledgeable user to vary conditions according to the specific virulence factors associated with the biological hazard being used and the hazards associated with the tasks being performed. For example, one might work on the open bench in a biosafety level 1 (BSL1) facility to create a viral vector using an *Escherichia coli* K-12 strain, an attenuated strain of *E. coli* used for recombinant DNA work. However, one might then move to a biological safety cabinet within a BSL2 facility to produce the viral particles in a cell culture system. Some institutions approve protocols at a biosafety level representing the highest risk; in the above example, the protocol would be approved at BSL2. Other institutions may designate different biosafety levels for each part of the work; in the above example, the protocol would specify BSL1 for the molecular cloning work and BSL2 for virus production.

THE BIOHAZARDOUS AGENT

Assessing the risk of work with biohazardous agents is not as straightforward as it is for chemical and physical hazards. Biohazardous microbial agents exist in a variety of environmental niches and can express different virulence factors in dynamic host-parasite interactions. For some agents of human disease, we do not even know the nutritional or host cell requirements that would allow their propagation. Although biohazardous agents do not fit into rigid categories, it is possible to assess the relative risk of an infectious microorganism and to place it into one of four risk groups (RGs). Variants, types, or strains of microbial species need to be assessed for the risk associated with variation from wild type. The RG is the classification of an organism based on known factors. It can be used as a starting point for evidence-based risk assessment.

RISK GROUP CLASSIFICATION

The World Health Organization (WHO) provides basic definitions for the classification of infective microorganisms by RG in their *Laboratory Biosafety Manual* (3).

Their classification is based on factors such as pathogenicity, transmission, and host range, as influenced by existing levels of immunity in the local population and the density and movement of hosts with that population. Other considerations in determining RG classification include the presence of appropriate vectors, standards of environmental hygiene, and local availability of effective preventive measures. These preventive measures could include sanitary precautions (e.g., food and water hygiene), control of animal reservoirs or arthropod vectors, movement of people or animals, importation of potentially infected animals or animal products, and prophylaxis by vaccination or antiserum. Such prophylaxis could include passive immunization and postexposure vaccination, as well as the use of antibiotics, antivirals, and chemotherapeutic agents, taking into consideration the possible emergence of resistant strains (3). Due to the many factors considered in determining RG classifications, lists will vary from country to country and should only be applied in the country of origin. Examples of countries that have established their own RG definitions are the United States, Australia, New Zealand, Belgium, Canada, the European Union, and the United Kingdom (4–11). A summary of these definitions is conveniently located on the ABSA International website, along with a searchable database (12).

The WHO has cautioned that simple reference to the RG is not sufficient for a risk assessment. Other factors need to be considered, including pathogenicity of the specific strain, infectious dose, potential outcome of an exposure, natural route of infection, laboratory route of infection, stability and concentration of the agent, volume to be manipulated, presence of a suitable host (human or animal), information from animal studies, reports on laboratory-acquired infections, types of laboratory activities planned (homogenization, sonication, centrifugation, aerosolization, etc.), types of genetic manipulation that could extend the host range or alter sensitivity to known effective treatment, and local availability of appropriate treatment or prophylaxis (3).

The prevailing conditions in the geographical area in which the microorganisms are handled must also be taken into account. Individual governments may decide to prohibit the handling or importation of certain pathogens except for diagnostic purposes. Competent authorities in Australia, Canada, the European Union, and the United States have placed biohazardous agents into four defined groups reflecting an increasing risk to the user and the environment. The WHO *Laboratory Biosafety Manual* provides general guidance on biosafety, taking into account factors that affect international application, such as difference in risk attributed to certain biological agents and the availability of appropriate laboratory facilities and trained staff to handle such agents (3).

Different countries and organizations around the world have their own definitions for RGs, and lists of microorganisms vary accordingly. A common theme among the definitions is the assessment of individual versus community risk as follows: RG 1 contains agents with low individual and low community risk; RG 2 contains agents with moderate individual risk and low community risk; RG 3 contains agents with high individual risk and low community risk; and RG 4 contains agents with high individual and high community risk.

Some countries consider the risk to animals, plants, and the environment when defining their RGs. The WHO, Australia, and Canada have included not only the risk to humans but also the risk to livestock and the environment due to economic concerns. Australia also added the risk to plants of economic importance. Canada took into consideration the economic impact on the environment, including plants, but did not include plant pathogens in their list of pathogens. Canada does not include a printed list of human and animal pathogens but rather maintains a more dynamic listing available online. The United States and European Union limit their lists of pathogens to those that could potentially cause disease in healthy adult humans and workers.

In the United States, the publication entitled *Biosafety in Microbiological and Biomedical Laboratories* (also known as BMBL) does not actually use RGs (8). Instead, this publication describes how agents are to be handled at each of four BSL containment levels. The protection of animals, plants, and the environment is not mentioned at BSL1 and BSL2, where the focus is on the protection of workers. Community risk is addressed at BSL3, with aerosol-transmitted agents to be handled under primary containment, which, when combined with the secondary containment provided by the facility, is meant to reduce or eliminate release to the surrounding community. BMBL describes four BSLs to provide actual containment criteria for agents known to have caused laboratory-acquired infections, those that could reasonably be expected to do so, and those that would have serious consequences.

The *NIH Guidelines* provide definitions for each of four RGs (2). However, these guidelines, as well as those from Australia, Canada, and the European Union, do not mention aerosols in the definition of an RG 3 agent. This important variation results in the inclusion of human immunodeficiency virus (HIV) in the RG 3 list, while containment conditions in BMBL range from BSL2 to BSL3. This subtle difference in risk management causes great consternation among those who would put infectious agents into rigid classification schemes. These variations cause concern because the list has been taken at face value without reading the accompanying document (10). In the European Union directive, introductory note 8 in annex III states, "Certain biological agents classified in group 3 which are indicated in the appended list by two asterisks (**), may present a limited risk of infection for workers because they are not normally infectious by the airborne route. Member States shall assess the containment measures to be applied to such agents, taking account of the nature of specific activities in question and of the quantity of the agent involved, with a view to determining whether, in particular circumstances, some of these measures may be dispensed with." Looking at the list of viral agents in RG 3 in that document, one finds HIV, human T-lymphotropic virus 1 and 2, hepatitis B virus, and hepatitis C virus, among others.

The BMBL handles the situation more directly. Indeed, although BMBL appears to be an outlier in not defining RGs, it does provide flexibility in risk management via the BSLs recommended in the agent summary statements. For example, since HIV has not been shown to be transmitted by the aerosol route, BSL2 containment is appropriate for clinical labs. Amplification of HIV by cocultivation increases the risk of exposure and leads to an increase in the level of containment to include BSL3 practices with no requirement for a BSL3 facility. The European Union and BMBL actually reflect two approaches to risk assessment that can reach the same conclusions for risk management. Several countries and organizations (namely, Australia, Canada, the European Union, and the WHO) define HIV as an RG 3 agent, but they agree with BMBL that it can be safely handled at either BSL2 or BSL3, according to the activity. This situation highlights the importance of doing not only an agent-based risk assessment but also, more importantly, a protocol-driven risk assessment. It also shows that the RG of an agent can be related to but does not have to equate with the biosafety containment level (3). Although this difference in listings of agents according to RG is to be expected when following the WHO guidelines, this lack of global uniformity has not been well received in the United States. The disparity must be interpreted appropriately by those who assess the risk of work with biohazardous agents at academic, government, and industrial sites in different countries. Those who do not understand the European Union directive assess the risk by using only the RG listed for the agent, thus erroneously assuming that BSL3 practices and facilities are required for all work with HIV in the European Union.

In assigning an agent to an RG, one must also take into account that "there are in all groups of microorganisms naturally occurring strains which vary in virulence, and may thus need to be handled at a higher or lower level of containment" (10). Exotic and restricted agents differ from one country to another. Thus, the lists of agents in the four RGs differ among countries. Agents that have not been assessed for risk or that have not been listed by the competent authority of a participating country should not automatically be considered nonhazardous. A default risk assessment of a minimum of RG 2 and the use of standard microbiological precautions can be found in the Australian, Canadian, and European Union guidelines mentioned above. In the United States, this would be the use of standard precautions (8) or universal precautions (13) for contact with blood and certain body fluids and substances, which translates as BSL2 in the laboratory.

Many of the agent-specific factors to be considered in a risk assessment were mentioned above. Some of these factors form the basis of RG definitions, while others are to be considered in the overall risk assessment. An initial agent-based risk estimate for work with a microorganism can be predicated upon information provided in guidelines available from government agencies, professional associations, academic institutions, or designated competent authorities (2, 3, 8–10, 14–18). Those countries providing a WHO RG classification have either published their lists of pathogens or made them available online. The list for the European Union is included in a directive issued for the protection of workers from exposure to biohazardous agents (10).

The agent summary statements in BMBL provide the risk assessment and containment recommendations for the use of agents that have been reported to cause laboratory-acquired infections or could be expected to have a serious outcome (8). Many potentially biohazardous agents are not included. For example, *Bacteroides fragilis*, *Enterobacter aerogenes*, *Haemophilus influenzae* type b, and *Staphylococcus aureus* are not covered by agent summary statements because they have not been shown to pose a serious hazard to a healthy adult as a laboratory-acquired infection. However, in performing any risk assessment, care must be taken to determine the true disease potential of an agent in the activity proposed. Information on human communicable diseases is provided in the *Control of Communicable Diseases Manual* from the American Public Health Association (17). The relative risk of isolates from human clinical samples is also available in the literature (16, 19). Lists of biohazardous agents that require special practices in handling, packaging, and transporting are provided by the CDC (20), as the government agency responsible for identifying biohazardous agents that are to be regulated in transport as well as the packaging requirements needed for the import and transport of such agents. The Department of Health and Human Services and the United States Department of Agriculture have provided lists of agents and toxins that are restricted under new regulations to prevent their use in bioterrorist activities (21).

Information collected in the risk assessment may confirm an increased virulence of the specific strain or serotype in use, in which case the risk assessment may be altered enough to require an increase in the level of containment. For example, organisms that have developed resistance to multiple therapeutic drugs, such as multidrug-resistant *Mycobacterium tuberculosis*, are considered to be of a higher risk due to the lack of treatment alternatives and are to be handled with more stringent precautions. *M. tuberculosis* is RG 3, but the extra precautions required for safe work with the multidrug-resistant strains would still not be expected to require higher than BSL3. Conversely, if the assessment indicates a lower level of virulence, a relaxation of some of the protective measures may be in order. For example, an avirulent strain of the same microbe, *M. tuberculosis* H37Ra, has been handled safely under BSL2 containment, even though the virulent strain H37Rv is to be handled at BSL3; staff should be made aware that seroconversion could happen in either case. *Streptococcus pneumoniae* is RG 2. However, strains exist that no longer possess the capsule, a virulence factor that allows this pathogen to evade the phagocytic arm of the host immune system. Therefore, these strains can be handled safely at BSL1. The need for a knowledgeable evaluator cannot be overemphasized.

Yersinia enterocolitica, a bacterial agent of enteric disease, is known to have specific invasion genes as virulence factors, but it is also known to have strains that are nonpathogenic (22–24). A rigid agent classification system would impose a higher level of containment than is necessary for work with the nonpathogenic strains of pathogenic organisms. There should be enough flexibility in the criteria to allow such strains to be handled and shipped as nonpathogens even though the wild-type strain belongs in an RG that requires a higher level of containment and special packaging.

Microorganisms that are attenuated for use as killed or live vaccines may no longer require the same containment as the wild-type parent organism (10). For example, some attenuated influenza virus vaccine strains may be handled at BSL1, while work with the parent virus is done at BSL2. Even higher levels of containment should be considered for highly pathogenic avian influenza virus or gain-of-function strains in order to protect the community from a pandemic. The strain of *Mycobacterium bovis* known as bacillus Calmette-Guérin (BCG) is handled at BSL2, but the wild-type RG 3 *M. bovis* strain is more virulent and usually requires BSL3 work practices and facilities. Similar assessments have led to exemptions from registration of vaccine strains or inactivated preparations of selected agents (25). Recent incidents in which live or more dangerous strains of *Bacillus anthracis* and influenza A virus were shipped unknowingly to recipients have identified a need for both suppliers and recipients to confirm attenuation or inactivation through appropriate testing (26).

A rigid classification of risk would inadvertently exclude the use of higher containment for situations in which the virulence of a pathogenic agent has been reassessed or significantly enhanced. The live poliovirus vaccine strains, which were given orally to children and adults, were once considered safe to handle at BSL1, although the wild-type strains of poliovirus were handled at BSL2. When vaccine-associated cases of polio began to occur more frequently than naturally occurring cases, the live oral polio vaccine began to be replaced by the inactivated polio vaccine on the recommendation of the Advisory Committee on Immunization Practices (27).

The WHO recommended a program in which stocks of poliovirus were to be destroyed in anticipation of global eradication of the virus by 2005. Due to ongoing cases appearing in countries where polio is endemic, namely Afghanistan and Pakistan, the timetable has been revised (see reports at https://www.cdc.gov/ and http://www.who.int/en/). At present, poliovirus is to be handled at BSL2/polio. Once the poliovirus is eradicated in the wild, laboratories wishing to work with wild poliovirus or infectious or potentially infectious materials will do so under BSL3/polio. The specific requirements for this level of containment are meant to be applied only to work with the poliovirus (30). At that time, the only source of poliovirus will be laboratories and vaccine production facilities. The relative risk of reintroduction of poliovirus from a variety of sources in such facilities has been assessed and is provided in Table 1. The most important source of virus for reintroduction is facility personnel, who should thus be the focus for management of the risk of release to the community (31). Eradication plans and the proposed changes in containment requirements must now take into account the fact that scientists have synthesized poliovirus (32).

TABLE 1.

Risk of poliovirus reintroduction from laboratory or vaccine production facilities[a]

Potential viral source	Relative risk
Facility personnel	
Infected	++++
Contaminated	++++
Liquid effluent	– to +++
Air effluent	+/–
Solid-waste disposal	+/–
Materials in transit	+/–
Laboratory animals	–

[a]From reference 31.

RISK ASSESSMENT OF UNKNOWNS

We continue to be challenged by emerging infectious diseases, such as Middle East respiratory syndrome virus, Zika virus, and highly pathogenic avian influenza A virus, which must be assessed for risk to prevent epidemics and pandemics, even as they are first being identified. Laboratory-associated infections reported with these agents suggest either that the risks were not assessed correctly or that the workers failed to use the laboratory practices needed to control those risks. We also continue to be faced with reemerging infections with agents that develop multiple antibiotic resistance, such as metronidazole-resistant *Clostridium difficile*, multidrug-resistant *M. tuberculosis*, and methicillin- and vancomycin-resistant *S. aureus*. The risk becomes greater when there are no effective treatments. In other situations, where a sample or agent has not been well characterized, there may not be enough data to allow the risk to be assessed with any confidence. The information that is available is used to develop a rational default process for handling such unknown agents. Highly pathogenic avian influenza A virus is known to extend its host range from avian to human and has a high case fatality in humans (over 50% for some strains, such as H5N1 and H7N9). Gain-of-function strains are complete unknowns as they are laboratory creations with possibly no existing cross-reactive immunity from previous epidemics. The existing influenza medications, namely the neuraminidase inhibitors Tamiflu and Relenza, may not be effective against these unknowns, and there are limitations on the effectiveness of these drugs even in the absence of resistance. Stockpiled vaccines may also be ineffective against these unknowns. The minimum default risk assessment would require BSL3 containment for diagnostic work, with enhanced precautions for research, depending on the exact nature of the work to be performed.

In another situation, soil samples are sent to a drug discovery facility from sites around the world, which may vary from a tropical rain forest to a pigeon-infested city park. All such samples are subject to a risk assessment because pathogenic agents ranging from exotic viruses to the spores of pathogenic fungi, such as *Histoplasma capsulatum*, may be present. It is prudent to consider such soil samples as potentially contaminated with organisms indigenous to the area, using BSL2 at a minimum (3) and processing the sample so as not to expose workers to such agents by minimizing the production of and exposure to splash, spatter, and true infectious aerosols (8).

ACTIVITY-BASED RISK ASSESSMENT

Information on the exposure potential associated with specific work practices and equipment helps to identify situations that need to be controlled before the work begins. Factors to be studied prior to the proposed work include (i) potential for generation of aerosols (including splash and splatter), (ii) quantity (volume, concentration/titer, infectious dose, etc.), and (iii) work proposed (*in vitro*, *in vivo*, aerosol challenge, or environmental release).

The protocols or standard operating procedures being developed for the specific tasks and equipment involving the etiologic agents of human disease can be assessed to identify the need for special containment practices or protective equipment. We are faced with the potential for exposures associated with new equipment that can produce unexpected aerosols due to the lack of sham (or surrogate) testing. We must assess work activities, such as centrifugation, homogenization, sonication, etc. A risk assessment is especially important for new procedures accompanying technological advances in related fields, such as synthetic biology and nanotechnology. Do chemists and molecular biologists assess and understand the risks? Are such scientists aware of the outcome of exposure? We must take these concerns into account in performing a risk assessment and identify the potential for exposure to be encountered in various work environments, from molecular biology to the testing of potential new antibiotics against the latest clinical isolates.

RISK ASSOCIATED WITH RECOMBINANT ACTIVITIES

Factors involved with recombinants include properties of the transgenes, such as those that code for virulence factors, toxins, host range, integration, replication, reversion to wild type, etc. (3). The actual process of producing a recombinant organism is not one of the factors of concern since the National Research Council (33) concluded that there is no evidence of any unique hazard posed by recombinant DNA techniques. They concluded that the risks associated with recombinant DNA are the same in kind as those associated with unmodified organisms or organisms modified by other means. The National Research Council recommended that risk assessment of work involving environmental release of recombinants be based on the nature of the organism and the environment into which it is introduced, and not on the method by which it is produced. This recommendation for risk assessment of the product rather than the process was also accepted by the U.S. Office of Science and Technology Policy (34) and the National Research Council (33). Guidelines for assessing the risk of recombinant work have been provided by the NIH (2), Canada (7), and the WHO (3). Although all possible scenarios cannot be addressed, the potential hazards may actually be novel and uncharacterized. The WHO (3) lists the following factors to be considered in the risk assessment: properties of the donor organism, nature of the genetic sequences that will be transferred, properties of the recipient organism, and properties of the environment. There is always the risk that a more virulent organism will evolve from the insertion of a gene not normally found in that host. For example, a more virulent ectromelia virus (mousepox) resulted from the insertion of the gene for interleukin 4 (35).

SCALE-UP AND LARGE-SCALE ACTIVITIES

Work with biohazards can be divided into research, diagnostic, or large scale, with the last usually referring to levels greater than 10 liters in the United States and Canada. Volume, however, is not the sole determinant, because the intent of the work can also determine the scale in countries such as the United Kingdom (28). The workplace can create unnatural situations that increase the risk of employee exposure to infectious, toxigenic, or allergenic agents or materials. In the research laboratory, work is usually limited to relatively few organisms that are known to the investigator, who can usually choose the agents of interest. The diagnostic or clinical laboratory handles unknowns in clinical samples, but there is general knowledge of the types of agents that can be isolated from each patient site in that specific geographic location and whether or not such agents are known to cause human disease (19). Regarding work with large volumes of agents, the agent summary statements in BMBL suggest considering an increase in the level of containment (8), without any mention of specific laboratory precautions that could be applicable to work done in bioreactors or fermenters (36). Many of the agents being used in large-scale production are not covered by agent summary statements. Each manufacturer using such agents must assess the situation and determine the appropriate containment. Appendix K of the *NIH Guidelines* (2) continues to provide assistance in developing specific practices for large-scale recombinant work.

AGENT-ACTIVITY INTERACTION

A thorough evaluation of the hazard potential of the work practices, procedures, and equipment to be used for proposed tasks is called a job safety analysis. In such an analysis, the supervisor or his or her designee analyzes the risk of employee exposure from each task in which a biohazard is involved. Ideally, those who develop the work procedures and those who will actually perform the work are involved in the analysis in order to recommend less risk-prone options wherever possible. The job is divided into steps that describe what is to be done. Instructions are reviewed or developed for each step, and key points are provided, such as warnings of a specific hazard or a potential accident. Methods of control for each potential hazard are then developed. An example of a job safety analysis for a sterilizer/autoclave is described by Songer (37).

A job safety analysis is to be done in advance of work with hazardous agents. Information on the best work practices and the safest equipment for handling biohazardous materials, along with other hazards associated with the tasks, can then be provided to the worker prior to carrying out the procedures. It is important to consider that animal experiments pose additional activity risks, such as biting, scratching, and escape. The supervisor is responsible for the safety orientation and the specific training required for the safe performance of the work. The OSHA standard for worker protection from blood-borne pathogens requires that information and training be provided prior to work with blood-borne pathogens and prior to offering the employee the hepatitis B virus vaccine, to allow informed consent (13). The requirement for advance information and training should ideally be applied to all work with biohazardous agents. Directives for the protection of workers from exposure to biohazardous agents have been issued in Europe (10). Although they are regulations abroad, they can provide adequate incentive to voluntarily protect our own workers.

EXPOSURE DETERMINATION

As a part of the risk assessment process, a definition of what constitutes an exposure to the agent should be determined in advance of an incident. This allows the identification of work activities or equipment that could pose a hazard to be assessed and addressed. The definition of an exposure also prevents unnecessary medical treatment and employee concern. Individuals who are present during an accident but are not exposed to the agent by one of the routes of infection do not require medical treatment, although such individuals may need counseling or further training on principles of infection.

BIOSAFETY MANUAL

Written documentation of the risk estimate and the actual protocols are to be placed in an infection control plan or biosafety manual. Regulatory requirements for the prevention of occupationally acquired blood-borne infectious diseases in the United States now mandate such documentation in an exposure control plan (10). The regulations for handling Select Agents also mandate such documentation (21).

HOST FACTORS: HEALTH STATUS

A more thorough risk evaluation of the agent-host-activity triad is required to develop appropriate containment for actual work with etiologic agents. After the agent and activity are assessed for risk, the remainder of the risk assessment involving the host is beyond the purview of the biosafety professional. Each institution should have its own policy on occupational medicine programs. In the United States, the *NIH Guidelines* require an occupational health program for large-scale research or production activities requiring BSL3 containment and for research involving RG 3 influenza viruses conducted at enhanced BSL3. The need to consider occupational health for lower containment levels cannot be emphasized enough in light of the death caused by an attenuated strain of *Yersinia pestis*, which was approved for BSL2 (38). This death was the first documented case of a laboratory-acquired infection resulting from enhanced virulence due to a host's hematochromatosis. Institutions performing animal research must provide assurance to the Association for Assessment and Accreditation of Laboratory Animal Care that at-risk personnel are evaluated under an occupational health and safety program in order to maintain accreditation.

The BMBL guidelines in the United States (8) presume that the worker is an immunocompetent adult. Declaration or identification of impaired host defense factors, such as immune deficiencies and extremes of age, is the responsibility of the employee. An example of a host condition that may affect susceptibility to infection is diabetes, which may cause immunosuppression or enhanced opportunity for transmission due to finger pricking. The risk assessment must include other individuals who may enter the work area, such as custodial and maintenance workers, temporary student helpers, and laboratory observers.

An evaluation of fitness for duty may need to be obtained from a physician, especially if there is a change in health status that could place that employee at increased risk of infection. The risk assessment needs to be kept current and relevant to the work in progress to control any potential increase in risk. The importance of evaluating the epidemiological triad (agent-host-activity) in establishing the containment level has been emphasized by the National Research Council Committee on Hazardous Biological Substances in the Laboratory (39).

Opportunistic pathogens and normal microbial flora that pose no or low risk to healthy adult coworkers can cause disease in immunocompromised or immunosuppressed adults. Frank pathogens usually pose a greater risk of more serious disease in such individuals, and the additional risk must be addressed. A similar approach is required for pregnant females, due to risk to the intrinsically immunocompromised fetus. BMBL (8) provides strong recommendations to preclude certain individuals from working with certain agents; e.g., serologically negative women of childbearing age should not work with *Toxoplasma*. Warnings of increased risk to the immunocompromised are also found in some of the agent summary statements (8). The use of potentially harmful biological agents needs to be evaluated on a case-by-case basis to prevent discrimination under the Americans with Disabilities Act (ADA) (29). The worker must be advised of the specific hazard and understand what can be done to offer protection through a reasonable accommodation, such as a special engineering design or personal protective equipment. The worker may even be asked to sign a document indicating that he or she has been informed of the special personal hazard.

It may be difficult for the biosafety professional to understand how to comply effectively with the seemingly mutually exclusive requirements of the ADA and BMBL. Because of this, and for reasons of confidentiality, these discussions and decisions are usually handled by the worker in consultation with the occupational health physician or personal physician. When such host factors are evaluated and a task-based risk analysis is completed, an appropriately trained professional can provide advice to reduce or prevent exposure to the biohazardous agent.

ACCEPTABILITY OF THE RISK OF WORK WITH BIOLOGICAL HAZARDS

How do we assess the acceptability of the risk of work with biological hazards? Some people accept more risk than others; they perceive and tolerate risk differently. Therefore, judging the acceptability of risk is a subjective process that involves personal, social, cultural, and possibly even religious judgments. Scientists cannot measure whether something is safe or not, but they are prepared to measure risk in terms of probability, which is a more objective process. We are more likely to judge something as unsafe if it is unfamiliar to us. A safe activity is one in which the risks are considered to be acceptable. The acceptability of a risk is constantly changing, due to changes in social values, even though the actual level of that risk might remain the same. Safety is not an intrinsic, absolute, measurable property of things. Songer cautioned us not to attempt to anticipate all of the possible risks that might occur in a work situation, such as a biomedical laboratory (37). He advised the use of some general, results-oriented guidelines to assess and control potential risks. For example, reduction in needlestick accidents could indicate success in a blood-borne pathogen program. An increase in needlesticks could signal a defect that should be addressed immediately. Using the framework provided by regulations and guidelines, we can measure

TABLE 2.

Risk prioritization

Probability of accident	RG 1	RG 2	RG 3	RG 4
Negligible	Very low	Low	Low	Medium
Low	Very low	Low	Medium	High
Medium	Low	Medium	High	Very high
High	Low	Medium	High	Very high

risks and make safety judgments regarding the acceptability of those risks based on observed results on a day-to-day basis. Codification of risks along with prescribed requirements can adversely affect a search for better work procedures.

RISK PRIORITIZATION

A facility with limited personnel and funds should direct attention to the risks with the greatest probability of occurrence and harm. One way to prioritize risks is to develop a matrix based on the severity of the consequences and the probability that an infection will occur under the conditions of use (Table 2). The RG of the agent, which reflects agent-based risk and the relative severity of the consequences of a laboratory-acquired infection, could be used as a surrogate for a severity assessment. The probability of an accident can be estimated from the known hazards associated with the protocol, including the use of aerosol-generating equipment, such as sonicators and homogenizers, as well as procedures that use sharps, such as injecting animals. The probability of an accident, based on an assessment of the specific situation, would be recorded as negligible, low, medium, or high. The risk matrix provides a priority for that risk that can be read from the table as very low through very high.

CONCLUSION

As we assess risk and attempt to translate guidelines, regulations, and standards into work practices, we should continue to search for better control methods. Rules and regulations cannot be written to encompass every possible set of circumstances. We must continue to assess the risk of biohazardous work based upon current information and common sense in order to recommend appropriate and realistic methods of containment for preventing worker exposure and environmental contamination. In the final analysis, we must be sure that the individuals working with biohazardous agents are trained to a level of competence that provides for their own safety and that of their coworkers and community.

References

1. **Motorcycle Safety Foundation.** 2014. *The Motorcycle Safety Foundation Basic Ridercourse Rider Handbook*, 1.0 ed. Motorcycle Safety Foundation, Irvine, CA. https://www.msf-usa.org/downloads/BRCHandbook.pdf.
2. **National Institutes of Health.** 2016. *NIH Guidelines for Research Involving Recombinant or Synthetic Nucleic Acid Molecules.* National Institutes of Health, Bethesda, MD. http://osp.od.nih.gov/sites/default/files/NIH_Guidelines.html#_Toc446948312.
3. **World Health Organization.** 2004. *Laboratory Biosafety Manual*, 3rd ed. World Health Organization, Geneva, Switzerland. http://www.who.int/csr/resources/publications/biosafety/WHO_CDS_CSR_LYO_2004_11/en/.
4. **Ministry of the Brussels-Capital Region.** 2002. [Laws, decrees, ordinances, and regulations.] (In French and Flemish.) http://www.biosafety.be/PDF/ArrRB02.pdf.
5. **Ministry of the Walloon Region.** 2002. [Laws, decrees, ordinances, and regulations.] (In French and Flemish.) http://www.biosafety.be/CU/ArrRW02_FR/ArrRW02FR_TC.html.
6. **Ministry of the Flemish Region.** 2004. [Laws, decrees, ordinances, and regulations.] (In French and Flemish.) http://www.biosafety.be/PDF/BesVG04_NL.pdf.
7. **Public Health Agency of Canada.** 2016. Canadian Biosafety Standard, 2nd ed. http://canadianbiosafetystandards.collaboration.gc.ca/cbh-gcb/index-eng.php#ch4.
8. **U.S. Department of Health and Human Services, Public Health Service, Centers for Disease Control and Prevention, National Institutes of Health.** 2009. *Biosafety in Microbiological and Biomedical Laboratories*, 5th ed. HHS Publication no. (CDC) 21-112. http://www.cdc.gov/biosafety/publications/bmbl5/BMBL.pdf.
9. **Standards New Zealand, Joint Technical Committee CH-026.** 2010. Australian/New Zealand Standard, Safety in laboratories, Part 3: Microbiological safety and containment, 6th ed. SAI Global Limited and Standards New Zealand, Sydney, Australia, and Wellington, New Zealand. https://law.resource.org/pub/nz/ibr/as-nzs.2243.3.2010.pdf.
10. **European Parliament and Council of the European Union.** 2000. Directive 2000/54/EC of the European Parliament and of the Council of 18 September 2000 on the protection of workers from risks related to exposure to biological agents at work (seventh individual directive within the meaning of Article 16(1) of Directive 89/391/EEC). Official Journal of the European Communities http://www.biosafety.be/PDF/2000_54.pdf.
11. **Health and Safety Executive, Advisory Committee on Dangerous Pathogens.** 2013. *The Approved List of Biological Agents*, 3rd ed. HSE Books, London, United Kingdom. http://www.hse.gov.uk/pubns/misc208.pdf.
12. **American Biological Safety Association International.** 2016. Risk Group Database. https://my.absa.org/tiki-index.php?page=Riskgroups.
13. **Occupational Safety and Health Administration.** 2012. Regulations (Standards - 29.CFR) Occupational Safety and Health Standards: Toxic and Hazardous Substances: Bloodborne pathogens. https://www.osha.gov/pls/oshaweb/owadisp.show_document?p_table=STANDARDS&p_id=10051.
14. **Health Canada.** 2004. *Laboratory Biosafety Guidelines*, 3rd ed. Minister of Health Population and Public Health Branch Centre for Emergency Preparedness and Response, Ottawa, Ontario, Canada.
15. **Fleming DO, Hunt DL (ed).** 2006. *Biological Safety: Principles and Practices*, 4th ed. ASM Press, Washington, DC.

16. **Brooks GF, Butel JS, Morse SA (ed).** 2004. *Jawetz, Melnick and Adelberg's Medical Microbiology*, 23rd ed. Lange Medical Books/McGraw Hill, New York, NY.
17. **Heymann DL (ed).** 2014. *Control of Communicable Diseases Manual*, 20th ed. APHA Press, Washington, DC.
18. **Kuenzi M, Assi F, Chmiel A, Collins CH, Donikian M, Dominguez JB, Financsek L, Fogarty LM, Frommer W, Hasko F, Hovland J, Houwink EH, Mahler JL, Sandvist A, Sargeant K, Sloover C, Muijs GT.** 1985. Safe biotechnology general considerations. A report prepared by the Safety in Biotechnology Working Party of the European Federation of Biotechnology. *Appl Microbiol Biotechnol* **21:**1–6.
19. **Isenberg HD, D'Amato RF.** 1995. Indigenous and pathogenic microorganisms of humans, p 5–18. *In* Murray PR, Baron EJ, Pfaller MA, Tenover FC, Yolken RH (ed), *Manual of Clinical Microbiology*, 6th ed. ASM Press, Washington, DC.
20. **Centers for Disease Control and Prevention, HHS.** 1997. Additional Requirements for Facilities Transferring or Receiving Select Agents. 42 CFR Part 72. https://grants.nih.gov/grants/policy/select_agent/42CFR_Additional_Requirements.pdf.
21. **Public Health Security and Bioterrorism Preparedness and Response Act Public Law 107–188.** 2002. https://www.congress.gov/107/plaws/publ188/PLAW-107publ188.pdf.
22. **Finlay BB, Falkow S.** 1997. Common themes in microbial pathogenicity revisited. *Microbiol Mol Biol Rev* **61:**136–169.
23. **Miller VL.** 1992. *Yersinia* invasion genes and their products. *ASM News* **58:**26–33.
24. **Miller VL, Falkow S.** 1988. Evidence for two genetic loci in Yersinia enterocolitica that can promote invasion of epithelial cells. *Infect Immun* **56:**1242–1248.
25. **Kroger AT, Atkinson WL, Marcuse EK, Pickering LK, Advisory Committee on Immunization Practices (ACIP) Centers for Disease Control and Prevention (CDC).** 2006. General recommendations on immunization: recommendations of the Advisory Committee on Immunization Practices (ACIP). *MMWR Recomm Rep* **55**(RR-15)**:**1–48.
26. **Kaiser J.** 2014. Lab incidents lead to safety crackdown at CDC. *Science* http://www.sciencemag.org/news/2014/07/lab-incidents-lead-safety-crackdown-cdc.
27. **Centers for Disease Control and Prevention.** 2016. Birth-18 years & "catch-up" immunization schedules. http://www.cdc.gov/vaccines/schedules/hcp/child-adolescent.html.
28. **Health and Safety Executive, Advisory Committee on Dangerous Pathogens.** 2005. Biological agents: managing the risks in laboratories and healthcare premises. HSE Books, Suffolk, United Kingdom. http://www.hse.gov.uk/biosafety/biologagents.pdf
29. **United States Department of Justice, Civil Rights Division.** 1990. Information and technical assistance on the Americans with Disabilities Act of 1990 (42 U.S.C. § 12101 et seq.). https://www.ada.gov/ada_intro.htm.
30. **World Health Organization.** 2004. *WHO Global Action Plan for Laboratory Containment of Wild Polioviruses*, 2nd ed. World Health Organization, Geneva, Switzerland.
31. **Wolff C, Fleming DO, Dowdle W.** 2005. Assessment and management of post-eradication poliovirus facility associated community risks. 48th Annual Biological Safety Conference, Vancouver, Canada.
32. **Cello J, Paul AV, Wimmer E.** 2002. Chemical synthesis of poliovirus cDNA: generation of infectious virus in the absence of natural template. *Science* **297:**1016–1018.
33. **National Research Council (NRC) Committee on the Introduction of Genetically Engineered Organisms into the Environment.** 1987. *Committee on the Introduction of Recombinant DNA-Engineered Organisms into the Environment: Key Issues.* National Academy Press, Washington, DC.
34. **U.S. Office of Science and Technology Policy.** 1986. Coordinated framework for regulation of biotechnology; announcement of policy; notice for public comment. *Fed Regist* **51:**23302–23350.
35. **Jackson RJ, Ramsay AJ, Christensen CD, Beaton S, Hall DF, Ramshaw IA.** 2001. Expression of mouse interleukin-4 by a recombinant ectromelia virus suppresses cytolytic lymphocyte responses and overcomes genetic resistance to mousepox. *J Virol* **75:**1205–1210.
36. **Cipriano ML.** 2002. *Cumitech 36, Biosafety Considerations for Large-Scale Production of Microorganisms.* ASM Press, Washington, DC.
37. **Songer JR.** 1995. Laboratory safety management and the assessment of risk, p 257–268. *In* Fleming DO, Richardson JH, Tulis JJ, Vesley D (ed), *Laboratory Safety: Principles and Practices*, 2nd ed. ASM Press, Washington, DC.
38. **Centers for Disease Control and Prevention (CDC).** 2011. Fatal laboratory-acquired infection with an attenuated Yersinia pestis strain—Chicago, Illinois, 2009. *MMWR Morb Mortal Wkly Rep* **60:**201–205.
39. **National Research Council (US) Committee on Hazardous Biological Substances in the Laboratory.** 1989. Biosafety, p 22–24. *In The Laboratory: Prudent Practices for the Handling and Disposal of Infectious Materials.* National Academies Press, Washington, DC.

6

Protozoa and Helminths

BARBARA L. HERWALDT

Parasitic diseases have public health and clinical importance throughout the world, not just in developing countries. Some parasitic diseases are endemic globally (e.g., toxoplasmosis and cryptosporidiosis), and even those that are endemic primarily in developing countries or in the tropics and subtropics are receiving increasing attention in developed countries, in part because of their importance in returning travelers and immigrants. As clinical interest in and laboratory research about parasitic diseases increase and as the numbers of infected patients increase, so do the numbers of persons working in settings in which they could be exposed to parasites.

Persons working in research and clinical laboratories, as well as health care workers providing patient care, are at risk of becoming infected with parasites through accidental exposures, which may or may not be recognized when they occur. Persons working in clinical settings often do not know that particular patients are, or might be, infected with parasites. Even persons, such as researchers, who realize they have had an accidental exposure to a particular parasite and are knowledgeable about parasitic diseases may not know whether they truly were exposed to viable organisms, what clinical manifestations may occur when natural modes of transmission are bypassed, how to monitor for infection postexposure, and whether to begin antimicrobial therapy before infection is documented. In part because of such uncertainties and the potential severity of some parasitic diseases even in immunocompetent persons, the first reactions to accidental exposures may include bewilderment, anxiety, fear, and shame. However, another common reaction is to conclude inappropriately that an exposure was inconsequential and not worthy of mention.

The main purpose of this review is to inform laboratorians, biosafety personnel, and health care workers about the potential hazards of working in settings in which exposures to viable parasites could occur. Table 1 provides information about parasites that have caused or could cause accidental infections in laboratorians and health care workers. Factors that affect whether exposures result in infection and disease are listed in Table 2.

Ideally, accurate counts of accidental exposures, as well as resultant cases of infection, would be available for the United States and other countries, as would information about the magnitudes of the risks per person-hour or

TABLE 1.

Parasites to which laboratorians and health care workers could be exposed[a]

Parasite[a]	Routes of exposure[a]	Infective stages	Protective measures[a]	Diagnostic approaches[a,b]	Clinical manifestations[c]
Blood and tissue protozoa					
Acanthamoeba spp.	Wound, eye (aerosol?) (needle?)	Trophozoite, cyst	Gloves, mask, gown, class II BSC; wound and needle precautions	Examinations of pertinent tissue specimens (e.g., brain, cornea, skin) with various techniques (also: brain scans; CSF examinations; serology)	Headache, neurologic impairment, skin lesion(s), pneumonitis; keratitis, conjunctivitis
Babesia spp.	Needle, wound, vector	Intraerythrocytic stages; sporozoite	Gloves; wound and needle precautions	Examinations of blood specimens with various techniques, including light-microscopic examinations of blood smears; serology	Fever, chills, other flu-like symptoms; hemolytic anemia
Balamuthia mandrillaris	Wound; needle (aerosol?)	Trophozoite, cyst	Gloves, mask, gown, class II BSC; wound and needle precautions	Examinations of pertinent tissue specimens (e.g., brain, skin) with various techniques (also: brain scans; CSF examinations; serology)	Headache, neurologic impairment, skin abscess (pneumonitis?)
Leishmania spp.	Needle, wound, transmucosal, vector	Amastigote (in tissue); promastigote (in culture and vector)	Gloves; wound, mucous membrane,[d] and needle precautions	Examinations of pertinent tissue specimens (e.g., skin, mucosa, bone marrow) with various techniques; for visceral leishmaniasis, also serology	Cutaneous: skin lesion(s) at or near exposure site; adenopathy; ascending lymphangitis Mucosal: naso-oropharyngeal/laryngeal lesion(s) Visceral: fever (early); hepatosplenomegaly and pancytopenia (later)
Naegleria fowleri	Transmucosal (nasopharynx), aerosol (needle?)	Trophozoite (flagellate?) (cyst?)	Gloves, mask, gown, class II BSC; wound and needle precautions	Examinations of CSF (brain tissue) with various techniques	Headache, stiff neck, neurologic impairment (including sense of smell), coma
Plasmodium spp.	Needle, wound, vector	Intraerythrocytic stages; sporozoite	Gloves; wound and needle precautions	Examinations of blood specimens with various techniques, including light-microscopic examinations of blood smears; antigen and antibody tests	Fever, chills, other flu-like symptoms; hemolytic anemia
Sarcocystis spp.	Oral (see text)	Sarcocyst; oocyst or sporocyst	Gloves, mask, hand washing	Stool examinations; examinations of muscle or cardiac biopsy specimens	Gastrointestinal symptoms; eosinophilic myositis
Toxoplasma gondii	Oral, needle, wound, transmucosal (aerosol?)	Oocyst, tachyzoite, bradyzoite	Gloves, hand washing; wound, mucous membrane,[d] and needle precautions	Serology; examinations of lymph node or other tissue specimens with various techniques	Adenopathy, fever, malaise, skin eruption
Trypanosoma cruzi (American trypanosomiasis)	Needle, wound, transmucosal, vector (aerosol?)	Trypomastigote	Gloves; wound, mucous membrane,[d] and needle precautions	Examinations of blood/buffy coat and tissue (if pertinent) specimens with various techniques; serology	Swelling and/or redness at exposure site, fever, skin eruption, adenopathy; electrocardiographic changes

(continued)

TABLE 1.

(Continued)

Parasite[a]	Routes of exposure[a]	Infective stages	Protective measures[a]	Diagnostic approaches[a,b]	Clinical manifestations[c]
Trypanosoma brucei rhodesiense and *T. brucei gambiense* (African trypanosomiasis)	Needle, wound, transmucosal, vector (aerosol?)	Trypanosome	Gloves; wound, mucous membrane,[d] and needle precautions	Examinations of blood/buffy coat, CSF, and tissue (if pertinent) specimens with various techniques	Swelling and/or redness at exposure site, fever, skin eruption, adenopathy, headache, fatigue, neurologic signs
Intestinal protozoa[e]					
Cryptosporidium spp.	Oral; transmucosal (aerosol?)	Oocyst (sporozoite)	Gloves, hand washing; mucous membrane precautions[d]	Stool examinations (see text)	Symptoms of gastroenteritis
Cyclospora cayetanensis	Oral	Oocyst (sporozoite)	Gloves, mask, hand washing	Stool examinations (see text)	Symptoms of gastroenteritis
Entamoeba histolytica	Oral	Cyst	Gloves, mask, hand washing	Stool examinations (see text); serology (for invasive infection)	Symptoms of gastroenteritis (stools may be bloody)
Giardia intestinalis (*Giardia lamblia*)	Oral (aerosol?)	Cyst	Gloves, mask, hand washing	Stool examinations (see text)	Symptoms of gastroenteritis
Cystoisospora belli (*Isospora belli*)	Oral	Oocyst (sporozoite)	Gloves, mask, hand washing	Stool examinations (see text)	Symptoms of gastroenteritis
Helminths[f]					
Ascaris lumbricoides	Oral	Egg	Gloves, mask, hand washing	Stool examinations	Cough, fever, pneumonitis; abdominal symptoms[g]; hypersensitivity reactions
Enterobius vermicularis	Oral	Egg	Gloves, mask, hand washing; nail cleaning	Scotch tape test	Perianal pruritus
Fasciola spp.	Oral	Metacercaria	Gloves, mask, hand washing	Examinations of stool or bile; serology	Right upper quadrant pain, biliary colic, obstructive jaundice; elevated transaminase levels
Hookworm	Percutaneous[h]	Larva	Gloves, gown, hand washing	Stool examinations	Animal species[i]: cutaneous larva migrans or creeping eruption (skin) Human species: abdominal symptoms; anemia[g]
Hymenolepis nana	Oral	Egg	Gloves, mask, hand washing	Stool examinations	Abdominal symptoms
Schistosoma spp.	Percutaneous[h]	Cercaria	Gloves, gown, hand washing	Stool examinations; serology	Acute schistosomiasis: dermatitis, fever, cough, hepatosplenomegaly, adenopathy
Strongyloides stercoralis	Percutaneous[h]	Larva	Gloves, gown, hand washing	Stool examinations (motile larvae may be seen in wet preparations); serology	Cough and chest pain, followed by abdominal symptoms[g]
Taenia solium	Oral	Egg, cysticercus	Gloves, mask, hand washing	Cysticercosis: serology; brain scans, soft tissue X rays Worm: stool examinations	Cysticercosis: neurologic symptoms and signs Worm: usually asymptomatic but may cause abdominal symptoms

(continued)

TABLE 1.

Parasites to which laboratorians and health care workers could be exposed[a] (*Continued*)

Parasite[a]	Routes of exposure[a]	Infective stages	Protective measures[a]	Diagnostic approaches[a,b]	Clinical manifestations[c]
Trichinella spp.	Oral	Larva	Gloves, mask, hand washing	Serology; examinations of muscle biopsy specimens	Abdominal symptoms; muscle pain[g]
Trichuris trichiura	Oral	Egg	Gloves, mask, hand washing	Stool examinations	Abdominal symptoms (e.g., tenesmus)[g]

[a]This table provides a general overview about selected parasites that have caused or could cause accidental infections in laboratorians and health care workers; additional parasites are discussed in the text. For each column in the table (e.g., routes of exposure, diagnostic approaches, clinical manifestations), refer to the text, other tables, and the website of CDC's Division of Parasitic Diseases and Malaria (http://www.cdc.gov/parasites) for additional details and perspective. BSL-2 guidelines apply to these parasites (4). Class II biological safety cabinets (BSC), other physical containment devices, and/or personal protective equipment (e.g., face shield) should be used during procedures that could create aerosols or droplets. In this table, the "needle" route signifies parenteral transmission (i.e., percutaneous transmission, via a contaminated sharp object, such as a needle); and the "wound" route signifies contamination (e.g., via a spill or splash) of a preexisting abrasion, cut, or break in the skin.

[b]With some exceptions, the information in this column focuses on the types of specimens to consider examining (if pertinent) rather than on the types of examinations to consider conducting, such as whether molecular approaches (227) are available (if so, what and where) and appropriate for the situation at hand. Diagnostic testing should be individualized, with expert consultation. CDC's Division of Parasitic Diseases and Malaria (http://www.cdc.gov/parasites) offers diagnostic guidance and assistance through DPDx (http://www.cdc.gov/dpdx).

[c]The clinical manifestations, if any, can be highly variable, depending in part on such factors as the species/strain of the parasite, the size of the inoculum, the stage/severity of the infection, and the host's immunologic status/competence (Table 2). The listed clinical features are not all-inclusive and do not necessarily include potentially life-threatening manifestations (e.g., cerebral malaria or myocarditis/encephalitis in persons with toxoplasmosis or Chagas' disease).

[d]Use of a class II BSC provides optimal protection against exposure of the mucous membranes of the eyes, nose, and mouth; see footnote *a*.

[e]The possibility of becoming infected via swallowing inhaled infectious aerosols/droplets has been raised for *Cryptosporidium* oocysts (192) and *Giardia* cysts (188). The same principle could apply to other intestinal protozoa. *Cryptosporidium* oocysts can bypass the gastrointestinal tract and directly establish a pulmonary infection.

[f]Eosinophilia is common for helminthic infections with an invasive tissue stage. See footnote *c* regarding the variability of clinical manifestations.

[g]Symptoms are unusual, unless the inoculum is large, which would be unlikely in most laboratory exposures.

[h]The parasite can penetrate intact skin.

[i]Cutaneous larva migrans usually is caused by animal hookworms (typically, *Ancylostoma* spp.) and sometimes by animal and human *Strongyloides* spp. and other species.

person-year of relevant work and of the risks associated with different types and severities of accidents. However, exposures and infections (even if clinically manifest) often are not recognized and, even if recognized, are not reported to local authorities or in the published literature; risk data, with few exceptions (Table 3), are unavailable (1).

Even so, much can be learned from the occupationally acquired cases that have been reported; the terminology "laboratory acquired" is used in this chapter in the generic sense, to encompass all of the cases of interest, including those in health care workers. The criteria for case selection are enumerated in Table 4 (e.g., only accidental exposures that resulted in infection are included). The 229 identified cases that met these criteria are tabulated in Table 5. The 115 cases counted by Pike (2) are tabulated separately in Table 5; Pike primarily provided summary, tabular data (2), comparing different types of microbes, rather than data or references about individual cases.

The 229 cases (Table 5) were ascertained through various means used to identify published and unpublished cases. Although literature searches were conducted, many of the published cases that were ascertained and counted here could easily have been missed; as is evident from the reference list for this chapter, many case reports were buried in publications that had unrevealing titles or that were not detectable through "searches," available electronically, or readily identifiable because of when (e.g., decades ago), where (e.g., in book chapters or obscure journals), or how (e.g., in foreign languages) they were written.

A substantial minority of the identified cases, as well as risk data (Table 3), were ascertained by means available through my work and contacts at the CDC and elsewhere. Such means included requests for teleconsultations with CDC staff about accidental exposures and occupationally acquired infection, diagnostic services, and antiparasitic drugs from the CDC Drug Service; review of CDC surveillance data for nationally notifiable diseases; ad hoc reporting of laboratory-acquired cases; and informal surveys of colleagues at the CDC and contacts elsewhere in the United States and other countries.

The case descriptions provided here focus on the known or likely type of exposure that resulted in the infection; the length of the incubation period from exposure (day 0) until clinical manifestations were noted; the manifestations that developed, especially those that were noteworthy or severe or that were apparent before the case was diagnosed; and the methods used to diagnose infection. Whether clinical manifestations versus posi-

TABLE 2.

Factors that affect whether accidental exposures to parasites cause infection and disease[a]

Factors related to the parasite[a,b]
• Species and strain • Stage of development • Laboratory handling or manipulation[b] • Viability, virulence,[b] and pathogenicity • Infective dose
Factors related to the exposure[a,c]
• Source (e.g., a culture, person, insect vector, or other nonhuman animal) • Route, exposure site, and penetration • Inoculum size
Factors related to the exposed person[a]
• Status of anatomic and physiologic barriers (e.g., whether the exposed skin was intact)[c] • Comorbidities and immunologic status/competence • Awareness/recognition of the exposure[c] • Type and timing of postexposure actions (e.g., wound care and/or presumptive antimicrobial therapy[d])

[a]Some factors are interrelated.
[b]Researchers should adhere to standard biosafety practices even when working with "attenuated" strains.
[c]Unrecognized and seemingly inconsequential exposures can result in infection.
[d]Considerations regarding whether to provide presumptive therapy after accidental exposures also have been discussed elsewhere (1). Expert consultation is encouraged, especially regarding persons who were or may have been exposed to blood or tissue protozoa.

TABLE 3.

Available data about rates of laboratory accidents and infections with particular parasites[a]

Toxoplasma gondii
Laboratory A in the United Kingdom (114) • Rate of recognized laboratory accidents per person-hour of relevant work: one accident per 9,300 person-hours (three accidents in 27,750 person-hours of "performing the dye test or demonstrating viable *T. gondii*") • Total number of probable laboratory-acquired infections: one, which occurred in a person who had been symptomatic but had not noted an accident and whose case was detected through a serosurvey
Laboratory B in the United States (1) • Number of person-years of work: ~48 person-years (average of two or three persons working at a time, over a 19-year period; not limited to hours of relevant work) • Rate of recognized laboratory accidents per person-year: one accident per 12 person-years (four accidents in 48 person-years) • Rate of infections per person-year: one infection per 24 person-years (two symptomatic seroconversions in 48 person-years; testing conducted at baseline and postexposure)
Trypanosoma cruzi
State of São Paulo, Brazil[b] • Number of person-years of work: 126.5 person-years over a period of ~17 years, including 91.5 person-years of relatively high-risk work (e.g., working with needles, preparing viable parasites, or working with tissue cultures with large numbers of parasites) by 21 persons • Rate of recognized laboratory accidents per high-risk person-year: one accident per 15 person-years (six accidents in 91.5 person-years) • Rate of infections per high-risk person-year: one infection per 46 person-years (two infections in 91.5 person-years)
Schistosoma mansoni
Laboratory C in the United States (1) • Rate of infections: four asymptomatic seroconversions, without recognized accidents, among ~20 persons, during the period from the late 1970s through mid-1999 (number of person-years of work not available); two of the four persons had positive stool specimens • Collective data from an unspecified number of laboratories that included "over 100 persons handling millions of cercariae for over 20 years" (225) • Number of symptomatic infections: none • Number of asymptomatic seroconversions: two

[a]See text for additional details. The extent to which these data are representative of research laboratories and laboratorians working with these parasites is unknown.
[b]Data from M Rabinovitch and R de Cassia Ruiz, personal communication (1).

tive test results were noted first and when they were noted depended on such factors as the person's level of self-awareness, the frequency of physical examination, and the type and frequency of diagnostic testing. In addition, the validity of the data about the incubation periods for the cases described here is highly dependent on accurate recall and reporting of the timing of the relevant exposures and clinical manifestations. The early manifestations often were mild or nonspecific and/or initially were overlooked or attributed to other etiologies (e.g., viral illnesses), which underscores the importance of reporting all accidental exposures to local authorities (e.g., supervisor and safety officer) and closely monitoring for clinical and laboratory evidence of infection.

Persons working with parasites that can cause infection detectable by serologic testing (3) should have serum specimens obtained preemployment; periodically thereafter (e.g., semiannually), to screen for asymptomatic infection (especially if detection of asymptomatic infection is medically relevant); after accidental exposures (i.e., immediately postexposure and periodically thereafter; see below); and if clinical manifestations suggestive of parasitic infection develop. The specimens obtained at the time of employment, periodically thereafter, and immediately after an accident are useful for comparison with subsequent postexposure specimens, particularly if follow-up specimens test positive. Freezing multiple aliquots of baseline specimens helps minimize repeated freezing and thawing of individual specimens, which might affect the performance of some assays.

TABLE 4.

Criteria for including occupationally acquired cases of parasitic infection in this chapter and examples of types of cases that were and were not included[a]

	Comments	
Criterion for cases	**Cases that were included**	**Cases that were not included**
Occurred in employees or students in laboratories (research or clinical) or clinical settings (e.g., medical or veterinary hospitals or clinics)	Cases in persons providing "direct patient care" and in ancillary staff (e.g., secretarial and janitorial)	Cases of parasitic infections in patients, even if nosocomial, and cases of *Cryptosporidium* infection in persons exposed to naturally infected animals[a]
Likely resulted from accidental exposure in a work setting	Cases in persons with unrecognized exposures (if possible, the likely route of exposure was induced)	Cases known or likely to have been acquired by natural means or intentional exposures (e.g., experimental inoculations)
Caused by a parasite	Symptomatic and asymptomatic cases	Cases caused by organisms no longer classified as parasites (e.g., microsporidia)
Caused by a pathogen	Cases caused by parasites previously not known to be pathogenic for humans (e.g., *Plasmodium cynomolgi*)	Cases caused by parasites not considered pathogens (e.g., *Chilomastix mesnili*)
Confirmed (e.g., by parasitologic or immunologic methods)	Cases for which details about the means of confirmation were not known or specified	Cases of accidental exposures that did not result in infection[a]
Reported (e.g., in published articles or via personal communications)	Cases reported by persons who requested anonymity	Cases simply noted "in passing" in articles, without any details (see text and other tables for exceptions)[a]

[a]Some publications that describe such cases are cited in the text.

Additional information about diagnosing and treating parasitic infections can be obtained from other reference materials—including the website of CDC's Division of Parasitic Diseases and Malaria (DPDM), at http://www.cdc.gov/parasites—and by consulting local experts or DPDM staff, at 404-718-4745. Questions about the availability of antiparasitic drugs in the United States can be directed to staff of the CDC Drug Service, at 404-639-3670. Although most parasitic diseases are treatable, some cases of treatable diseases may be difficult to treat because of their stage/severity (e.g., advanced cases), parasite factors (e.g., antimicrobial resistance), host factors (e.g., immunosuppression), and/or drug-associated toxicity. Despite therapy, some parasites (e.g., *Toxoplasma gondii*) can cause persistent infection that can reactivate if the host becomes immunocompromised.

Some of the accidental exposures that resulted in parasitic infections were linked directly to poor practices, such as recapping needles or working barehanded (Table 6). Clearly, preventing accidental exposures is preferable to managing their consequences. To minimize the risk for exposures to parasites, laboratorians should use the containment conditions known as biosafety level 2 (BSL-2) (4), which are based on standard microbiological practices and incorporate personal protective equipment and biological safety cabinets when appropriate. Animal containment guidelines specify practices for working safely with BSL-2 agents in the animal arena (4, 5). In clinical settings (e.g., microbiology laboratories), following Universal (Standard) Precautions when handling human specimens entails consistently using BSL-2 facilities and practices (6). The Occupational Safety and Health Administration (http://www.osha.gov) Bloodborne Pathogen Standard (29 CFR 1910.1030) includes regulations for occupational exposures to blood-borne pathogens (7, 8), and the CDC periodically updates guidelines for managing such exposures (9). The Clinical and Laboratory Standards Institute (formerly, NCCLS) develops voluntary consensus standards for various laboratory issues and practices, such as worker protection and verification of training for laboratory personnel (10). Requirements for interstate shipment of etiologic agents have been delineated (6). Additional information about such topics can be found in other chapters of this book and in other reference materials (4, 11).

This chapter, like its previous iterations (1, 12–15), is intended as a reference document, with the expectation that readers will focus on the sections relevant to their work. The blood and tissue protozoa, which are the focus of the chapter, are discussed first. Thereafter, intestinal protozoa and helminths (both intestinal and nonintestinal) are discussed.

INFECTIONS WITH PROTOZOA

Blood and Tissue Protozoa

Summary data

This section focuses on the protozoa that cause leishmaniasis, malaria, toxoplasmosis, Chagas' disease (American

TABLE 5.

Number of reported cases of occupationally acquired parasitic infections

Parasite[a]	No. of cases counted in this chapter[b] ($n = 229$)	No. of cases counted by Pike[b,c] ($n = 115$)
Blood and tissue protozoa		
Trypanosoma cruzi	67	17[c]
Plasmodium spp.	50	18[d]
Toxoplasma gondii	49	28[c]
Leishmania spp.	15	4[b]
Trypanosoma brucei subspp.	8	
Intestinal protozoa		
Cryptosporidium spp.	21	
Cystoisospora belli (*Isospora belli*)	3	5[e]
Giardia intestinalis (*G. lamblia*)	2	2
Entamoeba histolytica		23
Helminths		
Schistosoma spp.	8–10	1
Strongyloides spp.	4[f]	2[g]
Ancylostoma spp.	1[f]	
Ascaris lumbricoides		8
Enterobius vermicularis		1
Fasciola hepatica	1 possible case	1
Hookworm		2[g]

[a]Under each subheading (e.g., "blood and tissue protozoa"), the relevant parasites are ordered in descending frequency, according to the numbers of cases counted in this chapter.

[b]See Table 4 regarding the criteria for including cases in this chapter. In contrast, all 115 cases tallied by Pike (see footnote *c*) reportedly were symptomatic cases in persons working in research or clinical laboratories (2). However, the cases were not necessarily laboratory acquired (e.g., one of the four cases of leishmaniasis [228]) or associated with accidental exposures. He included four cases that resulted from intentional exposures but did not specify which cases these were; 38 (34.2%) of the other 111 cases were known to be associated with accidental exposures. Three of the 115 cases he counted are not listed in this table: 1 case of *Sarcocystis* infection (see text about the improbability of such a case), 1 case of infection with *Chilomastix* (not a pathogen), and 1 case of infection with a *Leukocytozoon* species (not known to infect humans).

[c]The 115 cases tabulated by Pike in his 1976 article (2) had "come to [his] attention" by December 1974, by various means (228–231). He primarily provided summary, tabular data, comparing different types of microbes (e.g., parasites, bacteria, and viruses), by such variables as type of accident and type of laboratory, rather than data or references about individual cases. He cited one reference about a case of Chagas' disease (154) and one about a fatal case of toxoplasmosis (17). In an article published in 1978 (229), he noted being aware of 116 (versus 115) cases of "laboratory-associated" parasitic infections and commented that only 74 (63.8%) had been "published."

[d]Pike commented that 8 of the 18 cases were caused by *P. cynomolgi* and cited two references (41, 42).

[e]Presumably, the etiologic agent of these cases of "coccidiosis" was *C. belli*.

[f]Cutaneous larva migrans (creeping eruption or "ground itch").

[g]Pike did not specify whether these were cases of cutaneous larva migrans or of intestinal infection.

trypanosomiasis), and African trypanosomiasis. Summary data regarding 189 occupationally acquired cases of infection with the protozoa that cause these diseases are provided in the text and tables; some tables focus on individual parasites (see Tables 10 to 16), whereas others facilitate comparisons among the parasites (such as Tables 7 to 9).

The 189 cases occurred in 186 persons: three U.S. men each had two laboratory-acquired cases of malaria; these men are counted twice in the analyses, such as in the following demographic data. Among the 126 case-patients whose sex was known, 72 (57.1%) were men. The median age of the 87 case-patients with available data was 31 years (range, 19 to 71). The case-patients included research and clinical laboratorians, health care workers, and ancillary staff (secretarial and janitorial personnel). They ranged from new employees, students, or trainees to persons with decades of experience, including principal investigators and an emeritus researcher. The work settings included insectaries, animal facilities, research laboratories (e.g., in universities, public health agencies, and pharmaceutical companies), clinical laboratories, hospital wards, and autopsy suites (two cases occurred in persons conducting postmortem examinations).

The case-patients worked in at least 30 countries, and the case reports were written in six languages (i.e., English, German, French, Spanish, Portuguese, and Dutch). Among the 146 persons for whom data were available, 63 (43.2%) worked in the United States (Table 8). The years when the cases occurred (if known) or were reported ranged from 1924 through 2012. For some parasites, marked decade-to-decade variability was noted for the numbers of cases, either overall (Table 7) or for particular species (e.g., *Plasmodium cynomolgi*) or for particular routes of transmission (e.g., accidental ingestion of *T. gondii* oocysts); see the text and Table 6. The data in Tables 7 and 8 should be interpreted with caution because they do not take into account the variability by time and place in the numbers of persons doing relevant work and in the likelihood that cases were reported, either in publications or via personal communications.

Because protozoa, in contrast to most helminths, multiply in the human host, even small inocula can cause illness. As described below, some case-patients either did not recall an accidental exposure or initially considered it trivial and remembered/reported it only after they developed clinical manifestations of infection. Even more of the exposures were unrecognized than is apparent from the data in the table regarding routes of transmission (Table 9) because, for some cases, the likely route could be determined (e.g., ingestion of *T. gondii* oocysts) even if a specific accident had not been recognized. Of the 129 case-patients with available data about the known or likely route of transmission, 62 (48.1%) had percutaneous exposure

TABLE 6.

Examples of practices and occurrences that have resulted in occupationally acquired cases of parasitic infections

Practice or occurrence	Generic examples	Specific examples
Made incorrect assumptions	An organism (whose life cycle was being investigated) was not hardy in the environment	Assumed that *T. gondii* oocysts in feline feces did not survive long
	An organism previously known to infect only nonhuman animals could not infect humans	Assumed that the simian parasite *P. cynomolgi* could not infect humans
	A stock contained a particular subspecies not infective for humans rather than one that is	Assumed that a stock contained *T. brucei brucei* rather than *T. brucei rhodesiense*
	An intermediate host no longer was shedding infective organisms	Assumed that snails had stopped shedding cercariae of *S. mansoni*
Was unprepared for movements (or presence) of persons, animals, or insect vectors	Patients	Had a needlestick injury when obtaining a blood specimen from a restless child
	Colleagues	Had a needlestick injury when standing between a colleague (with a contaminated needle) and a sharps container
	Laboratory animals	Had a needlestick injury when dropped a syringe/needle after being stunned by a sudden movement of an animal Had a needlestick injury when a mouse kicked a syringe/needle Was bitten by an infected animal (the person was infected via the bite or contamination of the wound) Was sprayed with droplets of an inoculum by a coughing or regurgitating animal
	Insect vectors	Was bitten by a mosquito that got loose in an insectary Was bitten after inadvertently placing an arm on a cage that had infected mosquitoes
Used improper laboratory procedures or disposal practices or used defective equipment	Needles/stylets (e.g., needlestick injuries)	Attempted to recap a needle or to remove it from a syringe Did not promptly dispose of a contaminated needle (e.g., set it aside, with its "point" facing up) "Crossed hands" while discarding a contaminated needle
	Capillary hematocrit tubes	Broke a tube while pressing it into clay sealant
	Test tubes/glassware	Unintentionally opened a test tube during disposal or handled contaminated glassware
	Syringes and associated paraphernalia	Used a syringe that had a faulty piston or that was attached to perforated tubing
Did not use protective laboratory garb or equipment	Gloves	Worked barehanded or wore torn gloves
	Clothing	Wore short- versus long-sleeved clothing
	Protection of mucous membranes	Did not consider the possibility of aerosols or mucosal exposures (see Table 1) Pipetted "by mouth"
Miscellaneous occurrences	Work behaviors	Worked unsupervised, before fully trained Worked too fast, without being "careful" Worked late at night, when tired

via a contaminated sharp (i.e., a needle or other sharp object), which is referred to here as parenteral transmission. Accidental puncture with a needle while working with animals (among researchers) or patients (among clinicians caring for patients with malaria) was particularly common. Even a small volume of blood can contain a large number of parasites, much larger than the infective dose. Under experimental conditions that simulated a needlestick injury (specifically, with a 22-gauge needle attached to a syringe containing 2 ml of blood), the mean volume

TABLE 7.

Number of reported cases of occupationally acquired infections caused by blood and tissue protozoa, by decade of occurrence (if known) or publication[a]

Decade	*Leishmania* spp. ($n = 15$)	*Plasmodium* spp. ($n = 50$)	*Toxoplasma gondii* ($n = 49$)	*Trypanosoma cruzi* ($n = 67$)	*Trypanosoma brucei* subspp. ($n = 8$)	Total no. (% of 189; % of 146[b])
1920s	0	1	0	0	0	1 (0.5; 0.7)
1930s	1	0	0	1	0	2 (1.1; 1.4)
1940s	1	0	4	0	1	6 (3.2; 4.1)
1950s	0	4	18	0	1	23 (12.2; 15.8)
1960s	0	7	9	7	0	23 (12.2; 15.8)
1970s	0	9	7	3	1	20 (10.6; 13.7)
1980s	7	13	6	4	2	32 (16.9; 21.9)
1990s	3	9	5	8	3	28 (14.8; 19.2)
2000s	3	4	0	2	0	9 (4.8; 6.2)
2010s (to date)		2				2 (1.1; 1.4)
Unknown	0	1	0	42[c]	0	43 (22.8; NA[d])

[a]The data represent cases, not rates, and do not account for the numbers of persons at risk during the various periods. A total of 189 cases are included in the table. For 35 (24.0%) of the 146 cases for which the decade is provided, the data are based on the decade of publication because the decade of occurrence was not known or specified.
[b]Percentages also are provided using the number of cases with available data as the denominator.
[c]Brener did not provide data for most of the cases he tallied (18, 150).
[d]NA, not applicable.

inoculated was 1.40 μl (range, 0 to 6.13 μl; 20 replicates) (16). Of note, 1 μl of blood from a patient with malaria who has a parasitemia level of 1% (and ~4 million erythrocytes/μl) contains ~40,000 parasites.

The reported cases ranged from asymptomatic (10 cases caused by *T. gondii* and 2 by *Trypanosoma cruzi*) to fatal (one case each caused by *T. gondii*, *T. cruzi*, and *Plasmodium falciparum*); nonfatal but severe cases of various diseases also were reported. Only one of the three reported fatal cases was described in a detailed case report, i.e., the case of toxoplasmosis, which was associated with symptoms of encephalitis that prompted admission to a psychiatric hospital (17). The fatal case of Chagas' disease was mentioned in passing, with few details, in a short report about many laboratory-acquired cases (18); the fatal case of malaria, which occurred in a physician whose case was not suspected or diagnosed until a postmortem examination was conducted, was described briefly in the equivalent of surveillance reports (19, 20).

Many of the cases described in publications were reported for reasons other than their severity; even the fact that they were laboratory acquired may have been incidental. For example, some cases served as opportunities to study immunologic or other responses to infection in nonimmune hosts with known dates of exposures; the mere occurrence of some cases may have increased scientific knowledge (e.g., the simian parasite *Plasmodium cynomolgi* can infect humans); and some cases were described in routine surveillance summaries or in articles that focused on acute, incident cases of infection, regardless of how or why they occurred.

Blood and tissue protozoa of potential relevance to laboratorians and health care workers are discussed below in alphabetical order by genus.

Acanthamoeba *spp.,* Balamuthia mandrillaris*,* Naegleria fowleri*, and* Sappinia pedata

Acanthamoeba spp., *B. mandrillaris*, *N. fowleri*, and *Sappinia pedata* are free-living amebae that cause life-threatening infection of the central nervous system (CNS) (21). Infection with *N. fowleri* typically is acquired by swimming in freshwater. This ameba invades the CNS through the nasal mucosa and cribriform plate and causes primary amebic meningoencephalitis, which typically is rapidly fatal. *Acanthamoeba* spp. and *B. mandrillaris* cause more subacute or chronic infection. Both cause granulomatous amebic encephalitis, which may result from hematogenous dissemination in the context of pulmonary or skin lesions, and *Acanthamoeba* spp. cause keratitis in persons who wear contact lenses or have corneal abrasions.

Relatively few laboratorians work with these amebae; to my knowledge, no laboratory-acquired infections have been reported. However, infection could result from inhalation of aerosols or droplets, exposure to mucous membranes (e.g., by splashes), or, potentially, via accidental needlestick injuries or preexisting microabrasions of the skin.

TABLE 8.

Number of reported cases of occupationally acquired infections caused by blood and tissue protozoa, by country or region of the world where the case occurred[a]

Geographic area	*Leishmania* spp. (*n* = 15)	*Plasmodium* spp. (*n* = 50)	*Toxoplasma gondii* (*n* = 49)	*Trypanosoma cruzi* (*n* = 67)	*Trypanosoma brucei* subspp. (*n* = 8)	Total no. (% of 189; % of 146[b])
United States	9	23	23	8	0	63 (33.3; 43.2)
Europe	1	23	20	3	5	52 (27.5; 35.6)
Latin America	3	0	2	17	0	22 (11.6; 15.1)
Asia	1	2	1	0	0	4 (2.1; 2.7)
Australia/ New Zealand	0	1	1	0	0	2 (1.1; 1.4)
Africa	0	1	0	0	1	2 (1.1; 1.4)
Canada	1	0	0	0	0	1 (0.5; 0.7)
Unknown	0	0	2	39[c]	2[d]	43 (22.8; NA[e])
Subtotals						
– United States	9	23	23	8	0	63 (33.3; 43.2)
– Other areas	6	27	24	20	6	83 (43.9; 56.8)
– Unknown	0	0	2	39[c]	2[d]	43 (22.8; NA)

[a]The data represent cases, not rates, and do not account for the numbers of persons at risk in the various regions (e.g., doing research on particular parasitic diseases) or the likelihood that cases would be recognized and reported. The geographic areas are listed by descending frequency (see last column). A total of 189 cases are included in the table.

[b]Percentages also are provided using the number of cases with available data as the denominator.

[c]Brener did not provide data for most of the cases he tallied (18, 150); see text.

[d]Europe was the probable region for at least one of these cases.

[e]NA, not applicable.

Babesia *spp.*

In nature, *Babesia* spp. are transmitted by the bite of infected ticks. Transmission via blood transfusion is well documented (22); congenital/perinatal transmission also has been reported. In the United States, the recognized etiologic agents of human cases of babesiosis include *Babesia microti* (for most reported cases), *B. duncani* (formerly, the WA1-type parasite), CA1-type parasites, and *B. divergens*-like organisms (23–26). Zoonotic *Babesia* cases in Europe have been attributed to *B. divergens sensu stricto*, the EU1 agent (also known as *B. venatorum*) (27), and *B. microti*. Persons who are asplenic, elderly, or otherwise immunocompromised are at increased risk for clinically manifest and severe cases of infection.

Although, to my knowledge, no occupationally acquired cases of babesiosis have been reported, such cases could be acquired via contact with infected ticks or blood from infected persons or animals. Because ticks can be controlled somewhat more easily than mosquitoes in laboratory settings, the risk of becoming infected through contact with ticks is relatively low.

If babesiosis is suspected, peripheral blood smears should be examined by light microscopy for intraerythrocytic parasites. Approaches for amplifying subpatent parasitemia (i.e., parasitemia too low to be detected on blood smears) include molecular testing (e.g., PCR) and animal inoculation; no *in vitro* culture technique has been developed for *B. microti*. Molecular methods also can be useful for species identification. Serologic testing (traditionally, with indirect fluorescent antibody assays [IFA]) also can be useful in some settings. Acute cases of babesiosis typically are treated either with clindamycin plus quinine (the standard of care for severe cases) or atovaquone plus azithromycin (28).

Leishmania *spp.*

General. Leishmaniasis is caused by diverse species of the genus *Leishmania*, which are transmitted in nature by the bite of infected female phlebotomine sand flies (29). Transmission also can occur via transfusion (30) and congenitally. The promastigote form of the parasite is found in the insect vector and axenic cultures, and the amastigote form is found in macrophages and other mononuclear phagocytic cells in mammalian hosts. The three main clinical syndromes in infected humans are visceral leishmaniasis, which affects internal organs (e.g., spleen and bone marrow) and can be life-threatening; cutaneous leishmaniasis, which causes skin lesions that can persist for months, sometimes years; and mucosal leishmaniasis, a metastatic complication of cutaneous infection with some New World leishmanial species, which involves the naso-oropharyngeal/laryngeal mucosa and can cause substantial morbidity (29).

In laboratory settings, leishmaniasis could be acquired through inadvertent contact with an infected sand fly; containment measures for infected flies should be strictly

TABLE 9.

Number of reported cases of occupationally acquired infections caused by blood and tissue protozoa, by known or likely route of exposure[a]

Route of exposure	*Leishmania* spp. (*n* = 15)	*Plasmodium* spp. (*n* = 50)	*Toxoplasma gondii* (*n* = 49)	*Trypanosoma cruzi* (*n* = 67)	*Trypanosoma brucei* subspp. (*n* = 8)	Total no. (% of 189; % of 149[b])
Parenteral[c]	9	19	15	13	6	62 (32.8; 41.6)
No available information			1	38	1	40 (21.2; NA[d])
Vector-borne transmission	1	24		2		27 (14.3; 18.1)
No accident recognized[e]	1		12	7		20 (10.6; 13.4)
Mucous membrane exposure[f]	1		9	3		13 (6.9; 8.7)
Other skin exposure (e.g., via a spill or splash)[g]						
– Nonintact skin[e,g]	1	7	1	2	1	12 (6.3; 8.1)
– Skin, other				1[h]		1 (0.5; 0.7)
Ingestion (presumptive mode)			9			9 (4.8; 6.0)
Bite (not necessarily the source of infection)[i]	2		1	1		4 (2.1; 2.7)
Aerosol transmission?[e]			1			1 (0.5; 0.7)

[a]The routes of exposure are listed by descending frequency (see last column). If there was uncertainty about the nature of the exposure or no accident was recognized but evidence suggested that a particular route was most likely, this route usually was presumed, for purposes of this table, to have been the mode of transmission. However, the threshold for doing so was subjective because the information available about the cases varied in quantity and quality. Similarly, the distinction between "no accident recognized" and "no available information" was not always clear in the case reports. See text and other tables about the individual parasites for caveats regarding various cases.

[b]Percentages also are provided using the number of cases with available data as the denominator. Cases without a recognized accident were kept in the denominator.

[c]Parenteral exposures involved needles or other sharp objects (e.g., glass coverslip, Pasteur pipette, or broken capillary hematocrit tube) that punctured, scratched, or grazed skin.

[d]NA, not applicable.

[e]Some of the laboratorians who did not recall discrete exposures might have had subtle exposures, such as contamination of unrecognized microabrasions or exposure through aerosolization or droplet spread.

[f]With the exception of the case described in footnote *h*, the exposure was assumed to have been mucosal if the person's face was splashed.

[g]This category includes a hodgepodge of nonparenteral skin exposures. Sometimes the report specified that the person had preexisting skin abrasions, cuts, or breaks (i.e., nonintact skin), whereas other times this was a presumption (e.g., someone who worked barehanded and did not recall parenteral exposures or someone who developed a chagoma at the site of a cuticle was assumed to have had transmission across nonintact skin).

[h]The laboratorian apparently got blood from an infected mouse on his face when a centrifuge tube broke (see text); whether this represented skin or mucosal contact or transmission by aerosol or droplets was unclear.

[i]All of the cases in persons who were bitten by animals are counted here to highlight the importance of this type of injury, even though contamination of the bite wound rather than the bite itself might have been the route of transmission for some of the cases.

followed. Transmission also could occur through contact with cultured parasites or specimens from infected persons or animals (e.g., through accidental needlestick injuries or via preexisting microabrasions of the skin); blood specimens also should be handled with care.

Laboratory-acquired cases. (A) SUMMARY DATA. Fifteen cases of laboratory-acquired infection, caused by eight *Leishmania* spp., have been reported (12, 14, 15, 31–40) (Table 10). The first such reported case occurred in 1930, in a researcher who described his own case (32); and the second occurred in 1948 (40). Among the 15 reported cases, over half (9 [60.0%]) occurred in the United States, one-fifth (3 [20.0%]) in Latin America, and one-fifth (3 [20.0%]) elsewhere. All 15 cases were acquired in research settings; one case was vector borne.

Although most of the infected persons developed cutaneous leishmaniasis, one person also developed mucosal leishmaniasis, and one person developed visceral leishmaniasis. In aggregate, among the 14 persons who recalled a discrete or potential exposure, the median incubation period was ~2.5 months (range, 2.5 weeks to 8 months) (Table 10).

(B) DESCRIPTION OF SIX CASES CAUSED BY ORGANISMS IN THE *LEISHMANIA DONOVANI* SPECIES COMPLEX. Among the six laboratorians known to have become infected with

TABLE 10.

Characteristics of the reported cases of occupationally acquired infection with *Leishmania* spp.[a]

Characteristic	No. (%) of cases (n = 15)
Species	
– *L. donovani* species complex	6 (40.0)[b]
– *L.* (*Viannia*) *braziliensis*	3 (20.0)
– *L. tropica*	2 (13.3)
– *L. major*	1 (6.7)
– *L.* (*V.*) *guyanensis*	1 (6.7)
– *L. mexicana*	1 (6.7)
– *L. amazonensis*	1 (6.7)
Decade of occurrence (if known) or publication	
– 1930s	1 (6.7)
– 1940s	1 (6.7)
– 1950s	0
– 1960s	0
– 1970s	0
– 1980s	7 (46.7)
– 1990s	3 (20.0)
– 2000s	3 (20.0)
– 2010s (to date)	0
Country or region of occurrence	
– United States	9 (60.0)
– Latin America	3 (20.0)
– Canada	1 (6.7)
– Europe	1 (6.7)
– Asia	1 (6.7)
Route of exposure	
– Parenteral	9 (60.0)
– Animal bite	2 (13.3)[c]
– Nonintact skin	1 (6.7)
– Mucous membrane?	1 (6.7)[d]
– Vector-borne transmission	1 (6.7)
– No accident recognized	1 (6.7)
Clinical manifestations	
– Symptomatic cases	15 (100)
– Severe cases	2 (13.3)[e]
– Fatal cases	0

[a]Among the 14 cases with sufficient data to determine or approximate the interval to clinical manifestations, the median incubation period was approximately 2.5 months (range, 2.5 weeks to 8 months). For the subset of nine cases caused by parenteral exposures, the median incubation period was 8 weeks (range, 3 weeks to 6 months). Overall, five cases reportedly had incubation periods ≤1 month: the vector-borne *L. major* case, the two cases caused by *L. tropica*, and two (of five) cases caused by *L. donovani*.

[b]Reportedly, five cases were caused by *L. donovani* (see text) and one was caused by *L. chagasi* (generally considered synonymous with *L. infantum*).

[c]For at least one of the cases, contamination of the bite wound rather than the bite itself was considered the likely means of transmission (31).

[d]The laboratorian had repeatedly contaminated his fingers and oral mucosa (apparently during mouth pipetting) with blood from infected squirrels and once had swallowed blood (32). For purposes of this table, the presumptive route was mucosal.

[e]The two cases classified as severe include one case of mucosal leishmaniasis (14, 15) and one case of visceral leishmaniasis (32).

organisms in the *L. donovani* species complex (12, 32, 34, 35, 40), five laboratorians reportedly were infected with *L. donovani* and one was infected with *L. chagasi* (generally considered synonymous with *L. infantum*). Although both *L. donovani* and *L. chagasi* typically are considered etiologic agents of visceral leishmaniasis, both also can cause cutaneous infection, with or without clinical or laboratory evidence of concomitant visceral infection.

Only one of the six laboratorians infected with parasites in the *L. donovani* species complex developed manifestations consistent with visceral involvement (e.g., fever, splenomegaly, and leukopenia). The affected laboratorian, who published his own case report, "hope[d] that the report of [his] case [would] at least serve as a warning to laboratory workers to safeguard themselves in handling *Leishmania donovani*" (32). His case, which occurred in China in 1930, was the first documented case of laboratory-acquired leishmaniasis; vector-borne transmission was considered unlikely. The possibility that the investigator was infected with *L. infantum* (vs. *L. donovani*) cannot be excluded in retrospect.

Apparently during mouth pipetting, "while making blood counts," he accidentally swallowed blood from an infected squirrel; he "sucked" ~30 to 40 µl of blood into his mouth but probably swallowed a smaller volume. He added: "Through neglect of precautions, contamination of the mouth cavity with infected blood subsequently occurred on many occasions. As it was often necessary to stop the hemorrhage from the infected squirrels following punctures of the superficial veins for blood counts, the fingers of the right hand were not infrequently contaminated due to the fact that the cotton sponges used to check the bleeding was [sic] often soaked with the infected blood. The fingers in turn contaminated the rubber tubes of the blood-counting pipettes" (32). The incubation period was a minimum of several months but could not be determined with certainty, in part because he repeatedly had mucosal and skin contact with contaminated blood. Alternative diagnoses that were considered included influenza and brucellosis. Ultimately, the parasite was isolated by cultures of blood and a "liver puncture."

One of the other five laboratorians infected with parasites in the *L. donovani* species complex—a woman with "mild gait and sensory deficits secondary to multiple sclerosis"—punctured the palm of her right hand, on the thenar eminence, with a needle containing 5×10^8 amastigotes/ml in a suspension of splenic tissue from a hamster (35). The *L. donovani* strain (Humera; L82) with which she was infected had been passaged in hamsters for 14 years. Three weeks postexposure, she developed intermittent erythema, swelling, joint pain, and stiffness in her entire thumb distal to the inoculation site. A nodule was noted at the site by week 7, and regional lymphadenopathy was detected by week 8. The parasite was isolated by culture of a skin biopsy specimen, and amastigotes were noted by histologic examination of the tissue. No parasites were detected by light-microscopic examination and cultures of bone marrow and peripheral-blood buffy coat, and she did not have clinical or laboratory evidence of systemic infection.

While recapping a needle, a physician accidentally inoculated himself with amastigotes from a hamster infected with a strain of *L. donovani* (MHOM/SU/00/S3) that had been maintained in laboratory animals for ~30 years (34). He noticed a nodule at the inoculation site 6 months later, but he did not develop lymphadenopathy or systemic symptoms. The prolonged incubation period might have been attributable to "reduced virulence of the isolate" (34). The parasite was isolated by culture of a skin biopsy specimen, and amastigotes were noted by histologic examination of the tissue; an accentuated lymphoproliferative response to leishmanial antigen also was noted.

Another case of *L. donovani* infection occurred in a laboratorian who had an accidental percutaneous exposure (HW Murray, C Tsai, and D Helfgott, personal communication) (12). This case was caused by an isolate from Sudan that was lethal for golden hamsters despite having been passaged in laboratory animals for decades. Approximately 90 min after harvesting parasites from infected hamsters, the laboratorian accidentally dropped the contaminated syringe/needle onto her thigh; the needle pierced her skin and caused some bleeding. Approximately 4 weeks postexposure, she noted a papular lesion at the site. Her physical examination otherwise was unremarkable (e.g., no fever, lymphadenopathy, or hepatosplenomegaly), as were the values of routine laboratory tests (e.g., complete blood count). The parasite was isolated by culture of a skin biopsy specimen; a blood specimen tested negative by culture.

A technician working with laboratory animals infected with *L. donovani* developed a swollen finger as well as epitrochlear and axillary lymphadenopathy (40). His fingers had been bitten several times "within the few months" before the clinical manifestations developed. Whether he became infected through subsequent contamination of the bite wounds is unknown. The parasite was isolated by culture of a biopsy specimen from a lymph node, and amastigotes were noted in an impression smear of the tissue. No parasites were detected in smears or a culture of bone marrow.

The one reported case of laboratory-acquired *L. chagasi* infection was associated with a needlestick injury, in a fatigued graduate student working late at night, who accidentally inoculated himself rather than a mouse (12). The mouse he was preparing to inoculate suddenly moved, such that he lost his grip on the syringe that contained the inoculum. Although the number of parasites he self-inoculated was not known, the syringe contained 10^7 stationary-phase promastigotes of a Brazilian isolate that had been serially passaged in hamsters for more than a decade. The student noted a "bump" at the injection site ~4 months postexposure, first considered the possibility of leishmaniasis ~1 month later, informed his advisor ~2 months thereafter, and declined evaluation by medical personnel until after he submitted his thesis the next month. When he was first evaluated ~8 months postexposure (~4 months after he first noted the lesion), a nodular, nonulcerative skin lesion, ~1.5 cm in diameter, was noted at the injection site. His infection was parasitologically confirmed by culturing *Leishmania* parasites from skin biopsy and blood specimens. Although he had a positive blood culture, he did not have clinical manifestations suggestive of visceral leishmaniasis.

(C) DESCRIPTION OF THREE CASES CAUSED BY *LEISHMANIA (VIANNIA) BRAZILIENSIS*. Of the three laboratorians infected with *L.* (*V.*) *braziliensis*, one was a student who, when unsupervised, passaged suspensions of amastigotes in hamsters barehanded. He did not recall a recent accident, but "spillage had occurred" (38). He ultimately developed an ulcerative lesion on a finger. Leishmaniasis was diagnosed by demonstrating amastigotes in an impression smear of a biopsy specimen from the lesion and by inoculating a hamster with biopsy material.

A student bitten by a hamster she was inoculating with *L.* (*V.*) *braziliensis* amastigotes from infected hamsters subsequently developed leishmaniasis (31). The bite wound was thought to have become contaminated with the inoculum, but details of the exposure were not specified. By 2 months after the bite, a papular lesion that had developed at the site at an unspecified time had evolved into an ulcerative nodule and ascending lymphangitis was noted. The diagnoses of erysipelas and sporotrichosis were considered. She ultimately developed numerous papular lesions, and leishmaniasis was diagnosed on the basis of histologic examination 10 months after the bite.

A laboratorian became infected with *L.* (*V.*) *braziliensis* (L1794 MHOM/VE/84[VE3]) by accidentally puncturing her thumb with a needle that "pierced its plastic hood" after she inoculated a hamster with an infective macerate containing ~2,000 amastigotes/µl (33). The inoculum was thought "to be low by experimental standards [but] likely high when compared with natural infections." Eight weeks postexposure, she developed an ulcerative skin lesion at the inoculation site; the results of PCR analysis of a blood specimen were positive. During week 18, amastigotes were detected in a skin biopsy specimen.

(D) DESCRIPTION OF TWO CASES CAUSED BY *LEISHMANIA TROPICA*. Of the two laboratorians reportedly infected with *L. tropica*, one was a graduate student who had a needlestick injury while passaging amastigotes (NIH strain 173) in mice (37). He noticed an erythematous, tender nodule at the inoculation site 4 weeks postexposure, which ulcerated 2 weeks later. A lymphoproliferative response to leishmanial antigen became detectable during week 5. However, no *Leishmania* parasites were detected by histologic examination or culture of a skin biopsy specimen obtained during week 12.

The other laboratorian became infected via accidental self-inoculation while injecting an animal and developed an inflammatory nodule at the injection site 3 weeks later (14, 15). The diagnosis was parasitologically confirmed.

(E) DESCRIPTION OF ONE CASE CAUSED BY *LEISHMANIA MAJOR*. A researcher was bitten by a *Lutzomyia longipalpis* sand fly—experimentally infected with strain WR2885 (39)—that had gotten loose in an insectary. The researcher developed a papule at the site of the bite ~2.5 weeks postexposure (RC Jochim and TE Nash, personal communication). On the basis of PCR analysis, both the sand fly (which had been dissected) and the researcher (from whom a skin biopsy specimen was obtained) were infected with *L. major*.

(F) DESCRIPTION OF ONE CASE CAUSED BY *LEISHMANIA* (*VIANNIA*) *GUYANENSIS*. A graduate student accidentally inoculated herself while preparing to inject mice with a strain that had been isolated from a patient 8 years earlier (14, 15). She noted pruritus at the inoculation site 3 months postexposure, and an ulcerative skin lesion developed over the next 2 months. A culture of a biopsy specimen was positive.

(G) DESCRIPTION OF ONE CASE CAUSED BY *LEISHMANIA MEXICANA*. A technician receiving immunosuppressive therapy for systemic lupus erythematosus became infected with *L. mexicana* (36). Several hours after she had cut a finger and dressed "the wound using a sticking plaster," "the plaster was soaked by culture medium containing about 8×10^7 *L. mexicana* amastigotes [sic] when a test tube was unintentionally opened during disposal." A papule developed at the exposure site 8 months later and ulcerated 3 months thereafter. Leishmaniasis was diagnosed by histologic examination, culture, and PCR analysis.

(H) DESCRIPTION OF ONE CASE CAUSED BY *LEISHMANIA AMAZONENSIS*. A laboratorian infected with the Maria strain ultimately developed mucosal leishmaniasis as a sequela of a laboratory-acquired case of cutaneous leishmaniasis (14, 15). She had developed a local erythematous nodule within 3 months of scratching herself with a needle contaminated with a suspension of amastigotes. Culture of a biopsy specimen was positive. She received antileishmanial therapy with what now would be considered an inadequate course of treatment with the pentavalent antimonial compound sodium stibogluconate. The skin lesion regressed but recurred. She again received pentavalent antimonial therapy, and the lesion also was treated with heat. Although the local lesion healed, she developed mucosal leishmaniasis several years later.

Postexposure management. Persons who have had accidental exposures to *Leishmania* spp. should be monitored for clinical and laboratory evidence of infection. In general, presumptive antimicrobial therapy is not recommended; however, decisions should be individualized with expert consultation. Lesions that develop at or near the site of exposure, as well as potential manifestations of mucosal or visceral infection, should be evaluated (29). Periodic serologic testing should be conducted, especially if the organism to which the person was exposed can cause visceral infection. In addition to a baseline preemployment specimen, serum should be collected immediately postexposure, at least monthly for 8 to 12 months or until seroconversion is noted, and whenever clinical manifestations suggestive of leishmaniasis are noted. If seroconversion is noted or clinical manifestations suggestive of visceral infection develop, further evaluation (e.g., examination of bone marrow) may be indicated. Although the numbers are small, the fact that only one of the six persons infected with parasites in the *L. donovani* species complex developed manifestations of visceral leishmaniasis is reassuring.

Plasmodium *spp.*

General. Malaria parasites are transmitted in nature by the bite of infected female anopheline mosquitoes. Transmission also can occur congenitally and via transfusion. In nature, human infection usually is caused by *P. falciparum*, *P. vivax*, *P. ovale*, and *P. malariae* but also can be caused by *P. knowlesi*.

A common means by which laboratorians have become infected is via inadvertent, unrecognized contact with an infected mosquito that escaped from a mosquito colony. Strict containment measures for infected mosquitoes should be followed. Light traps should be operative 24 hours per day, at various levels (e.g., high and low), in rooms where escaped mosquitoes could be present. Laboratorians who dissect mosquitoes could become infected via subcutaneous injection of sporozoites. Another means of transmission to laboratorians and health care workers is via contact with contaminated blood from persons or animals or with cultured parasites, thereby bypassing the hepatic stage of the life cycle.

TABLE 11.

Characteristics of the reported cases of occupationally acquired infection with *Plasmodium* spp.[a]

Characteristic	No. (%) of cases (*n* = 50)
Species	
– *P. falciparum*	27 (54.0)
– *P. cynomolgi*	12 (24.0)
– *P. vivax*	11 (22.0)
Decade of occurrence (if known) or publication	
– 1920s	1 (2.0)
– 1930s	0
– 1940s	0
– 1950s	4 (8.0)
– 1960s	7 (14.0)
– 1970s	9 (18.0)
– 1980s	13 (26.0)
– 1990s	9 (18.0)
– 2000s	4 (8.0)
– 2010s (to date)	2 (4.0)
– Unknown	1 (2.0)
Country or region of occurrence	
– United States	23 (46.0)
– Europe	23 (46.0)
– Asia	2 (4.0)
– New Zealand	1 (2.0)
– Africa	1 (2.0)
Route of exposure	
– Vector-borne transmission	24 (48.0)
– Parenteral	19 (38.0)
– Nonintact skin	7 (14.0)
Clinical manifestations	
– Symptomatic cases	50 (100)
– Severe cases (including fatal case)	10 (20.0)[b]
– Fatal case	1 (2.0)

[a]The median incubation period was 12.5 days (range, 4 to 18 days) for the 24 non-vector-borne cases with available data (21 caused by *P. falciparum* and 3 by *P. vivax*). The median incubation period was 12 days (range, 10 to 15 days) for the five vector-borne cases with available data (three caused by *P. vivax* and two by *P. cynomolgi*). See text regarding a laboratorian for whom the interval from his presumptive vector-borne exposure to diagnosis of *P. vivax* infection was at least 14 months.

[b]See text; all of the cases classified as severe were caused by *P. falciparum*.

Laboratory-acquired cases. (A) SUMMARY DATA. Fifty occupationally acquired cases of malaria (in 47 different persons) caused by three *Plasmodium* spp. (*P. falciparum*, *P. vivax*, and *P. cynomolgi*) have been reported (1, 12, 14, 15, 19, 20, 41–81) (Table 11). A cryptic case of *P. falciparum* infection in a nurse's aide who became ill 10 days after obtaining a blood specimen from a patient reportedly infected with *P. vivax* (82) was not counted because of the discordant species data; blood smears from the patient were not available for reexamination. The nurse's aide did not recall discrete exposures to the patient's blood.

The first documented occupationally acquired case of malaria was reported in 1924 (67); the year of occurrence was not specified. Other reported cases have occurred in each decade since the 1950s. Twenty-three (46.0%) of the 50 reported cases have occurred in the United States. In contrast to the cases of other parasitic diseases, almost half (24 [48.0%]) of the 50 reported cases of malaria were known, or presumed to be, vector borne; only 1 other reported case was in a researcher (specifically, in a person with a parenteral exposure to cultured parasites [68]). Half (25 [50.0%]) of the cases occurred in health care workers exposed to contaminated blood from patients (e.g., while obtaining a blood specimen, inserting an intravenous catheter, administering an injection or transfusion, preparing a blood smear, or conducting a postmortem examination).

Most persons with recognized exposures became ill ~1 to 2 weeks thereafter. Among the 29 persons with available data, the median incubation period was 12 days (range, 4 to 18 days). The medians were comparable (10 to 13 days) for data stratified by *Plasmodium* species or route of exposure. The incubation period was <7 days for only one person, who reportedly became ill 4 days after a nonparenteral exposure to *P. falciparum* and was hospitalized on day 6 (73).

At least 10 (38.5%) of the 26 non-vector-borne cases were classified as severe—i.e., were in persons who had at least one of the following: definite or possible cerebral malaria, renal insufficiency or failure, pulmonary edema, marked hypotension, or a parasitemia level ≥5%. All 10 such cases were caused by *P. falciparum*. Ten (45.4%) of the 22 non-vector-borne cases caused by *P. falciparum* were severe, often at least in part because of delayed diagnosis. One of the 10 severe cases was fatal and was diagnosed postmortem (19, 20).

Some cases were noteworthy for other reasons. Notable vector-borne cases include those caused by the simian parasite *P. cynomolgi*; two vector-borne cases, at different times, in each of three persons (one person was infected once with *P. cynomolgi* and once with *P. vivax*, and the other two persons were infected twice with *P. vivax*); and a case of *P. vivax* infection in a secretary (1). Notable non-vector-borne cases include a case in a pregnant woman (66), a case in a physician exposed to blood from a patient who became infected via transfusion (63), cases in persons who authored or coauthored their reports (52, 55, 68, 71), and a case in a nurse reported by the physician who was responsible for her needlestick exposure (79).

(B) DESCRIPTION OF 24 VECTOR-BORNE CASES. At least 24 reported cases (in 21 persons) were attributed to mosquito-borne (sporozoite-induced) transmission, including at least 12 cases of *P. cynomolgi* infection (1, 14, 15, 41, 42, 58, 62, 64, 72, 75), 7 cases of *P. vivax* infection (1, 14, 15, 60, 76), and 7 cases of *P. falciparum* infection (14, 15, 56, 59, 81).

Among the 12 reported *P. cynomolgi* cases, the first 6 occurred in 1960 in the United States (41, 42, 64, 75) and the last 2 occurred in 1977 and 1979 in the same laboratory in France (62). *P. cynomolgi*, which naturally infects Asian monkeys, was isolated in 1957 and brought to the United States in 1960 for research purposes (64). As expressed by some investigators (41), "up to 1960, the attitude among malariologists generally was: 'Monkey malaria is for monkeys, and human malaria is for humans.'" In short, ". . . it was thought [that] 'man could not be infected with monkey malaria.'" Therefore, some investigators "paid scant attention to the occasional mosquito that escaped into the room." In studies of rhesus monkeys intravenously inoculated with sporozoites of *P. cynomolgi bastianellii* (one of the B strains of *P. cynomolgi*), the infectious dose was 10 sporozoites (83). In 55 volunteers with experimentally induced *P. cynomolgi* infection (24 with sporozoite-induced infection and 31 with blood-induced infection), the mean prepatent period until parasites were noted on blood smears was 19 days (range, 15 to 37 days) (84).

The seven reported vector-borne cases of *P. vivax* infection ranged in time from 1950 through 2012 and occurred in five different persons. The case in 1950 occurred in a secretary who inadvertently had laid her arm on a cage containing infected mosquitoes (1). The case diagnosed in 2012 (60) occurred in a laboratorian who worked with *Plasmodium*-infected mosquitoes, and who had had another presumptive vector-borne case of *P. vivax* infection, caused by a different laboratory strain, diagnosed 8 years earlier (76). Back in 2004, he did not recall a mosquito bite but likely had been exposed 12 days before he became symptomatic (76). On day 0, he had been conducting routine duties related to maintenance of a mosquito colony, while other employees, including trainees, were harvesting sporozoites from highly infected mosquitoes; presumably, a mosquito got loose and bit him. He also did not recall any mosquito bites or other laboratory accidents/exposures to which he could attribute his second case of *P. vivax* infection. Of note, the last time he had worked with the strain (India XIII) that caused his second case of malaria was 14 months before that case was diagnosed in 2012. His clinical course and the course of his *P. vivax* infection might have been affected by the ciprofloxacin therapy he periodically had received in the interim, reportedly for urinary tract infections. For purposes of this chapter, the incubation period for his case that was diagnosed in 2012 is considered unknown.

(C) DESCRIPTION OF 26 NON-VECTOR-BORNE CASES. At least 26 non-vector-borne cases have been reported: 22 caused by *P. falciparum* (16 acquired by parenteral and 6 by nonparenteral routes) and 4 caused by *P. vivax* (3 parenteral and 1 nonparenteral).

(i) Sixteen cases of P. falciparum *infection associated with parenteral exposures.* Seven of these 16 cases were classified as severe, including a fatal case, which was diagnosed at necropsy. Information regarding the fatal case was gleaned from two short descriptions (19, 20), neither of which included clinical details. The case occurred in 1997, in a physician in "the first aid unit" of a hospital in Sicily who had a needlestick injury while evaluating a patient he did not know was in transit from Africa back home to England. The fact that she had malaria, which was diagnosed in England (parasitemia level, 30%), where she was hospitalized 4 days after her brief stop in Sicily, was not known by staff in Sicily until after he died. According to one report (19), he died 20 days postexposure; according to the other report (20), he "subsequently developed malaria, which was not diagnosed until he died, within two weeks."

One of the other six severe cases was the first reported case of occupationally acquired malaria, which was described in 1924 (67). This case occurred in an assistant who pricked a finger during a postmortem examination ("while sewing up the body") and developed a febrile illness on day 15. His symptoms were suggestive of cerebral malaria but initially were attributed to postinfluenza encephalitis. On day 24, the parasitemia level was very high, with "every second or third erythrocyte" infected.

The other five severe cases occurred in nurses:

- A nurse stuck herself with a needle used to place an intravenous catheter in an infant (61). She developed a febrile illness 10 days later and was hospitalized on day 17. Her case was complicated by coma (Glasgow score, 5), "pulmonary edema," and "renal failure" (dialyzed three times); the maximum documented parasitemia level was 5%.
- A nurse who punctured herself while recapping a needle used to obtain a blood specimen developed a febrile illness on day 8 (temperature, 40°C) (48). Her clinical course was complicated by intracerebral hemorrhage, in the context of severe thrombocytopenia, and coma (stage 1; not attributable to the hemorrhage). When her case of malaria finally was diagnosed, 10 days after her first febrile episode, the level of parasitemia was 25%.
- A nurse who had a needlestick injury while obtaining a blood specimen from a "very restless" child developed a febrile illness 1 week later (53). When hospitalized (number of days postexposure not specified), she

was "sleepy, disoriented and extremely dehydrated"; a parasitemia level in the range of 5% to 10% was documented the next day.

- A nurse who considered her exposure to a "broken, blood-contaminated, malaria diagnostic (QBC) test tube" too "trivial" to report (80) became symptomatic 1 week postexposure (maximum temperature, 39.5°C) and was hospitalized 1 week thereafter. Six hours after she was hospitalized, her blood pressure had dropped to 60/32 mm Hg and her parasitemia level had increased to 6%.
- An occupationally acquired case in a nurse led to the first described instance of transmission from a health care worker to a patient (43, 57): Transmission from patient A to the nurse occurred via a parenteral route (a needlestick injury while placing an intravenous catheter) and was followed by transmission from the nurse (3 days before her case of malaria was diagnosed) to patient B; the nurse recalled having dry, chapped hands, with "occasional bleeding fissures," when she cared for patient B (e.g., inserted and manipulated intravenous catheters). In retrospect, the febrile illness the nurse developed on day 10 postexposure (see figure in reference 43) might have been caused by malaria; a blood smear was not examined then. At the time, her illness was attributed to an unconfirmed urinary tract infection, and she was treated with trimethoprim-sulfamethoxazole and ciprofloxacin. Her malaria infection, which was not diagnosed until day 38, when she was hospitalized because of acute onset of a febrile illness, might have been suppressed temporarily by the antimicrobial therapy. For purposes of this chapter, the incubation period for this case is presumed to have been 10 days. On day 38, when intraerythrocytic parasites were noted on a blood smear (parasitemia level, 7%), the diagnosis of babesiosis was initially considered because she had traveled in babesiosis-endemic areas in New England and did not mention her occupational exposure to blood until later during her hospitalization.

One of the nine reported nonsevere cases occurred in a pregnant nurse (6 weeks' gestation), who developed a febrile illness 14 days after "accidentally stabb[ing] herself" with the needle used to collect a blood specimen (66). She "forgot the incident" until her case of malaria was diagnosed and treated. She "fully recovered."

Another of the nonsevere cases was the one reported non-vector-borne case in a researcher, who coauthored his case report (68). He was working with parasites that had been in continuous culture for almost 4 years, which were considered sensitive to chloroquine, on the basis of initial *in vitro* testing. He developed a "minor puncture wound" when he broke a capillary hematocrit tube he had been pressing into clay sealant; presumptively treated himself with chloroquine on days 0 and 8; "soon [forgot about] the incident"; developed a febrile illness on day 17; presumptively treated himself again on day 18; felt ill on days 21 and 22; and mentioned his laboratory exposure to the physician he consulted on day 23. Intraerythrocytic ring forms (initially considered suggestive of *Babesia* parasites) were noted on a blood smear. Subsequent *in vitro* testing demonstrated that the *P. falciparum* parasites with which he had been working had become resistant to chloroquine, despite having been cultured "without chloroquine pressure."

A medical student who coauthored his case report had a needlestick injury after collecting a specimen of arterial blood (52, 55). He developed a sore throat, myalgia, and fatigue on day 8 and "was treated with ampicillin, without effect." On day 15, he developed "drenching sweats," nausea, and vomiting, which prompted hospitalization. He was "lethargic" (no details suggestive of cerebral malaria were provided) and had a temperature of 38.9°C (duration not specified), "tonsillar enlargement and splenomegaly," and a parasitemia level of 50,000 parasites/mm^3 (~1.25%).

A physician who accidentally "stung" the back of a nurse's hand with the stylet used to insert an intravenous catheter described her case (79). She developed a febrile illness 18 days postexposure.

Five other nonsevere cases were associated with parenteral exposures:

- A "Senior House Officer in pathology" was hospitalized 14 days after accidentally stabbing his finger while preparing a blood smear; he had developed a febrile illness 36 h prehospitalization (45).
- A "newly registered" nurse developed a febrile illness 10 days after "a deep, blood-letting" needlestick injury (77, 78). After obtaining a blood specimen "with an 18-gauge, peripheral venous catheter that had no safety feature . . . [, she] stuck herself as she crossed her hands to discard the stylet in a sharps container."
- A nurse who "pricked herself with a needle while transfusing a patient" became ill 13 days later (44).
- A nurse who had an accidental needlestick injury while obtaining a blood specimen developed a febrile illness 13 days postexposure (54).
- A health care assistant who had a needlestick injury while resuscitating a patient became ill 7 days thereafter (65).

(ii) Six cases of P. falciparum *infection associated with nonparenteral ezposures.* Five of the six cases occurred in persons who, preexposure, had demonstrable skin excoriations, cuts, sores, or chapping. Three cases occurred in physicians and three in nurses.

At least three of the cases were classified as severe:

- A case of cerebral malaria occurred in an intern in a clinical laboratory who had "skin excoriations" and handled "a highly contaminated blood sample" (73). He developed chills and headache 4 days postexposure (assuming the relevant exposure was recognized) and was hospitalized on day 6. On day 7, malaria was diagnosed; reportedly, the parasitemia level was 5%. He developed "oliguria" and cerebral malaria, with "altered mental status and hallucinations."
- A case associated with renal failure occurred in a nurse with chapped hands who became febrile 8 days after spilling blood—collected barehanded from a patient at a ski resort—onto her hands (74). On day 15, she was hospitalized in an intensive care unit confused and "in a state of shock." She had a temperature of 40°C, a parasitemia level of 40%, and renal failure, which persisted for 3 months but did not require dialysis.
- A nurse who had a 3-mm-long cut on a finger that was contaminated with a patient's blood during venipuncture developed "high fever" 17 days later (69). Her clinical status deteriorated during 4 weeks of treatment with penicillin and gentamicin for "sepsis of unknown origin." The possibility of malaria was not considered until she mentioned the incident; her parasitemia level was 22%, and her hemoglobin level was 7.4 g/dl.

The other three cases are described below:

- A nurse who had sores on her fingers that were contaminated with several drops of a patient's blood developed a severe headache on day 12 and intense chills and high fever (40°C) on day 13 (47); the parasitemia level was not specified.
- A "senior house-officer . . . fell ill" ~10 days after "admit[ting]" a patient with malaria (e.g., obtaining blood specimens, making blood smears, and placing an intravenous catheter), shortly after cutting one of his fingers "to the quick" while trimming his nails (50, 51). On day 12, "a half-humorous suggestion that he . . . had malaria was acted on . . ." The route of exposure was recalled only "after a careful review of his doings up to the time of examining" the patient (50).
- A physician developed fatigue and rigors 2 weeks after spilling blood—from a patient with malaria attributed to transfusion—onto "her hand at the time of . . . venesection" (63); she "could not recall having any abrasions or cuts on her hands."

(iii) Four cases caused by P. vivax. Three of the four reported cases were associated with parenteral exposures and one with a nonparenteral exposure. One of the four cases occurred in a physician, who described her own case (71). She had pricked her finger with a needle while doing a venipuncture but had "attached little importance to the event." Although she "began to suffer from benign tertian malaria" 12 days later, the diagnosis was not suspected until 7 days thereafter, when "the periodicity of the rigors became established and their unpleasant and debilitating nature obvious."

The other three cases were in nurses. A nurse who had a needlestick injury while placing an intravenous catheter developed a febrile illness 14 days later (1), a nurse who pricked her finger with a contaminated needle while administering an injection became ill after an unspecified incubation period (49), and a nurse with "several small scratches on her fingertips (caused by peeling potatoes)" developed a febrile illness 13 days after performing a venipuncture barehanded (46).

(D) PATIENT-TO-PATIENT TRANSMISSION (NOT INCLUDED IN THE CASE COUNTS). In addition, nosocomial patient-to-patient transmission (e.g., via contamination of a multidose-heparin container) has been reported (60, 70, 85–105). Patient-to-patient transmission is beyond the scope of this chapter (Table 4), and such cases were not included in the case counts. A particularly noteworthy example of patient-to-patient transmission has been reported (90). A cluster of three nosocomial cases of *P. falciparum* infection, one of which was fatal, was linked by molecular analyses to a travel-associated case of malaria (T. C. Boswell, *Abstr 39th Intersci Conf Antimicrob Agents Chemother*, abstr 2081, p 650, 1999). The probable means of transmission was repeated use of a vial of saline—"for flushing intravenous catheters and reconstituting drugs" (96)—that became contaminated with blood from the patient with travel-associated malaria.

Postexposure management. The possibility of malaria should be considered for persons with unexplained flulike or febrile illnesses who might have been exposed to malaria parasites. In contrast to researchers, who know they are working with a particular parasite, even health care workers who know they were exposed to blood might not know that the source was a patient with malaria. Health care workers (as well as researchers) should be encouraged to report even seemingly trivial exposures to local authorities. If relevant (e.g., on the basis of information obtained about the source patient), the exposed person should be counseled about and monitored for infection with malaria parasites, not just bloodborne viruses.

Infected persons can become symptomatic before they develop patent parasitemia (i.e., before parasites are detectable by light-microscopic examination of thick and thin blood smears) and before antibody is detectable by serologic testing. If malaria is a possibility but no parasites are detected on the initial blood smears, more specimens should be examined (three sets of smears, each separated by 12–24 h). In addition, molecular methods (e.g., PCR) can be useful for detecting subpatent parasitemia. Persons infected with *P. cynomolgi* typically have low-level, subpatent parasitemia; the diagnosis can be confirmed by molecular methods or animal inoculation (i.e., by inoculating a monkey with a blood specimen and monitoring the monkey thereafter for development of parasitemia).

The magnitude of the risks associated with accidental exposures has not been well defined, but the prognosis generally is good if infection is diagnosed and treated promptly. In contrast, delayed diagnosis and treatment, especially of cases of *P. falciparum* infection, can result in severe, sometimes fatal, illness. Because of the potential risks associated with infection and the relative safety and ease of administration of available medications, presumptive treatment should be considered, particularly after moderate to high-risk exposures to *P. falciparum* (or to an unidentified species). The following are examples of effective approaches for chloroquine-resistant parasites (PM Arguin, personal communication; also refer to http://www.cdc.gov/malaria):

- After exposures to blood-stage parasites, provide a 3-day course of therapy with daily atovaquone-proguanil (Malarone) or with twice daily artemether-lumefantrine (Coartem).
- After exposures to live sporozoites, provide a 28-day course of therapy with daily atovaquone-proguanil or doxycycline.

Sarcocystis spp.

Various *Sarcocystis* spp. can infect humans. Humans serve as the definitive host (i.e., for the sexual stage) for *S. hominis* and *S. suihominis*, for which the intermediate hosts (i.e., for the asexual stage) are cattle and swine, respectively. Persons working with raw beef or pork should guard against accidental ingestion of sarcocysts (i.e., the asexual stage) via contaminated fingers. Persons infected with these species can be asymptomatic or have gastrointestinal symptoms. Infection is diagnosed by finding oocysts or sporocysts in stool. Humans sometimes serve as intermediate hosts for other *Sarcocystis* spp.; sarcocysts with unknown life cycles and unknown carnivorous definitive hosts have been found in biopsy specimens of human skeletal and cardiac muscle (106–108), sometimes in association with eosinophilic myositis (106, 109). No specific therapy has been identified for treating human sarcocystosis.

Whether laboratorians could become infected via accidental parenteral inoculation of *Sarcocystis* spp. is unknown. Although cell culture-derived merozoites of the classic *Sarcocystis* spp. of domestic animals do not cause disease when inoculated into other animals, culture-derived merozoites of *S. neurona* (an equine species) cause encephalitis after parenteral inoculation of immunosuppressed mice (110, 111).

Toxoplasma gondii

General. *T. gondii*, the etiologic agent of toxoplasmosis, is transmitted in nature to persons who ingest tissue cysts in undercooked meat or oocysts from feline feces that have sporulated and become infective; waterborne transmission of oocysts also can occur. The possibility of transmission via swallowing inhaled oocysts has been suggested (112). Congenital transmission is well documented, and transfusion/transplantation-associated cases have been reported. Symptomatic *Toxoplasma* infection can range from a syndrome of fever and lymphadenopathy to diffuse, life-threatening involvement of internal organs (e.g., myocarditis and encephalitis).

Laboratorians can become infected through ingestion of sporulated oocysts from feline fecal specimens or via skin or mucosal contact with either tachyzoites or bradyzoites in human or animal tissue or culture. All *Toxoplasma* isolates should be considered pathogenic for humans even if they are avirulent for mice (113). Procedures for separating oocysts from feline feces and for infecting mice have been described; fecal flotations should be performed before oocysts sporulate (113). Instruments and glassware contaminated with oocysts should be sterilized because oocysts are not readily killed by exposure to chemicals or the environment (113).

Risks for laboratory accidents and infection. The magnitude of the risk associated with laboratory work with *T. gondii* was assessed in a small case-control study in the United Kingdom (114). The seroprevalence rates were comparable for the three study groups (16 persons per group). Two seropositive persons were identified in laboratory A (Table 3), among "medical laboratory scientific officers with experience of working in the toxoplasma reference unit"; one of the two persons was seropositive before beginning this work. Among groups of age- and sex-matched controls from a routine microbiology laboratory and the general population, zero and three seropositive persons, respectively, were identified.

Among the staff working with *T. gondii* in laboratory A, three persons reported having had accidental exposures to suspensions of viable organisms (i.e., via needlestick injury, spillage onto skin, and splash into an eye), for a rate of three accidents per 27,750 person-hours of relevant work (i.e., working with viable parasites or performing the Sabin Feldman dye test, a serologic test that uses live tachyzoites) or one accident per 9,300 person-hours.

Two of the three persons received presumptive antimicrobial therapy, and none of the accidents resulted in seroconversion. However, one case of infection that was associated with seroconversion and probably was laboratory acquired, via ingestion of oocysts, was identified in the study in a person without a recognized accident (114); see below.

In laboratory B in the United States (Table 3), during the 19-year period from 1980 to 1999, ~30 to 40 persons worked directly with *T. gondii* (1). On average, two to three persons worked in the laboratory at a time (range, one to five), which translates into ~48 person-years of work (not limited to hours of relevant work). Serologic testing was conducted at the time of employment and after accidental exposures. Only one person already was seropositive when hired. Four persons had recognized laboratory accidents: three had percutaneous needlestick injuries, and one squirted a *Toxoplasma*-containing solution into his eye. None of the four persons chose to be treated presumptively, and seroconversion was documented for two of the four, both of whom had needlestick injuries; their cases are described below. The four accidents occurred among the last five persons who joined the laboratory; three of these five persons had accidents within a few months of starting to work, in the context of increased turnover among the staff.

Some risk data regarding work with oocysts are provided below.

Laboratory-acquired cases. (A) SUMMARY DATA. Forty-nine laboratory-acquired cases of *Toxoplasma* infection have been reported (1, 17, 114–138) (Table 12). An unconfirmed case of toxoplasmosis that might have been laboratory associated was not counted (139), nor was an undescribed case attributed to laboratory transmission (details not specified) in a person who served as a source of specimens for studying reactivity against cyst antigens (140).

Reported laboratory-acquired cases occurred in each decade from the 1940s through the 1990s; the highest proportion of the reported cases (36.7%) occurred (or were described) in the 1950s. About half (23 [48.9%]) of the 47 reported cases with available data occurred in the United States. A substantial minority (9 [25.0%]) of the 36 cases for which the mode of transmission was known or suspected probably were attributable to ingestion of oocysts. Parenteral, mucosal, and unrecognized exposures also were quite common.

The median incubation period for the 20 cases with available data, all of which were associated with exposure to tissue stages of the parasite, was 8.5 days (range, 3 days to 2 months); with the exception of 2 cases that had incubation periods of ~2 months, the intervals were ≤13 days (Table 12). The incubation periods were comparable for cases associated with parenteral and mucosal

TABLE 12.

Characteristics of the reported cases of occupationally acquired infection with *Toxoplasma gondii*[a]

Characteristic	No. (%) of cases (*n* = 49)
Decade of occurrence (if known) or publication	
– 1940s	4 (8.2)
– 1950s	18 (36.7)
– 1960s	9 (18.4)
– 1970s	7 (14.3)
– 1980s	6 (12.2)
– 1990s	5 (10.2)
– 2000s	0
– 2010s (to date)	0
Country or region of occurrence	
– United States	23 (46.9)
– Europe	20 (40.8)[b]
– Latin America	2 (4.1)
– Australia	1 (2.0)
– Asia	1 (2.0)
– Unknown	2 (4.1)
Route of exposure[c]	
– Parenteral	15 (30.6)
– No accident recognized	12 (24.5)[d]
– Ingestion (presumptive route)	9 (18.4)[e]
– Mucous membrane	9 (18.4)
– Nonintact skin	1 (2.0)
– Animal bite (see text)	1 (2.0)
– Aerosol transmission? (no evidence provided)	1 (2.0)
– No available information (during an autopsy)	1 (2.0)
Clinical manifestations	
– Asymptomatic cases	10 (20.4)
– Symptomatic cases	39 (79.6)[f]
– Severe cases (including fatal case)	4 (8.2)[g]
– Fatal case	1 (2.0)

[a]The median incubation periods were 8.5 days (range, 3 days to 2 months) for cases related to exposure to tissue stages of the parasite (20 cases with available data), 8 days (range, 3 to 13 days) for the subset of these cases attributed to parenteral exposures (*n* = 11), and 7 days (range, 3 days to 2 months) for the subset of cases attributed to mucosal exposures (*n* = 7 with available data).

[b]For five case-patients, Europe was the presumptive region of occurrence, on the basis of the little available information.

[c]If there was uncertainty about the nature of the exposure or no accident was recognized but evidence suggested that a particular route was most likely, this route usually was presumed, for purposes of this table, to have been the mode of transmission. However, the threshold for doing so was subjective. Similarly, the distinction between "no accident recognized" and "no available information" was not always clear in the case reports. See text for caveats about the cases.

[d]See footnote *c*. At least three persons who did not recall an accident had performed the dye test and, therefore, might have gotten tachyzoites on their skin. The person whose case was fatal (17) had not reported an accident and is presumed, for purposes of this table, not to have recognized a discrete accident.

[e]Eight persons are thought to have ingested oocysts, and one person who "often pipetted toxoplasma exudate" might have become infected *per os* (115).

[f]Having lymphadenopathy was classified as being symptomatic.

[g]Four persons had encephalitis, two of whom also had myocarditis; one person with both conditions died (17).

exposures. Among the four cases with 3-day incubation periods, two were associated with mucosal exposures (splashes) and two were associated with parenteral exposures. At one end of the spectrum, 10 infected persons (20.4%) were asymptomatic. At the other end of the spectrum, four infected persons (8.2%) developed encephalitis, two of whom also developed myocarditis. One of the two persons who developed both encephalitis and myocarditis died (17, 130).

(B) DESCRIPTION OF EIGHT CASES ATTRIBUTED TO INGESTION OF OOCYSTS. Eight cases associated with seroconversion were circumstantially attributed to ingestion of oocysts (114, 126). Seven of these cases were documented in several laboratories in the late 1960s and early 1970s, before oocysts were known to be extraordinarily hardy. The infected persons, who had worked mainly with the M-7741 strain, were essentially asymptomatic, although one person had midcervical lymphadenopathy and two persons had either a "mild 'flu-like' episode" or "slight fatigue and malaise" (126). Before seropositivity was documented, these laboratorians had worked with tissue stages of the parasite for 51 person-years (average, 10 years; range, 1 to 30 years) and with oocysts for 16 person-years (average, 2.3 person-years). Seven other laboratorians who had worked for 75 person-months but not with infective oocysts had not become infected.

The eighth case was documented in laboratory A, in the study described above (Table 3) (114). The case was reported in 1992, but the year it occurred was not specified. The laboratorian—who had been extracting oocysts from the feces of a cat infected with the RH strain—developed malaise, mild fever, and lymphadenopathy. Presumably, his hands occasionally were contaminated with oocysts, which he inadvertently ingested.

(C) RAWAL'S REVIEW OF 18 CASES. In 1959, Rawal described his own case of infection and reviewed 17 others (130), some of which had been described previously (17, 115, 117, 118, 124, 132–134, 136, 137). The probable mode of transmission was unknown for 10 of the 18 cases, including the author's. He suspected that organisms had contaminated his skin, particularly when he performed the dye test, which uses live tachyzoites. Four of the 18 persons had needlestick injuries. For example, one person pricked his finger with a clogged needle he had "set . . . aside point uppermost" (117). Three persons splashed infective material onto their face or into an eye. One person might have become infected via the bite of an infected rabbit (132); *T. gondii*, which can invade susceptible tissue cells throughout the body, has been isolated from rabbit and murine saliva (141). One person who did not recall an accident "often pipetted *Toxoplasma* exudate" and might have become infected *per os* (115). He became ill just 18 days after starting to work in the laboratory. A person who did not have detectable antibody by the dye test ~1 month after the onset of symptoms had positive results when next tested ~3 weeks later (118).

The most commonly reported clinical manifestations in Rawal's case series were fever, headache, malaise, skin eruptions, and lymphadenopathy; two persons were asymptomatic. Three persons developed signs of encephalitis, two of whom also developed myocarditis. One person who had not reported a laboratory accident developed both encephalitis and myocarditis and died (17, 130). This person's case, which occurred in 1951, was described more fully in a separate report (17), whose authors noted that "the handling of toxoplasma in the laboratory [had] not previously been regarded as hazardous." Six days before the patient died, she was admitted to a psychiatric hospital; she had had delusions and hallucinations intermittently for 3 days, flu-like symptoms and poor coordination 4 days before admission, and several months of fatigue, somnolence, and "lack of desire to do things." When hospitalized, she was febrile, had a maculopapular skin eruption, and was delirious. She "spoke frequently to imaginary characters in the room and indicated that she was going to die from toxoplasmosis." She progressively became sicker and was transferred to a medical service in another hospital 4 days after admission to the psychiatric hospital. Although the diagnosis of toxoplasmosis was suspected after she became "seriously ill," confirmatory laboratory results apparently did not become available until after she had died; optimal therapy for toxoplasmosis had not yet been identified when this case occurred.

(D) DESCRIPTION OF 23 OTHER CASES. Four laboratory-acquired cases were described in 1970 among persons in the same laboratory whose cases were diagnosed using the dye test and IFA (131). Since 1962, three other persons in that laboratory had various types of accidents (i.e., a needlestick injury, a bite from an infected rabbit, and a cut with a coverslip containing infected tissue culture cells) that had not resulted in infection; one of these persons had been treated presumptively with sulfadiazine and pyrimethamine, starting on day 0. The number of persons who had worked in the laboratory from 1962 to 1970 was not specified.

Among the four persons in the laboratory who became infected, two recalled accidents (i.e., a needle scratch or puncture [RH strain]) and two did not. The person who scratched herself with a contaminated needle developed cervical and supraclavicular lymphadenopathy, which was noted 10 days after the accident. On the same day, the first postexposure dye-test titer was 1:4,096. The person who punctured herself with a needle immediately began presumptive therapy with sulfadiazine and pyrimethamine and remained asymptomatic but had a rise in her dye-test titer, from 1:256 (the titer when multiple preexposure specimens were tested) to 1:4,096 (the titer from ~1 month postexposure until at least 1 year thereafter).

One of the two persons in the laboratory who became infected but did not recall an accident was a medical student who worked with the RH strain in tissue culture and mice. He developed marked malaise and prolonged lymphadenopathy of unknown etiology. "Although familiar with the adenopathy caused by toxoplasma, [he] at no time considered this as a possible cause of his disease and did not inform [the laboratory director] of his illness" (131). Serologic testing was conducted after the student mentioned that he worked in a particular laboratory, whose director then was called. The other person who did not recall a specific accident was asymptomatic. Her case was detected through the laboratory's routine serologic monitoring program, which entailed testing at baseline and at least yearly thereafter. Her job included performing the dye test, and she was thought to have become infected while preparing the test. The possibility that she became infected outside the laboratory could not be excluded.

A researcher who described his own case pierced his thumb with a needle previously used for intraperitoneal inoculation of mice with a swine strain that had been passaged in mice for 26 months and had become highly pathogenic to mice (138). He was febrile intermittently on days 13 to 29 postexposure and started therapy on day 30. He also had "slight respiratory involvement, malaise, and occasional profuse nighttime sweating." Seroconversion was noted, from negative IFA results before and soon after the accident to IFA titers of 1:64 (day 15) and 1:256 (day 34).

A technician who scratched a finger on her left hand with a contaminated needle (RH strain) became infected (120). The inoculum probably did not exceed 0.02 ml or from 1 to 100 mouse 50% lethal doses. She developed transient epigastric cramping on day 4 and fever, chills, and headache on day 5. On day 7, she was evaluated by a physician, who thought she had influenza. On day 8, she noted tenderness in her left axilla and a tender, erythematous lesion at the inoculation site, which prompted her to remember and report her accident. On physical examination, when she was hospitalized on day 9, she had a skin eruption on her upper body and bilateral axillary and cervical lymphadenopathy; the lesion on her finger was 3 mm in diameter, with a purulent center. The results of the dye test were positive, and *T. gondii* was isolated from a blood specimen obtained on day 9.

A researcher who scratched a finger on his left hand with a needle while inoculating mice with peritoneal exudate from infected animals (RH strain) developed parasitologically confirmed toxoplasmosis (125). His wound was superficial and did not bleed spontaneously. The accident occurred just 21 days after he started to work in the laboratory. He developed generalized myalgia on day 6 postexposure, malaise and headache on day 7, left axillary "swelling" on day 8, and fever on day 9. Ultimately, he also developed a petechial eruption on his chest, cervical and inguinal lymphadenopathy, a pulmonary infiltrate, anemia, and lymphocytosis, with some atypical lymphocytes. *T. gondii* was recovered by animal inoculation—with a blood specimen obtained on day 9 (the day he was hospitalized) and lymph node tissue excised on day 15. The results of the dye test were negative on day 9 but positive on day 11.

Two cases in laboratory B (see above and Table 3) were associated with needlestick injuries (1). A technician working with a concentrated solution of the RH strain from murine peritoneal exudate stuck one of her fingers with a needle while recapping it. Approximately 7 to 10 days later, she developed a severe headache, stiff neck, and perhaps fever and was hospitalized to determine whether she had meningitis. She developed ipsilateral axillary lymphadenopathy and *Toxoplasma*-specific antibody. Another technician in the laboratory stuck herself with a needle while injecting 100 mice with *T. gondii* (C56 strain); she attributed the accident to working too fast. She noted malaise and fatigue on day 13 postexposure, and seroconversion was detected 1 month postexposure (the previous blood specimen was from day 1). She also had had a needlestick injury ~2 years earlier, without subsequent seroconversion.

A technician who stuck a finger with a contaminated needle developed headache, fever, and lymphadenopathy at an unspecified time thereafter (122). Antibody was detectable by IFA and solid-phase indirect hemadsorption when serologic testing was first conducted 1 week postexposure and was detectable later by complement fixation and indirect hemagglutination. The same group of investigators reported two other cases, including one in a medical assistant who accidentally injected parasites (BK strain) into her thumb (123). Three days later, her hand was painful and regional lymphadenopathy was noted. Antibody became detectable on day 14. The other person, a laboratory assistant, became infected by squirting a mixture of saline and tachyzoites (BK strain) from a syringe with a defective piston into his left eye (123). On day 4, his left ear was tender. On day 9, he developed edema of the left eye and left side of his face. Mandibular lymphadenopathy was noted on day 11, and seropositivity was noted on day 15.

Two cases in laboratorians infected with the RH strain were described briefly in an article about "the avidity maturation of IgGs [immunoglobulin Gs] in human toxoplasmosis" (135). One laboratorian had a needlestick injury and remained asymptomatic. The other became infected "by accidental spillage in the eyes" and "showed high fever and adenopathy during approximately 1 week"; the incubation period and details about the "spillage" were not specified.

Two cases diagnosed using multiple serologic techniques, including the dye test, were described in one report (127). One of the two laboratorians stuck his hand with a needle containing murine exudate (R strain). He began presumptive therapy with a sulfa drug the next day and remained asymptomatic, but seroconversion occurred. The other person accidentally sprayed murine peritoneal exudate (BK strain) into his right eye. For 5 days, beginning on day 9 postexposure, he noted malaise, headache, and myalgia. Fever was noted on day 11, and lymphadenopathy at the right angle of his jaw was noted on day 17.

Three other persons became infected through splash-related exposures to their eyes:

- A laboratorian who splashed her eye while manipulating a *Toxoplasma* suspension reportedly developed "relapsing meningoencephalitis" (117) (Franceschetti A, Bamatter F, *First Latin Congr Ophthalmol*, p 344, 1953).
- An investigator splashed infective material into his left eye while harvesting tissue cultures (1). He was passing the cells through a 25-gauge needle—to disrupt the cells and thereby to release the parasite—and the needle might have become clogged. On day 7, he developed fever, conjunctivitis, and tender preauricular and cervical lymphadenopathy. After he dreamed that night about the accident, he realized that the clinical manifestations were attributable to the exposure. Serologic testing at an unspecified time demonstrated a high titer of *Toxoplasma*-specific IgM.
- A laboratory assistant aspirating peritoneal exudate from a mouse infected with the RH strain splashed a small amount of exudate onto the right side of her face; the accident was attributed to a defective syringe (119). Although she did not think that any exudate got into her eye, she was presumed to have had a conjunctival exposure. On day 9, her right eye was bloodshot and she had a headache, earache, sore throat, and painful ipsilateral cervical adenopathy. On day 12, she became febrile. She also developed malaise. Seroconversion was noted by the dye test at 2 weeks, when the first postexposure testing was conducted, and hemagglutinating antibody was detectable 2 weeks later.

Four other occupationally acquired cases have been reported:

- A laboratorian who accidentally spilled ascitic fluid from infected mice onto small scratches on his left hand developed fever and left axillary lymphadenopathy 10 days later (129). Seroconversion was noted by enzyme immunoassay (EIA) for *Toxoplasma*-specific IgG, IgM, and IgA on day 40, when the first postexposure testing was conducted.
- An animal technician who had worked with the RH strain developed a case of toxoplasmosis diagnosed using the dye test (116). Although his infection was attributed to inhalation of aerosolized organisms, evidence to support this route of transmission and details about his work were not provided. Clinical manifestations included fever, rigors, vomiting, headache, generalized aching, tiredness, lethargy, dysphagia, a macular skin eruption, lymphadenopathy (axillary, inguinal, and cervical), and hepatosplenomegaly.
- A pathologist who supervised the postmortem examination of someone with cerebral toxoplasmosis became acutely ill 2 months later (128); potential exposures were not specified. The pathologist's clinical manifestations included fever, chills, severe malaise, profound weakness, lethargy, lymphadenopathy, and hepatosplenomegaly. The adenopathy initially was noted in the anterior and posterior cervical areas and later became generalized. Infection was documented by serologic testing (dye test and complement fixation) and by intraperitoneal inoculation of a mouse with an emulsified lymph node.
- A laboratorian developed fever, headache, conjunctivitis, a maculopapular skin eruption on his face, and antibody to *Toxoplasma* (titer of 1:32 by complement fixation) (121). The case purportedly was laboratory acquired. However, no information about the laboratorian's work or exposures was provided; for purposes of this chapter, the laboratorian is presumed not to have recalled an accident. Although the results of serologic testing for Q fever, a rickettsial disease caused by *Coxiella burnetii*, also were low-grade positive, skin eruptions are uncommon with Q fever.

Postexposure management. Serologic, parasitologic (mouse inoculation or tissue cell culture), molecular, and histologic examinations can be useful for diagnosing *Toxoplasma* infection. IFA and EIA are the most widely used serologic methods for detecting *Toxoplasma*-specific IgG (142). A single test result demonstrating elevated IgG levels can reflect previous infection and, therefore, is not helpful for diagnosing acute infection. If acute infection is suspected and test results for IgG are positive, the laboratorian's baseline specimens also should be tested for IgG; testing by an IgM-capture EIA also can be helpful. A negative *Toxoplasma*-specific IgM result essentially excludes acute infection, if more than several weeks have elapsed since exposure. Although a high-titer IgM result suggests that infection was acquired during the several-month period before the specimen was obtained, detectable levels of IgM can persist for up to 18 months or longer (142). If the IgM result is positive, a *Toxoplasma*-specific IgG avidity assay can be used to help determine when infection was acquired; the presence of high-avidity IgG indicates that infection was acquired at least 4 months before the specimen was obtained.

For naturally acquired *Toxoplasma* infection in immunocompetent persons, typical practice is to treat persons who have organ involvement (e.g., myocarditis) or persistent severe symptoms with pyrimethamine and either sulfadiazine or trisulfapyrimidines, in conjunction with folinic acid, for at least 3 to 4 weeks. For persons with moderate to high-risk exposures, administration of a 2-week course of presumptive therapy with these drugs (or alternative regimens for sulfa-intolerant persons) should be considered while diagnostic testing is in progress; treatment is likely to be the most effective if started promptly. A cautionary tale has been published (143), in which presumptive therapy resulted in neutropenia, complicated by a probable case of *Listeria* meningitis. Even persons treated presumptively should be monitored serologically for several months postexposure or until seroconversion is noted (i.e., they should be tested immediately after the exposure, weekly for at least 1 month, and at least monthly thereafter). As noted above, seroconversion can occur despite presumptive therapy; although presumptive therapy typically prevents disease or at least substantial morbidity, it does not necessarily prevent infection.

Trypanosoma cruzi

General. *T. cruzi*, the etiologic agent of Chagas' disease (American trypanosomiasis), a disease endemic in parts of Latin America, is transmitted by triatomine bugs, when bug feces containing infective metacyclic trypomastigotes contaminate a wound (e.g., the bug's bite wound) or mucous membranes. Oral (e.g., foodborne), congenital, and transfusion/transplantation-associated transmission also has been reported (144–147). After the parasite invades host cells, it replicates as the amastigote stage and differentiates into trypomastigotes, which are released when infected host cells rupture. Circulating trypomastigotes can invade other host cells or be taken up by the vector. In humans, the acute phase of infection lasts ~1 to 2 months and often is asymptomatic. However, the acute phase can be associated with mild, nonspecific clinical manifestations or with life-threatening myocarditis or meningoencephalitis. Years to decades later, ~20% to 30% of infected persons develop cardiac or gastrointestinal manifestations of chronic Chagas' disease.

Laboratorians can become infected via exposure to the feces of infected triatomine bugs, by handling cultures or blood specimens from infected persons or animals, and, potentially, by inhaling aerosolized organisms (148). Although the epimastigote stage of the parasite usually predominates in axenic cultures, trypomastigotes (the infective stage) also are found; the proportion of cultured parasites that are trypomastigotes depends on such factors as the strain of the parasite and the age of the culture. *T. cruzi* can infect persons through needlestick injuries or preexisting microabrasions of the skin or by crossing intact mucous membranes; mice have been infected experimentally by applying parasites to the conjunctiva or oral mucosa (149). Safety precautions for work with *T. cruzi* have been outlined (150–152). *Ex vivo* trypanosomatid parasites are killed by common laboratory disinfectants and by heat (5 min at 50°C) (153).

Laboratory-acquired cases. (A) SUMMARY DATA. Sixty-seven cases of laboratory-acquired *T. cruzi* infection have been reported (1, 14, 15, 18, 150, 154–167) (Table 13). For 37 (55.2%) of these 67 cases, no information is available other than that they occurred (18, 150), and limited information is available about some of the other cases. A possible case of *T. cruzi* infection in a person who was exposed via needlestick to blood from a patient with acute Chagas' disease was not included in the case counts (168).

The first reported case occurred in 1938, and additional reported cases occurred during and since the 1960s. Over half (17 [60.7%]) of the 28 cases with available data about the region of occurrence occurred in South America, which likely in part reflects the amount of research on Chagas' disease conducted there. Of the 22 cases for which the route of transmission was known or induced, 13 (59.1%) were (or likely were) attributable to parenteral exposures. The median incubation period for the 11 cases with available data was 10 days (range, 2 to 24 days); for the subset of 7 cases attributed to parenteral exposures, the median was 12 days (range, 5 to 24 days). One person reportedly had clinical manifestations 2 days postexposure, which constituted erythema along an exposed cuticle; he developed fever and myalgia 4 days postexposure (160). Of the 28 persons with available data on clinical manifestations, 2 (7.1%) were asymptomatic, whereas 9 (33.3%) had evidence of cardiac or neurologic involvement, 1 of whom developed myocarditis and died (18).

(B) DATA ABOUT EIGHT CASES IN THE STATE OF SÃO PAULO, BRAZIL, AND INCIDENCE DATA ABOUT ACCIDENTS AND INFECTION. Some data are available for the State of São Paulo, Brazil,

TABLE 13.

Characteristics of the reported cases of occupationally acquired infection with *Trypanosoma cruzi*[a]

Characteristic	No. (%) of cases (n = 67)[b]
Decade of occurrence (if known) or publication	
– 1930s	1 (1.5; 4.0)
– 1940s	0
– 1950s	0
– 1960s	7 (10.4; 28.0)
– 1970s	3 (4.5; 12.0)
– 1980s	4 (6.0; 16.0)
– 1990s	8 (11.9; 32.0)
– 2000s	2 (3.0; 8.0)
– 2010s (to date)	0
– Unknown	42 (62.7; NA[c])
Country or region of occurrence	
– Latin America	17 (25.4; 60.7)
– United States	8 (11.9; 28.6)
– Europe	3 (4.5; 10.7)
– Unknown	39 (58.2; NA)
Route of exposure	
– Parenteral	13 (19.4; 44.8)[a]
– No accident recognized	7 (10.4; 24.1)
– Mucous membrane	3 (4.5; 10.3)
– Nonintact skin (includes cuticle)	2 (3.0; 6.9)
– Vector-borne transmission	2 (3.0; 6.9)[d]
– Animal bite	1 (1.5; 3.4)[e]
– Skin, other	1 (1.5; 3.4)[f]
– No available information	38 (56.7; NA)
Clinical manifestations	
– Asymptomatic cases	2 (3.0; 7.1)
– Symptomatic cases	26 (38.8; 92.9)
– Unknown clinical status	39 (58.2; NA)
– Severe cases (including fatal case)	10 (14.9; 35.7)[g]
– Fatal case	1 (1.5; 3.6)

[a]The median incubation periods were 10 days (range, 2 to 24 days) for the 11 cases of infection with available data and 12 days (range, 5 to 24 days) for the subset of 7 cases attributed to parenteral exposures. For several cases, the incubation period was not specified but was less than 3 to 4 weeks (155–157), potentially as short as days (156); see text.

[b]Percentages also are provided using the number of cases with available data as the denominator. These numbers are 25 for decade of occurrence (if known) or publication, 28 for country or region of occurrence, 29 for route of exposure, and 28 for clinical manifestations.

[c]NA, not applicable.

[d]For both cases, infection was attributed to exposure to metacyclic trypomastigotes from infected triatomine bugs (150). Whether the laboratorians were exposed to the bugs per se was not specified. The case of a laboratorian who had ocular mucosal contact with triatomine feces (158) was attributed to mucosal transmission; whether the laboratorian had contact with the bug or only its feces was not specified.

[e]The laboratorian was bitten by an uninfected mouse (see text). The relevance of this bite per se was unclear (e.g., whether he bled infected mice near the time of the bite and contaminated the wound).

[f]The laboratorian apparently got blood from an infected mouse on his face when a centrifuge tube broke (see text); whether this represented skin or mucosal contact or transmission by aerosol or droplets was unclear.

[g]Nine persons had signs of cardiac or neurologic involvement, one of whom died (18). A pregnant woman developed a "severe" case of Chagas' disease, with fever, hepatosplenomegaly, and "high" parasitemia (157).

regarding the numbers of documented laboratory accidents and cases of *T. cruzi* infection (MA Shikanai-Yasuda and ES Umezawa, personal communication) (1). As of 1999, an unknown number of persons in at least 15 institutions worked with *T. cruzi*. Eight laboratory-acquired cases of infection, which are discussed below, were documented from 1987 to 1998 in six institutions; presumably, others occurred but were not detected or reported. Of the eight reported cases, two were in asymptomatic persons and two others were in persons who did not recall specific accidents.

In addition, 37 other persons in seven Brazilian institutions reportedly had laboratory accidents from 1984 to 1999 that did not result in infection; 22 (59.5%) of these 37 accidents occurred from 1997 to 1999. The number of exposed persons who were treated presumptively is unknown.

Incidence data for accidents and cases of infection are available for one of the laboratories in the State of São Paulo (M Rabinovitch and R de Cassia Ruiz, personal communication) (1) (Table 3). The data are for a period of ~17 years during the 1980s and 1990s, with 126.5 person-years of observation, including 91.5 person-years for 21 persons doing relatively high-risk work (e.g., working with needles, preparing viable parasites, and working with tissue cultures containing many parasites). Four accidents that did not result in infection and two that did were documented, all of which were included in the tallies of accidents and cases in the two previous paragraphs and occurred among the 21 persons doing relatively high-risk work. The persons who did not develop demonstrable infection had been treated presumptively. The two cases of infection occurred in persons working with the CL strain (Shikanai-Yasuda and Umezawa, personal communication). One of them apparently had conjunctival exposure attributed to defective (perforated) tubing attached to a syringe. Clinical manifestations, which developed after an unspecified incubation period, included fever, petechiae, a pericardial effusion, and peripheral edema; parasites were noted in a blood specimen. Dengue fever and leukemia also had been considered as possible diagnoses. The other infected person did not recall a specific accident. Clinical manifestations included fever, arthritis, congestive heart failure, and reversible facial paralysis; parasites were noted in a smear of a bone marrow aspirate. Leukemia also had been considered as a possible diagnosis.

Another laboratory in the State of São Paulo contributed five laboratory accidents, all with needles, to the above tallies (EA Almeida et al., *Abstr Rev Soc Bras Med Trop*, vol 27 [suppl II], abstr 11, p 145–146, 1994). The accidents occurred in 1993 and 1994, and one involved the principal investigator. The laboratorians who had the accidents were experienced, and their accidents were attributed to not being "careful." Two of the five exposed persons became infected (Y strain); both were asymptomatic. One of the two infected persons had a positive blood smear on day 15 postexposure, and the other person did not have any parasites detected on a blood smear on day 10 but had detectable *T. cruzi*-specific IgG and IgM on day 30. The three exposed persons who did not develop demonstrable infection had been treated presumptively.

All of the other four documented laboratory-acquired cases in the State of São Paulo (of the total of eight mentioned above) occurred among persons working with the Y strain (Shikanai-Yasuda and Umezawa, personal communication). One of the four persons did not recall a specific accident, one apparently got contaminated murine blood on his face when a centrifuge tube broke (whether this constituted skin or mucosal contact or transmission via aerosol or droplets was unclear), one had a needlestick accident, and one cut his hand with a contaminated Pasteur pipette (10^7 trypomastigotes/ml). All four persons developed febrile illnesses and, at a minimum, had serologic evidence of infection; three had parasitologically confirmed cases.

The case that involved the contaminated pipette is particularly noteworthy (MA Shikanai-Yasuda et al., *Abstr Rev Soc Bras Med Trop*, vol 26 [suppl II], abstr 119, p 127, 1993). On day 14 postexposure, the laboratorian developed acute Chagas' disease, despite having received a 10-day course of presumptive therapy with benznidazole; whether therapy was started on day 0 was not specified. His clinical manifestations included fever, headache, mild hepatosplenomegaly, and lymphocytosis. Xenodiagnosis and mouse inoculation were conducted with blood collected on day 22 postexposure; 20 days later, parasites were demonstrable by both means. He ultimately received a second course of therapy with benznidazole, for 80 rather than 10 days.

(C) BRENER'S COMMENTARY ABOUT MORE THAN 50 LABORATORY-ACQUIRED CASES. In a published letter prompted by a laboratory-acquired case described below (162, 164), Brener reported being aware of >50 laboratory-acquired cases of Chagas' disease, including a fatal case in an untreated person with "unusually severe myocarditis" (details not specified) (18); Brener did not specify the exact number of cases (for purposes of the case tallies in this chapter, the number 51 is used). In an earlier publication (150), Brener stated that he was aware of 45 cases, which included 8 previously published cases (154, 161, 163, 165–167) that are described below. Brener provided few details about the 45 cases in aggregate or individually. He noted that they were distributed among 11 countries in North, Central, and South America and in Europe. Of these cases, 16 had been acquired in university laboratories, 14 in nonacademic research laboratories, 12 in laboratories in pharmaceutical companies, and 3 in public health laboratories. The most frequent type of accident reportedly was "accidental puncture with the needle used to infect animals" (150). Contaminated blood was the source of infection for 15 of the 20 cases with available data. Two persons were infected with tissue culture-derived trypomastigotes, two persons were infected with metacyclic trypomastigotes from triatomine bugs, and one person had pipetted and swallowed flagellates from acellular culture medium.

A biochemist whose case Brener described became infected with the Y strain while inoculating mice; a syringe containing 800,000 trypomastigotes in 0.4 ml of blood "dropped from his hands and capriciously fell on his foot in an upright position" (150). Fever, malaise, and crural lymphadenopathy were noted 12 days later. On day 16, a chagoma (the inflammatory primary skin lesion) was noted at the inoculation site and trypomastigotes were found "by fresh blood examination." Later, the results of xenodiagnosis also were positive.

(D) DESCRIPTION OF 17 CASES, INCLUDING 8 CITED BY BRENER. The described cases ranged from asymptomatic to severe; the exposures ranged from unrecognized, to seemingly inconsequential, to noteworthy accidents of various types.

Three symptomatic cases of infection with the Tulahuen strain were described in the same article (166). Only two of the three infected persons recalled a specific laboratory accident, both of whom had superficial needlestick injuries that resulted in severe cases of infection. For one of these persons, the injury occurred when a mouse he was inoculating suddenly moved; he became febrile on day 5 and developed swelling and redness at the inoculation site on day 8. The other person who recalled an accident was wearing short sleeves and was scratched on an uncovered part of her left forearm by a contaminated needle used for an inoculation. She noted a lesion at the site on day 5 and fever, chills, malaise, and left axillary pain on day 6. Both persons who recalled accidents developed manifestations suggestive of meningoencephalitis, which were particularly marked in the woman, who also developed manifestations suggestive of myocarditis. Other clinical features in the three persons included generalized maculopapular skin eruptions, splenomegaly, and facial edema. "The final diagnosis was made by finding the parasite in the blood."

Another case with cardiac manifestations was mentioned briefly in an article about acute Chagas' disease (155); the article's first author provided more details via a personal communication (1). In December, the laboratorian accidentally inoculated himself with parasites from an axenic culture. In January, he had persistent fever and myalgia despite analgesic therapy; the incubation period was not specified. In February, he was hospitalized, but the cause of his illness was not determined. In March, he reported his accident and was evaluated for *T. cruzi* infection. Additional clinical manifestations included myocarditis, pericardial effusion, and arrhythmias. No organisms were detected by light-microscopic examination of blood, hemoculture, xenodiagnosis, or mouse inoculation. However, amastigotes were detected by histologic examination of an endomyocardial biopsy specimen, and the results of serologic testing by EIA, IFA, and direct agglutination were positive.

A "severe" case in a pregnant woman (32 weeks' gestation) was mentioned briefly in an article about "acute maternal infection" (157). The only published information about her exposure was that she was "manipulating *T. cruzi* in a research laboratory"; for purposes of this chapter, the route of transmission is presumed to have been parenteral. At an unspecified time between her exposure and her 3-week postexposure delivery of a healthy baby, she developed a "severe" case of Chagas' disease, with fever, hepatosplenomegaly, and a "high" level of parasitemia "detectable through direct blood examination."

A technician became infected when he stuck his left thumb with a needle contaminated with blood from a mouse infected with the CL strain. He had been trying to remove the needle from a syringe "in a manner prohibited" by the guidelines of the laboratory (164). His acute case of infection had several notable features (162, 164). He was well until 24 days postexposure, when he developed fever and chills; he ultimately had high fever (up to 42°C), with relative bradycardia. He developed a chagoma between the first and second metacarpals of the dorsum of his left hand (i.e., proximal to rather than at the inoculation site), which initially was a confusing feature. He had multiple negative smears of "concentrated blood" and relatively late seroconversion. Mouse inoculation and serologic testing by EIA and IFA simultaneously yielded positive results nearly 5 weeks postexposure; the mouse had been inoculated with a blood specimen 1 week earlier, when specific antibody was not yet detected. Serum neuraminidase activity was detected on day 12, peaked on day 24, and had become undetectable when *T. cruzi*-specific antibodies were first demonstrable (162). Other clinical manifestations included malaise; lethargy; easy fatigability; anorexia; a generalized skin eruption, which initially was maculopapular and later consisted of erythematous blotches; left axillary lymphadenopathy; and T-cell lymphopenia (598 cells/μl, with a normal helper/suppressor cell ratio).

A graduate student injecting mice with trypomastigotes (Brazil strain) had a needlestick injury: a mouse kicked the syringe, which flew into the air, and the needle grazed his abdominal skin (14, 15). The wound was so superficial that it did not bleed and could not be found later that day. On day 10, while on vacation, he noted a "small erythematous pimple" on his abdomen, which gradually expanded to 5 to 7 cm in diameter. When hospitalized on day 18, he was febrile and had a headache. No organisms were detected by light-microscopic examination of blood, buffy coat, or an impression smear of a skin biopsy specimen; and the results of serologic testing by IFA were negative. However, xenodiagnosis, which was performed weeks later, yielded positive results.

A research veterinarian became infected by accidentally puncturing a finger with a needle being used to inject mice with a Brazil strain (159, 163); the estimated inoculum was 1,500 organisms. Approximately 16 to 18 days later, he developed swelling and discoloration of the finger, tender ipsilateral epitrochlear and axillary lymphadenopathy, fever, rigors, and malaise. When he was hospitalized on day 19, an erythematous, blotchy, indurated skin eruption was noted on his upper body. Serum specimens from days 7, 19, 40, 72, 100, 128, and 159 were tested for total IgM and IgG and, by IFA, for *T. cruzi*-specific IgM and IgG (163). The results with all four assays showed elevated levels on day 40 and on all testing days thereafter, with decreases in some of the levels on day 100 and thereafter. Whereas *T. cruzi* was not detected by direct examination of blood on day 20, the parasite was isolated by hemoculture, mouse inoculation, and xenodiagnosis (159).

A technician with more than 15 years of experience in a research laboratory became infected with the Y strain (clone YP3); she "auto-inoculat[ed]" trypomastigotes "on the dorsal surface of the left hand at the base of the thumb, which she did not consider important at the time" (156). She was evaluated 3 weeks postexposure, "complaining of low fever, fatigue, chronic headache, malaise and depression lasting approximately 21 days." Although she had "an inoculation chagoma" and ipsilateral axillary adenopathy, she had not considered the possibility of Chagas' disease, which thereafter was diagnosed by PCR analysis of blood specimens.

Accidents that did not involve needles also have been reported, including, among others, the accident associated with the first documented case of unintentional, laboratory-acquired Chagas' disease, which occurred in 1938. This xenodiagnosis-confirmed case resulted from ocular mucosal contact with triatomine feces (158). Thirteen days postexposure, the investigator developed pain and redness of the internal angle of her exposed eye. The next day she developed ipsilateral palpebral edema, dacryocystitis, and increased tearing; generalized malaise; and fever. Other manifestations in the ensuing days included headache, myalgia, edema of the ipsilateral cheek, lymphadenopathy, and splenomegaly.

A microbiologist who spilled a solution of trypomastigotes (Tulahuen strain) onto slightly abraded skin on his left hand became infected (154). He was hospitalized 1 week postexposure, with a 4- to 5-day history of headache, low backache, anorexia, fever, chills, and fatigue. When hospitalized, he was drowsy and intermittently delirious and had photophobia, fever, sinus tachycardia, palatal petechiae, and ipsilateral lymphadenopathy. On day 4 of hospitalization (1.5 weeks postexposure), he developed a maculopapular skin eruption on his trunk, arms, and thighs; he never developed a chagoma. Two days later, a systolic murmur, a pericardial friction rub, cardiomegaly, and nonspecific T-wave changes on an electrocardiogram were noted. Testing for evidence of *T. cruzi* infection was conducted daily, starting on day 2 of hospitalization; on day 11, trypomastigotes were seen in direct smears of the patient's blood and in blood from a mouse that had been inoculated with his blood on day 5. His course might have been affected by concomitant bacteremia and corticosteroid therapy begun on day 5.

A researcher emeritus whose work included bleeding infected mice recalled being bitten on his left index finger by a control mouse (14, 15). The relevance of this bite per se was unclear (e.g., whether he bled infected mice near the time of the bite and contaminated the wound); for purposes of this chapter, the incubation period is considered unknown. On the day after the bite, he noted fatigue (which he also had, in addition to headache, the day before he was bitten), anorexia, fever, and chills; and his finger became red, swollen, and tender. On day 2, a nontender left axillary lymph node was noted. On day 3, the lesion on his finger was lanced and a small amount of serosanguineous material was released, the culture of which was positive for *Enterobacter cloacae*. Amastigotes were detected in a biopsy specimen obtained from the lymph node on day 13.

A medical technician in a pharmaceutical company who operated barehanded in the peritoneal cavity of a mouse infected with the Tulahuen strain developed erythema along a cuticle 2 days later and fever and myalgia 2 days thereafter (160). When hospitalized 12 days postexposure, he had splenomegaly and generalized lymphadenopathy. Electrocardiographic findings were consistent with myocarditis, and trypomastigotes were noted in "blood smears."

Four infected persons, in addition to three mentioned above, did not recall specific exposures. For one of these persons—a laboratorian who worked barehanded with the blood of infected mice and with contaminated triatomine feces—the presumptive exposure date was determined (165). On that day, he developed a bruise on his hand (circumstances not specified). Four days later, he noted local erythema and swelling. Subsequent clinical manifestations included anorexia; fatigue; myalgia; headache; fever, with relative bradycardia; a skin eruption on his trunk, extremities, and face; conjunctivitis; left axillary lymphadenopathy; and splenomegaly. The diagnosis was confirmed by mouse inoculation (conducted on day 22) and xenodiagnosis (day 25); the results of serologic testing by complement fixation (day 22) also were positive.

The other three persons who did not recall exposures did not know how or when they became infected:

- A technician who had worked for ~20 years with the Tulahuen strain developed a chagoma on her thumb and subsequently developed weakness, headache, fever, night sweats, regional lymphadenopathy, transient pedal edema, intermittent tachycardia, and nonspecific T-wave changes on an electrocardiogram (167). *T. cruzi* was isolated via hemocultures obtained 7 days after the chagoma was noted, and seroconversion was noted thereafter by complement fixation and hemagglutination.
- A nonscientist whose job included collecting glassware used for culturing *T. cruzi* was evaluated for unexplained fever and was discovered fortuitously, through a positive hemoculture, to be infected (161).
- A technician who worked with the Brazil strain (e.g., maintaining cultures and working with animals) developed symptoms suggestive of a viral illness, trypomastigotes were found when wet mounts of buffy coat were examined, and an enlarged epitrochlear lymph node was noted subsequently (14, 15). *T. cruzi* PCR and serologic results also were positive.

Postexposure management. Persons who have had accidental exposures to *T. cruzi* should be monitored for clinical and laboratory evidence of infection, even if they are treated presumptively. An example of a monitoring protocol is provided in Table 14, which includes comments about adapting the protocol for persons treated presumptively; details about examining blood for *T. cruzi* are provided in Table 15.

Experts in the field generally recommend prompt initiation of presumptive therapy for persons who have had moderate to high-risk accidents (147). Whereas persons with documented infection typically are treated for up to several months (~60 days with benznidazole or ~90 days with nifurtimox [144, 145]), the duration of presumptive therapy typically is shorter (e.g., a 10-day course of benznidazole therapy [147]). The rationale for presumptive therapy is 2-fold: (i) both acute and chronic Chagas' disease can be life threatening, and (ii) therapy is thought to be more effective the earlier it is started. Although this rationale is strong, the efficacy and optimal duration of presumptive therapy have not been established, for obvious reasons, in controlled clinical trials. As described

TABLE 14.

Clinical and laboratory monitoring for *Trypanosoma cruzi* infection after accidental exposures

General comments

- The duration and frequency of monitoring should be individualized; see text.
- Various types of monitoring should be conducted, regardless of whether the person who had the accidental exposure is treated presumptively (see text).
- For persons treated presumptively, the protocol below should be adapted to include more intensive monitoring after the course of therapy has been completed, because therapy could be suppressive (i.e., early test results could be negative and later results could be positive).

Monitoring for clinical manifestations of infection

- Temperature should be monitored daily for 4 weeks, and unexplained febrile or flu-like illnesses during the 6-month (or longer) postexposure period should be evaluated. In addition, persons who develop any of the following should be evaluated: skin lesions (e.g., a chagoma), conjunctivitis (after mucosal exposures), swelling, or erythema at or near the site of exposure; localized or generalized lymphadenopathy or skin eruptions; hepatosplenomegaly; or clinical manifestations suggestive of myocarditis or meningoencephalitis.

Monitoring for parasitemia

- A suggested approach is to monitor blood for parasitemia at least weekly for at least 4 weeks, bimonthly for 1 to 2 months, monthly for at least 1 to 2 months, and whenever clinical manifestations suggestive of Chagas' disease are noted. See Table 15 for details about light-microscopic examination of whole blood and buffy coat specimens for motile trypomastigotes. PCR analysis may facilitate early detection of infection (232, 233). Conventional means for parasitologic confirmation of infection also include histologic examination, hemoculture, and animal inoculation; xenodiagnosis is an approach that was used in the past.

Monitoring for development of antibody to the parasite

- Pre-employment serum and/or serum obtained immediately post-exposure should be tested in parallel with subsequent specimens, especially if the results for postexposure specimens are positive. In newly infected persons, the interval from exposure to development of detectable antibodies to *T. cruzi* is variable; antibodies typically are detectable by approximately 6 to 8 weeks postexposure, but the interval may be shorter (e.g., seropositivity within several weeks of the exposure) or longer. Especially if PCR analysis can be conducted at the intervals described above (starting with at least weekly specimens/testing for at least 4 weeks), serologic testing could be deferred until approximately 6 to 8 weeks postexposure. If the serologic results are negative, consider obtaining/testing serum specimens monthly (or bimonthly) for at least the next 4 months (or until seroconversion is noted) and whenever clinical manifestations suggestive of Chagas' disease are noted.

TABLE 15.

Practical guide for detection of circulating *Trypanosoma cruzi* trypomastigotes by light microscopy[a]

Obtain whole, anticoagulated blood by venipuncture or fingerstick.

Process and examine the blood while it is fresh.

- Use sterile technique if specimens also will be cultured or inoculated into animals.[b]

Prepare both whole blood and buffy coat for examination.

- If the blood was obtained by venipuncture, remove approximately 1 ml of whole blood from the tube, before centrifugation, and place it in a small vial for examination as described below. Centrifuge the rest of the blood to separate the erythrocyte, leukocyte (buffy coat), and plasma layers. Pass a pipette through the plasma to the buffy coat layer. Carefully remove the buffy coat and place it in a small vial for examination as described below.
- If the blood was obtained by fingerstick, fill at least two microhematocrit tubes with blood. Leave one tube uncentrifuged, so whole blood can be examined as described below. Centrifuge the other tube to separate the various layers of cells. Break the tube just above the buffy coat layer, remove the buffy coat, and place it in a small vial for examination as described below.
- Prepare multiple slides for examination. To facilitate semiquantitative analysis (see below), if 12-mm-diam circular coverslips are used, dot 1.5-µl aliquots of blood and separate aliquots of buffy coat onto slides and place a coverslip over each dot; if 22- by 22-mm square coverslips are used, use 6.4-µl aliquots.

Examine slides of both whole blood and buffy coat, under high power, by light microscopy, preferably phase contrast, looking for motile trypomastigotes (length, ~15–25 µm), which often are first manifest by the resultant movement of the other cells on the slide. Stain positive slides with Giemsa.

- Specimens of whole blood can be examined more quickly than buffy coat because erythrocytes are homogeneous in size and color, whereas leukocytes and debris in buffy coat are translucent and heterogeneous in size. However, trypomastigotes are present in higher concentrations in buffy coat than in whole blood. Therefore, both whole blood and buffy coat should be examined.
- If these recommendations about the sizes of aliquots and coverslips are followed, examining 200 high-power fields (magnification, ×400) of whole blood is the equivalent of examining 0.48 µl of blood, and finding on average 1 parasite per high-power field indicates that the specimen contains ~400,000 parasites/ml.

[a]LV Kirchhoff was instrumental in the development of this table (1). A clot contraction technique for concentrating hemoflagellates has been described (234). Appropriate precautions should be used when handling specimens; see text and Table 1.

[b]Residual buffy coat and whole blood can be examined by other methods; see Table 14.

above, a clinically evident case of acute Chagas' disease developed despite short-course presumptive therapy. Another potential concern is that such therapy could suppress parasitemia and mask indicators of inadequately treated infection. However, recommending a 2- to 3-month-long course of presumptive therapy also is problematic because treatment can be associated with substantial toxicity of various types (e.g., dermatologic, hematologic, gastrointestinal, neurologic). Persons being treated for laboratory-confirmed infections might be more willing than persons being treated presumptively to continue therapy despite development of side effects that do not necessitate stopping therapy.

Of note, toxicity is not necessarily restricted to long-course therapy, as highlighted by the following cautionary tale (12). A laboratorian who had a relatively low-risk exposure to *T. cruzi* developed a severe hypersensitivity reaction to benznidazole, which was manifested by fever (38.3°C) and probable erythema multiforme, with apparent "sparing" of mucous membranes. On the 5th day of the 10-day course of presumptive therapy, she noted a "mild rash," which markedly worsened on the 9th and 10th days. She was hospitalized on the 10th (last) day of therapy, for a total of 5 days (4 nights), and received parenteral corticosteroid therapy. She had developed confluent erythema on her face and neck, with widespread, coalescing, erythematous papules and plaques on her trunk and upper extremities. Some of the lesions had targetoid centers suggestive of erythema multiforme. This cautionary tale underscores the importance of considering the risks associated with the exposure and with presumptive therapy.

Trypanosoma brucei rhodesiense *and* Trypanosoma brucei gambiense

General. *T. brucei rhodesiense* and *T. brucei gambiense*, the etiologic agents of East and West African trypanosomiasis, respectively, are transmitted in sub-Saharan Africa by tsetse flies (169). Transmission also could occur congenitally and via transfusion. African trypanosomes multiply in the bloodstream of mammalian hosts. East African trypanosomiasis typically follows a more acute course than the West African disease and is characterized by early invasion of the CNS. Cases of laboratory-acquired African trypanosomiasis can result from contact with blood or tissue from infected persons or animals.

Laboratory-acquired cases. (A) SUMMARY DATA. Eight laboratory-acquired cases have been reported (1, 170–175); six cases were caused by *T. brucei gambiense* and the other two by *T. brucei rhodesiense* (Table 16). Of the eight cases, five (62.5%) were reported in a total of four published articles (one case was described in two articles), and three (37.5%) were reported via personal communications from the same person (A Van Gompel) (13). The median incubation period for the eight cases was 7 days (range, 1 to 10 days). As described below, the one person who reportedly became ill 1 day postexposure had chills on days 1 and 2 and became febrile on day 3.

TABLE 16.

Characteristics of the reported cases of occupationally acquired infection with *Trypanosoma brucei* subspp.[a]

Characteristic	No. (%) of cases (n = 8)
Subspecies	
– *T. brucei gambiense*	6 (75.0)
– *T. brucei rhodesiense*	2 (25.0)
Decade of occurrence (if known) or publication	
– 1940s	1 (12.5)
– 1950s	1 (12.5)
– 1960s	0
– 1970s	1 (12.5)
– 1980s	2 (25.0)
– 1990s	3 (37.5)
– 2000s	0
– 2010s (to date)	0
Country or region of occurrence	
– Europe	5 (62.5)
– Africa	1 (12.5)
– Unknown	2 (25.0)[b]
Route of exposure	
– Parenteral	6 (75.0)
– Nonintact skin	1 (12.5)[c]
– No available information	1 (12.5)
Clinical manifestations	
– Symptomatic cases	8 (100)
– Fatal cases	0

[a] The median incubation periods were 7 days (range, 1 to 10 days) for the eight cases of infection and 7 days (range, 1 to 10 days) for the subset of six cases attributed to parenteral exposures. See text about the plausibility of a 1-day incubation period.

[b] Europe was the probable region for at least one of these cases.

[c] The laboratorian did not recall a discrete accident (see text and references 170 and 171).

(B) DESCRIPTION OF SIX CASES CAUSED BY *T. BRUCEI GAMBIENSE*. The first two reported cases were caused by the Yaoundé strain from Cameroon, which had been passaged in laboratory animals since 1934 (173, 174). For the case in 1949, details regarding the laboratory exposure were not specified. The case in 1959 was associated with a needlestick injury, which occurred while the laboratorian was attempting to inoculate a struggling rat. For both cases, the incubation period was 10 days.

One of the other four persons infected with *T. brucei gambiense* was a technician who scratched his arm with a contaminated needle (strain Gboko/80/Hom/NITR. Kad.) "during pre-experimental passaging of Wistar rats with . . . parasites" (172). He was thought to have been exposed to a "tiny inoculum, part of which must have been washed out with soap and water" (172). When evaluated 1 week later, he had a large chancre (the inflammatory primary skin lesion) at the inoculation site, fever, headache, anorexia, and fatigue. Whether he first noted the "large chancre" earlier than 1 week postexposure was not specified. "Numerous trypanosomes" were found in blood smears. The case report did not include the values of cerebrospinal fluid (CSF) parameters.

A technician who stuck her thumb after inoculating mice became infected with a strain of *T. brucei gambiense* (FEO ITMAP-1893) that had been isolated from a patient

31 years earlier and maintained through passage in mice (175). She became febrile (39°C) 8 days later and developed erythema, warmth, and swelling of the thenar region of her hand 2 days thereafter; a swollen axillary lymph node and splenomegaly were noted the next day. Laboratory abnormalities included leukopenia and thrombocytopenia. Trypanosomes were isolated from the chancre and were detected in a blood specimen passed through a DEAE-cellulose column. Seroconversion was noted on day 18 by IFA testing. The leukocyte count and protein level in CSF were normal, and a mouse inoculated with CSF did not become infected.

While handling a mouse infected with *T. brucei gambiense* (cloned antigenic variant LiTat 1.3, serodeme LiTAR 1), a technician stuck his left hand with a contaminated needle (Van Gompel, personal communication). On day 7, he had fever, headache, and erythematous swelling at the inoculation site. On day 10, the site was still swollen, with bluish red induration; he also had a red lymphangitic streak and swelling of the ipsilateral epitrochlear lymph node. On day 11, trypanosomes were noted in a Giemsa-stained thin blood smear and aminotransferase levels were 3 to 5 times the upper limits of normal. He was hospitalized on day 12 and had 5 lymphocytes/µl in a specimen of CSF.

Another technician had a similar laboratory accident; he stuck his left fifth finger with a needle while handling a mouse infected with the same strain of *T. brucei gambiense* (Van Gompel, personal communication). He had chills on the next 2 days. On day 3, he had fever (39 to 40°C), headache, sore throat, and dark (concentrated) urine. On day 4, he developed nausea and vomiting and became agitated and profoundly fatigued. A total of six trypanosomes were noted in 50 microscopic fields of a Giemsa-stained thick blood smear; the CSF examination was unremarkable, including a negative culture for trypanosomes. He had leukopenia and thrombocytopenia; his aminotransferase levels rose to 2 to 3 times the upper limits of normal on day 5.

(C) DESCRIPTION OF TWO CASES CAUSED BY *T. BRUCEI RHODESIENSE*. One of the two persons known to have become infected with *T. brucei rhodesiense* was a medical student doing a summer research project that involved infecting mice and rats with stabilates of various serotypes (i.e., variable antigen types); separating trypanosomes from animal blood by column chromatography on DEAE-cellulose, which resulted in a concentrated suspension of organisms (~10^8/ml); and inoculating the parasites into chickens (170, 171). His role in the direct work with live animals was supportive (e.g., he restrained chickens that were being inoculated).

The trypanosomes were derived from a stock (BUSOGA/60/EATRO/3) isolated 14 years earlier from tsetse flies in Uganda. The stock mistakenly had been thought to be *T. brucei brucei* and, therefore, not infective for humans. The laboratory had stabilates of 12 different serotypes (ETat 1 to 12), only 1 of which (ETat 10) was infective for humans. The student had used several serotypes in experiments. Retrospective serologic investigations after his case was diagnosed showed that he was infected with ETat 10 (170). He had worked with this serotype 8 and 5 days before he became ill. The relevant exposure might have occurred 5 days before he became ill, when he exsanguinated infected rodents and separated trypanosomes from their blood. Although he did not recall a discrete accident, the route of transmission could have been via the abrasions he had gotten on his hands while restraining chickens.

The student developed an erythematous, swollen area on a finger. Other clinical manifestations included arthralgia, hyperalgesia of thigh and calf muscles, fever, rigors, fatigue, vomiting, diarrhea, tinnitus, headache, confusion, disorientation, a generalized skin eruption, cervical lymphadenopathy, and splenomegaly. One trypanosome was noted on examination of 400 microscopic fields of a Giemsa-stained thick blood smear (magnification not specified). Later, his serum IgM level increased markedly. The case report did not include CSF findings.

The other documented case of *T. brucei rhodesiense* infection was in a technician who cut his left hand with a glass coverslip contaminated with *T. brucei rhodesiense* (cloned antigenic variant ETat 1.10, serodeme ETat 1 [Van Gompel, personal communication]). On days 7 to 11 postexposure, he noted chills and fever (39 to 40°C), myalgia, and a painful ipsilateral axillary lymph node, which became more swollen and painful in the ensuing days. On day 11, one motile trypanosome was noted on examination of 40 microscopic fields of a wet mount of blood (magnification, ×400). When hospitalized, he was febrile (38.1°C) and had a positive Giemsa-stained thick blood smear (1 trypanosome in 25 microscopic fields; magnification, ×1,000). The CSF examination was unremarkable, including a negative culture for trypanosomes. His leukocyte count was within normal limits, but he had mild thrombocytopenia.

Postexposure management. The diagnosis of African trypanosomiasis is parasitologically confirmed by detecting trypanosomes in peripheral blood, CSF, or an aspirate of a chancre, lymph node, or bone marrow. The ease of finding trypanosomes in various tissues and fluids depends on the infecting subspecies (*T. brucei gambiense* versus *T. brucei rhodesiense*) and the stage of infection (hemolymphatic versus CNS). Whereas *T. brucei rhodesiense* typically is relatively easy to find on a blood smear (at least for vector-borne cases), *T. brucei gambiense* is more difficult to detect. Concentration methods that facilitate detection include microhematocrit centrifugation, followed by examination of buffy coat, as is done for *T. cruzi* (Table 15), and the miniature anion-exchange centrifugation technique using DEAE-cellulose (169, 176). Animal inoculation (for *T. brucei rhodesiense*) and *in vitro* culture can be used to isolate the parasite. The sensitivity of the card agglutination test (Institute of Tropical Medicine, Antwerp, Belgium) is high in most, but not all, areas where Gambian trypanosomiasis is endemic (169, 177).

The choice of therapy also depends on the infecting subspecies and the stage of the infection. For *T. brucei gambiense*, the medications typically recommended for the hemolymphatic and CNS stages are pentamidine and eflornithine, respectively; eflornithine is effective both as monotherapy and in combination with nifurtimox. For *T. brucei rhodesiense*, the hemolymphatic stage typically is treated with suramin; melarsoprol is the only effective medication for the CNS stage. Decisions regarding whether to administer presumptive postexposure therapy before infection is documented should be individualized with expert consultation.

Intestinal Protozoa

General. Intestinal protozoa of potential concern to laboratorians include *Entamoeba histolytica* (which also can cause extraintestinal infection); *Giardia intestinalis* (*Giardia lamblia*); and coccidian parasites, particularly *Cryptosporidium* spp. (i.e., *Cryptosporidium parvum*, *C. hominis*, and potentially other *Cryptosporidium* spp.), *Cystoisospora belli* (*Isospora belli*), and *Cyclospora cayetanensis*. (See above regarding *Sarcocystis* spp.) Fecally excreted *Cystoisospora* and *Cyclospora* oocysts require an extrinsic maturation period to become infective (178), whereas *E. histolytica* cysts, *Giardia* cysts, and *Cryptosporidium* oocysts are infective when excreted. Because protozoa multiply in the host, even small inocula can cause illness (179–181).

Laboratory personnel should observe routine precautions for work with stool specimens and fecally contaminated material (e.g., careful hand washing). Even preserved specimens should be handled with care because parasites in inadequately preserved specimens could be viable. Commercially available iodine-containing disinfectants are effective against *E. histolytica* and *G. intestinalis*, when used as directed, as are high concentrations of chlorine (1 cup of full-strength commercial bleach [~5% chlorine] per gallon of water [1:16, vol/vol]).

Environmental contamination with *Cryptosporidium* oocysts is problematic, especially for persons working with infected calves; during the peak period of shedding (~5 to 12 days postexposure), infected calves shed billions of oocysts per day (M Arrowood, personal communication) (178, 182). Although *Cryptosporidium* oocysts are inactivated by freezing (e.g., 20°C for 24 h) and moist heat (55°C for 30 s or 70°C for 5 s) (183), they are highly resistant to chemical disinfection (184–186), as are *Cystoisospora* and *Cyclospora* oocysts. Solutions known to kill *Cryptosporidium* oocysts, if the contact times are sufficiently long, include 5% ammonia and 10% Formol saline (185), both of which are noxious, and undiluted 3% ("10 vol") commercial hydrogen peroxide. Although these solutions probably also kill *Cystoisospora* and *Cyclospora* oocysts, insufficient data are available in this regard.

For all of these coccidian parasites, contaminated skin should be thoroughly washed, and a conventional laboratory detergent/cleaner should be used to remove "contaminating matter" from surfaces (e.g., of bench tops and equipment). After the organic material has been removed, if the contaminant included *Cryptosporidium* spp., 3% hydrogen peroxide can be used to disinfect the surfaces (data for *Cystoisospora* and *Cyclospora* oocysts are not available). Therefore, dispensing bottles containing 3% hydrogen peroxide should be readily available in laboratories in which surfaces could become contaminated; affected surfaces should be flooded (i.e., completely covered) with hydrogen peroxide. If surfaces are contaminated by spillage of a large volume of liquid, to avoid diluting the hydrogen peroxide, absorb the bulk of the spillage with disposable paper towels. Dispense hydrogen peroxide repeatedly, as needed, to keep affected surfaces covered (i.e., wet/moist) for ~30 min. Absorb residual hydrogen peroxide with paper towels, and allow surfaces to dry thoroughly (10 to 30 min) before use. Contaminated paper towels and other disposable materials should be autoclaved or similarly disinfected before they are discarded. Reusable laboratory items can be disinfected and washed in a laboratory dishwasher by using the "sanitize" cycle and a detergent that contains chlorine. Alternatively, immerse contaminated items for ~1 h in a water bath preheated to 50°C; thereafter, wash them in a detergent/disinfectant solution.

Laboratory-acquired cases. SUMMARY DATA. Relatively few cases of laboratory-acquired infections with intestinal protozoa have been reported (Table 5), probably, in part, because of the comparative ease with which such infections can be diagnosed and treated and because the illness typically is gastrointestinal rather than systemic. Two cases of giardiasis, 3 cases of cystoisosporiasis (isosporiasis), and 21 cases of cryptosporidiosis are described below.

Description of two cases caused by G. intestinalis *and three cases caused by* Cystoisospora belli. A worker who "checked in several hundred stool survey specimens, stamping numbers and dates on report cards, many of which had been contaminated from leaky containers," became infected with *G. intestinalis*. The parasite was detected in the person's stool "after typical incubation period and course of disease" (187).

A "debilitating bout" of giardiasis thought to have represented patient-to-staff transmission was described in a case-report (188). The case was in an orthopedic surgeon who had two preschool-age patients with giardiasis. One of these patients was a 1-year-old child who had her plaster cast adjusted on 9 March and removed on 16 April; on both days, the cast was noticeably stained with moist and dry feces. The physician became ill in early May and later had a positive stool specimen. Because he typically washed his hands before and after changing casts but rarely wore a mask, the possibility that he inhaled and then swallowed plaster dust contaminated with *Giardia* cysts (average length, 11 to 12 μm) was raised.

One of the three cases of cystoisosporiasis occurred in a laboratory technician who examined numerous stool specimens from a patient infected with *C. belli*; the technician became ill ~1 week after examining the first specimens, and *C. belli* was detected in his stool (189). Two researchers who were feeding a rabbit a capsule containing ~400 *Cystoisospora* oocysts were sprayed on their faces with droplets of infective material when the rabbit regurgitated it and vigorously shook its head; the researchers became ill 11 and 12 days later (187, 190). A case of cystoisosporiasis that might have been laboratory acquired has been described (191) but was not included in the case counts (Table 5).

Description of 21 cases caused by Cryptospiridium *spp.* Although cryptosporidiosis is a well-recognized occupational hazard for persons exposed to naturally infected calves and other animals (235–241 [these cases were not included in the case counts; Table 4]), cases of cryptosporidiosis also have been reported among persons exposed to experimentally infected animals (192–194). Seven such cases were included in the tallies:

- Five veterinary students who had direct (four) or indirect (one) contact with experimentally infected calves became ill 6 to 7 days later and had diarrhea for a median of 5 days (range, 1 to 13 days) (194); one student was hospitalized. In addition, oocysts were found in a stool specimen from the spouse of an infected student.
- A researcher developed gastrointestinal symptoms 5 days after a rabbit exposed to oocysts via a gastric tube coughed droplets of inoculum onto his face as he was removing the tube (193). Oocysts were found in the researcher's stool, which was first obtained for testing on day 6.
- A veterinary scientist developed flu-like symptoms 7 days after sniffing for gastric odor to check the position of a gastric tube in an infected calf; she did not recall other potential exposures to *Cryptosporidium* oocysts (192). She developed gastrointestinal symptoms on day 10, and oocysts were found in a stool specimen on day 16 (presumably, the first specimen tested). Although airborne transmission of this small organism (average dimensions, 4.5 by 5 μm) is plausible, aerosolization of oocysts from the rumen of a calf is speculative.

The tallies also include 13 cases of nosocomial transmission of *Cryptosporidium* spp. from human patients to hospital staff, who became symptomatic and had positive stool specimens or, for one person, had serologic evidence of infection; presumptive cases without laboratory evidence of infection were not counted. The infected persons who had positive stool specimens included a nurse caring for an infected patient who had received a bone marrow transplant (195), a nurse on a ward where an infected 13-month-old boy was hospitalized (196), a nurse caring for infected patients before and after renal transplantation (197), five nurses caring for an infected patient with AIDS (198–200), and five persons (nurses and ancillary staff) linked to another HIV-coinfected patient (T Poissant, personal communication). The exception regarding stool positivity was a case in a symptomatic intern with serologic evidence of *Cryptosporidium* infection after exposure to an infected patient (201); the stool that tested negative was obtained on day 17 of the intern's illness.

Nosocomial patient-to-patient transmission of *Cryptosporidium* spp. also has been reported (197, 198, 202–210); the reports have varied regarding the strength of the evidence that infection was hospital acquired. Patient-to-patient transmission is beyond the scope of this chapter (Table 4), and such cases were not included in the case counts. Nosocomial cases of cryptosporidiosis could result from direct person-to-person transmission (e.g., via health care workers), contact with contaminated surfaces or objects (e.g., medical devices), or ingestion of contaminated food or water.

Postexposure diagnostic testing. Infections with intestinal protozoa are diagnosed by examining stool specimens. Because organisms can be excreted intermittently and in small numbers, multiple stools, obtained on different days, may need to be examined. Stools should be preserved in 10% formalin and in polyvinyl alcohol or alternative fixatives, and a concentration technique should be used. *Cryptosporidium*, *Cystoisospora*, and *Cyclospora* oocysts, all of which are acid fast, are distinguishable by size and shape; *Cystoisospora* and *Cyclospora* spp. demonstrate autofluorescence with UV fluorescence microscopy (178, 211). Antigen detection tests are available for *E. histolytica*, *G. intestinalis*, and *Cryptosporidium* spp. Molecular methods (e.g., PCR) also are available for various intestinal protozoa.

INFECTIONS WITH HELMINTHS

General Information and Laboratory-Acquired Cases

Few laboratory-acquired helminthic infections have been reported (Table 5). The scarcity of such reports might in part reflect the fact that helminthic infections generally are less likely than protozoan infections to be acquired in the laboratory. Even if laboratorians became infected by ingestion of infective eggs or via percutaneous exposure to infective larvae, they typically would have low worm burdens and few, if any, symptoms because most helminths do not multiply in humans.

Flukes (trematodes) and most tapeworms (cestodes) require further larval development in a nonhuman host. One possible laboratory-acquired case of fascioliasis and at least 8 (maybe 10) cases of schistosomiasis are described below. Because the eggs of most intestinal nematodes (e.g., *Ascaris lumbricoides* and *Trichuris trichiura*) require an extrinsic maturation period of days to weeks to become infective, persons in diagnostic laboratories are unlikely to become infected with these organisms if the stool specimens were obtained recently. However, even preserved specimens should be handled with care because some helminth eggs can develop and remain viable in formalin (212). Because ascarid eggs are sticky, contaminated laboratory surfaces and equipment must be cleaned thoroughly to prevent worker exposure. Laboratorians working with *Ascaris* spp. should be aware that hypersensitivity reactions to *Ascaris* antigens can develop, which can include respiratory, dermatologic, and gastrointestinal manifestations (213–217).

The eggs of *Enterobius vermicularis* (pinworm) and *Hymenolepis nana* (dwarf tapeworm)—parasites that do not require intermediate hosts—are unusual in that they are infective immediately or shortly after excretion in feces; *H. nana* eggs can be found in human and rodent feces. Therefore, persons working in diagnostic laboratories or with rodents could become infected by ingesting these organisms if routine precautions, such as the use of gloves and careful hand washing, are neglected. Similarly, laboratory personnel exposed to mature filariform larvae of *Strongyloides stercoralis*, which can penetrate intact skin, could become infected. Lugol's iodine kills infective larvae and should be sprayed on exposed skin and contaminated laboratory surfaces. Although the larvae shed in stool typically are noninfective rhabditiform larvae, a few infective filariform larvae could be present. Hyperinfected persons can shed large numbers of larvae in respiratory secretions as well as in feces, some of which might be infective. Cases of cutaneous larva migrans (creeping eruption or "ground itch") caused by skin contact with *Strongyloides* spp. (four cases) (218, 219) or *Ancylostoma* spp. (one case) (220) have been described. The latter case was in an animal caretaker who fed and cared for a cat infected with *Ancylostoma braziliense* and *Ancylostoma caninum*.

Humans can serve as both the intermediate and the definitive host of *Taenia solium* (pork tapeworm). Ingestion of eggs from a tapeworm carrier can result in the development of larval cysts (i.e., cysticercosis) in the brain and elsewhere.

Human infection with *Echinococcus* spp., which are cestodes, requires ingestion of eggs shed in the feces of definitive hosts. Various canid species serve as definitive hosts for *Echinococcus granulosis*, *E. multilocularis*, and *E. vogeli*, whereas various feline species are the definitive hosts for *E. oligarthrus*. Therefore, infection could be acquired by persons in veterinary diagnostic or research laboratories.

Trichinella spp. are the only tissue nematodes that pose substantial risk to laboratory personnel. Preparations of fresh tissue, as well as specimens digested with pepsin hydrochloride, can contain encysted *Trichinella* larvae that are infective if ingested. Because most laboratorians would have ingested few organisms, serologic testing would be a more sensitive means of diagnosing infection than histologic examination of muscle biopsy specimens. Filarial infections, which also are caused by tissue nematodes, could be acquired by laboratory personnel working with infected arthropods.

Laboratory-acquired cases of fascioliasis and schistosomiasis

Because flukes need to develop in an intermediate host to become infective, the presence of eggs in mammalian feces does not pose a risk to personnel in diagnostic laboratories. However, persons in research laboratories who handle snails that are competent intermediate hosts should exercise caution. Laboratorians working with aquaria containing snail intermediate hosts could become infected via ingestion of *Fasciola* metacercariae, which encyst on aquatic grasses or plants, or through skin penetration by schistosome cercariae, which swim freely (221); dissecting or crushing infected schistosome-infected snails could result in exposure to droplets that contain cercariae. Therefore, laboratorians doing such work should wear gloves. In addition, persons at risk for exposure to schistosome cercariae should minimize the amount of uncovered skin by wearing a long-sleeved gown or coat and shoes rather than sandals. Snails and cercariae in the water of laboratory aquaria should be killed by chemicals (e.g., ethanol, hypochlorites, or iodine) or by heat before discharge to the sewer.

One possible laboratory-acquired case of fascioliasis and at least 8 (possibly 10) cases of schistosomiasis in at least six persons have been reported. A technician who worked with *Fasciola hepatica* in a veterinary laboratory developed clinical manifestations consistent with fascioliasis (i.e., lassitude, fever, weight loss, slight tenderness at the right costal margin, and eosinophilia) (222, 223). Although he was thought to have become infected through his work, the nature of his job was not described. The conclusion that he was infected was based on positive serologic results with a double-diffusion precipitin test; the fact that the parasite was not found in stool specimens was attributed to testing early in the invasive stage of infection.

A laboratory assistant working with snails (*Biomphalaria pfeifferi*) from an area where *Schistosoma mansoni* infection is endemic developed schistosomiasis (224). She had stopped wearing gloves 3 weeks after beginning this work because she thought the snails no longer were infective. On day 31 after she began to work barehanded, she developed what likely was a mild case of Katayama fever, which lasted 5 days and was manifested by fever, headache, and fatigue. Eosinophilia was noted on day 54, when her report of 3 days of "digestive complaints" prompted examination of a blood specimen. The results of serologic testing by EIA were negative on days 54 and 82, weakly positive on day 101, and strongly positive on day 234. Stool specimens tested negative on day 94 but positive on days 101 and 103.

Several asymptomatic cases of *S. mansoni* infection were detected in a laboratory (laboratory C in Table 3) whose staff worked daily with *S. mansoni*-infected snails and antigen preparations and were monitored twice yearly by serologic testing (1). If seroconversion was documented, stool was tested. During the period from the late 1970s through mid-1999, seroconversion was noted in 4 of approximately 20 persons. None of the four persons recalled a discrete laboratory accident, and all had used standard precautions. Two of the four persons had positive stool specimens (<40 eggs/g); no eggs were noted after antimicrobial therapy.

Several cases of schistosomiasis in persons working with cercariae were mentioned briefly in two reports (225, 226). In one of the reports, an investigator noted that he had been infected three times with *S. mansoni* (226). In the other report, a researcher who had solicited information about laboratory accidents from other investigators commented that "no lab infections were reported for over 100 people handling millions of cercariae for over 20 years, though two technicians became seropositive without developing symptoms, probably through torn gloves" (225). Whether these two cases were two of the four mentioned in the previous paragraph is not known, although not all of the details match. No information was provided about whether and how staff in the various laboratories were monitored for infection (225).

CONCLUSIONS

Many of the key details about the occupationally acquired cases of parasitic infections described in this chapter were summarized in tables and in various summary sections in the text. Clearly, preventing accidental exposures is preferable to managing their consequences. The resulting infections can range from asymptomatic to severe. Three fatal cases of occupationally acquired infection have been reported: one in a person with myocarditis caused by acute Chagas' disease (18), one in a person with myocarditis and encephalitis caused by toxoplasmosis (17), and one in a physician with malaria whose case was diagnosed at necropsy (19, 20). Congenital transmission also is a potential risk for some of the blood and tissue protozoa; women of childbearing age should exercise caution. One of the occupationally acquired cases of malaria (66) and one of the cases of Chagas' disease (157) occurred in pregnant women; fortunately, congenital transmission was not documented. Each of three laboratorians had two vector-borne cases of malaria, one person had several cases of schistosomiasis, and multiple cases of particular parasitic diseases occurred in persons working in the same laboratory.

To decrease the likelihood of accidental exposures, persons who could be exposed to pathogenic parasites must be thoroughly instructed about safety precautions before they begin to work and through ongoing training programs. Protocols should be provided for handling specimens that could contain viable organisms, using protective laboratory clothing and equipment, dealing with spills of infective organisms, and responding to accidental exposures. Laboratorians who work with parasites should follow parasite-specific and general laboratory precautions (e.g., wear gloves, use mechanical pipettors, adequately restrain animals that will be bled or inoculated, restrict the use of sharps, use needleless systems or devices with safety features that reduce the risk for percutaneous injuries, decontaminate work surfaces, and use biological safety cabinets when appropriate), many of which also apply to health care workers. The fact that some of the persons who became infected did not recall discrete exposures suggests that subtle exposures (e.g., contamination of unrecognized microabrasions) can result in infection. The occurrence of laboratory-acquired cases of infection with species not previously known to be infective for humans (e.g., *P. cynomolgi*) or to be extraordinarily hardy in the environment (e.g., *T. gondii* oocysts) highlights the need for special vigilance when working with organisms that have not been characterized fully in such regards.

The following persons contributed in various ways to this chapter or to previous editions: Nestor Añez, Paul M.

Arguin, Michael J. Arrowood, William E. Collins, Jennifer R. Cope, J. P. Dubey, Mark L. Eberhard, Paul J. Edelson, Gregory A. Filice, Diane O. Fleming, Loreen A. Herwaldt, Warren D. Johnson, Jeffrey L. Jones, Dennis D. Juranek, Louis V. Kirchhoff, Diana L. Martin, Anne C. Moore, Douglas Nace, Theodore E. Nash, Franklin A. Neva, Phuc P. Nguyen-Dinh, Monica E. Parise, Malcolm R. Powell, Jack S. Remington, Scott W. Sorensen, Francis J. Steurer, Herbert B. Tanowitz, Charles W. Todd, Govinda S. Visvesvara, Mary E. Wilson, and Jonathan S. Yoder. Some persons who provided information about unpublished cases of parasitic diseases asked to remain anonymous.

The findings and conclusions in this review are those of the author and do not necessarily represent the official position of the Centers for Disease Control and Prevention. Use of trade names is for identification only and does not imply endorsement by the Public Health Service or by the U.S. Department of Health and Human Services.

References

1. **Herwaldt BL.** 2001. Laboratory-acquired parasitic infections from accidental exposures. *Clin Microbiol Rev* **14:**659–688.
2. **Pike RM.** 1976. Laboratory-associated infections: summary and analysis of 3921 cases. *Health Lab Sci* **13:**105–114.
3. **Wilson M, Schantz PM, Nutman T.** 2006. Molecular and immunological approaches to the diagnosis of parasitic infections, p 557–568. *In* Detrick B, Hamilton RG, Folds JD (ed), *Manual of Molecular and Clinical Laboratory Immunology*, 7th ed. ASM Press, Washington, DC.
4. **Centers for Disease Control and Prevention and National Institutes of Health.** 2009. *Biosafety in Microbiological and Biomedical Laboratories*, 5th ed. Chosewood LC, Wilson DE (ed). U.S. Department of Health and Human Services, Washington, DC. [Online.] http://www.cdc.gov/biosafety/publications/bmbl5/BMBL.pdf.
5. **Hankenson FC, Johnston NA, Weigler BJ, Di Giacomo RF.** 2003. Zoonoses of occupational health importance in contemporary laboratory animal research. *Comp Med* **53:**579–601.
6. **Miller JM, Astles R, Baszler T, Chapin K, Carey R, Garcia L, Gray L, Larone D, Pentella M, Pollock A, Shapiro DS, Weirich E, Wiedbrauk D, Biosafety Blue Ribbon Panel, Centers for Disease Control and Prevention (CDC).** 2012. Guidelines for safe work practices in human and animal medical diagnostic laboratories. Recommendations of a CDC-convened, Biosafety Blue Ribbon Panel. *MMWR Suppl* **61:**1–102.
7. **Occupational Safety and Health Administration.** 1991. Title 29 CFR Part 1910.1030. Protection from bloodborne pathogens. *Fed Regist* **56:**64175–64182.
8. **Occupational Safety and Health Administration.** 2001. Title 29 CFR Part 1910.1030. Occupational exposure to bloodborne pathogens: needlesticks and other sharp injuries; final rule. *Fed Regist* **66:**5317–5325.
9. **U.S. Public Health Service.** 2001. Updated U.S. Public Health Service guidelines for the management of occupational exposures to HBV, HCV, and HIV and recommendations for postexposure prophylaxis. *MMWR Recomm Rep* **50**(RR-11):1–52.
10. **Clinical and Laboratory Standards Institute.** 2014. *Protection of Laboratory Workers from Occupationally Acquired Infections: Approved Guideline, 4th ed. CLSI document M29-A4.* Clinical and Laboratory Standards Institute, Wayne, Pa.
11. **Sewell DL.** 1995. Laboratory-associated infections and biosafety. *Clin Microbiol Rev* **8:**389–405.
12. **Herwaldt BL.** 2006. Protozoa and helminths, p 115–161. *In* Fleming DO, Hunt DL (ed), *Biological Safety: Principles and Practices*, 4th ed. ASM Press, Washington, DC.
13. **Herwaldt BL.** 2000. Protozoa and helminths, p 89–110. *In* Fleming DO, Hunt DL (ed), *Biological Safety: Principles and Practices*, 3rd ed. ASM Press, Washington, DC.
14. **Herwaldt BL, Juranek DD.** 1995. Protozoa and helminths, p 77–91. *In* Fleming DO, Richardson JH, Tulis JI, Vesley D (ed), *Laboratory Safety: Principles and Practices*, 2nd ed. ASM Press, Washington, DC.
15. **Herwaldt BL, Juranek DD.** 1993. Laboratory-acquired malaria, leishmaniasis, trypanosomiasis, and toxoplasmosis. *Am J Trop Med Hyg* **48:**313–323.
16. **Napoli VM, McGowan JE Jr.** 1987. How much blood is in a needlestick? *J Infect Dis* **155:**828.
17. **Sexton RC Jr, Eyles DE, Dillman RE.** 1953. Adult toxoplasmosis. *Am J Med* **14:**366–377.
18. **Brener Z.** 1987. Laboratory-acquired Chagas disease: comment. *Trans R Soc Trop Med Hyg* **81:**527.
19. **CDSC.** 1997. Needlestick malaria with tragic consequences. *Commun Dis Rep CDR Wkly* **7:**247.
20. **Romi R, Boccolini D, Majori G.** 1999. Malaria surveillance in Italy: 1997 analysis and 1998 provisional data. *Euro Surveill* **4:**85–87.
21. **Visvesvara GS, Roy SL, Maguire JH.** 2011. Pathogenic and opportunistic free-living amoebae: *Acanthamoeba* spp., *Balamuthia mandrillaris, Naegleria fowleri*, and *Sappinia pedata*, p 707–713. *In* Guerrant RL, Walker DH, Weller PF (ed), *Tropical Infectious Diseases: Principles, Pathogens & Practice*, 3rd ed. Elsevier, Philadelphia, Pa.
22. **Herwaldt BL, Linden JV, Bosserman E, Young C, Olkowska D, Wilson M.** 2011. Transfusion-associated babesiosis in the United States: a description of cases. *Ann Intern Med* **155:**509–519.
23. **Conrad PA, Kjemtrup AM, Carreno RA, Thomford J, Wainwright K, Eberhard M, Quick R, Telford SR III, Herwaldt BL.** 2006. Description of *Babesia duncani* n.sp. (Apicomplexa: Babesiidae) from humans and its differentiation from other piroplasms. *Int J Parasitol* **36:**779–789.
24. **Persing DH, Herwaldt BL, Glaser C, Lane RS, Thomford JW, Mathiesen D, Krause PJ, Phillip DF, Conrad PA.** 1995. Infection with a *Babesia*-like organism in northern California. *N Engl J Med* **332:**298–303.
25. **Herwaldt BL, de Bruyn G, Pieniazek NJ, Homer M, Lofy KH, Slemenda SB, Fritsche TR, Persing DH, Limaye AP.** 2004. *Babesia divergens*-like infection, Washington State. *Emerg Infect Dis* **10:**622–629.
26. **Herwaldt BL, Persing DH, Précigout EA, Goff WL, Mathiesen DA, Taylor PW, Eberhard ML, Gorenflot AF.** 1996. A fatal case of babesiosis in Missouri: identification of another piroplasm that infects humans. *Ann Intern Med* **124:**643–650.
27. **Herwaldt BL, Cacciò S, Gherlinzoni F, Aspöck H, Slemenda SB, Piccaluga P, Martinelli G, Edelhofer R, Hollenstein U, Poletti G, Pampiglione S, Löschenberger K, Tura S, Pieniazek NJ.** 2003. Molecular characterization of a non-*Babesia divergens* organism causing zoonotic babesiosis in Europe. *Emerg Infect Dis* **9:**942–948.
28. **Wormser GP, Dattwyler RJ, Shapiro ED, Halperin JJ, Steere AC, Klempner MS, Krause PJ, Bakken JS, Strle F, Stanek G, Bockenstedt L, Fish D, Dumler JS, Nadelman RB.** 2006. The clinical assessment, treatment, and prevention of lyme disease, human granulocytic anaplasmosis, and babesiosis: clinical practice guidelines by the Infectious Diseases Society of America. *Clin Infect Dis* **43:**1089–1134.

29. **Herwaldt BL.** 1999. Leishmaniasis. *Lancet* **354:**1191–1199.
30. **Dey A, Singh S.** 2006. Transfusion transmitted leishmaniasis: a case report and review of literature. *Indian J Med Microbiol* **24:** 165–170.
31. **Dillon NL, Stolf HO, Yoshida EL, Marques ME.** 1993. [Accidental cutaneous leishmaniasis.] In Portuguese. *Rev Inst Med Trop Sao Paulo* **35:**385–387.
32. **Chung HL.** 1931. An early case of kala-azar, possibly an oral infection in the laboratory. *Natl Med J China* **17:**617–621.
33. **Delgado O, Guevara P, Silva S, Belfort E, Ramirez JL.** 1996. Follow-up of a human accidental infection by *Leishmania (Viannia) braziliensis* using conventional immunologic techniques and polymerase chain reaction. *Am J Trop Med Hyg* **55:**267–272.
34. **Evans TG, Pearson RD.** 1988. Clinical and immunological responses following accidental inoculation of *Leishmania donovani*. *Trans R Soc Trop Med Hyg* **82:**854–856.
35. **Freedman DO, MacLean JD, Viloria JB.** 1987. A case of laboratory acquired *Leishmania donovani* infection; evidence for primary lymphatic dissemination. *Trans R Soc Trop Med Hyg* **81:**118–119.
36. **Knobloch J, Demar M.** 1997. Accidental *Leishmania mexicana* infection in an immunosuppressed laboratory technician. *Trop Med Int Health* **2:**1152–1155.
37. **Sadick MD, Locksley RM, Raff HV.** 1984. Development of cellular immunity in cutaneous leishmaniasis due to *Leishmania tropica*. *J Infect Dis* **150:**135–138.
38. **Sampaio RN, de Lima LM, Vexenat A, Cuba CC, Barreto AC, Marsden PD.** 1983. A laboratory infection with *Leishmania braziliensis braziliensis*. *Trans R Soc Trop Med Hyg* **77:**274.
39. **Stamper LW, Patrick RL, Fay MP, Lawyer PG, Elnaiem DE, Secundino N, Debrabant A, Sacks DL, Peters NC.** 2011. Infection parameters in the sand fly vector that predict transmission of *Leishmania major*. *PLoS Negl Trop Dis* **5:**e1288.
40. **Terry LL, Lewis JL Jr, Sessoms SM.** 1950. Laboratory infection with *Leishmania donovani*; a case report. *Am J Trop Med Hyg* **30:**643–649.
41. **Coatney GR, Collins WE, Warren M, Contacos PG.** 1971. *The Primate Malarias*. U.S. Government Printing Office, Washington, DC.
42. **Eyles DE, Coatney GR, Getz ME.** 1960. *Vivax*-type malaria parasite of macaques transmissible to man. *Science* **131:**1812–1813.
43. **Alweis RL, DiRosario K, Conidi G, Kain KC, Olans R, Tully JL.** 2004. Serial nosocomial transmission of *Plasmodium falciparum* malaria from patient to nurse to patient. *Infect Control Hosp Epidemiol* **25:**55–59.
44. **Antunes F, Forte M, Tavares L, Botas J, Carvalho C, Carmona H, Araujo FC.** 1987. Malaria in Portugal 1977–1986. *Trans R Soc Trop Med Hyg* **81:**561–562.
45. **Bending MR, Maurice PD.** 1980. Malaria: a laboratory risk. *Postgrad Med J* **56:**344–345.
46. **Börsch G, Odendahl J, Sabin G, Ricken D.** 1982. Malaria transmission from patient to nurse. *Lancet* **2:**1212.
47. **Bourée P, Fouquet E.** 1978. [Malaria: direct interhuman contamination.] In French. *Nouv Presse Med* **7:**1865.
48. **Bouteille B, Darde ML, Weinbreck P, Voultoury JC, Gobeaux R, Pestre-Alexandre M.** 1990. [Pernicious malaria by accidental needle inoculation.] In French. *Bull Soc Fr Parasitol* **8:**69–72.
49. **Bruce-Chwatt LJ.** 1982. Imported malaria: an uninvited guest. *Br Med Bull* **38:**179–185.
50. **Burne JC.** 1970. Malaria by accidental inoculation. *Lancet* **2:**936.
51. **Burne JC, Draper CC.** 1971. Accidental self-inoculation with malaria. *Trans R Soc Trop Med Hyg* **65:**1.
52. **Cannon NJ Jr, Walker SP, Dismukes WE.** 1972. Malaria acquired by accidental needle puncture. *JAMA* **222:**1425.
53. **Carosi G, Maccabruni A, Castelli F, Viale P.** 1986. Accidental and transfusion malaria in Italy. *Trans R Soc Trop Med Hyg* **80:**667–668.
54. **Carrière J, Datry A, Hilmarsdottir I, Danis M, Gentilini M.** 1993. [Transmission of Plasmodium falciparum following an accidental sting.] In French. *Presse Med* **22:**1707.
55. **Center for Disease Control.** 1972. *Malaria Surveillance 1971 Annual Report. DHEW publication (HSM) 72–8152*. Center for Disease Control, Atlanta, Ga.
56. **Centers for Disease Control.** 1984. *Malaria Surveillance Annual Summary 1982*. Centers for Disease Control, Atlanta, Ga.
57. **Centers for Disease Control and Prevention.** 2002. Malaria surveillance—United States, 2000. *MMWR Surveill Summ* **51:**9–21.
58. **Cross JH, Hsu-Kuo MY, Lien JC.** 1973. Accidental human infection with *Plasmodium cynomolgi bastianellii*. *Southeast Asian J Trop Med Public Health* **4:**481–483.
59. **Cullen KA, Arguin PM, Division of Parasitic Diseases and Malaria, Center for Global Health, Centers for Disease Control and Prevention (CDC).** 2013. Malaria surveillance—United States, 2011. *MMWR Surveill Summ* **62:**1–17.
60. **Cullen KA, Arguin PM, Centers for Disease Control and Prevention (CDC).** 2014. Malaria surveillance—United States, 2012. *MMWR Surveill Summ* **63:**1–22.
61. **Daumal M, Verreman V, Daumal F, Manoury B, Colpart E.** 1996. [*Plasmodium falciparum* malaria acquired by accidental needle puncture.] In French. *Med Mal Infect* **26:**797–798.
62. **Druilhe P, Trape JF, Leroy JP, Godard C, Gentilini M.** 1980. [Two accidental human infections by *Plasmodium cynomolgi bastianellii*: a clinical and serological study.] In French. *Ann Soc Belg Med Trop* **60:**349–354.
63. **Freedman AM.** 1987. Unusual forms of malaria transmission. A report of 2 cases. *S Afr Med J* **71:**183–184.
64. **Garnham PC.** 1967. Malaria in mammals excluding man. *Adv Parasitol* **5:**139–204.
65. **Haworth FL, Cook GC.** 1995. Needlestick malaria. *Lancet* **346:**1361.
66. **Hira PR, Siboo R, Al-Kandari S, Behbehani K.** 1987. Induced malaria and antibody titres in acute infections and in blood donors in Kuwait. *Trans R Soc Trop Med Hyg* **81:**391–394.
67. **Holm K.** 1924. Über einen Fall von Infektion mit Malaria Tropica an der Leiche. In German. *Klin Wochenschr* **3:**1633–1634.
68. **Jensen JB, Capps TC, Carlin JM.** 1981. Clinical drug-resistant *falciparum* malaria acquired from cultured parasites. *Am J Trop Med Hyg* **30:**523–525.
69. **Kociecka W, Skoryna B.** 1987. *Falciparum* malaria probably acquired from infected skin-cut. *Lancet* **2:**220.
70. **Lettau LA.** 1991. Nosocomial transmission and infection control aspects of parasitic and ectoparasitic diseases. Part II. Blood and tissue parasites. *Infect Control Hosp Epidemiol* **12:**111–121.
71. **Lewis J.** 1970. Iatrogenic malaria. *N Z Med J* **71:**88–89.
72. **Most H.** 1973. *Plasmodium cynomolgi* malaria: accidental human infection. *Am J Trop Med Hyg* **22:**157–158.
73. **Petithory J, Lebeau G.** 1977. [A probable laboratory contamination with *Plasmodium falciparum*.] In French. *Bull Soc Pathol Exot Filiales* **70:**371–375.
74. **Raffenot D, Rogeaux O, De Goer B, Zerr B.** 1999. *Plasmodium falciparum* malaria acquired by accidental inoculation. *Eur J Clin Microbiol Infect Dis* **18:**680–681.
75. **Schmidt LH, Greenland R, Genther CS.** 1961. The transmission of *Plasmodium cynomolgi* to man. *Am J Trop Med Hyg* **10:**679–688.
76. **Skarbinski J, James EM, Causer LM, Barber AM, Mali S, Nguyen-Dinh P, Roberts JM, Parise ME, Slutsker L, Newman RD.** 2006. Malaria surveillance—United States, 2004. *MMWR Surveill Summ* **55:**23–37.
77. **Tarantola AP, Rachline AC, Konto C, Houzé S, Lariven S, Fichelle A, Ammar D, Sabah-Mondan C, Vrillon H, Bouchaud O, Pitard F, Bouvet E, and Groupe d'Etude des Risques d'Exposition des Soignants aux agents infectieux.** 2004.

Occupational malaria following needlestick injury. *Emerg Infect Dis* **10:** 1878–1880.

78. **Tarantola A, Rachline A, Konto C, Houzé S, Sabah-Mondan C, Vrillon H, Bouvet E; Group for the Prevention of Occupational Infections in Health Care Workers.** 2005. Occupational *Plasmodium falciparum* malaria following accidental blood exposure: a case, published reports and considerations for post-exposure prophylaxis. *Scand J Infect Dis* **37:**131–140.
79. **van Agtmael MA.** 1997. A most unfortunate needlestick injury: why the docter [sic] paid a taxi for the nurse. (letter). *BMJ* **314:**h.
80. **Vareil MO, Tandonnet O, Chemoul A, Bogreau H, Saint-Léger M, Micheau M, Millet P, Koeck JL, Boyer A, Rogier C, Malvy D.** 2011. Unusual transmission of *Plasmodium falciparum*, Bordeaux, France, 2009. *Emerg Infect Dis* **17:**248–250.
81. **Williams JL, Innis BT, Burkot TR, Hayes DE, Schneider I.** 1983. *Falciparum* malaria: accidental transmission to man by mosquitoes after infection with culture-derived gametocytes. *Am J Trop Med Hyg* **32:**657–659.
82. **Pasticier A, Mechali D, Saimot G, Coulaud JP, Payet M.** 1974. [Endemic malaria. Diagnostic problems]. In French. *Bull Soc Pathol Exot Filiales* **67:**57–64.
83. **Contacos PG, Collins WE.** 1973. Letter: malarial relapse mechanism. *Trans R Soc Trop Med Hyg* **67:**617–618.
84. **Collins WE.** 1982. Simian malaria, p 141–150. *In* Steele JH, Jacobs L, Arambulo P (ed), *Parasitic Zoonoses*, vol 1. CRC Press, Inc, Boca Raton, Fla.
85. **Abulrahi HA, Bohlega EA, Fontaine RE, al-Seghayer SM, al-Ruwais AA.** 1997. *Plasmodium falciparum* malaria transmitted in hospital through heparin locks. *Lancet* **349:**23–25.
86. **Al-Hamdan NA.** 2009. Hospital-acquired malaria associated with dispensing diluted heparin solution. *J Vector Borne Dis* **46:**313–314.
87. **Al-Saigul AM, Fontaine RE, Haddad Q.** 2000. Nosocomial malaria from contamination of a multidose heparin container with blood. *Infect Control Hosp Epidemiol* **21:**329–330.
88. **Bruce-Chwatt LJ.** 1972. Blood transfusion and tropical disease. *Trop Dis Bull* **69:**825–862.
89. **Centers for Disease Control and Prevention.** 2001. Malaria surveillance—United States, 1996. *MMWR Surveill Summ* **50:** 1–22.
90. **CDSC.** 1999. Hospital-acquired malaria in Nottingham. *Commun Dis Rep CDR Wkly* **9:**123.
91. **Chen KT, Chen CJ, Chang PY, Morse DL.** 1999. A nosocomial outbreak of malaria associated with contaminated catheters and contrast medium of a computed tomographic scanner. *Infect Control Hosp Epidemiol* **20:**22–25.
92. **de Oliveira CG, Freire F Jr.** 1948. [An epidemic of inoculated malaria]. In Portuguese. *Bol Clin Hosp Civis Lisb* **12:**375–404.
93. **Dziubek Z, Kajfasz P, Basiak W.** 1993. [Hospital infection of tropical malaria in Poland]. In Polish. *Wiad Lek* **46:**860–863.
94. **González L, Ochoa J, Franco L, Arroyave M, Restrepo E, Blair S, Maestre A.** 2005. Nosocomial *Plasmodium falciparum* infections confirmed by molecular typing in Medellín, Colombia. *Malar J* **4:**9.
95. **Jain SK, Persaud D, Perl TM, Pass MA, Murphy KM, Pisciotta JM, Scholl PF, Casella JF, Sullivan DJ.** 2005. Nosocomial malaria and saline flush. *Emerg Infect Dis* **11:**1097–1099.
96. **Jones JL.**2000. Recommendations from an external review following transmission of malaria in hospital. *Euro Surveill* 4(10):pii=1645. [Online.] http://www.eurosurveillance.org/ViewArticle.aspx?ArticleId=1645.
97. **Kim JY, Kim JS, Park MH, Kang YA, Kwon JW, Cho SH, Lee BC, Kim TS, Lee JK.** 2009. A locally acquired *falciparum* malaria via nosocomial transmission in Korea. *Korean J Parasitol* **47:**269–273.
98. **Kirchgatter K, Wunderlich G, Branquinho MS, Salles TM, Lian YC, Carneiro-Junior RA, Di Santi SM.** 2002. Molecular typing of *Plasmodium falciparum* from Giemsa-stained blood smears confirms nosocomial malaria transmission. *Acta Trop* **84:**199–203.
99. **Lee EH, Adams EH, Madison-Antenucci S, Lee L, Barnwell JW, Whitehouse J, Clement E, Bajwa W, Jones LE, Lutterloh E, Weiss D, Ackelsberg J.** 2016. Healthcare-associated transmission of *Plasmodium falciparum* in New York City. *Infect Control Hosp Epidemiol* **37:**113–115.
100. **Moro ML, Romi R, Severini C, Casadio GP, Sarta G, Tampieri G, Scardovi A, Pozzetti C, Malaria Outbreak Group.** 2002. Patient-to-patient transmission of nosocomial malaria in Italy. *Infect Control Hosp Epidemiol* **23:**338–341.
101. **Mortimer PP.** 1997. Nosocomial malaria. *Lancet* **349:**574.
102. **Navarro P, Betancurt A, Paublini H, Medina I, Núñez MJ, Domínguez M.** 1987. [Plasmodium falciparum malaria as a nosocomial infection]. In Spanish. *Bol Oficina Sanit Panam* **102:**476–482.
103. **Piro S, Sammud M, Badi S, Al Ssabi L.** 2001. Hospital-acquired malaria transmitted by contaminated gloves. *J Hosp Infect* **47:**156–158.
104. **Varma AJ.** 1982. Malaria acquired by accidental inoculation. *Can Med Assoc J* **126:**1419–1420.
105. **Winterberg DH, Wever PC, van Rheenen-Verberg C, Kempers O, Durand R, Bos AP, Teeuw AH, Spanjaard L, Dankert J.** 2005. A boy with nosocomial malaria tropica contracted in a Dutch hospital. *Pediatr Infect Dis J* **24:**89–91.
106. **Arness MK, Brown JD, Dubey JP, Neafie RC, Granstrom DE.** 1999. An outbreak of acute eosinophilic myositis attributed to human *Sarcocystis* parasitism. *Am J Trop Med Hyg* **61:**548–553.
107. **Beaver PC, Gadgil K, Morera P.** 1979. *Sarcocystis* in man: a review and report of five cases. *Am J Trop Med Hyg* **28:**819–844.
108. **Dubey JP, Speer CA, Fayer R.** 1989. *Sarcocystosis of Animals and Man.* CRC Press, Inc, Boca Raton, Fla.
109. **Esposito DH, Stich A, Epelboin L, Malvy D, Han PV, Bottieau E, da Silva A, Zanger P, Slesak G, van Genderen PJ, Rosenthal BM, Cramer JP, Visser LG, Muñoz J, Drew CP, Goldsmith CS, Steiner F, Wagner N, Grobusch MP, Plier DA, Tappe D, Sotir MJ, Brown C, Brunette GW, Fayer R, von Sonnenburg F, Neumayr A, Kozarsky PE; Tioman Island Sarcocystosis Investigation Team.** 2014. Acute muscular sarcocystosis: an international investigation among ill travelers returning from Tioman Island, Malaysia, 2011–2012. *Clin Infect Dis* **59:**1401–1410.
110. **Dubey JP, Lindsay DS.** 1998. Isolation in immunodeficient mice of *Sarcocystis neurona* from opossum (*Didelphis virginiana*) faeces, and its differentiation from *Sarcocystis falcatula*. *Int J Parasitol* **28:**1823–1828.
111. **Marsh AE, Barr BC, Lakritz J, Nordhausen R, Madigan JE, Conrad PA.** 1997. Experimental infection of nude mice as a model for *Sarcocystis neurona*-associated encephalitis. *Parasitol Res* **83:**706–711.
112. **Teutsch SM, Juranek DD, Sulzer A, Dubey JP, Sikes RK.** 1979. Epidemic toxoplasmosis associated with infected cats. *N Engl J Med* **300:**695–699.
113. **Dubey JP, Beattie CP.** 1988. *Toxoplasmosis of Animals and Man.* CRC Press, Inc, Boca Raton, Fla.
114. **Parker SL, Holliman RE.** 1992. Toxoplasmosis and laboratory workers: a case-control assessment of risk. *Med Lab Sci* **49:**103–106.
115. **Strom J.** 1951. Toxoplasmosis due to laboratory infection in two adults. *Acta Med Scand* **139:**244–252.
116. **Baker CC, Farthing CP, Ratnesar P.** 1984. Toxoplasmosis, an innocuous disease? *J Infect* **8:**67–69.
117. **Beverley JK, Skipper E, Marshall SC.** 1955. Acquired toxoplasmosis with a report of a case of laboratory infection. *BMJ* **1:** 577–578.
118. **Brown J, Jacobs L.** 1956. Adult toxoplasmosis; report of a case due to laboratory infection. *Ann Intern Med* **44:**565–572.

119. **Field PR, Moyle GG, Parnell PM.** 1972. The accidental infection of a laboratory worker with *Toxoplasma gondii*. *Med J Aust* **2:** 196–198.
120. **Frenkel JK, Weber RW, Lunde MN.** 1960. Acute toxoplasmosis. Effective treatment with pyrimethamine, sulfadiazine, leucovorin calcium, and yeast. *JAMA* **173:**1471–1476.
121. **Giroud P, Le Gac P, Roger F, Gaillard JA.** 1953. [Toxoplasmosis of the adult]. In French. *Sem Hop* **29:**4036–4039.
122. **Hermentin K, Picher O, Aspöck H, Auer H, Hassl A.** 1983. A solid-phase indirect haemadsorption assay (SPIHA) for detection of immunoglobulin M antibodies to *Toxoplasma gondii*: application to diagnosis of acute acquired toxoplasmosis. *Zentralbl Bakteriol Mikrobiol Hyg [A]* **255:**380–391.
123. **Hermentin K, Hassl A, Picher O, Aspöck H.** 1989. Comparison of different serotests for specific *Toxoplasma* IgM-antibodies (ISAGA, SPIHA, IFAT) and detection of circulating antigen in two cases of laboratory acquired *Toxoplasma* infection. *Zentralbl Bakteriol Mikrobiol Hyg [A]* **270:**534–541.
124. **Hörmann J.** 1955. [Laboratory infection with *Toxoplasma gondii*: contribution to the clinical picture of the adult toxoplasmosis]. In German. *Z Gesamte Inn Med* **10:**150–152.
125. **Kayhoe DE, Jacobs L, Beye HK, McCullough NB.** 1957. Acquired toxoplasmosis—observations on two parasitologically proved cases treated with pyrimethamine and triple sulfonamides. *N Engl J Med* **257:**1247–1254.
126. **Miller NL, Frenkel JK, Dubey JP.** 1972. Oral infections with *Toxoplasma* cysts and oocysts in felines, other mammals, and in birds. *J Parasitol* **58:**928–937.
127. **Müller WA, Färber I, Wachtel D.** 1972. [Relation between the indirect immunofluorescent reaction, the Sabin-Feldmann test, and the complement fixation test studying the titer level in laboratory infection with toxoplasmosis]. In German. *Dtsch Gesundheitsw* **27:**82–85.
128. **Neu HC.** 1967. Toxoplasmosis transmitted at autopsy. *JAMA* **202:**844–845.
129. **Partanen P, Turunen HJ, Paasivuo RT, Leinikki PO.** 1984. Immunoblot analysis of *Toxoplasma gondii* antigens by human immunoglobulins G, M, and A antibodies at different stages of infection. *J Clin Microbiol* **20:**133–135.
130. **Rawal BD.** 1959. Laboratory infection with *Toxoplasma*. *J Clin Pathol* **12:**59–61.
131. **Remington JS, Gentry LO.** 1970. Acquired toxoplasmosis: infection versus diseases. *Ann N Y Acad Sci* **174:**1006–1017.
132. **Sabin AB, Eichenwald H, Feldman HA, Jacobs L.** 1952. Present status of clinical manifestations of toxoplasmosis in man; indications and provisions for routine serologic diagnosis. *J Am Med Assoc* **150:**1063–1069.
133. **Thalhammer O.** 1954. [Two remarkable cases of fresh toxoplasma infection]. In German. *Osterr Z Kinderheilkd Kinderfuersorge* **10:**316–321.
134. **Van Soestbergen AA.** 1957. [The course of a laboratory infection with Toxoplasma gondii]. In Dutch. *Ned Tijdschr Geneeskd* **101:**1649–1651.
135. **Villavedra M, Battistoni J, Nieto A.** 1999. IgG recognizing 21–24 kDa and 30–33 kDa tachyzoite antigens show maximum avidity maturation during natural and accidental human toxoplasmosis. *Rev Inst Med Trop Sao Paulo* **41:**297–303.
136. **Wettingfeld RF, Rowe J, Eyles DE.** 1956. Treatment of toxoplasmosis with pyrimethamine (daraprim) and triple sulfonamide. *Ann Intern Med* **44:**557–564.
137. **Wright WH.** 1957. A summary of the newer knowledge of toxoplasmosis. *Am J Clin Pathol* **28:**1–17.
138. **Zimmerman WJ.** 1976. Prevalence of *Toxoplasma gondii* antibodies among veterinary college staff and students, Iowa State University. *Public Health Rep* **91:**526–532.
139. **Umdenstock R, Mandoul R, Pestre-Alexandre M.** 1965. [Laboratory accident caused by the bite of a Toxoplasma-infected mouse. Auto-observation]. In French. *Bull Soc Pathol Exot Filiales* **58:**207–209.
140. **Woodison G, Balfour AH, Smith JE.** 1993. Sequential reactivity of serum against cyst antigens in *Toxoplasma* infection. *J Clin Pathol* **46:**548–550.
141. **Jacobs L.** 1957. The interrelation of toxoplasmosis in swine, cattle, dogs, and man. *Public Health Rep* **72:**872–882.
142. **McAuley JB, Jones JL, Singh AK.** 2015. Toxoplasma, p 2373–2386. *In* Jorgensen JH, Pfaller MA, Carroll KC, Funke G, Landry ML, Richter SS, Warnock DW (ed), *Manual of Clinical Microbiology*, 11th ed. ASM Press, Washington, DC.
143. **Peters E, Seaton A.** 2005. Bio-hazards and drug reactions: a cautionary tale. *Scand J Infect Dis* **37:**312–313.
144. **Bern C, Montgomery SP, Herwaldt BL, Rassi A Jr, Marin-Neto JA, Dantas RO, Maguire JH, Acquatella H, Morillo C, Kirchhoff LV, Gilman RH, Reyes PA, Salvatella R, Moore AC.** 2007. Evaluation and treatment of chagas disease in the United States: a systematic review. *JAMA* **298:**2171–2181.
145. **Bern C.** 2015. Chagas' Disease. *N Engl J Med* **373:**456–466.
146. **Shikanai-Yasuda MA, Carvalho NB.** 2012. Oral transmission of Chagas disease. *Clin Infect Dis* **54:**845–852.
147. **WHO Expert Committee.** 2002. Control of Chagas disease. *World Health Organ Tech Rep Ser* **905:**i–vi, 1–109, back cover.
148. **Zeledón R.** 1974. Epidemiology, modes of transmission and reservoir hosts of Chagas' disease, p 51–85. *In* Ciba Foundation Symposium 20 (new series), *Trypanosomiasis and Leishmaniasis with Special Reference to Chagas' Disease*. Associated Scientific Publishers, Amsterdam, The Netherlands.
149. **Kirchhoff LV, Hoft DF.** 1990. Immunization and challenge of mice with insect-derived metacyclic trypomastigotes of *Trypanosoma cruzi*. *Parasite Immunol* **12:**65–74.
150. **Brener Z.** 1984. Laboratory-acquired Chagas' disease: an endemic disease among parasitologists? p 3–9. *In* Morel CM (ed), *Genes and Antigens of Parasites: a Laboratory Manual*, 2nd ed. Fundação Oswaldo Cruz, Rio de Janeiro, Brazil.
151. **Gutteridge WE, Cover B, Cooke AJ.** 1974. Safety precautions for work with *Trypanosoma cruzi*. *Trans R Soc Trop Med Hyg* **68:**161.
152. **Transactions of the Royal Society of Tropical Medicine and Hygiene.** 1983. Suggested guidelines for work with live *Trypanosoma cruzi*. *Trans R Soc Trop Med Hyg* **77:**416–419.
153. **Wang X, Jobe M, Tyler KM, Steverding D.** 2008. Efficacy of common laboratory disinfectants and heat on killing trypanosomatid parasites. *Parasit Vectors* **1:**35.
154. **Aronson PR.** 1962. Septicemia from concomitant infection with *Trypanosoma cruzi* and *Neisseria perflava*. First case of laboratory-acquired Chagas' disease in the United States. *Ann Intern Med* **57:** 994–1000.
155. **Añez N, Carrasco H, Parada H, Crisante G, Rojas A, Gonzalez N, Ramirez JL, Guevara P, Rivero C, Borges R, Scorza JV.** 1999. Acute Chagas' disease in western Venezuela: a clinical, seroparasitologic, and epidemiologic study. *Am J Trop Med Hyg* **60:**215–222.
156. **Kinoshita-Yanaga AT, Toledo MJ, Araújo SM, Vier BP, Gomes ML.** 2009. Accidental infection by *Trypanosoma cruzi* follow-up by the polymerase chain reaction: case report. *Rev Inst Med Trop Sao Paulo* **51:**295–298.
157. **Moretti E, Basso B, Castro I, Carrizo Paez M, Chaul M, Barbieri G, Canal Feijoo D, Sartori MJ, Carrizo Paez R.** 2005. Chagas' disease: study of congenital transmission in cases of acute maternal infection. *Rev Soc Bras Med Trop* **38:**53–55.
158. **Herr A, Brumpt L.** 1939. Un cas aigu de maladie de Chagas contractée accidentellement au contact de triatomes Mexicains: observation et courbe fébrile. In French. *Bull Soc Pathol Exot Filiales* **32:**565–571.
159. **Allain DS, Kagan IG.** 1974. Isolation of *Trypanosoma cruzi* in an acutely infected patient. *J Parasitol* **60:**526–527.

160. **Centers for Disease Control.** 1980. Chagas' disease—Michigan. *MMWR Morb Mortal Wkly Rep* **29:**147–148.
161. **Coudert J, Despeignes J, Battesti MR, Michel-Brun J.** 1964. [A case of Chagas' disease caused by accidental laboratory contamination by *T. cruzi*.] In French. *Bull Soc Pathol Exot Filiales* **57:**208–213.
162. **de Titto EH, Araujo FG.** 1988. Serum neuraminidase activity and hematological alterations in acute human Chagas' disease. *Clin Immunol Immunopathol* **46:**157–161.
163. **Hanson WL, Devlin RF, Roberson EL.** 1974. Immunoglobulin levels in a laboratory-acquired case of human Chagas' disease. *J Parasitol* **60:**532–533.
164. **Hofflin JM, Sadler RH, Araujo FG, Page WE, Remington JS.** 1987. Laboratory-acquired Chagas disease. *Trans R Soc Trop Med Hyg* **81:**437–440.
165. **Melzer H, Kollert W.** 1963. [Contribution on the clinical and therapeutic aspect of Chagas disease (South American trypanosomiasis)]. In German. *Dtsch Med Wochenschr* **88:**368–377.
166. **Pizzi T, Niedmann G, Jarpa A.** 1963. [Report of 3 cases of acute Chagas' disease produced by accidental laboratory infections]. In Spanish. *Bol Chil Parasitol* **18:**32–36.
167. **Western KA, Schultz MG, Farrar WE, Kagan IG.** 1969. Laboratory acquired Chagas' disease treated with Bay [sic] 2502. *Bol Chil Parasitol* **24:**94.
168. **Shikanai-Yasuda MA, Lopes MH, Tolezano JE, Umezawa E, Amato Neto V, Barreto AC, Higaki Y, Moreira AA, Funayama G, Barone AA, Duarte A, Odone V, Cerri GC, Sato M, Pozzi D, Shiroma M.** 1990. [Acute Chagas' disease: transmission routes, clinical aspects and response to specific therapy in diagnosed cases in an urban center]. In Portuguese. *Rev Inst Med Trop Sao Paulo* **32:**16–27.
169. **World Health Organization.** 2013. Control and surveillance of human African trypanosomiasis. *World Health Organ Tech Rep Ser* **984:**1–237.
170. **Herbert WJ, Parratt D, Van Meirvenne N, Lennox B.** 1980. An accidental laboratory infection with trypanosomes of a defined stock. II. Studies on the serological response of the patient and the identity of the infecting organism. *J Infect* **2:** 113–124.
171. **Robertson DH, Pickens S, Lawson JH, Lennox B.** 1980. An accidental laboratory infection with African trypanosomes of a defined stock. I. The clinical course of the infection. *J Infect* **2:**105–112.
172. **Emeribe AO.** 1988. *Gambiense* trypanosomiasis acquired from needle scratch. *Lancet* **1:**470–471.
173. **Fromentin H.** 1959. [Accidental human trypanosomiasis due to *Trypanosoma gambiense*. Considerations on the infective strain and on the response of the organism]. In French. *Bull Soc Pathol Exot Filiales* **52:**181–188.
174. **Nodenot L.** 1949. Note sur une infection accidentelle avec une souche de *Trypanosoma gambiense*. In French. *Bull Soc Pathol Exot Filiales* **42:**16–18.
175. **Receveur MC, LeBras M, Vincendeau P.** 1993. Laboratory-acquired Gambian trypanosomiasis. *N Engl J Med* **329:**209–210.
176. **Lumsden WH, Kimber CD, Evans DA, Doig SJ.** 1979. *Trypanosoma brucei*: Miniature anion-exchange centrifugation technique for detection of low parasitaemias: Adaptation for field use. *Trans R Soc Trop Med Hyg* **73:**312–317.
177. **Magnus E, Vervoort T, Van Meirvenne N.** 1978. A card-agglutination test with stained trypanosomes (C.A.T.T.) for the serological diagnosis of *T. B. gambiense* trypanosomiasis. *Ann Soc Belg Med Trop* **58:**169–176.
178. **Herwaldt BL.** 2000. *Cyclospora cayetanensis*: a review, focusing on the outbreaks of cyclosporiasis in the 1990s. *Clin Infect Dis* **31:**1040–1057.
179. **Messner MJ, Chappell CL, Okhuysen PC.** 2001. Risk assessment for *Cryptosporidium*: a hierarchical Bayesian analysis of human dose response data. *Water Res* **35:**3934–3940.
180. **Okhuysen PC, Chappell CL, Crabb JH, Sterling CR, DuPont HL.** 1999. Virulence of three distinct *Cryptosporidium parvum* isolates for healthy adults. *J Infect Dis* **180:**1275–1281.
181. **Rendtorff RC.** 1954. The experimental transmission of human intestinal protozoan parasites. II. *Giardia lamblia* cysts given in capsules. *Am J Hyg* **59:**209–220.
182. **Dorner SM, Huck PM, Slawson RM.** 2004. Estimating potential environmental loadings of *Cryptosporidium* spp. and *Campylobacter* spp. from livestock in the Grand River Watershed, Ontario, Canada. *Environ Sci Technol* **38:**3370–3380.
183. **Fujino T, Matsui T, Kobayashi F, Haruki K, Yoshino Y, Kajima J, Tsuji M.** 2002. The effect of heating against *Cryptosporidium* oocysts. *J Vet Med Sci* **64:**199–200.
184. **Blewett DA.** 1989, p 107–115. *In* Angus KW, Blewett DA (ed), *Cryptosporidiosis. Proceedings of the First International Workshop.* Animal Disease Research Association, Edinburgh, Scotland.
185. **Campbell I, Tzipori AS, Hutchison G, Angus KW.** 1982. Effect of disinfectants on survival of *cryptosporidium* oocysts. *Vet Rec* **111:**414–415.
186. **Pavlásek I.** 1984. [The effect of disinfectants on the infectivity of Cryptosporidium sp. oocysts]. In Czech. *Cesk Epidemiol Mikrobiol Imunol* **33:**97–101.
187. **Cook EB.** 1961. Safety in the public health laboratory. *Public Health Rep* **76:**51–56.
188. **Schuman SH, Arnold AT, Rowe JR.** 1982. Giardiasis by inhalation? *Lancet* **1:**53.
189. **McCracken AW.** 1972. Natural and laboratory-acquired infection by *Isospora belli. South Med J* **65:**800–818, passim.
190. **Henderson HE, Gillepsie GW, Kaplan P, Steber M.** 1963. The human *Isospora. Am J Hyg* **78:**302–309.
191. **Jeffery GM.** 1956. Human coccidiosis in South Carolina. *J Parasitol* **42:**491–495.
192. **Højlyng N, Holten-Andersen W, Jepsen S.** 1987. Cryptosporidiosis: a case of airborne transmission. *Lancet* **2:**271–272.
193. **Blagburn BL, Current WL.** 1983. Accidental infection of a researcher with human *Cryptosporidium. J Infect Dis* **148:**772–773.
194. **Pohjola S, Oksanen H, Jokipii L, Jokipii AM.** 1986. Outbreak of cryptosporidiosis among veterinary students. *Scand J Infect Dis* **18:**173–178.
195. **Dryjanski J, Gold JW, Ritchie MT, Kurtz RC, Lim SL, Armstrong D.** 1986. Cryptosporidiosis. Case report in a health team worker. *Am J Med* **80:**751–752.
196. **Baxby D, Hart CA, Taylor C.** 1983. Human cryptosporidiosis: a possible case of hospital cross infection. *Br Med J (Clin Res Ed)* **287:**1760–1761.
197. **Roncoroni AJ, Gomez MA, Mera J, Cagnoni P, Michel MD.** 1989. *Cryptosporidium* infection in renal transplant patients. *J Infect Dis* **160:**559.
198. **Casemore DP, Gardner CA, O'Mahony C.** 1994. Cryptosporidial infection, with special reference to nosocomial transmission of *Cryptosporidium parvum*: a review. *Folia Parasitol (Praha)* **41:** 17–21.
199. **Gardner C.** 1994. An outbreak of hospital-acquired cryptosporidiosis. *Br J Nurs* **3:**152, 154–158.
200. **O'Mahony C, Casemore DP.** 1992. Hospital-acquired cryptosporidiosis. *Commun Dis Rep CDR Rev* **2:**R18–R19.
201. **Koch KL, Phillips DJ, Aber RC, Current WL.** 1985. Cryptosporidiosis in hospital personnel. Evidence for person-to-person transmission. *Ann Intern Med* **102:**593–596.
202. **Arikan S, Ergüven S, Akyön Y, Günalp A.** 1999. Cryptosporidiosis in immunocompromised patients in a Turkish university hospital. *Acta Microbiol Immunol Hung* **46:**33–40.
203. **Foot AB, Oakhill A, Mott MG.** 1990. Cryptosporidiosis and acute leukaemia. *Arch Dis Child* **65:**236–237.
204. **Lettau LA.** 1991. Nosocomial transmission and infection control aspects of parasitic and ectoparasitic diseases: part I. Introduction/enteric parasites. *Infect Control Hosp Epidemiol* **12:**59–65.

205. **Martino P, Gentile G, Caprioli A, Baldassarri L, Donelli G, Arcese W, Fenu S, Micozzi A, Venditti M, Mandelli F.** 1988. Hospital-acquired cryptosporidiosis in a bone marrow transplantation unit. *J Infect Dis* **158:**647–648.
206. **Navarrete S, Stetler HC, Avila C, Garcia Aranda JA, Santos-Preciado JI.** 1991. An outbreak of *Cryptosporidium* diarrhea in a pediatric hospital. *Pediatr Infect Dis J* **10:**248–250.
207. **Neill MA, Rice SK, Ahmad NV, Flanigan TP.** 1996. Cryptosporidiosis: an unrecognized cause of diarrhea in elderly hospitalized patients. *Clin Infect Dis* **22:**168–170.
208. **Ravn P, Lundgren JD, Kjaeldgaard P, Holten-Anderson W, Højlyng N, Nielsen JO, Gaub J.** 1991. Nosocomial outbreak of cryptosporidiosis in AIDS patients. *BMJ* **302:**277–280.
209. **Sarabia-Arce S, Salazar-Lindo E, Gilman RH, Naranjo J, Miranda E.** 1990. Case-control study of *Cryptosporidium parvum* infection in Peruvian children hospitalized for diarrhea: possible association with malnutrition and nosocomial infection. *Pediatr Infect Dis J* **9:**627–631.
210. **Wittenberg DF, Miller NM, van den Ende J.** 1989. Spiramycin is not effective in treating *cryptosporidium* diarrhea in infants: results of a double-blind randomized trial. *J Infect Dis* **159:**131–132.
211. **Eberhard ML, Pieniazek NJ, Arrowood MJ.** 1997. Laboratory diagnosis of *Cyclospora* infections. *Arch Pathol Lab Med* **121:**792–797.
212. **Garcia LS.** 2016. *Diagnostic Medical Parasitology*, 6th ed. ASM Press, Washington, DC.
213. **Coles GC.** 1975. Letter: gastro-intestinal allergy to nematodes. *Trans R Soc Trop Med Hyg* **69:**364–365.
214. **Coles GC.** 1985. Allergy and immunopathology of ascariasis, p 167–184. *In* Crompton DWT, Nesheim MC, Pawlowski ZS (ed), *Ascariasis and its Public Health Significance*. Taylor and Francis, London, United Kingdom.
215. **Jones TL, Kingscote AA.** 1935. Observations on *Ascaris* sensitivity in man. *Am J Epidemiol* **22:**406–413.
216. **Sprent JF.** 1949. On the toxic and allergic manifestations produced by the tissues and fluids of *Ascaris*; effect of different tissues. *J Infect Dis* **84:**221–229.
217. **Turner KJ, Fisher EH, McWilliam AS.** 1980. Homology between roundworm (*Ascaris*) and hookworm (*N. americanus*) antigens detected by human IgE antibodies. *Aust J Exp Biol Med Sci* **58:**249–257.
218. **Maligin SA.** 1958. [A case of cutaneous form of strongyloidiasis caused by larvae of S. ransomi, S. westeri and S. papillosus]. In Russian. *Med Parazitol (Mosk)* **27:**446–447.
219. **Roeckel IE, Lyons ET.** 1977. Cutaneous larva migrans, an occupational disease. *Ann Clin Lab Sci* **7:**405–410.
220. **Stone OJ, Levy A.** 1967. Creeping eruption in an animal caretaker. *Lab Anim Care* **17:**479–482.
221. **Fölster-Holst R, Disko R, Röwert J, Böckeler W, Kreiselmaier I, Christophers E.** 2001. Cercarial dermatitis contracted via contact with an aquarium: case report and review. *Br J Dermatol* **145:**638–640.
222. **Ashton CR, Beresford OD.** 1974. Letter: fascioliasis. *BMJ* **2:**121.
223. **Beresford OD.** 1976. A case of fascioliasis in man. *Vet Rec* **98:**15.
224. **Van Gompel A, Van den Enden E, Van den Ende J, Geerts S.** 1993. Laboratory infection with *Schistosoma mansoni*. *Trans R Soc Trop Med Hyg* **87:**554.
225. **Elsevier.** 1998. Accidental infections. *Parasitol Today* **14:**55.
226. **Elsevier.** 1998. Schistosomiasis: symptoms of mild infections? *Parasitol Today* **14:**8.
227. **Vasoo S, Pritt BS.** 2013. Molecular diagnostics and parasitic disease. *Clin Lab Med* **33:**461–503.
228. **Pike RM.** 1979. Laboratory-associated infections: incidence, fatalities, causes, and prevention. *Annu Rev Microbiol* **33:**41–66.
229. **Pike RM.** 1978. Past and present hazards of working with infectious agents. *Arch Pathol Lab Med* **102:**333–336.
230. **Pike RM, Sulkin SE, Schulze ML.** 1965. Continuing importance of laboratory-acquired infections. *Am J Public Health Nations Health* **55:**190–199.
231. **Sulkin SE, Pike RM.** 1951. Survey of laboratory-acquired infections. *Am J Public Health Nations Health* **41:**769–781.
232. **Kirchhoff LV, Votava JR, Ochs DE, Moser DR.** 1996. Comparison of PCR and microscopic methods for detecting *Trypanosoma cruzi*. *J Clin Microbiol* **34:**1171–1175.
233. **Schijman AG, Bisio M, Orellana L, Sued M, Duffy T, Mejia Jaramillo AM, Cura C, Auter F, Veron V, Qvarnstrom Y, Deborggraeve S, Hijar G, Zulantay I, Lucero RH, Velazquez E, Tellez T, Sanchez Leon Z, Galvão L, Nolder D, Monje Rumi M, Levi JE, Ramirez JD, Zorrilla P, Flores M, Jercic MI, Crisante G, Añez N, De Castro AM, Gonzalez CI, Acosta Viana K, Yachelini P, Torrico F, Robello C, Diosque P, Triana Chavez O, Aznar C, Russomando G, Büscher P, Assal A, Guhl F, Sosa Estani S, DaSilva A, Britto C, Luquetti A, Ladzins J.** 2011. International study to evaluate PCR methods for detection of *Trypanosoma cruzi* DNA in blood samples from Chagas disease patients. *PLoS Negl Trop Dis* **5:**e931.
234. **Strout RG.** 1962. A method for concentrating hemoflagellates. *J Parasitol* **48:**100.
235. **Anderson BC, Donndelinger T, Wilkins RM, Smith J.** 1982. Cryptosporidiosis in a veterinary student. *J Am Vet Med Assoc* **180:**408–409.
236. **Current WL, Reese NC, Ernst JV, Bailey WS, Heyman MB, Weinstein WM.** 1983. Human cryptosporidiosis in immunocompetent and immunodeficient persons. Studies of an outbreak and experimental transmission. *N Engl J Med* **308:** 1252–1257.
237. **Konkle DM, Nelson KM, Lunn DP.** 1997. Nosocomial transmission of *Cryptosporidium* in a veterinary hospital. *J Vet Intern Med* **11:**340–343.
238. **Lengerich EJ, Addiss DG, Marx JJ, Ungar BL, Juranek DD.** 1993. Increased exposure to cryptosporidia among dairy farmers in Wisconsin. *J Infect Dis* **167:**1252–1255.
239. **Levine JF, Levy MG, Walker RL, Crittenden S.** 1988. Cryptosporidiosis in veterinary students. *J Am Vet Med Assoc* **193:**1413–1414.
240. **Reif JS, Wimmer L, Smith JA, Dargatz DA, Cheney JM.** 1989. Human cryptosporidiosis associated with an epizootic in calves. *Am J Public Health* **79:**1528–1530.
241. **Drinkard LN, Halbritter A, Nguyen GT, Sertich PL, King M, Bowman S, Huxta R, Guagenti M.** 2015. Outbreak of cryptosporidiosis among veterinary medicine students—Philadelphia, Pennsylvania, February 2015. *MMWR Morb Mortal Wkly Rep* **64:**773.

7 Mycotic Agents

WILEY A. SCHELL

The number of fungal species is conservatively estimated to be 1.5 million, and at least 98,000 have been described formally (1). Although more than 300 of these are documented as causing disease in humans, only about 100 are encountered regularly as pathogens of humans. Virulence among these fungi varies, as do the entry portals through which they cause disease in the host and the manner in which they subsequently could spread. These various differences provide a convenient basis for broadly categorizing the mycoses, and they also help in delineating biosafety measures needed for the safe handling and storage of the fungi involved.

First, there historically have been two main categories of fungi regarding virulence: frank pathogens and opportunistic pathogens. Frank pathogens possess elevated infectivity and virulence and are capable of causing disease in normal hosts, whereas opportunists are less infectious, less virulent, and cause disease mainly in hosts whose defenses have been compromised in some manner. Next, there are three main classes of fungal disease on the basis of the anatomic location involved, namely, cutaneous, subcutaneous, and pulmonary mycoses. Cutaneous mycoses are those that are limited to the dermis and its appendages (hair, nail). Most common are the dermatophytic diseases, which are caused by species of the three dermatophyte genera *Epidermophyton, Microsporum,* and *Trichophyton* (primarily *E. floccosum, M. canis, M. gypseum, T. rubrum,* and *T. tonsurans*). Several nondermatophytic fungi also can cause cutaneous afflictions. Subcutaneous mycoses are infections that normally require a breach of the cutaneous barrier by trauma (e.g., thorn, splinter, abrasion) and implantation of the fungus into host tissue. The most frequent of these in North America are sporotrichosis, caused by *Sporothrix schenckii*; phaeohyphomycosis, which is caused by species of *Alternaria, Exophiala,* and species of a large number of additional genera; and mucormycosis, which usually is caused by species of *Rhizopus* and *Mucor.* Pulmonary mycoses begin in the lungs and often can disseminate to other organs, including the skin. The best known in North America are cryptococcosis (caused by *Cryptococcus neoformans* and *C. gattii),* blastomycosis (caused by *Blastomyces dermatitidis*), coccidioidomycosis (caused by *Coccidioides immitis,* and *C. posadasii*), histoplasmosis (caused by *Histoplasma capsulatum* var. *capsulatum*), aspergillosis (caused mainly by *Aspergillus fumigatus* and

A. flavus), and mucormycosis (caused mainly by species of *Rhizopus* and *Mucor*).

These traditional distinctions about inherent virulence and portal of infection provide a useful starting point for considering the pathogenic fungi. However, advances in immunosuppression and other medical interventions during the past 25 years have altered important aspects of the population of those people who are particularly susceptible to fungal infection. As a result, new fungal pathogens have been reported and new aspects of previously well-characterized pathogens have been recognized. Some of these changes are so significant that strict adherence to traditional distinctions concerning pathogenic fungi can sometimes be a hindrance to an effective understanding of modern medical mycology and its related safety issues. For this reason, discussion of fungal agents of human disease in this chapter will be organism based rather than disease based or anatomically based. This discussion is not exhaustive for all documented fungal pathogens, but instead seeks to address those fungi that are seen with some regularity in clinical, veterinary, and environmental mycology laboratories. Species-based compilations of documented infections (2, 3), as well as diseased-based approaches to medical mycology (4–6), and approaches that address sexuality and phylogeny (2, 3, 7–9) can be found elsewhere.

INCIDENCE OF LABORATORY-ASSOCIATED FUNGAL INFECTIONS

The true incidence of laboratory-acquired mycoses is unknown and therefore can only be estimated. In North America, some U.S. states require reporting of certain mycoses such as coccidioidomycosis in California, Arizona, and others; blastomycosis in Louisiana, Wisconsin, and some provinces of Canada; and histoplasmosis in Kentucky and Wisconsin. Ohio requires reporting of blastomycosis, histoplasmosis, and sporotrichosis, but only when there is an outbreak or unusual incidence. Ontario, Canada recently has discontinued inclusion of blastomycosis as a notifiable disease (10).

There is no U.S. federal requirement for notification of fungal infections, whether laboratory acquired or otherwise, except for species covered by the Select Agent Program and recombinant organisms covered under the NIH Guidelines for Research Involving Recombinant or Synthetic Nucleic Acid Molecules (11–16). Four fungi (in the broad taxonomic sense) currently are included on the National Select Agent Registry, namely, *Synchytrium endobioticum, Sclerophthora rayssiae* var. *zeae, Phoma glycinicola* (also known as *Coniothyrium glycines, Pyrenochaeta glycines,* and *Phoma glycinicola*), and *Peronosclerospora philippinensis*. Two other fungi, *Coccidioides immitis* and *C. posadasii*, recently were removed from the Registry. Apart from the requirements of those two programs, coccidioidomycosis became the only mycosis on the list of nationally notifiable infectious diseases, effective 1995. However, reports of this infection by state agencies to the Centers for Disease Control and Prevention (CDC) are voluntary. Individual institutions have policies for reporting certain microbial exposures, but there is no means for tabulating exposures to pathogens that are not covered by state mandates. The experience with laboratory-acquired fungal infections as of about 1980 was thoroughly reviewed and summarized by Schwarz, and this publication probably characterizes the extent of such infections fairly accurately (17). Subsequent cases have been reported when circumstances or etiologic agents were unusual. However, because these incidents generally rely on the peer-review publication process, laboratory-acquired infections that are unremarkable are not likely to be regarded as worthy of publication in the scientific literature. As a result, information about the incidence of fungal infections in general and laboratory-related infections in particular is fragmentary and sometimes anecdotal (Table 1).

PREVENTION OF OCCUPATIONAL EXPOSURES

Biological safety measures for various fungi of medical concern are recommended by the CDC (18). Strict adherence to CDC-recommended biosafety guidelines for each of those fungi provides excellent protection against accidental exposure. The major risk for laboratory exposure is inadvertent release of mold spores (Fig. 1). A Class II biological safety cabinet (BSC) is the single most important device that can be used for containment of fungi that are infectious via the airborne route. Use of the BSC is required for work with particular mold species, is essential for safe work with unidentified molds, and is desirable for work involving any mold regardless of its potential infectivity. The effectiveness of this approach is reflected by a recent example from the mycology section of a large clinical laboratory, in which a bacterial culture plate had been referred for identification of a mold. The plate in question contained a faintly brown mold described as coming from "sinus." On the basis of the anatomic site description and on the brownish appearance of the mold, it was suspected by the mycologist to be a relatively harmless dematiaceous mold isolated from a paranasal sinus specimen. Instead, however, the mold proved to be an isolate of *C. immitis* that had been cultured from a cutaneous fistula (i.e., a sinus). *C. immitis* is one of the most infectious microbes known and, in this instance, laboratory policy that specified use of a BSC when opening any plate containing an unidentified mold

TABLE 1.

Laboratory-acquired mycoses

Fungus	BSL/ABSL	Route	Infectious potential	No. cases	References
Blastomyces dermatitidis	2/2	Transcutaneous	Serious/possibly fatal	12	124
					125
					17
					126
					62
		Inhalation	Serious /possibly fatal	2	62
					64
Candida albicans	2?[a]/2?[a]	Cutaneous	Minor	1[b]	28
Coccidioides posadasii[c]	3[d]/2	Inhalation	Serious/possibly fatal	Numerous[e]	17
		Transcutaneous	Serious/possibly fatal	4	127
		Spray	Serious	1	69
Coccidioides immitis[c]	3[d]/2	Inhalation	Serious/possibly fatal	Numerous[c]	17, 128, 129
Cryptococcus neoformans	2/2[f]	Transcutaneous	Serious	1	35
Dermatophytes	2/2	Cutaneous	Minor	Numerous	82
					17
					81
Penicillium marneffei[g]	2/2	Transcutaneous	Serious	1	100
		Inhalation (presumed)	Serious/possibly fatal	1	101
Histoplasma capsulatum	3[d]/2	Inhalation	Serious/possibly fatal	Numerous	17
		Transcutaneous	Serious	3	88
					89
					90
		Spray	Serious	1	91
Sporothrix schenckii	2/2	Transcutaneous	Serious	ca. 2 dozen[h]	17
					105
		Spray	Serious	3	17
					130

[a]See discussion under "Considerations for Specific Fungi" (subhead "Yeasts"), below.
[b]Two additional cases have been mentioned, but without citation or supportive evidence.
[c]Four additional laboratory exposures to *Coccidioides* species in the BSL3 setting were reported between 2004 and 2010, but it is not clear from the reports whether infections resulted (see Appendix D, A Review of Reported Incidents, Exposures and Infections in BSL-3AND BSL-4 Laboratory Facilities. *In* Final Supplementary Risk Assessment Report for the Boston University National Emerging Infectious Diseases Laboratories, National Institutes of Health. 2012. http://www.bu.edu/neidl/files/2013/01/SFEIR-Volume-III.pdf).
[d]BSL2 is considered acceptable for identification of clinical isolates of *Coccidioides* species and *Histoplasma* species, but BSL3 is recommended for further work with known isolates. BSL3 is recommended for work with fomites likely to harbor conidia. ABSL3 is recommended for studies using animal that are infected by any route other than injection.
[e]See discussion regarding recent taxonomic changes.
[f]See discussion regarding use of additional ABSL precautions for *Cryptococcus* species in "Biosafety Considerations When Using Animals."
[g]The natural history, epidemiology, and clinical characteristics of *Penicillium marneffei* bear significant similarities to those of *Histoplasma capsulatum*. It would be prudent to follow biosafety precautions similar to those for *H. capsulatum*.
[h]Includes veterinary clinic-acquired cases of sporotrichosis.

provided the necessary primary protection from infectious airborne spores.

Culture plates often are used in the mycology laboratory because they offer the advantage of greater surface area. Because prevention of accidental release of mold spores is the single most important element of mycology laboratory safety, culture plate lids must be taped at two points so they cannot be opened accidentally. Even this precaution will not always be adequate when certain fungi are involved because the plastic lid incorporates three tabs to keep the lid slightly raised from the plate bottom to facilitate air exchange. It is possible for spores to escape via this vented circumference. Thus, it is best to subculture molds to medium promptly in screw-cap tubes and to wrap the plates (parafilm or shrink-wrap) until they can be decontaminated.

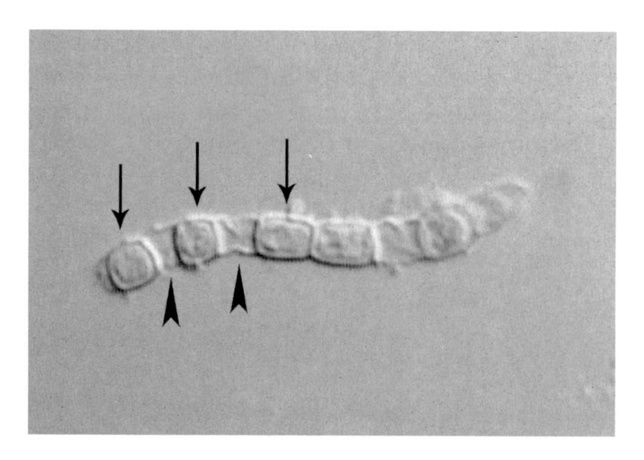

Figure 1: Spore formation provides an efficient mechanism for dispersal of molds. Spores from some species pose a risk of infection to laboratory personnel. In this image of *Coccidioides immitis,* cells of a vegetative hypha have transformed into spores (arrows) that will be liberated when walls of the adjacent cells, now dead and withered, become fractured (arrowheads).

A recent survey found that laboratory compliance with biosafety specifications generally is high (19). The main deficiency identified by the survey was a lack of annual certification of BSCs, which was documented for 27% of laboratories surveyed. The survey did not distinguish between cabinets with recently expired certification and those for which there was no evidence of a certification policy ever having been implemented. The need to assure that primary biocontainment equipment—mainly the BSC—has current certification, and that the certification was performed by an NSF International-accredited Class II BSC field certifier, is of paramount importance. However, it must be recognized that other issues can significantly diminish the biocontainment efficacy of this equipment, such as improper technique when using the BSC, or a suboptimal location of the BSC within the laboratory. The survey also determined a 16.7% failure rate for laboratories that were designed to have directional airflow into the laboratory. Directional airflow can be difficult to maintain consistently, and often is beyond the immediate control of the laboratory staff. However, laboratory staff can make efforts to monitor airflow by the use of periodic smoke checks and by mounting tissue strips or equivalent at the bottom of doors as visual indicators, in addition to use of built-in monitoring devices. All of these engineering controls should be evaluated for use based on the risk assessment for the laboratory.

Probably two of the most frequently overlooked personal protection measures are eye protection and gloves. Biosafety level 2 (BSL2) criteria call for eye protection in circumstances where splashes are possible, and BSL1 criteria specify eye protection when contact lenses are worn (18). These precautions are particularly important in the clinical laboratory when pipetting liquid suspensions of fungi (e.g., in conducting susceptibility testing) and in the research laboratory when pipetting suspensions or intravenously injecting animals with fungal suspensions. In the latter circumstance, an unsuccessful attempt to inject fungal inoculum into the tail vein of a mouse can lead to pressure at the injection site, sufficient to cause inoculum to spray back toward the researcher's face. Similarly, it should be emphasized that BSL2 criteria specify the use of gloves when hands might contact potentially infectious materials, contaminated surfaces, or equipment. BSL1 criteria call for use of gloves when skin is broken or there is a rash (18). Skin contact with *Candida albicans* rarely can cause dermatitis, contact with dermatophytes can cause dermatophytosis, and simple contact with *S. schenckii* is now recognized as an initiating event for sporotrichosis. Also, even though some molds are regarded as essentially harmless, it is recommended that all molds be manipulated within a BSC to contain the spread of spores that can be allergenic and which can pose an increased contamination risk to other cultures.

MOLDS OF SPECIAL CONCERN

Of the medically important fungi, only three fungal genera, *Blastomyces, Coccidioides,* and *Histoplasma,* contain species that require BSL3 biocontainment precautions and facilities (18). These requirements apply to known (i.e., identified) isolates and to environmental samples that are suspected to contain infectious spores. No fungi are classified as requiring BSL4 biocontainment precautions. *C. immitis* recently was split into two species on the basis of DNA sequence analysis (20). The two species, *C. immitis* and *C. posadasii,* have identical potential for causing infection, disease, and death in humans and other animals, and they require the same precautions for safe handling, storage and shipping. Apart from the species in those three genera, there are other molds of special concern. Working with these additional molds can be made safer by incorporating added precautions, in particular the use of a BSC for culture manipulations. These molds include *Cladophialophora bantiana, Paracoccidioides brasiliensis, Penicillium marneffei, Rhinocladiella mackenziei,* and *Ochroconis gallopava* (*Verruconis gallopava*). Accidental exposure to these molds can pose significant risk, even to immunocompetent hosts. In addition to the use of a BSC for handling all molds, culture vessels used for propagating particularly hazardous molds should be screw-cap tubes rather than Petri plates whenever possible. If glass culture tubes must be used, they can be wrapped with a transparent vinyl tape (e.g., 3M #471) to render them shatter resistant.

Use of disposable aerosol barrier pipette tips is another advisable precaution when working with suspensions of particularly hazardous molds. These aerosol barrier tips can prevent contamination of internal components of the pipette and subsequent inadvertent propagation or aerosolization of the fungus.

Culture storage for most fungal species needs no special labeling, but, if possible, molds of special concern ought to be clearly labeled by name and with the universal biohazard symbol so that spills or other mishaps involving these fungi can be recognized quickly and dealt with and documented appropriately. Many laboratories store fungal cultures on a long-term basis for research applications, teaching purposes, and as reference materials. Storage of hazardous molds at any temperature, including refrigeration, calls for clear labeling with the fungal name and the universal biohazard symbol, if possible. Long-term storage (e.g., at –70°C to –80°C) of fungal suspensions in 2-ml screw-cap polypropylene vials containing 15% glycerol is particularly convenient and has become increasingly common. In addition to the technical advantages of this method, an incidental advantage is that spilled conidial suspensions in 15% glycerol are relatively less likely to generate infectious aerosols compared to spills from other storage methods, and 15% glycerol also evaporates slowly, thereby facilitating effective disinfection of the contaminated area. The small size of these vials, however, can make full labeling difficult. In this case, use of miniature adhesive biohazard symbols, or coding the containers using a red marker, for example, could be an effective alternative.

Added caution is called for when discarding cultures of *C. immitis, C. posadasii, H. capsulatum, B. dermatitidis, C. bantiana, P. brasiliensis, P. marneffei, R. mackenziei*, and *O. gallopava (V. gallopava)*. When such cultures are to be discarded, they should be decontaminated without delay rather than being placed into a biohazard bin that might remain for several days or longer before being removed for decontamination. If this is not possible, cultures could be sealed in a plastic bag containing disinfectant, placed inside a second sealed bag, and then placed into the biohazard bin. In the case of fungi that are on the Select Agent Registry, destruction must be witnessed and documented, and the documentation must be retained permanently on site. Alternatively, these regulated agents may be shipped to a registered facility if storage of the isolate is desirable.

REGULATIONS

Select Agent Regulations

The Public Health Security and Bioterrorism Preparedness and Response Act of 2002 required establishment of regulations regarding the possession, use, and transfer of select biological agents and toxins. The Select Agent Program, as it commonly is known, specifies infectious agents and toxins that are regulated either by the Department of Health and Human Services (HHS) or the Department of Agriculture (USDA) or both (13, 14, 21, 22). Those agents regulated by both HHS and USDA are known as "overlap agents." The Select Agent list includes species in four genera of fungi (including the Chromista supergroup). All four are plant pathogens regulated exclusively by the USDA.

The Select Agent list should not necessarily be taken as the best indicator of hazard associated with a given fungal species. Specifically, *C. immitis* and *C. posadasii* were included on earlier versions of the Select Agent Registry ". . . because they are highly infectious when aerosolized and sporulate easily in culture," whereas *H. capsulatum* and *B. dermatitidis* were not included "... because they are difficult to cultivate and do not sporulate readily" (18, 21). However, it should be noted that *H. capsulatum* and *B. dermatitidis* are not difficult to propagate and, although they do grow and sporulate more slowly than the *Coccidioides* species, their spores can become airborne and cause serious infection under ordinary laboratory conditions. As noted above, the Select Agent list is subject to change.

Shipping Regulations

National and international shipping regulations for transporting infectious or potentially infectious materials have undergone significant changes in recent years, primarily to formulate requirements that are based on actual health risks to transport personnel. Infectious materials have undergone significant changes in recent years, primarily to formulate requirements that are based on actual health risks to transport personnel. Regulations for packaging and shipping of biological materials are covered in another chapter of this book. Certain fungi are especially hazardous and require significant precautions, such as *C. immitis*. However, most fungi (i.e., species of *Candida*) shipped to other laboratories pose low or no risk to the transporter or other people in the environment.

The most current international regulations provide two categories for shipment of infectious substances. Category A substances are defined as those that may cause permanent disability or a life-threatening or fatal disease in otherwise healthy humans and animals. Category B substances are those that do not meet the definition of Category A. The regulations allow the shipper to use discretion and professional judgment when deciding in which category a substance belongs. Although the packaging instructions for both categories require the use of triple packing for transport, there are significant differences in requirements for documentation, marking, and labeling the outer container, and the testing specifications of the package.

Shippers should avoid simply classifying all shipments as either Category A or B, because this is not the intent of the regulations. However, in instances where there is doubt about whether a substance meets the criteria of Category A or B, it should be classified as a Category A substance and shipped accordingly. Shipping regulations are revised every 2 years and should be consulted regularly for changes.

CONSIDERATIONS FOR SPECIFIC FUNGI

Yeasts

Candida species (*C. albicans, C. tropicalis,* and *C. glabrata*)

C. albicans is part of the normal microbiota in the gastrointestinal tract, vagina, and oropharynx, but can cause disease in both immunocompetent and immunodeficient hosts. Its frequency of isolation from these sites has been reported as 0–55%, 2.2–68%, and 1.9–41.4% respectively, and *C. albicans* is not readily isolated from healthy glabrous skin in immunocompetent hosts (23). *C. albicans* frequently causes esophageal candidiasis among acquired immunodeficiency syndrome (AIDS) patients and can cause invasive candidiasis in other immunodeficient hosts. Invasive candidiasis can begin by invasion from the gastrointestinal tract or from breaches of the skin caused by medical devices, such as indwelling intravenous catheters (24). Other at-risk persons include burn patients and patients who have received protracted treatment with broad-spectrum antibacterial drugs. Chronic candidiasis of the skin, nails, and mucosal membranes can occur in people who are genetically disposed or who have endocrine dysfunction. Candidiasis also can occur in healthy people under certain circumstances. These include vaginitis, nail infection, oral candidiasis in infants and the elderly, and infection of macerated skin. *Candida* species often have also been regarded as a leading cause of fungal sinusitis in healthy hosts, but a review of all published reports has cast doubt on this view (25).

Candida species do not form airborne spores as part of their life cycle and are safely handled using basic precautions. Canadian authorities specify BSL2 biocontainment precautions for *C. albicans,* but U.S. authorities have not yet specified a biosafety level (26). It has been suggested that BSL2 precautions for *C. albicans* might be unnecessary (27). Laboratory risk assessment for individual situations can determine suitable precautions. Eye protection is recommended in situations where a splash or spray is possible. The risk of infection for laboratory workers is very low and no documented report of laboratory-acquired *Candida* infection has been found. One source states that two laboratory-acquired infections with *C. albicans* have occurred, but no reference is provided (26). However, in one unpublished incident a medical student developed an itchy, erythematous rash and folliculitis 2 days after spilling a heavy suspension of *C. albicans* on her naked leg during a laboratory experiment. Initial self-directed therapy using topical nystatin was ineffective, but the condition resolved after a 5-day course of oral fluconazole (28).

Cryptococcus species

C. neoformans is widespread in nature and associated with the guano of pigeons and with decayed wood. There are four serotypes of the fungus originally known as *C. neoformans,* namely A, D, B, and C, and an AD hybrid. Changes in nomenclature have been proposed as follows, and are widely accepted. The B and C serotypes are known as *C. gattii,* the D serotype remains known as *C. neoformans* (var. *neoformans*), and the A serotype is known as *C. neoformans* var. *grubii.* Serotype A is ubiquitous, serotype D occurs worldwide but more in Europe, serotype hybrid AD is global but rare, and serotypes B and C are found mostly in subtropical to tropical climates (29). A dramatic exception for serotype B isolates (*C. gattii*) has been discovered during the past decade, in the form of about 200 infections that have occurred in temperate regions of British Columbia, Canada, and the Pacific Northwest of the United States. Many of these infections occurred in people who had a normal immune status (30, 31).

C. neoformans causes pulmonary infection and meningitis most often in people who have a compromised cell mediated immune response. Serologic studies have shown that exposure to the fungus is widespread in the human population (32, 33). Although natural resistance to infection is high, people with apparently normal immune systems have become infected in rare cases. The fungus exists in nature as a yeast. In the laboratory, yeast cells can be induced to form filamentous sporulating structures. It is believed that the mold form occurs in nature also and that airborne spores likely are formed and dispersed in air (34). Because *C. neoformans* grows as a yeast in the laboratory, BSL2 precautions are adequate for its handling (see "Biosafety Considerations When Using Animals," below).

The lungs are the portal of infection for *Cryptococcus* species. Yeasts cells grown in the laboratory on agar media will not become airborne and are too large to lodge in the alveoli. However, desiccated yeast cells (as from dehydrated culture plates, or guano submitted for culture) can be smaller in size and are regarded as a potential hazard. Basidiospores that can form under favorable research conditions (such as mating experiments between compatible strains) should be regarded as potentially serious airborne hazards as well.

Cryptococcosis resulting from a needle puncture to the thumb during blood collection from an AIDS patient with cryptococcal fungemia has been documented in a phlebotomy laboratory setting (35). Two additional instances of percutaneous cryptococcal inoculation from needlestick were reported. Both of these people received fluconazole prophylaxis, and no symptoms of infection developed (36). There have been two reports of eye infection related to surgical procedures (37, 38) and, accordingly, eye protection is advisable where splashes from *C. neoformans* are possible.

Other yeast species

Yeasts other than *Candida* and *Cryptococcus* species are relatively infrequent findings in the clinical laboratory. *Malassezia* species can cause cosmetic afflictions in healthy people and bloodstream infections in patients who are receiving intravenously administered lipids for nutrition (39). *Geotrichum capitatum* (formerly known as *Blastoschizomyces capitatus*) and *Trichosporon* species have caused serious and fatal infections in immune compromised hosts (40, 41). *Trichosporon* species are capable of causing minor dermal afflictions in healthy hosts. *Rhodotorula* species have caused bloodstream infections in patients with indwelling vascular catheters (42). All of these, and other relatively rare yeast species, are handled safely using BSL1 criteria. No laboratory acquired infections have been reported for any of these fungi.

Molds

Biocontainment and biosafety issues are more challenging for molds than for yeasts because molds have evolved the capacity to form airborne spores as a dispersal mechanism (see Fig. 1). For the most hazardous mold species mentioned previously, handling precautions focus on the infectious potential of these spores, which can be transmitted by air, traumatic inoculation, or casual contact depending on the particular fungus. Although spores from other mold species such as *A. fumigatus* are less infectious, they too require biocontainment measures because they are capable of causing infection in both normal and compromised hosts under certain circumstances.

Recently, an additional safety consideration for working with molds has gained attention. There are 300 or more metabolic compounds produced by fungi that are considered to be toxins. Much is known about the effects from ingesting the most hazardous of these mycotoxins, but relatively little is yet known about potential adverse effects from contact exposure to them (43–45). For some molds, mycotoxins are bound in the fungal cell wall of the spores. A partial list of species that can have mycotoxins in their spores includes *Alternaria alternata, A. fumigatus, Aspergillus flavus, Aspergillus parasiticus, Fusarium graminearum, Fusarium sporotrichioides,* and *Stachybotrys chartarum* (46). Experience so far suggests that exposure to very high concentrations of such spores is required to elicit a clear response (47). However, local tissue injury may be possible following deposition of spores having high levels of toxins (48). Exposure precautions are judged to be warranted until more can be learned about the potential effects of mycotoxins on human health via the respiratory tract or dermal exposure (45). Thus, for all molds regardless of species, incorporating the use of a BSC is recommended to contain spores and prevent their dispersal. Control of spores within the laboratory can prevent potential infection, potential exposure to mycotoxins, and allergic discomfort in sensitive personnel and can reduce the potential for contamination of other culture media.

Aspergillus species

Aspergillus species number more than 250 and are globally widespread in nature, mostly as saprobes but in a few cases as plant pathogens. Several are documented as causing invasive infection in humans, but only *A. fumigatus, A. flavus,* and perhaps *A. terreus* are routinely encountered as pathogens. The great majority of infections follow inhalation of conidia by hosts who have a poor cellular immune response, and most of these patients are profoundly immunocompromised. However, infection in immunocompetent people has been reported (49, 50). Inhalation of *A. fumigatus* conidia also causes a noninfectious hypersensitivity lung disease known as allergic bronchopulmonary aspergillosis, which affects asthmatic and cystic fibrosis patients (51). Nonpulmonary infection following direct introduction of spores or hyphal cells is possible, as in patients whose peritoneal dialysis catheter has become contaminated and among healthy individuals who suffer trauma to corneal or cutaneous tissue (52–54).

Disease also can occur in apparently normal hosts in the form of chronic sinusitis, which can be broadly divided into allergic fungal sinusitis (AFS) and chronic invasive fungal sinusitis (55). AFS is characterized by growth of the fungus within the paranasal sinus cavities but without invasion of living host tissue. AFS sometimes is caused by *Aspergillus* species but much more often is caused by various species of dematiaceous molds. Chronic invasive fungal sinusitis, as the name denotes, invades living tissue and it can occur in healthy or immune-compromised individuals (56, 57). The chronic invasive form can be fatal. These diseases presumably are preceded by inhalation of the fungal spores. No *Aspergillus* infections have been attributed to laboratory sources. *Aspergillus* species typically begin to sporulate within 1 day under laboratory conditions and sporulation becomes extremely heavy. Although these species are not known to cause disease consistently in healthy adults, BSL2 precautions often are appropriate for *Aspergillus* species due to their heavy formation of airborne spores, and culture manipulation should be conducted within a BSC.

Blastomyces dermatitidis

B. dermatitidis is one of the frank fungal pathogens (along with *Coccidioides, Histoplasma, Paracoccidioides, P. marneffei,* and *S. schenckii*) that routinely cause infection in healthy humans. Its endemic range is not as well delineated as those of *Coccidioides* and *Histoplasma* species because there is neither a useful skin test antigen nor an adequate serologic test for blastomycosis. *B. dermatitidis* is found mainly in the eastern half of North America and in Africa, but a few autochthonous cases have been confirmed in India, Europe, the Middle East, Central America, and Mexico. Its environmental niche is unclear, but evidence is accumulating that the mold grows as a saprobe in damp earth having a particularly high content of decaying wood and other organic matter (58).

Infection typically begins with the inhalation of spores, but transcutaneous infection is documented. Recommended biocontainment precautions for handling known isolates of this species recently have changed from BSL2 to those of BSL3 precautions and facilities (18). As with histoplasmosis and coccidioidomycosis, blastomycosis does occur in human immunodeficiency virus (HIV)-infected patients, but only several dozen cases have been reported in this population (59). Because no good skin test antigen exists, the extent to which exposure to this fungus can result in subclinical, self-limiting infections is unknown. Imported cases of blastomycosis have been reported (60, 61), which illustrates that laboratory personnel who work in regions where there are no endemic frank fungal pathogens still should use BSL2 with BSL3 biocontainment precautions when working with unidentified molds. These handling precautions also should be used for fomites that are suspected possibly to harbor the fungus (e.g., soil samples). Blastomycosis has been acquired in the laboratory as a result of transcutaneous inoculation with the yeast form (62, 63) and from inhalation of conidia (62, 64). Infections are seen often in dogs, and human infection acquired from tissue of infected animals has been reported (65).

Coccidioides species

C. immitis was divided into two species, *C. immitis* and *C. posadasii,* in 2002 on the basis of molecular data (20). The virulence of the two species is the same, and all handling precautions are the same. In addition to nucleotide differences in several genes, there are differences in the geographic distribution of the two species (20). These fungi are saprobes in certain desert soils of the southwestern United States, Mexico, and Central and South America. *C. immitis* occurs primarily in California and *C. posadasii* occurs primarily in Arizona, but also in Mexico and South America (66). The spores of these species are easily released and can be carried long distances in nature by wind currents. As frank pathogens, the fungi cause infection in healthy people, but in nearly all cases the infection is no more than a brief flu-like illness. In endemic areas, coccidioidomycosis is one of the most frequently seen infections in AIDS patients. In immunocompetent hosts, coccidioidomycosis can become symptomatic years after the initial infection. There have been occasional reports of laboratories in nonendemic continents encountering *Coccidioides* species in patients with a relevant travel history (67).

Biosafety level criteria are tiered for *Coccidioides* species. BSL2 precautions are deemed adequate until a culture is known to be *Coccidioides* species. At that point BSL3 precautions are specified for further work with the culture (18). This distinction is necessary because many clinical laboratories that are likely to recover isolates of *Coccidioides* species from medical or veterinary specimens are not BSL3 facilities. Fomites (soil) suspected to contain *Coccidioides* spores should be handled using BSL3 precautions. *Coccidioides* species grow and sporulate so rapidly in the laboratory that bacterial culture plates present a serious hazard if these fungi are present. Bacterial culture plates often are opened on the open benchtop for subsequent workup. These plates always should be evaluated for the presence of mold colonies before work on an open benchtop is undertaken. If a culture plate contains a mold, it should be opened only within a Class II BSC. Because their spores are highly infectious, *C. immitis* and *C. posadasii* have been responsible for the great majority of laboratory-acquired fungal infections, and *Coccidioides* has been identified as one of the 10 most frequently reported laboratory-associated infections worldwide (68). Nearly all of these infections were pulmonary following inhalation of spores (17). Rarely, transcutaneous inoculations and spray from a syringe have been reported (69). Even though *Coccidioides* also infects many animals species (mammals in all cases but one) (70), occupational infection in veterinary medicine has been reported only once. In that instance, infection of a veterinary assistant began following a bite from an infected cat (71).

Dematiaceous molds

Dematiaceous molds are globally ubiquitous as saprobes and as parasites of plants. At least 100 species have been documented as causing infection in humans. Dematiaceous molds cause infection when traumatically inoculated to skin, subcutaneous, or corneal tissue. Paranasal sinus, pulmonary, or cerebral infection following the inhalation of spores is possible as well. Infection caused by these fungi is called chromoblastomycosis, phaeohyphomycosis, or mycetoma depending on clinical features and the fungal morphology in tissue, and these distinctions are discussed elsewhere (72, 73).

Fonsecaea pedrosoi (including *F. monomorpha*) and *Phialophora verrucosa* (including *P. americana*) infections are rare in developed countries. Infections occur mainly in normal hosts and most are subcutaneous following traumatic inoculation from contaminated vegetation. However, pulmonary infection by *F. pedrosoi*, presumably following inhalation of spores, has been reported (74). Species of *Alternaria, Curvularia, Bipolaris, Phialophora,* and several other dematiaceous genera cause infection following transcutaneous implantation, and many of these also can cause sinusitis following inhalation of spores (72). These infections occur both in normal and compromised hosts.

Two other dematiaceous molds, namely *C. bantiana* and *R. mackenziei,* have been documented dozens of times from cases of cerebral infection in healthy hosts (75, 76). *O. gallopava (V. gallopava)* and *Chaetomium* species have caused similar infections in compromised hosts. Handling of these fungi should take place only within a Class II BSC.

C. bantiana has caused brain lesions in dozens of healthy people, and about half of these have been fatal. Even so, its infectivity probably is relatively low on the basis of experience in the teaching setting. Prior to 1985, a yearly medical mycology course included open benchtop manipulations of this species by dozens of students, without incident. By 1985, killed cultures were substituted in response to a growing concern that infections caused by this species might have a pulmonary rather than cutaneous origin (77). No laboratory-acquired infections from these dematiaceous fungi are known. However, several technicians in an Ohio medical research laboratory in 2004 possibly were exposed to *C. bantiana* spores when a culture plate was accidentally opened (M.A. Ghannoum 2004, personal communication). It was not certain that the plate contained *C. bantiana.* Although the risk of infection probably was very low, the outcome of any infection could have been grave, so oral terbinafine for 6 weeks was used as an antifungal prophylaxis regimen. There were no adverse outcomes from this incident (E.W. Davidson 2004, personal communication).

Pseudallescheria boydii, Scedosporium apiospermum, and *S. prolificans* have been regarded both as dematiaceous fungi and nondematiaceous fungi, and the taxonomy of these isolates is in flux. However, those distinctions are unimportant for the purposes this chapter (77, 78). These isolates are common in the environment and can cause infection in both immunocompetent and immunocompromised hosts following implantation or inhalation. As well, they often are encountered as a pulmonary colonizer of cystic fibrosis patients (79, 80). These fungi are highly resistant to antifungal drugs. No laboratory-acquired infections are known but spore biocontainment is recommended.

Dermatophytes

Dermatophyte species of the genera *Epidermophyton, Microsporum,* and *Trichophyton* number about three dozen, and can invade keratized structures such as the stratum corneum, hair, and nail. These keratinolytic molds can be transmitted from person to person, and they represent one of the most common transmissible infections known (81). Additional distinctions can be made regarding their nonhuman reservoirs. Some, such as *M. gypseum*, are geophilic and can be contracted from soil where they are found in conjunction with keratinized materials that were shed in the form of hair, skin scales, feathers, etc. Others have animal reservoirs, such as *M. canis* (cats, dogs), *T. verrucosum* (cattle), and *Trichophyton mentagrophytes* (cattle, horses), and these species can be contracted from infected animals even though the animals may be asymptomatic (81). Other dermatophytes such as *Microsporum audouinii* have evolved to the point that they are pathogens of humans and are rarely found elsewhere. Many of the dermatophytic molds readily form spores that can be present in clinical specimens and fomites (81), as well as other spore types that are formed in culture, all of which can become airborne. Thus there is potential for laboratory-acquired infection. One review concluded that *T. mentagrophytes* seemed to be the fungus most commonly associated with laboratory-acquired infection (82). However, these infections appear to be related to animal handling (83). More recent reports of infections acquired from laboratory animals have appeared (17, 81). One report of infection acquired from a fungal culture was found (81). No reports of infection contracted from dermatologic specimens sent for culture were found. BSL2 precautions are recommended for the dermatophytes.

Fusarium species

Fusarium species are globally widespread as plant pathogens and saprobes. Medically, they were first known as causes of keratitis following corneal trauma. By the early 1980s, they became recognized as colonizers of eschar in burn patients and, within a few years, *Fusarium* species emerged as frightening opportunistic pathogens of neutropenic patients. In neutopenic patients, hematogenous spread is extremely rapid and the outcome invariably is fatal if the patient's white cell production does not return. The mechanism of this rapid dissemination was shown to be the ability of *Fusarium* species to release a series of spores into the bloodstream following angioinvasion (84, 85).

Fusarium infection in a normal host is possible in some circumstances. One such circumstance is onychomycosis of toenails (especially the great toe), although it is still not clear under what conditions the infection originates. Another is corneal, cutaneous, and subcutaneous infections in healthy hosts, following trauma with contaminated material (86). Peritonitis in peritoneal dialysis patients has been reported numerous times (86). *Fusarium* species rarely can cause acute invasive sinusitis regardless of host immune status (86, 87). Several mycotoxins are associated with *Fusarium* species (44). Growth of *Fusarium* species under laboratory conditions is rapid and sporulation is heavy. No laboratory-acquired infections are known but biocontainment of spores and BSL2 precautions are recommended.

Histoplasma capsulatum

H. capsulatum currently accommodates three varieties, *H. capsulatum* var. *capsulatum*, *H. capsulatum* var. *duboisii*, and *H. capsulatum* var. *farciminosum*. The first two of these have been isolated from nature as well as from infected humans and animals, and they also exhibit a sexual portion of their life cycle that is classified in the genus *Ajellomyces*. *H. capsulatum* var. *farciminosum* is known only from disease in equine animals.

Handling precautions are the same for all three varieties because all are virulent and all are able to form airborne spores. Infections are pulmonary in origin, following inhalation of spores, and can disseminate to skin and internal organs. Cutaneous infection can arise from percutaneous inoculation wounds (88, 89). In the Western Hemisphere, only *H. capsulatum* var. *capsulatum* is encountered. It is one of the frank pathogens and, being endemic in six continents, has the widest global distribution. It is found in association with roosting sites that have been enriched with guano of various birds, particularly blackbirds and starlings. The birds are not infected, and their role in the spread of the fungus is inferred. Bat guano similarly supports growth of the fungus and, in contrast to birds, bats exhibit intestinal lesions of histoplasmosis. As with most molds, spores are formed readily under laboratory conditions. Biocontainment requirements are BSL2 for isolates until they are identified as *Histoplasma* spp. Propagation and further handling of isolates in culture, as well as handling of fomites such as soil and guano that are received for study, call for BSL3 biocontainment precautions and facilities. Tiered BSL biocontainment requirements are used because many medical laboratories that will encounter *H. capsulatum* are not BSL3 facilities.

Laboratory-acquired infections from *H. capsulatum* var. *capsulatum* have been numerous and are second only to those from *Coccidioides* species in clinical laboratories (17). Most have been pulmonary infection following inhalation of spores but transcutaneous infection has been reported from microbiology laboratories and from injury during an autopsy (88–90). A case of presumed conjunctival histoplasmosis from a syringe spray also has been reported (91).

Mucoraceous Fungi

Rhizopus oryzae and most other mucoraceous fungi are extremely common as saprobes in the environment and have evolved as very efficient primary colonizers of plant materials. Infection in humans by these fungi is well recognized and feared in certain settings, such as with ketoacidotic diabetic patients and with patients undergoing immunosuppressive therapy (92). These molds grow very rapidly and spore formation follows quickly. Most infections are the result of spore inhalation, but some have resulted from traumatic implantation and others from contact with macerated skin. There is no evidence that inhalation of spores is a danger to normal hosts, but spore biocontainment is advisable. As for all molds, cultures should be handled within a BSC for the purpose of spore biocontainment. No laboratory-acquired infections are known.

Paracoccidioides brasiliensis

P. brasiliensis is the etiologic agent of paracoccidioidomycosis, a mycosis endemic in South America (particularly Brazil) but not Chile, and a lesser extent in Mexico and Central America. Imported paracoccidioidomycosis has been reported (93).The fungus exists as a yeast in the host but grows as a mold at lower temperatures. Infection is believed to begin in the lungs following inhalation of airborne spores, although a natural reservoir for the fungus in nature has not been demonstrated convincingly. Spore formation in the culture vessel is scant or absent, but BSL2 with BSL3 biocontainment precautions (and BSC) are advisable. No laboratory-associated infections are known.

Paecilomyces lilacinus

P. lilacinus is a saprobe that is widespread in the environment. It also has been approved by the U.S. Environmental Protection Agency as an agricultural biocide that can be applied to control nematodes (94). As an opportunistic pathogen of humans, it has become a regular, if infrequent, finding in large medical laboratories. It causes soft tissue infections almost exclusively among compromised hosts, but can cause keratitis following corneal trauma in healthy people. Nosocomial outbreaks from contaminated ocular lens implants and contaminated skin lotion have occurred, possibly as a result of its relatively high resistance to disinfection processes (84, 95, 96). Additional infections in healthy people include sinusitis and onychomycosis (97–99). Virulence of this species is relatively slight and infections are slow to progress, but the species is highly resistant to antifungal drugs. No laboratory-acquired infections are known, but biocontainment of spores is recommended.

Penicillium marneffei

P. marneffei was discovered causing a liver lesion in a laboratory bamboo rat (100). It is known to occur in nature in conjunction with at least four species of these animals, presumably from a soil niche. The known natural range of this mold extends from Southeast Asia to India's easternmost state, and the fungus is widely encountered in those regions as a disseminated infection in AIDS patients. In addition, infection has occurred in dozens of people who are HIV-negative. The fungus has been encountered several times in people who have traveled to an endemic area. In 1996, the College of American Pathologists (CAP) shipped this mold to hundreds of medical laboratories in North America as an unidentified challenge isolate in a proficiency testing program. As a result of prompt concern by recipients, CAP sent out a facsimile letter revealing the identity of the fungus and directing all participants to cease work and destroy the isolate. This action was in response to concern for potential laboratory-acquired infection, but an additional, broader issue was the potential to expand the geographic range of this fungus via accidental release.

The spectrum of infection caused by *P. marneffei* is similar to that caused by *H. capsulatum*. Infection is presumed to begin in the respiratory tract in most cases. However, a laboratory-acquired infection by puncture wound was reported the same year the species was first described (100). A second case of what is believed to have been a laboratory-acquired infection was pulmonary in nature (101). Because all other *Penicillium* species are widely regarded as noninfectious, it is possible that laboratory personnel might be complacent in the handling of *Penicillium* species. Fortunately, *P. marneffei* forms a red-orange pigment that diffuses into the culture medium, which provides an early indication of its possible identity. *P. marneffei* should be handled at BSL2 using BLS3 precautions (18).

Sporothrix schenckii

S. schenckii is widespread in nature and can be isolated from a wide variety of senescent or dead vegetation. Historically, sporotrichosis has been regarding mainly as a cutaneous/subcutaneous infection that develops after traumatic implantation of the fungus into dermal or subdermal tissue. Infection in conjunction with AIDS has been reported for about 20 patients, and it appears that such patients are disposed to dissemination of the fungus, including to the central nervous system (102, 103).

Laboratory-acquired infections from inoculation wounds have been reported (17, 104, 105). A 1992 report noted that simple contact with the fungus had caused sporotrichosis on the finger of a researcher who had used a mortar and pestle to homogenize the mold (106). Molecular comparison between the isolate from the finger lesion and the isolate used in the experiment suggested that they were the same strain (107). This finding provided support for assertions that appeared from time to time in earlier literature that sporotrichosis is transmissible by casual contact.

Reports of person-to-person transmission date as far back as 1924 (108) and have appeared again as recently as 1990 (109). Abundant evidence for the transmissibility of sporotrichosis is also found in veterinary medicine, where human infection contracted from a cat was reported as early as 1909. Since then, at least 150 similar cases have been documented (110–112). In most of these, there was no associated trauma such as cat bite or scratch to introduce the fungus. On the basis of these experiences, recommendations for handling this mold now expressly call for the use of gloves (18). Also, it has been recommended that patients who are diagnosed with sporotrichosis be asked whether an infected pet may have been the source of infection, and that any such animals be investigated as a potential risk to the health of additional people (112).

Because pulmonary infection following inhalation of conidia is possible, it is advisable to take respiratory precautions when working with cultures. Roughly 100 cases of pulmonary infection have been reported, and, although most patients had underlying medical conditions, many appeared to be immunocompetent (102).

BIOSAFETY CONSIDERATIONS WHEN USING ANIMALS

Biocontainment precautions need to be taken into account when working with animal models of infectious diseases. Animal Biosafety Levels (ABSL) 1 through 4 have been established for this purpose, and specific recommendations are available for particular fungi (18). It is well known that some dermatophytoses have a zoonotic aspect and that dermatophytes can be transmitted readily from animal hosts to humans. Accordingly, contact precautions such as gloves and gown are in order when using an animal model of dermatophytosis (ABSL2). Sporotrichosis also can be easily transmitted from animals to humans by simple contact, as noted previously. Also, there are two well-documented cases of sporotrichosis initiated by a bite from a captured field mouse (113, 114). Sporotrichosis is a potentially serious infection, so careful precautions against skin contact are essential when using an animal model of this mycosis (ABSL2).

Coccidioides species, *H. capsulatum,* and *B. dermatitidis* can be excreted in the urine of systemically infected humans (115). Information about whether this occurs in animals, and therefore could pose a risk to investigators, is fragmentary. There is one report of *H. capsulatum* being present in urine of experimentally infected guinea pigs (116), and one report of *H. capsulatum* being present in nasal secretions of intravenously infected rabbits (117). A similar situation exists for *C. neoformans* where it has been shown that *C. neoformans* can be isolated from the cage bedding of mice infected by the nasal route but not by intravenous or intraperitoneal routes (118). It is thought that infected mice shed the fungal cells from their infected lungs, either by coughing or respiration. There have been no cases of coccidioidomycosis, histoplasmosis, blastomycosis, or cryptococcosis reported in humans as a result of an inhalation exposure during animal experimentation. ABSL2 has been recommended for animal work with these fungi. For animal models of coccidioidomycosis that use a nonparental route of inoculation, ABSL3 is recommended (18).

Risk assessment in individual circumstances may lead investigators to implement additional animal model precautions. For example, in the case of *C. neoformans,* serotypes B and C (now known as *C. gattii*), infections occur primarily in healthy humans and can be more difficult to manage medically as compared to infections with A, D, or AD serotypes (32). Consequently, implementation of steps to contain aerosolized bedding particles, such as using a HEPA-filtered chamber for bedding changes and animal transfer, and using biocontainment caging with filtered air exchange (individually ventilated caging system) to house the animals throughout the experiment may be prudent.

There are no known risks associated with animal models of other mycoses. Some of the most commonly used animal models are the murine models of candidiasis. A recent assessment of workplace biosafety in a murine model of intravenously initiated candidiasis concluded that viable *C. albicans* could be recovered from bedding, but only at rather low concentrations. There is no risk of infection presented by animal bedding, animal handling, associated work surfaces, and implements (27), and there have been no reports of candidiasis contracted from working with infected animals. ABSL2 biocontainment precautions are adequate for mouse models of candidiasis.

BIOSAFETY CONSIDERATIONS WHEN WORKING WITH FOMITES

Fomite samples can pose a potential hazard from two standpoints. First, they are a potential source of infection. In earlier years of medical mycology, fomite specimens typically consisted of bird or bat guano, sometimes mixed with soil. The fungi associated with these specimens were *H. capsulatum, C. neoformans,* and *Coccidioides* species. Few laboratories accepted such materials, and those laboratories tended to be research facilities that were well acquainted with the recommended handling guidelines for biocontainment of potentially infectious airborne fungal spores from soil and fomites (119). In more recent years, interest in health issues related to indoor air quality and water-damaged building materials has grown tremendously, and scores of commercial environmental microbiology laboratories now offer services dealing with processing of bulk materials for detection and enumeration of fungi. Consequently, fungal laboratory study of fomites has become commonplace. Of interest in these cases are molds that, rather than potentially being infectious, are a concern for potentially eliciting allergic responses or forming mycotoxins that can be integral to the cell wall of the fungal spores.

BIOSAFETY CONSIDERATIONS FOR MEDICAL SPECIMENS

Last, biocontainment precautions for specimens from human and veterinary medical practices warrant consideration. In contrast to fomite specimens, medical specimens pose little hazard of airborne fungal infection. Routine biocontainment practices for specimen handling and processing provide sound protection for laboratory personnel, as well as protection of specimens and cultures from extraneous contamination (43). Specimens from cases of dermatophytosis are somewhat unusual in that it is best not to transport skin scrapings and hair and nail samples in an airtight container. The reason for this is that increased relative humidity within the vessel can promote the growth of other microbes, especially fungal contaminants, which can hinder recovery of dermatophytes. Accordingly, it is common to receive skin flakes, hair, and nail specimens in paper or glassine enclosures. Care should be taken to ensure that these small fragments do not fall out of the packet and pose a possibility of transmitting a dermatophytic infection via skin contact.

EMPLOYEE HEALTH AND EXPOSURE MANAGEMENT ISSUES

In addition to biocontainment precautions described thus far, the laboratory safety manual and personnel training policies should provide for a specific plan to implement in response to an accidental release of pathogenic microorganisms in the laboratory setting. Informal annual or semiannual laboratory drills that simulate a release event are advisable. Annual inspection of the laboratory facilities by an institutional biosafety officer is recommended.

In cases of accidental exposure to dangerous fungi, there are no guidelines regarding prophylactic antifungal therapy. Even though risks from exposure to all of the various fungi are not clear, potential consequences from exposure to some species could be significant in terms of morbidity and, perhaps, mortality. Accordingly, and because there are several effective oral antifungal drugs available now that are relatively safe to administer, there have been some instances where laboratory personnel were treated prophylactically following accidental exposure (see "*Cryptococcus* species" and "Dematiaceous molds," above).

The additional precaution of immunization of medical laboratory workers is not an option because there are no FDA-approved fungal immunizing agents available at this time. However, there is continuing effort to develop an effective vaccine against coccidioidomycosis, which comprises the greatest fungal threat to microbiology laboratory personnel (120–121). Other fungi, such as *C. neoformans,* and *C. albicans,* also are being considered as vaccine candidates (122). Collection and storage of baseline serum has been recommended for laboratory, housekeeping, clerical, and other support personnel who will be working in space used in conjunction with fungal cultures such that, in the event of suspected laboratory exposure to fungi such as *C. immitis, C. posadasii, B. dermatitidis,* or *H.capsulatum,* reference sera will be available for comparison (123).

References

1. **Kirk PM, Canon PF, Minter DW, Stalpers JM (ed).** 2008. *Dictionary of the Fungi,* 10th ed. CAB International, Wallingford, U.K.
2. **de Hoog GS, Guarro J, Gené J, Figueras MJ (ed).** 2009. *Atlas of Clinical Fungi,* 3rd ed. Centraalbureau voor Schimmelcultures, Utrecht, The Netherlands.
3. **de Hoog GS, Guarro J, Gené J, Figueras MJ (ed).** 2000. *Atlas of Clinical Fungi,* 2nd ed. Centraalbureau voor Schimmelcultures, Utrecht, The Netherlands.
4. **Merz WG, Hay RJ (ed).** 2005. *Medical Mycology,* 10th ed. Hodder Arnold, London.
5. **Anaissie EJ, McGinnis MR (ed).** 2009. *Clinical Mycology,* 2nd ed. Churchill Livingstone, London.
6. **Kauffman CA, Pappas PG, Sobel JD, Dismukes WE (ed).** 2010. *Essentials of Clinical Mycology,* 2nd ed. Springer, New York.
7. **Howard DH (ed).** 2003. *Pathogenic Fungi in Humans and Animals,* 2nd ed. Marcel Dekker, Inc., New York.
8. **Gilgado F, Cano J, Gené J, Sutton DA, Guarro J.** 2008. Molecular and phenotypic data supporting distinct species statuses for *Scedosporium apiospermum* and *Pseudallescheria boydii* and the proposed new species *Scedosporium dehoogii. J Clin Microbiol* **46:**766–771.
9. **Park B, Park J, Cheong KC, Choi J, Jung K, Kim D, Lee YH, Ward TJ, O'Donnell K, Geiser DM, Kang S.** 2011. Cyber infrastructure for *Fusarium*: three integrated platforms supporting strain identification, phylogenetics, comparative genomics and knowledge sharing. *Nucleic Acids Res* **39**(Database)**:**D640–D646.
10. **Morris SK, Nguyen CK.** 2004. Blastomycosis. *Univ Toronto Med J* **81:**172–175.
11. **Department of Health and Human Services, National Institutes of Health.** 2009. NIH Guidelines for Research Involving Recombinant DNA Molecules. 2016 Guidelines http://osp.od.nih.gov/sites/default/files/NIH_Guidelines.html.
12. **Halde C, Valesco M, Flores M.** 1992. The need for a mycoses reporting system. *Curr Top Med Mycol* **4:**259–265.
13. **Department of Agriculture.** 2005. Part II. Agricultural Bioterrorism Protection Act of 2002; Possession, Use, and Transfer of Biological Agents and Toxins; Final Rule. 70. Title 7 CFR Part 331 and Title 9 CFR Part 121.
14. **Department of Health and Human Services.** 2005. Part III. Possession, use, and transfer of Select Agents and Toxins; Final Rule, 70. Title 42 CFR 72 and 73, Office of the Inspector General 42 CFR Part 1003.
15. **Chamberlain AT, Burnett LC, King JP, Whitney ES, Kaufman SG, Berkelman RL.** 2009. Biosafety training and incident-reporting practices in the united States: A 2008 Survey of biosafety professionals. *Appl Biosaf* **14:**135–143.
16. **Kimman TG, Smit E, Klein MR.** 2008. Evidence-based biosafety: a review of the principles and effectiveness of microbiological containment measures. *Clin Microbiol Rev* **21:**403–425.
17. **Schwarz J.** 1983. Laboratory infections with fungi, p 215–227. *In* Di Salvo AF (ed), *Occupational Mycoses.* Lea & Febiger, Philadelphia.
18. **Centers for Disease Control and Prevention and National Institutes of Health.** 2009. *Biosafety in Microbiological and Biomedical Laboratories,* 5th ed. U.S. Department of Health and Humans Services, Washington, DC.
19. **Zerwekh JT, Emery RJ, Waring SC, Lillibridge S.** 2004. Using the results of routine laboratory workplace surveillance activities to assess compliance with recommended biosafety guidelines. *Appl Biosaf* **9:**76–83.
20. **Fisher MC, Koenig GL, White TJ, Taylor JW.** 2002. Molecular and phenotypic description of *Coccidioides posadasii* sp. nov., previously recognized as the non-California population of *Coccidioides immitis. Mycologia* **94:**73–84.
21. **Department of Health and Human Services.** 2002. Part IV. Possession, use and transfer of select agents and toxins; Interim Final Rule, 67. Title 42 CFR Part 73 Title 42 CFR Part 1003.
22. **Department of Agriculture.** 2002. Agricultural Bioterrorism Protection Act of 2002. Possession, Use, and Transfer of Biological Agents and Toxins; Interim Final Rule, 67. 7 CFR Part 331 and 9 CFR Part 121.
23. **Odds FC.** 1988. *Candida and Candidosis,* 2nd ed. Baillière Tindall, London, United Kingdom.
24. **Ostrosky-Zeichner L, Sable C, Sobel J, Alexander BD, Donowitz G, Kan V, Kauffman CA, Kett D, Larsen RA,**

Morrison V, Nucci M, Pappas PG, Bradley ME, Major S, Zimmer L, Wallace D, Dismukes WE, Rex JH. 2007. Multicenter retrospective development and validation of a clinical prediction rule for nosocomial invasive candidiasis in the intensive care setting. *Eur J Clin Microbiol Infect Dis* **26:**271–276.
25. **Schell WA.** 2000. Histopathology of fungal rhinosinusitis. *Otolaryngol Clin North Am* **33:**251–276.
26. **Public Health Agency of Canada.** 2016. *Candida albicans*—pathogen safety data sheet. Public Health Agency of Canada, Ottawa. http://www.phacaspc.gc.ca/labbio/res/psdsftss/msds30eeng.php
27. **MacCallum DM, Odds FC.** 2004. Safety aspects of working with *Candida albicans*-infected mice. *Med Mycol* **42:**305–309.
28. **Perfect JR, Schell WA.** 2004. Laboratory-acquired *Candida albicans* skin infection. Personal Commnication.
29. **Litvintseva AP, Kestenbaum L, Vilgalys R, Mitchell TG.** 2005. Comparative analysis of environmental and clinical populations of *Cryptococcus neoformans*. *J Clin Microbiol* **43:** 556–564.
30. **Bartlett K, Byrnes III EJ, Duncan C, Fyfe M, Galanis E, Heitman J, Hoang L, Kidd S, MacDougall L, Mak S, Marr K.** 2011. The Emergence of *Cryptococcus gattii* infections on Vancouver Island and expansion in the Pacific Northwest, p 313–325. *In* Heitman J, Kozel TR, Kwon-Chung KJ, Perfect JR, Casadevall A (ed), *Cryptococcus: From Human Pathogen to Model Yeast.* ASM Press, Washington, DC.
31. **Harris J, Lockhart S, Chiller T.** 2012. *Cryptococcus gattii*: where do we go from here? *Med Mycol* **50:**113–129.
32. **Perfect JR, Casadevall A.** 2002. Cryptococcosis. *Infect Dis Clin North Am* 16:837–874, v–vi. PubMed PMID: 12512184.
33. **Goldman DL, Khine H, Abadi J, Lindenberg DJ, Pirofski L-a, Niang R, Casadevall A.** 2001. Serologic evidence for *Cryptococcus neoformans* infection in early childhood. *Pediatrics* **107:**e66.
34. **Hsueh Y, Lin X, Kwon-Chung KJ, Heitman J.** 2011. Sexual reproduction of *Cryptococcus*, p 81–96. *In* Heitman J, Kozel TR, Kwon-Chung KJ, Perfect JR, Casadevall A (ed), *Cryptococcus: From Human Pathogen to Model Yeast.* ASM Press, Washington, DC.
35. **Glaser JB, Garden A.** 1985. Inoculation of cryptococcosis without transmission of the acquired immunodeficiency syndrome. *N Engl J Med* **313:**266.
36. **Casadevall A, Mukherjee J, Yuan R, Perfect J.** 1994. Management of injuries caused by *Cryptococcus neoformans*–contaminated needles. *Clin Infect Dis* **19:**951–953.
37. **Beyt BE Jr, Waltman SR.** 1978. Cryptococcal endophthalmitis after corneal transplantation. *N Engl J Med* **298:**825–826.
38. **Perry HD, Donnenfeld ED.** 1990. Cryptococcal keratitis after keratoplasty. *Am J Ophthalmol* **110:**320–321.
39. **Tragiannidis A, Bisping G, Koehler G, Groll AH.** 2010. Minireview: *Malassezia* infections in immunocompromised patients. *Mycoses* **53:**187–195.
40. **Martino R, Salavert M, Parody R, Tomás JF, de la Cámara R, Vázquez L, Jarque I, Prieto E, Sastre JL, Gadea I, Pemán J, Sierra J.** 2004. *Blastoschizomyces capitatus* infection in patients with leukemia: report of 26 cases. *Clin Infect Dis* **38:** 335–341.
41. **Pfaller MA, Diekema DJ.** 2004. Rare and emerging opportunistic fungal pathogens: concern for resistance beyond *Candida albicans* and *Aspergillus fumigatus*. *J Clin Microbiol* **42:**4419–4431.
42. **Zaas AK, Boyce M, Schell W, Lodge BA, Miller JL, Perfect JR.** 2003. Risk of fungemia due to *Rhodotorula* and antifungal susceptibility testing of *Rhodotorula* isolates. *J Clin Microbiol* **41:**5233–5235.
43. **Miller JD, Rand TG, Jarvis BB.** 2003. *Stachybotrys chartarum*: cause of human disease or media darling? *Med Mycol* **41:** 271–291.
44. **DeVries JW, Trucksess MW, Jackson LS (ed).** 2002. *Mycotoxins and Food Safety.* Kluwer Academic/Plenum Publishers, New York, N.Y.
45. **Committee on Damp Indoor Spaces and Health,** Board on Health Promotion and Disease Prevention. 2004. *Damp Indoor Spaces and Health*, p. 12–13. The National Academies Press, Washington, DC.
46. **Sorenson WG.** 2001 Occupational respiratory disease: organic dust toxic syndrome, p 145–153. *In* Flannigan B, Samson RA, Miller JD (ed), *Microorganisms in Home and Indoor Work Environments.* Taylor & Francis, London, England.
47. **Jarvis BB.** 2002; Chemistry and toxicology of moulds isolated from water-damaged buildings, p 43–52. *In* DeVries JW, Trucksess MW, Jackson LS (ed), *Mycotoxins and Food Safety.* Kluwer Academic/Plenum Publishers, New York, NY.
48. **Pestka JJ, Yike I, Dearborn DG, Ward MD, Harkema JR.** 2008. *Stachybotrys chartarum*, trichothecene mycotoxins, and damp building-related illness: new insights into a public health enigma. *Toxicol Sci* **104:**4–26.
49. **Clancy CJ, Nguyen MH.** 1998. Acute community-acquired pneumonia due to *Aspergillus* in presumably immunocompetent hosts: clues for recognition of a rare but fatal disease. *Chest* **114:**629–634.
50. **Patterson TF, Kirkpatrick WR, White M, Hiemenz JW, Wingard JR, Dupont B, Rinaldi MG, Stevens DA, Graybill JR.** 2000. Invasive aspergillosis. Disease spectrum, treatment practices, and outcomes. I3 *Aspergillus* Study Group. *Medicine (Baltimore)* **79:**250–260.
51. **Knutsen AP, Slavin RG.** 2011. Allergic bronchopulmonary aspergillosis in asthma and cystic fibrosis. *Clin Dev Immunol* **2011:**843763.
52. **Anderson LL, Giandoni MB, Keller RA, Grabski WJ.** 1995. Surgical wound healing complicated by *Aspergillus* infection in a nonimmunocompromised host. *Dermatol Surg* **21:**799–801.
53. **Sawyer RG, Schenk WG III, Adams RB, Pruett TL.** 1992. *Aspergillus* flavus wound infection following repair of a ruptured duodenum in a non-immunocompromised host. *Scand J Infect Dis* **24:**805–809.
54. **Hope WW, Walsh TJ, Denning DW.** 2005. The invasive and saprophytic syndromes due to *Aspergillus* spp. *Med Mycol* **43**(Suppl 1) :207–238.
55. **Ferguson BJ.** 2000. Definitions of fungal rhinosinusitis. *Otolaryngol Clin North Am* **33:**227–235.
56. **Deshazo RD.** 2009. Syndromes of invasive fungal sinusitis. *Med Mycol* **47**(Suppl 1)**:**S309–S314.
57. **Stringer SP, Ryan MW.** 2000. Chronic invasive fungal rhinosinusitis. *Otolaryngol Clin North Am* **33:**375–387.
58. **Burgess JW, Schwan WR, Volk TJ.** 2006. PCR-based detection of DNA from the human pathogen *Blastomyces dermatitidis* from natural soil samples. *Med Mycol* **44:**741–748.
59. **Lortholary O, Dupont B.** 2010. Fungal Infections Among Patients with AIDS, p 525. *In* Kauffman CA, Pappas PG, Sobel J, Dismukes WE (ed), *Essentials of Clinical Mycology*, 2nd ed. Springer, New York.
60. **Velázquez R, Muñoz-Hernández B, Arenas R, Taylor ML, Hernández-Hernández F, Manjarrez ME, López-Martínez R.** 2003. An imported case of *Blastomyces dermatitidis* infection in Mexico. *Mycopathologia* **156:**263–267.
61. **Marty P, Brun S, Gari-Toussaint M.** 2000. [Systemic tropical mycoses]. *Med Trop (Mars)* **60:**281–290.
62. **Denton JF, Di Salvo AF, Hirsch ML.** 1967. Laboratory-acquired North American blastomycosis. *JAMA* **199:**935–936.
63. **Harrell ER, Curtis AC.** 1959. North American blastomycosis. *Am J Med* **27:**750–766.
64. **Baum GL, Lerner PI.** 1970. Primary pulmonary blastomycosis: a laboratory-acquired infection. *Ann Intern Med* **73:**263–265.
65. **Larsh HW, Schwarz J.** 1977. Accidental inoculation blastomycosis. *Cutis* **19:**334–335, 337.

66. **Barker BM, Jewell KA, Kroken S, Orbach MJ.** 2007. The population biology of *coccidioides*: epidemiologic implications for disease outbreaks. *Ann N Y Acad Sci* **1111:**147–163.
67. **Verghese S, Arjundas D, Krishnakumar KC, Padmaja P, Elizabeth D, Padhye AA, Warnock DW.** 2002. Coccidioidomycosis in India: report of a second imported case. *Med Mycol* **40:**307–309.
68. **Singh K.** 2009. Laboratory-acquired infections. *Clin Infect Dis* **49:**142–147.
69. **Trimble JR, Doucette J.** 1956. Primary cutaneous coccidioidomycosis; report of a case of a laboratory infection. *AMA Arch Derm* **74:**405–410.
70. **Pappagianis D.** 2005. Coccidioidomycosis, p 502–518. *In* Merz WG, Hay RJ (ed), *Medical Mycology*. Hodder Arnold, London, UK.
71. **Gaidici A, Saubolle MA.** 2009. Transmission of coccidioidomycosis to a human via a cat bite. *J Clin Microbiol* **47:**505–506.
72. **Schell WA.** 2003. Dematiaceous Hyphomycetes, p 565–636. *In* Howard DH (ed), *Pathogenic Fungi in Humans and Animals*. Marcel Dekker, New York, N.Y.
73. **Mendoza N, Arora A, Arias C, Hernandez C, Madkam V, Tyring S.** 2009. Cutaneous and subcutaneous mycoses, p 509–523. *In* Anaissie EJ, McGinnis MR, Pfaller MA (ed), *Clinical Mycology*, 2nd ed Elsevier, Inc.
74. **Morris A, Schell WA, McDonagh D, Chaffee S, Perfect JR.** 1995. Pneumonia due to *Fonsecaea pedrosoi* and cerebral abscesses due to *Emericella nidulans* in a bone marrow transplant recipient. *Clin Infect Dis* **21:**1346–1348.
75. **Jabeen K, Farooqi J, Zafar A, Jamil B, Mahmood SF, Ali F, Saeed N, Barakzai A, Ahmed A, Khan E, Brandt ME, Hasan R.** 2011. *Rhinocladiella mackenziei* as an emerging cause of cerebral phaeohyphomycosis in Pakistan: a case series. *Clin Infect Dis* **52:**213–217.
76. **Horré R, De Hoog GS.** 1999. Ecology and evolution of black yeasts and their relatives. *Studies Mycol* **43:**176–193.
77. **Schell WA, Salkin IF, McGinnis MR.** 2003. *Bipolaris, Exophiala, Scedosporium, Sporothrix,* and other dematiaceous fungi, p 825–846. *In* Murray P et al (ed), *Manual of Clinical Microbiology. 2*, 8th ed. ASM Press, Washington, D.C.
78. **Sigler L.** 2003. Miscellaneous opportunistic fungi: Microascaceae and other ascomycetes, hyphomycetes, coelomycetes and basidiomycetes, p 637–676. *In* Howard DH (ed), *Pathogenic Fungi in Humans and Animals*, 2nd ed. Marcel Dekker, New York.
79. **Defontaine A, Zouhair R, Cimon B, Carrère J, Bailly E, Symoens F, Diouri M, Hallet JN, Bouchara JP.** 2002. Genotyping study of *Scedosporium apiospermum* isolates from patients with cystic fibrosis. *J Clin Microbiol* **40:**2108–2114.
80. **Williamson EC, Speers D, Arthur IH, Harnett G, Ryan G, Inglis TJ.** 2001. Molecular epidemiology of Scedosporium apiospermum infection determined by PCR amplification of ribosomal intergenic spacer sequences in patients with chronic lung disease. *J Clin Microbiol* **39:**47–50.
81. **Kane JR, Summerbell R, Sigler L, Krajden S, Land G.** 1997. *Laboratory Handbook of Dermatophytes: A Clinical Guide and Laboratory Manual of Dermatophytes and Other Filamentous Fungi from Skin, Hair, and Nails*. Star Publishing Co., Belmont, CA.
82. **Collins CH, Kennedy DA.** 1999. *Laboratory-Acquired Infections: History, Incidence, Causes and Prevention*, 4th ed. Butterworth Heinemann, Oxford, UK.
83. **Sewell DL.** 1995. Laboratory-associated infections and biosafety. *Clin Microbiol Rev* **8:**389–405.
84. **Schell WA.** 1995. New aspects of emerging fungal pathogens. A multifaceted challenge. *Clin Lab Med* **15:**365–387.
85. **Liu K, Howell DN, Perfect JR, Schell WA.** 1998. Morphologic criteria for the preliminary identification of *Fusarium, Paecilomyces*, and *Acremonium* species by histopathology. *Am J Clin Pathol* **109:**45–54.
86. **Dignani MC, Anaissie E.** 2004. Human fusariosis. *Clin Microbiol Infect* **10**(Suppl 1):67–75.
87. **Schell WA.** 2000. Unusual fungal pathogens in fungal rhinosinusitis. *Otolaryngol Clin North Am* **33:**367–373.
88. **Tesh RB, Schneidau JD Jr.** 1966. Primary cutaneous histoplasmosis. *N Engl J Med* **275:**597–599.
89. **Tosh FE, Balhuizen J, Yates JL, Brasher CA.** 1964. Primary cutaneous histoplasmosis. Report of a case. *Arch Intern Med* **114:**118–119.
90. **Buitrago MJ, Gonzalo-Jimenez N, Navarro M, Rodriguez-Tudela JL, Cuenca-Estrella M.** 2011. A case of primary cutaneous histoplasmosis acquired in the laboratory. *Mycoses* **54:**e859–e861.
91. **Spicknall CG, Ryan RW, Cain A.** 1956. Laboratory-acquired histoplasmosis. *N Engl J Med* **254:**210–214.
92. **Sun HY, Singh N.** 2011. Mucormycosis: its contemporary face and management strategies. *Lancet Infect Dis* **11:**301–311.
93. **Van Damme PA, Bierenbroodspot F, Telgtt DSC, Kwakman JM, De Wilde PCM, Meis JFGM.** 2006. A case of imported paracoccidioidomycosis: an awkward infection in The Netherlands. *Med Mycol* **44:**13–18.
94. **Madsen AM.** 2011. Occupational exposure to microorganisms used as biocontrol agents in plant production. *Front Biosci (Schol Ed)* **3:**606–620.
95. **Orth B, Frei R, Itin PH, Rinaldi MG, Speck B, Gratwohl A, Widmer AF.** 1996. Outbreak of invasive mycoses caused by *Paecilomyces lilacinus* from a contaminated skin lotion. *Ann Intern Med* **125:**799–806.
96. **Castro LG, Salebian A, Sotto MN.** 1990. Hyalohyphomycosis by *Paecilomyces lilacinus* in a renal transplant patient and a review of human Paecilomyces species infections. *J Med Vet Mycol* **28:**15–26.
97. **Innocenti P, Pagani E, Vigl D, Höpfl R, Huemer HP, Larcher C.** 2011. Persisting *Paecilomyces lilacinus* nail infection following pregnancy. *Mycoses* **54:**e880–e882.
98. **Rockhill RC, Klein MD.** 1980. *Paecilomyces lilacinus* as the cause of chronic maxillary sinusitis. *J Clin Microbiol* **11:**737–739.
99. **Fletcher CL, Hay RJ, Midgley G, Moore M.** 1998. Onychomycosis caused by infection with *Paecilomyces lilacinus*. *Br J Dermatol* **139:**1133–1135.
100. **Segretain G.** 1959. *Penicillium marneffei* n. sp., agent d'une mycose du systeme reticulo-endothelial. *Mycopathol Mycol Appl* **11:**327–353.
101. **Hilmarsdottir I, Coutellier A, Elbaz J, Klein JM, Datry A, Guého E, Herson S.** 1994. A French case of laboratory-acquired disseminated *Penicillium marneffei* infection in a patient with AIDS. *Clin Infect Dis* **19:**357–358.
102. **Kauffman CA.** 1999. Sporotrichosis. *Clin Infect Dis* **29:**231–236, quiz 237.
103. **Galhardo MCG, Silva MTT, Lima MA, Nunes EP, Schettini LEC, de Freitas RF, Paes RA, Neves ES, do Valle ACF.** 2010. *Sporothrix schenckii* meningitis in AIDS during immune reconstitution syndrome. *J Neurol Neurosurg Psychiatry* **81:**696–699.
104. **Harrell ER.** 1964. *Occupational Diseases Acquired from Animals*. The University of Michigan School of Public Health, Ann Arbor, MI.
105. **Dunstan RW, Reimann KA, Langham RF.** 1986. Feline sporotrichosis. *J Am Vet Med Assoc* **189:**880–883.
106. **Cooper CR, Dixon DM, Salkin IF.** 1992. Laboratory-acquired sporotrichosis. *J Med Vet Mycol* **30:**169–171.
107. **Cooper CR Jr, Breslin BJ, Dixon DM, Salkin IF.** 1992. DNA typing of isolates associated with the 1988 sporotrichosis epidemic. *J Clin Microbiol* **30:**1631–1635.
108. **Forester HR.** 1924. Sporotrichosis. *Am J Med Sci* **167:**55–76.
109. **Jin XZ, Zhang HD, Hiruma M, Yamamoto I.** 1990. Mother-and-child cases of sporotrichosis infection. *Mycoses* **33:**33–36.

110. **Barros MB, Schubach TP, Coll JO, Gremião ID, Wanke B, Schubach A.** 2010. [Sporotrichosis: development and challenges of an epidemic]. *Rev Panam Salud Publica* **27:**455–460.
111. **Barros MB, Schubach AO, do Valle AC, Gutierrez Galhardo MC, Conceição-Silva F, Schubach TM, Reis RS, Wanke B, Marzochi KB, Conceição MJ.** 2004. Cat-transmitted sporotrichosis epidemic in Rio de Janeiro, Brazil: description of a series of cases. *Clin Infect Dis* **38:**529–535.
112. **Arenas R.** 2005. Sporotrichosis, p 367–384. *In* Merz WG, Hay RJ (ed), *Topley & Wilson's Microbiology and Microbial Infecctions. Medical Mycology.* Hodder Arnold, London, UK.
113. **Frean JA, Isaäcson M, Miller GB, Mistry BD, Heney C.** 1991. Sporotrichosis following a rodent bite. A case report. *Mycopathologia* **116:**5–8.
114. **Moore JJ, Davis DJ.** 1918. Sporotrichosis following mouse bite with certain immunologic data. *J Infect Dis* **23:**252–265.
115. **Kwon-Chung KJ, Bennett JE.** 1992. *Medical Mycology.* Lea & Febiger, Philadelphia, PA.
116. **Reid JD, Scherer JH, Herbut PA, Irving H.** 1942. Systemic histoplasmosis. *J Lab Clin Med* **27:**419–434.
117. **Daniels LS, Berliner MD, Campbell CC.** 1968. Varying virulence in rabbits infected with different filamentous types of *Histoplasma capsulatum. J Bacteriol* **96:**1535–1539.
118. **Nosanchuk JD, Mednick A, Shi L, Casadevall A.** 2003. Experimental murine cryptococcal infection results in contamination of bedding with *Cryptococcus neoformans. Contemp Top Lab Anim Sci* **42:**9–12.
119. **Ajello L, Weeks RJ.** 1983. Soil decontamination and other control measures, p 229–238. *In* DiSalvo AF (ed), *Occupational Mycoses.* Lea & Febiger, Philadelphia, PA.
120. **Cole GT, Xue JM, Okeke CN, Tarcha EJ, Basrur V, Schaller RA, Herr RA, Yu JJ, Hung CY.** 2004. A vaccine against coccidioidomycosis is justified and attainable. *Med Mycol* **42:**189–216.
121. **Xue J, Chen X, Selby D, Hung CY, Yu JJ, Cole GT.** 2009. A genetically engineered live attenuated vaccine of Coccidioides posadasii protects BALB/c mice against coccidioidomycosis. *Infect Immun* **77:**3196–3208.
122. **Mochon AB, Cutler JE.** 2005. Is a vaccine needed against *Candida albicans? Med Mycol* **43:**97–115.
123. **McGinnis MR.** 1980. *Laboratory Handbook of Medical Mycology.* Academic Press, New York, NY.
124. **Butka BJ, Bennett SR, Johnson AC.** 1984. Disseminated inoculation blastomycosis in a renal transplant recipient. *Am Rev Respir Dis* **130:**1180–1183.
125. **Ramsey FK, Carter GR.** 1952. Canine blastomycosis in the United States. *J Am Vet Med Assoc* **120:**93–98.
126. **Larson DM, Eckman MR, Alber RL, Goldschmidt VG.** 1983. Primary cutaneous (inoculation) blastomycosis: an occupational hazard to pathologists. *Am J Clin Pathol* **79:**253–255.
127. **Sorensen RH, Cheu SH.** 1964. Accidental cutaneous coccidioidal infection in an immune person. A case of an exogenous reinfection. *Calif Med* **100:**44–47.
128. **Subcommittee on Oversight and Investigations, Committee on Energy and Commerce, House of Representatives.** 2008. *Germs, Viruses, and Secrets: The Silent Proliferation of Bio-Laboratories in the United States.* Serial No. 110-70. U.S. Government Printing Office, Washington, DC.
129. **National Insitute of Allergy and Infectious Diseases, National Institutes of Health.** Recombinant DNA Incident Reports 1977–May 2010. Freedom of Information Act case 377372010.
130. **Wilder WH, McCollough CP.** 1914. Sporotrichosis of the eye. *J Am Med Assoc* **LXII:**1156–1160.

8

Bacterial Pathogens

TRAVIS R. McCARTHY, AMI A. PATEL, PAUL E. ANDERSON,
AND DEBORAH M. ANDERSON

BACTERIAL VIRULENCE STRATEGIES

Bacterial Endotoxins

Bacterial endotoxins are heat-stable components of the Gram-negative outer membrane that are released after a bacterial cell lyses and are also continuously shed by viable bacteria. Endotoxins cause systemic inflammatory responses and are associated with sepsis as well as chronic inflammation (1). The outer membrane of Gram-negative bacteria is covered up to 75% by lipopolysaccharides (LPSs); the remainder consists of proteins that primarily serve as channels for the entry and exit of molecules or as structures that mediate interactions with the environment (2). A LPS is composed of three parts—lipid A, core polysaccharide, and a glycan, typically O antigen. Lipid A is the ligand that stimulates inflammation by binding to receptors on the host cell surface as well as intracellular membranes and cytoplasm (3). Many bacterial pathogens, such as *Francisella, Yersinia, Brucella,* and *Coxiella,* synthesize an alternative form of lipid A that is less stimulatory in the mammalian host and significantly dampens the early host inflammatory responses, thus facilitating the establishment of disease.

Secretion of Exotoxins

Bacteria secrete proteins using specialized and dedicated protein machinery that is embedded in the cell membrane and spans the entire length of the envelope (peptidoglycan for Gram-positive bacteria; periplasm and outer membrane for Gram-negative bacteria). Nine secretion systems have been defined for the secretion of exotoxins or other factors that can be associated with virulence, survival in the environment, and/or motility (4). Eight are found in Gram-negative bacteria and one in Gram-positive bacteria. There is significant genetic conservation of genes encoding the secretion machineries amongst bacteria, and many of these are encoded on mobile genetic elements. Frequently associated with virulence of *Enterobacteriaceae* is the type III secretion system. *Salmonella, Shigella, Escherichia coli, Pseudomonas,* and *Yersinia,* among others, depend on the type III secretion system to inject proteins into the cytoplasm of host cells where they can reprogram cellular events and gene expression in a manner that favors virulence. Intracellular Gram-negative pathogens, such as *Legionella, Coxiella,* and *Brucella,* depend on type IV secretion systems, which can recognize and secrete more than 200 proteins from the bacteria-containing vacuole into the cytoplasm. These proteins can manipulate vacuole maturation, nutrient transport, vacuole membrane lysis, and programmed cell death, which together provide the pathogen with immune evasion and the establishment of a replicative niche (5). Toxin secretion is often accomplished through the type V secretion system. Gram-positive bacteria, such as *Mycobacterium*, *Staphylococcus*, and *Bacillus*, encode type VII secretion systems, which are associated with virulence. Secretion systems can also be used to assemble extracellular structures that remain attached to cells, such as pili, which are important for adherence to target cells and play a central role in pathogenesis.

Biofilms

Many bacterial species encode genes for the production of an extracellular polymeric substance (EPS) matrix that is produced in response to environmental conditions, such as starvation (6). The EPS facilitates adherence of bacterial cells to one another and to surfaces, as well as the cooperation of the population that enhances survival by trapping nutritional material. For bacteria such as *Streptococcus*, *Staphylococcus*, and *Pseudomonas*, the EPS is essential for adherence to tissues and provides protection from antibiotics. Biofilms promote chronic infections as well as resistance to phagocytosis and antimicrobial peptides. In polymicrobial environments, biofilms are ideal for facilitating close contact with heterologous species necessary for horizontal gene exchange, such as the transfer of antimicrobial resistance.

Dormancy

Sporulation and dormancy are strategies commonly applied by Gram-positive bacteria to survive in human and animal hosts. *Bacillus* and *Clostridium* encode elaborate sporulation systems that are developmental programs induced under starvation conditions and other environmental signals and result in the generation of a dormant spore that is able to withstand very harsh environmental conditions for nearly an indefinite period of time. Germination of the spore occurs when conditions become more favorable. *Bacillus anthracis* sporulates upon death of its animal host and can remain dormant in the soil for decades until inhaled by a susceptible host, where it will germinate in the lungs. For these organisms, the spore is the infectious form, but the vegetative cell that results from germination produces toxins and other virulence factors that cause disease.

Mycobacterium tuberculosis does not form a spore but can initiate a dormant state from the host lung environment (7). Dormant mycobacteria cause no inflammation or disease symptoms but when reactivated will cause acute tuberculosis, in a process that is not well understood.

Antibiotic Resistance

Antibiotic resistance in the modern era has been acquired in an alarming number of cases by many pathogens that commonly infect humans. *Staphylococcus*, *Enterococcus*, *Klebsiella*, *Bacteroides*, *Mycobacterium*, and *Salmonella*, for example, are so successful in the development of resistance that no antibiotic treatment options are available. Resistance is acquired through multiple mechanisms but develops as a genetically heritable trait in the pathogen that spreads along with transmission of infection. Many bacteria are naturally competent for internalizing DNA, and very often resistance is transmitted in polymicrobial environments, such as the mammalian digestive tract or biofilms, where DNA from live and dead cells can be taken up by recipient cells. For these reasons, care should be taken in the inactivation of bacterial samples, especially those carrying antibiotic resistance genes.

HOST RESPONSE TO BACTERIAL INFECTIONS

Inflammation

Production of proinflammatory cytokines leads to recruitment of neutrophils that combat extracellular bacterial infection. Uncontrolled production of inflammatory cytokines causes tissue destruction. Sepsis is defined as high levels of circulating inflammatory cytokines. Proinflammatory cytokines are expressed when membrane-bound or cytosolic pattern-recognition receptors (PRRs) are stimulated by pathogen-associated molecular patterns (PAMPs), such as LPS, peptidoglycan, and nucleotides. Stimulation of these PRRs activates expression of inflammatory genes as well as type I interferon (IFN), nitric oxide synthase, and other components of the innate immune response. Proinflammatory cytokines interleukin-1β (IL-1β) and IL-18 are expelled from cells that have activated inflammasomes, which are stimulated when they bind to bacterial products, such as lipid A, peptidoglycan, and small nucleotides, in the cytoplasm. For respiratory infections, unregulated inflammation quickly exacerbates disease as the lung alveoli become congested and the epithelium becomes necrotic. These events cause rapid decline of the host and also provide nutrients for the bacteria. Many bacterial pathogens, including *Brucella* and *Borrelia*, induce chronic inflammation that can cause arthritis, endocarditis, or other inflammatory diseases. Chronic inflammation occurs as a result of incomplete bacterial clearance, but can also occur even after the infection is cleared.

Type I IFN is an essential component of the antiviral response; however, when induced during bacterial infection, its effect can be detrimental (8). Type I IFN can activate inflammatory responses, but it is also associated with sepsis and the activation of inflammasome-mediated cell death. Intracellular PRRs activate type I IFN expression, and there are no known instances whereby bacteria completely prevent its expression. *Listeria*, *Mycobacterium*, *Staphylococcus*, *Salmonella*, and *Yersinia* are examples in which promotion of type I IFN leads to disease progression, whereas the list of bacterial infections in which type I IFN benefits the host is much smaller.

Release of nitric oxide accompanies the inflammatory response and is toxic to most microbial pathogens. However, facultative anaerobes can take advantage of nitric oxide in the environment to boost their anaerobic growth (9). As a result, in the gut or lung environment, inflammation alters the microbiome present and can favor the growth of pathogenic species such as *Enterobacteriaceae* and *Pseudomonas*.

Autophagy

Autophagy is the process by which cells recycle material from the cytoplasm and plays an important role in many processes, including bacterial clearance (10). Autophagy of bacteria is referred to as xenophagy. Membranes, typically derived from the endoplasmic reticulum, are recruited to bacteria that are residing in the cytosol, resulting in enclosure of the pathogen in a vacuole. This vacuole later fuses with the lysosome, where the bacteria are lysed and degraded. For many infections, autophagy plays a central role in innate immunity. Autophagy can also lead to programmed cell death as a mechanism to eliminate intracellular pathogens (11). Intracellular bacterial pathogens, such as *Legionella* and *Mycobacterium*, block autophagy, while others, such as *Coxiella*, *Brucella*, *Francisella*, and *Anaplasma*, exploit it as a survival mechanism to establish a replicative niche and/or for nutrient acquisition.

Programmed Host Cell Death

Bacterial infections are now known to induce programmed cell death through apoptosis, pyroptosis, necroptosis, and other regulated necrosis pathways in addition to toxin-mediated cell lysis. While essential to a successful immune response to the invading pathogen, the possible outcomes of host cell death, including replication of the pathogen as well as inflammation, can be additive toward tissue destruction, dissemination of the infection, and the progression of disease. In some cases, this response is so acute that the efficacy of antibiotic therapy in preventing disease begins to wane shortly after infection. Bacterial secretion systems are common inducers of programmed cell death in a manner that promotes the infection. Below we briefly summarize programmed cell death pathways known to be involved in the host response to bacterial infection.

Apoptosis

Programmed cell death via apoptosis is used by eukaryotic organisms to recycle cellular material and remove damaged cells. For the mammalian immune system, apoptosis of neutrophils occurs after they successfully defend against bacterial infection in order to signal the downregulation of the inflammatory response (12). Macrophages are the primary host cell that removes apoptotic cells, a process termed efferocytosis. In addition to inducing an anti-inflammatory state, efferocytosis of neutrophils carrying live bacteria can result in spread of infection, a property used by some bacterial pathogens, such as *Mycobacterium* and *Anaplasma*. Some pathogens, such as *Listeria monocytogenes*, induce apoptosis as a mechanism to eliminate effector cells of the innate immune system. Others, such as *Legionella pneumophila*, induce apoptosis following intracellular replication as a mechanism to spread to neighboring cells. *M. tuberculosis* inhibits apoptosis of macrophages as a virulence strategy (13). The process of apoptosis involves an initial phase in which the cell membrane inverts and blebs, a property that provides the signal for efferocytosis, the so-called "eat me" signal. Enzymatically driven damage to host cell DNA causes DNA fragmentation. This phase is promptly followed by disruption in cell integrity and cell death.

Pyroptosis

Lysis of host cells via pyroptosis sends alarms to the innate immune system, inducing a potent proinflammatory response driven by IL-1β, which is associated with fever, and the activation of neutrophils for clearance of intracellular pathogens (14). Pyroptosis is induced by a multiprotein complex termed the inflammasome, which responds to perturbations of the cytoplasm. Rupture of the host cell membrane by toxins or other pore-forming proteins produced by bacterial pathogens is recognized by more types of inflammasomes and activates a protease-driven pathway that disrupts cell integrity and results in the release of mature IL-1β and IL-18. In recent years, it has become appreciated that pyroptosis can be both beneficial and harmful to host cells. Bacterial virulence factors, especially those involving the type III secretion systems, such as *Salmonella*, *Yersinia*, and *E. coli*, induce pyroptosis as a mechanism to eliminate immune cells, penetrate protective epithelium or mucosa, facilitate spread to neighboring cells, or induce clinical responses that promote the transmission of the pathogen (11).

Necroptosis

Necroptosis is another form of inflammatory host cell death that involves the activation of signaling pathways at the host cell plasma membrane that induce the rupture of the membrane and the release of "alarmins," which activate inflammation and neutrophil recruitment (15). Necroptosis is defined as programmed necrosis that is dependent on the RIP1 and RIP3 kinases. Bacterial toxins, such as staphylococcal toxins, can induce necroptosis, as will the activation of so-called "death receptors," including Fas ligand and tumor necrosis factor-α receptors. Although this process can be protective due to the activation of inflammation, like pyroptosis, it may also be activated by virulence factors to promote the infection.

Nonnecroptosis forms of regulated necrosis

NETosis occurs in neutrophils and results in the controlled release of neutrophil extracellular traps (NETs), which are antibacterial (16). Poly(ADP-ribose) polymerase (PARP) proteins, such as PARP1, modify target proteins to deplete cellular NAD^+, which causes necrosis, a form of cell death known as parthanatos. PARP1 is proteolytically inactivated during apoptosis to preserve the energy stores necessary to carry out apoptosis, but PARP1 causes necrosis when overactive.

BIOHAZARD CONTAINMENT: LAWS, GUIDELINES, AND RISK GROUP ASSIGNMENTS

In the United States, federal guidelines published by the Occupational Safety and Health Administration (OSHA), the Centers for Disease Control and Prevention (CDC), the National Institutes of Health (NIH), the U.S. Department of Agriculture (USDA), and others provide information for individuals and institutions seeking to work with biological hazards. State, county, and municipal public health departments have their own regulations governing biohazard use within their jurisdiction and in some cases have their own biosafety committee to review proposed biohazard use. Globally, the World Health Organization (WHO) publishes guidelines for work with biological agents, and many federal governments that support funded research with biological agents print their own guidelines with information pertinent to work within that country's borders. Each of these guidelines is designed to assist institutions, researchers, and biosafety professionals to determine how to safely handle the agents in their possession by providing standardized information from a trusted source. It is the responsibility of everyone working with biohazards to be aware of and familiar with these guidelines and regulations. Although each of these guidelines will be described further in other chapters of this book, examples and information pertinent to this chapter are provided below.

Risk Groups

Risk Groups (RGs) classify biological agents on the basis of hazard. In the United States, RG classifications of infectious agents are published by the NIH and included in the *NIH Guidelines for Research Involving Recombinant and Synthetic Nucleic Acid Molecules* (*NIH Guidelines*). Most bacterial pathogens fall into the RG2 category (agents that typically cause self-limiting disease in healthy humans and are transmitted through direct contact), with some in the RG3 category (agents that cause severe, treatable disease in healthy adults and can be transmitted through aerosol exposure in addition to direct contact). There are no naturally occurring bacterial agents that are RG4 (agents that cause serious lethal and sometimes untreatable disease in humans). RG classifications focus primarily on the communicability and severity of the disease in healthy individuals and availability of effective treatment strategies. Many other governments and organizations publish RG recommendations for agents used within their borders.

Biosafety Levels

The biosafety levels (BSLs) describe the level of physical containment necessary to provide adequate personnel, environment, and community protection from the agent in use. Like RGs, there are four BSLs representing increasing amounts of containment as levels get higher. BSL1 is to be used with agents posing little to no risk to healthy adults, whereas BSL4 is for agents causing significant morbidity and mortality with high risk to the community. Bacterial pathogens fall into BSL2 or BSL3 because they are capable of causing moderate to severe morbidity and mortality, but treatments are typically available. BSL3 is chosen when work with the agent poses a significant aerosol risk to the scientists. It is important to note that the listing of an organism within a certain RG does not necessarily mean it must be handled under the corresponding containment level. The BSL at which an agent is worked requires a thorough risk assessment performed by a biosafety professional, taking into account the risks to the laboratory worker and surrounding environment based on the pathogenic properties of the bacteria, as well as the activities planned with the organism and the training and skill levels of the laboratorian.

Biological Select Agents and Toxins

The U.S. Departments of Health and Human Services and Agriculture have designated a set of biological agents and toxins that they deem to pose a significant threat to public health and safety, animal health, animal products, and/or plant health. Use of these agents, Biological Select Agents and Toxins (BSAT), is overseen by the Federal Select Agent Program (FSAP). Companies and institutions that work with BSAT must register with the FSAP to be able to possess, use, or transfer any of the agents on this list. FSAP regulations focus on the physical security of BSAT and the safety of the individuals working with the agents and toxins. The published rule can be found in the *Federal Register* (42 C.F.R. Part 73, 7 C.F.R. Part 331, 9 C.F.R. Part 121). Entity inspections use these rules as well as guidelines published in the BMBL and *NIH Guidelines* to judge containment and biosafety procedures. There are significant punitive and criminal penalties for violation of the Select Agent rule. Many of the bacterial agents described in this section are classified as Select Agents.

Dual Use

As technology has advanced, it has become easier, faster, and safer to make recombinant pathogenic organisms. Although this research is intended to be used for peaceful purposes, including development of novel therapeutics and vaccines, it is clear that a subset involving agents with severe individual and public health outcomes, including many of the bacteria discussed later in this chapter, may also be used to create strains that could be used to cause harm. Unlike the Select Agent regulations, which were developed to ensure the physical security of Select Agents and Toxins, the U.S. Government Policy for Institutional Oversight of Life Sciences Dual Use Research of Concern was created to protect knowledge, information, products, or technologies developed by American scientists from being misused against the public. All Tier 1 Select Agents, which include *B. anthracis*, *Burkholderia pseudomallei*, *Francisella tularensis*, and *Yersinia pestis*, are subject to Dual Use oversight.

GRAM-POSITIVE EXTRACELLULAR BACTERIAL PATHOGENS

Staphylococcus

Staphylococci are ubiquitously present, Gram-positive, human pathogens that cause infections of the skin and soft tissue, as well as invasive severe disease. A common feature of staphylococcal infections is their recurrence, as natural immunity essentially does not develop (17). Recent emergence of antibiotic resistance in *Staphylococcus aureus*, the so-called methicillin-resistant *S. aureus* (MRSA), and the evolution of community-acquired MRSA (CA-MRSA) are variants that highlight the genetic plasticity of *S. aureus*, with horizontal acquisition of virulence genes as well as antibiotic resistance genes. Humans commonly carry *S. aureus* in the nares; it is also prevalent on the skin. CA-MRSA is more invasive and is a prevalent cause of staphylococcal pneumonia, osteomyelitis, bloodstream infection, and endocarditis. *S. aureus* produces numerous toxins and secreted immune evasive proteins that are associated with virulence and form biofilms that adhere to medical devices. Staphylococcal toxins can also cause gastrointestinal disease. *Staphylococcus* can survive in neutrophils, a trait that is important to its characteristic abscess formation. These bacteria are thought to survive well in aerosols and on surfaces and are generally spread by hand contact (18).

Age, indwelling catheters, diabetes, or other immunocompromising conditions are risk factors for invasive staphylococcal diseases. There are no licensed vaccines. *Staphylococcus* requires BSL2 containment to protect workers. Cultures of large volumes of MRSA or vancomycin-resistant *S. aureus* are recommended for handling using BSL3 practices, including biosafety cabinet and respiratory protection, especially for aerosol-generating activities.

Streptococcus

The genus *Streptococcus* is a group of non-spore-forming, Gram-positive bacteria important to medicine, industry, and the normal microbial flora of animals and humans. Streptococci can be categorized both according to their hemolytic properties on blood agar as well as their serogroup identification of cell wall components. Alpha-hemolytic species oxidize iron in hemoglobin molecules within red blood cells, giving it a greenish color on blood agar. Beta-hemolytic species cause complete lysis of red blood cells, resulting in wide areas clear of blood cells surrounding bacterial colonies. Gamma-hemolytic species cause no hemolysis. The group-specific references to streptococci (A through T) refer to their differentiation by serological reactivity according to the Lancefield classification of cell wall carbohydrate antigens and fermentation patterns (19). Group A streptococci, particularly *S. pyogenes*, cause strep throat, scarlet fever, and impetigo as well as nonsuppurative sequelae, such as rheumatic fever, pneumonia, and glomerulonephritis (20–22). *S. agalactiae*, a group B streptococcus, causes neonatal sepsis and meningitis in infants. Other groups of streptococci cause diseases ranging from dental caries (*S. mutans*) to endocarditis (group D streptococci) to abscesses and gangrene (23).

The most common streptococci associated with human disease are the group A streptococci (GAS). Acute disease is most often a respiratory infection (pharyngitis or tonsillitis) or a skin infection (pyoderma). However, late immunologic sequelae not directly attributable to dissemination of bacteria are also medically significant. Because the cytoplasmic membrane of *S. pyogenes* has antigens similar to those of human cardiac, skeletal, and smooth muscle as well as heart valve fibroblasts and neuronal tissues, molecular mimicry can result where the body develops an immune response against its own tissues, as seen in acute rheumatic fever (ARF), rheumatic heart disease, acute glomerulonephritis, and pediatric neuropsychiatric disorder associated with streptococcal infection (PANDAS) (24, 25). Work with *Streptococcus* is recommended at BSL2.

Pneumococcus

The term "pneumococci" refers to a group of over 90 known serotypes of *Streptococcus pneumoniae*, lancet-shaped, Gram-positive, facultative, anaerobic bacteria that remain the leading cause of community-acquired pneumonia, meningitis, and bacteremia in children and adults (26). Pneumococci are common inhabitants of the respiratory tract and may be isolated from the nasopharynx of 5% to 90% of healthy persons, depending on the population and setting. It is estimated that about 900,000 Americans get pneumococcal pneumonia each year and about 5% to 7% die from it (27, 28). When pneumococci invade parts of the body that are normally free from germs, infections are considered "invasive." For example, pneumococcal bacteria can invade the bloodstream, causing bacteremia, and the tissues and fluids surrounding the brain and spinal cord, causing meningitis. Children under 2 years of age and adults over 65 years of age, as well individuals with certain underlying medical conditions, such as cardiovascular and pulmonary disease, immunosuppression, leukemia, and systemic corticosteroid use, are at higher risk (29). Although antimicrobial-resistant pneumococcal infections have been documented as early as 1912, it was not until the latter half of the 20th century that antibiotic-resistant and multidrug-resistant (MDR) clinical isolates became common (3, 30, 31). Since the year 2000, aggressive vaccination of at-risk groups with pneumococcal conjugate vaccines (PCV13) as well as pneumococcal polysaccharide vaccine (PPSV23) has been very effective at reducing disease in children, although efficacy in the elderly remains the subject of debate (32, 33).

Bacillus anthracis

B. anthracis is the causative agent of the severe acute zoonosis called anthrax. Like other *Bacillus* spp., *B. anthracis* forms spores capable of surviving for extended periods under extreme environmental conditions (34). Anthrax is found commonly in the western United States associated with disease in cattle and other livestock. Humans come into contact with *B. anthracis* spores while working the soil, handling infected animals and their products, or eating improperly prepared meat. The vegetative form is not infectious. Historically, the most common forms of anthrax are cutaneous, pneumonic, and gastrointestinal; however, recent years have seen an increase in transmission of anthrax among intravenous drug users, leading to cases of injectional anthrax (35). The most common presentation, cutaneous anthrax, is typically associated with a local necrotic lesion at the exposure site, leading to a painless eschar that is treatable with antibiotics.

The most severe form of the disease, pneumonic or inhalational anthrax, is associated with high morbidity and mortality, even with aggressive antibiotic treatment. Inhaled spores are transported by macrophages to the mediastinal and peribronchial lymph nodes. From there, vegetative bacteria, hidden from the immune system by a poly-D-glutamic acid capsule, begin disseminating through the bloodstream, where they begin to release anthrax toxin. Anthrax toxin is a tripartite AB toxin consisting of PA, or protective antigen, that is responsible for cell binding and trafficking of both EF (edema factor) and LF (lethal factor) (36). EF is an adenosine cyclase that increases the intracellular cAMP, leading to edema. LF is a zinc-dependent metalloprotease that is cytotoxic to host cells. The resulting intoxication ultimately leads to rapid onset of severe symptoms, deterioration, and death. Pneumonic anthrax is often characterized by a prodromal phase wherein a patient momentarily recovers before rapid onset of severe symptoms and death.

Currently licensed vaccines to prevent anthrax are based on neutralizing antibodies to PA (37). AVA (anthrax vaccine absorbed) is a filtered supernatant that contains PA and small amounts of LF. AVA has been used in humans for 60 years. The vaccine is very reactogenic, however, likely due to the presence of LF, and is only recommended in situations of potential occupational exposure. Antibodies to PA have been licensed as post-exposure prophylactic treatment for use in combination with antibiotics. In situations with potential for exposure to aerosolized *B. anthracis* spores, BSL3 is recommended. *B. anthracis* is classified as a Tier 1 Select Agent due to the potential for genetically engineered antibiotic-resistant strains, aerosol transmission, and the environmental stability of spores.

Erysipelothrix rhusiopathiae

E. rhusiopathiae is a Gram-positive, non-spore-forming, facultative anaerobe found in land and sea animals worldwide. The organism is ubiquitous in land and marine environments, is able to persist for long periods of time in the environment, and infects a broad range of animals, including humans. Infection is typically through the skin, and the resulting disease ranges from localized cutaneous lesions (erysipeloids) to sepsis and endocarditis (38). The infection usually occurs as a result of occupational exposure during contact with contaminated animals, waste products, or soil, and the bacteria can be present in laboratory animals. Single or multiple skin lesions (usually on the hands, but can spread) are raised and painful, but characteristically free of pus, and are self-limiting over 3 weeks. Sepsis rarely develops from the localized infections, but when it develops there is a high degree of endocarditis and a high mortality rate, usually due to congestive heart failure. The bacteria express an antiphagocytic capsule and multiple cell-surface proteins involved in adhesion to host cells. Multiple serotypes of *E. rhusiopathiae* have been identified, and two major serotypes are associated with disease. Diagnosis can be difficult, and the bacteria are naturally resistant to many antibiotics, including kanamycin, neomycin, and vancomycin. Natural immunity to *E. rhusiopathiae* does not appear to occur, but the bacteria are sensitive to antibiotics and are readily killed by disinfectants. Penicillin G is the drug of choice for treatment. BSL2 precautions are recommended.

PATHOGENS OF THE GASTROINTESTINAL TRACT

Bacteroides spp.

Bacteroides spp. are Gram-negative anaerobes and are normal residents of the human gut microbiota, accounting for more than 5% of the intestinal microbial population (39). Occasionally, *Bacteroides*, especially *B. fragilis*, is associated with invasive infections that can be fatal. *B. fragilis* synthesizes a polysaccharide capsule that is antiphagocytic but stimulates $CD4^+$ T-cell development in the gut mucosa, resulting in the production of the anti-inflammatory cytokine IL10. During disease, the polysaccharides aid in immune evasion, stimulate abscess formation, and are subject to antigenic variation. Diseases in adults that are associated with *Bacteroides* include intraabdominal sepsis, perforated and gangrenous appendicitis, gynecological infections, skin and soft tissue infections, endocarditis and pericarditis, bacteremia, sepsis, and even brain abscess and meningitis.

Bacteroides can carry antibiotic resistance determinants and is an emerging MDR pathogen, encoding resistances to penicillin, cephalosporins, and tetracyclines as well as carbapenem and 5-nitroimidazole antibiotics, which are currently used for treatment. Consequently, infection with MDR *Bacteroides* has a high mortality rate (40). Even though *Bacteroides* is a resident of the normal microbiota, work in the laboratory with *Bacteroides* is recommended to be performed at BSL2 due to its pathogenic properties.

Clostridium difficile

C. difficile is a Gram-positive, spore-forming bacterium that typically causes enterocolitis and is the leading infection in hospitals (41). Age is a risk factor for inflammatory bowel disease, which can be lethal. Susceptibility is influenced by the microbiota of the intestine, with an increase in spore germination associated with antibiotic treatment that reduces that microflora population. *C. difficile* is transmitted by ingestion of spores, which can be airborne as well as long-lived. Work in the laboratory with *C. difficile* is recommended to be performed at BSL2.

Enteric Bacteria

Vibrionaceae

Vibrio spp. are Gram-negative bacteria that naturally occur in aquatic environments and include some species that infect humans. The most common pathogenic species, *V. cholerae*, grows in a biofilm in the environment that enables these bacteria to resist starvation for extended periods of time (42). Pathogenesis of *V. cholerae* in humans is strongly associated with strains that produce the cholera toxin, which is encoded on an encrypted bacteriophage, as well as the toxin-coregulated pilus (TCP), which is necessary for the bacteria to adhere in the small intestine. The production of the toxin is induced in the environment of the intestine. The toxin-induced severe watery diarrhea causes death by creating electrolyte and fluid imbalances within hours to days postinfection, especially in young children. Humans are the only known natural hosts of *V. cholerae*. More than 200 serotypes of *V. cholerae* exist in nature, only a few of which (primarily O1 and O139) are associated with human diseases. Genetic factors, including blood type, and vitamin A deficiency are associated with susceptibility to disease. The human infectious dose is very high due to the sensitivity of the bacteria to stomach acid and is believed to be especially high from laboratory-grown strains. The infectious dose is significantly reduced when exposure originates from patient samples. *V. cholerae* can be transmitted via contaminated food or water or in fomites. Supportive liquid replacement therapy to restore electrolyte imbalance is usually sufficient to contain disease. Preventive vaccines against the O139 serotype are available (43).

Twelve of the 100 known *Vibrio* species are pathogenic in humans. *V. parahaemolyticus* and *V. vulnificus* are the most common after *V. cholerae*. Diseases caused by vibrios are usually self-limiting with the primary symptom of diarrhea, but these pathogens can also invade through open wounds, which can lead to septicemia and death within 48 hours (44). Between 2004 and 2013, 13 cases of septicemia caused by noncholera vibrios were reported in Germany with 5 deaths. Bacteria are transmitted through contaminated food and are shed in human fecal material. *V. parahaemolyticus* is the most common cause of illness from undercooked fish worldwide, whereas *V. vulnificus* is commonly acquired through eating contaminated raw oysters. Individuals with malignancy, adrenal insufficiency, liver cirrhosis, or diabetes or who have high circulating iron levels are at greater risk for sepsis and death from *Vibrio* infection. Invasive *Vibrio* infections are treatable with antibiotics, including fluoroquinolones and cephalosporins (45). Vibrios are recommended for handling at BSL2.

Campylobacter

Campylobacter is a Gram-negative anaerobe (or microaerobe) that causes zoonotic infections. The primary intestinal *Campylobacter* species associated with human disease are *C. jejuni* and *C. coli*. The bacteria can infect numerous animal species, including sheep and chickens (46). Typically, exposure to *Campylobacter* occurs from contaminated meat, especially poultry, but the infection can also be spread through raw or unpasteurized milk and possibly environmental sources such as water. Young children and elderly and immunocompromised individuals are more at risk for serious complications associated with campylobacteriosis, including bacteremia, hepatitis and pancreatitis. The disease is usually a self-limiting diarrhea, with symptom onset approximately 2 to 5 days after exposure. Supportive fluid replacement therapy to restore electrolyte imbalance is usually sufficient to contain disease. Invasive infections are treated with antibiotics, including tetracyclines, quinolones, and azithromycin (47). Antibiotic-resistant strains have emerged due to the overuse of antibiotics in the poultry industry. Infected individuals can be asymptomatic while shedding bacteria. *Campylobacter* is recommended for handling at BSL2.

Helicobacter pylori

H. pylori is a Gram-negative bacterium that colonizes the harsh environment of the stomach and is estimated to affect half the world's population. Oral transmission between family members during childhood is believed the primary route of transmission, and once infected, the individual remains infected for life (48). The majority of people experience chronic mild gastritis. In about 10% of cases, the infection progresses to peptic ulcers, mucosal-associated lymphoid tissue lymphoma, or gastric adenocarcinoma. *H. pylori* produces a toxin that induces apoptosis in epithelial cells and contributes to disease (49). There is a significant amount of variability in the toxin, with those with increased activity associated with greater risk of severe disease in humans. Combination antibiotic treatment along with proton pump inhibitors is usually successful, but antibiotic-resistant strains are emerging (50). There is currently no licensed vaccine. BSL2 precautions are appropriate for handling *H. pylori* in the laboratory.

Enterobacteriaceae

The family *Enterobacteriaceae* contains the genera *Escherichia, Shigella, Salmonella, Proteus, Klebsiella, Enterobacter, Serratia, Citrobacter,* and *Yersinia*. Many of these, including *Klebsiella, Escherichia, Citrobacter*, and *Yersinia*, can also infect other tissues, including the lungs, where they can lead to severe pneumonia and sepsis.

Escherichia coli

Shiga toxin-producing *E. coli* strains (STEC), including O157:H7 (also called verocytotoxin-producing *E. coli* or VTEC), are zoonotic pathogens and account for an estimated 2.8 million foodborne infections per year worldwide with an approximately 0.01% fatality rate (51). It is a major cause of bloody diarrhea, and more than 10% of STEC infections develop into hemolytic uremic syndrome (HUS), which has a fatality rate of 3% to 5%. Symptoms of HUS include hemolytic anemia, acute kidney failure, and low platelet count. STEC is a subset of the enterohemorrhagic *E. coli* (EHEC). One or more Shiga toxins are produced by these strains that are encoded on a cryptic bacteriophage (52). Shiga toxins block ribosomal protein synthesis, which leads to cytotoxicity. In addition, EHEC and enteropathogenic *E. coli* (EPEC) use a secretion system to promote their adherence to epithelial cells and the subsequent disruption of the microvilli on the intestinal epithelium. Treatment for EHEC is supportive, as the use of antibiotics results in bacterial lysis and the release of stored toxins that promote the development of HUS. Carrier states of greater than 6 months have been reported wherein affected individuals can transmit disease (53). There is currently no licensed vaccine.

EPEC causes persistent diarrhea in children and adults, with outbreaks found in developing countries (54). There are more than 200 serotypes of EPEC. In addition to the secretion system that mediates attachment and disruption of epithelial microvilli, EPEC encodes for pili that contribute to attachment (55). Secreted EPEC proteins inhibit phagocytosis and induce host inflammatory responses that also promote disease.

Enterotoxigenic *E. coli* (ETEC) infection causes noninflammatory watery diarrhea and is exclusively found in humans. Of the *E. coli* strains, ETEC is the most common cause of diarrhea (56). Disease is caused by the production of one or two toxins, known as heat-labile toxin (LT) and heat-stable toxin (ST). Adherence to the intestinal epithelium is due to the expression of at least 20 adhesins, and this permits bacterial colonization of the epithelium and subsequent LT/ST-induced disease. There is currently no licensed vaccine for ETEC due to the diversity of ETEC strains and poor immunogenicity of ST. Limited efficacy is provided by vaccines against cholera toxin, due to the high degree of similarity between LT and cholera toxin. All *E. coli* strains are recommended for handling at BSL2.

Salmonella

S. enterica is responsible for nontyphoidal and typhoidal fever, invasive infections that are currently the leading cause of morbidity in southern Africa and parts of Asia. Moreover, multidrug resistance is becoming a significant concern, with reports of 30% to 75% MDR strains (57). Fluoroquinolones and cephalosporins are less effective on these strains, and resistant strains account for more than 50% of cases in Africa (58). Invasive nontyphoidal *S. enterica* (iNTS) strains (exemplified by strain STS313) cause blood infection with a high mortality rate (10% to 30%), especially in young children. Malaria and HIV infection, as well as sickle cell disease and malnutrition, increase the risk of mortality from iNTS. Evolution of the iNTS strains and of the *S. enterica* subspecies Typhi includes genome degradation, with loss-of-function mutations especially prevalent in genes required for colonization of the gastrointestinal tract (59). Typhoid fever is characterized by a nondescript fever that leads to septicemia and metastatic purulent infections (60). *S.* Typhi is exclusively transmitted by humans, with no animal or environmental reservoir. Infection can lead to a carrier state, wherein infected individuals continuously shed bacteria in their stool and urine for many years (61).

Salmonella is transmitted by the handling of contaminated food or water. Some strains exclusively infect humans, while others may also infect animals. NTS infection in humans is typically a self-limiting enterocolitis associated with eating contaminated food. Licensed vaccines for *S.* Typhi are available, with killed whole-cell vaccines as well as polysaccharide-based versions shown to be effective (62). There are no licensed vaccines that protect against nontyphoidal strains of *S. enterica*, and current development efforts are focused on live attenuated platforms. Prompt antibiotic treatment is recommended for NTS infection. *Salmonella* accounts for a significant portion of laboratory-acquired infections (LAIs) and is recommended for handling in the laboratory at BSL2.

Shigella

Shigellosis affects only humans, especially young children. *Shigella* is Gram-negative and is endemic worldwide; however, epidemics and disease are primarily limited to developing countries with poor public health practices (56). Shigellosis is an acute intestinal infection with a range of symptoms, from watery diarrhea to severe inflammatory bacillary dysentery with blood and mucus in the stool. The infectious dose for *Shigella* is believed to be very low, approximately 10 CFU. During infection, *Shigella* uses a secretion system to promote its internalization by epithelial cells. Intracellular *Shigella* escapes the vacuole and grows in the cytoplasm, where the bacterium polymerizes and depolymerizes cytoskeletal actin to promote spread into neighboring cells. There is currently no licensed vaccine for shigellosis. Antibiotic treatment is used for severe infections. *Shigella* is recommended for handling at BSL2.

Yersinia

Y. enterocolitica and Y. pseudotuberculosis

All *Yersinia* spp. commonly associated with human disease target lymphoid tissues and are carried by animals. Yersiniae are Gram-negative and susceptible to many broad-spectrum antibiotics. *Y. pseudotuberculosis* and *Y. enterocolitica* cause yersiniosis, a typically self-limiting gut-associated disease involving febrile gastroenteritis that resembles appendicitis, diarrhea, mesenteric lymphadenitis, and terminal ileitis (63). Septicemia can develop, especially in patients with conditions that cause iron overload. *Y. enterocolitica* is very heterogeneous, with 60 serotypes currently identified, 11 of which are associated with human disease. Comparatively fewer *Y. pseudotuberculosis* strains are associated with human disease. *Y. pseudotuberculosis* and *Y. enterocolitica* are typically not associated with lethal disease and are recommended for handling at BSL2. The incubation period in humans is 5 to 10 days. Immunocompromised patients are more susceptible to severe disease, including sepsis and meningitis, which can be fatal. *Y. pseudotuberculosis* is the closest relative of the plague bacterium *Y. pestis*, which evolved less than 5,000 years ago (64).

Y. pestis

Y. pestis causes bubonic, pneumonic, and septicemic plague, and, unlike *Y. pseudotuberculosis*, is a flea-borne zoonotic infection. Historically, plague has been responsible for three worldwide pandemics due to the zoonotic and vector life cycles of the pathogen combined with its extreme virulence. In the 21st century, plague is endemic in four continents with more than 11,000 cases worldwide, including the United States, during the period 2000–2010 (65). Pneumonic plague can be spread between individuals through inhalation of *Y. pestis* aerosol and is typically lethal within 7 days. Early antibiotic treatment is usually protective, but the disease rapidly becomes untreatable, often before a diagnosis can be made (66). Plague, especially the pneumonic form, remains a challenge for treatment, even in areas of endemicity, and antibiotic-resistant strains have been isolated from human plague patients (67–69).

In humans, there is a short lag phase following inhalation of *Y. pestis* where the patient is asymptomatic (70).

Bacteria are thought to invade the alveoli and replicate, and this eventually elicits a robust but ineffective inflammatory response. The resulting tissue damage caused by bacterial and host responses permits dissemination of *Y. pestis* through the vasculature, causing severe sepsis. Patients typically succumb from pneumonia and/or sepsis along with thrombocytopenia (71). A hallmark of pneumonic plague in humans is the development of a high fever (>101.5°F, 38.6°C). Once fever develops, and often this is the first symptom, successful antibiotic treatment becomes difficult.

There is currently no plague vaccine licensed for humans (72). Considerable effort has recently been put toward a subunit vaccine based on two dominant protein antigens, CaF1 and LcrV (73, 74). Several delivery platforms have been moved forward that involve these two antigens, and it is likely that one or more will ultimately be licensed for human use. A medical surveillance program is recommended for laboratory workers, military personnel, or others that are at risk for exposure. There have been only a few cases of laboratory-acquired plague, including a recent case of lethal septicemic plague in a laboratory worker. This is the only known lethal human infection caused by the so-called nonpigmented mutant *Y. pestis* strain, which lacks the capability to scavenge iron from the host. Hereditary hemochromatosis is now known to cause increased susceptibility to nonpigmented *Y. pestis*, due at least in part to elevated iron levels in the blood (75, 76). If accidental or suspected exposure to *Y. pestis* occurs in the laboratory, exposed persons should receive 7 to 10 days of antibiotic therapy. BSL3 precautions are required for wild-type strains of *Y. pestis*. Attenuated strains, including the nonpigmented mutant, can be safely handled at BSL2. *Y. pestis* is classified as a Tier 1 Select Agent.

PSEUDOMONADS

Pseudomonas

Among the genus *Pseudomonas*, the opportunistic human pathogen *P. aeruginosa* is the most recognized and medically relevant species. *P. aeruginosa* is a rod-shaped, Gram-negative, facultative anaerobic bacterium with high relevance to human health. *Pseudomonas* produces a number of toxins that are critical to virulence. Immunocompromised patients and patients with cystic fibrosis (CF), AIDS, and burn wounds are at high risk for severe disease with a high mortality rate (77–79).

Respiratory infections with *P. aeruginosa* can be categorized into transient and persistent manifestations. Transient infections are generally found in patients in intensive care units or with mild lung injury and can be eliminated with antibiotics, whereas persistent or chronic infections are generally found in patients suffering from CF and may last for decades (80, 81). *P. aeruginosa* commonly occurs in the environment, has been found on the bedrails, floors, and sinks of hospitals, and is known to transmit through fomites and vectors. Due to its ability to form biofilms on living and nonliving surfaces alike and its presence on surgical instruments and scopes, even after decontamination processes, there is an increased risk of transmission during hospital visits. Hospital-acquired infections (HAIs) are prevalent all across the globe.

The pathogen's ability to acquire MDR genes, form biofilms, and rapidly evolve multidrug efflux systems makes it increasingly resistant to antibiotic treatment, which has significant clinical relevance (82–84). More recently, highly divergent and possibly hypervirulent strains have been isolated from hospital patients around the world, which pose a growing threat to human health (85). Current antibiotic treatment of *P. aeruginosa* infections includes fluoroquinolones, anti-pseudomonad β-lactams, and aminoglycosides. For MDR *P. aeruginosa* isolates that are resistant to these drugs, use of colistin has been reported to treat the infection (86). *P. aeruginosa* is recommended for handling at BSL2.

Burkholderia spp.

B. mallei and *B. pseudomallei* are Gram-negative bacterial pathogens and the causative agents of the zoonotic diseases glanders and melioidosis, respectively. Both bacteria grow well on artificial media, but in the host they are intracellular pathogens that survive by escaping the phagosomal compartment and hijacking the host actin filament polymerization system to spread from cell to cell. Cell-to-cell movement is also facilitated by formation of multinucleated giant cells (87). Both agents can be transmitted by aerosol or contact with broken skin.

Glanders is a disease of horses, mules, and donkeys that is transmissible to humans. Horses develop prominent pulmonary infections with ulcerative skin lesions and lymphatic thickening, followed by systemic disease. Humans develop ulcerative lesions on the skin and/or mucous membranes, followed by lymphadenitis and sepsis, which are often fatal if untreated. Inhalation of *B. mallei* produces primary pneumonia.

Melioidosis is a glanders-like disease that may be acute, subacute, or chronic. In acute infections, the incubation period is short, lasting only 2 to 3 days. When chronic or latent infections ensue, the disease may last for years (88). Sheep, goats, swine, horses, and other animals in Southeast Asia and northern Australia experience epizootics caused by this bacterium, but these animals do not appear to be the natural reservoir. In humans, initial localized infections at the site of exposure may give way to bacteremia. Pulmonary pneumonia is common after inhalation of the bacterium from cultures or aerosolized infected tissues and body fluids, accompanied by fever and leukocytosis leading to consolidation in the upper lobes of the lung. Chronic infections without fever develop, and the disease may be reactivated following immunosuppression or after other infections. Patients with diabetes are at higher risk for severe disease. If untreated, this disease has a high mortality rate. Relapses are common, even after antibiotic treatment.

No vaccine is available for either glanders or melioidosis. BSL3 containment is recommended. Both *B. mallei* and *B. pseudomallei* are registered under the U.S. Departments of Health and Human Services and Agriculture Select Agent programs as Overlap Agents and are subject to restrictions for possession, handling, and transport.

OTHER GRAM-NEGATIVE EXTRACELLULAR PATHOGENS

Acinetobacter

A. baumannii is a Gram-negative opportunistic pathogen that causes skin and soft tissue infections as well as pneumonia in immunocompromised individuals (89). It survives well in the soil and water and resists disinfectants. MDR strains that resist all commonly used antibiotics have been isolated (90). *Acinetobacter* is recommended for use at BSL2.

Bordetella pertussis

B. pertussis, a reemerging bacterial pathogen that causes acute respiratory infection, has become a major public health problem in recent years. *B. pertussis* is a Gram-negative, pleomorphic, aerobic coccobacillus that causes pertussis (whooping cough) in infants, children, and adults. Other *Bordetella* species, *B. parapertussis* and *B. holmesii*, are also known to cause whooping cough-like disease, but with milder symptoms in humans (91, 92). Symptoms of whooping cough include low-grade fever, runny nose, and apnea (in infants) that lasts 1 to 2 weeks followed by violent and rapid coughs with the characteristic whoop sound, exhaustion, and vomiting. The recovery is slow and symptoms can last for up to 10 weeks. Pertussis is highly contagious and is transmitted from person to person via aerosols (93). *B. pertussis* infects ciliated epithelium and macrophages in the respiratory tract, where bacteria secrete toxins and other virulence factors, resulting in local and systemic pathogenesis (94, 95). Due to the highly contagious nature of the disease, it is recommended that affected individuals be isolated to prevent an epidemic.

Although pertussis is an endemic disease in many parts of the world, estimates vary widely from year to year due to limited surveillance infrastructure in some countries (96). Due to potential serious adverse events linked with whole-cell pertussis vaccines, acellular pertussis vaccine is administered routinely in infants and children and, if required, adolescent and adults (97). Therapeutics such as erythromycin, azithromycin, clarithromycin, or trimethoprim-sulfamethoxazole can be used to treat pertussis. *B. pertussis* is recommended for handling at BSL2.

Klebsiella pneumoniae

K. pneumoniae is a Gram-negative opportunistic pathogen that can cause pneumonia, bacteremia, liver abscess, and urinary tract infections. It resides in the environment and can be found in soil and water and on medical devices (98). In humans, *K. pneumoniae* is transmitted via oral and respiratory routes, colonizing mucosal tissues and resisting the innate immune defenses of the host. Recently, carbapenem-resistant *K. pneumoniae* strains have emerged, especially in hospitals, some of which have acquired additional resistances such that no currently available antibiotics are effective (90). In addition, hypervirulent strains have been isolated from patients with severe disease. *K. pneumoniae* is recommended for handling at BSL2.

Neisseria spp.

There are two species in the genus *Neisseria* that are pathogenic for humans, *N. gonorrhoeae* and *N. meningitidis*. Both species are highly similar genetically (they share about 80% DNA homology) but inhabit distinctly different niches within their human host (99). *N. gonorrhoeae* (gonococcus) is the causative agent of gonorrhea, pelvic inflammatory disease (PID) in women, and epididymitis in men. Gonococci are obligate human pathogens and do not survive in the environment or produce disease in experimental animals. The organism is able to evade immune recognition by varying its antigenic profile during infection. *N. gonorrhoeae* can be carried in both males and females as a chronic, asymptomatic infection; however, even subclinical infections may result in severe reproductive consequences and should be treated promptly upon discovery. In addition to infection of the urogenital tract, *N. gonorrhoeae* causes ocular infections. Penicillin- and quinolone-resistant strains of gonococcus have become prevalent globally and throughout the United States (100, 101).

N. meningitidis, also called meningococcus, causes meningitis and septicemia. As much as 5% to 10% of the healthy population is a carrier of *N. meningitidis* during nonendemic times (102, 103). Meningococcus is protected by a capsule and possesses a potent hemolysin, which upon reaching the bloodstream causes disseminated intravascular coagulation and central nervous system infections, especially in young children. Numerous LAIs have been reported with *N. meningitidis* (104). *N. meningitidis* and *N. gonorrhoeae* should be handled under BSL2 conditions.

Pasteurella

Zoonotic species of *Pasteurella*, which are Gram-negative, animal pathogens, produce a range of human diseases. *P. multocida* is widely present in farm, domestic, and feral animals around the world where it can cause hemorrhagic septicemia. It is the most common cause of wound infection in cat and dog bites in humans. Mild human infections also occur when the pathogen is transmitted to humans through broken skin. The bacterium can establish itself in the normal human flora and may be transmitted from human to human. The disease caused by *Pasteurella* spp. occurs a few hours to a few days after exposure and typically is a local swelling at the infection site with tenderness, erythema, and significant pain. In a few patients, the disease progresses to a low-grade fever and regional lymphadenopathy, followed by a course of self-limiting arthritis, tenosynovitis, and osteomyelitis. Workers handling laboratory cats (infection rate, 50% to 90%) and dogs (50%) are at greater risk for contracting *P. multocida* infection (105).

P. haemolytica is present in the upper respiratory tract of infected farm and domestic animals and fowl. This bacterium produces epidemics in cattle and sheep and causes fowl cholera in chickens and turkeys. *P. pneumotropica* occurs frequently in the respiratory and gastrointestinal tracts of rats and mice and can cause pneumonia and/or sepsis in stressed rodents. Humans bitten by infected animals may develop mild disease. *P. ureae* resides in animals and is associated with mixed respiratory infections that linger as a chronic disease in humans.

No vaccine is available for these infections in humans. Penicillin is the drug of choice in treating *P. multocida* and can be used prophylactically to prevent infection after cat scratches. BSL2 is recommended for handling these organisms or working with animals likely infected with them.

SPIROCHETES

Borrelia spp.

B. recurrentis causes epidemic relapsing fever and is transmitted to humans from animals by the body louse. Endemic relapsing fever infections are caused by *B. hermsii*, which is transmitted by the tick *Ornithodoros hermsi*. In the United States, only the endemic form is found. *Borrelia* spp. are spirochetes that can survive in refrigerated blood for months. Tick-borne *Borrelia* can be transmitted to other ticks as well as to humans, but *B. recurrentis* cannot pass from louse to louse. Antibodies form in humans in response to *Borrelia* infections, but the bacterium frequently changes its antigen structure such that it evades adaptive immunity. Protective immunity only develops after 3 to 10 episodes of recurring infections and is often short-lived. In fatal cases, borreliae are found in the spleen, liver, and other organs and can also infect the cerebrospinal fluid to cause meningitis. *Borrelia* spp. disappear from the blood at the end of each recurring episode and may reside dormant in the brain. The incubation period is 3 to 10 days after initial infection, followed by chills, fever, myalgia, arthralgia, headache, and spiking temperature as well as septicemia. Headaches and recurring fever occur, with each attack caused by a new variant of the original infecting bacterium, selected by antibodies appearing in response to the previous infection. Mortality in endemic episodes is typically low, but can be as high as 30% in epidemics of infection with *B. recurrentis*. No vaccine is available. More than 45 cases of laboratory-acquired relapsing fever have been reported, making it the seventh most common LAI caused by a bacterium (106). Short-term treatment with tetracycline, erythromycin, or penicillin is usually effective in breaking the infection. BSL2 containment and biosafety practices are required for work with *Borrelia* spp. Infection in the laboratory by contamination of the mucous membranes of the eye, nose, and mouth has been reported. Ectoparasites infected with *Borrelia* spp. can also transmit to laboratory workers.

B. burgdorferi produces a fever and other debilitating symptoms, first described in the initial classic outbreak that occurred in Lyme, CT. *B. burgdorferi* is transmitted from infected animals by the deer tick *Ixodes*. Lyme disease manifests as distinctive skin lesions accompanied by flu-like symptoms that can be mild to severe. Late in infection, arthralgia and arthritis are common. Spirochetes are present in blood, cerebrospinal fluid, and skin lesions. Serially laboratory-passaged *B. burgdorferi* loses virulence rapidly. The organism is more easily cultured from tick saliva than from human tissues on artificial medium. Various strains isolated from humans exhibit heterogeneity in DNA homology, plasmids, and morphology. Infected ticks must be attached to the human for at least 24 h to ensure transmission. Other rodents and birds may also serve as reservoirs.

The slowly expanding skin lesion that occurs after 3 days to 4 weeks following the tick bite in humans is a classic diagnostic sign. These skin lesions are annular and flat, with distinctive reddened areas, termed erythema chronicum migrans. The bacterium reproduces in these lesions to high titer (stage 1) and spreads to the blood, regional lymph nodes, musculoskeletal sites, and organs (stage 2) over the next several weeks or months. Stage 3 of the persistent infection occurs over the following months or years with chronic neurologic, arthritic, and other manifestations that range from arthralgia to meningitis with fever, facial nerve palsy, painful radiculoneuritis, and accompanying cardiac damage. Tetracycline, doxycycline, amoxicillin, or penicillin G treatment of early symptoms when annular skin lesions appear is effective, but such treatment can leave the patient susceptible to reinfection when the disease is arrested. About 50% of patients receiving early treatment with penicillin or erythromycin still experience minor complications, including headache and joint pain. A vaccine (Lymerix, SmithKline Beecham) has been approved for human use. LAIs with *B. burgdorferi* have not been reported. BSL2 precautions are recommended.

Leptospira

Leptospira, a genus of spirochete bacteria, has both pathogenic and free-living saprophytic species. *L. interrogans*, a Gram-negative obligate spirochete, is a pathogenic species with over 230 serovars, which causes leptospirosis. Leptospirosis is an important zoonotic disease spread worldwide, especially in areas with temperate and tropical climate (107). It has been recently classified as an emerging infectious disease due to widespread outbreaks in China, India, Brazil, Nicaragua, and United States amongst the inner-city homeless populations, most of which are associated with heavy rainfall and flooding (108–110). Leptospirosis has also become a widespread problem in urban slums due to lack of or inadequate sanitation, which produces conditions favorable for transmission to humans. Leptospirae require enzootic circulation within an animal reservoir, such as rats, mice, guinea pigs, cattle, and dogs (111, 112). They colonize in the renal tubules and are shed into lakes and ponds. Leptospirae are known to survive in contaminated water, soil, and vegetation for several weeks. Humans are accidental hosts and become infected through abraded skin or mucous membranes or by consuming contaminated water. Those suffering from the disease also shed potentially infectious bacteria in their urine for several weeks (111). Upon entry into the host cells, leptospirae translocate across polarized cell monolayers and reside in phagosomal compartments of nonphagocytic host cells, thereby evading host immune response and killing. They can be isolated from the bloodstream within minutes after inoculation and can spread to multiple organs within 72 h of infection. Leptospirosis is a biphasic illness, and disease symptoms range from subclinical infection to potentially lethal pulmonary illness. The initial acute phase is defined by fever, headache, chills, nausea, vomiting, and severe myalgia that lasts for a week. Fulminant disease arises in 5% to 10% of patients and can be fatal due to renal, cardiac, or respiratory failure. Pulmonary manifestations of leptospirosis are on the rise and have a high fatality rate due to multiorgan failure following acute respiratory distress syndrome (111, 112).

Surface proteins mediate interaction between the bacteria and host cells and are known virulence factors. Secreted proteins are also thought to play a role in virulence, but this has not yet been proven. LPS-specific antibodies are protective (111). Antibiotic treatment with penicillin and tetracycline early in infection attenuates the disease, while weekly treatment with doxycycline provides prophylactic protection to personnel at risk from exposure to contaminated water, soil, and infected animals. Whole leptospirae-based vaccines have been used to vaccinate livestock and humans; however, there are concerns about their use because immunity is serovar specific and adverse reactions to the vaccine have been documented (107, 113). Several cases of laboratory-acquired leptospirosis have also been reported. Work with leptospirae is recommended at BSL2.

INTRACELLULAR PATHOGENS

Mycobacterium spp.

Tuberculosis (TB) is caused by any member of the *M. tuberculosis* complex of bacteria, including *M. tuberculosis*, *M. bovis*, *M. microti*, *M. africanum*, *M. caprae*, and *M. pinnipedii*. *M. tuberculosis*, the most common cause of the disease, is a facultative intracellular pathogen that persists inside macrophages and other phagocytic cells by interrupting the normal phagolysosomal maturation pathway (114, 115). In acute disease, the organism disseminates throughout the body before being controlled by the natural immune process in caseous granulomas. During the ensuing latent stage of infection, the organism survives with little or no metabolic activity in the harsh environment inside the granuloma. Secondary tuberculosis develops following reactivation of latent bacilli, commonly associated with the weakening of the host immune system due to advancing age or HIV infection. Tuberculosis can remain latent for many years before reactivation. It is estimated that as many as one-third of the world's population may have been exposed to *M. tuberculosis* and are latently infected. Healthy, immunocompetent individuals who become infected with tuberculosis have a 10% lifetime chance of developing secondary disease. This risk is increased in HIV-infected individuals (116).

A live vaccine strain called bacillus Calmette-Guérin (BCG) was developed in the early 1900s at the Pasteur

Institute through repeated passage of *M. bovis in vitro*. Outside of the United States, many countries administer the vaccine to children to reduce the likelihood of development of childhood tuberculosis, but efficacy wanes with age. Despite being the most widely administered vaccine in the world (117), BCG provides little protection against development of active disease (116).

M. tuberculosis is transmitted primarily from person to person through respiratory droplets from close contacts, but may also be contracted through ingestion or percutaneous exposure. Because *M. tuberculosis* also infects nonhuman primates (NHPs) with high efficiency, it is possible for researchers and animal caretakers to transmit the disease to the animals they work with and vice versa. Individuals in clinical and laboratory settings working with members of the *M. tuberculosis* complex or NHPs should be included in an occupational exposure program. All laboratory work involving the manipulation or propagation of *M. tuberculosis* should be done under BSL3 or animal BSL3 (ABSL3) containment.

Leprosy, an ancient disease, results from infection with *M. leprae*. Despite a considerable push to eradicate leprosy worldwide, the disease remains a major global public health problem (118). Of the two forms of leprosy, tuberculous and lepromatous, only the lepromatous form is transmissible. Humans are the primary reservoir and host for *M. leprae*, although armadillos may serve as a reservoir in the United States. The bacilli can only be cultured in armadillo and mouse footpads and grow extremely slowly, with a proposed doubling time as long as 14 days. Primary hazards for work with *M. leprae* involve contact with infected animals and percutaneous exposure. BSL2 or ABSL2 containment and procedures are necessary for *M. leprae* laboratory work.

Brucella spp.

Human brucellosis (undulant fever or Malta fever) is a zoonotic disease caused by any of a number of closely related Gram-negative, obligate, intracellular bacteria. Of the currently characterized *Brucella* species, five are known to cause disease in humans. These include *B. suis* (swine), *B. abortus* (cattle), *B. canis* (dog), *B. melitensis* (sheep and goats), and *B. maris* (marine mammals). Brucellosis is transmitted from infected animals to humans via consumption of contaminated milk, through direct contact with contaminated tissues, through contact with feces and urine of infected animals, and through aerosols. Humans are not the natural host for brucellae and as such do not display typical disease manifestations. Specifically, the incubation period in humans is typically longer, there are fewer asymptomatic infections, and abortion due to infection is not observed. Although human-to-human transmission is not common, sexual transmission has been reported. *B. melitensis* is the most pathogenic species for humans, producing the most serious form of the disease.

Most cases of brucellosis are subclinical and thus go unreported. Following infection, the incubation period typically lasts from 1 to 6 weeks before clinical symptoms begin to appear. Initial symptoms are generally flu-like with insidious onset of malaise, fever, weakness, aches, and sweats. The disease is oftentimes called undulant fever, referring to the characteristic rising and falling of the fever in waves. Other symptoms include gastrointestinal and central nervous system symptoms, lymph node enlargement, and splenomegaly, as well as hepatitis, jaundice, and vertebral osteomyelitis. Symptoms resolve after several weeks or months in most patients; however, orchitis, endocarditis, meningitis, and septic arthritis can occur and are often accompanied by a distinctive itching or burning rash caused by a hypersensitivity response to the endotoxin. Chronic brucellosis is characterized by recurring aches and pains, low-grade fever, nervousness, and psychoneurotic symptoms, and often *Brucella* cannot be isolated from these patients. Death is rare (<2% of infected individuals), even without treatment. The placentas and fetal membranes of animals harboring the bacterium contain erythritol, a required growth factor for the organism; the bacterium causes septic abortion in these species. Significantly, the human placenta and fetus lack this growth factor and the bacterium is unable to injure the human fetus.

The infectious dose of *Brucella* is extremely low, and brucellosis remains the world's leading cause of zoonosis. Occupational exposures, including infectious aerosols, have made *Brucella* the leading cause of bacterial infections in the workplace (119). Individuals working with *Brucella* spp. in the laboratory are far more likely to become infected than are individuals in the general population. Twenty-two cases of infection with *B. melitensis* by airborne transmission among laboratory workers were reported from one laboratory (120). Infection typically results in long-term immunity. No licensed vaccine is currently available. BSL3 precautions are essential to protect laboratory workers from this highly infectious bacterium. *B. abortus*, *B. melitensis*, and *B. suis* are registered under the U.S. Departments of Health and Human Services and Agriculture Select Agent programs as Overlap Agents and are subject to restrictions for possession, handling, and transport.

Chlamydiae

The chlamydiae are obligate, intracellular, Gram-negative pathogens capable of causing latent or long-lasting subclinical infections in humans. The most common human-associated pathogens are *Chlamydia psittaci* (also called

Chlamydophila psittaci), *C. trachomatis*, and *C. pneumoniae* (also called *Chlamydophila pneumoniae*). *C. psittaci* is a zoonotic pathogen transmitted to humans through the respiratory secretions, urine, or feces of infected birds. Because the chlamydiae do not survive outside of host cells, they were thought to be viruses until the development of electron microscopy in 1965. *C. trachomatis* and *C. pneumoniae* are human-only pathogens. *C. trachomatis* is the leading cause of infectious sexually transmitted disease in America and the leading cause of preventable blindness globally (121). Chlamydiae transition between three distinct morphologies (EB, elementary body; RB, reticulate body; IB, intermediate body) as they infect human cells and modulate the immune response to survive inside nonacidified vacuoles called inclusions. Inside the vacuole, the infectious EB transitions into the metabolically active RB, which propagates inside the inclusion, transitions back into an EB, and eventually lyses the cell (122). Persistent infections are not well understood, but play an important role in survival and transmission of chlamydial infection (123).

Pike reported that *C. psittaci* was responsible for 10 deaths from LAIs prior to 1976 (119). This was ranked third among bacterial pathogens. Most LAIs are associated with exposure of workers to infectious aerosols. The infectious dose of chlamydiae is unknown. Work with *C. psittaci* requires BSL3 containment conditions and procedures.

Bartonella spp.

Of the greater than 20 species of *Bartonella*, only three regularly cause human infections. The most common of these, *B. henselae*, causes the benign, self-limiting zoonosis cat scratch disease (CSD). *B. henselae* is a Gram-negative rod estimated to be carried by one-third of all domestic cats at one time or another. Humans get CSD through close contact with an infected cat. The disease presents with the onset of low-grade fever and lymphadenopathy some 2 weeks after the affected person receives a scratch, lick, or bite from an infected cat. Cat fleas are believed important to transmission between cats; however, their involvement in human transmission is minimal. Skin papules or pustules appear at the injury site 3 to 10 days after exposure. Headaches, sore throat, and conjunctivitis also are common. The swollen regional lymphatics draining the bite site are often tender. Sterile pus is frequently discharged, but symptoms are self-limiting and disappear in weeks to months. The CDC reports that more than 20,000 cases of CSD occur annually in the United States. The self-limiting nature of CSD infections makes determining good treatment options difficult; however, antibiotic therapy may reduce disease severity and improve recovery times. In immunosuppressed patients, *B. quintana* and *B. bacilliformis* cause the diseases commonly known as trench fever and Carrion's disease, respectively. Unlike *B. henselae*, humans appear to be the natural reservoir for these agents. *Bartonella* species survive inside erythrocytes in the circulatory system of infected hosts, facilitating transmission via arthropod vectors, including ticks, sandflies, and fleas.

Immunosuppressed patients, especially those with HIV infections, develop bacillary angiomatosis following infection with *B. quintana*. The disease primarily targets the skin, but may be found in the subendothelial tissue of nearly every organ. The characteristic lesion in infected endothelial tissue where *B. quintana* is detected appears as a cranberry-red papule with surrounding erythema. The area ulcerates and forms a fibromyxoid matrix surrounding cystic, blood-filled spaces. Antibiotic therapy is very effective, and lesions begin to resolve following treatment, due in part to the continuous need for bacterial presence to maintain the lesion. BSL2 containment and practices are recommended for work with *Bartonella* spp.

Listeria monocytogenes

L. monocytogenes is a Gram-positive, facultative anaerobe that grows in the cytoplasm of host cells and is the causative agent of listeriosis. The *Listeria* genus consists of six species, with only *L. monocytogenes* reported to cause human illness, and has at least 13 serotypes. It can cause disease in both animals and humans. Listeriosis is a foodborne infection caused by consumption of contaminated food, such as soft cheese, processed meat, and even vegetables tainted with animal waste. *L. monocytogenes* survives and grows at lower temperatures and survives food-processing methods that rely on low pH or high salinity (124, 125). *Listeria* infection is more common in immunocompromised patients and pregnant women and can cross the placenta and cause serious infection in the fetus, with an approximate mortality rate of 30% (126). *L. monocytogenes* causes invasive and noninvasive gastrointestinal listeriosis, which is usually self-limiting. Invasive listeriosis can manifest as meningoencephalitis with symptoms of fever, nausea, vomiting, and intense headache or septicemia and can be fatal. *L. monocytogenes* can also show a variety of focal infections, including liver and brain abscess, peritonitis, joint infection, and myocarditis (124, 127). Diagnostic tools include culture isolation from blood, amniotic fluid, cerebrospinal fluid, or meconium on blood agar plates by PCR. *L. monocytogenes* infections can be treated with ampicillin or penicillin alone or in combination with gentamicin, erythromycin, or trimethoprim-sulfamethoxazole (128). There is no vaccine currently available for listeriosis. However, live-attenuated *L. monocytogenes* is being evaluated as a vaccine delivery platform currently in clinical trials. Gamma-irradiated *L. monocytogenes* was also able to induce a protective T-cell response in mice (129, 130). Work with *L. monocytogenes* is recommended at BSL2, with additional precautions necessary for at-risk populations, such as women who are pregnant.

Rickettsia

Rickettsia spp. that infect humans are usually transmitted by fleas, lice, ticks, and mites from animal reservoirs. Endemic typhus caused by *R. prowazekii* is uniquely transmitted by the body louse. It is found in humans and flying squirrels. *Rickettsia* species typically maintain a close relationship with arthropod vectors that may serve to transmit them to mammalian hosts. The arthropod vector transmitting various *Rickettsia* species to humans is specific for each species of *Rickettsia*. In some arthropods such as fleas, *Rickettsia* can be transmitted transovarially.

Rickettsiae are small, Gram-negative bacteria about 1/10 the size of *E. coli*. They are obligate, intracellular parasites, grow poorly on artificial laboratory media, and are restricted to slow reproduction, with a generation time of 8 to 10 h in the cytoplasm of host mammalian cells. Some rickettsiae can grow in the nuclei of many types of mammalian cells. Most rickettsiae are very fragile outside of mammalian cells, but *R. prowazekii* can survive for weeks in the dried feces of infected lice.

Rickettsia spp. rely heavily on their hosts for nutrients to supplement essential pathways, including basic metabolism and nucleic acid and amino acid production. Rickettsiae invade and multiply in endothelial cells of the small blood vessels. *In vitro*, *Rickettsia* spp. are not restricted to endothelial cells and can invade many mammalian cell types. *In vivo*, disseminated intravascular coagulation occurs, and the brain, heart, and other organs may develop aggregations of immunocytes called typhus nodules.

Infections caused by rickettsiae are divided into three groups—typhus, spotted fever, and scrub typhus. The two rickettsial species most often associated with human disease are *R. prowazekii* (epidemic typhus) and *R. rickettsii* (Rocky Mountain spotted fever, RMSF). Due to their severe disease manifestations, high morbidity when left untreated, and the possibility of aerosol dissemination, the utility of these organisms as a biological weapon has been explored in the past. *R. prowazekii* and *R. rickettsii* are classified as Select Agents.

Endemic typhus caused by *R. typhi* is a milder form of epidemic typhus and is rarely fatal. The incubation period in laboratory-acquired typhus varies from 4 to 14 days after inoculation or exposure. Aerosol transmission in the laboratory has been most common, but other routes of transmission, including punctures with sharps and splashes to the eye, have been reported for personnel working with cultured organisms. Research on *Rickettsiae* is recommended at BSL3.

Brill-Zinsser disease is caused by latent infection in individuals previously infected with *R. prowazekii* and is characterized by rash, fever, and headaches. Epidemic typhus caused by *R. prowazekii* is a severe disease with a high fatality rate (10% to 60%) characterized by a fever that lasts for 2 weeks. Age is a risk factor, as patients over the age of 40 have a substantially higher fatality rate from typhus. Symptoms include high fever, severe headache, muscle pain, cough, rash, and mental confusion.

Humans appear to be the primary host and reservoir for *R. prowazekii*, although other sources, including flying squirrels, have been identified recently. Transmission occurs following contact with the arthropod's feces. Scratching of the bite site generates a portal of entry through the breaks in the skin and can also generate aerosols that are inhaled by the host. *R. prowazekii* survives in dried louse feces and is efficiently transmitted via aerosol. The incubation period for the bacteria after infection is 7 to 14 days. The bacteria enter lymphatic tissue and the bloodstream and spread throughout the body. *R. prowazekii* targets epithelial cells and attaches to the membrane via adhesion proteins. Upon reaching high numbers within the cells, the bacteria induce cell lysis by disruption of the plasma membrane, which facilitates spread into surrounding cells. This results in widespread vasculitis, humoral inflammation, pulmonary edema, and the characteristic rashes on the skin.

Few methods, such as quantitative real-time PCR, immunofluorescence, and plate microagglutination, are successful in detecting *R. prowazekii*-specific infection in humans (131). Antibiotic therapy using doxycycline, tetracycline, or chloramphenicol can be used to treat the disease. Several vaccines were used successfully in the 20th century; however, availability of efficient antibiotic treatment and a limited market for a vaccine have made the prospects of vaccine development dim. More than 100 cases of LAIs with *R. prowazekii* have been reported (132, 133). Aerosol transmission, as well as punctures with sharps and splashes to the eye, have been reported for personnel working with cultured organisms.

Spotted fevers have a characteristic maculopapular and petechial rash that appears first on the extremities, moves centripetally, and eventually involves the soles of the feet and palms of the hands. RMSF is caused by *R. rickettsii* and is transmitted by ticks that will transmit the infection transovarially to their eggs. RMSF has an incubation period in humans of 1 to 8 days, followed by the classic symptoms, although the characteristic rash may not always be present. LAIs of *R. rickettsii* are not uncommon and can be associated with fatality (119). LAIs from handling infected eggs, tissue culture, or ticks have been reported. Cases have also occurred in persons simply entering the laboratory where *R. rickettsii* was in use, suggesting a low infectious dose in humans. The respiratory route, mucous membrane contamination, and direct inoculation with contaminated needles have been the most often implicated causes of LAI. Killed RMSF vaccine from yolk sacs has been used for laboratory personnel, but direct-challenge studies have shown it to be ineffective.

Laboratory testing by PCR, culture, and immunofluorescence is used for diagnostics and helps determine the specificity of *R. rickettsii* in infected individuals (134). Antibiotic treatment includes tetracycline and chloramphenicol, with delay in treatment increasing the risk of fatal outcome. Despite developing and testing few vaccines, currently there are no effective vaccines against RMSF.

Besides RMSF, there are other spotted fevers in mammals that are transmitted through tick bites. Among them, Mediterranean spotted fever in humans is caused by an infection with *R. conorii* upon inoculation of the mucous membrane from crushed ticks. This disease is found mainly in North Africa, Asia, the Indian subcontinent, the Middle East, and southern Europe, and the disease manifestations are similar to other spotted fevers caused by *Rickettsia* species (135). Prophylactic treatment is similar to RMSF, and there are no effective vaccines available currently.

Orientia tsutsugamushi

Scrub typhus is endemic in the West Pacific and South and East Asia and is caused by the rickettsial species *O. tsutsugamushi*, with a reservoir in the rodent mite. Scrub typhus affects 1 million people each year worldwide. The incubation period in humans is 1 to 3 weeks, followed by rapid onset of chills and fever, headache, myalgia, and a nonproductive cough. Scrub typhus features blackened scabbed lesions (called eschars), where mites have fed on the skin. Lymphadenopathy and lymphocytosis are common in scrub typhus, and heart and brain involvement are often severe as well. LAIs have been reported for *O. tsutsugamushi* (119). Transmission by droplet aerosol, direct contact, mucous membranes, and rat and mite bites accounts for the majority of the LAIs. No vaccine is available for scrub typhus.

The bacteria enter at the site of feeding and multiply, resulting in development of the characteristic eschar (136). The bacterium spreads in the body via lymphatic tissues and the bloodstream. Oftentimes, scrub typhus is a febrile illness; however, if left untreated, it can result in multiorgan failure, and the mortality rates can reach up to 30%. Laboratory testing by indirect immunofluorescence, culture, and PCR are very specific; however, they are not readily available in underdeveloped countries where the disease is more prevalent. Antibiotics such as tetracycline and chloramphenicol are the choice of therapy to treat scrub typhus, and the symptoms usually subside within 24 h after treatment. Due to high antigenic variation in *O. tsutsugamushi* strains, there are no vaccines currently effective against scrub typhus. Prevention is the primary way to avoid contracting the disease. *O. tsutsugamushi* is recommended for handling at BSL3.

Anaplasma and *Ehrlichia*

Most rickettsiae escape the membrane-enclosed vacuole of their host cell and replicate in the cytoplasm. *Anaplasma* and *Ehrlichia* do not and instead have a developmental life cycle inside the host cell. The dense core (DC) cell is the infectious form that enters professional phagocytes, typically neutrophils and macrophages. The phagosome is modified, and the DC develops into the vegetative form, which is the replicative form. The modified phagosome, called a morula, is a dense particle that can be easily seen by microscopy. Following bacterial replication, the DC develops and is released. Intracellular *Anaplasma* and *Ehrlichia* block apoptosis of their host cells to preserve their replicative niche.

Anaplasma phagocytophilum infects neutrophils and is the causative agent of human granulocytic anaplasmosis (137). Invasion of neutrophils is facilitated by numerous bacterial adhesins. The host cell autophagy pathway is exploited to generate the replicative niche. *A. phagocytophilum* infects a large range of animal hosts and is transmitted by ticks. The bacteria are endemic in the United States in all areas where Lyme disease is found. In 2015, there were approximately 2,600 cases of the tick-borne disease (138). Symptoms include fever and malaise, accompanied by leukopenia, thrombocytopenia, and/or increased transaminases. Severe complications requiring hospitalization are common, and the disease can be fatal. Immunocompromised individuals and the elderly are at greater risk for severe diseases, especially if antibiotic therapy is delayed. Diagnosis is by blood smear and PCR. Antibody testing is generally not recommended because antibodies are not always present. Seroconversion does not indicate active infection. There are currently no licensed vaccines to prevent anaplasmosis, but doxycycline treatment is usually very effective. *A. phagocytophilum* is recommended for handling at BSL2.

Ehrlichia chaffeensis is the causative agent of human monocytic ehrlichiosis, a disease that resembles RMSF, except that there is no rash (139). Ehrlichiosis is an emerging disease, with cases on the rise in the United States. *E. chaffeensis* lacks LPS, pili, and capsules, but instead has many surface-exposed proteins that are highly immunogenic and the focus of vaccine development. Upon invasion of monocytes or neutrophils, *E. chaffeensis* resides in a vacuolar compartment and avoids fusion of lysosomes to form its replicative niche. The bacteria require a type IVa secretion system to survive intracellularly. Like *Anaplasma, E. chaffeensis* also has a broad host range in animals, including humans, and is a tick-borne infection. In humans, the infection has a broad range of outcomes, from seroconversion to death, with elderly and immunocompromised individuals at greater risk for severe disease. Treatment with doxycycline is usually effective, and there is currently no licensed vaccine. *E. chaffeensis* is recommended for handling at BSL2.

Coxiella burnetii

Q fever is a bacterial zoonosis caused by the obligate, intracellular, Gram-negative bacterium *C. burnetii. Coxiella* is a member of the *Gammaproteobacteria* family consisting of *Legionella, Francisella,* and *Rickettsiae.* Humans, along with sheep, goats, cattle, dogs, birds, rodents, and other domestic animals, are infected by this bacterium. Q fever is spread worldwide except in New Zealand. Humans may become infected with *C. burnetii* following direct contact with infected animals or animal fluids or by inhaling infectious aerosols. Although not considered a significant route of transmission to humans, *C. burnetii* survives inside tick vectors that may serve to transmit the disease in the wild. *C. burnetii* is resistant to killing in the environment and may survive for months on surfaces.

Symptoms of Q fever usually develop between 1 and 3 weeks from the time of infection. Q fever can manifest either as an acute or chronic form of disease. Clinical symptoms of acute Q fever include chest pain while breathing, cough, fever, headache, and shortness of breath, while symptoms of chronic Q fever include chills, fatigue, night sweats, and prolonged fever. Oftentimes, low-dose infection results in an asymptomatic disease that can develop months or even years later (140). PCR-based assay or immunofluorescence assays that have a high rate of false positives are still used for clinical diagnosis. Q fever is often treated by antibiotics, such as doxycycline and other tetracyclines. Antibiotic therapy can last from 3 weeks for acute illness to 18 months for chronic illness. Vaccines have been used in humans with some success, and Q-Vax, consisting of formalin-inactivated *C. burnetii,* is approved for use in Australia. However, there are many side effects from this vaccine, ranging from tenderness at the site of injection to flu-like symptoms. There is currently no vaccine licensed for human use in the United States. Q fever has been classified as a Select Agent due to its low infectious dose by the aerosol route and its resistance to physical stresses. BSL3 precautions are recommended while working in a laboratory setting (141).

C. burnetii readily converts to a small colony variant, known as Phase II, which differs from Phase I in its LPS structure. Phase II bacteria are unable to revert to Phase I and are unable to cause disease in humans. Antigens expressed by Phase II bacteria can elicit antibody responses that have been associated with protection to the Phase I pathogen. *C. burnetii* is highly infectious by both aerosol and percutaneous routes of exposure. The estimated human infectious dose is less than 10 viable organisms. The majority of infections are asymptomatic or subclinical. The incubation period and severity of the disease are closely related to infectious dose and can occur in as little as 1 day (142). Acute disease is rarely fatal (1% to 2% if left untreated); however, endocarditis resulting from the chronic infection leads to estimated mortality rates as high as 65%. Antibiotic treatment is most successful if initiated within the first few days of illness. Treatment of chronic Q fever requires extended treatment with combination drug therapies up to 4 years in length.

Q fever is the second most commonly reported laboratory infection, and outbreaks involving 15 or more persons have been documented with a low fatality rate (119). In one report detailing 50 laboratory-acquired cases of Q fever at an army biological laboratory between 1950 and 1965, only 5 of the cases resulted from known exposures (142). Sixteen cases occurred in researchers with close contacts but no known incident, while 28 of the cases resulted in individuals employed in the laboratory or who visited the laboratory with no known exposure to the agent. At-risk individuals, including those that work with livestock and sheep and laboratory workers, are recommended to receive the Australian vaccine.

Francisella tularensis

F. tularensis is a fastidious, facultative, intracellular Gram-negative pathogen found naturally throughout much of the Northern Hemisphere. Pathogenic subspecies of *F. tularensis* are uniquely distributed with the more pathogenic subsp. *tularensis* (Type A) restricted primarily to North America, while its close relatives subsp. *holarctica* (Type B) and subsp. *mediasiatica* are found throughout the Northern Hemisphere or Asia, respectively. A second, nonpathogenic *Francisella* species, *F. novicida,* previously had been considered another subspecies of *tularensis;* however, recent data indicate it represents a separate species (143). Type A strains rank among the most infectious bacterial pathogens known, with a human infectious dose via aerosol or percutaneous exposure of less than 10 organisms. The resulting zoonosis, called tularemia or rabbit fever, is a fulminant disease with rapid onset of high fever, chills, and general flu-like symptoms, particularly following inhalational exposure. Direct contact through a cut, animal bite, or bite from an infected arthropod leads to development of ulceroglandular tularemia. The ulceroglandular form results in development of a local lesion at the site of infection and, if left untreated, may result in dissemination to local lymph nodes and possibly to the lung that may lead to development of more severe disease such as pulmonary tularemia. Pulmonary disease is fatal in 30% to 60% of cases if left untreated; however, with treatment, the fatality rate is reduced to less than 2%. Humans acquire tularemia from infected rabbits, muskrats, or other wild animals; from dogs or cats that have had contact with an infected animal; or from bites by infected ticks or deerflies. Occasionally, the source is contaminated water or food. Ocular infections may also occur following direct contact of *F. tularensis* with the eye. Tularemia is the third most common cause of bacterial LAI (119).

A vaccine derived from a live, attenuated, type B strain (live vaccine strain, LVS) through repeated serial passage was developed in Russia and gifted to the United States in the late 1950s (144). Despite its effectiveness against subcutaneous and low-dose aerosol infection, the strain's utility has been limited by its inability to protect against high-dose aerosol challenge and a predilection for causing disease in susceptible individuals following aerosol exposure. LVS remains unlicensed in the United States, and subsequent efforts to produce an improved vaccine have proven difficult (145). Because of its high level of infectivity, rapid disease onset, high morbidity and mortality, and ease of dissemination, *F. tularensis* has long been considered a potential bioweapon and, with the exception of the LVS strain, is classified as a Tier 1 Select Agent. Safe handling of *F. tularensis* in the laboratory requires strict adherence to BSL3 standards, including limiting aerosol formation, use of respiratory protection, and liberal use of biosafety cabinets. Safe sharps handling techniques must be employed at all times.

Legionella pneumophila

L. pneumophila, a Gram-negative bacillus, is known for the outbreak of Legionnaires' disease that caused severe pneumonia in hundreds of attendees and killed 29 people at an American Legion Convention in Philadelphia, PA in 1976 (146). Of more than 50 species of *Legionella*, at least 24 species have been known to cause disease in humans. However, ~90% of the disease is caused by *L. pneumophila*, which has at least 15 serotypes. Legionellae are found in freshwater environments that are either natural or man-made habitats, such as lakes, streams, fountains, and air-conditioning cooling towers. The bacteria multiply inside host cells such as amoebae, with which they form parasitic or commensal relationships. *L. pneumophila* can be resistant to many antibiotics as well as acids and thermal and osmotic stresses. *L. pneumophila* makes use of a type IVb secretion system to translocate more than 200 different effector proteins into the host cell cytoplasm to establish and maintain its replicative niche (147).

Symptoms of Legionnaires' disease are similar to pneumonia and include high fever, chills, and fatigue. Several laboratory tests, such as PCR, urinary antigen assay, culture, and serological testing, are used to identify and confirm cases of Legionnaires' disease. Suggested treatments include use of antibacterial macrolides and fluoroquinolones, such as azithromycin, clarithromycin, roxithromycin, and rifampicin. Recurrence of Legionnaires' disease is rare and there is currently no approved vaccine (148). Work with *L. pneumophila* is recommended to be performed at BSL2.

CONCLUSIONS

In this chapter, we have provided an overview of what is currently known about bacterial pathogenesis strategies and the innate immune responses of their hosts. We outlined summary risk assessments of many bacterial pathogens that commonly infect humans in the modern world. In the context of the evolutionary relationship between humans and their bacterial pathogens, this constitutes only a snapshot in time and will not be the final chapter. Bacteria will continue to evolve new strategies to improve their chance of survival, including the acquisition of antibiotic resistance, virulence, or transmission traits that may change the risk severity to individuals and the environment. We must continue to discover new methods for diagnosis, prevention, and treatment to keep the odds in our favor. The strategies outlined here should provide a framework for the evaluation of risks not only for those pathogens specifically discussed, but also for other bacteria with known or unknown potential to infect humans.

We acknowledge and thank Dr. Joseph Coggins for his contributions to the original edition of this chapter. Portions of the original chapter have been incorporated into this edition.

References

1. **Zielen S, Trischler J, Schubert R.** 2015. Lipopolysaccharide challenge: immunological effects and safety in humans. *Expert Rev Clin Immunol* **11:**409–418.
2. **Molinaro A, Holst O, Di Lorenzo F, Callaghan M, Nurisso A, D'Errico G, Zamyatina A, Peri F, Berisio R, Jerala R, Jiménez-Barbero J, Silipo A, Martín-Santamaría S.** 2015. Chemistry of lipid A: at the heart of innate immunity. *Chemistry* **21:**500–519.
3. **Morgenroth J, Kaufmann M.** 1912. Arzneifestigkeit bei Bakterien (Pneumokokken). *Z Immunit Exp Ther* **15:**610
4. **Costa TR, Felisberto-Rodrigues C, Meir A, Prevost MS, Redzej A, Trokter M, Waksman G.** 2015. Secretion systems in

Gram-negative bacteria: structural and mechanistic insights. *Nat Rev Microbiol* **13:**343–359.
5. **Personnic N, Bärlocher K, Finsel I, Hilbi H.** 2016. Subversion of retrograde trafficking by translocated pathogen effectors. *Trends Microbiol* **24:**450–462.
6. **Valentini M, Filloux A.** 2016. Biofilms and cyclic di-GMP (c-di-GMP) signaling: lessons from *Pseudomonas aeruginosa* and other bacteria. *J Biol Chem* **291:**12547–12555.
7. **Latorre I, Domínguez J.** 2015. Dormancy antigens as biomarkers of latent tuberculosis infection. *EBioMedicine* **2:**790–791.
8. **Dhariwala MO, Anderson DM.** 2014. Bacterial programming of host responses: coordination between type I interferon and cell death. *Front Microbiol* **5:**545.
9. **Scales BS, Dickson RP, Huffnagle GB.** 2016. A tale of two sites: how inflammation can reshape the microbiomes of the gut and lungs. *J Leukoc Biol* **100:**943–950.
10. **Winchell CG, Steele S, Kawula T, Voth DE.** 2016. Dining in: intracellular bacterial pathogen interplay with autophagy. *Curr Opin Microbiol* **29:**9–14.
11. **Lai XH, Xu Y, Chen XM, Ren Y.** 2015. Macrophage cell death upon intracellular bacterial infection. *Macrophage Houst* **2:**e779.
12. **Ucker DS.** 2016. Exploiting death: apoptotic immunity in microbial pathogenesis. *Cell Death Differ* **23:**990–996.
13. **Abebe M, Kim L, Rook G, Aseffa A, Wassie L, Zewdie M, Zumla A, Engers H, Andersen P, Doherty TM.** 2011. Modulation of cell death by *M. tuberculosis* as a strategy for pathogen survival. *Clin Dev Immunol* **2011:**678570.
14. **Chow SH, Deo P, Naderer T.** 2016. Macrophage cell death in microbial infections. *Cell Microbiol* **18:**466–474.
15. **Wallach D, Kang TB, Dillon CP, Green DR.** 2016. Programmed necrosis in inflammation: toward identification of the effector molecules. *Science* **352:**aaf2154.
16. **Vanden Berghe T, Linkermann A, Jouan-Lanhouet S, Walczak H, Vandenabeele P.** 2014. Regulated necrosis: the expanding network of non-apoptotic cell death pathways. *Nat Rev Mol Cell Biol* **15:**135–147.
17. **Thammavongsa V, Kim HK, Missiakas D, Schneewind O.** 2015. Staphylococcal manipulation of host immune responses. *Nat Rev Microbiol* **13:**529–543.
18. **Thompson KA, Bennett AM, Walker JT.** 2011. Aerosol survival of *Staphylococcus epidermidis*. *J Hosp Infect* **78:**216–220.
19. **Lancefield RC.** 1933. A serological differentiation of human and other groups of hemolytic streptococci. *J Exp Med* **57:**571–595.
20. **Dillon HC Jr.** 1979. Post-streptococcal glomerulonephritis following pyoderma. *Rev Infect Dis* **1:**935–945.
21. **Centor RM, Meier FA, Dalton HP.** 1986. Diagnostic decision: throat cultures and rapid tests for diagnosis of group A streptococcal pharyngitis. *Ann Intern Med* **105:**892–899.
22. **Quinn RW.** 1989. Comprehensive review of morbidity and mortality trends for rheumatic fever, streptococcal disease, and scarlet fever: the decline of rheumatic fever. *Rev Infect Dis* **11:**928–953.
23. **Venezio FR, Gullberg RM, Westenfelder GO, Phair JP, Cook FV.** 1986. Group G streptococcal endocarditis and bacteremia. *Am J Med* **81:**29–34.
24. **Patterson M.** 1996. Chapter 13, *Streptococcus*. *In Medical Microbiology*, 4th ed. University of Texas Medical Branch at Galveston, Galveston, TX.
25. **Orefici G, Cardona F, Cox C, Cunningham M.** 2016. *Pediatric autoimmune neuropsychiatric disorders associated with Streptococcal infections (PANDAS)*. University of Oklahoma Health Sciences Center, Oklahoma City, OK.
26. **Lynch JP III, Zhanel GG.** 2009. *Streptococcus pneumoniae*: epidemiology, risk factors, and strategies for prevention. *Semin Respir Crit Care Med* **30:**189–209.
27. **Huang SS, Johnson KM, Ray GT, Wroe P, Lieu TA, Moore MR, Zell ER, Linder JA, Grijalva CG, Metlay JP, Finkelstein JA.** 2011. Healthcare utilization and cost of pneumococcal disease in the United States. *Vaccine* **29:**3398–3412.
28. **Centers for Disease Control and Prevention.** 2012. *Epidemiology and prevention of vaccine-preventable diseases*, vol 2. Public Health Foundation, Washington, DC.
29. **Immunization Practice Advisory Committee.** 1997. Prevention of pneumococcal disease: recommendation of the advisory committee on immunization practices (ACIP). *MMWR Morb Mortal Wkly Rep* **46:**1–24.
30. **Jacobs MR, Koornhof HJ, Robins-Browne RM, Stevenson CM, Vermaak ZA, Freiman I, Miller GB, Witcomb MA, Isaäcson M, Ward JI, Austrian R.** 1978. Emergence of multiply resistant pneumococci. *N Engl J Med* **299:**735–740.
31. **Pérez JL, Linares J, Bosch J, López de Goicoechea MJ, Martín R.** 1987. Antibiotic resistance of *Streptococcus pneumoniae* in childhood carriers. *J Antimicrob Chemother* **19:**278–280.
32. **Conklin L, Loo JD, Kirk J, Fleming-Dutra KE, Deloria Knoll M, Park DE, Goldblatt D, O'Brien KL, Whitney CG.** 2014. Systematic review of the effect of pneumococcal conjugate vaccine dosing schedules on vaccine-type invasive pneumococcal disease among young children. *Pediatr Infect Dis J* **33**(Suppl 2)**:**S109–S118.
33. **Hochman M, Cohen PA.** 2015. Reconsidering guidelines on the use of pneumococcal vaccines in adults 65 years or older. *JAMA Intern Med* **175:**1895–1896.
34. **Lechner S, Mayr R, Francis KP, Prüss BM, Kaplan T, Wiessner-Gunkel E, Stewart GS, Scherer S.** 1998. *Bacillus weihenstephanensis sp. nov.* is a new psychrotolerant species of the *Bacillus cereus* group. *Int J Syst Evol Microbiol* **48:**1373–1382.
35. **Berger T, Kassirer M, Aran AA.** 2014. Injectional anthrax—new presentation of an old disease. *Euro Surveill* **19:**20877.
36. **Friebe S, van der Goot FG, Bürgi J.** 2016. The ins and outs of anthrax toxin. *Toxins (Basel)* **8:**69.
37. **Williamson ED, Dyson EH.** 2015. Anthrax prophylaxis: recent advances and future directions. *Front Microbiol* **6:**1009.
38. **Wang Q, Chang BJ, Riley TV.** 2010. *Erysipelothrix rhusiopathiae*. *Vet Microbiol* **140:**405–417.
39. **Wexler HM.** 2007. *Bacteroides*: the good, the bad, and the nitty-gritty. *Clin Microbiol Rev* **20:**593–621.
40. **Sóki J, Hedberg M, Patrick S, Bálint B, Herczeg R, Nagy I, Hecht DW, Nagy E, Urbán E.** 2016. Emergence and evolution of an international cluster of MDR *Bacteroides fragilis* isolates. *J Antimicrob Chemother* **71:**2441–2448.
41. **Rodriguez C, Van Broeck J, Taminiau B, Delmée M, Daube G.** 2016. *Clostridium difficile* infection: early history, diagnosis and molecular strain typing methods. *Microb Pathog* **97:**59–78.
42. **Nelson EJ, Harris JB, Morris JG Jr, Calderwood SB, Camilli A.** 2009. Cholera transmission: the host, pathogen and bacteriophage dynamic. *Nat Rev Microbiol* **7:**693–702.
43. **Bishop AL, Camilli A.** 2011. *Vibrio cholerae*: lessons for mucosal vaccine design. *Expert Rev Vaccines* **10:**79–94.
44. **Huehn S, Eichhorn C, Urmersbach S, Breidenbach J, Bechlars S, Bier N, Alter T, Bartelt E, Frank C, Oberheitmann B, Gunzer F, Brennholt N, Böer S, Appel B, Dieckmann R, Strauch E.** 2014. Pathogenic vibrios in environmental, seafood and clinical sources in Germany. *Int J Med Microbiol* **304:**843–850.
45. **Tang HJ, Chen CC, Lai CC, Zhang CC, Weng TC, Chiu YH, Toh HS, Chiang SR, Yu WL, Ko WC, Chuang YC.** *In vitro* and *in vivo* antibacterial activity of tigecycline against *Vibrio vulnificus*. *J Microbiol Immunol Infect*, in press.
46. **Gölz G, Rosner B, Hofreuter D, Josenhans C, Kreienbrock L, Löwenstein A, Schielke A, Stark K, Suerbaum S, Wieler LH, Alter T.** 2014. Relevance of *Campylobacter* to public health—the need for a One Health approach. *Int J Med Microbiol* **304:**817–823.
47. **Wieczorek K, Osek J.** 2013. Antimicrobial resistance mechanisms among *Campylobacter*. *BioMed Res Int* **2013:**340605.
48. **Keilberg D, Ottemann KM.** 2016. How *Helicobacter pylori* senses, targets and interacts with the gastric epithelium. *Environ Microbiol* **18:**791–806.

49. **Kim IJ, Blanke SR.** 2012. Remodeling the host environment: modulation of the gastric epithelium by the *Helicobacter pylori* vacuolating toxin (VacA). *Front Cell Infect Microbiol* **2:**37.
50. **Safavi M, Sabourian R, Foroumadi A.** 2016. Treatment of *Helicobacter pylori* infection: current and future insights. *World J Clin Cases* **4:**5–19.
51. **Majowicz SE, Scallan E, Jones-Bitton A, Sargeant JM, Stapleton J, Angulo FJ, Yeung DH, Kirk MD.** 2014. Global incidence of human Shiga toxin-producing *Escherichia coli* infections and deaths: a systematic review and knowledge synthesis. *Foodborne Pathog Dis* **11:**447–455.
52. **Rahal EA, Fadlallah SM, Nassar FJ, Kazzi N, Matar GM.** 2015. Approaches to treatment of emerging Shiga toxin-producing *Escherichia coli* infections highlighting the O104:H4 serotype. *Front Cell Infect Microbiol* **5:**24.
53. **Agger M, Scheutz F, Villumsen S, Mølbak K, Petersen AM.** 2015. Antibiotic treatment of verocytotoxin-producing *Escherichia coli* (VTEC) infection: a systematic review and a proposal. *J Antimicrob Chemother* **70:**2440–2446.
54. **Franzin FM, Sircili MP.** 2015. Locus of enterocyte effacement: a pathogenicity island involved in the virulence of enteropathogenic and enterohemorragic *Escherichia coli* subjected to a complex network of gene regulation. *BioMed Res Int* **2015:** 534738.
55. **Ochoa TJ, Contreras CA.** 2011. Enteropathogenic *E. coli* (EPEC) infection in children. *Curr Opin Infect Dis* **24:**478–483.
56. **O'Ryan M, Vidal R, del Canto F, Carlos Salazar J, Montero D.** 2015. Vaccines for viral and bacterial pathogens causing acute gastroenteritis: Part II: Vaccines for *Shigella*, *Salmonella*, enterotoxigenic *E. coli* (ETEC) enterohemorragic *E. coli* (EHEC) and *Campylobacter jejuni*. *Hum Vaccin Immunother* **11:**601–619.
57. **Kariuki S, Gordon MA, Feasey N, Parry CM.** 2015. Antimicrobial resistance and management of invasive *Salmonella* disease. *Vaccine* **33**(Suppl 3)**:**C21–C29.
58. **Ao TT, Feasey NA, Gordon MA, Keddy KH, Angulo FJ, Crump JA.** 2015. Global burden of invasive nontyphoidal *Salmonella* disease, 2010(1). *Emerg Infect Dis* **21:**941–949.
59. **Kariuki S, Onsare RS.** 2015. Epidemiology and genomics of invasive nontyphoidal *Salmonella* infections in Kenya. *Clin Infect Dis* **61**(Suppl 4)**:**S317–S324.
60. **Andrews JR, Ryan ET.** 2015. Diagnostics for invasive *Salmonella* infections: current challenges and future directions. *Vaccine* **33**(Suppl 3)**:**C8–C15.
61. **Watson CH, Edmunds WJ.** 2015. A review of typhoid fever transmission dynamic models and economic evaluations of vaccination. *Vaccine* **33**(Suppl 3)**:**C42–C54.
62. **Tennant SM, Levine MM.** 2015. Live attenuated vaccines for invasive *Salmonella* infections. *Vaccine* **33**(Suppl 3)**:**C36–C41.
63. **Valentin-Weigand P, Heesemann J, Dersch P.** 2014. Unique virulence properties of *Yersinia enterocolitica* O:3—an emerging zoonotic pathogen using pigs as preferred reservoir host. *Int J Med Microbiol* **304:**824–834.
64. **Cui Y, Yu C, Yan Y, Li D, Li Y, Jombart T, Weinert LA, Wang Z, Guo Z, Xu L, Zhang Y, Zheng H, Qin N, Xiao X, Wu M, Wang X, Zhou D, Qi Z, Du Z, Wu H, Yang X, Cao H, Wang H, Wang J, Yao S, Rakin A, Li Y, Falush D, Balloux F, Achtman M, Song Y, Wang J, Yang R.** 2013. Historical variations in mutation rate in an epidemic pathogen, *Yersinia pestis*. *Proc Natl Acad Sci USA* **110:** 577–582.
65. **Butler T.** 2013. Plague gives surprises in the first decade of the 21st century in the United States and worldwide. *Am J Trop Med Hyg* **89:**788–793.
66. **Wang H, Cui Y, Wang Z, Wang X, Guo Z, Yan Y, Li C, Cui B, Xiao X, Yang Y, Qi Z, Wang G, Wei B, Yu S, He D, Chen H, Chen G, Song Y, Yang R.** 2011. A dog-associated primary pneumonic plague in Qinghai Province, China. *Clin Infect Dis* **52:**185–190.
67. **Organization WH.** 2016. Plague around the world, 2010–2015. *Wkly Epidemiol Rec* **91:**89–93.
68. **Galimand M, Guiyoule A, Gerbaud G, Rasoamanana B, Chanteau S, Carniel E, Courvalin P.** 1997. Multidrug resistance in *Yersinia pestis* mediated by a transferable plasmid. *N Engl J Med* **337:**677–681.
69. **Guiyoule A, Gerbaud G, Buchrieser C, Galimand M, Rahalison L, Chanteau S, Courvalin P, Carniel E.** 2001. Transferable plasmid-mediated resistance to streptomycin in a clinical isolate of *Yersinia pestis*. *Emerg Infect Dis* **7:**43–48.
70. **Pollitzer R.** 1954. *Plague*. World Health Organization, Geneva, Switzerland.
71. **Li YF, Li DB, Shao HS, Li HJ, Han YD.** 2016. Plague in China 2014-All sporadic case report of pneumonic plague. *BMC Infect Dis* **16:**85.
72. **Feodorova VA, Motin VL.** 2012. Plague vaccines: current developments and future perspectives. *Emerg Microbes Infect* **1:**e36.
73. **Williamson ED, Eley SM, Griffin KF, Green M, Russell P, Leary SE, Oyston PC, Easterbrook T, Reddin KM, Robinson A, Titball R.** 1995. A new improved sub-unit vaccine for plague: the basis of protection. *FEMS Immunol Med Microbiol* **12:**223–230.
74. **Heath DG, Anderson GW Jr, Mauro JM, Welkos SL, Andrews GP, Adamovicz J, Friedlander AM.** 1998. Protection against experimental bubonic and pneumonic plague by a recombinant capsular F1-V antigen fusion protein vaccine. *Vaccine* **16:**1131–1137.
75. **Frank KM, Schneewind O, Shieh WJ.** 2011. Investigation of a researcher's death due to septicemic plague. *N Engl J Med* **364:** 2563–2564.
76. **Quenee LE, Hermanas TM, Ciletti N, Louvel H, Miller NC, Elli D, Blaylock B, Mitchell A, Schroeder J, Krausz T, Kanabrocki J, Schneewind O.** 2012. Hereditary hemochromatosis restores the virulence of plague vaccine strains. *J Infect Dis* **206:**1050–1058.
77. **Franzetti F, Cernuschi M, Esposito R, Moroni M.** 1992. *Pseudomonas* infections in patients with AIDS and AIDS-related complex. *J Intern Med* **231:**437–443.
78. **Lyczak JB, Cannon CL, Pier GB.** 2000. Establishment of *Pseudomonas aeruginosa* infection: lessons from a versatile opportunist. *Microbes Infect* **2:**1051–1060.
79. **Williams BJ, Dehnbostel J, Blackwell TS.** 2010. *Pseudomonas aeruginosa*: host defence in lung diseases. *Respirology* **15:**1037–1056.
80. **Safdar N, Crnich CJ, Maki DG.** 2005. The pathogenesis of ventilator-associated pneumonia: its relevance to developing effective strategies for prevention. *Respir Care* **50:**725–739, discussion 739–741.
81. **Bragonzi A, Paroni M, Nonis A, Cramer N, Montanari S, Rejman J, Di Serio C, Döring G, Tümmler B.** 2009. *Pseudomonas aeruginosa* microevolution during cystic fibrosis lung infection establishes clones with adapted virulence. *Am J Respir Crit Care Med* **180:**138–145.
82. **Hirakata Y, Srikumar R, Poole K, Gotoh N, Suematsu T, Kohno S, Kamihira S, Hancock RE, Speert DP.** 2002. Multidrug efflux systems play an important role in the invasiveness of *Pseudomonas aeruginosa*. *J Exp Med* **196:**109–118.
83. **Zhang L, Mah TF.** 2008. Involvement of a novel efflux system in biofilm-specific resistance to antibiotics. *J Bacteriol* **190:**4447–4452.
84. **Fricks-Lima J, Hendrickson CM, Allgaier M, Zhuo H, Wiener-Kronish JP, Lynch SV, Yang K.** 2011. Differences in biofilm formation and antimicrobial resistance of *Pseudomonas aeruginosa* isolated from airways of mechanically ventilated patients and cystic fibrosis patients. *Int J Antimicrob Agents* **37:**309–315.
85. **Huber P, Basso P, Reboud E, Attrée I.** *Pseudomonas aeruginosa* renews its virulence factors. *Environ Microbiol Rep*, in press.
86. **Falagas ME, Bliziotis IA.** 2007. Pandrug-resistant Gram-negative bacteria: the dawn of the post-antibiotic era? *Int J Antimicrob Agents* **29:**630–636.

87. **Galyov EE, Brett PJ, DeShazer D.** 2010. Molecular insights into *Burkholderia pseudomallei* and *Burkholderia mallei* pathogenesis. *Annu Rev Microbiol* **64:**495–517.
88. **Ngauy V, Lemeshev Y, Sadkowski L, Crawford G.** 2005. Cutaneous melioidosis in a man who was taken as a prisoner of war by the Japanese during World War II. *J Clin Microbiol* **43:**970–972.
89. **Yan Z, Yang J, Hu R, Hu X, Chen K.** 2016. *Acinetobacter baumannii* infection and IL-17 mediated immunity. *Mediators Inflamm* **2016:**9834020.
90. **Rice LB.** 2009. The clinical consequences of antimicrobial resistance. *Curr Opin Microbiol* **12:**476–481.
91. **He Q, Viljanen MK, Arvilommi H, Aittanen B, Mertsola J.** 1998. Whooping cough caused by *Bordetella pertussis* and *Bordetella parapertussis* in an immunized population. *JAMA* **280:** 635–637.
92. **Mooi FR, Bruisten S, Linde I, Reubsaet F, Heuvelman K, van der Lee S, King AJ.** 2012. Characterization of *Bordetella holmesii* isolates from patients with pertussis-like illness in The Netherlands. *FEMS Immunol Med Microbiol* **64:**289–291.
93. **Yesmin K, Mamun K, Shamsazzaman S, Chowdhury A, Khatun K, Alam J.** 2010. Isolation of potential pathogenic bacteria from nasopharynx from patients having cough for more than two weeks. *Bangladesh J Med Microbiol* **4:**13–18
94. **Prasad SM, Yin Y, Rodzinski E, Tuomanen EI, Masure HR.** 1993. Identification of a carbohydrate recognition domain in filamentous hemagglutinin from *Bordetella pertussis. Infect Immun* **61:**2780–2785.
95. **Hewlett EL, Burns DL, Cotter PA, Harvill ET, Merkel TJ, Quinn CP, Stibitz ES.** 2014. Pertussis pathogenesis—what we know and what we don't know. *J Infect Dis* **209:**982–985.
96. **Carbonetti NH.** 2016. *Bordetella pertussis:* new concepts in pathogenesis and treatment. *Curr Opin Infect Dis* **29:**287–294.
97. **Kilgore PE, Salim AM, Zervos MJ, Schmitt HJ.** 2016. Pertussis: microbiology, disease, treatment, and prevention. *Clin Microbiol Rev* **29:**449–486.
98. **Paczosa MK, Mecsas J.** 2016. *Klebsiella pneumoniae*: going on the offense with a strong defense. *Microbiol Mol Biol Rev* **80:** 629–661.
99. **Schielke S, Frosch M, Kurzai O.** 2010. Virulence determinants involved in differential host niche adaptation of *Neisseria meningitidis* and *Neisseria gonorrhoeae. Med Microbiol Immunol (Berl)* **199:**185–196.
100. **Dillon JA, Yeung KH.** 1989. β-Lactamase plasmids and chromosomally mediated antibiotic resistance in pathogenic *Neisseria* species. *Clin Microbiol Rev* **2**(Suppl):S125–S133.
101. **Tapsall JW.** 2009. *Neisseria gonorrhoeae* and emerging resistance to extended spectrum cephalosporins. *Curr Opin Infect Dis* **22:**87–91.
102. **Cartwright KA, Stuart JM, Jones DM, Noah ND.** 1987. The Stonehouse survey: nasopharyngeal carriage of meningococci and *Neisseria lactamica. Epidemiol Infect* **99:**591–601.
103. **Stephens DS.** 1999. Uncloaking the meningococcus: dynamics of carriage and disease. *Lancet* **353:**941–942.
104. **Singh K.** 2009. Laboratory-acquired infections. *Clin Infect Dis* **49:**142–147.
105. **Talan DA, Citron DM, Abrahamian FM, Moran GJ, Goldstein EJ, Emergency Medicine Animal Bite Infection Study Group.** 1999. Bacteriologic analysis of infected dog and cat bites. *N Engl J Med* **340:**85–92.
106. **Dworkin MS, Anderson DE Jr, Schwan TG, Shoemaker PC, Banerjee SN, Kassen BO, Burgdorfer W.** 1998. Tick-borne relapsing fever in the northwestern United States and southwestern Canada. *Clin Infect Dis* **26:**122–131.
107. **Vinetz JM.** 2001. Leptospirosis. *Curr Opin Infect Dis* **14:**527–538.
108. **Vijayachari P, Sugunan AP, Shriram AN.** 2008. Leptospirosis: an emerging global public health problem. *J Biosci* **33:**557–569.
109. **Ko AI, Galvão Reis M, Ribeiro Dourado CM, Johnson WD Jr, Riley LW, Salvador Leptospirosis Study Group.** 1999. Urban epidemic of severe leptospirosis in Brazil. *Lancet* **354:**820–825.
110. **Vinetz JM, Glass GE, Flexner CE, Mueller P, Kaslow DC.** 1996. Sporadic urban leptospirosis. *Ann Intern Med* **125:**794–798.
111. **Ko AI, Goarant C, Picardeau M.** 2009. *Leptospira*: the dawn of the molecular genetics era for an emerging zoonotic pathogen. *Nat Rev Microbiol* **7:**736–747.
112. **Dolhnikoff M, Mauad T, Bethlem EP, Carvalho CR.** 2007. Pathology and pathophysiology of pulmonary manifestations in leptospirosis. *Braz J Infect Dis* **11:**142–148.
113. **Levett PN.** 2001. Leptospirosis. *Clin Microbiol Rev* **14:**296–326.
114. **Deretic V, Vergne I, Chua J, Master S, Singh SB, Fazio JA, Kyei G.** 2004. Endosomal membrane traffic: convergence point targeted by *Mycobacterium tuberculosis* and HIV. *Cell Microbiol* **6:**999–1009.
115. **Kumar D, Rao KV.** 2011. Regulation between survival, persistence, and elimination of intracellular *mycobacteria*: a nested equilibrium of delicate balances. *Microbes Infect* **13:**121–133.
116. **Kaufmann SH, McMichael AJ.** 2005. Annulling a dangerous liaison: vaccination strategies against AIDS and tuberculosis. *Nat Med* **11**(Suppl):S33–S44.
117. **Trunz BB, Fine P, Dye C.** 2006. Effect of BCG vaccination on childhood tuberculous meningitis and miliary tuberculosis worldwide: a meta-analysis and assessment of cost-effectiveness. *Lancet* **367:**1173–1180.
118. **Rodrigues LC, Lockwood DN.** 2011. Leprosy now: epidemiology, progress, challenges, and research gaps. *Lancet Infect Dis* **11:**464–470.
119. **Pike RM.** 1976. Laboratory-associated infections: summary and analysis of 3921 cases. *Health Lab Sci* **13:**105–114.
120. **Ollé-Goig JE, Canela-Soler J.** 1987. An outbreak of *Brucella melitensis* infection by airborne transmission among laboratory workers. *Am J Public Health* **77:**335–338.
121. **Thylefors B, Négrel AD, Pararajasegaram R, Dadzie KY.** 1995. Global data on blindness. *Bull World Health Organ* **73:**115–121.
122. **Harkinezhad T, Geens T, Vanrompay D.** 2009. *Chlamydophila psittaci* infections in birds: a review with emphasis on zoonotic consequences. *Vet Microbiol* **135:**68–77.
123. **Hogan RJ, Mathews SA, Mukhopadhyay S, Summersgill JT, Timms P.** 2004. Chlamydial persistence: beyond the biphasic paradigm. *Infect Immun* **72:**1843–1855.
124. **Allerberger F, Wagner M.** 2010. Listeriosis: a resurgent foodborne infection. *Clin Microbiol Infect* **16:**16–23.
125. **Hoffman AD, Gall KL, Norton DM, Wiedmann M.** 2003. *Listeria monocytogenes* contamination patterns for the smoked fish processing environment and for raw fish. *J Food Prot* **66:**52–60.
126. **Smith B, Kemp M, Ethelberg S, Schiellerup P, Bruun BG, Gerner-Smidt P, Christensen JJ.** 2009. *Listeria monocytogenes*: maternal-foetal infections in Denmark 1994–2005. *Scand J Infect Dis* **41:**21–25.
127. **Cone LA, Somero MS, Qureshi FJ, Kerkar S, Byrd RG, Hirschberg JM, Gauto AR.** 2008. Unusual infections due to *Listeria monocytogenes* in the Southern California Desert. *Int J Infect Dis* **12:**578–581.
128. **Hof H.** 2003. Therapeutic options. *FEMS Immunol Med Microbiol* **35:**203–205.
129. **Yoshimura K, Jain A, Allen HE, Laird LS, Chia CY, Ravi S, Brockstedt DG, Giedlin MA, Bahjat KS, Leong ML, Slansky JE, Cook DN, Dubensky TW, Pardoll DM, Schulick RD.** 2006. Selective targeting of antitumor immune responses with engineered live-attenuated *Listeria monocytogenes. Cancer Res* **66:**1096–1104.
130. **Datta SK, Okamoto S, Hayashi T, Shin SS, Mihajlov I, Fermin A, Guiney DG, Fierer J, Raz E.** 2006. Vaccination with irradiated *Listeria* induces protective T cell immunity. *Immunity* **25:**143–152.

131. **Bechah Y, Capo C, Raoult D, Mege JL.** 2008. Infection of endothelial cells with virulent *Rickettsia prowazekii* increases the transmigration of leukocytes. *J Infect Dis* **197:**142–147.
132. **Johnson JE III, Kadull PJ.** 1967. Rocky Mountain spotted fever acquired in a laboratory. *N Engl J Med* **277:**842–847.
133. **Oster CN, Burke DS, Kenyon RH, Ascher MS, Harber P, Pedersen CE Jr.** 1977. Laboratory-acquired Rocky Mountain spotted fever. The hazard of aerosol transmission. *N Engl J Med* **297:** 859–863.
134. **Demma LJ, Traeger MS, Nicholson WL, Paddock CD, Blau DM, Eremeeva ME, Dasch GA, Levin ML, Singleton J Jr, Zaki SR, Cheek JE, Swerdlow DL, McQuiston JH.** 2005. Rocky Mountain spotted fever from an unexpected tick vector in Arizona. *N Engl J Med* **353:**587–594.
135. **Nicholson WL, Allen KE, McQuiston JH, Breitschwerdt EB, Little SE.** 2010. The increasing recognition of rickettsial pathogens in dogs and people. *Trends Parasitol* **26:**205–212.
136. **Watt G, Parola P.** 2003. Scrub typhus and tropical rickettsioses. *Curr Opin Infect Dis* **16:**429–436.
137. **Truchan HK, Seidman D, Carlyon JA.** 2013. Breaking in and grabbing a meal: *Anaplasma phagocytophilum* cellular invasion, nutrient acquisition, and promising tools for their study. *Microbes Infect* **15:**1017–1025.
138. **Sanchez E, Vannier E, Wormser GP, Hu LT.** 2016. Diagnosis, treatment, and prevention of Lyme disease, human granulocytic anaplasmosis, and babesiosis: a review. *JAMA* **315:**1767–1777.
139. **Rikihisa Y.** 2015. Molecular pathogenesis of *Ehrlichia chaffeenis* infection. *Annu Rev Microbiol* **69:**283–304.
140. **Maurin M, Raoult D.** 1999. Q fever. *Clin Microbiol Rev* **12:**518–553.
141. **Oyston PC, Davies C.** 2011. Q fever: the neglected biothreat agent. *J Med Microbiol* **60:**9–21.
142. **Johnson JE III, Kadull PJ.** 1966. Laboratory-acquired Q fever. A report of fifty cases. *Am J Med* **41:**391–403.
143. **Keim P, Johansson A, Wagner DM.** 2007. Molecular epidemiology, evolution, and ecology of *Francisella*. *Ann N Y Acad Sci* **1105:**30–66.
144. **Eigelsbach HT, Downs CM.** 1961. Prophylactic effectiveness of live and killed tularemia vaccines. I. Production of vaccine and evaluation in the white mouse and guinea pig. *J Immunol* **87:**415–425.
145. **Pechous RD, McCarthy TR, Zahrt TC.** 2009. Working toward the future: insights into *Francisella tularensis* pathogenesis and vaccine development. *Microbiol Mol Biol Rev* **73:**684–711.
146. **Brenner DJ, Steigerwalt AG, McDade JE.** 1979. Classification of the Legionnaires' disease bacterium: *Legionella pneumophila*, genus novum, species nova, of the family Legionellaceae, familia nova. *Ann Intern Med* **90:**656–658.
147. **Newton HJ, Ang DK, van Driel IR, Hartland EL.** 2010. Molecular pathogenesis of infections caused by *Legionella pneumophila*. *Clin Microbiol Rev* **23:**274–298.
148. **Amsden GW.** 2005. Treatment of Legionnaires' disease. *Drugs* **65:**605–614.

Viral Agents of Human Disease: Biosafety Concerns

9

MICHELLE ROZO, JAMES LAWLER, AND JASON PARAGAS

This chapter discusses the unique hazards of handling biosamples that contain viruses and the illnesses that can result from occupational exposure and infection. Basic concepts of virology and clinical syndromes of viral infections are reviewed, with specific attention dedicated to viral pathogens associated with a risk of work-acquired infections. The last section addresses laboratory safety challenges presented by newer biotechnologies. Armed with the knowledge presented in this chapter, the laboratorian will be able to define a risk management matrix based on the risk of a virus in a specimen from a patient manifesting a specific illness, the types of procedures used in the laboratory, and the potential outcome of a laboratory-acquired infection (LAI).

Since the initial discovery of a virus as the cause of yellow fever (YF) in the early 1900s, viruses have been demonstrated to be the cause of major epidemics and pandemics of human disease (1). An examination of the etiology of the majority of emerging infections in the 21st century reveals newly recognized viruses that either have evolved, have been previously unrecognized, or have emerged as a result of the encroachment by humans into the environment where viruses are endemic; this point is exemplified by the recent, and previously obscure, Zika virus outbreak (2–5). Viruses are unmistakably important human pathogens and pose a significant risk for the laboratorian working with clinical specimens or the research virologist.

In most cases, an acute viral infection results in the production of an asymptomatic or subclinical infection; however, human infections caused by a variety of pathogens can be complicated by fever, pneumonia, severe multiorgan illness, encephalitis, and hemorrhage. The classical or severe manifestation of a viral infection is a result of a number of viral and host factors, such as viral virulence, the amount of virus in the exposure, the route of viral entry, and the host's immune status, genetic makeup, and age. These viral and host factors also define the risks to the laboratory worker for an occupational infection.

Previous publications on laboratory-associated infections demonstrate that viruses pose a significant risk in the laboratory environment (Table 1). An early survey documented 222 viral infections, of which 21 were fatal (case-fatality ratio of 11%), 12% were associated with a laboratory accident, and more than 30% were associated with the handling of infected animals and tissues (6–8).

TABLE 1.

Summary of previously published laboratory-associated viral infections[a]

Family	Genus	Virus	No. of cases	No. of deaths	Recommended biosafety level[b]
Adenoviridae	*Aviadenovirus*	Fowl plague virus	1	0	2
	Mastadenovirus	Adenovirus	10	0	2
Arenaviridae	*Arenavirus*	Lassa virus	3	0	4
		Lymphocytic choriomeningitis (LCM) virus	102	9	2 for work with infectious material or infected animal, 3 for activities with high potential for aerosol production or production of large quantities of material, or work with infected hamsters
		Junin virus	15	1	4 (level 3 if immunized)
		Machupo virus	7	1	4
		Sabia virus	2	0	4
		SPH114202 virus	1	0	4
Bunyaviridae	*Nairovirus*	Crimean-Congo hemorrhagic fever virus (CCHFV)	5	0	4
		Dugbe virus	1	0	3
		Nairobi sheep disease virus	3	0	3
	Orthobunyavirus	Apeu virus	2	0	2
		Marituba virus	2	0	2
		Oriboca virus	2	0	2
		Bunyamwera virus	8	0	2
		Germiston virus	5	0	3
		Oropouche virus	5	0	3
		Guaroa virus	1	0	2
		Ossa virus	1	0	2
	Phlebovirus	Rift Valley fever (RVF) virus	103	4	3
	Hantavirus	Hantaan virus	226	0	2 for infected tissue samples using level 3 practices, 3 for cell culture propagation, 4 for large-scale growth and viral concentration
Coronoviridae	*Coronavirus*	Middle East respiratory syndrome coronavirus (MERS-CoV)	0	0	2; 3 for cell culture propagation
		Severe acute respiratory syndrome coronavirus (SARS-CoV)	6	0; but a researcher in China subsequently infected her mother, who succumbed to the infection	2; 3 for cell culture propagation
Filoviridae	*Marburgvirus*	Marburg virus	31	9	4
	Ebolavirus	Ebola virus	4	1	4
Flaviviridae	*Flavivirus*	Dengue virus	14	0	2
		Japanese encephalitis virus	2	0	3
		Kunjin virus	4	0	2
		Kyasanur Forest disease (KFD) virus	132	0	4

(continued)

TABLE 1.

(Continued)

Family	Genus	Virus	No. of cases	No. of deaths	Recommended biosafety level[b]
		Louping-ill virus	67	0	3
		Omsk hemorrhagic fever virus	9	0	4
		Powassan virus	2	0	3
		Rio Bravo virus	12	0	2
		St. Louis encephalitis virus	3	0	3
		Spondweni virus	4	0	3
		Tick-borne encephalitis (TBE) virus	39	2	3
		Wesselsbron virus	10	0	3
		West Nile virus	27	0	3
		Yellow fever (YF) virus	140	24	3
		Zika virus	4	0	2
		Negishi virus	1	0	3
Hepadnaviridae	*Orthohepadnavirus*	Hepatitis B virus (HBV)	19	0	2; 3 for activities with potential for droplet or aerosol production and viral concentration
Herpesviridae	*Lymphocryptovirus*	Epstein-Barr virus	2	0	2; 3 for producing, purifying, and concentrating
	Simplexvirus	Herpes B virus	37	20	2 for tissue specimens from macaques; 3 for suspected infected tissue or *in vitro* propagation, 4 for cultures with high-titer virus
		Herpes simplex virus	2	0	2; 3 for producing, purifying, and concentrating
	Varicellovirus	Pseudorabies virus	2	0	2
		Bovine encephalomyelitis virus	1	0	Unclassified
		Varicella-zoster virus	3	0	2
Orthomyxoviridae	Unspecified	Influenza virus	22	1	2; 3 for highly pathogenic avian influenza viruses (HPAI) and the 1918 influenza virus strain
Paramyxoviridae	*Rubulavirus*	Newcastle disease virus	77	0	Restricted animal pathogen
		Mumps virus	8	0	2
	Morbillivirus	Measles virus	2	0	2
	Pneumovirus	Respiratory syncytial virus (RSV)	1	0	2
	Respirovirus	Sendai virus	1	0	2
Parvoviridae	*Erythrovirus*	B19 virus	10	0	2
Picornaviridae	*Aphthovirus*	Foot-and-mouth disease virus	2	0	Restricted animal pathogen
	Cariovirus	Mengovirus encephalomyocarditis	2	0	
	Enterovirus	Coxsackie virus	39	0	
		Poliomyelitis virus	21	5	2
		Swine vesicular disease	1	0	Restricted animal pathogen
		Echovirus	3	0	2
	Hepatovirus	Hepatitis A virus (HAV)	5	0	2

(continued)

TABLE 1.

(Continued)

Family	Genus	Virus	No. of cases	No. of deaths	Recommended biosafety level[b]
Poxviridae	*Orthopoxvirus*	Vaccinia and smallpox viruses	38	0	2 for facilities with three practices (if immunized, smallpox an exception; work on live variola can only be conducted within two approved BSL4/ABSL4 facilities: CDC, Atlanta, GA, and the State Research Center of Virology and Biotechnology, VECTOR, in Koltsovo, Russia)
	Yatapoxvirus	Yaba and Tana viruses	24	0	2 (if immunized)
	Parapoxvirus	Orf virus	2	0	2
Reovirus	*Coltivirus*	Colorado tick fever virus	19	0	2
Retroviridae	*Lentivirus*	Human immunodeficiency virus (HIV)	45	0	2; 3 for industrial scale or high concentrations of virus
		Simian immunodeficiency virus (SIV)	3	0	2; 3 for industrial scale or high concentrations of virus
	Oncornavirinae	Simian type D retrovirus	2	0	
	Spumavirus	Simian foamy virus	13	0	2
Rhabdoviridae	*Lyssavirus*	Rabies virus	2	0	2; 3 for aerosol-producing activities or high concentrations of virus
	Vesiculovirus	Vesicular stomatitis virus	78	0	3 (exception for laboratory-adapted strains, level 2)
		Piry virus	10	0	3
Togaviridae	*Alphavirus*	Chikungunya virus	33	0	3
		Eastern equine encephalitis virus	7	0	2 (level 3 and immunized if working with infection of newly hatched chickens)
		Mayaro virus	6	0	3
		Mucambo virus	4	0	3
		Venezuelan equine encephalitis (VEE) virus	187	2	3
		Western equine encephalitis (WEE) virus	16	4	2 (level 3 if immunized and working with infection of newly hatched chickens)
	Rubivirus	Rubella virus	7	0	
Unspecified	Unspecified	Hepatitis virus	360	2	2
Unspecified	Unspecified	Viral diarrhea virus	2	0	
Unspecified	Unspecified	Hemorrhagic nephrosonephritis virus	1	0	

[a]Data from references 6–9, 46, 66, 81, 82, 126, 157, 212–214, 227–253.
[b]Data from reference 185.

The majority of the viral agents in this early survey were accounted for by the viruses causing YF, Rift Valley fever (RVF), Venezuelan equine encephalomyelitis (VEE), and lymphocytic choriomeningitis (LCM). Of the LAIs described, only 27 (12%) had a recognized accident and cause of exposure; the routes of exposure for the rest of the cases were unknown. Of course, there were likely many LAIs that were never reported or detected. In a follow-up survey conducted in 1951, 1,342 LAIs were documented, of which 39 resulted in deaths, an overall case-fatality ratio of 3% (8). Viral diseases were associated with a higher case-fatality ratio of 4.5% in this series.

In a worldwide survey of laboratory-associated infections conducted in 1974, 3,921 infections were documented, of which 164 were fatal (case-fatality ratio of 4%) (6). Overall, viruses were responsible for 27% of laboratory-associated infections with defined causes. As seen from these early surveys, arboviruses pose a particular risk to the laboratorian. In a survey of overt laboratory-acquired arboviral infection, 428 infections were documented, with 16 fatal (case-fatality ratio of 3.7%) (9). Exposure to infectious aerosols was the most common source of infection in this series, with the most common viruses being VEE virus, Kyasanur Forest disease (KFD) virus, YF virus, vesicular stomatitis virus (VSV), RVF virus, tick-borne encephalitis virus (TBEV), louping-ill virus, and chikungunya virus. Fatalities were caused most commonly by exposure to YF virus, followed by Western equine encephalitis (WEE) virus, VEE virus, TBE virus, Junin virus, Machupo virus, and RVF virus. The probable sources of infection in this series, in declining order of frequency, were experimentally infected animals, aerosol, agent handling, and accidents. Other less frequent sources of infection were preparation of vaccine or antigens, experimentally infected chicken embryos, and discarded glassware. Table 1 summarizes the published literature on laboratory-associated viral infections and the viruses' families, genera, and recommended laboratory biosafety levels (BSLs). Comparison of the more historic accounts of LAIs to those that are recently acquired shows the addition of other pathogens to the list—Ebola virus, West Nile virus (WNV), severe acute respiratory syndrome coronavirus (SARS-CoV), and vaccinia virus. Shifting research priorities may account for the differences, as well as the reemergence of pathogens or the discovery, and/or creation, of new pathogens with high risk.

VIRUSES: BASIC CONCEPTS

Taxonomy and Classification

Taxonomy and classification of viruses are overseen by the International Committee on Taxonomy of Viruses (ICTV), and viruses are grouped broadly into orders, families, subfamilies, genera, and species based on characteristics (10). Viruses are classified based on the type and organization of the viral genome (double-stranded DNA, single-stranded DNA, RNA and DNA reverse transcribing, double-stranded RNA, negative-sense single-stranded RNA, positive-sense single-stranded RNA, and subviral agents), the strategy of viral replication, and the structure of the virion (10). Discriminators used to determine virus species are the relatedness of the genome sequence, natural host range, cell and tissue tropism, pathogenicity and cytopathology, mode of transmission, physicochemical properties, and antigenic properties of the viral proteins. Currently the complete virus taxonomy consists of 7 orders, 111 families, 27 subfamilies, 609 genera, and 3,704 species. Reports are periodically updated by the ICTV, and a ready reference can be found on its website (http://www.ictvonline.org/), which contains all its published information and a universal virus database. The ICTV database contains the symptom data for viruses that infect humans and is consistent with the World Health Organization International Code of Diseases (ICD-10) and found on its website (http://www.who.int/classifications/icd/en/) (10).

Virology and Viral Epidemiology

Viruses, the smallest of the replicating organisms, are obligate intracellular parasites, requiring the cellular host's machinery for reproduction. They are essentially packages of genetic material containing a core of DNA or RNA, packaged in a protein coat, or capsid. They may be enveloped, meaning that the capsid is covered by an outer lipid membrane coat derived from the host cell nuclear membrane, Golgi membrane, or the outer membrane during budding. The proteins embedded in the outer membrane coat function as ligands or cellular receptors, which are required for cellular attachment and viral entry into the host cell. The life cycle of viruses is similar. They require attachment to their targeted host cell and entry of the virus through the cellular membrane. After cellular entry, there is release of the viral genetic material, transport of the genetic information within the cell, and replication, which employs the host cell machinery to copy the viral genome. Some viruses have the ability to integrate their genetic material into the host cell, as a stage of viral replication, or as an inactive viral infection, which then allows for future replications of the virus during host cell replication. Then, viral protein is produced, followed by assembly of the virus, and budding or cellular lysis to release mature virions. Viruses have evolved a number of unique sophisticated methods in order to evade the host's defenses to facilitate this process of replication and propagation.

Viral Diversity

RNA viruses are uniquely situated in nature to rapidly evolve and adapt to differing environments and hosts. The same is not true for DNA viruses, which contain a DNA polymerase with editing exonuclease activity that corrects mutational events that may occur during replication. Unlike the DNA viruses, the RNA viruses lack this enzyme and thus are highly susceptible to point mutations. Point mutations occur in the range of 10^{-4} to 10^{-5}

substitutions per nucleotide copied (11). During replication, homologous and heterologous recombination, gene reassortments, and the formation of quasispecies may occur among the RNA viruses (12). The high mutation and recombination rates, the resultant competition among mutant genomes, and the natural selection of strains best fit to adapt to the host environment provide for the diversity seen among the RNA viruses (13). The end result is an RNA virus that undergoes rapid evolution to become highly adaptable to the host and environment. An emerging new virus causing an epidemic will more than likely be an RNA virus, and this has been the case in many recent outbreaks—both Ebola and Zika are RNA viruses.

Arboviruses, a group of viruses transmitted by arthropod vectors, exemplify the diversity present in virus replication and disease. The arboviruses all contain RNA as their genetic code but differ in their vectors of transmission, epidemiology, pathogenesis, and clinical manifestation. Although these viruses differ in the degree of clinical illness that they produce, they share the ability to produce direct cellular activation of immune cells, cell damage, cell death, and derangements in the host's coagulation and complement pathways. An example of the diversity of the arboviruses can be seen in the family *Bunyaviridae*, which includes various viruses such as RVF virus (genus *Phlebovirus*); Crimean-Congo hemorrhagic fever virus (CCHFV) (genus *Nairovirus*); and Hantaan, Dobrava, Saaremaa, Seoul, and Puumala hantaviruses causing hemorrhagic fever with renal syndrome (HFRS) (genus *Hantavirus*). These viruses all cause severe human disease, have wide geographic distribution, and have emerged as major pathogens. They share similar morphological features: a spherical virion and a size between 80 and 120 nm (14). Their lipid envelope contains two or three glycoproteins that determine cell tropism and host pathogenicity and are sites for viral neutralization by antibody (15–18). Their genetic information consists of a single negative strand of RNA organized into three segments termed large, medium, and small, which code for the virus nucleocapsid, glycoproteins, and polymerase proteins, respectively (19, 20).

Viral factors associated with human disease are illustrated by CCHFV. The M-segment-encoded polyprotein contains a mucin-like domain and a furin cleavage site (21). This polyprotein has been implicated in endothelial damage, cellular cytotoxicity, and interferon antagonism. These are seen in other hemorrhagic fever viruses, such as in Ebola (22). The effect on host gene regulation by hantavirus illustrates the ability of these viruses to up- or downregulate host genes as a mechanism of producing severe disease. One difference between the pathogenic and nonpathogenic strains is that the pathogenic strains suppress early cellular interferon responses that are activated by nonpathogenic strains (23).

Arboviral Epidemiology

The complex epidemiology and large disease burden associated with the arboviruses can be illustrated by the dengue and Zika viruses (family *Flaviviridae*, genus *Flavivirus*). Dengue is the most common arboviral infection of tropical and subtropical regions of the world, with over a million infections per year (24). The dengue viruses consist of four serologically and genetically distinct viruses, serotypes 1 to 4 (25). Each serotype can produce human infection, which results in lifelong protective immunity but does not protect against the other three serotypes. The first infection is termed a primary dengue virus infection, which can be followed by a secondary infection.

Epidemiological data suggest that there are serotype differences among the dengue viruses in their ability to produce large outbreaks of human disease and in their ability to produce severe clinical disease (26). A study of dengue virus in Thailand over a 26-year period suggested that dengue 3 produced large epidemics of disease and all serotypes were associated with dengue hemorrhagic fever (DHF) (26). Genetic characterization of the dengue 2 strains circulating in Thailand revealed that many different virus variants circulate simultaneously, reflecting the quasispecies nature of these viruses (27). Distinct genotypic groups are associated with dengue fever and DHF patients, implying both a common progenitor and that DHF-producing dengue 2 viruses segregate into only one genotypic group that has evolved independently in Southeast Asia (27).

Dengue disease and DHF are primarily childhood diseases in regions of endemicity, although all ages are at risk of dengue virus infection and severe dengue disease. The reason for the continuous transmission of dengue in a population is largely due to the transmission of different dengue virus serotypes that become predominant in any given year. Since immunity is conferred from the infecting dengue virus serotype but not to other dengue virus serotypes, a person may be infected by two or more dengue virus serotypes during a lifetime. This was demonstrated in a prospective school-based study of dengue virus transmission and severe disease (28, 29). Despite the proximity of the schools studied, there was marked spatial and temporal clustering of transmission of each dengue virus serotype; one or two serotypes were responsible for an outbreak in a given school, with each serotype alternating every year. The constant influx of susceptible children and the continuous spread of differing dengue virus serotypes and intrinsic population dynamics provide an environment of continuous dengue transmission without the development of herd immunity in the population (30).

Zika virus was first discovered in 1947 in the Zika Forest in Uganda; the first human cases were reported in 1952, and outbreaks were reported in tropical Africa, Southeast Asia, and the Pacific islands (31). Before 2007, only 14 cases of Zika were documented, although many were probably undiagnosed given the antigenic and symptomatic similarities to dengue virus. However, the virus has undergone significant changes in protein and nucleotide sequence in the last half century, likely contributing to the outbreaks in Micronesia (2007), French Polynesia (2013), and currently in the Americas, and the novel associations with Guillain-Barré syndrome and microcephaly (32). A virus being able to cross the barrier of the placenta is not unique. Varicella-zoster virus (VZV, or chicken pox), cytomegalovirus (CMV), rubella virus, herpes simplex virus (HSV), human immunodeficiency virus (HIV), among others, all can spread from the mother to the developing fetus; however, Zika is the first of the arboviruses to demonstrate this ability. As of January 2016, the WHO declared Zika as a Public Health Emergency of International Concern and forecasted that four million individuals would be affected through the course of the outbreak (33).

This brief discussion on some of the arboviruses illustrates the diversity among these viruses in their mode of transmission, virulence factors, geographic distribution, and burden of disease in a population. Understanding the basic epidemiology of viruses and the viral and host factors that interact to produce clinical disease will provide a fundamental knowledge base for the laboratorian to estimate the likelihood for a particular pathogen to cause a severe illness based on the history of the clinical specimen, clinical presentation of the patient, and the risk for potential infections in the laboratory.

Infectious and Lethal Doses

The infectious dose of a virus is the amount of virus that can produce infection *in vivo* or *in vitro*, while the lethal dose is the amount of virus necessary to cause death of the host. Infectious dose can be large (10^5 organisms for *Escherichia coli*) to very small (less than 10 organisms for *Mycobacterium tuberculosis*). Infection in a cell culture system (*in vitro*) consists of cellular entry and viral replication with the production of mature virions that can further infect cells and continue to replicate. Infection in an animal model or human host (*in vivo*) is demonstrated by viral replication in the animal or human host and is confirmed by isolating viable virus from blood or tissue and/or serological evidence of viral replication by the production of virus-specific immunoglobulin. A host may be infected and not develop clinical disease.

The amount of virus required to produce infection in a cell culture system is determined by serial dilution of virus followed by inoculation into an appropriate cell culture system with a defined end point of infection, such as cytopathic effect. In an animal model the end point is death. Virulence is typically described quantitatively by utilizing either the 50% infectious dose (ID_{50}) or the 50% lethal dose (LD_{50}), the number of microbes necessary to cause infection or death, respectively, in half of the infected population. The lower the ID_{50}/LD_{50}, the more virulent the organism (34).

The infectious dose is one of the main criteria in defining the BSL necessary to work with viruses or other organisms. Agents or toxins are characterized by (i) their effect on human health, (ii) the degree of contagiousness (infectious dose/lethal dose) and the methods of transmission to humans, (iii) the availability of medical countermeasures, and (iv) any other criteria including the needs of vulnerable populations. The infectious dose of viruses should be taken into account by laboratory workers as they characterize their risk.

RISK MATRIX

Risk is a function of likelihood and consequences. To be able to adequately assess risk, laboratorians must understand the underlying hazard posed by live viral agents in the laboratory environment as well as the consequence of infection with various viruses. Figure 1 provides a potential risk matrix that can be used by the laboratorian to assess risk. The hazard probability is an estimate based on the type of procedures used in the laboratory and the likelihood for an accidental release of virus as a function of the consequences to the individual and beyond should the virus be released.

The laboratorian should use what is known about a specimen, virus, and the type of procedures being used to assess the likelihood component of risk. For example, a viral pathogen with a low ID_{50} used in high concentrations (i.e., 20 to 50 ID_{50}s) would pose a hazard if the virus is released from a centrifugation accident or a spill from a culture plate. A viral pathogen with a high ID_{50} used in low concentrations (0.1 to 0.5 ID_{50}, for example) would represent a lower hazard to laboratory workers. Additionally, lab practices can be associated with risk likelihood; for example, open centrifugation could be expected to release the virus into the work environment, but centrifugation in a class III glove box or the use of secondary barriers could reduce the risk.

Equally important is the severity of clinical illnesses associated with the viral pathogen. A virus with a high ID_{50} and 100% mortality would be assessed at a different risk

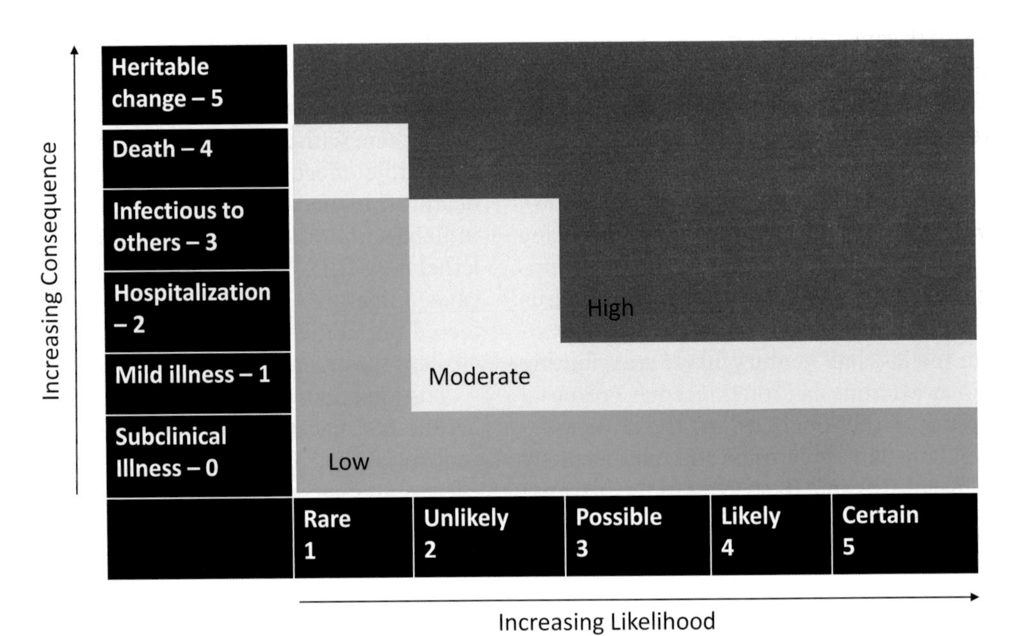

Figure 1: Risk matrix that may serve as a framework for making decisions for planned experiments and for analyzing the impact of experimental and engineering controls used to manage risk. The consequences, of increasing severity, are noted on the *y* axis. The *x* axis represents the probability of an event occurring. Stoplight colors are used to indicate regions of the matrix that may represent areas of differing risks. The eclipse is a representative of the wide scope that genetically engineered viruses may represent in the matrix.

than a pathogen that has a low ID_{50} but is not known to cause a severe illness. A significant proportion of risk with viral agents is driven by the consequences of infection, which ranges from subclinical disease to severe illness resulting in hospitalization to death. Clinical severity is based on a number of incompletely understood factors, including pathogen virulence and host susceptibility. Immunosuppression or advanced age generally imparts greater risk for a laboratory worker compared to a younger adult with a competent immune system, although there may be exceptions to this rule. A vaccinated worker would likely have a lower clinical severity associated with occupational infection; however, other confounding medical conditions may impact the effectiveness of vaccination and clinical severity score (35). It needs to be reiterated that vaccination is not an acceptable primary biosafety barrier.

Using this risk matrix, work can be categorized as high risk to low risk and individualized to the laboratory worker and the procedures being used. A risk assessment matrix is presented as a paradigm to determine risk in the laboratory based on the viral pathogen, risk of the biosample or laboratory procedure, the appropriate BSL of potential viral pathogens, and the severity of resultant clinical illness. The risk matrix is intended to frame a discussion before an experiment is performed with the appropriate professionals. Experiments with high risk can be performed and in some cases need to be performed but will require the addition of risk mitigation. The risk matrix is not intended as an analytic tool but as conceptual framework. It is possible that procedures may contain multiple consequences and differing likelihoods. It will rest on the investigator to balance the risks with potential benefits toward advancing the scientific frontiers for society and apply risk mitigation practices wherever possible. Mitigation can be in the form of engineering controls, changes to standard operating procedures, and/or the addition of robotics/automation to replace human operation in higher-risk situations.

The risk matrix can also be utilized when evaluating gene-editing technologies (see discussion in "Genetically Engineered Viruses," below). There are two important considerations: (i) the use of a viral vector to deliver clustered regularly interspaced short palindromic repeat (CRISPR)/Cas9-mediated transgene insertions, and (ii) the modification of virus genomes using gene editing. Neither transgenes with viral vectors nor genetically engineered viruses are novel; however, the efficiency and specificity of CRISPR/Cas9-based approaches have significantly lowered the barriers for genome modifications in somatic as well as germ line cells. Additionally, gene drive is a derivative technology that circumvents rules of Mendelian inheritance, and thus creates a novel dimension for biosafety considerations. Consequences of LAIs for the laboratory worker may affect offspring of the infected individual or populations when working with constructs with gene-editing potential. It is important to consider that CRISPR/Cas9 technology may not simply be an additive risk but may have synergistic effects when in combination with the unique attributes of viruses. With the maturing of technologies, like the CRISPR/Cas9 family of gene-editing technologies, it is advisable for the laboratory to frequently reevaluate risks.

CLINICAL MANIFESTATION OF VIRAL DISEASE

Knowledge of the clinical consequences of viral infections is essential to determining the potential risk of laboratory work with known viruses and unknown clinical specimens. Understanding the clinical manifestations of these infections may be critical to recognizing a laboratory exposure and mitigating such risk. Diagnostic specimens from an undiagnosed patient pose a particular risk for the laboratorian, and workers and supervisors in laboratories that handle such specimens should be especially vigilant for clinical syndromes associated with viral infection.

Constant vigilance for potential symptoms of occupational infection is vital. Although the incubation period for most viruses is short, ranging from 3 to 15 days following exposure, it can be considerably longer for certain viruses, such as hepatitis B and C viruses (HBV and HCV, respectively) or rabies virus. The incubation period can also vary with the amount of virus inoculated; in general, the more virus inoculated, the shorter the incubation period. For example, the incubation period of CCHFV can be a short as 2 days following a large viral exposure, as seen in nosocomial infections (36).

Clinical Manifestations by Syndrome

Fever is an early presenting feature of clinical illness from a viral infection, and every laboratorian working with infectious agents should be evaluated for any fever as a potential LAI. At presentation, fever is often associated with an array of generalized or systemic symptoms such as myalgia, arthralgia, malaise, and headache during the early clinical period. For many viral infections, classic differentiating signs and symptoms, such as bleeding or encephalitis, tend to occur in the middle to late clinical course. Grouping viral diseases into major clinical syndromes is a helpful approach in assessing evidence of occupational exposure.

Fever and rash

Fever with rash is a hallmark of many viral infections. Viral exanthems are frequently nondescript, but the timing or morphology can often suggest particular etiologies, and in some cases is highly suggestive. Macular or maculopapular "classic" viral exanthems are seen with enteroviruses, adenovirus, rubeola virus (measles), rubella virus (German measles), human herpesvirus 6 (HHV-6), and parvovirus B-19. Vesicular or pustular rashes are associated with HSV-1 or -2, VZV, hand-foot-and-mouth disease (HFMD)-associated enteroviruses, and poxviruses. Papular eruptions can be seen with molluscom contagiosum virus. The Gianotti-Crosti syndrome is a popular pruritic eruption usually sparing the trunk that is associated with a number of viruses, including Epstein-Barr virus (EBV), CMV, coxsackie virus, parvovirus B19, hepatitis A and B viruses, and influenza A (37).

Most of the common arboviruses produce fever and rash prior to the more classic signs that define these viruses, such as arthritis or central nervous system (CNS) disease. In addition, many of the hemorrhagic fever virus infections are associated with transient macular rash early on in the course of illness. These rashes may be especially difficult to appreciate in darker-skinned individuals. Thus, meticulous examination for rash should be undertaken in any laboratorian with fever.

Acute neurologic disorders

Common neurologic manifestations of viral infections include meningitis and encephalitis. A number of viral infections can result in meningoencephalitis, which is a mixed presentation of the two and is characterized by fever, headache, photophobia, mental status changes, focal neurologic deficits, and neck stiffness.

Pure viral meningitis generally involves fever, headache, and neck stiffness, but less commonly alterations of mental status or more serious neurological signs. Meningitis is more frequently associated with nonpolio enteroviruses (including coxsackie and echoviruses), HSV-2, lymphocytic choriomeningitis virus (LCMV), acute HIV, and EBV. The clinical course of viral meningitis is generally self-limited in immunocompetent adults, with the vast majority of patients recovering full neurologic functioning without clinical intervention.

Most of the arboviruses can cause a mixed meningoencephalitis picture. These viruses include the California encephalitis group of bunyaviruses (La Crosse, Jamestown Canyon, Snowshoe hare), the CNS-tropic flaviviruses (St. Louis encephalitis virus, WNV), Colorado tick fever virus (an orbivirus), sandfly fever virus, VEE, WEE, and Eastern equine encephalitis (EEE). Other viruses causing meningoencephalitis include mumps virus, measles virus, poliovirus, rubella virus, Nipah virus, parvovirus B19, adenovirus, influenza virus, and parainfluenza virus.

Encephalitis is due to viral infection of the brain parenchyma and is associated with fever and earlier and more pronounced alterations of consciousness. Naturally occurring pure encephalitis syndrome is most commonly encountered with rabiesvirus, HSV-1, and VZV infection, but many other viruses such as enterovirus, arboviral encephalidites (WEE, EEE, VEE), and other arboviruses can be a cause. Herpes B virus (Cercopithecine herpesvirus) is a potential cause of encephalitis among laboratorians working with Old World macaques. Encephalitis syndromes are generally more severe, with high fatality and neurologic sequelae in those that do survive.

In addition to syndromes involving the brain and meninges, neurological manifestations of viral infection can also localize to the spinal cord. The most recognized of these syndromes is acute flaccid paralysis (AFP), defined by the sudden loss of motor function to the extremities and trunk muscles that may lead to loss of respiratory function. While AFP and myelitis are classically seen with poliovirus infection, related syndromes can be seen with many other viral agents such as enteroviruses, WNV, Japanese encephalitis virus (JEV), TBE virus, and human T-cell lymphotropic virus (HTLV).

Respiratory

Respiratory illnesses associated with viral infections include upper respiratory syndromes, such as bronchitis or tracheitis, pleuritis, and lower respiratory infection involving the lung parenchyma (pneumonia). Upper respiratory infections are characterized by fever, sore throat, nasal congestion, and dry nonproductive cough. Lower respiratory infections are characterized by cough (occasionally productive), tachypnea, and pleuritis. Typically, chest X ray demonstrates an atypical pneumonia with an interstitial pattern. Severe cases of viral pneumonia may present with an acute respiratory distress syndrome-like picture with diffuse fluffy infiltrates. Examples of viruses that produce a respiratory illness are rhinovirus, adenovirus, respiratory syncytial virus (RSV), influenza virus, coronavirus, including SARS and Middle East respiratory syndrome (MERS), parainfluenza virus, and hantavirus.

Hepatitis

Classic viral hepatitis can be divided into four clinical phases: an incubation period, a prodromal period, an icteric phase, and a convalescent period. The prodromal period is characterized by fever, loss of appetite, fatigue, malaise, muscle pain, nausea, and vomiting. The icteric phase of hepatitis is characterized by the appearance of dark urine, pale stools, and yellow discoloration of the mucous membranes, conjunctivae, and skin. Laboratory findings demonstrate an increase in liver enzymes and total bilirubin during the icteric phase. Fever usually subsides after the first few days of the icteric phase. Examples of viruses that produce hepatitis are HAV, HBV, HCV, hepatitis delta virus (HDV), and HEV.

Acute gastroenteritis

Acute gastroenteritis is characterized by the acute onset of nausea, vomiting, abdominal pain or cramps, anorexia, and diarrhea. Fever may be present, and other constitutional symptoms such as headache, chills, myalgia, and sore throat may occur, with dehydration a major complication without appropriate fluid replacement. Acute gastroenteritis is usually self-limited, lasting on average 2 to 4 days. Examples of viruses that produce acute gastroenteritis are Norwalk viruses (norovirus), astrovirus, rotavirus, and adenovirus.

Hemorrhagic fever

Viral hemorrhagic fever is characterized by fever, severe systemic symptoms, capillary leak syndrome, and various degrees of coagulopathy. Coagulopathy is manifested by hemorrhage into the skin as petechiae or ecchymoses, oozing at a puncture site, epistaxis, gingival bleeding, hemorrhagic conjunctivitis, hematemesis, melena, or severe vaginal bleeding. Although laboratory indicators of coagulopathy can accompany early disease, overt bleeding disorder generally occurs late in disease. Significant bleeding events are generally associated with more severe infections and poor prognosis; however, significant blood loss and hemorrhagic shock are usually not major contributors to patient demise. Cardiovascular collapse and shock syndrome can occur through intravascular plasma leakage into the extravascular space or other less-understood mechanisms. Viruses producing hemorrhagic fever syndrome are members of four families: *Flaviviridae* (dengue, YF, KFD virus, Omsk hemorrhagic fever viruses, Alkumra), *Bunyaviridae* (hantavirus, CCHFV, and RVF virus), *Arenaviridae* (Lassa, Junin, Machupo, Sabia viruses, and *Filoviridae* (Ebola virus, Marburg virus).

Mononucleosis syndrome

Acute infection with EBV is the prototypical mononucleosis syndrome and presents typically as a flu-like syndrome with fever, rash, pharyngitis, fatigue, and swollen lymph nodes. Other symptoms and signs include myalgias, arthralgias, diarrhea, nausea, vomiting, headache, enlarged liver and spleen, weight loss, and neurologic symptoms. The duration of the acute presentation can be as long as 3 weeks, with a convalescent phase associated with fatigue and malaise that may persist for months. Examples of viruses that commonly cause a mononucleosis syndrome are EBV, CMV, and HIV.

Viremia and clinical illness day

The clinical illness day predicts the virus burden in a biospecimen, which is important for the laboratorian to know when assessing risk in the laboratory. Peak viremia can correspond to the height of clinical illness in some diseases, but it may precede also more significant clinical manifestations (such as with CCHF or severe dengue).

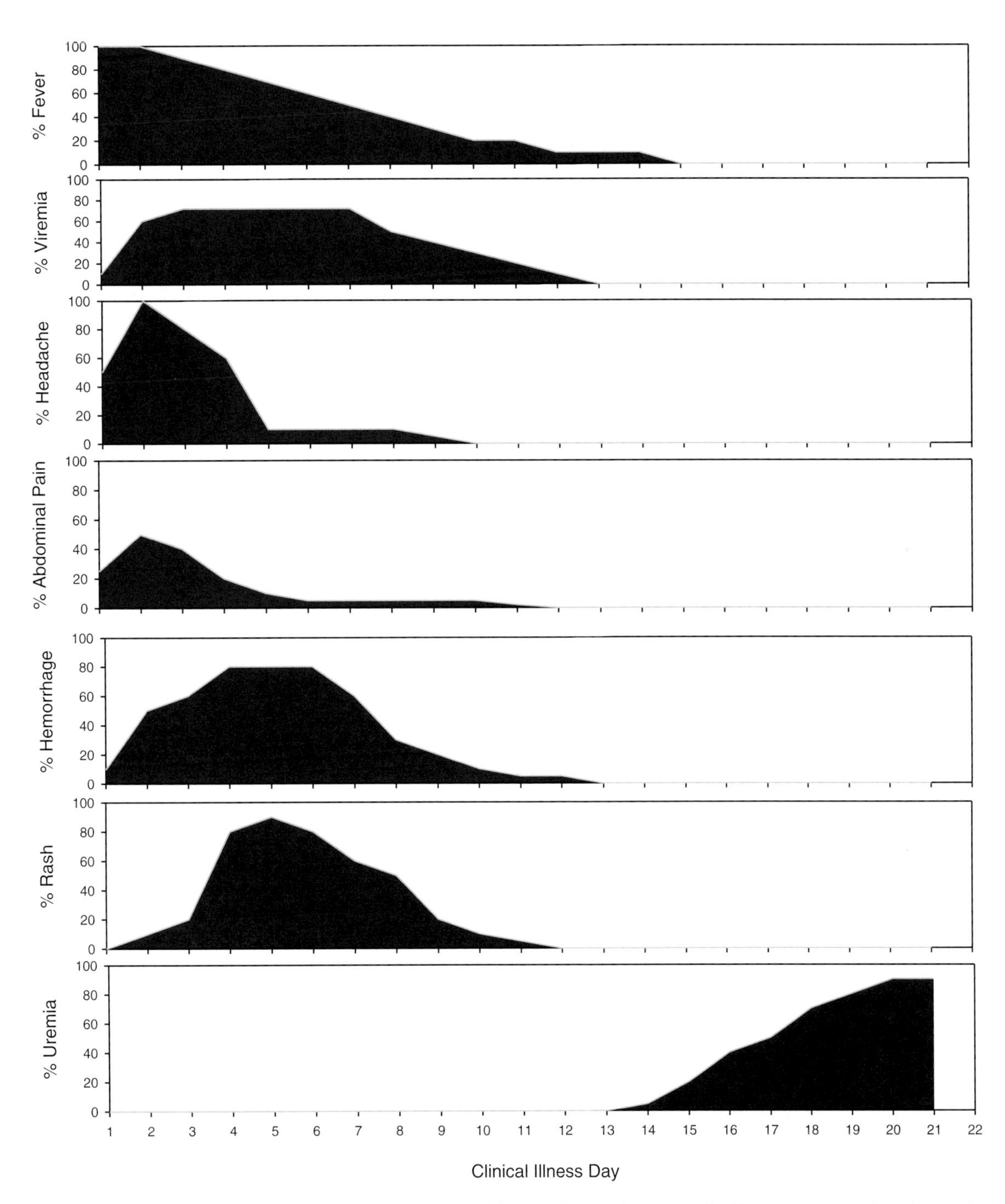

Figure 2: Chart representing the course of differing aspects of CCHFV infection in human. Incidences are noted on the *y* axis and the clinical day is noted on the *x* axis. (Adapted with permission from reference 38.)

This results in a potential exposure to health care providers and laboratorians working from specimens obtained during the early clinical period and before a diagnosis was made. Peak viremia represents the highest-risk period for a laboratory exposure from a clinical specimen. Figure 2 represents the presence of symptoms and signs and viremia by clinical illness day in patients with CCHF (38). This figure demonstrates that viremia in CCHF occurs very early in the infection, reaching its peak on days 2 to 8. The associated symptoms occurring during peak viremia are protean and consist of headache and abdominal pain. The presentation of rash and hemorrhage does not occur until days 4 to 7 of clinical illness. More severe disease, such as renal failure and uremia, in contrast, occurs during the latter part of the clinical course after viremia has diminished.

POSTACCIDENT MANAGEMENT

Postaccident management of a viral exposure should be part of a carefully planned contingency that is specific for each laboratory. Despite engineering controls, training, and safety programs, laboratory workers can still become infected. Prior to working in the virology laboratory, individual workers should be medically assessed for their ability to perform the laboratory tasks and their susceptibility to occupational viral infections. For example, immunocompromised individuals are at greater risk than immunocompetent laboratory workers. An exposed laboratory worker living with or caring for an immunocompromised person or young children puts those individuals at risk. Pregnant laboratory workers may put at risk a successful outcome of their pregnancy by working with particular viruses. Establishment of the preexposure health status of the individual enhances the medical response to the exposures and may identify a need for additional precautions. For example, a person for whom vaccination against orthopoxviruses is contraindicated may work with the agent with supplemental respiratory protection. Another example would be an individual with manual dexterity problems due to arthritis of the hands who would be unable to perform inoculations safely. To ameliorate that risk, modified duty may be prescribed to preclude the use of needles or other sharps. As important as it is for the laboratory worker to know the viruses being used and the corresponding symptoms, it is imperative that the medical caregivers be told of the agents to which these workers could be exposed. For example, a person performing orthopoxvirus experiments with non-human primates would be at risk for herpes B virus infection as well as orthopoxvirus infection. Many non-human primates are infected with herpes B virus, which is benign in non-human primates but highly lethal in humans. Guidelines for handling non-human primate wounds are available, and arrangements should be made in advance (39). A stock of acyclovir and Dryvax may be recommended as part of a postexposure contingency (39).

The recognition of an exposure may occur either at the time of the incident (i.e., needlestick) or weeks later when symptoms of the virus become apparent (Sabia virus infection described above). Since viral exposures may occur through unconventional means, the development of symptoms may be the first indication of infection. Laboratory workers must be familiar with the disease symptoms even though many viral diseases manifest themselves in ways similar to common colds and influenza virus infections. The postexposure procedure should be outlined in advance and familiar to all involved. The primary concern should be for the health and welfare of the infected worker and the community.

SELECTED VIRAL PATHOGENS PRODUCING LAIS

Laboratory-acquired viral infections are not only harmful to the individual worker, but due to the possibility of secondary transmissions, represent a potential threat to the community and to global public health. The LAIs in 2003 of SARS-CoV threatened to restart the epidemic, even though there were no naturally occurring infections of humans at that time (40–42). Human and mechanical failures resulting in LAIs are an inherent risk of virology diagnostics and research. Viral laboratory exposures occur through mechanical inoculations (i.e., needlesticks, cuts, or bites), aerosols (including splashes), and fomite contamination from direct or indirect contact. These events often happen through human failure to follow recommended precautions. Many LAIs are subclinical or are commonly attributed to community-acquired infections (see discussion of Sabia virus under "*Arenaviridae*," below). Both human error and mechanical failure can be controlled through robust safety and training programs, routine servicing of instrumentation, and the insistence of the principal investigator that safety be an aspect of successful experimentation. Although a virus may not normally infect an individual by certain routes, the virology laboratory may present a novel risk for infection by nontraditional means. The following section describes examples of LAIs representing nontraditional routes of exposure that would only occur in the laboratory or health care setting. The laboratory worker must be aware of the mechanisms for viral exposures for each experimental task. Below are selected viral pathogens that have been known to cause LAIs, as summarized in Table 1.

Adenoviridae

General

The *Adenoviridae* are DNA-containing viruses and classified into two genera, *Mastadenovirus* and *Aviadenovirus*. Adenoviruses within the *Aviadenovirus* genus are limited to viruses of birds; the *Mastadenovirus* genus contains human, simian, murine, porcine, ovine, equine, bovine, canine, and opossum viruses. Among the human adenoviruses, there are 49 serotypes that produce a variety of clinical illnesses. Adenoviruses are transmitted to susceptible hosts by direct contact with infectious oral, nasopharyngeal, or conjunctival secretions. Depending on the serotype, a variety of clinical illnesses can occur, including acute pharyngitis, respiratory disease, pneumonia, keratoconjunctivitis, hemorrhagic cystitis, gastroenteritis, meningoencephalitis, hepatitis, and myocarditis.

LAI

From previous surveys of LAIs due to the *Adenoviridae*, one LAI has been associated with the genus *Aviadenovirus*, the fowl plague virus, and 10 LAIs have been associated with human adenovirus, genus *Mastadenovirus*. In 2011, a novel adenovoirus, titi monkey adenovirus (TMAdV), was found to be responsible for infecting 65 titi monkeys at the California National Primate Research Center (CNPRC) (43). The researcher in closest contact with the animals, as well as two close family members with no contact with CNPRC, developed an acute respiratory illness. Convalescent serum samples from the researcher and a family member were seropositive samples for TMAdV, suggesting a zoonotic infection and the potential for human-to-human transmission of the virus. It is likely that more human infections occur but are either not reported or are attributed to community-acquired adenoviral infection.

Postaccident management

The approach to postaccident management of potential laboratory exposures to the *Adenoviridae* is similar to those already discussed and should include a thorough physical exam, close clinical follow-up, and diagnostic assays to establish the exposure. An effective vaccine to serotypes 4 and 7 is licensed in the United States and is currently used by the U.S. military to vaccinate high-risk individuals such as basic training participants.

Arenaviridae

General

The *Arenaviridae* are enveloped viruses containing two single-stranded RNA segments designated small and large. The *Arenaviridae* family contains a single genus, *Arenavirus*, consisting of 18 species; there are two antigenic groups within the genus, the Old World group and the New World (Tacaribe complex) group. The arenaviruses cause a chronic infection in rodents that are indigenous to Europe, Africa, and the Americas. Humans become incidentally infected when in contact with infected animals in their natural habitat or elsewhere. Human infections with the arenaviruses are characterized by the gradual onset of fever and muscle pain. A viral hemorrhagic fever may result, with severe constitutional symptoms and bleeding that result from coagulopathy. CNS infection with meningoencephalitis may also occur. LCMV is the prototype virus of the family *Arenaviridae* (44). Other members of the family that cause important zoonoses include Lassa fever virus, as well as Argentine (Junin), Bolivian (Machupo), Venezuelan (Guanarito), and Brazilian (Sabia) hemorrhagic fever viruses and Lujo virus.

LAI

There are reports of human LCMV infection associated with laboratory animal and pet contact, particularly with mice and hamsters (44–48). LCMV is widely distributed among the wild mouse population throughout most of the world, presenting a zoonotic hazard (49–51). Mice, hamsters, guinea pigs, non-human primates, swine, and dogs are among the laboratory animal hosts that sustain natural infections, with reports of infectious hepatitis in marmosets and tamarins in zoological parks in both the United States and England (52–56). Rodent (mouse) infestations of these zoos, and/or the supplementation of the diets of tamarins and marmosets with suckling mice, are potential sources for LCMV infection and infection of laboratory workers (46, 57, 58). Immunodeficient mouse strains may pose a special risk for harboring silent, chronic infections (46). Bedding material and other fomites contaminated by LCMV-infected animals can be an important source of LAI (46, 59). The experimental passage of tumors and cell lines contaminated with LCMV has been documented and is a potential threat for the introduction of LCMV into animal facilities (46, 60, 61). Infection in humans may occur by parenteral inoculation, inhalation, or contamination of mucous membranes or broken skin with infectious tissues or fluids from infected animals. Airborne transmission is well documented and plays an important role in human infections (62, 63). Humans usually develop a flu-like illness following an incubation period of 1 to 3 weeks. Some patients can develop more serious manifestations of the disease, including maculopapular rash, lymphadenopathy, meningoencephalitis, and, rarely, orchitis, arthritis, and epicarditis (64). CNS involvement has resulted in death in several cases. Infections during pregnancy pose a risk of infection for the human fetus (65).

Another example of the hazards of the arenaviruses in the laboratory is the case of a laboratory-acquired Sabia virus infection of an investigator at the Yale University Arbovirus Laboratory in 1994 (66, 67). In this case, the worker, wearing a surgical mask, gown, and gloves, was clarifying a preparation of Sabia virus using high-speed centrifugation in a BSL3 laboratory. On completion of the run, he opened the centrifuge to discover that one of the centrifuge bottles was wet and that infectious cell culture fluid had gathered in the bucket of the rotor. He proceeded to clean the spill with sodium hypochlorite before continuing with the planned experiments. On day 8 postexposure, he experienced myalgias, headache, stiff neck, and fever. He treated himself with ibuprofen for 48 h before seeking medical attention. Initially the case presented as a relapse of *Plasmodium vivax*. He reported no laboratory accidents. A negative smear ruled out malaria. Further interviews with the patient revealed the laboratory spill 10 days prior. Following immediate hospitalization, he was treated with intravenous ribavirin. Fortunately, the worker survived despite being symptomatic for Sabia virus infection. The Sabia virus infection was confirmed by reverse transcription-PCR testing. The contribution of ribavirin is unclear, and there are to date no efficacy studies to support the use of ribavirin for Sabia virus infection. Due to the potential for infection by the aerosol route and the high mortality rates associated with infection, all arenaviruses are now classified as BSL4 pathogens.

Postaccident management

The exposed worker should seek immediate medical consultation. Intravenous ribavirin therapy may reduce mortality in patients infected with Lassa fever virus and may be of some benefit in patients with severe LCMV infections (68). However, use of ribavirin should only occur in consultation with an experienced infectious disease clinician, as it has been shown to have numerous off-target effects and substantial toxicity. Favipiravir is a promising RNA antiviral shown to disrupt early stages of viral replication to inhibit arenaviruses (69). However, these antiviral effects against LCMV have only been tested in cell culture to date.

Bunyaviridae

General

The *Bunyaviridae* are single-stranded RNA viruses that contain three genome segments designated large, medium, and small. This family encompasses a large group of arthropod-borne viruses sharing similar structural and antigenic properties. The *Bunyaviridae* family contains four genera of viruses that infect animals—*Bunyavirus, Hantavirus, Nairovirus,* and *Phlebovirus*—and one genus of viruses that infect plants—*Tospovirus*. Two significant zoonotic bunyaviruses are CCHFV and hantaviruses. Work with CCHFV in the United States is restricted to BSL4 facilities. However, the virus is endemic in Central Asia and parts of Africa and can be encountered in clinical or lower-containment research laboratories. Typically, CCHFV is transmitted by the bite of an infected tick or fluid exposure from infected animals (70). However, nosocomial infections are common. Furthermore, laboratory workers have been infected by suspected aerosol, fluid, and needlestick exposures. The infected individual presents with a high fever, chills, headache, dizziness, and back and abdominal pain. More severe cases may have hemorrhagic manifestations such as petechiae and ecchymosis. Mortality ranges from 20 to 80%. Barrier nursing practices are sufficient to prevent spread in most situations (71). However, the potential for aerosol exposure may require additional precautions.

The hantaviruses are widely distributed in nature among wild rodent reservoirs, but unlike other members of the family, they are usually not transmitted by insect vectors (72). New World hantaviruses typically cause hantavirus pulmonary syndrome (HPS), while hantaviruses from other regions are associated with HFRS. An outbreak of hantavirus infection resulting in numerous deaths in adults was first recognized in the United States (72–74). Since this initial outbreak, cases of HPS have been reported from 30 states, and about three-quarters of the patients have been from rural areas. Rodents from the genera *Apodemus, Clethrionomys, Mus, Rattus, Pitymys,* and *Microtus* have been implicated in foreign outbreaks of the disease. In the United States, serological surveys have detected evidence of hantavirus infection in urban and rural areas involving *Rattus norvegicus, Peromyscus maniculatus, Peromyscus leucopus, Microtus pennsylvanicus, Tamias* spp., *Sigmodon hispidus, Reithrodontomys megalotis, Oryzomys palustris,* and *Neotoma* (72, 75, 76). Numerous cases of hantavirus infection have occurred among laboratory animal facility personnel from exposure to infected rats (*Rattus*), including outbreaks in Korea, Japan, Belgium, France, and England (77). No cases of hantavirus infection associated with laboratory animals have been reported in the United States. In the United States, HPS in humans has been associated with outdoor activities and occupations that place them into close proximity with wild infected rodents and their excrement (72, 78, 79). Several cases have involved individuals from academic institutions involved in field studies. Epidemiological data indicate that cats may become infected through rodent contact and serve as a potential reservoir (80).

Hantavirus infection is transmitted through the inhalation of infectious aerosols, and extremely brief exposure times (5 min) have resulted in human infection. Rodents shed the virus copiously in respiratory secretions, saliva, urine, and feces for months (75). Transmission of the infection may result from an animal bite or when dried materials contaminated with rodent excreta are disturbed, allowing wound contamination, conjunctival exposure, or ingestion to occur. Recent cases have occurred in the laboratory animal facility environment involving infected laboratory rats (81). Person-to-person transmission apparently is a very rare feature of hantavirus infection (72). The clinical signs vary in severity and expression according to the hantavirus involved. In the recent cases of HPS in the United States, patients had fever, thrombocytopenia, and leukocytosis similar to HFRS. Patients progress rapidly to respiratory failure due to capillary leakage into the lungs, followed by shock and cardiac complications. The form of the disease that has been noted following laboratory animal exposure fits the classical pattern for HFRS characterized by fever, headache, myalgia, petechiae, and other hemorrhagic manifestations, including anemia and gastrointestinal bleeding, oliguria, hematuria, severe electrolyte abnormalities, and shock (82). Animal BSL4 (ABSL4) guidelines are recommended for animal studies involving hantavirus infections in permissive hosts such as *P. maniculatus* and in wild-caught rodents brought into the laboratory that are susceptible to hantaviruses producing HPS or HFRS (83). ABSL2 practices are sufficient for handling rodent strains known not to excrete the virus.

LAI

For both hantavirus and CCHFV, the laboratory worker is at risk for infection by fluid contact, aerosol exposure, and needlesticks. When hantavirus is worked with at BSL3, the laboratorian must be aware of procedures that may generate aerosols. Currently there are no approved vaccines or chemotherapeutics for either CCHFV or hantavirus. The CCHFV laboratory worker is less at risk for aerosol and fluid exposure because of the engineering solutions of BSL4 laboratories but not for the risk of needlestick exposures. The hantavirus laboratory worker should look to prevent aerosol exposures by using positive-pressure breathing apparatus and biocontainment cabinets.

Postaccident management

There are no FDA-approved treatments for bunyaviruses. Anecdotal reports suggest that ribavirin may be effective in some cases, and a single clinical trial demonstrated probable efficacy against Hantaan virus (causative agent of Korean hemorrhagic fever) (84). A recent meta-analysis confirmed conventional wisdom that HPS-causing virus infections do not appear to benefit from ribavirin (85). While observational studies suggest that ribavirin may improve outcomes with CCHF (86, 87), a randomized clinical trial investigating its intravenous use in treatment of CCHFV has yet to be done.

In the event of occupational infection, CCHFV is capable of human-to-human transmission, requiring barrier nursing practices. Furthermore, individuals in close personal contact with the infected individual should also be monitored and quarantined from the general public. The incubation time is approximately 1 to 5 days depending on the mode of transmission. The CDC and the State Health Department should be advised of any potential CCHFV exposure. While most hantaviruses do not spread from human to human, there is growing evidence for nosocomial and household transmission of Andes hantavirus (88).

Coronaviridae

General

Coronaviruses are in the order *Nidovirales,* family *Coronaviridae,* and are single-stranded, positive-sense RNA viruses. The *Coronaviridae* contain many important animal pathogens, including feline coronavirus and porcine epidemic diarrhea virus. Human pathogens include the human coronaviruses SARS-CoV and MERS-CoV. The human coronaviruses cause an upper respiratory infection of adults and children and are similar to the rhinoviruses in producing the common cold. SARS-CoV is a recently recognized emerging viral agent that caused significant mortality and morbidity from November 2002 to June 2003. The virus quickly spread throughout the globe, mediated by airline travel. Asia and Canada were particularly affected. During the epidemic, no vaccines or chemotherapeutic formulations were available for prophylaxis or treatment. The crisis spurred a vigorous response by the research community to develop effective medical countermeasures against SARS. The laboratory isolates were distributed by a number of agencies to virology laboratories throughout the world. SARS-CoV was expected to reemerge the following November, but despite predictions, has not yet naturally reemerged.

Instead, a different SARS-like infection emerged in Saudi Arabia in 2012, caused by the genetically distinct MERS-CoV (89). MERS presents similarly to SARS; patients can be asymptomatic or can develop fever, cough, shortness of breath, and the potential for pneumonia. The fatality rate of MERS-CoV appears to be much higher than with SARS-CoV infection, with 36% of the reported MERS patients succumbing to the illness (90). The route of transmission is not well understood, but camels and nosocomial exposures are likely to be the major sources for MERS-CoV infection of humans. While the potential for human-to-human transmission still appears to be relatively low, there are concerns that it may mutate to a strain that does permit human transmission (for more discussion, see "Genetically Engineered Viruses," below). MERS-CoV appears to be circulating throughout the Arabian Peninsula, although 27 countries to date have reported cases to the WHO, with 1,733 laboratory-confirmed infections since 2012. A total of 628 deaths have been associated with MERS-CoV in that time (89).

LAI

A number of LAIs from SARS-CoV threatened to reignite the epidemic observed in 2003 (41, 42, 91, 92). These accidents occurred mostly through human error and resulted from a large cadre of inexperienced laboratory workers, working with an environmentally stable virus capable of aerosol and fomite transmission. Global health was potentially threatened by a failure to recognize that expedited increases in worldwide BSL3 operations occurred without having the appropriate controls in place. Several members of the community were infected and developed disease. These exposures and the resulting consequences emphasize that careful training in biosafety, experience, and active supervision by principal investigators are integral to safe viral laboratory operations and global health. There have been no reports of MERS-CoV LAIs to date, although health care workers are infected at high rates (~14%).

Postaccident management

SARS-CoV and MERS-CoV LAIs have the potential to become an epidemic despite the lack of naturally occurring infections. Laboratory workers suspected of an LAI or having symptoms consistent with SARS/MERS disease should immediately report the incident to the institutional biosafety officer and follow postexposure protocols. Infected workers should quarantine themselves to reduce the likelihood of introduction of the virus into the general population. Currently, there are no FDA-approved vaccines and/or antiviral products for the treatment of SARS or MERS. Interferon alfacon-1 was evaluated in SARS virus-infected patients in a pilot experiment. Other type 1 interferons have been evaluated in experimentally infected non-human primates and tissue culture models of SARS-CoV infection. These studies have suggested efficacy, but to date there are no studies demonstrating efficacy in humans. The first clinical trial of a vaccine for MERS-CoV, GLS-3000, a DNA plasmid vaccine, began phase I enrollment in January 2016 (registered at ClinicalTrials.gov under NCT02670187). The current recommendations focus on supportive care and containment of infected patients.

Filoviridae

General

Filoviruses are negative-stranded RNA viruses that are classified in the order *Mononegavirales,* family *Filoviridae,* and are divided into three genera—*Ebolavirus, Marburgvirus,* and the recently discovered *Cuevavirus* (93). Only one species of *Marburgvirus* exists but with two subtypes, Marburg virus (MARV) and Ravn virus (RAVV). By contrast, *Ebolavirus* is divided into five separate species: Ebola virus (EBOV, *Zaire ebolavirus*), Sudan virus (SUDV, *Sudan ebolavirus*), Bundibugyo virus (BDBV, *Budibugyo ebolavirus*), Tai Forest virus (TAFV, formerly *Cote d'Ivoire ebolavirus*), and Reston virus (RESTV, *Reston ebolavirus*). *Cuevavirus* contains one single virus, Lloviu virus (LLOV), which was discovered in 2002 during the investigation of massive bat die-offs in Cueva del Lloviu, Spain, in 2002 (94).

Strikingly, RESTV and LLOV have not been shown to cause symptomatic human disease, and TAFV has been documented in only a single nonlethal human infection (95). However, the remaining filoviruses can cause human disease that is among the most severe of all hemorrhagic fevers. In general, filovirus infection presents with abrupt onset of fever, chills, malaise, myalgia, and gastrointestinal symptoms. Hemorrhage is manifested by a coagulopathy as petechiae, bruising, mucosal hemorrhages, and internal bleeding (96). During 2013–2016, West Africa experienced the largest filovirus outbreak in history with over 15,000 confirmed cases of Ebola virus disease (EVD) and over 11,000 confirmed deaths (97). In this outbreak of EVD caused by EBOV, fever, headache, malaise, weakness, and gastrointestinal symptoms predominated, often with voluminous diarrhea reminiscent of cholera (98, 99). Depending on the specific species, filovirus infection has been associated with fatality in 20% to over 80% of cases.

LAI

MARV was first described and isolated during an outbreak of hemorrhagic fever in laboratory workers that occurred in Marburg and Frankfurt, Germany, and Belgrade, Yugoslavia (100). The source of the infection in all three cities was African green vervet monkeys (*Cercopithecus aethiops*) captured in the Lake Kyoga area in central Uganda. A total of 31 patients developed clinical illness, with seven deaths. Most of these individuals had handled fresh tissues or primary cell cultures from the monkeys. Secondary infection occurred in six persons who had contact with those patients who had originally been exposed through blood, tissue, or cell cultures.

EVD was first discovered in an outbreak in Zaire (currently Democratic Republic of the Congo) in 1976 (101). Human cases of EVD have been confined to or originated from the continent of Africa, and primary cases are often traced to forest and bush animal exposure, with at least one case resulting from contact with infected chimpanzees (95, 102).

The vast majority of naturally acquired EVD results from direct contact with blood and body fluids of infected patients (103). Although aerosol transmission does not appear to be an important epidemiological phenomenon in natural outbreaks, it may occur and is certainly a more significant concern for LAI (104). In the initial RESTV outbreak in a quarantine facility in Reston, Virginia, cynomolgus monkeys imported from the Philippines appeared to infect several animal caretakers by an unknown route, although no clinical illness developed (105). Risk factors for transmission were examined during an epidemic of EVD (SUDV) in Uganda during 2000 (106). The magnitude of the outbreak was largely determined by secondary transmission, with three generations from three primary index cases documented during this outbreak (106). The source of infection in the primary cases could not be determined. Secondary contact occurred either within the family or in the hospital; contact with patient's body fluids was the strongest risk factor for transmission. The outbreak of EBOV in 2014 in West Africa was the largest and most complex EVD epidemic to date and provided even more insight into possible modes of transmission and clinical features of EVD. Virus has now been detected in body fluids and immune-privileged reservoirs, like breast milk, semen, and intraocular fluid, for weeks or months after infection (107). Unconventional transmission, including sexual transmission from Ebola survivors and a mother who may have transmitted Ebola to her baby via breastfeeding, is thought to have been responsible for some of the cases at the tail end of the epidemic. In one study, EBOV was still detectable in the semen of survivors over 270 days after infection (108). Recently, a case of EVD relapse in the CNS was documented 9 months after initial presentation (109). Studies on these Ebola survivors are ongoing, as many suffer from prolonged mental health issues and issues with pregnancy, arthritis, and arthralgia, which may be attributed to long-term effects of the infection.

LAIs of Ebola are fortunately quite rare, and only two open published cases of research LAIs exist (110, 111). Additional high-risk exposures, i.e., needlestick injuries during animal experiments, have occurred without documented infection (112, 113). It is possible that laboratory exposures in the West Africa epidemic resulted in occupational infection; however, the nature of operations and extent of risks at many of these sites were sufficiently complicated as to preclude definitive conclusions (114).

Laboratory research with filoviruses should be restricted to BSL4, which is designed to engineer out possible exposures by aerosol and fluid contacts (see chapter 27). However, the ever-present danger of laboratory sharps and animal bites cannot be fully controlled, even in BSL4 laboratory operations—both needlestick injuries described previously were sustained under BSL4 conditions (112, 113). The addition of the personal protective suit prevents accidental aerosol and fluid exposures but does not provide protection from sharps. For example, a laboratory worker at the Vektor Laboratory in the Russian Federation became infected via a needlestick injury while working with virus-infected guinea pigs (111). The worker was apparently treated aggressively by local physicians but did not survive. All biosamples containing potentially infectious virus must be handled with extreme care. Biosamples may be handled under reduced containment if the virus is inactivated. For example, a Western blot does not have to be done under BSL4 conditions. Treatment with chemicals, heat, enzymes, or gamma irradiation is needed to inactivate the virus. Recently, researchers published a series of inactivation procedures for EBOV, demonstrating numerous ways to kill the virus in biologic specimens (115). Although the authors indicate these chemical inactivation methods may be effective against other negative-sense RNA viruses, it is still important for the investigator to validate inactivation by safety testing through attempts to culture virus from the inactivated samples. Samples are passaged multiple times through sensitive cell lines or animals and then tested for viral nucleic acids, or by plaque assay or immunofluorescence assay.

Postaccident management

Medical consultation with a subject matter expert should be sought immediately after a suspected exposure. The virus is transmitted by direct contact with contaminated fluids to mucous membranes, nonintact skin, or through percutaneous introduction and, at least in a laboratory setting, potentially through aerosolized infected fluids. Currently, no studies have examined the efficacy of emergency decontamination procedures for filovirus-contaminated wounds or mucous membranes. Standard emergency decontamination procedures developed for potential HIV exposures are a prudent guide.

While several experimental products hold promise, no licensed products for postexposure prophylaxis or therapy are currently available. During the West Africa outbreak, multiple experimental products were given to a number of symptomatically infected health care workers repatriated to Europe or the United States (116). The most commonly administered were (untitered) convalescent plasma, zMapp, favipiravir, brincidofovir, and TKM-Ebola (TKM-100802), but products were generally given outside of structured research protocols, so no conclusions regarding effectiveness were possible. In addition, at least 10 health care workers judged to have significant exposures were given postexposure prophylaxis with favipiravir, rVSV-ZEBOV, or TKM-100802 upon evacuation to the United States or United Kingdom (117, 118). Although none of these treated workers developed clinical EVD, the low numbers were insufficient to assess statistical relevance of these data.

The most compelling data for postexposure prophylaxis come from a cluster-randomized ring-vaccination trial of a recombinant VSV vector expressing EBOV in Guinea (119). Between April and July of 2015, 2,014 contacts in 48 clusters received immediate vaccination with rVSV-ZEBOV, while 2,380 contacts in randomly assigned 42 clusters received vaccine 21 days later. The immediate group experienced no cases of symptomatic EVD beyond 10 days postrandomization, while the delayed group experienced 16 cases in the same timeframe. This contact vaccination exhibited 100% efficacy for primary outcome in this preliminary analysis; however, one should note that a total of 13 vaccinated individuals developed EVD within 10 days after vaccination. Nevertheless, this was the first vaccine to demonstrate human efficacy against filovirus infection, providing hope that effective pre- and postexposure prophylaxis may eventually become available.

The rapidly evolving array of experimental therapeutics and vaccines for filoviruses will likely render any current recommendation for postexposure prophylaxis obsolete in the near future. In addition to the products mentioned above, recombinant adenovirus-vectored vaccines and MVA-vectored vaccines for EBOV are undergoing clinical trials, as is the nucleotide analogue GS-5734 that was administered to a single patient on a compassionate use provision during the 2014 outbreak (117). Decisions regarding postexposure prophylaxis or treatment for an LAI should be discussed extensively with the CDC, FDA, and subject matter experts in the care of patients with EVD.

If clinical EVD does result from exposure, the incubation time ranges 2 to 21 days. During this time, the individual should be closely monitored for the onset of flu-like symptoms. The appropriate use and level of quarantine during this period is a subject of intense debate. However, patients must be isolated if they develop symptoms consistent with EVD. As of 2015, the CDC had designated 55 Ebola treatment centers in the United States; however, only five U.S. hospitals have managed cases of EVD, and only Emory University Hospital, University of Nebraska Medical Center, and the NIH Clinical Center managed more than one case. Vigorous supportive care, including fluid and electrolyte management, are the mainstays of recommended treatment options (116). This aggressive care may result in significant survival benefit, and, if possible, patients should be referred to hospitals with clinical experience in management of EVD.

Flaviviridae

General

The *Flaviviridae* are positive-stranded RNA viruses, contain many arthropod-borne human pathogens, and comprise three genera: *Flavivirus*, *Pestivirus*, and *Hepacivirus*. Members of the *Flavivirus* genus are grouped by vector: tick, mosquito, and unknown. The tick-borne viruses include KFD, Langat, Omsk hemorrhagic fever, Powassan, Karshi, TBE, and louping-ill viruses. The mosquito-borne flaviviruses include the Aroa virus group (Aroa, Bussuquara, Iguape, and Naranjal viruses), the dengue group (dengue virus serotypes 1 to 4 and Kedougou virus), the Japanese encephalitis group (Cacipacore, Koutango, Japanese encephalitis, Murray Valley, St. Louis, West Nile, and Yaounde viruses), the Kokobera group (Kokobera virus), the Ntaya group (Bagaza, Ilhéus, and Tembusu viruses), the Spondweni group (Zika virus), and the YF group (Banzi, Sepik, Uganda S, Wesselsbron, and YF viruses). Other important flaviviruses without a known vector include Entebbe bat and Rio Bravo viruses. The genus *Pestivirus* contains many important animal pathogens and includes bovine viral diarrhea virus, classical swine fever virus (also known as hog cholera virus), and border disease virus of sheep. The genus *Hepacivirus* contains one virus, HCV.

The arthropod-borne flaviviruses produce a range of clinical illnesses from a mild febrile illness to a severe hemorrhagic fever or encephalitis. They do not appear to be a risk for human-to-human transmission, with the notable exception of the sexual transmission of Zika virus (120). Hepatitis C is transmitted by contact with infected blood through a needlestick injury, the use of contaminated injection equipment or poor parenteral practices, blood transfusion, sexual activity, or during childbirth. HCV infection results in an acute and chronic hepatitis. In addition to the typical routes of arthropod-borne transmission, the Zika virus has demonstrated sexual transmission and potentially also transfusion and perinatal transmission (121). While most infected individuals present with a mild and self-limiting febrile disease, associations with the much more serious Guillain-Barré syndrome and with affected pregnant women later giving birth to babies with neurological defects have caused international concern (122). The amount of laboratory research on Zika virus has increased dramatically in an attempt to determine its cause of action as well as potential countermeasures, therefore increasing the likelihood of LAIs.

LAI

The *Flaviviridae* represent a family of viruses that have been a major cause of LAIs, with 464 documented cases. Among the flaviviruses, KFD and YF viruses were responsible for the majority of infections, followed by louping-ill, TBE, and West Nile viruses. With the exception of HCV and dengue virus, the flaviviruses in general have demonstrated the ability to infect through aerosol transmission and thus are a potential hazard in the laboratory environment. HCV is a blood-borne pathogen that is an occupational hazard among health care workers who sustain a percutaneous exposure from infected patients. Infection from aerosol or contaminated body fluids or stools from exposures other than percutaneous ones has not been established; however, there have been two case reports of infection following a blood splash to the conjunctiva (123). The incidence of HCV infection following a needlestick injury ranges from 1.8 to 10% (124, 125).

Mosquito sampling of insects infected with Zika led to LAIs in two American scientists while working in

Senegal in August 2008 (126). Both men became ill upon returning home to Colorado, presenting with fatigue, headaches, swelling and arthralgia in joints, and in the case of one patient, hematospermia. This patient's wife later developed similar clinical symptoms, implicating a sexually transmitted infection. Needlestick injury has also been linked to Zika transmission in a female researcher in Pennsylvania in June 2016 (127). Until the link between the Zika virus infection and fetal malformations is better understood, pregnancy should be considered a significant factor in the risk assessment matrix for researchers working with Zika virus. Workers who are pregnant or have partners who are pregnant should minimize their involvement in studies with the virus as much as possible.

Postaccident management

Postaccident management of potential laboratory exposures to or LAIs with the *Flaviviridae* should include a thorough physical exam, close clinical follow-up, and diagnostic assays to establish the exposure. Human-to-human infection does not occur among the majority of the *Flaviviridae*. Hepatitis C and potentially Zika are transmitted from person to person sexually, to a newborn infant during delivery, and through contaminated blood products or needles. Therefore, contact tracing from an infected laboratorian is required for HCV or Zika LAIs. Laboratory workers who are pregnant or have pregnant partners should immediately inform their physician in the event of a Zika LAI. At present, the only vaccines available for human use are YF, Japanese encephalitis, tick-borne encephalitis, and since 2015, the dengue vaccine; workers potentially exposed to these viruses should be vaccinated (254). At least 18 agencies around the world are currently attempting to develop a Zika vaccine; a phase I clinical study was approved in June 2016 by the FDA for GLS-5700, a synthetic DNA plasmid vaccine developed by GeneOne Life Science Inc., in collaboration with Inovio Pharmaceuticals (128, 129). HCV exposure should result in consultation with infectious disease or hepatology specialists with experience managing such infections, as therapeutic options have greatly expanded with the recent advent of effective anti-HCV nucleotide analogues (130).

Hepadnaviridae

General

HBV is the prototype virus for a family of DNA viruses called the *Hepadnaviridae*. HBV is an enveloped DNA virus that causes acute and chronic hepatitis, cirrhosis, and hepatocellular carcinoma (131, 132). The virus is transmitted by percutaneous exposure to contaminated serum, blood, or body fluids. The primary infection ranges from mild or subclinical hepatitis to severe acute hepatitis. The virus can establish a persistent infection without liver disease. Chronic hepatitis is a more severe form of the infection that can lead to macronodular cirrhosis. Both chronic and acute HBV infections present with hepatocellular necrosis, inflammatory response, lymphocytic infiltration, and liver cell regeneration. Liver cirrhosis is described as having regenerative nodules and diffuse fibrosis. Infection of the young is generally less severe than infection of older adults. There are anecdotal reports that infection is less severe in immunologically impaired adults, suggesting a possible immunological role in disease progression. The virus itself is sufficient to cause disease; HBV also serves as a helper virus for HDV. HDV is a defective RNA virus that cannot replicate without the helper functions of HBV,. which provides the virion envelope polypeptide.

LAI

Those who handle HBV are at risk for exposures through self-inoculations and contact with contaminated fluids (133). The persistent danger of sharps pervades operations at BSL2 to BSL4, and to date there are limited effective solutions to preventing these types of injuries and exposures. Fluid exposures to HBV-infected biosamples can be reduced by proper use of barriers, shields, and biosafety cabinets and the routine decontamination of the laboratory surfaces and waste materials. Although there are no reported cases of aerosol exposure resulting in a confirmed human infection, precautions should be used to prevent such exposures. Before allowing an employee to work with the virus, the employer is required by federal regulations to offer the hepatitis B vaccine, which the worker is advised to take. Although the vaccine is not a primary barrier for exposure prevention, it may help to lessen the impact of an accidental exposure.

Postaccident management

Following a suspected laboratory exposure, the worker should immediately seek predetermined medical assistance (134). There are effective vaccines, antiviral drugs, and hyperimmune serum for the treatment of HBV infection (135, 136). Recent advances in antiviral therapeutics for HBV have revolutionized the clinical management of chronic HBV infection (137). Potential interventions for postexposure prophylaxis or management of chronic infection should be made in consultation with a hepatologist or infectious disease specialist with experience managing HBV infection. The worker should be closely monitored for signs of viral hepatitis and refrain from donating blood. Family members are not at risk of acquiring the infection unless exposed to body fluids. The availability of medical countermeasures for HBV may lessen the consequences of an LAI.

Herpesviridae

General

Herpes simplex viruses (HSV) (family *Herpesviridae*, genus *Simplexvirus*) are members of a family of viruses with genomes that consist of a single, large, double-stranded DNA molecule (138). The virus is composed of an electrondense core containing DNA, an icosadeltahedral capsid, a layer of proteins named tegument that surrounds the capsid, and an outer envelope. Natural HSV infection of humans is characterized by a localized primary lesion, latency, and a tendency to localized recurrence. The distinct clinical syndromes that are characteristic of HSV types 1 and 2 depend upon the portal of entry (139). Transmission is dependent on oral or genital contact of susceptible individuals with someone actively excreting HSV (140). HSV has a specific tissue tropism, with replication in the genital, perigenital, or anal skin sites and seeding and colonization of the sacral ganglia (141). Most HSV infections result in viral latency in the dorsal root ganglia. Active viral replication can lead to life-threatening CNS infection (herpes encephalitis) or recurrent oral and/or genital ulceration. In the immunosuppressed, in neonates, or during pregnancy, severe disseminated multiorgan involvement can occur. Mucosal infection and resultant ulceration are a result of virally mediated cell death and host inflammatory response. Cellular changes include ballooning of infected cells and the formation of condensed chromatin within the nuclei of cells (138). Cells form multinucleated giant cells, and with cell lysis, a clear vesicular fluid appears between the epidermis and dermal layer, forming a blister. Natural infection does not typically manifest as disease in the hand or digits. In the laboratory, however, needlestick injuries directly into a digit with high-titer preparations have been reported (142, 143) that resulted in a cutaneous lesion called herpetic whitlow (144).

There are many herpesviruses described for non-human primates and other research animal species. *Saimiriine herpesvirus* replicates in human tissues and is classified as an oncogenic virus by the National Cancer Institute, and herpesvirus *tamarinus* has been shown to produce skin pustules, fever, and nonfatal encephalitis in humans (58, 145, 146). Cercopithecine herpesvirus 1 is a serious zoonosis with potential for lethal infection in humans. B virus was first described in 1933 as causing a life-threatening neurologic disease of humans, and human cases have continued to occur within the past decade (39, 147–149). This virus is still widely distributed among colonies of macaques used in biomedical research. B virus infection in the macaque generally produces a mild clinical disease similar to human HSV infection. During primary infection, macaques develop lingual and/or labial vesicles or ulcers that generally heal within a 1- to 2-week period; keratoconjunctivitis or corneal ulcer also may be noted, with virus becoming latent in the trigeminal and genital ganglia of the macaque. Viral shedding may be reactivated in asymptomatic animals by physical or psychological stressors or immunosuppressive therapy (150). Transmission between macaques occurs through close contact involving the oral, conjunctival, and genital mucous membranes (151). In colonies with endemic B virus infection, there is an age-related increase in the incidence of B virus infection during adolescence, with all animals becoming infected by their first breeding season (152). B virus should be considered endemic among Asian monkeys of the genus *Macaca* unless these animals have been obtained from specific breeding colonies known to be B virus free. Only macaques are known to harbor B virus naturally, but several species of New World monkeys and other Old World monkeys are known to be susceptible to fatal B virus infection (153, 154).

LAI

B virus transmission to humans primarily occurs via exposure to contaminated saliva through bites and scratches. Airborne exposure may have played a role in several human cases, and exposure of ocular mucous membranes to biological material, possibly fecal, has been confirmed in a recent human fatality (39, 155). Needlestick injury, exposure to infected non-human primate tissues, and human-to-human transmission have produced human infections (147, 156–158). The incubation period of human B virus infection ranges from 2 days to 2 to 5 weeks, with one reported case of development of clinical disease 10 years after exposure (156). Following a bite, scratch, or other local trauma, humans may develop a herpetiform vesicle at the site of inoculation. In ocular exposures, a swollen, painful orbit with conjunctivitis may develop (39). Other clinical signs of B virus include myalgia, fever, headache, and fatigue, followed by progressive neurologic disease characterized by numbness, hyperesthesia, paresthesia, diplopia, ataxia, confusion, urinary retention, convulsions, dysphagia, and an ascending flaccid paralysis.

Guidelines for the prevention of B virus infection in animal handlers have been developed in response to the 1987 outbreak in monkey handlers (39, 58, 147, 153). These recommendations emphasize the need for non-human primate handlers to use personal protective equipment and a program based upon a thorough hazard assessment of all work procedures, potential routes of exposure, and adverse health outcomes (39). Protective clothing consists of leather gloves or full-length-sleeve garments, goggles for splash protection, and a mask to protect mucous membranes from exposure to infectious secretions. The use of a face shield alone is not considered a sufficient method for protection against ocular exposure. Droplet splashes to the head may run down into the eyes and infectious materials may enter via the gap along the margins of the shield.

Postaccident management

CDC recommendations specify that institutions should be prepared to handle patients with a suspected exposure and that patients should have direct and immediate access to a local medical consultant knowledgeable about B virus. The wound should be cleansed thoroughly, and serum samples and cultures should be obtained for serology and viral isolation from both the patient and the monkey. The initiation of antiviral therapy with valacyclovir or other antiviral agents may be warranted based on the risk of exposure. The management of antiviral therapy in B virus-infected patients is controversial because increasing antibody titer has been demonstrated for a patient following the discontinuation of acyclovir therapy (153). Physicians should consult the Viral Exanthems and Herpesvirus Branch, Division of Viral Diseases, CDC, Atlanta, GA, for assistance in case management. Additional information about B virus diagnostic resources is available through the B Virus Research and Resource Laboratory, Georgia State University, Atlanta, GA.

Orthomyxoviridae

General

The family *Orthomyxoviridae* consists of enveloped viruses containing a segmented single-stranded, negative-stranded RNA genome. The *Orthomyxoviridae* family contains four genera: *Influenzavirus A*, *Influenzavirus B*, *Influenzavirus C, and Influenzavirus D* (also known as *Thogotovirus*). Influenza A, B, and C viruses are known human respiratory pathogens, causing significant mortality and morbidity each year. These viruses are spread by aerosols and fomites. The influenza viruses undergo two types of evolution: genetic shift and genetic drift. Genetic shift results from the wholesale exchange of a genomic segment with that of a different strain of influenza virus (reassortment). Genetic drift is the genetic change over time caused by immune pressure and mutations (polymerase errors). The dominant strain of influenza virus may change from year to year. Therefore, the vaccine compositions must be reconsidered every year. Humans are considered the reservoir for human influenza virus infections. However, influenza virus infections from different antigenic strains occur naturally in many animals, including avian species, swine, horses, mink, and seals (159). Animal reservoirs are thought to contribute to the emergence of new human strains of influenza virus by the passage of avian influenza viruses through pigs that act as the intermediate host (160). In the laboratory, ferrets are highly susceptible to human influenza virus and often are used as experimental models of influenza virus infection (159). The transmission of animal influenza virus strains from animals to humans is an uncommon occurrence. However, a study has shown that pigs experimentally infected with influenza virus in the laboratory can directly and readily spread this agent to persons working with these animals (161, 162). The colloquially named swine flu epidemic of 2009 was shown to be genetically similar to swine H1Ni; however, investigations of initial human cases did not identify exposures to pigs (163). It quickly became apparent that a novel reassortment of the influenza virus was circulating among humans; the global infection rate climbed to between 11% and 21% of the world's population in 2009 (164). More recently, highly pathogenic avian strains of influenza A virus (H5N1), although currently not transmissible from human to human, have been sporadically infecting humans; 850 cases have been reported to the WHO by May 2016, with 449 deaths (165). Mutations permitting mammalian transmission have been identified, which critics fear could be coupled with technology to generate influenza viruses entirely from cDNA in order to engineer biological weapons (see discussion in "Genetically Engineered Viruses," below).

LAI

Laboratory workers are at risk for an LAI by influenza virus by aerosol and fomite transmissions. The prototypical laboratory strains are not suspected of producing disease in humans. However, work with primary human isolates poses a risk of LAI for laboratory workers (166). All workers should be vaccinated against influenza virus on an annual basis (166, 167). In 2007, the FDA licensed the first vaccine for avian H5N1, manufactured by Sanofi Pasteur. The vaccine has been added to the National Stockpile in the event that the H5N1 influenza virus develops the capacity to spread efficiently from human to human.

Postaccident management

Following a suspected exposure, the laboratory worker should seek immediate medical attention and inform the institutional biosafety office. Currently there are two types of antivirals available for the treatment of influenza virus infections: the neuraminidase inhibitors and the M2 ion channel blockers (168, 169). Since these antivirals are most effective if promptly administered within 48 h of exposure, it is advisable to seek medical treatment immediately following an exposure. Influenza virus is highly contagious. The exposed individual should not have contact with small children, immunosuppressed people, or the elderly. Although prototypical laboratory strains represent a limited threat, primary human and avian isolates are a serious concern to both the exposed worker and the general population. Work with recombinant viruses may pose an added threat depending on the types of genetic modifications that were engineered and the genetic background of the virus (laboratory versus primary human and/or avian). This information must be provided to the physician overseeing the medical management of the exposed laboratory worker.

Paramyxoviridae

General

The *Paramyxoviridae* are enveloped, negative-stranded RNA viruses that share similar properties with two other important families of viruses, the *Orthomyxoviridae* and the *Rhabdoviridae*. The family *Paramyxoviridae* comprises two subfamilies, the *Paramyxovirinae* and the *Pneumovirinae*. The subfamily *Paramyxovirinae* contains three genera, *Respirovirus*, *Rubulavirus*, and *Morbillivirus*. The subfamily *Pneumovirinae* contains the genera *Pneumovirus* and *Metapneumovirus*. The *Paramyxoviridae* contain viruses that are important animal and human pathogens and include the animal viruses Newcastle disease virus and rinderpest virus and the human viruses measles virus, RSV, and parainfluenza and mumps viruses. The *Paramyxoviridae* also contain several newly emerging human and animal pathogens, including Hendra virus (from horses and humans in Australia) and Nipah virus (from pigs and humans in Malaysia and, most recently, Bangladesh).

LAI

The paramyxoviruses have been a frequent cause of LAIs. Of the animal paramyxoviruses, Newcastle disease virus has been the most common cause of LAIs, with over 70 associated with this virus over time (Table 1) (170). Newcastle disease is seen among wild and domestic birds, with transmission among the bird population caused by contaminated food and water (171). Illness in birds is characterized by anorexia and respiratory disease in adult birds and neurologic signs in young birds. Newcastle disease spreads to humans by aerosol transmission and produces a follicular conjunctivitis with a mild fever and cough. Respiratory involvement can range from bronchiolitis to pneumonia in more severe cases. A less frequent cause of LAI among the animal paramyxoviruses is Sendai virus, mouse parainfluenza virus type 1, with one reported LAI.

Among the human paramyxoviruses, LAIs have been associated with mumps and measles viruses and RSV. Mumps virus infection results in fever, salivary gland swelling (especially the parotid glands), and submaxillary gland enlargement. Virus is present in the saliva for up to 5 days after the onset of clinical disease. Measles virus infection produces fever, malaise, anorexia with cough, coryza, and conjunctivitis. Amaculopapular rash appears 3 to 4 days after the onset of symptoms. In the majority of cases, measles virus infection results in the classic rash followed by a gradual clearance and convalescence. Complications can occur from measles virus infection and include interstitial pneumonitis, hepatitis, myocardial disease, corneal ulceration with blindness, and encephalitis. Measles in the immunocompromised host can be particularly devastating, with the development of giant cell pneumonia and encephalitis. RSV is primarily a disease of infants causing a low-grade fever, cough, and rhinorrhea. In more severe cases, pneumonia may develop with wheezing, tachypnea, and severe coughing requiring ventilator support in infants severely compromised. Adults may develop an RSV infection characterized by cough, bronchitis, rhinorrhea, fatigue, and headache. In older individuals, severe pneumonia may develop, leading to adult respiratory distress syndrome requiring ventilator support. Like measles, RSV infection can be particularly devastating to the immunocompromised host, producing giant cell pneumonitis.

Postaccident management

Newcastle disease can be prevented from spreading in the laboratory environment by immunizing susceptible birds for this disease or obtaining birds from flocks known to be free of this agent. Adequate face and respiratory personal protective equipment should be worn by personnel working with infected birds. Mumps and measles are diseases preventable by vaccination; personnel working with these agents or potentially exposed should have their antibody titers measured and boosted with vaccine as needed. There is currently no licensed vaccine against RSV, although a medication called palivizumab may prevent RSV infections. Potential LAIs from RSV should be confirmed, and symptomatic personnel should be removed from the laboratory environment and monitored clinically. They should avoid coming in contact with infants, the elderly, and the immunocompromised during the symptomatic phase of their illness.

Picornaviridae

General

The family *Picornaviridae* is composed of small, nonenveloped viruses with a single-stranded RNA and consists of six genera: *Aphthovirus*, *Cardiovirus*, *Enterovirus*, *Hepatovirus*, *Parechovirus*, and *Rhinovirus*. Important human pathogens within this family include poliovirus, coxsackievirus, echovirus, and enterovirus (genus *Enterovirus*); HAV (genus *Hepatovirus*), and rhinovirus (genus *Rhinovirus*). The enteroviruses and hepatovirus are transmitted orally from contaminated water or through the fecal-oral route, and rhinovirus is transmitted by aerosol droplet. In the majority of cases, enterovirus infection results in an asymptomatic infection. Clinical disease results in a spectrum of illness ranging from a mild febrile illness to diarrhea. In more severe cases, meningitis and poliomyelitis are found. Pneumonia, bronchiolitis, hemorrhagic conjunctivitis, and HFMD may also occur. Rhinovirus produces the common cold, with sneezing, nasal discharge, sore throat, headache, cough, and malaise. HAV produces an acute viral hepatitis, with elevation in the liver enzymes and jaundice.

LAI

Previous surveys have demonstrated that the *Picornaviridae* are responsible for a large number of LAIs. Mengo encephalomyocarditis virus was responsible for two laboratory infections; coxsackievirus was responsible for 39; HAV was responsible for five; and poliomyelitis virus was responsible for 21, with five fatalities. Of the picornaviruses that are associated with animal disease, there were two cases of human LAIs from foot-and-mouth disease virus and one from swine vesicular disease virus. HAV is unique among the *Picornaviridae* for causing infection in laboratory non-human primates, with more than 100 human cases of HAV infection associated with newly imported chimpanzees (172).

Postaccident management

Postaccident management of potential laboratory exposures to or LAIs with the *Picornaviridae* should include a thorough physical exam, close clinical follow-up, and diagnostic assays to establish the exposure. Personnel working with poliovirus or HAV should be vaccinated with periodic booster vaccines. If their vaccination is not up to date or is unknown, all potentially exposed personnel should be vaccinated immediately after exposure. Specific immune serum globulin also can be used for protection and should be given in the recommended dose of 0.02 ml/kg of body weight before experimental animal HAV infection studies begin or at a postexposure dose of 0.02 ml/kg within 2 weeks of exposure (58, 173).

Poxviridae

General

The *Poxviridae* family is composed of viruses that contain single linear double-stranded DNA and enzymes that synthesize mRNA. The *Poxviridae* are divided into two subfamilies, the insect viruses *Entomopoxvirinae*, which include three genera, and the animal viruses *Chordopoxvirinae*, with eight genera: *Orthopoxvirus*, *Parapoxvirus*, *Avipoxvirus*, *Capripoxvirus*, *Leporipoxvirus*, *Suipoxvirus*, *Molluscipoxvirus*, and *Yatapoxvirus*. The prototype virus of the *Chordopoxvirinae* is vaccinia virus, which has an unknown origin and no known natural host. Vaccinia virus infection of immunized and nonimmunized laboratory workers has been reported. The recent fear of bioweapons has reinvigorated orthopoxvirus research. Many laboratories are conducting experiments with vaccinia virus, which is very similar to the vaccine (Dryvax) used to prevent variola virus infection. The poxviruses involved in zoonotic transmission in the laboratory animal facility represent three genera—*Orthopoxvirus*, *Parapoxvirus*, and *Yatapoxvirus*. The non-human primate serves as a host for the majority of the zoonotic poxviral species (174). In humans, these infections usually are characterized by the development of cutaneous or subcutaneous lesions.

Monkeypox, an orthopoxvirus, is similar to smallpox in clinical presentation and its ability to produce a sustained cycle of person-to-person infection (175, 176). Monkeypox causes human disease in Africa, and natural outbreaks have been recorded for non-human primates in the wild and in the laboratory (174, 177). Non-human primates develop clinical disease similar to that in humans, with fever followed in 4 to 5 days by cutaneous eruptions distributed on the limbs, trunk, face, lips, and buccal cavity. In humans, infection is characterized by fever, malaise, headache, severe backache, prostration, occasional abdominal pain, lymphadenopathy, and a maculopustular rash (178). Human-to-human transmission occurs through close contact with active lesions, contaminated fomites, or respiratory secretions (174). In addition to non-human primates, squirrels of the genera *Funisciurus* and *Heliosciurus* have been identified as hosts and significant reservoirs of the disease (179). Recently monkeypox has been implicated in an outbreak within the United States from imported Gambian rats which spread to prairie dogs and to humans (180). Smallpox vaccination will protect against monkeypox in humans and has been used for the control of this disease in monkeys also.

Tanapox virus is a yatapoxvirus that has been zoonotic in the laboratory environment (181). Tanapox is endemic in regions of Africa, and cases of disease in humans have been detected in Africa during the course of surveillance for monkeypox (176). Tanapox affects monkeys of the genus *Presbytis* in Africa and captive macaques in the United States (182). The disease is spread rapidly among non-human primates in gang cages, suggesting direct transmission (183). Infections in animal handlers have been attributed to contamination of skin abrasions. In both humans and non-human primates, tanapox is characterized by the development of circumscribed, oval to circular, elevated red lesions on the eyelids, face, body, or genitalia. Humans may also experience headache, backache, prostration, and intense itching at the site of lesion development (184). These lesions regress spontaneously in 4 to 6 weeks. The barrier protections defined for vertebrate ABSL2 should be sufficient to prevent the zoonotic transmission of tanapox virus (185).

Non-human infections with Yaba monkey tumorvirus, a yatapoxvirus, were reported initially for rhesus monkeys in Yaba, Nigeria (186). Subsequent outbreaks have been documented for laboratory-housed non-human primates (187). Other species that can be infected include the pigtail macaques (*Macaca nemestrina*), stumptail macaques (*Macaca arctoides*), and cynomolgus (*Macaca fascicularis*), African green (*Chlorocebus aethiops*), sooty mangabey (*Cercocebus atys*), and patas monkeys (*Erythrocebus patas*) (188, 189). New World non-human primate species are resistant to infection (188). Experimental studies have demonstrated that the virus can be spread by aerosol transmission and must be considered a potential hazard to humans, although no human cases of natural disease have been associated with non-human primate contact (190). Monkeys infected with Yaba virus develop subcutaneous benign histiocytomas appearing as palpable pink nodules, reaching a maximum size 6 weeks postinoculation and regressing approximately 3 weeks thereafter, conferring immunity to reinfection (191). The surgical removal of a Yaba tumor in a baboon prior to natural tumor regression was associated with subsequent susceptibility and reinfection with Yaba virus (192). Six human volunteers have been inoculated experimentally with Yaba virus and developed tumors similar to, but smaller than, those seen in monkeys; tumor regression also was earlier. Yaba tumor induction also has been recorded as a result of accidental self-inoculation (needlestick) in a laboratory worker using the agent tumor, and complete tumor resection was curative (193).

Orf is a parapoxvirus disease of sheep, goats, and wild ungulates that continues to be prevalent in the United States and worldwide and causes human infections associated with occupational exposures, including research animal contact. The disease affects all age groups, although young animals are most frequently and most severely affected. In sheep, orf virus infection does not reliably confer protection against reinfection, aiding in viral persistence within a population (194). Orf virus produces proliferative, pustular encrustations on the lips, nostrils, and mucous membranes of the oral cavity and urogenital orifices of infected animals. Orf virus is transmitted to humans by direct contact with exudates from virus-laden lesions. Transmission of this agent by fomites or other animals contaminated with the virus also is possible due to the environmental persistence of this virus for up to 12 years in dried crusts (195). Vaccines have been developed to induce protective immunity in animals (196). Orf in humans is characterized by the development of a solitary lesion located on the hands, arms, or face. The lesion is maculopapular or pustular initially and progresses to a weeping proliferative nodule with central umbilication. Occasionally, several nodules are present, each measuring up to 3 cm in diameter and persisting for 3 to 6 weeks, followed by spontaneous regression with minimal residual scarring. Although rare, more extensive disease involvement includes regional adenitis, lymphangitis, erythema multiforme, bacterial superinfection, and blindness in infected eyes (197). The use of personal protective equipment and clothing as defined for ABSL2 (185) is recommended for the prevention of orf virus transmission to humans.

LAI

Vaccinia virus infections in laboratory workers have been documented. For example, a laboratory worker immunized in childhood (26 years old at time of infection) had a needlestick injury with a syringe containing vaccinia virus (198). The individual developed a severe localized lesion at the injection site caused by vaccinia virus. A combination of high titers and a waning immunity to vaccinia virus were the causes of the severe pathology of a relatively attenuated live vaccine. In another example, a nonimmunized laboratory worker developed generalized vaccinia after working with vaccinia virus (199). This confirmed infection with vaccinia virus resembled a typical adverse event associated with vaccinia virus immunization. In both cases, a comprehensive risk assessment should have been performed before initiating studies with vaccinia virus. These examples illustrate that proper evaluation and vaccination (booster vaccination every 3 years) may have reduced the severity of disease in the first case; in the second case, proper evaluation should have shown that the laboratory worker was at risk and either should not have worked with the virus, should have been immunized, or should have taken added precautions against fluid and aerosol exposures. LAIs with orthopoxviruses are potentially serious events; however, the consequences of accidental infection can be reduced by prior vaccination. The laboratory worker is at risk for aerosol, fomite, fluid, and needlestick exposures. Engineering controls and safe work practices must be used. Prior to working with orthopoxviruses, the laboratory worker should be evaluated for possible complications from a live viral vaccine. If vaccination is contraindicated, the laboratory worker should consider additional precautions (i.e., positive air breathing apparatus) for working with orthopox viruses.

Postaccident management

Effective vaccines for orthopoxvirus infection exist. Currently, the principal vaccine in the United States is ACAM2000, a live vaccina virus derived from a clone of the New York City Board of Health strain and produced using modern cell culture technology. A Modified Vaccinia Ankara (MVA) vaccine is undergoing clinical trials as an alternative to the ACAM2000, as it may offer a less-severe adverse event profile. After exposure, the worker should seek immediate medical consultation and inform the biosafety officer. Early postexposure prophylaxis with vaccine is the most proven historical intervention for preventing clinical orthopoxvirus disease. For patients with significant immune impairment, relative contraindications to vaccination, or other concerns, additional options exist. Tecovirimat (ST-246) is an orthopox-specific antiviral compound that is currently held in the U.S. Strategic National Stockpile and is available for emergency Investigational New Drug (IND) use through the CDC (200). In addition, vaccinia immune globulin (VIG) may also be obtained through the CDC. Management of such exposures should occur in close consultation with CDC and subject matter experts. In general, lab workers with exposure risk should have been vaccinated prior to work (within the last 3 years); the physician may choose to revaccinate after exposure. If the individual was not previously vaccinated, the vaccine should be effective at preventing disease if given within 4 days of exposure. Because the artificially high titers of virus to which a worker may be exposed may compress the window of effective therapeutic vaccination, the exposed worker should be treated immediately after exposure. After vaccination, the worker must follow the same restrictions as other vaccinated individuals.

Retroviridae

General

The family *Retroviridae* comprises a large group of viruses that are unique in having RNA as their genome, which is transcribed into DNA upon host cell entry. The viral DNA is integrated into the host chromosomal DNA, forming a provirus that serves as the template for production of viral proteins. The integration of viral DNA into the host chromosomal DNA confers a powerful ability to the virus to maintain a persistent infection within the host as well as vertical transmission. The *Retroviridae* contain important animal and human viruses and are classified into seven genera: *Alpharetrovirus*, *Betaretrovirus*, *Gammaretrovirus*, *Deltaretrovirus*, *Epsilonretrovirus*, *Lentivirus*, and *Spumavirus*. Examples for each genus are avian leukosis virus and RSV for alpharetrovirus; mouse mammary tumor virus for betaretrovirus; murine leukemia and feline leukemia virus for gammaretrovirus; HTLV-1 and HTLV-2 for deltaretrovirus; HIV types 1 and 2 and simian immunodeficiency virus (SIV) for lentivirus; and human foamy virus for spumavirus. HTLV-1 and -2 result in adult T-cell leukemia in the majority of infections, although this association is less clear for HTLV-2. HTLV-1 may also result in a slowly progressive neurologic disorder termed HTLV-associated myelopathy or tropical spastic paraparesis. Acute HIV infection results in the onset of fever and "flu-like" symptoms consisting of myalgias, arthralgias, headache, and weight loss. Opportunistic infections may occur during this acute period. Chronic HIV infection results in a progressive loss of CD4 cells and subsequent immunosuppression and opportunistic infections. The human retroviruses are transmitted from person to person by exposure to infected blood and fluid secretions through sexual transmission, by contaminated blood products, by needlestick injury, or by injectional drug use.

LAI

HIV is the virus most associated with occupationally acquired infections among the retroviruses. As of 2010, 143 U.S. workers had been reported to the CDC's national surveillance system for occupationally acquired HIV infection (201, 202). Three of these infections were from exposure to concentrated virus in the laboratory, and the remaining were among health care workers with occupational exposure to HIV-infected blood from percutaneous and/or mucocutaneous exposure. Studies of health care workers have estimated the risk of HIV transmission via infected blood after percutaneous exposure as approximately 0.3% and that after a mucous membrane exposure as 0.09% (203, 204). In a study of exposure of health care workers in the United Kingdom between 1997 and 2000, 293 health care workers were documented as having been exposed to HIV, with one transmission occurring despite postexposure prophylaxis (205). Of the health care workers exposed, eight were laboratory workers. Factors associated with the risk of HIV transmission after percutaneous exposure include a deep injury, a needle or device that is visibly contaminated with the patient's blood, procedures involving a needle placed directly in the patient's vein or artery, and a source from an AIDS patient who died within 60 days of the percutaneous exposure (206). The risk of transmission from other body fluids and from aerosol transmission appears to be negligible (207, 208).

Other retroviruses with potential for causing LAIs include SIV and simian foamy virus. SIV is a lentivirus that infects Old World non-human primates and produces an immunodeficiency syndrome equivalent to HIV infection in humans. The seroprevalence of SIV in Asian macaques is low, but that among wild-caught African non-human primate species is much higher (209). Following a known exposure incident, there have been two human cases of seroconversion to SIV and possible other infections based on serological surveys (210–213). Of the two known infections, the first occurred following a puncture of the skin with a needle contaminated by the blood of an infected macaque. The second case occurred in an individual who had dermatitis on the hands and forearms and handled SIV-infected blood specimens without wearing gloves. In both cases, no symptoms or evidence of immunodeficiency occurred. Simian foamy viruses have been isolated from a number of New and Old World non-human primates and have been reported as causing infection in humans accidentally exposed to infected non-human primates (214, 215). In a serological survey of 231 individuals, 4 were positive (214). No clinical illness was associated with infection.

Postaccident management

Postaccident management of potential laboratory exposures to HIV or other retroviruses begins with thorough cleaning or flushing of exposed wounds or mucous membranes. Chemoprophylaxis is available for HIV exposures and should be discussed with workers with a known risk for HIV transmission (201). The decision to initiate antiretroviral postexposure prophylaxis (PEP) is not straightforward, and one must weigh the significance of exposure and the potential adverse effects of antiretroviral drugs, which are often prohibitive. However, if PEP is to be pursued, it should be done as quickly as possible. Initiation of PEP within 2 to 3 hours should be the goal, as earlier initiation is thought to be more beneficial; after 72 hours, it is likely ineffective. A three-drug regimen is standard, and decisions on the specific drugs should be made in consultation with an experienced HIV physician, if possible. The University of California, San Francisco runs a highly regarded clinical consultation service PEP Hotline (1-888-448-4911) that can be extremely helpful in addressing PEP questions.

In diagnosing HIV infection, the time from infection to the development of antibody and a positive test can range from as early as 2 to 3 weeks to as long as 12 months following exposure (216). Nucleic acid detection assays can be positive within a week and should be used for early detection, particularly in the presence of symptoms. The majority of infections become antibody positive within 6 months following exposure. Thus, laboratorians potentially exposed to HIV will need long-term follow-up and testing for at least 6 months following exposure. Potentially exposed workers should be counseled to refrain from donating blood or other tissues during the follow-up period and to refrain from breast-feeding and to use condoms during sexual intercourse.

LAIs with the other retroviruses should be managed similarly to HIV LAI. The utility of using HIV antiretroviral drug prophylaxis for these viruses has not been established. Appropriate follow-up for LAIs with other retroviruses that can cause human infection includes a thorough physical exam, close clinical follow-up, and diagnostic assays to establish the exposure where possible.

Rhabdoviridae

General

The *Rhabdoviridae* are enveloped, nonsegmented, negative-stranded RNA viruses in the order *Mononegavirales*, family *Rhabdoviridae*. They are classified into two genera: *Vesiculovirus* and *Lyssavirus*. *Vesiculovirus* contains VSV, a zoonotic pathogen that primarily affects cattle, horses, and swine, causing severe economic losses. *Lyssavirus* contains rabies virus, an important animal and human pathogen. Rabies virus is spread through the bite of an infected animal and by aerosol transmission and results in acute encephalitis with nearly 100% mortality in humans (217).

LAI

Rabies virus LAI can occur through puncture wounds, through aerosol exposures, and through the bite of an experimentally infected animal. An effective vaccine is available against rabies virus, and all laboratory workers should be vaccinated before working with the virus (218). Vaccination of laboratory workers is not intended as a primary barrier from viral infection but is one of several layers of protection against LAI.

Postaccident management

Postaccident management of potential laboratory exposures to or LAIs with rabies virus should include a comprehensive physical exam, close clinical follow-up, and diagnostic assays to establish the exposure. Laboratorians with a suspected exposure should undergo postexposure rabies treatment consisting of a combination of vaccination and treatment with equine rabies immunoglobulin (ERIG) or human rabies immunoglobulin (HRIG) (219, 220). It has been suggested that rabies vaccine failures are significant if the vaccine is not given in combination with ERIG or HRIG (221). The infected laboratory worker is not at risk for spreading the virus to others. However, postexposure rabies treatment should be administered as soon as possible after a suspected exposure.

Togaviridae

General

The *Togaviridae* are small, lipid-enveloped, positive-stranded RNA viruses. They consist of two genera: *Alphavirus* and *Rubivirus*. The alphaviruses consist of 22 separate species grouped into seven antigenic complexes. *Rubivirus* contains only one virus, rubella virus. Both alphaviruses and rubella virus are important causes of human disease. The alphaviruses include Barmah Forest, chikungunya, EEE, Getah, Mayaro, o'nyong-nyong, Ross River, Sindbis, Semliki Forest, VEE, and WEE viruses. The alphaviruses are important arboviruses and produce clinical illnesses ranging from a mild febrile illness with rash, severe arthralgias, and arthritis to encephalitis. Chikungunya virus has been associated with the production of hemorrhagic fever. Rubella virus infection is by droplet spread or direct contact with nasopharyngeal secretions and causes a common childhood illness consisting of fever, conjunctivitis, sore throat, and arthralgias. A rash appears along the trunk and limbs. The most serious complication of rubella is a postinfectious encephalopathy and a congenital rubella syndrome consisting of hearing loss, congenital heart disease, and retardation.

LAI

The *Togaviridae* have been responsible for over 253 reported LAIs and six deaths as documented by previous laboratory surveys (see Table 1), thus demonstrating the propensity of these viruses to aerosolize and cause severe infection. The vast majority of LAIs were caused by the alphaviruses, with VEE virus responsible for 186 LAIs and two deaths. Rubella virus was responsible for only one LAI by previous surveys and no deaths.

Postaccident management

Postaccident management of potential laboratory exposures to or LAIs with the *Togaviridae* should include a thorough physical exam, close clinical follow-up, and diagnostic assays to establish the exposure. Human-to-human infection does not occur among the alphaviruses as it does for rubella virus. Therefore, contact tracing from an infected laboratorian should be limited to only those with rubella virus infection. There are two IND vaccines, C-84 strain and TC-83, that are indicated only for use in the military and research positions that risk contracting the virus (222). An effective vaccine is available for rubella as part of the measles, mumps, and rubella vaccine, and all workers working with this virus should have antibody titers checked and should be vaccinated for low or lack of antibodies to rubella virus.

GENETICALLY ENGINEERED VIRUSES

General

Recently, it has become possible to generate viruses entirely from cDNA. This technology allows for genetically engineered viruses to be readily generated. Recombinant viruses may be altered to carry foreign genes, generate reassortant viruses (segment exchange between different viruses), and alter viral factors affecting replication and pathogenesis, which may increase their transmissiblity. Substantial consideration must be made regarding the BSL of such gain-of-function (GOF) recombinant viruses. Although it is unlikely that recombinant viruses generated from prototypical laboratory strains would be a risk to human health, it is possible. Work with genetically engineered viruses derived from primary human isolates is a larger concern. Therefore, work with recombinant viruses may require higher levels of biocontainment depending on the composition of the recombinant virus.

Gain-of-Function Viruses

GOF influenza research may help in understanding the pandemic potential of the virus by identifying mutations that would make it more transmissible, for use in biosurveillance and vaccine development. Experiments performed in 2011, but not published until 2012 due to intense controversy, demonstrated mutations permissive for mammal-to-mammal transmission of the highly pathogenic avian H5N1 influenza virus (223, 224). Concerns have been raised about the risk of these studies not only to the individual researchers, but also to the community due to their dual-use potential and the possibility for accidental or deliberate release of extremely virulent pathogens. In October 2014, as result of these debates, the U.S. Government paused federal funding for GOF influenza research, as well as for SARS-CoV and MERS-CoV research, and called for scientists to voluntarily halt all work on these viruses until a comprehensive risk assessment could be performed (225).

In May 2016, after holding eight meetings (six committee meetings and two large National Academy of Sciences workshops) and contracting Gryphon Scientific to perform a 1,000-page risk assessment (http://www.gryphonscientific.com/gain-of-function/), the National Institutes of Health's National Science Advisory Board for Biosecurity (NSABB) approved a final proposal on the process for approving what it now calls Gain of Function Research of Concern (GOFROC) (225). One finding was that only a small subset of GOF research is GOFROC, and it is only this subset that warrants additional oversight. In addition, the NSABB recommended that the U.S. Government should consider developing a system to collect and analyze data about LAIs, as well as undertake broad efforts to strengthen laboratory biosafety and biosecurity (226). Supporters of the policy suggest that it will not be difficult to implement, as it is similar to policies already in place for dual-use research of concern. It is unclear, however, how much longer the moratorium on GOF research will remain. Additionally, what BSL level and biosafety standards and procedures will be required for researchers working with GOF/GOFROC is unclear and needs further investigation.

Viral-Vectored CRISPR/CAs9

Viral-vectored genome editing is an emerging technology that adds a new dimensionality to biosafety risk management. Genome-editing technology has applications in the direct editing of a viral genome. It is possible to directly edit RNA- and DNA-based genomes using CRISPR/Cas9 technology. Although the existing frameworks that are currently employed for genetically engineering viruses already accomplish these feats, CRISPR/Cas9 does it more efficiently, specifically, and with much less difficulty for the researcher. These parameters virtually ensure that increasing numbers of scientists and even "DIY bio" will be utilizing the technology. However, the CRISPR/Cas9 system, once placed on a viral vector, requires special attention because these constructs could possess the capability for germ line changes. Additionally, the derivative technology, gene drive, circumvents Mendelian inheritance rules and by doing so creates novel implications for LAI. Therefore, it is possible to have effects that are passed to the children of the affected workers. The pace of advancement and adoption of gene-editing technology has been unprecedented. The U.S. National Academies and other global scientific organizations are providing leadership on the societal impacts of gene editing technology (http://nationalacademies.org/gene-editing/index.htm). The laboratory worker will need to stay abreast beyond the confines of this text.

CONCLUSION

Viral infection in the laboratory is a serious event that can have severe, even irreversible, consequences for the individual and the community. However, a coordinated program can be developed to manage the risk using concepts described in this chapter. A thorough risk assessment, the appropriate risk management, and a postexposure action plan should provide an environment that is safe for the laboratorian and coworkers.

Disclaimer

Copyright Statement

References

1. **Monath TP.** 1991. Yellow fever: Victor, Victoria? Conqueror, conquest? Epidemics and research in the last forty years and prospects for the future. *Am J Trop Med Hyg* **45:**1–43.
2. **Holland DJ.** 1998. Emerging viruses. *Curr Opin Pediatr* **10:**34–40.
3. **Marwick C.** 1989. Scientists ponder when, why of emerging viruses. *JAMA* **262:**16.
4. **Nedry M, Mahon CR.** 2003. West Nile virus: an emerging virus in North America. *Clin Lab Sci* **16:**43–49.
5. **Ríos Olivares E.** 1997. The investigation of emerging and re-emerging viral diseases: a paradign. *Bol Asoc Med P R* **89:**127–133.
6. **Pike RM.** 1976. Laboratory-associated infections: summary and analysis of 3921 cases. *Health Lab Sci* **13:**105–114.
7. **Sulkin SE, Pike RM.** 1949. Viral infections contracted in the laboratory. *N Engl J Med* **241:**205–213.
8. **Sulkin SE, Pike RM.** 1951. Survey of laboratory-acquired infections. *Am J Public Health Nations Health* **41:**769–781.
9. **Hanson RP, Sulkin SE, Beuscher EL, Hammon WM, McKinney RW, Work TH.** 1967. Arbovirus infections of laboratory workers. Extent of problem emphasizes the need for more effective measures to reduce hazards. *Science* **158:**1283–1286.
10. **Büchen-Osmond C.** 2003. Taxonomy and classification of viruses, p 1217–1226. *In* Murray P, Baron E, Jorgensen J, Pfaller M, Yolken R (ed), *Manual of Clinical Microbiology*, 8th ed, vol 2. ASM Press, Washington, DC.
11. **Domingo E, Escarmís C, Sevilla N, Moya A, Elena SF, Quer J, Novella IS, Holland JJ.** 1996. Basic concepts in RNA virus evolution. *FASEB J* **10:**859–864.
12. **Duarte EA, Novella IS, Weaver SC, Domingo E, Wain-Hobson S, Clarke DK, Moya A, Elena SF, de la Torre JC, Holland JJ.** 1994. RNA virus quasispecies: significance for viral disease and epidemiology. *Infect Agents Dis* **3:**201–214.
13. **Domingo E, Menéndez-Arias L, Holland JJ.** 1997. RNA virus fitness. *Rev Med Virol* **7:**87–96.
14. **Martin ML, Lindsey-Regnery H, Sasso DR, McCormick JB, Palmer E.** 1985. Distinction between Bunyaviridae genera by surface structure and comparison with Hantaan virus using negative stain electron microscopy. *Arch Virol* **86:**17–28.
15. **Arikawa J, Schmaljohn AL, Dalrymple JM, Schmaljohn CS.** 1989. Characterization of Hantaan virus envelope glycoprotein antigenic determinants defined by monoclonal antibodies. *J Gen Virol* **70:**615–624.
16. **Foulke RS, Rosato RR, French GR.** 1981. Structural polypeptides of Hazara virus. *J Gen Virol* **53:**169–172.
17. **Pekosz A, Griot C, Nathanson N, Gonzalez-Scarano F.** 1995. Tropism of bunyaviruses: evidence for a G1 glycoprotein-mediated entry pathway common to the California serogroup. *Virology* **214:**339–348.
18. **Sanchez AJ, Vincent MJ, Nichol ST.** 2002. Characterization of the glycoproteins of Crimean-Congo hemorrhagic fever virus. *J Virol* **76:**7263–7275.

19. **Bishop DH.** 1979. Genetic potential of bunyaviruses. *Curr Top Microbiol Immunol* **86:**1–33.
20. **Hooper JW, Larsen T, Custer DM, Schmaljohn CS.** 2001. A lethal disease model for hantavirus pulmonary syndrome. *Virology* **289:**6–14.
21. **Vincent MJ, Sanchez AJ, Erickson BR, Basak A, Chretien M, Seidah NG, Nichol ST.** 2003. Crimean-Congo hemorrhagic fever virus glycoprotein proteolytic processing by subtilase SKI-1. *J Virol* **77:**8640–8649.
22. **Yang ZY, Duckers HJ, Sullivan NJ, Sanchez A, Nabel EG, Nabel GJ.** 2000. Identification of the Ebola virus glycoprotein as the main viral determinant of vascular cell cytotoxicity and injury. *Nat Med* **6:**886–889.
23. **Geimonen E, Neff S, Raymond T, Kocer SS, Gavrilovskaya IN, Mackow ER.** 2002. Pathogenic and nonpathogenic hantaviruses differentially regulate endothelial cell responses. *Proc Natl Acad Sci USA* **99:**13837–13842.
24. **Gubler DJ, Trent DW.** 1993. Emergence of epidemic dengue/dengue hemorrhagic fever as a public health problem in the Americas. *Infect Agents Dis* **2:**383–393.
25. **Calisher CH, Karabatsos N, Dalrymple JM, Shope RE, Porterfield JS, Westaway EG, Brandt WE.** 1989. Antigenic relationships between flaviviruses as determined by cross-neutralization tests with polyclonal antisera. *J Gen Virol* **70:**37–43.
26. **Nisalak A, Endy TP, Nimmannitya S, Kalayanarooj S, Thisayakorn U, Scott RM, Burke DS, Hoke CH, Innis BL, Vaughn DW.** 2003. Serotype-specific dengue virus circulation and dengue disease in Bangkok, Thailand from 1973 to 1999. *Am J Trop Med Hyg* **68:**191–202.
27. **Rico-Hesse R, Harrison LM, Nisalak A, Vaughn DW, Kalayanarooj S, Green S, Rothman AL, Ennis FA.** 1998. Molecular evolution of dengue type 2 virus in Thailand. *Am J Trop Med Hyg* **58:**96–101.
28. **Endy TP, Chunsuttiwat S, Nisalak A, Libraty DH, Green S, Rothman AL, Vaughn DW, Ennis FA.** 2002. Epidemiology of inapparent and symptomatic acute dengue virus infection: a prospective study of primary school children in Kamphaeng Phet, Thailand. *Am J Epidemiol* **156:**40–51.
29. **Endy TP, Nisalak A, Chunsuttiwat S, Libraty DH, Green S, Rothman AL, Vaughn DW, Ennis FA.** 2002. Spatial and temporal circulation of dengue virus serotypes: a prospective study of primary school children in Kamphaeng Phet, Thailand. *Am J Epidemiol* **156:**52–59.
30. **Hay SI, Myers MF, Burke DS, Vaughn DW, Endy T, Ananda N, Shanks GD, Snow RW, Rogers DJ.** 2000. Etiology of interepidemic periods of mosquito-borne disease. *Proc Natl Acad Sci USA* **97:**9335–9339.
31. **Wikan N, Smith DR.** 2016. Zika virus: history of a newly emerging arbovirus. *Lancet Infect Dis* **16:**e119–e126.
32. **Wang L, Valderramos SG, Wu A, Ouyang S, Li C, Brasil P, Bonaldo M, Coates T, Nielsen-Saines K, Jiang T, Aliyari R, Cheng G.** 2016. From mosquitos to humans: genetic evolution of Zika virus. *Cell Host Microbe* **19:**561–565.
33. **World Health Organization.** 2016. *WHO Director-General Summarizes the Outcome of the Emergency Committee Regarding Clusters of Microcephaly and Guillain-Barré Syndrome.* World Health Organization, Geneva, Switzerland.
34. **Mahy BWJ, Kangro HO.** 1996. *Virology Methods Manual.* Academic Press, London, United Kingdom.
35. **Centers for Disease Control and Prevention (CDC).** 2011. Fatal laboratory-acquired infection with an attenuated Yersinia pestis Strain—Chicago, Illinois, 2009. *MMWR Morb Mortal Wkly Rep* **60:**201–205.
36. **Altaf A, Luby S, Ahmed AJ, Zaidi N, Khan AJ, Mirza S, McCormick J, Fisher-Hoch S.** 1998. Outbreak of Crimean-Congo haemorrhagic fever in Quetta, Pakistan: contact tracing and risk assessment. *Trop Med Int Health* **3:**878–882.
37. **Biesbroeck L, Sidbury R.** 2013. Viral exanthems: an update. *Dermatol Ther (Heidelb)* **26:**433–438.
38. **Swanepoel R, Shepherd AJ, Leman PA, Shepherd SP, McGillivray GM, Erasmus MJ, Searle LA, Gill DE.** 1987. Epidemiologic and clinical features of Crimean-Congo hemorrhagic fever in southern Africa. *Am J Trop Med Hyg* **36:**120–132.
39. **Centers for Disease Control and Prevention (CDC).** 1998. Fatal Cercopithecine herpesvirus 1 (B virus) infection following a mucocutaneous exposure and interim recommendations for worker protection. *MMWR Morb Mortal Wkly Rep* **47:**1073–1076, 1083.
40. **Normile D.** 2003. Infectious diseases. SARS experts want labs to improve safety practices. *Science* **302:**31.
41. **Orellana C.** 2004. Laboratory-acquired SARS raises worries on biosafety. *Lancet Infect Dis* **4:**64.
42. **Ryder RW, Gandsman EJ.** 1995. Laboratory-acquired Sabiá virus infection. *N Engl J Med* **333:**1716.
43. **Chen EC, Yagi S, Kelly KR, Mendoza SP, Tarara RP, Canfield DR, Maninger N, Rosenthal A, Spinner A, Bales KL, Schnurr DP, Lerche NW, Chiu CY.** 2011. Cross-species transmission of a novel adenovirus associated with a fulminant pneumonia outbreak in a new world monkey colony. *PLoS Pathog* 7:e1002155.
44. **Jahrling PB, Peters CJ.** 1992. Lymphocytic choriomeningitis virus. A neglected pathogen of man. *Arch Pathol Lab Med* **116:**486–488.
45. **Bowen GS, Calisher CH, Winkler WG, Kraus AL, Fowler EH, Garman RH, Fraser DW, Hinman AR.** 1975. Laboratory studies of a lymphocytic choriomeningitis virus outbreak in man and laboratory animals. *Am J Epidemiol* **102:**233–240.
46. **Dykewicz CA, Dato VM, Fisher-Hoch SP, Howarth MV, Perez-Oronoz GI, Ostroff SM, Gary H Jr, Schonberger LB, McCormick JB.** 1992. Lymphocytic choriomeningitis outbreak associated with nude mice in a research institute. *JAMA* **267:**1349–1353.
47. **Lehmann-Grube F, Ibscher B, Bugislaus E, Kallay M.** 1979. [A serological study concerning the role of the golden hamster (*Mesocricetus auratus*) in transmitting lymphocytic choriomeningitis virus to humans (author's transl)]. *Med Microbiol Immunol (Berl)* **167:**205–210.
48. **Rousseau MC, Saron MF, Brouqui P, Bourgeade A.** 1997. Lymphocytic choriomeningitis virus in southern France: four case reports and a review of the literature. *Eur J Epidemiol* **13:**817–823.
49. **Childs JE, Glass GE, Korch GW, Ksiazek TG, Leduc JW.** 1992. Lymphocytic choriomeningitis virus infection and house mouse (*Mus musculus*) distribution in urban Baltimore. *Am J Trop Med Hyg* **47:**27–34.
50. **Morita C, Tsuchiya K, Ueno H, Muramatsu Y, Kojimahara A, Suzuki H, Miyashita N, Moriwaki K, Jin ML, Wu XL, Wang FS.** 1996. Seroepidemiological survey of lymphocytic choriomeningitis virus in wild house mice in China with particular reference to their subspecies. *Microbiol Immunol* **40:**313–315.
51. **Smith AL, Singleton GR, Hansen GM, Shellam G.** 1993. A serologic survey for viruses and *Mycoplasma pulmonis* among wild house mice (*Mus domesticus*) in southeastern Australia. *J Wildl Dis* **29:**219–229.
52. **Lucke VM, Bennett AM.** 1982. An outbreak of hepatitis in marmosets in a zoological collection. *Lab Anim* **16:**73–77.
53. **Montali RJ, Ramsay EC, Stephensen CB, Worley M, Davis JA, Holmes KV.** 1989. A new transmissible viral hepatitis of marmosets and tamarins. *J Infect Dis* **160:**759–765.
54. **Stephensen CB, Jacob JR, Montali RJ, Holmes KV, Muchmore E, Compans RW, Arms ED, Buchmeier MJ, Lanford RE.** 1991. Isolation of an arenavirus from a marmoset with callitrichid hepatitis and its serologic association with disease. *J Virol* **65:**3995–4000.
55. **Stephensen CB, Montali RJ, Ramsay EC, Holmes KV.** 1990. Identification, using sera from exposed animals, of putative viral antigens in livers of primates with callitrichid hepatitis. *J Virol* **64:**6349–6354.
56. **Stephensen CB, Park JY, Blount SR.** 1995. cDNA sequence analysis confirms that the etiologic agent of callitrichid hepatitis is lymphocytic choriomeningitis virus. *J Virol* **69:**1349–1352.

57. **Richter C, Lehner N, Henrickson R.** 1984. Primates, p 297–383. *In* Fox JG, Cohen BJ, Loew FM (ed), *Laboratory Animal Medicine*. Academic Press, Orlando, FL.
58. **Adams SR.** 1995. Zoonoses, biohazards and other health risks, p 391–412. *In* Bennett BT, Abee CR, Henrickson R (ed), *Nonhuman Primates in Biomedical Research*. Academic Press, San Diego, CA.
59. **Lehmann-Grube F.** 1982. Lymphocytic choriomeningitis virus, p 231–266. *In* Foster H, Small J, Fox J (ed), *The Mouse in Biomedical Resesarch*. Academic Press, New York, NY.
60. **Bhatt PN, Jacoby RO, Barthold SW.** 1986. Contamination of transplantable murine tumors with lymphocytic choriomeningitis virus. *Lab Anim Sci* **36:**136–139.
61. **Nicklas W, Kraft V, Meyer B.** 1993. Contamination of transplantable tumors, cell lines, and monoclonal antibodies with rodent viruses. *Lab Anim Sci* **43:**296–300.
62. **Biggar RJ, Woodall JP, Walter PD, Haughie GE.** 1975. Lymphocytic choriomeningitis outbreak associated with pet hamsters. Fifty-seven cases from New York State. *JAMA* **232:**494–500.
63. **Hinman AR, Fraser DW, Douglas RG, Bowen GS, Kraus AL, Winkler WG, Rhodes WW.** 1975. Outbreak of lymphocytic choriomeningitis virus infections in medical center personnel. *Am J Epidemiol* **101:**103–110.
64. **Johnson KM.** 1990. Lymphocytic choriomeningitis virus, lassa virus (lassa fever) and other arenaviruses, p 1329–1334. In Mandell GL, Gordon DR, Bennett JE (ed), *Principles and Practices of Infectious Diseases*. Churchill Livingstone Inc, New York, NY.
65. **Wright R, Johnson D, Neumann M, Ksiazek TG, Rollin P, Keech RV, Bonthius DJ, Hitchon P, Grose CF, Bell WE, Bale JF Jr.** 1997. Congenital lymphocytic choriomeningitis virus syndrome: a disease that mimics congenital toxoplasmosis or cytomegalovirus infection. *Pediatrics* **100:**E9.
66. **Barry M, Russi M, Armstrong L, Geller D, Tesh R, Dembry L, Gonzalez JP, Khan AS, Peters CJ.** 1995. Brief report: treatment of a laboratory-acquired Sabiá virus infection. *N Engl J Med* **333:**294–296.
67. **Gandsman EJ, Aaslestad HG, Ouimet TC, Rupp WD.** 1997. Sabia virus incident at Yale University. *Am Ind Hyg Assoc J* **58:**51–53.
68. **Andrei G, De Clercq E.** 1993. Molecular approaches for the treatment of hemorrhagic fever virus infections. *Antiviral Res* **22:**45–75.
69. **Mendenhall M, Russell A, Juelich T, Messina EL, Smee DF, Freiberg AN, Holbrook MR, Furuta Y, de la Torre JC, Nunberg JH, Gowen BB.** 2011. T-705 (favipiravir) inhibition of arenavirus replication in cell culture. *Antimicrob Agents Chemother* **55:**782–787.
70. **Whitehouse CA.** 2004. Crimean-Congo hemorrhagic fever. *Antiviral Res* **64:**145–160.
71. **Swanepoel R, Gill DE, Shepherd AJ, Leman PA, Mynhardt JH, Harvey S.** 1989. The clinical pathology of Crimean-Congo hemorrhagic fever. *Rev Infect Dis* **11**(Suppl 4)**:**S794–S800.
72. **Schmaljohn C, Hjelle B.** 1997. Hantaviruses: a global disease problem. *Emerg Infect Dis* **3:**95–104.
73. **Childs JE, Kaufmann AF, Peters CJ, Ehrenberg RL, Centers for Disease Control and Prevention.** 1993. Hantavirus infection—southwestern United States: interim recommendations for risk reduction. *MMWR Recomm Rep* **42**(RR-11)**:**1–13.
74. **Centers for Disease Control and Prevention (CDC).** 1993. Update: outbreak of hantavirus infection—southwestern United States, 1993. *MMWR Morb Mortal Wkly Rep* **42:**477–479.
75. **Tsai TF.** 1987. Hemorrhagic fever with renal syndrome: mode of transmission to humans. *Lab Anim Sci* **37:**428–430.
76. **Tsai TF, Bauer SP, Sasso DR, Whitfield SG, McCormick JB, Caraway TC, McFarland L, Bradford H, Kurata T.** 1985. Serological and virological evidence of a Hantaan virus-related enzootic in the United States. *J Infect Dis* **152:**126–136.
77. **LeDuc JW.** 1987. Epidemiology of Hantaan and related viruses. *Lab Anim Sci* **37:**413–418.
78. **Hjelle B, Tórrez-Martínez N, Koster FT, Jay M, Ascher MS, Brown T, Reynolds P, Ettestad P, Voorhees RE, Sarisky J, Enscore RE, Sands L, Mosley DG, Kioski C, Bryan RT, Sewell CM.** 1996. Epidemiologic linkage of rodent and human hantavirus genomic sequences in case investigations of hantavirus pulmonary syndrome. *J Infect Dis* **173:**781–786.
79. **Jay M, Hjelle B, Davis R, Ascher M, Baylies HN, Reilly K, Vugia D.** 1996. Occupational exposure leading to hantavirus pulmonary syndrome in a utility company employee. *Clin Infect Dis* **22:**841–844.
80. **Xu ZY, Tang YW, Kan LY, Tsai TF.** 1987. Cats—source of protection or infection? A case-control study of hemorrhagic fever with renal syndrome. *Am J Epidemiol* **126:**942–948.
81. **Kawamata J, Yamanouchi T, Dohmae K, Miyamoto H, Takahaski M, Yamanishi K, Kurata T, Lee HW.** 1987. Control of laboratory acquired hemorrhagic fever with renal syndrome (HFRS) in Japan. *Lab Anim Sci* **37:**431–436.
82. **Lee HW, Johnson KM.** 1982. Laboratory-acquired infections with Hantaan virus, the etiologic agent of Korean hemorrhagic fever. *J Infect Dis* **146:**645–651.
83. **Centers for Disease Control and Prevention.** 1994. Laboratory management of agents associated with hantavirus pulmonary syndrome: interim biosafety guidelines. *MMWR Recomm Rep* **43**(RR-7)**:**1–7.
84. **Huggins JW, Hsiang CM, Cosgriff TM, Guang MY, Smith JI, Wu ZO, LeDuc JW, Zheng ZM, Meegan JM, Wang QN, Oland DD, Gui XE, Gibbs PH, Yuan GH, Zhang TM.** 1991. Prospective, double-blind, concurrent, placebo-controlled clinical trial of intravenous ribavirin therapy of hemorrhagic fever with renal syndrome. *J Infect Dis* **164:**1119–1127.
85. **Moreli ML, Marques-Silva AC, Pimentel VA, da Costa VG.** 2014. Effectiveness of the ribavirin in treatment of hantavirus infections in the Americas and Eurasia: a meta-analysis. *Virusdisease* **25:**385–389.
86. **Dokuzoguz B, Celikbas AK, Gök SE, Baykam N, Eroglu MN, Ergönül Ö.** 2013. Severity scoring index for Crimean-Congo hemorrhagic fever and the impact of ribavirin and corticosteroids on fatality. *Clin Infect Dis* **57:**1270–1274.
87. **Ozbey SB, Kader Ç, Erbay A, Ergönül Ö.** 2014. Early use of ribavirin is beneficial in Crimean-Congo hemorrhagic fever. *Vector Borne Zoonotic Dis* **14:**300–302.
88. **Martinez-Valdebenito C, Calvo M, Vial C, Mansilla R, Marco C, Palma RE, Vial PA, Valdivieso F, Mertz G, Ferrés M.** 2014. Person-to-person household and nosocomial transmission of andes hantavirus, Southern Chile, 2011. *Emerg Infect Dis* **20:**1629–1636.
89. **World Health Organization.** 2016. Middle East respiratory syndrome coronavirus (MERS-CoV), *on* World Health Organization. http://who.int/emergencies/mers-cov/en/.
90. **Zumla A, Hui DS, Perlman S.** 2015. Middle East respiratory syndrome. *Lancet* **386:**995–1007.
91. **Li RW, Leung KW, Sun FC, Samaranayake LP.** 2004. Severe acute respiratory syndrome (SARS) and the GDP. Part I: Epidemiology, virology, pathology and general health issues. *Br Dent J* **197:**77–80.
92. **Senio K.** 2003. Recent Singapore SARS case a laboratory accident. *Lancet Infect Dis* **3:**679.
93. **Kuhn JH, et al.** 2014. Virus nomenclature below the species level: a standardized nomenclature for filovirus strains and variants rescued from cDNA. *Arch Virol* **159:**1229–1237.
94. **Negredo A, Palacios G, Vázquez-Morón S, González F, Dopazo H, Molero F, Juste J, Quetglas J, Savji N, de la Cruz Martínez M, Herrera JE, Pizarro M, Hutchison SK, Echevarría JE, Lipkin WI, Tenorio A.** 2011. Discovery of an ebolavirus-like filovirus in europe. *PLoS Pathog* **7:**e1002304.
95. **Formenty P, Hatz C, Le Guenno B, Stoll A, Rogenmoser P, Widmer A.** 1999. Human infection due to Ebola virus, subtype Côte d'Ivoire: clinical and biologic presentation. *J Infect Dis* **179**(Suppl 1)**:**S48–S53.
96. **Geisbert TW, Hensley LE.** 2004. Ebola virus: new insights into disease aetiopathology and possible therapeutic interventions. *Expert Rev Mol Med* **6:**1–24.

97. **Centers for Disease Control and Prevention.** 2016. 2014 Ebola Outbreak in West Africa. https://www.cdc.gov/vhf/ebola/outbreaks/2014-west-africa/.
98. **Schieffelin JS, Shaffer JG, Goba A, Gbakie M, Gire SK, Colubri A, Sealfon RS, Kanneh L, Moigboi A, Momoh M, Fullah M, Moses LM, Brown BL, Andersen KG, Winnicki S, Schaffner SF, Park DJ, Yozwiak NL, Jiang P-P, Kargbo D, Jalloh S, Fonnie M, Sinnah V, French I, Kovoma A, Kamara FK, Tucker V, Konuwa E, Sellu J, Mustapha I, Foday M, Yillah M, Kanneh F, Saffa S, Massally JL, Boisen ML, Branco LM, Vandi MA, Grant DS, Happi C, Gevao SM, Fletcher TE, Fowler RA, Bausch DG, Sabeti PC, Khan SH, Garry RF, Program KGHLF, Viral Hemorrhagic Fever C, Team WHOCR, KGH Lassa Fever Program, Viral Hemorrhagic Fever Consortium, WHO Clinical Response Team.** 2014. Clinical illness and outcomes in patients with Ebola in Sierra Leone. *N Engl J Med* **371:**2092–2100.
99. **WHO Ebola Response Team.** 2014. Ebola virus disease in West Africa—the first 9 months of the epidemic and forward projections. *N Engl J Med* **371:**1481–1495.
100. **Isaacson M.** 1988. Marburg and Ebola virus infections, p 185–197. In Gear JHS (ed), *Handbook of Viral and Rickettsial Hemorrhagic Fevers.* CRC Press, Boca Raton, FL.
101. **Breman JG, Johnson KM.** 2014. Ebola then and now. *N Engl J Med* **371:**1663–1666.
102. **Formenty P, Boesch C, Wyers M, Steiner C, Donati F, Dind F, Walker F, Le Guenno B.** 1999. Ebola virus outbreak among wild chimpanzees living in a rain forest of Côte d'Ivoire. *J Infect Dis* **179**(Suppl 1)**:**S120–S126.
103. **Dallatomasina S, Crestani R, Sylvester Squire J, Declerk H, Caleo GM, Wolz A, Stinson K, Patten G, Brechard R, Gbabai OB, Spreicher A, Van Herp M, Zachariah R.** 2015. Ebola outbreak in rural West Africa: epidemiology, clinical features and outcomes. *Trop Med Int Health* **20:**448–454.
104. **Osterholm MT, Moore KA, Kelley NS, Brosseau LM, Wong G, Murphy FA, Peters CJ, LeDuc JW, Russell PK, Van Herp M, Kapetshi J, Muyembe JJ, Ilunga BK, Strong JE, Grolla A, Wolz A, Kargbo B, Kargbo DK, Sanders DA, Kobinger GP.** 2015. Correction for Osterholm et al., Transmission of Ebola viruses: what we know and what we do not know. *MBio* **6:**e01154.
105. **Jahrling PB, Geisbert TW, Johnson ED, Peters CJ, Dalgard DW, Hall WC.** 1990. Preliminary report: isolation of Ebola virus from monkeys imported to USA. *Lancet* **335:**502–505.
106. **Francesconi P, Yoti Z, Declich S, Onek PA, Fabiani M, Olango J, Andraghetti R, Rollin PE, Opira C, Greco D, Salmaso S.** 2003. Ebola hemorrhagic fever transmission and risk factors of contacts, Uganda. *Emerg Infect Dis* **9:**1430–1437.
107. **Chughtai AA, Barnes M, Macintyre CR.** 2016. Persistence of Ebola virus in various body fluids during convalescence: evidence and implications for disease transmission and control. *Epidemiol Infect* **144:**1652–1660.
108. **Sow MS, Etard JF, Baize S, Magassouba N, Faye O, Msellati P, Toure AI, Savane I, Barry M, Delaporte E, for the Postebogui Study Group.** 2016. New evidence of long-lasting persistence of Ebola virus genetic material in semen of survivors. *J Infect Dis* **214:**1475–1476.
109. **Jacobs M, Rodger A, Bell DJ, Bhagani S, Cropley I, Filipe A, Gifford RJ, Hopkins S, Hughes J, Jabeen F, Johannessen I, Karageorgopoulos D, Lackenby A, Lester R, Liu RS, MacConnachie A, Mahungu T, Martin D, Marshall N, Mepham S, Orton R, Palmarini M, Patel M, Perry C, Peters SE, Porter D, Ritchie D, Ritchie ND, Seaton RA, Sreenu VB, Templeton K, Warren S, Wilkie GS, Zambon M, Gopal R, Thomson EC.** 2016. Late Ebola virus relapse causing meningoencephalitis: a case report. *Lancet* **388:**498–503.
110. **Emond RT, Evans B, Bowen ET, Lloyd G.** 1977. A case of Ebola virus infection. *BMJ* **2:**541–544.
111. **Miller J.** 2004. Russian scientist dies in Ebola accident at former weapons lab. *In* The New York Times. New York, NY. http://www.nytimes.com/2004/05/25/world/russian-scientist-dies-in-ebola-accident-at-former-weapons-lab.html?_r=0.
112. **Günther S, Feldmann H, Geisbert TW, Hensley LE, Rollin PE, Nichol ST, Ströher U, Artsob H, Peters CJ, Ksiazek TG, Becker S, ter Meulen J, Olschläger S, Schmidt-Chanasit J, Sudeck H, Burchard GD, Schmiedel S.** 2011. Management of accidental exposure to Ebola virus in the biosafety level 4 laboratory, Hamburg, Germany. *J Infect Dis* **204**(Suppl 3)**:**S785–S790.
113. **Kortepeter MG, Martin JW, Rusnak JM, Cieslak TJ, Warfield KL, Anderson EL, Ranadive MV.** 2008. Managing potential laboratory exposure to ebola virus by using a patient biocontainment care unit. *Emerg Infect Dis* **14:**881–887.
114. **Olu O, Kargbo B, Kamara S, Wurie AH, Amone J, Ganda L, Ntsama B, Poy A, Kuti-George F, Engedashet E, Worku N, Cormican M, Okot C, Yoti Z, Kamara KB, Chitala K, Chimbaru A, Kasolo F.** 2015. Epidemiology of Ebola virus disease transmission among health care workers in Sierra Leone, May to December 2014: a retrospective descriptive study. *BMC Infect Dis* **15:**416.
115. **Haddock E, Feldmann F, Feldmann H.** 2016. Effective chemical inactivation of Ebola virus. *Emerg Infect Dis* **22:**1292–1294.
116. **Uyeki TM, Mehta AK, Davey RT Jr, Liddell AM, Wolf T, Vetter P, Schmiedel S, Grünewald T, Jacobs M, Arribas JR, Evans L, Hewlett AL, Brantsaeter AB, Ippolito G, Rapp C, Hoepelman AI, Gutman J, Working Group of the U.S.–European Clinical Network on Clinical Management of Ebola Virus Disease Patients in the U.S. and Europe.** 2016. Clinical management of Ebola virus disease in the United States and Europe. *N Engl J Med* **374:**636–646.
117. **Jacobs M, Aarons E, Bhagani S, Buchanan R, Cropley I, Hopkins S, Lester R, Martin D, Marshall N, Mepham S, Warren S, Rodger A.** 2015. Post-exposure prophylaxis against Ebola virus disease with experimental antiviral agents: a case-series of health-care workers. *Lancet Infect Dis* **15:**1300–1304.
118. **Wong KK, Davey RT Jr, Hewlett AL, Kraft CS, Mehta AK, Mulligan MJ, Beck A, Dorman W, Kratochvil CJ, Lai L, Palmore TN, Rogers S, Smith PW, Suffredini AF, Wolcott M, Ströher U, Uyeki TM.** 2016. Use of postexposure prophylaxis after occupational exposure to Zaire ebolavirus. *Clin Infect Dis* **63:**376–379.
119. **Henao-Restrepo AM, Longini IM, Egger M, Dean NE, Edmunds WJ, Camacho A, Carroll MW, Doumbia M, Draguez B, Duraffour S, Enwere G, Grais R, Gunther S, Hossmann S, Kondé MK, Kone S, Kuisma E, Levine MM, Mandal S, Norheim G, Riveros X, Soumah A, Trelle S, Vicari AS, Watson CH, Këita S, Kieny MP, Røttingen JA.** 2015. Efficacy and effectiveness of an rVSV-vectored vaccine expressing Ebola surface glycoprotein: interim results from the Guinea ring vaccination cluster-randomised trial. *Lancet* **386:**857–866.
120. **Frank C, Cadar D, Schlaphof A, Neddersen N, Günther S, Schmidt-Chanasit J, Tappe D.** 2016. Sexual transmission of Zika virus in Germany, April 2016. *Euro Surveill* **21:**21.
121. **Musso D, Roche C, Robin E, Nhan T, Teissier A, Cao-Lormeau VM.** 2015. Potential sexual transmission of Zika virus. *Emerg Infect Dis* **21:**359–361.
122. **Panchaud A, Stojanov M, Ammerdorffer A, Vouga M, Baud D.** 2016. Emerging role of Zika virus in adverse fetal and neonatal outcomes. *Clin Microbiol Rev* **29:**659–694.
123. **Sartori M, La Terra G, Aglietta M, Manzin A, Navino C, Verzetti G.** 1993. Transmission of hepatitis C via blood splash into conjunctiva. *Scand J Infect Dis* **25:**270–271.
124. **Alter MJ.** 1997. The epidemiology of acute and chronic hepatitis C. *Clin Liver Dis* **1:**559–568, vi–vii.
125. **Mitsui T, Iwano K, Masuko K, Yamazaki C, Okamoto H, Tsuda F, Tanaka T, Mishiro S.** 1992. Hepatitis C virus infection in medical personnel after needlestick accident. *Hepatology* **16:**1109–1114.
126. **Foy BD, Kobylinski KC, Chilson Foy JL, Blitvich BJ, Travassos da Rosa A, Haddow AD, Lanciotti RS, Tesh RB.** 2011. Probable non-vector-borne transmission of Zika virus, Colorado, USA. *Emerg Infect Dis* **17:**880–882.

127. **Anonymous.** 2016. Zika virus: laboratory acquired case reported in pittsburgh area. *Outbreak News Today*, June 9, 2016.
128. **Hayden EC.** 2016. The race is on to develop Zika vaccine. *Nature* doi:10.1038/nature.2016.19634. http://www.nature.com/news/the-race-is-on-to-develop-zika-vaccine-1.19634
129. **Schnirring L.** 2016. FDA paves way for first human Zika vaccine trial. *CIDRAP News*, June 20, 2016.
130. **Hull MW, Yoshida EM, Montaner JS.** 2016. Update on current evidence for hepatitis C therapeutic options in HCV mono-infected patients. *Curr Infect Dis Rep* **18:**22.
131. **Lee JY, Locarnini S.** 2004. Hepatitis B virus: pathogenesis, viral intermediates, and viral replication. *Clin Liver Dis* **8:**301–320.
132. **Locarnini S.** 2004. Molecular virology of hepatitis B virus. *Semin Liver Dis* **24**(Suppl 1):3–10.
133. **Bouvet E, Tarantola A.** 1998. [Protection of hospital personnel against risks of exposure to blood]. *Rev Prat* **48:**1558–1562.
134. **Cavalieri J.** 2001. Responding rapidly to occupational blood and body-fluid exposures. *JAAPA* 14:22–24, 27–30, 33–25.
135. **Westland CE, Yang H, Delaney WE IV, Wulfsohn M, Lama N, Gibbs CS, Miller MD, Fry J, Brosgart CL, Schiff ER, Xiong S.** 2005. Activity of adefovir dipivoxil against all patterns of lamivudine-resistant hepatitis B viruses in patients. *J Viral Hepat* **12:**67–73.
136. **Zoulim F.** 2004. Antiviral therapy of chronic hepatitis B: can we clear the virus and prevent drug resistance? *Antivir Chem Chemother* **15:**299–305.
137. **Lok AS, McMahon BJ, Brown RS Jr, Wong JB, Ahmed AT, Farah W, Almasri J, Alahdab F, Benkhadra K, Mouchli MA, Singh S, Mohamed EA, Abu Dabrh AM, Prokop LJ, Wang Z, Murad MH, Mohammed K.** 2016. Antiviral therapy for chronic hepatitis B viral infection in adults: a systematic review and meta-analysis. *Hepatology* **63:**284–306.
138. **Roizman B, Sears AE.** 1993. Herpes simplex viruses and their replication, p 1795–1841. *In* Roizman B, Whitley RJ, Lopez C (ed), *The Human Herpesviruses*. Raven Press, New York, NY.
139. **Heymann JB, Conway JF, Steven AC.** 2004. Molecular dynamics of protein complexes from four-dimensional cryo-electron microscopy. *J Struct Biol* **147:**291–301.
140. **Spruance SL, Overall JC Jr, Kern ER, Krueger GG, Pliam V, Miller W.** 1977. The natural history of recurrent herpes simplex labialis: implications for antiviral therapy. *N Engl J Med* **297:**69–75.
141. **Baringer JR, Swoveland P.** 1973. Recovery of herpes-simplex virus from human trigeminal ganglions. *N Engl J Med* **288:**648–650.
142. **Douglas MW, Walters JL, Currie BJ.** 2002. Occupational infection with herpes simplex virus type 1 after a needlestick injury. *Med J Aust* **176:**240.
143. **Manian FA.** 2000. Potential role of famciclovir for prevention of herpetic whitlow in the health care setting. *Clin Infect Dis* **31:**E18–E19.
144. **Rosato FE, Rosato EF, Plotkin SA.** 1970. Herpetic paronychia—an occupational hazard of medical personnel. *N Engl J Med* **283:**804–805.
145. **Hunt RD, Carlton WW, King NW.** 1978. Viral diseases, p 1313. *In* Benirschke K, Garner FM, Jones TC (ed), *Pathology of Laboratory Animals*. Springer-Verlag, New York, NY.
146. **Mansfield K, King N.** 1998. Viral diseases, p 1–57. *In* Bennett BT, Abee CR, Henrickson R (ed), *Nonhuman Primates in Biomedical Research*. Academic Press, San Diego, CA.
147. **Centers for Disease Control (CDC).** 1987. Guidelines for prevention of Herpesvirus simiae (B virus) infection in monkey handlers. *MMWR Morb Mortal Wkly Rep* **36:**680–682, 687–689.
148. **Centers for Disease Control (CDC).** 1989. B virus infections in humans—Michigan. *MMWR Morb Mortal Wkly Rep* **38:**453–454.
149. **Gay F, Holden M.** 1933. The herpes encephalitis problem, II. *J Infect Dis* **53:**287–303.
150. **Zwartouw HT, Boulter EA.** 1984. Excretion of B virus in monkeys and evidence of genital infection. *Lab Anim* **18:**65–70.
151. **Weigler BJ, Scinicariello F, Hilliard JK.** 1995. Risk of venereal B virus (cercopithecine herpesvirus 1) transmission in rhesus monkeys using molecular epidemiology. *J Infect Dis* **171:**1139–1143.
152. **Weigler BJ, Hird DW, Hilliard JK, Lerche NW, Roberts JA, Scott LM, Weigler BJ, Hird DW, Hilliard JK, Lerche NW, Roberts JA, Scott LM.** 1993. Epidemiology of cercopithecine herpesvirus 1 (B virus) infection and shedding in a large breeding cohort of rhesus macaques. *J Infect Dis* **167:**257–263.
153. **Holmes GP, Chapman LE, Stewart JA, Straus SE, Hilliard JK, Davenport DS.** 1995. Guidelines for the prevention and treatment of B-virus infections in exposed persons. The B virus Working Group. *Clin Infect Dis* **20:**421–439.
154. **Kalter SS, Heberling RL, Cooke AW, Barry JD, Tian PY, Northam WJ.** 1997. Viral infections of nonhuman primates. *Lab Anim Sci* **47:**461–467.
155. **Palmer AE.** 1987. B virus, Herpesvirus simiae: historical perspective. *J Med Primatol* **16:**99–130.
156. **Benson PM, Malane SL, Banks R, Hicks CB, Hilliard J.** 1989. B virus (Herpesvirus simiae) and human infection. *Arch Dermatol* **125:**1247–1248.
157. **Holmes GP, et al.** 1990. B virus (*Herpesvirus simiae*) infection in humans: epidemiologic investigation of a cluster. *Ann Intern Med* **112:**833–839.
158. **Wells DL, Lipper SL, Hilliard JK, Stewart JA, Holmes GP, Herrmann KL, Kiley MP, Schonberger LB.** 1989. Herpesvirus simiae contamination of primary rhesus monkey kidney cell cultures. CDC recommendations to minimize risks to laboratory personnel. *Diagn Microbiol Infect Dis* **12:**333–335.
159. **Benenson AS.** 1995. *Control of Communicable Diseases Manual*, 16th ed. American Public Health Association, Washington, DC.
160. **Webster RG.** 1997. Influenza virus: transmission between species and relevance to emergence of the next human pandemic. *Arch Virol Suppl* **13:**105–113.
161. **Marini RP, Adkins JA, Fox JG.** 1989. Proven or potential zoonotic diseases of ferrets. *J Am Vet Med Assoc* **195:**990–994.
162. **Wentworth DE, McGregor MW, Macklin MD, Neumann V, Hinshaw VS.** 1997. Transmission of swine influenza virus to humans after exposure to experimentally infected pigs. *J Infect Dis* **175:**7–15.
163. **Cohen J.** 2009. Flu researchers train sights on novel tricks of novel N1H1. *Science* 324:870–871.
164. **Kelly H, Peck HA, Laurie KL, Wu P, Nishiura H, Cowling BJ.** 2011. The age-specific cumulative incidence of infection with pandemic influenza H1N1 2009 was similar in various countries prior to vaccination. *PLoS One* **6:**e21828.
165. **World Health Organization.** 2016. *Cumulative Number of Confirmed Human Cases for Avian Influenza A (H5N1) Reported to WHO*. World Health Organization, Geneva, Switzerland.
166. **Ruef C.** 2004. Immunization for hospital staff. *Curr Opin Infect Dis* **17:**335–339.
167. **von Hoersten B, Sharland M.** 2004. RSV and influenza. Treatment and prevention. *Adv Exp Med Biol* **549:**169–175.
168. **De Clercq E.** 2004. Antiviral drugs in current clinical use. *J Clin Virol* **30:**115–133.
169. **Schmidt AC.** 2004. Antiviral therapy for influenza: a clinical and economic comparative review. *Drugs* **64:**2031–2046.
170. **Barkley WE, Richardson JH.** 1984. Control of biohazards associated with the use of experimental animals, p 595–602. *In* Fox JG, Cohen BE, Loew FM (ed), *Laboratory Animal Medicine. American College of Laboratory Animal Medicine*. Academic Press, San Diego, CA.
171. **Mufson MA.** 1989. Parainfluenza viruses, mumps, and Newcastle disease virus, p 669–691. *In* Schmidt N, Emmons R, *Diagnostic Procedures for Viral, Rickettsial, and Chlamydial Infections*. American Public Health Association, Washington, DC.
172. **Centers for Disease Control.** 1971. Hepatitis surveillance. Report no. 34, p 10–14. Centers for Disease Control, Atlanta, GA.
173. **Anonymous.** 1991. Update on adult immunization. Recommendations of the Immunization Practices Advisory Committee (ACIP). *MMWR Recomm Rep* **40**(RR-12):1–94.

174. **Fenner F.** 1990. Wallace P. Rowe lecture. Poxviruses of laboratory animals. *Lab Anim Sci* **40:**469–480.
175. **Breman JG, Kalisa-Ruti, Steniowski MV, Zanotto E, Gromyko AI, Arita I.** 1980. Human monkeypox, 1970–79. *Bull World Health Organ* **58:**165–182.
176. **Jezek Z, Arita I, Mutombo M, Dunn C, Nakano JH, Szczeniowski M.** 1986. Four generations of probable person-to-person transmission of human monkeypox. *Am J Epidemiol* **123:**1004–1012.
177. **Soave O.** 1981. Viral infections common to human and nonhuman primates. *J Am Vet Med Assoc* **179:**1385–1388.
178. **Jezek Z, Gromyko AI, Szczeniowski MV.** 1983. Human monkeypox. *J Hyg Epidemiol Microbiol Immunol* **27:**13–28.
179. **Jezek Z, Fenner F.** 1988. Epidemiology of human monkeypox, p 81–110, *In* Jezek Z, Fenner F, *Human Monkeypox, Monographs in Virology*, vol 17. Karger, Basel, Switzerland.
180. **Fleischauer AT, Kile JC, Davidson M, Fischer M, Karem KL, Teclaw R, Messersmith H, Pontones P, Beard BA, Braden ZH, Cono J, Sejvar JJ, Khan AS, Damon I, Kuehnert MJ.** 2005. Evaluation of human-to-human transmission of monkeypox from infected patients to health care workers. *Clin Infect Dis* **40:**689–694.
181. **McNulty WP Jr, Lobitz WC Jr, Hu F, Maruffo CA, Hall AS.** 1968. A pox disease in monkeys transmitted to man. Clinical and histological features. *Arch Dermatol* **97:**286–293.
182. **Espana C.** 1971. Apox disease of monkeys tranmissible to man, p 694–708. *In* Goldsmith E, Moor-Jankowski J (ed), *Medical Primatology*. Karger, Basel, Switzerland.
183. **Hall AS, McNulty WP Jr.** 1967. A contagious pox disease in monkeys. *J Am Vet Med Assoc* **151:**833–838.
184. **Nakano JH, Esposito JJ.** 1989. Poxviruses, p 453–511. *In* Schmidt NJ, Emmons RW (ed), *Diagnostic Procedures for Viral, Rickettsial and Chlamydial Infections*, 6th ed. American Public Health Association, Washington, DC.
185. **U.S. Department of Health and Human Services, Public Health Service, Centers for Disease Control and Prevention, National Institutes of Health.** 2009. *Biosafety in Microbiological and Biomedical Laboratories*, 5th ed. HHS Publication no. (CDC) 21-112. http://www.cdc.gov/biosafety/publications/bmbl5/BMBL.pdf.
186. **Bearcroft WG, Jamieson MF.** 1958. An outbreak of subcutaneous tumours in rhesus monkeys. *Nature* **182:**195–196.
187. **Walker DH, Voelker FA, McKee AE Jr, Nakano JH.** 1985. Diagnostic exercise: tumors in a baboon. *Lab Anim Sci* **35:**627–628.
188. **Ambrus JL, Strandström HV.** 1966. Susceptibility of Old World monkeys to Yaba virus. *Nature* **211:**876.
189. **Ambrus JL, Strandstrom HV, Kawinski W.** 1969. 'Spontaneous' occurrence of Yaba tumor in a monkey colony. *Experientia* **25:**64–65.
190. **Wolfe LG, Griesemer RA, Farrell RL.** 1968. Experimental aerosol transmission of Yaba virus in monkeys. *J Natl Cancer Inst* **41:**1175–1195.
191. **Niven JS, Armstrong JA, Andrewes CH, Pereira HG, Valentine RC.** 1961. Subcutaneous "growths" in monkeys produced by a poxvirus. *J Pathol Bacteriol* **81:**1–14.
192. **Bruestle ME, Golden JG, Hall A III, Banknieder AR.** 1981. Naturally occurring Yaba tumor in a baboon (Papio papio). *Lab Anim Sci* **31:**292–294.
193. **Grace JT Jr, Mirand EA, Millian SJ, Metzgar RS.** 1962. Experimental studies of human tumors. *Fed Proc* **21:**32–36.
194. **Haig DM, McInnes C, Deane D, Reid H, Mercer A.** 1997. The immune and inflammatory response to orf virus. *Comp Immunol Microbiol Infect Dis* **20:**197–204.
195. **Gibbs EPJ.** 1998. Contagious ecthyma, p 619–620. *In* Aiello SA (ed), *The Merck Veterinary Manual*, 8th ed. Merck & Co, Whitehouse Station, NJ.
196. **Mercer A, Fleming S, Robinson A, Nettleton P, Reid H.** 1997. Molecular genetic analyses of parapoxviruses pathogenic for humans. *Arch Virol Suppl* **13:**25–34.
197. **Johannessen JV, Krogh H-K, Solberg I, Dalen A, van Wijngaarden H, Johansen B.** 1975. Human orf. *J Cutan Pathol* **2:**265–283.
198. **Moussatché N, Tuyama M, Kato SE, Castro AP, Njaine B, Peralta RH, Peralta JM, Damaso CR, Barroso PF.** 2003. Accidental infection of laboratory worker with vaccinia virus. *Emerg Infect Dis* **9:**724–726.
199. **Wlodaver CG, Palumbo GJ, Waner JL.** 2004. Laboratory-acquired vaccinia infection. *J Clin Virol* **29:**167–170.
200. **Grosenbach DW, Jordan R, Hruby DE.** 2011. Development of the small-molecule antiviral ST-246 as a smallpox therapeutic. *Future Virol* **6:**653–671.
201. **Beltrami EM, Williams IT, Shapiro CN, Chamberland ME.** 2000. Risk and management of blood-borne infections in health care workers. *Clin Microbiol Rev* **13:**385–407.
202. **Centers for Disease Control and Prevention.** 2011. Surveillance of occupationally acquired HIV/AIDS in healthcare personnel, as of December 2010. https://www.cdc.gov/hai/organisms/hiv/surveillance-occupationally-acquired-hiv-aids.html.
203. **Bell DM.** 1997. Occupational risk of human immunodeficiency virus infection in healthcare workers: an overview. *Am J Med* **102**(5B):9–15.
204. **Ippolito G, Puro V, De Carli G.** 1993. The risk of occupational human immunodeficiency virus infection in health care workers. Italian Multicenter Study. The Italian Study Group on Occupational Risk of HIV infection. *Arch Intern Med* **153:**1451–1458.
205. **Evans B, Duggan W, Baker J, Ramsay M, Abiteboul D.** 2001. Exposure of healthcare workers in England, Wales, and Northern Ireland to bloodborne viruses between July 1997 and June 2000: analysis of surveillance data. *BMJ* **322:**397–398.
206. **Cardo DM, Culver DH, Ciesielski CA, Srivastava PU, Marcus R, Abiteboul D, Heptonstall J, Ippolito G, Lot F, McKibben PS, Bell DM, Centers for Disease Control and Prevention Needlestick Surveillance Group.** 1997. A case-control study of HIV seroconversion in health care workers after percutaneous exposure. *N Engl J Med* **337:**1485–1490.
207. **Bell DM.** 1991. Human immunodeficiency virus transmission in health care settings: risk and risk reduction. *Am J Med* **91**(3B)**:**S294–S300.
208. **Fahey BJ, Koziol DE, Banks SM, Henderson DK.** 1991. Frequency of nonparenteral occupational exposures to blood and body fluids before and after universal precautions training. *Am J Med* **90:**145–153.
209. **Hayami M, Ido E, Miura T.** 1994. Survey of simian immunodeficiency virus among nonhuman primate populations. *Curr Top Microbiol Immunol* **188:**1–20.
210. **Centers for Disease Control and Prevention(CDC).** 1992. Anonymous survey for simian immunodeficiency virus (SIV) seropositivity in SIV-laboratory researchers—United States, 1992. *MMWR Morb Mortal Wkly Rep* **41:**814–815.
211. **Centers for Disease Control and Prevention (CDC).** 1992. Seroconversion to simian immunodeficiency virus in two laboratory workers. *MMWR Morb Mortal Wkly Rep* **41:**678–681.
212. **Khabbaz RF, Heneine W, George JR, Parekh B, Rowe T, Woods T, Switzer WM, McClure HM, Murphey-Corb M, Folks TM.** 1994. Brief report: infection of a laboratory worker with simian immunodeficiency virus. *N Engl J Med* **330:**172–177.
213. **Khabbaz RF, Rowe T, Heneine WM, Kaplan JE, Folks TM, Schable CA, George JR, Pau C, Parekh BS, Curran JW, Schochetman G, Lairmore MD, Murphey-Corb M.** 1992. Simian immunodeficiency virus needlestick accident in a laboratory worker. *Lancet* **340:**271–273.
214. **Heneine W, Switzer WM, Sandstrom P, Brown J, Vedapuri S, Schable CA, Khan AS, Lerche NW, Schweizer M, Neumann-Haefelin D, Chapman LE, Folks TM.** 1998. Identification of a human population infected with simian foamy viruses. *Nat Med* **4:**403–407.
215. **Neumann-Haefelin D, Fleps U, Renne R, Schweizer M.** 1993. Foamy viruses. *Intervirology* **35:**196–207.

216. **Busch MP, Satten GA.** 1997. Time course of viremia and antibody seroconversion following human immunodeficiency virus exposure. *Am J Med* 102:117–124; discussion 125–126.
217. **Jackson AC.** 2003. Rabies virus infection: an update. *J Neurovirol* **9:**253–258.
218. **Rupprecht CE, Gibbons RV.** 2004. Clinical practice. Prophylaxis against rabies. *N Engl J Med* **351:**2626–2635.
219. **Anderson LJ, Sikes RK, Langkop CW, Mann JM, Smith JS, Winkler WG, Deitch MW.** 1980. Postexposure trial of a human diploid cell strain rabies vaccine. *J Infect Dis* **142:**133–138.
220. **Suntharasamai P, Warrell MJ, Warrell DA, Viravan C, Looareesuwan S, Supanaranond W, Chanthavanich P, Supapochana A, Tepsumethanon W, Pouradier-Duteil X.** 1986. New purified Vero-cell vaccine prevents rabies in patients bitten by rabid animals. *Lancet* **2:**129–131.
221. **Servat A, Lutsch C, Delore V, Lang J, Veitch K, Cliquet F.** 2003. Efficacy of rabies immunoglobulins in an experimental post-exposure prophylaxis rodent model. *Vaccine* **22:**244–249.
222. **Pratt WD, Davis NL, Johnston RE, Smith JF.** 2003. Genetically engineered, live attenuated vaccines for Venezuelan equine encephalitis: testing in animal models. *Vaccine* **21:**3854–3862.
223. **Herfst S, Schrauwen EJ, Linster M, Chutinimitkul S, de Wit E, Munster VJ, Sorrell EM, Bestebroer TM, Burke DF, Smith DJ, Rimmelzwaan GF, Osterhaus AD, Fouchier RA.** 2012. Airborne transmission of influenza A/H5N1 virus between ferrets. *Science* **336:**1534–1541.
224. **Imai M, Watanabe T, Hatta M, Das SC, Ozawa M, Shinya K, Zhong G, Hanson A, Katsura H, Watanabe S, Li C, Kawakami E, Yamada S, Kiso M, Suzuki Y, Maher EA, Neumann G, Kawaoka Y.** 2012. Experimental adaptation of an influenza H5 HA confers respiratory droplet transmission to a reassortant H5 HA/H1N1 virus in ferrets. *Nature* **486:**420–428.
225. **Kaiser J.** 2015. NIH moving ahead with review of risky virology studies. *Sci News* http://www.sciencemag.org/news/2015/02/nih-moving-ahead-review-risky-virology-studies.
226. **National Science Advisory Board for Biosecurity.** 2016. *Recommendations for the Evaluation and Oversight of Proposed Gain-of-Function Research.* Office of Science Policy, National Institutes of Health, Bethesda, MD.
227. **Anderson RA, Woodfield DG.** 1982. Hepatitis B virus infections in laboratory staff. *N Z Med J* **95:**69–71.
228. **Artenstein AW, Hicks CB, Goodwin BS Jr, Hilliard JK.** 1991. Human infection with B virus following a needlestick injury. *Rev Infect Dis* **13:**288–291.
229. **Cohen BJ, Couroucé AM, Schwarz TF, Okochi K, Kurtzman GJ.** 1988. Laboratory infection with parvovirus B19. *J Clin Pathol* **41:**1027–1028.
230. **Lisieux T, et al.** 1994. New arenavirus isolated in Brazil. *Lancet* **343:**391–392.
231. **Davenport DS, Johnson DR, Holmes GP, Jewett DA, Ross SC, Hilliard JK.** 1994. Diagnosis and management of human B virus (Herpesvirus simiae) infections in Michigan. *Clin Infect Dis* **19:**33–41.
232. **Desmyter J, Johnson KM, Deckers C, LeDuc JW, Brasseur F, van Ypersele de Strihou C.** 1983. Laboratory rat associated outbreak of haemorrhagic fever with renal syndrome due to Hantaan-like virus in Belgium. *Lancet* 32**2:**1445–1448.
233. **Douron E, Moriniere B, Matheron S, Girard PM, Gonzalez JP, Hirsch F, McCormick JB.** 1984. HFRS after a wild rodent bite in the Haute-Savoie—and risk of exposure to Hantaan-like virus in a Paris laboratory. *Lancet* **1:**676–677.
234. **Freifeld AG, Hilliard J, Southers J, Murray M, Savarese B, Schmitt JM, Straus SE.** 1995. A controlled seroprevalence survey of primate handlers for evidence of asymptomatic herpes B virus infection. *J Infect Dis* **171:**1031–1034.
235. **Grist NR.** 1983. Infections in British clinical laboratories 1980–81. *J Clin Pathol* **36:**121–126.
236. **Grist NR, Emslie J.** 1985. Infections in British clinical laboratories, 1982–3. *J Clin Pathol* **38:**721–725.
237. **Grist NR, Emslie JA.** 1987. Infections in British clinical laboratories, 1984–5. *J Clin Pathol* **40:**826–829.
238. **Ippolito G, Puro V, Heptonstall J, Jagger J, De Carli G, Petrosillo N.** 1999. Occupational human immunodeficiency virus infection in health care workers: worldwide cases through September 1997. *Clin Infect Dis* **28:**365–383.
239. **Lloyd G, Jones N.** 1986. Infection of laboratory workers with hantavirus acquired from immunocytomas propagated in laboratory rats. *J Infect* **12:**117–125.
240. **Masuda T, Isokawa T.** 1991. [Biohazard in clinical laboratories in Japan]. *Kansenshogaku Zasshi* **65:**209–215.
241. **Moore DM, MacKenzie WF, Doepel F, Hansen TN.** 1983. Contagious ecthyma in lambs and laboratory personnel. *Lab Anim Sci* **33:**473–475.
242. **Schweizer M, Turek R, Hahn H, Schliephake A, Netzer KO, Eder G, Reinhardt M, Rethwilm A, Neumann-Haefelin D.** 1995. Markers of foamy virus infections in monkeys, apes, and accidentally infected humans: appropriate testing fails to confirm suspected foamy virus prevalence in humans. *AIDS Res Hum Retroviruses* **11:**161–170.
243. **Shiraishi H, Sasaki T, Nakamura M, Yaegashi N, Sugamura K.** 1991. Laboratory infection with human parvovirus B19. *J Infect* **22:**308–310.
244. **Umenai T, Woo Lee P, Toyoda T, Yoshinaga K, Horiuchi T, Wang Lee H, Saito T, Hongo M, Ishida N.** 1979. Korean haemorrhagic fever in staff in an animal laboratory. *Lancet* **313:** 1314–1316.
245. **Vasconcelos PF, Travassos da Rosa AP, Rodrigues SG, Tesh R, Travassos da Rosa JF, Travassos da Rosa ES.** 1993. [Laboratory-acquired human infection with SP H 114202 virus (Arenavirus: Arenaviridae family): clinical and laboratory aspects]. *Rev Inst Med Trop Sao Paulo* **35:**521–525.
246. **Wong TW, Chan YC, Yap EH, Joo YG, Lee HW, Lee PW, Yanagihara R, Gibbs CJ Jr, Gajdusek DC.** 1988. Serological evidence of hantavirus infection in laboratory rats and personnel. *Int J Epidemiol* **17:**887–890.
247. **Lerche NW, Switzer WM, Yee JL, Shanmugam V, Rosenthal AN, Chapman LE, Folks TM, Heneine W.** 2001. Evidence of infection with simian type D retrovirus in persons occupationally exposed to nonhuman primates. *J Virol* **75:**1783–1789.
248. **Heymann DL, Aylward RB, Wolff C.** 2004. Dangerous pathogens in the laboratory: from smallpox to today's SARS setbacks and tomorrow's polio-free world. *Lancet* **363:**1566–1568.
249. **Rusnak JM, Kortepeter MG, Aldis J, Boudreau E.** 2004. Experience in the medical management of potential laboratory exposures to agents of bioterrorism on the basis of risk assessment at the United States Army Medical Research Institute of Infectious Diseases (USAMRIID). *J Occup Environ Med* **46:**801–811.
250. **MacNeil A, Reynolds MG, Damon IK.** 2009. Risks associated with vaccinia virus in the laboratory. *Virology* **385:**1–4.
251. **Wurtz N, Papa A, Hukic M, Di Caro A, Leparc-Goffart I, Leroy E, Landini MP, Sekeyova Z, Dumler JS, Bădescu D, Busquets N, Calistri A, Parolin C, Palù G, Christova I, Maurin M, La Scola B, Raoult D.** 2016. Survey of laboratory-acquired infections around the world in biosafety level 3 and 4 laboratories. *Eur J Clin Microbiol Infect Dis* **35:**1247–1258.
252. **Harding A, Byers K.** 2006. Epidemiology of laboratory-associated infections, p 53–77. *In* Fleming D, Hunt D (ed), *Biological Safety*, vol 4. ASM Press, Washington, DC.
253. **Hsu CH, Farland J, Winters T, Gunn J, Caron D, Evans J, Osadebe L, Bethune L, McCollum AM, Patel N, Wilkins K, Davidson W, Petersen B, Barry MA.** 2015. Laboratory-acquired vaccinia virus infection in a recently immunized person—Massachusetts, 2013. *MMWR Morb Mortal Wkly Rep* **64:**435–438.
254. **Wong SS, Poon RW, Wong SC.** 2016. Zika virus infection—the next wave after dengue? *J Formos Med Assoc* **115:**226–242.

Emerging Considerations in Virus-Based Gene Transfer Systems

10

J. PATRICK CONDREAY, THOMAS A. KOST, AND CLAUDIA A. MICKELSON

Accompanying the development of molecular biological tools for identifying gene sequences and functions has been the development of novel gene transfer vectors for shuttling gene sequences between different organisms. As illustrated in Fig. 1, a search of the PubMed database reveals a large number of publications describing the application of frequently used recombinant viruses. The use of these vector systems has continued to increase significantly since the publication of the 4th edition of this book in 2006. This increase has been driven by several factors, including the proliferation of gene therapy protocols (Fig. 2), the increased recognition of the general utility of viral vectors as gene transfer agents for elucidating and studying gene function, and the increased commercial access of the reagents required to produce these viral vectors. Additionally, our understanding of the molecular biology of a wide variety of virus families has increased to the point where novel chimeric viruses, containing the unique properties of two or more viruses, are being constructed routinely. Due to their unique nature, these viruses pose special challenges in risk assessment.

This chapter will provide a general overview of the construction, safety features, and suggested containment level of viral gene transfer vectors in common laboratory use or under development. In addition, we will discuss the challenges facing Institutional Biosafety Committees (IBCs) in conducting viral vector risk assessments. The discussion of each vector will focus on similarities of the biosafety features and issues arising from the construction and use of these gene transfer systems. This effort was driven in part by the recognition that different IBCs may assess the use of similar vector-transgene constructs at different containment levels and that developments in the field of novel virus-mediated gene transfer vectors continue to progress at a rapid rate. The safety assessments and suggested containment levels presented in this chapter are the opinions of the authors only. The ideas and resources mentioned in this chapter can be used to inform the discussions of local IBCs and assist in the assessment of the risks inherent in the use of these vectors, thereby increasing the effectiveness of the research oversight process.

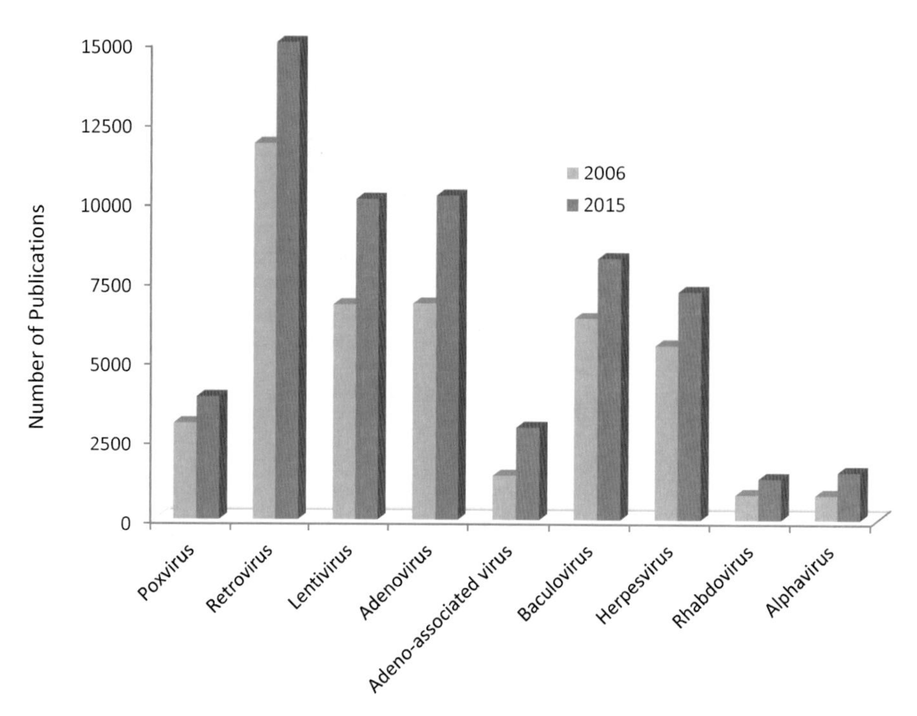

Figure 1: Number of PubMed citations found for recombinant viral vectors.

VIRAL VECTOR TECHNOLOGY

Viruses are desirable systems for recombinant gene delivery platforms due to their ability to condense, package, and deliver nucleic acids to new cells. This property of reproducing and disseminating their nucleic acid payload throughout a population of cells, and the host's response to that spread, is mainly responsible for the pathogenicity of a viral infection in a host. Therefore, to exploit the gene delivery capability of a virus, it is often

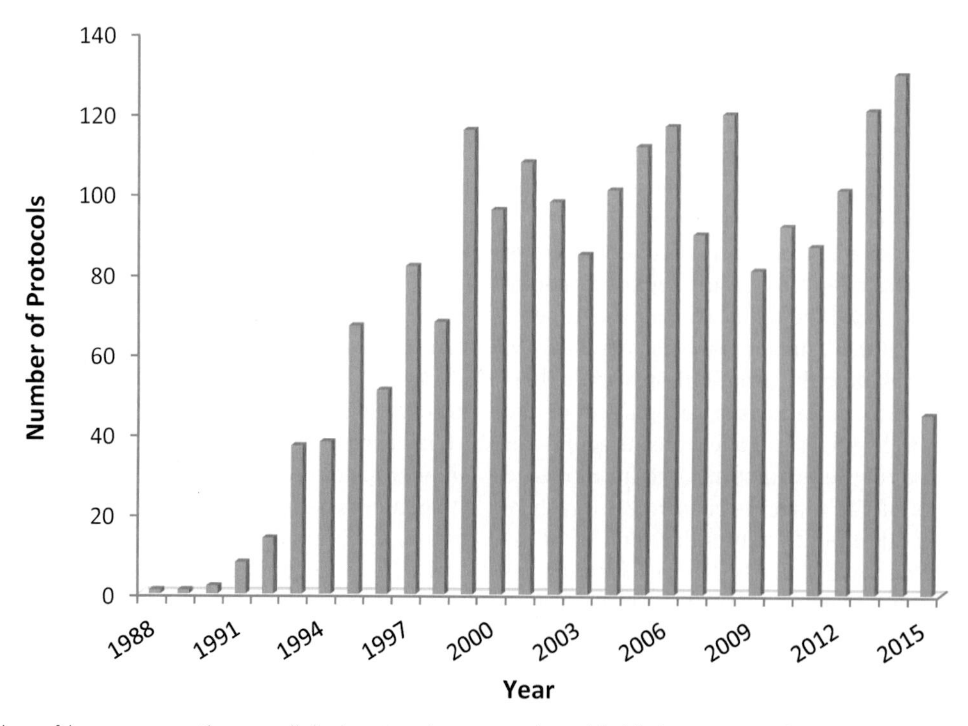

Figure 2: Number of human gene therapy clinical protocols approved worldwide by year as of July 2015. Source data are from reference 182.

necessary first to disable its ability to replicate in a host, thus reducing its capability to disseminate and cause pathogenic responses.

General Viral Genetic Structure

The vast diversity that exists in the virus world makes generalizations difficult; however, viruses typically consist of viral nucleic acid densely packed in a protein capsid structure, usually referred to as a nucleocapsid. Although capsids can vary widely in their structure and shape, they all function to lend protection to the nucleic acid genome. For certain virus families (enveloped viruses, e.g. retroviruses, poxviruses), the nucleocapsid is surrounded by a membrane called the viral envelope derived from the infected host cell. Viral envelopes are studded with one or more virus-encoded proteins that are involved in cell recognition/binding/entry. Other virus types (nonenveloped viruses, e.g., adenovirus, adeno-associated virus) only have a rigid capsid structure that serves to protect the viral nucleic acid.

Being obligate intracellular parasites, viruses rely on the host cell for much of the metabolic and biosynthetic machinery they require; however, they need to encode sufficient gene functions to adapt the cellular environment to their needs for completion of their infection cycle. These genes can be divided into three broad classes: (i) genes whose products are involved in the replication and amplification of the viral genome, (ii) structural genes whose products are needed to assemble viral particles, and (iii) accessory genes whose products are involved in regulation of viral processes to commandeer or block host cell processes. In addition to gene sequences, viral genomes also encode a number of regulatory sequences that are crucial to processes such as regulation of viral gene expression or replication of the genome. The ends of viral genomes often contain specific sequences that are important for proper processing of progeny nucleic acids and can act as signal to ensure that a viral nucleic acid is properly packaged into a virus particle.

The genetic components of a viral system can be divided into two broad categories: *trans*-acting and *cis*-acting functions. The term "*trans*-acting" refers to components that are encoded on one piece of nucleic acid but are able to act on other pieces of nucleic acid. These would generally be the genes in a viral genome that encode the various proteins that are required to synthesize and package new viral genomes. These genes do not need to be contained on an input viral genome to be able to generate new copies of the genome. The fact that they can be provided on different nucleic acid constructs is central to the ability to engineer viruses as vectors. Other genetic elements are referred to as *cis*-acting elements and are functional only on the nucleic acid upon which they reside. An example of these would be a sequence in the nucleic acid that targets that particular nucleic acid for packaging into a viral particle. In general, these *cis*-acting elements are regions of the viral genome that must remain in the genome even when it is converted into a gene delivery vector. A list of *trans*- and *cis*-acting viral elements is given in Table 1. As will be discussed below, generally the *trans*-acting factors are the elements that one attempts to remove from a viral genome to turn it into a gene transfer vector, whereas the *cis*-acting elements need to remain in the vector genome.

TABLE 1.

Viral vector *trans*- and *cis*-acting elements

trans-Acting elements	*cis*-Acting elements
Replication-related proteins	Nucleic acid ends
• Polymerases	• Inverted terminal repeats
• Replicases	• Long terminal repeats
• Proteases	• Internal ribosome entry site (IRES) elements
• Integrases	Packaging signals
Accessory proteins	Regulatory sequences
• Transcription regulators	• Promoters
• Nuclear export factors	• Polyadenylation signals
• Pathogenicity factors	• Splice donor/acceptor sites
Structural proteins	• Replication origins
• Capsid components	• Accessory protein-binding sites
• Nucleocapsid components	
• Matrix proteins	
• Envelope proteins	

Disabling Virus Pathogenicity

Attenuation of the pathogenicity of a virus is necessary for it to be useful as a gene transfer tool. One method to achieve this is to start with viral systems that exhibit reduced pathogenicity in relevant hosts. Baculovirus vectors used for mammalian cell gene transfer (BacMam viruses) represent one extreme of this concept. The virus is replication competent in insect cells, yet it is inherently replication incompetent in mammalian target cells (1) and thus would be unlikely to exhibit pathogenicity in humans. Similarly avipox vectors replicate efficiently in avian cell lines but are replication incompetent in mammalian systems (2). Vaccinia virus is an example of a virus that is replication competent and can be grown efficiently in cultured cells yet has reduced ability to grow in human hosts, either naturally or through deliberate derivation of attenuated strains. For example, the highly attenuated modified vaccinia Ankara (MVA) isolate grows efficiently in chicken embryo fibroblasts, yet does not replicate efficiently in primate cells.

Many of the virus systems in common use as gene transfer vectors are genetically modified to disable their ability to make progeny virions. Some of these viruses are minimally altered such that they constitute replicon-type systems, i.e., they are replication competent in some respects, yet are rendered unable to produce progeny virions. An example of this would be the alphaviruses in which the structural genes encoding the nucleocapsid and envelope glycoproteins of the virus are replaced with a desired transgene (3). To package the vector genome into viral particles, the structural genes are provided on a separate nucleic acid than the vector genome. The *trans*-acting structural proteins will package the vector genomes giving rise to defective vector virions. When these are used to infect a target cell, the vector genome is able to replicate itself and the transgene is expressed. However, due to the absence of structural proteins, the vector genomes are unable to be packaged into new viral particles.

By far the more common approach to making a viral vector is to remove sufficient gene functions (*trans*-acting functions) so as to disable fully its ability to replicate in a host, such that the delivery of a nucleic acid payload is a terminal event and new viral particles cannot be generated (4). This approach can span a wide spectrum as to how much of the viral genome is deleted. Early generations of adenovirus and herpesvirus vectors are deleted of a few gene products that are expressed early in infection and control the expression of later genes that remain in the vector genome. At the other extreme are retroviral, lentiviral, and adeno-associated viral vectors in which often all of the viral gene products have been removed from the vector genomes. Deletion of all viral gene products from the vector genome is certainly more desirable because it precludes inadvertent expression of viral gene products in the vector-transduced host, which has led to problems in *in vivo* gene transfer trials. In addition, removal of all viral genes reduces the probability of reconstituting viral replication ability through recombination as will be discussed below.

Vector Development and Production

The first step in producing a viral vector is to generate a viral vector genome construct that contains a transgene cassette and is capable of being replicated, amplified, and packaged into virus particles. Vector genome constructs may also be referred to as transfer plasmids. Deletion of nucleic acid from a viral genome to serve as a vector is done for two main reasons: to disable the replication of the virus and to provide space for transgene constructs. In general, nonenveloped viruses such as adenovirus or adeno-associated virus (AAV) have rigid limitations in the amount of nucleic acid that can be efficiently packaged into a viral particle, whereas the enveloped viruses offer more flexibility. In its most minimal form (i.e., with all of the viral genes removed), a viral vector genome must contain at least the *cis*-acting sequences necessary for replication, processing, and packaging of the genome. Transgene cassettes may be inserted into the vector construct using standard molecular biology techniques for construction of recombinant plasmids. As the vector genome increases in complexity, such as those for viruses that have large genomes with many genes, the vector genome may still contain many viral *trans*-acting genes in addition to the *cis*-acting sequences. Often these vector genome constructs are quite large and are not easily manipulated in plasmids using recombinant DNA techniques. In these cases, the transgene cassette may be introduced into the vector genome through natural recombination processes in an appropriate host cell.

Once a vector genome with transgene has been obtained, the next step is to introduce the construct into a cell that has been suitably modified with necessary complementing *trans*-acting functions to generate progeny genomes and virion structural elements to package those genomes. There are two general methods to accomplish this. The first involves stable introduction of the required viral genes into a cell line to complement the defects engineered into the virus to make it a vector. Some of the most common examples of these complementing cell lines are the adenovirus E1-expressing cell lines used to propagate E1-deleted adenoviral vectors (5, 6). The various packaging cell lines for retroviral vectors engineered to express all of the viral genes stably are also examples of complementing cell lines (7). The second method to generate viral particles of a disabled vector genome is to introduce multiple plasmid DNAs transiently that contain the vector genome and the required *trans*-acting factors into a suitable cell line followed by collection of the resulting viral particles. Each of these systems has advantages and disadvantages in terms of ease of use. Complementing cell lines generally require transfection of only a single plasmid DNA, whereas the transient transfection systems can require simultaneous introduction of several plasmids, which can be technically challenging. These different methods can have biosafety implications for the vector system, as will be discussed below.

Maintenance of Replication Deficiency

Possibly the most important characteristic of any viral vector system, or the method used to generate vector particles, is maintenance of the replication defect in the vector genomes that are replicated, amplified, and packaged. The joining together of two pieces of nucleic acid, which can happen through a variety of mechanisms, is termed recombination. Recombination events between defective viral vector genomes and the DNA constructs encoding the *trans*-acting functions used in their production can lead to reconstitution of a replication-competent viral genome. Complementation of defects by coinfection of cells with the vector and related replication-competent virus can also be a concern. There are a variety of modifications that can be made to vector systems to engineer their safety and maintain replication incompetence. Some of these features can be specific for particular viral systems; however, there are a number of them that are fairly generic and will be described below.

The viral genes (*trans*-acting functions) that need to be expressed require *cis*-acting functions, such as promoters and polyadenylation sites, for their expression. However, they do not require the entire *cis*-acting elements from the virus; some portions can be replaced with heterologous elements. This substitution has the effect of reducing regions of identity between the nucleic acid constructs containing the packaging/replication functions and the viral vector genome, with the goal of reducing the probability of recombination events between them.

Recombination events have a certain probability of occurrence, depending on design of the system. One method to reduce the probability of restoring packaging/replication functions to a viral vector genome is to split these functions onto two or more plasmid constructs. These are sometimes referred to as split genomes and have the effect of increasing the number of recombination events needed to reacquire all of the necessary genes for replication competence. Thus, the probability of several independent recombination events is the product of all the probabilities, making recombination more unlikely. An example of this concept is the packaging plasmid mixes used for lentiviral vectors, which, depending upon the system used, can contain up to five plasmids to deliver the necessary *trans*-acting functions to make vector particles.

Early methods for construction of retroviral vectors employed packaging cell lines that had been stably transfected with the requisite packaging/replication genes. This allowed investigators to then stably introduce the viral vector genome and create a producer cell line that could be propagated continually in culture, constantly producing viral vector particles (8, 9). This practice keeps the defective vector genome in proximity with the packaging/replication genes for multiple cell passages, thus increasing the opportunity for recombination events between these nucleic acids. Furthermore, if a recombination event does occur, it can be propagated in culture. This complication can be avoided by using packaging cell lines constructed with the packaging/replication functions resident on split genomes as mentioned above, transiently transfecting the vector genome into the cells, and then collecting the vector particles produced. Alternatively, single-batch transfections of multiplasmid mixtures into cells can be performed to collect the virus made from a single transfection batch.

The choice of a host cell for amplification and packaging of a viral vector can have important implications for recombination or complementation of the vector genome. As mentioned previously, removal of viral *cis*-acting functions from viral genes needed for vector replication is done to reduce opportunities for recombination. The complementing HEK293 cell line is commonly used to amplify defective E1-deleted adenoviral vectors. This line was constructed by stably transfecting sheared pieces of adenoviral DNA into cells followed by selection of a clone that would support growth of the defective virus (5). Thus, there are regions of homology including viral *cis*-acting functions between the E1 region insert in the cell line and the defective viral genome that can lead to recombination events resulting in replication-competent adenovirus. More recently derived cell lines, such as PER. C6 cells, have been developed that eliminate these regions of overlap and reduce the risk of recombination events (6).

Complementation of endogenous viruses in cell lines by viral vector packaging functions can also be a concern that influences the choice of host cells for vector generation. Certain murine (10) and avian (11) cell lines are known to carry retroviral genomes in their genetic complement. Although these endogenous viruses may be defective for replication, their genomes are capable of being packaged into virus particles by packaging functions used to generate retroviral vector particles. These mobilized endogenous viral genomes are then capable of being delivered to naïve cells. Use of cell lines derived from species that do not harbor endogenous viral genomes reduces the risk of this type of complementation. This coupled with removal of overlap regions between packaging function constructs and viral vector genomes, as mentioned above, has led to development of safer cell lines for vector production (6, 7). This concept of complementation of a defective virus can go in the other direction also; a defective viral vector can be complemented by coinfection with a wild-type, replication-competent virus. This can be a concern especially for *in vivo* gene transfer experiments and is one reason for the development of self-inactivating (SIN) retroviral and lentiviral vectors that will be described in the sections "Retrovirus Vectors" and "Lentivirus Vectors," below.

One final general concept is that of reducing the number of viral genes present in a packaging mix. Completely deleting some of the normal genetic complement of a virus from the packaging system means that a wild-type virus cannot be reconstituted by recombination between a defective vector genome and the plasmids encoding the packaging functions. This is best illustrated in the lentivirus vector system based on human immunodeficiency virus (HIV). Of the six additional genes other than *gag*, *pol*, and *env* encoded by the virus, the four accessory genes (*nef*, *vpr*, *vif*, *vpu*) are not necessary for viral replication. Thus, these genes are not included in vector packaging mixes, meaning that wild-type HIV cannot be reconstituted from these systems. Pseudotyping of viral vectors can be thought of as a variation on this theme. Enveloped viruses contain surface proteins in their membranes that are responsible for cell recognition, binding, and entry. In some cases, the surface protein of one virus can be exchanged with the surface protein of a different virus, which is referred to as pseudotyping. Most retroviral and lentiviral vectors are pseudotyped with the G glycoprotein from vesicular stomatitis virus (VSV). This process broadens the host range and increases stability (12), two modifications that increase risk. However, in some cases, as with the envelope protein from group A avian sarcoma and leukosis virus and its cognate receptor tumor virus receptor A (TVA), which is not found on mammalian cells, pseudotyping can be used to limit the host range of a virus (13). Furthermore, pseudotyping a vector removes one of the normal viral genes from a packaging mix so that a wild-type virus cannot be reconstituted from the system.

COMMON VIRAL VECTORS

This section discusses some details of the biology of specific virus families that are in common use for a variety of *in vitro* and *in vivo* gene delivery applications. A list of some of the characteristics of the parental viruses is given in Table 2. Descriptions of how each virus has been made into a vector will be given along with risk assessment considerations particular to each family, with a focus on the features designed to maintain any replication defects in the system. As much as possible, recent reviews have been identified as sources for this information. These serve as excellent supplemental information to this chapter and can direct the reader to the original literature if desired. Comprehensive reviews related to this section are available (4, 14–16). The nature of the transgene to be expressed is a risk assessment factor common to all recombinant DNA (rDNA) work and will be dealt with in "Containment Levels for Viral Vectors," below.

Poxvirus Vectors

Biology and recombinant vectors

The poxviruses constitute a large family of DNA viruses that replicate in the cytoplasm of infected vertebrate and invertebrate cells. The most well-known member of this family, variola virus, is the etiologic agent of smallpox, a contagious disease with a mortality rate of approximately 30% (17, 18). Routine vaccination for smallpox was discontinued in the United States in 1971, and this dreaded disease was declared eradicated in 1980 by the World Health Organization after a successful worldwide immunization program. In response to the potential bioterrorism threat posed by smallpox infection, a select group of individuals in the United States were offered vaccination in 2003 (19, 20). A comprehensive description of the adverse events associated with this vaccination program has been published by Casey et al. (21).

In addition to variola virus, three other species of *Orthopoxvirus* can cause human infections. These include monkeypox virus, cowpox virus, and vaccinia virus. The poxviruses contain a linear double-stranded DNA genome that varies between virus types from 130 to 300 kilobases (kb) and replicates in the cytoplasm of infected cells. Vaccinia virus serves as the prototype poxvirus and has been used widely as the basis of the human smallpox vaccine. The molecular virology of vaccinia virus has been studied extensively. It has a wide host range in cultured cells, replicates to high titers, and is easily quantified by plaque assay (22).

The development of recombinant poxvirus vectors was first described in 1982 (23, 24). Recombinant poxviruses provide a number of advantages as viral vectors (2, 25–27). A large proportion of the viral DNA is dispensable for viral replication and therefore can be replaced by

TABLE 2.

Characteristics of viral systems commonly used for recombinant gene transfer[a]

Virus type	Nucleic acid	Enveloped	Route of transmission	Genome size (kb)[a]	Risk group	Persistence of nucleic acid	Integration
Poxvirus	dsDNA	Y	Aerosol/direct contact	192	2	N	N
Retrovirus	Diploid ssRNA	Y	Blood-borne	8.3	1/2	Y	Y
Lentivirus	Diploid ssRNA	Y	Blood-borne	9.7	2/3	Y	Y
Adenovirus	dsDNA	N	Aerosol	36	2	N	N
Adeno-associated virus	ssDNA	N	Aerosol	4.5	1	Y	Y/N[b]
Baculovirus	dsDNA	Y	NA	134	1	N	N
Herpesvirus	dsDNA	Y	Direct contact	152	2	Y	N

[a]Abbreviations: kb, kilobases; ds, double-stranded; ss, single-stranded; NA, not applicable.
[b]Wild-type viruses expressing *rep* functions integrate; vector genomes do not integrate.

inserted foreign gene sequences that can reach lengths greater than 20 kb. Viral DNA is not inserted into the host cell genome, and relatively high levels of foreign gene expression can be obtained. In addition, the development and selection of recombinant vectors is relatively easy. Most methods for producing recombinant poxviruses employ plasmid transfer vectors that contain an expression cassette consisting of a poxvirus promoter with adjacent restriction endonuclease sites for foreign gene insertion, flanked by poxvirus sequences that direct recombination to the desired viral region (28). Recombinant viruses can be readily identified and isolated using a variety of selection procedures (29). The interest in poxvirus vectors focuses primarily on their potential use as vaccines and anticancer agents (30–32). Although studies have been published describing their use for the general production of recombinant proteins, this application has not been adopted widely (33–36).

Risk assessment

In most studies, vaccinia virus has been used for deriving recombinant vectors. The Western Reserve strain of vaccinia virus is commonly used because high yields of virus can be obtained, discrete viral plaques are formed, and the strain is well adapted to laboratory animals. The viruses are quite stable and can persist at room temperature when dried. Virus can be inactivated by 50% ethanol within 1 hour and by 1% phenol over 24 hours. A recent article by de Oliviera et al. (37) describes the susceptibility of vaccinia virus to a variety of chemical disinfectants. Unlike recombinant adenoviruses and retroviruses, most recombinant vaccinia viruses are not designed to be replication defective. Thus, infection with such a recombinant virus can lead to viral replication and shedding. Percutaneous inoculation of a nonimmune individual with vaccinia virus typically results in the development of a papule that becomes vesicular and pustular, reaching a maximum size in 8 to 10 days. The pustule dries and separates within 14 to 21 days. For more detailed information on the vaccination process, see Evans and Lesnaw (38), Evans (39), and Walsh and Dolin (30). The secondary transmission of vaccinia has been reviewed by Sepkowitz (40). A significant number of laboratory-acquired exposures to vaccinia virus vectors have been reported (41–49). Thus, laboratory workers should be particularly careful when working with these vectors (50–53). In addition to laboratory coats and gloves, eye protection is recommended when handling vaccinia virus vectors, and all possible precautions should be employed when handling sharps. In the United States, the Advisory Committee on Immunization Practices recommended that laboratory work with non-highly attenuated vaccinia virus strains be conducted at biosafety level 2 (BSL2) containment and work practices (19, 54). Vaccination is recommended for laboratory workers who directly handle cultures or animals contaminated or infected with non-highly attenuated vaccinia virus, recombinant vaccinia viruses derived from non-highly attenuated vaccinia strains, or other orthopoxviruses that infect humans (e.g., monkeypox virus, cowpox virus, and variola virus) (54, 55). It is important to inform potential vaccine recipients of the benefits and potential adverse events that may be associated with vaccination (21, 56). Interestingly, a short report by Benzekri et al. (52) suggests that the main barrier to vaccination among laboratory workers may be fear associated with possible vaccine adverse effects and a willingness to risk accidental exposure rather than be vaccinated. Laboratory and other health care personnel who work with highly attenuated strains of vaccinia virus (e.g., MVA and NYVAC strains) do not require routine vaccinia vaccination. In addition, laboratory and other health care personnel who work with the avipoxvirus strains ALVAC and TROVAC do not require routine vaccination. Immunosuppressed individuals should not be vaccinated (54, 55). In the United States, the CDC is the only source of vaccine for civilians; thus, requests for vaccine should be referred to the CDC (55).

A number of approaches have been taken to attenuate vaccinia virus and thereby enhance its safety profile. A highly modified MVA has been developed with significantly reduced pathogenicity. At the NIH, laboratory studies with MVA are permitted at BSL1 without vaccination in laboratories where no other vaccinia viruses are being manipulated (25). The NYVAC strain derived from vaccinia virus (Copenhagen), TROVAC strain derived from fowlpox virus, and ALVAC strain derived from canarypox virus are also considered low-risk strains. In 1993, the Recombinant DNA Advisory Committee (RAC) of the NIH reduced the biosafety level for these viruses to BSL1 (57, Appendix D in reference 56). However, in assigning an appropriate biosafety level to recombinant viruses derived from these strains, it is always important to consider the properties of the gene product one wishes to produce.

Retrovirus Vectors

Biology and recombinant vectors

Retroviruses comprise a family of enveloped viruses with positive-strand RNA genomes that range from 7 to 10 kb in length. These viruses infect a variety of species, resulting in a range of pathogenic effects in their hosts. One major difference between the simple retroviruses and lentiviruses (see below) is that the simple retroviruses only replicate in dividing cell populations. Many of the retroviral vectors in use are derived from murine leukemia viruses. Maetzig et al. (58) have recently published an excellent review of gamma retroviral vectors. The infection cycle is quite well understood and does not result in cell lysis; rather the virus sets up a chronic infection in which infected cells are capable of continuous production of virions. There are two unique features of the retroviral life cycle. The RNA genome is replicated through a double-stranded DNA intermediate called the proviral genome. This reaction is performed by an RNA, DNA-directed DNA polymerase referred to as a reverse transcriptase. Additionally the proviral DNA must enter the nucleus of the infected cell and integrate into the host cell genome so that progeny genomes (as well as viral mRNAs) can be transcribed from the integrated provirus (16).

The genome of retroviruses consists of three genes (*gag*, *pol*, and *env*) that encode all of the *trans*-acting functions required for the viral life cycle, bounded by *cis*-acting sequences referred to as long terminal repeats (LTRs). The LTRs contain the signals necessary for transcription initiation and termination and packaging of progeny nucleic acids. The viruses are converted to vectors by removal of the three viral genes, leaving the LTRs in the vector genome and providing up to 8 kb of space for transgene cassettes. The viral gene products are provided in *trans* to replicate and package vector genomes into virion particles. The packaging functions can be stably introduced into appropriate cells to construct complementing cell lines called packaging cell lines (16). The vector genome can then be stably introduced into packaging cells to yield a producer cell line capable of producing recombinant vector particles (8, 9). From a large-scale or clinical material production perspective, this method has particular advantages (59). Alternatively, the vector genome can be transfected into packaging cells, or the vector genome can be transfected with plasmids containing the packaging functions into naïve cells. Either of these alternatives is then used to transiently produce a single batch of virus (60).

Retroviral vectors have found utility in a number of *in vitro* and *in vivo* applications. They remain a popular choice for gene therapy approaches (61), although their restriction to transduction of dividing cell populations limits their use to therapies for hematopoietic and tumor cells (62). The ability to integrate their genome into that of a host cell makes them a popular choice for gene transfer applications where long-term expression of a transgene is desirable. Retroviral vectors are able to infect a wide variety of cells from different species, with the range of species being controlled by the particular *env* gene used for packaging. Often vectors are pseudotyped with the VSV G protein, as mentioned previously, which extends the range of cell types the viruses can transduce. This modification also makes the virions more stable, helping to overcome the low yields of virus and sensitivity to purification methods that are characteristic of these vectors (58).

Due to safety concerns arising from the integration of viral nucleic acid into the genome of host cells (discussed below), several types of nonintegrating vectors based on retroviruses have been developed. Inactivation of the viral integrase enzyme by mutation gives rise to vectors that mediate transgene expression from proviral DNA maintained as episomal elements in the host nucleus. Introduction of mutations in the virus that interfere with the ability to reverse transcribe the RNA genome results in vectors in which the transgene is translated directly off the entering vector genome. Vectors have also been described that do not package a vector genome; they are virus-like particles that contain transgene products as fusion proteins with retroviral proteins that are part of the virion particle. As these proteins are processed by the viral protease, the foreign proteins are released so they can be delivered to an infected cell. These new technologies and their limitations have been reviewed recently by Schott et al. (63).

Risk assessment

Murine leukemia viruses are classified as Risk Group 1 agents because they are not known to cause disease in humans. However, as with all viral vectors, the activity of the particular transgene product to be expressed is a major determinant in considering appropriate containment levels for protocols. When considering risks of inadvertent exposures, it is an advantage that there are no viral genes remaining in the vectors, minimizing concerns for immune responses to expression of viral gene products in transduced cells. Furthermore, the viruses will not transduce nondividing cell populations and they are sensitive to inactivation by human complement (9). The host range of vectors can be restricted to cells of murine origin by packaging with an ecotropic *env* protein, although this also limits the utility of the vector. Being enveloped viruses, they are not resistant to drying and contaminated surfaces can be decontaminated with detergents or 70% ethanol solutions (38, 39).

Recombinant retroviral genomes integrate into the genome of host cells. Thus, the risk of systemic exposure raises the concern that an integrated provirus can cause disruptions in normal cellular gene expression. This can happen by internal insertion into a gene, thus inactivating expression of that gene, or, because the proviral LTR sequences have promoter activity, integration can activate transcription of host genes downstream from an insertion site. Either of these events could have deleterious consequences for the host. To address this concern, SIN retroviral vectors have been developed in which the promoter function of the 3′ LTR is disabled by making deletions of the U3 sequence of the vector genome. When this U3-deleted vector genome enters a host, the defective region is duplicated to the 5′ LTR of the proviral DNA during reverse transcription. Thus, the promoter function of both 5′ and 3′ LTRs is disabled in the integrated proviral DNA. Disabling the 3′ LTR promoter function reduces the risk of activation of cellular genes located downstream of the integration site. Reduction of transcription from the 5′ LTR has two effects that enhance the safety of these vectors. First, it precludes the use of the 5′ LTR promoter to drive transgene expression, necessitating the addition of another promoter in the transgene cassette. Second, transcription from the 5′ LTR of an integrated provirus in a normal infection is responsible for generation of full-length viral progeny RNAs. Thus, SIN vectors are unable to generate those transcripts, meaning that they are not able to package vector genomes if a transduced cell is coinfected by a replication-competent retrovirus (58). Insertional mutagenesis is a risk that should be acknowledged in risk assessments for any integrating viral vector. The implications or insertional mutagenesis for human gene transfer protocols will be discussed in the section "Vector Characteristics and Safety Issues."

A major biosafety concern for any replication-defective vector system is the possibility of generation of replication-competent virus through recombination. Thus, it is important to understand the features of the vector construct and packaging system that are designed to decrease the opportunities for recombination events. As retroviral packaging cell lines or transient plasmid systems have developed, there are a number of modifications that have been made for this reason, many of which have been discussed in the previous section on viral vector technology. The only *trans*-acting functions that are required are the *gag*, *pol*, and *env* genes. They are generally present on two different nucleic acid constructs (*gag-pol* on one, *env* on the other) as well as having viral *cis*-acting sequences, such as transcriptional control and packaging signals, removed. Additionally viral vector genomes have been trimmed to remove regions of overlap with packaging sequences, and different cell lines have been developed to avoid complementation with endogenous retroviruses (9, 64).

Lentivirus Vectors

Biology and recombinant vectors

Lentiviruses are a class of retroviruses with more complex genomes than the murine leukemia viruses. Examples of viruses in this family include HIV-1 and HIV-2, feline and bovine immunodeficiency viruses (FIV, BIV), equine infectious anemia virus (EIAV), and caprine arthritis encephalitis virus (CAEV), all of which have been converted into vector systems (62). Many of the vectors in common use are based on the HIV-1 genome (16). In addition to the *gag*, *pol*, and *env* genes found in the simple retroviruses, HIV-1 encodes two regulatory genes (*tat* and *rev*) and four accessory genes (*vpr*, *vpu*, *vif*, and *nef*). Lentiviruses also infect both dividing and nondividing cell populations, giving vectors derived from them a distinct advantage for many *in vivo* and *in vitro* applications (16, 62).

Packaging constructs for lentiviral vectors have undergone several iterations to engineer safety into the system. The major development has been the ability to limit the viral genes required in the packaging mix to *gag*, *pol*, *rev*, and in certain cases *tat*. In most cases the envelope protein used is the VSV G protein, for the same reasons mentioned previously (12). However, an active area of vector development is determining other surface glycoproteins that can be used to expand the clinical utility of lentiviral vectors (65). Excellent discussions of the development of lentivral packaging systems are provided in reviews by Cockrell and Kafri (62) and Pauwels et al. (66). The requirement for *rev* as a *trans*-acting factor is due to the Rev protein's role in binding to the Rev-responsive element (RRE), a viral *cis*-acting element that is still present in vector genomes. The Rev-RRE interaction is required for efficient nuclear export of full-length vector RNAs for packaging (12). The product of the *tat* gene is required for high-level transcription directed from the viral LTR. If a vector genome construct is dependent upon the LTR to express progeny RNA genomes for packaging, then the *tat* gene must be included in the *trans*-acting packaging functions. However, this requirement can be avoided by replacing the *tat*-dependent promoter of the 5′ LTR with a constitutive promoter, such as the cytomegalovirus promoter (62).

Production of vector virions is commonly carried out by transient transfection of a cell line (generally a HEK293 derivative) with plasmids containing the vector genome and the requisite *trans*-acting genes, followed by collection of the virus batch. One reason for using transient packaging rather than packaging cell lines is that some of the required gene products are deleterious to cells, making it challenging to derive stable cell lines that express them constitutively. However, the need to easily scale up production of these vectors for human gene therapy applications has necessitated the development of packaging cell lines based on conditional expression of the packaging genes (59, 62). A recent report describes construction of a lentivirus vector packaging cell line that continually expresses packaging functions and can be used to derive stable producer cell lines (67).

Risk assessment

Most of the elements of risk assessment for lentiviral vectors are identical to those of the retroviral vectors. There are the same concerns about the transgene to be expressed, the possibility of insertional mutagenesis, and the generation of replication-competent virus through recombination. Lentiviral vector genomes deleted of all viral genes have been made self-inactivating similar to retroviruses to reduce risks of vector genome mobilization and activation of downstream genes. Pseudotyping vectors with the VSV G protein, as is commonly done, confers the same characteristics of broader host range and virion stability, as mentioned previously, plus there is the added factor that these vectors transduce both dividing and nondividing cell populations. The packaging gene mixes that are used are delivered on multiple plasmid constructs to maximize the number of recombination events that would be required to reconstitute a replicative virus. Furthermore, several of the wild-type virus genes are not present in the packaging mix, thus preventing reconstitution of the starting virus. For a detailed review of risk assessment and biosafety of these vectors, see Pauwels et al. (66).

Adenovirus Vectors

Biology and recombinant vectors

Human adenovirus was first isolated in 1953 from cultured degenerating human adenoid tissue (68). The name adenovirus was proposed by Enders et al. (69). Fifty-seven serotypes of human adenoviruses have been identified, with adenovirus type 2 (Ad2) and Ad5 being the most well characterized. (70). The viruses are icosahedral, nonenveloped, double-stranded DNA viruses with a linear genome of 36 kb and a capsid diameter ranging from 70 to 90 nm. Viral replication and assembly occur in the nucleus of infected cells. The viruses exhibit a broad host range both *in vitro* and *in vivo* and can infect both dividing and nondividing cell types. Adenoviruses are easy to grow and manipulate in cell culture, highly infective, and can be obtained in high titers, e.g., 10^{10}–10^{12} PFU/ml (70, 71).

Recombinant adenoviruses are frequently used as gene delivery vectors in cell culture experiments and *in vivo* animal experiments to study gene function (72–75). Vector systems based on adenovirus, as well as recombinant viruses, are readily available commercially for use as gene delivery tools. They are also in development as potential gene therapy therapeutics, vaccine delivery vectors, and anticancer agents (76–81). Recombinant adenovirus vectors possess a number of advantages as gene delivery vectors. The molecular biology of adenoviruses is well understood and genetic modifications of the viral genome can be made relatively easily to accommodate foreign gene sequences up to 30 kb. The viruses can infect a wide variety of cell types, including nondividing cells such as hepatocytes, and can be grown to high titers. A variety of systems have been developed for generating replication-deficient adenovirus (RDA) viruses carrying the gene of interest. In most instances, recombinant vectors are derived from two components: a viral DNA vector and a packaging cell line, most commonly 293 cells, a human kidney cell line that has been stably transfected with a fragment of the adenovirus genome containing the early region 1A (E1A) region. This allows the vector to be made and replicated within the 293 cell. Vectors prepared from the cell line will lack the E1A region and thus remain replication defective. PER.C6 cells, engineered human retinoblastoma cells containing a more defined E1 region, have also been used for packaging virus (6). Kovesdi and Hedley (82) have reviewed the available producer cell lines. As with any viral vector system generated using a packaging cell line, one must ensure that the vector produced is free of wild-type virus contamination. Theoretically, infection with replication-defective adenoviruses should not lead to the production and shedding of infectious virus (83, 84). A number of assays have been described for detecting the breakthrough of repletion-competent adenovirus in specimens (85–90). Adenoviral DNA does not usually integrate into host cell DNA, and only transient expression of the foreign gene is observed. However, a report by Harui et al. (91) demonstrated that E1-substituted and helper-dependent vectors achieved integration efficiencies ranging from 10^{-3} to 10^{-5} per cell. Thus, although adenoviral DNA does not pose the level of concern for insertional mutagenesis as do retroviruses, a low level of integration *in vivo* may be possible (92, 93). A comparison of the fate of adenovirus variants administered to animals in several different ways indicated that the probability of excretion of E1 and E2 gene-deleted vectors is very low (94). A review by Tiesjema et al. (95) focuses on the distribution and shedding of replication-deficient human Ad5.

More recent versions of adenovirus vectors have been developed in which all viral gene coding sequences have been deleted (77). These viruses are referred to by a variety of names, including third-generation, helper-dependent, and gutless vectors. These vectors offer a number of advantages; however, the need for a helper virus for production requires that special precautions be taken to ensure that the presence of helper virus is minimized in the preparation of high-titer virus stocks. Due to the need for large quantities of adenovirus vectors for preclinical and clinical testing, large-scale production methods have been developed and described (96).

Risk assessment

Adenoviruses have been isolated from a large number of vertebrate species and at least five avian species. The viruses can infect and replicate at various sites of the respiratory tract, eye, gastrointestinal tract, liver, and urinary bladder. Most adenovirus infections in humans are subclinical; however, clinical pathology can range from acute febrile respiratory infections to keratoconjunctivitis, gastroenteritis, and acute hemorrhagic cystitis (70). Virus shedding can continue for years after infection (97). Adenoviruses have been isolated from immunocompromised patients and have contributed to morbidity and mortality (38, 70). Human Ad12 has been shown to be oncogenic in rodents (98); however, to date, adenoviruses have not been associated with human malignancies (70, 99).

Adenovirus is transmitted by droplets or close personal contact. It also may be transmitted by fecal-oral contact, especially in children. The virus is stable and can persist on surfaces for long periods. A recent study in which the level of adenovirus contamination on laboratory surfaces was surveyed identified centrifuges as hot spots for contamination (100). The virus is not inactivated by soap and water, alcohol, or chlorohexidine gluconate. Sodium hypochlorite (1–10% household bleach solution) can be used as an effective disinfectant for fluids containing virus. Contaminated solid waste should be appropriately packaged, labeled and autoclaved or incinerated.

Wild-type adenoviruses have been classified as Risk Group 2 agents, and laboratory protocols involving the introduction of recombinant DNA into adenoviruses should be reviewed by the IBC. Experiments can usually be conducted at BSL2; however, it is important to conduct a thorough risk assessment of the proposed activities to identify specific concerns that may warrant additional precautions. Important elements to consider in conducting the risk assessment are described in the CDC/NIH publication *Biosafety in Microbiological and Biomedical Laboratories*, commonly referred to as the BMBL (101) and by Knudsen (102). For example, the nature of the foreign gene sequence inserted into the virus, altered viral tropism, or volume of virus may be of sufficient concern that a BSL2+ or BSL3 containment level is more appropriate. In practice, such determinations are not always straightforward, and it is important that the laboratory director or principal investigator work closely with the IBC to implement the appropriate BSL for conducting the proposed experiments. It is also important that the individuals conducting experiments with recombinant adenoviruses have had specific training relevant to adenovirus biology and the handling of infectious agents.

Adeno-Associated Virus Vectors

Biology and recombinant vectors

AAV is a small, helper virus-dependent parvovirus with a single-stranded, linear DNA genome of 4.7 kb. The virus was originally identified as a contaminant of adenovirus-infected cultures, and like adenovirus its route of infection is thought to be either respiratory or gastrointestinal. Due to its small size and the simplicity of its genome, the virus requires a coinfecting helper virus such as adenovirus or herpesvirus for productive replication. In the absence of a helper virus, wild-type AAV integrates into the host cell genome, where it resides in a latent state until infection with a helper virus occurs, at which point the proviral DNA excises and replicates. Interestingly, wild-type AAV is often inserted into a single site on chromosome 19q13. AAV has a broad host range and can infect both dividing and quiescent cells. Epidemiologic studies have shown that AAV infection is common in the general population, with 50% to 80% of tested individuals having serologic evidence of AAV infection (103, 104).

Recombinant AAV has attracted significant interest as a vector for gene transfer and gene therapy (105, 106). The vectors lack all viral genes, are not immunogenic, infect dividing and quiescent cells, and can give rise to long-term gene expression *in vivo* after a single virus injection. Importantly, despite the ubiquitous nature of AAV, the virus had for many years been considered to be nonpathogenic because it was not associated with any human disease. Genetic manipulation of AAV became possible when the viral genome containing intact palindromic terminal repeats was cloned into bacterial plasmids. Recombinant AAV can be generated by various routes. One approach requires coinfection with helper adenovirus (103), whereas more recently developed methods use helper-free packaging systems, thus eliminating the need for concurrent helper adenovirus infection (107). The majority of AAV vectors in use are based on the AAV2 serotype, but other natural serotypes exhibit different tissue tropisms that make them of interest for clinical applications (108). Additionally, an active area of development is to use rational design or directed evolution strategies to engineer the AAV capsid to improve its properties to target specific cell populations and transduce them more efficiently (109).

Risk assessment

As mentioned above, until recently, AAV infection has not been associated with any known clinical disease. However, a recent report found clonal integrations of regions of the wild-type AAV2 genome associated with several proto-oncogenes in human hepatocellular carcinomas, suggesting that AAV2 insertional mutagenesis may have led to tumor development (110). These insertions presumably act as promoters or enhancers, leading to overexpression of the genes to which they are linked. Although a region of the AAV genome present in these insertional events is not present in AAV vectors, Russell and Grompe (111) speculate that these findings could have implications for use of AAV vectors in gene therapy applications. Due to the widespread interest in AAV as a gene therapy vector, numerous preclinical studies have been conducted that provide data regarding the safety of various recombinant

AAVs in animal models. Removal of the AAV genes from vector genomes takes away the specific chromosomal integration seen with wild-type AAV; instead, the vectors achieve long-term gene expression by maintaining their genomes as extrachromosomal DNA in the host. At a low frequency, this leads to chromosomal integration events (112). Thus, the presence of strong enhancer-promoter elements in vectors coupled with the high doses of vector particles often used warrants increased consideration of vector and study design in AAV trials (111). AAV is still considered a Risk Group 1 agent; however, in light of these recent findings, it seems prudent to work with these vectors at BSL2. As with all recombinant virus experimentation it is also important to consider the gene product(s) one is attempting to express and other potential risk factors in deciding on an appropriate BSL.

Herpesvirus Vectors

Biology and recombinant vectors

Herpes simplex virus 1 (HSV-1) is an enveloped virus with a linear, double-stranded DNA genome of approximately 152 kb. The genome has an interesting structure consisting of two unique regions called U_L and U_S that are each flanked by distinct inverted repeat sequences. These inverted repeats contain *cis*-acting sequences involved in genome processing and packaging as well as *trans*-acting functions that are required for the viral replication cycle. Many of the approximately 90 viral genes are nonessential for growth of the virus in cultured cells, a characteristic that is exploited in construction of some herpesvirus vectors. The virion consists of a tegument, a matrix of virus-encoded proteins located between the nucleocapsid and the viral envelope that enters the cell along with the nucleocapsid. Replication of viral DNA and its assembly into progeny nucleocapsids occurs in the nucleus of infected cells. The virus is spread by direct contact and replicates lytically in epithelial cells. However, HSV is able to infect peripheral neurons, where it can establish a latent infection with its nucleic acid persisting in the cell in a quiescent state for years. The virus can remain dormant in neuronal cells, but it can also be reactivated by environmental factors to undergo a lytic infection.

During the lytic phase of growth, HSV gene expression proceeds in a cascading manner. Transcription of the immediate-early, or α, genes is activated by one of the tegument proteins, which in turn are responsible for activating expression of the early (β) genes. The early gene products are involved in DNA synthesis, leading to production of long concatamers of the viral genome. Once DNA replication is initiated, the late (γ) genes are transcribed, giving rise to the viral structural proteins. Genomes are processed and packaged into nucleocapsids that bud through the nuclear membrane and are subsequently transported to the cell surface for release.

In latent infection, the lytic cascade of gene expression is somehow avoided and the viral DNA is maintained episomally in the nucleus. The only viral gene transcripts detected in latency are those from the *LAT* gene, which is controlled by two neuronal-specific promoters called LAP1 and LAP2. The major RNA found is a 2-kb molecule with a lariat structure whose role in establishment of, maintenance of, or reactivation from the latent state is unclear. Latent virus can become reactivated through unknown mechanisms, initiating a productive infection. More details about aspects of HSV replication outlined above can be found in Pellott and Roizman (113) and Roizman et al. (114).

HSV infects a broad range of cell types, but its ability to mediate gene expression long term in neuronal cells makes it particularly attractive as a recombinant vector for gene therapy for neuronal disorders (115). It has also found utility as a vaccine vector and for anti-cancer therapy applications (116–119). Recently, the Food and Drug Administration (FDA) approved a genetically modified herpesvirus for the treatment of advanced melanoma (120). Three types of vectors have been derived from herpesviruses: amplicons, replication-defective, and replication-competent (118). Early methods for construction of recombinant HSV vectors relied on recombination between viral DNA and plasmids containing transgenes (121); however, a variety of different methods have been devised to move transgene cassettes into the viral vector genome (4, 122, 123).

Amplicons were derived from studying the makeup of defective HSV particles that are able to replicate in the presence of helper virus (124). Amplicon plasmids contain two *cis*-acting sequences from HSV, an origin of replication and a packaging/processing signal; they allow for up to 150 kb of transgene DNA to be packaged into HSV virions. *trans*-acting functions from HSV-derived helper constructs replicate the amplicon plasmid into long concatamers that are then processed into genome-length (~152 kb) molecules and packaged; thus, for small transgene cassettes, several copies of the payload can be contained in a single particle. The nature of the helper construct has been a major area of development for the amplicon system. Early methods utilized helper viruses to provide *trans*-acting packaging functions, but helper virus-free systems have been developed that deliver the required gene functions on cosmids or bacterial artificial chromosomes (BACs) (125). Amplicons are useful vectors for many of the same applications as other HSV-derived vectors because they retain the same tropism and host range; however, the large payload capacity provides opportunities for delivery of regulated transgene systems and construction of hybrid viral amplicons that will allow persistent maintenance of amplicon DNA (125). Furthermore, the ability of amplicon vectors to transduce antigen-presenting cells without concern for expression of HSV genes that interfere with the function of the immune response makes them good candidates for vaccine vectors (126).

The design of replication-defective HSV vectors exploits the fact that early and late viral gene expression is dependent upon expression of immediate early gene products (4). Deletion of the *ICP4* and *ICP27* genes precludes later gene expression and blocks the lytic cycle of the virus. Vector genomes can be amplified and packaged in complementing cell lines. Other nonessential immediate early genes have been deleted from vector genomes because their products are cytotoxic and to provide space for transgenes. Addition of transgene cassettes controlled by constitutive promoters results in vectors that are able to mediate transient, high-level expression in many cell types, although placing transgenes under control of the LAP promoters yields vectors that can mediate expression in neurons for many weeks. Another type of defective HSV vector that has been used mainly for vaccine applications is disabled infection single-cycle (DISC) viruses (118). These vectors are essentially replicons; they are deficient in structural genes such that the cascade of viral gene expression and DNA replication takes place (thought to aid in the antigenicity of the vector to aid immune responses), yet they are unable to make progeny virions unless the structural genes are provided in *trans*.

Replication-competent HSV vectors are under development for antitumor applications, so-called oncolytic viruses (4, 116, 118, 119). These vectors are based on observations that certain HSV gene products are nonessential for growth of the virus in rapidly growing cell culture populations, such as tumor cells, yet the loss of these genes severely attenuates virus propagation in quiescent cell populations, such as neurons. The genes that have been deleted to derive these vectors are involved in aspects of nucleotide metabolism, for example, thymidine kinase (*tk*) and ribonucleotide reductase (*ICP6*), or they are involved in various mechanisms the virus has evolved to modify the cellular environment for its own benefit, such as *ICP34.5* and *vhs*. These vectors have undergone iterations of single or combinations of mutants to identify viruses that are sufficiently attenuated in neuronal cells while retaining their ability to grow in tumor cells. A further stage in development of oncolytic vectors is the inclusion of transgenes whose expression will increase the toxicity of the vector or stimulate immune responses to the infected tumor mass.

Risk assessment

HSV-1 is a Risk Group 2 agent and is readily transmissible by direct contact with epithelial or mucosal surfaces; it is able to propagate in a wide variety of dividing and nondividing cells. It is an enveloped virus, and work surfaces can be decontaminated with 70% ethanol solution or ionic detergents (38). A significant percentage of the population is seropositive for HSV-1, and approximately 65% of adults in the United States are estimated to harbor latent infections of the trigeminal nerve (14). This becomes a concern for use of vectors derived from this virus because maintenance and reactivation from latency are poorly understood. Infection by herpesvirus vectors of latently infected cells potentially could reactivate the wild-type virus, or alternatively, spontaneous reactivation of a latent infection could create a situation where replication-defective vectors could be propagated (127).

Amplicon vectors deliver no viral genes and are completely replication defective and so would appear to pose minimal risk, although the nature of the transgene payload remains a major consideration in the risk assessment. Additionally, potential for contamination of amplicon preparations with helper virus from the system used for packaging must be considered. Improvements in this area include development of multiple cosmid systems, specially engineered helper viruses, and BACs to provide the necessary *trans*-acting functions (125).

Continued improvement of the biosafety of HSV-derived replication-defective and replication-competent vectors is driven largely by active research to develop these pathogenic viruses into clinically useful therapeutics (118). One biosafety advantage for HSV vectors is the availability of clinical therapeutics in case of exposures or adverse events; however, the susceptibility of HSV to these drugs is dependent upon an active *tk* gene. Issues of cytotoxicity of vectors have been addressed by deletion of several immediate early gene products. Removal or modification of some nonessential gene products has been performed to improve transgene expression or to affect the cell type specificity of the virus. Consideration must be given to the nature of these modifications (i.e., mutation or deletion of the genes) because this relates to the stability of the defect during propagation. Additionally the potential for recombination with complementing genes in packaging systems should be assessed.

Baculovirus Vectors

Biology and recombinant vectors

Baculoviruses are a large family of lytic viruses that infect a variety of arthropod species. Individual members of the family generally exhibit narrow species specificity, but by all accounts they are unable to replicate in mammalian cells and do not cause any disease in humans. The two baculoviruses used most commonly as gene transfer vectors are the *Autographa californica* multiple nucleopolyhedrosis virus (AcMNPV) and the *Bombyx mori* nucleopolyhedrosis virus (BmNPV). For many years, vectors derived from these viruses have been used extensively to overexpress hundreds of proteins in insect larvae or cultured insect cells (128), and in recent years they have attracted interest as a production system for other viral vectors (107). Baculovirus virions are found in two morphogenetic forms that are important for different stages of the virus life cycle and are both used as recombinant vectors. Occlusion-derived virus, a form in which viral nucleocapsids are embedded in a protein matrix, is responsible for *de novo* natural infection of insect larvae and is used in the laboratory for infection of larvae for protein production. The budded virus (BV) form consists of a single viral nucleocapsid surrounded by a cell-derived membrane and is responsible for cell-to-cell spread of virus within an infected larva. The BV form is used in the laboratory for production of proteins in insect or mammalian cells.

Use of these viruses as gene transfer vectors in insect systems is predicated on promoters for viral gene products that are expressed at very high levels late in the infection process. These promoters are recognized only by the viral RNA polymerase. Therefore, placing a transgene under their control in a recombinant virus will give rise to a virus that transiently directs high-level expression of the desired protein between 24 and 48 hours postinfection of a susceptible insect cell line. Methods for construction of recombinant viral DNA along with other aspects of baculovirus biology have been recently reviewed by van Oers (1). Additionally, an online book by Rohrmann (129) is available that provides detailed information on many aspects of baculovirus biology.

Although baculoviruses will not replicate in mammalian cells, they have been shown to deliver their DNA to these cells. Because the viral promoters are not active in mammalian cells, it is necessary to introduce an expression cassette controlled by a mammalian cell-active promoter to enable transgene expression from these vectors (130), which have been commonly termed "BacMam" viruses. Since its introduction, this system has found several uses as a delivery system for mammalian cell-based assay development (131). Uses that have appeared recently include tissue engineering (132, 133), delivery of genes to generate induced pluripotent stem cells (134), and antigen delivery vehicles for vaccine development (135, 136). Additionally, many of the characteristics of BacMam vectors make this a desirable system to explore for *ex vivo* and *in vivo* gene therapy applications (16, 137).

Risk assessment

Recombinant baculoviruses have a favorable biosafety profile, making them an attractive choice for expression of recombinant proteins in mammalian cells (138). The viruses are inherently unable to replicate in mammalian cells. As mentioned above, the only difference between BacMam vectors and recombinant baculoviruses used in insect cells is the choice of promoter that controls expression of the transgene. Baculoviruses do not cause any disease in humans; thus they are classified as Risk Group 1 agents, although the expression of particular transgenes could raise concerns (Table 3). Additionally, the BV form routinely used in the laboratory is noninfectious for the natural host, thus decreasing the environmental risk from inadvertent release (139). Virions are enveloped and so work surfaces can be decontaminated with a 70% ethanol solution.

Baculoviruses are rapidly inactivated by human complement. This limits their use for *in vivo* applications, but presents an advantage in terms of biosafety. However, there have been reports in the literature of *in vivo* gene transfer by BacMam vectors, and methods to overcome complement inactivation remain an active area of investigation for these vectors (140). Viruses pseudotyped by addition of VSV G glycoprotein to the envelope have been demonstrated to increase resistance to inactivation by animal sera (141). These pseudotyped viruses also demonstrate improved transduction *in vitro*. To allow efficient delivery to a broader range of cell types, a second generation of BacMam vector has recently become commercially available that contains this modification (BacMam pCMV-DEST from Invitrogen). In addition, this second-generation BacMam vector contains the woodchuck hepatitis virus posttranscriptional regulatory element (142) that has been used in a number of viral vectors to increase transgene expression. These pseudotyped BacMam viruses could represent a slightly elevated risk to users if introduced systemically due to resistance to complement inactivation and thus minimization of sharps for use with these vectors is warranted.

Other Viral Vectors

In addition to the more commonly used viral vectors, researchers continue to develop gene delivery systems using other viruses. These include, but are not limited to, measles virus (143), alphaviruses (144), coxsackievirus B3 (145), simian virus 40 (SV40) (146), flavivirus (147), foamy virus (148), VSV (149, 150), and rabies virus (151). Undoubtedly, this group will continue to expand and include further development of novel chimeric viruses designed to incorporate the most desirable characteristics of various virus families (152, 153). A detailed description of all of these viral vector systems is beyond the scope of this chapter and highlights the rapid developments in this field. Another area that is beginning to evolve is the melding of chemistry and engineered biomaterials with viral

TABLE 3.

Viral vectors and transgene containment

Gene transfer vector[a]	Host range[b]	Insert or gene function[c]	Laboratory containment level[d]
MMLV based, *gag, pol, env* deleted	Ecotropic	S, E, M, G, CC, T, MP, DR, R, TX, O_v, O_c	BSL1*
	Amphotropic, VSV-G pseudotyped	S, E, M, MP, DR, T, G,	BSL2
		O_v, O_c, R, CC	BSL2+
		TX	BSL3
Herpes virus based, nonlytic	Broad host range, nervous system	S, E, M, MP, DR, T, G	BSL2
		O_v, O_c, R, CC	BSL2+
		TX	BSL3
Lentivirus based, HIV, SIV, EIAV, FIV, etc.; *gag, pol, env, nef, vpr* deleted	Ecotropic, amphotropic, VSV-G pseudotyped	S, E, M, MP, DR, T, G	BSL2
		O_c, O_v, R, CC	BSL2+
		TX	BSL3
Adenovirus based, serotypes 2, 5, 7; E1 and E3 or E4 deleted	Broad host range, infective for many cell types	S, E, M, T, MP, DR, R, G, CC	BSL2
		O_v, O_c,	BSL2+
		TX	BSL3
Rabies virus, SAD B19 strain (veterinary vaccine strain), G protein deleted	Broad mammalian host range, nervous system specific	Marker proteins such as EGFP	BSL2
		O_v, O_c,	BSL2+
		TX	BSL3
Baculovirus based	Broad mammalian host cell range	S, E, M, T, MP, DR, R, G, CC	BSL1*
		O_v, O_c,	BSL2
		TX	BSL2+/BSL3
AAV based, *rep, cap* defective	Broad host range, infective for many cell types including neurons	S, E, M, T, MP, DR, G	BSL2
		O_v, O_c, R, CC	BSL2+
		TX	BSL2+/BSL3
Poxvirus based, canarypox, vaccinia[e]	Broad host range	S, E, M, T, DR, MP, CC, R, G,	BSL2
		O_v, O_c,	BSL2+
		TX	BSL3

[a] Refers to the parental or wild type virus and some of the common deletions used in viral vectors.

[b] Refers to ability of vector to infect cells from a range of species. Ecotropic generally means able to infect only cells of species originally isolated from or identified in. Please note that the ecotropic host for HIV and HSV would be human cells, but the ecotropic host for MMLV would be murine cells. Amphotropic and VSV-G pseudotyped virus host range includes human cells.

[c] These are general categories of marker or cellular genes and functions. Please note that there are differences in the containment level for the same gene class, depending on whether the viral vector integrates into the recipient genome at high rate. The general categories are: EGFP, enhanced green fluorescent protein; S, structural proteins, actin, myosin, etc.; E, enzymatic proteins, serum proteases, transferases, oxidases, phosphatases etc.; M, metabolic enzymes, amino acid metabolism, nucleotide synthesis, etc.; G, cell growth, housekeeping; CC, cell cycle, cell division; DR, DNA replication, chromosome segregation, mitosis, meiosis; MP, membrane proteins, ion channels, G-coupled protein receptors, transporters, etc.; T, tracking genes such as GFP, luciferases, photoreactive genes; TX, active subunit genes for toxins such as ricin, botulinum toxin, Shiga and Shiga-like toxins; R, regulatory genes, transcription and cell activators such as cytokines, lymphokines, tumor suppressors; O_v and O_c, oncogenes identified via transforming potential of viral and cellular analogs, or mutations in tumor suppressor genes resulting in a protein that inhibits/moderates the normal cellular wild-type protein. This does not include SV40 T antigen. SV40 T antigen-containing cells should not be considered more hazardous that the intact virus. SV40 is considered a Risk Level 1 agent (the lowest level) according to the *NIH Guidelines* (Appendix B in reference 57). The prevalence of SV40 infection in the U.S. population due to contaminated polio vaccine does not seem to have caused a statistically significant increase in the rate of cancers.

[d] This is a general assessment of containment levels for laboratory construction and use of these vectors for nonproduction quantities only based on the 2009 edition of BMBL (101). Please note this table cannot cover every potential use within a research or laboratory setting, as information is gained risk assessments and containment levels may be changed. Local IBCs should use all available information and their best judgment to determine appropriate containment levels. BSL1* refers to the containment level based on parent virus risk group; however, most procedures involving the handling and manipulation of the viral vectors are done at BSL2 to protect cell cultures and viral stocks from contamination.

[e] Certain specific strains of poxviruses, such as MVA, NYVAC, ALVAC, and TROVAC, are considered low-risk agents and can be handled at BSL1 in certain cases (see "Poxvirus Vectors" section).

vector development. Physical or chemical modification may be used to modify tropism, stability, and other properties of vectors (154).

There is increasing interest in clinical use of viruses as anticancer treatments. Some of these oncolytic viruses are unmodified agents, such as Newcastle disease virus (155), that replicate in tumor cells but not in normal human tissues. However, a number of agents are recombinant viruses that have been engineered to exploit a particular cellular environment to allow them to replicate and destroy tumor cells (156, 157) and thus fall under the purview of IBC oversight. The therapeutic effects of these treatments are not solely due to virus-mediated oncolysis, but are also due to immunogenic mechanisms that follow oncolysis. For this reason, many of these oncolytic viruses have been engineered to express immunostimulatory gene products, such as granulocyte-macrophage colony-stimulating factor (GM-CSF) (155, 158).

As the capability to develop and produce novel viral vectors continues to expand, it will prove increasingly challenging for IBCs to evaluate and recommend the appropriate BSLs for such experimentation. In particular, careful risk assessments need to be conducted for animal experimentation in which a thorough knowledge of viral transmission and shedding may not be known. It is important that IBCs are informed regarding advances in molecular virology and have access to the appropriate knowledge experts to make knowledgeable recommendations and decisions.

GENE TRANSFER VECTOR USE IN ANIMALS

The majority of gene transfer vectors discussed in this chapter are replication defective and in wide use within the research community either to create transgenic animals or as transgene/RNA interference (RNAi) delivery vehicle in various animal models of disease. The use of viral vectors in animals opens up a number of additional safety concerns that fall into the three general areas, as follows. (i) Does complementation or rescue of the replication-defective viral vector occur *in vivo*? (ii) How long does the viral vector persist in the animal and where? (iii) Is the viral vector shed, excreted, or exhaled during the experiment? This information determines the risk of exposure of animal care staff and researchers. Unfortunately, most biodistribution and pharmacology/toxicology studies for viral vectors are conducted in uninfected animals so that animal models of rescue and complementation can only be assessed for viral vectors that are related to remnants of ancient endogenous retroviruses embedded in the host genome. However, some human clinical trials offer unique data where viral vectors are used in participants previously infected with the wild-type parent virus (the presence of the wild-type virus is not part of the participant exclusion criteria). Although the number is comparatively small, some gene transfer trials have been registered with the National Institutes of Health (NIH) Office of Biotechnology Activities (OBA) and reviewed by the NIH RAC in which lentiviral vectors have been given to HIV-infected individuals (see the NIH OBA Protocol list at http://osp.od.nih.gov/office-biotechnology-activities/biomedical-technology-assessment-recombinant-dna-advisory-committee/human-gene-transfer-protocols-registered-oba). Recent data discussed at the NIH OBA Policy Conference on RNA Oligonucleotides: Emerging Clinical Applications held in December of 2011 (http://osp.od.nih.gov/office-biotechnology-activities/event/2011-12-15-131500-2011-12-16-201500/rna-oligonucleotides-emerging-clinical-applications) showed no indication of lentiviral vector rescue or complementation of the RNAi-lentiviral vector in these participants.

Where information is lacking, the IBC and Biosafety Officer (BSO) should collaborate with investigators to determine if data should be generated so that appropriate containment and animal handling procedures can be implemented. BSL2 is recommended for laboratory generation and characterization of viral vector preparations, but once the vector is injected into the animal, the dosed animal may not have to be held at animal BSL2 (ABL2) for the duration of the experiment. The animal may serve as the primary containment if the vector is (i) well characterized and the dose is below the level that might contain one replication-competent particle, (ii) the wild-type parent virus is not known to be excreted, exhaled, or secreted from that tissue or in that animal, and (iii) there is data that rescue and complementation are not detectable in that system. In cases in which the IBC, investigators, BSO, and IACUC all agree with the data and that the standard animal facility and care practices and procedures are sufficiently stringent, viral vector-dosed animals do not require ABL2 containment for the duration of the experiment. They may be handled as if uninfected once the injection is completed and ideally once circulating viral vector cannot be detected (if this can be determined). This should be decided on a case-by-case basis, on the basis of experimental data or published literature, and may vary depending on the injection site and dose (159). In terms of overall facility and research convenience and efficiency, reductions in containment usually coincide with a cage change.

THE IBC AND RISK ASSESSMENT

The *NIH Guidelines for Research Involving Recombinant or Synthetic Nucleic Acid Molecules* (*NIH Guidelines*), first published in 1976, established the need for IBCs. Institutions that receive NIH monies for any recombinant DNA research are required to establish an IBC. Initially the scope was narrowly defined and limited to rDNA-based research, but over time, the scope of the *NIH Guidelines* and IBCs was expanded to include use of rDNA technologies in animals, plants, and humans. Most recently, the *NIH Guidelines* were amended to include synthetic nucleic acid technology. Although the IBC is considered the cornerstone of institutional oversight programs for rDNA research, the range of today's biological research has far outstripped the NIH-mandated scope of the IBC. Some institutions have retained the limited rDNA focus of their IBC, whereas others have recognized the value of expanding the responsibilities of the IBC to include all of the biological research conducted at their institution. In these latter instances, the purview of IBCs has expanded to include not just rDNA research, but also use of infectious agents, vaccine development and testing, nanotechnology-based gene delivery, use of human materials and cell lines, synthetic biology, Select Agents, biosafety in animal research, and biosecurity. This far exceeds the original mandate of the IBC, but it does meet the public expectation that institutions that receive public monies to conduct biological research have a coherent and consistent oversight program that encompasses all biological research conducted at a particular institution or entity. The institution via the IBC is expected to ensure the safe and responsible conduct of all the biological research at their institution, not just rDNA-based work.

This increasing scope and workload is a strain on the IBC oversight process, and assessing new or rapidly changing technologies can be problematic. Finding the right balance between asking for more information on specific experiments while not unduly obstructing the pace of research is challenging. By the very nature of new or rapidly changing technologies, there may not be much information on which to base a risk assessment. The lack of information or understanding of the technology often leads IBCs to make conservative risk assessments and require higher containment levels. These risk assessments may be modified over time as more data become available, and it is extremely helpful if investigators and the IBC collaborate to design experiments to address the safety concerns. If the safety-related data can be published, this collaboration eventually benefits the entire research and IBC communities. The more relevant data being made available to IBCs allow for more informed and realistic risk assessments and containment level recommendations.

An example of a rapidly changing field and the need for collaboration between researchers and IBCs is the question of whether local IBCs should require routine testing of lentiviral vector preparations for replication-competent lentivirus (RCL). The current assays—antibody-based p24, reverse transcriptase (RT) assays, reverse transcriptase PCR (RT-PCR)—for detection of replication-competent particles in lentiviral vector preps can be time-consuming and/or technically challenging (160, 161). The latest generation of four to five multiplasmid packaging systems for SIN lentiviral vectors makes the risk of generation of RCL extremely low, although not zero. From the IBC viewpoint, it would be good to know that there were no RCL detectable in a given number of vector particles for each vector preparation. The other side of this safety issue is the need for a scientifically rigorous positive control (in this instance infectious HIV) for determining the limits of detection of the assay.

Balancing the need for good scientific design against requiring investigators to handle HIV to validate defective vector preparations creates an issue for both the researcher and the IBC. These gene transfer vectors have been highly engineered to be as safe and debilitated as feasible. This has created some reluctance on the part of IBCs to ask for RCL testing because the handling of HIV is not a trivial risk. IBCs have considered allowing use of lentiviruses other than HIV to establish the limits of detection of the assay (161). This is a partial solution, but it is difficult to publish these experiments because the positive control is a surrogate. There should be some effort made by the research community to create a specifically attenuated HIV that could be used as a routine positive control for RCL assays without incurring undue risk to the investigator. There has certainly been progress in the refinement of assays for RCL, but the need for quick, accurate, and less technically challenging assays remains. Some of these issues were discussed at the March and June 2006 meetings of the NIH RAC and are reiterated in the 2006 RAC Guidance document on Lentiviral Vector Containment. (This guidance document can be found on the NIH OBA web page, http://osp.od.nih.gov/office-biotechnology-activities.) At the end of their deliberation, the NIH RAC left the decision about requiring RCL testing for laboratory-generated viral vector preparations up to the IBCs.

The development and refinement of viral vectors continues to be a fast-growing area of research. As their use increases, viral vectors are constantly modified in an effort to improve cell tropism and gene delivery efficacy and expression and to enhance safety. The ultimate aim in many instances is to develop a well-characterized, safe, and effective gene transfer vehicle suitable for clinical use. Viral vectors continue to be the predominant delivery method in human gene therapy clinical trials (Fig. 3) with the result that there is now a body of peer-reviewed literature containing safety data, pharmacology/toxicology information, and outcomes from a portion of these clinical

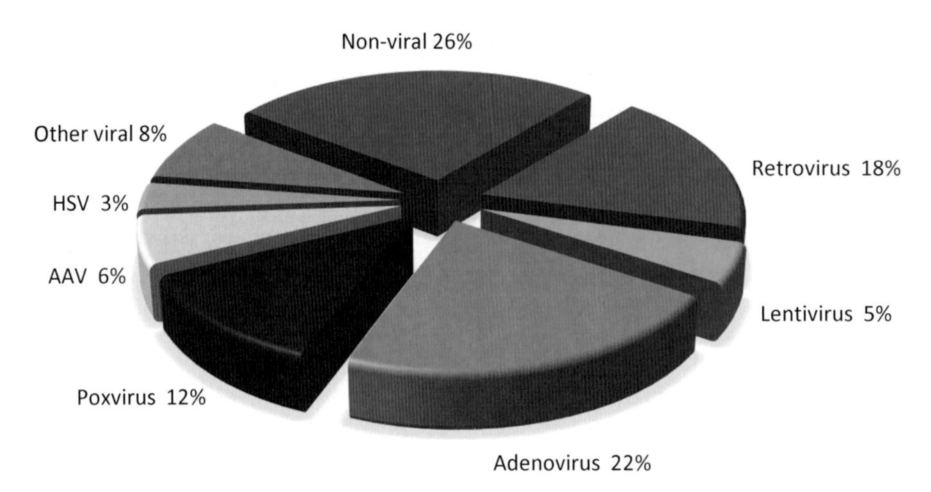

Figure 3: Summary of gene transfer clinical trial delivery methods worldwide through July 2015. Source data are from reference 182.

trials (162) as well as preclinical trials (159, 163). Additional information about outcomes, safety data, and pharmacology/toxicology studies is available to the research community and the public via the NIH Genetic Modification Clinical Research Information System (GeMCRIS) database posted on the NIH OBA website and the National Gene Vector Biorepository (NGVB) databases at the NGVB website (https://ngvbcc.org/Home.action). The FDA Center for Biologics Evaluation and Research has also published many guidance documents on its website addressing safety issues in gene therapy clinical trials in the United States (164). The availability of this growing body of published data can only help biosafety professionals and IBCs make more informed assessments.

VECTOR CHARACTERISTICS AND SAFETY ISSUES

Many of the safety concerns surrounding viral vectors are derived from the very attributes of viruses that make them so useful in a research setting. For example, the ability to integrate into the host genome is a desirable trait for creating transgenic animals and ensuring transgene persistence. On the other hand, vector integration also raises the specter of possible insertional mutagenesis and activation of downstream cellular genes. This has been clearly seen in recent human gene transfer trials.

The undesirable side effects of retroviral vector integration were observed in three separate human gene therapy clinical trials for inherited immunodeficiencies. Two of the three clinical trials were for treatment of X chromosome-linked severe combined immunodeficiency (SCID-X1). The third trial was for X-linked chronic granulomatous disease (CGD) (165, 166). Five of the 20 participants enrolled in the two SCID-X1 gene therapy trials developed T-cell leukemia with one death. The remaining four children responded to chemotherapy for their leukemia. Detailed mapping of the retroviral insertion sites in the expanded T-cell clones in the five leukemic patients indicated retroviral vector insertion sites near the *LMO2* proto-oncogene. The integration of the retroviral vector near the *LMO2* locus brought the cellular gene under the control of the strong gamma retroviral LTR promoter, resulting in continuous expression of the *LMO2* gene and a selective growth advantage of these particular T-cell clones. The selective growth advantage conferred by continued expression of the *LMO2* gene resulted in the uncontrolled expansion of transduced stem cells carrying the integrated retrovirus near the *LMO2* locus. The nature of the disease, the very young age of the participants, the characteristics of the gene transfer viral vector and possible role of the transgene, the *ex vivo* transduction methodology, and the vector insertion sites all played a role in these serious adverse events (167–169). Although the serious adverse events have dominated this discussion so far, it should be noted that 17 out of the total of 20 participants in the two SCID-X1 trials benefited from the gene therapy with nearly all patients' T-cell counts reaching normal levels and functionally competent. Nearly 10 years of follow-up of the earliest SCID-X1 patients have shown retention of T-cell levels and diverse function.

In the CGD trial, there was no clinical improvement in any of the three trial participants. However, two of the three patients showed persistent activation of genes near the vector insertion sites with clonal expansion of transduced myeloid cells. Analysis of the expanded clones indicated that the viral vector was present but there was no expression of the transgene. The promoter for the transgene was silenced in the recipients. Yet again, as in the earlier SCID-X1 trials, the retroviral vector LTRs activated expression of various downstream genes. The clonal expansions were created by the continuous expression of the downstream genes, unrelated to any expression of the transgene. In all three trials, the vector insertion site and the strong viral LTR promoters created the serious adverse events.

The SCID-X1 trials were the first gene therapy clinical trials to show clinical improvements in recipients and also the first to show the safety issues associated with integrating retroviral gene transfer vectors. Unfortunately, insertional mutagenesis and downstream activation have in one way or another been associated with other gene therapy trials (170).

The outcomes of the three clinical trials discussed above led to a series of questions for the research community that are relevant to IBCs charged with review and oversight of gene transfer research. How should preclinical animal testing models be modified to reflect the human disease and predict safety issues better? What types of assays should be developed to identify insertion sites and nearby genes clearly, and possible outcomes of vector insertion such as activation of proto-oncogenes? What improvements should be made to viral vectors to prevent insertional mutagenesis and downstream gene activation? What modifications should be made to gene transfer vectors to make transgene expression better mimic the more natural temporal patterns of expression? Many of the modifications made to both retroviral and lentiviral gene transfer vectors that are discussed in the earlier sections of this chapter came about as a result of the questions raised by these particular studies and the regulators response to those issues. In particular, investigators focused on the development of the specific deletions in the viral LTRs resulting in loss of LTR promoter activity and inability of the integrated vector to be reactivated if a secondary viral infection were to occur (SIN vectors), the use of tissue-specific or responsive promoters to allow a more physiological expression pattern, as well as the development of chimeric vectors (171, 172). Since 2006, the FDA requires assessment of all retroviral integration sites for any gene therapy trial employing integrating vector systems and that all recipients of integrating vector systems are followed for the onset of any serious adverse event for 10 years. All of these changes are intended to address some of the safety issues observed in these clinical trials.

The present inability to control vector integration locations is a significant safety issue. There is a growing body of data that suggests that retroviral and lentiviral vector integration is not completely random (63, 173, 174). Even though retroviral and lentiviral vectors now carry multiple deletions and modifications intended to improve their safety profile, the preference for insertion near actively expressed genes continues to raise concerns.

The cellular tropism, or tissue specificity, of a recombinant virus is a function of the vector envelope and can be manipulated by investigators to suit the needs of their research. At the moment, there is no viral vector that is single cell type or organ specific. Viral receptors can be ubiquitous, e.g., the CAR, coxsackie-adenovirus receptor of adenovirus, or a virus may be able to use multiple receptors to infect a range of cell types and organs, e.g., HIV. If a large number of possible cell targets exist within an animal or a human recipient, either a very large number of vector particles must be used in an attempt to saturate all possible receptor-positive cell types, or the vector must be introduced in such a way that most of the viral vector only reaches the particular target cells or organ of interest. Giving high doses of viral vectors to humans carries risks, not just from the viral vector but from the heterogeneous nature of any vector preparation, which contains empty capsids and fragments, etc. Purification processes are improving, but no viral vector preparation process leads to a product that would meet the standards set for the purity and homogeneity of a new drug. *Ex vivo* transduction of participant autologous cells sidesteps many of these issues and allows both the manipulation and enrichment of specific cell types, the growth of patients' cells, and better control of the transduction process in general. Control of the transduction process means that lower viral vector-to-target cell ratios can be attained and that the numbers of successfully transduced cells can be increased through subsequent rounds of cell culture growth and testing for insertion site locations. These steps reduce the risks associated with the introduction and integration of many vector cDNA copies into each cell and increase the odds that the larger numbers of transduced cells may have a therapeutic effect when reintroduced into the patient.

CONTAINMENT LEVELS FOR VIRAL VECTORS

Determining the appropriate level of containment for a viral gene transfer vector and its experimental use is a complex process that includes consideration of virulence, pathogenicity, dose, route of administration or exposure, environmental stability, and host range (57, 101). In certain instances, gene insert properties, such as physiological activity, toxicity, oncogenic properties, and allergenicity, might also be important. Using the information on mode of transmission, virulence, vector design, and safety features in the "Biology and recombinant vectors" subsections for individual virus families in "Common Viral Vectors" above, a general assessment of the appropriate containment level for viral vector construction and laboratory research use can be created. It must be said that only a general assessment can be made, because it would be impossible to cover every potential viral vector application and transgene. The BSLs suggested in Table 3 are only intended as a rough guideline. As experience is gained with various constructs, and vector safety and packaging cell line systems continue to improve, containment levels may be altered.

This table was originally constructed in 1999 in an effort to help address the many requests to the NIH RAC for assistance from IBCs in conducting viral vector risk assessments. It was meant to assist IBCs in their discussions, but not preclude IBCs from conducting their own independent risk assessments. Eventually the NIH RAC published its document "Biosafety Considerations for Research with Lentiviral Vectors" (175), which essentially agrees with the containment levels outlined in Table 3 for lentiviral and retroviral vectors and supports the original assessment paradigm shown here. The table in this latest revision of this chapter no longer includes information for alphaviral vectors, as these seem not to be a "vector of choice" currently. Rabies virus-based vectors have been included instead. The use of rabies viral vectors is on the rise because of their ability to trace neural networks (176). It is important to note that the rabies virus vector is based on the SAD B19 strain (176, 177), which is an attenuated veterinary vaccine strain.

The criteria for assessing intrinsic vector and insert safety attempts take into account (i) the risk group of the parent unmodified virus; (ii) the extent and type of modifications used to generate the defective vector; (iii) some indication about the function of the inactivated viral genes, e.g., genes involved in DNA replication or deletion of late capsid genes; (iv) vector pseudotyping and host range; (v) whether the viral vector integrates into the host genome with high efficiency; and (vi) transgene function.

The risk assessment paradigm begins with the risk group and containment level for the parent virus. In most cases, the lowest initial containment level for the recombinant viral vector is usually equivalent to the parent wild-type virus as given in the *NIH Guidelines* (57) and the 5th edition of the BMBL (101). This initial containment level recognizes the possibility of the presence of replication-competent contaminants. This may not be necessary as improvements in vector systems are made.

Beyond the viral vector background characteristics, the second part of the assessment identifies categories of transgenes and the possible effect of different categories of transgenes on the potential risks in vector construction and use. Categories of transgene inserts were created according to function and an estimation of biological effect. The largest group consisted of genes coding for proteins with no known oncogenic potential that are either marker proteins such as enhanced green fluorescent protein (EGFP), enzymes, or are involved in the function and maintenance of a cell. These are considered to present the lowest risk from a biological safety standpoint. Members of this category would include the types of transgenes used in most human gene transfer clinical trials for monogenic diseases. The second category contains transgenes encoding proteins that might be involved in the regulation of expression or in growth control might have the ability to activate various immune cells, or are known oncogenes. Viral vectors containing this type of transgene are placed at a somewhat higher containment level due to the enhanced biological effects of the transgenes and the increase in risk associated with working with these vector-transgene constructs. Small quantities of these gene products introduced into a cell may have a profound effect on the normal function of a particular cellular pathway. In all cases, the cloning and expression of genes encoding the active portion of a toxin in mammalian cells are put at the highest containment level, as these would carry the highest risk of damage or cell death for any recipient.

These general transgene categories are fairly basic, but it is important to remember that proteins do not always have a single function. For example, some highly conserved and common proteins function as both ion transporters and viral receptors (178, 179). Although one may consider a particular gene product as fairly innocuous, elevated or nonregulated expression of a protein may lead to unanticipated physiological effects. As mentioned in the introduction to this section, these suggested containment levels should be used only as a starting point for classification and as a basis for IBC discussion and independent risk assessment.

CONCLUSION

Recombinant nucleic acid and gene transfer technologies have dramatically increased the pace of biological research and drastically altered fundamental concepts in biology, genetics, cloning, and medicine. This increased pace, coupled with new recombinant technologies, has shortened the time from lab bench experimentation to potential human therapy. Part of this rapid change is due to the refinement of and continuing developments in gene transfer technology and the application of this technology to new questions, problems, and diseases. For example, several trials have featured an immunological approach to cancer treatment by modifying T cells with chimeric antigen receptors (CARs) engineered to the T cells in response to binding tumor cell-surface antigens (CAR T-cell therapy). In these trials, T-cell populations collected from patients are virally transduced *ex vivo* before being infused into the patient (180). Additionally, similar *ex vivo* cell modification by viral transduction trials are used in treatments for primary immune deficiencies (181). As viral vector applications are extended into new areas, vectors are being genetically modified and novel vectors are being constructed to provide increased efficiencies on all levels. One recent example of this is the PVS-RIPO recombinant oncolytic virus in which the internal ribosomal entry site (IRES) of poliovirus is replaced with the IRES from a human rhinovirus (157). This results in a poliovirus derivative that does not exhibit cytotoxicity for normal central nervous system cells, but grows lytically in tumor cells that display the poliovirus receptor. The impact of this rapid rate of gene transfer vector development has given rise to a wide range of vectors with very different characteristics but with little coordinated or assembled information on which to base containment and safety assessments. This lack of coordinated information has made it challenging for IBCs and biosafety professionals to deal with the emerging field of gene transfer technologies.

The safety assessments and suggested containment levels presented in this chapter are the opinions of the authors only, but can be used to inform the discussion of local IBCs, investigators, and biosafety professionals, and assist in the assessment of risks inherent in the use of these particular viral vectors. As experience is gained and data accumulated on the *in vivo* fate and persistence of the different vectors, containment requirements may change and risk assessments may be altered. Where information on the outcome of human exposures and infections is known, it should be taken into account.

As investigators derive new vectors, it might be prudent for IBCs and IACUCs to ask investigators to develop "safety profiles" as part of a vector characterization process and use in the laboratory and in animals. These are important pieces of information and help determine the appropriate containment procedures for future work with a particular vector. The appropriate containment level can then be determined within the context of the proposed experimental procedures and the biological activity of the transgene or insert. Publication of the results of laboratory and animal safety profile studies would help both the research and biosafety communities. Unless this information is available, local oversight committees are forced to make more conservative estimates of the risks inherent in the research than may be warranted. It is to everyone's advantage—IBCs, biosafety professionals, and researchers—to generate safety information at the early stages of vector development and to publish biodistribution and pharmacology/toxicology data as well as the clinical trial outcomes as soon as feasible.

References

1. **van Oers MM.** 2011. Opportunities and challenges for the baculovirus expression system. *J Invertebr Pathol* **107**(Suppl):S3–S15.
2. **Vanderplasschen A, Pastoret PP.** 2003. The uses of poxviruses as vectors. *Curr Gene Ther* **3**:583–595.
3. **Lundstrom K.** 2003. Virus-based vectors for gene expression in mammalian cells: Semliki Forest Virus, p 207–230. *In* Makrides SC (ed), *Gene Transfer and Expression in Mammalian Cells*. Elsevier Science B. V, Amsterdam, The Netherlands.
4. **Shen Y, Post L.** 2007. Viral vectors and their applications, p 539–564. *In* Knipe DM, Howley PM (ed), *Fields Virology*, 5th ed. Lippincott Williams & Wilkins, Philadelphia, PA.
5. **Graham FL, Smiley J, Russell WC, Nairn R.** 1977. Characteristics of a human cell line transformed by DNA from human adenovirus type 5. *J Gen Virol* **36**:59–72.
6. **Fallaux FJ, Bout A, van der Velde I, van den Wollenberg DJ, Hehir KM, Keegan J, Auger C, Cramer SJ, van Ormondt H, van der Eb AJ, Valerio D, Hoeben RC.** 1998. New helper cells and matched early region 1-deleted adenovirus vectors prevent generation of replication-competent adenoviruses. *Hum Gene Ther* **9**:1909–1917.
7. **Rigg RJ, Chen J, Dando JS, Forestell SP, Plavec I, Böhnlein E.** 1996. A novel human amphotropic packaging cell line: high titer, complement resistance, and improved safety. *Virology* **218**: 290–295.
8. **Miller AD, Trauber DR, Buttimore C.** 1986. Factors involved in production of helper virus-free retrovirus vectors. *Somat Cell Mol Genet* **12**:175–183.
9. **Merten OW.** 2004. State-of-the-art of the production of retroviral vectors. *J Gene Med* **6**(Suppl 1):S105–S124.
10. **Scadden DT, Fuller B, Cunningham JM.** 1990. Human cells infected with retrovirus vectors acquire an endogenous murine provirus. *J Virol* **64**:424–427.
11. **Ronfort C, Girod A, Cosset FL, Legras C, Nigon VM, Chebloune Y, Verdier G.** 1995. Defective retroviral endogenous RNA is efficiently transmitted by infectious particles produced on an avian retroviral vector packaging cell line. *Virology* **207**: 271–275.
12. **Dull T, Zufferey R, Kelly M, Mandel RJ, Nguyen M, Trono D, Naldini L.** 1998. A third-generation lentivirus vector with a conditional packaging system. *J Virol* **72**:8463–8471.
13. **Wickersham IR, Lyon DC, Barnard RJO, Mori T, Finke S, Conzelmann K-K, Young JAT, Callaway EM.** 2007. Monosynaptic restriction of transsynaptic tracing from single, genetically targeted neurons. *Neuron* **53**:639–647.
14. **Braun A.** 2006. Biosafety in handling gene transfer vectors. *Curr Protoc Hum Genet* 50:12.1.1–12.1.18.
15. **Kamimura K, Suda T, Zhang G, Liu D.** 2011. Advances in gene delivery systems. *Pharmaceut Med* **25**:293–306.
16. **Warnock JN, Daigre C, Al-Rubeai M.** 2011. Introduction to viral vectors. *Methods Mol Biol* **737**:1–25.
17. **Alcami A, Moss B.** 2010. Smallpox vaccines, p 1–15. *In* Khan AS, Smith GL (ed), *Scientific Review of Variola Research, 1999–2010*. Khan, AS. World Health Organization, Geneva, Switzerland. http://apps.who.int/iris/bitstream/10665/70508/1/WHO_HSE_GAR_BDP_2010.3_eng.pdf.
18. **Damon IK.** 2013. Poxviruses, p 2160–2184. *In* Knipe DM, Howley PM (ed), *Fields Virology*, 6th ed. Lippincott Williams and Wilkins, Philadelphia, PA.
19. **Wharton M, Strikas RA, Harpaz R, Rotz LD, Schwartz B, Casey CG, Pearson ML, Anderson LJ, Advisory Committee on Immunization Practices, Healthcare Infection Control Practices Advisory Committee.** 2003. Recommendations for using smallpox vaccine in a pre-event vaccination program. Supplemental recommendations of the Advisory Committee on Immunization Practices (ACIP) and the Healthcare Infection

Control Practices Advisory Committee (HICPAC). *MMWR Recomm Rep* **52**(RR-7):1–16.
20. **Lofquist JM, Weimert NA, Hayney MS.** 2003. Smallpox: a review of clinical disease and vaccination. *Am J Health Syst Pharm* **60:**749–756, quiz 757–758.
21. **Casey CG, Iskander JK, Roper MH, Mast EE, Wen XJ, Török TJ, Chapman LE, Swerdlow DL, Morgan J, Heffelfinger JD, Vitek C, Reef SE, Hasbrouck LM, Damon I, Neff L, Vellozzi C, McCauley M, Strikas RA, Mootrey G.** 2005. Adverse events associated with smallpox vaccination in the United States, January-October 2003. *JAMA* **294:**2734–2743.
22. **Moss B.** 2013. Poxviridiae, p 2129–2159. *In* Knipe DM, Howley PM (ed), *Fields Virology*, 6th ed. Lippincott Williams and Wilkins, Philadelphia, PA.
23. **Mackett M, Smith GL, Moss B.** 1982. Vaccinia virus: a selectable eukaryotic cloning and expression vector. *Proc Natl Acad Sci USA* **79:**7415–7419.
24. **Panicali D, Paoletti E.** 1982. Construction of poxviruses as cloning vectors: insertion of the thymidine kinase gene from herpes simplex virus into the DNA of infectious vaccinia virus. *Proc Natl Acad Sci USA* **79:**4927–493.
25. **Moss B.** 1996. Genetically engineered poxviruses for recombinant gene expression, vaccination, and safety. *Proc Natl Acad Sci USA* **93:**11341–11348.
26. **Paoletti E.** 1996. Applications of pox virus vectors to vaccination: an update. *Proc Natl Acad Sci USA* **93:**11349–11353.
27. **Guo ZS, Bartlett DL.** 2004. Vaccinia as a vector for gene delivery. *Expert Opin Biol Ther* **4:**901–917.
28. **Wyatt LS, Earl PL, Moss B.** 2015. Generation of recombinant vaccinia viruses. *Curr Protoc Microbiol* **39:**1–18.
29. **Issacs SN (ed).** 2004. Vaccinia Virus and Poxvirology: Methods and Protocols. *Methods Mol Biol* vol 269. Humana Press, Totowa, NJ.
30. **Walsh SR, Dolin R.** 2011. Vaccinia viruses: vaccines against smallpox and vectors against infectious diseases and tumors. *Expert Rev Vaccines* **10:**1221–1240.
31. **Thorne SH.** 2012. Next-generation oncolytic vaccinia vectors. *Methods Mol Biol* **797:**205–215.
32. **Sánchez-Sampedro L, Perdiguero B, Mejías-Pérez E, García-Arriaza J, Di Pilato M, Esteban M.** 2015. The evolution of poxvirus vaccines. *Viruses* **7:**1726–1803.
33. **Bleckwenn NA, Golding H, Bentley WE, Shiloach J.** 2005. Production of recombinant proteins by vaccinia virus in a microcarrier based mammalian cell perfusion bioreactor. *Biotechnol Bioeng* **90:**663–674.
34. **Hebben M, Brants J, Birck C, Samama JP, Wasylyk B, Spehner D, Pradeau K, Domi A, Moss B, Schultz P, Drillien R.** 2007. High level protein expression in mammalian cells using a safe viral vector: modified vaccinia virus Ankara. *Protein Expr Purif* **56:**269–278.
35. **Jester BC, Drillien R, Ruff M, Florentz C.** 2011. Using Vaccinia's innate ability to introduce DNA into mammalian cells for production of recombinant proteins. *J Biotechnol* **156:**211–213.
36. **Guo W, Cleveland B, Davenport TM, Lee KK, Hu SL.** 2013. Purification of recombinant vaccinia virus-expressed monomeric HIV-1 gp120 to apparent homogeneity. *Protein Expr Purif* **90:**34–39.
37. **de Oliveira TM, Rehfeld IS, Coelho Guedes MI, Ferreira JM, Kroon EG, Lobato ZI.** 2011. Susceptibility of vaccinia virus to chemical disinfection. *Am J Trop Hyg* **85:**152–157.
38. **Evans ME, Lesnaw JA.** 2002. Infection control for gene therapy: a busy physician's primer. *Clin Infect Dis* **35:**597–605.
39. **Evans ME.** 2003. Gene therapy and infection control, p 262–278. *In* Wenzel RP (ed), *Prevention and Control of Nosocomial Infections*, 4th ed. Lippincott Williams & Wilkins, Philadelphia, PA.
40. **Sepkowitz KA.** 2003. How contagious is vaccinia? *N Engl J Med* **348:**439–446.
41. **Jones L, Ristow S, Yilma T, Moss B.** 1986. Accidental human vaccination with vaccinia virus expressing nucleoprotein gene. *Nature* **319:**543.
42. **Openshaw PJ, Alwan WH, Cherrie AH, Record FM.** 1991. Accidental infection of laboratory worker with recombinant vaccinia virus. *Lancet* **338:**459.
43. **Rupprecht CE, Blass L, Smith K, Orciari LA, Niezgoda M, Whitfield SG, Gibbons RV, Guerra M, Hanlon CA.** 2001. Human infection due to recombinant vaccinia-rabies glycoprotein virus. *N Engl J Med* **345:**582–586.
44. **Loeb M, Zando I, Orvidas MC, Bialachowski A, Groves D, Mahoney J.** 2003. Laboratory-acquired vaccinia infection. *Can Commun Dis Rep* **29:**134–136.
45. **Mempel M, Isa G, Klugbauer N, Meyer H, Wildi G, Ring J, Hofmann F, Hofmann H.** 2003. Laboratory acquired infection with recombinant vaccinia virus containing an immunomodulating construct. *J Invest Dermatol* **120:**356–358.
46. **Moussatché N, Tuyama M, Kato SE, Castro AP, Njaine B, Peralta RH, Peralta JM, Damaso CR, Barroso PF.** 2003. Accidental infection of laboratory worker with vaccinia virus. *Emerg Infect Dis* **9:**724–726.
47. **Lewis FM, Chernak E, Goldman E, Li Y, Karem K, Damon IK, Henkel R, Newbern EC, Ross P, Johnson CC.** 2006. Ocular vaccinia infection in laboratory worker, Philadelphia, 2004. *Emerg Infect Dis* **12:**134–137.
48. **Centers for Disease Control and Prevention (CDC).** 2009. Laboratory-acquired vaccinia virus infection—Virginia, 2008. *MMWR Morb Mortal Wkly Rep* **58:**797–800.
49. **Centers for Disease Control and Prevention (CDC).** 2009. Human vaccinia infection after contact with a raccoon rabies vaccine bait—Pennsylvania, 2009. *MMWR Morb Mortal Wkly Rep* **58:**1204–1207.
50. **Isaacs SN.** 2004. Working safely with vaccinia virus: laboratory technique and the role of vaccinia vaccination. *Methods Mol Biol* **269:**1–14.
51. **Byers KB.** 2005. Biosafety tips. See subhead, Biosafety issues in laboratory experiments using vaccinia virus vectors. *Appl Biosaf* **10:**118–122.
52. **Benzekri N, Goldman E, Lewis F, Johnson CC, Reynolds SM, Reynolds MG, Damon IK.** 2010. Laboratory worker knowledge, attitudes and practices towards smallpox vaccine. *Occup Med (Lond)* **60:**75–77.
53. **MacNeil A, Reynolds MG, Damon IK.** 2009. Risks associated with vaccinia virus in the laboratory. *Virology* **385:**1–4.
54. **Rotz LD, Dotson DA, Damon IK, Becher JA, Advisory Committee on Immunization Practices.** 2001. Vaccinia (smallpox) vaccine: recommendations of the Advisory Committee on Immunization Practices (ACIP), 2001. *MMWR Recomm Rep* **50**(RR-10):1–25, quiz CE1–CE7.
55. **Petersen BW, Harms TJ, Reynolds MG, Harrison LH, Centers for Disease Control and Prevention.** 2016. Use of vaccinia virus smallpox vaccine in laboratory and health care personnel at risk for occupational exposure to orthopoxviruses—recommendations of the Advisory Committee on Immunization Practices (ACIP), 2015. *MMWR Morb Mortal Wkly Rep* **65:** 257–262.
56. **Baggs J, Chen RT, Damon IK, Rotz L, Allen C, Fullerton KE, Casey C, Nordenberg D, Mootrey G.** 2005. Safety profile of smallpox vaccine: insights from the laboratory worker smallpox vaccination program. *Clin Infect Dis* **40:**1133–1140.
57. **National Institutes of Health.** 2013. NIH Guidelines for Research Involving Recombinant or Synthetic Nucleic Acid Molecules (NIH Guidelines) 78 FR 66751 (November 6, 2013). The current version of the NIH Guidelines can be accessed at http://osp.od.nih.gov /office-biotechnology-activities/biosafety/nih-guidelines.
58. **Maetzig T, Galla M, Baum C, Schambach A.** 2011. Gammaretroviral vectors: biology, technology and application. *Viruses* **3:** 677–713.
59. **Schweizer M, Merten OW.** 2010. Large-scale production means for the manufacturing of lentiviral vectors. *Curr Gene Ther* **10:**474–486.

60. **Parolin C, Palù G.** 2003. Virus-based vectors for gene expression in mammalian cells: retrovirus, p 231–250. *In* Makrides SC (ed), *Gene Transfer and Expression in Mammalian Cells.* Elsevier Science B. V, Amsterdam, The Netherlands.
61. **Miller AD.** 2014. Retroviral vectors: from cancer viruses to therapeutic tools. *Hum Gene Ther* **25:**989–994.
62. **Cockrell AS, Kafri T.** 2007. Gene delivery by lentivirus vectors. *Mol Biotechnol* **36:**184–204.
63. **Schott JW, Hoffmann D, Schambach A.** 2015. Retrovirus-based vectors for transient and permanent cell modification. *Curr Opin Pharmacol* **24:**135–146.
64. **Yi Y, Noh MJ, Lee KH.** 2011. Current advances in retroviral gene therapy. *Curr Gene Ther* **11:**218–228.
65. **Lévy C, Verhoeyen E, Cosset F-L.** 2015. Surface engineering of lentiviral vectors for gene transfer into gene therapy target cells. *Curr Opin Pharmacol* **24:**79–85.
66. **Pauwels K, Gijsbers R, Toelen J, Schambach A, Willard-Gallo K, Verheust C, Debyser Z, Herman P.** 2009. State-of-the-art lentiviral vectors for research use: risk assessment and biosafety recommendations. *Curr Gene Ther* **9:**459–474.
67. **Sanber KS, Knight SB, Stephen SL, Bailey R, Escors D, Minshull J, Santilli G, Thrasher AJ, Collins MK, Takeuchi Y.** 2015. Construction of stable packaging cell lines for clinical lentiviral vector production. *Sci Rep* **5:**9021.
68. **Rowe WP, Huebner RJ, Gilmore LK, Parrott RH, Ward TG.** 1953. Isolation of a cytopathogenic agent from human adenoids undergoing spontaneous degeneration in tissue culture. *Proc Soc Exp Biol Med* **84:**570–573.
69. **Enders JF, Bell JA, Dingle JH, Francis T Jr, Hilleman MR, Huebner RJ, Payne AMM.** 1956. Adenoviruses: group name proposed for new respiratory-tract viruses. *Science* **124:**119–120.
70. **Wold WSM, Ison MG.** 2013. Adenoviruses, p 1732–1767. *In* Knipe DM, Howley PM (ed), *Fields Virology*, 6th ed. Lippincott Williams and Wilkins, Philadelphia, PA.
71. **Berk AJ.** 2013. Adenoviridiae, p. 1704–1731. *In* Knipe DM, Howley PM (ed.), *Fields Virology*, 6th ed. Lippincott Williams and Wilkins, Philadelphia, PA.
72. **Wang I, Huang I.** 2000. Adenovirus technology for gene manipulation and functional studies. *Drug Discov Today* **5:**10–16.
73. **Bourbeau D, Zeng Y, Massie B.** 2003. Virus-based vectors for gene expression in mammalian cells: adenovirus, p 109–124. *In* Makrides SC (ed), *Gene Transfer and Expression in Mammalian Cells.* Elsevier Science B. V, Amsterdam, The Netherlands
74. **McVey D, Zuber M, Brough DE, Kovesdi I.** 2003. Adenovirus vector library: an approach to the discovery of gene and protein function. *J Gen Virol* **84:**3417–3422.
75. **Ames RS, Lu Q.** 2009. Viral-mediated gene delivery for cell-based assays in drug discovery. *Expert Opin Drug Discov* **4:**243–256.
76. **Hackett NR, Crystal RG.** 2009. Adenoviruses for gene therapy, p. 39–68. *In* Templeton NS (ed), *Gene and Cell Therapy,* 3rd ed. CRC Press, Taylor & Francis Group, Boca Raton, FL.
77. **Brunetti-Pierri N, Ng P.** 2009. Helper-dependent adenoviral vectors for gene therapy, p. 87–114. *In* Templeton NS (ed), *Gene and Cell Therapy*, 3rd ed. CRC Press, Taylor & Francis Group, Boca Raton, FL.
78. **Toth K, Dhar D, Wold WS.** 2010. Oncolytic (replication-competent) adenoviruses as anticancer agents. *Expert Opin Biol Ther* **10:**353–368.
79. **Crystal RG.** 2014. Adenovirus: the first effective *in vivo* gene delivery vector. *Hum Gene Ther* **25:**3–11.
80. **Appaiahgari MB, Vrati S.** 2015. Adenoviruses as gene/vaccine delivery vectors: promises and pitfalls. *Expert Opin Biol Ther* **15:** 337–351.
81. **Zhang C, Zhou D.** 2016. Adenoviral vector-based strategies against infectious disease and cancer. *Hum Vaccin Immunother* **22:**1–11. ePub.
82. **Kovesdi I, Hedley SJ.** 2010. Adenoviral producer cells. *Viruses* **2:**1681–1703.
83. **Hitt MM, Parks RJ, Graham FL.** 1999. Structure and genetic organization of adenovirus vectors, p 61–86. *In* Friedmann T (ed), *The Development of Human Gene Therapy.* Cold Spring Harbor Laboratory, Cold Spring Harbor, NY.
84. **Wivel NA, Gao GP, Wilson JM.** 1999. Adenovirus vectors, p 87–110. *In* Friedmann T (ed), *The Development of Human Gene Therapy.* Cold Spring Harbor Laboratory, Cold Spring Harbor, NY.
85. **Dion LD, Fang J, Garver RI Jr.** 1996. Supernatant rescue assay vs. polymerase chain reaction for detection of wild type adenovirus-contaminating recombinant adenovirus stocks. *J Virol Methods* **56:**99–107.
86. **Ishii-Watabe A, Uchida E, Iwata A, Nagata R, Satoh K, Fan K, Murata M, Mizuguchi H, Kawasaki N, Kawanishi T, Yamaguchi T, Hayakawa T.** 2003. Detection of replication-competent adenoviruses spiked into recombinant adenovirus vector products by infectivity PCR. *Mol Ther* **8:**1009–1016.
87. **Wang F, Patel DK, Antonello JM, Washabaugh MW, Kaslow DC, Shiver JW, Chirmule N.** 2003. Development of an adenovirus-shedding assay for the detection of adenoviral vector-based vaccine and gene therapy products in clinical specimens. *Hum Gene Ther* **14:**25–36.
88. **Lichtenstein DL, Wold WSM.** 2004. Experimental infections of humans with wild-type adenoviruses and with replication-competent adenovirus vectors: replication, safety, and transmission. *Cancer Gene Ther* **11:**819–829.
89. **Murakami P, Havenga M, Fawaz F, Vogels R, Marzio G, Pungor E, Files J, Do L, Goudsmit J, McCaman M.** 2004. Common structure of rare replication-deficient E1-positive particles in adenoviral vector batches. *J Virol* **78:**6200–6208.
90. **Schalk JA, de Vries CG, Orzechowski TJ, Rots MG.** 2007. A rapid and sensitive assay for detection of replication-competent adenoviruses by a combination of microcarrier cell culture and quantitative PCR. *J Virol Methods* **145:**89–95.
91. **Harui A, Suzuki S, Kochanek S, Mitani K.** 1999. Frequency and stability of chromosomal integration of adenovirus vectors. *J Virol* **73:**6141–6146.
92. **Stephen SL, Sivanandam VG, Kochanek S.** 2008. Homologous and heterologous recombination between adenovirus vector DNA and chromosomal DNA. *J Gene Med* **10:**1176–1189.
93. **Stephen SL, Montini E, Sivanandam VG, Al-Dhalimy M, Kestler HA, Finegold M, Grompe M, Kochanek S.** 2010. Chromosomal integration of adenoviral vector DNA *in vivo. J Virol* **84:**9987–9994.
94. **Oualikene W, Gonin P, Eloit M.** 1994. Short and long term dissemination of deletion mutants of adenovirus in permissive (cotton rat) and non-permissive (mouse) species. *J Gen Virol* **75:** 2765–2768.
95. **Tiesjema B, Hermsen HP, van Eijkeren JC, Brandon EF.** 2010. Effect of administration route on the biodistribution and shedding of replication-deficient HAdV-5: a qualitative modelling approach. *Curr Gene Ther* **10:**107–127.
96. **Kallel H, Kamen AA.** 2015. Large-scale adenovirus and poxvirus-vectored vaccine manufacturing to enable clinical trials. *Biotechnol J* **10:**741–747.
97. **Fox JP, Hall CE, Cooney MK.** 1977. The Seattle Virus Watch. VII. Observations of adenovirus infections. *Am J Epidemiol* **105:** 362–386.
98. **Trentin JJ, Yabe Y, Taylor G.** 1962. The quest for human cancer viruses. *Science* **137:**835–841.
99. **Green M, Wold WSM, Brackmann KH.** 1980. Human adenovirus transforming genes: group relationships, integration, expression in transformed cells and analysis of human cancers and tonsils, p 373–397. *In* Essex M, Todaro G, zur Hausen H (ed), *Cold Spring Harbor Conference on Cell Proliferation and Viruses in Naturally Occurring Tumors.* Cold Spring Harbor Laboratory, Cold Spring Harbor, NY.
100. **Bagutti C, Alt M, Schmidlin M, Vogel G, Vögeli U, Brodmann P.** 2011. Detection of adeno- and lentiviral (HIV1) contaminations

on laboratory surfaces as a tool for the surveillance of biosafety standards. *J Appl Microbiol* **111:**70–82.

101. **U.S. Department of Health and Human Services, Public Health Service, Centers for Disease Control and Prevention, National Institutes of Health.** 2009. *Biosafety in Microbiological and Biomedical Laboratories,* 5th ed. HHS Publication no. (CDC) 21-112. http://www.cdc.gov/biosafety/publications/bmbl5/BMBL.pdf.
102. **Knudsen RC.** 1998. Risk assessment for biological agents in the laboratory. *In* Richmond JY (ed), *Rational Basis for Biocontainment: Proceedings of the Fifth National Symposium on Biosafety.* American Biological Safety Association, Mundelein, IL.
103. **Samulski RJ, Sally M, Muzyczka N.** 1999. Adeno-associated viral vectors, p 131–172. *In* Friedmann T (ed), *The Development of Gene Therapy.* Cold Spring Harbor Laboratory, Cold Spring Harbor, NY.
104. **Berns KI, Parrish CR.** 2013. Parvoviridiae, p 1768–1791. *In* Knipe DM, Howley PM (ed), *Fields Virology,* 6th ed. Lippincott Williams and Wilkins, Philadelphia, PA.
105. **Carter BJ, Burstein H, Peluso RW.** 2009. Adeno-associated virus and AAV vectors for gene delivery, p 115–158. *In* Templeton NS (ed), *Gene and Cell Therapy,* 3rd ed. CRC Press, Taylor & Francis Group, Boca Raton, FL.
106. **Weitzman MD, Linden RM.** 2011. Adeno-associated virus biology. *Methods Mol Biol* **807:**1–23.
107. **Kotin RM.** 2011. Large-scale recombinant adeno-associated virus production. *Hum Mol Genet* **20**(R1)**:**R2–R6.
108. **Lisowski L, Tay SS, Alexander IE.** 2015. Adeno-associated virus serotypes for gene therapeutics. *Curr Opin Pharmacol* **24:**59–67.
109. **Büning H, Huber A, Zhang L, Meumann N, Hacker U.** 2015. Engineering the AAV capsid to optimize vector-host-interactions. *Curr Opin Pharmacol* **24:**94–104.
110. **Nault JC, Datta S, Imbeaud S, Franconi A, Mallet M, Couchy G, Letouzé E, Pilati C, Verret B, Blanc JF, Balabaud C, Calderaro J, Laurent A, Letexier M, Bioulac-Sage P, Calvo F, Zucman-Rossi J.** 2015. Recurrent AAV2-related insertional mutagenesis in human hepatocellular carcinomas. *Nat Genet* **47:** 1187–1193.
111. **Russell DW, Grompe M.** 2015. Adeno-associated virus finds its disease. *Nat Genet* **47:**1104–1105.
112. **Schultz BR, Chamberlain JS.** 2008. Recombinant adeno-associated virus transduction and integration. *Mol Ther* **16:**1189–1199.
113. **Pellet PE, Roizman B.** 2013. Herpesviridiae, p 1802–1822. *In* Knipe DM, Howley PM (ed), *Fields Virology,* 6th ed. Lippincott Williams & Wilkins, Philadelphia, PA.
114. **Roizman B, Knipe DM, Whitely RJ.** 2013. Herpes simplex viruses, p 1823–1897. *In* Knipe DM, Howley PM (ed), *Fields Virology,* 6th ed. Lippincott Williams and Wilkins, Philadelphia, PA.
115. **Glorioso JC.** 2014. Herpes simplex viral vectors: late bloomers with big potential. *Hum Gene Ther* **25:**83–91.
116. **Parker JN, Bauer DF, Cody JJ, Markert JM.** 2009. Oncolytic viral therapy of malignant glioma. *Neurotherapeutics* **6:**558–569.
117. **Yeomans DC, Wilson SP.** 2009. Herpes virus-based recombinant herpes vectors: gene therapy for pain and molecular tool for pain science. *Gene Ther* **16:**502–508.
118. **Manservigi R, Argnani R, Marconi P.** 2010. HSV recombinant vectors for gene therapy. *Open Virol J* **4:**123–156.
119. **Zemp FJ, Corredor JC, Lun X, Muruve DA, Forsyth PA.** 2010. Oncolytic viruses as experimental treatments for malignant gliomas: using a scourge to treat a devil. *Cytokine Growth Factor Rev* **21:**103–117.
120. **Ledford H.** 2015. Cancer-fighting viruses win approval. *Nature* **526:**622–623.
121. **Post LE, Roizman B.** 1981. A generalized technique for deletion of specific genes in large genomes: a gene 22 of herpes simplex virus 1 is not essential for growth. *Cell* **25:**227–232.
122. **Parker JN, Zheng X, Luckett W Jr, Markert JM, Cassady KA.** 2011. Strategies for the rapid construction of conditionally-replicating HSV-1 vectors expressing foreign genes as anticancer therapeutic agents. *Mol Pharm* **8:**44–49.
123. **Goins WF, Huang S, Cohen JB, Glorioso JC.** 2014. Engineering HSV-1 vectors for gene therapy. *Methods Mol Biol* **1144:** 63–79.
124. **Spaete RR, Frenkel N.** 1982. The herpes simplex virus amplicon: a new eucaryotic defective-virus cloning-amplifying vector. *Cell* **30:**295–304.
125. **Oehmig A, Fraefel C, Breakefield XO.** 2004. Update on herpesvirus amplicon vectors. *Mol Ther* **10:**630–643.
126. **Santos K, Duke CMP, Dewhurst S.** 2006. Amplicons as vaccine vectors. *Curr Gene Ther* **6:**383–392.
127. **Lim F, Khalique H, Ventosa M, Baldo A.** 2013. Biosafety of gene therapy vectors derived from herpes simplex virus type 1. *Curr Gene Ther* **13:**478–491.
128. **Condreay JP, Kost TA.** 2007. Baculovirus expression vectors for insect and mammalian cells. *Curr Drug Targets* **8:**1126–1131.
129. **Rohrmann GF.** 2013. *Baculovirus Molecular Biology,* 3rd ed. Bookshelf ID: NBK114593. National Center for Biotechnology Information, Bethesda, MD. http://www.ncbi.nlm.nih.gov/books/NBK114593/.
130. **Boyce FM, Bucher NL.** 1996. Baculovirus-mediated gene transfer into mammalian cells. *Proc Natl Acad Sci USA* **93:** 2348–2352.
131. **Kost TA, Condreay JP, Ames RS.** 2010. Baculovirus gene delivery: a flexible assay development tool. *Curr Gene Ther* **10:**168–173.
132. **Chen C-Y, Wu H-H, Chen C-P, Chern S-R, Hwang S-M, Huang S-F, Lo W-H, Chen G-Y, Hu Y-C.** 2011. Biosafety assessment of human mesenchymal stem cells engineered by hybrid baculovirus vectors. *Mol Pharm* **8:**1505–1514.
133. **Hitchman RB, Murguía-Meca F, Locanto E, Danquah J, King LA.** 2011. Baculovirus as vectors for human cells and applications in organ transplantation. *J Invertebr Pathol* **107** (Suppl)**:**S49–S58.
134. **Takata Y, Kishine H, Sone T, Andoh T, Nozaki M, Poderycki M, Chesnut JD, Imamoto F.** 2011. Generation of iPS cells using a BacMam multigene expression system. *Cell Struct Funct* **36:** 209–222.
135. **Zhang J, Chen XW, Tong TZ, Ye Y, Liao M, Fan HY.** 2014. BacMam virus-based surface display of the infectious bronchitis virus (IBV) S1 glycoprotein confers strong protection against virulent IBV challenge in chickens. *Vaccine* **32:**664–670.
136. **Keil GM, Pollin R, Müller C, Giesow K, Schirrmeier H.** 2016. BacMam platform for vaccine antigen delivery. *Methods Mol Biol* **1349:**105–119.
137. **Hu YC.** 2008. Baculoviral vectors for gene delivery: a review. *Curr Gene Ther* **8:**54–65.
138. **Kost TA, Condreay JP.** 2002. Innovations—biotechnology: Baculovirus vectors as gene transfer vectors for mammalian cells: biosafety considerations. *Appl Biosaf* **7:**167–169.
139. **O'Reilly DR, Miller LK, Luckow VA.** 1994. *Baculovirus Expression Vectors: A Laboratory Manual.* Oxford University Press, New York, NY.
140. **Kaikkonen MU, Ylä-Herttuala S, Airenne KJ.** 2011. How to avoid complement attack in baculovirus-mediated gene delivery. *J Invertebr Pathol* **107**(Suppl)**:**S71–S79.
141. **Tani H, Limn CK, Yap CC, Onishi M, Nozaki M, Nishimune Y, Okahashi N, Kitagawa Y, Watanabe R, Mochizuki R, Moriishi K, Matsuura Y.** 2003. *In vitro* and *in vivo* gene delivery by recombinant baculoviruses. *J Virol* **77:**9799–9808.
142. **Donello JE, Loeb JE, Hope TJ.** 1998. Woodchuck hepatitis virus contains a tripartite posttranscriptional regulatory element. *J Virol* **72:**5085–5092.
143. **Russell SJ, Peng KW.** 2009. Measles virus for cancer therapy. *Curr Top Microbiol Immunol* **330:**213–241.

144. **Ehrengruber MU, Schlesinger S, Lundstrom K.** 2011. Alphaviruses: semliki forest virus and sindbis virus vectors for gene transfer into neurons. *Curr Prot Neurosci* 57:4.22.1–4.22.27.
145. **Kim DS, Nam JH.** 2011. Application of attenuated coxsackievirus B3 as a viral vector system for vaccines and gene therapy. *Hum Vaccin* **7:**410–416.
146. **Louboutin JP, Marusich E, Fisher-Perkins J, Dufour JP, Bunnell BA, Strayer DS.** 2011. Gene transfer to the rhesus monkey brain using SV40-derived vectors is durable and safe. *Gene Ther* **18:**682–691.
147. **Reynard O, Mokhonov V, Mokhonova E, Leung J, Page A, Mateo M, Pyankova O, Georges-Courbot MC, Raoul H, Khromykh AA, Volchkov VE.** 2011. Kunjin virus replicon-based vaccines expressing Ebola virus glycoprotein GP protect the guinea pig against lethal Ebola virus infection. *J Infect Dis* **204** (Suppl 3)**:**S1060–S1065.
148. **Lindemann D, Rethwilm A.** 2011. Foamy virus biology and its application for vector development. *Viruses* **3:**561–585.
149. **Ayala-Breton C, Barber GN, Russell SJ, Peng KW.** 2012. Retargeting vesicular stomatitis virus using measles virus envelope glycoproteins. *Hum Gene Ther* **23:**484–491.
150. **Tani H, Morikawa S, Matsuura Y.** 2012. Development and applications of VSV vectors based on cell tropism. *Front Microbiol* **2:**272.
151. **Gomme EA, Wanjalla CN, Wirblich C, Schnell MJ.** 2011. Rabies virus as a research tool and viral vaccine vector. *Adv Virus Res* **79:** 139–164.
152. **Ramsey JD, Vu HN, Pack DW.** 2010. A top-down approach for construction of hybrid polymer-virus gene delivery vectors. *J Control Release* **144:**39–45.
153. **Jurgens CK, Young KR, Madden VJ, Johnson PR, Johnston RE.** 2012. A novel self-replicating chimeric lentivirus-like particle. *J Virol* **86:**246–261.
154. **Jang JH, Schaffer DV, Shea LD.** 2011. Engineering biomaterial systems to enhance viral vector gene delivery. *Mol Ther* **19:** 1407–1415.
155. **Zamarin D, Palese P.** 2012. Oncolytic Newcastle disease virus for cancer therapy: old challenges and new directions. *Future Microbiol* **7:**347–367.
156. **Fueyo J, Gomez-Manzano C, Alemany R, Lee PSY, McDonnell TJ, Mitlianga P, Shi Y-X, Levin VA, Yung WKA, Kyritsis AP.** 2000. A mutant oncolytic adenovirus targeting the Rb pathway produces anti-glioma effect *in vivo*. *Oncogene* **19:** 2–12.
157. **Goetz C, Dobrikova E, Shveygert M, Dobrikov M, Gromeier M.** 2011. Oncolytic poliovirus against malignant glioma. *Future Virol* **6:**1045–1058.
158. **Bartlett DL, Liu Z, Sathaiah M, Ravindranathan R, Guo Z, He Y, Guo ZS.** 2013. Oncolytic viruses as therapeutic cancer vaccines. *Mol Cancer* **12:**103.
159. **Wolfe D, Niranjan A, Trichel A, Wiley C, Ozuer A, Kanal E, Kondziolka D, Krisky D, Goss J, Deluca N, Murphey-Corb M, Glorioso JC.** 2004. Safety and biodistribution studies of an HSV multigene vector following intracranial delivery to non-human primates. *Gene Ther* **11:**1675–1684.
160. **Sastry L, Cornetta K.** 2009. Detection of replication competent retrovirus and lentivirus. *Methods Mol Biol* **506:**243–263.
161. **Cornetta K, Yao J, Jasti A, Koop S, Douglas M, Hsu D, Couture LA, Hawkins T, Duffy L.** 2011. Replication-competent lentivirus analysis of clinical grade vector products. *Mol Ther* **19:** 557–566.
162. **Schenk-Braat EA, van Mierlo MM, Wagemaker G, Bangma CH, Kaptein LCM.** 2007. An inventory of shedding data from clinical gene therapy trials. *J Gene Med* **9:**910–921.
163. **Gonin P, Gaillard C.** 2004. Gene transfer vector biodistribution: pivotal safety studies in clinical gene therapy development. *Gene Ther* **11**(Suppl 1)**:**S98–S108.
164. **Wilson CA, Cichutek K.** 2009. The US and EU regulatory perspectives on the clinical use of hematopoietic stem/progenitor cells genetically modified ex vivo by retroviral vectors. *Methods Mol Biol* **506:**477–488.
165. **Qasim W, Gaspar HB, Thrasher AJ.** 2009. Progress and prospects: gene therapy for inherited immunodeficiencies. *Gene Ther* **16:**1285–1291.
166. **Aiuti A, Roncarolo MG.** 2009. Ten years of gene therapy for primary immune deficiencies. *Hematology (Am Soc Hematol Educ Program)* **2009:**682–689.
167. **Gene Therapy Expert Group of the Committee for Proprietary Medicinal Products (CPMP), European Agency for the Evaluation of Medicinal Products–June 2003 Meeting.** 2004. Insertional mutagenesis and oncogenesis: update from non-clinical and clinical studies. *J Gene Med* **6:**127–129.
168. **Ginn SL, Liao SH, Dane AP, Hu M, Hyman J, Finnie JW, Zheng M, Cavazzana-Calvo M, Alexander SI, Thrasher AJ, Alexander IE.** 2010. Lymphomagenesis in SCID-X1 mice following lentivirus-mediated phenotype correction independent of insertional mutagenesis and gammac overexpression. *Mol Ther* **18:**965–976.
169. **Woods NB, Bottero V, Schmidt M, von Kalle C, Verma IM.** 2006. Gene therapy: therapeutic gene causing lymphoma. *Nature* **440:**1123.
170. **Cavazzana-Calvo M, Payen E, Negre O, Wang G, Hehir K, Fusil F, Down J, Denaro M, Brady T, Westerman K, Cavallesco R, Gillet-Legrand B, Caccavelli L, Sgarra R, Maouche-Chrétien L, Bernaudin F, Girot R, Dorazio R, Mulder GJ, Polack A, Bank A, Soulier J, Larghero J, Kabbara N, Dalle B, Gourmel B, Socie G, Chrétien S, Cartier N, Aubourg P, Fischer A, Cornetta K, Galacteros F, Beuzard Y, Gluckman E, Bushman F, Hacein-Bey-Abina S, Leboulch P.** 2010. Transfusion independence and HMGA2 activation after gene therapy of human β-thalassaemia. *Nature* **467:**318–322.
171. **Mátrai J, Chuah MK, VandenDriessche T.** 2010. Recent advances in lentiviral vector development and applications. *Mol Ther* **18:**477–490.
172. **Zhou S, Mody D, DeRavin SS, Hauer J, Lu T, Ma Z, Hacein-Bey Abina S, Gray JT, Greene MR, Cavazzana-Calvo M, Malech HL, Sorrentino BP.** 2010. A self-inactivating lentiviral vector for SCID-X1 gene therapy that does not activate LMO2 expression in human T cells. *Blood* **116:**900–908.
173. **Ciuffi A.** 2008. Mechanisms governing lentivirus integration site selection. *Curr Gene Ther* **8:**419–429.
174. **Biffi A, Bartolomae CC, Cesana D, Cartier N, Aubourg P, Ranzani M, Cesani M, Benedicenti F, Plati T, Rubagotti E, Merella S, Capotondo A, Sgualdino J, Zanetti G, von Kalle C, Schmidt M, Naldini L, Montini E.** 2011. Lentiviral vector common integration sites in preclinical models and a clinical trial reflect a benign integration bias and not oncogenic selection. *Blood* **117:**5332–5339.
175. **National Institutes of Health Recombinant DNA Advisory Committee.** 2006. Biosafety considerations for research with lentiviral vectors. http://osp.od.nih.gov/office-biotechnology-activities/biosafety-guidance-institutional-biosafety-committees/guidance-biosafety-considerations-research-lentiviral-vectors-0.
176. **Wickersham IR, Finke S, Conzelmann KK, Callaway EM.** 2007. Retrograde neuronal tracing with a deletion-mutant rabies virus. *Nat Methods* **4:**47–49.
177. **Wickersham IR, Sullivan HA, Seung HS.** 2010. Production of glycoprotein-deleted rabies viruses for monosynaptic tracing and high-level gene expression in neurons. *Nat Protoc* **5:** 595–606.
178. **Tufaro F.** 1997. Virus entry: two receptors are better than one. *Trends Microbiol* **5:**257–258, discussion 258–259.
179. **Johann SV, Gibbons JJ, O'Hara B.** 1992. GLVR1, a receptor for gibbon ape leukemia virus, is homologous to a phosphate

permease of Neurospora crassa and is expressed at high levels in the brain and thymus. *J Virol* **66:**1635–1640.
180. **Almåsbak H, Aarvak T, Vemuri MC.** 2016. CAR T cell therapy: a game changer in cancer treatment. *J Immunol Res* **2016:** 5474602.
181. **Kuo CY, Kohn DB.** 2016. Gene therapy for the treatment of primary immune deficiencies. *Curr Allergy Asthma Rep* **16:**39.
182. **Journal of Gene Medicine.** 2015. Gene therapy clinical trials worldwide. John Wiley and Sons Ltd., Hoboken, NJ. http://www.abedia.com/wiley/index.html.

11

Biological Toxins: Safety and Science

JOSEPH P. KOZLOVAC AND ROBERT J. HAWLEY

INTRODUCTION

Biological toxins are poisonous by-products of microorganisms, plants, and animals that produce adverse clinical effects in humans, animals, or plants. A toxin is defined as "a poisonous substance that is a specific product of the metabolic activities of a living organism and is usually very unstable, notably toxic when introduced into the tissues, and typically capable of inducing antibody formation" (Merriam-Webster OnLine, http://www.merriam-webster.com/dictionary/toxin). Biological toxins include metabolites of living organisms, degradation products of dead organisms, and materials rendered toxic by the metabolic activity of microorganisms. Some toxins can also be produced by bacterial or fungal fermentation, the use of recombinant DNA technology, or chemical synthesis of low-molecular-weight toxins. Because they exert their adverse health effects through intoxication, the toxic effect is analogous to chemical poisoning rather than to a traditional biological infection.

Biological toxins may be classified according to the class of organism from which the toxin is derived: bacterial, fungal, algal, plant, or animal. Toxins may also be classified according to their mode of action. A list of the sources of some toxins and venoms and their modes of action are provided in Table 1. Many toxins are used in research to study biochemical processes; therefore, biosafety professionals should become familiar with the risks associated with the toxins to be used in the laboratory. Several venoms and toxins of natural origin have drastic pharmacological effects and are thus of great interest to the scientific community. Toxins have frequently been used to elucidate physiological mechanisms, such as the classic work by Claude Bernard on curare in the 1850s (1). Toxins and venoms have also been used to develop therapeutic drugs. Snake venoms and their isolated enzymes have been used to study and treat blood coagulation problems. The venom of the jararacussu (*Bothrops jararacussu*), a snake native to South America, is used to make the antihypertensive drug captopril (Capoten), which is effective against heart disease and hypertension and has angiotensin-converting enzyme inhibitor activity. Capoten, which gained U.S. Food and Drug Administration (FDA) approval in 1981, is used the world over and has saved millions of lives (2).

TABLE 1.

Sources and mechanisms for various toxins and venoms (16)

Toxin or class	Source(s)	Mechanism of action
Small molecules		
Tetrodotoxin	Puffer fish, octopus, salamander	Na^+ channel blocker
Saxitoxin	Shellfish contaminated with dinoflagellates	Na^+ channel blocker
Ciguatoxin	Large tropical fish contaminated with dinoflagellates	Actions on Na^+ channel
Cardiac glycosides	Toad skin	Adenosine triphosphatase (ATPase) inhibitor
Batrachotoxin	Frog skin	Central nervous system toxin
Palytoxin	Sea anemone	Lanophore
Proteins and polypeptides		
α-Bungarotoxin	Elapid snakes (kraits)	Nicotinic receptor blocker
β-Bungarotoxin	Elapid snakes (kraits)	Presynaptic cholinergic nerves
α-Conotoxin	Cone shells	Nicotinic acetylcholine receptor ligand
μ-Conotoxin	Cone shells	Skeletal muscle Na^+ channel blocker
w-Conotoxin	Cone shells	N-type Ca^{2+} antagonist
Cardiotoxin	Elapid snakes	Direct-acting cardiotoxin
Phospholipases	Many snakes	Cell membrane destruction
Bacterial toxins		
Botulinum toxin	*Clostridium botulinum*	Synaptin in nerve endings
Cholera toxin	*Vibrio cholerae*	Activation of G_s protein
Pertussis toxin	*Bordetella pertussis*	Inactivates G_o/G_s protein
Endotoxin	Gram-negative bacteria	Cell membranes
Tetanus toxin	*Clostridium tetani*	Cell membrane ionophore
Staphylococcal toxin	*Staphylococcus* spp.	Enterotoxin
Pseudomonas exotoxin A	*Pseudomonas aeruginosa*	Inhibits protein synthesis
Diphtheria toxin	*Corynebacterium diphtheriae*	Adenosine diphosphate (ADP)-ribosylation of elongation factor 2

Studies are being conducted with the venom of the mamba snake (*Dendroaspis*), which contains a homologous protein class, the dendrotoxins, that facilitate and help regulate the release of neurotransmitters (3, 4). Continued work with this venom may lead to therapeutics for neurodegenerative disorders, including Alzheimer's disease and associated memory disorders (5). Another example of a toxin that can be used as a treatment is botulinum toxin, one of the most lethal substances known, which has proven to be a useful treatment for localized muscle spasms (6, 7).

The spectrum of naturally occurring toxins is too broad to cover in this chapter; therefore, the discussion is limited to a general review of some of the biological toxins of concern to biomedical laboratory workers. For a detailed discussion of the mechanism of toxicity, the reader is referred to general toxinology/toxicology references (8–13) or reviews of medical microbiology (14, 15).

With an increase in the use of biological toxins in biomedical research, there is a growing need for information on working safely with these materials. This chapter is intended to serve as a guide for laboratory personnel and biosafety professionals for work with diagnostic and research laboratory quantities of biological toxins. Its contents may not be applicable to an industrial setting where large quantities of toxin are being produced.

ROUTES OF EXPOSURE

In the laboratory setting, the routes of exposure for biological toxins and venoms, including ingestion, inhalation, and absorption (dermal, percutaneous, or ocular), are similar to those for infectious agents. In a laboratory or biomedical setting, needlesticks are a primary concern for exposure to toxin, and procedures that use hollow-bore needles to inject or aspirate toxins are among the most hazardous. The skin is an effective barrier to most water-soluble toxins and venoms, but not to fat-soluble substances (16). Although very few toxins (e.g., trichothecene mycotoxins) represent a dermal exposure threat, cutaneous absorption is a potential risk if the toxin is solubilized in a diluent such as dimethyl sulfoxide, which can readily transport the toxin through the skin (17, 18). Dermal exposure may result in localized inflammation; however, a systemic response through percutaneous penetration is also possible. In general, toxins are a health hazard if ingested. However, the toxins of greatest concern in the laboratory differ from those that are most often reported as causes of food-borne outbreaks in the general population (19). Ingestion of even minute amounts of *Clostridium botulinum* toxin can result in death, as opposed to the more common food poisoning outbreaks, usually caused by staphylococcal enterotoxin (20). Another unique exposure is by direct envenomation due to bites from venomous animals during their capture, either from the wild, from the vivarium, or during venom extraction.

All toxins should be considered hazardous by inhalation, although biological toxins are not intrinsically volatile. Aerosols of toxins are usually generated through mechanical agitations, for example, those associated with high-energy laboratory operations such as vortexing, vigorous pipette operations, and mixing. Generation of aerosols is increased when working with the dry form of toxins, especially lyophilized powders; thus, solutions should be used whenever possible. The use of good laboratory practices, proper use of primary containment devices, and availability of appropriate personal protective equipment (PPE) will significantly reduce the hazard potential of common laboratory procedures (21).

DEGREE OF TOXICITY

The degree of toxicity of a hazardous toxin is commonly expressed in terms of a lethal dose to a specific animal species, reported in terms of the dose (LD_{50}) per kilogram of body weight that will kill 50% of the test animals via a specific route of exposure. Other expressions include the minimum lethal dose (MLD) or 100% lethal dose (LD_{100}) per kilogram of body weight of test species for a specific exposure route. The MLD is often assumed to be about twice the LD_{50}; however, this is not always the case (22–24). It is important also to note that toxicity testing in animal species cannot always be readily extrapolated to human exposure. Table 2 contains toxicity data for selected biological toxins.

With an increase in the use of biological toxins in biomedical research, there is a growing need for information on working safely with these materials. This chapter is intended to serve as a guide for laboratory personnel and biosafety professionals for work with diagnostic and research laboratory quantities of biological toxins. Its contents may not be applicable to an industrial setting where large quantities of toxin are being produced, such as for vaccine (toxoid) production.

BACTERIAL TOXINS

Bacterial toxins are "soluble substances that alter the normal metabolism of host cells with deleterious effects on the host" (25) and are the primary virulence factors for a variety of pathogenic bacteria (26). Bacterial toxins are classified as either endotoxins or exotoxins, the characteristics of which are listed in Table 3.

Exotoxins

The concept that a symptom or specific pathological condition may have its origin in the activity of a single compound elaborated by a single organism is best exemplified by the exotoxins produced by certain bacteria (27). Exotoxins are cellular products excreted from viable organisms or released when an organism disintegrates by autolysis. Although most exotoxins are produced by Gram-positive bacteria, such as *Staphylococcus aureus* and *C. botulinum*, some Gram-negative bacteria, such as *Pseudomonas aeruginosa*, also produce exotoxins (28). Several of the exotoxins can be of concern to laboratory workers if proper work practices are not followed. *Clostridium* spp., including *C. tetani*, *C. botulinum*, and *C. perfringens*, are all known to cause human disease as a direct result of exotoxin release (29). One important characteristic of exotoxins is the ease of conversion to the nontoxic, immunogenic toxoids used for immunization, for example, against tetanus and diphtheria (30).

Several other species of bacteria produce a specific type of exotoxin, usually referred to as enterotoxins, which are associated with food-borne and waterborne outbreaks of diarrheal diseases. These include, but are not limited to, *S. aureus*, *Vibrio cholerae*, and *Pasteurella*, *Bordetella*, and *Shigella* spp. All of these are known to cause significant human morbidity and mortality but are not considered to be great threats to laboratory workers if basic biosafety practices are followed while working with these organisms.

Endotoxins

Endotoxins are derived from the lipopolysaccharide (LPS) cell wall of Gram-negative organisms, such as *Escherichia coli* and *Shigella dysenteriae*, and are released from the cell upon death or autolysis. Purified endotoxin is referred to as a LPS, with the lipid A subunit being the biologically active component. Endotoxins produce symptoms in the host that range from endotoxin fever to disseminated intravascular coagulation, and death (see Table 3). Acute

TABLE 2.

Toxicity data for Select Toxins

			Toxic dose[a]				
Toxin type	**Toxin source**	**Toxin name**	**Mice**	**Guinea pigs**	**Rabbits**	**Monkeys**	**Humans**
Bacterial toxins	*Bacillus anthracis*	LFPA	1.25 µg i.v.				
	Clostridium botulinum		2.4 µg i.v. (rat)				
	Type A	Neurotoxin A	1.2 ng i.p.	(0.6 ng)[b]	(0.5 ng)[b]	(0.5–0.7 ng)[b]	(ca. 1 ng)[b]
	Type B	Neurotoxin B (proteolytically activated)	0.5 ng i.v.	0.6 ng i.p.			
	Type C	Neurotoxin C1	1.1 ng i.v.	(ca. 1.1 ng)[b]	(ca. 0.15ng)[b]	(ca. 0.4 ng)[b]	
	Type D	Neurotoxin D	0.4 ng	0.1 ng[b]	0.08 ng[b]	40 ng[b]	
	Type E	Neurotoxin E	(1.1 ng)	0.6 ng[b]	1.1 ng[b]	1.1 ng[b]	
	Type F	Neurotoxin F	2.5 ng i.p.				
	Clostridium tetani	Tetanus toxin	(0.1 ng)	(ca. 0.3 ng)[b]	(0.5–5 ng)[b]		(<2.5 ng)[b]
	Corynebacterium diphtheriae	Diphtheria toxin	(1.6 mg s.c.)				(<100 ng s.c.)
	Escherichia coli	Enterotoxin	250 µg i.v.				
	Staphylococcus aureus	Alpha toxin	40–46 µg i.v.		1.3 µg		
	Vibrio cholerae	Cholera toxin	250 µg				
	Yersinia pestis	Murine toxin	0.50 µg i.p.				
Plant toxins	Rosary pea plant	Abrin	0.04 µg				
	Castor bean plant	Ricin	3.0 µg				
Marine toxins	Cone snail	Conotoxin	5.0 µg				
	Marine dinoflagellates	Saxitoxin	10.0 µg				
	Fish/marine dinoflagellates	Ciguatoxin	0.4 µg				
	Puffer fish	Tetrodotoxin	8.0 µg				
Mycotoxin	Various fungi	T-2 mycotoxin	1,210 µg				

[a]Values are LD_{50}s except for those in parentheses, which are MLDs.
[b]Values are calculated from mouse toxicities (23, 79, 70).
i.v., intravenous; i.p., intraperitoneal; s.c., subcutaneous; LD_{50}, median lethal dose; MLD, minimum lethal dose.

exposure to endotoxin-containing dusts can cause chest tightness, organic dust toxic syndrome, and byssinosis, and long-term exposure is associated with an accelerated decline in lung function and chronic respiratory disease. There is a growing body of knowledge that is supportive of the hypothesis that endotoxin exposure is possibly beneficial with respect to reducing lung cancer risk (31). Also termed cell-associated toxins, endotoxins have a chemical composition and toxic properties distinct from those of bacterial exotoxins. Endotoxins are heat stable and have molecular weights between 3,000 and several million (29). Unlike exotoxins, they are not tissue specific.

Agricultural workers and processors of vegetable fibers are most likely to be at risk from endotoxins. Airborne concentrations of endotoxin in excess of 50 ng/m^3 have been reported in several occupational settings. They include swine and poultry confinement buildings (32), grain storage facilities (33), cotton and flax mills (34), and poultry houses (35). At least one study has proposed thresholds for acute pulmonary toxicity in a range of 10 to 33 ng/m^3 and recommended that consideration be given to limiting exposure to airborne endotoxin in work environments (36).

Mode of Action

Bacterial toxins can also be categorized by their mode of action as seen in Tables 1 and 4. Toxins may damage cell membranes, inhibit protein synthesis, hinder neurotransmitter release, or activate the host immune response. Table 4 lists bacterial toxins according to mode of action, the target of the toxin, characteristics of the disease produced, and whether toxin has been implicated in that disease.

TABLE 3.

Primary features of bacterial exotoxins and endotoxins (14)

Exotoxins	Endotoxins
Excreted by living cells; high concentrations in liquid medium	Integral part of the cell wall of Gram-negative bacteria; released during bacterial death and in part during growth; may not need to be released to have biological activity
Produced by both Gram-positive and Gram-negative bacteria	Produced only by Gram-negative bacteria
Polypeptides	Lipopolysaccharide complexes
Relatively unstable; toxicity often destroyed rapidly at temperatures above 60°C	Relatively stable; withstand heating at temperatures above 60°C for hours without loss of activity
Highly antigenic; stimulate formation of high-titer antibodies	Weakly immunogenic; relationship between antibody titers and protection from disease is less clear than with exotoxins
Converted to antigenic toxoids when inactivated by treatment with formalin, acid, heat, etc. Toxoids are used for preventive immunization (i.e., tetanus toxoid)	Not converted to toxoids
Highly toxic; fatal to animals in a dose of microgram quantities or less	Moderately toxic; fatal to animals in a dose of tens to hundreds of micrograms
Usually bind to specific receptors on cells	Specific receptors not found on cells
Usually do not produce fever in the host	Usually produce fever in the host through the release of interleukin-1 and other mediators
Frequently controlled by extrachromosomal genes (i.e., plasmids)	Synthesis directed by chromosomal genes

Anthrax toxin

Anthrax toxin is an exotoxin produced by the Gram-positive, spore-forming *Bacillus anthracis*, which occurs worldwide. It exists in the soil as a spore that germinates into a vegetative cell when environmental conditions are favorable. The anthrax exotoxin is composed of three proteins: protective antigen (PA, 83 kDa), lethal factor (LF, 90 kDa), and edema factor (EF, 89 kDa). None of these proteins are toxic individually. The three proteins act in binary combinations of PA plus LF (lethal toxin) and PA plus EF (edema toxin) that exert their effects on the host defenses and eventually kill the host. Edema toxin causes the extreme tissue swelling associated with cutaneous anthrax; lethal toxin is responsible for death of the host. Production of LF and EF requires the plasmid pX01. PA serves as the receptor-binding moiety that delivers LF and EF into the cell. PA is cleaved by cell-surface proteases, releasing a 20-kDa fragment from the amino-terminal end of PA. Upon release of the 20-kDa fragment, LF or EF binds competitively to the exposed site. The complex undergoes receptor-mediated endocytosis, allowing translocation to the cell cytosol to exert its toxic effects. Although these toxins play a very important role in the pathophysiology of anthrax, virulence is also dependent on the presence of the bacterial capsule (37). If the bacillus loses the capsule, it loses the ability to cause disease.

Infection with *B. anthracis* can usually be treated successfully with antibiotics if the disease is in the cutaneous form (38); if the disease is in the inhalation form (39), treatment may be possible only if started early (40). Treatment or postexposure prophylaxis while on antibiotic treatment can also involve administration of the vaccine (41). Although there is at present no available prophylaxis or treatment beyond supportive therapy once the toxicity has progressed, studies have been conducted for efficient postexposure therapy for *B. anthracis* infection. One approach has been passive immunization through the administration of monoclonal antibodies that mitigate the biological action of anthrax toxin (39, 42, 43).

Clostridium toxins

There are more than 100 species in the genus *Clostridium*, Gram-positive, spore-forming, catalase-negative, anaerobic bacilli that are distributed widely in nature, in soil, as well as in freshwater and marine sediments throughout the world. Only those species for which toxin formation is of primary concern are discussed here.

Botulinum neurotoxins

Strains of *C. botulinum* produce seven serologically distinct but related botulinum neurotoxins (A, B, C1, D, E, F, and G) (44), with types A, B, E, and F most often implicated in human disease. The botulinum neurotoxin is the most potent toxin known, but the serotypes vary in toxicity in different animal species (45). The primary concern with *C. botulinum* is the life-threatening neuromuscular paralysis, clinically known as botulism, which follows ingestion of toxin-containing food (19). The most common

TABLE 4.

Characteristics of bacterial toxins[a] (26)

Organism/toxin	Mode of action	Target(s)	Disease(s)	Toxin implicated in disease[b]
Damage membranes				
Aeromonas hydrophila/ aerolysin	Pore former	Glycophorin	Diarrhea	(Yes)
Clostridium perfringens/ perfringolysin O	Pore former	Cholesterol	Gas gangrene[c]	?
Escherichia coli/hemolysin[d]	Pore former	Plasma membrane	UTIs	(Yes)
Listeria monocytogenes/ listeriolysin O	Pore former	Cholesterol	Food-borne systemic illness, meningitis	(Yes)
Staphylococcus aureus/ alpha toxin	Pore former	Plasma membrane	Abscesses[c]	(Yes)
Streptococcus pneumoniae/ pneumolysin	Pore former	Cholesterol	Pneumonia[c]	(Yes)
Streptococcus pyogenes/ streptolysin O	Pore former	Cholesterol	Strep throat, SF[c]	?
Inhibit protein synthesis				
Corynebacterium diphtheriae/ diphtheria toxin	Adenosine disphosphate (ADP)-ribosyltransferase	Elongation factor 2	Diphtheria	Yes
Escherichia coli and *Shigella dysenteriae*/Shiga toxins	*N*-Glycosidase	28S RNA	HC and HUS	Yes
Pseudomonas aeruginosa/ exotoxin A	ADP-ribosyltransferase	Elongation factor 2	Pneumonia[c]	(Yes)
Activate second messenger pathways				
Escherichia coli				
CNF	Deamidase	Rho G proteins	UTIs	?
LT	ADP-ribosyltransferase	G proteins	Diarrhea	Yes
ST[d]	Stimulates guanylate cyclase	Guanylate cyclase receptor	Diarrhea	Yes
CLDT[d]	G2 block	Unknown	Diarrhea	(Yes)
EAST	ST-like?	Unknown	Diarrhea	?
Bacillus anthracis/EF	Adenylate cyclase	Adenosine triphosphate (ATP)	Anthrax	Yes
Bordetella pertussis/ dermonecrotic toxin	Deamidase	Rho G proteins	Rhinitis	(Yes)
Bordetella pertussis/ pertussis toxin	ADP-ribosyltransferase	G protein(s)	Pertussis	Yes
Clostridium botulinum/C2 toxin	ADP-ribosyltransferase	Monomeric G-actin	Botulism	?
Clostridium botulinum/C3 toxin	ADP-ribosyltransferase	Rho G protein	Botulism	?
Clostridium difficile/toxin A	Glucosyltransferase	Rho G protein(s)	Diarrhea/AAPC	(Yes)
Clostridium difficile/toxin B	Glucosyltransferase	Rho G protein(s)	Diarrhea/AAPC	?
Vibrio cholerae/cholera toxin	ADP-ribosyltransferase	G protein(s)	Cholera	Yes
Activate immune response				
Staphylococcus aureus/ enterotoxins	Superantigen	TCR and MHC II	Food poisoning[c]	Yes
Staphylococcus aureus/ exfoliative toxins	Superantigen (and serine protease?)	TCR and MHC II	SSS[c]	Yes
Staphylococcus aureus/ toxic shock toxin	Superantigen	TCR and MHC II	TSS[c]	Yes

(continued)

TABLE 4.

(Continued)

Organism/toxin	Mode of action	Target(s)	Disease(s)	Toxin implicated in disease[b]
Streptococcus pyogenes/ pyrogenic exotoxins	Superantigen	TCR and MHC II	SF/TSS[c]	Yes
Protease				
Bacillus anthracis/LF	Metalloprotease	MAPKK1/MAPKK2	Anthrax	Yes
Clostridium botulinum/ neurotoxins A–G	Zinc metalloprotease	VAMP/synaptobrevin, SNAP-25, syntaxin	Botulism	Yes
Clostridium tetani/tetanus toxin	Zinc metalloprotease	VAMP/synaptobrevin	Tetanus	Yes

[a]Abbreviations: AAPC, antibiotic-associated pseudomembranous colitis; CLDT, cytolethal distending toxin; CNF, cytotoxic necrotizing factor; EAST, enteroaggregative *E. coli* heat-stable toxin; HC, hemorrhagic colitis; HUS, hemolytic-uremic syndrome; LT, heat-labile toxin; MAPKK, mitogen-activated protein kinase kinase; MHC II, major histocompatibility complex class II; SNAP-25, synaptosomal associated protein; SF, scarlet fever; SSS, scalded skin syndrome; ST, heat-stable toxin; TCR, T-cell receptor; TSS, toxic shock syndrome; UTI, urinary tract infection; VAMP, vesicle-associated membrane protein.
[b]Yes, strong causal relationship between toxin and disease; (yes), role in pathogenesis has been shown in animal model or appropriate cell culture; ?, unknown.
[c]Other diseases are also associated with the organism.
[d]Toxin is also produced by other genera of bacteria.

offenders are spiced, smoked, vacuum-packed, or canned alkaline foods that are eaten without sufficient cooking. Home canning of vegetables, meats, and smoked fish provides the anaerobic conditions needed for the germination of spores into vegetative forms that produce botulinum toxin (46). *C. botulinum* has also been implicated in rare cases of wound infection (47, 48).

Botulinum toxin acts by blocking release of acetylcholine at nerve synapses and neuromuscular junctions. The toxin consists of two polypeptide subunits. The B subunit binds to a receptor on the axons of motor neurons. The toxin is taken into the axon, where the A chain exerts its cytotoxic effect, preventing release of acetylcholine and neuromuscular transmission (presynaptic inhibition). Symptoms begin 18 to 24 h after ingestion of the toxic food, with visual disturbances, inability to swallow, and speech difficulty; signs of bulbar paralysis are progressive, and death occurs from respiratory paralysis or cardiac arrest. Gastrointestinal symptoms are not usually prominent, and there is no fever. The patient remains fully conscious until shortly before death. Use of a mechanical respirator and other supportive measures has reduced the mortality rate from 65% to below 35%, if diagnosis is made early. Patients who recover do not develop protective antibodies.

In addition to supportive care, botulinum intoxication may be treated with heptavalent equine antitoxin, which may decrease disease progression if given before onset of clinical signs. However, substantial serious adverse effects may occur following administration. The antitoxin is available as an Investigational New Drug from the Centers for Disease Control and Prevention (CDC), Atlanta, GA (http://www.cdc.gov/laboratory/drugservice/formulary.html#ia; telephone, 404-639-3311). Human botulism immune globulin for the treatment of infant botulism caused by type A or B botulinum toxin has been licensed (49) and has avoided many years of hospital stay and many millions of dollars of hospital costs.

An investigational pentavalent vaccine effective against botulinum toxin serotypes A, B, C, D, and E is available for use as a prophylaxis against botulinum toxin. When tested in several thousand volunteers and at-risk laboratory workers, it induced serum antitoxin levels that correspond to protective levels in animals. The vaccine, when given at 0, 2, and 12 weeks, followed by a booster at 1 year, results in a protective titer in >90% of vaccinees after 1 year. Administration of this vaccine is recommended for personnel working with cultures of *C. botulinum* or its toxins (50). The vaccine was shown to be effective against aerosol challenge of animals (51). The disadvantages of the vaccine are that it is a formalin-inactivated product, it is costly to produce, and it is in limited supply. The Department of Defense is currently developing a recombinant vaccine that is homogeneous in composition, easy and affordable to produce, and effective in inducing long-term protective immunity (52).

Decontamination of toxin-contaminated articles or materials can be accomplished using soap and water to remove the toxin burden and treating with a 0.5% sodium hypochlorite solution in water for a contact time of 10 to 15 min (53–55).

C. tetani

Tetanus is caused by a toxin released by *C. tetani*, a ubiquitous bacterium found in soil, especially soil rich in fecal matter. Agricultural workers are at the greatest risk of exposure (56). Tetanus toxin, tetanospasmin, is a potent neurotoxin that inhibits transmitter release from nerve terminals, causing local or generalized paralysis, and is responsible for the signs and symptoms of tetanus (57). Diagnosis is by history and examination. The only condition that closely resembles tetanus is strychnine poisoning (58). Human disease can be prevented by immunization with tetanus toxoid (59, 60).

Escherichia coli

Laboratory-acquired *E. coli* infection, presumably due to failure to follow standard biosafety procedures, has been documented (61, 62). Several different types of heat-labile and heat-stable enterotoxins have been identified among human, porcine, and bovine *E. coli* strains (63–65). In recent years, an especially toxic form of *E. coli* (O157:H7) has been implicated in severe disease, including renal failure and hemolytic-uremic syndrome (66).

Shigella species

Clinical laboratory workers have been reported to be at increased risk of exposure to *Shigella* infections (67, 68). Upon autolysis, all shigellae release their toxic LPS endotoxin (69), which probably contributes to the irritation of the bowel wall. *S. dysenteriae* type 1 (Shiga bacillus) also produces a heat-labile exotoxin, a protein that is antigenic and lethal for experimental animals, affecting both the gut and the central nervous system (70). Acting as an enterotoxin, it produces diarrhea, as does the *E. coli* verotoxin, perhaps by the same mechanism. In humans, the exotoxin also inhibits sugar and amino acid absorption in the small intestine. Acting as a "neurotoxin," this material may contribute to the extreme severity and fatal nature of *S. dysenteriae* infections and to the central nervous system reactions observed (i.e., meningismus and coma). Patients with *S. flexneri* or *S. sonnei* infections develop antitoxin that neutralizes *S. dysenteriae* exotoxin *in vitro*. The toxic activity is distinct from the invasive property of shigellae. The two may act in sequence, the toxin producing an early nonbloody, voluminous diarrhea followed by the invasion of the large intestine resulting in dysentery with blood and pus in stools (14).

Staphylococcus species

There are several soluble staphylococcal enterotoxins produced by nearly 50% of *S. aureus* strains, including SEA, SEB, SEC1, SEC2, SEC3, SED, SEE, and toxic shock syndrome toxin 1, which is the same as enterotoxin F. These enterotoxins are heat stable (resist boiling for 30 min) and resistant to the action of intestinal enzymes. As important causes of food poisoning, enterotoxins are produced when *S. aureus* grows in carbohydrate and protein foods (71). The gene for enterotoxin production may be on the bacterial chromosome (72), but a plasmid may carry a protein that regulates active toxin production (73). SEA, SEB, and SEC1 are potent mitogens (74, 75).

Staphylococcus spp. produce several toxins that are lethal for animals on injection, cause necrosis in skin, and contain soluble hemolysins, which can be separated by electrophoresis. The alpha toxin is a heterogeneous protein that can lyse erythrocytes and damage platelets and is probably identical to the lethal and dermonecrotic factors of exotoxin. Alpha toxin also has a powerful action on vascular smooth muscle (76). Beta toxin degrades sphingomyelin and is toxic for many kinds of cells, including human red blood cells (14). These toxins and two others, the gamma and delta toxins, are antigenically distinct and bear no relationship to streptococcal lysins (77). Exotoxin treated with formalin gives a nonpoisonous but antigenic toxoid, but it is not useful clinically (14).

SEB enterotoxin is of concern due to its potential for use in bioterrorist activities. Ocular exposure resulting in symptoms in three laboratory workers was reported for the first time, along with reports of 16 laboratory-acquired inhalational SEB intoxications (78). Within 1 to 6 h postexposure, all three individuals developed localized cutaneous swelling and conjunctivitis. Gastrointestinal symptoms occurred in two of the three. The emetic response is allegedly due to stimulation of nerve centers in the gut (78). This case study illustrates that toxin can be transferred by contaminated hands to the mouth or mucous membranes and emphasizes the necessity of using appropriate engineering controls and PPE when working with SEB.

The ingestion of as little as 25 μg of enterotoxin B by humans or monkeys results in vomiting and diarrhea (79). SEB intoxication is a severely incapacitating illness with an onset of 3 to 4 h after ingestion. Symptoms vary, but include fever, malaise, headache, respiratory changes, gastrointestinal involvement, and leukocytosis lasting for 3 to 4 days. There is also significant morbidity and potential mortality when these toxins are inhaled. In one episode, which occurred in 1964, at least nine laboratory workers were exposed to aerosolized SEB following what was described as an accident in a laboratory (74). The ensuing illness was heralded by rigors and fever with readings as high as 106°F. The onset of the fever averaged 12 h after the exposure (range, 8 to 20 h), and the febrile period lasted from 12 to 76 h (mean duration, 50 h). The fever was associated with muscle aches and headache. Respiratory symptoms, in the form of a nonproductive cough, began about the same time as the fever and muscle pain. Of the nine workers, five had an abnormal lung exam associated with shortness of breath, and three had shortness of breath at rest. One of these had "profound" shortness of breath for the first 12 h of symptoms and exertional shortness of breath for 10 days. Chest X rays obtained during this sublethal exposure were abnormal. Chest pain, described as moderately intense, also occurred with the respiratory symptoms, lasting an average of 1 day (range, 4 h to 4 days). Vomiting and loss of appetite developed in most, with a mean onset of 17 h (range, 8 to 24 h), with the anorexia lasting several days and the vomiting limited to a mean of 9 h (range, 4 to 20 h).

Although there is no specific therapy or prophylaxis, vaccine development is in progress (80–82). Studies addressing protection with monoclonal antibodies suggest they represent excellent therapeutic candidates for further preclinical and clinical development (83). In studies involving nine mildly intoxicated patients, supportive therapy, including aspirin or acetaminophen, antihistamines, fluid and electrolyte replacement, and cough suppressants, appeared to be adequate in reducing toxicity in these patients (84).

Yersinia pestis

The toxin produced by *Y. pestis*, the plague bacillus, often referred to as murine toxin, rapidly kills mice and rats, especially at low temperatures (85). It may act as an antagonist of epinephrine, but the biochemical mechanism is not known. Previously, inactivated plague vaccine was recommended for laboratory workers with frequent exposure to *Y. pestis* (86). Currently there is no U.S. licensed plague vaccine that could be used for prophylaxis during laboratory and fieldwork with *Y. pestis* (87). However, there is continued progress in alternate protective strategies, including the development of subunit, DNA, and live carrier platform delivery using bacterial and viral vectors, as well as other approaches for controlled attenuation of virulent strains of *Y. pestis* (88).

FUNGAL TOXINS

The hazards associated with eating poisonous mushrooms have been recognized for centuries (89); however, the hazards associated with certain filamentous fungi have only recently been documented (90). The single-cell fungi, known as yeasts, do not produce toxins of concern to human health.

Mycotoxins are low-molecular-weight, secondary metabolites of various species of terrestrial filamentous microfungi. They contaminate a wide array of foods, including beans, cereals, coconuts, milk, peanuts, and sweet potatoes, and pose a significant threat to human food supplies (91). The mycotoxins of greatest importance are the aflatoxins and tricothecenes, produced by *Aspergillus*, *Fusarium*, and *Stachybotrys* spp. (92).

Aflatoxin

Aflatoxins have been associated with acute liver damage and liver cirrhosis, tumor development, teratogenesis immunosuppression, and altered protein energy metabolism and hemoglobin levels. They include aflatoxins B1, B2, G1, and G2, produced by the fungi *Aspergillus flavus*, *Aspergillus parasiticus*, and *Aspergillus nomius*. The toxins may be found in peanuts, maize, and cottonseed. Aflatoxins have been shown to be carcinogenic in animal studies, and epidemiological studies suggest an association between aflatoxin consumption and liver cancer in humans (93, 94). At least one study also suggests that exposure to aerosols of aflatoxin might be related to bronchial carcinoma, colon cancer, and liver cancer (95). Treatment is supportive, as there is no specific antidote or therapy for aflatoxin intoxication (96, 97).

Trichothecene Mycotoxins

The trichothecene mycotoxins, which include T-2 mycotoxin, diacetoxyscirpenol, nivalenol, and 4-deoxynivalenol, are a diverse group of more than 40 low-molecular-weight toxins produced by at least seven genera of fungi (*Trichothecium*, *Fusarium*, *Stachybotrys*, *Cephalosporium*, *Myrothecium*, *Gibberella*, and *Trichoderma*). They are stable under various environmental conditions and are implicated in diseases in animals and humans caused by the consumption of moldy grains (92, 98). Onset of symptoms can occur within 10 to 30 min of exposure to tricothecene mycotoxins. Following ingestion of mycotoxins, depending upon whether the exposure is acute or chronic, there is weight loss, vomiting, bloody diarrhea, diffuse hemorrhage, skin inflammation, and possibly death (99). The clinical signs of acute trichothecene mycotoxicosis are similar to those of radiation toxicity (100).

The T-2 toxin is toxic by inhalation, ingestion, and dermal contact. The LD_{50} by inhalation is 50 to 100 μg/kg, whereas the LD_{50} by oral ingestion is 5 to 10 mg/kg. Erythema and dermal necrosis are caused by nanogram and microgram quantities. The major threat of T-2 toxin is its dermal activity. The toxin may also affect clotting factors and lead to hemorrhage. There is no specific antidote or therapy for T-2 mycotoxin-induced mycotoxicosis. Treatment is supportive (96, 101). However, vesicant creams offer protection against T-2 mycotoxin exposure (102). Washing with soap and water within 1 h after exposure to T-2 toxin effectively prevents dermal toxicity (53, 103). Polyethylene glycol 300 was shown in rat studies to be more effective than a soap-and-water wash at removing high doses (100 μg) of T-2 mycotoxins (104).

ANIMAL TOXINS (ZOOTOXINS)

Arthropods, snakes, snails, fish, and other marine animals synthesize and secrete or excrete toxins; however, the toxins are not considered to be of major concern for laboratory workers. Research workers using animal specimens would typically have a higher risk of an exposure than individuals who work with venom or toxin preparations, but both groups of workers should focus on prevention measures (105). These include the use of PPE, including gloves and a fully buttoned laboratory coat or other protective garment. Only individuals who have been specifically and adequately trained on the handling of a particular venomous organism should be permitted to perform research with venomous animal specimens (106). Most animal toxin work can be safely performed using biosafety level 2 (BSL2) practices and procedures.

Arthropods

Many arthropods inject their toxin by a tail stinger or other spike or mouth parts. The usual route of exposure for envenomation is through physical trauma to the skin with the release or injection of venom. A less frequent exposure route is to the oral mucosa through a bite or sting (107). In addition to toxins, venom also contains proteins that can cause sensitization (108, 109). Anaphylaxis is the most severe allergic response to insect stings and requires immediate treatment. Arthropod venoms can cause a severe skin rash, facial swelling, and constriction of the throat, resulting in difficulty in breathing, nausea, cramps, anxiety, lowered blood pressure, unconsciousness, shock, and even death (110).

Snake Venom

Snake venoms and their isolated enzymes have been used to study and treat blood coagulation problems (111). There are more than a thousand species of snakes, 400 of which are venomous, and these cause about 3,000,000 bites per year worldwide, with over 100,000 deaths. Snake toxins differ in their mechanisms of action (112, 113), but all contain components to immobilize prey and aid in digestion.

Marine Animal Toxins

Marine toxins may be found as contaminants of seafood or as venom used for hunting prey or defense from predators (114). Most of the marine toxins are heat stable and represent an increased hazard to the consumer because toxin-contaminated seafood looks, smells, and tastes normal (115). Marine animals known to produce or contain toxins include stingrays, sea urchins, octopi, cone snails, weaver fish, zebra fish, scorpion fish, wasp fish, devilfish, lumpfish, catfish, and puffer fish. Paralytic shellfish poisoning is a potentially fatal syndrome associated with the consumption of shellfish containing saxitoxin-producing (STX^+) cyanobacteria. STX is produced by these microscopic marine dinoflagellate algae. Studies showed the coordination of the expression of foreign and native genes in the common ancestor of STX^+ cyanobacteria (116). Some of these toxins have been shown to be produced by microorganisms present in the animals' flora, indicating that bacterial genes encoding antibacterial effectors may have been horizontally transferred to diverse eukaryotes (117). Further information on marine toxins may be found in reference 118 and at http://www.emedicine.com.

Conotoxins

Conotoxins are peptide neurotoxins from *Conus geographus*, the fish-hunting cone snail (119). They comprise 15 to 40 amino acids held in tight conformations by multiple disulfide bridges. Each of the 500 species of cone snail generates roughly 50 to 100 distinct conotoxins, which are used to immobilize prey (120). Conotoxin A is currently being investigated for use as a treatment for chronic pain (121–123). Tests on rats suggest that it may be 10,000 times more potent than morphine, is nonaddictive, and lacks side effects (124).

The onset of symptoms almost immediately follows the injection of the conotoxin. Symptoms in nonfatal cases include burning pain, swelling of the affected part of the body, and local numbness that rapidly spreads to involve the entire body, with some cardiac and respiratory distress. There is neither a rapid diagnostic assay nor an antidote. Treatment is limited to supportive care, which may involve artificial respiration and treatment of symptoms (125). Despite their toxicity, conotoxins are valuable tools in medical research.

Tetrodotoxin

Tetrodotoxin is an organic molecule assimilated into the tissues of the puffer fish or into the modified salivary glands of the blue-ringed octopus. Tetrodotoxin is thought to be synthesized by bacteria such as *Vibrio*, *Pseudomonas*, and *Photobacterium phosphoreum* (126) or by dinoflagellates associated with puffer fish. It is a potent blocker of sodium ion channels, has become a useful tool for physiological studies, and was recently synthesized (127). A rapid onset of symptoms occurs within 15 min to several hours following the ingestion of tetrodotoxin-containing food (128). Symptoms include numbness in the mouth and lips followed by numbness of the face and extremities. Acute respiratory failure, through paralysis of the respiratory musculature, may result in death within 4 to 6 h. Although there are no antidotes available, the outcome is usually favorable with supportive care (118, 129, 130).

TABLE 5.

Dinoflagellate toxin-associated diseases: public health strategies[a]

Clinical syndrome	Organism(s)	Toxins	Exposure vehicle	Public health response
Paralytic shellfish poisoning	*Alexandrium* spp. and related organisms	Saxitoxins	Consumption of bioaccumulated toxins in shellfish	Phytoplankton monitoring (*Alexandrium* is a large and distinctive dinoflagellate); shellfish bed closures based on mouse bioassay for saxitoxins in shellfish
Amnesic shellfish poisoning	*Pseudo-nitzschia* spp.	Domoic acid	Consumption of bioaccumulated toxins in shellfish, particularly mussels	Phytoplankton and shellfish domoic acid content monitoring by HPLC[b] (difficult to distinguish toxic from nontoxic diatoms)
Neurotoxic shellfish poisoning	*Gymnodinium breve*	Brevetoxins	Consumption of bioaccumulated toxins; inhalation of toxin-containing aerosols generated by wave and wind action	Phytoplankton monitoring; shellfish bed (and beachfronts, for onshore blooms) closure based on counts (>5,000 cells/liter = closure)
Ciguatera fish poisoning	Benthic coral reef dinoflagellate species (*Gambierdiscus toxicus*, *Prorocentrum* spp., *Amphidinium carterae*, and others)	Cigatoxins/ maitotoxins	Consumption of bioaccumulated toxins in tropical reef fish, in particular predators such as barracuda, grouper, and snapper	No feasible method currently available for screening of individual fish; toxins are heat stable; common-sense measures (avoid consumption of high-risk fish species from high-risk locales)

[a]Used with permission from reference 178.
[b]HPLC, high-performance liquid chromatography.

Pfiesteria toxins

The toxin produced by the dinoflagellate *Pfiesteria piscicida* has been associated with massive fish kills in North Carolina and Maryland (131) and is thought to be responsible for neurocognitive disorders in exposed humans (132, 133). Illness has been reported among researchers working with fish that have been exposed to *Pfiesteria*. Much of the early work to define the nature of this toxin was conducted at BSL3 (134). See Table 5 for additional dinoflagellate toxin-associated diseases.

PLANT TOXINS

Phytotoxins are poisonous substances produced by plants for protection and transport of compounds (135, 136). The most notorious plant toxins are the lectins, which include ricin and abrin. Other categories of phytotoxins include cyanogenic glycosides, alkaloids, oxalates, coumarins, and phenols.

Ricin

Ricin is the water-extractable toxin of the castor bean, *Ricinus communis*, which is commonly naturalized in many parts of the world and is the source of castor oil (137). Worldwide, over 1,000,000 tons of castor beans are processed annually for industrial uses; the waste mash is 3% to 5% ricin. The mechanism of action of ricin, a globular glycoprotein with a molecular mass of 66 kDa, is inhibition of cellular protein synthesis (138). The ricin molecule consists of A and B subunits joined by a disulfide bond. At pH 7.8, the toxin is heat stable at 80°C for 10 min and at 50°C for about 1 h. Ricin is marginally toxic (LD_{50}, 3 to 30 μg/kg) compared to botulinum neurotoxin and SEB (see Table 2). The latent period for the appearance of symptoms in humans is about 8 to 10 h after ingestion. Symptoms include nausea, vomiting, abdominal cramps, severe diarrhea, and vascular collapse. Ricin is severely toxic to the pulmonary system when inhaled. Following inhalation there is a latent period of 18 to 24 h before development of nonspecific symptoms, which include fever, chest tightness, cough, dyspnea, and nausea, followed by hypothermia and pulmonary edema. Death may occur as early as 3 days, but probably varies with the route of exposure.

Currently there is no specific therapy or prophylaxis. However, an engineered vaccine is being developed that lacks the active site of the toxin molecule. Mice immunized with this vaccine are protected against injected ricin at 10 times the LD_{50}, with no side effects. Studies are in progress to challenge mice by other routes of exposure

(139). Investigators plan to immunize volunteers to learn if human antiserum will protect mice against ricin, the rationale being that if the vaccine protects passively, it will protect actively (140). Two recombinant vaccines and neutralizing monoclonal antibodies that passively protect animals are being evaluated for their passive postexposure protection (141). A thermostable, aluminum-adjuvant–containing formulation of a ricin vaccine having two mutations was recently shown to protect rhesus macaques from a lethal dose of aerosolized ricin toxin (142). The safety and immunogenicity of the ricin vaccine RVEc in a phase 1 multiple-dose, open-label, non-placebo-controlled, dose-escalating (20, 50, and 100 μg), single-center study was recently reported (143). That study revealed that the RVEc vaccine was well tolerated and immunogenic at 20-μg and 50-μg dose levels. A single boost vaccination of RVEc greatly enhanced immunogenicity of the vaccine in human subjects.

The purpose for the development of a vaccine against ricin toxin was to enable its use as a prophylactic countermeasure. Passive transfer studies will be considered in the near future as proof of concept that human antibodies could protect against ricin intoxication. The efficacy of such antibodies must be demonstrated in animal models to obtain information about the usefulness of ricin-specific antibodies in the form of human ricin vaccine immune plasma or human ricin vaccine immune globulin.

OCCUPATIONAL HEALTH

The availability of preventive treatment (toxoid vaccines), postexposure prophylaxis (antitoxins), and postexposure supportive care for biological toxins is extremely variable (144). These issues should be addressed with occupational medicine and health care providers in conjunction with supervisors and safety professionals prior to starting work with biological toxins as part of the overall project risk assessment. If a vaccine is available for the toxin being handled, at-risk individuals should be knowledgeable of the efficacy of the vaccine and any potential adverse reactions. Although antitoxin or antivenin is available for several toxins, postexposure treatment for many toxins is limited to supportive therapy. In preparation for an off-hours emergency, the occupational medicine staff should communicate with the local hospital or emergency clinic so the staff is aware of the toxins in use at the facility. The at-risk personnel should be trained in advance so they will know to report to the local hospital for initial treatment. This coordinated approach ensures that the treating facility will have the applicable postexposure prophylaxis available when needed.

JOB HAZARD ANALYSIS (JOB SAFETY ANALYSIS) AND RISK MANAGEMENT

When planning work with toxins or toxin-containing materials, it is essential to be familiar with the toxin and the procedures that will be used. Information can be found in biological agent summary statements (50), material safety data sheets, toxicology and microbiology texts, and so forth. Investigators should develop a job hazard analysis with the assistance of a biological safety professional or safety committee as a part of the risk assessment process. A job hazard analysis is a technique that focuses on job tasks as a way to identify hazards before they occur. It focuses on the relationship between the worker, the task, the tools, and the work environment. Ideally, after you identify uncontrolled hazards, you will take steps to eliminate or reduce them to an acceptable risk level (145). This analysis is the genesis of a written standard operating procedure (SOP) to be followed while working with the material. Each procedure or step is then analyzed using an operational risk management process, such as in Risk Management (Field Manual 100-14), developed by the U.S. Army (146). This is a five-step process that includes: (i) identifying the hazards, (ii) assessing the hazards, (iii) developing controls and making a risk decision, (iv) implementing controls, and (v) supervising and evaluating the controls. The process is not static but changes continually as the situation changes. Another process is based on the ATOM (Active Threat and Opportunity Management) technique (147), which could be applied to the laboratory environment. These types of analyses can be incorporated as part of the qualitative risk assessment processes employed by and favored by many life science institutions.

Investigators are encouraged to employ a simulant, or less toxic material, for initial training on the SOPs. Practice runs or sham operations have been used as training tools for generations. The risk management process can be integrated into the practice run to assist an investigator in identifying and subsequently mitigating any hazardous procedures within the SOP. This approach has been used successfully by one of the authors (J.P.K.) of this chapter in conjunction with an Institutional Biosafety Committee (IBC) as part of an initial risk assessment. In this case, the researchers used a surrogate agent that fluoresces when exposed to black light while conducting a number of routine manipulations and various degrees of equipment failure. These exercises clearly demonstrated where the hazards existed in the SOPs and where modifications were necessary. This type of training and subsequent documentation of proficiency of laboratory staff by the principal investigator or laboratory supervisor is rapidly becoming the norm, and many institutions have adopted more formal training and mentoring programs for individuals new to the laboratory regardless of past education and experience. A potential driver for this renewed focus on training and documentation of proficiency includes the recent highly publicized laboratory incidents and case law related to laboratory incidents, such as the University of California (UC)–Los Angeles District Attorney (LADA) Agreement (148) (see https://www.depts.ttu.edu/vpr/integrity/csb-response/downloads/UC-agreement.pdf) and the June, 2014 Deferred Prosecution Agreement between Patrick Haran and the LADA (149) (see http://newsroom.ucla.edu/releases/agreement-resolves-charges-against-ucla-organic-chemistry-professor). One could interpret the stipulations in Appendix A of the UC-LADA agreement to be a new due diligence standard for research institutions, principal investigators and safety staff regarding:

- Principal Investigator training requirements
- Laboratory staff training and documentation
- Documentation change management (formalized review and approval of laboratory specific SOPs and other risk-associated documentation).

LABORATORY FACILITIES AND SAFETY EQUIPMENT

In general, most of the biological toxins used in research projects can be safely handled by strict adherence to BSL2 work practices in a laboratory designed to meet BSL2 facility recommendations and equipped with appropriate engineering controls. In addition to the BSL2 facility recommendations as specified in *Biosafety in Microbiological and Biomedical Laboratories* (BMBL) (50), the U.S. Department of Agriculture (USDA) Animal, Plant Health Inspection Service (APHIS) states "the minimum air change rate for laboratory space is 8 air changes per hour regardless of space cooling load" and "airflow will be from areas of lower hazard to higher hazard" (150).

It is prudent that laboratories be equipped with a single-pass ventilation system that provides 8 to 10 air changes per hour with directional airflow from areas of lower hazard potential to areas of higher hazard potential (151). An inward airflow rate of 50 to 100 ft^3 per min is recommended. In some cases, a risk assessment of the proposed work may reveal a need for BSL3 practices and procedures in a BSL2 facility or a need to use a BSL3 facility. Variables such as the toxin under study, the physical state of the toxin (liquid or dry form), the volume of the material manipulated, the type of study, and equipment used (aerosol challenges, etc.) will determine if the work needs to be conducted at a higher level of containment. If other hazards, such as infectious agents, are used in the same laboratory, the risk associated with these materials must also be considered when making decisions on laboratory containment (50).

The toxin laboratory should be equipped with an emergency drench shower and eyewash that meet the current American National Standards Institute (ANSI) standard (152, 153). An eyewash station should be located within "10 seconds" (preferably in every room) of a splash potential with a toxin or other hazardous chemical. Each toxin laboratory should have a sink for hand washing. If the laboratory is equipped with a vacuum system, it should be protected by a high-efficiency particulate air (HEPA) filter or equivalent at each point of use to avoid contaminating associated piping. The surfaces of the laboratory, including casework, should have finishes that allow for ease of cleaning and that are compatible with laboratory disinfectants. Countertops should be impervious to water and resistant to acids, alkalis, organic solvents, and moderate heat (50, 154). Large-scale production (a volume of 10 liters or greater) (155, 156) of toxin or activities with a high potential for aerosol or droplet generation should be conducted using BSL3 practices, procedures, and equipment as defined in the BMBL. If a recombinant organism is involved, the BL2 large-scale criteria defined in Appendix K of the *NIH Guidelines for Research Involving Recombinant or Synthetic Nucleic Acid Molecules* (*NIH Guidelines*) (157) will be applicable.

It is very important to prevent contamination of work surfaces, clothing, and skin when working with biological toxins, because accidental ingestion is the primary route of exposure for most toxins. All operations that can generate aerosols or droplets must be conducted within appropriately ventilated engineering controls, such as a Class II or Class III biological safety cabinet (BSC), glove box, or in some cases a chemical fume hood with an associated HEPA filter exhaust. Laboratory-scale operations (less than 10 liters) involving liquid toxin may be conducted in a Class II BSC. For large-scale operations (more than 10 liters), additional containment equipment is required. Detailed information on the types of BSCs as well as the selection, use, and installation of these engineering controls is available from a variety of sources, such as the ANSI/National Sanitation Foundation (158) and the CDC/NIH (159). When volatiles are used in conjunction with biological toxins, Class II, Type B1 or B2, or Class III cabinets, which are designed to be used with small amounts of volatiles and radionuclides, are recommended. For operations involving dry-form toxins or powders, glove boxes or Class III BSCs should be used. A less desirable option would be to use a glove bag within a Class II BSC. For experiments involving intentional generation of aerosols, we concur with the use of a "box in a box" concept (18). The exposure chamber, the nose-only exposure apparatus (i.e., Henderson apparatus), and the generation system are all to be contained within a Class

III BSC or a glove box equipped with a HEPA-filtered supply and exhaust.

In addition to the engineering controls found in the research laboratory, there are a number of devices that have been designed to reduce the occupational hazards associated with handling venomous animals, primarily snakes (160). There is a broad "potato masher" pinning device that is easier to use and provides much more control of the head, which is a considerable improvement over the "forked stick" device. Hide boxes with detachable doors aid the transfer of snakes to "hoop bags," without any direct handling by the herpetologist. Once transferred into the bag, the snake can be milked for venom or examined with relative safety (161). The availability of tongs, tubes, and other handling devices renders direct contact between investigators and the head or neck of a venomous snake unnecessary. There are few, if any, legitimate circumstances where venomous snakes must be handled with the investigator's bare hand at the head or neck. Those handling venomous snakes or lizards should be knowledgeable concerning the proper methods and tools for handling these animals. A training plan should be in place that emphasizes safe procedures and responsibility (162).

Personal Protective Equipment

PPE to be used should be chosen following a careful risk assessment and should be appropriate to the toxin, as well as to any other hazardous material(s) in the experiment.

Protective clothing

Laboratory gowns, preferably back-closing, should be worn to protect street clothing. If gowns are reusable, they should be restricted to the toxin work area, and the facility should have a means to decontaminate and launder them. It is recommended that laboratory clothing not be taken home (163).

Eye and face protection

Appropriate eye protection, such as face shield, goggles, and safety glasses with side shields, should be provided and required to be worn during laboratory or animal operations involving biological toxins. The level of protection should be based upon the exposure potential of the specific operation (159).

Gloves

The glove material must be impervious to the toxin and the medium being used. For example, latex gloves are inappropriate for handling solutions of toxin in ethanol, as alcohol degrades latex. Nonstatic gloves are needed for handling the dry form of toxins (21). When conducting work with dermally active toxins, a full-face shield and additional arm and hand coverings (double gloving) are recommended in addition to the basic protective clothing.

Respiratory protection

Respiratory protection, such as the use of a full-face respirator with HEPA and/or combination cartridges, may be necessary if adequate engineering controls are not available (21). Consult with a knowledgeable safety professional to ensure that appropriate respiratory protective equipment is provided and that the individual receiving the respirator has been medically cleared, trained, and fit-tested in its use. In the United States, these individuals should be participating in the institution's respiratory protection program in compliance with the Occupational Safety and Health Administration (OSHA) Respiratory Protection Standard 29 Code of Federal Regulation 1910.134 (see https://www.osha.gov/pls/oshaweb/owadisp.show_document?p_table=STANDARDS&p_id=12716).

WORK PRACTICES

Many of the standard practices described for BSL2 containment (50) should be incorporated into the SOPs for the toxin laboratory. On the basis of the risk assessment of the work to be conducted, BSL3 practices may also be incorporated into the SOP. Individuals working with toxins should be formally trained on the laboratory's SOPs, and this training should be documented. The laboratory director must ensure that individuals are proficient in all laboratory operations and knowledgeable about the work practices specific to the toxin laboratory, such as the following:

- When work with toxins is being conducted, the room shall be posted to indicate the BSL of the facility and of the practices being employed, special entry requirements, and emergency contact information.
- Each laboratory should develop a biosafety or chemical hygiene plan specific for the toxin being used with detailed information on the toxin and emergency procedures for spills and exposures.
- When performing high-risk operations, such as manipulating dry-form toxins, injecting animals, or using hollow-bore needles in conjunction with an estimated lethal dose for a human, at least two knowledgeable individuals wearing appropriate PPE shall be present and maintain a direct line of sight of each other. Work with toxins, toxin stock solutions, and dry-form toxins should be conducted using engineering controls as identified by a risk assessment. Individuals should have specific training on the proper use and limitations of the safety equipment and engineering controls.

TABLE 6.

Chemical inactivation of toxins[a] (166)

Toxin	2.5% NaOCl + 0.25 N NaOH	2.5% NaOCl	1% NaOCl	0.1% NaOCl
T-2 mycotoxin	Yes	No	No	No
Brevetoxin	Yes	Yes	No	No
Microcystin	Yes	Yes	Yes	No
Tetrodotoxin	Yes	Yes	Yes	No
Saxitoxin	Yes	Yes	Yes	Yes
Palytoxin	Yes	Yes	Yes	Yes
Ricin	Yes	Yes	Yes	Yes
Botulinum	Yes	Yes	Yes	Yes
Staphylococcal enterotoxin	Yes (?)	Yes (?)	Yes (?)	Yes (?)

[a]Inactivation methods consisted of 30 min of exposure to various concentrations of sodium hypochlorite (NaOCl) with and without sodium hydroxide (NaOH).
Key: yes, complete inactivation; yes (?), assumed inactivation.

DECONTAMINATION

In the toxin laboratory, a safe work environment is maintained through stringent housekeeping procedures, frequent decontamination of potentially contaminated surfaces and equipment, and appropriate decontamination and disposal of toxin-contaminated waste (164). The decontamination method of choice varies with the toxin being manipulated (see Table 6). A dilute solution (0.25 to 0.5%) of sodium hypochlorite (NaOCl) with an adequate contact time is recommended for surface and equipment decontamination for many bacterial toxins (165). However, this procedure would be inadequate for aflatoxin or T-2 mycotoxin, which requires a solution of 2.5% NaOCl and 0.25 N sodium hydroxide (NaOH) with a 30-min contact time (166). Table 6 shows a comparison of toxin-inactivation efficacies of various concentrations of NaOCl. Glassware or other items grossly contaminated with mycotoxins should be soaked in a 2.5% NaOCl plus 0.25 N NaOH solution for 2 to 8 h. Not only is aflatoxin B1 ineffectively detoxified by NaOCl alone, but also its use can lead to the formation of the potent carcinogen and mutagen aflatoxin B1-2,3-dichloride (167). To eliminate this carcinogen, the treated solution may be diluted to approximately 1 to 1.5% NaOCl by volume, followed by the addition of acetone to give a final concentration of 5% (vol/vol) (168). Aflatoxin B1 may also be decontaminated with a solution of potassium permanganate in sulfuric acid or NaOH (169).

Temperatures in excess of 121°C for a minimum of 20 min are considered adequate for inactivation of many biological toxins. The use of a properly operating autoclave will inactivate the proteinaceous bacterial toxins, but this method should not be used to inactivate heat-stable, low-molecular-weight toxins, such as mycotoxins, snake venoms, and marine toxins (166, 170). The inactivation efficacy of autoclaving is compared with those of a 10-min exposure to various temperatures of dry heat in Table 7.

TABLE 7.

Heat inactivation of toxins[a] (166)

Toxin	Autoclaving	200°F	500°F	1,000°F	1,500°F
T-2 mycotoxin	No	No	No	No	Yes
Brevetoxin	No	No	No	No	Yes
Microcystin	No	No	Yes	Yes	Yes
Tetrodotoxin	No	No	Yes	Yes	Yes
Saxitoxin	No	No	Yes	Yes	Yes
Palytoxin	No	No	Yes	Yes	Yes
Ricin	Yes	Yes	Yes	Yes	Yes
Botulinum	Yes	Yes	Yes	Yes	Yes
Staphylococcal enterotoxin	Yes (?)	Yes (?)	Yes (?)	Yes (?)	Yes (?)

[a]Inactivation methods consisted of autoclaving or 10 min of exposure to dry heat at various temperatures.
Key: yes, complete inactivation; yes (?), assumed inactivation.

Incineration in an appropriately licensed medical waste incinerator is an excellent method of inactivating all biological toxins. Regardless of the method used, all waste associated with toxin work should be decontaminated and disposed of according to federal, state, and local laws and regulations.

REGULATORY ISSUES

Biological Toxin Registration

Registration and review of research with biohazardous agents, including recombinant DNA (rDNA), etiologic agents of human disease, and zoonotic agents, by an IBC and/or a biosafety professional are an essential part of any biosafety management program (171). Although many institutions have had long-established registration programs for experiments involving rDNA molecules and pathogenic microorganisms, only a minority in the United States required formal registration of work involving toxins of biological origin prior to the promulgation of the Select Agent regulations (172, 173), which included a number of toxins as both CDC Select Agents and CDC/USDA Overlap Agents.

A registration program for toxins of biological origin is an essential component of the institutional biological safety program and protects the scientist by legitimizing possession and use of biohazardous agents. This is especially true for scientists in the United States since the U.S. Patriot Act of 2001 (174) amended Section 175 of the U.S. Criminal Code (175) to allow prosecution of individuals who knowingly possess any biological agent, toxin, or delivery system of a type or in a quantity not reasonably justified by prophylactic, preventive, bona fide research, or other peaceful purpose. The registration document is essentially a tool that can provide the necessary information for an IBC and/or a safety professional to perform a risk assessment. The registration document should provide, at a minimum, answers to the following questions:

- Who is responsible?
- What is the agent?
- Where will the research be conducted?
- What types of agent manipulations are planned?
- What is the volume?
- What types of engineering controls are available?
- Where will material be stored?
- Who will perform the work?
- How will the agent be safely disposed of?

The registration document must be accessible to the entire research community, preferably through a website or in a downloadable form. Registration should be mandatory for researchers who wish to use toxins.

Security of Toxins

Regardless of whether a toxin is subject to the Select Agent regulations (172, 173, 176, 177), each laboratory working with toxins should establish the following minimal security requirements as a prudent practice:

- Toxins should be secured in a locked refrigerator, freezer, or storage cabinet when not in use.
- An accurate inventory of material should be maintained. A physical inventory of all toxin stocks should be conducted on a regular basis.
- Access to laboratories where toxins are stored or where work with toxins is conducted should be restricted to those individuals whose work assignments require access. When work with the toxin is being conducted, access to the laboratory should be restricted to individuals who have been advised of the potential hazards and meet all laboratory entry requirements.

If an institution(s) possesses, uses, and transfers any of the Select Toxins listed in Table 8 in amounts above the permissible levels published in 42 C.F.R. Part 73 listed in Table 9, they must register with Federal Select Agent Program and comply with all of the relevant requirements as stated in 42 C.F.R. Part 73. Facilities in possession of botulinum neurotoxin, which is currently the only Tier 1 Select Toxin, must employ additional security requirements such as:

- Establish procedures that limit access to only those individuals who have been approved by Health and Human Services (HHS) following a Security Risk Assessment by the Attorney General, have had an entity-conducted preaccess suitability assessment, and are subject to the entity's procedures for ongoing suitability assessment.
- Establish procedures that limit access to laboratory and storage facilities outside normal business hours to only those specifically approved by the Responsible Official or designated alternate.
- Establish procedures for allowing visitors, their property, and vehicles at the entry and exit points to the registered space, or at other designated points of entry to the building, facility, or compound, that are based on the entity's site-specific risk assessment.
- A minimum of three security barriers are present, where each security barrier adds to the delay in reaching secured areas where Select Agents and Toxins are used or stored. One of the security barriers must be monitored in such a way as to detect intentional and

TABLE 8.

Select Toxins list[a]

HHS Select Toxins	Exclusions (attenuated strains)
Botulinum neurotoxins[b]	Botulinum neurotoxin type C atoxic derivative (BoNT/C ad) Fusion proteins of the heavy-chain domain of BoNT/translocation domain of diphtheria toxin Recombinant catalytically inactive botulinum A1 holoprotein (ciBoNT/A1 HP) BoNT purified protein (BoNT/A1 atoxic derivative, ad, E224A/Y366A) Recombinant Botulinum neurotoxin serotype A (R362A, Y365F)
Conotoxins[c] (short, paralytic alpha conotoxins containing the amino acid sequence) X1CCX2PACGX3X4X5X6CX7)[d]	Conotoxins (nonshort, paralytic alpha conotoxins)
Diacetoxyscirpenol	N/A
Ricin	N/A
Saxitoxin	N/A
Staphylococcal enterotoxins A, B, C, D, E subtypes	SEA triple mutant (L48R, D70R, and Y92A) SEB triple mutant (L45R, Y89A, Y94A) SEC double mutant (N23A and Y94A)
T-2 toxin	
Tetrodotoxin	Anhydrotetrodotoxin, a derivative of wild-type tetrodotoxin

[a]Sources: http://www.selectagents.gov/SelectAgentsandToxinsList.html and http://www.selectagents.gov/exclusions-hhs.html.
[b]Denotes Tier 1 Agent.
[c]Proposed for removal based on the January 2016 Notice of Proposed Rule Making.
[d]C, cysteine residues are all present as disulfides, with the 1st and 3rd cysteine, and the 2nd and 4th cysteine forming specific disulfide bridges; the consensus sequence includes known toxins α-MI and α-GI (shown above) as well as α-GIA, Ac1.1a, α-CnIA, α-CnIB. X1, any amino acid(s) or Des-X; X2, asparagine or histidine; P, proline; A, alanine; G, glycine; X3, arginine or lysine; X4, asparagine, histidine, lysine, arginine, tyrosine, phenylalanine or tryptophan; X5, tyrosine, phenylalanine, or tryptophan; X6, serine, threonine, glutamate, aspartate, glutamine, or asparagine; X7, any amino acid(s) or Des X; "Des X", "an amino acid does not have to be present at this position." For example if a peptide sequence were XCCHPA then the related peptide CCHPA would be designated as Des-X.

unintentional circumventing of established access control measures under all conditions (day/night, severe weather, etc.). The final barrier must limit access to the Select Agent or Toxin to personnel who have been approved to access Tier 1 Select Agent or Toxin as described above.

TABLE 9.

Permissible toxin amounts[a]

HHS toxins [§73.3(d)(3)]	Current amount	Proposed amount
Abrin	100 mg	1,000 mg
Botulinum neurotoxins (BoNT)	0.5 mg	1 mg
Short, paralytic alpha conotoxins	100 mg	—
Diacetoxyscirpenol (DAS)	1,000 mg	10,000 mg
Ricin	100 mg	1,000 mg
Saxitoxin	100 mg	500 mg
Staphylococcal enterotoxins (subtypes A, B, C, D, and E)	5 mg	100 mg
T-2 toxin	1,000 mg	10,000 mg
Tetrodotoxin	100 mg	500 mg

[a]Sources: http://www.selectagents.gov/PermissibleToxinAmounts.html and https://www.federalregister.gov/articles/2016/01/19/2016-00758/possession-use-and-transfer-of-select-agents-and-toxins-biennial-review-of-the-list-of-select-agents#h-22

- All registered space or areas that reasonably afford access to the registered space must be protected by an intrusion detection system (IDS) unless physically occupied, and staff monitoring these systems must be capable of evaluating and interpreting the alarm and alerting a security response force or local law enforcement.
- Establish procedures to ensure that security is maintained in the event of a powered access control system failure due to a power disruption affecting Tier 1 agent registered space.
- Ensure that response time for security response force or local law enforcement does not exceed 15 minutes from the time of initial alarm or report of a security incident and that sufficient security barriers are provided to delay unauthorized access until a security force or law enforcement arrives to prevent theft, intentional release, or unauthorized access.

CONCLUSION

Biological toxins are chemical compounds of great diversity, produced by microorganisms, plants, and animals. The effect of these toxins on humans and animals ranges from discomfort to incapacitating or life-threatening events. Medical treatment for intoxication also varies, ranging from administration of antidotes (antivenin or antitoxin) or vaccines (i.e., toxoids) to supportive therapy. Those who plan research with toxins should be aware of the potency and significant characteristics of the toxins, and of the work practices, PPE, and possible regulations required to study them. Armed with this knowledge, laboratorians can conduct work with toxins safely and with confidence that adverse consequences can be minimized.

The authors gratefully acknowledge Evelyn M. Hawley, Dr. David R. Franz, and Dr. Steve Kappes for their gracious contributions and critical review of the manuscript.

References

1. **Bernard C, Dumas JB, Bert P.** 1878. *La science experimentale, par Claude Bernard*. J. B. Bailliaere & fils, Paris, France.
2. **Cushman DW, Ondetti MA.** 1991. History of the design of captopril and related inhibitors of angiotensin converting enzyme. *Hypertension* **17**:589–592.
3. **Hider RC, Karlsson E, Namiranian S.** 1991. Separation and purification of toxins from snake venoms, p 1–34. *In* Harvey A (ed), *Snake Toxins*. Pergamon Press, New York.
4. **Hollecker M, Marshall DL, Harvey AL.** 1993. Structural features important for the biological activity of the potassium channel blocking dendrotoxins. *Br J Pharmacol* **110**:790–794.
5. **Wulff H, Zhorov BS.** 2008. K+ channel modulators for the treatment of neurological disorders and autoimmune diseases. *Chem Rev* **108**:1744–1773.
6. **Schwartz BS, Mitchell CS, Weaver VM, Cloeren M.** 1994. Bacteria, p 318–381. *In* Wald PH, Stave GM (ed), *Physical and Biological Hazards of the Workplace*. Van Nostrand Reinhold, New York, N.Y.
7. **Royal MA.** 2003. Botulinum toxins in pain management. *Phys Med Rehabil Clin N Am* **14**:805–820.
8. **Cuatrecasas P (ed).** 1977. *The Specificity and Action of Animal, Bacterial and Plant Toxins (Receptors and Recognition), series B*, vol 1. John Wiley & Sons, Inc, New York.
9. **Menez A (ed).** 2002. *Perspectives in Molecular Toxinology*. John Wiley and Sons, Ltd, West Sussex, England.
10. **Oehme FW, Keyler DE.** 2007. Plant and animal toxins, p 983–1050. *In* Hayes AW (ed), *Principles and Methods of Toxinology*, 5th ed. Informa Healthcare USA, Inc, New York, N.Y.
11. **Mackessy SP (ed).** 2009. *Handbook of Venoms and Toxins of Reptiles*. CRC Press, Boca Raton, FL.
12. **Klaassen C.** 2013. *Casarett & Doull's Toxicology: The Basic Science of Poisons*, 8th ed. McGraw-Hill Company, New York, N.Y.
13. **Derelanko MJ, Auletta CS.** 2014. *Handbook of Toxicology*, 3rd ed. CRC Press, Boca Raton, FL.
14. **Brooks G, Butel J, Morse S (ed).** 2004. *Jawetz, Melnick & Adelberg's Medical Microbiology*, 23rd ed. McGraw Hill Companies, New York.
15. **Levinson W.** 2014. *Review of Medical Microbiology and Immunology*, 13th ed. McGraw-Hill Education / Medical, Columbus, OH.
16. **Walker M.** 1997. Toxins and poisons, p 523–537. *In* Curtis M, Sutter M (ed), *Integrated Pharmacology*. C. V. Mosby, Chicago, IL.
17. **Kemppainen BW, Pace JG, Riley RT.** 1987. Comparison of in vivo and in vitro percutaneous absorption of T-2 toxin in guinea pigs. *Toxicon* **25**:1153–1162.
18. **Johnson B, Mastnjak R, Resnick IG.** 2000. Safety and health considerations for conducting work with biological toxins, p 88–111. *In* Richmond J (ed), *Anthology of Biosafety II: Facility Design Considerations*. American Biological Safety Association, Mundelein, IL.
19. **Ramanathan H.** 2010. Food poisoning by *Clostridium botulinum*, p 10–15. *In Food Poisoning—A Threat to Humans*. Marsland Press, Richmond Hill, NY. http://www.sciencepub.net/book/041_1349book.pdf.
20. **Centers for Disease Control and Prevention (CDC).** 2013. Outbreak of staphylococcal food poisoning from a military unit lunch party—United States, July 2012. *MMWR Morb Mortal Wkly Rep* **62**:1026–1028. http://www.cdc.gov/mmwr/pdf/wk/mm6250.pdf.
21. **Wilson DE, Chosewood LC.** 2009. Guidelines for work with toxins of biological origin, p 385–393. *In Biosafety in Microbiological and Biomedical Laboratories*. 5th ed., HHS Publication No. (CDC) 21-1112. http://www.cdc.gov/biosafety/publications/bmbl5/bmbl5_appendixi.pdf.
22. **Van Heyningen WE, Mellanby J.** 1971. Tetanus toxin, p 69–108. *In* Kadis S, Montie T, Ajl S (ed), *Microbiological Toxins,* vol. 2A. Academic Press, New York, NY.
23. **Fodstad O, Johannessen JV, Schjerven L, Pihl A.** 1979. Toxicity of abrin and ricin in mice and dogs. *J Toxicol Environ Health* **5**:1073–1084.
24. **Ezzell JW, Ivins BE, Leppla SH.** 1984. Immunoelectrophoretic analysis, toxicity, and kinetics of in vitro production of the protective antigen and lethal factor components of Bacillus anthracis toxin. *Infect Immun* **45**:761–767.
25. **Schlessinger D, Schaechter M.** 1993. Bacterial toxins, p 162–175. *In* Schaechter M, Medoff G, Eisenstein BI (ed), *Mechanisms of Microbial Disease*, 2nd ed. Williams and Wilkins, Baltimore, MD.
26. **Schmitt CK, Meysick KC, O'Brien AD.** 1999. Bacterial toxins: friends or foes? *Emerg Infect Dis* **5**:224–234.
27. **Lamanna C.** 1959. The most poisonous poison: what do we know about the toxin of botulism? What are the problems to be solved?. *Science* **130**:763–772.
28. **Ramachandran G.** 2014. Gram-positive and gram-negative bacterial toxins in sepsis: a brief review. *Virulence* **5**:213–218., http://www.tandfonline.com/doi/pdf/10.4161/viru.27024
29. **Mahon CR, Flaws ML.** 2011. Host-parasite interaction, B. Pathogenesis of infection, p 33. *In* Mahon CR, Lehman DC, Manuselis G Jr (ed), *Textbook of Diagnostic Microbiology*, 4th ed. W.B. Saunders, Maryland Heights, MO.
30. **Schoenbach EB, Jezukawicz JJ, Mueller JH.** 1943. Conversion of hydrolysate tetanus toxin to toxoid. *J Clin Invest* **22**: 319–320.
31. **Lenters V, Basinas I, Beane-Freeman L, Boffetta P, Checkoway H, Coggon D, Portengen L, Sim M, Wouters IM, Heederik D, Vermeulen R.** 2010. Endotoxin exposure and lung cancer risk: a systematic review and meta-analysis of the published literature on agriculture and cotton textile workers. *Cancer Causes Control* **21**:523–555.
32. **Attwood P, Brouwer R, Ruigewaard P, Versloot P, de Wit R, Heederik D, Boleij JS.** 1987. A study of the relationship between airborne contaminants and environmental factors in Dutch swine confinement buildings. *Am Ind Hyg Assoc J* **48**:745–751.
33. **DeLucca AJ II, Palmgren MS.** 1987. Seasonal variation in aerobic bacterial populations and endotoxin concentrations in grain dusts. *Am Ind Hyg Assoc J* **48**:106–110.
34. **Rylander R, Morey P.** 1982. Airborne endotoxin in industries processing vegetable fibers. *Am Ind Hyg Assoc J* **43**:811–812.
35. **Jones W, Morring K, Olenchock SA, Williams T, Hickey J.** 1984. Environmental study of poultry confinement buildings. *Am Ind Hyg Assoc J* **45**:760–766.
36. **Castellan RM, Olenchock SA, Kinsley KB, Hankinson JL.** 1987. Inhaled endotoxin and decreased spirometric values. An exposure-response relation for cotton dust. *N Engl J Med* **317**:605–610.

37. **Goossens PL, Tournier J-N.** 2015. Crossing of the epithelial barriers by *Bacillus anthracis*: the known and the unknown. *Front Microbiol* **6:**1122.
38. **Stevens DL, Bisno AL, Chambers HF, Patchen Dellinger E, Goldstein EJC, Gorbach LSL, Hirschmann JV, Kaplan SL, Montoya JG, Wade JC.** 2014. Practice guidelines for the diagnosis and management of skin and soft tissue infections: 2014 Update by the Infectious Diseases Society of America. *Clin Infect Dis* 59:e10–e52. http://cid.oxfordjournals.org/content/early/2014/06/14/cid.ciu296.full.pdf.
39. **Schneemann A, Manchester M.** 2009. Anti-toxin antibodies in prophylaxis and treatment of inhalation anthrax. *Future Microbiol* **4:**35–43.
40. **Centers for Disease Control and Prevention (CDC).** 2001. Update: Investigation of Bioterrorism-Related Anthrax and Interim Guidelines for Exposure Management and Antimicrobial Therapy. *MMWR Morb Mortal Wkly Rep* **50:**909–919.
41. **Wright JG, Quinn CP, Shadomy S, Messonnier N; Centers for Disease Control and Prevention (CDC).** 2010. Use of anthrax vaccine in the United States. Recommendations of the Advisory Committee on Immunization Practices (ACIP). *MMWR Recomm Rep* 59(RR-6)**:**1–30. http://www.cdc.gov/mmwr/pdf/rr/rr5906.pdf
42. **Albrecht MT, Li H, Williamson ED, LeButt CS, Flick-Smith HC, Quinn CP, Westra H, Galloway D, Mateczun A, Goldman S, Groen H, Baillie LWJ.** 2007. Human monoclonal antibodies against anthrax lethal factor and protective antigen act independently to protect against *Bacillus anthracis* infection and enhance endogenous immunity to anthrax. *Infect Immun* **75:**5425–5433.
43. **Chen Z, Moayeri M, Purcell R.** 2011. Monoclonal antibody therapies against anthrax. *Toxins (Basel)* **3:**1004–1019.
44. **Hill KK, Smith TJ.** 2013. Genetic diversity within *Clostridium botulinum* serotypes, botulinum neurotoxin gene clusters and toxin subtypes, p 1–20. *In* Rummel A, Binz T (ed), *Botulinum Neurotoxins*. Springer, New York.
45. **Hambleton P.** 1992. *Clostridium botulinum* toxins: a general review of involvement in disease, structure, mode of action and preparation for clinical use. *J Neurol* **239:**16–20.
46. **Sobel J, Tucker N, Sulka A, McLaughlin J, Maslanka S.** 2004. Foodborne botulism in the United States, 1990–2000. *Emerg Infect Dis* **10:**1606–1611.
47. **FitzGerald S, Lyons R, Ryan J, Hall W, Gallagher C.** 2003. Botulism as a cause of respiratory failure in injecting drug users. *Ir J Med Sci* **172:**143–144.
48. **Maslanka SE.** 2014. Botulism as a disease of humans, p 259–289. *In* Foster KA (ed), *Molecular Aspects of Botulinum Toxin*, vol 4. Springer, New York.
49. **Arnon SS.** 2007. Creation and development of the public service orphan drug Human Botulism Immune Globulin. *Pediatrics* **119:**785–789.
50. **Wilson DE, Chosewood LC.** 2009. *Biosafety in Microbiological and Biomedical Laboratories*. 5th ed., HHS Publication No. (CDC) 21-1112. http://www.cdc.gov/biosafety/publications/bmbl5/bmbl.pdf.
51. **Rusnak JM, Boudreau EF, Hepburn MJ, Martin JW, Bavari S.** 2007. Medical Countermeasures, p 495. *In* Dembek Z (ed), *Medical Aspects of Biological Warfare, Defense Dept., Army*. Office of the Surgeon General, Borden Institute, Washington, DC.
52. **Dux MP, Huang J, Barent R, Inan M, Swanson ST, Sinha J, Ross JT, Smith LA, Smith TJ, Henderson I, Meagher MM.** 2011. Purification of a recombinant heavy chain fragment C vaccine candidate against botulinum serotype C neurotoxin [rBoNTC(H(c))] expressed in Pichia pastoris. *Protein Expr Purif* **75:**177–185.
53. **Hawley RJ, Kozlovac J.** 2004. Decontamination, p 333–348. *In* Lindler LE, Lebeda FJ, Korch GW (ed), *Biological Weapons Defense: Infectious Diseases and Counterbioterrorism*. Humana Press, Inc, Totowa, NJ.
54. **Kortepeter M.** 2001. Decontamination, p 118–129. *In* Kortepeter M (ed), *Medical Management of Biological Casualties Handbook*. U.S. Army Medical Research Institute of Infectious Diseases, Fort Detrick, MD.
55. **Dembek Z.** 2011. Decontamination, p 165–167. *In* Dembek Z (ed), *Medical Management of Biological Casualties Handbook*, 7th ed. U.S. Army Medical Research Institute of Infectious Diseases, Fort Detrick, MD.
56. **Simon HB, Swartz MN.** 1992. Pathophysiology of fever and fever of undetermined origin, p 8–12. *In Scientific American Medicine*. WebMD Professional Publishing, New York.
57. **Critchley DR, Nelson PG, Habig WH, Fishman PH.** 1985. Fate of tetanus toxin bound to the surface of primary neurons in culture: evidence for rapid internalization. *J Cell Biol* **100:**1499–1507.
58. **Bleck TP.** 1991. Tetanus: pathophysiology, management, and prophylaxis. *Dis Mon* **37:**551–603.
59. **Guilfoile P, Babcock H.** 2008. How is tetanus prevented?, p 55–63. *In* Guilfoile P, Babcock H (ed), *Tetanus*. Chelsea House, New York.
60. **National Center for Immunization and Respiratory Diseases.** 2011. General recommendations on immunization—Recommendations of the Advisory Committee on Immunization Practices (ACIP). *MMWR Recomm Rep* 60**:**1–64.
61. **Parry SH, Abraham SN, Feavers IM, Lee M, Jones MR, Bint AJ, Sussman M.** 1981. Urinary tract infection due to laboratory-acquired *Escherichia coli*: relation to virulence. *Br Med J (Clin Res Ed)* **282:**949–950.
62. **Wilson ML, Reller B.** 2014. Chapter 22, Clinical laboratory-acquired infections, p 320–328. *In* Jarvis WR (ed), *Bennett and Brachman's Hospital Infections*, 6th ed. Lippincott, Williams & Wilkins, Philadelphia, PA.
63. **Staples SJ, Asher SE, Giannella RA.** 1980. Purification and characterization of heat-stable enterotoxin produced by a strain of *E. coli* pathogenic for man. *J Biol Chem* **255:**4716–4721.
64. **Saeed AMK, Magnuson NS, Sriranganathan N, Burger D, Cosand W.** 1984. Molecular homogeneity of heat-stable enterotoxins produced by bovine enterotoxigenic *Escherichia coli*. *Infect Immun* **45:**242–247.
65. **Erume J, Berberov EM, Kachman SD, Scott MA, Zhou Y, Francis DH, Moxley RA.** 2008. Comparison of the contributions of heat-labile enterotoxin and heat-stable enterotoxin b to the virulence of enterotoxigenic *Escherichia coli* in F4ac receptor-positive young pigs. *Infect Immun* **76:**3141–3149.
66. **Pavia AT, Nichols CR, Green DP, Tauxe RV, Mottice S, Greene KD, Wells JG, Siegler RL, Brewer ED, Hannon D, Blake PA.** 1990. Hemolytic-uremic syndrome during an outbreak of *Escherichia coli* O157:H7 infections in institutions for mentally retarded persons: clinical and epidemiologic observations. *J Pediatr* **116:**544–551.
67. **Grist NR, Emslie JAN.** 1987. Infections in British clinical laboratories, 1984–5. *J Clin Pathol* **40:**826–829.
68. **Grist NR, Emslie JAN.** 1991. Infections in British clinical laboratories, 1988–1989. *J Clin Pathol* **44:**667–669.
69. **Sandvig K, Bergan J, Dyve A-B, Skotland T, Torgersen ML.** 2010. Endocytosis and retrograde transport of Shiga toxin. *Toxicon* **56:**1181–1185.
70. **Gill DM.** 1987. Bacterial toxins: lethal amounts, p 127–135. *In* Laskin A, Lechevalier HA (ed), *CRC Handbook of Microbiology*, 2nd ed, vol VIII. CRC Press, Boca Raton, FL.
71. **Pinchuk IV, Beswick EJ, Reyes VE.** 2010. Staphylococcal enterotoxins. *Toxins (Basel)* **2:**2177–2197.
72. **Mallonee DH, Glatz BA, Pattee PA.** 1982. Chromosomal mapping of a gene affecting enterotoxin A production in *Staphylococcus aureus*. *Appl Environ Microbiol* **43:**397–402.
73. **Cunha MLRS, Calsolari RAO.** 2008. Toxigenicity in *Staphylococcus aureus* and coagulase-negative staphylococci: Epidemiological and molecular aspects. *Microbiol Insights* **1:**13–24.
74. **Ulrich RG, Sidell S, Taylor TJ, Wilhelmsen C, Franz DR.** 1997. Staphylococcal enterotoxin B and related pyrogenic toxins,

p 621–630. *In* Zajtchuk R, Bellamy RF (ed), *Medical Aspects of Chemical and Biological Warfare.* Borden Institute, Washington, DC.

75. **Kortepeter M.** 2011b. Staphylococcal enterotoxin B, p 138–145. *In* Kortepeter M (ed), *Medical Management of Biological Casualties Handbook.* U.S. Army Medical Research Institute of Infectious Diseases, Fort Detrick, MD.
76. **Bhakdi S, Tranum-Jensen J.** 1991. Alpha-toxin of Staphylococcus aureus. *Microbiol Rev* **55:**733–751.
77. **Nilsson I-M, Hartford O, Foster T, Tarkowski A.** 1999. Alpha-toxin and gamma-toxin jointly promote *Staphylococcus aureus* virulence in murine septic arthritis. *Infect Immun* **67:**1045–1049.
78. **Rusnak JM, Kortepeter M, Ulrich R, Poli M, Boudreau E.** 2004. Laboratory exposures to staphylococcal enterotoxin B. *Emerg Infect Dis* **10:**1544–1549.
79. **Franz DR.** 1997. *Defense Against Toxin Weapons.* U.S. Army Medical Research Institute of Infection Diseases, Fort Detrick, MD.
80. **Stiles BG, Garza AR, Ulrich RG, Boles JW.** 2001. Mucosal vaccination with recombinantly attenuated staphylococcal enterotoxin B and protection in a murine model. *Infect Immun* **69:**2031–2036.
81. **Boles JW, Pitt MLM, LeClaire RD, Gibbs PH, Torres E, Dyas B, Ulrich RG, Bavari S.** 2003. Generation of protective immunity by inactivated recombinant staphylococcal enterotoxin B vaccine in nonhuman primates and identification of correlates of immunity. *Clin Immunol* **108:**51–59.
82. **Morefield GL, Tammariello RF, Purcell BK, Worsham PL, Chapman J, Smith LA, Alarcon JB, Mikszta JA, Ulrich RG.** 2008. An alternative approach to combination vaccines: intradermal administration of isolated components for control of anthrax, botulism, plague and staphylococcal toxic shock. *J Immune Based Ther Vaccines* **6:**5.
83. **Karauzum H, Chen G, Abaandou L, Mahmoudieh M, Boroun AR, Shulenin S, Devi VS, Stavale E, Warfield KL, Zeitlin L, Roy CJ..** 2012. Synthetic human monoclonal antibodies toward staphylococcal enterotoxin B (SEB) protective against toxic shock syndrome. *J Biol Chem* **287:**25203–25215.
84. **Franz DR, Jahrling PB, Friedlander AM, McClain DJ, Hoover DL, Bryne WR, Pavlin JA, Christopher GW, Eitzen EM Jr.** 1997. Clinical recognition and management of patients exposed to biological warfare agents. *JAMA* **278:**399–411.
85. **Hinnebusch BJ, Rudolph AE, Cherepanov P, Dixon JE, Schwan TG, Forsberg A.** 2002. Role of Yersinia murine toxin in survival of *Yersinia pestis* in the midgut of the flea vector. *Science* **296:**733–735.
86. **Butler T.** 1990. *Yersinia* species (including plague), p 1748–1756. *In* Mandell GL, Douglas GR, Bennett JE (ed), *Principles and Practices of Infectious Diseases,* 3rd ed. Churchill Livingstone, New York.
87. **Powell BS, Andrews GP, Enama JT, Jendrek S, Bolt C, Worsham P, Pullen JK, Ribot W, Hines H, Smith L, Heath DG, Adamovicz JJ.** 2005. Design and testing for a nontagged F1-V fusion protein as vaccine antigen against bubonic and pneumonic plague. *Biotechnol Prog* **21:**1490–1510.
88. **Feodorova VA, Motin VL.** 2012. Plague vaccines: current developments and future perspectives. *Emerg Microbes Infect* **1:**e36.
89. **Clarke D, Crews C.** 2014. Natural toxicants—mushrooms and toadstools, p 269–276. *In* Motarjemi Y, Moy G, Todd E (ed), *Encyclopedia of Food Safety,* vol 2. Academic Press, San Diego, CA.
90. **Yamaguchi MU, Rampazzo RCP, Yamada-Ogatta SF, Nakamura CV, Ueda-Nakamura T, Filho BP.** 2007. Yeasts and filamentous fungi in bottled mineral water and tap water from municipal supplies. *Braz Arch Biol Technol* **50:**1–9.
91. **Philp RB.** 2008. Mycotoxins and other toxins from unicellular organisms, p 283–300. *In* Philp RB (ed), *Ecosystems and Human Health—Toxicology and Environmental Hazards,* 3rd ed. CRC Press, Boca Raton, FL.
92. **Bennett JW, Klich M.** 2003a. Mycotoxins. *Clin Microbiol Rev* **16:**497–516.
93. **Stoloff L.** 1977. Aflatoxins: an overview, p 7–28. *In* Rodricks J, Hesseltine C, Mehlmann M (ed), *Mycotoxins in Human and Animal Health.* Pathotox Publishers, Inc, Park Forest South, Ill.
94. **World Health Organization International Agency For Research On Cancer (WHO).** 2002. Alfatoxins, p 193–194. *In* IARC Working Group (ed.), *IARC Monographs On The Evaluation Of Carcinogenic Risks To Humans,* vol. 82. *Some Traditional Herbal Medicines, Some Mycotoxins, Naphthalene and Styrene.* IARC Press, Lyon, France.
95. **Shotwell OL, Burg W.**1982. Aflatoxin in corn: potential hazard to agricultural workers, p. 69–86. *In* Kelly W (ed.), *Agricultural Respiratory Hazards (Annals of the American Conference of Governmental Industrial Hygienists).* American Conference of Governmental Industrial Hygienists, Cincinnati, OH.
96. **Bennett JW, Klich M.** 2003b. Mycotoxins. *Clin Microbiol Rev* **16:**497–516.
97. **Heymann DL.** 2015. Aspergillosis, p 58–61. *In* Heymann DL (ed), *Control of Communicable Diseases Manual,* 20th ed. American Public Health Association, Washington, DC.
98. **Wannemacher RW, Wiener SL.** 1997. Trichothecene mycotoxins, p 655–676. *In* Zajtchuk R, Bellamy RF (ed), *Medical Aspects of Chemical and Biological Warfare.* Borden Institute, Washington, DC.
99. **Huebner KD, Wannemacher RW, Stiles BG, Popoff MR, Poli MA.** 2007. Additional toxins of clinical concern, p 359–361. *In* Dembek Z (ed), *Medical Aspects of Biological Warfare.* Office of the Surgeon General, Borden Institute, Washington, DC.
100. **Fung F, Clark R, Williams S.** 1998. *Stachybotrys,* a mycotoxin-producing fungus of increasing toxicologic importance. *J Toxicol Clin Toxicol* **36:**79–86.
101. **Huebner KD, Wannemacher RW, Stiles BG, Popoff MR, Poli MA.** 2007. Additional toxins of clinical concern, p 361–365. *In* Dembek Z (ed), *Medical Aspects of Biological Warfare.* Office of the Surgeon General, Borden Institute, Washington, DC.
102. **Hospenthal DR.** 2004. Mycotoxins, p 194. *In* Roy MJ (ed), *Physician's Guide to Terrorist Attack.* Humana Press, Totowa, NJ.
103. **Kortepeter M.** 2011. T-2 mycotoxins, p 146–165. *In* Kortepeter M (ed), *Medical Management of Biological Casualties Handbook.* U.S. Army Medical Research Institute of Infectious Diseases, Fort Detrick, MD.
104. **Fairhurst S, Maxwell SA, Scawin JW, Swanston DW.** 1987. Skin effects of trichothecenes and their amelioration by decontamination. *Toxicology* **46:**307–319.
105. **Gwaltney-Brant S, Dunayer E, Youssef H.** 2012. Terrestrial zootoxins, p 969–992. *In* Gupta RC (ed), *Veterinary Toxicology Basic and Clinical Principles,* 2nd ed. Academic Press, Waltham, MA.
106. **Doucet ME, DiTada I, Martori R, Abalos A.** 1978. Security standards in a serpentarium, p 467–470. *In* Rosenberg P (ed), *Toxins, Animal, Plant and Microbial.* Pergamon Press, New York.
107. **Palmier JA, Palmier C.** 2002. Envenomations, p 581–603. *In* Wald PH, Stave GM (ed), *Physical and Biological Hazards of the Workplace,* 2nd ed. Wiley-Interscience, New York.
108. **Keele CA.** 1967. The chemistry of pain production. *Proc R Soc Med* **60:**419–422.
109. **Voght JT, Kozlovac JP.** 2006. Safety considerations for handling imported fire ants (*Solenopsis spp.*) in the laboratory and field. *Appl Biosaf* **11:**88–97.
110. **Goddard J.** 2007. Signs and symptoms of arthropod-borne diseases, p 107–114. *In Physician's Guide to Arthropods of Medical Importance,* 5th ed. CRC Press, Boca Raton, FL.
111. **Joseph B, Raj SJ, Edwin BT, Sankarganesh P.** 2011. Pharmacognostic and biochemical properties of certain biomarkers in snake venom. *Asian J. Biol. Sci.* **4:**317–324.
112. **Tsetlin VI, Hucho F.** 2004. Snake and snail toxins acting on nicotinic acetylcholine receptors: fundamental aspects and medical applications. *FEBS Lett* **557:**9–13.

113. **Kang TS, Georgieva D, Genov N, Murakami MT, Sinha M, Kumar RP, Kaur P, Kumar S, Dey S, Sharma S, Vrielink A, Betzel C, Takeda S, Arni RK, Singh TP, Kini RM.** 2011. Enzymatic toxins from snake venom: structural characterization and mechanism of catalysis. *FEBS J* **278:**4544–4576.
114. **Whittle K, Gallacher S.** 2000. Marine toxins. *Br Med Bull* **56:** 236–253.
115. **Park DL, Guzman-Perez SE, Lopez-Garcia R.** 1999. Aquatic biotoxins: design and implementation of seafood safety monitoring programs. *Rev Environ Contam Toxicol* **161:**157–200.
116. **Moustafa A, Loram JE, Hackett JD, Anderson DM, Plumley FG, Bhattacharya D.** 2009. Origin of saxitoxin biosynthetic genes in cyanobacteria. *PLoS One* **4:**e5758. http://journals.plos.org/plosone/article?id=10.1371/journal.pone.0005758.
117. **Chou S, Daugherty MD, Peterson SB, Biboy J, Yang Y, Jutras BL, Fritz-Laylin LK, Ferrin MA, Harding BN, Jacobs-Wagner C, Yang XF, Vollmer W, Malik HS, Mougous JD.** 2015. Transferred interbacterial antagonism genes augment eukaryotic innate immune function. *Nature* **518:**98–101.
118. **Baden DG, Fleming LE, Bean JA.** 1995. Marine toxins, p 141–175. *In* deWolf FA (ed), *Handbook of Clinical Neurology: Intoxications of the Nervous System Part H. Natural Toxins and Drugs.* Elsevier Press, Amsterdam, The Netherlands.
119. **Gray WR, Olivera BM, Cruz LJ.** 1988. Peptide toxins from venomous Conus snails. *Annu Rev Biochem* **57:**665–700.
120. **Becker S, Terlau H.** 2008. Toxins from cone snails: properties, applications and biotechnological production. *Appl Microbiol Biotechnol* **79:**1–9.
121. **Satkunanathan N, Livett B, Gayler K, Sandall D, Down J, Khalil Z.** 2005. Alpha-conotoxin Vc1.1 alleviates neuropathic pain and accelerates functional recovery of injured neurones. *Brain Res* **1059:**149–158.
122. **Essack M, Bajic VB, Archer JAC.** 2012. Conotoxins that confer therapeutic possibilities. *Mar Drugs* **10:**1244–1265.
123. **Hannon HE, Atchison WD.** 2013. Omega-conotoxins as experimental tools and therapeutics in pain management. *Mar Drugs* **11:**680–699.
124. **Layer RT, McIntosh JM.** 2006. Conotoxins: therapeutic potential and application. *Mar Drugs* **4:**119–142.
125. **Chand P.** 2009. Marine envenomations, p 454–459. *In* Dobbs MR (ed), *Clinical Neurotoxicology: Syndromes, Substances, Environments.* Saunders Elsevier, Philadelphia, PA.
126. **Liu F, Fu Y, Shih DY.** 2004. Occurrence of tetrodotoxin poisoning in *Nassarius papillosis* alectrion and *Nassarius gruneri* niotha. *J Food Drug Anal* **12:**189–192.
127. **Taber DF, Storck PH.** 2003. Synthesis of (–)-tetrodotoxin: preparation of an advanced cyclohexenone intermediate. *J Org Chem* **68:**7768–7771.
128. **Kheifets J, Rozhavsky B, Girsh Solomonovich Z, Marianna R, Soroksky A.** 2012. Severe tetrodotoxin poisoning after consumption of *Lagocephalus sceleratus* (pufferfish, fugu) fished in Mediterranean Sea, treated with cholinesterase inhibitor. *Case Rep Crit Care* **2012:**782507.
129. **Benzer TI.**2005. Toxicity, tetrodotoxin. *E-Medicine* [Online.] http://www.emedicine.com/emerg/topic576.htm.
130. **Osterbauer PJ, Dobbs MR.** 2009. Neurobiological weapons, p 631–645. *In* Dobbs MR (ed), *Clinical Neurotoxicology: Syndromes, Substances, Environments.* Saunders Elsevier, Philadelphia, PA.
131. **Vogelbein WK, Lovko VJ, Reece KS.** 2008. Pfiesteria, p 297–330. *In* Walsh PJ, Smith SL, Fleming LE, Solo-Gabriele HM, Gerwick WH (ed), *Oceans and Human Health—Risks and Remedies from the Seas.* Academic Press, Burlington, MA.
132. **Peterson JS.** 2000. Pfiesteria, p 273–280. *In* Fleming DO, Hunt DL (ed), *Biological Safety: Principles and Practices,* 3rd ed. ASM Press, Washington, DC.
133. **Morris JG Jr.** 2001. Human health effects and *Pfiesteria* exposure: a synthesis of available clinical data. *Environ Health Perspect* **109**(Suppl 5)**:**787–790.
134. **Burkholder J.**1998. *Pfiesteria piscicida.* Eagleson Lecture, American Biological Safety Association Annual Conference, 26 October, Orlando, FL.
135. **Amusa NA.** 2006. Microbially produced phytotoxins and plant disease. *Afr J Biotechnol* **5:**405–414.
136. **Duke SO, Dayan FE.** 2011. Modes of action of microbially-produced phytotoxins. *Toxins (Basel)* **3:**1038–1064.
137. **Roels S, Coopman V, Vanhaelen P, Cordonnier J.** 2010. Lethal ricin intoxication in two adult dogs: toxicologic and histopathologic findings. *J Vet Diagn Invest* **22:**466–468.
138. **Al-Tamimi FA, Hegazi AE.** 2008. A case of castor bean poisoning. *Sultan Qaboos Univ Med J* **8:**83–87.
139. **Smallshaw JE, Vitetta ES.** 2010. A lyophilized formulation of RiVax, a recombinant ricin subunit vaccine, retains immunogenicity. *Vaccine* **28:**2428–2435.
140. **Smallshaw JE, Firan A, Fulmer JR, Ruback SL, Ghetie V, Vitetta ES.** 2002. A novel recombinant vaccine which protects mice against ricin intoxication. *Vaccine* **20:**3422–3427.
141. **Smallshaw JE, Vitetta ES.** 2012. Ricin vaccine development, p. 259–272. *In* Mantis N (ed.), *Ricin and Shiga Toxins.* Springer, Berlin, Heidelberg.
142. **Roy CJ, Brey RN, Mantis NJ, Mapes K, Pop IV, Pop LM, Ruback S, Killeen SZ, Doyle-Meyers L, Vinet-Oliphant HS, Didier PJ, Vitetta ES.** 2015. Thermostable ricin vaccine protects rhesus macaques against aerosolized ricin: epitope-specific neutralizing antibodies correlate with protection. *Proc Natl Acad Sci USA* **112:**3782–3787.
143. **Pittman PR, Reislerb RB, Lindsey CY, Güereña F, Rivard R, Clizbe DP, Chambers M, Norris S, Smith LA.** 2015. Safety and immunogenicity of ricin vaccine, RVEc™, in a Phase 1 clinical trial. *Vaccine* 33**:**7299–7306.
144. **Wilson DE, Chosewood LC.** 2009. Neurotoxin-producing Clostridia species, p 135. *In Biosafety in Microbiological and Biomedical Laboratories.* 5th ed, HHS Publication No. (CDC) 21-1112. http://www.cdc.gov/biosafety/publications/bmbl5/bmbl5_sect_viii.pdf.
145. **Occupational Safety and Health Administration.** 2015. OSHAcademy Course 706 Study Guide—Job Hazard Analysis. http://www.oshatrain.org/courses/studyguides/706studyguide.pdf.
146. **Headquarters, Department of the Army.** 1998. *Risk Management.* Field manual 100-14, p 2-0–2-21. Headquarters, Department of the Army, Washington, DC.
147. **Hillson D, Simon P.** 2012. Making it Work, p 9–20. *In* Hillson D, Simon P (ed), *Practical Project Risk Management—the ATOM Methodology,* 2nd ed. Management Conceptpress, Tysons Corner, VA.
148. **Case No.: BA392069 Prosecution Enforcement Agreement**, Administrative Enforcement Terms and Conditions, Penal Code 1385. State of California v. The Regents of the University of California, a Public Corporation, and Patrick Harran, Superior Court of the State of California for the County of Los Angeles. July 2012.
149. **Case No.: BA392069 Deferred Prosecution Agreement**, State of California v. Patrick Harran. Superior Court of the State of California for the County of Los Angeles. June 2014.
150. **United States Department of Agriculture Marketing and Regulatory Programs, Animal and Plant Health Inspection Service (APHIS)**. 2010. Laboratory Ventilation Management, p 13–14. Administrative Notice APHIS 11-3.
151. **United States Department of Agriculture, Agricultural Research Service.** 2012. ARS Facilities Design Standards. *Manual* **242:**247.
152. **United States Department of Defense (DOD).** 2010. Safety Standards for Microbiological and Biomedical Laboratories. Department of Defense Manual 6055.18-M. *Enclosure* **8:**50.
153. **ANSI/ISEA.** 2014. American National Standard for Emergency Eyewash and Shower Equipment; Z358.1. Washington, DC.

154. **Headquarters, Department of the Army.** 1993. *Biological Defense Safety Program.* Pamphlet 385–69. Headquarters, Department of the Army, Washington, DC.
155. **Cipriano M.** 2006. Large-scale production of microorganisms, p 561–577. *In* Fleming D, Hunt D (ed), *Biological Safety: Principles and Practices.* ASM Press, Washington, DC.
156. **Canadian Biosafety Standards and Guidelines (CBSG).** 2015. Chapter 14.Large scale work, p xxii. *In Canadian Biosafety Standard,* 2nd ed. Public Health Agency of Canada, Ottawa, Canada. http://canadianbiosafetystandards.collaboration.gc.ca/cbs-ncb/assets/pdf/cbsg-nldcb-eng.pdf.
157. **National Institutes of Health.** 2013. *NIH Guidelines for Research Involving Recombinant DNA Molecules (NIH Guidelines),* 59 FR 34496 (July 5, 1994), as amended. http://osp.od.nih.gov/sites/default/files/NIH_Guidelines_0.pdf.
158. **NSF International Standard/American National Standard (NSF/ANSI).** 2014. *Biosafety Cabinetry: Design, Construction, Performance, and Field Certification. NSF/ANSI standard 49-2014.* NSF International, Ann Arbor, MI.
159. **Wilson DE, Chosewood LC.** 2009. Primary containment for biohazards: selection, installation and use of biological safety cabinets, p 290–325. *In Biosafety in Microbiological and Biomedical Laboratories.* 5th ed, HHS Publication No. (CDC) 21-1112. http://www.cdc.gov/biosafety/publications/bmbl5/bmbl5_appendixa.pdf.
160. **World Health Organization (WHO).** 2010. WHO Guidelines for the Production, Control and Regulation of Snake Antivenom Immunoglobulins. World Health Organization, Geneva, Switzerland. http://www.who.int/bloodproducts/snake_antivenoms/SnakeAntivenomGuideline.pdf.
161. **Pearn JH, Covacevich J, Charles N, Richardson P.** 1994. Snakebite in herpetologists. *Med J Aust* **161:**706–708.
162. **Herpetological Animal Care and Use Committee (HACC), Beaupre SJ et al (ed).** 2004. Guidelines for Use of Live Amphibians and Reptiles in Field and Laboratory Research. American Society of Ichthyologists and Herpetologists, Lawrence, KS.
163. **Wilson DE, Chosewood LC.** 2009. Section IV-Laboratory biosafety level criteria, laboratory biosafety level criteria: BSL-2, C. Safety equipment (primary barriers and personal protective equipment), p 36. *In Biosafety in Microbiological and Biomedical Laboratories.* 5th ed, HHS Publication No. (CDC) 21-1112. http://www.cdc.gov/biosafety/publications/bmbl5/bmbl.pdf.
164. **Canadian Biosafety Standards and Guidelines (CBSG).** 2015. Chapter 4.8. Decontamination and waste management, p 75–77. In *Canadian Biosafety Standard,* 2nd ed. Public Health Agency of Canada, Ottawa, Canada. http://canadianbiosafetystandards.collaboration.gc.ca/cbs-ncb/assets/pdf/cbsg-nldcb-eng.pdf.
165. **Seto Y.** 2009. [Decontamination of Chemical and Biological Warfare Agents.] *Yakugaku Zasshi* **129:**53–69.
166. **Wannemacher RW.**1989. Procedures for inactivation and safety containment of toxins, p 115–122. In *Proceedings of Symposium on Agents of Biological Origin.* U.S. Army Research, Development and Engineering Center, Aberdeen Proving Ground, MD.
167. **Suzuki T, Noro T, Kawamura Y, Fukunaga K, Watanabe M, Ohta M, Sugiue H, Sato Y, Kohno M, Hotta K.** 2002. Decontamination of aflatoxin-forming fungus and elimination of aflatoxin mutagenicity with electrolyzed NaCl anode solution. *J Agric Food Chem* **50:**633–641.
168. **Castegnaro M, Friesen M, Michelon J, Walker EA.** 1981. Problems related to the use of sodium hypochlorite in the detoxification of aflatoxin B1. *Am Ind Hyg Assoc J* **42:**398–401.
169. **Lunn G, Sansone EB.** 1994. Aflatoxin, p 23–30. *In* Lunn G, Sansone EB (ed), *Destruction of Hazardous Chemicals in the Laboratory,* 2nd ed. John Wiley and Sons, Inc, New York.
170. **Poli MA.** 1988. Laboratory procedures for detoxification of equipment and waste contaminated with brevetoxins PbTx-2 and PbTx-3. *J Assoc Off Anal Chem* **71:**1000–1002.
171. **Gilpin RW.** 2000. Elements of a biosafety program, p 443–462. *In* Fleming DO, Hunt DL (ed), *Biological Safety: Principles and Practices,* 3rd ed. ASM Press, Washington, DC.
172. **Centers for Disease Control and Prevention and Office of the Inspector General, U.S. Department of Health and Human Services.** 2005. Possession, use and transfer of select agents and toxins; final rule (42 CFR Part 73). *Fed Regist* **70:**13316–13325. http://www.gpo.gov/fdsys/pkg/FR-2005-03-18/pdf/05-5216.pdf#page=23.
173. **Animal and Plant Health Inspection Service, U.S. Department of Agriculture.** 2005. Agricultural Bioterrorism Protection Act of 2002: possession, use and transfer of biological agents and toxins; final rule (7 CFR Part 331; 9 CFR Part 121). *Fed Regist* **70:**13278–13292. http://www.gpo.gov/fdsys/pkg/FR-2005-03-18/pdf/05-5063.pdf#page=37.
174. **107th Congress.** 2001.USA PATRIOT Act. *Public Law* 107–56 https://www.gpo.gov/fdsys/pkg/FR-2005-03-18/pdf/05-5063.pdf
175. **Title 18 United States Code (U.S.C.).** 2006. Prohibitions with respect to biological weapons, p. 40–41. Section 175, Chapter 10. http://www.gpo.gov/fdsys/pkg/USCODE-2011-title18/pdf/USCODE-2011-title18-partI-chap10-sec175.pdf.
176. **Centers for Disease Control and Prevention (CDC) Division of Select Agents and Toxins Animal and Plant Health Inspection Service (APHIS) Agriculture Select Agent Program (CDC/APHIS).** 2013. Security Guidance for Select Agent or Toxin Facilities. http://www.selectagents.gov/resources/Security_Guidance_v3-English.pdf.
177. **Wilhelm K.R.** 2015. Biosecurity countermeasures by major agency regulation or guidance, p 142–146. *In* Wilhelm K (ed), *Biological Laboratory Applied Biosecurity & Biorisk Management Guide.* Alfa-Graphics, Westlake, OH.
178. **Oldach D, Brown E, Rublee P.** 1998. Strategies for environmental monitoring of toxin producing phantom dinoflagellates in the Chesapeake. *Md Med J* **47:**113–119.

12

Molecular Agents

DAWN P. WOOLEY

INTRODUCTION

Molecular agents are some of the most challenging types of agents encountered in biosafety in terms of assessing risk and determining the appropriate containment levels. Molecular agents are often found on the cutting edge of science. As such, there are many unknowns, and they do not fit into discrete risk group categories. This chapter discusses some of the major categories of molecular agents that may be encountered in biological safety. Various types of nucleic acids are described, including recombinant, nonrecombinant, oncogenic, and pathogenic DNA. Synthetic, naked, and free nucleic acids are also discussed. Gene transfer techniques are mentioned, and a review of RNA technologies is provided, including a discussion of different types of RNA interference, such as small interfering RNA (siRNA) and microRNA (miRNA). Innovative molecular tools for genome editing and cancer immunotherapy are described, namely zinc finger nucleases, transcription activator-like effector nucleases (TALENs), meganucleases, clustered regularly interspersed short palindromic repeats (CRISPRs), chimeric antigen receptors (CARs), and engineered T-cell receptors (TCRs). Nanotechnology and how this new field relates to biosafety are discussed. Finally, a review of biosafety issues related to prion diseases is given. For each of the topics, a general description of the technology is provided and biosafety considerations for risk assessment and containment are presented.

NUCLEIC ACIDS

Regulation of genetic engineering varies around the world, with some countries policing activities through the enactment of laws and others controlling activities through the issuance of national guidelines. State and local ordinances may place additional and more stringent regulations over the federal requirements. In the United States, both recombinant and synthetic DNA are covered extensively by the *NIH Guidelines for Research Involving Recombinant or Synthetic Nucleic Acid Molecules* (1), hereafter referred to as the *NIH Guidelines*. These guidelines were originally issued on June 23, 1976, and amended on March 5, 2013, to include synthetic DNA. Institutions accepting research funding from the NIH are subject to these guidelines, and all researchers at these institutions are subject to the guidelines, even if their research is not specifically funded by the NIH. Whether or not an institution is required to follow the guidelines, they serve as a national standard for the United States. Other countries may choose to consider these recommendations as a model in developing their own regulations.

The *NIH Guidelines* came about as a result of public concern over the new science of recombinant DNA that emerged in the early 1970s. In response to this concern, the Director of the NIH created the Recombinant DNA Molecule Program Advisory Committee (now called the Recombinant DNA Advisory Committee, or the RAC). In 1975 scientists also convened the Asilomar Conference on Recombinant DNA Molecules to address public apprehension. Proceedings from the Asilomar Conference and inputs from the RAC over the years have shaped the guidelines into what they are today. The guidelines are constantly evolving as new technology emerges. Amendments are made to the guidelines and posted to the U.S. Federal Register. Readers are encouraged to check the latest version, which is readily available online (1).

There are many different types of nucleic acids encountered in biosafety and some of the main categories are defined here:

Recombinant DNA. Generally defined as the joining of at least two DNA molecules or "any technique for manipulating DNA or RNA" (2). More specifically defined as "molecules that a) are constructed by joining nucleic acid molecules and b) that can replicate in a living cell" (1).
Nonrecombinant DNA. DNA that exists naturally.
Synthetic nucleic acids. DNA and RNA that are "chemically or by other means synthesized or amplified, including those that are chemically or otherwise modified but can base pair with naturally occurring nucleic acid molecules" (1).
Naked nucleic acids. DNA and RNA produced in the laboratory; histone-free.
Free nucleic acids. DNA and RNA produced in the laboratory and released into the environment.
Oncogenic DNA. DNA that can induce tumors in experimental animals or cause transformation of cells *in vitro.*
Pathogenic DNA. DNA that can cause disease in experimental animals or cause cytopathic effects *in vitro.*

Nucleic acids can be synthesized in different ways. Solid-phase DNA synthesis is a procedure whereby a DNA molecule is built nucleotide-by-nucleotide onto a solid support system using a programmed sequence; this is chemical synthesis. Techniques such as PCR or *in vitro* transcription use proteins known as polymerases and primers to copy an existing molecule; this is biological synthesis.

Naked nucleic acids are used in research, industry, and medicine. They have a large size range from about 20 to millions of bases in length. Naked DNAs often contain antibiotic resistance markers, transgenes, and xenobiotics. They take the form of plasmids, transposons, cDNA of RNA viral genomes, artificial vectors, artificial chromosomes, DNA vaccines, PCR-amplified sequences, and oligonucleotides. Naked RNAs include molecules such as viral genomes, antisense RNA, ribozymes, RNA vaccines, self-replicating RNA, and RNA/DNA hybrids.

Assessing risk for recombinant or synthetic DNA research requires careful deliberation. It is important to determine whether nucleic acids will be introduced into humans, animals, or plants; these special situations are addressed in specific appendixes of the *NIH Guidelines*. Other biosafety considerations include whether the molecular constructs express any drug or immunological resistance genes, oncogenic or pathogenic genes, toxins, microorganisms, or prions. Each molecular construct should be described in detail, including information such as the name, provider, and function of the transgene. The name, type, species, and strain of the vector are also important, along with information about the inclusion of control elements, such as promoters and enhancers that effect gene expression. A schematic diagram of the molecular construct can be helpful, if available. The host cell, strain, and host range of the recombinant or synthetic molecule are essential knowledge, together with information about the highest concentration and largest production volume to be used; large-scale work is described in specific sections of the *NIH Guidelines*. It is important to know where the construct was made or synthesized in order to determine whether the appropriate material transfer agreements are in place and whether the material will be transported in a safe manner. In some cases, materials may need to be verified upon arrival to ensure that a given material does not represent or is not contaminated with a more dangerous agent. If the vector is a genome, e.g., a viral vector, it is important to know the percentage of the genome that has been deleted or substituted.

An important consideration for recombinant or synthetic DNA work is the level of review required before the work begins. Section III of the *NIH Guidelines* spells out six categories of experiments, ranging from exempt to those that require review by three separate entities (Institutional Biosafety Committee, RAC, and NIH Director) prior to initiating experiments.

Gene Transfer

The beginnings of gene transfer date back to 1944 with the famous experiments of Oswald Avery, Colin Munro

MacLeod, and Maclyn McCarty in which they isolated DNA from a virulent strain of *Pneumococcus* type III and successfully used it to transform a nonvirulent strain (3). Transformation refers to the alteration of the genetic makeup of a cell as a result of the uptake of exogenous genetic material. This term generally refers to bacterial cells and has a special meaning with regard to animal cells, indicating progression to a cancerous state. Thus, the term is avoided when describing the uptake of genetic material into animal cells.

Different types of vectors can be used to transfer recombinant DNA into cells, including plasmids, phagemids, expression vectors, shuttle vectors, transposons, and viral vectors. Plasmids are the most common vehicle; they are small (~2 to 20 kbp) circular DNA molecules that replicate independently of a bacterial chromosome. In nature, plasmids express antibiotic resistance genes and other genes that perform specific functions for a bacterial cell. Phagemids are plasmids that contain an origin of replication from a bacteriophage, or phage (a virus that infects bacterial cells), allowing it to be expressed as a plasmid or packaged into a virus particle. Expression vectors are designed to contain elements, such as promoters and termination signals, that allow a particular gene sequence to be expressed as a protein. Shuttle vectors are constructed in a way that allows replication in at least two different host species, such as prokaryotic and eukaryotic. Transposons are small (1 to 4 kbp) DNA sequences that can move throughout a bacterial genome, and they can be engineered to carry recombinant DNA. Finally, viral vectors are viral genomes modified to deliver foreign genetic material into a cell. Viral vectors may be contained in and/or expressed by another vector, such as a bacterial plasmid or other virus, e.g., phage. As such, characteristics of both the viral vector and secondary carrier vector must be considered in a biosafety risk assessment. Host range is an important consideration for assessing the risk of plasmids and transposons, while tropism (preference for a particular cell type) is important for viral vectors and phage.

There are many different methods to introduce (or transfect) nucleic acids into cells, including the use of chemicals (such as DEAE, $CaPO_4$, and DMSO) (4–6), liposomes, electroporation, microinjection, gene gun, and viral vectors. The term "transduction" has its origins in bacteriology and refers to the transfer of genetic materials from one bacterial cell to another by phage. However, the term transduction has been broadened in modern times to refer to the use of viral vectors to transfer genes. Generally, all of the nonviral methods are less than 50% efficient for transduction. The viral methods can approach 90% efficiency, which is why the use of viral vectors has become so popular as gene transfer agents. Efficiency of transduction depends heavily on the cell type.

Whatever method is used to transfer nucleic acid, its fate depends on many factors. Initially, a nucleic acid must survive nucleases in the environment. Nucleases are enzymes that destroy nucleic acids, and RNA is more susceptible to nucleases than DNA because of its hydroxyl group on the second carbon position of the ribose sugar. Next, the nucleic acids must adsorb to and enter cells. Once inside the cell, nucleic acids are often destroyed by cellular defense mechanisms. If the nucleic acid is RNA, it must be expressed in the cytoplasm. If the nucleic acid is DNA, it must find its way to the nucleus where it may be expressed or ligated to other DNA and become integrated into the genome, potentially causing insertional mutagenesis. During this process, the DNA often becomes mutated, which raises additional biosafety concerns (7). DNA may become integrated through a homologous recombination process in which cellular repair proteins are used, or it may become integrated through an illegitimate process mediated by nonhomologous end-joining proteins (8). This latter process can be more efficient as compared to homologous recombination because there are more sites available. Illegitimate recombination is associated with open chromatin and enhanced by DNA damage (8).

Oncogenic DNA

An oncogene is a gene that can induce tumors in experimental animals or cause transformation of cells *in vitro*. Oncogenes were originally discovered in retroviruses as mutated cellular genes incorporated into viral genomes and carried into cells. Oncogenes encode proteins that control cellular growth, such as growth factors and their receptors, signal transducers, and transcription factors. The first oncogene discovered was v-*src* (9, 10), which encodes an enzyme that transfers a phosphate group to the tyrosine amino acid on a protein. The letter "v" in front of the gene name stands for the viral version of the gene. The cellular version of the gene (also referred to as the proto-oncogene) is called c-*src*. Other oncogenes follow the same naming scheme.

An important question in biosafety is whether cellular DNA can be oncogenic, and there is evidence that it can. There are three possible mechanisms by which such DNA could cause cancer through insertional mutagenesis: (i) insertion of an activated oncogene into a cell; (ii) inactivation of a tumor-suppressor gene; and (iii) activation of a cellular oncogene via proximal insertion of a promotor or enhancer. The risk of transformation by cellular DNA was calculated using a theoretical model employing a statistical Poisson distribution approach in which assumptions were made about oncogene size, frequency, biological integrity, and transformation efficiency under optimal conditions (11, 12). The *in vitro* risk was estimated to be 10^{-6} (1 in a million) with a 10-pg contaminant and 100 oncogenes per cell, and the *in vivo* risk was estimated to be 10^{-9} (1 in a billion) with a 1,000-pg contaminant and 100 oncogenes per cell (12). Experimental data gathered *in vivo* showed that a single application of 10 µg of plasmid DNA containing the human T24 H-*ras* oncogene (1.1×10^{12} molecules) caused cancer when applied to scarified mouse skin in the absence of further promotional stimuli (13). Injection of plasmid DNA containing the human T24 H-*ras* oncogene and c-*myc* proto-oncogene (12.5 mg each) into mice caused cancer in 20% to 80% of the animals in the absence of further promotional stimuli (14). Injection of BK virus DNA (5 mg) into the brains of newborn hamsters caused cancer in 5% of the animals in the absence of further promotional stimuli (15). Injection of plasmid DNA (2 mg) containing the BK virus early region and the human T24 H-*ras* oncogene into newborn hamsters caused cancer in 73% of the animals in the absence of further promotional stimuli (16, 17). Approximately 2 µg of cloned v-*src* (2.5×10^{11} molecules) was found to be oncogenic in chickens (18, 19). Thus, it has been shown that oncogenic DNA can cause cancer in experimental animals, including application to nonintact skin.

Pathogenic DNA

Pathogenic DNA is that which can cause disease in experimental animals or cause cytopathic effects *in vitro*. When nucleic acid is extracted from a pathogenic microorganism, it is important to assess whether the preparation may still contain any of the intact parental microorganism. It is also important to determine whether the naked DNA or RNA is infectious. For example, the positive-stranded RNA genomes of viruses are usually infectious because they resemble an mRNA and can be translated immediately in the cytoplasm of a cell. Some DNA genomes of viruses are infectious. If it is determined that naked nucleic acid is infectious, then containment should be the same as that for the microorganism from which it was isolated.

The infectious dose to 50% of individuals (ID_{50}) of viral DNA in biologicals has been determined for some viruses, including *Simian immunodeficiency virus* (SIV) in monkeys (38 mg) (20, 21), murine retrovirus in mice (2.5 mg) (22), and polyomavirus in mice and hamsters (0.004 mg) (23). On the basis of SIV risk data, the likelihood of an infection resulting from intramuscular injection of 1 mg of cellular DNA containing a single viral genome per cell (about 150,000 cells worth of DNA) is ~2.5×10^{-8} (or 1 in 40 million) (20, 21). For a product that contains 1 mg of viral DNA and 100 viral genomes per cell, the risk would be 1 in 400,000 if each individual were to receive one dose (20, 21). Currently, the WHO and U.S. FDA recommend a limit of 10 ng and 200 base pairs for residual DNA in a final product dose (24). The fact that there are limits on the amount of cellular DNA allowed in biological products implies that there is some risk.

Overall, the risks associated with handling naked DNA and RNA are low. However, the risk is not zero. RNA is more easily destroyed in the environment as compared to DNA. Sharps should be avoided in manipulative procedure, and caution should be used when DNA and RNA are used in conjunction with solvents that can penetrate the skin. Gloves should be worn with consideration to the chemicals being used. Proper disposal should be performed to avoid environmental contamination. Nonrecombinant DNA should be considered in risk assessments, i.e., it should not be ignored.

RNAs

There are two main categories of RNA, coding and noncoding. Coding RNA takes the form of mRNA. mRNA has a plus-strand polarity that matches the sequence of the coding strand of double-stranded DNA, with uracil replacing thymine as one of the bases. In eukaryotic systems, mRNA has a chemical structure known as a "cap" at the 5′ (phosphate) end and a tail of adenine bases known as a "poly(A) tail" at the 3′ (hydroxyl) end.

Within the noncoding RNA category, there are several classes. Translational RNAs consist of transfer RNA (tRNA) and ribosomal RNA (rRNA), which assist in the translation of mRNA into protein. Molecules involved in RNA splicing and other processing are known as small nuclear RNA (snRNA) and small cytoplasmic RNA (scRNA). Small noncoding RNAs include siRNA, miRNA, Piwi-interacting RNA (piRNA), and short hairpain RNA (shRNA). Long noncoding RNA (lncRNA) is generally defined as a molecule longer than 200 bases that does not encode for protein. lncRNAs play a role in gene expression and cellular processes. RNA molecules that perform an enzymatic function are known as ribozymes.

Interfering RNAs are used for gene silencing, and RNAi is a blanket term for RNA interference in general. RNA silencing appears to have evolved, at least in part, as an antiviral and antitransposon defense mechanism (25, 26). siRNA is found naturally in plants and lower animals. It is expressed by the same genes that it regulates and is formed from shRNA or from long synthetic RNA. A set of proteins that form an RNA-induced silencing complex (RISC) are capable of recognizing siRNA and cleaving mRNA that is complementary to the siRNA, effectively silencing gene expression (27). Recent research indicates that direct translational repression of protein expression precedes mRNA degradation and is a prerequisite for it (28). Endonuclease-prepared siRNA (esiRNA) is a mixture of siRNAs resulting from the cleavage of a long strand of double-stranded RNA. shRNAs are encoded by DNA and consist of two complementary RNA sequences connected by a linker such that the molecule can fold back on itself, also known as DNA-directed RNA interference (ddRNAi). miRNA is similar to siRNA with the exception that it does not typically silence its own expression. A new form of noncoding RNA known as circular RNA (circRNA) has recently been identified as a possible regulator of miRNA, transcription, and splicing (29, 30). piRNA was discovered in *Drosophila* and shown to be involved in germ line development. The mouse ortholog is Miwi, and the human ortholog is Hiwi. Repeat-associated small interfering RNA (rasiRNA) is a specific form of piRNA (26). In animals, these RNAs are expressed mainly in testes where they trigger gene silencing as part of the normal process of spermatogenesis. They are also found in somatic cells of animals and in ovarian somatic and neuronal cells of invertebrates. In general, silencing RNAs are being exploited as tools in research and as potential therapies for a variety of diseases and medical conditions (27).

The number of different types of small RNA molecules and their acronyms can be overwhelming. However, from a risk assessment perspective, all of these interfering, silencing, or enzymatic RNAs are of similar concern with regard to inappropriate expression and off-target activity that could lead to unwanted toxicities. The off-target effects are species-specific. Therefore, small animal studies may not predict all of the negative side effects in humans. However, there are ways to mitigate the risks, such as by using redundancy and rescue designs (31). Redundancy involves targeting more than one sequence within the same gene, while rescue involves providing a functional target gene that is resistant to silencing. Database searches and *in vitro* experiments in cell culture may also help to identify negative side effects. Examples of public databases that could be searched for off-target effects are http://genomernai.org/ and http://www.ncbi.nlm.nih.gov/pcassay. Interfering and small RNA molecules can be delivered by various DNA constructs, including viral vectors, in which case they have the potential to spread, contaminate the germ line, or be continually produced in cells through constitutive expression. Overexpression of any of these RNA molecules could be potentially toxic, and preclinical testing should be performed to rule out possible deleterious consequences prior to a clinical trial.

GENOME EDITING

In recent years, genome editing has become a popular way of making specific changes to a DNA sequence, and some of these technologies are making their way into clinical trials (clinicaltrials.gov and www.gemcris.od.nih.gov) (32). There are four major types of editing technologies: zinc fingers, TALENs, CRISPRs, and meganucleases. All of them rely on making double-stranded cuts in the DNA sequence. Each technology differs in the way it targets a specific sequence, with some having more specificity than others.

Zinc fingers were the first of the genome-editing technologies to become popular. Each finger consists of approximately 30 amino acids that recognize a 3-base-pair stretch of DNA. Typically three to six zinc fingers are put together to create a 9- to 18-base-pair recognition sequence on the DNA strand, and the array is fused to the cleavage domain of the FokI restriction nuclease. Since the FokI restriction enzyme must be dimerized, zinc fingers are designed to bind the complementary strands of DNA. This feature, along with the fact that there is a required spacing between the zinc finger and the FokI endonuclease, creates specificity in binding to the DNA sequence (33).

TALEN is an acronym for transcription activator-like effector nuclease. TALENs are similar to zinc finger nucleases in that they have a DNA-binding protein (the TAL effector) fused to a restriction endonuclease, usually FokI. The TAL effector proteins were originally discovered in plants (34) and differ from those of zinc fingers by having a one-to-one recognition of nucleotides (35). TALENs contain a 34-amino-acid repeat domain with hypervariable amino acids at positions 12 and 13 that recognize a specific nucleotide according to a specific code (36). Because one TAL is needed for each nucleotide, TALENs are larger than zinc fingers and are more difficult to deliver.

Meganucleases are similar to naturally occurring restriction enzymes, except they have much larger recognition sites on the DNA molecule, in the range of 14 to 40 bp (37, 38). Because the recognition sites are large, the number of sites in any given genome is low, allowing for specificity of cutting (39). The number of naturally occurring meganucleases is extremely limited, so research efforts are being made to modify the binding sites of existing meganucleases, thus making designer meganucleases (40, 41).

CRISPRs are the newest genome-editing tools and are generating much excitement due to the potential for greater specificity. Unlike the other genome-editing technologies described above, CRISPRs are guided by RNA rather than protein. CRISPR is an acronym that stands for clustered regularly interspaced short palindromic repeats, which were discovered as part of a bacterial immune system (42). CRISPR-associated proteins, known as Cas proteins, are the endonucleases that make the double-stranded cuts in the DNA, analogous to the FokI enzyme described above. Thus, there are two components to the system, which is why it is often referred to as a CRISPR/Cas system. There are three different types of systems, and the type II CRISPR/Cas9 system is the most popular one used in genetic engineering (42). In nature, foreign DNA is incorporated into a CRISPR locus in a bacterial genome. This locus is transcribed by the bacterial cell and guide RNAs, called trans-activating CRISPR RNA (tracrRNA) and CRISPR RNA (crRNA), are created to identify and bind the foreign DNA. The Cas9 endonuclease is then recruited to degrade the invading DNA (43). Researchers exploit this technology to create synthetic guide RNAs (gRNAS) for targeting specific regions of DNA genomes.

Despite having different targeting and DNA-binding strategies, all of these genome-editing technologies ultimately result in the creation of a double-stranded cut in the DNA. One of two repairs is then made by the cell to alter the genome. Nonhomologous end joining is one mechanism in which the DNA is ligated at the breakpoint. During this process, small insertions and deletions are created that result in a mutated or truncated protein product when the gene is expressed. This mechanism is used to disrupt genes and create knockouts. Homologous repair is the other mechanism in which exogenous sequences may be inserted at the breakpoint (33).

Similar to the small RNA technologies, the main concern with genome editing is off-target effects. Zinc fingers are the most clinically advanced of all genome-editing technologies, and some zinc fingers have already been associated with cytotoxicity, presumably due to cleavage at nontargeted sites. Design features that limit homodimerization of zinc fingers may help to improve their safety (33). Strategies to increase specificity of zinc fingers include oligomerized pool engineering and context-dependent assembly (44–46). CRISPR/Cas9 systems have been made more specific by truncating the 5′ end of the gRNA, combining different platforms (e.g., fusing the FokI nuclease to a catalytically inactive Cas9), and engineering Cas9 to create a nick rather than a double-stranded break (47). Another strategy for improving safety is to direct genome editing to "safe harbors" within the genome that have minimal risk for insertional mutagenesis (47). Bioinformatics tools such as PROGNOS (http://bao.rice.edu/Research/BioinformaticTools/prognos.html) may help to predict off-target sites (47), and preclinical testing should include both *in vitro* cell culture testing and *in vivo* animal testing whenever possible. Genome-editing molecules can be delivered by various DNA constructs, including viral vectors, in which case they have the potential to spread, contaminate the germ line, or be continually produced in cells through constitutive expression. Ethical issues include germ line alteration, nontherapeutic genome enhancement, and animal chimeras for organ transplant.

CHIMERIC ANTIGEN RECEPTORS AND T-CELL RECEPTORS

A promising approach in cancer immunotherapy has been the use of gene-modified T cells to seek out and destroy cancer cells. Two main strategies have been used to genetically modify T cells—engineered TCRs and CARs.

Natural TCRs are composed of two protein chains (α and β) that each have a constant and variable region, like an antibody molecule. The variable region gives the molecule its specificity. The α and β chains are associated with three transmembrane signaling molecules that form what is called a CD3 complex. The TCR recognizes its target antigen in the context of a major histocompatibility complex (MHC) on the target cell. The MHC presents a small piece of a foreign antigen (peptide) for binding to the TCR. Simultaneous binding of a coreceptor (like CD8 for a killer T cell) is required for full activation. When TCRs are modified for cancer immunotherapy, sequence changes are made to encode new TCR α and β chains with different peptide specificity. Changes are also made to the transmembrane regions to avoid interaction with endogenous TCRs (48).

The CAR has some similarities to a TCR, but it is simpler in terms of binding to the target. It is constructed in a way that links an extracellular antibody fragment to an intracellular signaling domain that directs killing action. The antibody fragment consists of what is called a single-chain variable fragment (scFv). The scFv is found at the tip of the Y-shaped antibody molecule, and it is very specific for a particular antigen, like a lock and key (49). The scFv can bind directly to an antigen on a target cell; it does not need to be presented by an MHC molecule nor does it need a coreceptor such as CD8.

The variable portions of the heavy and light chains of an scFv must be held together by a linker because the rest of the antibody molecule has been removed. In a CAR construct, the antibody fragment is connected to a CD3ζ intracellular signaling domain by a hinge and transmembrane region. This construct represents the first generation of CAR. Since it was not possible to achieve complete T-cell activation with the first generation, a costimulatory domain (either CD28 or 4-1BB) was added just inside the cell between the transmembrane region and the CD3ζ signaling domain to create the second generation of CAR. To further improve T-cell activation, a third-generation CAR was engineered that expresses two costimulatory domains at the same time—a combination of CD27, CD28, 4-1BB, inducible costimulator (ICOS), or OX40 (50). In a biosafety risk assessment, it would be important to identify which generation of CAR is being used and which costimulatory domains have been incorporated in order to predict or correlate possible adverse reactions in human subjects or exposed workers.

Many antigens have been targeted by TCR and CAR constructs, and some have reached clinical trials (48, 51, 52). When used in immunotherapies, a patient's T cells are removed from the body and transduced *ex vivo* with the TCR or CAR. Retroviral or lentiviral constructs are generally used as vectors for the constructs because they are efficient and integrate naturally into the genome for stable expression. Therefore, all of the biosafety issues surrounding the delivery vehicle would need to be considered in a risk assessment, such as contamination with replicating virus, insertional mutagenesis, germ line integration, and horizontal and vertical transmission.

Before the genetically modified cells are infused, the patient is given chemotherapy to reduce the number of circulating T cells in order to give an advantage to the infused cells. This treatment could promote an expansion of the infused cells. Once the cells are infused, there are three types of major adverse effects that have been observed: on-target off-tumor activity, off-target reactivity, and cytokine-release syndromes (48). On-target off-tumor effects occur when normal tissues express the antigen that was targeted, even when expressed at low levels. Thus, an autoimmune reaction occurs. Off-target effects occur when there is cross-reactivity, and there is more chance for this with TCRs because they recognize peptides. Cytokine release syndrome (aka cytokine storm) occurs when large numbers of immune cells become activated and release inflammatory cytokines. The syndrome is characterized by fever and low blood pressure. Life-threatening complications include heart and lung problems, neurologic toxicity, kidney or liver failure, and blood clots (53). Treatment for the syndrome often involves steroids, vasopressors, and supportive therapy in intensive care (52). *In silico, in vitro,* and *in vivo* testing should be performed before clinical trials take place to rule out all possible deleterious effects in humans, including screens against self-peptides (51). As part of the biosafety risk assessment, consequences to researchers involved in preparing the product and to medical professionals involved in administering the product should be considered.

NANOPARTICLES

Nanotechnology is a relatively new field at the cutting edge of commercial and medical product development. Nanoparticles are generally defined as having at least one dimension between 1 and 100 nm. They have a large surface-to-volume size ratio as compared to the bulk form of the same substance; a given weight of nanoparticles has a much larger surface area than the same weight of bigger particles made of the same material. Due to their small size and unique reactivity, nanomaterials do not obey the traditional laws of physics but rather the laws of quantum physics (54, 55). For these reasons, they are more chemically reactive and may exhibit unpredictable behavior, making the risk assessment more challenging.

There is a lot of confusion with regard to nanoparticles and biological safety. Some people are under the impression that because a nanoparticle is small like a virus it automatically falls under the purview of biological safety. Indeed, there is overlap in size with viruses that range from 18 nm for the *Parvoviridae* (56) to over 700 nm for the *Pandoraviridae* (57). Viral capsid proteins are being used to synthesize "virus-like particles" in the size range of nanoparticles, which serve as empty shells to carry payloads of drugs and biological materials (58, 59), and some of these virus-like particles are referred to as nanoparticles, even though they may fall outside the size range for nanoparticles. Different types of materials are used to make nanoparticles, including carbon and assorted metals, such as silver, gold, platinum, cadmium, zinc, copper, iron, and titanium. Among metals, silver is known for its antimicrobial properties, both in its ionic and nanoparticle form (60, 61). The carbon- and metal-based nanoparticles can be "functionalized" with a variety of biological molecules, such as DNA, protein, and antibodies, to create bionanomaterials (62, 63).

Nanoparticles are made in a variety of ways. They are chemically synthesized by either a top-down or bottom-up approach. The top-down approach begins with a larger portion of material that is somehow processed into smaller pieces (through grinding, cutting, etching, etc.). The bottom-up approach builds the nanomaterial piece-by-piece, molecule-by-molecule. The size and shape of nanoparticles can be controlled by adding other chemical agents in a specific fashion using specific ratios under certain sets of conditions (time, temperature, etc.) (64–67). Nanoparticles can vary in size within a given batch. Metallic nanoparticles are classified as "monodisperse" if they differ by less than 15% from the average particle size, and they are classified as having a "narrow size distribution" if the standard deviation is less than 15% to 20% of the mean particle diameter. Stabilizers can be added to disperse nanoparticles and prevent agglomeration (64, 67–69).

In addition to chemical synthesis methods, nanoparticles can also be made in biological systems, such as microorganisms and plants (70), and they are quite frequently used in the fields of biology and medicine. Thus, although a given type of nanoparticle might be composed of metal, which falls under the purview of chemical safety, it might be made or used in a biological system. Biological safety is reaching into the fields of engineering, physics, and chemistry, among others. Nanoparticles are used in drug and gene delivery, biosensing, bioimaging, antimicrobials, antipesticides, cosmetics, medical devices, cancer therapy, cell labeling, and many other novel applications (70, 71).

In a risk assessment for nanoparticles, one should first determine whether the material is actually a biohazardous agent, which is defined in this text as an infectious agent or other substance produced by a living organism that causes disease in another living organism. Many nanoparticles are purely chemical hazards; size does not define a biohazardous agent. Some nanoparticles are conjugated to biological agents, while others are synthesized in a biological system. Some nanoparticles carry payloads of biological agents. In these cases, the risk assessment should focus on the type of biological agent being used in conjunction with the nanoparticle, keeping in mind the unpredictable nature and unknown behavior of materials in the nanoparticle size range. Nanoparticles composed of viral capsids need to be assessed for tropism (preference for a particular cell type) or toxic effects, such as inflammatory or autoimmune reactions. Nanomaterials are still relatively novel agents; thus they must be assessed on a case-by-case basis. Recently, the National Research Council performed a study to develop a framework for assessing the safety of engineered nanomaterials (72).

For those nanoparticles that fall into the biohazardous category, biocontainment will depend on the nature of the biohazardous agent(s) used in conjunction with the nanoparticle and the types of procedures to be performed with the nanomaterial. If aerosol-generating procedures are employed, then a biological safety cabinet should be used for containment. It is important to note that the upper end of the nanoparticle size range overlaps with the particle size that is least efficiently removed by HEPA filters based on historical data, which is 100 to 300 nm (0.1–0.3 μm). Since nanoparticles do not obey the traditional laws of physics, they may not follow the same diffusion patterns as other small particles studied to date. Thus, their efficiency of capture in the HEPA filter may not be predictable based on extrapolation. Respirators should be used for procedures not contained within a biological safety cabinet. Research showed that the most penetrating nanoparticle size for an N95 respirator was about 50 nm (73). Sonicators are often used to disperse nanoparticles, and it can be difficult to achieve a closed system because the sonicator probe must be placed into the nanomaterial. For this reason, sonicators should be placed within a chemical fume hood or biological safety cabinet (depending on whether a biohazardous agent is present and whether product protection is required), and hearing protection should be worn by all persons in the vicinity. When the HEPA filters of biological safety cabinets are changed, workers should be made aware if nanomaterials have been used. The typical decontamination methods for a biological safety cabinet would not necessarily inactivate all types of nanomaterials. Thus, workers removing the HEPA filters should wear respiratory protection, and the HEPA filters should be bagged upon removal and disposed of appropriately.

PRIONS

Prions are transmissible pathogenic agents that cause a group of invariably fatal neurodegenerative diseases, including scrapie (the prototype prion disease in sheep and goats), bovine spongiform encephalopathy (BSE) in cattle, chronic wasting disease (CWD) in deer and elk, and Creutzfeldt-Jakob disease (CJD) in humans. A complete list of currently known animal and human prion diseases (also known as transmissible spongiform encephalopathies) is provided in Table 1. The newly identified α-synuclein prion is a the probable cause of multiple system atrophy, a neurodegenerative disorder that presents with symptoms of Parkinson's disease or cerebellar dysfunction (74). Recently, the first plant prion protein, named Luminidependens, has been identified as being involved in the plant flowering process (75).

Prions are unprecedented in nature because, unlike other infectious pathogens, they do not contain any nucleic acids. Rather, they are "infectious proteins" composed of abnormally folded proteins. They transmit the "infection" by contacting the normal form of the protein and causing it to fold improperly. The normal host protein is called the prion protein (PrP). The abnormal isoform is designated with a superscript, such as PrP^{Sc} for the "scrapie-like" isoform of PrP. Prion diseases are disorders of protein conformation involving template-assisted replication, resulting in accumulation of abnormal protein in the brain, which causes neuronal dysfunction, degeneration, and death. Prion diseases represent a truly novel pathogenic mechanism (76–78).

A distinctive feature of prion diseases is their ability to manifest as infectious, inherited, and spontaneous illnesses (79). Familial CJD, Gerstmann-Straüssler-Scheinker syndrome, and fatal familial insomnia are all dominantly inherited prion diseases linked to mutations in the gene encoding the prion protein, whereas sporadic CJD (sCJD) is thought to be due to a spontaneous conversion of the protein (80). Yet in all three manifestations of prion disease, infectious prions are generated in the brains of afflicted individuals, and these prions are composed of disease-causing molecules with the amino acid sequence encoded by the gene of the affected host. When prions pass into the brain of a new host species, a "species barrier," primarily related to differences in protein sequence, is responsible for inefficient infection (81–85). However, if interspecies transmission does occur, then the prions generated in the brain of the new host carry the amino

TABLE 1.

Prion diseases

Prion diseases	Mode of transmission
Prion disease/host	
Scrapie/sheep	Unknown, genetic susceptibility
Bovine spongiform encephalopathy (BSE)/cattle	Oral, foodborne
Transmissible mink encephalopathy (TME)/mink	Oral, foodborne
Chronic wasting disease (CWD)/mule deer, elk	Unknown
Feline spongiform encephalopathy (FSE)/cats	Oral, foodborne
Exotic ungulate encephalopathy (EUE)/greater kudu, nyala, oryx	Oral, foodborne
Human prion diseases	
Kuru[a] (extinct)	Ritualistic endocannibalism
Somatic Creutzfeldt-Jakob disease (sCJD)	Somatic mutation or spontaneous
Variant Creutzfeldt-Jakob disease (vCJD)	Oral, foodborne, transfusion, human-to-human
Familial Creutzfeldt-Jakob disease (fCJD)	Germ line mutation
Iatrogenic Creutzfeldt-Jakob disease (iCJD)	Therapeutic agents, devices, and procedures
Fatal sporadic insomnia (FSI)	Somatic mutation or spontaneous
Fatal familial insomnia (FFI)	Germ line mutation
Gerstmann–Sträussler–Scheinker syndrome (GSS)	Germ line mutation
Multiple system atrophy (MSA)	Spontaneous

[a]Restricted to the Fore people of Papua New Guinea.

acid sequence encoded by the new host and not the original host; in other words, the prions replicating in the new host brain are not the same as those that initiated replication. This scenario is profoundly different from what happens during a viral infection.

The biosafety issues related to prions have been addressed in several guidelines published by health authorities over the years, with recommendations to limit the potential risk associated with prion contamination in laboratory studies as well as in foods and medicinal products. These biosafety issues have taken on a heightened concern over the past decade, due to the potential transmission of scrapie, CWD, and BSE to animals and humans through consumption of contaminated foodstuffs. Prions in meat and bone meal, most likely derived from scrapie-infected sheep offal, are believed to be the cause of the BSE epidemic in the United Kingdom (86, 87). The transmission of a similar disease to cats is suspected to be due to prion-contaminated cat food (88). Demonstration of a link between BSE and a new variant of CJD (vCJD) has provoked a profound reassessment of public health policy worldwide on prion-associated risks to the human population (83, 89, 90). In addition, a further potential hazard to human health was recognized in the administration of biological as well as medicinal products derived from or associated with human or animal tissues potentially contaminated with prions. Professional, public, and political reaction to the rising number of CJD cases in recipients of cadaveric dura mater grafts or of pituitary-extracted human growth hormone from donor cadavers with undiagnosed CJD has led to global concerns about potential prion contamination of products derived from human tissues and about surgical instruments or medical devices exposed to such tissues (91). These concerns have spread to questions regarding the safety of human blood and blood products as well.

Physical Properties of Prions

The smallest infectious prion particle is probably a dimer or small oligomer of prion protein. This estimate is consistent with an ionizing radiation target size of 55 ± 9 kDa (92). A recent study designed to evaluate the relationship between infectivity and the size of prion protein aggregates showed that the most infectious particles consisted of oligomers of 14 to 28 molecules in the 17- to 27-nm (300- to 600-kDa) range (93). Therefore, prions may not be retained by all of the filters that efficiently eliminate bacteria and viruses. Additionally, prions aggregate into particles of nonuniform size that can affect their filtration capacity, and they cannot be solubilized by detergents, except under denaturing conditions where infectivity is lost (94, 95). Prions resist inactivation by nucleases (96), UV irradiation at 254 nm (97, 98), and treatment with psoralens

(99), divalent cations, metal ion chelators, acids (between pH 3 and 7), hydroxylamine, formalin, boiling, or proteases (100, 101).

Laboratory Exposure to Prions

Human prions and those propagated in apes and monkeys are generally handled at biosafety level 2 (BSL2) or BSL3, depending on the studies being conducted (102). BSE prions are likewise handled at BSL2 or BSL3, due to the fact that BSE prions have been transmitted to humans in the United Kingdom, France, and elsewhere (102, 103). All other animal prions are generally handled at BSL2.

Clinical Exposure to Human Prions

The highest concentrations of prions are found in the central nervous system and its coverings. Based on data from animal studies, it is presumed that high concentrations of prions may be found in the spleen, thymus, and lymph nodes. Moreover, in vCJD, prions are routinely detected in lymphoid tissue of tonsil, spleen, and appendix (104–107). In addition, prions have been found in muscle tissue of experimentally infected mice (108) and patients who died of sCJD (109).

In the care of patients dying of human prion disease, those precautions used for patients with AIDS or hepatitis are certainly adequate. In contrast to these viral illnesses, the human prion diseases are not contagious (110). There is no evidence of contact or aerosol transmission of prions from one human to another. However, prions are infectious under some circumstances, such as ritualistic cannibalism, administration of prion-contaminated growth hormone, transplantation of prion-contaminated dura mater grafts, and blood transfusion (111–115).

Surgical procedures on patients with a diagnosis of prion disease should be minimized. It is thought that CJD was spread from a CJD patient who underwent neurosurgical procedures to two other patients shortly thereafter in the same operating theater (116). Sterilization of reusable instruments and decontamination of surfaces should be performed in accordance with CDC (http://www.cdc.gov/) and WHO guidelines (117). Routine autopsies and the processing of small amounts of formalin-fixed tissues containing human prions require BSL2 precautions. The absence of any known effective treatment for prion disease demands caution in the manipulation of potentially infectious tissues. Because it is important to establish a definitive diagnosis of a human prion disease and to distinguish between sporadic or familial cases and those acquired by infection as a result of medical procedures or from prion-contaminated food products, unfixed brain tissue should be obtained. Unfixed samples of brain, spinal cord, and other tissues containing human prions should be processed with extreme care at a minimum of BSL2 with BSL3 practices (102). The main precaution to be taken when working with prion-infected or -contaminated material is to avoid puncture of the skin (110). The prosector should wear cut-resistant gloves if possible. If accidental contamination of skin occurs, the area should be washed with detergent and abundant quantities of warm water with a brief exposure (1 minute) to 1 N NaOH or a 1:10 dilution of bleach for maximum safety (102).

Animal Prions

BSE is the most worrisome of all the animal prion diseases from a biosafety standpoint. It is widely thought to be a man-made epidemic, caused by a form of industrial cannibalism in which cattle were fed meat and bone meal produced through faulty industrial processes from prion-contaminated cattle and sheep offal (86, 87, 118, 119). Epidemiologically, BSE has shown a disquieting propensity to cross species barriers through oral consumption of prion-contaminated bovine foodstuffs (88, 120, 121), and this propensity has been confirmed experimentally in laboratory transmission studies employing several routes of inoculation, including the oral and intravenous routes (122–129). Most alarmingly, there is now solid epidemiological and experimental evidence that BSE is responsible for vCJD primarily in the United Kingdom, but also elsewhere (83, 90, 105, 128,130–132). The first U.S. case of BSE in cows was discovered in 2003 (133), and the most recent case was reported in 2012 (134).

Scrapie is transmitted between farms by movement of infected animals. Once introduced into a flock, both vertical and horizontal routes of transmission can occur (135). To date, there is no epidemiological evidence that scrapie is transmitted to humans (136). However, recent experiments in humanized mice show that the scrapie proteins have zoonotic potential (137). There is no epidemiological evidence that any of the other naturally occurring animal prion diseases—CWD, transmissible mink encephalopathy (TME), feline spongiform encephalopathy (FSE), and exotic ungulate encephalopathy (EUE)—are transmissible to humans. Recent experiments using transgenic mice expressing human or ovine prion protein show that CWD prion protein can transmit to mice expressing ovine but not human prion protein (138–141).

Inactivation of Prions

Prions are characterized by extreme resistance to conventional inactivation procedures, including irradiation, boiling, dry heat, and chemicals (formalin, β-propiolactone, and alcohols). However, they are inactivated by 1 N NaOH, 4.0 M guanidinium hydrochloride or isocyanate, sodium hypochlorite (≥2% free chlorine), and steam autoclaving at 132°C for 4.5 h (142–145). It is recommended that dry waste be autoclaved at 132°C for 4.5 h or incinerated. Large volumes of infectious liquid waste containing high titers of prions can be completely sterilized by treatment with 1 N NaOH (final concentration) followed by autoclaving at 132°C for 4.5 h. Disposable plasticware, which can be discarded as a dry waste, is highly recommended. Biological safety cabinet work surfaces must be decontaminated with 1 N NaOH, followed by 1 N HCl, and rinsed with water. The paraformaldehyde or vaporous hydrogen peroxide decontamination procedures routinely used to decontaminate biosafety cabinet filters do not diminish prion titers; the HEPA filters should be autoclaved and incinerated.

Prion infectivity is diminished by prolonged digestion with proteases and other treatments such as boiling in sodium dodecyl sulfate (SDS). Sterilization of rodent brain extracts with high titers of prions requires autoclaving at 132°C for 4.5 h, denaturing organic solvents such as phenol (1:1), chaotropic agents such as guanidine isocyanate or hydrochloride (>4 M), or alkali such as NaOH for 24 h (142–145). Although there is no evidence to suggest that aerosol transmission occurs in the natural disease, it is prudent to avoid the generation of aerosols or droplets during the manipulation of tissues or fluids and during the necropsy of experimental animals. It is further strongly recommended that gloves be worn for activities that provide the opportunity for skin contact with infectious tissues and fluids. Formaldehyde-fixed and paraffin-embedded tissues, especially of the brain, remain infectious.

Recent fears of the potential spread of human prion disease, particularly vCJD, through contaminated surgical instruments, as well as other medical and dental equipment, prompted investigators to research novel, less corrosive methods for prion disinfection. Some of these consisted of modifications of conventional cleaning reagents, which were reported to have utility in diminishing prion titers (146, 147). A recent major breakthrough, however, was reported in an investigation of the inactivation of prions by SDS in weak acid (148). Using SDS combined with autoclaving for 15 min completely inactivated sCJD prions bound to stainless steel wires, providing the basis for a noncorrosive system suitable for inactivating prions on surgical instruments and other medical and dental equipment.

CONCLUSION

In assessing risk for molecular agents in biosafety, it is important to understand the basic science behind the technology in order to help identify the hazards that will be encountered. For each of the topics in this chapter, a general description of the agent or new technology was provided and current papers were cited for more in-depth, follow-up reading. Some of the known safety issues were discussed to assist safety professionals in asking the necessary questions for determining containment levels and protective equipment. Many of the molecular agents described in this chapter are headed for or are already being used in clinical trials. In consideration of the administrative changes to the NIH RAC, which no longer require each gene therapy protocol be evaluated, members of local Institutional Biosafety Committees and Institutional Review Boards have an increased burden and responsibility to be vigilant in their reviews and seek outside expertise when needed. Recombinant DNA has come a long way in the 40-plus years since the Asilomar Conference was convened in 1975 to address concerns over the first genetic engineering experiments. Emerging fields like nanotechnology show us that the field of biosafety will be constantly evolving to keep pace with new molecular inventions.

Excerpts from the chapter on prions written by Henry Baron and Stanley B. Prusiner from the 4th edition of this book were included in this chapter and updated for the current edition.

References

1. **National Institutes of Health.** 2016. *NIH Guidelines for Research Involving Recombinant or Synthetic Nucleic Acid Molecules.* National Institutes of Health, Bethesda, MD. http://osp.od.nih.gov/sites/default/files/NIH_Guidelines.html#_Toc446948312.
2. **Watson JD, Gilman M, Witkowski J, Zoller M.** 1998. *Recombinant DNA*, 2nd ed. Scientific American Books, New York, NY.
3. **Avery OT, Macleod CM, McCarty M.** 1944. Studies on the chemical nature of the substance inducing transformation of pneumococcal types: induction of transformation by a desoxyribonucleic acid fraction isolated from Pneumococcus type III. *J Exp Med* **79:**137–158.
4. **Graham FL, van der Eb AJ.** 1973. A new technique for the assay of infectivity of human adenovirus 5 DNA. *Virology* **52:**456–467.
5. **Kawai S, Nishizawa M.** 1984. New procedure for DNA transfection with polycation and dimethyl sulfoxide. *Mol Cell Biol* **4:**1172–1174.
6. **McCutchan JH, Pagano JS.** 1968. Enchancement of the infectivity of simian virus 40 deoxyribonucleic acid with diethylaminoethyl-dextran. *J Natl Cancer Inst* **41:**351–357.
7. **Laakso MM, Sutton RE.** 2006. Replicative fidelity of lentiviral vectors produced by transient transfection. *Virology* **348:**-406–417.
8. **Lee SH, Oshige M, Durant ST, Rasila KK, Williamson EA, Ramsey H, Kwan L, Nickoloff JA, Hromas R.** 2005. The SET domain protein Metnase mediates foreign DNA integration and links integration to nonhomologous end-joining repair. *Proc Natl Acad Sci USA* **102:**18075–18080.
9. **Czernilofsky AP, Levinson AD, Varmus HE, Bishop JM, Tischer E, Goodman HM.** 1980. Nucleotide sequence of an avian

sarcoma virus oncogene (*src*) and proposed amino acid sequence for gene product. *Nature* **287:**198–203.
10. **Parker RC, Varmus HE, Bishop JM.** 1981. Cellular homologue (c-src) of the transforming gene of Rous sarcoma virus: isolation, mapping, and transcriptional analysis of c-src and flanking regions. *Proc Natl Acad Sci USA* **78:**5842–5846.
11. **Löwer J.** 1990. Risk of tumor induction in vivo by residual cellular DNA: quantitative considerations. *J Med Virol* **31:**50–53.
12. **Petricciani JC, Regan PJ.** 1987. Risk of neoplastic transformation from cellular DNA: calculations using the oncogene model. *Dev Biol Stand* **68:**43–49.
13. **Burns PA, Jack A, Neilson F, Haddow S, Balmain A.** 1991. Transformation of mouse skin endothelial cells in vivo by direct application of plasmid DNA encoding the human T24 H-ras oncogene. *Oncogene* **6:**1973–1978.
14. **Sheng L, Cai F, Zhu Y, Pal A, Athanasiou M, Orrison B, Blair DG, Hughes SH, Coffin JM, Lewis AM, Peden K.** 2008. Oncogenicity of DNA *in vivo*: tumor induction with expression plasmids for activated H-*ras* and c-*myc*. *Biologicals* **36:**184–197.
15. **Corallini A, Altavilla G, Carra L, Grossi MP, Federspil G, Caputo A, Negrini M, Barbanti-Brodano G.** 1982. Oncogenity of BK virus for immunosuppressed hamsters. *Arch Virol* **73:**243–253.
16. **Corallini A, Pagnani M, Caputo A, Negrini M, Altavilla G, Catozzi L, Barbanti-Brodano G.** 1988. Cooperation in oncogenesis between BK virus early region gene and the activated human c-Harvey *ras* oncogene. *J Gen Virol* **69:**2671–2679.
17. **Corallini A, Pagnani M, Viadana P, Camellin P, Caputo A, Reschiglian P, Rossi S, Altavilla G, Selvatici R, Barbanti-Brodano G.** 1987. Induction of malignant subcutaneous sarcomas in hamsters by a recombinant DNA containing BK virus early region and the activated human c-Harvey-*ras* oncogene. *Cancer Res* **47:**6671–6677.
18. **Fung YK, Crittenden LB, Fadly AM, Kung HJ.** 1983. Tumor induction by direct injection of cloned v-src DNA into chickens. *Proc Natl Acad Sci USA* **80:**353–357.
19. **Halpern MS, Ewert DL, England JM.** 1990. Wing web or intravenous inoculation of chickens with v-*src* DNA induces visceral sarcomas. *Virology* **175:**328–331.
20. **Krause PR, Lewis AM Jr.** 1998. Safety of viral DNA in biological products. *Biologicals* **26:**317–320.
21. **Letvin NL, Lord CI, King NW, Wyand MS, Myrick KV, Haseltine WA.** 1991. Risks of handling HIV. *Nature* **349:**573.
22. **Portis JL, McAtee FJ, Kayman SC.** 1992. Infectivity of retroviral DNA in vivo. *J Acquir Immune Defic Syndr* **5:**1272–1273.
23. **Israel MA, Chan HW, Hourihan SL, Rowe WP, Martin MA.** 1979. Biological activity of polyoma viral DNA in mice and hamsters. *J Virol* **29:**990–996.
24. **Yang H.** 2013. Establishing acceptable limits of residual DNA. *PDA J Pharm Sci Technol* **67:**155–163.
25. **Meister G, Tuschl T.** 2004. Mechanisms of gene silencing by double-stranded RNA. *Nature* **431:**343–349.
26. **Pélisson A, Sarot E, Payen-Groschêne G, Bucheton A.** 2007. A novel repeat-associated small interfering RNA-mediated silencing pathway downregulates complementary sense gypsy transcripts in somatic cells of the *Drosophila* ovary. *J Virol* **81:**1951–1960.
27. **Kanasty R, Dorkin JR, Vegas A, Anderson D.** 2013. Delivery materials for siRNA therapeutics. *Nat Mater* **12:**967–977.
28. **Wilczynska A, Bushell M.** 2015. The complexity of miRNA-mediated repression. *Cell Death Differ* **22:**22–33.
29. **Chen LL.** 2016. The biogenesis and emerging roles of circular RNAs. *Nat Rev Mol Cell Biol* **17:**205–211.
30. **Jeck WR, Sharpless NE.** 2014. Detecting and characterizing circular RNAs. *Nat Biotechnol* **32:**453–461.
31. **Jackson AL, Linsley PS.** 2010. Recognizing and avoiding siRNA off-target effects for target identification and therapeutic application. *Nat Rev Drug Discov* **9:**57–67.
32. **Reardon S.** 2016. First CRISPR clinical trial gets green light from US panel. *Nature News*, June 22.
33. **Carlson DF, Fahrenkrug SC, Hackett PB.** 2012. Targeting DNA with fingers and TALENs. *Mol Ther Nucleic Acids* **1:**e3.
34. **Bogdanove AJ, Voytas DF.** 2011. TAL effectors: customizable proteins for DNA targeting. *Science* **333:**1843–1846.
35. **Li T, Yang B.** 2013. TAL effector nuclease (TALEN) engineering. *Methods Mol Biol* **978:**63–72.
36. **Boch J, Scholze H, Schornack S, Landgraf A, Hahn S, Kay S, Lahaye T, Nickstadt A, Bonas U.** 2009. Breaking the code of DNA binding specificity of TAL-type III effectors. *Science* **326:**1509–1512.
37. **Arnould S, Delenda C, Grizot S, Desseaux C, Pâques F, Silva GH, Smith J.** 2011. The I-CreI meganuclease and its engineered derivatives: applications from cell modification to gene therapy. *Protein Eng Des Sel* **24:**27–31.
38. **Daboussi F, Zaslavskiy M, Poirot L, Loperfido M, Gouble A, Guyot V, Leduc S, Galetto R, Grizot S, Oficjalska D, Perez C, Delacôte F, Dupuy A, Chion-Sotinel I, Le Clerre D, Lebuhotel C, Danos O, Lemaire F, Oussedik K, Cédrone F, Epinat JC, Smith J, Yáñez-Muñoz RJ, Dickson G, Popplewell L, Koo T, VandenDriessche T, Chuah MK, Duclert A, Duchateau P, Pâques F.** 2012. Chromosomal context and epigenetic mechanisms control the efficacy of genome editing by rare-cutting designer endonucleases. *Nucleic Acids Res* **40:**6367–6379.
39. **Molina R, Montoya G, Prieto J.** 2011. *Meganucleases and Their Biomedical Applications*. Wiley Online Library.
40. **Ashworth J, Havranek JJ, Duarte CM, Sussman D, Monnat RJ Jr, Stoddard BL, Baker D.** 2006. Computational redesign of endonuclease DNA binding and cleavage specificity. *Nature* **441:**656–659.
41. **Zaslavskiy M, Bertonati C, Duchateau P, Duclert A, Silva GH.** 2014. Efficient design of meganucleases using a machine learning approach. *BMC Bioinformatics* **15:**191.
42. **Jinek M, Chylinski K, Fonfara I, Hauer M, Doudna JA, Charpentier E.** 2012. A programmable dual-RNA-guided DNA endonuclease in adaptive bacterial immunity. *Science* **337:**816–821.
43. **Reis A, Hornblower B.** 2014. CRISPR/Cas9 and targeted genome editing: a new era in molecular biology. *NEB Expressions* **1:**3–6. https://www.neb.com/tools-and-resources/feature-articles/crispr-cas9-and-targeted-genome-editing-a-new-era-in-molecular-biology?device=pdf.
44. **Maeder ML, Thibodeau-Beganny S, Osiak A, Wright DA, Anthony RM, Eichtinger M, Jiang T, Foley JE, Winfrey RJ, Townsend JA, Unger-Wallace E, Sander JD, Müller-Lerch F, Fu F, Pearlberg J, Göbel C, Dassie JP, Pruett-Miller SM, Porteus MH, Sgroi DC, Iafrate AJ, Dobbs D, McCray PB Jr, Cathomen T, Voytas DF, Joung JK.** 2008. Rapid "open-source" engineering of customized zinc-finger nucleases for highly efficient gene modification. *Mol Cell* **31:**294–301.
45. **Sander JD, Dahlborg EJ, Goodwin MJ, Cade L, Zhang F, Cifuentes D, Curtin SJ, Blackburn JS, Thibodeau-Beganny S, Qi Y, Pierick CJ, Hoffman E, Maeder ML, Khayter C, Reyon D, Dobbs D, Langenau DM, Stupar RM, Giraldez AJ, Voytas DF, Peterson RT, Yeh JR, Joung JK.** 2011. Selection-free zinc-finger-nuclease engineering by context-dependent assembly (CoDA). *Nat Methods* **8:**67–69.
46. **Sander JD, Reyon D, Maeder ML, Foley JE, Thibodeau-Beganny S, Li X, Regan MR, Dahlborg EJ, Goodwin MJ, Fu F, Voytas DF, Joung JK, Dobbs D.** 2010. Predicting success of oligomerized pool engineering (OPEN) for zinc finger target site sequences. *BMC Bioinformatics* **11:**543.
47. **Corrigan-Curay J, O'Reilly M, Kohn DB, Cannon PM, Bao G, Bushman FD, Carroll D, Cathomen T, Joung JK, Roth D, Sadelain M, Scharenberg AM, von Kalle C, Zhang F, Jambou R, Rosenthal E, Hassani M, Singh A, Porteus MH.** 2015. Genome editing technologies: defining a path to clinic. *Mol Ther* **23:**796–806.

48. **Sharpe M, Mount N.** 2015. Genetically modified T cells in cancer therapy: opportunities and challenges. *Dis Model Mech* **8:**337–350.
49. **Ahmad ZA, Yeap SK, Ali AM, Ho WY, Alitheen NB, Hamid M.** 2012. scFv antibody: principles and clinical application. *Clin Dev Immunol* **2012:**980250.
50. **Maude SL, Teachey DT, Porter DL, Grupp SA.** 2015. CD19-targeted chimeric antigen receptor T-cell therapy for acute lymphoblastic leukemia. *Blood* **125:**4017–4023.
51. **Debets R, Donnadieu E, Chouaib S, Coukos G.** 2016. TCR-engineered T cells to treat tumors: seeing but not touching? *Semin Immunol* **28:**10–21.
52. **Sadelain M, Brentjens R, Rivière I.** 2013. The basic principles of chimeric antigen receptor design. *Cancer Discov* **3:**388–398.
53. **Lee DW, Gardner R, Porter DL, Louis CU, Ahmed N, Jensen M, Grupp SA, Mackall CL.** 2014. Current concepts in the diagnosis and management of cytokine release syndrome. *Blood* **124:**188–195.
54. **Della Torre E, Bennett LH, Watson RE.** 2005. Extension of the BLOCH *T*(3/2) law to magnetic nanostructures: Bose-Einstein condensation. *Phys Rev Lett* **94:**147210.
55. **Gieseler J, Quidant R, Dellago C, Novotny L.** 2014. Dynamic relaxation of a levitated nanoparticle from a non-equilibrium steady state. *Nat Nanotechnol* **9:**358–364.
56. **Pattison JR, Patou G.** 1996. Parvoviruses. *In* Baron S (ed), *Medical Microbiology*. University of Texas Medical Branch at Galveston, Galveston, TX.
57. **Philippe N, Legendre M, Doutre G, Couté Y, Poirot O, Lescot M, Arslan D, Seltzer V, Bertaux L, Bruley C, Garin J, Claverie JM, Abergel C.** 2013. Pandoraviruses: amoeba viruses with genomes up to 2.5 Mb reaching that of parasitic eukaryotes. *Science* **341:**281–286.
58. **Hernandez-Garcia A, Kraft DJ, Janssen AF, Bomans PH, Sommerdijk NA, Thies-Weesie DM, Favretto ME, Brock R, de Wolf FA, Werten MW, van der Schoot P, Stuart MC, de Vries R.** 2014. Design and self-assembly of simple coat proteins for artificial viruses. *Nat Nanotechnol* **9:**698–702.
59. **Lu Y, Chan W, Ko BY, VanLang CC, Swartz JR.** 2015. Assessing sequence plasticity of a virus-like nanoparticle by evolution toward a versatile scaffold for vaccines and drug delivery. *Proc Natl Acad Sci USA* **112:**12360–12365.
60. **Lara HH, Garza-Treviño EN, Ixtepan-Turrent L, Singh DK.** 2011. Silver nanoparticles are broad-spectrum bactericidal and virucidal compounds. *J Nanobiotechnology* **9:**30.
61. **Silvestry-Rodriguez N, Sicairos-Ruelas EE, Gerba CP, Bright KR.** 2007. Silver as a disinfectant. *Rev Environ Contam Toxicol* **191:**23–45.
62. **Honek JF.** 2013. Bionanotechnology and bionanomaterials: John Honek explains the good things that can come in very small packages. *BMC Biochem* **14:**29.
63. **Sapsford KE, Algar WR, Berti L, Gemmill KB, Casey BJ, Oh E, Stewart MH, Medintz IL.** 2013. Functionalizing nanoparticles with biological molecules: developing chemistries that facilitate nanotechnology. *Chem Rev* **113:**1904–2074.
64. **Bajpai SK, Mohan YM, Bajpai M, Tankhiwale R, Thomas V.** 2007. Synthesis of polymer stabilized silver and gold nanostructures. *J Nanosci Nanotechnol* **7:**2994–3010.
65. **Barnickel P, Wokun A, Sager M, Eicke E-F.** 1992. Size-tailoring of silver colloids by reduction in W/O microemulsions. *J Colloid Interface* **148:**80–90.
66. **Chen S, Carroll DL.** 2002. Synthesis and characterization of truncated triangular silver nanoplates. *Nano Lett* **2:**1003–1007.
67. **Cushing BL, Kolesnichenko VL, O'Connor CJ.** 2004. Recent advances in the liquid-phase syntheses of inorganic nanoparticles. *Chem Rev* **104:**3893–3946.
68. **Luo C, Zhang Y, Zeng X, Zeng Y, Wang Y.** 2005. The role of poly(ethylene glycol) in the formation of silver nanoparticles. *J Colloid Interface Sci* **288:**444–448.
69. **Radziuk D, Skirtach A, Sukhorukov G, Mohwald H.** 2007. Stabilization of silver nanoparticles by polyelectrolytes and poly(ethylene glycol). *Macromol Rapid Commun* **28:**848–855.
70. **Singh P, Kim YJ, Zhang D, Yang DC.** 2016. Biological synthesis of nanoparticles from plants and microorganisms. *Trends Biotechnol* **34:**588–599.
71. **De M, Ghosh PS, Rotello VM.** 2008. Applications of nanoparticles in biology. *Adv Mater* **20:**4225–4241.
72. **National Research Council.** 2012. *A Research Strategy for Environmental, Health, and Safety Aspects of Engineered Nanomaterials*. Committee to Develop a Research Strategy for Environmental, Health, and Safety Aspects of Engineered Nanomaterials, Washington, DC.
73. **Rengasamy S, Eimer BC.** 2011. Total inward leakage of nanoparticles through filtering facepiece respirators. *Ann Occup Hyg* **55:**253–263.
74. **Prusiner SB, Woerman AL, Mordes DA, Watts JC, Rampersaud R, Berry DB, Patel S, Oehler A, Lowe JK, Kravitz SN, Geschwind DH, Glidden DV, Halliday GM, Middleton LT, Gentleman SM, Grinberg LT, Giles K.** 2015. Evidence for α-synuclein prions causing multiple system atrophy in humans with parkinsonism. *Proc Natl Acad Sci USA* **112:**E5308–E5317.
75. **Chakrabortee S, Kayatekin C, Newby GA, Mendillo ML, Lancaster A, Lindquist S.** 2016. Luminidependens (LD) is an Arabidopsis protein with prion behavior. *Proc Natl Acad Sci USA* **113:**6065–6070.
76. **Prusiner SB.** 1998. Prions. *Proc Natl Acad Sci USA* **95:**13363–13383.
77. **Prusiner SB, Scott MR, DeArmond SJ.** 2004. Transmission and replication of prions, p 187–242. *In* Prusiner SB (ed), *Prion Biology and Diseases*. Cold Spring Harbor Laboratory Press, Cold Spring Harbor, NY.
78. **Weissmann C, Enari M, Klöhn PC, Rossi D, Flechsig E.** 2002. Transmission of prions. *Proc Natl Acad Sci USA* **99**(Suppl 4): 16378–16383.
79. **Huang WJ, Chen WW, Zhang X.** 2015. Prions mediated neurodegenerative disorders. *Eur Rev Med Pharmacol Sci* **19:**4028–4034.
80. **Bishop MT, Will RG, Manson JC.** 2010. Defining sporadic Creutzfeldt-Jakob disease strains and their transmission properties. *Proc Natl Acad Sci USA* **107:**12005–12010.
81. **Pattison IH.** 1965. Experiments with scrapie with special reference to the nature of the agent and the pathology of the disease, p 249–257. *In* Gajdusek DC, Gibbs CJ Jr, Alpers MP (ed), *Slow, Latent and Temperate Virus Infections NINDB Monograph 2*. US Government Printing Office, Washington, DC.
82. **Scott M, Foster D, Mirenda C, Serban D, Coufal F, Wälchli M, Torchia M, Groth D, Carlson G, DeArmond SJ, Westaway D, Prusiner SB.** 1989. Transgenic mice expressing hamster prion protein produce species-specific scrapie infectivity and amyloid plaques. *Cell* **59:**847–857.
83. **Scott MR, Will R, Ironside J, Nguyen HO, Tremblay P, DeArmond SJ, Prusiner SB.** 1999. Compelling transgenetic evidence for transmission of bovine spongiform encephalopathy prions to humans. *Proc Natl Acad Sci USA* **96:**15137–15142.
84. **Telling GC, Scott M, Mastrianni J, Gabizon R, Torchia M, Cohen FE, DeArmond SJ, Prusiner SB.** 1995. Prion propagation in mice expressing human and chimeric PrP transgenes implicates the interaction of cellular PrP with another protein. *Cell* **83:**79–90.
85. **Asante EA, Linehan JM, Desbruslais M, Joiner S, Gowland I, Wood AL, Welch J, Hill AF, Lloyd SE, Wadsworth JD, Collinge J.** 2002. BSE prions propagate as either variant CJD-like or sporadic CJD-like prion strains in transgenic mice expressing human prion protein. *EMBO J* **21:**6358–6366.
86. **Wilesmith JW, Ryan JB, Atkinson MJ.** 1991. Bovine spongiform encephalopathy: epidemiological studies on the origin. *Vet Rec* **128:**199–203.

87. **Pattison J.** 1998. The emergence of bovine spongiform encephalopathy and related diseases. *Emerg Infect Dis* **4:**390–394.
88. **Wyatt JM, Pearson GR, Smerdon TN, Gruffydd-Jones TJ, Wells GA, Wilesmith JW.** 1991. Naturally occurring scrapie-like spongiform encephalopathy in five domestic cats. *Vet Rec* **129:**233–236.
89. **Hill AF, Desbruslais M, Joiner S, Sidle KC, Gowland I, Collinge J, Doey LJ, Lantos P.** 1997. The same prion strain causes vCJD and BSE. *Nature* **389:**448–450, 526.
90. **Will RG, Ironside JW, Zeidler M, Cousens SN, Estibeiro K, Alperovitch A, Poser S, Pocchiari M, Hofman A, Smith PG.** 1996. A new variant of Creutzfeldt-Jakob disease in the UK. *Lancet* **347:**921–925.
91. **Jaunmuktane Z, Mead S, Ellis M, Wadsworth JD, Nicoll AJ, Kenny J, Launchbury F, Linehan J, Richard-Loendt A, Walker AS, Rudge P, Collinge J, Brandner S.** 2015. Evidence for human transmission of amyloid-β pathology and cerebral amyloid angiopathy. *Nature* **525:**247–250.
92. **Bellinger-Kawahara CG, Kempner E, Groth D, Gabizon R, Prusiner SB.** 1988. Scrapie prion liposomes and rods exhibit target sizes of 55,000 Da. *Virology* **164:**537–541.
93. **Silveira JR, Raymond GJ, Hughson AG, Race RE, Sim VL, Hayes SF, Caughey B.** 2005. The most infectious prion protein particles. *Nature* **437:**257–261.
94. **Gabizon R, Prusiner SB.** 1990. Prion liposomes. *Biochem J* **266:**1–14.
95. **Safar J, Ceroni M, Piccardo P, Liberski PP, Miyazaki M, Gajdusek DC, Gibbs CJ Jr.** 1990. Subcellular distribution and physicochemical properties of scrapie-associated precursor protein and relationship with scrapie agent. *Neurology* **40:**503–508.
96. **Bellinger-Kawahara C, Diener TO, McKinley MP, Groth DF, Smith DR, Prusiner SB.** 1987. Purified scrapie prions resist inactivation by procedures that hydrolyze, modify, or shear nucleic acids. *Virology* **160:**271–274.
97. **Alpers M.** 1987. Epidemiology and clinical aspects of kuru, p 451–465. *In* Prusiner SB, McKinley MP (ed), *Prions—Novel Infectious Pathogens Causing Scrapie and Creutzfeldt-Jakob Disease.* Academic Press, Orlando, FL.
98. **Bellinger-Kawahara C, Cleaver JE, Diener TO, Prusiner SB.** 1987. Purified scrapie prions resist inactivation by UV irradiation. *J Virol* **61:**159–166.
99. **McKinley MP, Masiarz FR, Isaacs ST, Hearst JE, Prusiner SB.** 1983. Resistance of the scrapie agent to inactivation by psoralens. *Photochem Photobiol* **37:**539–545.
100. **Brown P, Wolff A, Gajdusek DC.** 1990. A simple and effective method for inactivating virus infectivity in formalin-fixed tissue samples from patients with Creutzfeldt-Jakob disease. *Neurology* **40:**887–890.
101. **Prusiner SB.** 1982. Novel proteinaceous infectious particles cause scrapie. *Science* **216:**136–144.
102. **U.S. Department of Health and Human Services, Public Health Service, Centers for Disease Control and Prevention, National Institutes of Health.** 2009. *Biosafety in Microbiological and Biomedical Laboratories,* 5th ed. HHS Publication no. (CDC) 21-112. http://www.cdc.gov/biosafety/publications/bmbl5/BMBL.pdf.
103. **Will RG.** 1996. Incidence of Creutzfeldt-Jakob disease in the European Community, p 364–374. *In* Gibbs CJ Jr (ed), *Bovine Spongiform Encephalopathy: the BSE Dilemma.* Springer-Verlag, New York, NY.
104. **Bruce ME, McConnell I, Will RG, Ironside JW.** 2001. Detection of variant Creutzfeldt-Jakob disease infectivity in extraneural tissues. *Lancet* **358:**208–209.
105. **Hill AF, Zeidler M, Ironside J, Collinge J.** 1997. Diagnosis of new variant Creutzfeldt-Jakob disease by tonsil biopsy. *Lancet* **349:**99–100.
106. **Hilton DA, Ghani AC, Conyers L, Edwards P, McCardle L, Ritchie D, Penney M, Hegazy D, Ironside JW.** 2004. Prevalence of lymphoreticular prion protein accumulation in UK tissue samples. *J Pathol* **203:**733–739.
107. **Wadsworth JD, Joiner S, Hill AF, Campbell TA, Desbruslais M, Luthert PJ, Collinge J.** 2001. Tissue distribution of protease resistant prion protein in variant Creutzfeldt-Jakob disease using a highly sensitive immunoblotting assay. *Lancet* **358:** 171–180.
108. **Bosque PJ, Ryou C, Telling G, Peretz D, Legname G, DeArmond SJ, Prusiner SB.** 2002. Prions in skeletal muscle. *Proc Natl Acad Sci USA* **99:**3812–3817.
109. **Glatzel M, Abela E, Maissen M, Aguzzi A.** 2003. Extraneural pathologic prion protein in sporadic Creutzfeldt-Jakob disease. *N Engl J Med* **349:**1812–1820.
110. **Ridley RM, Baker HF.** 1993. Occupational risk of Creutzfeldt-Jakob disease. *Lancet* **341:**641–642.
111. **Centers for Disease Control and Prevention.** 1985. Fatal degenerative neurologic disease in patients who received pituitary-derived human growth hormone. *MMWR Morb Mortal Wkly Rep* **34:**359–360, 365–356.
112. **Centers for Disease Control and Prevention.** 1997. Creutzfeldt-Jakob disease associated with cadaveric dura mater grafts—Japan, January 1979-May 1996. *MMWR Morb Mortal Wkly Rep* **46:**1066–1069.
113. **Public Health Service Interagency Coordinating Committee.** 1997. Report on human growth hormone and Creutzfeldt-Jakob disease. U.S. Department of Health and Human Services, Washington, DC.
114. **Dietz K, Raddatz G, Wallis J, Müller N, Zerr I, Duerr HP, Lefèvre H, Seifried E, Löwer J.** 2007. Blood transfusion and spread of variant Creutzfeldt-Jakob disease. *Emerg Infect Dis* **13:**89–96.
115. **Gajdusek DC.** 1977. Unconventional viruses and the origin and disappearance of kuru. *Science* **197:**943–960.
116. **Brown P, Preece MA, Will RG.** 1992. "Friendly fire" in medicine: hormones, homografts, and Creutzfeldt-Jakob disease. *Lancet* **340:**24–27.
117. **World Health Organization.** 1999. WHO infection control guidelines for transmissible spongiform encephalopathies. Report of a WHO consultation, Geneva, Switzerland, March 23–26, 1999. Geneva, Switzerland.
118. **Anderson RM, Donnelly CA, Ferguson NM, Woolhouse ME, Watt CJ, Udy HJ, MaWhinney S, Dunstan SP, Southwood TR, Wilesmith JW, Ryan JB, Hoinville LJ, Hillerton JE, Austin AR, Wells GA.** 1996. Transmission dynamics and epidemiology of BSE in British cattle. *Nature* **382:**779–788.
119. **Prusiner SB.** 1997. Prion diseases and the BSE crisis. *Science* **278:**245–251.
120. **Kirkwood JK, Wells GA, Wilesmith JW, Cunningham AA, Jackson SI.** 1990. Spongiform encephalopathy in an arabian oryx (Oryx leucoryx) and a greater kudu (Tragelaphus strepsiceros). *Vet Rec* **127:**418–420.
121. **Willoughby K, Kelly DF, Lyon DG, Wells GA.** 1992. Spongiform encephalopathy in a captive puma (Felis concolor). *Vet Rec* **131:**431–434.
122. **Baker HF, Ridley RM, Wells GA.** 1993. Experimental transmission of BSE and scrapie to the common marmoset. *Vet Rec* **132:**403–406.
123. **Barlow RM, Middleton DJ.** 1990. Dietary transmission of bovine spongiform encephalopathy to mice. *Vet Rec* **126:**111–112.
124. **Dawson M, Wells GA, Parker BN, Scott AC.** 1990. Primary parenteral transmission of bovine spongiform encephalopathy to the pig. *Vet Rec* **127:**338.
125. **Foster JD, Bruce M, McConnell I, Chree A, Fraser H.** 1996. Detection of BSE infectivity in brain and spleen of experimentally infected sheep. *Vet Rec* **138:**546–548.
126. **Fraser H, Bruce ME, Chree A, McConnell I, Wells GA.** 1992. Transmission of bovine spongiform encephalopathy and scrapie to mice. *J Gen Virol* **73:**1891–1897.

127. **Hunter N, Foster J, Chong A, McCutcheon S, Parnham D, Eaton S, MacKenzie C, Houston F.** 2002. Transmission of prion diseases by blood transfusion. *J Gen Virol* **83:**2897–2905.
128. **Lasmézas CI, Deslys JP, Demaimay R, Adjou KT, Lamoury F, Dormont D, Robain O, Ironside J, Hauw JJ.** 1996. BSE transmission to macaques. *Nature* **381:**743–744.
129. **Lasmézas CI, Fournier JG, Nouvel V, Boe H, Marcé D, Lamoury F, Kopp N, Hauw JJ, Ironside J, Bruce M, Dormont D, Deslys JP.** 2001. Adaptation of the bovine spongiform encephalopathy agent to primates and comparison with Creutzfeldt-Jakob disease: implications for human health. *Proc Natl Acad Sci USA* **98:**4142–4147.
130. **Bruce ME, Will RG, Ironside JW, McConnell I, Drummond D, Suttie A, McCardle L, Chree A, Hope J, Birkett C, Cousens S, Fraser H, Bostock CJ.** 1997. Transmissions to mice indicate that 'new variant' CJD is caused by the BSE agent. *Nature* **389:**498–501.
131. **Collinge J, Sidle KC, Meads J, Ironside J, Hill AF.** 1996. Molecular analysis of prion strain variation and the aetiology of 'new variant' CJD. *Nature* **383:**685–690.
132. **Zeidler M, Stewart GE, Barraclough CR, Bateman DE, Bates D, Burn DJ, Colchester AC, Durward W, Fletcher NA, Hawkins SA, Mackenzie JM, Will RG.** 1997. New variant Creutzfeldt-Jakob disease: neurological features and diagnostic tests. *Lancet* **350:**903–907.
133. **Centers for Disease Control and Prevention (CDC).** 2004. Bovine spongiform encephalopathy in a dairy cow—Washington state, 2003. *MMWR Morb Mortal Wkly Rep* **52:**1280–1285.
134. **U.S. Department of Agriculture, Animal and Plant Health Inspection Service, Veterinary Services.** 2012. Summary Report. California bovine spongiform encephalopathy case investigation. https://www.aphis.usda.gov/animal_health/animal_diseases/bse/downloads/BSE_Summary_Report.pdf.
135. **Dexter G, Tongue SC, Heasman L, Bellworthy SJ, Davis A, Moore SJ, Simmons MM, Sayers AR, Simmons HA, Matthews D.** 2009. The evaluation of exposure risks for natural transmission of scrapie within an infected flock. *BMC Vet Res* **5:**38.
136. **Chatelain J, Cathala F, Brown P, Raharison S, Court L, Gajdusek DC.** 1981. Epidemiologic comparisons between Creutzfeldt-Jakob disease and scrapie in France during the 12-year period 1968–1979. *J Neurol Sci* **51:**329–337.
137. **Cassard H, Torres JM, Lacroux C, Douet JY, Benestad SL, Lantier F, Lugan S, Lantier I, Costes P, Aron N, Reine F, Herzog L, Espinosa JC, Beringue V, Andréoletti O.** 2014. Evidence for zoonotic potential of ovine scrapie prions. *Nat Commun* **5:**5821.
138. **Barria MA, Balachandran A, Morita M, Kitamoto T, Barron R, Manson J, Knight R, Ironside JW, Head MW.** 2014. Molecular barriers to zoonotic transmission of prions. *Emerg Infect Dis* **20:**88–97.
139. **Béringue V, Herzog L, Jaumain E, Reine F, Sibille P, Le Dur A, Vilotte JL, Laude H.** 2012. Facilitated cross-species transmission of prions in extraneural tissue. *Science* **335:**472–475.
140. **Collinge J.** 2012. Cell biology. The risk of prion zoonoses. *Science* **335:**411–413.
141. **Sandberg MK, Al-Doujaily H, Sigurdson CJ, Glatzel M, O'Malley C, Powell C, Asante EA, Linehan JM, Brandner S, Wadsworth JD, Collinge J.** 2010. Chronic wasting disease prions are not transmissible to transgenic mice overexpressing human prion protein. *J Gen Virol* **91:**2651–2657.
142. **Prusiner SB, Groth D, Serban A, Stahl N, Gabizon R.** 1993. Attempts to restore scrapie prion infectivity after exposure to protein denaturants. *Proc Natl Acad Sci USA* **90:**2793–2797.
143. **Prusiner SB, McKinley MP, Bolton DC, Bowman KA, Groth DF, Cochran SP, Hennessey EM, Braunfeld MB, Baringer JR, Chatigny MA.** 1984. Prions: methods for assay, purification and characterization, p 294–345. *In* Maramorosch K, Koprowski H (ed), *Methods in Virology*. Academic Press, New York, NY.
144. **Taylor DM, Woodgate SL, Atkinson MJ.** 1995. Inactivation of the bovine spongiform encephalopathy agent by rendering procedures. *Vet Rec* **137:**605–610.
145. **Taylor DM, Woodgate SL, Fleetwood AJ, Cawthorne RJ.** 1997. Effect of rendering procedures on the scrapie agent. *Vet Rec* **141:**643–649.
146. **Fichet G, Comoy E, Duval C, Antloga K, Dehen C, Charbonnier A, McDonnell G, Brown P, Lasmézas CI, Deslys JP.** 2004. Novel methods for disinfection of prion-contaminated medical devices. *Lancet* **364:**521–526.
147. **Race RE, Raymond GJ.** 2004. Inactivation of transmissible spongiform encephalopathy (prion) agents by environ LpH. *J Virol* **78:**2164–2165.
148. **Peretz D, Supattapone S, Giles K, Vergara J, Freyman Y, Lessard P, Safar JG, Glidden DV, McCulloch C, Nguyen HO, Scott M, Dearmond SJ, Prusiner SB.** 2006. Inactivation of prions by acidic sodium dodecyl sulfate. *J Virol* **80:**322–331.

13 Biosafety for Microorganisms Transmitted by the Airborne Route

MICHAEL A. PENTELLA

For some pathogenic microorganisms, the airborne route is the predominant means of transmission to humans. These agents, which may be transmitted from humans, animals, and the environment, i.e., soil and water, include certain pathogenic viruses, bacteria, and fungi. Although certain species of fungi and mycobacteria share the airborne route of transmission, they are very different in substantive elements of transmission, including their natural habitats and reservoirs. It is only with a clear knowledge of these differences that the laboratorian can adequately perform a risk assessment and implement appropriate safety protocols to abate those hazards in the laboratory.

The term "laboratory" in its broadest application involves all types of situations, including the collection and evaluation of materials in the external environment. In this instance, the laboratorian may be exposed to winds or air currents containing fungal spores. Of particular concern are fungal spores that are known to naturally infect human hosts, especially those that produce systemic mycoses. In North American soils, these are primarily *Coccidioides immitis* in the desert Southwest and *Blastomyces dermatitidis* and *Histoplasma capsulatum*, both found in the Southeast and Midwest. The human host who becomes infected with these agents is not normally a source of infection to social contacts or hospital caregivers. This is reflected in the lack of any recommendation for isolation precautions for such patients by the Centers for Disease Control and Prevention (CDC) guidelines for isolation precautions in hospitals (1).

The absence of person-to-person transmission is due to unique properties of the fungi. These organisms are dimorphic. The form in which the organisms are transmissible is the hyphal form that occurs in nature and, under certain circumstances, in the laboratory. The hyphal forms of these fungi produce conidia that are readily transmissible by the airborne route. The *in vivo* tissue forms of these fungi are yeasts or spherules and are not readily transmissible to other humans, either by direct contact or by the airborne route. However, there are isolated case reports of percutaneous transmission in laboratorians or health care workers. Infection with these agents has occurred when infectious material from the patient was introduced via trauma, for example, at autopsy, into the subcutaneous tissues of the laboratorian. Moreover, there is one report in which a draining wound from a patient with coccidioidomycosis apparently propagated infectious arthroconidia on the surface of a cast where body fluids were deposited and not washed away. But these are the exception.

The primary source of infectious hazards due to these organisms in the health care environment is in the laboratory via the airborne route. It is in the laboratory manipulation of the organisms that the infectious form of the agent, the hyphal stage, is propagated. To isolate and accurately identify the agents of systemic mycoses, the laboratorian incubates the cultures at a temperature that is permissive for the production of the more hazardous, airborne form. This laboratory circumstance should be the focus of laboratory safety with the systemic dimorphic fungi.

The airborne transmission of the agent *Mycobacterium tuberculosis* is the primary focus where laboratory safety is concerned. With *M. tuberculosis*, the natural transmission to humans is almost exclusively from another human by infectious droplet nuclei, since the human is the only natural reservoir. The mechanism of transmission from human to human is analogous to the most common means of acquisition of the organism in the laboratory setting. In both cases, the primary means of transmission is via droplet nuclei. Efficient means of production of droplet nuclei in nature are sneezing, coughing, and vibration of the larynx, all of which introduce energy that subdivides fluids into tiny droplets. Any procedure that imparts energy into a microbial system produces aerosols. Procedures such as pipetting, the inoculation of biochemicals, flame sterilization of a wire loop, vortexing, pouring fluids, centrifugation, or even lifting the lid of a petri dish have the potential to aerosolize pathogens. These types of procedures are ubiquitous in the daily activities of the laboratory environment. These aerosols usually remain unnoticed, yet they are are pervasive in the lab environment, placing all in the vicinity at risk. Hence, simply being in the same room when the aerosol-generating procedure is performed may be sufficient to cause infection. Therefore, with tuberculosis it is useful to understand natural transmission as a paradigm for accidental laboratory acquisition.

The extent of laboratory-acquired tuberculosis occurring in laboratory workers is unknown because neither laboratory-acquired infections nor skin test conversions are reportable conditions. The lack of reporting deprives us of the opportunity to learn from those experiences and determine best practices. The frequency of infection for staff who manipulate *M. tuberculosis* in a laboratory is estimated to be 100 times greater than for the general population (2). When laboratorians who work with *M. tuberculosis* are compared to laboratorians who do not, the estimates are that tuberculosis among persons who work with *M. tuberculosis* is three to five times greater than among those who do not. It is estimated that between 8% and 30% of laboratorians may experience tuberculin conversions (3). In a review of laboratory-acquired infections, tuberculosis ranked fourth among pathogens commonly identified as responsible for laboratory-acquired infections. Only brucellosis, Q fever, and typhoid fever ranked higher (4). Kubica described 15 separate incidents in which 80 of 291 (27%) exposed lab staff developed a positive tuberculin skin test (TST) following specific incidents. Eight of the incidents, the majority, involved poor directional airflow in the laboratory (3). Another five cases were associated with failures of the biological safety cabinets (BSCs). An additional case was associated with an autoclave failure, and another case was due to equipment failure. Using Workers' Compensation Data from 1996–2000 in the State of Washington (5), it was determined that overall health care worker TST conversion was 2.3 per 10,000 full-time equivalent employees (FTEs) in nonhospital settings. TST reactivity claims of 3.7 per 10,000 FTEs were highest for physician offices. Medical labs ranked second with 2.6 per 10,000 FTEs. With the emergence of multidrug-resistant (MDR) and extensively drug-resistant (XDR) strains of *M. tuberculosis*, tuberculosis can no longer be considered curable as it once was. The risk of acquiring tuberculosis spans past the walls of the clinical microbiology lab to include other areas of the clinical lab, such as surgical pathology, the autopsy suite, and the cytology lab (6). Outside of the clinical lab, biosafety practices in the research lab environment are also of concern.

Other bacterial and fungal agents that are not typically hazardous by the airborne route may represent a risk in the laboratory. A number of circumstances dictate that transmission in the artificial laboratory setting may be enhanced over that in the natural setting. The first of these involves the high concentration of microorganisms that is often manipulated in the laboratory. Culture amplification of microorganisms increases their numbers by many orders of magnitude over the numbers commonly found in human clinical material. With amplified cultures, the inoculum that is necessary to produce an infection by the airborne route might be easily achieved by modest manipulations that produce minor aerosols. The second of these involves the efficiency of production of aerosols that may be available in the laboratory. Unlike clinical disease affiliated with tuberculosis that produces a cough as a mechanism for spread of the mycobacterium, not all infectious processes invoke such an efficient transmitting host response. Thus, it is only in the laboratory where the organism may be manipulated with such an efficient means of aerosol production that the agent is transmissible by an airborne route. Manipulations conducted within the laboratory with infectious organisms can result in airborne droplets that can rapidly dry into droplet nuclei of 1 to 5 μm in size. These airborne droplet nuclei can remain on air currents for extended periods of time and can subsequently reach the pulmonary alveoli, as can small-droplet particles (>5 μm) that settle more quickly within 3 to 6 ft of the source. In the third circumstance, the laboratorian may work with large volumes of the fluid cultures in which organisms are amplified and also may manipulate the fluids. Therefore, the net amount of aerosol produced may be much larger because of the large volume of fluid manipulated, even with an inefficient means of aerosol production. In all three circumstances, organisms effectively become more hazardous than their counterparts in clinical specimens. These agents are listed in Table 1. For the most part, the air does not pose a significant health risk from airborne bacterial and fungal species in a clinical microbiology biosafety level 2 (BSL2) environment. This was demonstrated by testing the air in a busy laboratory and finding 30 bacterial and 28 fungal isolates, none of which would pose a hazard to an immunocompetent host (7).

DEVELOPMENT OF A SAFETY POLICY

Several organizations, such as the Occupational Safety and Health Administration (OSHA), the CDC, College of American Pathologists, Clinical and Laboratory Standards Institute (CLSI), and Joint Committee on Accreditation of Healthcare Organizations (JCAHO), provide guidelines for biosafety practices. Each laboratory is charged with developing specific written policies that establish biosafety practices for the individual facility. Furthermore, staff must be trained to understand and follow the policies and practices. It is the responsibility of management to monitor and ensure compliance. The CDC has published competencies for working with biologic agents

TABLE 1.

Airborne pathogen containment for work with agents usually handled at BSL2 in diagnostic quantities[a]

Organism	Containment level[b]	
	Droplet/aerosols	High titer/large volume
Bacteria		
Bacillus anthracis	BSL3	BSL3
Bordetella pertussis	BSL3 E	BSL3
Burkholderia pseudomallei	BSL3	BSL3
Chlamydia pneumoniae	BSL3	BSL3
Chlamydia psittaci	BSL3	BSL3
Chlamydia trachomatis	BSL3	BSL3
Clostridium botulinum (toxin)	BSL3	BSL3
Legionella pneumophila	BSL3 P, E	BSL3 P, E
Neisseria gonorrhoeae	BSL3 P, E	BSL3 P, E
Neisseria meningitidis	BSL3 P, E	BSL3 P, E
Salmonella enterica serovar Typhi	BSL3 P	BSL3 P
Yersinia pestis	BSL3 P, E	BSL3 P, E
Fungi		
Cryptococcus neoformans (contaminated environmental samples)	BSL3 E	
Viruses		
Hepatitis B and C viruses	BSL3 P, E	BSL3 P, E
HIV and SIV[c]	BSL3 P, E	BSL3
Human herpesviruses	BSL3 P, E	BSL3 P, E
Lymphocytic choriomeningitis virus	BSL3 P, E	BSL3 P, E
Poxviruses (vaccinia, cowpox, monkeypox)		BSL3 E
Rabies virus	BSL3 P, E	BSL3 P, E
Prions (human)	BSL3 P, E	

[a]Adapted from agent summary statements in reference 12.
[b]P, personal practices and procedures; E, primary containment equipment, i.e., BSC.
[c]HIV and SIV, human and simian immunodeficiency viruses, respectively.

in BSL2–4 that serve to define the essential skills, knowledge, and abilities (8). The competencies are tiered to a worker's experience and position in the facility, whether entry-level, experienced, or management positions. These competencies include many critical task-level details. In 2015, the CDC published a companion document that is a comprehensive set of competencies for laboratory professionals (9). These competencies cover multiple topic areas including biosafety. The 2015 competencies are based on the 2011 guidelines, and the content has been revised and restructured.

To establish facility-specific policies and practices, each laboratory must perform a risk assessment that relies on factual information to define the health effects of exposure to workers from exposure to hazardous materials and situations (10). Data to conduct risk assessment for the biological laboratory may be limited due to underreporting or even a unique laboratory setting, resulting in a subjective approach of risk management that incorporates hazard probability and severity to provide recommended actions to ensure a safe environment. Integral to the risk management process is knowledge of the biological agent, facility, safety equipment, and work processes (11). Qualitative risk assessment is encompassed in the BSLs established in *Biosafety in Microbiological and Biomedical Laboratories* (12).

SELECTION OF APPROPRIATE BIOSAFETY PRACTICES

Determination of BSL practices to be implemented for manipulations with airborne pathogens encompasses several evaluation tools and resources to determine the appropriate level. Selection of the appropriate BSL practice includes evaluation of resources for the recommended BSL, such as agent summary statements by the CDC and National Institutes of Health (NIH) (12), Risk Group classifications (13), review of procedures and processes for potential aerosol and droplet generation and/or genetic modification (in particular if the procedure is atypical or novel), volume, expertise of staff, facilities, and recent literature.

The current laboratory safety-oriented biosafety classification of microorganisms reflects what is known regarding the tendency for an organism to be transmitted by an airborne route, if there is effective treatment available for infection, and whether a vaccine is available. The infectious dose and impact on public health are also considered. BSL1 organisms are not known to cause disease in healthy adults, and practices with BSL1 organisms require no containment of aerosols. BSL2 organisms are transmitted primarily by percutaneous exposure, mucous membrane exposure, or ingestion. BSL2 practices call for containment in a BSC when aerosol-generating procedures are carried out with culture-amplified organisms if there is any risk of aerosol transmission (see Table 1). BSL3 organisms are those with a pronounced tendency to be transmitted by aerosols. Affiliated practices for BSL3 organisms call for containment in a BSC whenever culture-amplified materials are manipulated or when efficient aerosol-generating activities are carried out on nonamplified body fluids thought likely to contain certain organisms that are efficient in producing infections at low concentrations. BSL3 also requires a number of engineering controls to protect those not directly working with the microbe in the event of a spill or accident. Organisms that require BSL4 practices may be transmitted by either route (aerosol or mucous membrane exposure) and must be contained more fastidiously because they have a greater, or unknown, tendency to pose a high risk of lethality. These organisms require absolute containment of all materials in more specialized containment cabinets afforded by BSL4 practices. The selection of appropriate BSL practices may not always be clear-cut. For example, a good portion of the work with microorganisms in clinical laboratories involves the identification of unknown agents isolated from patients with diseases of unknown etiology. Later in this chapter, guidance is provided on ways in which lab practices may be employed that involve knowledge of the agents and their hazards.

SELECTION OF PERSONAL PROTECTIVE EQUIPMENT

Selection of personal protective equipment (PPE) begins with a review of the appropriate regulations and literature for emerging technology. The CDC currently recommends laboratory staff use a powered air-purifying respirator when working with XDR *M. tuberculosis* strains (14). Specific OSHA regulations that apply to the laboratory environment are included in Bloodborne Pathogens 29 C.F.R. 1910.1030 and Occupational Exposure to Hazardous Chemicals 29 C.F.R. 1910.1450. Additionally, OSHA specifies general PPE requirements in Personal Protective Equipment 29 C.F.R. 1910.132 and specific PPE regulations in the Respiratory Protection standard 29 C.F.R. 1910.134, Eye and Face Protection standard 29 C.F.R. 1910.133, and Hand Protection standard 29 C.F.R. 1910.138. To ensure protection of the employee, a PPE program begins with the risk assessment and includes activities, facility, and equipment to identify the type of PPE necessary. After selection, the employer must provide the employee with the PPE and ensure that he or she is trained in appropriate use and maintenance of this equipment. A successful program includes cooperation of the employee to wear and maintain equipment and to report ineffective equipment.

DISCARD OF WASTE

Laboratories utilizing BSL1 and BSL2 practices should develop strategies to handle microbial cultures safely for disposal. The strategies may include on-site inactivation (e.g., steam sterilization, incineration, or alternative treatment technology) or packaging and shipping untreated wastes to an off-site facility for inactivation and disposal (15). A BSL3 or BSL4 laboratory must inactivate microbial cultures on site by an approved inactivation method, for example, autoclaving or incineration, before transport to and disposal in a sanitary landfill.

INCIDENT REPORTING: REVIEW OF PRACTICES

Required reporting of safety incidents is limited to specific situations involving laboratory activities conducted in areas such as those described in the *NIH Guidelines for Research Involving Recombinant DNA Molecules* (*NIH Guidelines*) (13) and Possession, Use, and Transfer of Select Agents and Toxins 42 C.F.R. Part 73. The lack of reporting requirements allows for incidents of laboratory-acquired infection to remain underreported and contributes to insufficient data in addressing any changes needed in practices or equipment. Recent reports of laboratory-acquired infections have elevated concern and discussion regarding incident reporting. An employer that embraces incident reporting beyond that of the minimal requirements encourages development of a work environment where all employees actively participate in safety. Even reports of a "near miss" can lead to improvements in work practices and equipment that could prevent occupational exposure. The ideal safety culture is composed of employees who do not fear retaliation; employees and employers who view reporting an incident as a positive response; employers who respond appropriately to address safety equipment, facility, or behavior changes; and programs that include analysis of near misses or employee suspicion that an action could result in an exposure or incident (8, 16).

SPECIFIC HAZARDS ASSOCIATED WITH *M. TUBERCULOSIS* NATURAL TRANSMISSION: DROPLET NUCLEI

Studies of airborne transmission of tuberculosis conducted during the first half of the last century (17) led to the framing of the concept of the "droplet nucleus."

TABLE 2.

Evaporation time and falling distance of droplets based on size[a]

Diameter of droplet (μm)	Evaporation time (s)	Distance (ft) fallen before evaporation
200	5.2	21.7
100	1.3	1.4
50	0.31	0.085
25	0.08	0.0053

[a]Adapted from reference 17.

Photographs of sneezes and coughs revealed the discharge of thousands of droplets ranging in size from a few micrometers to several hundred micrometers in diameter. Further studies with more sophisticated photographic techniques and physical modeling revealed that particles that are very small when they are discharged evaporate very readily (see Table 2), transforming within hundredths of a second to a dehydrated mass containing the previously dissolved solutes of the discharged solution and any particulates that were carried with the droplet. The fineness of the division of the discharged particle determines its ultimate fate. The larger particles drop within seconds to the floor, where they most often form an aggregate of dust that is not readily redispersed into the air. It can be seen from Table 2 that for droplets discharged from a height of 6 ft, those larger than some 140 μm would tend to fall to the ground before they evaporated and those smaller than that size would be more likely to evaporate before contacting the ground. The aerodynamic properties of the residues of these evaporated droplets are such that they are thence carried aloft for very long periods of time. These particles have been designated "droplet nuclei."

An understanding of the formation and life cycle of droplet nuclei has allowed us to understand the means to prevent transmission of tuberculosis. A tissue held over the mouth of an individual who is coughing collects all the tiny droplets as they are discharged and before they have time to evaporate and minimizes the production of droplet nuclei, because the tiniest droplets have no time to evaporate before they coalesce as a mass of fluid in the tissue. This mass will ultimately dry, but the coalesced material cannot be converted into droplet nuclei without the intervention of powerful physical forces. Thus, the tissue may be safely disposed of into the wastebasket without generating harm to others under normal circumstances. Therefore, a patient who is compliant in using tissues when coughing can reduce the load of microorganisms introduced into the air of the room.

The laboratorian who is exposed to patients with tuberculosis is at little excess risk if the patient contains the coughing with a simple mask or tissue. Prevention of the formation of droplet nuclei is key in preventing transmission. Once a droplet nucleus has been allowed to form, its small size means that it can penetrate the fibers of a tissue or a routine surgical mask, and these products do not represent adequate physical barriers to the aerosol transmission of organisms. The appropriate barrier is a well-fitted respirator device that does not allow leakage of air around its edges and blocks passage of microorganisms in the fibers or pores through which the air is inspired. Although a simple surgical mask applied to the noncompliant patient who must be transported outside the isolation room will prevent the dispersal of the organisms as droplet nuclei, such a mask will not provide adequate protection to the individual who must breathe air containing droplet nuclei. An N95 particulate respirator is recommended for health care workers caring for a patient with suspected or confirmed active tuberculosis (18).

Transmission of Mycobacteria in the Laboratory

The transmission of the agent *M. tuberculosis* is the primary focus of laboratory safety with regard to processing specimens for mycobacterial testing. In natural infections, *M. tuberculosis* bacteria are inhaled into the alveoli in the lower respiratory tract, where they are ingested by alveolar macrophages, a necessary step in the first phase of establishing a productive infection. To reach the deep lung, the bacteria are aerosolized in the form of droplet nuclei—particles of 1- to 5-μm diameter that can remain suspended for long periods of time (19). The settling of droplet nuclei is a function of size, time, and evaporation and is influenced by both host and environmental factors, such as relative humidity (17). Efficient means of production of droplet nuclei in humans are sneezing, coughing, and other vibrations of the larynx, which introduce energy that subdivides fluids into tiny droplets.

Risk Assessment

In the laboratory, the risk of acquiring infection from *M. tuberculosis* is via the aerosol route and by percutaneous inoculation. Inhalation of droplet nuclei is also the most common means of infection in the laboratory setting. Any procedure that imparts energy into a microbial system produces aerosols. Laboratory procedures such as pipetting, the inoculation of biochemical reagents into cultures, flame sterilization of a wire loop, vortex mixing, pouring off supernatants, centrifugation, or lifting the lid of a petri dish have the potential to aerosolize bacteria.

Precautions against percutaneous inoculation are the same as for other pathogens. These include wearing appropriate PPE, such as gloves and lab coat, and the elimination of the use of sharps whenever possible for prevention and control. Ideally, one would also limit the potential for broken glass; however, glass is commonly used in media bottles and slides, so there should be appropriate procedures and supplies in place for clearing up broken glass incidents. The focus of this chapter is addressing aerosol-acquired infections. A detailed risk assessment of the specific laboratory layout, specimen types processed, and activities should be taken so that appropriate controls and procedures can be implemented to prevent exposures. Due to the increase in the prevalence of MDR and XDR strains of *M. tuberculosis*, the potential for receiving these types of specimens in the laboratory should be evaluated and addressed. The laboratory director, safety officer, infection control, and any other involved parties should be included in the assessment. Recommendations on conducting the risk assessment and on laboratory biosafety have been published by the CDC (18), the National Institute for Occupational Safety and Health (NIOSH, tuberculosis), and OSHA (OSHA, tuberculosis) and provide excellent additional resources. The Association of Public Health Laboratories created a detailed tool to assess a laboratory that performs *M. tuberculosis* testing (20). Since it is designed as a quality assessment, it goes beyond the risk assessment categories to cover related topics of quality laboratory testing.

Laboratories that handle specimens that potentially contain *M. tuberculosis* should use BSL2 practices and facilities, with limited access and under negative pressure (12). The risk assessment will determine the necessary PPE. At a minimum, personnel should wear a protective garment such as a laboratory coat, eye protection, and disposable gloves. All specimens should be contained in sealed containers and manipulations should occur in a BSC, Class I or II, unless the specimen has been rendered inactive by treatment with bleach or other tuberculocidal agent (18). The use of an N95 respirator or equivalent is highly recommended for processing and manipulating specimens or *M. tuberculosis* cultures since no BSC is 100% effective (16). Growth of *M. tuberculosis* from clinical specimens and subsequent handling of these cultures imparts additional risks that must be addressed by enhanced processes, such as use of an N95 or similar respirator and work being conducted in a BSL3 facility. BSL3 practices and facilities are required for culture of XDR *M. tuberculosis*. Other situations that incur additional risk include certain research activities such as genetic manipulation of bacterial strains, large-scale production of mycobacteria for vaccine manufacture, and studies involving laboratory animals.

Laboratory biosafety must address all areas involving facility layout, engineering controls such as appropriate equipment specifications and maintenance, and work practice. Without addressing all aspects and providing appropriate training, there will be gaps in the system.

The risk assessment should include the process for documenting competency, detailed decontamination protocols, and spill response and waste handling procedures. Laboratorians should be included in a tuberculosis surveillance program. If wearing a respirator, they should be fit tested and enrolled in a respirator program.

Laboratory Practice

The types of specimens that may contain *M. tuberculosis* include a variety of lower respiratory specimens, including sputa and bronchial alveolar lavage, in addition to pleural fluid, cerebrospinal fluid, gastric aspirates, blood, and tissues. The main goal in processing any of these specimen types is to prevent the formation of aerosols. Because aerosols can be formed any time that energy is added to a liquid specimen by such mechanisms as pipetting, vortex mixing, inoculation of media, pouring off supernatant, placing a drop of liquid onto a flat surface, etc., these procedures must be undertaken with appropriate environmental and procedural practices to prevent exposure.

Particular attention should be paid to the manipulation of liquid cultures, which pose a greater risk than manipulation of colonies on solid media. Mycobacteria are extremely hydrophobic, and liquid culture medium often contains surfactants such as Tween 80 to allow for more homogeneous growth. Mycobacteria suspended in Tween 80 and/or subjected to physical forces, such as shaking or rotation during incubation to disperse the clumps into single organisms, represent the greatest risk (21). The droplets that are aerosolized by the manipulations of the liquid culture become droplet nuclei if they dry before they land on a horizontal surface. If the organisms are not clumped, they are more likely to have effective diameters less than 5 µm and thus are able to reach the lung alveoli (17). To prevent the formation of droplet nuclei, the formation of aerosols must be mitigated. A typical risk assessment based on task performance and mitigation of aerosol risk is presented in Table 3. Although not all mycobacteria present a risk of aerosol infection, it is recommended that practices equivalent to those for *M. tuberculosis* are utilized for working up all specimens being evaluated for acid-fast bacilli. Even with preliminary evidence to rule out *M. tuberculosis*, there can be mixed infections, where a fast-growing mycobacterial strain is initially cultured and *M. tuberculosis* is later discovered.

Some recommended practices for prevention of aerosol spread are as follows.

- The work area inside the BSC must have a tuberculocidal disinfectant-soaked pad to capture any drops or splatter that may result from manipulation of the specimen, pipettes, loops, tubes, slides, etc.
- Use aerosol-resistant pipette tips that contain a hydrophobic microporous filter bonded onto the walls of the pipette tip.
- Incineration of needles or loops is particularly prone to aerosol formation, so the conventional flame should be replaced with an electric incinerator or a disposable loop.
- Centrifugation should occur in aerosol-proof, sealed rotor cups that are only opened in the BSC. The O-rings on these containers should be inspected for cracks and regularly lubricated or replaced as needed.
- Following vortex mixing of liquids, allow the tubes to stand for 5 minutes before opening inside of the BSC.
- Make sure that all tubes and containers are completely sealed and wiped off or submerged in disinfectant before they are removed from the BSC.
- Molecular methods are becoming increasingly common. They employ nucleic acid extraction steps that typically kill tubercle bacilli so that they can be worked with outside of the BSC. Empirically verify that that the bacteria are killed by the treatment before putting that process into action.
- Killing by tuberculocidal disinfectant type and contact time should be appropriate for the matrix of contaminated material.
- Use a splash-proof container when decanting fluids in the BSC. Splash-proof containers need to have disinfectant added to them prior to use. If a funnel is used, rinse it with disinfectant after use.

BACTERIA (OTHER THAN *M. TUBERCULOSIS*) TRANSMITTED BY THE AIRBORNE ROUTE

The many bacteria that have caused laboratory-acquired infections in humans include *Brucella*, *Francisella tularensis*, and *Burkholderia pseudomallei* (22). Infections due to these agents have often occurred when they were being manipulated as unknown organisms in a clinical bacteriology setting operating routinely as a BSL2 laboratory. But in fact some aerosol-prone procedures have not always been contained in the BSC, as required by BSL2 practices. For example, it is not uncommon for the laboratorian to dispense with concern for a microorganism once it is suspended in a fluid. The use of automated instruments for identification and susceptibility testing further places the laboratorian at risk because most of the operations involve manipulation of organisms suspended in fluids. Some microbiologists practice the sniffing of open culture plates to assist in identification of isolates; this can pose a risk of airborne exposure, although it is considered a low risk (23). During the risk assessment process, every step of the procedures should be evaluated to detect aerosol-prone conditions. Since it is impractical

TABLE 3.

Stratification of tasks according to risk of aerosol spread of tuberculosis

Task and risk assessment	Practice	Special instructions
Manipulation of body fluids potentially containing *M. tuberculosis* or *M. bovis* BSL2 task; agents are present in lower numbers in patient body fluids Respirator program or a mask not required	All work with open vessels is conducted in a BSC Centrifugation outside the BSC is conducted in sealed, break-proof containers	1. Pour into splash-proof container and rinse funnel with disinfectant. 2. Pipette over disinfectant-soaked pad. Do not "blow out" pipette. 3. Immerse used pipettes in disinfectant or seal discard container before removing from the BSC. 4. Vortex tightly sealed tubes and allow to stand for 30 min before opening. 5. Open safety centrifuge cups in the BSC. Inspect surface of tubes for leakage; disinfect cups, if contaminated, before reuse. 6. Safety centrifuge cups are part of a routine preventive maintenance program; replace O-rings to ensure adequate seal. 7. Submerge all empty vessels in disinfectant or seal tightly before removal from the BSC for transport to autoclave.
Manipulation of colonies of *M. tuberculosis* or *M. bovis* BSL4 task; organism numbers are amplified by culture; risk is less than with culture-amplified organisms suspended in fluids Respirator program not necessary	Plates are sealed with tape or shrink-sealed and opened only in the BSC. All manipulations of opened tubes or plates are conducted only in the BSC.	1. Transfer colonies from solid medium to solid medium, with streaking of plates restricted to those plates with a smooth surface. 2. Disinfect loops in a safety bacticinerator, or remove microorganisms from the loop by a phenol sand trap before incineration in a flame. (See text for precautions on use of gas burners in BSCs.) Immerse disposable loops in disinfectant before removal from the BSC. 3. Seal plates and tubes for discard with an aerosol-proof seal before removal from the BSC for transport to autoclave.
Manipulation of fluids containing culture-amplified *M. tuberculosis* or *M. bovis* BSL3 task; volume and extent of potential aerosolization determine whether a BSL3 facility is required or whether BSL3 practices in a BSL2 facility will suffice.	1. Pipetting or aspirating fluid from sealed bottles 2. Vortexing 3. Centrifuging 4. Sonicating 5. Blending	A respiratory protection program and negative room air are recommended, especially for items 2 to 5 below, and particularly when organisms are well dispersed, for example, using Tween 80. 1. Work over disinfectant-soaked pad. Do not blow out pipette/syringe. Immerse used devices in disinfectant or seal discard container before removing from the BSC. 2. Vortex tightly sealed tubes and allow to stand for 30 min before opening. 3. Centrifuge in sealed centrifuge cups; open in the BSC. If the centrifuge is installed in a BSC, evaluate for any interference with BSC operation. 4. Always sonicate in BSC, even if using a closed container, to protect from organisms introduced onto external surfaces or from accidentally opened tubes. 5. Use special containment blenders or total-containment BSC (Class III) when blending.

to perform all of the work in the BSC, it is likely that many of the aerosol-prone procedures can be better contained by engineering controls. For example, all bacteriology laboratories should employ microincinerators for loop sterilization. Staff must be trained to suspect a pathogen that can be transmitted by the aerosol route. It is judged that when alert staff employ appropriate safety measures instituted in bacteriology laboratories, such as working within the BSC, the occasional encounter with agents such as *Brucella*, although it is classified as BSL3, will not result in infection.

In one reported experience, despite stringent infection control practices in the laboratory, the risk of laboratory-acquired brucellosis could not be eliminated when brucellosis was hyperendemic in the population (24). In this case, the large number of infected specimens handled by the laboratory resulted in seven cases over a 10-year period. Although *Brucella* is the most commonly reported laboratory-acquired bacterial pathogen (4), the fact that *Brucella* has a propensity to stain as Gram positive in young blood broth cultures makes it difficult for the microbiologist to suspect the organism from a Gram stain of the blood culture broth and then manipulate the cultures in a BSC. Therefore, it is most practical to handle all blood culture vials that are detected as positive by an automated blood culture system inside a BSC until the suspect agent is deemed of sufficient low risk of pathogenicity that work on the bench in a BSL2 environment is not a risk to laboratory workers. After overnight incubation, further growth characteristics of *Brucella* would guide the microbiologist to perform subsequent manipulations inside a BSC. Ideally, the fact that brucellosis is the suspected diagnosis would be communicated to the laboratory before receipt of the specimen. When a laboratory is evaluating a culture for suspected *Brucella*, besides performing all of the work inside a BSC, access to the laboratory should be restricted and potential fluctuations of the air-handling system should be avoided by keeping external doors and windows closed. If *Brucella* is identified in a culture, then all workers with exposure must be evaluated. To determine whether postexposure prophylaxis (PEP) is indicated, the CDC recommends classifying exposures as either high risk or low risk (25). For example, a high-risk exposure is described as a laboratorian working with cultures on the open bench in a BSL2 environment. A low-risk exposure would be another laboratorian working in close proximity during manipulation of the exposure.

As with brucellosis, it is important that the laboratory be informed prior to receipt of the specimens whenever there is a clinical suspicion of tularemia. One laboratory reported that 12 microbiology laboratory employees were exposed to *F. tularensis* from a rapidly fatal case of pulmonary tularemia (26). As a result of the exposure, 13 staff members received prophylactic doxycycline due to concerns about transmission. None developed signs or symptoms of tularemia.

N. meningitidis poses a risk to microbiologists for laboratory-acquired meningococcal disease from the manipulation of culture isolates on the bench without respiratory protection. In a review of 16 cases of meningococcal disease in laboratorians that occurred between 1985 and 2001, it was found that microbiologists experience this disease at a higher incidence (attack rate of 13/100,000) than adults in the United States in general (attack rate of 0.2/100,000) (27). As a consequence of this risk, it is recommended that all work in clinical or research laboratories with cultures that contain isolates of *N. meningitidis* be manipulated only when using a BSC. All workers who have the potential to handle meningococcal isolates should seriously consider vaccination with the quadrivalent meningococcal vaccine or quadrivalent meningococcal conjugate vaccine, which includes serogroup A, C, Y, and W-135 capsular polysaccharides, and vaccination for *N. meningitidis* serogroup B. *N. meningitidis* serogroup B is available in two vaccines—MenB-4C, which consists of three recombinant proteins and outer membrane vesicles containing outer membrane protein PorA serosubtype P1.4, and MenB-FHbp, which consists of two purified recombinant FHbp antigens.

Other candidate organisms for aerosol transmission include *Legionella pneumophila, Bartonella henselae*, and *Bartonella quintana*. Laboratory work with these organisms involves the prolonged incubation of media, often inoculated with specimens from patients for whom suspected tuberculosis is included in their differential diagnosis. It has been documented that *M. tuberculosis* will grow on the media and under the incubation conditions employed for the isolation of these agents (28). In fact, routine blood and chocolate agar support the development of microscopic colonies of *M. tuberculosis* within a week of incubation. Moreover, brucella agar and pertussis media are particularly good in supporting the growth of this agent. Microscopic appearance of colonies on these media precedes the microscopic appearance of *M. tuberculosis* colonies on Middlebrook 7H11 medium. Mature colonies of several weeks' growth are always smaller on routine media than they are on mycobacterial media. Although they may not progress to form macroscopically visible colonies, they will be present in an amplified, but undetected, form. On the basis of this information, the laboratory practices of the past that have included "hot looping" agar should be discouraged. For example, touching the surface of agar without apparent colonies on it to cool a loop might produce a substantial aerosol from a macroscopically invisible colony of *M. tuberculosis*. The first clinical laboratory that was successful in the isolation of *B. henselae* reports that it is not infrequently receiving referred organisms for confirmation of *Bartonella* identity that prove to be *M. tuberculosis* that was previously unsuspected by the referring laboratory (D.E. Welch, University of Oklahoma Health Science Center, personal communication). Thus, bacteriology laboratories are advised to review each procedure employed in the laboratory and consider whether it might be necessary to use a BSC when examining plates subjected to prolonged incubation.

Work with organisms grown or suspended in fluids to high concentrations, particularly those dispersed by detergent or mechanical action, should be considered of the greatest risk, and each element of the work should be evaluated individually to determine whether it is necessary. If deemed necessary, each step should be assessed carefully, and containment protocols should be developed. The case of a laboratorian who had been purifying proteins from cells infected with *Orientia tsutsugamushi*, the causative agent of scrub typhus, and subsequently developed scrub typhus pneumonitis is an example of work practices that required further review (29). This laboratorian used an ultrasonication method that typically generates significant aerosols, and the choice of this method of cellular disruption, without appropriate safety practices, led to infection.

VIRUSES TRANSMITTED BY THE AIRBORNE ROUTE

Before working with specimens for the detection of viruses, a risk assessment must be performed. If the procedures involve simple steps that are not expected to generate aerosols, then those procedures can be performed on the benchtop using splash guard protection. If the procedure is more complex, such as vortexing or other steps that could generate aerosols, then a BSC must be used (16). This precaution is applicable to both human and veterinary diagnostic specimens. There are a number of viruses to which this applies, including influenza viruses types A, B, and C. Work with specimens potentially containing rabies virus should always be performed in a BSC. Staff working with specimens that could contain rabies virus must be vaccinated and have detectable immunity as demonstrated by laboratory testing prior to initial training.

FUNGI TRANSMITTED BY THE AIRBORNE ROUTE

Certain fungi can readily be aerosolized from the cell mass that is cultured on solid media in the laboratory with little apparent physical intervention. In nature, most fungal hyphae develop structures intended for dispersal in air, either conidia on specialized aerial fruiting bodies or hyphal elements that mature into transmissible subsegments (arthroconidia). As these transmissible elements are intended in nature to disperse the organisms via air currents, the dispersing bodies are engineered to resist desiccation and UV light. Moreover, conidial forms are constructed to be readily discharged into the air and to remain aloft for long periods. When inhaled by a susceptible host, the conidia multiply and develop the alternate tissue form known as the yeast phase. *Histoplasma capsulatum* is present in tissues as the yeast form and only with rare exceptions is detected in its hyphal form. *C. immitis* is found in tissues as an endosporulating spherule or as individual endospores from a ruptured spherule. *B. dermatitidis* occurs in tissues as broad-based budding yeast. When cultivated in the laboratory, the systemic dimorphic fungi tend to form hyphal structures when exposed to certain conditions of growth on artificial media at lower temperatures (25 to 30°C). The hyphal structures ultimately elaborate specialized conidial forms. In culture, *C. immitis* fungal hyphae transition within a few days to segments known as arthroconidia that are readily dispersed into the air. The other two fungal agents generally require a more prolonged incubation period before the infectious conidia are elaborated. Once conidial structures are present, the laboratory culture represents a hazard if the containers are opened, allowing the conidia to be lofted into the air. Of 4,000 laboratory-acquired infections, approximately 9% were due to fungi (22). Although there are undoubtedly many laboratory-acquired fungal infections that are unreported, mycotic morbidity is a well-recognized occupational risk for mycologists (30).

For work with known isolates of the systemic dimorphic fungal organisms, hazard assessment is relatively simple and precautions are straightforward (12). As airborne hazards, these agents must be handled in the laboratory with BSL3 practices. Because the infectious conidia are autoairborne, the mere lifting of a culture plate lid is sufficient to produce a substantial infectious dose if inhaled. Thus, with all such organisms grown in the laboratory at temperatures of 25 to 30°C, all containers that have been culture amplified are sealed with tape and opened only in the BSC. It is also prudent to contain materials that were grown at more elevated temperatures (35 to 37°C) because the agents have a tendency to convert to the infectious form unless there is rigorous control of the conditions of incubation and evaluation. Rational protocols for work with such organisms in culture call for the implementation of further safety considerations. Laboratories should have a protocol to follow when there is an accident involving a container that breaks outside of the BSC (31). After an exposure from opening the lid of a plate containing *C. immitis*, the lab should be evacuated immediately, doors and windows should be closed, and the biosafety officer should be contacted to direct cleanup and postexposure management (32). Exposed individuals may be given PEP. There should be provisions in the procedure manual that prohibit the making of slide cultures with agents whose conidial structures have not developed sufficiently to rule out the systemic dimorphic fungi (33). Until confirmation of the identity of an agent other than a systemic dimorphic pathogen, the assessment of the hyphal structures should be done with a tease mount rather than a transparent tape preparation (33). Moreover, there should be administrative controls in place to make certain that cultures of the organisms are not disseminated to untrained outsiders for use in science fairs or research in uncontrolled conditions. Finally, there should be strict adherence to the rules for shipping samples of the organisms (34).

For work with unknown fungal isolates, in particular in the clinical laboratory, hazard assessment and abatement must be proactive. For all situations in which fungal cultures are ordered, culture plates should be taped and culture tubes should have secure screw caps, both of which should be opened only in the BSC. In no case should early growth or pigment production be considered evidence against the prospect that a fungal isolate is one of the systemic dimorphic fungi. For example, *C. immitis* often grows within a few days and produces pigmentation that may range from pink to green. The infectious arthroconidia may be absent when certain culture media are used, for example, cycloheximide media, but may be present on other media. The absence of characteristic conidia should not be considered evidence against identification of the unknown isolate as *H. capsulatum*. It is common to encounter hyphal forms that are slow to produce characteristic tuberculate macroconidia. Whenever such sterile hyphae are encountered, particularly when they are small in cross section and able to grow on cycloheximide agar, the unknown isolate should be considered a strong candidate for an infectious hazard until proven otherwise. The most problematic setting for laboratory safety with dimorphic fungi is not the mycology laboratory, where their growth is anticipated and safety procedures are fully implemented. Rather, other circumstances may offer more risk; for example, a pharmaceutical lab using soil samples for drug discovery must use safety practices when processing soil potentially contaminated with dimorphic fungi. In the bacteriology laboratory, it is not unusual for a laboratorian to open a plate and encounter a "fuzzy" colony.

There are three basic protocols to prevent exposure that should be instituted in the bacteriology laboratory (32). First, no culture of an unknown mold should be opened outside a BSC. Second, when culture plates inoculated with clinical specimens are incubated or held for 3 or more days, the lids should be taped or sealed or the technologist should be forewarned not to open the plate without first examining the culture plate surface for evidence of hyphal growth. Third, when a plate with evidence of hyphal growth is accidentally opened before the growth is discovered, the technologist should quickly close the plate and perform all further workup of the organisms in the BSC. Moreover, the hyphal growth should be evaluated to rule out systemic dimorphic fungi, even if the protocol does not call for the full identification of the organism in question. If there are sterile hyphae present and the growth does not readily suggest the identity of the organism, the supervisor should be contacted for an assessment of whether to refer the isolate to a mycology laboratory for identification. The presence or absence of conidial structures at the time of exposure should be evaluated so that a medical decision can be made regarding the potential amount of exposure, should the organism prove to be one of the hazardous agents. Unknown fungal growth referred to another laboratory as a potential systemic dimorphic fungus should be shipped or transported according to packaging instructions (34) for the dimorphic fungi, because the infectious conidia may develop in transit and place at risk those who transport or open the container.

Disposal of culture plates may present a danger of fungal infection to the laboratorian. For example, many microbiology laboratories in clinical settings save one or more culture plates from each patient sample for a week or more before discarding them. This is a valuable asset for revisiting culture results and obtaining cultures for rechecking identities or susceptibility to antibiotics. But during that week of exposure to room temperature, hyphal forms may have grown and sporulated. When such plates are thrown blindly into a plastic bag, they will open and release spores. Discarded plates should be taped before discard. Ideally, all plates should be taped in a stack during the week of holding so that none are opened without appropriate precautions in a BSC.

The greatest problem encountered in the laboratory safety arena is generally not those involving decision patterns for known problems. Rather, it is more often the unknown safety precautions required for a particular agent that causes the laboratorian to fear infection. Thus, it is good to mention instances in which organisms are not known to represent an infectious hazard in the laboratory. *Cryptococcus neoformans* and *Sporothrix schenckii*, although potentially pathogenic if inoculated subcutaneously or splashed into the eye (35), do not appear to constitute a laboratory hazard with regard to the respiratory route under usual clinical or investigative conditions. There is a suggestion that the dematiaceous organisms may represent a hazard via the respiratory route when growing in their hyphal form (12); therefore, these organisms should be manipulated in the BSC. Although many of the opportunistic and saprophytic fungi do not represent a substantial infection risk to the immunocompetent laboratorian (7), the dispersal of their conidia in the laboratory is problematic because there may be allergic reactions as well as contamination of media. Therefore, opening the culture containers only in the BSC should minimize gross exposures.

CONCLUSION

The respiratory route of transmission poses a significant risk to the laboratorian in multiple laboratory settings. Bacteria, viruses, and fungi all represent a hazard through airborne transmission. The many methods used in the laboratory that can generate aerosols must be evaluated in performing the risk assessment. While the lack of reporting laboratory-acquired infections hampers our ability to identify adequately all the risks that have resulted in laboratory-acquired infections in the past, there are sufficient laboratory-acquired infections reported in the literature to recognize that the risk is present and that mitigation measures must be taken. The risk assessment dictates appropriate prevention efforts and is the foundation of a sound biosafety program when accompanied by the selection and use of mitigation measures such as biosafety practices, PPE, and policies on discarding waste and incident reporting.

References

1. **Garner JS.** 1996. Guideline for isolation precautions in hospitals. Part I. Evolution of isolation practices, Hospital Infection Control Practices Advisory Committee. *Am J Infect Control* **24:** 24–31.
2. **Reid DD.** 1957. Incidence of tuberculosis among workers in medical laboratories. *BMJ* **2:**10–14.
3. **Kubica GP.** 1990. Your tuberculosis laboratory: are you really safe from infection? *Clin Microbiol Newsl* **12:**85–87.
4. **Sewell DL.** 1995. Laboratory-associated infections and biosafety. *Clin Microbiol Rev* **8:**389–405.
5. **Shah SM, Ross AG, Chotani R, Arif AA, Neudorf C.** 2006. Tuberculin reactivity among health care workers in nonhospital settings. *Am J Infect Control* **34:**338–342.
6. **Nolte KB, Taylor DG, Richmond JY.** 2002. Biosafety considerations for autopsy. *Am J Forensic Med Pathol* **23:**107–122.
7. **Nagano Y, Walker J, Loughrey A, Millar C, Goldsmith C, Rooney P, Elborn S, Moore J.** 2009. Identification of airborne bacterial and fungal species in the clinical microbiology laboratory of a university teaching hospital employing ribosomal DNA (rDNA) PCR and gene sequencing techniques. *Int J Environ Health Res* **19:**187–199.
8. **Delany J, Rodriguez J, Holmes D, Pentella M, Baxley K, Shah K.** 2011. Guidelines for biosafety laboratory competency: CDC and the Association of Public Health Laboratories. *MMWR Suppl* **60:**1–23.
9. **Ned-Sykes R, Johnson C, Ridderhof JC, Perlman E, Pollock A, DeBoy JM, Centers for Disease Control and Prevention (CDC).** 2015. Competency guidelines for public health laboratory professionals: CDC and the Association of Public Health Laboratories. *MMWR Suppl* **64**(MMWR Suppl)**:**1–81.
10. **Boa E, Lynch J, Lilliquist DR.** 2000. *Risk Assessment Resources.* American Industrial Hygiene Association, Fairfax, VA.
11. **Ryan TJ.** 2003. Biohazards in the work environment, p 363–393. *In* DiNardi SR (ed), *The Occupational Environment: Its Evaluation, Control, and Management*, 2nd ed. American Industrial Hygiene Association, Fairfax, VA.
12. **U.S. Department of Health and Human Services, Public Health Service, Centers for Disease Control and Prevention, National Institutes of Health.** 2009. *Biosafety in Microbiological and Biomedical Laboratories*, 5th ed. HHS Publication no. (CDC) 21-112. http://www.cdc.gov/biosafety/publications/bmbl5/BMBL.pdf.
13. **National Institutes of Health.** 2002. *NIH Guidelines for Research Involving Recombinant DNA Molecules (NIH Guidelines)*, 59 FR 34496 (July 5, 1994), as amended. The current amended version can be accessed at http://osp.od.nih.gov/sites/default/files/resources/NIH_Guidelines_PRN_2-sided.pdf.
14. **Centers for Disease Control and Prevention.** July 2010. Interim laboratory biosafety guidance for extensively drug-resistant (XDR) *Mycobacterium tuberculosis* strains. http://www.cdc.gov/tb/topic/laboratory/biosafetyguidance_xdrtb.htm
15. **Centers for Disease Control and Prevention and the Healthcare Infection Control Advisory Committee.** 2003. Guidelines for environmental infection control in health care facilities: recommendation of CDC and the Healthcare Infection Control Advisory Committee. *MMWR Morb Mortal Wkly Rep* **52**(RR10)**:**1–42.
16. **Miller JM, Astles R, Baszler T, Chapin K, Carey R, Garcia L, Gray L, Larone D, Pentella M, Pollock A, Shapiro DS, Weirich E, Wiedbrauk D, Biosafety Blue Ribbon Panel, Centers for Disease Control and Prevention (CDC).** 2012. Guidelines for safe work practices in human and animal medical diagnostic laboratories. Recommendations of a CDC-convened, Biosafety Blue Ribbon Panel. *MMWR Suppl* **61**(Suppl)**:**1–102.
17. **Wells WF.** 1934. On air-borne infection. II. Droplets and droplet nuclei. *Am J Hyg* **20:**611–618.
18. **Jensen PA, Lambert LA, Iademarco MF, Ridzon R, CDC.** 2005. Guidelines for preventing the transmission of *Mycobacterium tuberculosis* in health-care settings, 2005. *MMWR Recomm Rep* **54**(RR-17)**:**1–141.
19. **Gralton J, Tovey E, McLaws ML, Rawlinson WD.** 2011. The role of particle size in aerosolised pathogen transmission: a review. *J Infect* **62:**1–13.
20. **Association of Public Health Laboratories.** 2013. *Mycobacterium tuberculosis*: assessing your laboratory. http://www.aphl.org/AboutAPHL/publications/Documents/ID_2013Aug_Mycobacterium-Tuberculosis-Assessing-Your-Laboratory.pdf#search=TB%20Assessing. Accessed December 23, 2015.
21. **Collins CH.** 1993. *Laboratory Acquired Infections*, 3rd ed. Butterworth/Heinemann, Oxford, United Kingdom.
22. **Pike RM.** 1976. Laboratory-associated infections: summary and analysis of 3921 cases. *Health Lab Sci* **13:**105–114.
23. **Barkham T, Taylor MB.** 2002. Sniffing bacterial cultures on agar plates: a useful tool or a safety hazard? *J Clin Microbiol* **40:**3877.
24. **Memish ZA, Mah MW.** 2001. Brucellosis in laboratory workers at a Saudi Arabian hospital. *Am J Infect Control* **29:**48–52.
25. **Centers for Disease Control and Prevention.** 2008. Laboratory-acquired brucellosis—Indiana and Minnesota, 2006. *MMWR Morb Mortal Wkly Rep* 57:39–42. http://www.cdc.gov/mmwr/preview/mmwrhtml/mm5702a3.htm.
26. **Shapiro DS, Schwartz DR.** 2002. Exposure of laboratory workers to *Francisella tularensis* despite a bioterrorism procedure. *J Clin Microbiol* **40:**2278–2281.
27. **Sejvar JJ, Johnson D, Popovic T, Miller JM, Downes F, Somsel P, Weyant R, Stephens DS, Perkins BA, Rosenstein NE.** 2005. Assessing the risk of laboratory-acquired meningococcal disease. *J Clin Microbiol* **43:**4811–4814.
28. **Shaw CH, Gilchrist MJR, Guruswamy AP, Welch DF.** 1994. Culture of mycobacteria: microcolony method, p 3.6.b.1–3.6.b.6. *In* Isenberg HD (ed), *Clinical Microbiology Procedures Handbook.* ASM Press, Washington, DC.
29. **Oh M, Kim N, Huh M, Choi C, Lee E, Kim I, Choe K.** 2001. Scrub typhus pneumonitis acquired through the respiratory tract in a laboratory worker. *Infection* **29:**54–56.
30. **DiSalvo AF.** 1987. Mycotic morbidity—an occupational risk for mycologists. *Mycopathologia* **99:**147–153.

31. **McGinnis MR.** 1980. *Laboratory Handbook of Medical Mycology.* Academic Press, New York, NY.
32. **Stevens DA, Clemons KV, Levine HB, Pappagianis D, Baron EJ, Hamilton JR, Deresinski SC, Johnson N.** 2009. Expert opinion: what to do when there is *Coccidioides* exposure in a laboratory. *Clin Infect Dis* **49:**919–923.
33. **Haley LD, Callaway CS.** 1979. *Laboratory Methods in Medical Mycology.* U.S. Government Printing Office, Washington, DC.
34. **International Air Transport Association.** 2006. *Infectious Substances Shipping Guidelines,* 7th ed. Ref. No. 9052–07. International Air Transport Association, Montreal, Quebec, Canada. https://www.sujb.cz/fileadmin/sujb/docs/zakaz-zbrani/Infectious-Substances-Shipping-Guidelines.pdf. Accessed March 1, 2016.
35. **Thompson DW, Kaplan W.** 1977. Laboratory-acquired sporotrichosis. *Sabouraudia* **15:**167–170.

14

Cell Lines: Applications and Biosafety

GLYN N. STACEY AND J. ROSS HAWKINS

ANIMAL CELLS AS SUBSTRATE AND PRODUCTION SYSTEM

Animal cells have been used in biotechnology since the early 1950s. The Salk polio vaccine, licensed in 1954, was the first product produced on animal cells as a substrate, and for many years the only products produced using animal cells were viral vaccines. Primary animal cells were used for many years for vaccine production and are still used in certain cases. These vaccines have generally proven to be acceptable and safe, but there are notable exceptions that have directed manufacturers and regulatory bodies to be very cautious in their assessment of new cell substrates. The earliest cell lines used to manufacture biological products were human diploid fibroblast finite cell lines. WI-38 and MRC-5, two of the best-known examples, have been used in the manufacture of a number of licensed products (Table 1). The early use of continuous cell lines (CCLs) for the manufacture of biological products is represented by the manufacture of foot-and-mouth disease vaccine in the Syrian hamster cell line BHK, the production of interferon from the B-lymphoblastoid cell line Namalwa, and the introduction of monoclonal antibodies from hybridoma cells in the 1990s (for a general reference, see reference 1).

More recently, the use of animal CCLs took a significant step forward with the acceptance of a Chinese hamster ovary (CHO) cell line in the production of Activase (tissue plasminogen activator) (1). This was the first therapeutic protein manufactured from a transfected mammalian cell line to be marketed. Today, a wide range of potential diagnostic and therapeutic products are being developed in CHO and BHK cell expression systems (e.g., hormones, interleukins, erythropoietin, tumor necrosis factor, interferons), and an ever-expanding range of cell substrates are being worked with as candidate production cells, including myeloma cell lines (e.g., NS0, SP2/0). However, all will be subject to rigorous safety testing and validation before the products they are used to make can be licensed.

From a biochemical perspective, animal cells are currently indispensable when it comes to the manufacture of human therapeutics in the form of complex proteins. In certain cases, complex glycoprotein structures may be required to provide the necessary biological activity and an acceptable half-life *in vivo*, and recombinant microorganisms, such as *Escherichia coli* and yeasts, may not yet

TABLE 1.

Cell culture applications

Year	Application
1949	Virus in cell culture (Enders, Weller, and Robins)
1954	Salk polio vaccine on monkey kidney cells
1955	Sabin polio vaccine on monkey kidney cells
1963	Measles vaccine on chicken embryo cells
1964	Rabies vaccine on WI-38 cells
1967	Mumps vaccine on WI-38 cells
1969	Rubella vaccine (WI-38 cells)
1974	Varicella, cytomegalovirus, and tick-borne encephalitis vaccines
1975	Hybridoma technology developed by Kohler and Milstein
1980	Interferon from Namalwa cells at 8000-L scale
1981	Antibody diagnostics kit
1981	Mouse embryonic stem cells isolated
1982	Recombinant insulin from *Escherichia coli*
1986	Gamma interferon from lymphoblasts licensed
1986	Polio and rabies vaccine on Vero cells
1988	Recombinant tissue plasminogen activator, Genentech licensed
1989	Recombinant erythropoietin (EPO) licensed
1998	Human embryonic stem cell lines isolated
2007	Human induced pluripotent stem cells isolated

[a]Modified from references 1, 182.

provide the necessary posttranslational modifications required for effective biological activity. Large-scale animal culture resulted in annual sales of diagnostic and therapeutic products of over $5 billion, with more than 100 candidate drugs in phase I, II, and III clinical trials, by the 1990s (2). Ongoing commentaries have indicated that for the early part of the 2000s the private sector activity in the world for the regenerative medicine industry alone (although including non-cell-based medicines) is approaching $2.5 billion (3, 4), and more recent analysis has given evidence that overall this is increasing significantly (5). However, the path to commercially successful cell therapies is challenging (6). More recently, it has been possible to assess experiences in the early stages of products in the area of cell-based medicines (7), and (at least at this point in time) it seems that there is renewed interest and investment in such advanced medicines, driven in some part by the notable successes in chimeric antigen receptor therapy (CAR T) approaches (see discussion under "*Ex Vivo* Cell Proliferation for Cancer Adoptive Immunotherapy," below).

A number of cell lines have been used and studied extensively, and many industrial cell lines can be traced back to these origins (Table 2).

ACCEPTANCE OF CELL LINES FOR PHARMACEUTICAL PRODUCTION

Of the only cells accepted for biopharmaceutical applications, including vaccine production, up to 1967 were primary cultures, e.g., monkey cells (8). Safety issues were debated by members of the International Association of Microbiological Societies (9) over the use of human diploid cells (HDCs). But HDCs could be proven to lack detectable viruses and to be nontumorigenic. At the time, CCLs were not really considered an option. HDCs, with their finite life span, were just gaining acceptance. The production by Wellcome of interferon for clinical trials from the continuous lymphoblastoid cell line Namalwa was discussed in 1978 at Lake Placid, NY. The concept of its use was supported. bioMérieux then used another CCL, Vero, to produce an improved polio vaccine backed by World Health Organization (WHO) decisions. In a 1985 conference, the discussion moved to the risk associated with three potential contaminants: viruses, host cell DNA, and transforming proteins, with a major focus on potentially oncogenic DNA. A provisional limit of 10 pg of host cell DNA per dose was suggested and developed into a nominal standard. That standard was challenged, because the contaminating DNA was whole-cell DNA and not highly purified oncogenic sequence, as was the assumption in the first place. In 1987, a report of the WHO Study Group on Biologicals concluded that there was no reason to exclude CCLs as substrates for the production of biologicals. The emphasis was now shifted to viral contaminants, and 100 pg of DNA per dose was cited as an acceptable level in certain circumstances; it was understood that higher levels of DNA contamination may be acceptable also under certain circumstances. However, it should be emphasized that acceptability was still linked to case-by-case scientific evaluation of the product and the process, which must be shown to be capable of removing or inactivating potential contaminating organisms to an acceptable level. A unit for blood transfusion will carry substantial amounts of cellular DNA (75 to 450 μg) with no reported ill effects. Thus, 500 ng of DNA per dose for biopharmaceuticals may be acceptable (8).

A joint conference of the International Association of Biological Standardization (IABS), the Cell Culture Committee of the IABS, WHO, and the European Society of Animal Cell Technology was held in 1988 in Arlington, VA. This conference addressed concerns related to the carriage and expression in CCLs of endogenous retroviruses that might be similar to tumorigenic retroviruses, and WHO guidance on cell substrates for vaccine production was updated in the same year (10). In more recent years, WHO established a Cell Substrates Study Group, which has progressed discussion on a number of challenging issues for cell substrates (e.g., DNA contami-

TABLE 2.

Common cell lines[a]

Cell line	Use	Description	Origin
BHK-21 (baby hamster kidney)	Veterinary vaccines (e.g., foot-and-mouth disease) and recombinant proteins	Originally fibroblast-like, anchorage dependent; later also suspension	Kidney cells from 1-day-old baby hamsters (1963)
CHO-K1 (Chinese hamster ovary)	rRNA gene products	Epithelium-like, anchorage and suspension growth; clone K1 requires proline	Ovary cells from adult Chinese hamsters (1957)
HeLa	Early experimental vaccines	Epithelium-like, suspension growth	Human (Henrietta Lacks) cervix carcinoma (1952)
McCoy	Diagnosis of chlamydial infections	Fibroblasts	Mouse
MDCK (Madin-Darby canine kidney epithelial cells)	Veterinary vaccines and recently as a substrate for human influenza vaccines	Polarized epithelial cell line	Canine kidney cells isolated by Madin-Darby from normal tissue
Mouse L cells	*In vitro* studies	Fibroblasts; can be grown in suspension	Connective tissue from a 100-day-old mouse (1943)
MRC-5 (Medical Research Council)	Human vaccine production and virus detection	Diploid finite cell line of limited life span, produces collagen	Human embryonic lung tissue (1966)
Namalwa	Interferon production	B lymphoblastoid, suspension	Human (Namalwa) Burkitt's lymphoma
NS0	Recombinant protein production for human vaccines	Continuous mouse myeloma cell line	Derived from tumor cells derived from the MOPC-31 mouse strain
PerC6	Proposed as a substrate	Weakly adherent	Human embryonic retinoblasts transformed with adenovirus 5 (Ad5) E1A and E1B sequences
3T3	*In vitro* studies of cell transformation by oncogenic viruses	Fibroblasts, anchorage dependent	Fibroblasts from mouse embryos (1963)
Vero	Polio vaccine and other human vaccines	Contact-inhibited fibroblasts, anchorage dependent	Kidney from an adult African green monkey (1962)
WI-38 (Wistar Institute)	Vaccine production	Finite diploid fibroblast line of limited life span; produces collagen	Human female embryonic lung tissue

[a]Data modified from reference 183.

nation of products, need for tumorigenicity, and animal testing) (11), culminating in new guidance on the evaluation of cell substrates for the manufacture of biologicals in general, including animal cell-derived recombinant therapeutics (12).

The range of animal cell substrates being proposed for the manufacture of biological products is expanding, e.g., Madin-Darby canine kidney (MDCK) cells. The first licensed vaccine to be developed using MDCK cells was Optaflu (Novartis) in 2007 (13). The recombinant cell line Per.C6 has been used as a substrate, particularly for human manufacture of influenza vaccine (14, 15), recombinant therapeutic proteins (e.g., references 16, 17), and gene therapy vectors (18). As each product is presented for product licensing, it is vital that it receives the closest scrutiny and undergoes appropriate validation and safety testing (see discussion under "Safety Testing of Cell Lines," below and reference 10).

A major change to the application of cell lines over the past decade has been the widespread introduction of cell culture-based therapies, such as monoclonal antibodies, hormones, and cytokines. The range and significance of biological drugs are set to expand greatly. The worldwide sales of biological drugs reached $130 billion in 2009. It has been predicted that biologics sales will grow annually by more than 10% (20), and in spite of the 2009 recession, global growth in total health care spending is anticipated to grow by more than 4% in 2015–2019 (see Deloitte Global Life Sciences Outlook, 2016 at https://www2.deloitte.com/). While the production of biologic generics or "biosimilars" will impact drug company profits as patents expire, the threat of out-of-patent sales of competing biosimilars is much less than that of generics to the chemical drug market and so encourages the development of new biologics. The quality and regulatory aspects of biosimilar production are still developing, however. How minor differences in molecular structure affect efficacy, toxicity, and immune tolerance has yet to be understood.

Biosimilars made from cell culture will still require careful consideration of the substrates used for manufacture to ensure consistency, safety, and efficacy of the final product. International recommendations from the WHO have been established (11, 12).

APPLICATION OF ANIMAL CELLS AS CELLULAR THERAPIES AND TEST SYSTEMS

Ex Vivo Cell Proliferation for Cancer Adoptive Immunotherapy

Adoptive cell therapy, consisting of the *ex vivo* activation and expansion of tumor-reactive immune cells, has achieved promising results in clinical trials of the treatment of highly immunogenic tumors, such as malignant melanoma and virus-associated malignancies (21, 22). Tumor-infiltrating lymphocytes can be recovered *ex vivo* from tumors, selected by cytokine and cell-surface marker expression, and cultured and rapidly expanded for therapeutic transfer back to the patient. In future, the genetic modification of adoptively transferred T cells, such as the redirection of T-cell antigen specificity with transgenic T-cell receptors or expression of transgenic cytokines or other receptors, may reduce the ability of tumors to evade the immune system. The genetic modification of natural killer (NK) cells to express CARs may also enhance the immune response against the expression of low levels of class I major histocompatibility complex molecules by tumor cells.

In 2010, the Provenge (sipuleucel-T) prostate cancer treatment produced by Dendreon Corporation became the first form of autologous cell therapy to gain U.S. Food and Drug Administration (FDA) approval for clinical use. Used for the treatment of asymptomatic or minimally symptomatic metastatic hormone-refractory prostate cancer, the sipuleucel-T process involved the harvesting of dendritic cell precursors from the patient's blood and stimulation of these cells with a prostatic acid phosphatase (tumor antigen)/granulocyte-macrophage colony stimulating factor (cytokine) fusion protein. In culture, the cells were differentiated into mature dendritic cells, which, when infused back into the patients, present tumor antigen to T cells, thus priming a cytotoxic T-cell-mediated immune response. A range of other cell-based medical products have been established and their progress has been reviewed by Maziarz and Driscoll (6).

Treatments targeting T cells to leukemic cell populations by means of expression of chimeric antibodies on T cells (CAR T cell) have been in development since the 1980s, with the first clinical trial in 2006 (23). Presently there are thought to be more than 50 ongoing clinical trials (24, 25). Approaches are also being developed that involve replacing the antibody domain with peptides and proteins. The modified receptors comprise the active targeting component (antibody or peptide), an extracellular spacer (which may be key to the efficacy of these therapies, as longer spacers give better access to epitopes), and a transmembrane component. Current antibodies used may cross-react with nontarget tissues, and so "on target/off tissue" toxicity could therefore be a consequence of CAR T-cell therapy. New approaches are being developed to enable treatment of solid tumors. However, for all the variants of this kind of therapy, there still remain significant regulatory science challenges, including the need for a better understanding of dose-effect relationships and improved cryopreservation methods.

Because each batch of cells produced in the laboratory is patient specific, autologous immunotherapy creates its own set of manufacturing problems. It remains critical that samples remain sterile and free of cross-contamination, but little time is available for appropriate testing. Scheduling of production can also be problematic due to the uneven rate of sample acquisition and the short shelf life of the incoming blood sample and outgoing cell product. Because scale-up is limited, the costs of processing multiple small batches will likely remain high.

Ex Vivo Cell Proliferation in Gene Therapy

In recent years, the interest in new approaches to transplantation of cell populations and whole tissues or organs has developed rapidly. An increasing demand for transplants for the treatment of hepatic failure, leukemia, and skin burns and the problem of graft rejection have revealed the shortcomings of traditional donor-to-recipient transplantation methods. One of the most advanced approaches is the removal, genetic modification, *ex vivo* expansion, and subsequent reimplantation of patient cells or tissues.

A range of packaging cell lines (26, 27) are used in the production of gene therapy vectors (e.g., retroviral shuttle vectors, adenoviral vectors, adeno-associated viral vectors, herpes simplex viral vectors), and these host cells should be treated in the same way as cell substrates used for the manufacture of biological products. Specific safety guidelines have been drawn up for the preparation and testing of gene therapy products, and these also refer in detail to the requirements for packaging cell lines (28–31). It is important to note that the cell and gene therapy fields are highly dynamic and that new guidances and updates to the regulation are emerging. Readers are guided to their respective regulators for relevant information and to the following websites for current updates from the U.S. FDA (http://www.fda.gov/BiologicsBloodVaccines

/GuidanceComplianceRegulatoryInformation/Guidances/CellularandGeneTherapy/) and the European Medicines Agency (EMA) (http://www.ema.europa.eu/ema/index.jsp?curl=pages/regulation/landing/human_medicines_regulatory.jsp&mid=).

A number of approaches have been developed to enable corrective gene editing of cells from patients with genetic disease (32). One of these techniques, called clustered regularly interspaced short palindromic repeats (CRISPR), was only recently discovered (33) but presents a much simpler approach that is now widely used and is even being taken forward toward clinical trials. However, it will be important to address the potential for a number of unintended genetic changes that may occur during genetic editing, including off-target and frameshift mutations. More robust technology is being developed all the time, and in the United States the Defense Advanced Research Projects Agency (DARPA) has launched a national program to reduce potential risks (www.darpa.mil/). It is worth noting that these technologies create changes in the genome, and thus some significant ethical issues will need to be addressed (34).

Cell Culture *Ex Vivo* for Cell Therapy

A number of different tissues can be removed from the patient or donor, expanded *ex vivo*, and used for transplantation, and in the case of stem cell-based therapies it appears that there are now more than 1,500 clinical trials in development around the world (35). Bone marrow transplant of hematopoietic stem cells (HSCs) has been established for more than 40 years, and the repair of skin tissue is another long-established cell therapy (36, 37). For the latter treatment, keratinocytes from small pieces of skin are grown into sheets of epithelium that can be used for wound treatment. In such processes, the protection of the graft from adventitious agents during manipulations is of utmost importance. In some cases, feeder cells are used which require careful safety evaluation. In the case of skin engraftment, the mouse cell line NIH 3T3 (clone J12) is used to provide feeder cells. In the United Kingdom, a central bank of these cells has been established at the National Institute for Biological Standards and Control. In a process almost identical to that described for *ex vivo* generation of skin grafts from keratinocytes, preparations of healthy limbus (a tissue structure located at the perimeter of the cornea) are cultured on feeder layers of 3T3 cells, and limbal stem cells are expanded and reimplanted to repair damaged areas of the corneal epithelium. In the near future, it is likely that synthetic matrices are likely to replace feeder cells (e.g., reference 38).

Stem cells isolated from bone marrow, cord blood, or peripheral blood may be cultured and selected for particular stem cell types prior to transplantation. Significant numbers of clinical trials have been established with such cells using gene therapy (35). In addition, important advances have been made toward development of fetal and stem cell-based treatments for neurological diseases, such as diabetes and Parkinson's disease (39, 40), and cell-based human therapies are also in development (41). Many clinical trials have been based on the use of bone marrow- and blood-derived HSCs, largely based on their track record in the traditional therapeutic application in the treatment of leukemias. However, for regenerative medicine in general, it will be important to establish a clear rationale and proposed mode of action for each new application. Cord blood is a well-established source of HSCs and has become an important source of these cells for treating hematologic malignancies. Use of umbilical cord blood (UCB) was shown to reduce graft-versus-host disease (42) and has been developed to provide broader therapeutic options for treatment where unrelated donors are the only source of HSCs for treatment (43, 44). Therapy with UCB and peripheral blood stem cells is often constrained by limitations on the numbers of cells that can be obtained from individual donations. However, methods have been developed whereby such cells may be cultured *in vitro* using growth factors to stimulate expansion and also may be cultured in bioreactors or stromal culture systems (45). A challenge for such systems is to maintain optimal therapeutic cell populations while avoiding the development of abnormal cells. Such populations have been reported by a number of research groups studying *in vitro* UCB cultures, but the significance of these findings has yet to be determined, as they do not appear to be tumorigenic by *in vitro* assays for tumorigenic potential (46, 47). These concerns should also be considered in the perspective of traditional bone marrow transplantation, where there is known to be a residual risk of donor-associated leukemia in recipient patients (48).

Mesenchymal stromal cells (sometimes called "mesenchymal stem cells") are a significant and readily isolated progenitor cell type that can be isolated from numerous sites in the body, including blood, umbilical cord endothelium, bone marrow, and adipose tissue (49–51). These cells are capable of differentiating into a range of potentially useful cells for therapy (52–55), and microvesicles derived from them have also been proposed as therapeutic products (56).

In the 1980s, embryonic stem cells (ES cells) were first isolated from mouse blastocysts (57), and in 1998 the first report of human embryonic stem cell (hESC) lines isolated from the inner cell mass of a human blastocyst was published (58). These cells have the capacity to generate cells representative of the three germ layers required to produce all the tissues of the body (i.e., are pluripotent) and can be passaged indefinitely while also retaining a diploid karyotype and the capacity to differentiate. These cells offer the exciting potential for the preparation of differentiated cells and tissues for regenerative medicine techniques.

Immunological mismatch between hESCs and the patient can mean that therapies derived from hESCs may involve long-term immune suppression using cytotoxic drugs to inhibit rejection, which in some cases could create a health risk to the patient that must be balanced against the benefit to the patient and the severity of his or her disease.

Proposals have been made for panels of human pluripotent stem cell (hPSC) lines that would permit delivery of therapy for the broad patient community based on 10 to 100 hESCs (59, 60). However, other challenging immunological barriers remain to be addressed, including polymorphic cell-surface proteins other than major histocompatibility complex molecules (61).

In 2007, the development of human induced pluripotent stem cell (hiPSC) lines, derived by reprogramming donor somatic cells into hESC-like cultures, was reported (62). This clearly opened up exciting possibilities to generate patient-specific stem cultures for therapy that had the potential to avoid immune rejection. hiPSCs also have the advantage of not requiring the destruction of a human embryo for their creation. However, hiPSCs have been shown to acquire mutations during the culture process (63, 64) and to be capable of provoking an immune response (65), although this may not apply to differentiated iPSCs (66). The rapid development of new reprogramming methods may provide more acceptable cells for therapy; however, until hiPSCs can be unequivocally confirmed as indistinguishable from hESCs, it may not be possible for them to achieve the same level of regulatory acceptability. In conclusion, clinical products based on the use of iPSCs will require the development of robust methods for reprogramming that avoid permanent genetic modification of the cells and provide consistent and effective reprogramming irrespective of the parental cells, but do not present risk due to the viral or other genes used to elicit reprogramming.

A major concern in the clinical application of pluripotent stem cells (both hESCs and hiPSCs) surrounds their potential to generate tumors that may be a benign expression of pluripotency (i.e., teratomas) or potentially cancer forming (i.e., teratocarcinomas). Teratomas may not form in immunologically competent recipients, but there is a theoretical risk that they could cause patient problems if impure populations of differentiated cells contaminated with pluripotent cells are transplanted into immune-suppressed patients. Ensuring the safety of therapies based on such cells will require the development (i) of methods to remove pluripotent cells efficiently and (ii) of sensitive analytical approaches to exclude their presence in final therapy products.

A principal advantage in the development of hPSC lines for therapy is their ability to be expanded into cell numbers that can facilitate preparation of cell therapies for multiple patients. However, this scale-up process brings a key challenge for hPSC lines, as they may exhibit phenotypic and genotypic instability *in vitro*. The well-established cut-and-paste method for scale-up of hESC and hiPSC colonies is highly labor intensive, and resultant colonies are prone to undergo uncontrolled differentiation, which can potentially waste a large proportion of the original cells as they will not be useful for generation of target cell types in the final differentiation protocols. Passaging hPSC lines *in vitro* has also been associated with genetic changes (67) and even the generation of abnormal and potentially tumorigenic cell types (e.g., references 63, 68). Another challenge to the delivery of therapies based on PSC lines is the identification of cell lines suitable for the development of specific therapies. Key elements of this suitability will include full traceability of the history of the cells, including (i) traceability to sources of cells with fully informed consent from donors for therapeutic application and (ii) traceability and documentation of all procedures, critical reagents, and other ancillary materials used in the isolation, culture, preservation, and storage of the cells.

Cells for use in clinical trials will need to meet the requirements of Good Manufacturing Practices (GMP), the principles of which are expressed in Good Tissue Practice (69), the European Tissues and Cells Directive (70), and WHO guidance (71). Early commercial initiatives utilized PSC lines derived in research laboratories.

Following successful studies in rodents (72), the first clinical trial for hESC-based therapy was approved in 2010. This phase I trial by Geron Corporation involved the use of hESCs differentiated into myelin-producing oligodendrocyte progenitor cells for the treatment of spinal cord injury. Although this therapy failed to show any adverse effects in patients, the clinical trial was halted in 2011 due to commercial reasons. In 2011, clinical trials were approved for hESC-based therapy of juvenile and age-related macular degeneration. This is a particularly promising area of clinical research, as the retina consists of a relatively small number of cells and is readily accessible for surgical cell replacement and transplanted cells can be easily monitored should adverse developments arise.

A variety of tissues in addition to blood and bone marrow have now proved to support somatic stem cell populations. In particular, stem cell populations appear to have been isolated from the human brain, liver, pancreas, intestine, corneal limbus, lung, and teeth. These may all prove to have importance in future therapies, either for transplantation or through development of therapies to mobilize endogenous stem cells by use of growth factors to stimulate tissue repair.

Safety Issues for *Ex Vivo*-Manufactured Cell Therapies

Significant assumptions are made regarding the safety of autologous treatments. Given that these involve returning the manipulated cells to the original donor, it has been assumed that virological testing is not necessary. However, there are hazards to address in high-throughput and complex production systems, including those associated with potential microbiological contamination arising from reagents and materials of animal origin and cross-contamination between patients' cell preparations. Rapid testing for microbiological contamination is crucial for cell preparations with very short shelf lives. Regulatory guidance is available (70; European Pharmacopoeia, General Chapter 2.6.27, Microbiological examination of cell-based preparations, in press; available at http://online.edqm.eu/EN/entry.htm).

Altered phenotype of cells on *in vitro* culture and genetic instability are both key issues that must be monitored carefully, and procedures for in-process monitoring during the manufacturing process should also be considered. Consideration should also be given to understanding the characteristics of apparently nontherapeutic (sometimes called "contaminant") cell populations. For general references on considerations in the manufacture of cell therapies see Hayakawa et al. (73) and Petricciani et al. (submitted for publication). For general approaches to characterization of cell therapy preparations and their use, see references 74–77.

Cell and Tissue *Ex Vivo* Cultures as Test Systems

Cell and tissue culture is becoming increasingly important for pharmacological, toxicological, and growth factor screening studies, providing an alternative to animal models. Some test systems utilizing animal cells have been standardized by the establishment of national pharmacopoeia protocols. Standardization of such tests is generally based on reference control preparations of active compounds; however, it is also important to use an appropriately qualified and validated stock of cells. For more accurate models of the *in vivo* human response, more sophisticated *in vitro* models may prove valuable, such as cocultures of different cell types, three-dimensional culture techniques, and induced differentiation of cell lines.

A recent significant opportunity for the development of human cell-based *in vitro* systems has been the possibility of using hESC and iPSC lines to establish "tissue in a dish" (78). The establishment of iPSC technology has been especially important in this respect, as it enables generation of pluripotent stem cell lines from a broad range of somatic tissues and individuals with important genetic disease traits, offering as yet untested potential in toxicology, product safety testing, and drug discovery. Such applications are subject to regulation of protocols for safety testing of products. The use of tissue and cell cultures may also introduce potentially significant additional variation in data from *in vitro* assays, especially where differentiated stem cell lines are used (79). Accordingly, Good Cell Culture Practice has been established to support these approaches (79) and is under development to accommodate hiPSC-based assays and three-dimensional systems (80).

RISKS ASSOCIATED WITH HANDLING OF ANIMAL CELL CULTURES

Potential hazards associated with the handling of animal cell cultures involve mainly the contamination of cells or media with pathogenic agents and/or the tumorigenicity for cells used in therapeutic applications. Long-standing experience has shown that contamination with pathogenic agents is the most important hazard and merits careful assessment of safety precautions both for laboratory use (81–83) and manufacture of biologicals (12) and cell therapies (85, 188).

Contaminating Microbial Agents

Around 20 documented cases of laboratory workers infected while handling primary cell cultures have been reported in the early history of cell culture (86, 87, 188). As with all other activities in research and production, it is advisable to evaluate the risks by systematic and reliable risk analysis methods (82, 88) to establish the appropriate protective measures, according to four internationally accepted risk groups of pathogens based on severity of infection and availability of preventive measures and treatments for the infection. The definitions of the European Federation Biotechnology (81) risk groups (based on definitions established by WHO) are as follows:

- Risk Group 1: Microorganisms that have never been identified as causative agents of disease in humans and that offer no threat to the environment.
- Risk Group 2: Microorganisms that may cause disease in humans and might therefore offer a hazard to laboratory workers. They are unlikely to spread into the environment. Prophylactics are available and treatment is effective.
- Risk Group 3: Microorganisms that offer a severe threat to the health of laboratory workers but a comparatively small risk to the population at large. Prophylactics are available and treatment is effective.
- Risk Group 4: Microorganisms that cause severe illness in humans and offer a serious hazard to laboratory workers and people at large. In general, effective

prophylactics are not available and no effective treatment is known.

These classes are closely linked to the concept of containment categories. With increasing risk, additional organizational measures and specialized laboratory equipment and design have to be implemented. Appropriate guidelines have been developed by most countries and are based on guidelines issued by international organizations such as the Organisation for Economic Co-operation and Development and the WHO.

Viruses

Viruses are of particular concern when handling animal cells. In some cases, viral infections may produce no cytopathic effect or may be latent and thus particularly hard to detect. Viral contamination of cell cultures may stem from the donor or from contamination by the operator or material used in the cell culture process (e.g., enzymes, serum, proteins, fetal extracts, hormones, and growth factors).

Retroviruses

Retroviruses are RNA viruses that comprise both endogenous viruses, normally harmless to the host, and exogenous viruses, which include a number of highly pathogenic species. An enzyme unique to retroviruses, reverse transcriptase (RT), enables the virus to transcribe its RNA genome into double-stranded DNA, which can then be integrated into the host genome as a provirus and does not require cell division to achieve this state. This ability for genomic integration enables the virus to persist indefinitely in the host. Outside the host, retroviruses tend to be labile and can be easily inactivated. Normally retroviruses will not infect hosts through intact skin, and only under certain circumstances, such as very high virus loads or long exposure, will retroviruses infect via intact mucosa.

Endogenous retrovirus sequences are present in many species, including humans, and are normally considered to be of very low risk. These seem to be remnants of historical infections in humans and animals and have limited function. The only endogenous retroviruses demonstrated to be pathogenic to their host have been observed in highly inbred immunocompromised nude mice. However, endogenous retrovirus-like sequences have been associated with certain human diseases—e.g., glomerulonephritis in some cases of systemic lupus erythematosus, some autoimmune diseases, and some forms of rheumatoid arthritis—and may yet prove to have some causative effect (89). Expression of incomplete endogenous retroviral RNA has been demonstrated in cancer cell lines (90), but this does not appear to present a hazard for laboratory workers or for vaccinees in the case of vaccines containing endogenous retroviral DNA (91).

C-type particles are retrovirus-like particles and have been found in a number of cell lines, such as CHO cells and hybridomas. C-type-like particles have been considered a potential hazard in cell culture-derived biopharmaceuticals (92) but have not yet been associated with any laboratory-acquired infections. Endogenous avian retroviral particles have been identified in most sources of embryonated hens' eggs and chicken embryo fibroblast cultures used in production of a number of vaccines, including influenza vaccines, but do not appear to represent a hazard in these biologicals (93). Approaches are now being developed that will enable all cell substrates used for vaccine manufacture to be screened for expression of such agents (94). Similarly, type D retrovirus-like particles have been observed in several cell lines, such as HBL-100 and Namalwa, and have also not been associated with any laboratory-acquired infections.

General viral contamination

It is generally assumed that human cell lines are the most likely cultures to be contaminated with highly pathogenic viruses, such as hepatitis B virus or human retroviruses such as the AIDS virus (human immunodeficiency virus, HIV). The risk of contamination with human pathogens is not limited to cells of human origin. Many other mammalian cells, both of primate origin and nonprimate origin, may contain viruses with broader host range. Primate cells may contain simian retroviruses or other simian viruses associated with severe human disease (e.g., herpesvirus simiae or Marburg virus). Rodents carrying lymphocytic choriomeningitis virus, Reo-3 virus, and hantavirus have led to human disease (81) and fatalities in some historical cases (95).

Exogenous retroviruses can be found in the onco-, spuma-, and lentiviruses. The lentiviruses comprise highly pathogenic viruses such HIV, and oncoviruses include human T-cell leukemia virus (HTLV-I) and hairy cell leukemia virus (HTLV-II). The spumavirus group also contains "foamy" viruses found in primates, which are often identified by the foamy cytopathic effect they produce in primary cell cultures. The role of spumaviruses as human pathogens is unclear, but laboratory workers and others with close primate contact frequently demonstrate an antibody response to these viruses (96–98). A precautionary approach has therefore been taken in consideration of risk from spumavirus infection, particularly as novel human pathogens could arise from infection in humans (99, 100).

Oncogenic viruses

In addition to the oncogenic viruses mentioned above, viral and cellular oncogenes have been identified that are able to transform cells into malignant forms. Other viruses are known to be oncogenic for primates, such as feline sarcoma virus, Epstein-Barr virus, hepatitis B virus, and human papillomavirus. Avian and murine leucosis viruses and polyomaviruses are considered to be of low risk to humans, although persistent BK and JC virus infections have been identified in transplant patients (e.g., references 101–103). Malignant transformation is thought to be a multistep process and therefore often difficult to quantify in an animal model. Also, the validity of the animal model and the possible *in vivo* amplification of the (viral) oncogene have to be considered (104).

Bacteria and Fungi

Under most circumstances, bacterial growth can be readily detected in cell cultures, as these organisms will overgrow the culture and be readily observed by shifted medium pH and increased turbidity. However, it should be borne in mind that numerous microorganisms (e.g., *Leptospira*, *Mycobacterium*) may fail to grow under the standardized conditions of pharmocopeial methods for sterility testing conditions, and others may survive as intracellular parasitic infections in cell cultures, such as *Mycobacterium avium-intrcellulare* (105). In recent years there has also been concern regarding the putative microorganisms called nanobacteria that may arise in cell culture (106); however, whether the nanoparticles which appear on *in vitro* cell culture of affected patient tissue are in fact viable organisms remains controversial (107, 108). Other contaminants, such as *Achromobacter*, have also been identified that can cause significant problems in cell culture due to broad antimicrobial resistance (109).

Mycoplasmas

Mycoplasmas are small (0.2- to 2-μm) prokaryotes lacking a cell wall. Mycoplasmas most commonly identified in cell culture include *Mycoplasma arginini*, *M. fermentans*, *M. orale*, *M. hyorhinis*, *M. hominis*, and *Acholeplasma laidlawii*. Historically, mycoplasma has been commonly detected in research laboratories and continues to be an issue today (110). The presence of these organisms in cell culture is not well known as a source of infection for the operator. However, they are considered potentially hazardous in cell culture products due to the possibility of persistence of mycoplasma antigens and other biologically active mycoplasma products. These microorganisms can proliferate in the culture without increasing the turbidity of the culture medium and they tolerate antibiotics; thus, they may go undetected for many passages. In addition, they can cause a wide range of genetic and phenotypic changes, including cell transformation, chromosomal abnormalities, and other physiological changes (111). Careful testing is necessary to detect these organisms (112), and any positive cultures should be discarded unless there is no alternative source of the cells. Eradication may be achieved using antibiotics, such as ciprofloxacin, but this is not always effective and it may alter the characteristics of the cells. While a range of methods of eradication have been reported (113), very careful validation is required to demonstrate that contamination will not reappear over time. Routine testing of cells for mycoplasmas is recommended, particularly if the culture has been put through an eradication procedure.

Parasites

Parasites, including malaria and Chagas' disease organisms, may be of concern in freshly prepared primary cell cultures or organ cultures if the donor is known to be infected or if there is a high risk of infection with a specific parasite. Because of the specific pathways of infection and proliferation of parasites, the risk to the operator may be considered minor. However, a number of species of *Onchocerca*, *Brugia*, *Leishmania*, *Trypanosoma*, *Acanthamoeba*, and *Pneumocystis* may be sustained *in vitro* under cell culture conditions. If there is a high likelihood that such contamination could arise in cell culture work, such cultures should be contained by use of an appropriate biological safety cabinet (BSC), aseptic technique should be used when culturing the cells, and procedures for effective decontamination of laboratory waste should be implemented. If cell cultures or tissue cultures are used to cultivate parasites, the containment measures must obviously be adequate for the parasite. Risk assessment has to take into account the host range, the routes of infection, and the infectious and noninfectious forms of the parasite during its developmental cycle(s).

Prions

Prions were first described in connection with a group of fatal neurodegenerative diseases in animals and humans known as transmissible spongiform encephalopathies (TSEs), of which there are a wide variety such as Creutzfeldt-Jakob disease (CJD), new variant CJD, and kuru in humans; scrapie in sheep; bovine spongiform encephalopathy (BSE) in cattle; and neural degenerative diseases in other animals such as deer and mink.

These transmissible diseases became the focus of attention during the British BSE outbreak. In 1994, 55% of milking herds in Britain contained cases of BSE and 850 cases were reported per week. Subsequently, increased incidence of BSE has been reported in a number of other countries, while incidence of both BSE in cattle and its counterpart in humans, variant CJD (vCJD), has decreased. Scrapie is endemic among sheep in many parts of the world, and efforts are under way to eliminate it using selective breeding programs.

The most widely accepted hypothesis for the nature of the causative agent for TSEs is the protein-only hypothesis, which postulates the absence of nucleic acids in the infectious particles (114–116). A number of animal and human diseases are now linked to prions (117–119). While methods of detection of prions have been established, a sensitive and reliable detection method has yet to be developed. Risk of prion infection in animal-derived materials (120) can be addressed through a risk assessment process. For example, see the European Medicines Agency guidance document (121) and resources available through the U.S. Center for Biologics Evaluation and Research (http://www.fda.gov/BiologicsBloodVaccines/GuidanceComplianceRegulatoryInformation/Guidances/Blood/ucm074089.htm). In particular, bovine serum used in cell cultures providing products for human use should be sourced from countries without BSE and from animals less than 30 months old, by which time the disease would have been evident. Risk assessment procedures for tissues have been established by WHO (122; see also WHO Tables on Tissue Infectivity Distribution in Transmissible Spongiform Encephalopathies, updated 2010, at http://www.who.int/bloodproducts/tablestissueinfectivity.pdf). Special considerations for risk assessment of prion and general microbial contamination of donor cells for cell therapies have also been considered under European Union (EU) regulation (123).

Tumorigencity and Oncogenic Potential

Tumorigenicity of CCLs has been considered a minor hazard for healthy individuals receiving highly purified cell-derived products, even when the cells are directly inoculated into humans. Only one case has been reported of a laboratory worker who developed a tumor from an accidental needlestick transmission of a human tumor cell line, a human colonic adenocarcinoma (124). In the 1950s, tumor cells were deliberately inoculated into volunteers. No incidence of malignant disease was reported as a result of these experiments. However, the use of tumorigenic cell lines for the manufacture of biologicals intended for human therapy has raised safety concerns. The potential hazards of tumor cells for use in the manufacture of vaccines were considered at a meeting hosted by the Center for Biologics Evaluation and Research of the FDA in September 1999 in Rockville, MD (125), and more recently guidance on evaluation of cell substrates has been published by the WHO (12). The question of tumorigenicity was discussed in detail and was concluded to be a consideration during early evaluation of cell substrates, but it was not considered a major issue for biologicals derived from animal cell culture where the products are cell free and highly purified. DNA oncogenicity has been addressed on a case-by-case basis with appropriate recommended limits set for different product types. WHO has now produced new guidance on the evaluation of cell substrates for the manufacture of vaccines and biotherapeutics (12), and similar guidance is also published by the FDA. Such guidance, as a consensus among Europe, Japan, and the United States, has also been published by the International Conference on Harmonisation (126). New data on DNA oncogenicity in a mouse model (127) have indicated that the risks of oncogenic DNA contamination, while low, may need further investigation and indicate the need to retain careful oncogenicity testing for cell-derived products.

Tumorigenicity of cells used in cell therapy products is a very challenging issue for which there are no generic assays available so far. Assessment of genetic stability and how it relates to development of potentially malignant cells are also aspects of the development of safety assessment for cell therapies yet to be resolved (see, e.g., reference 85).

Medium Components

Any component of animal origin in growth media—e.g., growth medium, growth factors, serum, or purification or isolation reagents—that comes into contact with the cells is a potential source of contamination with microorganisms. These contaminants may proliferate in the exposed cell culture and are a real issue for the manufacture of biologicals from cell culture (128). The source, means of preparation, and quality control testing of any such reagents should be assessed to identify the level of risk they represent to the cells and the operator.

The most common sources of viral contamination in general cell culture are bovine serum and trypsin. Commercial bovine serum may be contaminated with multiple viruses (129), most frequently bovine viral diarrhea virus (BVDV) (130). More recently, newly emerging related strains of pestivirus have also been identified in commercial sources of bovine serum (131). Noncytopathic strains of BVDV can establish persistent infections in cell lines of bovine origin, and cells from other species may also become infected (132). While such viral contamination is not necessarily a direct hazard to the operator, it is of concern in the production of biologicals and may also influence virology studies using BVDV-contaminated cells (130, 133). If bovine medium additives are imported from overseas, it may be necessary to also screen for diseases common to the country of origin and for emerging strains. In addition, reagents such as monoclonal antibodies may also harbor animal viruses that may be a hazard to laboratory workers (134). It is also important to remember that human blood-derived products screened for the more serious blood-borne pathogens are not necessarily free of viral contamination (135).

The more stringent level of qualification that must be applied for such reagents, when sourcing components for growth of cells used in the manufacture of biologicals for use in humans, has been particularly affected by the United Kingdom outbreak of vCJD in the 1990s, raising concerns regarding materials of both human and bovine origin. These concerns can be addressed by risk mitigation as described above (see discussion under "Prions") and by using protein-free growth media and recombinant proteins. However, in the manufacture of human biologicals for human treatment, even these approaches may not meet the demands of regulatory scrutiny, since protein-free media may contain components of animal origin, such as tallow and lactose, and the bacteriological growth media used to produce recombinant products may contain bovine proteins. For the purposes of cellular therapies, special guidance on selection of raw materials has been published for Europe (189), Japan (190), and the United States (191) and should be consulted by manufacturers.

RISK ASSESSMENT

Cell Line Source

The source of the cell line is an important element in risk assessment; the risk of significant infectious agents will increase the closer the genetic relationship of the cell line species is to humans. In addition, the likely contaminants of the tissue of origin should also be considered. These factors are generally true both for contaminating biological agents and for the risks inherent to the cell line, such as tumorigenic potential.

The tissues or cell types of origin, in order of decreasing risk, are hematogenous cells and tissues (e.g., blood and lymphoid tissue); neural tissue, endothelium, and gut mucosa; and epithelial cells and fibroblasts (81). Culture types, in order of decreasing perceived risk, are primary cell cultures; CCLs; and intensively characterized cells, including human diploid fibroblasts (e.g., WI-38, MRC-5, and IM90). When dealing with primary human cells, risk assessments should also include the organ or tissue of origin, the quantity of cells per specimen, the number of specimens from different individuals, and the level of risk represented by the population or cohort from which specimens are obtained.

Laboratory technicians should never manipulate their own cells in the laboratory (particularly when this involves Epstein-Barr virus transformation or genetic manipulation), as such modified cells may not be recognized by the operator's immune system and theoretically could result in cancer if accidentally injected.

It is important to understand that these are generalizations and there may remain a significant risk from unexpected sources, e.g., hemorrhagic fever viruses in insect cells and hantavirus and lymphocytic choriomeningitis virus in rodent cells. In addition, there have been a number of cases of serious diseases arising from organisms that appear to have crossed a species barrier from the normal host, e.g., H5N1 influenza in Hong Kong and Q fever and BSE in the United Kingdom.

Acquired Properties of Cell Lines

Recombinant cell lines

The risk assessment of recombinant cell lines should primarily take into account the nature of the expressed product and the risk associated with the host cell line, as described above. In some cases, recombinant cells may have to be handled under higher containment than the host cells if the properties of the vector used for cell transformation or the insert will confer additional risk. When using viral sequences, the transfer of pathogenic functions, or virulence factors, the risk of transactivation of endogenous viruses should also be considered. Although some viral expression systems, e.g., adenoviral and retroviral, are disabled by replication deficiency, replication-competent virus may arise in cultures and must be monitored.

Most countries have detailed national guidelines on the risk evaluation of recombinant organisms, including cell cultures. Similar information may be provided by companies marketing standard cloning systems. A number of international type culture collections also provide risk assessment or safety recommendations. Some of these institutions maintain excellent online information sites (see additional information at the end of this chapter). These evaluations can help researchers conduct the risk assessment of their constructs and recombinant cell lines. However, researchers should also be familiar with their own local and national safety rules and regulations.

Culture conditions that can change cell line properties

Cell biologists are becoming much more sophisticated in the way they grow and modify cells, e.g., induction of differentiation and modification of cell cycle. When "normal" cells are grown or treated in a new way (e.g., low serum, low temperature, "microgravity," novel growth surfaces, and medium supplements), a number of changes can be induced in the existing cell line. These changes could have effects of significance to their safe handling, such as altered expression of oncogenes and proto-oncogenes, expression of endogenous viruses, and interactions between recombinant viruses and endogenous genomic provirus. Such conditions may also lead to expansion of genetically altered clones that may take over the culture and alter its characteristics. Scientists and laboratory staff performing such experiments should be aware of the possibility that such events may occur and should be vigilant for them.

DNA methylation at CpG island promoters in transformed cell lines has been demonstrated to differ from that in the parental tissue of origin and is accompanied by an increase in CpG island methylation with immortalization and passage number (136–138). However, *de novo* hypermethylation is much less common in nontransformed primary cell lines. This hypermethylation has been hypothesized to result from the loss of demethylase activity in culture. It has also been demonstrated that within 3 days of initiation of cell culture there is a global erasure of 5-hydroxymethylcytosine, suggesting rapid loss of methylcytosine dioxygenase activity. The observed epigenetic and transcriptional reprogramming has significant implications for the use of cell lines as faithful mimics of *in vivo* epigenetic and physiological processes (139).

OPERATOR PROTECTION

Most national guidelines recommend that human and other primate cells be handled using biosafety level 2 (BSL2) practices and containment and that all work be performed in a BSC. There is general guidance on Good Cell Culture Practice (80, 140), which is an important element in safe laboratory practices. Because of the capacity of cell cultures to harbor persistent infection with viruses, it is recommended that all cultures be treated as potentially infectious and appropriate containment be applied. Before handling a cell line for the first time, a risk assessment (see discussion under “Risk Assessment,” above) should be carried out to identify the likely infectious potential of the culture and how the risks might be affected by the particular culture and processing procedures that will be applied in the laboratory. All waste cell culture material should be decontaminated by incineration, autoclaving, or appropriate disinfection before discard.

Use of Well-Characterized and Tested Material and Prevention of Contamination

Using well-characterized cell lines or cell lines from controlled sources is a sensible safety measure. For primary cells, the animal colony from which they originate should be screened for key pathogens. In some closely controlled systems, the animals will be certified as “specified-pathogen-free” animals. Some culture collections also now routinely test cell lines for human pathogens. Even if the sources are well defined and the cell lines are tested for adventitious agents, contamination may be introduced through handling (the operator) or medium sources (e.g., serum or trypsin), and it is not practicable to cover all potential contaminants in such testing. Therefore, preventive measures are vital to ensure protection of both the operator and the cell line at all times, as described below.

Practical Quarantine Procedures

Notwithstanding any containment requirements that may be required for particular cells as mentioned above, there are certain fundamental quarantine procedures that are advisable when any new cell line is received in the laboratory. The most immediate concern is contamination with bacteria, fungi, and mycoplasmas, since these organisms will survive in the tissue culture environment and readily contaminate cell cultures. However, viruses released from cultures may also survive to some extent on work surfaces (e.g., herpesviruses, enteroviruses). A number of practical procedures can be adopted to prevent transfer of such agents between cultures, including the following:

- Provide operator training in aseptic technique.
- Treat each culture as potentially infectious; clean up any culture fluid spills immediately.
- Work with one cell line in the BSC at a time and disinfect the work surfaces between cell lines.
- Aliquot growth medium so that the same vessel is not used for more than one cell line.
- Avoid pouring actions, which are a potential source of cross-contamination.
- Turn on BSCs for a period before and after use and thoroughly disinfect BSC surfaces at the end of each working session.
- Do not clutter the BSC with materials that could disrupt the airflow and result in introduction of contaminated laboratory air into the cabinet.
- Restrict the use of antibiotics in growth media. Antimicrobial agents may mask, but not eliminate, contamination, and resistant organisms may arise. Antibiotics should be specifically excluded, except for backup cultures, when preparing cryopreserved cell stocks.
- Quarantine new cell cultures, where facilities allow, to a dedicated BSC or a separate laboratory operating at negative pressure to other laboratory areas. This containment should be maintained until the culture has at least (i) been shown to be negative in sterility tests for bacteria and fungi, (ii) been shown to be negative in tests for mycoplasmas, and (iii) shows no evidence of cytopathic effects due to viral contamination.
- Handle cell cultures from undefined or unscreened sources as Risk Group 2 agents. If there is a reasonable likelihood of adventitious agents of a higher-risk class,

e.g., cell lines from patients at risk of HIV, the cell line should be handled under the appropriate containment level until tests have proven the cell line's safety.

Should microbial contamination arise in the laboratory, there is a risk it may spread quickly, especially in multiuser labs, and rapid response may be required to manage such situations and limit the potential damage to experimental and manufacturing programs. For examples of how to prevent and manage such situations in the research laboratory, see references 83, 141.

Practical Safety Measures

The most important safety measures are aimed at reducing direct contamination through splashes and direct contact or aerosol formation. In general, procedures that might lead to penetration injuries (e.g., a needlestick, glass laceration), abrasive injuries, or exposure to aerosols (e.g., microcentrifugation, rubber-stoppered vessels) should be avoided or carefully controlled. Aseptic technique, while generally very effective as a barrier to infection, may not be able to protect against such "high-energy" processes such as ultracentrifuges, microcentrifuges, and vortex mixers, which may need additional containment measures to prevent transmission of infection.

Good Microbial Technique

All safety measures are based on some fundamental rules, and these are embodied in the good microbial techniques for the safe handling of microorganisms with risk potential or the principles of good occupational safety and hygiene. Best practice measures for the safe handling of microorganisms with risk potential include the following:

- The operators should have basic knowledge of microbiology. Spread of pathogens (for example, via contaminated surfaces, hands, or clothes) should be controlled using a combination of procedural and physical control and containment. All workers should be aware of the risks of cultivated pathogens to people in the vicinity and should be trained in emergency procedures. Entry to the working place should be confined to persons who are aware of these risks.
- Protective laboratory clothes and any other prescribed personal protective equipment must be worn by personnel to prevent infection and spreading of pathogens. Protective clothing should not be worn out of the work area.
- Eating, drinking, and mouth pipetting must be prohibited in the laboratory.
- Hands should be washed after each laboratory work session or after a spill of infectious material and upon leaving the laboratory.
- Working surfaces and tables should be disinfected after normal work as well as after spills of infectious material. In the case of a spill of infectious material, affected work surfaces and floors should be cleaned and disinfected.
- Activities that may produce aerosols should be eliminated or contained in the workplace. Blending and filling of infectious or potentially infectious material should be carried out in a BSC or under other appropriate containment approved under the local biological safety rules (see below).
- Equipment used should be reliable and effective for its purpose and monitored by staff for safe operation and correct function.
- Infectious waste should be placed in sealable containers, the outside of which should be disinfected before transport to the autoclave or incinerator. Special procedures may be required for waste contaminated with certain pathogens, such as those in Risk Group 3 and above.
- Heat or chemical sterilization processes should be investigated beforehand to ensure that the expected microbial load can be rendered safe, i.e., the required killing rate is obtained.
- In case of accidents, an emergency procedure with details of first aid, cleaning, and disinfection should be available, and staff should be trained accordingly. All laboratory workers should be familiar with local safety rules and have appropriate access to safety manuals and local safety advice.
- Procedures for safe handling, disinfection, waste disposal, and emergency should be documented and all staff should be aware of the correct procedures.

In addition to these recommendations, a risk assessment of each new procedure should be performed prior to initiating the work so that any necessary procedures can be put in place.

In 1991 the European Centre for the Validation of Alternative Methods initiated a program to establish best practice with special reference to cell culture, which was published as guidance on Good Cell Culture Practice in 2005 by Coecke et al. (140). This guidance lays out the principles of best practice in cell and tissue culture under six key principles:

1. Establishment and maintenance of a sufficient understanding of the *in vitro* system and of the relevant factors that could affect it.
2. Assurance of the quality of all materials and methods, and of their use and application, in order to maintain the integrity, validity, and reproducibility of any work conducted.
3. Documentation of the information necessary to track the materials and methods used, to permit the

repetition of the work, and to enable the target audience to understand and evaluate the work.

4. Establishment and maintenance of adequate measures to protect individuals and the environment from any potential hazards.
5. Compliance with relevant laws and regulations and with ethical principles.
6. Provision of relevant and adequate education and training for all personnel, to promote high-quality work and safety.

This guidance also identified approaches to achieve these practices and templates for quality control and reporting cell culture systems, thus providing a comprehensive guide for generation and use of cell cultures in almost any laboratory setting. At the time of writing, this guidance is being updated to include best practice principles for stem cell lines and three-dimensional culture (80).

Correct Installation, Use, and Maintenance of BSCs

To prevent the contamination of tissue cultures and to allow procedures to be carried out free of the effects of contamination, it is vital that the BSC in use be installed, used, and maintained appropriately according to local and national rules. The BSC should be maintained to ensure that the typical fine particles released within the cabinet cannot escape through the front of the cabinet (operator protection tests) and that the filters providing sterile air into the cabinet function correctly. Some of the specific issues in the use of BSCs for cell culture have been raised above.

Shipping

Typically cell lines are shipped cryopreserved in vials in solid carbon dioxide ("dry ice" or "cardice"), which gives reliable and stable transportation, provided appropriate arrangements are made with shipping agents (such as topping up on long-distance shipments). Occasionally instances have been reported of vials exploding due to expansion of liquid nitrogen within the vial following removal of shipped vials from cardice. Care should be taken with all vials that have been recently in liquid nitrogen, using appropriate protective masks and gloves. Cell lines may also be shipped successfully as growing cultures; while in most cases these will survive in good condition for up to 5 days, the impact of such treatment on sensitive cultures such as stem cell lines should be considered. To avoid frothing of the growth medium, which would apply damaging shear stress to the cells, the culture is usually prepared in a plastic tissue culture flask or other screw-top vessel that is completely filled with growth medium. Survival of rapidly growing cultures may benefit from reduced serum (i.e., 2% to 5%) in the transport medium to prevent overgrowth and excessive cell death. Sealing and secondary containment of flasks (with inclusion of sufficient adsorbent materials around the flask to hold all liquid contents) are important safety measures and will also help to prevent contamination of cultures during air freight. Such contaminants may arise from the flask seal being broken by vibration and pressure changes in transit. For air freight, cell cultures are classified as "diagnostic specimens" under the International Air Transport Association (IATA) regulations. Up-to-date information on packaging and labeling can be found on the IATA website (http://www.iata.org).

International shipment of cell lines of non-human origin (including animal and plant) may also be subject to the international Convention on Biodiversity, under which numerous countries are obliged to complete the requirements of the Nagoya Protocol; details can be found at https://www.cbd.int/abs/.

SAFETY CONSIDERATIONS IN GENE THERAPY AND GRAFT GENERATION

Gene therapy protocols have been much debated regarding patient safety, due to immunological reactions in the recipient and potential reactivation of recombinant competent viruses from the gene vectors used. In many cases, gene therapy concepts are based on viral vectors derived from adenoviruses, adeno-associated viruses, retroviruses, and herpes simplex virus (for reviews see references 142, 143). By engineering the vector to be replication deficient, e.g., by deletion of early genes of adenoviruses, the pathogenic potential and spreading *in vivo* can be reduced. If vectors do not integrate into the genome, the risk of mutagenesis is minimized. Various clinical trials have been established. Up-to-date information on U.S. trials can be found at http://www.genetherapynet.com/clinicaltrialsgov.html; a review of gene therapy trials worldwide was published in 2007 (144).

In recent years, many packaging lines have been developed to produce new recombinant viruses (27). The packaging cell lines require careful characterization similar to that required for other cells used to make biologicals (12). The testing regimens for packaging cells would also include characterization of integration sites (e.g., by fluorescent *in situ* hybridization) and vector and helper virus sequences (e.g., mRNA analysis; cellular DNA analysis by Southern blot and DNA). In the case of retroviral and adenoviral vector producer cell lines, a critical safety test for these cells lines is confirmation of the absence of replication-competent virus.

Gene therapy is also being tested for applications in the field of cancer treatment, for example, the therapy of melanomas by suicide genes inserted directly into melanoma cells or the insertion of the herpes simplex virus thymidine kinase gene into cancer cells to elicit ganciclovir sensitivity by direct injection of the plasmid (145). For a review of this aspect of gene therapy applications, see Lo et al. (146).

SAFETY MEASURES FOR LARGE-SCALE PRODUCTION OF BIOLOGICALS

Biotechnology products have an exceptional record with respect to process safety concerns. Before biopharmaceuticals produced from animal cells are approved for use, they must be scrutinized for the possibility of transmission of agents or other potential hazardous contaminants to patients. The required level of quality assurance is achieved by determining the quality of raw materials, the production process, and the downstream procedures, including purification. Characterized cell banks, well-designed and validated processes, and sophisticated analytical technology are the key elements for success (12).

Cell banks guarantee a reserve of original, consistent, and reproducible starting material. The typical cell banking systems consist of a master cell bank and a working cell bank (derived from a vial of the master) that is used for the seeding of production runs. In addition, regulatory authorities will seek information on the stability of the production cell line beyond the population doublings defined in the production process. Thus, extended cultivation and retesting are required for further characterization beyond the passage levels attained in production runs. Detailed documentation and environmental monitoring are required to ensure protection from contamination with dangerous substances or cross-contamination with previously handled cells. Safety testing of cell banks, including sterility tests, mycoplasma testing, and identity/stability tests, is required to be in compliance with Good Laboratory Practice (147, 148). For detailed guidance on the evaluation and characterization of cell substrates see references 12, 149.

To control potential contamination of biological products, it is necessary to address the specific issues of cell substrates, raw materials, purification process validation, and final product testing. Cell substrates must be tested at different stages of the process, as outlined in reference 12. This describes the tests normally applicable to the early seed stocks, master cell bank, working cell bank, and any extended cell banks.

Fermentation processes must be fully contained to ensure the safety of the product and to exclude adventitious agents from the cells; that is, the manufacturer must show that the fermentation was run under aseptic conditions (150). The art of aseptic design has developed rapidly in recent years. However, the importance of other aspects of hygienic design is sometimes underestimated. Surface finish, "dead" spaces, alignment of piping, and many other criteria are important in maintaining a high standard of cleanability and in avoiding the potential for buildup of contaminating materials. Appropriate engineering has also been developed for "cleaning in place" and "sterilization in place," which reduce the need to dismantle production equipment for disinfection between production runs (150).

The production process must be run under reproducible, validated conditions, since changes in the process have been reported to influence product safety in vaccine production, e.g., cell densities influencing accumulation of revertants in poliovirus vaccine (151). For a review of the elements of system and process validation see reference 88.

Purification processes must be validated to prove that they are capable of removing certain impurities to an acceptable level. Special consideration has been given to the capacity of the downstream procedure to remove the following:

- Components originating from the host cell (e.g., protein, DNA, or endogenous viruses)
- Impurities caused by medium components or substances used during downstream processing (nutrients, buffer components, stabilizers, chromatography media, etc.);
- Potential external contamination by adventitious agents (e.g., viruses, mycoplasmas, bacteria, or fungi) that should not be present throughout the process but could contaminate the culture by accident.

A number of physical techniques have been developed for the inactivation or removal of viruses. Those that are approved for use in the manufacture of biologicals include virus inactivation by pH extremes, heat, radiation, chromatography, and filtration (84, 152). A number of specific virus inactivation steps have been used by manufacturers, and these include the following:

- Formaldehyde (used in vaccine production)
- Solvent or detergent treatment (which is effective against many enveloped viruses but is not generally effective against enveloped viruses)
- Caprylate (effective against enveloped viruses)
- β-Propiolactone (directed against the viral genome)
- Gamma irradiation (directed against the viral genome).

Even if approved for one specific product, procedures have to be validated on a case-by-case basis if applied to a different process or production cell line. Manufacturers have to validate their purification systems to demonstrate inactivation (77, 84) and/or removal of viruses, nucleic acids, mycoplasmas, and scrapie-like agents. This validation process is extremely costly since it is time-consuming and requires expertise in handling of adventitious agents and analytical procedures.

The downstream processing and inactivation procedures are required to remove potential contamination with adventitious agents to such extremely low levels that they would not generally be detectable with standard sample sizes. However, the potential for infection in patients receiving such material would remain. Thus, the ability of downstream processes to achieve sufficient viral removal must be demonstrated by spiking the purification process with model viruses. Each part of the downstream procedures has to be spiked with model contaminants to evaluate the inactivation or removal capacity of each step with special reference to viruses. These spiking tests are performed on a model scale, and therefore sound scale-up strategies have to be used to guarantee equivalent contaminant clearance at the production scale. Validation studies are necessary to quantitate the inactivation and/or removal of relevant model agents by using a scaled-down model of the actual purification process, as it is used in the manufacturing procedure. The selection of the model contaminants should be based on the following:

- Model agents with a range of biophysical and structural features (see Table 3)
- Similarity of the model agents to contaminants to be expected considering the cell line origin and raw materials, such as the murine retroviruses commonly found in murine myeloma cell lines and hybridomas
- Availability of a sensitive and reliable assay for the model contaminant
- Biosafety of the model contaminant.

The European Commission prepared a note for guidance on the validation of virus removal and inactivation procedures (153, 169). This document made recommendations on validation processes, candidate model viruses, and a number of physiochemical properties for the selection of model viruses. Alternative viruses provide nonpathogenic models for serious pathogens (e.g., BVDV as a model for HCV). Many viruses have been used for validation studies, and examples of these are given in by Roberts (84) and are available through companies offering virus removal validation services. The early use of wild-type poliovirus as a model picornavirus is inconsistent with programs for eradication of this virus in populations internationally.

CHARACTERIZATION OF CELL LINES

Growth

An important element in characterization of a cell line is its growth characteristics, including growth rate, adherence, and maximum cell densities. Cell quantification is classically performed by dye exclusion from viable cells (e.g., with trypan blue or Evan's blue) and using a hemocytometer (microscopic counting chamber) or an electronic cell counter. Cell mass can be estimated by measuring dry mass, protein, or DNA. Numbers of adherent cells can be estimated by using indirect measurements such as oxygen or glucose consumption rates and the quantification of lactate dehydrogenase released by dead or dying cells. A wide variety of other biochemical tests can be used to estimate cell number and viability (see Table 4). However, these are based on widely differing parameters (e.g., membrane integrity, activation of apoptosis, enzyme activity, subcellular organelle function) and must be interpreted with caution in light of their biochemical basis.

Loss of adherence of anchorage-dependent cell lines or adherence of suspension-type cells can also indicate shifts in cell metabolism or indicate contamination or toxic effects. Failure to achieve typical population doubling rates and maximum cell densities may also be indicative of such problems.

Usually the maximum number of allowable population doublings has to be defined for the production of biopharmaceuticals in a standard production run. This addresses the concern of genetic and phenotypic drift or viral expression that may occur during extended serial subculture. Diploid cell lines become senescent (fail to replicate), and the limit for population doublings for use of such cell lines in production is set before the point of decrease in cell doubling rate (11, 154). For heteroploid cells, limits have to be defined but not as stringently controlled. A change in cell growth characteristics at standard culture conditions may also help to identify contamination or toxic effects of medium components or materials.

Quantifying growth efficiency

Growth curves are typically established by seeding a culture at low cell concentration and then measuring cell numbers (per milliliter or centimeter squared) and plotting the $\log^{10}$ value over time. This curve typically produces a linear phase (exponential phase), at which point cells overall are in the most rapid replication, and the period over which cell numbers double is given as the population doubling time. It is important to note that while this figure tends to be characteristic of the cell type, it may be affected by the culture conditions.

Plating efficiency can also be used to characterize cell cultures or determine cytopathic effects. Cells that show capacity for sustained proliferation and formation of colonies of at least 50 cells after five or six doublings are called "clonogenic." By relating the number of colonies formed to the number of cells plated, the colony-forming efficiency, or plating efficiency, can be calculated. It is important to remember that the growth medium and the quality of serum may affect the plating efficiency and will need to be controlled, typically by parallel culture of a reference cell line such as MRC-5 fibroblasts.

Morphology

Microscopic evaluation of cell morphology can yield valuable information on the cell line status, such as differentiation, dedifferentiation, proliferation, growth inhibition, and pathological effects, including viral cytopathic effect. Morphological evaluation can be greatly assisted by the use of histological staining methods (e.g., Giemsa or hematoxylin and eosin) that can reveal details of intracellular structure.

Isoenzyme Patterns

Isoenzyme analysis makes use of the fact that cells contain enzymes with the same substrate specificity but different molecular characteristics due to differences in gene expression or posttranslational modification. Where enzymes are used that are known to be monomorphic within species or strain but polymorphic between species, the resulting electrophoretic mobility patterns may be characteristic of the species or strain of origin. In many cases, sufficient information to identify the species of origin can be obtained from analyzing two sets of isoenzymes, i.e., lactate dehydrogenase and glucose-6-phosphate dehydrogenase. Specific identification of a cell line from a human individual by this method is not a realistic approach, and DNA profiling techniques are much more efficient for this purpose. The availability of reliable molecular tests (see discussion under "Safety Testing of Cell Lines," below) is now replacing this technique for cell line speciation. However, many pharmaceutical companies have established the use of isoenzyme analysis for authentication of cell lines of non-human origin (155).

DNA Barcoding

Using PCR amplification and DNA sequencing technology, it is possible to directly visualize DNA polymorphisms. Until recently it has been necessary to perform this type of work in a species-specific manner. However, the characterization of the mitochondrial cytochrome *c* oxidase I (COI) gene in many species has revealed a rapidly evolving region of DNA flanked by highly conserved regions. This enabled the design of primer pairs that should allow the PCR amplification of the COI in most, if not all, plant and animal species (156). The PCR product (648 bp in most species) is then sequenced to yield a "barcode" specific to the species being tested. The method has been widely suggested as panacea for molecular taxonomy (157). Therefore, it is likely that the method will become widely used to identify the species of cell cultures. Databases of DNA barcodes are maintained by organizations such as the Barcode of Life (www.boldsystems.org).

Immunological Characterization

Immunological characterization of cells by the use of species-specific antiserum and a fluorescence-labeled antibody directed against the antibody is used as a verification method for the donor species. Labeled cells can be directly examined by fluorescent light microscopy or fluorescence-activated cell sorting. This technique provides very high sensitivity compared to isoenzymology and nucleic acid probes. Cross-contamination of up to 1 in 10,000 cells can be detected. However, under the current principles to refine, reduce, and replace the use of animals, production of polyclonal sera in animals is not a favored procedure. Thus, for many laboratories this technique may be of historical interest only.

Molecular Characterization

A wide variety of techniques are now available for the characterization of RNA and protein expression profiles in cell cultures. These may be based on quantitative PCR for specific RNA species or microarray techniques that can analyze expression of thousands of sequences at any one time. In addition, proteomic approaches can reveal protein expression patterns. These techniques will find important applications in the future for the quality control and safety testing of cell cultures used in manufacture and cell therapy, but at this time they remain research tools until efficient and robust data mining and knowledge regarding the expression patterns can be established. Pathway-specific microarray and multiple quantitative PCR array tests for expression of well-characterized molecules may, however, become valuable in routine quality control in the near future.

Cytogenetic Analysis (Karyology)

Cell line characterization may include cytogenetic analysis by staining and examination by microscopy of the mitotic chromosomes. Each cell population has a unique modal number of copies of normal chromosomes and may also carry "marker" chromosomes. Frequently used chromosome staining methods include Giemsa, quinacrine fluorescence, constitutive heterochromatin, and reverse Giemsa stains. A full description and discussion of methods can be found in reference 158.

Karyology by Geimsa staining has been regarded as a key tool in cell line identification and stability testing,

TABLE 3.

Viruses commonly used for validation studies

Virus	Family	Natural host	Genome	Envelope (nm)	Size	Shape	Resistance
Poliovirus, Sabin type 1[a]	*Picornaviridae*	Humans	RNA	No	25–30	Isohedral	Medium
Reovirus 3	*Reoviridae*	Various	RNA	No	60–80	Spherical	High
Simian virus 40	*Polyomaviridae*	Monkey	DNA	No	45	Isohedral	High
Murine leukemia virus	*Retroviridae*	Mice	RNA	Yes	80–110	Spherical	Low
Human immuno-deficiency virus	*Retroviridae*	Humans	RNA	Yes	80–100	Spherical	Low
Vesicular stomatitis virus	*Rhabdoviridae*	Bovine	RNA	Yes	80–100	Bullet shape	Low
Parainfluenza virus	*Paramyxoviridae*	Various	RNA	Yes	150–300	Spherical to pleomorphic	Low
Pseudorabies virus	*Herpesviridae*	Swine	DNA	Yes	120–200	Spherical	Medium

[a]In view of the imminent eradication of polio and likely cessation of vaccination programs, an alternative for polio Sabin type 1 will have to be identified.

although the validity of this method for such testing has been questioned because of the inherent drift of chromosomal complement and a degree of variation in Geimsa-banded karyotypes between individuals of the same species. Nevertheless, where a particular cell line carries an aberrant and unique marker chromosome, identification can be straightforward and accurate.

Molecular Cytogenetic Analysis

A variety of new cytogenetic tools have been developed for genetics research that can give a range of information on specific mutations or larger-scale genetic changes, such as chromosomal translocation, inversion, and aneuploidy. Microarray comparative genome hybridization (aCGH) techniques have been developed that can provide data that are almost equivalent to those obtained with traditional karyology, but with greater resolution to detect much smaller genetic changes. Hybrid aCGH/single-nucleotide polymorphism microarrays can also detect (diploid) regions showing loss of heterozygosity, which was previously very difficult to do. This approach is rapidly being taken up as an alternative to preparing karyotypes. Other cytogenetic tools, such as "spectral" karyotyping, fluorescent *in situ* hybridization (FISH), and flow cytometry of chromosomes, give valuable data on chromosomal structure but are more typically research tools rather than routine quality control techniques.

DNA Fingerprinting, PCR, and Restriction Fragment Length Polymorphism

DNA fingerprinting identifies a cell line by visualization of the structure of the extremely variable repetitive component of genomic DNA and has certain advantages over other traditional identification techniques. Multi-locus DNA fingerprinting using M13 phage (159) or human minisatellite sequences (160, 161) has been used for the detection of cross-contamination and identification of cell lines (162–164). Although these methods are accurate for cell identification across a range of species, they are time-consuming and laborious. DNA fingerprinting and profiling methods have been des-

TABLE 4.

Examples of techniques used for the measurement of cell growth and viability

Method	Basic comments
Trypan blue dye exclusion	Exclusion of dye from cell indicates a functional intact membrane (apoptotic cell fragments may retain this function).
Fluorescein diacetate	Functional membrane esterase activity cleaves fluorescein diacetate to yield fluorescence within viable cells (184) (damaged tissues may also show a positive reaction).
MTT[a] assay	MTT is converted to an insoluble dark blue formazan product by biochemically active cells (185) (cell activation is measured, not cell proliferation).
Neutral red assay	3-Amino-7-dimethyl-2-methylphenazine hydrochloride (neutral red) is a supravital dye that accumulates in the lysosomes of viable cells and may be detected by spectrophotometric determination of neutral red in washed and fixed cells (186); the method often must be optimized for different cell types.

[a]MTT, 3-(4,5-dimethylthiazol-2-yl)-2,5-diphenyl tetrazolium bromide.

cribed as multilocus (where test conditions allow cross-hybridization of the DNA probe with related sequences) or single-locus (where test conditions generally permit a particular DNA probe to detect only one sequence that is its precise complementary sequence). These technologies are now not commonly used due to the advent of mutilocus PCR profiling, i.e., short tandem repeat (STR) profiling.

Single-locus testing analyzes variable numbers of tandem repeats at specific loci, producing simple DNA profiles. This technique can utilize PCR for amplification. Single-locus tests can be very specific when several primer sets are combined in multiplex PCRs for several loci. However, such primers often are designed for specific species (usually human) and may not be useful for other species. Producing and correctly interpreting DNA fingerprints require considerable care and experience. Purity of labeled probes, use of internal molecular weight markers and standard DNA samples, and reproducibility of electrophoresis and Southern blotting are essential. PCR is now commonly used to identify cell lines and to detect both cross-contamination by other cell lines and contamination by adventitious agents or in combination with other methods, as outlined above.

Restriction fragment length polymorphisms in specific genes and random amplified polymorphic DNA (RAPD) techniques have been used in the past to characterize the cell genome, but have not found useful broad application. However, an alternative technique has been developed that utilizes conserved intron sequences flanking polymorphic genes (165); it has been applied in the identification of cell lines (166) and has been used by some biopharmaceutical companies.

Product

The ability of a (recombinant) cell line to produce significant amounts of a highly specific protein can be used to characterize that cell line. However, changes in product characteristics and structure (such as glycosylation) can arise due to changes in the cell culture conditions (e.g., antibodies secreted by hybridoma cells, recombinant therapeutic proteins expressed by CHO cells). For reviews, see references 167, 168.

SAFETY TESTING OF CELL LINES

Cell lines should be tested for the most common environmental contaminants and the contaminants most likely to arise from the tissue and species of origin. A range of tests have been developed for the detection of mycoplasmas, bacteria, fungi, and viruses.

Bacteria and Fungi

It is important to test cell banks for the presence of bacteria and fungi and to repeat such testing when cells are maintained for extended periods with antibiotic-containing medium. Although many bacterial or fungal contaminants will become apparent through increased turbidity, abnormal shift in medium pH, and increased oxygen demand, their growth may be obscured by cell debris, slow growth, and antibiotics.

Examination of cultures by microscopy should be performed routinely. Phase-contrast optics and high magnification (e.g., oil immersion) may enhance detection. However, examination by microscopy will only enable the detection of fairly pronounced contamination. For increased sensitivity, culture tests using bacteriological growth media are performed. A panel of test media (e.g., blood agar, thioglycolate broth, tryptone soy broth, brain heart infusion broth, Sabouraud broth, YM broth, and 2% yeast broth) and incubation temperatures (26°C to 37°C) can be used to cover a wide spectrum of possible contaminants. It should be clear that culture tests will usually fail to detect viable organisms that will not grow under the sterility test conditions or bacteria and fungi with extremely low growth rates or long lag phases. Some bacteria are known to be detectable only weeks after inoculation. However, many contaminants should be detected by a system based on media such as tryptone soy broth and thioglycolate broth for bacteria and Sabouraud's medium for fungi, with incubation temperatures of 25°C and 35°C. It is important to note that standard sterility tests defined in national pharmacopeia are intended to expose general breaches in aseptic processing through the detection of readily culturable organisms. They should, therefore, not be used to provide a categorical statement that a cell culture is free of all bacterial and fungal species.

Mycoplasmas

Mycoplasmas can be detected by using bacteriological culture methods, indicator cell lines, DNA staining, PCR, and immunological or biochemical methods. A combination of these methods is necessary to enhance detection in important samples (e.g., master and working cell banks). Culture methods and the use of indicator cell lines, such as Vero cells, have the same drawback as discussed for bacterial and fungal contaminants. DNA staining is a commonly used method to detect the DNA in mycoplasmas around and on cells, using Hoechst stain 33258 or 4′,6′-diamidino-2-phenylindole (DAPI). Whatever method is used, it should allow detection of those *Mycoplasma* and *Acholeplasma* species known to occur in cell culture. The method should also be sensitive enough to detect contaminants when the mycoplasma growth may be inhibited by antibiotics.

Several kits are available for PCR-based detection of *Mycoplasma* spp., mostly utilizing PCR primers that will detect sequences found in all or most common contaminating species. Real-time detection kits are now available providing high sensitivity and specificity. Despite this advance, the culture method is still considered the "gold standard" in terms of sensitivity and wide range of species detected. The culture method, however, is often not the method of choice, as several weeks of culture may be required and the growth of *Mycoplasma* organisms presents an additional contamination threat to the cell culture laboratory.

As alluded to above, in relation to sterility testing, there are organisms that will grow in the rich nutritional environment of cell culture but not on standard bacteriological media used in standard sterility test methods. This means that some organisms, such as mycoplasmas, mycobacteria, and other organisms, may grow undetected in cell cultures. Identification of such contaminants will require isolation techniques tailored for fastidious organisms that may require both rich nutrition and an atmosphere enriched in CO_2. These organisms may not cause turbidity or other overt signs of contamination, but can cause difficulties where the infection may only become apparent under the different environmental conditions of a bioassay or bioreactor.

Viruses

Viruses may be detected by observation of cytopathic effects, such as syncytium formation, cell rounding, or vacuolation on indicator cell lines; by detection of viral proteins such as coat protein or enzymes; by detection of viral particles by electron microscopy; or by detection of nucleic acid sequences. General requirements for virological safety of biotechnology products have been published by the International Conference on Harmonisation (169).

Prions

Prions can be detected by a technique known as the protein misfolding cyclic amplification (PMCA) assay, in which the sample protein is mixed with brain homogenate from transgenic mice expressing human prion protein. The prion-positive material acts as an oligomeric substrate for PMCA in which misfolded monomeric prion protein is enabled to develop into polymers. If misfolded prions are present in the test sample, after repeated cycles of PMCA, the resultant polymers are detected by Western blotting (170).

Specific Tests

PCR

PCR provides a powerful tool for the detection of nucleic acid sequences present at very low levels. However, it may be inhibited by substances in the test samples and may give nonspecific reactions. Inhibition of PCR can be identified by running test samples in parallel with the same sample spiked with a positive control DNA at the lower limits of detection in addition to a separate positive control PCR. Failure to detect the spiked sample, alongside a satisfactory positive control result, may indicate inhibition of the PCR. PCR primers exist for many viruses, and it is necessary to identify the priorities for testing based on the likelihood of viral contamination (see discussion in "Risk Assessment," above). Primer sequences should be validated for each PCR method to ensure that there is no cross-reaction with cell DNA-specific genomic viral sequences in cells or with contaminating microorganisms.

Similar concerns apply with reverse transcriptase PCR (RT-PCR), in which mRNA is isolated from the test sample, transcribed into DNA by RT, and then subjected to PCR. The greater number of technical steps means that there are more risks involved that might cause the test to fail. The method of preparing the RNA, storage of RNA, and so forth can affect the sensitivity of detection, and the influence of such variations should be checked before a validated methodology is altered. One of the most common known contaminants in fetal calf serum is BVDV (171). Appropriate RT-PCR methods are available to screen new sources of serum for BVDV (131, 172). Sources of serum that are positive for BVDV by RT-PCR may not be infectious if the RNA is inactivated, but it would be wise to have evidence for viral inactivation before using such sera.

When using any DNA amplification technique, it is vital to have specialized equipment or isolated laboratories for the different stages of the PCR process to avoid contamination of new reactions with PCR products. Once cross-contamination becomes a problem, it can be very difficult to eradicate. A recent review of viral contamination risk for porcine trypsin and bovine serum by Marcus-Sekura et al. (173) has identified that testing regimes may need to be updated when screening for viral contamination in such materials.

Nonspecific Tests

Reverse transcriptase

Retroviruses can be detected using the RT assays from ultracentrifuged samples (e.g., 125,000 × *g* for 1 h) in the presence of magnesium or manganese. An inducing agent such as bromodeoxyuridine, iododeoxyuridine, or dexamethasone may enable the detection of latent virus.

Highly sensitive detection methods, commonly referred to as PCR-enhanced RT, are available. They are based on spiking test samples with specific template DNA for the RT activity, followed by PCR of resulting DNA (174, 175). However, such methods require very careful controls to enable identification of nonspecific reactions.

DNA sequencing

The sequencing reaction in "next-generation" sequencing methods is not primed by a specific primer but by a ligated linker molecule. As such, DNA molecules in a soup of molecules may be sequenced in a massively parallel manner. It is therefore possible to use next-generation DNA sequencing to identify contaminants in a DNA sample. In an exemplar of this approach, Victoria et al. (176) derived DNA sequence data from eight live attenuated viral vaccines. In one, a rotavirus vaccine, they detected significant contamination with porcine circovirus-1, which is likely to have derived from typsin treatment of the cell substrate. With the price of DNA sequencing dropping rapidly, it is likely that this approach will become commonplace in the safety testing of vaccines.

Cell coculture

Virus tests can be performed *in vitro* and *in vivo*. *In vitro* tests are carried out by the inoculation of a test sample into various susceptible indicator cell cultures capable of detecting a wide range of murine, human, bovine, and other possible animal viruses. The choice of cell line is dependent on the origin of the production cell line. Useful indicator cell lines include human diploid cell lines such as MRC-5, monkey kidney cell lines, and CCLs of various origins (12). The current WHO guidance recommends use of a panel of cell lines that include representatives of the following:

- Cultures (primary cells or CCLs) of the same species and tissue type as that used for production
- Cultures of a human diploid cell line. WHO also recommends that where the production cell line is of human origin, a simian kidney cell line should also be used (note that national guidance may also require a third cell line from a different species to be used).

For cell substrates new to application in manufacturing (e.g., insect cell lines, duck embryo cell line), further cell lines may be required to detect viruses known to be potentially harmful to humans.

Inoculated cultures are then observed regularly over an extended period for cytopathic effects. They may be finally screened by a variety of methods—e.g., electron microscopy, hemagglutination, or hemadsorption—to determine the presence of noncytopathic virus. Sensitivity for the detection of cell-associated viruses may be enhanced by applying cycles of freezing and thawing to the cells being tested. However, this effect may not be consistent for different cell-associated viruses, and virus viability may significantly decrease with multiple freeze-thaw cycles. The WHO guidance (12) includes recommendations on components of the testing. Where retroviral contamination is identified as a specific risk, then highly susceptible cells, such as *Mus dunni* cells, may be used.

In vivo assays

Where justified in terms of public health, *in vivo* assays may be performed by intramuscular injection of the sample into animals, including suckling mice, adult mice (these are also inoculated intracerebrally to test for the presence of lymphocytic choriomeningitis virus), guinea pigs, and embryonated eggs. The health of the animal is then monitored; if abnormalities occur, they are investigated to find their cause. Species-specific viruses can be detected by inoculating virus-free animals and monitoring the serum antibody level as in the mouse antibody production (MAP) test.

Electron microscopy

Retroviruses can also be detected by electron microscopy of cells before and after induction with chemicals such as 5-bromodeoxyuridine. Both transmission electron microscopy and negative stain scanning electron microscopy have been used (177), and new technologies are now enabling a broader range of high-resolution studies, including analysis of hydrated samples (178). In addition, electron microscopy has proven valuable to distinguish between viral and nonviral cytopathic effects sometimes seen in cell culture inoculation assays (179).

TUMORIGENICITY STUDIES

The ability of a particular cell substrate to cause tumors has been a cause for concern for many years. Assays are performed by inoculation of animals, but the appearance of tumors has not generally been associated with any real hazard, especially where the final product is highly purified and cell free. In the case of ES cells, formation of teratocarcinomas in mice that show evidence of the three primary germ layers of the embryo (endoderm, epiderm, and mesoderm) is used as an indication of their pluripotency and potential efficacy for providing a broad range of cell types for therapy, rather than a measure of risk. However, the risks represented by residual ES cell populations or transformed teratocarcinoma cells in such differentiated products for transplantation suggest that sensitive methods for detection of tumorigenic cells will be needed (85).

QUALITY STANDARDS

Safety testing of cells used in the manufacture of biologicals is usually performed with procedures accredited to Good Laboratory Practice (147, 148). Organizations using cell lines for services and production may also gain accreditation to ISO 9000, and where monitoring and testing are critical features, cell culture procedures may require the ISO Guide 25 (EN45001) quality standard. The preparation of a pharmaceutical product will need to be carried out in a current Good Manufacturing Practices (cGMP)-compliant facility. Any part of the process of manufacture may need to meet these and other requirements of cGMP and will apply to cell culture procedures used in production (180). Cell-based assays used for diagnostic purposes may need to comply with the international standard ISO 13485 and the In Vitro Diagnostics Directive in Europe. For cell therapy, requirements in the United States can be found at the U.S. FDA site (http://www.fda.gov/biologicsbloodvaccines/guidancecomplianceregulatoryinformation/guidances/cellularandgenetherapy/default.htm). In Europe, such requirements are covered under the European Tissues and Cells Directive (70) and Advanced Therapies Medicinal Products regulations (187). Additional guidance can be found in FDA and EMA advice on somatic cell therapies in the sections on cellular therapies (31, 181).

ADDITIONAL INFORMATION

Web Resources

Information on cell cultures

- ATCC (United States): http://www.atcc.org
- Coriell Cell Repositories, Coriell Institute for Medical Research (United States): https://www.coriell.org
- DSMZ (Germany): http://www.dsmz.de
- European Collection of Authenticated Cell Cultures: https://www.phe-culturecollections.org.uk/collections/ecacc.aspx
- European Medicines Agency: http://www.ema.europa.eu
- World Data Centre for Microorganisms (Japan): http://www.wdcm.org/
- Riken (Japan): http://en.brc.riken.jp
- Japanese Collection of Research Bioresources (Japan): http://cellbank.nibiohn.go.jp/english/
- Interlab Cell Line Collection (Italy): http://wwwsql.iclc.it
- Development Studies Hybridoma Bank: http://dshb.biology.uiowa.edu
- CABRI Europe: http://www.cabri.org
- UK Stem Cell Bank: http://www.nibsc.org/ukstemcellbank

Biosafety guidelines and online training

- United States: **U.S. Department of Health and Human Services, Public Health Service, Centers for Disease Control and Prevention, National Institutes of Health.** 2009. *Biosafety in Microbiological and Biomedical Laboratories*, 5th ed. HHS Publication no. (CDC) 21-112. http://www.cdc.gov/biosafety/publications/bmbl5/BMBL.pdf.
- Guidance on Good Cell Culture Practice. A report of the second ECVAM Task Force on Good Cell Culture Practice (Coecke et al., 2005). http://www.atla.org.uk
- WHO Laboratory Safety Manual, 3rd ed. http://www.who.int/csr/resources/publications/biosafety/WHO_CDS_CSR_LYO_2004_11/en

We thank Otto Doblhoff-Dier for his contributions to earlier forms of this manuscript.

References

1. **Griffiths JB.** 2007. The development of animal cell products: history and overview, p 1–14. *In* Stacey G, Davis J (ed), *Medicines from Animal Cells*. John Wiley & Sons, Chichester, United Kingdom.
2. **Cooney CL.** 1995. Are we prepared for animal cell technology in the 21st century? *Cytotechnology* **18:**3–8.
3. **Lysaght MJ, Jaklenec A, Deweerd E.** 2008. Great expectations: private sector activity in tissue engineering, regenerative medicine, and stem cell therapeutics. *Tissue Eng Part A* **14:** 305–315.
4. **Nerem, R.M.** 2010. Regenerative medicine: the emergence of an industry. *J R Soc Interface* **7**(Suppl 6)**:**S771–S775.
5. **Jaklenec A, Stamp A, Deweerd E, Sherwin A, Langer R.** 2012. Progress in the tissue engineering and stem cell industry: "Are we there yet?". *Tissue Eng Part B Rev* **18:**155–166.
6. **Maziarz RT, Driscoll D.** 2011. Hematopoietic stem cell transplantation and implications for cell therapy reimbursement. *Cell Stem Cell* **8:**609–612.
7. **Abou-El-Enein M, Elsanhoury A, Reinke P.** 2016. overcoming challenges facing advanced therapies in the EU market. *Cell Stem Cell* **19:**293–297.
8. **Petricciani JC.** 1995. The acceptability of continuous cell lines: a personal & historical perspective. *Cytotechnology* **18:**9–13.
9. **International Association of Microbiological Societies.** 1963. *Proceedings: Symposium on the Uses of Human Diploid Cell Strains.* Blasnikova-Tiskarna, Lubliana, Zagreb, Croatia.
10. **WHO Expert Committee on Biological Standardization and Executive Board.** 1998. *Requirements for the Use of Animal Cells as In vitro Substrates for the Production of Biologicals.* WHO Technical Report Series no. 878. World Health Organization, Geneva, Switzerland.

11. **Knezevic I, Stacey G, Petricciani J, Sheets R, WHO Study Group on Cell Substrates.** 2010. Evaluation of cell substrates for the production of biologicals: Revision of WHO recommendations. Report of the WHO Study Group on Cell Substrates for the Production of Biologicals, 22–23 April 2009, Bethesda, MD. *Biologicals* 38:162–169.
12. **WHO Cell Substrate Study Group.** 2010. Recommendations for the evaluation of animal cell cultures as substrates for the manufacture of biological medicinal products and for the characterization of cell banks. Proposed replacement of TRS 878, Annex 1. http://www.who.int/biologicals/BS2132_CS_Recommendations_CLEAN_19_July_2010.pdf.
13. **Doroshenko A, Halperin SA.** 2009. Trivalent MDCK cell culture-derived influenza vaccine Optaflu (Novartis Vaccines). *Expert Rev Vaccines* 8:679–688.
14. **Cox RJ, Madhun AS, Hauge S, Sjursen H, Major D, Kuhne M, Höschler K, Saville M, Vogel FR, Barclay W, Donatelli I, Zambon M, Wood J, Haaheim LR.** 2009. A phase I clinical trial of a PER.C6 cell grown influenza H7 virus vaccine. *Vaccine* 27:1889–1897.
15. **Genzel Y, Reichl U.** 2009. Continuous cell lines as a production system for influenza vaccines. *Expert Rev Vaccines* 8:1681–1692.
16. **Kuczewski M, Schirmer E, Lain B, Zarbis-Papastoitsis G.** 2011. A single-use purification process for the production of a monoclonal antibody produced in a PER.C6 human cell line. *Biotechnol J* **6:**56–65
17. **Ross D, Brown T, Harper R, Pamarthi M, Nixon J, Bromirski J, Li CM, Ghali R, Xie H, Medvedeff G, Li H, Scuderi P, Arora V, Hunt J, Barnett T.** 2012. Production and characterization of a novel human recombinant alpha-1-antitrypsin in PER.C6 cells. *J Biotechnol* **162:**262–273.
18. **Sakhuja K, Reddy PS, Ganesh S, Cantaniag F, Pattison S, Limbach P, Kayda DB, Kadan MJ, Kaleko M, Connelly S.** 2003. Optimization of the generation and propagation of gutless adenoviral vectors. *Hum Gene Ther* 14:243–254.
19. **European Medicines Agency.** 2011. Reflection paper on design modifications of gene therapy medicinal products during development. EMA/CAT/GTWP/44236/2009 (14 December 2011). http://www.ema.europa.eu/docs/en_GB/document_library/Scientific_guideline/2012/02/WC500122743.pdf.
20. **Damle B, White R, Wang HF.** 2015. Considerations for clinical pharmacology studies for biologics in emerging markets. *J Clin Pharmacol* **55**(Suppl 3)**:**S116–S122.
21. **Rosenberg SA, Restifo NP, Yang JC, Morgan RA, Dudley ME.** 2008. Adoptive cell transfer: a clinical path to effective cancer immunotherapy. *Nat Rev Cancer* **8:**299–308.
22. **Leen AM, Rooney CM, Foster AE.** 2007. Improving T cell therapy for cancer. *Annu Rev Immunol* **25:**243–265.
23. **Kershaw MH, Westwood JA, Darcy PK.** 2013. Gene-engineered T cells for cancer therapy. *Nat Rev Cancer* **13:**525–541.
24. **Gilham DE, Anderson J, Bridgeman JS, Hawkins RE, Exley MA, Stauss H, Maher J, Pule M, Sewell AK, Bendle G, Lee S, Qasim W, Thrasher A, Morris E.** 2015. Adoptive T-cell therapy for cancer in the United kingdom: a review of activity for the British Society of Gene and Cell Therapy annual meeting 2015. *Hum Gene Ther* **26:**276–285.
25. **Jackson HJ, Rafiq S, Brentjens RJ.** 2016. Driving CAR T-cells forward. *Nature Rev Clin Oncol* 13:370–383.
26. **Miller AD.** 1990. Retrovirus packaging cells. *Hum Gene Ther* **1:**5–14.
27. **Stacey GN, Merten O-W.** 2010. Host cells and cell banking, p 45–88. *In* Al-Rubeai M, Merten O-W (ed), *Viral Vectors for Gene Therapy: Methods and Protocols.* (Methods in Molecular Biology 1:737). Springer Science+Business Media LLC, New York, NY.
28. **Epstein S, U.S. Food and Drug Administration.** 1996. Addendum to the points to consider in human somatic cell and gene therapy (1991). *Hum Gene Ther* **7:**1181–1190.
29. **U.S. Food and Drug Administration.** 2011. *Guidance for Industry Potency Tests for Cellular and Gene Therapy Products.* U.S. Department of Health and Human Services, Food and Drug Administration, Center for Biologics Evaluation and Research, Silver Spring, MD.
30. **European Medicines Agency.** 2009. Committee for Medicinal Products for Human Use (CHMP), Doc.EMA/CHMP/GTWP/212377/2008. http://www.ema.europa.eu/docs/en_GB/document_library/Scientific_guideline/2010/01/WC500059111.pdf.
31. **European Medicines Agency.** 2001. Points to consider on the manufacture and quality control of hum somatic cell therapy medicinal products. CPMP/BMP/41450/98. https://www.old.health.gov.il/download/forms/a39_cimmittee.pdf.
32. **Esvelt KM, Wang HH.** 2013. Genome-scale engineering for systems and synthetic biology. *Mol Syst Biol* 9:641.
33. **Horvath P, Barrangou R.** 2010. CRISPR/Cas, the immune system of bacteria and archaea. *Science* **327:**167–170.
34. **Isasi R, Kleiderman E, Knoppers BM.** 2016. Editing policy to fit the genome? Framing genome editing policy requires setting thresholds of acceptability. *Science* **351:**337–339.
35. **Li MD, Atkins H, Bubela T.** 2014. The global landscape of stem cell clinical trials. *Regen Med* **9:**27–39.
36. **Green H, Kehinde O, Thomas J.** 1979. Growth of cultured human epidermal cells into multiple epithelia suitable for grafting. *Proc Natl Acad Sci USA* **76:**5665–5668.
37. **Navsaria HA, Myers SR, Leigh IM, McKay IA.** 1995. Culturing skin *in vitro* for wound therapy. *Trends Biotechnol* **13:**91–100.
38. **Wu J, Du Y, Watkins SC, Funderburgh JL, Wagner WR.** 2011. The engineering of organized human corneal tissue through the spatial guidance of corneal stromal stem cells. *Biomaterials* **33:**1343–1352.
39. **Ende N, Chen R.** 2002. Parkinson's disease mice and human umbilical cord blood. *J Med* **33:**173–180.
40. **Ende N, Chen R, Reddi AS.** 2004. Transplantation of human umbilical cord blood cells improves glycemia and glomerular hypertrophy in type 2 diabetic mice. *Biochem Biophys Res Commun* **321:**168–171.
41. **Barker RA, Mason SL, Harrower TP, Swain RA, Ho AK, Sahakian BJ, Mathur R, Elneil S, Thornton S, Hurrelbrink C, Armstrong RJ, Tyers P, Smith E, Carpenter A, Piccini P, Tai YF, Brooks DJ, Pavese N, Watts C, Pickard JD, Rosser AE, Dunnett SB, Simpson S, Moore J, Morrison P, Esmonde T, Chada N, Craufurd D, Snowdon J, Thompson J, Harper P, Glew R, Harper R, NEST-UK Collaboration.** 2013. The long-term safety and efficacy of bilateral transplantation of human fetal striatal tissue in patients with mild to moderate Huntington's disease. *J Neurol Neurosurg Psychiatry* **84:**657–665.
42. **Rocha V, Wagner JE Jr, Sobocinski KA, Klein JP, Zhang MJ, Horowitz MM, Gluckman E.** 2000. Graft-versus-host disease in children who have received a cord-blood or bone marrow transplant from an HLA-identical sibling. Eurocord and International Bone Marrow Transplant Registry Working Committee on Alternative Donor and Stem Cell Sources. *N Engl J Med* **342:**1846–1854.
43. **Garcia J.** 2010. Allogeneic unrelated cord blood banking worldwide: an update. *Transfus Apheresis Sci* **42:**257–263.
44. **Rubinstein P.** 2009. Cord blood banking for clinical transplantation. *Bone Marrow Transplant* **44:**635–642.
45. **Hofmeister CCJ, Zhang J, Knight KL, Le P, Stiff PJ.** 2007. Ex vivo expansion of umbilical cord blood stem cells for transplantation: growing knowledge from the hematopoietic niche. *Bone Marrow Transplant* **39:**11–23.
46. **Ge J, Cai H, Tan WS.** 2011. Chromosomal stability during ex vivo expansion of UCB CD34(+) cells. Cell Prolif **44:**550–557.
47. **Corselli M, Parodi A, Mogni M, Sessarego N, Kunkl A, Dagna-Bricarelli F, Ibatici A, Pozzi S, Bacigalupo A, Frassoni F, Piaggio G.** 2008. Clinical scale ex vivo expansion of cord blood-derived

outgrowth endothelial progenitor cells is associated with high incidence of karyotype aberrations. *Exp Hematol* **36:**340–349.
48. **Crow J, Youens K, Michalowski S, Perrine G, Emhart C, Johnson F, Gerling A, Kurtzberg J, Goodman BK, Sebastian S, Rehder CW, Datto MB.** 2010. Donor cell leukemia in umbilical cord blood transplant patients. A case study and literature review highlighting the importance of molecular engraftment analysis. *J Mol Diagn* **12:**530–537.
49. **Kassem M, Kristiansen M, Abdallah BM.** 2004. Mesenchymal stem cells: cell biology and potential use in therapy. *Basic Clin Pharmacol Toxicol* **95:**209–214.
50. **Calloni R, Cordero EA, Henriques JA, Bonatto D.** 2013. Reviewing and updating the major molecular markers for stem cells. *Stem Cells Devel* **22:**1455–1476.
51. **Nombela-Arrieta C, Ritz J, Silberstein LE.** 2011. The elusive nature and function of mesenchymal stem cells. *Nat Rev Mol Cell Biol* **12:**126–131.
52. **Horwitz EM, Gordon PL, Koo WK, Marx JC, Neel MD, McNall RY, Muul L, Hofmann T.** 2002. Isolated allogeneic bone marrow-derived mesenchymal cells engraft and stimulate growth in children with osteogenesis imperfecta: implications for cell therapy of bone. *Proc Natl Acad Sci USA* **99:**8932–8937.
53. **Ishikane S, Ohnishi S, Yamahara K, Sada M, Harada K, Mishima K, Iwasaki K, Fujiwara M, Kitamura S, Nagaya N, Ikeda T.** 2008. Allogeneic injection of fetal membrane-derived mesenchymal stem cells induces therapeutic angiogenesis in a rat model of hind limb ischemia. *Stem Cells* **26:**2625–2633.
54. **Nakajima H, Uchida K, Guerrero AR, Watanabe S, Sugita D, Takeura N, Yoshida A, Long G, Wright KT, Johnson WE, Baba H.** 2012. Transplantation of mesenchymal stem cells promotes the alternative pathway of macrophage activation and functional recovery after spinal cord injury. *J Neurotrauma* **29:**1614–1625.
55. **Puglisi MA, Tesori V, Lattanzi W, Piscaglia AC, Gasbarrini GB, D'Ugo DM, Gasbarrini A.** 2011. Therapeutic implications of mesenchymal stem cells in liver injury. *J Biomed Biotechnol* **2011:** 860578.
56. **Biancone L, Bruno S, Deregibus MC, Tetta C, Camussi G.** 2012. Therapeutic potential of mesenchymal stem cell-derived microvesicles. *Nephrol Dial Transplant* **27:**3037–3042.
57. **Evans MJ, Kaufman MH.** 1981. Establishment in culture of pluripotential cells from mouse embryos. *Nature* **292:**154–156.
58. **Thomson JA, Itskovitz-Eldor J, Shapiro SS, Waknitz MA, Swiergiel JJ, Marshall VS, Jones JM.** 1998. Embryonic stem cell lines derived from human blastocysts. *Science* **282:** 1145–1147.
59. **Taylor CJ, Bolton EM, Bradley JA.** 2011. Immunological considerations for embryonic and induced pluripotent stem cell banking. *Philos Trans R Soc Lond B Biol Sci* **366:**2312–2322.
60. **Barry J, Hyllner J, Stacey G, Taylor CJ, Turner M.** 2015. Setting up a haplobank: issues and solutions. *Curr Stem Cell Rep* **1:** 110–117.
61. **Fairchild PJ.** 2010. The challenge of immunogenicity in the quest for induced pluripotency. *Nat Rev Immunol* **10:**868–875.
62. **Takahashi K, Tanabe K, Ohnuki M, Narita M, Ichisaka T, Tomoda K, Yamanaka S.** 2007. Induction of pluripotent stem cells from adult human fibroblasts by defined factors. *Cell* **131:** 861–872.
63. **Gore A, Li Z, Fung HL, Young JE, Agarwal S, Antosiewicz-Bourget J, Canto I, Giorgetti A, Israel MA, Kiskinis E, Lee JH.** 2011. Somatic coding mutations in human induced pluripotent stem cells. *Nature* **471:**63–67.
64. **Lund RJ, Närvä E, Lahesmaa R.** 2012. Genetic and epigenetic stability of human pluripotent stem cells. *Nat Rev Genet* **13:** 732–744.
65. **Zhao T, Zhang ZN, Rong Z, Xu Y.** 2011. Immunogenicity of induced pluripotent stem cells. *Nature* **474:**212–215
66. **Guha P, Morgan JW, Mostoslavsky G, Rodrigues NP, Boyd AS.** 2013. Lack of immune response to differentiated cells derived from syngeneic induced pluripotent stem cells. *Cell Stem Cell* **12:**407–412.
67. **International Stem Cell Initiative.** 2011. Screening ethnically diverse human embryonic stem cells identifies a chromosome 20 minimal amplicon conferring growth advantage. *Nat Biotechnol* **29:**1132–1144.
68. **Baker DE, Harrison NJ, Maltby E, Smith K, Moore HD, Shaw PJ, Heath PR, Holden H, Andrews PW.** 2007. Adaptation to culture of human embryonic stem cells and oncogenesis in vivo. *Nat. Biotechnol* **25:**207–221.
69. **U.S. Food and Drug Administration.** 2007. *Guidance for Industry: Regulation of Human Cells, Tissues, and Cellular and Tissue-Based Products (HCT/Ps)—Small Entity Compliance Guide.* U.S. Department of Health and Human Services, Food and Drug Administration, Center for Biologics Evaluation and Research, Silver Spring, MD. http://www.fda.gov/BiologicsBloodVaccines/GuidanceComplianceRegulatoryInformation/Guidances/Tissue/ucm073366.htm.
70. **EUTCD.** 2004. Directive 2004/23/EC of the European Parliament and of the Council of 31 March 2004: Setting standards of quality and safety for the donation, procurement, testing, processing, preservation, storage and distribution of human tissue and cells. L102/48, Official Journal of the European Union, 7.4.2004.
71. **Global Programme for Vaccines and Immunization.** 1997. *The WHO Guide to Good Manufacturing Practice (cGMP) Requirements.* World Health Organization, Geneva, Switzerland.
72. **Keirstead HS, Nistor G, Bernal G, Totoiu M, Cloutier F, Sharp K, Steward O.** 2005. Human embryonic stem cell-derived oligodendrocyte progenitor cell transplants remyelinate and restore locomotion after spinal cord injury. *J Neurosci* **25:** 4694–4705.
73. **Hayakawa T, Aoi T, Bravery C, Hoogendoorn K, Knezevic I, Koga J, Maeda D, Matsuyama A, McBlane J, Morio T, Petricciani J, Rao M, Ridgway A, Sato S, Sato Y, Stacey G, Trouvin J-H, Umezawa A, Yamato M, Yano K, Yokote H, Yoshimatsu K, Zorzi-More P.** 2015. Report of the international conference on regulatory endeavors towards the sound development of human cell therapy products. *Biologicals* **43:**283–297.
74. **British Standards Institute Regenerative Medicines Committee.** 2011. *PAS 93:2011: Characterization of Human Cells for Clinical Applications. Guide.* British Standards Institute, London, United Kingdom.
75. **Sheridan B, Stacey G, Wilson A, Ginty P, Bravery C, Marshall D.** 2012. Standards can help bring products to market. *Bioprocess Int* **10:**18–20.
76. **Williams DJ, Archer R, Archibald P, Bantounas I, Baptista R, Barker R, Barry J, Bietrix F, Blair N, Braybrook J, Campbell J.** 2016. Comparability: manufacturing, characterization and controls, report of a UK Regenerative Medicine Platform Pluripotent Stem Cell Platform Workshop, Trinity Hall, Cambridge, 14–15 September 2015. *Regenerative Med* **11:**483–492.
77. **U.S. Food and Drug Administration.** 1998. *Guidance for Industry Guidance for Human Somatic Cell Therapy and Gene Therapy.* U.S. Department of Health and Human Services, Food and Drug Administration, Center for Biologics Evaluation and Research, Silver Spring, MD. http://www.fda.gov/BiologicsBloodVaccines/GuidanceComplianceRegulatoryInformation/Guidances/CellularandGeneTherapy/ucm072987.htm.
78. **Wobus AM, Löser P.** 2011. Present state and future perspectives of using pluripotent stem cells in toxicology research. *Arch Toxicol* **85:**79–117.
79. **Stacey GN, Coecke S, Price A, Healy L, Jennings P, Wilmes A, Pinset C, Sundstrom M, Myatt G.** 2016. Ensuring the quality of stem cell derived models for toxicity testing, p 259–297. *In* Eskes C, Whelan M (ed), *Validation of Alternative Methods for Toxicity Testing.* Springer, New York, NY.
80. **Pamies D, Bal-Price A, Simeonov A, Tagle D, Allen D, Gerhold D, Yin D, Pistollato F, Inutsuka T, Sullivan K, Stacey G.** 2016.

Good Cell Culture Practice for stem cells and stem-cell-derived models. *ALTEX* Online August 23, 2016, version 2. http://dx.doi.org/10.14573/altex.1607121.

81. **Frommer W, Archer L, Boon B, Brunius G, Collins CH, Crooy P, Doblhoff-Dier O, Donikian R, Economidis J, Frontali C, Gaal T, Hamp S, Haymerle H, Houwink EH, Küenzi MT, Krämer P, Lelieveld HLM, Logtenberg MT, Lupker J, Lund S, Mahler JL, Mosgaard C, Normand-Plessier F, Rudan F, Simon R, Tuijenburg Muijs G, Vranch SP, Werner RG.** 1993. Safe biotechnology (5). Recommendations for safe work with animal and human cell cultures concerning potential human pathogens. *Appl Microbiol Biotechnol* **39:**141–147.
82. **Jank B, Haymerle H, Doblhoff-Dier O.** 1996. Zurich hazard analysis in biotechnology. *Nat Biotechnol* **14:**894–896.
83. **Geraghty RJ, Capes-Davis A, Davis JM, Downward J, Freshney RI, Knezevic I, Lovell-Badge R, Masters JRW, Meredith J, Stacey GN, Thraves P, Vias M, Cancer Research UK.** 2014. Guidelines for the use of cell lines in biomedical research. *Br J Cancer* **111:**1021–1046.
84. **Roberts PL.** 2007. Virus safety of cell-derived biological products, p 371–392. *In* Stacey G, Davis J (ed), *Medicines from Animal Cells.* John Wiley & Sons, Chichester, United Kingdom.
85. **Andrews PW, et al.** 2015. Points to consider in the development of seed stocks of pluripotent stem cells for clinical applications: International Stem Cell Banking Initiative (ISCBI). *Regen Med* **10**(Suppl)**:**1–44.
86. **Davidson WL, Hummeler K.** 1960. B virus infection in man. *Ann N Y Acad Sci* **85:**970–979.
87. **National Research Council.** 1989. *Prudent Practices for the Handling and Disposal of Infectious Material*, p 13–33. National Academy Press, Washington, DC.
88. **Chesterton N.** 2007. System and process validation, p 285–302. *In* Stacey G, Davis J (ed), *Medicines from Animal Cells.* John Wiley & Sons, Chichester, United Kingdom.
89. **Urnovitz HB, Murphy WH.** 1996. Human endogenous retroviruses: nature, occurrence, and clinical implications in human disease. *Clin Microbiol Rev* **9:**72–99.
90. **Patzke S, Lindeskog M, Munthe E, Aasheim HC.** 2002. Characterization of a novel human endogenous retrovirus, HERV-H/F, expressed in human leukemia cell lines. *Virology* **303:**164–173.
91. **Weiss RA.** 2001. Adventitious viral genomes in vaccines but not in vaccinees. *Emerg Infect Dis* **7:**153–154.
92. **Adamson SR.** 1998. Experiences of virus, retrovirus and retrovirus-like particles in Chinese hamster ovary (CHO) and hybridoma cells used for production of protein therapeutics. *Dev Biol Stand* **93:**89–96.
93. **Weissmahr RN, Schüpbach J, Böni J.** 1997. Reverse transcriptase activity in chicken embryo fibroblast culture supernatants is associated with particles containing endogenous avian retrovirus EAV-0 RNA. *J Virol* **71:**3005–3012.
94. **Ma H, Khan AS.** 2011. Detection of latent retroviruses in vaccine-related cell substrates: investigation of RT activity produced by chemical induction of Vero cells. *PDA J Pharm Sci Technol* **65:**685–689.
95. **Lloyd G, Jones N.** 1986. Infection of laboratory workers with hantavirus acquired from immunocytomas propagated in laboratory rats. *J Infect* **12:**117–125.
96. **Saïb A, Périès J, de Thé H.** 1995. Recent insights into the biology of the human foamy virus. *Trends Microbiol* **3:**173–178.
97. **Schweizer M, Falcone V, Gänge J, Turek R, Neumann-Haefelin D.** 1997. Simian foamy virus isolated from an accidentally infected human individual. *J Virol* **71:**4821–4824.
98. **Huang F, Wang H, Jing S, Zeng W.** 2012. Simian foamy virus prevalence in *Macaca mulatta* and zookeepers. *AIDS Res Hum Retrovir* **28:**591–593.
99. **Khan AS.** 2009. Simian foamy virus infection in humans: prevalence and management. *Expert Rev Anti Infect Ther* **7:**569–580.
100. **Locatelli S, Peeters M.** 2012. Cross-species transmission of simian retroviruses: how and why they could lead to the emergence of new diseases in the human population. *AIDS* **26:**659–673.
101. **Kusne S, Vilchez RA, Zanwar P, Quiroz J, Mazur MJ, Heilman RL, Mulligan D, Butel JS.** 2012. Polyomavirus JC urinary shedding in kidney and liver transplant recipients associated with reduced creatinine clearance. *J Infect Dis* **206:**875–880.
102. **Sood P, Senanayake S, Sujeet K, Medipalli R, Zhu YR, Johnson CP, Hariharan S.** 2012. Management and outcome of BK viremia in renal transplant recipients: a prospective single-center study. *Transplantation* **94:**814–821.
103. **van Aalderen MC, Heutinck KM, Huisman C, ten Berge IJ.** 2012. BK virus infection in transplant recipients: clinical manifestations, treatment options and the immune response. *Neth J Med* **70:**172–183.
104. **Löwer J.** 1995. Acceptability of continuous cell lines for the production of biologicals. *Cytotechnology* **18:**15–20.
105. **Lelong-Rebel IH, Piemont Y, Fabre M, Rebel G.** 2009. *Mycobacterium avium-intracellulare* contamination of mammalian cell cultures. *In Vitro Cell Dev Biol Anim* **45:**75–90.
106. **Kajander EO, Ciftçioglu N.** 1998. Nanobacteria: an alternative mechanism for pathogenic intra- and extracellular calcification and stone formation. *Proc Natl Acad Sci USA* **95:**8274–8279.
107. **Shiekh FA.** 2012. Do calcifying nanoparticles really contain 16S rDNA? *Int J Nanomedicine* **7:**5051–5052.
108. **Kumon H, Matsumoto A, Uehara S, Abarzua F, Araki M, Tsutsui K, Tomochika K.** 2011. Detection and isolation of nanobacteria-like particles from urinary stones: long-withheld data. *Int J Urol* **18:**458–465.
109. **Gray JS, Birmingham JM, Fenton JI.** 2009. Got black swimming dots in your cell culture? Identification of Achromobacter as a novel cell culture contaminant. *Biologicals* **38:**273–277.
110. **Shannon M, Capes-Davis A, Eggington E, Georghiou R, Huschtscha LI, Moy E, Power M, Reddel RR, Arthur JW.** 2016. Is cell culture a risky business? Risk analysis based on scientist survey data. *Int J Cancer* **138:**664–670.
111. **Rottem S, Naot Y.** 1998. Subversion and exploitation of host cells by mycoplasmas. *Trends Microbiol* **6:**436–440.
112. **Uphoff CC, Drexler HG.** 2014. Detection of mycoplasma contamination in cell cultures. *Curr Protoc Mol Biol* **106:**1–14.
113. **Uphoff CC, Denkmann SA, Drexler HG.** 2012. Treatment of mycoplasma contamination in cell cultures with Plasmocin. *J Biomed Biotechnol* **2012:**267678.
114. **Alper T, Cramp WA, Haig DA, Clarke MC.** 1967. Does the agent of scrapie replicate without nucleic acid? *Nature* **214:**764–766.
115. **Griffith JS.** 1967. Self-replication and scrapie. *Nature* **215:**1043–1044.
116. **Prusiner SB.** 1982. Novel proteinaceous infectious particles cause scrapie. *Science* **216:**136–144.
117. **Aguzzi A, Heikenwalder M.** 2006. Pathogenesis of prion diseases: current status and future outlook. *Nat Rev Microbiol* **4:**765–775.
118. **Prusiner SB, Collinge J, Powell J, Anderton B (ed).** 1992. *Prion Diseases of Humans and Animals.* Ellis Horwood, London, United Kingdom.
119. **Sigurdson CJ, Miller MW.** 2003. Other animal prion diseases. *Br Med Bull* **66:**199–212.
120. **Kovacs GG, Budka H.** 2008. Prion diseases: from protein to cell pathology. *Am J Pathol* **172:**555–565.
121. **European Commission.** 2011. European Medicines Agency guidance document. Note for guidance on minimising the risk of transmitting animal spongiform encephalopathy agents via human and veterinary medicinal products (EMA/410/01 rev.3). *J European Union* http://www.emea.europa.eu/docs/en_GB/document_library/Scientific_guideline/2009/09/WC500003700.pdf.
122. **World Health Organization.** 2010. *WHO Guidelines on Tissue Infectivity Distribution in Transmissible Spongiform Encephalop-*

athies. World Health Organization, Geneva, Switzerland. http://www.who.int/bloodproducts/tse/WHO%20TSE%20Guidelines%20FINAL-22%20JuneupdatedNL.pdf.

123. **United Kingdom Department of Health.** 2014. Donation of starting material for cell-based advanced therapies: a SaBTO review. U.K. Department of Health. https://www.gov.uk/government/publications/donation-of-starting-material-for-advanced-cell-based-therapies.
124. **Gugel EA, Sanders ME.** 1986. Needle-stick transmission of human colonic adenocarcinoma. *N Engl J Med* **315:**1487.
125. **Centers for Disease Control and Prevention and National Institutes of Health.** 1999. *Biosafety in Microbiological and Biomedical Laboratories*, 4th ed. U.S. Government Printing Office, Washington, DC.
126. **Human Medicines Evaluation Unit.** 1997. *ICH Topic Q 5 D—Quality of Biotechnological Products: Derivation and Characterisation of Cell Substrates used for Production of Biotechnological/Biological Products*. European Agency for the Evaluation of Medicinal Products, ICH Technical Co-ordination, London, United Kingdom.
127. **Sheng-Fowler L, Cai F, Fu H, Zhu Y, Orrison B, Foseh G, Blair DG, Hughes SH, Coffin JM, Lewis AM Jr, Peden K.** 2010. Tumors induced in mice by direct inoculation of plasmid DNA expressing both activated H-*ras* and c-*myc*. *Int J Biol Sci* **6:**151–162.
128. **Garnick RL.** 1998. Raw materials as a source of contamination in large-scale cell culture. *Dev Biol Stand* **93:**21–29.
129. **Erickson GA, Bolin SR, Landgraf JG.** 1991. Viral contamination of fetal bovine serum used for tissue culture: risks and concerns. *Dev Biol Stand* **75:**173–175.
130. **Yanagi M, Bukh J, Emerson SU, Purcell RH.** 1996. Contamination of commercially available fetal bovine sera with bovine viral diarrhea virus genomes: implications for the study of hepatitis C virus in cell cultures. *J Infect Dis* **174:**1324–1327.
131. **Xia H, Vijayaraghavan B, Belák S, Liu L.** 2011. Detection and identification of the atypical bovine pestiviruses in commercial foetal bovine serum batches. *PLoS One* **6:**e28553.
132. **Onyekaba C, Fahrmann J, Bueon L, King P, Goyal SM.** 1987. Comparison of five cell lines for the propagation of bovine viral diarrhea and infectious bovine rhinotracheitis viruses. *Microbiologica* **10:**311–315.
133. **Nakamura S, Shimazaki T, Sakamoto K, Fukusho A, Inoue Y, Ogawa N.** 1995. Enhanced replication of orbiviruses in bovine testicle cells infected with bovine viral diarrhoea virus. *J Vet Med Sci* **57:**677–681.
134. **Nicklas W, Kraft V, Meyer B.** 1993. Contamination of transplantable tumors, cell lines, and monoclonal antibodies with rodent viruses. *Lab Anim Sci* **43:**296–300.
135. **Zhang W, Li L, Deng X, Blümel J, Nübling CM, Hunfeld A, Baylis SA, Delwart E.** 2016. Viral nucleic acids in human plasma pools. *Transfusion* **56:**2248–2255.
136. **Jones PA, Wolkowicz MJ, Rideout WM III, Gonzales FA, Marziasz CM, Coetzee GA, Tapscott SJ.** 1990. De novo methylation of the MyoD1 CpG island during the establishment of immortal cell lines. *Proc Natl Acad Sci USA* **87:**6117–6121.
137. **Antequera F, Boyes J, Bird A.** 1990. High levels of de novo methylation and altered chromatin structure at CpG islands in cell lines. *Cell* **62:**503–514.
138. **Wilson VL, Jones PA.** 1983. DNA methylation decreases in aging but not in immortal cells. *Science* **220:**1055–1057.
139. **Nestor CE, Ottaviano R, Reinhardt D, Cruickshanks HA, Mjoseng HK, McPherson RC, Lentini A, Thomson JP, Dunican DS, Pennings S, Anderton SM, Benson M, Meehan RR.** 2015. Rapid reprogramming of epigenetic and transcriptional profiles in mammalian culture systems. *Genome Biol* **16:**11.
140. **Coecke S, Balls M, Bowe G, Davis J, Gstraunthaler G, Hartung T, Hay R, Merten O-W, Price A, Shechtman L, Stacey GN, Stokes W.** 2005. Guidance on Good Cell Culture Practice. A report of the second ECVAM Task Force on Good Cell Culture Practice. ATLA **33:**1–27. http://www.atla.org.uk/guidance-on-good-cell-culture-practice
141. **Stacey GN.** 2010. Cell culture contamination, p 77–91. *In* Cree IA (ed), *Cancer Cell Culture: Methods and Protocols*. Springer Science+Business media LLC, New York, NY.
142. **Meager A.** 2007. Gene transfer vectors for clinical applications, p 125–142. *In* Stacey G, Davis J (ed), *Medicines from Animal Cells*. John Wiley & Sons, Chichester, United Kingdom.
143. **Kotterman MA, Chalberg TW, Schaffer DV.** 2015. Viral vectors for gene therapy: translational and clinical outlook. *Annu Rev Biomed Eng* **17:**63–89.
144. **Edelstein ML, Abedi MR, Wixon J.** 2007. Gene therapy clinical trials worldwide to 2007—an update. *J Gene Med* **9:**833–842.
145. **Calvez V, Rixe O, Wang P, Mouawad R, Soubrane C, Ghoumari A, Verola O, Khayat D, Colbère-Garapin F.** 1996. Virus-free transfer of the herpes simplex virus thymidine kinase gene followed by ganciclovir treatment induces tumor cell death. *Clin Cancer Res* **2:**47–51.
146. **Lo HW, Day CP, Hung MC.** 2005. Cancer-specific gene therapy. *Adv Genet* **54:**235–255.
147. **OECD Expert Group on Good Laboratory Practice.** 1982. *Good Laboratory Practice in the Testing of Chemicals*. Organisation of Economic Co-operation and Development, Paris, France.
148. **OECD.** 1998. OECD series on principles of good laboratory practice and compliance monitoring. Number 1. OECD principles on good laboratory practice. ENV/MC/CHEM(98)17. *Guideline* **33:**1–172.
149. **CBER.** 2010. *Characterization and Qualification of Cell Substrates and Other Biological Materials Used in the Production of Viral Vaccines for Infectious Disease Indications*. Food and Drug Administration, Bethesda, MD. http://www.fda.gov/BiologicsBloodVaccines/GuidanceComplianceRegulatoryInformation/Guidances/Vaccines/default.htm.
150. **Stoll TS.** 2007. Services and associated equipment for upstream processing, p 245–285. *In* Stacey G, Davis J (ed), *Medicines from Animal Cells*. John Wiley & Sons, Chichester, United Kingdom.
151. **Dragunsky E, Chumakov K, Norwood L, Parker M, Lu Z, Ran Y.** 1993. Live polio vaccine reversion: impact of cell density. *In Vitro* **3:**123A.
152. **Shukla AA, Hubbard B, Tressel T, Guhan S, Low D.** 2007. Downstream processing of monoclonal antibodies—application of platform approaches. *J Chromatogr B Analyt Technol Biomed Life Sci* **848:**28–39.
153. **Committee for Proprietary Medicinal Products Ad Hoc Working Party on Biotechnology/Pharmacy Working Party on Safety Medicines, Committee for Proprietary Medicin.** 1991. EEC regulatory document. Note for guidance: validation of virus removal and inactivation procedures. *Biologicals* **19:**247–251.
154. **Wood DJ, Minor PD.** 1990. Meeting report: use of diploid cells in production. *Biologicals* **18:**143–146.
155. **Nims RW, Shoemaker AP, Bauernschub MA, Rec LJ, Harbell JW.** 1998. Sensitivity of isoenzyme analysis for the detection of interspecies cell line cross-contamination. *In Vitro Cell Dev Biol Anim* **34:**35–39.
156. **Hebert PDN, Ratnasingham S, deWaard JR.** 2003. Barcoding animal life: cytochrome c oxidase subunit 1 divergences among closely related species. *Proc Biol Sci* **270**(Suppl 1)**:**S96–S99.
157. **Hebert PDN, Cywinska A, Ball SL, deWaard JR.** 2003. Biological identifications through DNA barcodes. *Proc Biol Sci* **270:**313–321.
158. **Chen TC.** 1998. Identity testing, authentication, karyology, p. 9A:1.1–9A:1.20. *In* Doyle A, Griffiths JB, Newell DG (ed.), *Cell & Tissue Culture: Laboratory Procedures*. John Wiley & Sons, Ltd., Chichester, United Kingdom.
159. **Vassart G, Georges M, Monsieur R, Brocas H, Lequarre AS, Christophe D.** 1987. A sequence in M13 phage detects hypervariable minisatellites in human and animal DNA. *Science* **235:**683–684.

160. **Jeffreys AJ, Wilson V, Thein SL.** 1985. Hypervariable 'minisatellite' regions in human DNA. *Nature* **314:**67–73.
161. **Jeffreys AJ, Wilson V, Thein SL.** 1985. Individual-specific 'fingerprints' of human DNA. *Nature* **316:**76–79.
162. **Stacey GN, Bolton BJ, Doyle A.** 1991. The quality control of cell banks using DNA fingerprinting. *EXS* **58:**361–370.
163. **Webb MB, Debenham PG.** 1992. Cell line characterisation by DNA fingerprinting; a review. *Dev Biol Stand* **76:**39–42.
164. **Gilbert DA, Reid YA, Gail MH, Pee D, White C, Hay RJ, O'Brien SJ.** 1990. Application of DNA fingerprints for cell-line individualization. *Am J Hum Genet* **47:**499–514.
165. **Lessa EP, Applebaum G.** 1993. Screening techniques for detecting allelic variation in DNA sequences. *Mol Ecol* **2:**119–129.
166. **Stacey GN, Hoelzl H, Stephenson JR, Doyle A.** 1997. Authentication of animal cell cultures by direct visualization of repetitive DNA, aldolase gene PCR and isoenzyme analysis. *Biologicals* **25:**75–85.
167. **Tarelli E.** 2007. Glycosylation of medicinal products, p 479–490. *In* Stacey G, Davis J (ed), *Medicines from Animal Cells.* John Wiley & Sons, Chichester, United Kingdom.
168. **Ghaderi D, Zhang M, Hurtado-Ziola N, Varki A.** 2012. Production platforms for biotherapeutic glycoproteins. Occurrence, impact, and challenges of non-human sialylation. *Biotechnol Genet Eng Rev* **28:**147–175.
169. **International Conference on Harmonisation.** 1999. ICH harmonised tripartite guideline: viral safety evaluation of biotechnology products derived from cell lines of human or animal origin Q5A(R1) Current Step 4 version dated 23 September. http://www.ich.org/products/guidelines/quality/article/quality-guidelines.html
170. **Moda F, Gambetti P, Notari S, Concha-Marambio L, Catania M, Park K-W, Maderna E, Suardi S, Haïk S, Brandel J-P, Ironside J, Knight R, Tagliavini F, Soto C.** 2014. Prions in the urine of patients with variant Creutzfeldt-Jakob disease. *N Engl J Med* **371:**530–539.
171. **Uryvaev LV, Dedova AV, Dedova LV, Ionova KS, Parasjuk NA, Selivanova TK, Bunkova NI, Gushina EA, Grebennikova TV, Podchernjaeva RJ.** 2012. Contamination of cell cultures with bovine viral diarrhea virus (BVDV). *Bull Exp Biol Med* **153:**77–81.
172. **Vilček S, Herring AJ, Herring JA, Nettleton PF, Lowings JP, Paton DJ.** 1994. Pestiviruses isolated from pigs, cattle and sheep can be allocated into at least three genogroups using polymerase chain reaction and restriction endonuclease analysis. *Arch Virol* **136:**309–323.
173. **Marcus-Sekura C, Richardson JC, Harston RK, Sane N, Sheets RL.** 2011. Evaluation of the human host range of bovine and porcine viruses that may contaminate bovine serum and porcine trypsin used in the manufacture of biological products. *Biologicals* **39:**359–369.
174. **Pyra H, Böni J, Schüpbach J.** 1994. Ultrasensitive retrovirus detection by a reverse transcriptase assay based on product enhancement. *Proc Natl Acad Sci USA* **91:**1544–1548.
175. **Lovatt A, Black J, Galbraith D, Doherty I, Moran MW, Shepherd AJ, Griffen A, Bailey A, Wilson N, Smith KT.** 1999. High throughput detection of retrovirus-associated reverse transcriptase using an improved fluorescent product enhanced reverse transcriptase assay and its comparison to conventional detection methods. *J Virol Methods* **82:**185–200.
176. **Victoria JG, Wang C, Jones MS, Jaing C, McLoughlin K, Gardner S, Delwart EL.** 2010. Viral nucleic acids in live-attenuated vaccines: detection of minority variants and an adventitious virus. *J Virol* **84:**6033–6040.
177. **Poiley JA, Bierley ST, Hillesund T, Nelson RE, Monticello TM, Raineri R.** 1994. Methods for estimating retroviral burden. *Biopharm Manuf* **7:**32–35.
178. **Goldsmith CS, Miller SE.** 2009. Modern uses of electron microscopy for detection of viruses. *Clin Microbiol Rev* **22:**552–563.
179. **Hendricks LC, Jordan J, Yang TY, Driesprong P, Haan GJ, Viebahn M, Mikosch T, Van Drunen H, Lubiniecki AS.** 2010. Apparent virus contamination in biopharmaceutical product at centocor. *PDA J Pharm Sci Technol* **64:**471–480.
180. **Sheu J, Klassen H, Bauer G.** 2014. Cellular manufacturing for clinical applications. *Dev Ophthalmol* **53:**178–188.
181. **U.S. Food and Drug Administration.** 2008. Guidance for FDA Reviewers and Sponsors: Content and Review of Chemistry, Manufacturing, and Control (CMC): Information for Human Somatic Cell Therapy Investigational New Drug Applications (INDs). http://www.fda.gov/BiologicsBloodVaccines/GuidanceComplianceRegulatoryInformation/Guidances/Xenotransplant.
182. **Griffiths B.** 1991. Products from animal cells, p 207–235. *In* Butler M (ed), *Mammalian Cell Biotechnology: A Practical Approach.* IRL Press, Oxford, United Kingdom.
183. **Butler M.** 1991. The characteristics and growth of cultured cells, p 1–25. *In* Butler M (ed), *Mammalian Cell Biotechnology: A Practical Approach.* IRL Press, Oxford, United Kingdom.
184. **Widholm JM.** 1972. The use of fluorescein diacetate and phenosafranine for determining viability of cultured plant cells. *Stain Technol* **47:**189–194.
185. **Mosmann T.** 1983. Rapid colorimetric assay for cellular growth and survival: application to proliferation and cytotoxicity assays. *J Immunol Methods* **65:**55–63.
186. **Bulychev A, Trouet A, Tulkens P.** 1978. Uptake and intracellular distribution of neutral red in cultured fibroblasts. *Exp Cell Res* **115:**343–355.
187. **European Parliament and Council of the European Union.** 2007. Regulation (EC) No 1394/2007 of the European Parliament and of the Council of 13 November 2007 on advanced therapy medicinal products and amending. Directive 2001/83/EC and regulation (EC) no 726/2004 (text with EEA relevance). *J European Union* **324:**121–137. http://eur-lex.europa.eu/LexUriServ/LexUriServ.do?uri=OJ:L:2007:324:0121:0137:en:PDF.
188. **Stacey GN.** 2007. Risk assessment of cell culture procedures, p 569–588. *In* Stacey GN, Davis JM (ed), *Medicines from Animal Cells.* J Wiley & Sons, Chichester, United Kingdom.
189. **European Pharmacopoeia.** 2014. 5.2.12 Raw materials for the production of cell-based and gene therapy medicinal products (PA/PH/Exp. RCG/T (14) 5 ANP). Pharmeuropa 2014, 26.4 (October 1, 2014).
190. **Ministry of Health, Labour and Welfare.** 2014. Notification No. 375 Standard for biological ingredients. http://www.pmda.go.jp/operations/shonin/info/saisei-iryou/pdf/H260926-kijun.pdf
191. **U.S. Pharmacopeia.** 2014. Chapter 1043: Ancillary materials for cell, gene and tissue-engineered products. http://www.pharmacopeia.cn/v29240/usp29nf24s0_c1043.html

Allergens of Animal and Biological Systems

15

WANDA PHIPATANAKUL AND ROBERT A. WOOD

INTRODUCTION

Allergy to laboratory animals is a significant occupational hazard and among the most common conditions affecting the health of workers involved in the care and use of research animals. At least 90,000 workers in the United States have direct contact with animals in research or industrial facilities, although some sources estimate 40,000 to 125,000 (1, 2). Workers who are in regular contact with furred animals often develop sensitivity to these animals. This sensitivity accounts for the high prevalence of laboratory animal allergy (LAA) in animal workers, and cross-sectional studies have estimated that as high as 44% of individuals working with laboratory animals report work-related symptoms (3, 4). Veterinarians are also at risk with similar levels of allergy development with symptoms (5). Of these symptomatic workers, up to 25% may eventually develop occupational asthma that persists even after the exposure ceases (6). This high prevalence rate has major medical and economic implications. When employees develop LAA, it often results in significant morbidity, at times necessitating a change in occupation. In addition, it may lead to reduced productivity, increased workloads for others, and increased health and worker's compensation costs for the employer. Due to recent awareness and increased surveillance and monitoring, recent studies have actually seen a decline in occupational asthma over the last 25 years (7), but recent studies suggest vigilance is necessary (8). Familiarity with LAA, including its clinical characteristics, etiology, pathophysiology, treatment, and preventative measures, can be vital in reducing the economic and physical impact of this important occupational hazard.

DISEASE PATHOGENESIS, CLINICAL SYMPTOMS, AND DIAGNOSIS

As with most environmental allergy, symptoms in LAA occur when an allergen contacts the skin or respiratory mucosa. With repeated allergen exposure, susceptible individuals develop antibodies known as immunoglobulin E (IgE) that are directed against the specific allergen. Subsequent exposures to that allergen cause cells known as mast cells and basophils, which have these IgE molecules on their surface, to degranulate and release a variety of chemical mediators, such as histamine, leukotrienes, and kinins. These mediators cause mucus production, swelling, and inflammation, which lead to the clinical symptoms of allergy. People who develop these symptoms and have demonstrated IgE antibodies against certain allergens are considered allergic or atopic.

Sensitization often occurs within the first 3 years of employment in an animal facility. Risk factors include personal or family history of atopy or allergy, preexisting allergies, high occupational allergen exposure, and environmental tobacco smoke exposure (1). The most important risk factor in the development of LAA is the level of exposure to laboratory animal allergens. Methods have been developed to allow quantitative estimates of the level of exposure (9–12), and studies suggest that higher levels of allergen exposure correlate well with both the development and severity of symptoms (13, 14). Questions exist as to whether individuals with coexisting allergies to allergens outside of the laboratory have an increased susceptibility to developing LAA, but most of the reported studies suggest that it is an important risk factor (2, 15, 16). Individual susceptibility has been examined carefully and multiple objective methods may be used to predict risk in a workplace setting.

Symptoms in laboratory animal allergy range from mild rashes to severe asthma. Overall, the most common symptom is allergic rhinoconjunctivitis, which consists of nasal congestion, clear nasal discharge, sneezing, and itchy, watery eyes. These symptoms have been reported in up to 80% of workers with LAA. Skin reactions occur in about 40% of affected individuals and include contact hives or a chronic, itchy rash known as eczema. Finally, certain individuals develop chest symptoms associated with asthma, such as wheezing, cough, and chest tightness upon exposure to such allergens. Asthma may develop in up to 20–30% of sensitive individuals. Anaphylaxis, a severe and sometimes fatal allergic reaction, can occur, although it is uncommon. Animal bites, more than other forms of exposure, pose the greatest risk (2, 16, 17).

The diagnosis of suspected LAA can be confirmed with the use of skin tests or radioallergosorbent tests (RASTs), which test for the presence of IgE antibodies to specific allergens. Skin tests are typically done with the prick or puncture method. With this procedure, a drop of allergen extract, such as animal pelt, is placed on the surface of the skin, and the skin is then pricked with a lancet or other device. If the patient is allergic to the substance, IgE antibodies on the surface of mast cells in the skin bind with the allergen and cause the release of mediators forming a small hive or wheal with surrounding redness. These reactions are measured after 10 to 15 minutes to determine the degree of reactivity and compared with histamine as a positive control and saline as a negative control. ImmunoCAP allergen-specific IgE is an *in vitro* method of measuring allergen-specific IgE, which generally correlates with skin test results. Skin tests and allergen specific ImmunoCAP can be used alone or together. In addition, physicians may use tests such as pulmonary function testing or methacholine challenge testing to assess for the presence of asthma. Other objective measures, such as nasal lavage, induced sputum, or acoustic rhiniometry, may help provide integrated evidence of when intervention is required, as in complete avoidance (18). By looking at the clinical history in conjunction with appropriate testing, one can determine who has likely been clinically affected by LAA or who may be at high risk of developing symptoms in the future.

ETIOLOGY

Allergens

Most laboratory animal allergens have been identified and characterized (Table 1) (6). The most common causes of LAA are rats and mice. This is primarily due to the fact that these animals are used more often than others, not because the other animals are necessarily less allergenic.

Three mouse allergens have been identified. Mouse urinary protein (Mus m 1 or MUP) has a molecular weight of 19 kDa. This allergen is produced in liver cells and found in hair follicles, dander, and urine. It is found at a level four times higher in males than females because its gene expression is testosterone dependent. A second allergen, Mus m 2, is a glycoprotein with a molecular weight of 16 kDa that is found in hair and dander but not in urine. A final allergen is albumin, which is allergenic in about 30% of mice-sensitive individuals (19–22). Furthermore, recent work in mouse research facilities has found co-exposures with levels of mouse and endotoxin (23, 24). This suggests that multiple exposures may be important in these settings.

Two rat allergens have been identified in urine, saliva, and pelt. Rat n 1A has a molecular weight of 20–21 kDa and Rat n 1B has a molecular weight of 16–17 kDa. Both allergens are variants of α_{2u}-globulin and are found in hair, dander, urine, and saliva. Rat n 1B is also produced in the liver and is androgen dependent. It can also be produced in the salivary, mammary, and other exocrine glands (12). Like mouse, rat albumin is also allergenic in about 24% of rat-allergic individuals (10, 25–27). There has been recent suggestion that dual sensitization to both rat and mouse allergens may be related to some cross-reactivity between the molecules (28).

In guinea pigs, two antigenic fragments termed Cav p 1 and Cav p 2 have been characterized and found in urine, hair, and dander (29). Although rabbit allergens have not been well characterized, two have been identified, Ory c 1

TABLE 1.

Laboratory animal allergens

Animal	Allergen	Molecular mass (kDa)	Source
Mouse (*Mus musculus*) Lipocalin family	Mus m 1 (prealbumin) Mus m 2 Albumin	19 16	Hair, dander, urine Hair, dander Serum
Rat (*Rattus norvegicus*) Lipocalin family	Rat n 1A/Rat n 1 B (α_{2u}-globulin)	16–21	Hair, dander, urine, saliva
Guinea pig (*Cavia porcellus*) Lipocalin family	Cav p 1 Cav p 2		Hair, dander, urine Hair, dander, urine
Rabbit (*Oryctolagus cuniculus*) Lipocalin family	Ory c 1 Ory c2	17 to 18	Urine, dander, saliva Hair, dander, urine
Cat (*Felis domesticus*)	Fel d 1 Albumin	38	Hair, dander saliva Serum
Dog (*Canis familiaris*)	Can f 1 Albumin	25	Hair, dander, saliva Serum
Cow (*Bos domesticus*)	Bos d 2		Hair, dander, saliva
Horse (*Equinus cavis*)	Equ c 1 Equ c 2 Equ c 4 Equ c 5	22 16 18.7 16.7	Hair, dander, saliva

and Ory c 2. Ory c 1 is a 17- to 18-kDa protein found in saliva, urine, and dander, whereas Ory c 2 is found in hair, dander, and urine (30). Recent observations have found that these allergens are in the lipocalin family, which is the same family of proteins as mouse and rat allergens (26).

Although cats and dogs are more often encountered as domestic pets rather than laboratory animals, they are common in laboratory environments as well (31). Although 12 allergenic cat proteins have been identified, the major cat allergen, Fel d 1, is the most important (32, 33). It is a tetrameric polypeptide with a molecular weight of 38 kDa. It is found in cat skin and saliva, and males produce higher levels of this allergen than females (33). The most important dog allergen is Can f 1, which is produced in hair, dander, and saliva, and has a molecular weight of 25 kDa. Dog albumin also has been described as a distinct allergen (34).

Other animals used in laboratories, including gerbils, hamsters, rabbits, cows, and sheep, may also occasionally cause reactions. The major allergens from some of these species, such as cow allergen (Bos d 2), have been identified as a member of the lipocalin family (35, 36). Horses can also be a potent source of allergen. The major horse allergen Equ c 1 is also from the lipocalin family (37). Species-specific data suggest that rates of allergy symptoms are higher in response to rabbit exposures because rabbits shed more fur and spray their urine (3, 38).

Even though primates are used in research facilities, few cases of sensitivity have been documented. There have been cases reported of allergy to bush baby (Galago) and the cottontop tamarin monkeys. These allergens were identified in the animal's dander (39).

Environmental Distribution

The aerodynamic and environmental properties of many of these allergens have been well characterized. Rodent allergens are found in a wide range of particle sizes, and it has been shown that small and large particles can migrate throughout a facility. For example, previous studies have characterized mouse allergen in public areas of an animal facility and found that rooms connected to the animal facility, but not actually containing mice, had detectable allergen on particles ranging in size from 0.4 to 3.3 μm (40). In freestanding, independently ventilated areas, such as a cafeteria not connected to a mouse facility, the allergen was predominantly greater than 10 μm in size (40, 41). This suggests that animal allergens can be carried substantial distances in animal facilities.

Airborne rat allergens are carried on particles ranging from 1 to 20 μm, with the majority on particles less than 7 μm. These allergens can remain airborne for 60 or more minutes after disturbance. Allergen levels have been studied in different settings, and the level of exposure has been shown to be primarily dependent on activity, with the highest exposures occurring among cage changers, room cleaners, and animal feeders (13). Levels are also increased with higher animal density and decreased relative humidity (25, 42, 43).

Much less is known or understood about the other laboratory animal allergens. Guinea pig allergen has been measured by RAST inhibition, and a high percentage of this allergen is found on particles less than 0.8 μm in diameter that will remain airborne for long periods (11, 44).

With regard to cat and dog allergens, they have best been studied in home settings. Cat allergen has been well characterized and shown to be on particles ranging from 1 to 20 μm in diameter. At least 15% of this allergen is carried on particles less than 5 μm in diameter (45). Although less is known about dog allergen, it appears to be distributed much like cat allergen, with about 20% of the airborne allergen carried on small particles (31, 46).

It is unclear what specific levels of exposure would be expected to induce symptoms. Data on the clinical relevance of airborne allergen levels are only available for rat and cat. In one study, rat allergen levels causing nasal symptoms ranged from 1.5 to 310 ng/m^3 (47). In a follow-up study, a dose-response was seen with greater symptoms at higher levels, although responses were so variable it was impossible to determine what level of exposure could be deemed "safe" (13). Likewise, studies on cat allergen have been inconclusive as to what level of allergen is the lowest capable of causing clinical symptoms (48–50).

PREVENTION MEASURES AND INTERVENTIONS—RISK FACTORS

Individual Susceptibility

The first step in prevention is identifying which workers may be more susceptible to the development of LAA. Preplacement screening evaluations may be helpful in identifying individuals who are at risk for developing LAA or asthma. A simple questionnaire may be a good starting point. Those with preexisting allergies or asthma not related to laboratory animal exposure are clearly at increased risk of developing LAA (4, 51). Studies have even suggested that some individuals have certain genes that predispose them to LAA, and a family history of asthma or allergies may help identify those at risk (52, 53). A history of allergy to domestic pets and tobacco smoking may also be important risk factors (12, 52). These preliminary assessments cannot be used legally to preclude employment, but may be helpful in determining job placement, for example, assigning tasks with less exposure to laboratory animal allergens. An example of a possible screening questionnaire is illustrated in Appendix A (17).

The evaluation of high-risk individuals identified from screening questionnaires may be supplemented by formal testing to detect specific IgE antibodies to animal or other allergens. Such tests can also be used as a baseline in high-risk individuals who might later develop LAA. By identifying individuals with susceptibility to allergy using the clinical history along with objective testing, one can implement preventive strategies by proper education and prevention measures (2, 4, 16, 52, 54).

As part of the medical surveillance, annual or semiannual evaluation using questionnaires may be conducted (Appendix A). If the worker starts to demonstrate signs and symptoms of LAA, further evaluation may be necessary (4).

Exposure Level

Epidemiologic studies have also shown that the greater the exposure to animal allergens, the more likely one is to become sensitized and have symptoms related to work (55, 56). Animal handlers and caretakers develop allergic symptoms more frequently than those who do not work in direct contact with the animals (14, 27). Therefore, identifying those with increased exposure is important in estimating risk and implementing measures for prevention.

Different job descriptions are associated with vastly different exposures to animal allergens. The highest exposures typically occur in handlers who are responsible for cage cleaning and feeding of the animals. Users are persons involved in daily experimental use of the animals, such as technicians, students, and investigators. These people have intermittent contact and therefore lower levels of exposure. Unexposed workers are secretaries and administrators with no direct contact with the animals. In looking at specific tasks, cleaning cages or manipulating active animals are activities associated with significantly higher levels of allergen exposure (57). Furthermore, it has been shown that symptomatic inflammatory responses in sensitized workers were correlated with airborne allergen concentrations, and that more symptoms occurred with active cage cleaning than quiet activity (13, 47).

Another interesting observation is that even those who do not have direct contact with animals can have work-related symptoms. Work-related symptoms were reported in one study in 56% of workers with no direct contact with animals (13, 47, 58, 59). Furthermore, while direct occupational exposure and atopy demonstrated more severe allergic symptoms than indirect exposure, several recent studies suggest that indirect exposure may be just as risky for allergy sensitization as direct exposure (60–62). This suggests that any exposure in environments where animals are present may induce disease.

Immunologic markers have recently been studied in the pathogenesis of occupational animal allergy. Cellular responses such as interleukin-4 (IL-4) cytokine release has been associated with the development of allergy in occupational workers sensitized to rat whereas nonspecific IL-10 and interferon-γ (IFN-γ) responses have not shown a protective effect (63). However, some literature has suggested that the development of mouse allergen immunoglobulin G (IgG; not IgE) antibodies may offer some markers of clinical tolerance (64). Furthermore, there have been some data suggesting certain gene-environmental markers may be useful (65), and perhaps future understanding of exposure and immune surveillance may help us understand who is at risk and what interventions could potentially prevent and decrease disease development and morbidity.

SURVEILLANCE AND MONITORING

After identifying those at risk through preliminary screening and testing, and identifying those with increased exposure by job description, surveillance and monitoring is also advisable at annual intervals. This can be accomplished through an annual questionnaire, employee records showing an abnormal pattern of illness, reports to the occupational health staff about symptoms of allergy, supplementary skin tests or allergen specific ImmunoCAP IgE, and tests of pulmonary function if indicated. Newer techniques of multiallergen surveillance may also help identify in a battery of tests allergen sensitization to the common exposure related to laboratory animal allergy and may improve prediction (66).

Health surveillance protects the health of the workers and assists in the evaluation of control measures. One should identify a risk-based approach to reducing exposure, and consider guidelines provided by the European Union (67). Self-reported symptoms may be monitored on a routine basis (3, 68). Newer, more sophisticated diagnostic models are being evaluated to improve questionnaire surveys (69). Early detection may enable one to implement precautionary measures, which may include limiting hours of exposure, withdrawing the individual from those procedures most likely to put him/her at risk, using respiratory protection and other personal protection equipment, using a safety cabinet where possible, increasing periodic monitoring to assess the efficacy of protective measures, monitoring any possible progression of the disease, and periodic assessment to determine continuing fitness for work (4, 70, 71).

Facility Design and Equipment

Attention to facility design and equipment may be helpful in reducing the incidence of laboratory animal allergy. The allergen load is dependent on the rate of allergen production and the rate of allergen removal from the air. These are a function of animal density and ventilation. To achieve a substantial reduction in allergen exposure in an area heavily populated with laboratory animals, frequent contact should be reduced. Effective respiratory protection devices may also be employed. Other approaches include evaluating building and ventilation design, altering work routines, and using personal protective measures.

With regard to building design, one possible approach is to isolate the laboratory animals strictly from the rest of the research facility in an attempt to reduce exposure and the spread of allergen. However, as mentioned earlier, the airborne characteristics of the allergens and practicality of this concept may make this very difficult or impossible. Building ventilation designed to reduce airborne allergen may be beneficial, but the clinical relevance of this approach has not been studied. Systems should allow adequate ventilation and airflow along with filtration systems to keep allergen levels as low as possible. Negative pressure in hallways may also be beneficial (52). Care also should be taken to reduce exposure when workers are not directly working with the animals. Adequate washing and showering facilities should be available in the animal units. In addition, changing rooms with ventilated lockers can help reduce contamination of personal clothing, and having employees use different clothes for direct contact with animals may be helpful (72). Recent studies have also suggested controlling allergens in animal rooms by using cage curtains to prevent spread from cages to workers (73).

Personal protection comes in two categories: the general protection of hygienic practices, such as hand washing and showering, and specific equipment, such as face masks, overalls, gloves, and shoe covers. Many facilities require personnel to use special clothing while working with animals to prevent microbiological contamination. Protective clothing may include a gown, cap, mask, gloves, and shoe covers. Furthermore, clothing should be handled by appropriate laundry facilities, or disposable protective clothing may be used that obviates the need for further handling of contaminated materials (72).

Ventilated face masks and helmets may also be considered as protective devices. Helmets may be bulky and cumbersome, and disposable face masks may be a more practical way of reducing airborne allergen exposure as long as the mask is of adequate quality (74).

Finally, work routines should be planned so that allergen contamination is kept to a minimum. Routines should be designed to prevent the spread of allergens to adjacent areas such as corridors, offices, and eating areas. Cages should be emptied in a separate cleaning area where dedicated equipment is used. Soiled cages should be cleaned and covered with cloth or plastic before transport. Furthermore, animal rooms should be kept clean using dust-reducing methods and frequent washing. Soiled bedding

TABLE 2.

Preventative measures and interventions

Method	Use	Advantage	Disadvantage
I. Screening and surveillance programs			
1. Questionnaires	Determine presence of nonwork-related allergic disease; determine existence of prior laboratory animal sensitization; assist in task assignment	Inexpensive	Accuracy of self-reporting
2. Skin testing or serologic assays for specific IgE antibodies to laboratory animal and other allergens	Determine presence of preexisting nonwork-related sensitization; determine baseline existence (or lack therof) of sensitization to laboratory animals	Ability to determine occupational relationship of sensitization or symptoms; early detection of sensitization	Cost and availability; invasive
3. Pulmonary function tests [peak expiratory flow rate (PEFR), spirometry]	Assess airway function; detect presence of reversible airway obstruction (asthma); required if patient is using effective respiratory protective gear	Early detection of asthma	Cost and availability
II. Facility design and equipment			
1. Ventilation systems (HEPA filters)	Decrease airborne levels	Effective but not proven to prevent or reduce symptoms	Very expensive
2. Ventilated cage/rack systems	Decrease airborne allergen levels	Effective but not proven to prevent or reduce symptoms	Expensive
3. Increase humidity in facility	Decrease airborne allergen levels	Inexpensive; not proven to reduce or prevent symptoms	May not be tolerated by animals or humans
4. Work stations for cage emptying/ cleaning	Decrease airborne allergen levels	Relatively inexpensive	May not totally eliminate high-level exposure
III. Work practices			
1. Education programs	Increase employee awareness of risks	Inexpensive	Time consuming
2. Job assignment	Reduce exposure in individuals at risk	Inexpensive	Validity not proven
IV. Personal protective respiratory equipment			
1. Respiratory protective gear	Reduce airborne allergen exposure	Efficient respirators effective in reducing symptoms	Requires motivated employee, medical supervision
V. Evaluation of the worker allergic to animals			
1. Referral to physician	Properly diagnose and treat affected individual	Improve employee health	Requires knowledgeable physician
VI. Emergency procedures			
1. Self-administered epinephrine	Prevent severe allergic reactions	Potentially life-saving	

and waste disposal containers should be emptied and carefully cleaned (72). A recent study evaluating a 10-year LAA prevention program showed that the annual incidence of primary LAA was reduced from 3.6% to 0% in the first 5 years and did not rise above 1.2% over the remaining years (75). This suggests that surveillance and preventive strategies can be quite effective in reducing the incidence of LAA. Table 2 summarizes these preventive measures and interventions and discusses the advantages and disadvantages of each (17).

Prevention

In line with surveillance and monitoring, prevention strategies are desirable. A recent study performed surveillance in the context of heightened awareness and prevention and still identified sensitization to laboratory animals ranged between 11.8% and 14.8% according to work seniority. Sixteen subjects (5%) reported asthma and 25 (7.9%) reported rhinitis when working with laboratory animals (76). The study revealed a low level of sensitization and symptoms of allergy to laboratory animals as a result of the preventive measures adopted to reduce exposure, but there is nonetheless a need to improve prevention so as to avoid the onset of LAA completely in students and workers. A strategy of wearing protective respiratory equipment was also effective in reducing prevalence of allergy in those with <5 years exposure, suggesting early identification of risk and reducing exposure is helpful in prevention (77).

MANAGEMENT OF WORKERS ALLERGIC TO ANIMALS

When a worker is suspected of having an animal allergy, intervention should occur in a timely fashion, including consultation with appropriate specialists who can make an accurate diagnosis and recommend appropriate medical management. For an animal facility worker with suspected animal allergy, the diagnosis is largely made on the history of clinical symptoms associated with exposure. Diagnosis is then confirmed with specific tests, such as skin tests or allergen specific ImmunoCAP tests to determine if the patient has IgE antibodies to the allergen.

One limitation in the use of these tests in the diagnosis of LAA is a lack of well-standardized allergen extracts for testing. However, recent studies have been promising in employing skin prick tests and measures of specific IgE in actual allergy to mouse in laboratory workers (78). Although there are many standardized allergen extracts, to date the only standardized animal extract is for cat. Therefore, the clinical history, skin testing, and RASTs may all be used in conjunction to make an accurate diagnosis. When skin testing for laboratory animal allergens, it is reasonable to include tests with extracts from several different animals whether or not the person is aware of exposure, such as those of rat, mouse, guinea pig, hamster, gerbil, and rabbit.

If asthma due to LAA is suspected, lung function measurements through serial peak flow meters or spirometry both in and out of the workplace may be helpful in documenting causality and management of symptoms. Serial exhaled nitric oxide measurements (a measure of inflammation used in research to evaluate asthma) conducted by having workers blow into a machine have also been considered to be helpful in research settings (79). Recent studies have also been evaluating the role of extremely high exposures and the development of IgG and IgG4 antibodies in those who tolerated, suggesting high-dose immunotherapy may help attenuation of risk at the highest of exposure (80), although the data are mixed in the full understanding of these mechanisms (81).

Ideally, medical management should begin with attempts to reduce exposure. Appropriate allergy and asthma medications should be administered to control symptoms; however, the highly sensitive individual who has continued symptoms despite reduction in exposure may require complete and absolute avoidance of the animal allergen.

Immunotherapy, or allergy shots, is a form of treatment in which increasing doses of allergen are injected subcutaneously, leading to an increased tolerance to the allergen(s). This treatment has been studied for animals such as cats and dogs with variable degrees of success. The benefits appear most applicable to intermittent exposure and not chronically exposed laboratory workers. The use of immunotherapy to protect workers from further symptoms, particularly for laboratory animals, has not been fully studied although it is used on occasion (2). Long-term effects include persistence of sensitization and even progression of lung function impairment over time regardless of allergy, further indicating the importance of this problem (82).

Emergency Procedures

Occasionally an allergic worker may experience anaphylaxis, a life-threatening allergic reaction from an animal bite or needle stick contaminated by animal allergens (83). Because of the rapid progression and potentially fatal consequences of these reactions, physicians may recommend that a worker carry a medication called epinephrine (e.g., Epi-Pen, or Ana-Kit), which can be lifesaving in such instances. In these cases, it is useful for coworkers to learn emergency procedures, such as cardiopulmonary resuscitation.

CONCLUSION

This chapter deals with laboratory animal allergy as a common and important occupational hazard. The symptoms of LAA can range from mild skin rashes to severe asthma. The allergens have been, for the most part, identified and purified, and risk of sensitization often occurs in the first 3 years of employment. Risk factors include a personal or family history of allergy, history of other nonwork-related allergies, and significant exposure. The allergens can be carried on small airborne particles and can remain airborne for long periods. Inhalation is the most common route of exposure, followed by skin and eye. Certain job descriptions are associated with higher exposure levels that can predispose susceptible individuals to develop sensitivity and eventually symptoms. Preliminary screening, ongoing surveillance, and intervention for affected workers are important to control this occupational hazard. Reducing exposure is the mainstay of both prevention and treatment. When workers develop symptoms, adequate intervention and management are necessary. This may at times require removal of the affected person from all exposure, and treatment and management by occupational health professionals. By understanding the etiology, pathophysiology, prevention, and management of LAA, hopefully the necessary measures can be implemented to control and prevent this disease.

References

1. **Acton D, McCauley L.** 2007. Laboratory animal allergy: an occupational hazard. *AAOHN J* **55:**241–244.
2. **Bush RK.** 2001. Mechanism and epidemiology of laboratory animal allergy. *ILAR J* **42:**4–11.
3. **Elliott L, Heederik D, Marshall S, Peden D, Loomis D.** 2005. Progression of self-reported symptoms in laboratory animal allergy. *J Allergy Clin Immunol* **116:**127–132.
4. **Seward JP.** 2001. Medical surveillance of allergy in laboratory animal handlers. *ILAR J* **42:**47–54. Review. 29 refs.
5. **Moghtaderi M, Farjadian S, Abbaszadeh Hasiri M.** 2014. Animal allergen sensitization in veterinarians and laboratory animal workers. *Occup Med (Lond)* **64:**516–520.
6. **Bush RK, Stave GM.** 2003. Laboratory animal allergy: an update. *ILAR J* **44:**28–51.
7. **Folletti I, Forcina A, Marabini A, Bussetti A, Siracusa A.** 2008. Have the prevalence and incidence of occupational asthma and rhinitis because of laboratory animals declined in the last 25 years? *Allergy* **63:**834–841.
8. **Muzembo BA, Eitoku M, Inaoka Y, Oogiku M, Kawakubo M, Tai R, Takechi M, Hirabayashi K, Yoshida N, Ngatu NR, Hirota R, Sandjaya B, Suganuma N.** 2014. Prevalence of occupational allergy in medical researchers exposed to laboratory animals. *Ind Health* **52:**256–261.
9. **Edwards RG, Beeson MF, Dewdney JM.** 1983. Laboratory animal allergy: the measurement of airborne urinary allergens and the effects of different environmental conditions. *Lab Anim* **17:**235–239.
10. **Cullinan P, Cook A, Gordon S, Nieuwenhuijsen MJ, Tee RD, Venables KM, McDonald JC, Newman Taylor AJ.** 1999. Allergen exposure, atopy and smoking as determinants of allergy to rats in a cohort of laboratory employees. *Eur Respir J* **13:**1139–1143.
11. **Harrison DJ.** 2001. Controlling exposure to laboratory animal allergens. *ILAR J* **42:**17–36.
12. **Gordon S.** 2001. Laboratory animal allergy: a British perspective on a global problem. *ILAR J* **42:**37–46.
13. **Eggleston PA, Ansari AA, Adkinson NF Jr, Wood RA.** 1995. Environmental challenge studies in laboratory animal allergy. Effect of different airborne allergen concentrations. *Am J Respir Crit Care Med* **151:**640–646.
14. **Hollander A, Van Run P, Spithoven J, Heederik D, Doekes G.** 1997. Exposure of laboratory animal workers to airborne rat and mouse urinary allergens. *Clin Exp Allergy* **27:**617–626.
15. **Wolfle TL, Bush RK.** 2001. The science and pervasiveness of laboratory animal allergy. *ILAR J* **42:**1–3.
16. **Bush RK.** 2001. Assessment and treatment of laboratory animal allergy. *ILAR J* **42:**55–64.
17. **Bush RK, Wood RA, Eggleston PA.** 1998. Laboratory animal allergy. *J Allergy Clin Immunol* **102:**99–112.
18. **Nguyen SB, Castano R, Labrecque M.** 2012. Integrated approach to diagnosis of associated occupational asthma and rhinitis. *Can Respir J* **19:**385–387.
19. **Schumacher MJ.** 1980. Characterization of allergens from urine and pelts of laboratory mice. *Mol Immunol* **17:**1087–1095.
20. **Schumacher MJ.** 1987. Clinically relevant allergens from laboratory and domestic small animals. *N Engl Reg Allergy Proc* **8:** 225–231.
21. **Longbottom JL, Price JA.** 1987. Allergy to laboratory animals: characterization and source of two major mouse allergens, Ag 1 and Ag 3. *Int Arch Allergy Appl Immunol* **82:**450–452.
22. **Phipatanakul W.** 2002. Rodent allergens. *Curr Allergy Asthma Rep* **2:**412–416.
23. **Pacheco KA, McCammon C, Thorne PS, O'Neill ME, Liu AH, Martyny JW, Vandyke M, Newman LS, Rose CS.** 2006. Characterization of endotoxin and mouse allergen exposures in mouse facilities and research laboratories. *Ann Occup Hyg* **50:**563–572.
24. **Park JH, Gold DR, Spiegelman DL, Burge HA, Milton DK.** 2001. House dust endotoxin and wheeze in the first year of life. *Am J Respir Crit Care Med* **163:**322–328. comment.
25. **Gordon S, Fisher SW, Raymond RH.** 2001. Elimination of mouse allergens in the working environment: assessment of individually ventilated cage systems and ventilated cabinets in the containment of mouse allergens. *J Allergy Clin Immunol* **108:**288–294.
26. **Baker J, Berry A, Boscato LM, Gordon S, Walsh BJ, Stuart MC.** 2001. Identification of some rabbit allergens as lipocalins. *Clin Exp Allergy* **31:**303–312.
27. **Hollander A, Heederik D, Doekes G.** 1997. Respiratory allergy to rats: exposure-response relationships in laboratory animal workers. *Am J Respir Crit Care Med* **155:**562–567.
28. **Jeal H, Harris J, Draper A, Newman Taylor A, Cullinan P, Jones M.** 2009. Dual sensitization to rat and mouse urinary allergens reflects cross-reactive molecules rather than atopy. *Allergy* **64:**855–861.
29. **Walls AF, Newman Taylor AJ, Longbottom JL.** 1985. Allergy to guinea pigs: II Identification of specific allergens in guinea pig dust by crossed radio-immunoelectrophoresis and investigation of the possible origin. *Clin Allergy* **15:**535–546.
30. **Warner JA, Longbottom JL.** 1991. Allergy to rabbits. III. Further identification and characterisation of rabbit allergens. *Allergy* **46:**481–491.
31. **Phipatanakul W.** 2001. Animal allergens and their control. *Curr Allergy Asthma Rep* **1:**461–465.
32. **Chapman MD, Aalberse RC, Brown MJ, Platts-Mills TA.** 1988. Monoclonal antibodies to the major feline allergen Fel d I. II. Single step affinity purification of Fel d I, N-terminal sequence analysis, and development of a sensitive two-site immunoassay to assess Fel d I exposure. *J Immunol* **140:**812–818.
33. **Anderson MC, Baer H, Ohman JL Jr.** 1985. A comparative study of the allergens of cat urine, serum, saliva, and pelt. *J Allergy Clin Immunol* **76:**563–569.
34. **Spitzauer S, Schweiger C, Anrather J, Ebner C, Scheiner O, Kraft D, Rumpold H.** 1993. Characterisation of dog allergens by means of immunoblotting. *Int Arch Allergy Immunol* **100:**60–67.
35. **Rautiainen J, Auriola S, Konttinen A, Virtanen T, Rytkönen-Nissinen M, Zeiler T, Mäntyjärvi R.** 2001. Two new variants of the lipocalin allergen Bos d 2. *J Chromatogr B Biomed Sci Appl* **763:**91–98.
36. **Ruoppi P, Virtanen T, Zeiler T, Rytkönen-Nissinen M, Rautiainen J, Nuutinen J, Taivainen A.** 2001. In vitro and in vivo responses to the recombinant bovine dander allergen Bos d 2 and its fragments. *Clin Exp Allergy* **31:**915–919.

37. **Goubran Botros H, Poncet P, Rabillon J, Fontaine T, Laval JM, David B.** 2001. Biochemical characterization and surfactant properties of horse allergens. *Eur J Biochem* **268:**3126–3136.
38. **Ooms TG, Artwohl JE, Conroy LM, Schoonover TM, Fortman JD.** 2008. Concentration and emission of airborne contaminants in a laboratory animal facility housing rabbits. *J Am Assoc Lab Anim Sci* **47:**39–48.
39. **Petry RW, Voss MJ, Kroutil LA, Crowley W, Bush RK, Busse WW.** 1985. Monkey dander asthma. *J Allergy Clin Immunol* **75:**268–271.
40. **Ohman JL Jr, Hagberg K, MacDonald MR, Jones RR Jr, Paigen BJ, Kacergis JB.** 1994. Distribution of airborne mouse allergen in a major mouse breeding facility. *J Allergy Clin Immunol* **94:**810–817.
41. **Gordon S, Kiernan LA, Nieuwenhuijsen MJ, Cook AD, Tee RD, Newman Taylor AJ.** 1997. Measurement of exposure to mouse urinary proteins in an epidemiological study. *Occup Environ Med* **54:**135–140.
42. **Gordon S, Tee RD, Newman Taylor AJ.** 1996. Analysis of the allergenic composition of rat dust. *Clin Exp Allergy* **26:**533–541.
43. **Gordon S, Tee RD, Stuart MC, Newman Taylor AJ.** 2001. Analysis of allergens in rat fur and saliva. *Allergy* **56:**563–567.
44. **Swanson MC, Agarwal MK, Reed CE.** 1985. An immunochemical approach to indoor aeroallergen quantitation with a new volumetric air sampler: studies with mite, roach, cat, mouse, and guinea pig antigens. *J Allergy Clin Immunol* **76:**724–729.
45. **Wood RA, Laheri AN, Eggleston PA.** 1993. The aerodynamic characteristics of cat allergen. *Clin Exp Allergy* **23:**733–739.
46. **Custovic A, Simpson A, Woodcock A.** 1998. Importance of indoor allergens in the induction of allergy and elicitation of allergic disease. *J Allergy Clin Immunol* **53(48 Suppl):**115–120.
47. **Eggleston PA, Ansari AA, Ziemann B, Adkinson NF Jr, Corn M.** 1990. Occupational challenge studies with laboratory workers allergic to rats. *J Allergy Clin Immunol* **86:**63–72.
48. **Gulbahar O, Sin A, Mete N, Kokuludag A, Kirmaz C, Sebik F.** 2003. Sensitization to cat allergens in non-cat owner patients with respiratory allergy. *Ann Allergy Asthma Immunol* **90:**635–639.
49. **Platts-Mills TA.** 2003. Allergen avoidance in the treatment of asthma and rhinitis. *N Engl J Med* **349:**207–208.
50. **Platts-Mills TA, Blumenthal K, Perzanowski M, Woodfolk JA.** 2000. Determinants of clinical allergic disease. The relevance of indoor allergens to the increase in asthma. *Am J Respir Crit Care Med* **162**(supplement_2)**:**S128–S133.
51. **Suarthana E, Malo JL, Heederik D, Ghezzo H, L'Archevêque J, Gautrin D.** 2009. Which tools best predict the incidence of work-related sensitisation and symptoms. *Occup Environ Med* **66:**111–117.
52. **Gordon S, Preece R.** 2003. Prevention of laboratory animal allergy. *Occup Med (Lond)* **53:**371–377. Review. 49 refs.
53. **Oxelius VA, Sjöstedt L, Willers S, Löw B.** 1996. Development of allergy to laboratory animals is associated with particular Gm and HLA genes. *Int Arch Allergy Immunol* **110:**73–78.
54. **Krakowiak A, Wiszniewska M, Krawczyk P, Szulc B, Wittczak T, Walusiak J, Pałczynski C.** 2007. Risk factors associated with airway allergic diseases from exposure to laboratory animal allergens among veterinarians. *Int Arch Occup Environ Health* **80:**465–475.
55. **Aoyama K, Ueda A, Manda F, Matsushita T, Ueda T, Yamauchi C.** 1992. Allergy to laboratory animals: an epidemiological study. *Br J Ind Med* **49:**41–47.
56. **Fuortes LJ, Weih L, Jones ML, Burmeister LF, Thorne PS, Pollen S, Merchant JA.** 1996. Epidemiologic assessment of laboratory animal allergy among university employees. *Am J Ind Med* **29:**67–74.
57. **Gordon S, Wallace J, Cook A, Tee RD, Newman Taylor AJ.** 1997. Reduction of exposure to laboratory animal allergens in the workplace. *Clin Exp Allergy* **27:**744–751.
58. **Venables KM, Tee RD, Hawkins ER, Gordon DJ, Wale CJ, Farrer NM, Lam TH, Baxter PJ, Newman Taylor AJ.** 1988. Laboratory animal allergy in a pharmaceutical company. *Br J Ind Med* **45:**660–666.
59. **Krakowiak A, Szulc B, Pałczyński C, Górski P.** 1996. [Laboratory animals as a cause of occupational allergy]. *Med Pr* **47:**523–531.
60. **Jang JH, Kim DW, Kim SW, Kim DY, Seong WK, Son TJ, Rhee CS.** 2009. Allergic rhinitis in laboratory animal workers and its risk factors. *Ann Allergy Asthma Immunol* **102:**373–377.
61. **Curtin-Brosnan J, Paigen B, Hagberg KA, Langley S, O'Neil EA, Krevans M, Eggleston PA, Matsui EC.** 2010. Occupational mouse allergen exposure among non-mouse handlers. *J Occup Environ Hyg* **7:**726–734.
62. **Krop EJ, Doekes G, Stone MJ, Aalberse RC, van der Zee JS.** 2007. Spreading of occupational allergens: laboratory animal allergens on hair-covering caps and in mattress dust of laboratory animal workers. *Occup Environ Med* **64:**267–272.
63. **Krop EJ, van de Pol MA, Lutter R, Heederik DJ, Aalberse RC, van der Zee JS.** 2010. Dynamics in cytokine responses during the development of occupational sensitization to rats. *Allergy* **65:** 1227–1233.
64. **Matsui EC, Diette GB, Krop EJ, Aalberse RC, Smith AL, Curtin-Brosnan J, Eggleston PA.** 2005. Mouse allergen-specific immunoglobulin G and immunoglobulin G4 and allergic symptoms in immunoglobulin E-sensitized laboratory animal workers. *Clin Exp Allergy* **35:**1347–1353.
65. **Pacheco KA, Rose CS, Silveira LJ, Van Dyke MV, Goelz K, MacPhail K, Maier LA.** 2010. Gene-environment interactions influence airways function in laboratory animal workers. *J Allergy Clin Immunol* **126:**232–240.
66. **Caballero ML, Ordaz E, Bermejo M, Rodriguez-Perez R, Alday E, Maqueda J, Moneo I.** 2012. Characterization of occupational sensitization by multiallergen immunoblotting in workers exposed to laboratory animals. *Ann Allergy Asthma Immunol* **108:**178–181.
67. **Westall L, Graham IR, Bussell J.** 2015. A risk-based approach to reducing exposure of staff to laboratory animal allergens. *Lab Anim (NY)* **44:**32–38.
68. **Nicholson PJ, Mayho GV, Roomes D, Swann AB, Blackburn BS.** 2010. Health surveillance of workers exposed to laboratory animal allergens. *Occup Med (Lond)* **60:**591–597.
69. **Suarthana E, Meijer E, Heederik D, Ghezzo H, Malo JL, Gautrin D.** 2009. The Dutch diagnostic model for laboratory animal allergen sensitization was generalizable in Canadian apprentices. *J Clin Epidemiol* **62:**542–549.
70. **Schmid K, Jüngert B, Hager M, Drexler H.** 2009. Is there a need for special preventive medical check-ups in employees exposed to experimental animal dust? *Int Arch Occup Environ Health* **82:**319–327.
71. **Tarlo SM, Liss GM.** 2001. Can medical surveillance measures improve the outcome of occupational asthma? *J Allergy Clin Immunol* **107:**583–585. comment.
72. **Fisher R, Saunders WB, Murray SJ, Stave GM.** 1998. Prevention of laboratory animal allergy. *J Occup Environ Med* **40:**609–613.
73. **Krohn TC, Itter G, Fosse R, Hansen AK.** 2006. Controlling allergens in animal rooms by using curtains. *J Am Assoc Lab Anim Sci* **45:**51–53.
74. **Perfetti L, Cartier A, Ghezzo H, Gautrin D, Malo JL.** 1998. Follow-up of occupational asthma after removal from or diminution of exposure to the responsible agent: relevance of the length of the interval from cessation of exposure. *Chest* **114:** 398–403.
75. **Goodno LE, Stave GM.** 2002. Primary and secondary allergies to laboratory animals. *J Occup Environ Med* **44:**1143–1152.
76. **Cauz P, Bovenzi M, Filon FL.** 2014. [Laboratory animal allergy: follow-up in a research centre]. *Med Lav* **105:**30–36.

77. **Jones M, Schofield S, Jeal H, Cullinan P.** 2014. Respiratory protective equipment reduces occurrence of sensitization to laboratory animals. *Occup Med (Lond)* **64:**104–108.
78. **Sharma HP, Wood RA, Bravo AR, Matsui EC.** 2008. A comparison of skin prick tests, intradermal skin tests, and specific IgE in the diagnosis of mouse allergy. *J Allergy Clin Immunol* **121:** 933–939.
79. **Hewitt RS, Smith AD, Cowan JO, Schofield JC, Herbison GP, Taylor DR.** 2008. Serial exhaled nitric oxide measurements in the assessment of laboratory animal allergy. *J Asthma* **45:** 101–107.
80. **Jones M, Jeal H, Schofield S, Harris JM, Shamji MH, Francis JN, Durham SR, Cullinan P.** 2014. Rat-specific IgG and IgG_4 antibodies associated with inhibition of IgE-allergen complex binding in laboratory animal workers. *Occup Environ Med* **71:** 619–623.
81. **Krop EJ, Doekes G, Heederik DJ, Aalberse RC, van der Zee JS.** 2011. IgG4 antibodies against rodents in laboratory animal workers do not protect against allergic sensitization. *Allergy* **66:** 517–522.
82. **Palmberg L, Sundblad BM, Lindberg A, Kupczyk M, Sahlander K, Larsson K.** 2015. Long term effect and allergic sensitization in newly employed workers in laboratory animal facilities. *Respir Med* **109:**1164–1173.
83. **Watt AD, McSharry CP.** 1996. Laboratory animal allergy: anaphylaxis from a needle injury. *Occup Environ Med* **53:**573–574.

APPENDIX A

LABORATORY ANIMAL ALLERGY QUESTIONNAIRE

Date ___________

Name: ____

Supervisor: ____

Department: ____

Age: ______________ Sex: ❑ Male ❑ Female

OCCUPATIONAL HISTORY Answer these questions about your present job:

1. Job title: ____
2. Number of years employed at this facility: ______________________ years
3. How many months/years at your present position? ______________________
4. Description of duties (briefly): ______________________

____

____

5. Do you work with laboratory animals? ❑ Yes ❑ No

   If yes, complete the following.

Animal	Yes	No	Approximate contact hours/day
Rats	❑	❑	______________
Mice	❑	❑	______________
Rabbits	❑	❑	______________
Guinea pigs	❑	❑	______________
Monkeys	❑	❑	______________
Cattle	❑	❑	______________
Dogs	❑	❑	______________
Cats	❑	❑	______________
Other	❑	❑	______________

6. Do you feel that you are allergic to any of these animals? ❑ Yes ❑ No

   ❑ Rats ❑ Mice ❑ Rabbits ❑ Dogs ❑ Other ❑ Cats ❑ Monkeys ❑ Cattle ❑ Guinea pigs

7. Did you work with laboratory animals prior to employment at this facility? ❑ Yes ❑ No

   If yes, how long? _____ years What type of animals? ______________________

8. Do you use or wear any of the following items when working with animals?

Protective eyeglasses	❑ Yes	❑ No
Mask/respiratory	❑ Yes	❑ No
Lab coat	❑ Yes	❑ No
Gloves	❑ Yes	❑ No

HOME ENVIRONMENT INFORMATION

9. Do you have any indoor pets? ❑ Yes ❑ No

If yes, which animals and for how long?

Animal	1–2 years	2–3 years	3–4 years	Over 4 years
Dogs	❑	❑	❑	❑
Cats	❑	❑	❑	❑
Other (type)______	❑	❑	❑	❑
____________	❑	❑	❑	❑

10. Do you regularly have any of the following symptoms? ❑ Yes ❑ No

Please indicate if the symptom is present and the year of onset. Also check in what location or time "period" the symptoms(s) is/are present.

			Symptoms are present			
Symptom	Yes/no present	Year of onset	At work	At home	On vacation	No difference
Cough	❑	____	❑	❑	❑	❑
Sputum production	❑	____	❑	❑	❑	❑
Shortness of breath	❑	____	❑	❑	❑	❑
Wheezing	❑	____	❑	❑	❑	❑
Chest tightness	❑	____	❑	❑	❑	❑
Asthma	❑	____	❑	❑	❑	❑
Nose congestion	❑	____	❑	❑	❑	❑
Runny nose	❑	____	❑	❑	❑	❑
Sneezing	❑	____	❑	❑	❑	❑
Itchy eyes	❑	____	❑	❑	❑	❑
Sinus problem	❑	____	❑	❑	❑	❑
Hay fever	❑	____	❑	❑	❑	❑
Frequent colds	❑	____	❑	❑	❑	❑
Hives	❑	____	❑	❑	❑	❑
Skin rash	❑	____	❑	❑	❑	❑
Swelling of eyes or lips	❑	____	❑	❑	❑	❑
Eczema	❑	____	❑	❑	❑	❑
Difficulty in swallowing	❑	____	❑	❑	❑	❑

11. Were you ever told by a doctor that you had allergies? ❑ Yes ❑ No

12. Have you ever been skin tested for allergies? ❑ Yes ❑ No

If yes, what substances were you found to be allergic to or sensitized to?

❑ Ragweed ❑ Grass ❑ Trees ❑ Mold ❑ Dust ❑ Cat ❑ Dog ❑ Mice

❑ Other ____________________________________

13. Have you ever received allergy (desensitization/ immunotherapy) shots? ❑ Yes ❑ No

14. Has a doctor ever said you have asthma? ❑ Yes ❑ No

If yes, when did your asthma start? __________ (year)

Are you currently taking medication (either over the counter or by prescription) to control your asthma? ❑ Yes ❑ No

15. Has a doctor ever told you that you have a medical condition caused by your working conditions? Yes ❑ No ❑

16. Do any of your blood relatives (grandparents, parents, brothers/sisters) have allergies or asthma? Yes ❑ No ❑

17. Are you under a doctor's care for any other illnesses? Yes ❑ No ❑

    If yes, please list illnesses: ______________________________

18. Do you take blood pressure medication(s)? ❑ Yes ❑ No

19. Do you regularly use "over-the-counter" (nonprescription) nose drops or nose sprays, e.g. Afrin, Neo-Synephrine? ❑ Yes ❑ No

20. Do you smoke cigarettes? ❑ Yes ❑ No

    If yes, how many cigarettes per day? How many years? _______

    If not presently smoking, did you ever smoke? ❑ Yes ❑ No

    If yes, when did you stop smoking cigarettes? _________ (year)

    How many years did you smoke? _________ years

    Comments: ____

    ____

    ____

    ____

    ____

Reviewed by: _______________

Date: _____________________

SECTION

III

Hazard Control

Design of Biomedical Laboratory and Specialized Biocontainment Facilities

16

JONATHAN T. CRANE AND JONATHAN Y. RICHMOND

World circumstances have changed the small, simple biocontainment facilities of the past into larger, more complex facilities with difficult design decisions. There is not "one way" to design any laboratory; therefore this chapter provides both laboratory users and designers with relevant information to assist in making choices appropriate for the needs of specific projects. If the architect and engineers make decisions without local input and informed consent, it is unlikely that the completed laboratory will be satisfactory. The design of biomedical research laboratories, particularly biocontainment laboratories, is an exercise in making choices that are often between competing ideas and needs. However, if the potential users become an active, integral part of the process and an experienced design team is engaged, the facility will likely meet current needs and future requirements. Competent professional assistance is a necessity in this design process.

This chapter deals with basic biomedical and clinical laboratories at biosafety level 2 (BSL2) and with containment laboratories, with the main emphasis on BSL3 and their enhancements. In addition, it provides an introduction to BSL4 laboratories and the decisions and issues faced as a part of a BSL4 design team. Additional information on BSL3 and BSL4 facilities can be found in the literature (1–3). In addition, containment criteria for research involving recombinant DNA can be found in the guidelines from NIH (4). The chapter also addresses specialized containment facilities such as BSL3 autopsy suites and patient biocontainment units (PBUs).

APPROACH AND PROCESS

Laboratories are specialized facilities in which clinical, research, and developmental work with hazardous materials can be performed safely. An assessment of the hazards expected to be present in each laboratory is a necessary part of the design process. The assessment of the risk of working with the hazardous material must come from the user of the laboratory. Engineering out the risk of such work is a major component of the thinking that goes into the design of the laboratory.

Most successful laboratory designs are based on simple, commonsense solutions to technological challenges. The biggest challenge to the laboratory design team is to keep the design simple and not to overdesign and make it too complex for the systems to work (5). The team must also keep from implementing new, untested technologies for the sake of technology innovation alone. It is the nature of architects and engineers to seek new and inventive solutions and to use the latest, most complex technology available. Everyone wants to design the "state-of-the-art laboratory." Improvements in laboratory design have generally evolved from recognition of basic needs and are consistent with proven principles. If a low-technology solution provides equivalent performance, it should be seriously considered before a high-technology solution.

A consensus on the level of systems performance must be reached before designing the facility. Rarely are systems designed for 100% performance except in life-critical situations (BSL4 would require the highest performance). For example, standard air-conditioning design criteria provide comfort for 80% of the population. To provide a system that will make 100% of the people comfortable, the designer must know, and the user must be prepared to bear, the (possibly extraordinary) additional costs of this level of design. Even taking it to the 100% level does not ensure that everyone will be satisfied all the time. The low-technology solution is a sweater for the person who likes it hot and shirtsleeves for the person who likes it cold. There are some laboratories in which Tyvek suits or air-supplied suits are worn routinely. These laboratories need cooler air for comfort than a routine laboratory does.

The level of complexity also must be considered. Complex systems may offer energy savings, better control of conditions, and more responsive systems. However, complex systems demand higher maintenance to retain their reliability, have higher initial design and construction costs, and take longer to construct and commission. A mistake often made in laboratory projects is to design complex systems and then back off on control and monitoring capabilities when the costs come in. Complex systems with inaccurate controls rarely work well. In general, as the need for system reliability increases, complexity should decrease.

Evolving Needs

A biomedical laboratory facility is constantly changing and evolving as the people, technology, and projects within it change. It should be structured to allow for adaptation. User needs should be evaluated on an ongoing basis, with building modifications made as required. The design of the facility should anticipate the ease of future modifications, balancing short-term needs with long-term goals. Unless unusual circumstances dictate, a laboratory should not be designed to meet only specific short-term needs; it should be generic enough to meet long-term needs as well.

Laboratory work requires stability to ensure accurate diagnostic results or success of long-term research projects. The facilities that house them must be reliable. Reliability can be achieved in many ways. Systems can be made simple for easy maintenance and operations. Duplicate or redundant systems can be provided to ensure continuance of critical services and to replace components in failure as required, especially for the higher containment levels. Systems can be overlapped as primary and secondary systems to provide a cumulative approach to reliability. Also, critical systems should be designed to fail in a position that minimizes the threat to life or property. Obviously, as BSLs increase from BSL1 through BSL4, reliability must also increase, sometimes leading to a requirement for redundant equipment, such as fans and freezers.

It is advisable to plan and budget some "forgiveness" into the project, because there is a gap between theoretical design and installed construction. Architecturally, this may mean leaving a little latitude in the space around equipment. Space allotment is especially important for biological safety cabinets (BSCs), which have outside dimensions larger than the work-surface dimensions by which they are described. It may also mean making rooms a little more generous than necessary and not planning to put 8 ft of base cabinets in an 8-ft-wide room. (For example, with acceptable construction tolerances, the room might end up 7 ft by 11 ft 3/4 in.) Mechanically, this might mean installing an exhaust fan that can be adjusted to move more air than the design calls for. This can allow for changes during the construction process in duct configuration due to space conditions, higher filter loadings, and changes in equipment operating specifications that vary from manufacturer to manufacturer. This will also provide more flexibility for fine-tuning the air balancing of the facility. Plan a facility that will allow the easy resolution of problems that will occur; allow for additional equipment requiring additional supply or exhaust air.

Function, both current and future, is the basis for laboratory design. An operational systems engineering approach should be taken in the design of the facility (6). Determine the movement of people and materials into and out of the facility and from room to room. Detail the steps of how supplies move into the facility, how they are used, and how they need to be handled as waste. What are storage and handling requirements along the way? What are alternative options? Analyze how the facility will be used at the date of occupancy and into the future. As Winston Churchill once said, "We shape our buildings, and then they shape us." The facility should not dictate the method of operation.

Sustainable (Green) Design

It is important to highlight the energy usage of biomedical laboratory facilities. The Environmental Protection Agency has instituted "Labs for the 21st Century," a program with a focus on energy efficiency and use of renewable resources, to encourage and measure design that can reduce the nation's dependence on energy and benefit the building owner with reduced operating costs. Because laboratories are inherently high energy users with high electrical power use, bright lighting, and 100% exhaust systems operating 24 h per day, they are also expensive to operate. Selection of highly sophisticated, energy-efficient equipment is one means of long-term cost reduction; however, there are some laboratories in which the systems require such major modifications after the initial installation that energy and cost savings are lost. Other laboratories do not maintain pressure relationships and safe ventilation patterns that form the original reason for building a new containment facility. Still others require more specialized maintenance personnel than are normally found in a laboratory facility.

All efforts should be made to provide energy savings that come with good basic commonsense design before deciding to develop costly and complex systems for energy reduction. Room volume might be minimized when room volume is the driving force in energy consumption. Appropriate insulation and solar barriers should be provided to reduce heat loss or gain in conditioned spaces. Natural light should be used where possible to augment artificial lighting. Other examples might include reduced air change rates for laboratories that use low volumes of chemicals, because air change rates have little impact on biological safety (7); use of heat recovery systems in lower-hazard laboratories; and use of materials with low inherent energy use in their manufacture. A great deal of information on sustainable-design opportunities for laboratories is available at http://www.labs21century.gov. In addition, the vast majority of architectural and engineering firms now make the use of sustainable technologies a standard part of their practice. In considering sustainable technologies for laboratories, particularly BSL3 and BSL4 laboratories, the laboratory function and systems performance must be the first priority.

Lighting has been a high contribution to laboratory energy usage. Historically, lighting contributed significant additional heat into the space beyond people, equipment, and processes. This heat requires additional cooling in the space, and this requires added system capacity. Light-emitting diode (LED) lighting systems can provide the same light quality and intensity with significantly reduced energy usage. Another added benefit is significantly longer bulb life, reducing operational impact from changing light bulbs.

Visualization

It is difficult to visualize the look and feel of the laboratory while reviewing the architectural plans. As the facility planning begins, some of the following techniques can be used to allow the entire design team to visualize the final project and to assess the needs of the laboratory.

Do a walk-through of the facility in use, even if the current facility is totally inadequate, as a source of information for the design team. Discuss the advantages and disadvantages of the layout, size, casework, lighting, and noise. Discuss the current operation and the proposed operation. Identify equipment to be moved to the new laboratory. Identify the way hazards are currently handled and decide if they should be handled differently in the new facility.

Key members of the project team should tour comparable facilities to learn quickly about the state of the art and gain a common basis for understanding the vocabulary used. This also can give the design team an understanding of some of the options in laboratory design.

Mock-ups of proposed or alternate designs can be invaluable in fine-tuning the design of the facility. These can be simple, for example, using masking tape to outline the size and shape of the room, with cardboard cutouts representing casework and equipment. This allows all parties to see the layout of the room, ensure that equipment will fit, and determine if room size will be comfortable. More complex mock-ups using actual construction to simulate the final project can be expensive, but it is the best way to make sure the design is correct to the smallest detail. The design team can simulate various operations in the mock-up to identify any weakness in the design. This also can be helpful when heating, ventilation, and air-conditioning (HVAC) systems need to be verified. Smoke testing can be performed to ensure proper airflow. This approach can be very cost-effective when several identical rooms, such as animal rooms and laboratory or containment modules, are planned.

Scaled physical models can be built to show buildings or detailed components. These models can be useful; however, they have been largely replaced by Building Information Modeling (BIM) systems that are now routinely used on laboratory projects by architectural and engineering firms. This computer modeling of the building, laboratories and laboratory systems can accurately show a three-dimensional viewable model of the spaces and systems. These models can include equipment and furnishings in the laboratory, including, for example, the details of the drawers and cabinets in the laboratory casework. Equipment can be placed at actual scale to demonstrate fit. HVAC and plumbing systems can be viewed to ensure accessibility for maintenance. It is much better to review and move a valve or damper on a drawing than to live without the ability to access these items for the life of the facility. This type of modeling also allows easy exploration of a variety of options in facility layout during early planning stages of a project.

Questioning

The entire design process is one of searching for answers and questioning those answers until the team is comfortable. Questions include: What are the customer's real needs? What has the customer not identified that is necessary? Will the design work? How long will it work? Where has it worked before? Are current methods of operating the best? Does the design reflect the protocols and practices that will be used? What is the answer for this project?

PREPLANNING

Program

Space requirements are determined before beginning laboratory design. The following laboratory requirements are the focus of this chapter.

Basic research laboratories

Basic laboratories should be generic and straightforward so that they allow occupancy by many programs with minimal changes while still usable. The biosafety facility design requirements of basic laboratories at BSL2 are described in *Biosafety in Microbiological and Biomedical Laboratories* (BMBL) (8, 9). The key elements for laboratory design are as follows:

- Facilities that house Select Agents must have doors that lock.
- New laboratories should be located away from public areas.
- Each laboratory should contain a sink for hand washing. Foot-, knee-, and automatically operated sinks are recommended. The sink should be located where protocols dictate the removal of gloves.
- The laboratory should be designed so that it can be easily cleaned. Carpets and rugs are inappropriate in laboratories.
- Benchtops should be impervious to water and resistant to moderate heat and the organic solvents, acids, alkalis, and chemicals used to decontaminate the work surfaces and equipment.
- Laboratory furniture should be capable of supporting anticipated loading and uses. Spaces between benches, cabinets, and equipment should be accessible for cleaning. Chairs and other furniture used in laboratory work should be covered with nonfabric material that can be easily decontaminated.
- BSCs must be installed in such a manner that fluctuations of the room supply and exhaust air do not cause them to operate outside their parameters for containment. Locate BSCs away from doors, from windows that can be opened, from heavily traveled areas, and from other potentially disruptive equipment so as to maintain the BSCs' parameters for containment. See chapter 18.
- An eyewash station must be readily available.
- Illumination should be adequate for all activities and should avoid reflections and glare that could impede vision.
- Windows that open to the exterior should be fitted with fly screens.

Although there are no specific ventilation requirements for BSL2, planning of new facilities should consider mechanical ventilation systems that provide an inward flow of air without recirculation to spaces outside the laboratory. Both the Centers for Disease Control and Prevention (CDC) and the National Institutes of Health (NIH) recommend directional, inward airflow at BSL2 (8, 9) to provide control for fumes as well as bioaerosols.

Containment research laboratories

Although containment laboratories should be designed to handle the specific needs of the program to be housed, they should also allow for as many different programs as possible to occupy them with minimal changes during their functional life. The facility design requirements of containment laboratories at BSL3 as described in BMBL include the following:

- The laboratory is separated from areas that are open to unrestricted traffic flow within the building, and access to the laboratory is restricted. Passage through a series of two self-closing doors is the basic requirement for entry into the laboratory from access corridors. Doors are lockable. A room for changing clothes may be included in the passageway.
- Each laboratory contains a hands-free or automatically operated sink for hand washing that is located near the laboratory exit door. The sink should be located where protocols dictate the removal of gloves.
- The interior surfaces of walls, floors, and ceilings of a BSL3 laboratory are constructed for easy cleaning and decontamination. Seams, if present, must be sealed. Walls, ceilings, and floors should be smooth, impermeable to liquids, and resistant to the chemicals and disinfectants normally used in the laboratory. Floors should be monolithic and slip resistant. Consideration should be given to the use of coved floor coverings. Penetrations in floors, walls, and ceiling surfaces are sealed.

Openings, such as those around ducts and the spaces between doors and frames, are capable of being sealed to facilitate decontamination.

- Benchtops are as for BSL2 above.
- Laboratory furniture is as for BSL2 above.
- All windows in the laboratory are closed and sealed.
- A method for decontaminating all laboratory wastes is available in the facility and utilized, preferably within the laboratory (i.e., autoclave, chemical disinfection, incineration, or other approved decontamination method). Consideration should be given to providing a decontamination air lock for decontaminating equipment.
- BSCs are required and are located away from doors, from room supply louvers, and from heavily traveled laboratory areas.
- A ducted exhaust air ventilation system is provided. This system creates directional airflow that draws air into the laboratory from "clean" areas into the laboratory toward "contaminated" areas. The exhaust air is not recirculated to any other area of the building. Filtration and other treatments of the exhaust air are not required but may be considered on the basis of the site requirements and specific agent manipulations and use conditions. The outside exhaust must be dispersed away from occupied areas and air intakes, or the exhaust must be HEPA filtered. Laboratory personnel must verify that the direction of the airflow (into the laboratory) is proper, i.e., by providing a visual monitoring device that indicates and confirms directional airflow into the laboratory at the laboratory entry. Consideration should be given to installing an HVAC control system to prevent sustained positive pressurization of the laboratory. Audible alarms should be considered to notify personnel of HVAC system failure.
- HEPA-filtered exhaust air from a Class II BSC can be recirculated into the laboratory if the cabinet is tested and certified at least annually. When exhaust air from Class II cabinets is to be discharged to the outside through the building exhaust air system, the cabinets must be connected to this system in a manner that avoids any interference with the air balance of the cabinets or the building exhaust system (e.g., an air gap between the cabinet exhaust and the exhaust duct). When Class III BSCs are used, they should be connected directly to the exhaust system. If the Class III cabinets are connected to the supply system, it is done in such a manner that prevents positive pressurization of the cabinets.
- Vacuum lines are protected with liquid disinfectant traps and HEPA filters, or their equivalent. Filters must be replaced as needed. An alternative is to use portable vacuum pumps (also protected with traps and filters).
- An eyewash station is available as for BSL2.
- Illumination is supplied as for BSL2.
- The BSL3 facility design and operational features must be documented. The facility must be tested for verification that the design and operational parameters have been met prior to operation. Facilities must be reverified, at least annually, against these procedures as modified by operational experience.
- Additional environmental protection (e.g., personnel showers, HEPA filtration of exhaust air, containment of other piped services, and provision of effluent decontamination) should be considered if recommended by the agent summary statement, or as determined by risk assessment, the site conditions, or applicable federal, state, or local regulations.

Maximum-containment laboratories

BSL4 laboratories are expensive, representing a quantum leap in investment, not only in the cost of construction but also in the cost of operation, maintenance, training, oversight, and community relations. In addition, these labs are time-consuming to enter and exit and difficult to work in, leading to a loss of productivity. Designing and building a BSL4 laboratory require careful consideration. There are agents and procedures that require BSL4 containment and thus a BSL4 laboratory. Visits to existing BSL4 laboratories to discuss the issues and challenges with those who have had experience are highly recommended.

There are two models for BSL4 containment, but either model, or a combination of the two models, must be located in the same rigorously designed type of facility (8, 9). In BSL4 cabinet laboratories, the focus is on enhanced primary containment by working with viable agents in a Class III BSC. In BSL4 suit laboratories, the agents are generally handled within primary containment, such as a Class II BSC, and redundant personnel protection is provided by positively pressurized suits supplied with breathable air. In these laboratories, the secondary containment is generally, in practice, a rigorously designed and constructed containment barrier. For routine diagnostics or antigen production at BSL4, a BSL4 cabinet laboratory might be the best choice, as such laboratories are less expensive to build, operate, and maintain. For complex research requiring more than a few animals for the research, or for work with nonhuman primates, a BSL4 suit laboratory would certainly be the preference of experts in this area. For more detail on BSL4 laboratory design and operations see Richmond (10).

Clinical laboratories

Clinical or diagnostic laboratories should be designed for the processing of biological samples from humans, animals, or the environment. Human specimen collection may occur in the field, in an area near the lab, in a physician's office, or at bedside. Risk assessments for activities related to handling human blood or other body fluids generally result in assigning BSL2 for facility design (11). Certain laboratories that process aerosol-transmissible agents (e.g., a tuberculosis laboratory) usually require BSL3 (12). Clinical laboratories are assigned one or more of the following functions—hematology, immunology, clinical chemistry, urinalysis, microbiology (bacteriology, virology, mycology, mycobacteriology, or parasitology), anatomic pathology, cytology, and blood banking. The main requirements for clinical laboratories as outlined by *Guidelines for Construction and Equipment of Hospital and Medical Facilities* (13) are as follows:

- Laboratory work counters provide space for microscopes and other equipment.
- Work areas include sinks with water for hand washing and disposal of nontoxic materials. Access to vacuum, gas, and electrical services is included as needed.
- Refrigerated blood storage facilities for transfusions are equipped with temperature monitoring and alarms.
- Storage facilities, including refrigeration, are provided for reagents, standards, supplies, stained specimen microscope slides, etc.
- Specimen (blood, urine, and feces) collection areas are provided but may be located outside the laboratory suite. Blood collection areas have work counters, space for patient seating, and hand washing facilities. The urine and feces collection room is equipped with a water closet and lavatory and may be located outside the laboratory suite.
- Safety provisions include emergency shower, eye-flushing devices, and appropriate storage for agents and chemical hazards, such as flammable liquids.
- Facilities and equipment for terminal sterilization (autoclave or electric oven) are provided for contaminated specimens before transport. (Terminal sterilization is not required for specimens that are incinerated on-site.)
- Requirements of authorities having jurisdiction should be verified.
- Administrative areas, including offices as well as space for clerical work, filing, and record maintenance, are provided.
- Lounge, locker, and toilet facilities are conveniently located for male and female laboratory staff. These may be outside the laboratory area and shared with other departments.
- Local (capture) ventilation systems are provided where appropriate to protect the worker from exposure to biohazardous agents or hazardous chemicals when it is not practical to place this equipment or perform these procedures in a BSC or fume hood (14, 15).

The small clinical laboratory

Many clinical laboratories in physicians' offices lack the benefit of appropriate laboratory air supply and exhaust systems due to HVAC systems that recirculate air in the building and have minimum capabilities for supply and exhaust air. Although these laboratories generally work at BSL2 and almost exclusively handle human tissues and bodily fluids, many of the recommendations for clinical laboratory design apply. Particular care should be taken in use and placement of aerosol- or fume-generating equipment and procedures to ensure that aerosols are not introduced in the work area or into air recirculated in the building. Appropriate containment should be provided along with required specimen and biomedical waste storage, including sharps containers.

Budget or Cost Constraints

Each project should be approached with a realistic budget; constraints on design resulting from that budget need to be reconciled early in the process. The more complex and specific the laboratory is, the more it will cost. Specialized systems, such as emergency power generators, central purified water systems, or effluent decontamination systems, will have a major impact on the budget.

The percentages of building construction cost for the major components of a new laboratory building are shown in Fig. 1. Note the high percentage of cost that typically is allocated for mechanical and electrical systems. Factors affecting the range of costs include complexity, size, and geographic location. Aesthetics of a laboratory have minimal impact on the overall costs of a laboratory facility. A high-quality design generally represents less than 5% of

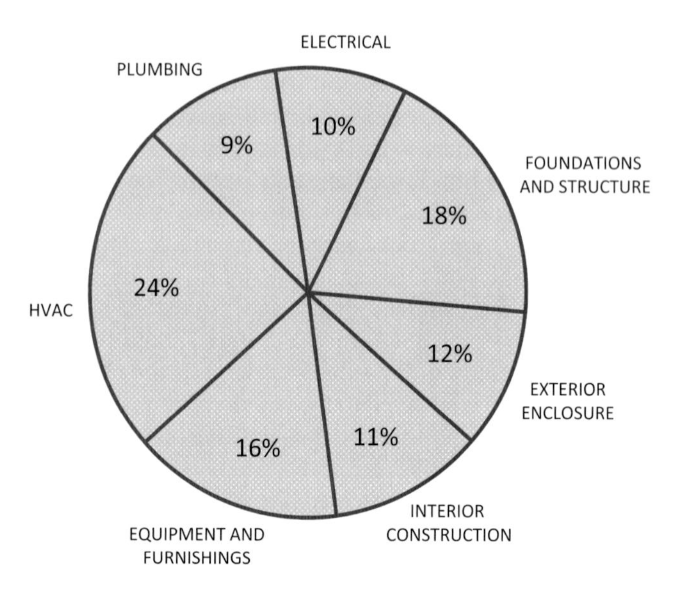

Figure 1: Percentage of building construction cost per new laboratory component.

the construction cost of a laboratory facility. Many psychological benefits from high-quality design are translated into actual benefits in the operation of the facility.

Schedule or Time Constraints

Laboratories take more time to plan, design, and construct than most other facilities. The design and construction process needs to fit the required schedule, but care should be taken not to sacrifice major needs of a long-term facility to meet short-range schedule requirements. Approximate time frames for design and construction of various types of projects are as follows: minor renovations, 3 to 6 months; major renovations, 1 to 2 years; minor new projects, 1 to 2 years; and major new projects, 3 to 5 years.

Operational Issues

Operational goals and constraints should be identified for both the long and short terms. Operational protocols, as could impact laboratory design, should be developed for normal laboratory support services; janitorial, maintenance, and waste handling operations; and emergency situations that may be anticipated.

Maintenance

The maintenance capabilities of the organization must be assessed to match the design of the facility with the support capabilities. One should not design and install systems that cannot be maintained because of budgetary or staffing reasons. Organizational experience with certain products, availability of supplies and parts, staff training, and maintenance budgeting should be evaluated prior to the design phase.

In summary, preplanning can provide a solid foundation for the development of the design of the facility. It reduces the potential for delays and cost increases that occur with changes in scope or approach to the project. At the end of preplanning, the entire design team should understand the goals and limitations of the project. The scope and quality of the project (what is to be built), the schedule for the project (how quickly it is to be built), and the budget for the project (funds available for the project) should be clearly defined and balanced before beginning the design.

DESIGN: DEVELOPING SAFE OPERATIONAL SYSTEMS

The design team should develop a space validation package for each room in the facility composed of the following:

- Review of applicable codes and regulations
- Development and analysis of hazardous-material use data
- Development and analysis of program data
- Code classification
- Design recommendations and review
- Development of room data sheets for recording requirements, including the casework and fixed equipment data, equipment listing and data, and functional layout and diagram
- Diagrams of proximity relationships to other rooms.

The BIM may be updated to include the additional level of information noted above so the room data can be viewed in conjunction with the three-dimensional models of the spaces. This provides a comprehensive analysis and record of requirements that will produce the following benefits:

- Rooms will meet codes and guidelines.
- Equipment will fit and have the proper services.
- The requirements and layout of each room will be understood.
- Hazardous materials will be identified and appropriate safeguards will be planned.
- Quality control reviews will verify that all requirements are met.

Biomedical laboratories should be designed to address safety concerns inherent or anticipated in such facilities. The potential for spread of contamination from the laboratories to other areas throughout the building needs to be minimized. The relationship of air handling within the building is critical (there should be no air movement from laboratories or containment or animal care facilities to other spaces).

Engineering controls can minimize hazards within the laboratory. Correct airflow and primary containment equipment allow investigators to perform their operations safely and may reduce contamination. Methods for handling exhaust air, waste, and hazardous by-products should be selected to disperse these wastes properly into the environment and to provide adequate protection for the worker and the community. Systems should be easy to maintain and safe for support personnel.

Primary barriers are specialized items designed for capture or containment of biological agents, e.g., BSCs, chemical fume hoods (CFHs), and animal cage dump stations (see chapters 15 through 17). Safe laboratories use an appropriately layered approach combining primary and secondary barriers to provide for personnel and environmental occupational safety and health.

Secondary barriers, the focus of this chapter, are facility-related design features that separate the laboratory from nonlaboratory areas or from the outside. Many of these barriers are physical in nature (e.g., walls and doors) and others result from mechanical devices, such as air handling systems.

Architectural and engineering features in a laboratory provide a secondary barrier system for protection that complements the primary barrier. These systems are designed to move the hazard away from the laboratory workers, restrict the hazard to a specific area, treat the hazard if necessary, and allow for easy cleanup in a confined area.

The primary tool for laboratory safety involves the practices and procedures used by the laboratory worker. The effectiveness of the laboratory design will be greatly increased if these protocols for use, cleanup, and maintenance of the facility are available to the design team during the project preplanning phase. Care should be taken to use a realistic approach to the development of protocols that will affect the laboratory design. It is as much a mistake to develop a laboratory protocol that is too stringent as it is to develop one that is too lax. Design teams also need to consider the practices and procedures used by the laboratory to ensure that necessary support facilities are provided. Conversely, a realistic approach needs to be taken relative to final practices so that engineering designs are not too complex. A protocol is only effective if it is followed. If the building makes it difficult to follow protocol or is designed around unnecessary procedures, it is unlikely that the protocol will be followed.

Although this book is focused on biohazards, the hazards that occur in biomedical laboratories that the design team must address are as follows.

- Chemical, such as flammables, carcinogens, toxins, and compressed gases
- Biological, which are known infectious agents and materials that may contain infectious agents
- Radioactive, including radionuclides and equipment that produces ionizing radiation
- Physical, comprising laser, magnetic, high voltage, UV, high noise, and extreme heat or cold.

How these hazards are to be manipulated is vital to laboratory design. Therefore, the risk of each hazard must be assessed individually. Appropriate measures must be taken for proper storage, handling, and disposal of all chemicals and biological and radioactive material. The threat to life and property from these hazards has caused the development of codes, regulations, and guidelines governing facilities and practices to minimize potential problems. Physical barriers, interlocked room access devices, and noise abatement strategies must be addressed. An approach to addressing these hazards in combination can be found in Crane and Riley (16) and in Riley et al. (17).

Regulations, codes, and guidelines in the design of laboratories fall into two main areas: (i) building and life safety codes that are adopted and administered on the local level and (ii) laboratory safety codes, which may be local, state, federal, or private association in nature. Determine the edition of code adopted by the authority having jurisdiction and governing the project, as requirements may differ greatly among various editions. In some instances, local jurisdictions have adopted even more restrictive codes than state or federal regulations.

Many states and the federal government have adopted facility design standards to allow access and use by disabled persons. Wheelchair turn radii, required clearances at doors, handicap-accessible workstations, and removal of obstructions are examples of types of requirements in these codes. These items have a large impact on size and layout of laboratory areas, providing adequate aisle width between benches and access to eyewash stations and emergency showers. Retrofits to accommodate disabled persons are expensive; however, in most cases they are required by law.

Offices and Conference and Administration Spaces

An administrative area, physically separated from all hazardous aspects of laboratory work, should be planned near the main entry to each building or floor. This area provides administrative support for the laboratories as well as acting to control access to the laboratory area.

Offices for scientific staff should be positioned as close as possible to each occupant's main nonoffice workspace. Research faculty offices typically range from 100 to 160 ft^2; thus a detailed layout should be developed to ensure that adequate space is provided for bookshelves, computers, desks, filing, and guest chairs. Each laboratory or office should ideally have a window to permit the occupant to look either into the corridor, into the laboratory, or outside. Interaction among scientific staff, particularly in multidisciplinary environments, is becoming increasingly important as a tool for sharing ideas and transferring knowledge. Informal spaces (e.g., alcoves with markerboards in corridors and seating areas along pathways) can enhance this interaction and should be considered in facility design.

A corridor, possibly fire rated, providing two means of exit from any point should service the laboratory block. Safety showers and spill control centers should be located in this corridor. If corridor widths are constructed to meet minimum fire code and equipment access requirements, it is impossible for corridors to be used as auxiliary laboratory work spaces, storage areas, or offices and break areas. If the corridors are wider than minimum requirements, these problems will inevitably occur.

Basic Research Laboratories

A standard laboratory module should be developed for flexibility (5). Laboratory space can be analyzed by the efficiency of workstations, generally measured by footage of usable bench and equipment space available. Typical laboratory modules vary in width from 9 ft 6 in. to 11 ft 6 in. Although the most often used widths are 10 to 11 ft, the jump from 10 to 11 ft can add 5% to the cost of the facility. Module depths generally range from 20 to 30 ft. A simple masking tape mock-up can provide insight for this critical decision.

To allow clearance for access of most equipment expected in a laboratory, laboratory doors should be a minimum of 3 ft 4 in. wide for single doors, or be a pair of doors 4 ft wide using an active leaf of 3 ft wide and an inactive leaf 1 ft wide. All active leaves should have closers. Most doors should open out from the laboratory, but doors opening into an exit corridor should be in a recessed pocket. Doors and windows in fire-rated walls should also be fire rated as required by code.

Flooring in basic laboratories can be sheet vinyl or vinyl composition tile with a standard rubber or vinyl base. Walls can be enamel paint. Ceilings in these laboratories should be lay-in acoustical tile to reduce noise and provide access to system components above the ceiling. See Fig. 2 for a photograph of a typical open BSL2 facility.

Basic Clinical Laboratories

Clinical laboratories should be set up to handle the volume of specimens anticipated in a manner that allows the specimens to flow from collection and receiving to processing, analysis, reporting, and storage in a logical manner. Much clinical laboratory work is now automated, and care must be taken in the design to provide space for the equipment to operate and be maintained. Heat load and data connections are main issues in the layout of this equipment. Record-keeping stations should be located at appropriate positions along the work flow. Although some components of clinical laboratories may be modular in nature, each individual area should be designed with its unique requirements in mind. Much has been written about clinical laboratories, including laboratory sizing and detailed design (18).

Basic Animal Facilities

The main consideration and components in the design of animal facilities outlined by the *Guide for the Care and Use of Laboratory Animals* (19) are as follows:

- Separation of animal facilities from personnel areas
- Separation of species
- Isolation of individual projects (when required by protocol)

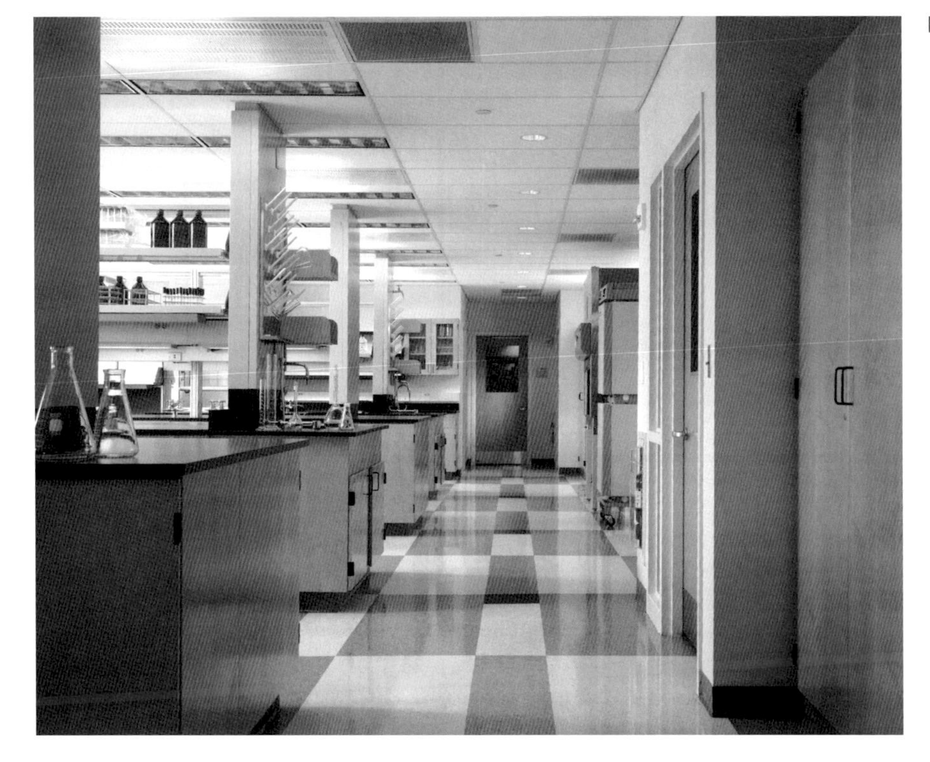

Figure 2: A typical BSL2 laboratory.

- Areas to receive, quarantine, and isolate animals
- Areas to house animals
- Specialized laboratories or individual areas contiguous with or near animal housing areas for such activities as surgery, intensive care, necropsy, radiography, preparation of special diets, experimental manipulation, treatment, and diagnostic laboratory procedures
- Containment facilities or equipment if hazardous biological, physical, or chemical agents are to be used
- Receiving and storage areas for food, bedding, pharmaceuticals and biologicals, and supplies
- Space for the administration, supervision, and direction of the facility
- Showers, sinks, lockers, and toilets for personnel
- An area separate from animal rooms for eating and drinking
- An area for washing and sterilizing equipment and supplies and, depending on the volume of work, machines for washing cages, bottles, glassware, racks, and waste cans; a utility sink; an autoclave for equipment, food, and bedding; and separate areas for holding soiled and clean equipment
- An area for repairing cages and equipment
- An area to store wastes prior to incineration or removal.

Additional specific requirements for construction can be found in the *Guide for the Care and Use of Laboratory Animals* (19) and BMBL (8, 9).

Traffic flow patterns are generally the determining factor in the layout of animal facilities, with consideration given to controlled access, functional conveniences, environmental control, and ease of movement of cages, waste, and personnel. Facilities are designed with standard corridors, clean/dirty corridor systems, or barrier systems depending on the species housed, the functional needs of the facility, and the type of housing unit selected. Isolation of animals either for biohazard containment or for animal protection (such as for severe combined immunodeficient mice) can be handled at the cage level with isolation cages, at the rack level with airflow modules, or at the room level with filtered, pressurized airflow in cubicle rooms.

Easy access and movement of cages and racks into and out of the cage-washing areas and storage space are important for a smoothly functioning facility. Storage space for cages becomes a larger factor when a variety of animals requiring differing cage types are held in the same facility. Consideration should be given to standardizing the animal care module for every major species grouping. Standardization provides uniform air requirements and consistency of airflow in modules and allows maintenance of the room HVAC systems. The size of the room should be based on rack layout, flexibility desired, and type of animal housed. Rooms that are over- or undersized are inefficient and waste energy. Material selection for rooms, flooring, and walls is important for accreditation and for ongoing U.S. Department of Agriculture inspections (19). Individually vented caging (IVC) systems have become a common practice for housing rodents. These rack-and-cage systems control cleanliness or contain biohazards at the cage level. This can provide an opportunity to reduce HVAC air change rates at the room level.

BSL3 CONTAINMENT FACILITIES

Layout requirements for containment facilities vary depending on their size and purpose. See Fig. 3, 4, and 5 for a layout of a small BSL3 laboratory, a typical tissue culture room, and a photograph of a BSL3 facility. An anteroom should be provided to create the "two doors in series" required by BMBL for entry into a BSL3 laboratory. Doors in a BSL3 facility normally allow air to flow into the laboratory through gaps along the side of and below the door. The anteroom should have adequate space for gowning and ungowning, along with space for storage and disposal of gowns, masks, gloves, etc. If protocols require the use of powered air-purifying respirators (PAPRs), space and electrical outlets for decontaminating and recharging these respirators should be provided. Hand wash-

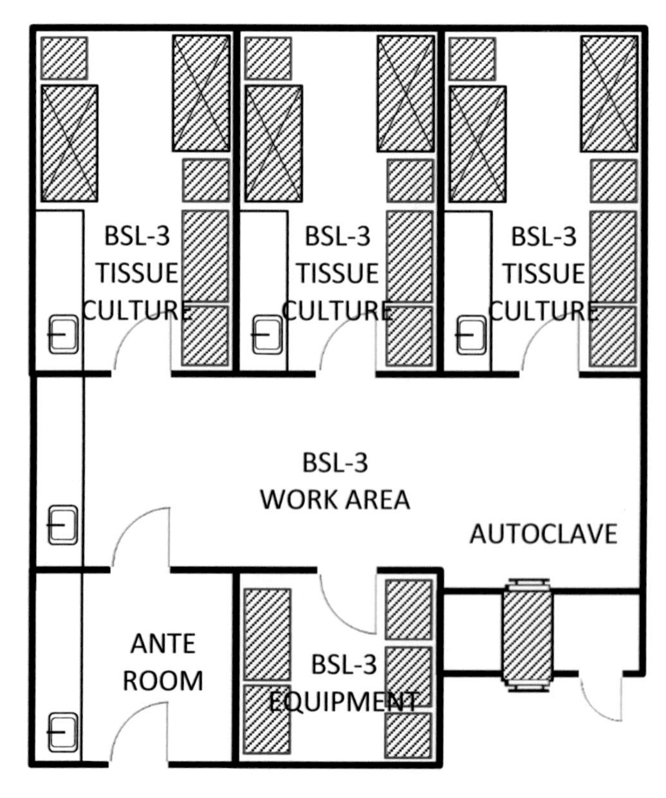

Figure 3: Example layout of a small BSL3 suite.

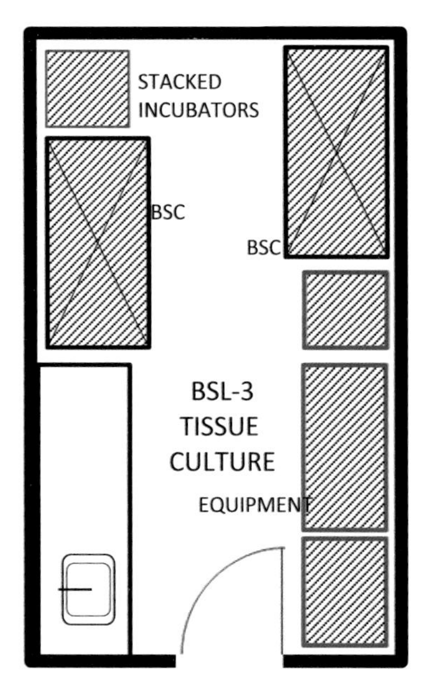

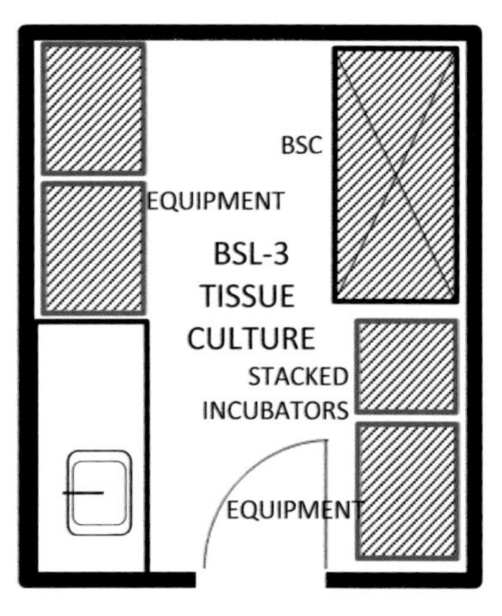

Figure 4: Example layouts of typical BSL3 tissue culture rooms.

ing facilities must be provided at locations where protocols dictate the removal of gloves. Space within the containment facility should be sufficient to handle the specific projects and personnel that are planned. Crowding in a containment facility can lead to unsafe situations.

Containment laboratories should have monolithic floors. Troweled epoxy or seamless sheet vinyl floors with an integral coved base are examples. If the laboratory includes animal holding with wet husbandry, epoxy may be the best choice; otherwise seamless vinyl is better due to its better comfort for work. Walls and ceilings of laboratory space are typically gypsum board and should be coated with easily cleanable paint, such as epoxy. Care should be taken in placing mechanical components requiring access above these ceilings to minimize service access in the containment laboratory and access panels in the

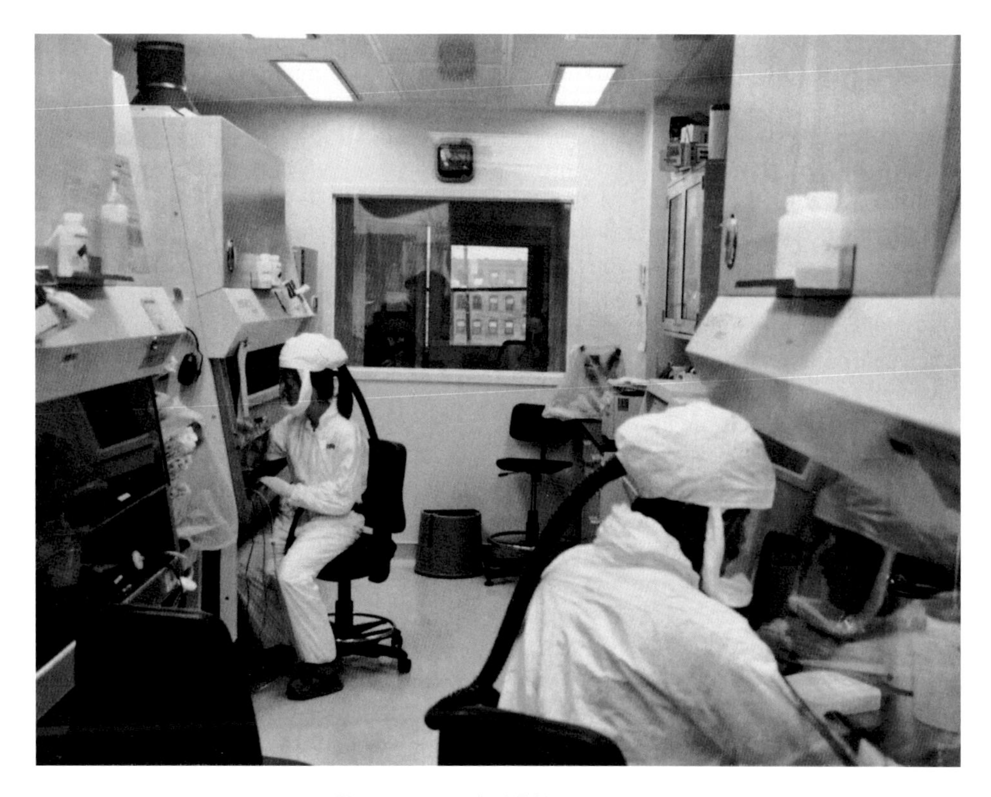

Figure 5: A typical BSL3 laboratory.

ceiling. Standard lay-in acoustical tile ceilings are not acceptable in BSL3 laboratories.

The BSL3 laboratory should be tightly sealed to facilitate decontamination with gas or vapor; however, the BSL3 facility is not required to withstand a pressure decay test. It is feasible to seal any openings such as supply and exhaust diffusers at the time of decontamination. If space decontamination will be a routine occurrence, consider remotely operated bubble tight dampers on these systems.

Enhanced BSL3 Containment Facilities

Risk assessment of the work at BSL3 might indicate increased secondary containment to protect the environment outside of the laboratory. Some of these enhancements are now found in many BSL3 facilities; however, it is important to note that they are not required unless a risk assessment identifies them as necessary for containment.

BMBL provides for use of an autoclave within the facility rather than in the room for BSL3 waste, if proper protocols are developed to transport it safely to the autoclave location. For agents where risk assessment indicates a higher environmental consequence, consider providing a pass-through double-door autoclave at the containment barrier of the laboratory. Provide interlocked doors to protect against the passage of potentially contaminated material through the chamber without a cycle being run. The autoclave should be located outside of the containment barrier for maintenance access and should have seals where it meets the containment barrier.

Gown-in, shower-out protocols may be considered to allow personnel a full-body shower after removing protective clothing prior to donning street clothes and leaving the facility. Ideally, the showers should be placed for direct pass-through upon exit from the facility and should be located between the inner change areas and the outer clean changing area to eliminate cross-contamination from dirty to clean clothing.

HEPA filtration may be provided on the exhaust system to prevent the environmental release of an agent. The filter units should allow access for maintenance and be designed to permit the scan testing of the filters in place after installation and to permit filter decontamination before removal. At BSL3, bag-in, bag-out filters might be considered. To reduce the length of and therefore the cost of potentially contaminated ductwork, HEPA filters should be located as near as possible to the containment barrier. Bioseal dampers should be provided in the exhaust ductwork on both sides of the filter housing to facilitate decontamination.

Supply HEPA filtration has been employed in some facilities; however, without an air pressure-resistant door, supply HEPA filtration does not have a significant impact on containment. In the event of a reversal in flow, air would flow out through the air gaps under the doors, making supply HEPA filtration ineffective. The air supply and exhaust systems must be interlocked to shut down the supply system in the event of exhaust failure to prevent reversal of the directional inward airflow if the containment space becomes positively pressurized.

For some high-consequence environmental pathogens, decontamination of liquid effluent from the facility may be required. Effluent decontamination systems are expensive to install and require significant oversight and maintenance, making them difficult to justify in small BSL3 facilities. With effective primary containment and carefully tailored protocols to eliminate effluent, the need for effluent decontamination may be reduced or eliminated. These systems have historically used heat treatment, although chemical systems can be considered and are particularly effective if the volume of waste is small. If shower waste must be decontaminated, it is important to consider carefully the number of personnel requiring showers, the length of showers, and the quantity of water flow. Four to six personnel per day can generate a significant amount of wastewater to be decontaminated, requiring either a large cook tank system or a high volume of chemical usage.

Animal BSL3 Holding

The facility requirements for animal holding at animal BSL3 (ABSL3) are similar to requirements for BSL3 facilities. A typical small facility layout is shown in Fig. 6. Construction materials for facilities where small and medium-sized animals are housed will be typical of normal animal facilities. Caging systems used as primary containment are an important consideration in the risk assessment and design of ABSL3 facilities. For large animals with certain agriculturally important agents, BSL3 Agricultural (BSL3Ag) facility requirements may apply as the room itself becomes the primary containment (see chapter 32).

Support

Most laboratories, including containment facilities, require support rooms. Large, expensive pieces of equipment (e.g., ultracentrifuges and scintillation counters) shared among laboratory programs require shared equipment rooms. For containment laboratories, isolating this equipment in a separate room can increase safety in the event of a malfunction, such as a rotor failure. These rooms also allow for noise-generating equipment to be housed away from the working laboratory. Separate rooms for freezers and ultralow-temperature freezers are recommended to minimize noise in laboratories and to address removal of the heat generated by the operation of this

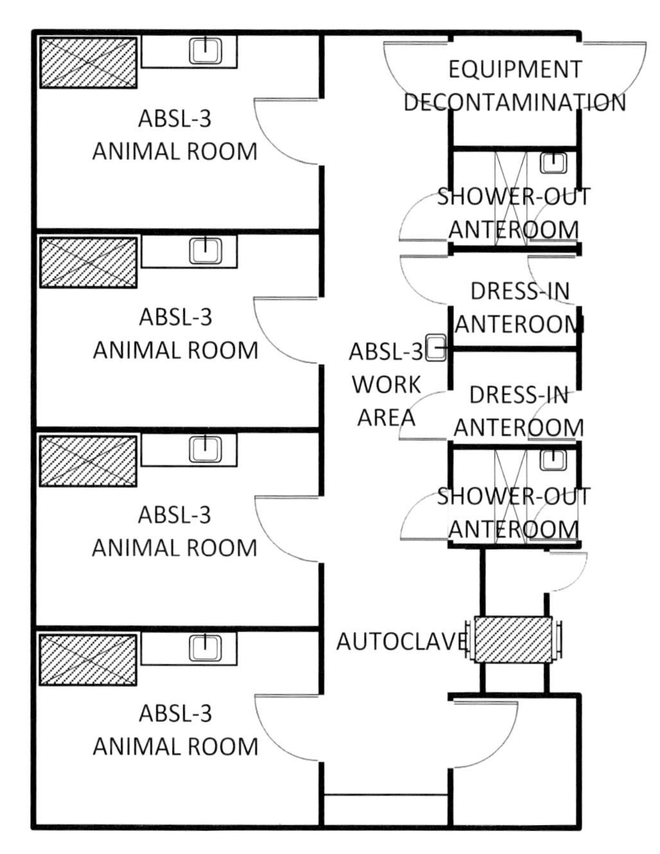

Figure 6: Example layout of an ABSL3 suite.

equipment adequately. Separating specimen storage areas may also allow for a higher level of security and access control.

Environmental rooms ranging from −20°C (freezers) to 4°C (cold rooms) to 37°C (warm rooms) are found in many laboratory facilities. These rooms are either for storage or work. Working rooms must be ventilated. Environmental rooms are complex systems that must be designed to the specific project requirements. Their costs vary greatly according to the accuracy and degree of temperature and humidity control required. Environmental rooms also require ramped access if they are not inset into the structure. As environmental rooms require a high degree of maintenance, careful consideration should be made prior to placing them in containment facilities. A proven alternative is to use smaller refrigerated enclosures, such as chromatography chambers or refrigerated storage units. These can be decontaminated and moved out of containment for any required maintenance.

Appropriate facilities located outside of containment need to be provided for preparation of culture media, which will involve an assessment of medium preparation and storage requirements. The unpleasant odors call for a well-ventilated, exhausted room. If primary containment is a requirement in making media, BSCs are recommended rather than laminar-flow clean benches.

Space for washing and drying some glassware, for ice machines, for dry ice storage, and for autoclaving, which produces heat and noise, should be provided in suitably ventilated support rooms outside of the laboratory. For higher containment levels, where protocol makes it difficult to enter and exit the laboratory, consider providing ice machines and dry ice holding inside containment. Providing ventilation above the doors of autoclaves and glassware washing and drying equipment helps remove heat, humidity, and odors. This equipment also requires adequate service space to minimize maintenance costs. Particular attention should be paid to flooring and floor drains in these areas to minimize damage due to leaks from equipment malfunction.

SPECIALIZED BIOCONTAINMENT FACILITIES

There are specialized facilities that require a different approach to biocontainment than typical BSL3 laboratories. The following examples of BSL3 autopsy suites and PBUs are provided to indicate how one might think about these facility types during the design process.

Contained Autopsy Facilities

In the 1990s, the emergence and re-emergence of highly pathogenic diseases with risk of aerosol transmission led to a reconsideration of morgue and autopsy facilities to safely handle the remains of those who died while infected with an infectious pathogen. This has resulted in upgrading the design and operational principles for these facilities. Autopsies of human remains pose special biocontainment risks beyond that found in most biocontainment facilities. For example, (i) there are potentially unknown risks from pathogenic microorganisms. The cause of death, say from an automobile crash, may have an underlying infection of a highly pathogenic infectious disease presenting risks above those immediately apparent. Even in the cases of infectious disease causing death, the actual organism might be different than expected, more virulent than normally encountered, or a drug-resistant strain. (ii) Primary containment is not practical with a human body. In addition, procedures such as opening the body, including cutting through bone and cartilage, may produce significant aerosols of infectious material. For these reasons, it is prudent to design a facility to prevent inadvertent infections in the prosectors, pathologists and technicians, or others who may be working in the facility and those outside the facility.

Approach to autopsy facility design and operations

As in the design of any biocontainment facility, as noted above, the approach to design involves primary containment,

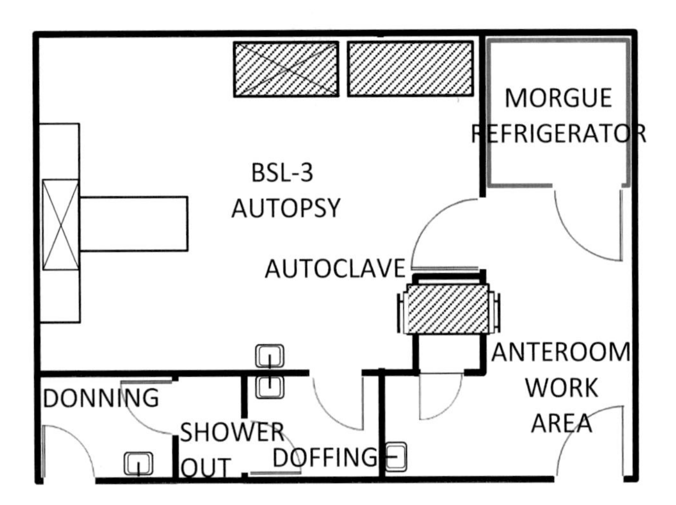

Figure 7: Example layout of a BSL3 autopsy suite.

secondary containment, and personal protective equipment (PPE). With the limitations on the use of primary containment identified above, PPE and secondary containment become more important design and operational considerations. Additional information can be found in previous editions of this publication (20) and other publications (21).

An important consideration is designing the flow and spaces to minimize potentially contaminated areas by locating the autopsy suite on a pathway from body intake that allows for decontamination after movement of the body. The goal is to create a self-contained suite with anterooms for both body movement and personnel entry/exit to the facility, and to provide the ability to decontaminate materials moving out of the facility, including liquid waste, and filtering infectious aerosols in the exhaust stream. Figure 7 shows a typical layout for a BSL3 autopsy suite.

Ergonomics and lighting are important features for minimizing incidents by the prosector and staff. Provide adequate space, including layout space for equipment and specimens. Provide BSCs for handling samples and specimens for further dissection. Consider providing a separate room for a BSC with access from the anteroom and pass-through from the autopsy room to minimize unnecessary personnel exposure to aerosols in the autopsy suite.

The use of a downdraft table can significantly reduce the aerosols while the autopsy is being performed; however, it is important to understand that downdraft tables are not as effective as BSCs and have only limited capability to contain aerosols. Wear appropriate PPE during autopsy procedures and room cleaning and disinfection. Consider the capability to decontaminate the space with vaporized hydrogen peroxide (VHP) or other chemical disinfectant. Penetrations into the rooms should be sealed to facilitate cleaning and decontamination.

Space requirements

A personnel locker and change room should be provided in the facility. This can be outside and away from the BSL3 autopsy suite. Provide an anteroom for personnel entry into the contained area and to act as a barrier against air movement out of the facility. This anteroom should be large enough to hold PPE required for the autopsy and other activities. After the autopsy, personnel should be provided the capability of showering out of the autopsy suite. Consider a layout to allow one-way personnel flow.

The autopsy room should be large enough to accommodate all planned equipment and allow adequate floor space for body movement into and out of the suite. Provide space to allow cleaning and disinfecting equipment and supplies in the room. Provide a pass-through autoclave to decontaminate materials, equipment, and supplies coming out of the autopsy suite. Consider a pass-through back into this area for gowning and other PPE that must be decontaminated after removal.

A separate anteroom should be provided for the movement of bodies and materials into the autopsy suite. This anteroom should be large enough for easy movement of gurneys and for transfer of bodies onto carts for use in the autopsy suite. This anteroom should contain a morgue refrigerator and the unload side of the pass-through autoclave. Last, consider providing a viewing space outside of containment with a window into the autopsy room for body identification without movement into the facility. The anteroom should have access to the autopsy room.

Handling air and effluents in the BSL3 autopsy suite

Like other BSL3 spaces, a BSL3 autopsy suite should have inward directional airflow to minimize the potential for organisms migrating from the suite back into the building through the movement of air. The autopsy room, being the highest biohazard area, should have no points where air flows out other than through controlled exhaust to create directional inward airflow. Consider locating exhaust low at the autopsy sinks to assist in limiting aerosols within the room. Air exhausted from the autopsy suite should be passed through a HEPA filter system that allows in-place decontamination of the filters.

There is significant liquid effluent produced during autopsies and cleaning of the autopsy facility. A liquid effluent decontamination system should be considered for this waste before discharge into the local sewer system.

Patient Biocontainment Units

The 2014 Ebola outbreak in West Africa and the spread of patient care issues to the United States and Europe have raised the awareness of the need for renewed efforts to care safely for patients with highly pathogenic infectious diseases. We learned that with these diseases there is a significant risk for health care workers. Through October 2014, in the 2014–2015 Ebola Zaire outbreak in West Africa, 3% to 5% of infections and deaths were in health care workers. There were over 800 health care workers infected with the virus and over 500 of those died. Care for those stricken is important; in the entire outbreak, over 27,000 persons were infected and over 11,000 died, per World Health Organization statistics.

While the risks of working in Africa and Western countries are different, there was a high rate of infection in Western personnel working with these patients both in Africa and in the United States. We also learned that modern critical care medicine can have a significant impact on reduction in mortality rates. If future outbreaks occur, it is likely that suspected or confirmed patients will be brought to the United States for care. It is important that we recognize, assess, and plan to address the risk in a rational and proactive manner. As observed in these events, a reactive, during-the-event approach creates problems.

Preparedness leads to a safer environment

Not all hospitals and health systems will want to have or already have the skills to treat patients with highly pathogenic infectious diseases such as Ebola; however, as the Texas Health Presbyterian Hospital Dallas experience demonstrated, all hospitals should be able to intake, hold, and care for a suspect patient safely until the patient can be transferred to an appropriate facility. In cases in which institutions do not want PBUs, appropriate measures in the emergency department should be considered to allow intake and holding until transfer can be made.

While a limited number of health systems in the United States and Europe actually cared for patients in the 2014 Ebola outbreak, awareness of the serious issues we face in preparing for emerging, re-emerging, and even purposefully engineered infectious disease was raised. Even for institutions that plan to operate PBUs for highly pathogenic infectious diseases, one size does not fit all from the perspective of design. For example, some institutions may face pressure from unions representing nursing or maintenance staff that may require highly engineered design solutions; others may have staff willing to implement more protocol-driven solutions. Some PBUs may be built in new facilities able to be designed for that specific use. Other PBUs may be placed in renovated space on upper floors of old buildings. In these cases, existing infrastructure, such as piping running down through the building, may be prone to leaks, creating risk in the facility. There are many similar issues that may be encountered. Each institution needs a facility tailored to their specific requirements and circumstances.

Despite these differences, there is a set of planning and design principles based on lessons learned from the institutions that have handled these patients, combined with experience from biocontainment facilities in health care and research that can provide guidance for the design of these facilities. These principles are outlined below.

Principle 1. There are unique risks in patient biocontainment

In fact, the risk to the worker is significantly higher in patient care facilities than in laboratories or facilities working with animal models of disease. Most PBUs are in the best case equivalent to BSL3 facilities. Laboratories at BSL3 typically know the agents they are working with, whereas hospitals are often dealing with unknown risks and agents. In laboratory facilities, including working with most animal models, primary containment protects both the worker and the environment. In health care it has proven extremely difficult to treat patients in primary containment isolators. They are not often used.

This means the worker is in the room with the patient, potentially exposed to infectious materials in the air and on surfaces. This exposure requires wearing complex PPE, including gowns, gloves, and respiratory protection that is difficult to don and doff. In a laboratory, due to the use of primary containment devices such as BSCs, PPE (other than gloves and sleeves) is not normally exposed to infectious material. When working with highly pathogenic infectious agents like Ebola in a laboratory, which would require BSL4, the highest biosafety level in laboratory facilities, a splash would require immediate disinfection. Even without a splash, the PPE would be completely decontaminated with chemicals before removal. In patient care areas, a worker's PPE may be exposed to aerosol-producing procedures, a high volume of infectious waste, and/or uncontrolled large-scale events, such as explosive vomiting and diarrhea. The worker then has to remove and decontaminate the PPE without exposing themselves.

Principle 2. Plan and prepare for the unexpected

Health care has little control over potential patients. With emerging, re-emerging, and evolving diseases, as well as increasing drug resistance and unknown agents that may create higher risks and consequences, a hospital never knows who might walk through the door requiring care. Patients may need medical care over and above normal infectious disease care, particularly in suspect cases. Plan the capability to provide unexpected medical care. A laboratory can choose if, when, and how to handle an infectious agent. A hospital may have no control of the same disease that may present itself in an unexpected manner. Plan your facilities to handle unexpected events and have proactive operational plans and contingencies in place.

Principle 3. Provide flexible patient care space

Four major lessons learned from the patient care activities during the care for Ebola cases in the United States during the 2014 outbreak were: (i) Patients were much sicker (higher acuity of disease) than typical infectious disease patients requiring a level of critical care not previously found with infectious diseases; kidney failure requiring high-risk procedures, such as dialysis, was one example. (ii) The acuity varied during a stay from healthy suspect cases to critically ill to recovering to a healthy state. (iii) The patients were required to stay in the PBUs much longer than anticipated. (iv) Current PBUs are not designed for all events and potential care required.

Patient care activities should accommodate diagnostics, procedures, labor, and delivery. Suspect patients as well as confirmed cases must be given appropriate medical care. Plan a facility that can accommodate both types of patients concurrently with maximum safety for both the patients and the staff. The patients may have existing comorbidities or conditions that would require unique diagnostics and treatments.

Patients may be pregnant and require labor, delivery, and mother-baby care capabilities. Equipment needs will change during the stay. Plan to allow equipment to enter and exit the room easily with appropriate decontamination. The rooms should be capable of providing most care without removal of patients from containment.

Principle 4. Prioritize engineering controls over protocols

Many accidents and unplanned events are a result of human error or the inability of staff to follow complex operational protocols consistently. Most containment laboratory incidents are a result of human error. Hospitals, in particular, have had a difficult time reducing error to acceptable levels. Engineering systems can be designed to accommodate failure scenarios. The facts show that humans are much more likely to make a mistake than will a well-engineered system fail.

In designing for biocontainment and biosafety, use engineering controls to replace protocols whenever practicable. An example would be to provide a pass-through autoclave to disinfect waste at the point of use rather than requiring bagging waste, disinfecting it, and moving it out of the facility to an autoclave. As the risk increases, engineering controls should increase, minimizing the need for protocols.

Principle 5. Integrate facility design with operational protocols

The facility design must be fully integrated with the planned operational models(s) to minimize the potential for adverse events. When facilities and operational protocols are mismatched, shortcuts and workarounds must be taken and the potential for an adverse event significantly increases. Design your facility to match planned operations, not the reverse.

Principle 6. Control contamination through separation

Separating contaminated areas from noncontaminated areas will minimize risk to patients, staff, and the community. First, look at how you can reduce the number of spaces that will potentially be contaminated to the absolute minimum required for operation. Where possible, achieve separation through primary containment. Take a particularly close look at eliminating clean and contaminated cross-flows. This would include the ability to eliminate the potential for cross-contamination between spaces serving confirmed and suspect patients as well as patients with different diseases.

Principle 7. Eliminate airborne spread of infectious agents

While it was generally believed that Ebola had limited potential for transmission through the aerosol route, that will not likely always be the case for future diseases. Plan flexibility in your facility to handle what may be required for diseases with higher aerosol risk. These provisions also provide safety with diseases such as Ebola because they will also limit airborne spread of virus to surfaces where it may be unexpectedly picked up through contact. Anterooms, filtration, and directional airflow would be important considerations for worker safety as well as preventing contamination outside the patient care space with airborne and aerosol risk. Consider HEPA filtration of exhaust and vent openings. Provide directional airflow from areas of lower risk to areas of higher risk. Recognize the limitations of directional airflow, and provide physical barriers such as anterooms that create a better air boundary because airflow must go through a minimum of two doors in series.

Principle 8. Surfaces and finishes must facilitate decontamination

In laboratories where the infectious agent is handled in a primary containment device, such as a BSC, there is rarely a need to decontaminate the room with chemicals that degrade finishes. However, due to the lack of primary containment, application of surface disinfectants will be required more often in PBUs than in comparable laboratory facilities. In addition, PBUs may not be able to be shut down for access for finish maintenance. This was a significant lesson learned from the facilities used during the 2014 outbreak. Most importantly, lack of finishes that can withstand these harsh chemicals will increase maintenance and downtime of these limited facilities.

Principle 9. Redundancy, reliability system isolation are critical considerations

When a lab has a system failure, you can shut down operations. However, with a patient in the room, all systems must function properly for safety. If a patient room supply or exhaust unit fails and needs maintenance during occupancy, you still must take care of the patient. A laboratory would shut down upon loss of airflow into the contained areas. A patient room that continues to operate will have increased risk for staff and other patients. Design to minimize the possibility of system failure. In the event of a system failure, separation of components by filtration will reduce the potential for exposure of maintenance personnel.

Principle 10. Define measuring containment success

A new facility type that has less definition is difficult to design, construct, and operate. Nothing is more frustrating than completing the design and construction of a biocontainment facility then finding there is disagreement on design or operational parameters. You can also find that you have excess alarm conditions that distract from operating the facility. Without defining ahead of time what specifically must achieve containment, there will be questions and disagreements from the team. You may not get the level of containment that you feel is appropriate.

Models of Patient Biocontainment Units

With lessons learned from the 2014 outbreak, new models for PBUs have been developing that combine lessons learned from BSL3 and BSL4 laboratory design with lessons learned from the initial operations of PBUs. The features of these models may include one-way flow to limit cross-contamination, laboratory testing, and decontamination of waste within the suite to reduce reliance on protocol and patient access to the suite from a restricted access corridor. See Fig. 8 for an example layout of the developing model for patient PBUs.

In the developing model, care activities utilize a clean support zone and corridor for patient care activities on the front side of the patient rooms combined with a potentially, but not routinely, contaminated access zone to the rear of the patient rooms. This area is kept clean by BSL3 entry and exit zones from patient care, decontamination of waste within the unit, and patient access from the non-clean side of the patient rooms. The staff enters from the clean corridor and dons PPE in a clean space. The staff then moves into the patient room to provide care. Upon exit from the patient room, the staff disposes of waste and then doffs their PPE before exiting through a shower into the clean side of the facility. A room is provided for moving equipment into the patient room and for decontaminating the equipment during removal.

The lab is entered and exited through the anteroom from the clean corridor. A BSC provides primary containment for protection of the worker and laboratory samples during analysis. The laboratory worker has the option of exiting the laboratory through the doffing and shower area.

This new model allows the ten principles of planning and design for PBUs to be implemented. This allows for improved patient care and increased safety for patients, staff, and the community.

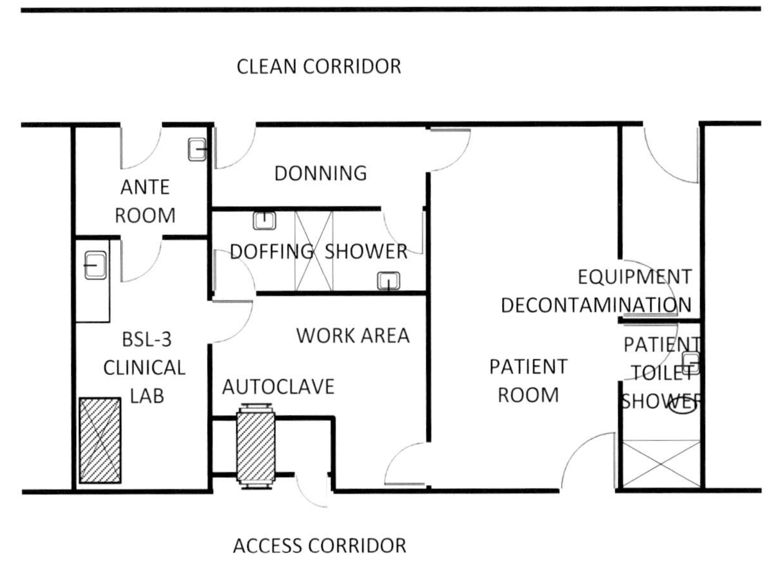

Figure 8: Example layout of a PBU.

Summary of Specialized Biocontainment Facilities

The above two examples provide an understanding of the processes and evaluation of issues involved when planning and designing novel biocontainment facilities. You must consider how principles of biocontainment apply to the operations involved and also look at how the operations and requirements change considerations for biocontainment.

DESIGNING SPECIFIC LABORATORY SYSTEMS

Heating, Ventilation, and Air Conditioning

Historically, laboratories had fairly simple static HVAC systems. A certain amount of air was supplied to the room and a proportional amount was exhausted (constant-volume systems). The temperature of the room was adjusted by varying the temperature of the air coming into the room (reheat systems). Pressure relationships were maintained by the proportion of the supply air exhausted. Once balanced, these systems with few moving parts were simple to maintain. Most problems came from dynamic changes in the system, such as static pressure change due to supply, exhaust, or ducted BSC filter loading, because these simple systems could not automatically respond to changes in conditions. Newer constant-volume systems using the technology of pressure-independent air valves allow an automatic response to dynamic changes in the system.

Over the past 25 years, the use of variable-volume systems that respond to dynamic changes in the laboratory environment have become the predominant model of systems in laboratories. These systems control temperatures in laboratories by supplying varying amounts of air to match the heat load produced in the space. Instead of a fixed damper, these systems use modulating dampers, or air valves, to control the amount of air supplied. Room-pressure relationships are maintained by modulating dampers or air valves in the exhaust system, varying the amount of air exhausted in proportion to the air being supplied. These valves and their controls add an additional level of design, construction coordination, and maintenance to the system. Although the individual components of the system have remarkable accuracy, when these individual components become interrelated, the design, construction, and operations of such systems can be difficult to control. When the airflow changes or a door opens in one laboratory, it may affect other laboratories or the entire system, and cumulative effects are difficult to predict. These systems must be well thought out and properly calibrated, commissioned, and maintained to work properly. For BSL3 facilities where air moves into and through the laboratory from areas of lowest to highest hazard, constant-volume systems have proven simple to operate. For BSL3Ag or BSL4, where the room is a sealed chamber, variable-volume systems have allowed the airflow to modulate as doors are opened and closed between sealed areas of the laboratory or animal holding.

The HVAC system is the most critical system in a laboratory to ensure worker comfort and safety. Early planning for this system is essential to allow proper sizing and placement of mechanical rooms with their large air-intake louvers and integration of the exhaust stacks into the design of the building. This is particularly critical in a renovation of, or conversion to, a laboratory where floor-to-floor heights begin to dictate system concepts. In a new building, floor-to-floor heights and location of chases for continuation of ducts through floors should be considered early.

The integration of architecture and engineering in a laboratory is critical to the cost, schedule, and quality control of a project. Module width for laboratories is critical where minimum air changes are a governing factor in air handling system design. For example, an 11-ft module adds an additional 10% to the air handling system requirements compared with a 10-ft module. This additional requirement has an impact on air handling systems, supply and exhaust duct sizes, chiller size, the cost of structural and architectural systems, and operating costs during the life of the building.

Ceiling height, and even whether to put a ceiling in the laboratories, must be decided early, because both decisions affect the volume of the room. When air changes are the governing factor, the lack of a ceiling can add about 33% to the size and costs of running the HVAC system. Floor-to-floor height is critical to properly operating HVAC systems. A floor-to-below-beam height of 12 to 13 ft is usually sufficient to allow duct systems to be installed with minimum offsets when planned properly. Offsets increase resistance to airflow in the duct and waste energy. In existing buildings where minimal floor-to-floor height is available, the entire system may have to be planned around possible duct routing.

Room size can be established early to maximize the match between the HVAC system and the room volume, and to eliminate duplication of systems such as fume hood exhaust and room exhaust.

Animal facility ventilation

Ventilation and airflow are the most critical factors in minimizing spread of aerosolized microorganisms, thereby protecting animals and personnel. Air movement is necessary for controlling odors in an animal facility. The air supply and exhaust systems should be independent and provide 100% outside air, totally exhausted with no recirculation. Systems should be designed for a high degree of reliability, providing constant temperature, air change rate, and humidity in the rooms. These requirements will vary with species of animals housed. The air pressure relationships between dirty areas and clean areas should be maintained at all times.

Integration of design

The air intake for the units should be on the building face, and the exhaust air should be discharged on the roof level. Air intakes must not pick up vehicle fumes from loading dock areas or discharges from the building exhaust stacks. Coordination and integration of the architectural treatment of the space with the HVAC systems are essential parts of design due to the large quantities of supply and exhaust air that must be moved.

Design parameters

Design parameters for HVAC systems are directional airflow, ventilation rates (air changes), pressure relationships, temperature, and humidity control (22). Fundamental guidelines are presented here, but conditions in each laboratory are different, and each laboratory should be engineered to its specific requirements.

Most research facilities at BSL2 or above are designed with no recirculation of air between laboratories. This minimizes the potential for spread of hazardous fumes or bioaerosols and reduces the chance of cross-contamination between laboratories. Air from the laboratories is exhausted directly to the outside. Air is usually introduced into the laboratory near the entry or desk working areas and sweeps the laboratory before being exhausted near the area of hazards. This reduces the area of the laboratory that becomes exposed to hazardous aerosols and is another good reason to put desk areas at the front of the laboratory. Care should be taken in the types of diffusers used to minimize drafts and cross-currents that might have an effect on experiments or upset the balance of safety devices. This is particularly important when air is introduced in the area of CFHs or BSCs. Laminar-flow air supply diffusers, which are large, perforated panels that allow air to ooze through, can be provided to minimize the velocity of air entering laboratories, thus reducing drafts that can affect hoods, cabinets, and laboratory bench work.

Air change rates vary depending on specific needs, types of HVAC systems, number of exhausted containment devices per laboratory, and cooling requirements of rooms. Typical infectious disease laboratories might have 6 to 10 air changes per hour, with animal facilities having 15 or more air changes per hour. As mentioned above in animal housing, the use of individually ventilated isolation cages in an animal facility can allow lowered air changes in animal rooms because the air change rate within the cage is high.

The rooms are then balanced so that air will flow from the corridor (or lower-hazard room) into the higher-hazard room, eventually being exhausted from the building (23). To have air moving from areas of low hazard to areas of higher hazard, the higher-hazard room is at a "negative" pressure relative to the lower-hazard room. Pressure relationships between rooms vary from 5% to 15%, depending on specific system requirements.

Airflow at low pressure differential is difficult to control, particularly with variable-volume systems, whose accuracy may be ±5%, which might allow the room pressure to be reversed in worst cases. With higher differentials, doors become difficult to open or are snatched open, depending on their direction of swing, and air infiltrates the room from any unsealed opening, such as electrical outlet boxes, door frames, or windows. The mechanical system design and level of architectural integrity of the laboratory must be balanced carefully. High negative pressures can also cause high infiltration of outside air through window joints and other locations that can bring in spores, pollen, fungi, and other pollutants or odors. This may disturb experiments, particularly in laboratories working with specimens similar to the pollutants, or contribute to contamination of tissue cultures or personnel problems (allergies).

The temperature in a biomedical research laboratory should be cool enough to provide comfort for workers wearing laboratory coats and gloves. Comfort levels are difficult to maintain unless sufficient cooling is provided to overcome the heat generated by equipment used in the laboratory. Humidity control is important to minimize condensation problems caused by high humidity or static electricity caused by low humidity.

Acoustical Considerations

Acoustical considerations in laboratories include vibration, which can affect sensitive equipment, such as balances, microscopes, microtomes, and electron microscopes, and noise, which can be detrimental to the health and comfort of the occupants. Most vibration in laboratory buildings comes either from the ground under the building or from mechanical equipment with moving parts in the building, such as chillers, fans, and air handling units. Proper design of these systems minimizes vibration transmission, and local dampening of equipment (e.g., balance tables) makes vibration a minimal problem in most laboratories. Noise, however, is a constant battle. Equipment noise and the noise inherently associated with large volumes of air movement are more difficult to control. CFHs, freezers, BSCs, centrifuges, blenders, and vacuum pumps produce noise. Where possible, items, such as freezers and centrifuges, that produce noise constantly for long periods should be located in separate rooms that are seldom occupied. Fume hood noise can be minimized by proper system design. Although most laboratories can meet

Occupational Safety and Health Administration (OSHA) noise level requirements of 85 dB over an 8-h period, they may not provide auditory comfort to the occupants. Sessler and Hoover (24) suggest that noise levels should not exceed 45 to 55 dB in research laboratories where telephone communication and creative thinking are done. However, these levels may be difficult to achieve in a laboratory; a realistic level would be 50 to 60 dB. Consideration of noise early in the design process will reduce potential problems.

Plumbing

Sanitary drainage, laboratory drainage, and vent systems

Laboratories often have two types of drainage systems—sanitary and laboratory waste. The sanitary drainage handles liquid waste from water closets, lavatories, drinking fountains, and other nonlaboratory sources. The laboratory waste system is acid resistant and serves sinks in laboratories and CFHs that may be contaminated with chemicals due to improper disposal or spills. Acid dilution tanks are often provided with laboratory waste systems. While regulations minimize the amount of waste that can be legally put into drainage systems, spills at CFHs can occur. Care must be taken to ensure that any wastes put into drainage systems meet local codes. Vent piping must be taken up through the laboratory to relieve sewer gases.

Domestic and laboratory water supply systems

Separate systems should be provided for potable and laboratory water in laboratory facilities. Backflow preventers or vacuum breakers are required on the laboratory water side to prevent contamination to potable water systems.

Eyewash stations should be provided at main laboratory sinks. Safety showers should be placed in corridors where required by code and ideally in each laboratory that contains a CFH. The safety showers should each have an eyewash associated with them. Considerations and local codes may govern whether these are on the laboratory or domestic water system and if the water should be tempered.

Specialized water systems (e.g., distilled, deionized, or microfiltered) may be required for water to be used in experiments. This water can be obtained from central or point-of-use systems.

Vacuum and Compressed-Air Systems

Central systems should be provided for vacuum and compressed air. The vacuum system from biosafety areas should have a disinfectant trap and HEPA filter on the line before entering the vacuum lines to keep the lines and collection systems from being contaminated and to keep biohazards from being discharged to the atmosphere. In-line air filters should be provided, where necessary, to keep oil from contaminating equipment or experiments. Alternatively, the use of oil-free compressors can alleviate the problem.

Laboratory Gases

Natural gas, carbon dioxide, and cryogenic and other specialty gases are normally found in laboratories. All gas cylinders in use or in storage must be restrained. Also, some hazardous specialty gases, such as hydrogen fluoride, may warrant additional safety support, such as showers or leak detection systems. High-volume-use gases are normally piped through central systems. It is also prudent to pipe gases into BSL3 laboratories. Cylinders should also be kept out of sensitive areas, such as tissue culture rooms, to minimize contamination from traffic. Sufficient storage areas and restraining devices must be provided for gas cylinders, both empty and full. Cylinders should be housed outside of BSL3 and BSL4 laboratories to minimize traffic into these laboratories and to avoid issues with decontaminating empty cylinders. Gases can be piped into the laboratories from the cylinder holding areas.

Fire Protection

Fire extinguishers

Fire extinguishers must be provided at locations required by the life safety codes and should be installed at other hazardous locations. Type ABC extinguishers are usually provided; other types (e.g., carbon dioxide or foam) are provided for specialized needs.

Sprinklers

The facility may not have to be equipped with automatic sprinklers to meet current National Fire Protection Association code, but consideration should be given to installing a sprinkler systen in the facility while it is being built. Sprinklers would provide the following benefits in the event of fire: increased life safety, minimized loss of experimental data, reduced possibility of breach of containment, and minimized loss of use of a unique facility. Containment of sprinkler discharge may be an issue in enhanced BSL3, BSL3Ag, and BSL4 facilities.

Other fire protection systems

Special electronic equipment or high-volume chemical transfer areas may require fire protection systems other than water sprinkler systems. CO_2, foam, and other systems are available for specialized needs but are not generally used in laboratories.

A fire alarm system should be provided to give early warning of fires or other life safety problems. The requirements for these systems are usually governed by the local codes. Visual systems must be provided for the hearing impaired and for those in high-noise areas or limited-access laboratories (BSL3) who may not hear the building alarm system.

Spill Control and Fire Stations

Spill control and fire stations should be set up in areas near hazards for quick access and response in the event of an emergency. Neutralization, absorption, and disinfectant materials to handle the hazards in the laboratory should be stored along with fire blankets, extinguishers, and first aid supplies. Limited-access laboratories (BSL3) should be stocked with sufficient suitable emergency response materials.

Electrical

Normal power supply and distribution

Laboratories have become high-volume power users, and use of electricity in laboratories is likely to increase. Normal power should be supplied generously to all laboratories, with many circuits feeding each laboratory. Also, particular equipment with unusual power requirements must be identified during the design phase so that appropriate power can be supplied. Additional empty conduits and junction boxes from the laboratory panels should be installed to allow easy wiring of future laboratory equipment. All levels of the laboratory power system, from the transformers to the individual laboratory panels, should allow for future growth.

Standby power

Consider installation of a back-up power generator to serve the life safety equipment, air handling, exhaust, biosafety systems, freezers, and incubators in the event of a loss of normal power. Establish in advance the critical equipment to be powered by the emergency generator to provide adequately for the power required. An uninterruptable power supply system can provide power continuity for equipment that cannot function with the momentary lapse in power during the start-up of the back-up power generator.

Lighting

Lighting in laboratories should be evenly distributed and task oriented. Lighting should be placed over benches in a manner to minimize glare, and shadows and should be of sufficient brightness to illuminate the workers as well as the work surfaces in the laboratory. Consider lighting under the shelves above countertops to provide task lighting to benches.

Instrument ground

Laboratory equipment may be sensitive to "electrical noise" from building systems and motors that may be distributed along the building grounding system. Care should be taken to minimize this problem by providing system separation between electronic noise producers and equipment that might be affected by the noise. Grounding systems can be developed to reduce this problem.

Access Control, Security, and Monitoring

Controlled access is becoming increasingly important in containment facilities. All parties have an interest in knowing that the facility is operating as designed. A computerized monitoring system can be installed to provide continuous monitoring of the conditions in the facility. Several systems have been developed and programmed for laboratory and animal facilities. Options for monitoring include the following:

- Card key access control to the facility
- Operation of critical equipment (e.g., HVAC, emergency generator, BSCs, CFHs, autoclaves, cold rooms, and freezers)
- Airflow, air changes, and pressurization
- Local environmental conditions (temperature, humidity, and lighting)
- Automatic 24-h notification of alarm systems
- Visual monitoring for security through closed-circuit TV

This system could be monitored in the facility, in the administrative area, or in the maintenance area. Some of the potential benefits are as follows:

- Controlled access to biological, chemical, and radioactive materials
- Advanced warning of system malfunction
- Record of entries and exits to the facility, including records of improper entry attempts
- Monitoring of the condition of incubators and freezers to prevent potential loss of research specimens
- Record of space conditions as backup for experimental result validation
- Higher level of public comfort concerning the facility

Communications

Telephone and data connections should be an integral part of the planning of all laboratory spaces. Cable trays should be provided above the ceiling for communication and computer wiring. This wiring should extend to all laboratory areas and a central point for connection to building-wide systems. Space should also be provided for data system racks and modems to connect to outside networks. Communication devices in BSL3 and BSL4 should be hands-free operation.

Laboratory Information Management Systems

Computer use is now integral to laboratory work. Laboratories use computer systems for accounting, data analysis, data acquisition, equipment operation, HVAC system operation, quality control, and system monitoring. Standalone and networked applications are being developed into laboratory information management systems, allowing for collection and analysis of data from a variety of laboratory sources. Most new laboratory equipment is computerized to some degree with data output into networks. Currently, much of this equipment has a personal computer-type computer and printer dedicated to it. A typical laboratory may have three or four computer-printer combinations dedicated for equipment use. Standards are being developed for systems to communicate through a network allowing one computer to handle multiple functions in each laboratory. This could reduce the need for up to 6 to 10 ft of bench space per laboratory.

Systems Distribution

A plan must be developed to provide a clear method for distribution of HVAC, plumbing, and electrical systems to the facility to allow for ease of operation and maintenance. Supply and exhaust ducts, pipes, conduits, and cable trays must be routed to minimize turns (which can reduce efficiency), crossings (which reduce ceiling heights or raise structure), and access locations (which may require access panels or drains). This routing must allow access for maintenance and repair. Because systems account for the most costs in a laboratory, the distribution scheme can either waste or save a lot of money. Evaluation should be made as to the best point to switch from main to branch distribution. Again, common sense should come into play, as systems should be located and developed to minimize the extent of the most expensive components. Systems serving BSL3 and BSL4 containment should have components requiring maintenance access located outside containment.

Consider providing interstitial space, such as a service floor just above the laboratory floor and above containment areas requiring HEPA filtration. This allows services to be distributed down into laboratories from above and HEPA filters to be located near the areas they serve. Interstitial floors provide a great deal of access for service and modifications, are very flexible, and minimize laboratory floor space required for service.

EQUIPMENT SELECTION

Casework

Casework type, construction, adaptability, and cost are important decisions in the design of any laboratory. Initial costs versus long-term cost must be considered, and casework tops and fixtures in relation to utility services are key components of design. Tops are available in several materials; those most commonly used in biomedical laboratories are stainless steel (easily cleanable and highly solvent resistant), epoxy resin (impact resistant and highly solvent and acid resistant), and plastic laminate (good chemical resistance and lower impact resistance). Cabinets for flammable and acid storage should be provided in each laboratory where these chemicals are used and stored.

Self-supporting (flexible) casework systems should be considered in laboratories where a high degree of change is anticipated. Clinical laboratories are one of the most successful users of flexible casework.

Benches should be designed with the appropriate mixture of cupboards and drawers for the use intended. An appropriate number of knee-space workstations should be provided to allow comfortable sit-down areas for working. Dry areas should be provided for working with paper and computers. The most common mistake in casework layout is putting too much casework into the laboratory during design. Floor space should be left for unanticipated large equipment. Casework can be added; it is seldom removed.

Vermin and Rodent Control

Vermin and rodents can be a nuisance and hazard in biomedical facilities, particularly animal and containment facilities. All penetrations, small holes, openings, and cracks should be sealed well. Materials and systems should be selected to minimize areas for vermin to hide. Adequate receiving areas, storage, and break rooms should be located away from laboratories to minimize these problems. An integrated pest control program as described in BMBL (8, 9) should be considered.

Waste Handling and Removal

Space for storage of wastes, flammable and solvent storage, and cold rooms or freezers for animal carcass storage should be an integral part of the facility. Provisions for the handling and storage of hazardous and nonhazardous wastes must be made during design. There is an increasing emphasis on waste management by institutions and regulatory agencies. Space must be provided if chemical recycling programs are to be carried out in the facility.

Space for general waste, biological waste, and noninfectious animal bedding should be provided along with appropriate pickup facilities. Space is also needed for radioactive waste to be held in the laboratory and to be picked up by the appropriate authorities. Freezer space for holding radioactive animal carcasses needs to be provided.

Providing an incinerator for biological waste should be considered if volume and local regulations warrant the expense and local regulations allow. Other methods of waste disposal (e.g., grinding and digestion) should be considered.

Decontamination

Consideration of how the facility will be decontaminated should be a part of the design. Provisions to allow separation of rooms from supply and exhaust systems should be made. Openings into the room should be sealed or be capable of being sealed, depending on the level of hazard (8, 9). Spill control, cleanup, and decontamination centers should be provided with easy access to all hazardous laboratories.

Signage and Information Systems

Five levels of signage should be considered for a laboratory project—general directional, informational, life safety, hazard identification, and system identification. General directional signage should direct persons to various locations in the facility. It should also include notification of access restrictions to laboratory areas.

Hazard identification signage has four categories. "Notice" states a policy related to safety of personnel or protection of property but is not for use with a physical hazard. "Caution" indicates a potentially hazardous situation that, if not avoided, may result in minor or moderate injury. "Warning" indicates a potentially hazardous situation that, if not avoided, will result in death or serious injury. "Danger" indicates an imminently hazardous situation that, if not avoided, will result in death or serious injury. There are specific sizes, shapes, colors, messages, and lettering requirements for each type of hazard identification signage.

Commissioning and Acceptance

As the laboratory is being constructed, it is important to ensure that the facility will operate per the intent of the design. This testing process is called commissioning. Ideally, commissioning will start during design so that the agent that will perform the testing can interact with the design, construction, and operations teams to be able to understand the intent fully. This early participation enhances the odds that the commissioning will go smoothly and the expectations of all stakeholders will be met.

Commissioning occurs at three levels: (i) testing of individual components of a system, (ii) testing of the system, and (iii) testing of all systems in an integrated fashion. An example of integrated systems testing would be testing how the supply and exhaust systems react, shut down, and restart upon loss of power, start up of the emergency generator, and transfer back to normal power. Included in this integrated test would be how the various alarms and notifications occur during this process.

For containment labs, commissioning would normally include testing the containment barrier for proper level of sealing, testing, and certification of BSCs and other primary containment equipment, testing of decontamination systems (autoclaves, effluent decontamination, and integrated gas or vapor systems), and testing of mechanical, electrical, and plumbing systems.

Validation of the containment laboratory should include verification that the written operational protocols match both the completed facility and the risk assessment of the work to be performed in the facility.

SUMMARY

Many considerations and decisions are involved in the design of biomedical research and specialized facilities, particularly facilities used for the containment of biohazards such as autopsy suites and PBUs. The users, administrators, and facility designers must provide adequate information so that correct decisions can be made at timely points in the design process. In the end, this will minimize lost time and added costs that occur when issues have to be revisited. Good communication among all parties is a key to the successful design of a biomedical laboratory or specialized facility. Each facility is unique; no method of design can provide a finished facility without adequate time, effort, and thought. The most successful facilities will provide for their unique requirements as well as the common elements required in every laboratory. Such a facility will meet the functional requirements for specific purposes and also be adaptable for future purposes.

References

1. **Crane JT, Riley JF.** 1999. Design of BSL3 laboratories, p 111–119. *In* Richmond JY (ed.), *Anthology of Biosafety*, vol. 1. *Perspectives on Laboratory Design*. American Biological Safety Association, Mundelein, IL.
2. **Crane JT, Bullock FC, Richmond JY.** 1999. Designing the BSL4 laboratory. *In* Richmond JY (ed.), *Anthology of Biosafety*, vol. 1. *Perspectives on Laboratory Design*. American Biological Safety Association, Mundelein, IL.

3. **Kuehne RW.** 1973. Biological containment facility for studying infectious disease. *Appl Microbiol* **26:**239–243.
4. **National Institutes of Health.** 2013. *NIH Guidelines for Research Involving Recombinant or Synthetic Nucleic Acid Molecules (NIH Guidelines)*, November 2013. http://osp.od.nih.gov/sites/default/files/NIH_Guidelines_0.pdf
5. **Dolan DC.** 1981. Design for biomedical research facilities: architectural features of biomedical design, p 75–86. *In* Fox DG (ed), *Design of Biomedical Research Facilities: Proceedings of a Cancer Research Safety Symposium, 1979.* NIH publication 81-2305. Frederick Cancer Research Center, National Institutes of Health, Bethesda, MD.
6. **West DL, Chatigny MA.** 1986. Design of microbiological and biomedical research facilities, p 124–137. *In* Miller BM (ed), *Laboratory Safety: Principles and Practices.* American Society for Microbiology, Washington, DC.
7. **Chatigny MA, West DL.** 1976. Laboratory ventilation rates: theoretical and practical considerations, p 71–100. *In Proceedings of the Symposium on Laboratory Ventilation for Hazard Control.* NIH Publication No. 82-1293. Frederick Cancer Research Center, Frederick, MD.
8. **Centers for Disease Control and Prevention and National Institutes of Health.** 1999. *Biosafety in Microbiological and Biomedical Laboratories*, 4th ed. U.S. Government Printing Office, Washington, DC.
9. **U.S. Department of Health and Human Services, Public Health Service, Centers for Disease Control and Prevention, National Institutes of Health.** 2009. *Biosafety in Microbiological and Biomedical Laboratories*, 5th ed. HHS Publication no. (CDC) 21-112. http://www.cdc.gov/biosafety/publications/bmbl5/BMBL.pdf.
10. **Richmond JY (ed).** 2002. *Anthology of Biosafety*, vol. 5. *BSL-4 Laboratories.* American Biological Safety Association, Mundelein, IL.
11. **Department of Labor.** 1999. *29 CFR Part 1910.1030, Bloodborne Pathogens, Final Rule. Occupational Safety and Health Administration.* U.S. Government Printing Office, Washington, DC.
12. **Centers for Disease Control and Prevention.** 1997. Goals for working safely with *Mycobacterium tuberculosis* in clinical, public health, and research laboratories. http://www.cdc.gov/od/ohs/tb/tbdoc2.htm.
13. **American Institute of Architects, Committee on Architecture for Health.** 2001. *Guidelines for Construction and Equipment of Hospital and Medical Facilities.* American Institute of Architects Press, Washington, DC.
14. **Department of Labor.** 1999. *29 CFR Part 1910.1450, Occupational Exposures to Hazardous Chemicals in Laboratories, Final Rule. Occupational Safety and Health Administration.* U.S. Government Printing Office, Washington, DC.
15. **Department of Labor.** 1999. *29 CFR Part 1990, Identification, Classification, and Regulation of Potential Occupational Carcinogens. Occupational Safety and Health Administration.* U.S. Government Printing Office, Washington, DC.
16. **Crane JT, Riley JF.** 1997. Design issues in the comprehensive BSL2 and BSL3 laboratory, p 63–114. *In* Richmond JY (ed), *Designing a Modern Microbiological/Biomedical Laboratory.* American Public Health Association, Washington, DC.
17. **Riley JF, Bullock FC, Crane JT.** 1999. Facility guidelines for BSL2 and BSL3 biological laboratories, p 99–109. *In* Richmond JY (ed.), *Anthology of Biosafety*, vol. 1. *Perspectives on Laboratory Design.* American Biological Safety Association, Mundelein, IL.
18. **College of American Pathologists.** 1985. *Medical Laboratory Planning and Design.* College of American Pathologists, Skokie, IL.
19. **National Research Council.** 1996. *Guide for the Care and Use of Laboratory Animals.* National Academies Press, Washington, DC.
20. **Nolte KB, Taylor DG, Richmond JY.** 2001. Autopsy biosafety, p 1–50, *In* Richmond JY (ed), *Anthology of Biosafety, IV. Issues in Public Health.* American Biological Safety Association (ABSA), Chicago, IL.
21. **Nolte KB, Taylor DG, Richmond JY.** 2002. Biosafety considerations for autopsy. *Am J Forensic Med Pathol* **23:**107–122.
22. **ASHRAE Technical Committee.** 1999. Industrial applications, laboratories, p 13.1–13.19. *In ASHRAE Handbook, Applications.* American Society of Heating, Refrigerating and Air-Conditioning Engineers, Inc., New York.
23. **National Fire Protection Association.** 2004. *NFPA 45 Fire Protection for Laboratories Using Chemicals.* National Fire Protection Agency, Quincy, MA.
24. **Sessler SM, Hoover RM.** 1983. *Laboratory Fume Hood Noise, Heating Piping and Air Conditioning.* Penton/PC Reinhold, Cleveland, OH.

17

Primary Barriers and Equipment-Associated Hazards

ELIZABETH GILMAN DUANE AND RICHARD C. FINK

Primary barriers are both techniques and equipment that guard against the release of biological material; they may also be referred to as primary containment. In general, they provide a physical barrier between the worker and/or the environment and the hazardous material. Primary barriers range from a basic laboratory coat to a biological safety cabinet (BSC). This chapter addresses some of the more common primary containment devices and personal protective equipment (PPE) and a variety of equipment-associated hazards. Other chapters cover respiratory protection, work practices, and BSCs that are more specific examples of primary containment.

The history of laboratory-acquired illnesses amply demonstrates how important primary barriers are and how equally important it is to select appropriate primary barriers and use them correctly. Examples from the literature are included whenever possible to illustrate this point. Other chapters contain information on laboratory-acquired infections.

PRIMARY CONTAINMENT DEVICES AND EQUIPMENT-ASSOCIATED HAZARDS

Animal Housing

The hazards from animals, other than bites and scratches, result chiefly from aerosolization of dried urine and feces (1). Thus, caging that provides protection from aerosols is important in research involving a wide variety of agents and viral vectors that can be excreted in urine and feces. A vivid example of the importance of containing aerosols is provided by the outbreak of lymphocytic choriomeningitis virus in researchers and visitors in 1972 to 1973 at the University of Rochester Medical Center. Forty-eight people became infected, 17 of whom did not have physical contact with the infected animals. Infection occurred not only from direct contact with infected hamsters but also from merely being present in the animal housing rooms (2).

There are a large variety of caging strategies available to minimize personnel exposure to aerosols generated by animals. There are more options in small-animal housing than in housing for large animals. For small animals, the housing options range from filter bonnets to laminar-airflow cubicles. The filter bonnet is effective in trapping aerosols created by movements of animals; however, once the bonnet is removed for animal care or research purposes, the protection is lost. Cages with filter bonnets can have an increase in humidity, carbon dioxide, and ammonia. To overcome these problems, there are ventilated cages and cage racks that are HEPA filtered. Some of these racks provide only cross-contamination protection and therefore exhaust unfiltered air into the room. Other rack designs are able to HEPA filter the air into and out of the cages (Fig. 1). Again, when the cages are removed from the racks and opened, personnel protection ends. Therefore, it is advisable to open cages within a BSC.

Cage racks can also be placed in HEPA-filtered, laminar-flow isolation rooms. These rooms can be permanent facilities, modular-cubicle, or portable cleanroom enclosures.

Figure 1: Super Mouse 750 Micro-Isolator (92140AR) high-density animal housing system equipped with EnviroGard environmental control supply and exhaust air units. (Courtesy of Lab Products, Inc.)

They may be maintained under either positive or negative pressure. Some of these units can be exhausted to the outside, resulting in the equivalent of a Class I BSC.

Centrifuge Safety

There are many opportunities to generate aerosols when working with a centrifuge. The action of filling and decanting centrifuge bottles or tubes creates an aerosol. Thus, in some situations, these procedures need to be performed in a BSC for personnel protection. When a centrifuge bottle or tube breaks during centrifugation, the size of the resulting aerosol can be very large (3). Centrifuges have been associated with hundreds of laboratory-acquired infections due to *Brucella*, *Coxiella burnetii*, Sabia virus, human immunodeficiency virus, *Pseudomonas mallei*, and other infectious agents (4–9).

In 1994, a virologist at Yale University apparently was exposed to an aerosol of Sabia virus when a centrifuge bottle developed a crack and tissue culture supernatant containing the virus leaked into a high-speed centrifuge. The virologist was working in a biosafety level 3 (BSL3) lab and cleaned the spilled material and the centrifuge while wearing a gown, surgical mask, and gloves. He subsequently developed the symptoms of illness, and a diagnosis of Sabia infection was confirmed by isolation of the virus from his blood. The virologist did, however, recover (6). This laboratory-acquired illness demonstrates not only that the appropriate PPE, namely, respiratory protection, was not used but also that centrifuges are capable of creating infectious aerosols.

Centrifuge safety cups provide a method of containment for centrifugation. Containers range from sealed tubes to larger screw-cap buckets and sealed rotors. The quality of the seal is important, due to the extremely high stresses of the process. There are several common devices that may be employed, such as gasket seals, which should be inspected regularly and replaced when compromised. To maintain personnel protection when working with infectious agents, centrifuge rotors must be loaded and unloaded in a BSC.

Rotor explosions have occurred in laboratories, and thus the possibility of a major accident when centrifuging infectious agents should be considered. In one such case, the explosion was due to the removal of the overspeed safety pin (10). As a result, the entire rotor and centrifuge were destroyed (Figs. 2 and 3). Equipment should always be used in the manner that it was designed to be used and should never be allowed to exceed design parameters. Proper preventative maintenance should be a part of the overall laboratory safety program.

Work with large volumes or titers of infectious agents may require placing the entire centrifuge in a specially

Figure 2: Ultracentrifuge rotor after the rotor exploded. (Courtesy of MIT Biosafety Office.)

designed ventilated enclosure such as a Class I BSC, which provides personnel and environmental protection. Alternatively, a Class II or Class III BSC (11) may be modified by the manufacturer to accommodate a centrifuge. The BSC provides protection for the user from aerosols that may be generated and possibly escape from the centrifuge.

Blenders, Sonicators, Homogenizers, Mixers, and Other Aerosol-Producing Equipment

Blenders and related mixing equipment are well known for creating aerosols. Covered blenders generated a detectable aerosol of *Serratia marcescens* during operation that averaged from 8.7 to 119.6 CFU per cubic foot of air sampled (12). Immediate removal of the top after blending resulted in a very large aerosol of 1,500 CFU per cubic foot of air sampled (3, 12).

Two examples of the dangers blenders can pose are illustrated by the following case reports of laboratory-acquired infections. A laboratory technician removed the lid of a blender in which he had homogenized egg yolk sacs infected with *Rickettsia prowazekii*. Shortly after this operation, the worker became ill with typhus fever (13). The second example occurred in the early 1970s when a laboratory worker homogenized 11 rabid goat brains using a kitchen-type blender. The blender had a typical loose-fitting plastic lid. The person subsequently developed a fatal case of rabies (14).

Commercially available autoclavable safety blender cups contain the aerosol generated during a blending operation. Care must still be taken to contain the aerosol generated when the lid is removed, and thus this procedure should be performed inside the BSC.

A probe sonicator (sonic oscillator) generated 6.3 CFU of *S. marcescens* per cubic foot of air sampled (12). Thus, in the absence of safety equipment, they too are capable of releasing infectious aerosols into the laboratory environment. In addition to the probe-type sonicator that must be placed within the material being sonicated, there are also horn-type sonicators. These are placed on the outside of the vessel being sonicated. As long as the vessel has containment, there is no release of aerosols.

Other mixing equipment, such as homogenizers, may generate infectious aerosols and must be placed into another means of primary containment, such as a BSC, if safety equipment is not available for the product. There are commercially available homogenizers that will contain the aerosols that are generated during the procedure. Homogenizer blenders that use a plastic bag to contain the contents and prevent the escape of aerosols are another type of primary containment device. These devices are frequently known as a stomacher, even though this is actually a trademarked name for a particular brand of blender. To operate this type of homogenizer blender as a primary containment device, the plastic bag has to be filled and emptied in a BSC. This contrasts with other

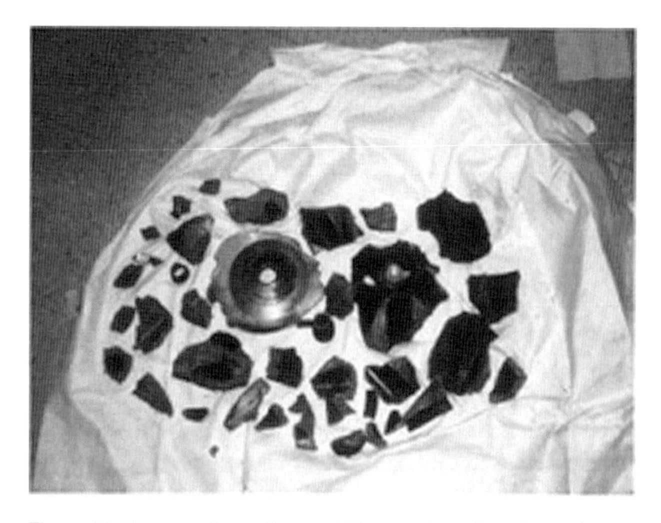

Figure 3: Pieces of an ultracentrifuge rotor after the rotor exploded. (Courtesy of MIT Biosafety Office.)

types of blenders and homogenizers that generate a large uncontrolled aerosol.

The equipment used for sorting of cells can produce aerosols. The International Society for Advancement of Cytometry has published cell sorter biosafety standards that provide guidance for containment levels and standard operating procedure (SOP) development (15). This guidance is particularly important when sorting unfixed cells, such as human cells or cells that are known to contain infectious agents. Aerosols are generated when the cell suspension is drawn up and pushed through a vibrating nozzle, and the resulting droplets fall between high voltage plates. High-speed sorters are capable of producing more small droplets compared to older equipment that operates at slower speeds, as high-speed sorters operate under greater pressure. Some cell sorters are designed to fit and operate within a BSC, which provides containment. Some sorters have an integral aerosol management system designed to capture the aerosols that are generated by the equipment (16).

Specimen Transport Containers

Closed-system carriers made of polycarbonate are a type of primary containment designed to facilitate the safe and efficient transport of tubes and samples containing infectious materials. Transport cases for biohazards are commercially available and include polysulfone clamps that securely hold the carrier closed and provide a leak-proof seal. Other appropriate secondary containers include plastic coolers with secured lids that have been properly labeled with a biohazard symbol. Secondary containers should be utilized for intralaboratory transport as well interlaboratory transport within a building or campus. For specimen shipping of biologicals that are infectious and require transport on public roadways and/or via air transport, country-specific regulations as applicable must be adhered to, as well as the requirements set forth by the International Air Transport Association (IATA) Dangerous Goods Regulations. Such regulations set forth criteria for specimen transport containers.

Vacuum Protectors and Vacuum Traps

Drawing a vacuum in the laboratory may result in the aerosolization of microorganisms and the subsequent contamination of the vacuum line, the pump, and the environment. To prevent this contamination, one can interpose any number of vacuum protectors. There are a number of commercially available in-line HEPA and microbial-grade pore filters. If a HEPA filter is used, a hydrophobic filter should be placed before it to protect the HEPA from wetting. Vacuum traps using vacuum flasks with a disinfectant can also be used to protect the vacuum system.

Positive-Pressure Suits

Positive-pressure personnel suits are necessary for entry into high-containment areas where infectious agents may not be adequately contained, such as in a Class II BSC. They isolate the worker by encapsulating him or her in a one-piece ventilated suit that features fresh air supplied from a breathing air supply. In a BSL4 laboratory, one-piece positive-pressure suits ventilated with a life support system can be used in conjunction with Class II BSCs. The air pressure in the suit is positive relative to the surrounding laboratory. Alternatively, all activities in a BSL4 laboratory may be contained in Class III BSCs, therefore not requiring the use of a positive-pressure suit.

Fermentors

Properly designed fermentors, from a simple shaker flask to a multi-thousand-liter stainless steel tank, can meet the criteria of a primary barrier. To make a shaker flask into a primary barrier, a tight-fitting plug must be added to the mouth of the flask. The plug can be of cotton, plastic foam, a polyfluorocarbon fiber closure, a gas-permeable film, or a microbial-grade filter. Each will either prevent the aerosol from exiting the flask or reduce the amount that exits.

A large fermentation tank can have multiple points where material can escape; thus, it must have multiple points of protection to qualify as a primary barrier. One of the most critical areas is the rotor shaft. If the shaft is at the bottom, a leak in the seals can result in a breach of containment. To prevent this breach, many fermentors use double mechanical seals or use a top-mounted agitator. Other possible breach points are exhaust gas vents and sampling ports. Exhaust gases can be passed through HEPA filters or an incinerator, and sampling ports can be fitted to a closed sampling system to avoid the generation of aerosols. The latter requires that personnel be trained in using the sampling ports to prevent the release of an aerosol during the connecting and disconnecting process. Air samples taken by the authors during sampling of a 1,500-liter fermenter showed that a detectable release of organisms could be found (the air sampler was located many feet from the sampling port).

PERSONAL PROTECTIVE EQUIPMENT

PPE includes all clothing and other work accessories designed to serve or be worn as a barrier against workplace hazards. Some common examples include such lab attire as gloves, head and shoe coverings, eye and face protection, and respiratory protection. PPE should not be used without first implementing the appropriate engineering, work practice, and administrative controls. Employers and employees must be conscious of the fact that PPE alone does not eliminate the hazard! If the primary containment fails or is insufficient, PPE often becomes an important barrier against exposure. An extensive evaluation of the PPE required to protect against the workplace hazards should be included in all workplace hazard assessments. For example, in 2014 in response to the Ebola virus outbreak in West Africa and the possibility of Ebola-infected individuals seeking care in U.S. hospitals, the Occupational Safety and Health Administration (OSHA) of the U.S. Department of Labor issued a fact sheet containing a PPE selection matrix (17). This matrix details a variety of work tasks and the recommended PPE elements, along with examples of workers who may require the particular level of protective equipment. This includes health care workers, maintenance and housekeeping personnel, and research and clinical laboratory workers.

A death in 2012 of a California laboratory worker due to infection with *Neisseria meningitidis* highlighted the need for both primary containment and PPE (18). The laboratory worker manipulated the *N. meningitidis* on the open bench instead of in a BSC. Investigation revealed multiple breaches in recommended biosafety practices, including the use of cloth laboratory coats that were not routinely decontaminated and lack of eye protection.

OSHA has issued a number of standards that address PPE requirements. For the laboratory environment, these include the following:

- PPE for general industry, 29 C.F.R. 1910.132 through 1910.138 (19)
- Respiratory protection standard, 29 C.F.R. 1910.134 (19)
- Hazard communication standard, 29 C.F.R. 1910.1200 (20)
- Bloodborne pathogens standard, 29 C.F.R. 1910.1030 (21)
- Laboratory standard, 29 C.F.R. 1910.1450 (22).

A useful publication known as the BMBL or *Biosafety in Microbiological and Biomedical Laboratories* (23) details the recommended PPE for BSL1 through BSL4. Table 1 illustrates the four BSL categories and their corresponding primary barriers.

The OSHA standards require that the employer furnish the appropriate PPE free of charge and that employees use the suitable PPE when there is the potential for injury or illness. The standards also outline the specific provisions for the various types of PPE. Equipment must be properly fitted and maintained in a clean and serviceable condition.

TABLE 1.

Biosafety levels and corresponding primary barriers[a]

BSL	Primary barriers
1	None required
2	Class I or II BSCs or other physical containment devices used for all manipulations that cause splashes or aerosols of infectious materials; laboratory coats, gloves, and face protection as needed
3	Class 1 or II BSCs or physical containment devices used for all open manipulations of agents; protective laboratory clothing, gloves, and respiratory protection as needed
4	All procedures conducted in Class III BSCs or Class I or II BSCs in combination with full-body, air-supplied, positive-pressure personnel suit

[a]BSL, biosafety level; BSC, biological safety cabinet.

Laboratory Clothing

There is a wide variety of commercially available laboratory clothing in different styles and sizes. This clothing is available in a range of fabrics, such as cotton, polyester, nylon, olefin, polyvinyl chloride, rubber, and specialty fabrics such as Tyvek, a spun-bonded olefin. The choice of style and fabric should be based on the job tasks to be performed and the material or hazards to which the wearer may be exposed. For example, clinical laboratory workers handling samples from patients with suspected or confirmed Ebola virus infection should wear fluid-resistant gowns per OSHA's PPE matrix (17).

Lab coats, scrubs, gowns, aprons, and coveralls

Laboratory coats and gowns are used to protect the worker from hazardous materials, such as infectious fluid. They can also provide protection for the material or product from contamination by the worker. Laboratory clothing gives various degrees of protection depending upon the material used. For example, a cotton lab coat may not be a suitable barrier should a lab worker be splashed with a large amount of liquid such as a chemical solvent.

In BSL1 and BSL2 laboratories, button-front lab coats may be acceptable. In a BSL3 lab, a wraparound or solid-front gown or coveralls must be used. The OSHA Bloodborne Pathogens Standard (21) specifies that the gown used must prevent blood, serum, etc., from soaking through to the worker's street clothing or skin.

There are data relating surgical gown construction (woven versus nonwoven), repellency and pore size to the microbial barrier effectiveness of the gown (24, 25). The type of material used in the lab clothing will also greatly affect its chemical and fire-resistance properties. When selecting lab clothing, it is important to balance worker comfort with risks due to chemical and biological permeation as well as fire resistance. Even the best lab clothing provides no protection if it is not worn appropriately and consistently. Another factor to consider is knit cuffs versus loose fit (no cuff). Knit cuffs make it easier to bring a glove cuff over the lab coat sleeve and minimize hanging sleeves that could knock over or come in contact with hazardous materials.

Because the lab coat protects the worker from contamination, it is important to realize that the coat can become contaminated. It is believed that in a 1900–1914 outbreak of anthrax in a woolen mill, workers' clothing disseminated anthrax spores to the workers' wives (26). In a report to the U.S. Congress, numerous chemical exposures were linked to contaminated work clothing. In addition, 12 cases of Q fever were believed to have been caused by contaminated work clothing (27). Whether one's work clothing includes overalls or a lab coat, prudent practice would dictate that potentially contaminated work clothing should not be brought home. Likewise, such clothing should not be worn in nonwork areas, such as restrooms, cafeterias, and offices.

Gloves

In 1987, four persons were admitted to hospitals with illnesses that were later confirmed to be caused by infection with *Cercopithecine herpesvirus* 1 (also known as *Herpesvirus simiae,* monkey B virus). Three were monkey handlers at the Naval Aerospace Medical Research Laboratory at the Pensacola Naval Air Station and the fourth was the wife of one of the three handlers. It was not certain if two of the three handlers were wearing gloves when bitten or scratched. The third handler was wearing only surgical gloves while holding a monkey. The wife of one of the infected handlers applied cortisone cream to her husband's skin lesions and subsequently applied the cream to an area of contact dermatitis on her own hands and thus spread the infection to herself (28). Protective gloves would have provided a primary barrier in this case.

There are a variety of disposable and nondisposable gloves available for a wide range of applications. These include vinyl, latex, and nitrile for disposable one-time use, as well as specialized gloves designed for hot and cold environments and those that provide various degrees of chemical and puncture resistance. Two excellent references for glove chemical resistance may be found at www.showagroup.com/products and http://www.ansellpro.com/download/Ansell_8thEditionChemicalResistanceGuide.pdf.

With the advent of the OSHA Bloodborne Pathogens Standard in 1991, the use of disposable gloves became mandatory for handling or contact with human blood or body fluids. Many health care workers were concerned about the use of vinyl and latex gloves and their integrity as barriers against blood-borne pathogens, such as human immunodeficiency virus and the hepatitis viruses. Experiments conducted to evaluate vinyl and latex examination gloves as barriers found that holes can be formed during procedures related to routine patient care (29). Under the conditions of the study, latex gloves appeared to be less susceptible to hole formation than vinyl gloves; additionally, vinyl gloves leaked the test phage more often than latex gloves. Of course, frequent hand washing, especially after removing gloves, is an essential work practice in the laboratory.

Cut-resistant gloves made of stainless steel mesh and Kevlar are useful when protection from cuts is desired, but they do not provide protection from needlestick injuries. Such gloves would be appropriate for work with a cryomacrocut, which is designed to slice tissue from a block of ice, or from bites and scratches from small animals. Newer fabrications offer resistance to needlesticks as tested in accordance with the ASTM Standard Test Method for Protective Clothing Material Resistance to Hypodermic Needle Puncture (30).

Head coverings

Tyvek and polypropylene hoods and bouffant caps are most frequently used when product protection is required, such as in the pharmaceutical industry, or when potential splashes to the head are anticipated. Other applications include use when entering and working in rooms where nonhuman primates are housed or as an integral part of a powered air-purifying respirator (PAPR).

Shoe coverings

Shoe coverings are generally not necessary in biomedical laboratories for routine laboratory procedures; however, they are useful for situations where shoe protection would be needed, such as cleaning a biological or chemical spill. In pharmaceutical plants and other work areas where product protection is important, polypropylene or Tyvek shoe coverings may be utilized. Open-toe shoes and sandals are not recommended as laboratory footwear because they do not provide sufficient protection against hazardous materials. Boots made of rubber or similar material may be necessary in work areas where there are large amounts of water, such as animal cage-washing rooms. Boots should be selected so that they offer slip-resistance to reduce the chance of slipping and falling on wet surfaces.

Eye and face protection

Numerous examples exist in the literature of laboratory-acquired infection in laboratory personnel due to facial exposures to infectious agents. For example, a case of fowl plague keratoconjunctivitis resulted when a laboratory technician splashed allantoic fluids containing fowl plague virus on the right side of her face. The accident occurred while she was opening the tops of infected eggs to harvest allantoic fluid (31). There was no mention of eye protection being worn while performing this task.

In another case, a technician dropped a moist rectal swab that was received during a national quality assurance exercise on the bench. She felt droplets falling on her gown and face. After replacing the swab in the transport medium, she removed her gown and washed her face and hands with water, followed by Hibiclens (chlorhexidine) and then Savlon (chlorhexidine plus cetrimide). About 54 hours later she developed fever, malaise, and dysentery that later required her to see a doctor. A stool sample culture yielded *Shigella flexneri*, which was confirmed as having the same serotype and antibiogram as the strain that had been isolated from the survey specimen (32).

In 1998, a technician at the Yerkes Regional Primate Research Center died due to an eye splash with a monkey's body fluid that contained herpes B virus. Herpes B virus is found in the blood, secretions, and tissues of macaque monkeys and can cause life-threatening central nervous system infections in humans. OSHA cited the center for not providing employees with appropriate eye and face protection against monkey body fluid splashes. The technician was not wearing eye protection while transferring a monkey from a transfer box into a cage (33).

SUMMARY

The selection and use of the appropriate primary barriers constitute just one important component of the overall laboratory safety program. Laboratory-acquired infections have occurred due to the lack of or misuse of primary containment devices and PPE. PPE should be chosen carefully and utilized appropriately.

References

1. **Wedum AG, Barkley WE, Hellman A.** 1972. Handling of infectious agents. *J Am Vet Med Assoc* **161:**1557–1567.
2. **Hinman AR, Fraser DW, Douglas RG, Bowen GS, Kraus AL, Winkler WG, Rhodes WW.** 1975. Outbreak of lymphocytic choriomeningitis virus infections in medical center personnel. *Am J Epidemiol* **101:**103–110.
3. **Reitman M, Wedum AG.** 1956. Microbiological safety. *Public Health Rep* **71:**659–665..
4. **Wedum AG.** 1964. Laboratory safety in research with infectious aerosols. *Public Health Rep* **79:**619–633.
5. **Pike RM.** 1978. Past and present hazards of working with infectious agents. *Arch Pathol Lab Med* **102:**333–336.
6. **Barry M, Russi M, Armstrong L, Geller D, Tesh R, Dembry L, Gonzalez JP, Khan AS, Peters CJ.** 1995. Brief report: treatment of a laboratory-acquired Sabiá virus infection. *N Engl J Med* **333:** 294–296.
7. **Ippolito G, Puro V, Heptonstall J, Jagger J, De Carli G, Petrosillo N.** 1999. Occupational human immunodeficiency virus infection in health care workers: worldwide cases through September 1997. *Clin Infect Dis* **28:**365–383.
8. **Centers for Disease Control.** 1988. Agent summary statement for human immunodeficiency virus and laboratory acquired infection with human immunodeficiency virus. *MMWR Morb Mortal Wkly Rep* **37:**1–22.
9. **Centers for Disease Control and Prevention.** 1994. Bolivian hemorrhagic fever—El Beni Department, Bolivia, 1994. *MMWR Morb Mortal Wkly Rep* **43:**943–946.
10. **Schaefer F, Liberman D, Fink R.** 1980. Decontamination of a centrifuge after a rotor explosion. *Public Health Rep* **95:**357–361.
11. **Chatigny MA, Dunn S, Ishimaru K, Eagleson JA, Prusiner SB.** 1979. Evaluation of a class III biological safety cabinet for enclosure of an ultracentrifuge. *Appl Environ Microbiol* **38:** 934–939.
12. **Kenny MT, Sabel FL.** 1968. Particle size distribution of *Serratia marcescens* aerosols created during common laboratory procedures and simulated laboratory accidents. *Appl Microbiol* **16:** 1146–1150.
13. **Wright LJ, Barker LF, Mickenberg ID, Wolff SM.** 1968. Laboratory-acquired typhus fevers. *Ann Intern Med* **69:**731–738.
14. **Winkler WG, Fashinell TR, Leffingwell L, Howard P, Conomy P.** 1973. Airborne rabies transmission in a laboratory worker. *JAMA* **226:**1219–1221.
15. **Holmes KL, Fontes B, Hogarth P, Konz R, Monard S, Pletcher CH Jr, Wadley RB, Schmid I, Perfetto SP.** 2014. International Society for the Advancement of Cytometry cell sorter biosafety standards. *Cytometry A* **85:**434–453.
16. **Byers KB.** 2008. Biosafety tips. *Appl Biosaf* **13:**57–59.
17. **Occupational Safety and Health Administration, U.S. Department of Labor.** 2014. OSHA fact sheet. PPE selection matrix for occupational exposure to Ebola virus. https://www.osha.gov/Publications/OSHA3761.pdf.
18. **Centers for Disease Control and Prevention.** 2014. Fatal meningococcal disease in a laboratory worker—California, 2012. *MMWR Morb Mortal Wkly Rep* **63:**770–772.
19. **Occupational Safety and Health Administration, U.S. Department of Labor.** Personal protective equipment. Title 29 CFR Subtitle B Ch. XVII Part 1910.132-138. https://www.osha.gov/pls/oshaweb/owadisp.show_document?p_table=STANDARDS&p_id=9777.
20. **Occupational Safety and Health Administration, U.S. Department of Labor.** Hazard communication. 29 CFR Subtitle B Ch. XVII Part 1910.1200. https://www.osha.gov/pls/oshaweb/owadisp.show_document?p_table=standards&p_id=10099
21. **Occupational Safety and Health Administration, U.S. Department of Labor.** Occupational exposure to bloodborne pathogens. 29 CFR Subtitle B Ch. XVII Part 1910.1030. https://www.osha.gov/pls/oshaweb/owadisp.show_document?p_table=STANDARDS&p_id=10051.
22. **Occupational Safety and Health Administration, U.S. Department of Labor.** Occupational exposure to hazardous chemicals in laboratories. 29 CFR Subtitle B Ch. XVII Part 1910.1450. https://www.osha.gov/pls/oshaweb/owadisp.show_document?p_table=STANDARDS&p_id=10106.
23. **U.S. Department of Health and Human Services, Public Health Service, Centers for Disease Control and Prevention, National Institutes of Health.** 2009. *Biosafety in Microbiological and Biomedical Laboratories*, 5th ed. HHS Publication no. (CDC) 21-112. http://www.cdc.gov/biosafety/publications/bmbl5/BMBL.pdf.
24. **Leonas KK, Jinkins RS.** 1997. The relationship of selected fabric characteristics and the barrier effectiveness of surgical gown fabrics. *Am J Infect Control* **25:**16–23.

25. **McCullough EA.** 1993. Methods for determining the barrier efficacy of surgical gowns. *Am J Infect Control* **21:**368–374.
26. **Carter T.** 2004. The dissemination of anthrax from imported wool: Kidderminster 1900–14. *Occup Environ Med* **61:**103–107.
27. **U.S. Department of Health and Human Services, Public Health Service, Centers for Disease Control and Prevention, National Institute for Occupational Safety and Health.** 1995. Report to Congress on Workers' Home Contamination Study Conducted Under the Workers' Family Protection Act (29 U.S.C. 671A). DHHS (NIOSH) Publication no. 95-123.
28. **Griffen DG, Sutton EW, Goodman PL, Zimmern WA, Bernstein ND, Bean TW, Ball MR, Schindler CM, Houghton JO, Brady JA, Rupert AH, Ward GS, Wilder MH, Hilliard JK, Buck RL, Trump DH.** 1987. Leads from the MMWR. B-virus infection in humans—Pensacola, Florida. *JAMA* **257:**3192–3193, 3198.
29. **Korniewicz DM, Laughon BE, Cyr WH, Lytle CD, Larson E.** 1990. Leakage of virus through used vinyl and latex examination gloves. *J Clin Microbiol* **28:**787–788.
30. **ASTM International.** 2010. ASTM F2878-10, Standard test method for protective clothing material resistance to hypodermic needle puncture. ASTM International, West Conshohocken, PA. http://www.astm.org/cgi-bin/resolver.cgi?F2878-10.
31. **Taylor HR, Turner AJ.** 1977. A case report of fowl plague keratoconjunctivitis. *Br J Ophthalmol* **61:**86–88.
32. **Ghosh HK.** 1982. Laboratory-acquired shigellosis. *Br Med J (Clin Res Ed)* **285:**695–696.
33. **U.S. Department of Labor, Occupational Safety and Health Administration.** 1998. Death of a technician at Georgia research center prompts OSHA citations and fines totaling $105,300. News release USDL 98-175.

18 Primary Barriers: Biological Safety Cabinets, Fume Hoods, and Glove Boxes

DAVID C. EAGLESON, KARA F. HELD, LANCE GAUDETTE, CHARLES W. QUINT, JR., AND DAVID G. STUART

Deadly disease outbreaks have become a frequent occurrence throughout the world with infectious agents such as human immunodeficiency virus (HIV), hepatitis B virus, severe acute respiratory syndrome-related coronavirus (SARS-CoV), Middle East respiratory syndrome coronavirus (MERS-CoV), Nipah virus of Malaysia, Hendra virus of Australia, hantavirus in the United States, and most recently, the Ebola virus in Africa (1). Additionally, new laboratory techniques have become reliant on the use of infectious agents for common procedures. Lentivirus, adenovirus, vaccinia virus, *Escherchia coli,* and human cancer cells are frequently found in research laboratories worldwide. With the increased exposure to these biological agents comes the greater risk of developing a laboratory-associated infection (LAI). Previous studies have accounted for 5,527 LAIs from 1930 to 2004, with 204 of these resulting in death (2–4). In a recent study of all LAIs from 1976 to 2010, it was found that there were 197 cases reported to the National Institutes of Health (NIH) due to exposure to specifically recombinant DNA-based materials (5). Unfortunately, most LAIs (82%) cannot be traced to a single incident to determine the cause of exposure (3, 5–7). Although good sterile and aseptic techniques are critical, virtually every activity in the laboratory gives rise to aerosols (8–10). Aerosols containing infectious agents, compounded by contact spread (11), could create an epidemic before any symptoms present. This underscores the need for protection from these agents, such as use of proper aseptic technique, personal protective equipment (PPE), and appropriate primary barriers.

The primary barrier provides a workspace within which hazardous material can be contained. Airflow sweeps aerosols and/or vapors away at the site of their generation. High-efficiency particulate air (HEPA) filtration, a leak-tight cabinet, and air inflow at the work opening prevent contaminants from escaping from the barrier unless they are carried out by contact. Containment is aided by established techniques, such as keeping everything in appropriate containers inside the barrier until the work is finished, decontaminating the containers before removing them from the barrier, removing outer gloves before exiting the barrier, and disposing properly of all used materials.

It is essential to understand that, by definition, HEPA filters must be at least 99.97% efficient in filtering out particles that are 0.3 μm in size (99.99% for those used in biological safety cabinets (BSCs) (12). Because 0.3 μm is the nominal, most penetrating particle size (most difficult to filter out) for HEPA filters, particles both larger and smaller than 0.3 μm are retained more efficiently (13). This includes viruses, many of which are much smaller than 0.3 μm. It is just as important to realize that gases and vapors readily pass through HEPA filters.

The use of primary containment for chemical reactions goes back centuries. The alchemist working at an enclosed, raised hearth was a common genre in the work of Dutch artists of the 17th century (14). A typical example is Adriaen Van Ostade's painting *An Alchemist* (1661). Enclosures for rearing germ-free animals were used as early as the 1800s (15). Although bioclean isolators have been commonly used for sterility testing, their use as enclosures for aseptic high-production, continuous, pharmaceutical filling lines has only recently been realized.

The early primary barriers used for biosafety were developed at Fort Detrick, MD during Arnold G. Wedum's biosafety program and were used with considerable success (16). Maximum-containment glove boxes, now called Class III BSCs, were developed during the 1940s (17), and partial-containment fume hood-like Class I BSCs made their appearance in the mid-1950s (18).

The forerunner of the present-day Class II BSC can be traced to a cabinet designed at the request of a pharmaceutical company in 1964. This cabinet provided both clean air in the work area and containment of powders using HEPA filter technology and a single blower. The Baker Company explored the need for a BSC based on this concept with Lewis Coriell of the Institute for Medical Research in Camden, NJ. In the meantime, Baker Company representatives had discussed a similar concept with the NIH, resulting in development of the forerunner to the Class II, Type A BSC (called a Type 1 cabinet at that time). Soon after this, cooperation with the National Cancer Institute (NCI) Division of Biological Safety, then under the direction of Emmett Barkley, led to a cabinet project in association with the Pitman Moore division of Dow Chemical Company. These activities resulted in the development and delivery to the NCI in 1967 of a version of what are now called Class II, Type B1 BSCs. The first publication of microbiological testing of the performance of "laminar flow biological safety cabinets," as they were originally called, appeared in 1968 (19, 20).

Primary containment is "the protection of personnel and the immediate laboratory from exposure to infectious agents" and "is provided by both good microbiological technique and the use of appropriate safety equipment" (21). It is absolutely essential that the laboratory worker understands and appreciates that primary barriers are not magic boxes that will take care of them, nor are they a substitute for good laboratory technique and practice.

Even though there are many similarities among primary barriers, there are also some very important differences. Primary barriers may provide personnel protection (containment), product protection (a virtually particle-free work area to help minimize contamination of cultures or other products), environmental protection (helps prevent contamination of the laboratory, the building, and the community), or some combination thereof. To be able to select and use primary barriers successfully, one must understand what they are, how they function, what their capabilities are, and what their limitations are so as not to confuse them with each other. It is important to select a primary barrier that will provide appropriate protection for each specific application.

SELECTION AND USE OF PRIMARY BARRIERS: RISK ASSESSMENT

The selection of the primary barrier and the practices to be followed while using it must be based on a thorough risk assessment as explained elsewhere in this book. Briefly, one should identify all of the potentially hazardous agents that will be used, determine the nature of the hazards (chemical, radiological, biological, physical, or a combination of these), determine the containment level for each hazard (biosafety level [BSL] and chemical safety level [CSL] of the agents), and match the overall risk with the BSL and CSL of the laboratory to be used (21, 22). Also to be considered is whether product protection is required and at what level. One should analyze the procedures and practices for handling these agents at this containment level, estimate the exposure that might be expected from this situation, and consider the dose response of each agent. By determining the level of performance offered by each of the primary barriers, one should choose the equipment that best suits the specific combination of hazardous agents, product protection, and practices being planned for the particular laboratory.

Care must be taken with hazardous chemicals that might vaporize in BSCs. Because gases and vapors pass through HEPA filters, they will recirculate in some types of BSCs. Therefore, the amounts of volatile chemicals used in these BSCs must be limited. Flammable vapors must not be allowed to reach their lower explosive limits, and BSCs are generally placarded against the use of such materials.

Amounts of chemical carcinogens and other toxic or hazardous materials, such as radionuclides, used in BSCs must also be limited if their vapor pressures are such that they will vaporize at room temperature. NSF/ANSI standard 49 (hereafter referred to simply as NSF) states that only "volatile chemicals and radionuclides required as adjuncts to microbiological studies" may be used in cer-

TABLE 1.

Overview of primary barrier application at various BSLs and with volatile hazards[a]

Primary barrier	BSL (12, 21)	Use of toxic chemicals and radio nuclides that may vaporize (12)
Fume hoods	No biohazards	Yes
BSCs		
Class I	1, 2, and 3 (4 with air-supplied positive-pressure suits)	Only when exhaust air is vented outdoors and in small amounts
Class II, Type A1	1, 2, and 3 (4 with suits)	No volatile hazards allowed
Class II, Type A2	1, 2, and 3 (4 with suits)	No volatile hazards allowed when vented to the room
Class II, Type A2 exhausted	1, 2, and 3 (4 with suits)	Very small amounts used to treat microorganisms as long as the materials used cause no harm when recirculated in the downflow air
Class II, Type B1	1, 2, and 3 (4 with suits)	Very small amounts used to treat microorganisms when working in the direcly exhausted back of the work area, or if the materials used cause no harm when recirculated in the downflow air
Class II, Type B2	1, 2, and 3 (4 with suits)	As required to treat agents in the course of microbiological work
Class III[b]	1, 2, 3, and 4	As required to treat agents in the course of microbiological work
Pharmacy glove box (isolator)	No biohazards	When appropriately HEPA filtered, under negative pressure, and exhausted outdoors (48)

[a]Because the information in the table is necessarily very abbreviated, be sure to consult the source documents.
[b]BSL2 or -3 agents may be upgraded to Class III cabinets on the basis of concentration of agent and/or aerosol generation potential.

tain types of BSCs (12). The term "adjuncts" means the actual use of the diluted agent during the study. Activities involving weighing out and diluting such materials should not be performed in the BSC. These activities are to be performed in appropriate equipment such as fume hoods or glove boxes. Because the actual amount of material required will vary for each particular procedure, the quantity considered an "adjunct" cannot be defined with specific units. For radionuclides, an example of "adjunct" levels is the amount of diluted radioactive material required to trace labeled substrates through biochemical pathways in microorganisms.

A representation of the relationship between the different types of BSCs and the use of hazardous materials is shown in Table 1. A safety professional should decide these issues.

FUME HOODS AND CLASS I BSCs

Most people associated with laboratories are familiar with fume hoods. Unfortunately, BSCs are often also called "hoods." In addition, Class I BSCs function much like fume hoods. Therefore, it is not uncommon for the two to be confused.

Fume Hoods

To prevent chemicals from reaching unacceptable levels, local ventilation with air exhausted outdoors is used. Fume hoods use this principle to provide primary containment of hazardous chemicals (23). Fume hoods also help prevent contact transmission of contaminants when operators use proper technique.

Figure 1 is a schematic design for a continuous-bypass fume hood. Contaminated air is separated from uncontaminated room air at the plane of the sash. An airfoil at the bottom of the work opening allows a smooth flow of air across the work surface even when the sash is closed. The airfoil also improves the airflow into the fume hood when the sash is open to a predetermined level and the hood is being used. A recessed work surface contains liquid spills. In some fume hoods, the bypass directs fresh air down the inside of the sash to dilute contamination in the roll of airflow within the work area and allows more inflow through the bypass as the sash is closed to keep the face velocity within acceptable limits. The standard fume hood has no bypass to help regulate face velocity variations as the sash is moved up and down. The rear baffle and the slots within a fume hood control the air leaving the work area. In many fume hoods these baffles and/or slots are adjustable.

Recent models of fume hoods have combination sashes. The combination sash has horizontally sliding sashes within the vertically sliding sash. When the combination sash is closed, the horizontally sliding sashes enable one to reach into the hood with an arm on each side of a sash, thus providing a physical barrier in front of the user's torso. Another advantage of using the horizontal option is that the maximum sash opening size is less than that of the fully open vertical sash, thus minimizing the airflow required to maintain acceptable face velocities.

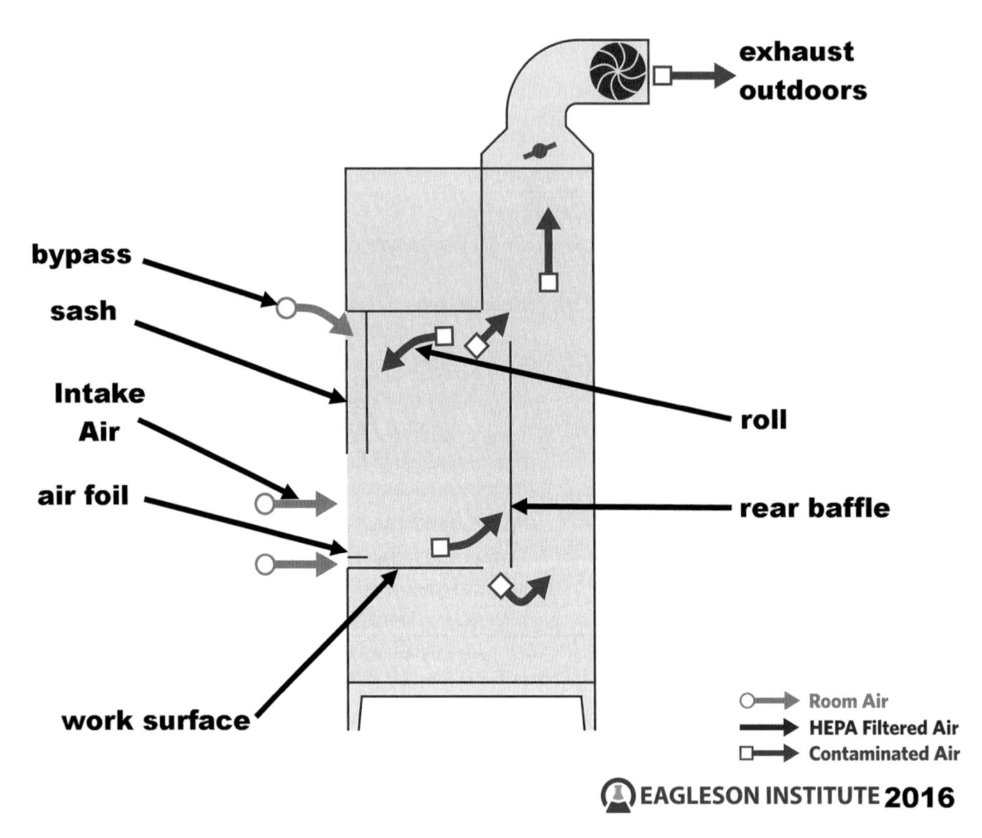

Figure 1: Fume hood basic design and airflow schematic.

The air outlet of the fume hood must be connected to an exhaust fan on the roof of the building for the hood to function. There should be a means of controlling the exhaust airflow rate via an adjustable speed fan and/or an inline damper in the exhaust duct. The amount of air leaving the room through the fume hood affects the amount of air that must be supplied to the room, the pressure in the room, and the operation of other equipment, such as BSCs that are vented to the outdoors (24). A 4-ft fume hood with the sash wide open and a face velocity of 100 ft/min (fpm) will require close to 1,000 ft^3/min (cfm) at about 0.5 inches (in.) of water column negative pressure.

The exhaust system used to pull the air through fume hoods may be operated with constant airflow that draws the same amount of air out of the room no matter how the fume hood is being used. To reduce the amount of conditioned air being vented to the outdoors, a variable airflow exhaust system may be used. The variable airflow system maintains constant face velocity, as the sash is moved up and down, by modulating the exhaust airflow rate as a function of sash height (25).

Further savings of initial exhaust system equipment costs can be realized by implementing laboratory exhaust diversity. It is highly unlikely that every fume hood in a building would be in use simultaneously, at its maximum exhaust level. Diversity is the percentage of total available exhaust capacity that is actually used (26). Thus, the size and total airflow of an exhaust system can be reduced by designing to the laboratory exhaust diversity.

The average face velocity of air flowing into the fume hood is measured to set and maintain the operation of the hood within the specified range. Traditionally, this was recommended at a range of 80 to 120 fpm (25); however more recently it has become common for higher-performing designs to operate in the range of 60 to 80 fpm (27). Visualizing the behavior of the airflow at the face of the hood by using smoke also indicates proper operation. The actual performance of a fume hood is measured with the American Society of Heating, Refrigeration and Air-Conditioning Engineers (ASHRAE) standard 110 sulfur hexafluoride tracer gas test (23, 28, 29). The industry consensus standard for fume hood containment is an average escape rate of not more than 0.1 ppm as used (25) under the conditions of the ASHRAE standard 110 test (28). These tests are often performed by manufacturers and by fume hood certifiers in the field.

Specialized Fume Hoods

Specialized hoods have to meet the operational and performance requirements of regular fume hoods. Some examples follow.

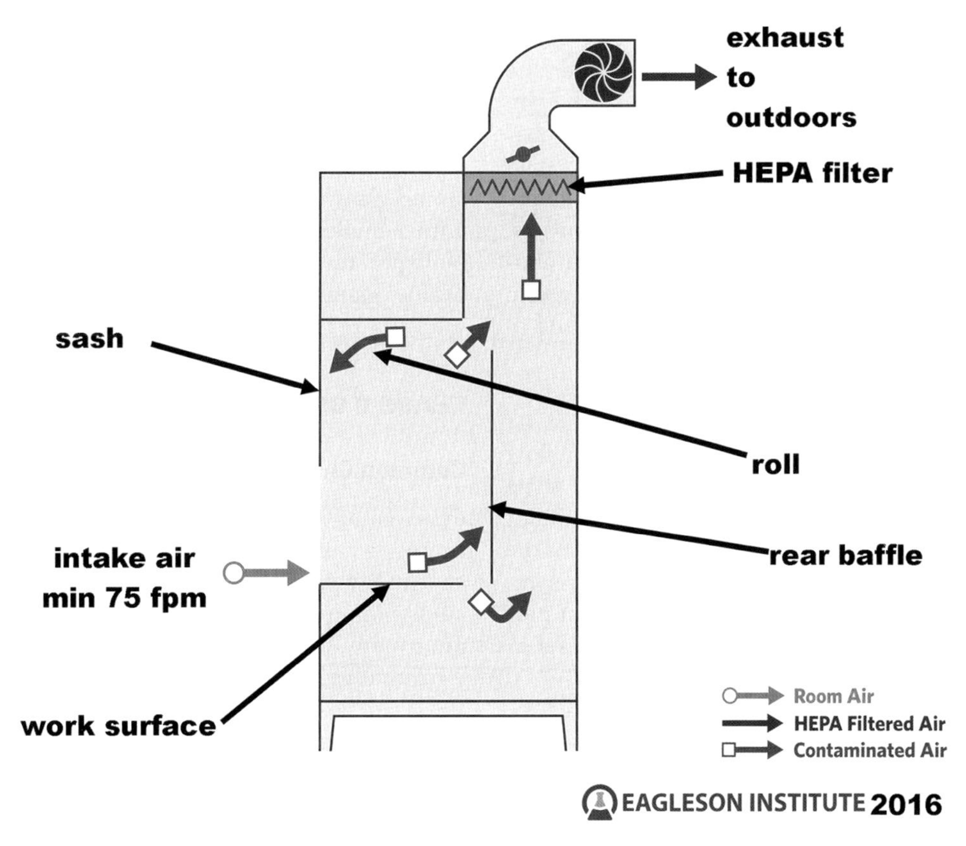

Figure 2: Class I BSC basic design and airflow schematic.

Radioisotope fume hoods have work surfaces made of continuous nonporous material that is easily cleaned and decontaminated (30). Work surfaces should be reinforced to support lead shielding. The hood should be labeled for radioactive work.

Perchloric acid fume hoods have exhaust ducts and fans that are constructed from acid-resistant materials. The hood has a watertight work surface and a spray system that will wash down the entire exhaust system and the hood to prevent buildup of explosive precipitates. The hood must be labeled appropriately (31).

Walk-in fume hoods are used when the required apparatus is too large to be confined in a regular hood. The walk-in hood is designed to stand on the floor and to have doors and/or sashes as required to allow access to the apparatus and still provide a closed physical barrier (25).

Recirculated (nonducted) hoods may also be found in the microbiological or biomedical laboratory. They recirculate air through filters that remove gases and vapors to some extent. Their use should be carefully monitored, because filter leaks or saturation will lead to fumes being released back into the laboratory. As such, these "ductless" fume hoods are preferred only when ducting is not practical and only low-risk chemicals are used. Their purpose is primarily for odor control.

Class I BSCs

The purpose of the Class I BSC is to provide primary containment (personnel and environmental protection) of microbiological hazards assigned to BSL1, -2, or -3. There is no attempt to keep the work area air clean, so there is no product protection. The Class I BSC functions much like a fume hood, with air pulled in through the work access opening and across the work area. Unlike the fume hood, the air is then drawn up through a HEPA exhaust filter prior to being vented to the outdoors. The complete definition can be found in the NSF standard (12). Some installations have had blowers mounted above the cabinet and HEPA-filtered air from the cabinet returned to the room. The use of toxic gas or vapors is not acceptable in such an installation.

The diagram in Fig. 2 depicts the design of a Class I BSC. The separation of the contaminated air inside the cabinet from the room air is at the plane of the sash. The airfoil at the bottom of the work opening does not allow air to flow under it as in the fume hood. This is because the BSC is designed and tested to perform at an 8- or 10-in. sash height and should never to be used with the sash at any other than the design height. The BSC design does not account for operation with sash heights varying from closed to halfway open or more, as the fume hood design does. There is no bypass because there will be no variation of intake airflow velocity with sash height when the BSC is used with the sash at the one designated height. There are no adjustable slots in the rear baffle, again because the cabinet will be used at only one sash opening height. The HEPA filter at the top of the cabinet filters out particulate contaminants, such as microorganisms, providing protection for the environment or for the laboratory if the exhaust air is circulated back into the room. Arm port panels with no gloves are sometimes used to improve containment performance.

Although the exhaust system for a 4-ft Class I BSC handles a lower rate of airflow than a 4-ft fume hood (about 220 cfm and 1,000 cfm, respectively), this flow will have much more negative pressure than the fume hood (about 1.5 in. of water column and 0.5 in. of water column, respectively). This is to account for pulling the air through a loading HEPA filter in the cabinet. It is important for the cabinet static pressure requirement plus that of the exhaust system to be factored into the selection of the exhaust fan.

The NSF standard thoroughly describes how to establish and maintain the proper operation of the Class I BSC (12). This is done by measuring intake air velocity (minimum of 75 fpm) and visualizing airflow patterns with smoke. In addition, the HEPA filter must be leak tested (no leak greater than 0.01% of the upstream challenge concentration). The performance of the BSC is measured using the microbiological aerosol tracer test for personnel protection (no more than 10 bacterial spores found in six all-glass impinger air samplers outside the cabinet with a challenge of 1×10^8 to 8×10^8 spores). This microbiological test may be performed on Class I BSCs by some cabinet manufacturers but not in the field.

CLASS II BSCs

Common Characteristics

Class II BSCs provide personnel protection, product protection, and environmental protection. Class II BSCs also utilize unidirectional downflow air to limit the possibility for cross contamination within the cabinet work area.

Characteristics that are shared by all types of Class II BSCs can be seen in Fig. 3. There are two sets of HEPA filters, one for supply air and another for exhaust air. Clean air descends through the work area entraining aerosols, and splits as it nears the work surface, with

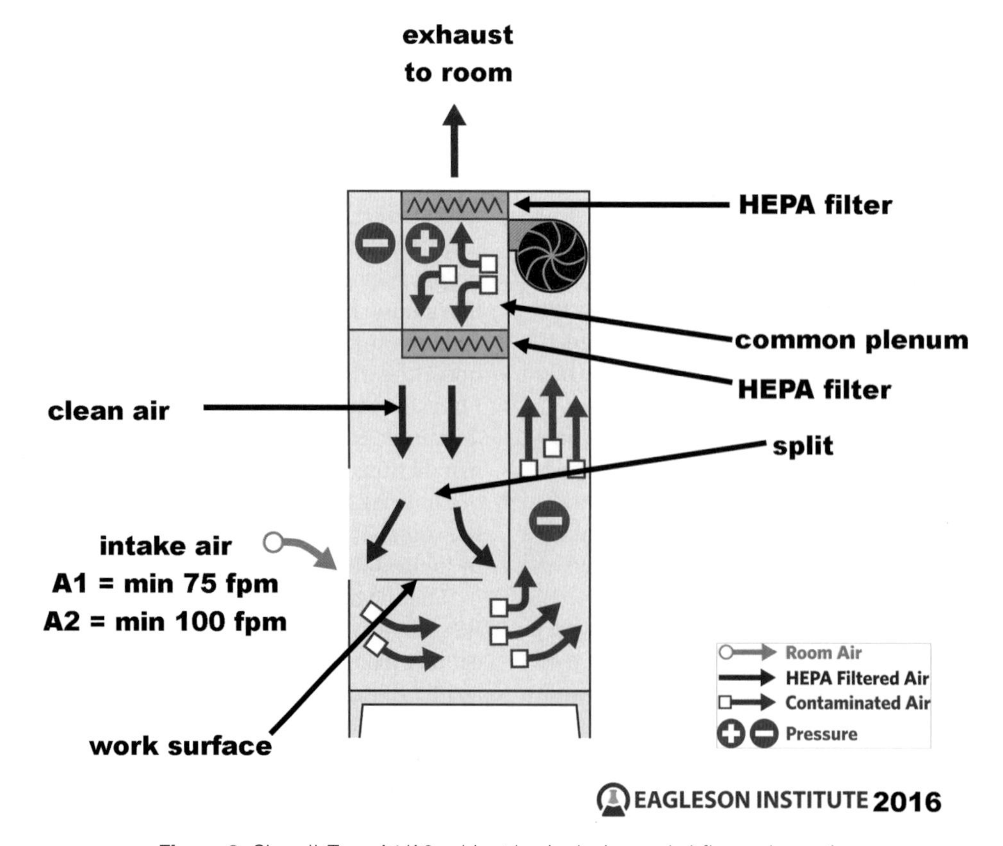

Figure 3: Class II, Type A1/A2 cabinet basic design and airflow schematic.

some air directed through the rear half of the front perforated grille and the rest to the rear grille of the cabinet. Intake air is drawn in and downward (much different from the straight-in flow of the fume hood) through the work access opening, filling only the front half of the front perforated grille. The HEPA-filtered downflow air is directed into the rear half of the front grille and prevents contaminants in the room air from entering the work area. At the same time, the intake air rushing into the front half of the grille helps prevent aerosols generated within the cabinet from escaping into the laboratory. All of the air leaving the work area is HEPA filtered by either the supply filter(s) or the exhaust filter(s). The exhaust filter(s) provides environmental protection by preventing particulates from escaping via the exhaust air duct. Leak-tight construction, and the use of negative pressure plenums, prevents escape of contaminants through joints, seams, or penetrations of the cabinet structure. It becomes readily apparent that proper balance between intake and downflow air is essential for a Class II BSC to perform as intended.

Proper operation of Class II BSCs is verified by measuring intake air and downflow air velocities to ensure proper airflow set point and balance, smoke pattern testing, and HEPA filter leak testing (12). Cabinet integrity testing, along with other testing, is usually done less often and is discussed below in the section Certification of Primary Barriers. For an in-depth discussion of BSCs refer to Eagleson (32). For issues involving the connection of primary barriers to laboratory ventilation systems, refer to Ghidoni (33).

Performance of Class II BSCs

Performance of Class II BSCs is determined by subjecting them to the microbiological aerosol tracer test described by NSF (12). This is a test in which ideal conditions are standardized and reproduced, rather than being an attempt to simulate field conditions with activity in the cabinet. Barkley developed a "containment factor" of 10^5 as a design criterion, which means that an aerosol of 100,000 organisms dispersed inside the BSC would be contained under the conditions of the test (34). This is essentially the containment performance criterion used by NSF. Microbiological aerosol tracer tests are also described for product protection and for cross contamination within the cabinet (12). A cabinet must pass each of three replicate runs of all three microbiological tests, in addition to many other tests, for it to be certified by NSF as a Class II BSC.

Intake and downflow air must be balanced within certain limits for a cabinet to pass the microbiological personnel and product protection tests. In fact, Class II cabinets have performance envelopes that are peculiar to the make, model, and size of cabinet (35). As can be seen in Fig. 4, the balance of intake to downflow air velocity must be within the performance envelope lines for the cabinet to pass the microbiological tests. It should be noted that the performance of the BSC declines sharply as the airflow settings move toward the edges of the performance envelope. A cabinet set well within the envelope will pass both the personnel and product protection tests, while a mix of passes and failures is observed as the settings approach the edge of the envelope. At airflow settings outside of the envelope only failing tests are seen. Thus, it is important that cabinet designers select a set point well within the performance envelope.

NSF requires that a given Class II BSC be balanced so that the airflows are within ±5 fpm of the set point that was used when a cabinet of the same make, model, and size passed the microbiological aerosol tracer tests conducted by NSF officials. In addition to microbiologically testing for personnel, product, and cross-contamination protection at nominal set point, NSF testing agents also perform personnel and product protection tests over an extended range of ±10 fpm from nominal set point. See the NSF test range as illustrated in Fig. 4. This provides data to show that the cabinet will pass these microbiological tests over a wider range than the certifier is allowed to set it (represented in Fig. 4 by the line at ±5 fpm from nominal set point). The section below on certification discusses the test methods to be used to ensure that a cabinet is operating properly as installed in the field.

Class II, Type A1 BSC (Formerly Called Type A)

The purpose of the Class II, Type A1 BSC is to provide personnel, product, and environmental protection from biohazards assigned to BSL1, -2, or -3. The Type A1 cabinet (Fig. 3) recirculates a portion (nominally 60–70%) of the total cabinet airflow (the mixed downflow and inflow air). This is acceptable when microorganisms are being used, because the HEPA filters will remove them. However, the Type A1 BSC, even when exhausted out of the laboratory, should not be used with hazardous chemicals that might vaporize.

Intake airflow must be at least 75 fpm. Supply air and exhaust air are mixed in a common plenum. The original design had biologically contaminated ducts that were under positive pressure relative to the room as shown in Fig. 5. If a leak occurred, contaminants could flow into the room. To ensure containment, this cabinet was pressurized to 2 in. of water column with a tracer gas and required to show leakage less than 5×10^{-7} cubic centimeters (ml) per second (s). This original design had fallen out of

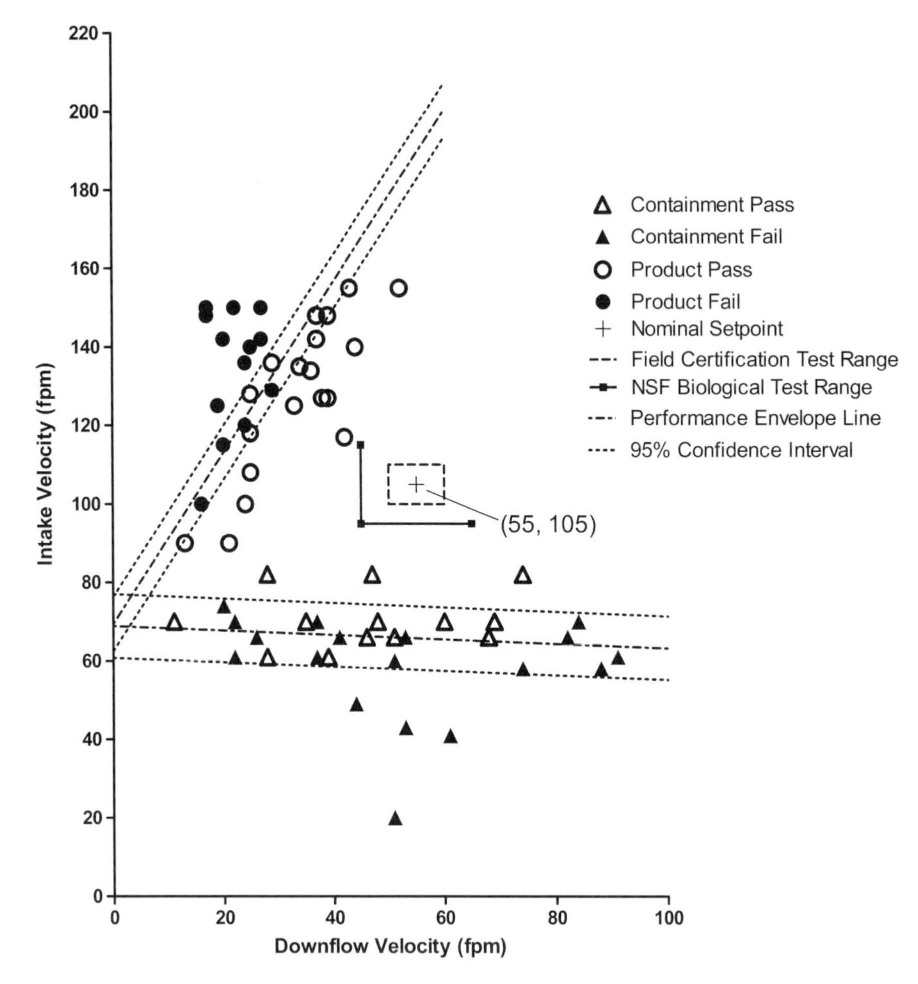

Figure 4: Representative performance envelope of a Class II, Type A2 BSC. The BSC passes the microbiological tests within the performance envelope.

favor due to the potential for leakage from pressurized contaminated ducts; NSF now requires Type A1 cabinets to have biologically contaminated plenums under negative pressure to the room, or to be surrounded by negative pressure plenums as shown in Fig. 3 (12). Thus, for all cabinets approved by NSF since this change in 2008, the only functional difference between Type A cabinets is a higher intake velocity on the Type A2.

The Type A1 cabinet is often operated with the exhaust air returned to the room, which has no impact on the facility's ventilation systems. This recirculation of air saves the expense and complications of venting the exhaust air to the outdoors. It also has the advantage of cleaning the room air as it recirculates through the cabinet's HEPA filters. Exhaust air from the Type A1 BSC can be vented to the outdoors, as is discussed in the next section.

Type A1 cabinets meet or exceed the requirements of the microbiological tests, but they are not suitable for use with volatile hazardous chemicals (12), as the lower intake airflow of the Type A1 BSC provides less dilution of volatile chemicals, and makes the cabinet more susceptible to escape of contaminants through the front access opening.

Class II, Type A2 BSC (Formerly Called A/B3)

A variation of the Class II, Type A1 BSC evolved that typically had an intake air velocity of 100 fpm and ducts that are under negative pressure relative to the room (Fig. 3). The purposes for this design are to ensure that potential leaks in the cabinet structure would be inward rather than outward into the room and to provide stronger performance by increasing the minimum intake air velocity from 75 to 100 fpm. This cabinet is now defined by NSF as a Class II, Type A2 BSC, and is the most commonly used type of BSC. The Type A2 cabinet may be used with volatile chemicals and radionuclides as an adjunct to microbiological studies, but must be exhausted to the outdoors

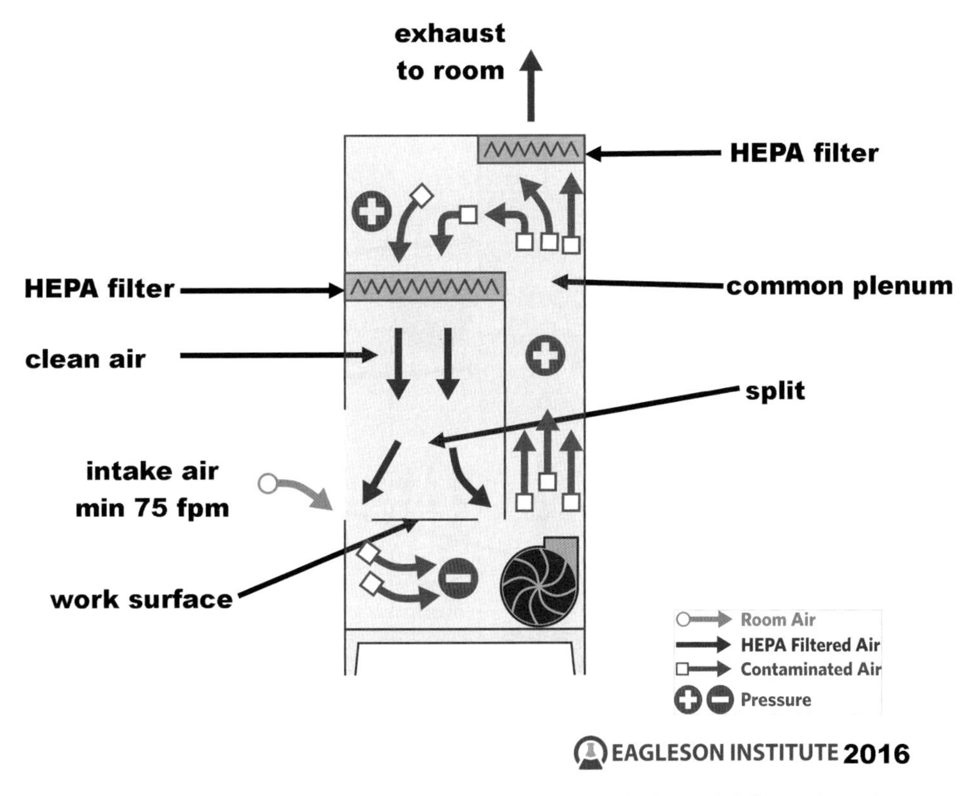

Figure 5: Class II, Type A1 (original Type A) cabinet basic design and airflow schematic.

when doing so (12). Prior to the 2002 revision of NSF/ANSI standard 49, the Type A2 cabinet exhausted to the outdoors was called a Type B3.

Class II, Type A2-Exhausted (Formerly Called Type B3) BSC

The purpose of the Class II, Type A2 cabinet exhausted outdoors (formerly called a Type B3, Fig. 6) is to allow use of very small amounts of toxic chemicals that might vaporize as an adjunct to the microbiological work. The vapor handling performance of the Type A2-exhausted BSC is similar to that of the Type B1 (as described below in the section, Class II, Type B1 BSC) when work is being carried out at the front of the work area of the Type B1 cabinet (Fig. 7).

As can be see in Fig. 6, a canopy connection must be used for exhausting a Type A1 or A2 cabinet outdoors (12), which introduces a carefully designed air gap between the cabinet and the building exhaust duct. If a canopy is designed properly, it is possible for the operation of a Type A2-exhausted cabinet to be virtually independent from the building exhaust system variations in rate of airflow (36).

When the building exhaust system pulls more air than the cabinet is exhausting, extra air comes in through the canopy gap, entraining all of the cabinet exhaust air. Traditional canopy designs operated at a nominal 20% above the cabinet exhaust flow; however, more recent designs commonly require 5% or less additional air (37). The canopy design should be tested with tracer gas to ensure that there is no leakage from the canopy gap. Should the building exhaust system be blocked off completely, and the canopy is properly designed, the cabinet continues running at a reduced intake air velocity, with HEPA-filtered cabinet exhaust air venting back into the room through the canopy gap (12). However, if the cabinet blower should fail, added static pressure may be applied to the building exhaust system, and product protection within the cabinet may be compromised.

In the past, many Type A cabinets (both A1- and A2-exhausted) had been installed as hard-ducted to the building exhaust, with no canopy gap. With a hard-duct installation, Type A cabinet operation is sensitive to building exhaust airflow variations. If the exhaust system fails in this situation, cabinet exhaust airflow is hindered and the cabinet blower continues to deliver the same total airflow. This results in cabinet intake airflow dropping while the downflow increases, creating the possibility for breach of containment at the work access opening. For this reason, the NSF standard has now disallowed the use of the hard-duct for Type A cabinets (12).

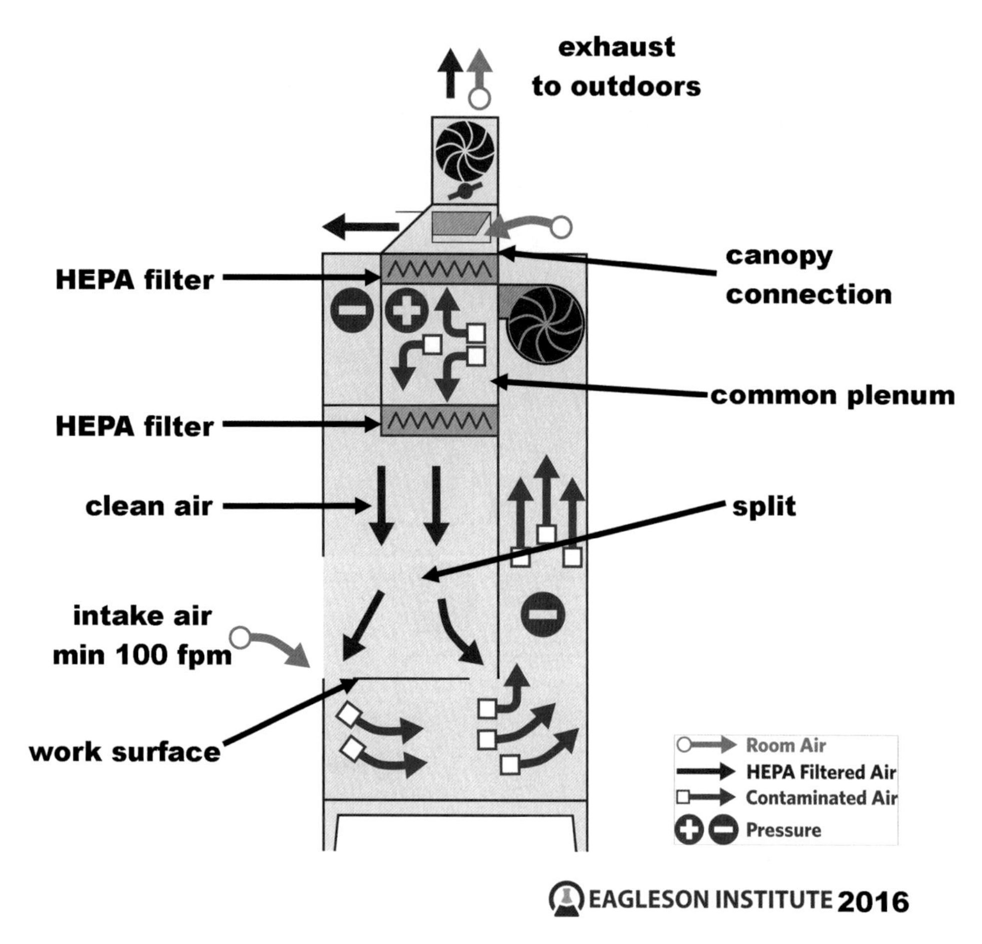

Figure 6: Class II, Type A2-exhausted cabinet basic design and airflow schematic.

Exhaust system requirements for a 4-ft Type A2-exhausted cabinet will range from about 300 to 400 cfm (cabinet exhaust, depending on the set point and additional flow required for the air gap) at approximately 0.1 in. of water column negative pressure. Type A2-exhausted cabinets are required to be supplied with an alarm that alerts the user in the event that the exhaust system fails (12).

The Type A2-exhausted BSC performs like the A2 cabinet on the microbiological tests. The Type A2-exhausted and Type B1 cabinets handle vapors similarly when work is carried out at the front of the work area of the Type B1 cabinet. However, the location of the work performed on the work surface has no effect on the vapor-handling performance of the Type A2-exhausted cabinet (Fig. 7), unlike that of the Type B1 cabinet.

Class II, Type B1 BSC (after the Original NCI Type B Design)

The purpose of the Class II, Type B1 BSC is to provide personnel, product, and environmental protection from biohazards assigned to BSL1, -2, or -3 containment and from small amounts of hazardous materials that may vaporize. The Type B1 BSC provides effective evacuation of vapors. In fact, Type B1 is unique among the Class II BSCs in that it is the only one that has 100% of the total airflow of the cabinet coming down through the work area. It also minimizes the escape of vapors into the room by maintaining a minimum 100-fpm intake air velocity and venting exhaust air outdoors (12).

Some of the differences in design, compared to the Type A1 or A2 BSC, can be seen in Fig. 8. Of the total cabinet airflow through the work area, roughly two-thirds is exhausted via a slot at the back of the work surface and is vented directly to the outdoors through a dedicated exhaust duct. The one-third of the downflow air that goes to the front of the work area is joined by intake air (to make up for the two-thirds that was exhausted out the back), and the total airflow of the cabinet is immediately filtered through a HEPA filter just below the work surface. This filter, present only on versions of the original Type B design, prevents both biological and chemical aerosol contamination of the inside of the cabinet, with only HEPA-filtered air passing through internal ducts and plenums (chemical vapors, while passing through HEPA

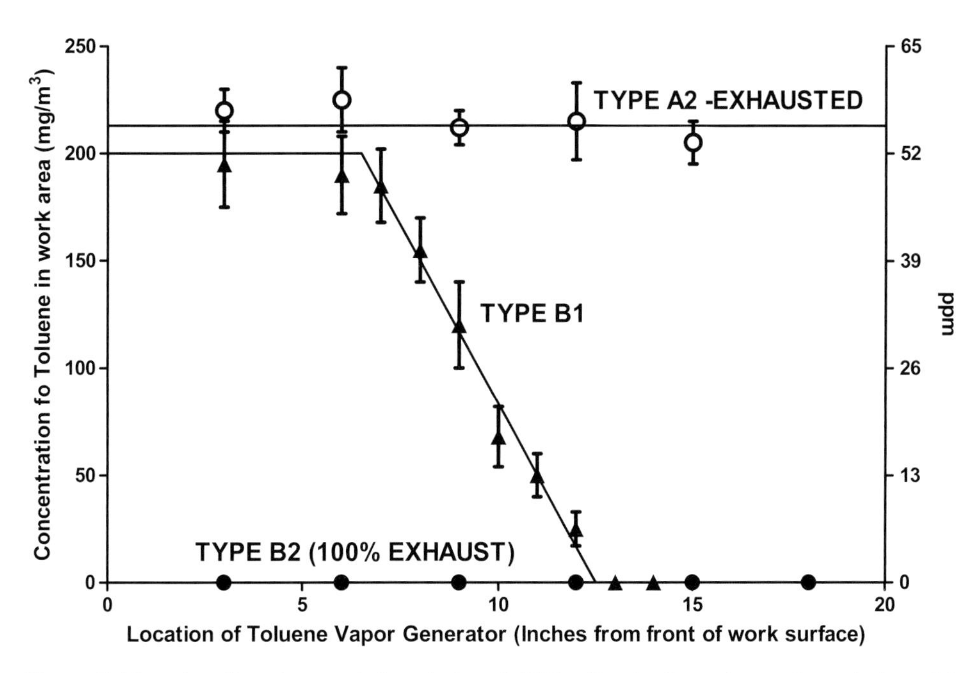

Figure 7: Vapor-handling characteristics of class II BSCs. Results from the mathematical model compared to observed concentration. Given an estimated rate of volatile generation in the cabinet, and measured cabinet airflows, the model will calculate a volatile concentration in the cabinet downflow air.

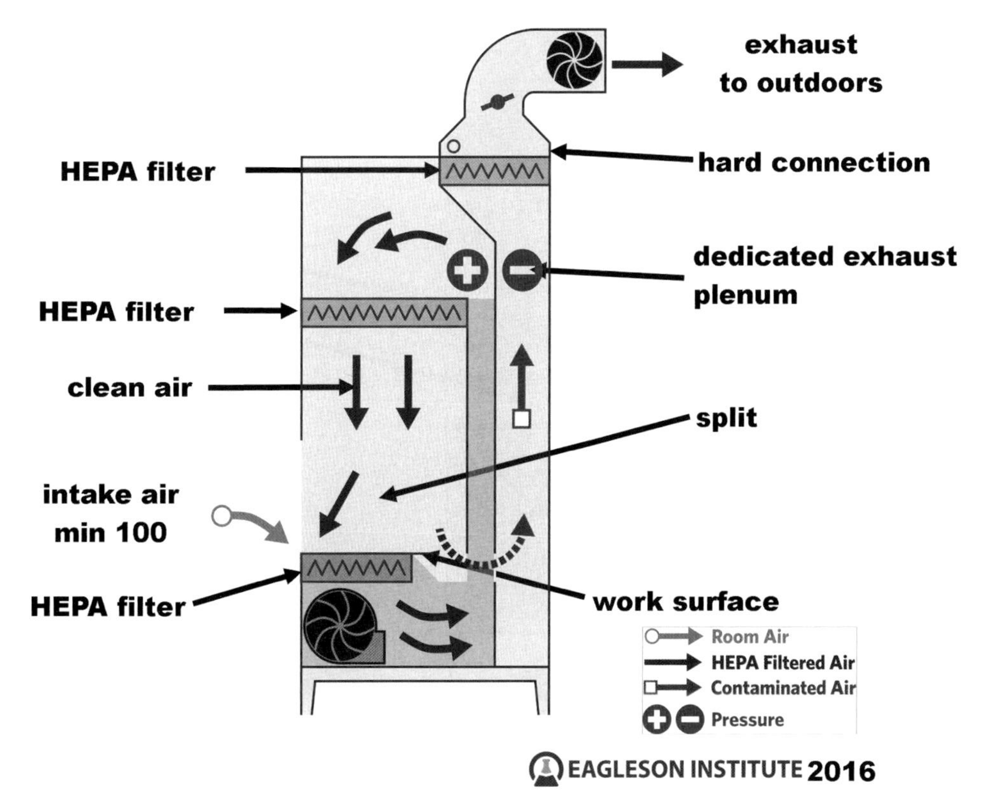

Figure 8: Class II, Type B1 (original Type B) basic design and airflow schematic.

filters, will eventually circulate and be exhausted to the outdoors). This is important when working with hazardous chemicals because they are not neutralized by gaseous decontamination of the cabinet.

The Class II, Type B1 BSC must be hard ducted to the outdoors by an exhaust system that will hold the airflow rate constant within ±5%. For safety purposes, there must be an interlock that turns off the cabinet supply fan and sounds an alarm within 15 s should the exhaust system airflow drop below 80% of the certified total flow (12). Since the cabinet fan only recirculates air within the cabinet, the exhaust fan has to pull air out through the slot in the back, through the cabinet exhaust duct and the exhaust HEPA filter before pulling it up through the building exhaust duct. For a 4-ft unit this means drawing approximately 270 cfm against 0.9 in. of water column (assuming a loaded filter) just to get the air out of the cabinet. Provisions must be made to provide the required supply air to the room, and possible interactions with other ventilation equipment must be taken into consideration.

The Class II, Type B1 BSC meets or exceeds the requirements of the microbiological performance tests. When vapors are released at the rear of the work area they are immediately exhausted through a HEPA filter to the outdoors (Fig. 8). However, vapors that are released at the front of the work area are removed from the work area via the front perforated grille. After being diluted by mixing with the intake air, the vapors recirculate back into the work area. Placing materials and working as far back in the cabinet as possible will take the best advantage of the Type B1 BSC's performance (38).

Class II, Type B1 BSC (NSF Definition)

The NSF definition for the Class II, Type B1 BSC (12) was written to include the original NCI Type B design, but does not require a HEPA filter immediately below the work surface. Therefore, some Class II, Type B1 BSCs will lack the HEPA filter below the work surface. In such cabinets the cabinet blower will likely be at the top of the cabinet (Fig. 9). Caution should be used when dealing with chemical aerosols in such cabinets. An aerosol is an aerial suspension of nongaseous particles that are so small that they have little tendency to fall out by gravity (colloidal). The particles may be solid, as in a smoke, or liquid droplets, as in a fog (39). HEPA filters filter out nongaseous particles. If aerosols of solid or liquid chemicals are generated in the work zone of the cabinet, a HEPA filter placed immediately below the

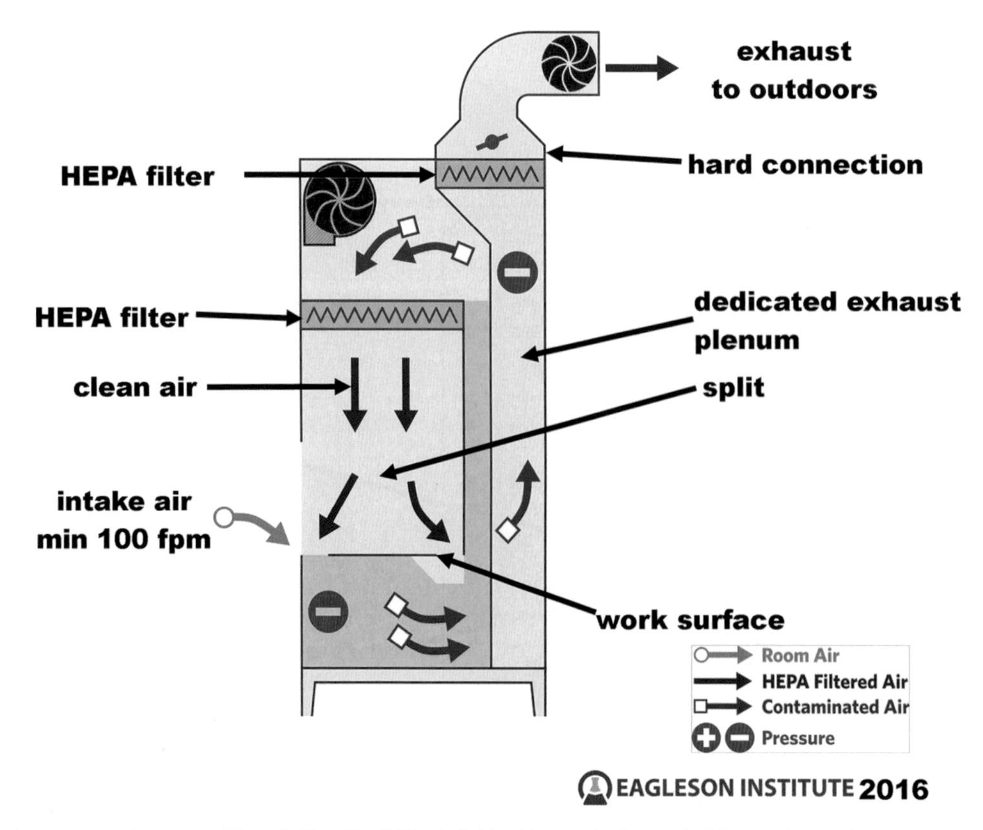

Figure 9: Class II, Type B1 (NSF definition) basic design and airflow schematic.

cabinet's work surface (as in the original Type B design) will stop these aerosols and prevent them from spreading throughout the inside of the cabinet. However, vapors from chemicals used in the cabinet will pass through the HEPA filter and keep on traveling with the airstream until they are exhausted outdoors. Hence, the HEPA filter below the work surface prevents chemical as well as biological contamination of the interior of the cabinet, however this filter is not present on all Type B1 cabinets.

Class II, Type B2 (100% Exhaust) BSC

The purpose of the Class II, Type B2 BSC is to provide personnel, product, and environmental protection from biohazards assigned to BSL1, -2, or -3 containment and to complete the spectrum of improved vapor handling. The minimum intake air velocity of the Class II, Type B2 BSC is 100 fpm, with no air recirculation, all plenums except the supply plenum (which carries only room air) are under negative pressure relative to the room, and exhausted air vents to the outdoors (12).

The simplicity of the airflow patterns in the Class II, Type B2 BSC can be seen in Fig. 10. Room air is pushed through the supply filter and down through the work area. For safety purposes, there must be an interlock that turns off the supply fan and sounds an alarm within 15 s should the exhaust system airflow drop by more than 20% (12). The building exhaust fan pulls the intake air into the cabinet and the total airflow of the cabinet out through the exhaust HEPA filter, through the exhaust duct, and to the outdoors.

The Class II, Type B2 BSC must be hard ducted to the outdoors, and the building exhaust system must have the capability of handling the airflow rate and static pressure (Table 2). The Class II, Type B2 BSC must meet or exceed the requirements of the microbiological performance tests. The difference between the four different types of Class II BSCs is not in their biosafety performance but in their vapor-handling performance. There is no question that the Type B2 is best of the Class II BSCs for handling vapors because the entire volumetric flow of the cabinet is immediately swept from the work area to the outdoors (Fig. 7). This does not make a BSC a fume hood, however. BSCs have a lower volumetric airflow rate through the work area than fume hoods do and may have air recirculating through the work area (excluding Type B2). Also, electrical outlets and fixtures in the work area, and fragile HEPA filters, may be damaged by the use of some chemicals. BSCs must be used with the sash at a designated height, which might be different from the sash height needed for a fume hood.

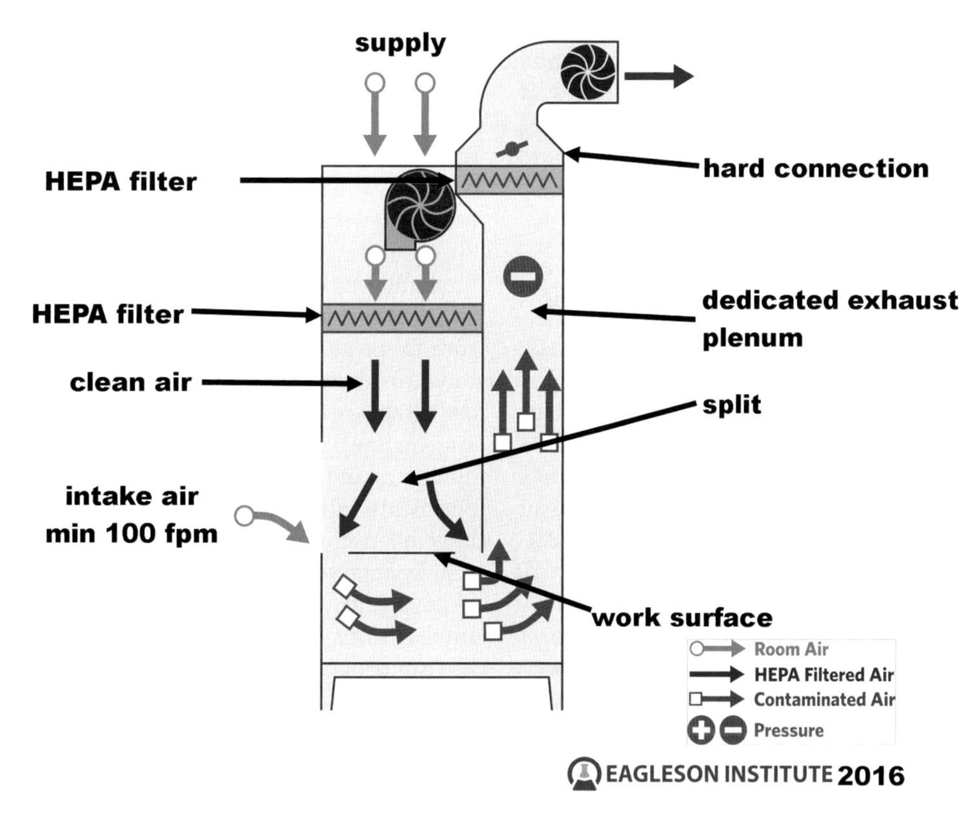

Figure 10: Class II, Type B2 cabinet basic design and airflow schematic.

TABLE 2.

Comparison of building exhaust airflow rates and negative static pressure requirements for various types of 4-ft primary barriers

Barrier type	Exhaust airflow rate (cfm)	Negative static pressure (ΔP, in. of water column)
Fume hood	1,000	0.5
BSC		
Class I	220[a]	1.5
Class II, Type A1	0[b]	0
Class II, Type A2	0[b]	0
Class II, Type A2-ex	320[a]	0.1
Class II, Type B1	280[a]	0.8
Class II, Type B2	750[a]	2.0
Class III	All are special cases	All are special cases
Glove box/isolator	All are special cases	All are special cases

[a]BSC cfm is nominal for 8-in. openings at a 105-fpm intake velocity.
[b]Types A1 and A2 when vented to the room.

It is clear from reading the use statements in NSF that the intent is to use only small diluted amounts of volatile toxic chemicals in BSCs as adjuncts to the biological work, not to weigh and dilute quantities of hazardous and/or volatile chemicals in the cabinet (12). BSCs have been designed to provide biosafety. Great care must be exercised not to overwhelm the airflows or damage HEPA filters by trying to use BSCs for activities for which they were not intended. Use of a BSC should be based on a thorough risk assessment performed by qualified people.

Type B2 Cabinets Modified for Fume Hood Activities

In some cases, modifications have been made to Type B2 BSCs to increase the safety of working with volatile chemicals. One such version places all electrical components outside the work area and provides remote shutoff valves for plumbed services. Taking this a step further, in some instances a BSC has been modified and called a "clean air fume hood." This approach takes a Class II, Type B2 BSC and places all electrical components outside the work area, removes the exhaust filter, and may use alternative materials for the work area (e.g., protective coatings over stainless steel). The supply HEPA filter thus provides a largely contamination-free work area; however, it is important with this so-called "clean air fume hood" to ban any biohazardous work. It is important to realize that without an exhaust HEPA filter for environmental protection, either on the cabinet or in the exhaust system, the hood is no longer a BSC. Even so, chemical work must not be carried out in this equipment unless a careful risk assessment shows that it would be safe.

Caution: BSC Look-Alikes

As research requirements and contamination control equipment designs evolve, some clean air equipment can appear to be a BSC at first glance. The vertical-flow clean bench (VFCB) is a good example. VFCBs were designed specifically for product protection. They do not provide personnel protection, as by design some amount of air will flow from the work area out toward the user, and there is no HEPA filter for the VFCB exhaust air (Fig. 11). It is very important to complete a careful risk assessment to determine whether use of a VFCB would be safe. A VFCB will have many features of a BSC: a stainless steel work area, unidirectional downflow of clean air in the work area, some air recirculation, a hinged view-screen, a perforated front grille, and an access opening to perform the work. However, after carefully reviewing the airflow diagram and cabinet specifications, it can be seen that a VFCB is not a BSC.

GLOVE BOXES AND CLASS III BSCs

Glove Boxes

The purpose of a glove box is to provide an enclosed work area with a controlled environment and to be a primary barrier against hazardous materials being handled inside, offering personnel, environmental, and/or product protection (40). This is accomplished by means of a physically sealed enclosure. If air is to flow through the enclosure, appropriate air-cleaning devices are used on both the intake and exhaust ports. The enclosure is operated under negative pressure for containment or under positive pressure for product protection.

Glove boxes are used in a myriad of applications. Figure 12 shows an example of an animal aerosol exposure system. Specialty Class III glove box systems protect the operators and experiment during aerosol inhalation experiments with animals. The purposeful aerosolization of large quantities of infectious agents violates basic industrial hygiene principles but is required to study the effects of the agents, particularly in the field of vaccine research. Increased air purge rates in the chamber enclosing the aerosol generation equipment will facilitate aerosol removal both during normal operation and in the case of an accidental release. These dual-sided main chambers enclose and integrate with the aerosol exposure equipment. Special care is required in defining ergonomic features required for performing the experiments and decontamination procedures due to the potentially high hazardous

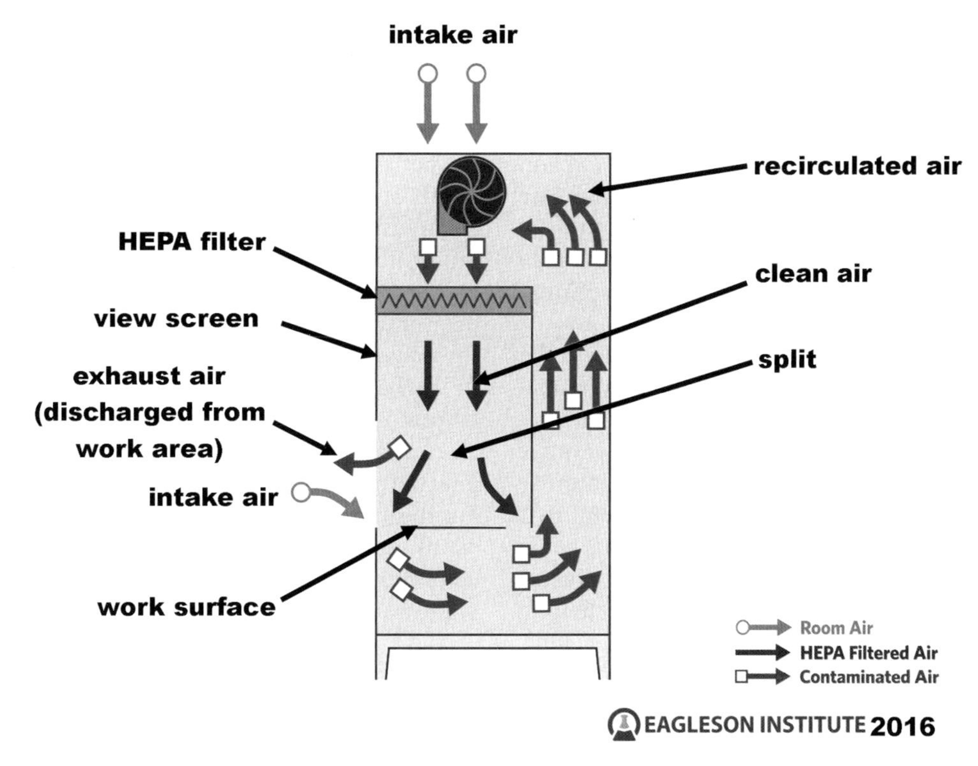

Figure 11: Basic design and airflow schematic for a VFCB, showing unfiltered outward airflow.

material burden. The other basic component to the system is the self-contained, mobile animal transfer cart. Usually the animals are housed in a location remote from the exposure room. The transfer cart provides: ventilation for respiration, through-the-wall docking connections, and door interlock components. It also requires ergonomic provisions for animal transfer, both at the aerosol lab and holding room, and for decontamination.

A glove box used to handle high-risk chemicals in the proposed CSL4 laboratory (22, 41) is a gastight stainless steel box operated under negative pressure with air flowing through it. Intake air is HEPA filtered and exhaust air is double HEPA filtered. The glove box is to be electrically grounded and have an internal fire extinguisher. All electrical power in the box is interlocked with exhaust airflow so as to kill power upon loss of exhaust. These glove boxes may have continuous hazardous-gas monitors to alarm for toxic or flammable fumes or gas (41). Exhaust air may have to be treated before it reaches the HEPA filter to protect the filter from chemicals that might degrade it. In addition, a pass-through that maintains the barrier while moving materials in or out of the glove box should be considered. This type of equipment has particular exhaust requirements, which will vary from none at all for still air boxes to sophisticated systems for specialized applications.

The performance of glove boxes is commonly measured with a pressure decay test (40), with the pressure being either positive or negative and varying from 1.5 to 10 in. of water column depending on the magnitude of the hazard involved. An industry guideline for a leak rate/pressure decay rate acceptance criterion is 0.5% of the box volume per hour (40). The pressure decay test

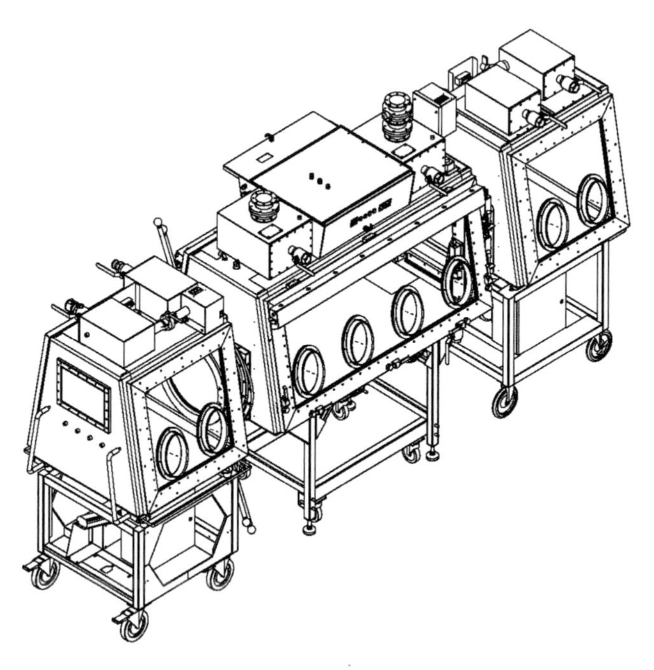

Figure 12: Basic design schematic of a glove box system to handle animals during inhalation exposure.

measures the total leakage of the entire system being tested. To ensure that most of the total leakage is not accounted for by a few relatively large leaks, rather than minute leaks spread throughout the system, a leak test using tracer gas under pressure is performed to find and quantify individual leaks. Use of a mass spectrometer leak meter to measure helium leaks is a common approach. Concentrations of helium within the box range from 0.5 to 100% under at least 2 in. of water column pressure.

The leak rate must be interpreted with respect to the concentration of gas being used. The entire individual leak rate resulting from a measurement of 10^{-4} ml/s using 100% helium in the box is 10^{-4} ml/s. However, if 0.5% helium is in use, the leak meter sees only the helium in the leak (0.5%) and none of the air (99.5%) that is passing through the leak. Therefore, a leak that measures 1×10^{-4} ml/s using 0.5% helium in the box is actually 2×10^{-2} ml/s after the viscosities of air and helium are taken into account (42, 43). While measured individual leak rate acceptance criteria run from 10^{-4} to 10^{-7} ml/s, the actual leak rate range is from 10^{-2} to 10^{-7} ml/s depending on the concentration of tracer gas used in the box.

Class III BSCs

Class III BSCs are specialized glove boxes. Their purpose is to provide maximum containment of high-risk microorganisms, thus protecting personnel and the environment. The objective is to attain as close to absolute containment as possible. This is accomplished by placing a leak-tight physical barrier between the laboratory worker and the agents being handled. The physical barrier is operated under at least 0.5 in. of water column negative pressure so that any leaks will be inward. In addition, if an agent is so dangerous that it is to be handled within a barrier such as this, one would not just open the door of the barrier, remove the agent, and carry it across the room. There must be controlled, safe access ports in the barrier that will prevent contamination of the laboratory when used. An interlocked double-door autoclave and/or a pass-through that can be sterilized or disinfected after each use are examples of these access ports. For the same reasons, Class III cabinets usually occur in "lines" or chains of interconnected cabinets that contain all of the equipment needed to carry out the work without removing the agent from the system (44). Reasonable product protection can be provided when the airflow within the Class III box resembles the mass displacement airflow of a Class II cabinet.

A Class III BSC can be seen in Fig. 13. Components include a stainless steel glove box with safety glass window, heavy gloves, inlet and exhaust HEPA filters, a ventilated pass-through, and a decontaminating dunk tank pass-through. All components are assembled gastight and operated under a minimum of 0.5 in. of water column negative pressure. Air exhausted from the cabinet must

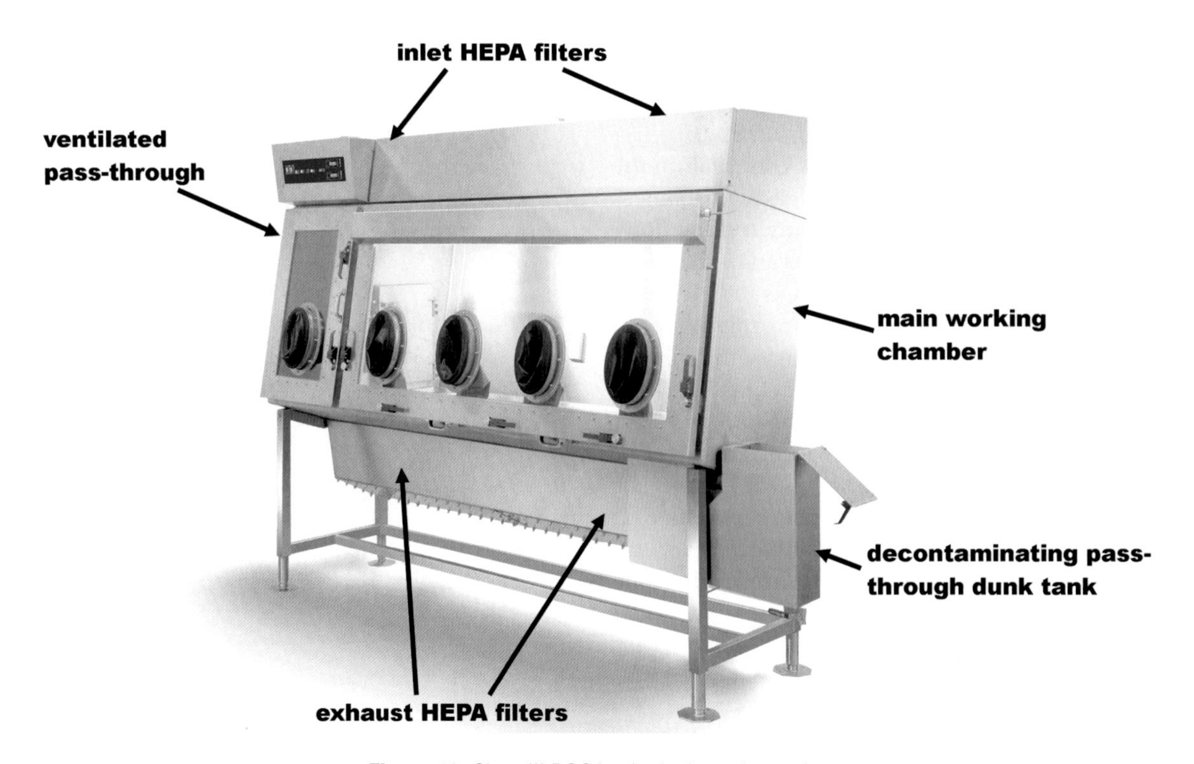

Figure 13: Class III BSC basic design schematic.

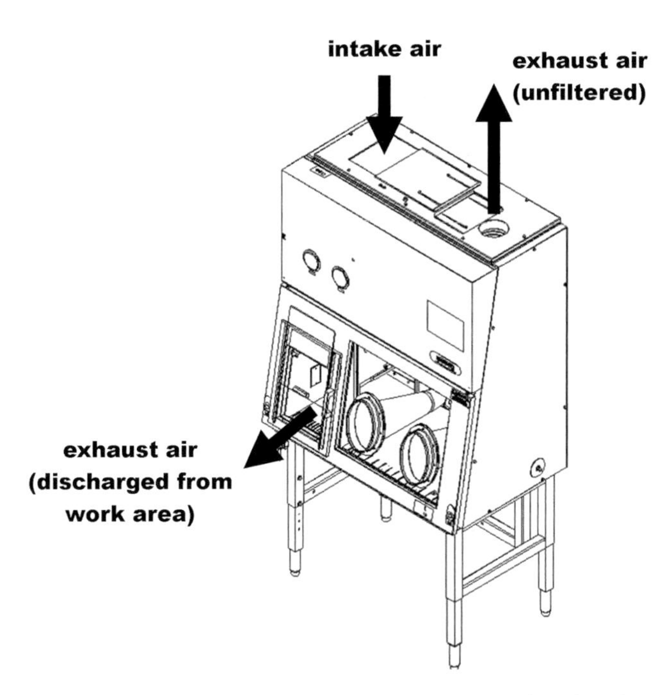

Figure 14: Basic design schematic of a pharmacy glove box for aseptic medication preparations.

be double HEPA filtered or HEPA filtered and incinerated. Should a glove come off, the inward velocity of air through the arm port must be at least 100 fpm. Class III BSCs are frequently custom designed for specific applications.

Class III systems require that building exhaust systems be balanced with room air supply systems for them to function. The exhaust air is double HEPA filtered, with both filters integral to the cabinet or one filter integral to the cabinet and another located close to the cabinet in the building exhaust system. Airflow rate and pressure requirements may vary from one Class III cabinet to another.

The performance of Class III systems depends on leak tightness, air cleaning, and specialized and thorough procedures and practices of maintenance and operation of the system that preclude any breach of containment. For more detailed information on Class III BSCs see references 43 to 45.

Caution: Class III Look-Alikes

Glove boxes are used for many applications, but not all glove boxes are Class III BSCs. While the physical isolation of the inside of the glove box from the laboratory may give the impression that the user is protected from hazards, this may not be the case.

Some systems may be designed as specialized glove boxes, or a series of interconnected gloveboxes, with adaptations incorporated that create a work area in which specific applications can be carried out under near-sterile aseptic conditions. These may have recirculated filtered air with an air handler providing conditioned air, as well as sterilize-in-place capability, such as vapor-phase hydrogen peroxide. The Trexler Isolator is one such example, and has been used to rear germ-free and specific pathogen-free mice (14).

Glove boxes are also being used more frequently in the clinical setting for the preparation of drugs to be administered to patients. Some pharmacy glove boxes (isolators) create a barrier by operating under positive pressure to isolate the product and reduce the risk of introducing contamination from the user or the room (46). However, these products may exhaust unfiltered air (Fig. 14). Other pharmacy glove boxes (isolators) (Fig. 15) operate under negative pressure and have been designed to also provide containment to protect pharmacy personnel from exposure to hazardous drugs (47). It is important to understand that these glove boxes are not designed to meet the performance requirements of a Class III BSC.

A safety professional should be consulted to perform a risk assessment to your particular application before selecting or using a glove box.

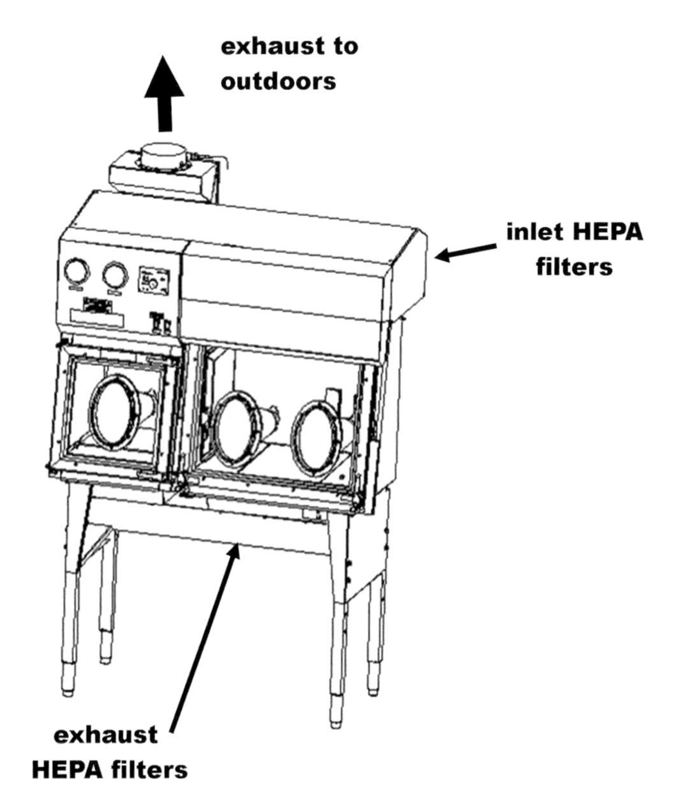

Figure 15: Basic design schematic of a pharmacy glove box for aseptic preparation of medications containing hazardous chemicals.

CERTIFICATION OF PRIMARY BARRIERS

Before any primary barriers are put into use, they must be located and installed properly. The equipment's operational integrity (including maintenance issues such as fan belts and dampers) must be demonstrated and documented at installation and at least annually thereafter (21, 25). This work should be performed by qualified personnel using equipment calibrated traceable to National Institute of Standards and Technology (21, 12). If the barrier has been used with hazardous material, the work area surface should be disinfected before testing and the interior of BSCs should be gas decontaminated before the contaminated areas of the cabinet are opened (48). The many tests performed on primary barriers are operational tests and performance tests. Operational tests measure how the equipment is functioning, for example, that the air is flowing properly. Performance tests measure whether the equipment is providing the personnel, environmental, and/or product protection it is designed to provide.

Fume Hoods and Class I BSCs

Operational testing

Operational tests include plumbing leak test, airflow rate (measured in the exhaust duct if possible), airflow face velocity, and airflow smoke pattern tests, as well as checking and documenting the calibration of all monitors and alarms (25, 27). These tests plus a HEPA filter leak test (12) are done on Class I BSCs.

Performance testing

The containment actually provided by the fume hood is measured using the ASHRAE standard 110 test (28). Four liters of SF6 tracer gas per minute is ejected into the work area of the fume hood under standard conditions. A manikin is placed in front of the fume hood in a standard location, with the probe of an SF6 detector in the manikin's breathing zone. The test is run for 5 min and the data are reported as a performance rating. The test may be run under as-manufactured (AM), as-installed (AI), or as-used (AU) conditions. The performance rating takes the form of "*AM yyy, AI yyy, or AU yyy,*" where the SF6 challenge rate is 4 liters/min and *yyy* represents the average concentration of SF6 at the breathing zone of the manikin in parts per million. ASHRAE gives no pass-fail criterion. Industrial hygienists at the user's site base decisions regarding the acceptance of the hood's performance on the performance rating obtained and the conditions under which it was measured. The Class I BSC is performance tested using the NSF personnel protection test. It may also be tested using the ASHRAE standard 110 test (28).

Class II BSCs

Operational testing

Operational tests include cabinet integrity leak test, plumbing leak test, HEPA filter leak test, intake air velocity, downflow air velocity, and airflow smoke pattern tests. Site assessment tests are also performed to verify proper cabinet location, check for harmful cross drafts, and confirm that alarms are functioning properly. Tests to help reduce worker fatigue and, in turn, help prevent accidents, spills, or mistakes that may breach containment or damage product may also be done. They include electrical tests (polarity, current leakage, and resistance to ground) and noise, lighting, and vibration (12).

Performance testing

Additional validation of BSC performance as used in the laboratory may be desired, particularly when larger instruments are to be placed within the BSC work area. BSC suppliers may be able to conduct these tests upon request for specific cabinet and instrument combinations; however, the microbiological aerosol tracer test mentioned earlier in the chapter is not suitable for performance testing of BSCs in the working laboratory because of the spores used as the tracer. These bacterial spores will survive many years in the laboratory as potential contaminants. The following sections describe alternate methods to test performance (beyond the operational tests).

One method of *in situ* performance testing for BSCs is called a KI discus test (49, 50), which is carried out by dropping potassium iodide (KI) on a spinning disk situated on the work surface. This flings droplets toward the work access opening. Air outside the work opening is sampled through a membrane filter, which is "developed," and the brown spots on the filter are counted using a magnifying glass. The acceptance criterion is a protection factor of at least 10^5. This test was developed and is most commonly used in the United Kingdom.

In the United States, a "bioanalog test" was developed for the measurement of BSC performance in the field (51). Both the personnel and the product protection tests from the bioanalog test mimic the microbiological tests closely, but use surrogates for the aerosolized spore suspension. A tracer gas (SF6) is used in the personnel protection test, while mineral oil is used for the product protection test. A protection factor is calculated from the data obtained. The results from both of these tests correlate well with those from the microbiological tests.

More recently a method using PCR amplification has been developed to assess BSC performance *in situ* (52). The PCR method has not been widely adopted for this use, but it does represent progress toward a more practical field test.

Glove Boxes

Operational testing

Operational tests include box leak test, plumbing leak test, pressure in box, volumetric flow rate and velocity through the box, and integrity test of air-cleaning devices (40). These tests, plus the HEPA filter leak test, are used to test the Class III BSC; an ultralow-penetrating air filter leak test is used for the bioclean system (53). HEPA filter leak tests and airflow tests are conducted for pharmacy glove boxes (isolators) (46).

Performance testing

Glove boxes with no special internal environment rely on system leak testing to measure containment performance (40). Measuring the ability to maintain a low oxygen concentration in a box intended to hold a nitrogen atmosphere is an example of performance testing of a special-environment box. Single-particle counting for air cleanliness, microbiological surface and air sampling, and media fills are examples of performance testing of bioclean isolators (14) and pharmacy glove boxes (isolators) (46).

SPECIAL DESIGNS AND MODIFICATIONS OF PRIMARY BARRIERS

If the risk assessment results in a requirement for a primary barrier that is not satisfied by standard models of available equipment, modifications of existing designs or special designs can often solve the problem. A few examples are given below.

Fume Hoods

A triple fume hood having two process modules and a service module was designed to provide personnel protection from aerosols and vapors during specialized animal inhalation exposure procedures (54).

Biosafety Cabinets

Class II BSCs have been modified for many different situations. They have been adapted to house microscopes, centrifuges, water baths, and cell-harvesting equipment and for bedding disposal (55). Modifications for special handling of waste within the cabinet are also common (56). Class III cabinets are often specially designed in interconnected lines; for an example, see the Georgia State University Class III line (45).

Glove Boxes and Isolators

It is common for glove boxes and isolators to be adapted to specific applications. Figure 16 shows a schematic of a glove box system that has been designed to receive and analyze potentially hazardous unknown samples. Class III systems are used due to the unknown nature of the hazard. Some of these systems are standalone, existing in their own separate laboratory. Some systems interface the chambers with a sealed wall pass-through allowing material to be passed from BSL3 to BSL4 lab space for further procedures, depending on the outcome of the initial analysis. Glove boxes have been designed to house an ultracentrifuge (57), as well as to enclose pharmaceutical filling lines (58).

Special-Purpose Primary Barriers

As the needs of end users have changed, new applications for primary barriers have surfaced. This selection of special-purpose primary barriers represents some common applications seen in laboratory use.

Low-flow fume hoods

Increasing concern about energy utilization has led to the search for fume hood designs to operate at lower exhaust flow rates (collectively referred to as "low-flow" or "high-performance" fume hoods). This is accomplished by introducing features that enable performance at lower face velocities and/or by decreasing the open area at the fume hood face (25). There is no accepted standard for face velocity on low-flow fume hoods; however, some designs claim that their hoods pass the ASHRAE standard 110 AM performance test at face velocities of 60 fpm or lower (27). Before considering the use of a low-flow fume hood, it is important to consult safety and industrial hygiene personnel and conduct a thorough risk assessment. In addition, the benefits of a low-flow device should be considered in the context of the overall heating, ventilation, and air-conditioning system for the laboratory, as certain minimum exhaust flows may be needed to provide the air balancing and air change requirements for the room.

Large-scale fume hoods

To accommodate larger-scale equipment and procedures, such as those in a pharmaceutical production pilot plant, standard benchtop fume hoods may not be suitable. To meet these specialized requirements, larger "walk-in" fume hoods or "island" fume hoods have been designed

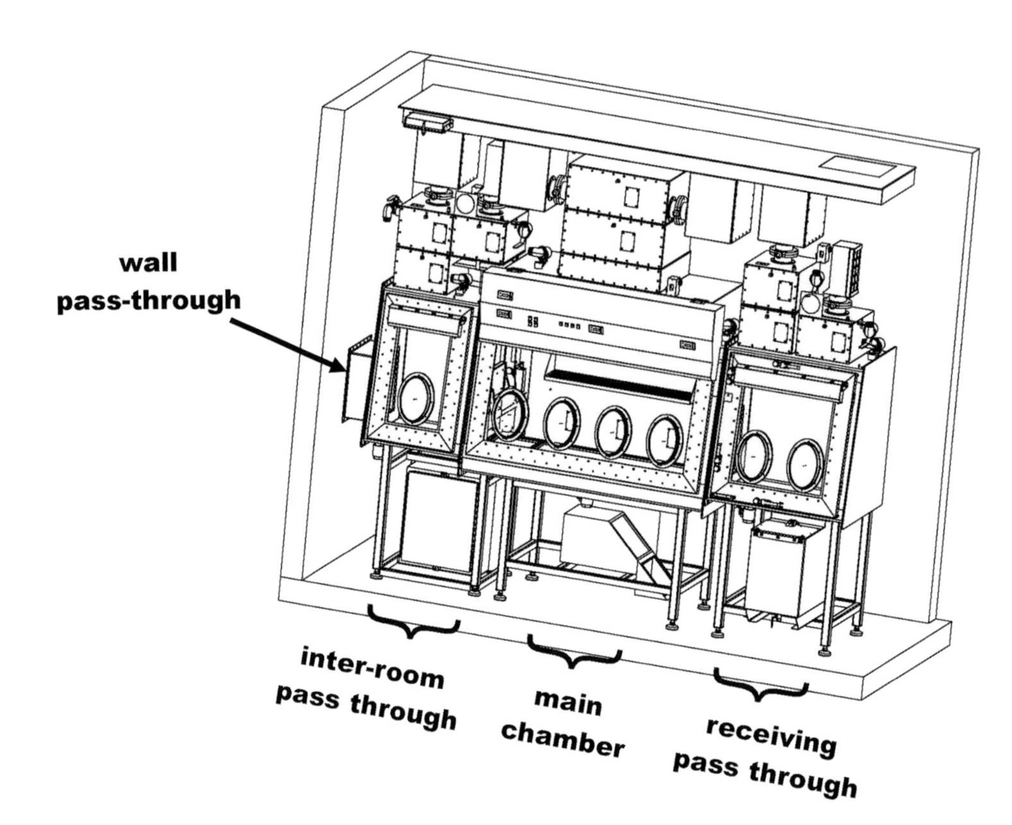

Figure 16: Basic design schematic of a through-the-wall Class III system for handling potentially hazardous unknown terrorism samples.

and manufactured. These hoods rest on the floor and may have recessed drain pans and/or floor grates. Access into such hoods is gained through large doors that allow users to move equipment in and out of the hood.

Adaptations for animal research

Standard BSCs, and other primary containment devices, are sometimes modified to accommodate procedures involving animal subjects. Larger-access openings allow operators to move animal cages into and out of the enclosure without raising the sash and also provide better ergonomics for cage changing or other procedures. Prefilters (non-HEPA/ultra-low particulate air [ULPA]) may be added to prevent larger materials shed by the animals from depositing in cabinet plenums or blowers. Postfilters (usually carbon) may be placed downstream of the cabinet exhaust filter to remediate odors. BSCs may also be equipped for necropsy procedures by including wash-down hoses and work surfaces that slope to a drain. Cabinets are also modified to allow direct transfer of soiled bedding to waste receptacles within the containment enclosure.

Enclosures for automated equipment

In recent years, the use of robotics and other automated equipment, such as automatic pipettors, cell culture feeders, cell sorters/counters, and bioprinters, has become commonplace in the laboratory. This has been driven by the desire for increased throughput, better accuracy, and reduction of repetitive strain injuries in laboratory personnel. Just as with the manual laboratory procedures that this equipment replaces, this work may require barriers for protection of the product and/or the laboratory personnel.

Specialized containment enclosures have been designed to accommodate automated equipment and associated support services (maintenance access, controls, and utilities). Some enclosures function as a larger fume hood, with air exhausted and (possibly) filtered; others, such as the enclosure in Fig. 17, are designed to function in a manner similar to a Class II BSC by providing product, personnel, and environmental protection. However, some enclosures for robotics are designed simply to provide clean air to the work area for product protection. While these enclosures have physical barriers between the laboratory and the equipment, they exhaust unfiltered air back to the room and should not be used for work with any potentially hazardous agents. Flow cytometry is an application that has drawn quite a bit of attention, as best practice guidelines have been written calling for primary containment for instruments used for sorting unfixed cells (59). Cell sorters (and indeed any instru-

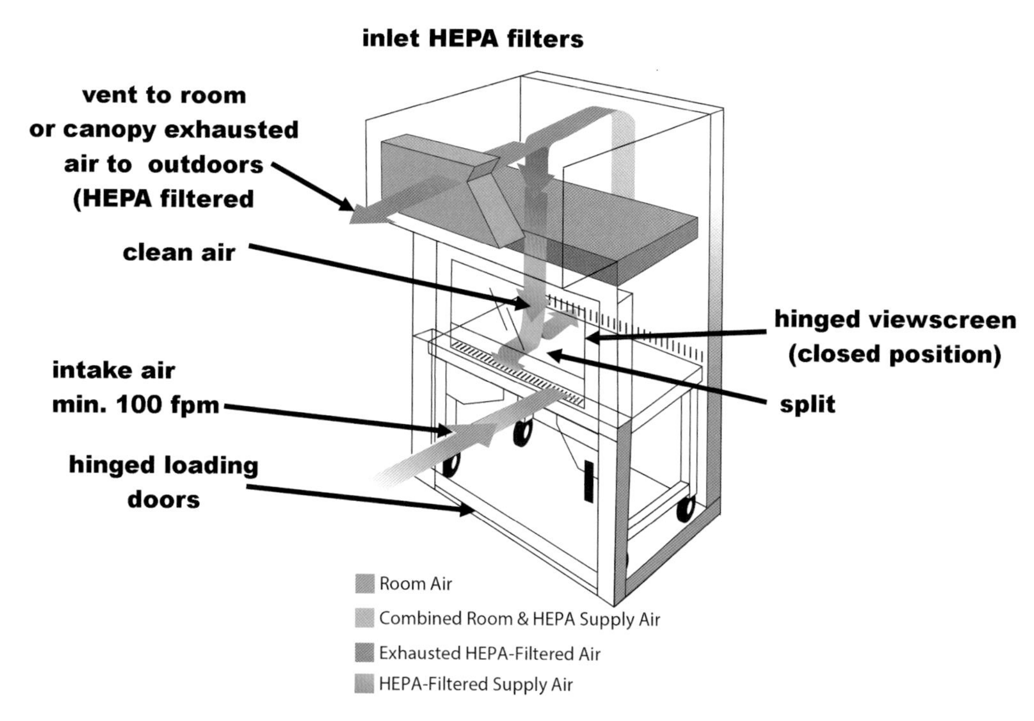

Figure 17: Basic design schematic of a Class II-style enclosure for automated laboratory equipment.

ment placed within primary containment) are of sufficient size that disruption to the flow in the enclosure is likely. Additionally, the heat generated by lasers and other energy-intensive components creates convection effects that can alter enclosure performance. For these reasons, it is critical that all specially designed enclosures be validated to provide the necessary performance (containment and/or product protection) with the instrument in place.

Equipment for mobile laboratories

Small BSCs and fume hoods have been designed and manufactured to fit into miniature laboratories that are housed in trucks, vans, or shipping containers. Samples can be passed into the lab module and then studied in an appropriate primary containment enclosure. A common application for these labs is in terrorism response, to handle unknown samples safely (suspected nuclear, biological, and/or chemical materials) in Class III BSCs. Such labs are also used to provide modern scientific capabilities in remote locations.

USE OF PRIMARY BARRIERS

To achieve and maintain biological safety, it is important to base selection and use of primary barriers on an understanding of their purpose, design, and performance. Together with a risk assessment, this will enable the use of appropriate equipment for the particular work that is to be carried out. Locating the primary barrier in the room and developing work practices should be based on the capabilities and limitations of the equipment.

Location

Fume hoods and BSCs should be located in the room so that there is comfortable access by the operator during its use and access for the certifier to test and service the equipment. Open-front hoods and cabinets are sensitive to cross drafts (29, 55) and must be located in low-traffic areas and more than 10 ft away from a doorway with no cross drafts in excess of one-half the face/intake velocity (25). The NIH design manual provides specific guidelines for placement of Class II BSCs relative to doorways, other BSCs, and personnel traffic (60).

Glove boxes have no special requirements for location, but there has to be room to work comfortably through the gloves; they should not be located in high-traffic areas (25).

General Work Practices in Open-Fronted Barriers

1. Read the operator's manual and follow the manufacturer's instructions and recommendations.
2. Make sure the barrier is up to date on certification and/or routine operational checks.
3. Respond appropriately to all warnings and alarms.

4. Check gauges and/or monitors to make sure that the equipment is functioning properly.
5. Prepare a written checklist and follow it to ensure that everything needed is in the unit to minimize arm movements in and out of the barrier.
6. Wear lab coat, gloves, forearm sleeves, and other appropriate personal protective equipment.
7. If it is not kept running, turn on the hood/cabinet at least 3 to 4 min before starting work.
8. Keep your head out of the hood/cabinet work area.
9. Move smoothly and deliberately in and out of the unit, with arms oriented perpendicularly to the plane of the sash.
10. Clean and disinfect the work area before and after each use.
11. Keep the sash at the specified opening height (lowest possible on fume hood) while working.
12. Be sure the work area is clean and work surfaces are disinfected before raising or removing the sash for setup.
13. Place plastic-backed absorbent material on the work surface to reduce splatter.
14. Place materials and equipment in the work area so as not to block airflow slots or grilles.
15. Put, and keep, only materials actively in use inside the work area.
16. Work should be conducted within the unobstructed clean airstream exiting the supply HEPA filter (sometimes referred to as the "first air").
17. Write, keep available, and use emergency spill protocols.
18. Safely discard all waste and contaminated personal protective equipment.
19. Avoid exposure to germicidal light if UV lamps are present.

These are general work practices for generic types of primary barriers. Work practices specific for the equipment used and the activities involved should be written for every individual laboratory. More specific information on work practices is available via publications from the Centers for Disease Control and Prevention (CDC) (20) and from the American Industrial Hygiene Association (AIHA) (25).

CONCLUSIONS

There are many variations of hazards associated with work conducted in microbiological and biomedical laboratories. There are many different kinds of primary barriers to help minimize the risk of working with those hazards. Safety requirements determined from a thorough risk assessment can be matched with primary barriers (based on an understanding of their purpose, design, and performance) to provide the personnel, environmental, and/or product protection required for the specific situation. When detailed work practices (based on thorough knowledge of the risk assessment and the selected primary barriers) are written, implemented, and followed, biological safety can be achieved.

References

1. **Choffnes ER, Mack A.** 2015. Emerging viral diseases—the one health connection: workshop Summary (2015). *Appl Biosaf* **20:**61.
2. **Collins CH, Kennedy DA (ed).** 1999. *Laboratory-Acquired Infections: History, Incidences, Causes and Preventions.* Butterworth-Heinemann, Oxford, U.K.
3. **Harding L, Byers K.** 2006. Epidemiology of laboratory-associated infections, p 53–77. *In* Fleming DO, Hunt, DL (ed)., *Biological Safety: Principles and Practices,* 4th ed. ASM Press, Washington, DC.
4. **Pike RM.** 1978. Past and present hazards of working with infectious agents. *Arch Pathol Lab Med* **102:**333–336.
5. **Campbell MJ.** 2015. Characterizing accidents, exposures, and laboratory-acquired infections reported to the National Institutes of Health's Office of Biotechnology Activities (NIH/OBA) Division Under the NIH Guidelines for Work with Recombinant DNA Materials from 1976–2010. *Appl Biosaf* **20:**12–26.
6. **Pike RM.** 1976. Laboratory-associated infections: summary and analysis of 3921 cases. *Health Lab Sci* **13:**105–114.
7. **Pike RM.** 1979. Laboratory-associated infections: incidence, fatalities, causes, and prevention. *Annu Rev Microbiol* **33:** 41–66.
8. **Bennett A, Parks S.** 2006. Microbial aerosol generation during laboratory accidents and subsequent risk assessment. *J Appl Microbiol* **100:**658–663.
9. **Chatigny MA, Clinger DI.** 1969. Contamination control in aerobiology, p 194–263. *In* Dimmick RL, Akers AB (ed), *An Introduction to Experimental Aerobiology.* John Wiley & Sons, Inc, New York.
10. **Pottage T, Jhutty A, Parks SR, Walker JT, Bennett AM.** 2014. Quantification of microbial aerosol generation during standard laboratory procedures. *Appl Biosaf* **19:**124–131.
11. **Sansone EB, Losikoff AM.** 1977. A note on the chemical contamination resulting from the transfer of solid and liquid materials in hoods. *Am Ind Hyg Assoc J* **38:**489–491.
12. **NSF International.** 2014. Class II (Laminar Flow) Biosafety Cabinetry. NSF/ANSI standard 49-2014. NSF International, Ann Arbor, MI.
13. **First MW.** 1998. HEPA filters. *J Am Biol Saf Assoc* **3:**33–42.
14. **Corbett JR.** 2006. Convention and change, p 249–271. *In* Wamberg J (ed), *Art & Alchemy.* Museum Tusculanum Press, Copenhagen, Denmark.
15. **Wagner CM, Akers JE (ed).** 1995. *Isolator Technology.* Interpharm Press, Inc, Buffalo Grove, Ill.
16. **Wedum AG.** 1957. Biological safety program at Camp Detrick—1 July 1953 to 30 June 1954. Technical report ABL-S-261. Army Biological Laboratories, Frederick, MD. Abstract at BiblioLine http://www.nisc. com.
17. **Barbeito MS.** 2002. The evolution of biosafety from the U.S. biological warfare program (1941–1972), p 1–28. *In* Richmond JY (ed), *Anthology of Biosafety V. BSL-4 Laboratories.* American Biological Safety Association, Mundelein, Ill.
18. **Kruse RH, Puckett WH, Richardson JH.** 1991. Biological safety cabinetry. *Clin Microbiol Rev* **4:**207–241.

19. **McDade JJ, Sabel FL, Akers RL, Walker RJ.** 1968. Microbiological studies on the performance of a laminar airflow biological cabinet. *Appl Microbiol* **16:**1086–1092.
20. **Coriell LL, McGarrity GJ.** 1968. Biohazard hood to prevent infection during microbiological procedures. *Appl Microbiol* **16:**1895–1900.
21. **Centers for Disease Control and Prevention and National Institutes of Health (CDC/NIH).** 2009. *Biosafety in Microbiological and Biomedical Laboratories*, 5th ed. U.S. Government Printing Office, Washington, D.C.
22. **American Chemical Society (ACS).** 2013. *Identifying and Evaluating Hazards in Research Laboratories.* American Chemical Society, Washington, D.C.
23. **Maupins K, Hitchings DT.** 1998. Reducing employee exposure potential using the ANSI/ASHRAE 110 Method of Testing Performance of Laboratory Fume Hoods as a diagnostic tool. *Am Ind Hyg Assoc J* **59:**133–138.
24. **Ghidoni, D. A., and R. L. Jones, Jr.** 1994. Methods of exhausting a biological safety cabinet (BSC) to an exhaust system containing a VAV component. ASHRAE Trans 100(part 1)**:**1275–1281.
25. **American Industrial Hygiene Association (AIHA).** 2003. American National Standard for Laboratory Ventilation. ANSI/AIHA standard Z9.5-2003. American Industrial Hygiene Association, Fairfax, VA.
26. **Hitchings DT, Shull RS.** 1993. Measuring and calculating laboratory exhaust diversity—three case studies. ASHRAE Trans 99(Part 2):1059–1071.
27. **Scientific Equipment & Furniture Association (SEFA).** 2014. *Laboratory Fume Hoods, Recommended Practices. SEFA I-2010.* Scientific Equipment & Furniture Association, Garden City, NY.
28. **American Society of Heating, Refrigeration and Air-Conditioning Engineers (ASHRAE).** 1995. American National Standard: Method of Testing Performance of Laboratory Fume Hoods. ANSI/ASHRAE standard 110. American Society of Heating, Refrigeration and Air-Conditioning Engineers, Atlanta, GA.
29. **Altemose BA, Flynn MR, Sprankle J.** 1998. Application of a tracer gas challenge with a human subject to investigate factors affecting the performance of laboratory fume hoods. *Am Ind Hyg Assoc J* **59:**321–327.
30. **National Fire Protection Association (NFPA).** 2014. *Recommended Fire Protection Practice for Facilities Handling Radioactive Materials. NFPA 801.* National Fire Protection Association, Quincy, Mass.
31. **National Fire Protection Association (NFPA).** 2015. *Standard on Fire Protection for Laboratories Using Chemicals. NFPA 45.* National Fire Protection Association, Quincy, Mass.
32. **Eagleson D.** 1990. Biological safety cabinets, p 303–331. *In* Ruys T (ed), *Handbook of Facilities Planning*, vol. I. *Laboratory Facilities.* Van Nostrand, New York.
33. **Ghidoni DA.** 1999. HVAC issues in secondary biocontainment, p 63–72. *In* Richmond JY (ed), *Anthology of Biosafety I. Perspectives on Laboratory Design.* American Biological Safety Association, Mundelein, Ill.
34. **Barkley WE.** 1972. Evaluation and development of controlled airflow systems for environmental safety in biomedical research. Ph.D. thesis. University of Minnesota, Minneapolis.
35. **Jones RL Jr, Stuart DG, Eagleson D, Greenier TJ, Eagleson JM Jr.** 1990. The effects of changing intake and supply airflow on biological safety cabinet performance. *Appl Occup Environ Hyg* **5:**370–377.
36. **Jones RL Jr, Tepper B, Greenier T, Stuart D, Large S, Eagleson D.** 1989. Abstr 32nd Biol Safety Conf, p. 28–29.
37. **Lloyd R, Eagleson D, Eagleson DC. 2012.** Biological safety canopy exhaust connection saves energy and improves overall safety performance. Acumen **10** (3). The Baker Co., Inc., Sanford, ME.
38. **Stuart DG, First MW, Jones RL Jr, Eagleson JM Jr.** 1983. Comparison of chemical vapor handling by three types of class II biological safety cabinets. *Particul Microb Control* **2:**18–24.
39. **Hinds WC.** 1999. *Aerosol Technology: Properties, Behavior, and Measurement of Airborne Particles*, 2nd ed. John Wiley & Sons, New York.
40. **American Glovebox Society (AGS).** 2007. *Guideline for Gloveboxes*, 3rd ed. American Glovebox Society, Denver, CO.
41. **Hill RH Jr, Gaunce JA, Whitehead P.** 1999. Chemical safety levels (CSLs): a proposal for chemical safety practices in microbiological and biomedical laboratories. *Chem Health Saf* **6:** 6–14.
42. **Stuart D, Ghidoni D, Eagleson D.** 1997. Helium as a replacement for dichlorodifluoromethane in class II biological safety cabinet integrity testing. *J Am Biol Saf Assoc* **2:**22–29.
43. **Stuart DG, Eagleson DC, Lloyd R, Hersey C, Eagleson D.** 2012. Analysis of the Class III Biological Safety Cabinet Integrity Test. *Appl Biosaf* **17:**128–131.
44. **Stuart DG, Kiley MP, Ghidoni DA, Zarembo M.** 2004. The class III biological safety cabinet, p 57–72. *In* Richmond JY (ed), *Anthology of Biosafety VII. Biosafety Level 3.* American Biological Safety Association, Mundelein, IL.
45. **Stuart D, Hilliard J, Henkel R, Kelley J, Richmond J.** 1999. Role of the class III cabinet in achieving BSL-4, p 149–160. *In* Richmond JY (ed), *Anthology of Biosafety I. Perspectives on Laboratory Design.* American Biological Safety Association, Mundelein, IL.
46. **United States Pharmacopeial Convention (USP).** 2015. Pharmaceutical compounding sterile preparations, p 2350–2370. *In* USP 39-NF 34. United States Pharmacopeial Convention, Rockville, MD.
47. **National Institute for Occupational Safety and Health (NIOSH).** 2004. *Preventing Occupational Exposures to Antineoplastic and Other Hazardous Drugs in Healthcare Settings.* National Institute for Occupational Safety and Health, Cincinnati, OH.
48. **Fink R, Liberman DF, Murphy K, Lupo D, Israeli E.** 1988. Biological safety cabinets, decontamination or sterilization with paraformaldehyde. *Am Ind Hyg Assoc J* **49:**277–279.
49. **Osborne R, Durkin T, Shannon H, Dornan E, Hughes C.** 1999. Performance of open-fronted microbiological safety cabinets: the value of operator protection tests during routine servicing. *J Appl Microbiol* **86:**962–970.
50. **CEN (European Committee for Standardization).** 2000. *Performance criteria for microbiological safety cabinets.* Central Secretariat, Brussels.
51. **Jones RL Jr, Ghidoni DA, Eagleson D.**1997. The bio-analog test for field validation of biosafety cabinet performance. Acumen **4**(1). The Baker Co., Inc., Sanford, ME.
52. **Fontaine CP, Ryan T, Coschigano PW, Colvin RA.** 2010. Novel testing of a biological safety cabinet using PCR. *Appl Biosaf* **15:** 186–196.
53. **Institute of Environmental Sciences and Technology (IEST).** 2009. *HEPA and ULPA Filter Leak Tests. IEST-RPCC034.3.* Institute of Environmental Sciences and Technology, Arlington Heights, IL.
54. **Colby CL, Stuart DG.** 2000. Primary containment devices for toxicological research and chemical process laboratories, p 114–128. *In* Richmond JY (ed), *Anthology of Biosafety II. Facility Design Considerations.* American Biological Safety Association, Mundelein, Ill.
55. **Rake BW.** 1979. Microbiological evaluation of a biological safety cabinet modified for bedding disposal. *Lab Anim Sci* **29:**625–632.
56. **Stimpfel TM, Gershey EL.** 1991. Design modifications of a class II biological safety cabinet and user guidelines for enhancing safety. *Am Ind Hyg Assoc J* **52:**1–5.

57. **Chatigny MA, Dunn S, Ishimaru K, Eagleson JA Jr, Prusiner SB.** 1979. Evaluation of a class III biological safety cabinet for enclosure of an ultracentrifuge. *Appl Environ Microbiol* **38:**934–939.
58. **Stuart D.** 1999. Primary containment devices, p 45–61. *In* Richmond JY (ed), *Anthology of Biosafety I. Perspectives on Laboratory Design*. American Biological Safety Association, Mundelein, IL.
59. **Holmes KL, Fontes B, Hogarth P, Konz R, Monard S, Pletcher CH Jr, Wadley RB, Schmid I, Perfetto SP, International Society for Advancement of Cytometry (ISAC).** 2014. International Society for the Advancement of Cytometry cell sorter biosafety standards. *Cytometry A* **85:**434–453.
60. **National Institutes of Health (NIH).** 2008. *Office of Research Facilities Design Requirements Manual*. National Institutes of Health, Bethesda, MD.

19

Arthropod Vector Biocontainment

DANA L. VANLANDINGHAM, STEPHEN HIGGS, AND YAN-JANG S. HUANG

INTRODUCTION

The recent emergence and reemergence of arthropod-borne viruses (arboviruses) such as chikungunya and Zika viruses, which are transmitted by mosquitoes, highlights the need to increase the capacity to conduct research on these pathogens and the vectors that are involved in the transmission cycles. Over the past two decades, the United States has had several arboviruses introduced, highlighting the need for more facilities and researchers to study these viruses. The introduction of West Nile virus (WNV) into the United States in 1999 (1–4) revealed a lack of suitably trained entomologists/virologists who could conduct critical surveillance operations and fieldwork essential for effective targeting of vector control programs. It also highlighted the erosion of training and educational material for entomology related to public health (5). Following the introduction of WNV, the capacity for surveillance and research increased in some areas; however, there is a need to continue support for these programs to quickly control new introductions of diseases transmitted by arthropods.

New biocontainment facilities that are suitable for research on arthropods are needed throughout the world as the spread of diseases associated with various arthropods continues. The recent introduction of Zika virus into islands in the Pacific (6, 7) and more recently into South, Central, and North America and the Caribbean Islands (8) is one example that highlights the need for arthropod research capacity. Other viruses, for example, WNV, continue to spread to new areas, such as Europe, which has the vectors required to establish a local transmission cycle (9). Emergence and reemergence of these vector-borne viruses is largely fueled by international travel, for example, when tourists return home after becoming infected during visits to countries currently experiencing local transmission of an arbovirus. Other factors that are contributing to the spread of pathogens include the introduction of vectors such as the Asian tiger mosquito, *Aedes albopictus*, into new areas (10, 11), viral mutations that alter the infectivity for an arthropod vector (12), and the implications of climate change on pathogen emergence. Essential requirements to prepare for new introductions of these agents include building physical infrastructure and having well-trained scientists who are able to conduct research efficiently in facilities that maintain field-collected and colony arthropods for study.

About This Chapter

In this chapter, we will refer to facilities that are designed for rearing, housing, and studying arthropods and the agents with which they can be infected and transmit as arthropod containment laboratories or insectaries. The term "insectary" is commonly used, even though, in reality, some arthropods that are reared and used in these facilities, for example, ticks, are not insects. The rearing of flying insects, such as mosquitoes, midges, and sandflies, have different requirements in an arthropod containment laboratory than what is needed to raise ticks; however, the general guidelines in this chapter can be used for both situations unless specifically identified as appropriate for one particular group.

Additionally, for the purposes of this chapter, we are focusing on arthropods of medical importance, specifically arthropod vectors that can transmit human and animal pathogens and therefore will require different levels of biocontainment. We are not considering research on, for example, fruit flies, which are commonly used in classroom settings or rearing of Lepidoptera by the amateur entomologist. Furthermore, this chapter is written as a general introduction to guide researchers and others to resources pertaining to the safe and secure operation of an arthropod containment laboratory and use of arthropods in research on pathogens that they vector. Due to the diverse biology of different types of arthropods, the sometimes species-specific requirements for successful maintenance under laboratory conditions, and the diverse range of pathogens they transmit, it is not practical to cover all of the varied scenarios to enable research with arthropod vectors. Rather, we direct the reader to published works so they can obtain the information required for their specific needs.

FACILITY DESIGN

Suitable facilities for rearing, maintaining, and infecting arthropods are essential for conducting research with arboviruses and tick-borne pathogens. Proper facility design encompasses more than the room where the arthropod containment laboratories will be housed. For example, the physical space includes the building design and location, the individual rooms, the choice of equipment, the placement of work surfaces, storage availability, and how the room design enables arthropod containment. Before beginning the facility design for the insectary, an assessment of the scope of work and purpose of the facility should be made.

Defining the purpose and intent of the arthropod containment laboratory is the essential first step. Critical questions to examine are: What types of arthropod will be housed?; What is the purpose of the facility (i.e., rearing, infecting, or analysis)?; and What is the scale of use? The answers to the first two questions will define the biosafety level (BSL) of operation (BSL1, BSL2, BSL3, or BSL4). The level of biosafety required will determine design and operation parameters for the arthropod containment laboratory. A summary of the arthropod containment level (ACL) recommended for different circumstances is provided in Table 1.

If the intent is to use the arthropods for infection experiments, the well-defined BSL categories for different pathogens as specified in *Biosafety in Microbiological and Biomedical Laboratories* (BMBL) (13) will aid in determining the minimum standard at which an insectary can be built and operated. Infection studies involving an agent such as Sindbis virus, which can be used in a BSL2 facility, indicate the insectary must be built to meet BSL2 standards at a minimum. Infection experiments with an agent such as chikungunya virus necessitate at least a BSL3 insectary (see "Biosafety, Regulations, and Insectary Design," below).

Once the type of arthropod(s), the research to be conducted, and the level of biosafety required are determined, it is important to examine the planned scope of use. Some questions that should be examined could include: How many species will be used?; How many of each species will be needed at any one time?; and Will there be several different arthropods in the same facility? The scope of the planned research will dictate the overall size and design of the insectary. If there will be, for example, mosquitoes and ticks, the design of the individual rooms will need to be modified for the different containment issues specific to the various species.

General Insectary Design Criteria

In many ways, an insectary may be regarded as a laboratory with a very specialized function. In addition to meeting well-established design criteria that are available from general sources such as the BMBL and Richmond (14), the insectary must meet the additional criteria needed for the safe and secure rearing and maintenance of uninfected and perhaps infected arthropods. To accomplish these latter specialized functions, arthropod containment laboratories should have a vestibule between the main entrance door and the rearing/laboratory space. At ACL2, this vestibule may be separated from the main room by, for example, cloth or screened curtains, although self-closing solid doors would be preferable. For ACL3, two self-closing doors with appropriate locks are mandated (see BMBL). Standard operating procedures (SOPs) used in the insectary are designed to prevent escape; however, the SOPs and facility design should also enable rapid detection of an escaped arthropod so that it can be safely

TABLE 1.

Summary of arthropod containment levels[a]

Determinant	Arthropod containment level 1		2	3	4
Arthropod distribution, escaped arthropod fate	Exotic, inviable, or transient	Indigenous	Exotic with establishment, indigenous, and transgenic		
Infectious status	Uninfected or infected with nonpathogen		Up to BSL2	Up to BSL3	BSL4
Active VBD cycling	No	Irrelevant	Irrelevant	Irrelevant	Irrelevant
Practices	ACL1 standard arthropod-handling practices		ACL1 plus more rigorous disposal, signage, and limited access	ACL2 with more highly restricted access, training, and record keeping	ACL3 with high access restriction, extensive training, and full isolation
Primary barriers	Species-appropriate containers		Species-appropriate containers	Escape-proof arthropod containers, glove boxes, BSC	Escape-proof arthropod containers handled in BSC or suit laboratory
Secondary barriers	NA		Separated from laboratories, double doors, sealed electrical/plumbing openings; breeding containers and harborages minimized	BSL3	BSL4

[a]BSL, biosafety level; VBD, vector-borne disease; ACL, arthropod containment level; NA, not applicable; BSC, biosafety cabinet.

captured or killed. Depending on the type of arthropods being used, surfaces in the insectary, including floors, walls, benches, and furniture, should be of a color that provides sufficient contrast to allow an escaped arthropod to be easily seen. Typically, white or light colors are used. Equipment in the facility and storage of items on open shelving should be minimized so that an escaped arthropod cannot hide. If possible, ceilings should be low so an insect such as an adult mosquito cannot fly out of reach, and ceilings should not have spaces, for example, between panels, that allow a mosquito to become inaccessible. Ceiling height may be regulated by building codes, and one must also take into account height of equipment such as biosafety cabinets (BSCs). Vents and other conduits to the outside, including light fixtures, should be modified as necessary to prevent escape, perhaps by use of insect-proof screening. Similarly, escape into drains must be prevented either by design features or by protocols.

Once the general design of the insectary is determined, a close examination of the needs of the specific species should be made to ensure the design will meet the needs of the researcher. The successful maintenance of arthropods in captivity can be highly species-specific. Procedures appropriate for ticks, lice, triatomid bugs, fleas, mosquitoes, biting midges, black flies, sand flies, and tsetse flies are described in detail in different chapters of Section VII: Special Methods As Applied to Vectors of Disease Agents in *Biology of Diseases Vectors* (15). Generally, arthropod containment laboratories for rearing uninfected arthropods may use two basic approaches relating to the biology of arthropods that often demand relatively warm and humid conditions. One approach is to control the environment of the whole room to conditions that are conducive to arthropod development. Although this has been successfully employed for large-scale rearing of mosquitoes, the maintenance of these conditions can be problematic. First, entrance and egress of personnel can cause fluctuations when cold air rushes into the room, especially if the room has negative airflow. Second, the humidity can be difficult to maintain, especially if high air-exchange rates are needed, as in BSL3 facilities, and also high humidity can be associated with mold growth and damage of electrical equipment. An alternative approach to provide warm humid conditions is the use of environmental chambers. Depending on space, these can be used to rear relatively large numbers of arthropods and provide the opportunity to use different settings for optimal rearing of different species in the same room. In deciding which design and how many chambers will be placed in a room, electrical needs should be determined as well as air flow, because like refrigerators, the chambers can produce quite a lot of heat during operation. Use of emergency power and appropriate monitors is recommended so that local power failures do not result in temperature fluctuations that could lead to a loss of valuable arthropod colonies. Some environmental chambers also provide the ability to control light carefully, which for some arthropods is critical for captive breeding. With regard to rearing ticks, maintenance between active feeding can typically be accomplished in desiccation chambers held in environmental chambers (15). As such, the maintenance of tick colonies may not demand as much space as needed for other arthropods.

Resources for Insectary Design

When designing an insectary, be it a new facility or remodeling of an existing room, there are a number of invaluable sources of information that can provide guidance from concept to operation. The Arthropod Containment Guidelines (17) are freely downloadable (https://www.cdc.gov/biosafety/publications/bmbl5/bmbl5_appendixe.pdf). This publication provides a comprehensive review of risk assessment procedures to determine the ACL that is recommended for different situations, with details of pertinent procedures, special practices, safety equipment (primary barriers), and facilities (secondary barriers) for each of the ACLs (ACL1, ACL2, ACL3, and ACL4). Additionally, information on transportation for arthropods is discussed to guide the investigator in the procurement and distribution of arthropods between different facilities.

The design of various arthropod containment laboratories can be fairly simple to complex, depending on the needs of the user. Toward the end of this chapter, we have added a discussion of how to set up a basic insectary to house uninfected mosquitoes. Basic insectary design features were described in 1980 (18) and remain applicable to this day. Recommendations by the Subcommittee on Arbovirus Laboratory Safety (SALS) state that "Normal arthropods for vector competence studies are reared within an insectary that is physically separated from rooms holding infected laboratory animals by at least a solid wall and four screened or solid doors opening inwards and closing automatically. Space between doors must be sufficient for a person entering so that each door is closed before another is opened. If effectiveness is documented with the arthropod being maintained, an air curtain or cloth can be substituted for one door." Duthu et al. (19) discuss how to select the location for an insectary, access, construction materials for walls, ceilings, floors, doors and door frames, heating, ventilation and air conditioning (HVAC), mechanical and electrical systems, lighting, and plumbing. Integration of these different systems is obviously critical. In addition to these basic design features, Duthu et al. discuss considerations for choosing options for equipping the room, for example, room humidification, furniture, and laboratory equipment, and include images of an operational insectary. Higgs et al. (20) provides an overview of arthropod containment, including images of the arthropod containment guidelines. Design of BSL3 arthropod containment laboratories is described by Crane and Mottet (21). Figures 1 and 2 illustrate the basic features of an ACL2 and an ACL3 insectary, respectively.

Biosafety, Regulations, and Insectary Design

At the earliest stages of planning work with arthropod vectors, it is recommended that researchers engage in discussion with institution regulatory personnel. The Arthropod Containment Guidelines define minimum standards that are recommended but not necessarily mandated. However, in some institutions, internal review committees, for example Institutional Biosafety Committees (IBC) and Institutional Animal Care and Use Committees (IACUC), may take an approach that the presence of a pathogen in an arthropod automatically elevates the perceived risk to a higher level. It is not unknown for an IBC to require researchers to use BSL3 standards (or ACL3) for an agent that would typically be used in a BSL2 facility once the agent is in an arthropod. We have never heard of an IBC requirement that elevates the biosafety level required to conduct research from BSL3 to BSL4 once the agent is in an arthropod, but there are several instances where IBCs have required pathogens that would normally be researched at BSL2 to be elevated to BSL3 in the arthropod. One cannot deny that a virus in a sealed flask of cells does not require identical handling procedures to the same virus in a mosquito that can fly; however, whether these differences automatically justify raising the BSL is debatable. With appropriate training, and SOP development, it should be acceptable to work with agents that would normally be manipulated at BSL2 at BSL2/ACL2 after these agents are in the arthropod. Although this should be established during discussions between the investigator and the appropriate review committees, it is important for the investigator to be aware of the possibility they may suddenly be in a situation where they cannot perform the planned experiments in arthropods at the same BSL as agents that are outside of an arthropod.

The determination of insectary design features when working with uninfected arthropods is largely established on the basis of the species of arthropod, taking into consideration the geographic location of the insectary. The use of uninfected arthropods cannot be regarded as low risk and can require use of a containment level above the most basic ACL1. As shown in Table 1, an exotic arthropod, meaning a species that is not native to an area but which might become established in an area if it were to escape from the insectary, should be contained at least to ACL2 standards.

Arthropods infected with pathogens require further planning to ensure the design features of the laboratory will meet required containment conditions. The initial determination of containment conditions and SOPs

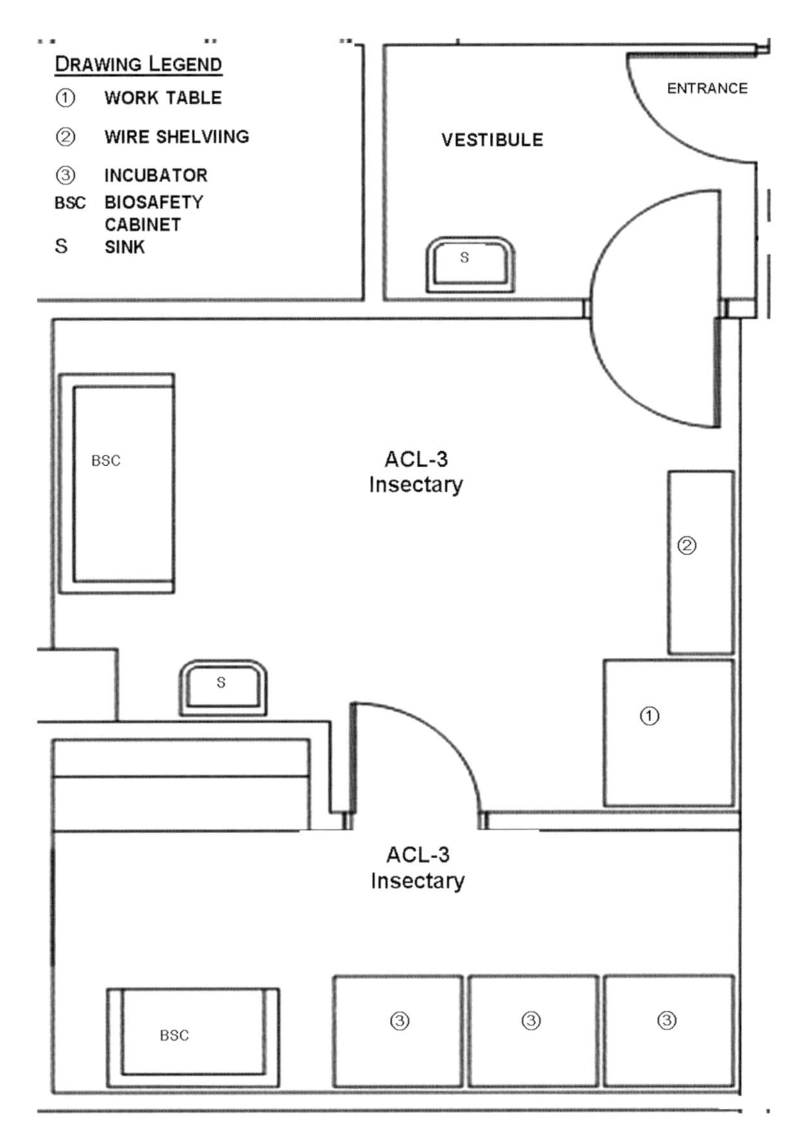

Figure 1: Example of ACL2 insectary layout. Modified with permission from Duthu et al. (19).

can be made on the basis of the BSL of the pathogen. Design of the insectary must then take into account and fulfill all of the requirements that apply to working with these agents. These requirements can vary between countries and may further be complicated by designation of pathogens within different BSL categories. For example, in the United States, the designation of pathogens as Select Agents is associated with regulations that cover, for example, storage, record keeping, and personnel qualifications. The importance of consultation with biosafety staff and research administrators cannot be overstated, and this consultation must be done at the earliest stages of research planning.

Determining if there is a need for a BSC for the planned research in the insectary should be considered early in the planning stages. In a typical research laboratory, even at BSL2, it is increasingly common, indeed even expected, that all procedures with infectious agents are performed in BSCs. These cabinets perform an essential safety function as secondary containment of the pathogen, protect the worker from aerosols, and also help to keep a relatively sterile environment to minimize contamination of, for example, cell cultures. For research with arthropods, however, BSCs can introduce risks that must be evaluated in the context of not only the pathogen but also of the vector. Anyone who has worked with mosquitoes in a BSC has probably experienced concern when these light bodies are placed close to a relatively strong air current. When in a primary container, such as a carton with a tightly fitted mesh top, this is not a problem. However, if the cold-anesthetized arthropods are counted onto a refrigerated "chill table," as is commonly used when manipulating them, the airflow of the BSC can be a major risk for escape simply by blowing the arthropods off the table. For this

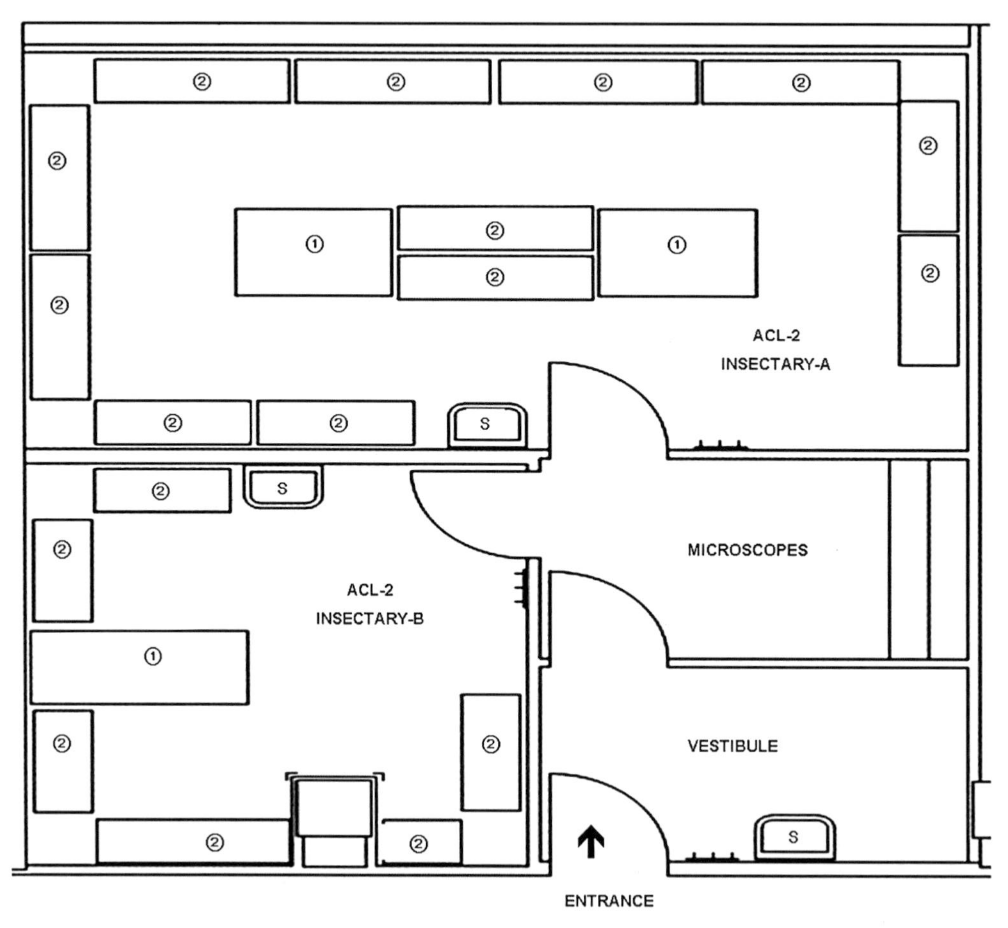

Figure 2: Example of ACL3 insectary layout. Modified with permission from Duthu et al. (19).

reason, it may be far safer and less of a risk for escape not to conduct the manipulation in a BSC, but instead use a purpose-designed glove box or, if permissible and approved, to work on the bench. With appropriately trained personnel and knowledge of the arthropods involved, SOPs can be developed that can effectively reduce risk of escape to a minimum and still protect the worker from infection. The term "minimum" is used here to acknowledge that zero risk may not be achievable, but risk can be reduced to such a highly unlikely level that it is almost zero. An important aspect of attaining this level of confidence is the training of workers to perform each procedure with uninfected arthropods to develop the necessary skills before working with pathogens. Manipulations, for example, intrathoracic inoculation, membrane feeding, feeding on vertebrates, and dissections of arthropods, must all be practiced using noninfectious material. A training plan for insectary staff should be developed and implemented, and there should be measurable criteria and standards that must be met prior to approval to work with exotic species and infectious material.

It must be emphasized that for some agents it is mandated that they must be handled in a BSC. Researchers should refer to relevant documents, for example, the BMBL, and consult with regulatory authorities if necessary. There are numerous recent examples highlighting how failure to comply with regulations, even if perhaps due to ignorance or misplaced confidence in the effectiveness of procedures such as inventory tracking or inactivation, can have severe penalties for individuals and institutions.

Other Considerations for Insectary Design

It is important to know the scope of work that is planned when developing the insectary design. For information about working with infected arthropods, a good resource is the fourth publication in the series *Anthology of Biosafety*, which is dedicated to arthropod-borne diseases and includes several chapters that discuss details needed for working with infected arthropods, including infected mosquitoes (22) and infected midges (23, 24).

Genetically Engineered Arthropods

Another area of research that may require further consideration when designing the insectary and determining the biosafety level needed is research on genetically engineered arthropods. In 2003, when the Guidelines for Genetically Engineered Arthropods (Guidelines) were published, the genetic manipulation of arthropods was achievable but not readily available. Nevertheless, given the interest to, for example, engineer mosquitoes to reduce vector competence, a discussion on containment of engineered vectors was included in these Guidelines. It was recommended that transgenic mosquitoes should require at least ACL2 practices, and, depending on circumstance, perhaps higher-level containment. The Guidelines provide a list of questions to help with the decision as to which category should be used.

Excluded from the Guidelines were discussions on fieldwork and use of human subjects for research on arthropod vectors. This omission has recently been addressed by Achee et al. (25). In a laboratory/insectary setting, the relevance of this document relates to the potential use of people to establish or sustain mosquito colonies by allowing mosquitoes to feed upon them, or by infecting vectors by allowing them to feed upon infected people. Both of these protocols may be justifiable and be permissible; however, the protocols will require approval by the appropriate authorities such as academic IBCs. In the United States, the protection of human research subjects is primarily regulated under the Common Rule (26).

Of relevance perhaps to genetic manipulation of vectors, full genomic sequences have been determined for the major mosquito species of medical importance (*Anopheles gambiae, Aedes aegypti, Aedes albopictus, Culex pipiens*). These mosquito species are responsible for transmission of several pathogens causing significant diseases of humans (malaria, chikungunya, dengue, yellow fever, Zika, Japanese encephalitis). To date, this knowledge of genomic sequences has not been translated into disruption of vector competence. However, technological advances mean that mosquitoes can now be routinely manipulated, with genetically modified mosquitoes being released in several countries.

The technology to engineer and develop so-called release of dominant lethal (RIDL) mosquitoes for *Ae. aegypti* population suppression has been described in numerous publications (27) and has been carefully evaluated and approved for use as part of campaigns to reduce the incidence of mosquito-borne viruses, specifically chikungunya, dengue, and Zika. In 2013, Higgs reviewed alternative approaches to control dengue and chikungunya and compared genetically engineered RIDL mosquitoes with nonengineered *Wolbachia*-infected mosquitoes (28).

The relatively recent development of a gene-drive capability based around CRISPR/Cas9 technology (29–31) has moved at a remarkable pace from discovery to widely available commercialization of inexpensive kits that have great potential for use in short-generation organisms such as arthropods. Indeed, CRISPR-engineered mosquitoes have already been produced (32, 33), and a report discussing gene-drive technology and its application to arthropods has been published recently (34). The report uses several case studies, including *Ae. aegypti* and *Ae. albopictus* to manage dengue, *Anopheles gambiae* to combat human malaria, and *Culex quinquefasciatus* to combat avian malaria in Hawaii. The chapter, "Phased Testing and Scientific Approaches to Reducing Potential Harms of Gene Drive," includes a discussion on containment, confinement, and mitigation strategies. Another chapter, "Governing Gene Drive Research and Applications," includes a discussion of biosecurity considerations. It is recommended that to reduce harm from the unintentional release of a genetically manipulated arthropod, particularly mosquitoes, experiments should be conducted at ACL2 at a minimum. ACL2 practices for genetically modified arthropods include conducting experiments in a separate room from other mosquitoes or in a BSC, having appropriate directional airflow in the room, using appropriate disposal techniques of the modified arthropods such as autoclaving or freezing, and using proper personal protective equipment (PPE). These issues are discussed in more detail in the Arthropod Containment Guidelines and Akbari et al. (17, 20, 35).

For engineered arthropods with a gene drive, IBCs are a key part of the approval and oversight process in an academic setting. In the United States, if the research is supported by National Institutes of Health (NIH) funding, these IBCs are accountable and must follow the *NIH Guidelines for Research Involving Recombinant or Synthetic Nucleic Acid Molecules* (36). For certain plant pests, research may be subject to regulations under the Biotechnology Regulatory Services of the Animal and Plant Health Inspection Agency (APHIS) of the U.S. Department of Agriculture (USDA). Also, the movement of insects may require transport permits through appropriate agencies. At present, gene-drive arthropods do not seem to be regulated by the Food and Drug Administration (FDA);

however, the FDA has been involved in the discussion of genetically engineered RIDL mosquitoes, presumably because a drug is used as part of the selection process. The National Academy concluded that in the case studies that were proposed, it was unclear as to which of several agencies would have jurisdiction within the United States. Depending on circumstances, the FDA, USDA, and Environmental Protection Agency (EPA) could potentially be involved, and in one case legislations related to the Endangered Species Act could apply.

Wolbachia-Infected Mosquitoes

Although not genetically engineered, mosquitoes infected with *Wolbachia* are now also being reared on a relatively large scale and released to reduce the incidence of certain diseases, not only because of the potential for population suppression through the phenomenon known as cytoplasmic incompatibility, but also because the presence of *Wolbachia* can reduce mosquito susceptibility to infection with some arboviruses and therefore can result in reduced transmission. The determination of the appropriate ACL and SOPs for experiments using modified arthropods should be taken into consideration to ensure that the insectary design meets the needs of the researcher.

Use of Live Vertebrates for Maintenance of Arthropods

Another consideration for insectary design that should be considered while still in the planning stages of the project is the use of blood meals. Blood meals are essential for reproduction of arthropod vectors and can add a unique requirement in an insectary that is not typical of a research laboratory. Methods to infect, maintain, and analyze arthropods are described in many publications (15, 37). Approaches range from relatively natural methods, such as feeding on infected vertebrates and feeding using artificial blood meals, to intrathoracic inoculation and delivery of pathogens via enema. Depending on the numbers of arthropods being maintained, blood may be presented to arthropods in purpose-built equipment, for example, the Hemotek feeding apparatus. However, large-scale rearing may be more efficient if live animals are used, and for some types of arthropods, for example, ticks, the use of live vertebrates is essential with no viable alternative. If it is decided that live vertebrates will be needed for maintenance of the arthropods, consideration should be given to building appropriate animal housing that is easily accessible from the insectary. Consultation with IACUC members and other appropriately qualified personnel, for example, veterinarians, is essential, so that housing and SOPs can be developed and approvals obtained in advance of procuring the arthropods. Insectary design and SOPs need to take into account how blood will be presented to the arthropods, especially if live vertebrates must be used and housed in or close to the insectary.

POINTS TO CONSIDER BEFORE BEGINNING

Planning and Designing an Insectary

To give a clearer idea of the biocontainment issues that should be addressed to develop an operational insectary, we discuss specific design parameters that should be considered when building an insectary to house uninfected mosquitoes. For this example, we discuss how to remodel a current laboratory to create a working insectary. In this scenario, we will assume that the building location and design are suitable for the level of containment needed for uninfected arthropods, either BSL1 or BSL2. When choosing a room to remodel into an insectary, items such as appropriate plumbing (sink, drain for ice machine, no open water sources), electricity supply (is there sufficient power to supply rearing chambers, centrifuge, refrigerator, freezer, autoclave), location and distance to outside doors, and overall space of the room should be considered. Other considerations are the ability to contain the arthropods in the room, i.e., if there were to be an escape, are the vents, cracks, and other holes in the room able to be covered or sealed so that the mosquito cannot leave the room? Is there room to build an anteroom for entry and exit from the insectary? Can the ceiling be dropped down so that if there were to be an escape, the insectary workers can reach the ceiling if needed?

The next questions to be addressed prior to remodeling are: What are the types of arthropods that will be used and for what purpose? In this example, we will assume that this insectary is being built for colony mosquitoes to be reared for experiments conducted in a higher-containment insectary located elsewhere. The number of mosquitoes to be reared at any one time will give an idea of the type of equipment and the overall room design that will be needed. For a relatively small-scale insectary, we have found that using growth chambers to house the mosquitoes works well. However, if there will be a lot of mosquitoes that will be needed at any one time, the room itself may have to be modified to allow for the proper temperature, lighting, and humidity to enable the mosquitoes to be reared on shelves. Modification of a room to enable higher-than-normal temperatures and humidity causes additional complexities. For example, the use of a humidifier may give the mosquitoes a good place to lay eggs if they escape, and the paint and equipment need to hold up to the hotter and more humid environment. Also, will only having primary containment during rearing cause arthropods to escape more frequently than if they were to be housed in a growth chamber? For this example, we will assume the use of rearing chambers that provide proper temperature, humidity, and light cycles for rearing mosquitoes. Another advantage of the growth chamber is the added layer of containment it provides as compared to having rearing pans or cages placed directly on a shelf in the room.

Other considerations when designing the insectary are the furniture and work surfaces, which should be minimal and a light color so that any escaped arthropod would have minimal places to hide and would be easy to spot against the light-colored surfaces. Storage is also an important consideration; only essential items are left out to reduce clutter and enable any arthropods to be easily spotted. The primary function of the room design is to help keep the arthropods contained.

Once issues specific to the physical design of the room and contents have been addressed, the next step to help ensure that the arthropods are contained is the development of SOPs, approvals from regulatory bodies, training programs for personnel, and development of matrixes to verify understanding of the SOPs and technical competence.

Infection and Manipulation of Infected Arthropods

Although the arthropod species studied can vary significantly among different laboratories, one of the important objectives of arthropod containment facilities is to infect the arthropods with human, animal, or zoonotic pathogens to determine vector competence of particular species of arthropods and to characterize the interactions among vectors, pathogens, and vertebrate hosts. Therefore, the design of the facilities, equipment, and procedures can be highly project specific but share several common principles. Regardless of the pathogens used, the facility and equipment must be designed to prevent the accidental release of arthropods. Scientists are required to treat each arthropod as if infected and capable of transmitting the pathogens, regardless of the status of infection. Use of multiple layers of secured barriers is a common approach to provide redundancy in preventing the escape of arthropods. One must keep in mind that the addition of containment on the arthropod species should not change or replace the existing good microbiological practice, PPE, and disinfecting/inactivating procedures designated for specific pathogens. However, due to the variation of endemic status and the availability of vaccines and prophylactic treatments in different regions, the biosafety level of pathogens may differ from one facility to the other. For example, research on virulent strains of Japanese encephalitis virus (JEV) in the United States is designated at BSL3. Manipulation of JEV-infected arthropods requires ACL3 laboratories. Similar studies will be designated at BSL2 and ACL2 in several countries in Asia due to the implementation of vaccination programs in the area.

Infection of arthropods requires the use of pathogens and arthropod species at the designated biosafety and arthropod-containment levels. The presence of pathogens requiring containment at BSL2 or above often requires the use of a BSC. It is common and preferred that the design of arthropod containment laboratories includes BSCs. If a BSC is absent, preparation of blood meals or inoculum containing infectious viruses should be performed in appropriate containment laboratories and transported in secured containers to arthropod containment laboratories. In addition to the protection offered during manipulation of pathogens, it is also acceptable for scientists to administer blood meals to arthropod species housed in cartons with tightly secured mesh on the top. However, manipulation of arthropods using cold anesthetization and dissection in a BSC is generally discouraged due to the presence of strong airflow that might increase the chance of losing arthropods during manipulation. An alternative approach to create a barrier to prevent the escape of arthropods is the use of glove boxes constructed of plastic or specimen-handling cages constructed with mesh or similar materials.

Infection can be done with different techniques at multiple stages of life cycle. However, one must consider if the approach taken is scientifically sound. For instance, to determine vector competence of arthropod species for specific pathogens, the use of intrathoracic inoculation is considered artificial and inappropriate because of the destruction of anatomic barriers and increase of infection and dissemination in arthropod species tested. The most commonly used method is to ingest viremic meals orally either by infected vertebrate hosts or blood products mixed with pathogens, which resembles the process of infection in nature.

Because of the limitation of space and the general principle of designing laboratories with the minimal amount of equipment and supplies, housing of vertebrate hosts before and after the challenge of pathogens generally does not take place in arthropod containment laboratories. With careful calculation, vertebrate hosts should be anesthetized with drugs at appropriate dosages prior to the exposure to arthropod species. Arthropods can be allowed to engorge through mesh in secure containers. Expected variation in titers of pathogens inoculated often exists in challenged animals. Although the use of infected vertebrate hosts might provide the necessary odorant and thermal cues for engorgement, it is technically challenging to control the quantities of pathogens administered through oral ingestion due to the variation in the titers of pathogens in the blood. An alternative that offers better control of the quantity ingested by individual arthropods is to deliver blood meals through other artificially heated apparatuses. Although some arthropods, such as several species of *Culex* mosquitoes, can be infected by feeding through cotton pledgets, the presence of chemical and physical cues like odors and heat generally increases the efficiency of blood feeding. Normally, the odorant cues for engorgement can be provided by the use of the skin of appropriate animal species. Temperature of blood meals can also be maintained by a heating device or heated liquid, such as water.

Glass feeders with water jackets filled with warm water were developed to infect hematophagous arthropods orally. Infectious blood meals containing specific pathogens are kept in the reservoir structure surrounded by the water jacket structure. A layer of animal skin or other membranous structure is secured to allow the feeding of arthropods. Because of the use of warm water to maintain the blood meals, a circulatory water bath that can be set at elevated temperature is normally required. As several arthropod species with high human and veterinary public health significance naturally lay eggs in water bodies, appropriate barriers should be placed to restrict the access to bodies of water in the worst-case scenario of unintentional release of engorged arthropods. An additional concern that should be addressed is the use of glassware, especially in containment facilities. Procedures for handling broken glassware and the required tools and protective equipment should be provided in the laboratories.

An improved format of administering blood meals is to provide blood meals through a heating device connected to feeders made of metal reservoirs, which replace glass feeders. A similar rationale in attracting hematophagous species to engorge infectious blood meals is achieved by mounting animal skins or membranous structures over the reservoir in the metal feeder. Significant improvements from glass feeders include the absence of glassware and a water body in arthropod containment laboratories.

In the event that high infection rates of arthropods must be obtained for experimental purposes, inoculation can be used for infection. Pathogens can generally be delivered through inoculation with glass capillary needles, as in the intrathoracic inoculation of mosquitoes. Procedures and protective equipment to mitigate the risks associated with aerosolization of pathogens during inoculation must be developed for the safety of laboratory personnel.

It is inevitable that arthropods must be immobilized by cold anesthetization prior to manipulation such as inoculation. A cold anesthetization procedure is also needed to isolate the engorged arthropods after feeding and to process arthropods collected in the experiments. Chill tables and ice are commonly used to create refrigeration temperatures. Glass petri dishes or other similar glassware generally provide the smooth surface and superior conductivity to maintain a low temperature in comparison to plastic ware.

After the completion of infection, the next critical step is to maintain the arthropod species to allow the replication and dissemination of pathogens. Housing of the infected arthropods shares the general requirement of rearing adult arthropods and requires further secondary barriers. In arthropod containment laboratories, a common challenge is the desiccation created by the high frequency of air exchange. Therefore, maintenance of temperature, humidity, and photo regimen must be achieved by environmental chambers or incubators. A crucial aspect in housing infected arthropods is to create several barriers to prevent the escape of arthropods created by any compromised structures in primary containers. Although well-constructed cartons designed to house infected arthropods should be sufficient to prevent the escape, it is generally advisable to place the cartons in transparent secondary containers to prevent any unexpected leakage of cartons without disrupting the supply of air to the arthropod species. Such containers are expected to be resistant to breakage caused by mechanical force and sufficiently transparent to see any potentially escaped arthropods from the primary container.

Sampling of arthropods exposed to pathogens from cartons should be performed by mechanical aspiration. Several models of mechanical aspirators adapted from flashlights or other electric devices are available from several suppliers. Similar to the concept to create a barrier to prevent the escape of arthropods, mechanical aspiration of arthropods and subsequent manipulations performed in a glove box or specimen-handling cage can substantially reduce the likelihood of escape. After being aspirated in a secure container, arthropods can be cold anesthetized by placing them temporarily on ice or at refrigerated temperature. Immobilized arthropods can subsequently be processed on ice or on a chill table.

TRAINING

Personnel Training and Development of Standard Operating Procedures

Following the analysis of the type of research to be conducted, the BSL required, and analysis of the institutional regulatory bodies that should be involved in the early planning, the next step is the development of SOPs and training of staff. Safety and security should take into consideration the safety of the laboratory staff, the facility, and the environment. The laboratory staff should be well trained to follow standardized protocols that are specific to the facility, the arthropods being used, and the pathogen. The development of SOPs is the first step to insure that staff follow procedures that will ensure that insects will not escape from containment within the insectary and will therefore not escape the insectary itself.

The top priority when designing, building, and conducting research in an insectary is to ensure that arthropods do not escape. This is not only important for arthropods that are infected with pathogens, but equally important with regard to uninfected arthropods. There should be zero tolerance for escape, even of species that are locally sourced and indigenous to the area where the insectary is located. Escape of these arthropods could still alter the potential for pathogen transmission or increase nuisance biting in the area. A zero tolerance policy also indicates a professional attitude of insectary staff and the home institute. An escape of a foreign arthropod, infected with a human or animal pathogen, is totally unacceptable because it represents obvious risks to the general population and the environment as the species could become established in some areas. Insectary protocols must be based on the premise that all arthropods are treated as though they are infected, dangerous, and trying to escape. It should be assumed that the arthropods have the potential capacity to establish in the environment and could initiate a pathogen transmission cycle. With this approach, there should be little risk of an accidental release of an arthropod due to a misguided attitude that "some species would not do any harm if they did escape." The "no escape is acceptable under any circumstances" philosophy of the staff will address one of the main risks of arthropod escape, which is human error.

Another potential pitfall related to arthropod containment is that efforts should primarily be targeted toward the adult stage. Just because the adult stage is seemingly more versatile with respect to escape potential, one must consider all stages when training staff, developing protocols, and designing the insectary. Adult mosquitoes can obviously move quickly, hide in corners, and fly from the insectary if a door is opened. However, eggs and larvae also have high escape potential. Embryonated eggs of Culicine mosquitoes are highly tolerant to desiccation, can survive for months, and are so small and dark that, if appropriate SOPs are not developed and strictly adhered to, they may escape by being discarded in trash or washed into the drains. When arthropods are being used for experiments, it is important to keep a record of numbers throughout the duration of the experiment, from the moment that the experiment begins until its termination; however, in the rearing rooms, it is rarely if ever possible to know accurately the numbers of arthropods in different stages that are present. It is simply impractical to count how many eggs are laid, how many larvae are in pans, and even how many adults are present in a colony cage. Unlike the experimental situation, where one must keep a running inventory of arthropods, in the rearing room where one does not know how many eggs for example are present, one cannot therefore detect an escape by checking the numbers present against the starting numbers.

CONCLUSION

There is a considerable amount of information available on the design and operation of arthropod containment laboratories, the rearing of most types of arthropod vectors, and the procedures and protocols that will enable safe, secure, and responsible research with most arthropod vectors and most arthropod-borne pathogens. On the basis of this information, situation-specific SOPs and staff education and training can be developed that will advance our understanding of the complex interactions between the arthropod vectors and the pathogens they transmit. Such knowledge and the commitment to develop and sustain a workforce with the interdisciplinary skills to work with arthropod vectors is a key need for combatting the constant threat of vector-borne diseases.

References

1. **Campbell GL, Marfin AA, Lanciotti RS, Gubler DJ.** 2002. West Nile virus. *Lancet Infect Dis* **2**:519–529.
2. **Granwehr BP, Lillibridge KM, Higgs S, Mason PW, Aronson JF, Campbell GA, Barrett ADT.** 2004. West Nile virus: where are we now? *Lancet Infect Dis* **4**:547–556.
3. **Nash D, Mostashari F, Fine A, Miller J, O'Leary D, Murray K, Huang A, Rosenberg A, Greenberg A, Sherman M, Wong S, Campbell GL, Roehrig JT, Gubler DJ, Shieh W-J, Zaki S, Smith P, Layton M, for the West Nile Outbreak Response Working G, 1999 West Nile Outbreak Response Working Group.** 2001. The outbreak of West Nile virus infection in the New York City area in 1999. *N Engl J Med* **344**:1807–1814.
4. **O'Leary DR, Marfin AA, Montgomery SP, Kipp AM, Lehman JA, Biggerstaff BJ, Elko VL, Collins PD, Jones JE, Campbell GL.** 2004. The epidemic of West Nile virus in the United States, 2002. *Vector Borne Zoonotic Dis* **4**:61–70.
5. **Spielman A, Pollack RJ, Kiszewski AE, Telford SR III.** 2001. Issues in public health entomology. *Vector Borne Zoonotic Dis* **1**:3–19.
6. **Duffy MR, Chen TH, Hancock WT, Powers AM, Kool JL, Lanciotti RS, Pretrick M, Marfel M, Holzbauer S, Dubray C, Guillaumot L, Griggs A, Bel M, Lambert AJ, Laven J, Kosoy O, Panella A, Biggerstaff BJ, Fischer M, Hayes EB.** 2009. Zika virus outbreak on Yap Island, Federated States of Micronesia. *N Engl J Med* **360**:2536–2543.
7. **Musso D, Nilles EJ, Cao-Lormeau VM.** 2014. Rapid spread of emerging Zika virus in the Pacific area. *Clin Microbiol Infect* **20**: O595–O596.
8. **Bogoch II, Brady OJ, Kraemer MU, German M, Creatore MI, Kulkarni MA, Brownstein JS, Mekaru SR, Hay SI, Groot E,**

Watts A, Khan K. 2016. Anticipating the international spread of Zika virus from Brazil. *Lancet* **387:**335–336.

9. **Hubálek Z, Halouzka J.** 1999. West Nile fever—a reemerging mosquito-borne viral disease in Europe. *Emerg Infect Dis* **5:** 643–650.
10. **Benedict MQ, Levine RS, Hawley WA, Lounibos LP.** 2007. Spread of the tiger: global risk of invasion by the mosquito *Aedes albopictus*. *Vector Borne Zoonotic Dis* **7:**76–85.
11. **Vanlandingham DL, Higgs S, Huang YS.** 2016. *Aedes albopictus* (Diptera: Culicidae) and mosquito-borne viruses in the United States. *J Med Entomol* **2016:**tjw025.
12. **Tsetsarkin KA, Vanlandingham DL, McGee CE, Higgs S.** 2007. A single mutation in chikungunya virus affects vector specificity and epidemic potential. *PLoS Pathog* **3:**e201.
13. **U.S. Department of Health and Human Services, Public Health Service, Centers for Disease Control and Prevention, National Institutes of Health.** 2009. *Biosafety in Microbiological and Biomedical Laboratories*, 5th ed. HHS Publication no. (CDC) 21-112. http://www.cdc.gov/biosafety/publications/bmbl5/BMBL.pdf.
14. **Richmond JY.** 1997. *Designing a Modern Microbiological/Biomedical Laboratory*. American Public Health Association, Washington, DC.
15. **Marquardt WC, Black WC, Freier JE, Hagedorn H, Moore C, Hemingway J, Higgs S, James A, Kondratieff B (ed).** 2004. *Biology of Disease Vectors*, 2nd ed. Elsevier Academic Press.
16. **Bouchard KR, Wikel SK.** 2004. Care, maintenance, and experimental infestation of ticks in the laboratory setting, p 705–711. *In* Marquardt WC, Kondratieff B, Moore CG, Freier J, Hagedorn HH, Black W III, James AA, Hemingway J, Higgs S (ed), *Biology of Disease Vectors*, 2nd ed. Elsevier Academic Press.
17. **Benedict MQ, Tabachnick WJ, Higgs S, American Committee of Medical Entomology, American Society of Tropical Medicine and Hygiene.** 2003. Arthropod containment guidelines. A project of the American Committee of Medical Entomology and American Society of Tropical Medicine and Hygiene. *Vector Borne Zoonotic Dis* **3:**61–98.
18. **The Subcommittee on Arbovirus Laboratory Safety of the American Committee on Arthropod-Borne Viruses.** 1980. Laboratory safety for arboviruses and certain other viruses of vertebrates. *Am J Trop Med Hyg* **29:**1359–1381.
19. **Duthu DB, Higgs S, Beets RL Jr, McGlade TJ.** 2001. Design issues for insectaries, p 227–244. *In* Richmond JY (ed), *Anthology of Biosafety IV: Issues in Public Health*. American Biological Safety Association, Mundelein, IL.
20. **Higgs S, Benedict MQ, Tabachnick WJ.** 2003. Arthropod containment guidelines, p 73–84. *In* Richmond JY (ed), *Anthology of Biosafety: VI Arthropod Borne Diseases*. American Biological Safety Association, Mundelein, IL.
21. **Crane J, Mottet M.** 2004. BSL-3 Insectary Facilities, p 29–34. *In* Richmond JY (ed), *Anthology of Biosafety VII: Biosafety Level 3*. American Biological Safety Association, Mundelein, IL.
22. **Olson K, Larson RE, Ellis RP.** 2003. Biosafety issues and solutions for working with infected mosquitoes, p. 25–38. *In* Richmond JY (ed), *Anthology of Biosafety VI: Arthropod Borne Diseases*. American Biological Safety Association. Mundelein, IL.
23. **Hunt GJ, Schmidtmann ET.** 2003. Safe and secure handling of virus-exposed biting midges within a BSL-3-AG containment facility, p 85–98. *In* Richmond JY (ed), *Anthology of Biosafety VI: Arthropod Borne Diseases*. American Biological Safety Association, Mundelein, IL.
24. **Drolet B, Campbell C, Mecham J.** 2003. Protect yourself and your sample: processing arbovirus-infected biting midges for viral detection assays and differential expression studies, p. 53–62. *In* Richmond JY (ed), *Anthology of Biosafety VI: Arthropod Borne Diseases*. American Biological Safety Association, Mundelein, IL.
25. **Achee NL, Youngblood L, Bangs MJ, Lavery JV, James S.** 2015. Considerations for the use of human participants in vector biology research: a tool for investigators and regulators. *Vector Borne Zoonotic Dis* **15:**89–102.
26. **Department of Health and Human Services.** 2009. CFR 45 Public welfare, Part 46: Protection of human subjects. Washington, DC. http://www.hhs.gov/ohrp/regulations-and-policy/regulations/45-cfr-46.
27. **Alphey L, Benedict M, Bellini R, Clark GG, Dame DA, Service MW, Dobson SL.** 2010. Sterile-insect methods for control of mosquito-borne diseases: an analysis. *Vector Borne Zoonotic Dis* **10:**295–311.
28. **Higgs S.** 2013. Alternative approaches to control dengue and chikungunya: transgenic mosquitoes. *Public Health* **24:**35–42.
29. **Barrangou R, Fremaux C, Deveau H, Richards M, Boyaval P, Moineau S, Romero DA, Horvath P.** 2007. CRISPR provides acquired resistance against viruses in prokaryotes. *Science* **315:** 1709–1712.
30. **Hale CR, Zhao P, Olson S, Duff MO, Graveley BR, Wells L, Terns RM, Terns MP.** 2009. RNA-guided RNA cleavage by a CRISPR RNA-Cas protein complex. *Cell* **139:**945–956.
31. **Sternberg SH, Doudna JA.** 2015. Expanding the biologist's toolkit with CRISPR-Cas9. *Mol Cell* **58:**568–574.
32. **Gantz VM, Jasinskiene N, Tatarenkova O, Fazekas A, Macias VM, Bier E, James AA.** 2015. Highly efficient Cas9-mediated gene drive for population modification of the malaria vector mosquito *Anopheles stephensi*. *Proc Natl Acad Sci* 112:E6736–E6743.
33. **Hammond A, Galizi R, Kyrou K, Simoni A, Siniscalchi C, Katsanos D, Gribble M, Baker D, Marois E, Russell S, Burt A, Windbichler N, Crisanti A, Nolan T.** 2016. A CRISPR-Cas9 gene drive system targeting female reproduction in the malaria mosquito vector *Anopheles gambiae*. *Nat Biotechnol* **34:**78–83.
34. **Committee on Gene Drive Research in Non-Human Organisms: Recommendations for Responsible Conduct; Board on Life Sciences; Division on Earth and Life Studies; National Academies of Sciences, Engineering, and Medicine.** 2016. *Gene Drives on the Horizon: Advancing Science, Navigating Uncertainty, and Aligning Research with Public Values*. National Academies Press, Washington, DC.
35. **Akbari OS, Bellen HJ, Bier E, Bullock SL, Burt A, Church GM, Cook KR, Duchek P, Edwards OR, Esvelt KM, Gantz VM, Golic KG, Gratz SJ, Harrison MM, Hayes KR, James AA, Kaufman TC, Knoblich J, Malik HS, Matthews KA, O'Connor-Giles KM, Parks AL, Perrimon N, Port F, Russell S, Ueda R, Wildonger J.** 2015. Safeguarding gene drive experiments in the laboratory. *Science* **349:**927–929.
36. **Office of Science Policy, National Institutes of Health.** 2016. *NIH Guidelines for Research Involving Recombinant or Synthetic Nucleic Acid Molecules (NIH Guidelines)*. http://osp.od.nih.gov/office-biotechnology-activities/biosafety/nih-guidelines.
37. **Higgs S, Olson KE, Kamrud KI, Powers AM, Beaty BJ.** 1997. Viral expression systems and viral infections in insects, p 459–483. *In* Crampton JM, Beard CB, Louis C (ed), *The Molecular Biology of Disease Vectors: A Methods Manual*. Chapman and Hall, UK.

20

Aerosols in the Microbiology Laboratory

CLARE SHIEBER, SIMON PARKS, AND ALLAN BENNETT

The control of microbial aerosols is the major driver in the design of microbiological containment laboratories. The provision of a negative-pressure laboratory area with a high efficiency particulate air (HEPA) filtered exhausted ventilation system is intended to prevent the escape of infectious aerosols from the facility. The use of directional airflow within open-fronted safety cabinetry is designed to prevent the release of any aerosols from the working area of the cabinets. Class III safety cabinets and isolator systems provide physical barriers between the operator and activity while maintaining negative pressure and high airflows, with HEPA filtration to prevent the release of aerosols. As a last resort, respiratory protection is used to prevent the exposed worker from inhaling the infectious agent. Yet, the average microbiologist may have only a limited understanding of the processes that generate aerosols in the laboratory and may have little knowledge of how effective preventative equipment and processes are.

The history of laboratory-acquired infection has shown that transmission by the aerosol route occurs and can affect many workers within, and, on some occasions, outside, the laboratory. However, many of the well-characterized incidents occurred before the use of safety cabinets and negative-pressure laboratories, sealed centrifuge rotors, and pipetting aids became standard practice. The current role of aerosols in laboratory-acquired infection is therefore unclear. Certain procedures in the laboratory are deemed to be aerosol generating, but evidence may be insufficient and the knowledge of aerosol physics may be lacking. Has modern microbiology prevented aerosol transmission of infection in the laboratory?

The objective of this chapter is to give a brief introduction to aerobiology—the study of microbial aerosols, and then to discuss the potential role of aerosols in laboratory infection and how they can be contained in the laboratory. Finally, a risk-assessment framework is provided to allow the risk of microbial laboratory processes to be determined in order to assess whether containment methods used are adequate and to assess potential worker exposure in the case of an accident.

AN INTRODUCTION TO AEROBIOLOGY

Aerobiology is the study of how microorganisms are aerosolized, how they behave in the aerosol state, and how they are deposited into the human respiratory tract. Having a basic knowledge of this science allows the recognition of aerosol hazards in the laboratory, the identification of suitable control measures, and the assessment of the remaining aerosol exposure risk. These factors can be split into basic aerosol physics, which determines the behavior of all aerosol particles, and microbial factors, which are dependent on aspects of the aerosolized microorganism.

Aerosol Physics

An aerosol is a suspension of a liquid or solid in a gas. In this chapter an aerosol is defined as particles of less than 10 μm in diameter for reasons that will be explained below. Larger particles are defined as droplets or splashes. The behavior of aerosol particles can be fully characterized by the laws of physics, and their movements can be mapped and measured. In recent years the availability of high power computers has allowed measurements to be accurately carried out using computer fluid dynamics, but such modeling is highly dependent on a comprehensive understanding of the systems and real-world working practices.

Generation

In the microbiology laboratory the main sources of material for aerosolization are fungal spores, liquid cultures, and lyophilized cultures. With fungal spores, the risks of aerosolization are clear and are a part of the natural cycle of the agent; however, with bacteria or viruses this is not the case. In order to produce an aerosol from a liquid, energy must be applied to break the forces holding the fluid together in order to form small aerosol particles. When aerosols are being generated intentionally, this energy can be supplied using high-pressure airflows, vibrating plates, spinning discs, and other high-energy processes (1). In the laboratory, as we shall see later, aerosols can be produced by equipment that applies high energy to liquids (centrifuges, homogenizers), accidents (dropping of flasks), and manual processes (pipetting, plating). If no energy is applied to a liquid, then no aerosol is generated. In the case of lyophilized materials, a lower energy is required to create an aerosol due to the state of the material.

Evaporation

When an aerosol is generated from a liquid, it is exposed to the air in the laboratory. This causes an immediate drying process to occur. The rapidity of this process depends on the relative humidity of the laboratory, but the period of time is less than a second. Drying forms so-called "droplet nuclei" which are smaller than the original droplet and are predominantly made up of solids. In the microbiology laboratory, these particles consist of the microorganisms and the dried remains of the suspension fluid (broth, buffer, blood etc.). Unless the microorganisms are aerosolized from a pure water or solvent suspension, without other impurities, the particle size of the aerosol is not the same as the microorganism size. This is important to recognize in the case of viruses, where the particle size primarily depends on the solid content of the fluid, not the agent itself. It is also important to recognize that aerosol particles may contain more than one microorganism, and the bigger an aerosol particle, the more microorganisms it is likely to contain due to its increasing volume, i.e., a 2 μm-diameter particle will contain eight times more agent than a 1 μm-diameter particle.

Particle size

The particle size is the key feature of a generated aerosol. It governs how long it remains in the aerosol state, where it is deposited in the human respiratory tract, how many microorganisms are likely to be in a particle, and potentially whether the microorganisms are likely to survive the aerosolization process.

Deposition

Deposition of an aerosol from air is defined by Stokes' Law, which states

$$u = \frac{\rho dp^2 g}{18\,\mu}$$

where u is deposition velocity (cm/s), ρ is density of particle (g/cm^3), μ is viscosity of air (g/cm s^{-1}), g is gravity (cm/s^2) and dp is particle diameter (cm).

Since g and μ are constants, the deposition velocity is directly proportional to the particle density (density of the microorganism and dried suspending fluid) and, more importantly, the particle diameter squared. The relationship is shown in Fig. 1.

Therefore, a 2-μm particle deposits at a rate of 0.012 cm/s in still air, whereas a 20-μm particle deposits at 1.2 cm/s. As such, small particles present a higher risk of inhalation due to their longer residence time in the air. Because these small airborne particles have irregular shapes, the term aerodynamic particle diameter is used to describe their size. This concept expresses the particle as a perfect sphere of unit density, and this is what is measured by size-fractioning air samples and is used to define inhalation criteria, which will be discussed later within this section.

The Respiratory Tract

The part of the respiratory tract a particle is able to access is dependent on the particle size. The particle size range that can access different parts of the respiratory

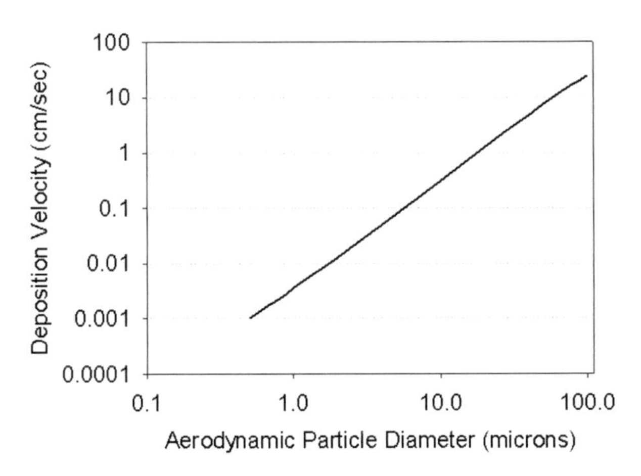

Figure 1: Relationship between deposition velocity and aerodynamic particle diameter. (Crown copyright.)

tract can be split into three fractions: respirable (red), thoracic (blue), and inhalable (green), as indicated in Fig. 2. The particle sizes that correspond with these areas are shown in Fig. 3. It can therefore be seen that only particles of 10 μm and below can access the lungs; however, particles of up to 30 μm can access the upper respiratory tract, and any particles can access the nose and mouth.

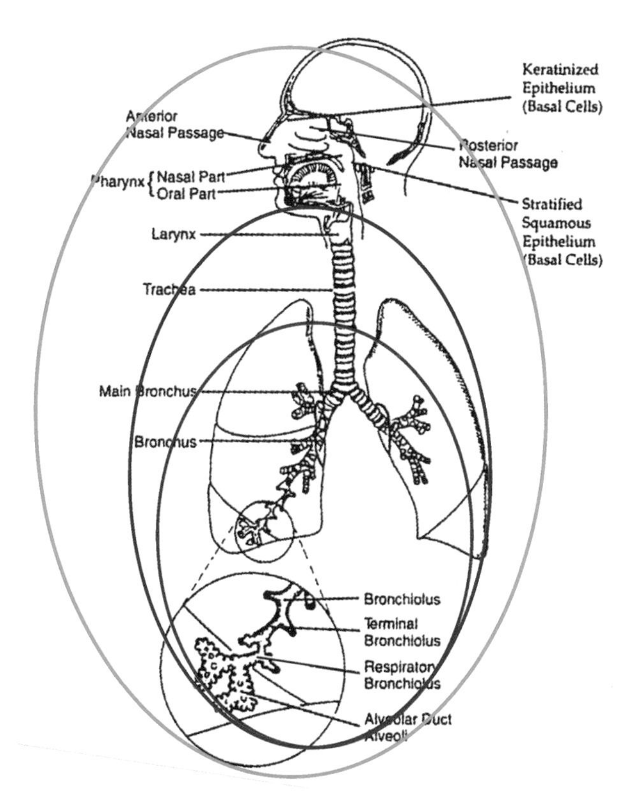

Figure 2: Demonstration of particle size and accessibility of different areas within the human respiratory tract. Adapted from the Human Respiratory Tract Model (34).

Survival of Microorganisms in the Air

When a microorganism is aerosolized, it is subjected to a number of stresses, including rapid drying, rapid cooling, formation of free radicals, and exposure to UV radiation; these stresses can permanently inactivate the microorganism, thus negating any risk of human exposure. Some microorganisms are extremely resistant to the stresses of aerosolization, and the diseases caused by these microorganisms are transmitted by the aerosol route in nature (tuberculosis, foot-and-mouth disease, measles). However, other microorganisms may be less resistant to the stresses of aerosolization. Whereas information is often unavailable about microbial aerosol stability, there are reviews that can be accessed which consider the factors that may affect survival and also the stability of several microorganisms (2–4).

Air Sampling

There are numerous devices that can be used to monitor the concentration of microbial aerosols (5). They operate by capturing airborne microorganisms on agar plates, on filters, or into a liquid. Air sampling can be used to assess the potential of aerosol generation during a process by using a microbial tracer or to assess the effectiveness of a control measure. However, while biological air samplers may, in the first instance, seem to be simple devices, their effectiveness can be highly variable, with many commercial hand-held samplers showing poor efficacy below 2 μm, making some unsuitable for evaluating laboratory processes. Furthermore, impaction samplers only record the number of bacteria-laden particles, but, as is discussed above, particles may contain more than one viable bacterium. Thus, impaction samplers can underestimate the total bacterial load.

Infectious Dose

Infection does not occur every time a microorganism is inhaled. Some diseases have a high infectious dose by the

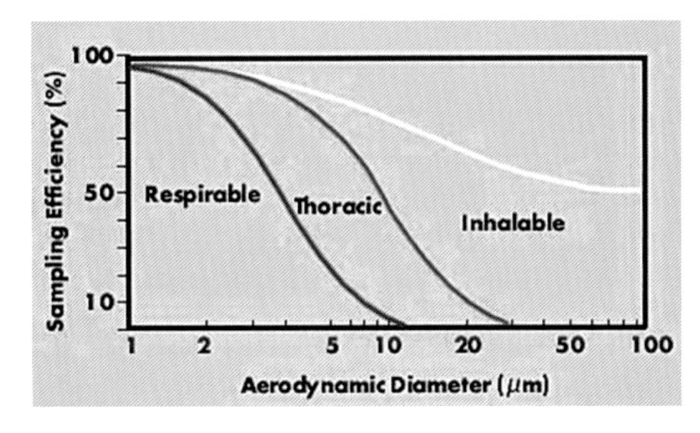

Figure 3: Aerosol particle sizes which can enter distinct areas within the human respiratory tract. (Data from reference 35.)

aerosol route. For example, the 50% infectious dose of *Bacillus anthracis* is thought to be 11,000 spores (6). However, other agents are thought to have extremely low infectious doses closer to 1 (*Mycobacterium tuberculosis*, measles virus [7, 8]). It must also be recognized that agents not known to be naturally transmitted through aerosolization have been transmitted this way in the laboratory, an example being rabies (9). The combination of high titers and the use of aerosol-generating equipment can create unnatural routes of transmission.

AEROSOL GENERATION IN THE MICROBIOLOGY LABORATORY

Historical Context

Sulkin and Pike's 1951 publication was one of the first reviews of laboratory-acquired infections. It featured a number of cases linked to aerosol generation of pathogens within the laboratory (10). Laboratory-acquired infections reported include an incident in which 94 cases of brucellosis were linked to the centrifugation of *Brucella melitensis* in the stairwell of a veterinary college in Michigan. Other incidents were linked to large-scale vaccine production of agents such as *Coxiella burnetii* which involved aerosol-generating procedures, such as open homogenization. Experiments involving the intentional aerosol infection of animals also caused laboratory-acquired infections. These observations led to the publication of several studies reviewing the potential for aerosol generation by quantifying the numbers of infectious particles released following certain microbiological procedures. Additionally, the generation of aerosols from laboratory animals was a concern for both experimental control and human exposure and was therefore measured by researchers in a bid to identify the routes of entry and potential for exposure of laboratory workers. As a result, significant developments in controls and practices were put forward (11).

Quantification of Aerosols for General Microbiological Procedures

In 1964, Wedum published data exploring the significance of airborne transmission in the development of laboratory-acquired infection and determining the levels of aerosol generation following standard laboratory techniques (12). This study concluded that exposure to the infectious particles most often occurs while working at the open bench and performing routine microbiological procedures, with accidents and incidents therefore not responsible for the greatest number of infections. This work influenced the increasing use of microbiological safety cabinets for handling pathogens.

In 1968 Kenny and Sabel measured the aerosols generated by then-common laboratory practices, using *Serratia marcescens* as a tracer (13) and an Andersen sampler (14), which allowed for determination of the particle-size distribution following a set procedure or after an accident. Kenny and Sabel found that the majority of the laboratory practices generated particle sizes below 5 µm, which is in the respirable range. In the early 1970s, Dimmick et al. carried out additional studies quantifying aerosol generation from laboratory accidents (15), and found the majority of aerosols generated were in a 2.5- to 3.5-µm range. Dimmick also introduced the concept of spray factors as a ratio of aerosolized microbial concentration to the suspension concentration (16). This critical step allows experimental data from simulated accidents to be used for risk assessment and to determine the possible exposure to aerosols. These early studies, though valuable, have limited relevance to current laboratory work because either the techniques monitored are now uncommon in modern laboratories or the equipment used is not representative.

However, for the purposes of illustration and to allow for comparison to more recent studies, Table 1 outlines some of the findings from the Kenny and Sabel paper, with spray factors calculated from the published data.

Much of this work was undertaken within small chambers or modified safety cabinets, in which a very high proportion of any aerosol generated would be captured; hence, as shown in Table 2, the exposure assessments within Dimmick's paper factor in room volumes, air change rates, and distance from the source. While this may, at first, seem to be a viable approach, the application of dilution factors, due to room volumes or distance from source, assumes perfect mixing within a space and hence could underestimate the possible exposure. Therefore,

TABLE 1.

Number of viable aerosolized organisms counted following different microbiological practices[a]

Procedure (suspension concentration)	Viable micro-organisms / m^3 air sampled	Spray factor ml m^{-3}
Mixing culture with pipette (1.2×10^9)	233	1.9×10^{-7}
Spilling culture from a pipette (1.2×10^9)	95	7.9×10^{-8}
Use of blender, top opened after operation (3.5×10^9)	52,972	1.51×10^{-5}
Use of sonic oscillator (3.6×10^9)	67.1	1.86×10^{-8}
Dropping flask containing 200 ml of culture (7.1×10^9)	54,773	7.71×10^{-6}

[a]Data from Kenny and Sabel (13).

TABLE 2.

Examples of aerosol generation studies and associated spray factors to allow quantification of risk[a]

Procedure	Spray factor (ft^{-3})
Mixing a liquid culture with a pipette	6.0×10^{-5}
Use of a sonic homogenizer	5.0×10^{-7}
Dropping single drops of liquid culture	1.0×10^{-7}
Dropping a flask of liquid culture	8.0×10^{-6}

[a]Data from Dimmick et al. (15).

TABLE 3.

Findings by Bennett and Parks with regard to aerosol production following simulated laboratory accidents

Procedure	Spray factor ml m^{-3}
Smashed/dropped flask—small volume (50 ml)	5.2×10^{-7}
Smashed/dropped flask—large volume (300 ml, glass tissue culture bottle)	6.85×10^{-6}
Spill 15 ml from bench to floor (0.9 m drop)	1.04×10^{-6}
Dropping three 50 ml bottles, each with 15 ml suspension	1.99×10^{-6}
Blocked tubing with peristaltic pump, causing connection to burst	2.59×10^{-6}
Blocked syringe filter (leading to filter disconnect under pressure)	8.85×10^{-6}
Centrifuge rotor, internal spill with seal removed	4.6×10^{-6}
Centrifuge bucket, internal spill with seal removed	1.7×10^{-7}

the risk assessment approach described later in this section is based on a simpler model, looking at maximum possible exposures.

During the 1970s and 1980s, great progress was made in improving the design of laboratory equipment to prevent aerosol release (centrifuges, analyzers, and homogenizers). A series of papers by Harper (17, 18), Druett (19), and others showed that with good design principles equipment could be made to contain aerosols. The work undertaken by Harper on centrifuge rotors was developed into a standard test method, which is the basis of that currently used for sealed centrifuge rotors, as defined by the IEC standard (1010-2-020). These developments allowed such equipment to be used outside primary containment and significantly reduced the possibility of contamination. However, as is discussed below, this placed a heavy reliance on the correct use by the operators. Whereas much of this work had been related to standard laboratory practices, Ashcroft and Pomeroy (1983) undertook studies looking at the aerosol risks associated with small-scale fermentation systems and demonstrated the potential for significant aerosol generation from accidents and misuse (20).

Bennett and Parks (2006) (21) carried out an updated version of Dimmick's study in which they measured aerosol generation during a series of laboratory procedures and accidents. These tests covered a range of possible accidents and were primarily undertaken with a test chamber more closely representing a small laboratory. The results are shown in Table 3. Furthermore, they also evaluated the effects of source concentration on aerosol potential using a range of spore suspensions of *Bacillus atrophaeus*. These tests showed a direct linear relationship between the suspension concentration and the aerosol generated by a given accident. This suggests that the spray factor could be considered a constant, regardless of titer, allowing for its scalable use within risk assessments. Examples of how spray factors can be used in risk assessment are given in the risk assessment section.

The centrifuge data above relate to sealed rotors and buckets, which had been shown to be aerosol tight, but which had had the primary O-ring removed. The removal of this one simple component led to significant aerosol generation and illustrates that the failure to follow operating instructions can lead to significant risks.

The introduction of new technologies can give rise to new hazards, and these can become more significant as the pathogenicity of the agents being handled increases. A good example of a new technology in which aerosol generation may occur is the fluorescence-activated cell sorter (FACS). Initial guidelines for the use of FACS were produced in 1997, with updates published in 2007 (22). The information provided in these standards was supported by studies undertaken using various tracers, both biological and radiological, to determine aerosol generation from the equipment in normal operating mode and also in failure mode (23, 24). These studies demonstrated that a significant aerosol risk may be present, especially if the instrument is in failure mode and has thus informed the use of primary and secondary containment to protect operators.

The Impact of Training and Experience on Aerosol Generation

As stated in *Biosafety in Microbiological and Biomedical Laboratories*, 5th ed. (BMBL), the technique of the operator can also influence the level of aerosol generation during a procedure (25). In a recent study, Pottage et al. measured the aerosols generated during basic laboratory procedures, such as serial dilution and plating of suspensions, comparing experienced workers with less experienced staff (26). Additionally, those who had undertaken training to work at high-containment levels (biosafety level 3 [BSL3] and above) were also compared with those whose work was limited to lower containment microbiology. A core part of the specialist high containment training was a full appreciation of the risks associated with poor working practices, aimed at raising the participant's awareness of the working methods. While the study showed that experienced staff produced significantly lower amounts of aerosol when performing serial dilutions compared to inexperienced staff, there was no significant difference between the two groups for the plating-out procedure. When levels of surface and personal contamination were evaluated, again there was little difference between the two groups. However, individuals who had undertaken high-containment training again produced significantly fewer aerosols than those who had not, irrespective of their length of laboratory career. This suggested that experience was not, of itself, an indicator of performance but that a clear working knowledge of the risk was just as important.

The study also found that, even though a high-titer suspension (1×10^9 CFU/ml) was used, aerosol generation was limited to a maximum of 203 particles within a 5-minute sampling period with an average of 9.9 CFU/m^3 being generated during the serial dilution study and 40.1 CFU/m^3 being produced while plating out the diluted suspension. This allows spray factors to be calculated for serial dilution and plating out of 9.9×10^{-9} and 4.01×10^{-8}, respectively.

AEROSOLS AND LABORATORY-ACQUIRED INFECTIONS IN MODERN MICROBIOLOGY LABORATORIES

Infection via inhalation is a well-documented primary route for many microorganisms. However, laboratory users need to consider the potential for infection via a nonstandard route due to the manipulations undertaken within the protocols used. Infections caused by microorganisms, that are usually transmitted through routes such as ingestion, cutaneous, or via mucosal membranes, can occur through the inhalational route if the microorganism is aerosolized in high concentrations due to the manipulation undertaken in the laboratory. Furthermore, there is potential for a laboratory infection caused by a noninhalational route to be misclassified as aerosol transmitted. In addition to transmission via an unnatural route, this may also result in an altered pathology of disease, thus adversely affecting diagnosis. Brucellosis is a classic example of a disease often transmitted by aerosol in the laboratory while, in nature, its primary mode of transmission is gastrointestinal.

Aerosol transmission of infection is difficult to prove. In his 1979 review of laboratory-acquired infections, Pike reports that the source of ca. 13.3% of 3,921 laboratory infections is aerosols (27). However, this seems to be used as a catch-all for unrecognized sources. Pike also states that the overall figure may be higher because some laboratory-acquired infections may be misclassified and placed in one of the other source categories. A large contribution to this 13% are infections caused by aerosol generation during large scale production of agents in pre-1945 biological weapons programs, including a centrifuge incident that caused 94 cases of brucellosis (2.4% of all the infections Pike reported). This review was carried out before the widespread use of safety cabinets, sealed centrifuge rotors, automatic pipettes, etc., which would probably have reduced the risk of aerosol infection.

The most recent survey of laboratory-acquired infections was carried out by Harding and Byers, but no breakdown is given of aerosol contribution to laboratory-associated infections (28). Many countries are considering developing a reporting system which is overseen by a regulatory authority, although bringing all these data together into a comprehensive database is yet to be achieved (29). This would allow better identification of the role aerosol transmission currently plays.

Aerosol Control in the Microbiology Laboratory

Aerosol generation in a microbiology laboratory can be reduced by simple and basic training in good microbiological practice. Trainees should be taught how to reduce aerosol generation of infectious materials by using good techniques. There are many simple ways to minimize the production of aerosols, for example, replacing the practice of shaking a liquid culture to ensure even distribution of microorganisms with gentle pipetting to achieve a uniform suspension. However, there are circumstances in which aerosol generation cannot be completely controlled by adherence to a protocol alone, and it is at this stage where additional controls are introduced to protect laboratory workers.

As shown in Fig. 4, the first objective to consider is whether the hazard can be eliminated; due to the nature of the work required in microbiology laboratories, hazard elimination is rarely possible, and therefore further controls must be enforced. Substitution may be appropriate, for example, replacing a microorganism of higher pathogenicity with one from a lower risk group. The next step in the hierarchy is the use of engineering controls. These are widely promoted because they often provide a barrier between the infectious work and the operator. Additionally, engineering controls may afford protection to all those working in the laboratory, not just the operator. As stated in the BMBL, "A procedure's potential to release microorganisms into the air as aerosol droplets is the most important operational risk factor that supports the need for containment equipment and facility safeguards" (25).

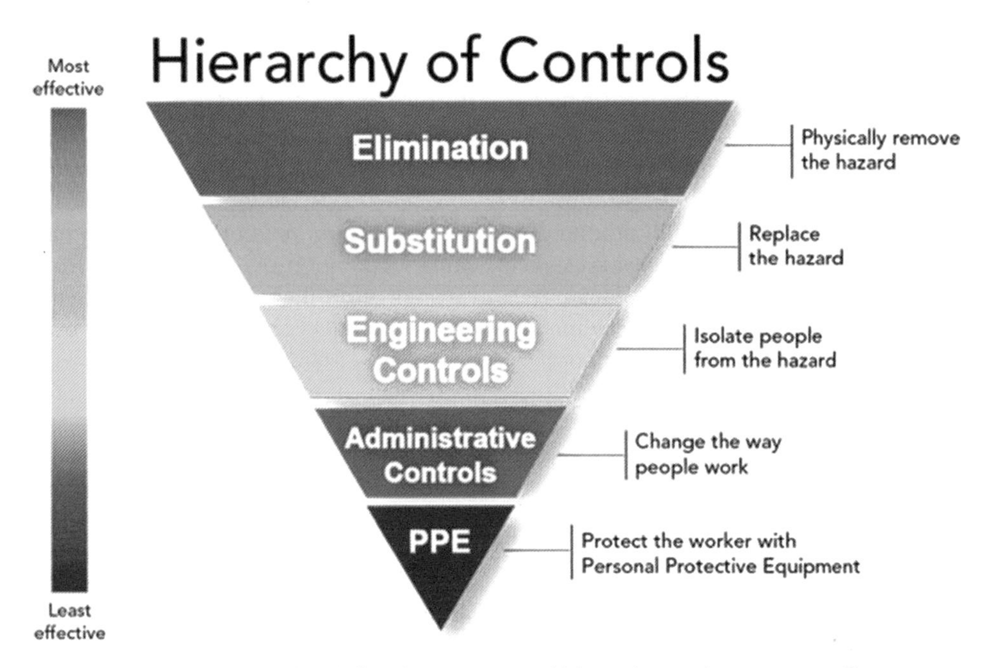

Figure 4: A hierarchy of controls is used to define the measures which can be used to protect staff. (Data from reference 36.)

Principles of Containment

Containment can be achieved through a range of mechanisms such as physical separation and directional airflow; these can often be combined to provide a higher level of assurance. However, containment strategies must also be seen in the wider context of good working practices and effective decontamination.

Containment strategies and systems can be viewed as two levels, Primary and Secondary. Often this can be seen as "Primary Barriers and Personal Protective Equipment" (Primary Containment) and "Facility Design/Construction" (Secondary Containment). However, within UK guidance, primary containment is considered to be not just the physical equipment but a combination of good microbiological practices and the use of containment devices, such as safety cabinets. This guidance recognizes that the role of primary containment to protect the workers and the immediate laboratory environment can be achieved only with good working practices; that is to say, engineering controls in themselves cannot give total protection, and staff training and good practice still play an integral role.

Furthermore, secondary containment, i.e., the protection of people and the environment outside the laboratory, should also be seen as a combination of the physical laboratory design and its operating practices, such as access control, air handling, and control of waste.

Within most laboratories the primary method of containment is the biological safety cabinet. Some devices have built-in primary equipment such as centrifuges (sealed rotors and buckets) and modern homogenizers, but other simple controls will also play an important part. Examples of engineering controls which may be used within a microbiology laboratory to protect against aerosols are discussed below.

High-efficiency filters

Most containment systems, whether they are biological safety cabinets, isolator systems, or high containment autoclaves, rely on filters to protect laboratory staff and the environment from infectious aerosols.

Within containment systems, filtration can be divided into two types, membrane filters and HEPA filters. Membrane filters work primarily by pore size and act like a sieve. Such filters are normally used for venting applications, such as fermenters and autoclaves. High grade membrane filters are suitable for applications involving very high humidity as they are not compromised by the presence of moisture; however, they are not well suited to applications requiring high airflow rates (containment cabinets) or low levels of resistance, such as respirator filters.

For most laboratory applications, HEPA filters are the first choice and are the basis of biological safety cabinets. There are several definitions of HEPA filter in national and international standards (e.g., DOE Standard 3020-2005 and EN1822:2009); however, the basic mode of operation is the same across the board. There is a commonly held misconception that HEPA filters behave like sieves, but the performance is a combination of factors, which come together to provide a very high level of efficiency over a wide range of particle sizes, including submicron particles.

The filter material within a HEPA filter is made of many densely packed fibers in a random pattern, and, as a particle passes through the material, there are three possible filtration mechanisms at work. The first is interception, where particles carried in the airflow around the filter fibers adhere to the filter. The particles must be within one radius of the filter fiber to be captured. Larger particles are often captured by the second mechanism, impaction. Due to their mass, these particles cannot adjust to sudden changes in the airflow patterns within the material and are embedded into a fiber.

The final mechanism is diffusion, which occurs due to Brownian motion, where small particles (<0.1 μm) move and interact with surrounding molecules in a random pattern, due to the action of surrounding molecules. This motion slows down a particle's path through the HEPA filter and increases the probability that the particle will be captured by either interception or impaction (30).

This combination of filtration mechanisms makes HEPA filters very efficient for both very large and very small particles, and, when the factors are combined, it has been shown that particles in the range of 0.2 to 0.3 μm are likely to be the most penetrating particle size, that is to say, the size range least likely to be captured. Hence, the efficacy for HEPA filters is defined as the performance against that most penetrating particle size and should be higher for particles that are larger or smaller.

Standards for HEPA filters vary across national boundaries. BMBL and NSF standards require HEPA filters of 99.99% efficiency (American National Standards Institute [ANSI] Z9.14) (2014). Europe uses filters with 99.995% efficiency (EN1822) as the standard for biocontainment. Because of the complex mechanisms involved, the stated efficiencies relate not only to the particle size used for testing, but also to the test conditions; flow rate has an impact on the stated results. In reality, the size of particles generated from microbial suspensions will normally be an order of magnitude higher than the most penetrating particle size and so the actual filter efficiency will be far greater than the test result implies.

An important aspect of filtration which can easily be overlooked is the effective and failsafe mounting of filters. Often the weak point in filtration systems is not the filter media but the seals and housings, which, if not appropriate, can lead to aerosols bypassing the filter, thereby reducing overall performance of the filter.

Biological safety cabinets

Biological safety cabinets are designed to provide operator protection from aerosols generated during working practices and potentially by laboratory equipment. The design of biological safety cabinets has evolved over many years and is clearly defined in international standards, such as NSF49 and EN12469. Designs can be split into three primary groups, Class I and II cabinets, which are open-fronted, and the fully enclosed Class III type cabinets primarily used for high-containment work.

While both Class I and II cabinets can provide a high level of operator protection from infectious aerosols, these cabinets do not provide protection from contamination through other means, such as splashes and spills. Open-fronted cabinets can also be affected by external airflows and other activities within the laboratory, which can lead to a drop or failure in performance. Cabinets should not be considered to be 100% protective and require careful installation by experienced engineers to ensure optimal performance. Furthermore, careful consideration must also be given to the type of equipment and work activities within open-fronted cabinets. Equipment, such as centrifuges, which generate strong air currents, can disrupt the airflow within the cabinets and compromise the containment provided. The installation of very large equipment within a Class I or II cabinet can also compromise the performance, and care needs to be taken to ensure that effective airflows are maintained and that adequate space is provided within the cabinet work zone for the operators to undertake their tasks.

The Class I biological safety cabinet

Class I cabinets (Fig. 5) are designed to provide operator protection only, with air drawn in through the front opening, across the work area, and vented through a HEPA filter to the atmosphere. The directional airflow prevents the release of aerosols but does not prevent possible contamination of the operators' hands and sleeves. This type of cabinet does not provide a clean working environment, but it does offer a simple, robust form of containment and is still widely used within diagnostic labs.

The Class II biological safety cabinet

As shown in Fig. 6, Class II cabinets are open-fronted like Class I cabinets but have additional internal HEPA filtration to provide a downflow of clean air over the work zone, thus providing additional protection for operator and product from external contamination but at the expense of a more complex cabinet and installation. Class II cabinets are the most widely used type and are split into a number of subgroups, reflecting different user requirements and installations, and these are described in a number of widely available publications, such as the BMBL 5th edition. However, in terms of their use and operator protection provided, the same caveats as above apply, and the limitations of open-fronted cabinets must still be considered.

The European standard for biological safety cabinets describes a test to determine operator protection or the

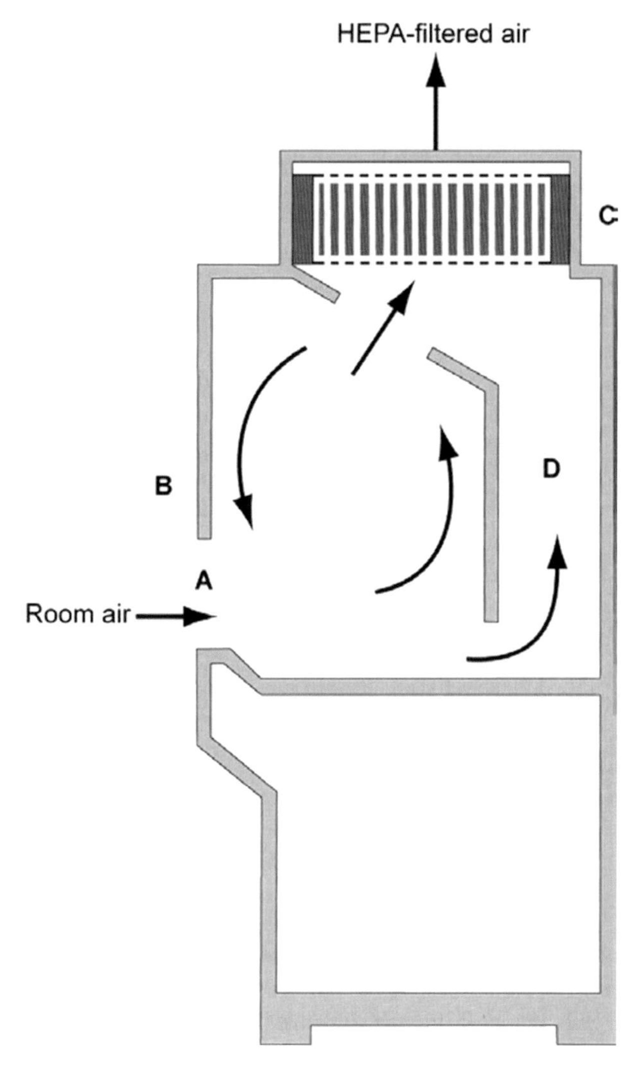

Figure 5: Example of a Class I biological safety cabinet. (Data from reference 25.)

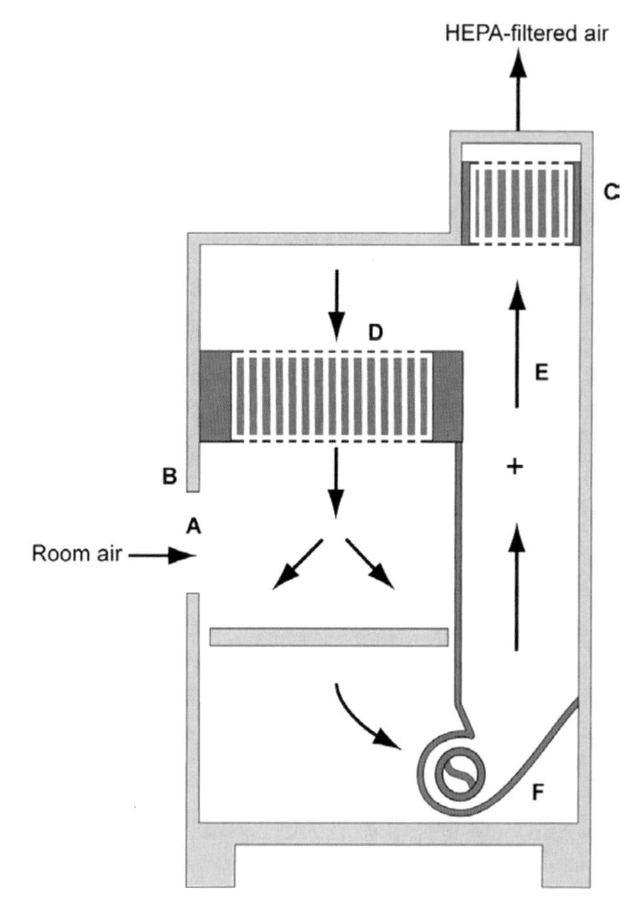

Figure 6: Example of a Class II biological safety cabinet. (Data from reference 25.)

ability to retain aerosols within the cabinet using aerosols of potassium iodide of particle size 7 μm (31). This test is routinely used within the UK for the *in situ* testing of Class I and II cabinets and provides an operator protection factor, which is the ratio of aerosol generated to that released from the cabinet. A correctly operating cabinet will give an operator protection factor of greater than 10^5 and often as high as 10^6. However, this test has repeatedly shown that poorly positioned cabinets, which appear through standard tests to be operating correctly, may have much lower protection factors and require additional adjustment to achieve an acceptable performance. This reduction in performance, often due to poorly considered installation or the impact of other airflows, such as room ventilation, may be happening with Class II cabinets used worldwide, but without detailed *in situ* testing it may not always be recognized. Hence, there is a need for highly trained service staff, who fully understand the operation of the equipment and the impact of the environment into which it is installed.

The Class III biological safety cabinet

Figure 7 provides an overview of Class III cabinets, often referred to as glove boxes, which are sealed enclosures. Air is drawn into the sealed cabinet via a HEPA filter giving turbulent flow within the cabinet, which is then exhausted via a second set of HEPA filters. Work is carried out using gloves or gauntlets mechanically attached to the cabinet, and samples are transferred in or out via pass-through ports or dunk tanks. The cabinet is maintained at negative pressure at all times, and there should be sufficient airflow to ensure that, in the event of a glove failure, inflow will prevent any escape of aerosols (>0.7 ms^{-1}).

These cabinets have historically been the preserve of the BSL4 high containment but are more widely used within the UK for BSL3 work when there is a high risk of aerosol production or large volumes of high-titer material are being handled. The cost and ergonomic limitations of such cabinets make them applicable only to very specialized applications, but the very high level of

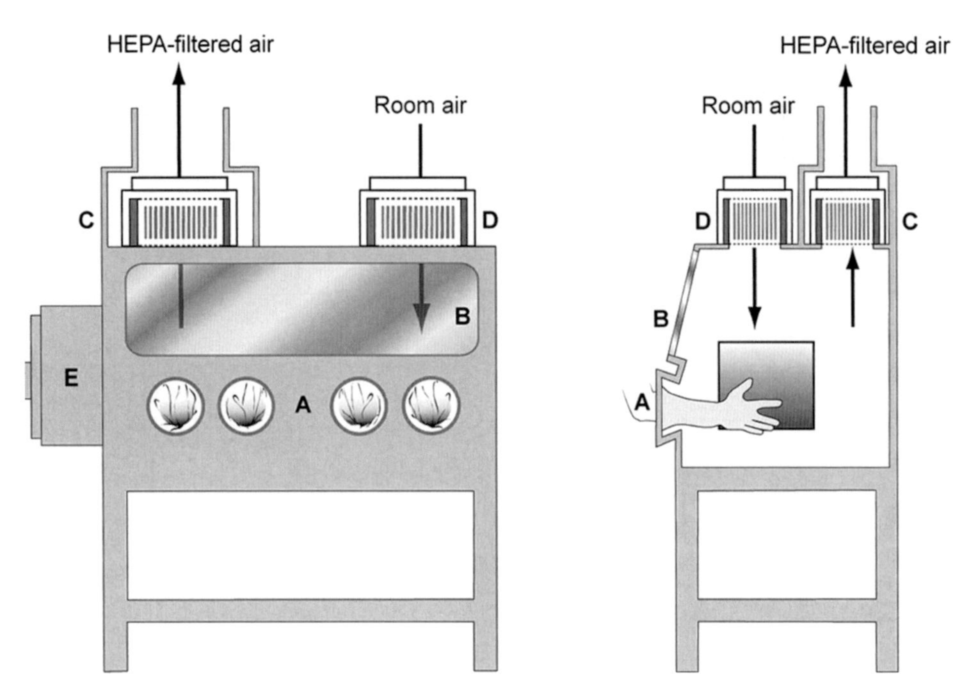

Figure 7: Example of a Class III biological safety cabinet. (Data from reference 25.)

performance assurance makes them an attractive alternative for high-hazard work and for containing complex or hazardous equipment.

Other forms of containment systems

Traditional safety cabinets provide a well-defined option for operator protection, but there are a number of other systems available that can also be utilized. The use of both film and ridged isolator systems within the pharmaceutical industry has become normal practice, and, with careful design, such systems can also be used to provide protection within the laboratory environment. HEPA-filtered, negative-pressure isolator systems are widely used and can be built around equipment or installations to give high levels of physical containment, while maintaining flexibility. Within animal facilities, the use of ventilated necropsy tables and individually ventilated cage systems is also commonplace, and, with the introduction of new working practices and equipment into laboratories, new or modified methods of containment are often needed.

Centrifuges

The use of centrifuges within the laboratory presents a number of possible issues. It is important to correctly position them within the laboratory as the strong air currents they produce can adversely affect the airflows of other equipment, such as biological safety cabinets. The combination of very high energy and infectious material makes centrifuges potential aerosol generators. In the event of an accident there is the possibility of generating a large aerosol of small respirable particles, so measures need to be taken to prevent this from happening. In the past, centrifuges were often used within a biological safety cabinet, but, over the last 30 years, aerosol-tight rotors and buckets have been developed for laboratory use. These sealed units prevent the release of an aerosol in the event that the primary container leaks or fails. However, because they only contain any leakage and aerosol and do not provide a decontamination step, they still need to be opened within a safety cabinet to prevent possible exposure. The design of sealed rotors and buckets has evolved, with designs now being produced that prevent any leakage from hitting the integral seal. However, it must be noted that, in most cases, the containment is still provided by a single O-ring. If the O-ring seals become damaged or are removed, containment is lost; training staff in the maintenance and inspection of these components is critical to their safe operation.

Secondary containment

Secondary containment is a combination of factors, a number of which are discussed below.

Directional airflow

High-containment microbiology laboratories establish a negative pressure by creating a directional airflow into the laboratory. Bennett et al. have demonstrated that the level of protection afforded by laboratories is related to

TABLE 4.

Number of minutes required for removal of airborne contaminants determined by air change rate

Air changes per hour	Percentage of airborne contaminants removed (%)		
	90	99	99.99
6	23	46	115
12	12	23	58
20	7	14	35
40	3	7	17

the magnitude of the airflow through the door of the laboratory and not to the degree of negative pressure (31).

The control of ventilation is critical to the correct operation of the laboratories, and the ability to maintain negative pressure, or, more pertinently, prevent positive pressurization is core to the design principles.

Whereas containment cabinets are designed with very high airflows to remove aerosols and provide rapid dilution of any aerosol release, laboratory ventilation is primarily driven by operator comfort and the heat loads generated within the working environment. However, knowledge of a laboratory's ventilation rate is still useful for both risk assessment and accident response. The UK ACDP guidance (32) gives details of the relationship between air change rates and dispersal of aerosols, which can be used in the event of an aerosol release. This information is shown in Table 4.

Anterooms

An anteroom (which may also be referred to as a lobby) allows separation between high-hazard work and lower-hazard work. Anterooms are routinely found prior to entry at BSL3 and are mandatory at BSL4, providing an additional layer of containment between the work being undertaken and the environment external to the laboratory. Anterooms are often used to provide a cascade in negative pressure which increases towards the high-containment laboratory. It has been demonstrated by Bennett et al. that anterooms may increase the protection afforded by a negative-pressure laboratory by a factor of 100 (31).

New equipment

When new equipment is introduced into a laboratory it is important to assess the potential for generation of microbial aerosols. Manufacturers may be able to provide information on testing undertaken to assure any aerosols are contained. However, often equipment may be developed for use with low-risk biological agents and may need additional containment when used with pathogenic agents.

Closed systems

Many new diagnostic or research instruments employ the use of a closed system, designed to protect both the operator and also the test samples from potential contamination. Automated diagnostic and research technologies often allow the user to place the sealed sample within a machine or robot, at which point the test is performed within the equipment, followed by a decontamination step prior to sample exit. In addition to enhanced quality control due to a decrease in the risk of external contamination, this system can also prevent the release of aerosols during manipulations that would normally result in energy transfer.

RISK ASSESSMENT

The term aerosol-generating procedure is often used in microbiology laboratories to indicate a hazardous manipulation or process. However, due to improvements in equipment and the use of containment devices, some procedures do not generate aerosols and others only generate low-concentration aerosols. While biosafety risk assessment often focuses on the agent category, the titer of agent is an extremely important factor in measuring the potential for aerosol generation and thus the potential for infection. If low-titer suspensions are to be used, then it is extremely unlikely that individuals will be exposed. A risk assessment model has been developed incorporating the spray-factor concept. This model can be used to assess whether biocontainment practices will protect against an aerosol risk from a process or accident and additionally to assess potential exposure of a person involved in an accident, which may be aerosol generating.

Risk Assessment Model

In order to model the risk of exposure to an aerosolized microbial agent, the potential inhaled dose of the exposed person needs to be calculated. This can be done by inputting information about the spray factor of a process, the titer of the suspension in use, the exposure time, and the person's breathing rate.

Aerosol concentration

As has previously been defined, spray factor (SF, ml/m^3) is equal to the aerosol concentration (AC, CFU/m^3) divided by the suspension concentration (SC, CFU/ml). Therefore the aerosol concentration (AC) generated by an accident/process equals the concentration of the suspension multiplied by the spray factor (SF × SC). Although this aerosol concentration normally decreases

TABLE 5.

Selected examples of the use of spray factor in risk assessment calculation

Process	Spray factor	Titer (CFU/ml)	Exposure time (min)	Protection used (factor)	Dose
Centrifuge leak	4.6×10^{-6}	10^9	10	None	690
		10^6	10	None	0.69
Centrifuge leak	4.6×10^{-6}	10^9	10	Safety cabinet (10^5)	0.007
		10^9	10	P95 (95)	34.5
Pipetting	9.9×10^{-9}	10^9	30	None	297
		10^6	30	None	0.297
Pipetting	9.9×10^{-9}	10^9	30	Safety cabinet (10^5)	0.003
		10^9	30	P95 (95)	14.9

with time due to deposition, dilution through ventilation, and agent inactivation, for simplicity, concentration can be used in the calculation of exposure, as a worst case.

Amount breathed

A person's exposure depends on the volume (V, liters) of the contaminated air they breathe, which is calculated from their breathing rate (BR) multiplied by their exposure time (T, minutes). A typical breathing rate for someone carrying out normal laboratory procedures is about 0.015 m^3/min (33).

Exposure

The exposure (CFU) is the amount of air breathed multiplied by the aerosol concentration.

$$\text{Exposure without containment} = BR \times T \times (SF \times SC)$$

If containment or respiratory protection equipment is used, there will be a protection factor (PF), which will reduce exposure, giving:

$$\text{Exposure} = BR \times T \times (SF \times SC) / PF$$

Model calculation

A laboratory worker, with a breathing rate of 0.015 m^3/min and a 15-minute exposure time period for a process with a spray factor of 10^{-6} on a 10^9 suspension of an agent carried on outside containment, is exposed to 225 organisms. If the titer is 10^6, the exposure would be 0.225. If the original suspension was used but the work was carried out in a safety cabinet with a PF of 10^5, the exposure would be 0.00225. This simple calculation can be used to assess potential exposure before beginning an operation or after an accident. Examples are provided in Table 5.

There are two major findings from this type of calculation. First, the use of a correctly performing safety cabinet protects from almost any foreseeable source of aerosol in the microbiology laboratory. Second, if a low-titer agent is being handled, then aerosol exposure is likely to be very low and only of concern for the organisms most infectious by the aerosol route.

CONCLUSIONS

The generation of microbial aerosols in the microbiology laboratory is a potential source of laboratory-acquired infection in those exposed and, potentially, those in the external environment. However, in the modern microbiology laboratory, the use of containment devices should greatly reduce the risk of aerosol exposure. With a good understanding of the mechanisms behind aerosol generation, working practices can be improved and an awareness of the risk, seen and unseen, can be raised. As an increased understanding of the role of aerosols within the laboratory develops, the importance of this information being disseminated to those working and managing these areas is crucial. Implications on risk assessment, equipment design and positioning, and training of staff can all be drawn from knowledge of aerobiology.

References

1. **Furr AK.** 2000. Laboratory facilities—design and equipment, p 195. *In CRC Handbook of Laboratory Safety, 5th ed.* CRC Press LLC, Boca Raton, FL.
2. **Cox CS.** 1995. Stability of airborne microbes and allergens, p 77–79. *In* Cox CS, Wathes CM (ed), *Bioaerosols Handbook.* CRC Press, Inc., Boca Raton, FL.
3. **Mitscherlich E.** 1984. Special influences of the environment, p 725–727. *In* Mitscherlich E, Marth EH (ed), *Microbial Survival in the Environment: Bacteria and Rickettsiae Important in Human and Animal Health.* Springer-Verlag, Berlin, Germany.
4. **Sinclair R, Boone SA, Greenberg D, Keim P, Gerba CP.** 2008. Persistence of category A select agents in the environment. *Appl Environ Microbiol* **74:**555–563.
5. **Griffiths WD, DeCosemo GAL.** 1994. The assessment of bioaerosols: a critical review. *J Aerosol Sci* **25:**1425–1458.
6. **Toth DJA, Gundlapalli AV, Schell WA, Bulmahn K, Walton TE, Woods CW, Coghill C, Gallegos F, Samore MH, Adler FR.** 2013. Quantitative models of the dose-response and time course of inhalational anthrax in humans. *PLoS Pathog* **9:**e1003555.
7. **Pfyffer GE.** 2007. *Mycobacterium*: general characteristics, laboratory detection, and staining procedures, p 543–572. *In* Murray PR et al (ed), *Manual of Clinical Microbiology*, 9th ed. ASM Press, Washington, DC.
8. **Knudsen RC.** 2001. Risk assessment for working with infectious agents in the biological laboratory. *Appl Biosaf* **6:**19–26.

9. **Winkler WG, Fashinell TR, Leffingwell L, Howard P, Conomy P.** 1973. Airborne rabies transmission in a laboratory worker. *JAMA* **226:**1219–1221.
10. **Sulkin SE, Pike RM.** 1951. Laboratory-acquired infections. *J Am Med Assoc* **147:**1740–1745.
11. **Phillips GB, Jemski JV.** 1963. Biological safety in the animal laboratory. *Lab Anim Care* **13:**13–20.
12. **Wedum AG.** 1964. Airborne infection in the laboratory. *Am J Public Health Nations Health* **54:**1669–1673.
13. **Kenny MT, Sabel FL.** 1968. Particle size distribution of *Serratia marcescens* aerosols created during common laboratory procedures and simulated laboratory accidents. *Appl Microbiol* **16:** 1146–1150.
14. **Andersen AA.** 1958. New sampler for the collection, sizing, and enumeration of viable airborne particles. *J Bacteriol* **76:**471–484.
15. **Dimmick RL, Vogl WF, Chatigny MA.** 1973. Potential for accidental microbial aerosol transmission in the biological laboratory, p 246–266. *In* Hellman A, Oxman MN, Pollack R (ed), *Biohazards in Biological Research*. Cold Spring Harbor Laboratory, Cold Spring Harbor, NY.
16. **Dimmick RL.** 1974. Laboratory hazards from accidentally produced airborne microbes. *Dev Ind Microbiol* **15:**44–47.
17. **Harper GJ.** 1984. Evaluation of sealed containers for use in centrifuges by a dynamic microbiological test method. *J Clin Pathol* **37:**1134–1139.
18. **Harper GJ.** 1984. An assessment of environmental contamination arising from the use of some automated equipment in microbiology. *J Clin Pathol* **37:**800–804.
19. **Druett HA, May KR.** 1952. A wind tunnel for the study of airborne infections. *J Hyg (Lond)* **50:**69–81.
20. **Ashcroft J, Pomeroy NP.** 1983. The generation of aerosols by accidents which may occur during plant-scale production of microorganisms. *J Hyg (Lond)* **91:**81–91.
21. **Bennett A, Parks S.** 2006. Microbial aerosol generation during laboratory accidents and subsequent risk assessment. *J Appl Microbiol* **100:**658–663.
22. **Schmid IC, Lambert D, Ambrozak D, Perfetto SP.** 2007. Standard safety practices for sorting of unfixed cells, supplement 39. *In Current Protocols in Cytometry,* Section 3.6.1–3.6.20. John Wiley and Sons Inc., Hoboken, NJ.
23. **Xie M, Waring MT.** 2015. Evaluation of cell sorting aerosols and containment by an optical airborne particle counter. *Cytometry A* **87:**784–789.
24. **Wallace RG, Aguila HL, Fomenko J, Price KW.** 2010. A method to assess leakage from aerosol containment systems: testing a fluorescence-activated cell sorter (FACS) containment system using the radionuclide technetium-99m. *Appl Biosaf* **15:**77–85.
25. **US Department for Health and Human Services, Public Health Service, Centers for Disease Control and Prevention, National Institutes of Health.** 2009. *Biosafety in Microbiological and Biomedical Laboratories,* 5th ed. http://www.cdc.gov/biosafety/publications/bmbl5/bmbl.pdf.
26. **Pottage T, Jhutty A, Parks S, Walker J, Bennett A.** 2014. Quantification of microbial aerosol generation during standard laboratory procedures. *Appl Biosaf* **19:**124–131.
27. **Pike RM.** 1979. Laboratory-associated infections: incidence, fatalities, causes, and prevention. *Annu Rev Microbiol* **33:** 41–66.
28. **Harding AL, Byers KB.** 2006. Epidemiology of laboratory-associated infections, p 53–77. *In* Fleming DO, Hunt DL (ed), *Biological Safety: Principles and Practices*, 4th ed. ASM Press, Washington, DC.
29. **Singh K.** 2011. It's time for a centralized registry of laboratory-acquired infections. *Nat Med* **17:**919.
30. **First MW.** 1998. HEPA filters. *Appl Biosaf* **3:**33–42.
31. **Bennett AM, Parks SR, Benbough JE.** 2005. Development of particle tracer techniques to measure the effectiveness of high containment laboratories. *Appl Biosaf* **10:**139–150.
32. **Health and Safety Executive.** 2001. The management, design and operation of microbiological containment laboratories. HSE Books, Surrey, United Kingdom. http://www.hse.gov.uk/pubns/priced/microbiologyiac.pdf.
33. **Heinsohn RJ.** 1991. *Industrial Ventilation: Principles and Practice.* Wiley, New York, NY.
34. **International Commission on Radiological Protection.** 1994. *Human Respiratory Tract Model for Radiological Protection: A Report of a Task Group of the International Commission on Radiological Protection.* Pergamon Press, Oxford, United Kingdom.
35. **Soderholm SC.** 1989. Proposed international conventions for particle size-selective sampling. *Ann Occup Hyg* **33:**301–320.
36. **Centers for Disease Control and Prevention, National Institute for Occupational Safety and Health (NIOSH).** 2015. *Hierarchy of Controls.* NIOSH, Washington, DC. http://www.cdc.gov/niosh/topics/hierarchy/.

21

Personal Respiratory Protection

NICOLE VARS McCULLOUGH

Respiratory protection is used when workplace air is unsuitable for breathing due to lack of oxygen or unsafe levels of contaminants. Respirators are designated as a last resort or temporary control measure to help reduce contaminant exposures in the workplace to acceptable levels or provide sufficient oxygen for breathing. In accordance with the industrial hygiene hierarchy of controls, available engineering and administrative controls should be implemented before considering personal respiratory protection as a control measure. When necessary, only respirators certified by the National Institute for Occupational Safety and Health (NIOSH) should be used in the United States. A full respiratory protection program administered by a trained individual as specified by the Occupational Safety and Health Administration (OSHA) must accompany any use of respirators in the workplace. A respiratory protection program is necessary to ensure safe and proper use of respirators and to help avoid misuse or injury or death to the respirator users. Important components of a program include written standard operating procedures (SOPs), medical evaluation, user training, respirator maintenance procedures, and properly fitting the respirator to the user. The program must have a designated and knowledgeable administrator, preferably someone trained in a field of occupational health and safety.

The objective of this chapter is to introduce the subject of respiratory protection to individuals working in the field of microbiology. It should not be used as an exclusive training tool for individuals who wish to administer a respiratory protection program. Persons charged with this responsibility should enroll in formal training and review the most current regulations and guidelines carefully.

TYPES OF RESPIRATORS

Respirators can be divided into two general classes: atmosphere supplying and air purifying. Atmosphere-supplying respirators utilize clean, breathable air from a gas cylinder or air compressor. A respirator that functions by drawing contaminated air through a filter or chemical cartridge before it reaches the wearer's respiratory system is an air-purifying respirator.

Atmosphere-Supplying Respirators

There are two types of atmosphere-supplying respirators: self-contained breathing apparatuses (SCBAs) and air-line respirators. SCBAs are similar to self-contained underwater breathing apparatuses (SCUBA) in that they utilize a breathable air cylinder worn on the back. SCBAs, which operate under positive pressure (called pressure demand), can be used in atmospheres containing concentrations of contaminants that may be immediately dangerous to life or health (IDLH) and for entering atmospheres that are oxygen deficient, such as in firefighting.

Air-line respirators, typically called supplied-air respirators (SARs), are supplied with air through a small hose connected to a cylinder or air compressor. There are several types of SARs, but the most used type is pressure demand, which supplies air to a face mask to keep it under positive pressure; the air quantity used is limited to that required during breathing. SARs can be coupled with air-purifying elements to allow movement into and out of the contaminated atmosphere before hooking up to the air-line, or for protection while switching between air-lines. They can also be coupled with an auxiliary air cylinder (referred to as an escape SCBA) that permits escape from the atmosphere if the air-line becomes nonfunctional. Only those air-line respirators combined with escape SCBA can be used in IDLH or oxygen-deficient atmospheres.

A special type of air-line respirator that is used in biosafety level 4 (BSL4) laboratories is a supplied-air suit (1). This type of suit consists of a full-body impermeable barrier supplied with air through an air-line and offers complete isolation from the work environment (Fig. 1). These suits are designed to be under positive pressure with regard to the room environment, so any leakage of air would be out of the suit. NIOSH does not have testing and certification criteria for these suits, and therefore they are not approved by NIOSH. (This is one case where a non-NIOSH-approved system may be acceptable for use.) However, the Los Alamos National Laboratory together with the Department of Energy has established guidelines that should be considered before use (2).

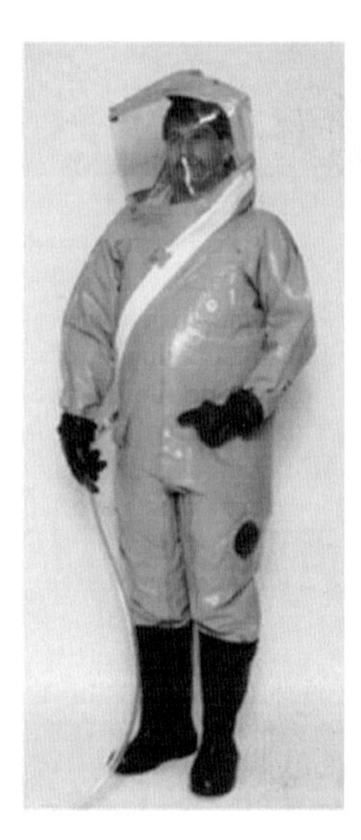

Figure 1: Supplied-air suit. (Courtesy of ILC Dover, Frederica, DE.)

Air-Purifying Respirators

Air-purifying respirators help reduce the concentration of contaminants in the air to an acceptable exposure level by passing the contaminated air through an air-purifying device such as a particulate filter or chemical cartridge. Air-purifying respirators can be further subdivided into two categories: (i) nonpowered and (ii) powered air-purifying respirators (PAPRs). Air-purifying respirators that are not powered rely on the wearer to draw air through the air-purifying element and into the facepiece. PAPRs employ a small motor which forces air through filtering elements and into the facepiece.

Respirators can be further described by the type of facepiece that is used. Respirator facepieces that fit tightly and form a seal with the face cover either half or all of the face. Half-mask respirators cover the nose and mouth but not the eyes. They seal below the chin, across the cheeks, and over the bridge of the nose. A subset of half-mask respirators are those that have the facepiece made entirely out of filtering material, referred to as filtering facepiece respirators (Fig. 2A and 2B). Full-facepiece respirators cover the nose, mouth, and eyes and seal below the chin, across the cheeks, and across the forehead (Fig. 2C). Hoods, helmets, and loose-fitting facepieces do not seal tightly to the face. Hoods cover the entire head and neck, incorporating a clear visor, with elastic forming a loose seal around the neck (Fig. 2D). Helmets also cover the head and neck but incorporate a hard-surfaced helmet to provide some head protection. Loose-fitting respirators typically cover only the face but do not form a tight seal between the face and the respirator (Fig. 2E). Table 1 indicates the typical components, facepieces, and clean-air sources for commonly used respirators.

The respirator design determines the degree of protection it will provide. The protection afforded by the different types of respirators is discussed under "Respirator Protection Factors," below.

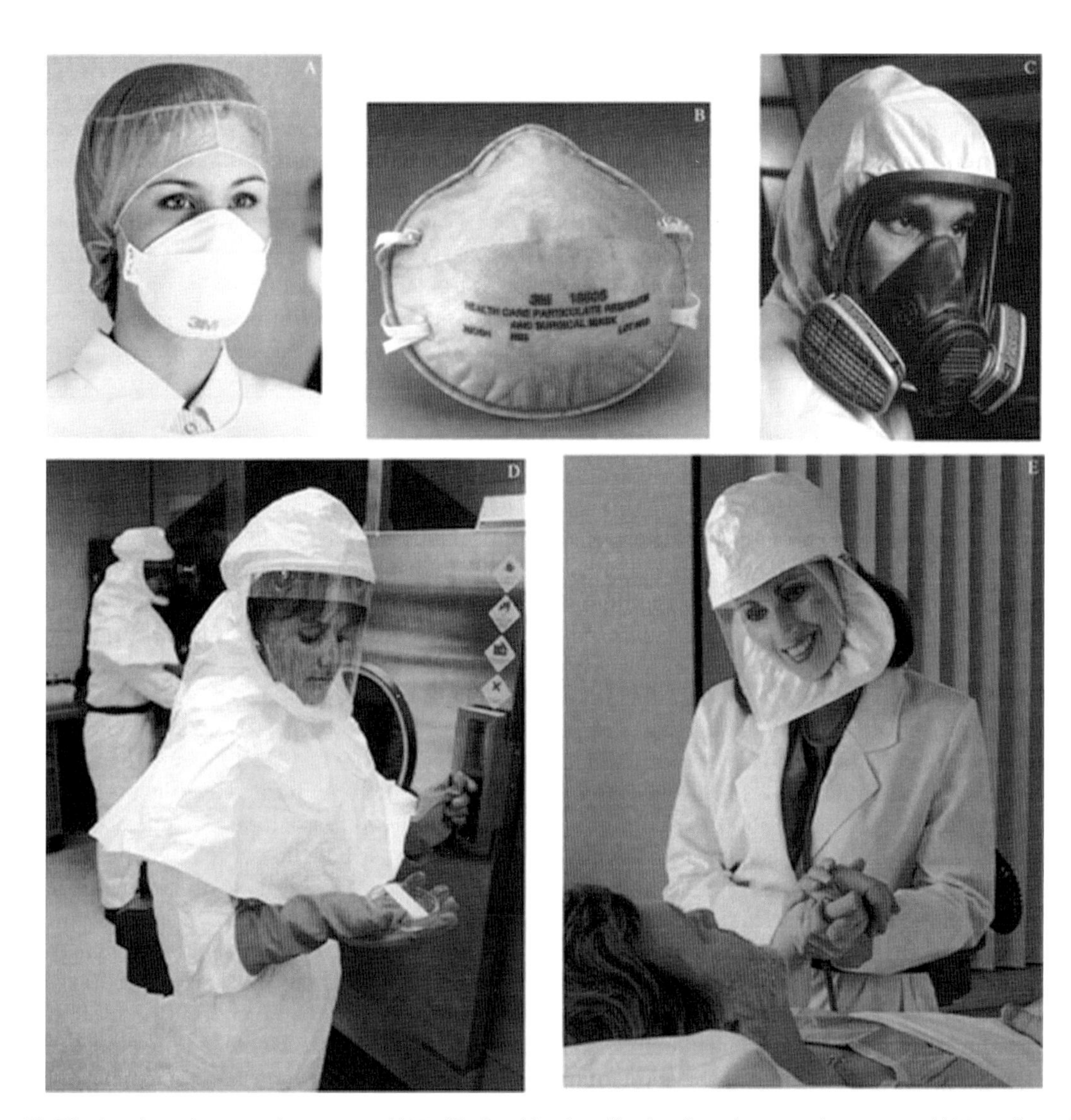

Figure 2: (A) Filtering facepiece respirator, type N95. (B) Combination filtering facepiece respirator, type N95, and surgical mask. (C) Full-facepiece respirator. (D) Powered air-purifying respirator (PAPR) with hood. (E) PAPR with loose-fitting facepiece. (Courtesy of 3M Company, St. Paul, MN.)

TABLE 1.

Typical components, facepieces, and clean-air sources for commonly used respirators[a]

Type of respirator	Components	Facepiece(s)	Source of clean air
Nonpowered air purifying	Facepiece and filter, cartridge, or canister	Half-mask, including filtering facepieces, full facepiece	Gas and vapor cartridges and canisters, particulate filters,[b] or combination cartridge and filter
PAPR	Facepiece and filter, cartridge, or canister; battery pack; breathing tube; blower and motor unit; harness	Half-mask, full facepiece, hood, helmet (loose-fitting)	Gas and vapor cartridges and canisters, HEPA particulate filters, or combination cartridge and filter
SAR	Facepiece, breathing tube, valve, hose, harness	Half-mask, full facepiece, hood, helmet (loose-fitting)	Air cylinder or air compressor
SCBA	Facepiece, breathing tube, cylinder, valve, harness	Full facepiece	Air cylinder

[a]All components must be available from the same manufacturer and approved by NIOSH as a system. PAPR, powered air-purifying respirator; HEPA, high-efficiency particulate air; SAR, supplied-air respirator; SCBA, self-contained breathing apparatus; NIOSH, National Institute for Occupational Safety and Health.

[b]The classes of particulate filters for nonpowered, air-purifying respirators are shown in Table 2.

PARTICLE FILTRATION AND GAS REMOVAL

Air-purifying respirators filter the gases, vapors, and particles from the contaminated air and deliver this "cleaned" air to the wearer's respiratory system. Understanding the basics of how the purifying mechanisms work is useful when selecting respirators and evaluating product claims.

Particles may be captured on a filter by both mechanical and electrostatic mechanisms. Four mechanical mechanisms contribute to particle deposition: interception, diffusion, gravity, and inertial impaction. Electrostatic attractive forces may also be responsible for considerable capture in some filters in which the filter fibers, particles, or both carry a charge. For every filter there is a particle size where none of the capture mechanisms is highly effective; this is referred to as the most penetrating particle size. This particle size, which ranges from 0.02 to 0.3 μm for most filters, represents the point at which the filter is the least efficient (3). In this range, particles are too large to be captured effectively by diffusion and too small to be captured effectively by interception, impaction, and gravity (4). Particles smaller than the most penetrating particle size, as well as those bigger than this size, will be captured more efficiently than particles in the most penetrating particle size range.

NIOSH approves filters for particulate air-purifying respirators according to test procedures found in 42 C.F.R. Part 84, which specifies that the test aerosol consist of particles in the most penetrating size range [U.S. Department of Health and Human Services (DHHS)] (5). These tests are designed so that the filter will exhibit the certified efficiency in most workplace conditions. Therefore, if a filter is demonstrated to have a certain efficiency using the most penetrating particle size, then it will be at least that efficient for all other particle sizes both larger and smaller given the same set of conditions (i.e., flow).

There are nine classes for particulate filters used with nonpowered respirators. These are based on the filter efficiency and whether the filter can be used in a work environment that contains airborne oil particles (Table 2). There are three efficiency levels of filters: 95%, 99%, and 99.97%. There are also three series (N, P, and R) that designate whether the filter can be used in an oil-containing atmosphere. This is important because oil particles may reduce the charge on some respirator filters that use electrostatic collection mechanisms. Some industrial workplaces, such as machine shops, have oil particles in their atmosphere. However, as most microbiology laboratories do not have oil particles in the air, an N series filter would typically be appropriate. Although the terminology in Table 2 refers to the filters only, filtering facepiece respirators are often mistakenly referred to by the filter designation, e.g., an N95 respirator. This terminology is not specific enough and could refer to any respirator with N95 filters. The proper way to refer to this type of respirator is as a "type N95 filtering facepiece respirator" or a "half-mask respirator with N95 filters."

TABLE 2.

Particulate filter classes for nonpowered, air-purifying respirators

Use in atmospheres containing oil aerosols	Filter class with indicated efficiency			
	Series	95%	99%	99.97%
Can be used for more than one shift	P	P95	P99	P100
Can be used for only one shift	R	R95	R99	R100
Cannot be used	N	N95	N99	N100

There is only one particle filter available for PAPRs, the high-efficiency (HE) filter. These filters are certified using test procedures similar to those for the P100 filter, which is certified to be 99.97% efficient in atmospheres containing all kinds of aerosols, even oil.

Particulate filters will not remove gases and vapors. Gases and vapors are filtered by drawing contaminated air through a chemical cartridge or canister. This cartridge or canister consists of a bed of sorption media that removes gases and vapors from breathed in air through physical adsorption or chemisorption. The sorption media of activated charcoal or resin may be treated to capture a specific chemical or to increase the capture of a given chemical. There is no one kind of sorption media that will capture all gases and vapors, and for this reason NIOSH has determined a set of chemicals for which cartridges can be tested and approved. In some cases, the cartridge is tested with and approved for a single chemical, such as chlorine or ammonia. In other cases, the cartridge is tested with a representative of a class of chemicals. For example, the approval test for the class of chemicals called "organic vapors" is performed using a single organic chemical. Because there are many more chemicals being used in the workplace than there are NIOSH approvals, it is important that the correct chemical cartridge be selected. If there is any question as to whether a cartridge is appropriate for a given contaminant, an industrial hygienist or the respirator manufacturer should be contacted. Chemical cartridges will not remove particles unless they are paired with a particulate filter.

SURGICAL MASKS

Health care providers routinely use surgical masks as part of their personal protective equipment (PPE).

Surgical masks were originally designed to protect the patient against large droplets expelled by the mask wearer. Traditional surgical masks have not been designed to protect the wearer from inhaling airborne particles created by infected patients; however, they have been mistakenly used by some hospitals for this purpose. Surgical masks do not provide adequate protection against inhalation of infectious aerosols, and personal respiratory protection should be employed. Surgical masks are not recommended to reduce exposures to *Mycobacterium tuberculosis*, Middle East respiratory syndrome coronavirus (MERS-CoV) (the virus that causes MERS), or other biological agents.

Surgical masks designed only to protect patients against large particles expelled by the wearer or the wearer from fluid splashes may not exhibit either high collection (filtration) efficiency or adequate face fit. Unlike respirators, surgical masks are not subject to standard filter certification tests. Each manufacturer is responsible for evaluating its claims. In addition, surgical masks are not required to fit the face or be tested for fit on an individual.

The overall efficiency, with regard to very small particles, of a surgical mask is only as good as the fit of the mask. Masks developed as barriers against expulsions from a wearer's nose and mouth or as a barrier against fluid splashes were designed to catch very large droplets. Therefore, tight fit was not a priority. Air will take the path of least resistance and travel through gaps between the mask and face. Small particles (such as bacteria and viruses) may be able to follow the airstream and by doing so enter the interior of the mask. Research has shown that face seal leakage can be a significant factor in lowering the filtration efficiency of surgical masks (6). Studies have found that the filtration media of surgical masks have a wide range of effectiveness when challenged with aerosols less than 1 μm in size (6–9). Whereas some surgical masks demonstrate filtration efficiencies below 50%, others have been found to be over 95% efficient. However, this effectiveness is of little importance when the lack of fit may allow significant particle leakage (6, 9).

Surgical respirators are NIOSH approved as respiratory protection devices as well as having clearance from the Food and Drug Administration (FDA) to be sold as a medical device. Use of these combination surgical masks/respirators must be included in a complete respiratory protection program that incorporates medical clearance and fit testing. These are the only types of surgical masks that can be relied upon to provide respiratory protection to the wearer. For example, a combination surgical mask/NIOSH-approved respirator, such as the one pictured in Fig. 2B, would meet the requirements for helping to control exposures to *M. tuberculosis*.

RESPIRATOR PROTECTION FACTORS

The level of protection provided by a certain type of respirator is called the protection factor. Assigned protection factors (APFs) are designated by type of respirator. The APF is defined as "the minimum expected workplace level of respiratory protection that would be provided by a properly functioning respirator or class of respirators, to a stated percentage of properly fitted and trained users" (10). In simplified terms, the APF is the factor by which a properly selected and fitted respirator will reduce contaminant exposures. If an APF is 10, then the concentration of contaminant that reaches the wearer's lungs will be reduced by a factor of 10.

Several groups, including the American National Standards Institute (ANSI) and NIOSH, have determined APFs; however, those designated by OSHA are the only ones with legal significance in the United States (11–14). In 2006 OSHA issued a revision to 29 C.F.R. 1910.134 which included a final rule on APFs. This rule clarified APFs for NIOSH-approved respirators (15). OSHA APFs for commonly used respirator types can be found in Table 3. APFs only apply if the respirator is properly

TABLE 3.

OSHA APFs for commonly used respirators[a]

Type of respirator	OSHA APF (OSHA, 2006) (43)
Air-purifying, nonpowered	
Half-mask	10
Full facepiece	50
PAPR	
Half-mask	50
Full facepiece	1,000
Loose-fitting facepiece	25
Hood or helmet	25/1,000[b]
Air-line respirator	
Continous flow	25/1,000[b]
Full facepiece	
Hood or helmet	
SCBA, pressure demand if quantitative fit testing is used	10,000

[a]OSHA, Occupational Safety and Health Administration; APF, assigned protection factor; PAPR, powered air-purifying respirator; SCBA, self-contained breathing apparatus; WPF, workplace protection factor; SWPF, simulated workplace protection factor; SAR, supplied-air respirator.

[b]The employer must have evidence provided by the respirator manufacturer that testing of these respirators demonstrates performance at a level of protection of 1,000 or greater to receive an APF of 1,000. This level of performance can best be demonstrated by performing a WPF or SWPF study or equivalent testing. Absent such testing, all other PAPRs and SARs with helmets/hoods are to be treated as loose-fitting facepiece respirators and receive an APF of 25.

selected and used within a respiratory protection program per 29 C.F.R. 1910.134. APFs are used in selecting a respirator that will afford protection in a given environment. The occupational exposure limit (OEL) and concentration (C_o) of a contaminant are used to determine the class of respirator needed to prevent overexposure through calculation of a hazard ratio (HR), as follows: C_o/OEL = HR. A respirator with an APF equal to or greater than the HR should be selected. If the HR lies between APFs, the most conservative (higher APF) class of respirators should be chosen. A lower APF may result in overexposure of the wearer.

OCCUPATIONAL EXPOSURE LIMITS

OELs are the concentrations of gases and vapors or particulates to which workers can be exposed during a typical workweek for a typical working lifetime, without adverse health effects. Several organizations set OELs in the United States. The American Conference of Governmental Industrial Hygienists (ACGIH) annually publishes OELs called threshold limit values (TLVs) (13). They consider epidemiological data as well as toxicology studies in setting these values. In 1970, OSHA adopted many of the 1968 threshold limit values for airborne substances and renamed them permissible exposure limits (PELs). These have been periodically reviewed and updated since. They can be found in 29 C.F.R. 1910.1000 (Table Z-2) and are enforceable by law.

OELs have not been set for biological aerosols, and there are no accepted safe levels of exposure. Therefore, respiratory protection can only reduce exposure to biological aerosols but cannot completely eliminate the risk of contracting a disease or infection.

REGULATIONS REGARDING RESPIRATOR USE

In the United States, two agencies hold responsibility for regulating the use of respirators in the workplace. NIOSH approves respirators for use in the workplace. This responsibility was shared with the Bureau of Mines and the Mine Safety and Health Administration until 1998, when it was transferred solely to NIOSH, with the exception of escape respirators used in mining. Respirators that are marketed for use in surgical settings, or are also designed as surgical masks, may be considered medical devices. In this case, respirators must be cleared for use with the FDA. OSHA has the responsibility for enforcing proper respirator use in the workplace, which includes industry, laboratories, and hospitals and other medical facilities.

Respirator Approval

All respirators used in the workplace to help protect individuals against exposure to airborne contaminants must be certified, in the United States, by NIOSH. Respirators are certified as entire systems, not as individual pieces. All of a respirator's components must be available through the manufacturer for use as a system (see Table 1 for components) and approved together by NIOSH. The one exception, for air-line respirators, is that the air compressor or gas cylinder does not have to be approved, but the air hose from the compressor or cylinder to the inlet valve on the respirator must be NIOSH approved with that respirator system.

Modern respirator performance requirements were first detailed in 30 C.F.R. Part 11 (16) and included four certification categories for nonpowered, air-purifying, particulate respirators: (i) dust, (ii) mist, (iii) fume, and (iv) high-efficiency particulate air (HEPA) filter. Two certification categories were available for PAPRs: dust/mist and HEPA. Air-purifying respirators certified under this standard can still be used, but manufacturers are no longer allowed to manufacture them.

In 1995, NIOSH adopted 48 C.F.R. Part 84. This standard included new certification procedures for nonpowered, air-purifying, particulate respirators (14) which created nine classes of particulate filters (for nonpowered, air-purifying respirators) as previously mentioned (see Table 2). This standard also eliminated the dust/mist category for PAPR filters.

OSHA 29 C.F.R. 1910.134: The General Respiratory Protection Standard

In 1971, OSHA adopted 29 C.F.R. 1910.134, the general respiratory protection standard. A revised standard became effective in April of 1998. This standard describes the requirements for using respiratory protection in all workplaces (including those with exposures to biological aerosols such as *M. tuberculosis*, Ebola virus, etc.). Respiratory protection is considered a temporary or last-resort solution to controlling exposures to airborne contaminants. Therefore, 29 C.F.R. 1910.134 requires that engineering controls be utilized first if they are available and feasible. It is required that, when necessary, employers provide the respiratory protection and be responsible for establishing a respirator program. Employees are required to use respirators in accordance with the training and instructions.

The employer must establish a written respirator protection program that is maintained, updated as necessary, and made available upon request. The program must detail the following nine procedures regarding respirator use in that workplace:

- Respirator selection for the specific workplace and hazards
- Medical evaluation of workers before being issued respiratory protection (Appendix C of 29 C.F.R. 1910.134 includes a questionnaire that must be reviewed by a licensed health care professional and retained by the employer)
- Proper use of respirators in the workplace for both routine and emergency situations, including a schedule of when to replace canisters and cartridges
- Initial and annual fit testing for respirators with tight-fitting facepieces
- Cleaning, storage, and maintenance of respirators
- Adequate breathing air for atmosphere-supplying respirators ensured, with the supplied air at least meeting the requirements for type 1 grade D breathing air and requirements for oxygen, hydrocarbon, carbon monoxide, and odor
- Training in hazards and proper use, including selection, emergency use, and the general requirements of the standard
- Evaluating the effectiveness of the program as necessary, including fit, selection, use, and maintenance

The program must designate a program administrator. This person must be qualified for the position through training or experience that is proportionate to the complexity of the program. The administrator supervises or administers the program and conducts the evaluations of program effectiveness. The training needed to become a qualified program administrator is available through courses offered by NIOSH education and resource centers, OSHA training institutions, organizations such as the American Industrial Hygiene Association (AIHA) and ACGIH, universities, and respirator manufacturers.

While the above paragraphs have highlighted the important parts of a respiratory protection program, 29 C.F.R. 1910.134 should be thoroughly reviewed before implementation of a respiratory protection program in a workplace. Additional resources that may be useful are listed at the end of this chapter and should be consulted as necessary.

RESPIRATOR FIT AND FIT TESTING

In general, respirators are designed to cover the nose and mouth of the wearer and to provide clean air to the respiratory system. How well the respirator seals to or "fits" on the wearer's face is a main component in ensuring that filtered or breathable air is delivered to the respiratory system. There are several types of facepieces that form different types of seals around the face and head. Two main categories are tight-fitting facepieces and those that do not fit tightly against the face, such as helmets, hoods, and loose-fitting facepieces.

Tight-fitting respirators must fit tightly against the face and form a seal between the edge of the respirator and the face of the wearer. This seal is a major factor in reducing the amount of air contaminants that can enter the respiratory system of the wearer. Gases, vapors, and very small particles can leak into a poorly fitting mask and dramatically reduce its protective capabilities. Even small amounts of facial hair on men have been shown to reduce the protective abilities of a tight-fitting respirator (17).

It has been recognized that a properly fitting facepiece is essential to the successful operation of a respirator. Several user requirements have been implemented to help ensure a well-fitting respirator. These requirements are for all respirators with tight-fitting facepieces and include testing the fit of the respirator (fit test) each time a new respirator model is assigned and at least annually thereafter, inspecting the respirator condition before each donning, and checking the seal each time the respirator is donned (user seal check).

The fit test is a method to determine if the respirator can achieve an adequate seal with the face of the respirator wearer. All persons issued NIOSH-approved respirators with tight-fitting facepieces must be fit tested prior to using them in the workplace. Before a respirator is issued, the worker must successfully pass a fit test or another respirator make, model, or size must be tried until the fit test is passed. Fit testing must be conducted prior to initial use; whenever a different make, model, or size of respirator is used and at least annually thereafter. Fit testing must be conducted more often if there are changes in the employee's physical condition that may affect fit (i.e., changes in weight, scarring, or cosmetic surgery). The only respirators that do not need to be fit tested are those with hoods, helmets, or loose-fitting facepieces.

Workers must be clean-shaven (within the last 24 hours) in the area where the respirator seals to the face prior to the fit test and every time the respirator is used. The only facial hair that is acceptable for use with a tight-fitting respirator is hair that does not come directly between the face and the edge of the respirator and does not interfere with valve performance. Fit testing and respirator use are prohibited if there is any hair growth between the skin and the respirator sealing surface, including stubble, beard, moustache, or sideburns. In addition, facial hair, such as a long beard, which may interfere with the respirator fit or function is not allowed with respirators with hoods, helmets, or loose-fitting facepieces.

There are two means by which to fit test respirators: quantitative and qualitative. Quantitative fit testing utilizes a probed respirator and a means to count the particles inside and outside of the respirator. Qualitative fit testing utilizes a substance that is sprayed around the respirator and that has a distinctive taste or smell, which is identified by the wearer if there is an inadequate seal. The two most common substances used for qualitative fit testing are sweet tasting (saccharin) and bitter tasting (Bitrex). Descriptions of these methods are included in 29 C.F.R. 1910.134. Records of fit testing must include identification of the employee; type of fit test; make, model, style, and size of the respirator; date of test; and test results. These records must be retained until the next fit test is administered.

RESPIRATOR SELECTION

Current Respirator Decision Logic

NIOSH has developed a process by which the proper respirator for the hazard and environment can be selected. This is referred to as the NIOSH Respirator Decision Logic and was developed for selecting respirators for exposures to industrial agents with OELs. It involves a systematic elimination of inappropriate respirators until an acceptable respirator can be chosen. The selection process consists of a series of steps that require the user to make choices based on the nature of the atmosphere, the concentration of the contaminant, and the work to be performed. The following section outlines the five steps of the decision process.

The first step in choosing a respirator involves determining the nature of the atmosphere and the work to be performed. If the atmosphere in which the respirator will be worn is oxygen deficient (<19.5% O_2) or if the work involves firefighting, the most protective respirator, a pressure-demand SCBA, is required.

The second step in choosing a respirator involves determining if the level of the contaminant is IDLH. Determining whether a workplace is IDLH depends on knowledge of the contaminant concentration, which can be directly measured in most cases or anticipated by examining the literature for published studies of similar workplaces and processes or by investigating historical air monitoring records from the workplace. The most accurate estimate of contaminant concentration is typically through direct measurement. If the employer cannot reasonably estimate the exposure, the atmosphere should be considered IDLH. If an IDLH condition exists, then a full-facepiece, pressure-demand SCBA with 30 min of use or a combination full-facepiece, pressure-demand SAR with an auxiliary self-contained air supply should be chosen.

The physical nature of the contaminant (gas, vapor, particulate, or mixture) is then identified (step 3).

The fourth step is selection of a respirator for the specific workplace atmosphere. This is done by determining the HR, which is a function of the workplace concentration and the OEL, as described earlier. A respirator with an APF equal to or greater than the HR is selected. In addition to APFs, the contaminant, work, and worker factors are considered when selecting a particular respirator. These include overall worker health, eye irritation, interference with work practices, comfort, wearing time, and battery life.

If an air-purifying respirator is chosen, the respirator must be equipped with the appropriate air-purifying elements, which is the fifth step of the selection process. Filters that remove particulates, canisters and cartridges that remove gases and vapors, and a combination are available. As mentioned earlier, particulate filters are classified by efficiency and the filter's ability to be used in workplace atmospheres containing oil. The necessary efficiency level is chosen first. If OSHA requires that a 99.97% efficient filter (or a HEPA filter) be used for a specific substance, then a 100-level filter should be used. If there is no substance-specific recommendation, as with most biological hazards, then a 95- or 99-level filter can be used. Where they exist, substance-specific standards should be reviewed for respirator recommendations. Next it must be determined if the atmosphere contains oil. If it does not, any of the filters can be used. If it does, then an R or P series filter must be used. Filters contain time use limitations, and the manufacturer's instructions should be reviewed when setting a filter change-out schedule.

When selecting chemical cartridges or canisters for capture of gases and vapors, a NIOSH-approved canister, cartridge, or combination canister-filter or cartridge-filter should be selected that has been designed to remove that chemical or class of chemicals. If it is not clear whether a given cartridge or canister will remove a particular chemical, the manufacturer should be consulted. Several manufacturers offer guides for selecting chemical cartridges, and at the time of this writing, OSHA was offering a program (the Respiratory Protection Advisor) through its website to instruct users on the proper selection of respiratory protection and the development of change schedules for gas/vapor cartridges.

Cartridges (and canisters) for gases and vapors used to be replaced when the user began to taste or smell the chemical or experienced sensory irritation under the previous standard (prior to 1998). However, OSHA has ruled that this is no longer an acceptable practice. In the current version of 29 C.F.R. 1910.134, one of the mandatory pieces of a respiratory protection program is development of a cartridge change-out schedule that is not based on the wearer's sensory information. OSHA stipulates that the change-out schedule for gas and vapor cartridges must be based on objective information and data that ensure that the cartridges are replaced before the end of their service life. One method of determining service life is to use a mathematical model. There are several of these available to the public, including the OSHA Respiratory Protection Advisor and others available through manufacturers. Very short change-out times indicate that atmosphere-supplying respirators or engineering controls may be a better solution than an air-purifying respirator for that particular gas and vapor exposure.

CALIFORNIA OSHA AEROSOL-TRANSMISSIBLE DISEASE STANDARD

In 2009 the California Occupational Safety and Health Administration published the first U.S. standard applying specifically to workplace exposures to aerosol-transmissible diseases (ATDs) (18). This standard applies to multiple settings, including health care institutions, first responders, and laboratories that may handle aerosol-transmissible pathogens. Aerosol-transmissible pathogens include those for which *Biosafety in Microbiological and Biomedical Laboratories* (BMBL) (1) or the site biosafety officer recommends a BSL3 and those in Appendix D of the standard. The standard is comprehensive and should be thoroughly reviewed by everyone in California potentially working with aerosol-transmissible pathogens or patients with ATDs. Although this standard is not enforceable outside of the state of California, it is considered a best practice in the United States and may be used by anyone working in the field.

With regard to respiratory protection, the standard requires that engineering and administrative controls be implemented first. The standard states that PPE, including respiratory protection, shall be provided by the employer when the potential of exposure remains after implementation of other control methods. It states that control methods shall be consistent with BMBL.

When performing care and tasks for suspected or confirmed airborne infectious disease cases or cadavers, respirators must be used that are at least as protective as an NIOSH-approved "N95" respirator (APF = 10) unless the biosafety officer decides that a more protective respirator is necessary. In 2010, the standard was updated to specify that a PAPR with a HE filter is necessary for those employees who perform high-hazard procedures on suspected or confirmed airborne infectious disease cases or cadavers unless the employer determines that this type of respirator could interfere with work tasks. There are exceptions for certain situations (e.g., medical care in a helicopter), and the full standard should be reviewed for a comprehensive list.

All respirator use in the California OSHA ATD standard must be in accordance with the full specifications in the standard, which are generally in line with the federal OSHA respiratory protection standard 29 C.F.R. 1910.134.

RECOMMENDATIONS PERTAINING TO MICROBIOLOGICAL LABORATORIES

The 5th edition of BMBL (1) recognizes that aerosols are a source of laboratory-acquired infections (LAI). Procedures such as pipetting, blenders, centrifuges, vortex mixers, and sonicators are known sources of aerosols. Good practices and engineering controls should always be the first line of defense to help reduce the generation of aerosols and workers' exposure to them.

Respiratory protection is considered a primary piece of safety equipment at BSL2 and BSL3. Respirators are also to be used at BSL2, BSL3, and BSL4 in rooms containing infected animals (as determined by the risk assessment). All workers issued respiratory protection must be a part of an appropriate respiratory protection program.

Respirators are included in the list of safety equipment to be considered, where appropriate, when working outside of a biological safety cabinet (BSC) at BSL3. Respiratory protection is specifically called for when the worker is in a room containing an infected animal. When working at vertebrate animal BSL2 (ABSL2), respirators should be considered whenever procedures with a high potential for creating aerosols (i.e., centrifugation, grinding, blending, vigorous shaking or mixing, sonic disruption, necropsy, intranasal inoculation, and harvesting infected tissues) are conducted. At ABSL3, BMBL calls for respiratory protection to be worn by all personnel entering animal rooms and for all work that is not done within a primary barrier. Guidance for respirator selection is not included.

Personnel working at BSL4 or ABSL4 may be provided with a one-piece positive-pressure suit that is ventilated by a life support system protected by HEPA filtration (1). The life support system should include redundant breathing air compressors, alarms, and backup breathing air cylinders. The supplied air should meet the requirements for grade D air (19). Anyone using these suits should be included in a full respiratory protection program and be given adequate time to adjust to working in the system.

In 2012 the CDC convened a special panel to review biosafety in laboratories. They made several references to respiratory protection. They suggested that fit-tested N95 respirators or other appropriate respiratory protection be worn by workers in mycobacteriology and virology laboratories, where organisms are manipulated. A risk assessment should be done to determine the appropriate level of respiratory protection and other additional safeguards (20).

SPECIFIC RECOMMENDATIONS FOR EXPOSURES TO AIRBORNE MICROORGANISMS

There are few recommendations for use of respiratory protection to control exposures to specific aerosolized microorganisms, or bioaerosols as they are sometimes called. These recommendations are typically very sparse and only include the minimum acceptable respirator with no explanation of the selection process. An accepted premise across agencies and academic institutions is that aerosolized microorganisms are particles and will be removed by filters with at least the same efficiency as nonbiological particles. Therefore, if an air-purifying respirator is selected, a particulate filter will remove the organism with at least the efficiency it is certified for (e.g., 95% for an N95 filter) (21, 22). Although properly used and fitted respirators can help reduce exposures to airborne microorganisms, they do not eliminate all risk and may not prevent disease, illness, or death.

Table 4 presents a general summary of many of the existing respirator recommendations for bioaerosols. Seven of these recommendations are discussed in detail below. This chapter only contains recommendations regarding respirators; therefore, the references should be consulted for other PPE (e.g., eye, face, and skin protection) recommendations and special procedures before work with these organisms is conducted. Because most of these recommendations are not made in regulatory documents, the reader should be aware that these recommendations may change or new ones may be issued at any time with little publicity. It is recommended that the current literature be reviewed periodically.

In the United States, any time an employer requires a worker to wear a respirator, all aspects of the OSHA General Respiratory Protection standard (29 C.F.R. 1910.134) must be followed. Therefore, the following recommendations must be implemented within a complete respiratory protection program.

Severe Acute Respiratory Syndrome

Several cases of severe acute respiratory syndrome coronavirus (SARS-CoV) disease have been reported for workers in research laboratories where SARS-CoV was being propagated. The CDC has issued "Public Health Guidance for Community-Level Preparedness and Response to Severe Acute Respiratory Syndrome (SARS) Version 2" (23). With respect to respiratory protection, the CDC currently recommends that manipulation of specimens such as respiratory secretions, stool, or tissues for procedures performed in laboratories be conducted at BSL2 and in a class II BSC. They recommend that SARS-CoV propagation in cell culture and initial characterization of viral agents recovered in cell cultures of SARS specimens be conducted in a BSL3 facility using BSL3 work practices. If the procedure cannot be conducted in a BSC, then PPE should be worn, including respiratory protection. The CDC lists acceptable methods of respiratory protection as including a properly fit-tested, NIOSH-approved "filter respirator (N95 or higher) or a powered air-purifying respirator (PAPR) equipped with HEPA filters" (24). Therefore, the minimum acceptable level of respiratory protection is a NIOSH-approved half-facepiece respirator equipped with particulate filters. However, higher levels of respiratory protection (e.g., PAPR) should be considered as appropriate.

Respiratory protection for other workers is also addressed in the CDC's guidance, including for health care workers. In addition, new laboratory guidelines may be introduced or revised at any time. Therefore, any worker who may be exposed to SARS-CoV should review the most current guidances from the CDC and other appropriate agencies before exposure.

Bacillus anthracis

The 2001 incidents involving *Bacillus anthracis* spores, or suspected *B. anthracis* spores, deliberately spread through the U.S. Mail heightened awareness regarding the potential need for respiratory protection when dealing with this bacterium in a laboratory setting. Prior to these incidents, the CDC had issued "Basic Laboratory Protocol for the Presumptive Identification of *Bacillus anthracis*" (25). This document states that respiratory protection should be considered when using materials or conducting analytical procedures determined to be potentially hazardous outside of a BSC. In addition, personnel involved in cleaning up a spill with the potential for aerosolization should consider respiratory protection.

In April of 2002, the CDC issued "Comprehensive Procedures for Collecting Environmental Samples for Culturing *Bacillus anthracis*" (26). In addition to specifying the types of laboratory facilities necessary for sample analysis, this document stresses that safety considerations for laboratory personnel are paramount and that proper procedures, such as those outlined in the 4th edition of BMBL, must be followed to reduce exposures during analysis. Regarding anthrax, the 4th edition of BMBL states that BSL2 practices are appropriate for clinical materials and diagnosis of cultures. ABSL2 is appropriate for infected laboratory rodents, and BSL3 should be used for work involving production quantities or concentrations of cultures and for activities with a high potential for aerosol production.

TABLE 4.

Recommendations relating to respirator use for biosafety applications[a]

Reference	Agent(s)	Activity or type of worker	Minimum respirator recommendation
CDC/NIH, 2009 (1)	Those used in microbiological laboratories	Working outside of BSC at BSL3	General respiratory protection (no specific respirator identified)
		Working at ABSL2 (vertebrates) when performing procedures with a high potential for creating aerosols	
		Working at ABSL3 (vertebrates) when entering animal rooms and working outside of a primary barrier	
CDC/NIH, 2009 (1)	Arboviruses and arenaviruses	Working with those assigned to BSL3 when enhancements to BSL3 practices are necessary	"Appropriate respiratory protection"
CDC, 2001 (44)	*Bacillus anthracis*	Working with or near machinery capable of generating aerosolized particles (e.g., electronic mail sorters) or at other work sites where such particles may be present	NIOSH-approved respirators at least as protective as an N95 respirator
NIOSH, 2001 (45)	*Bacillus anthracis*	Conducting environmental sampling	PAPRs with high-efficiency filters
CDC, 1999 (46)	*Blastomyces dermatitidis*	Disturbing the soil during prairie dog relocation	Half-mask respirator with N95 filters
CDC, 1997 (43, 47); CDC/NIH, 2009 (20)	*Chlamydia psittaci*	Working at risk of exposure to *C. psittaci* (handling infected birds, performing necropsies, cleaning cages)	Half-mask respirator with N95 filters
United States Army (48) Environmental Hygiene Agency, 1992 (48)	*Cryptococcus neoformans*, *Histoplasma capsulatum*	Cleaning and removing bird and bat droppings	Full-facepiece respirator with HEPA filters (100-level filters would also be appropriate) or full-facepiece air-line respirator
CDC, 2015 (35)	Ebola virus	Laboratory testing in a Class I or II BSC	Respiratory protection optional
CDC (36)		Employees collecting samples from patients	Respiratory protection including an N95 or PAPR
CDC, 1993, 1995 (32, 49, 50)	*Hantavirus*	Removing organs or obtaining blood from rodents potentially infected with hantavirus at a field processing site	Half-mask respirator (no filter identified) or PAPR
		Working with organism in the laboratory	Follow BMBL guidelines
		Cleaning in homes of persons with hantavirus infection or buildings with heavy rodent infestation	Half-mask respirator with HEPA filters (100-level filters would also be appropriate) or PAPR with HEPA filters
CDC, NIOSH, NCID (1997) (51)	*H. capsulatum*	During removal of an accumulation of bat or bird manure from an enclosed area such as an attic	A NIOSH-approved respirator with HEPA filters or any 42 C.F.R. Part 84 particulate filter
		In dusty conditions where spores are present	A full-face NIOSH-approved respirator
		During site surveys of bird roosts, collecting soil samples, or maintenance on filters of earth-moving equipment	Disposable or elastomeric half-facepiece respirators
	C. neoformans	When cleaning chimneys, working in attics, and in poultry houses	Respirators with higher APFs
		Same activities as for *H. capsulatum*	Same PPE recommendations as for *H. capsulatum*
CDC (1)	Novel influenza viruses: 1818 strain or HPAI	Any work	Negative pressure respirators with high efficiency (~100 level) filters or PAPRs
CDC (34)	Influenza A(H7N9)	Work at BSL3 and whenever aerosols may be generated; processing samples and/or performing non-culture-based diagnostic testing	Respiratory protection such as an N95 or PAPR

(continued)

TABLE 4.

Recommendations relating to respirator use for biosafety applications[a] (*continued*)

Reference	Agent(s)	Activity or type of worker	Minimum respirator recommendation
Wisconsin Department of Health and Social Services, 1987 (52)	*Legionella* spp.	Cleaning cooling towers and related equipment	Full-facepiece or half-mask air-purifying respirator with HEPA filters and chemical cartridge for chlorine, if used (100-level filters would also be appropriate)
CDC, 2015 (37)	MERS virus	Laboratory workers that may be handling potentially infectious MERS-CoV specimens	NIOSH-approved respirators
CDC, 1994 (27)	*Mycobacterium tuberculosis*	Workers caring for and interacting with patients with known or suspected tuberculosis cases in health care, correctional, long-term care, homeless, and drug treatment facilities	Half-mask air-purifying respirators with N95 filters
CDC/NIH, 1997 (28)	*M. tuberculosis complex*	Collecting sputum samples from patients not enclosed in a booth	Half-mask air-purifying respirators with N95 filters or PAPR
		Workers manipulating *M. tuberculosis* cultures in BSL3 laboratories	Half-mask air-purifying respirators with N95 filters
CDC/NIH, 2009 (1)	Prions	During autopsies of patients with suspected prion disease	"Appropriate respiratory protection"; PAPRs given as an example
CDC, 2004 (23)	SARS-CoV	Laboratory procedures conducted on SARS-CoV that cannot be conducted in a BSC	NIOSH-approved respiratory protection
CDC, 2003 (24)	SARS-CoV	Health care workers caring for patients with SARS	NIOSH-certified respirators
New York City Department of Health, 1993 (53)	*Stachybotrys*	Remediation of low levels of contaminant (<30 ft^2)	Half-facepiece respirator (no filter identified)
		Remediation of high levels of contaminant (<30 ft^2) or of HVAC systems	Full-facepiece respirator with HEPA filters or PAPR

[a]BSC, biosafety cabinet; BSL, biosafety level; ABSL, animal biosafety level; NIOSH, National Institute for Occupational Safety and Health; PAPR, powered air-purifying respirator; NCID, National Center for Infectious Diseases; NIOSH, National Institute for Occupational Safety and Health; HEPA, high-efficiency particulate air; APF, assigned protection factor; PPE, personal protective equipment; HPAI, highly pathogenic avian influenza; MERS-CoV, Middle East respiratory syndrome coronavirus; SARS, severe acute respiratory syndrome coronavirus; HVAC, heating, ventilation, and air-conditioning.

Mycobacterium tuberculosis

In 1994, the CDC issued "Guidelines for Preventing the Transmission of *Mycobacterium tuberculosis* in Health Care Facilities" (27). It specified that respiratory protection should be used by personnel entering rooms in which patients with known or suspected infectious tuberculosis are being isolated, those present during cough-inducing or aerosol-generating procedures, and those in other settings where other control methods may not be adequate. These other settings may include patient transport or surgical or dental care. The guidelines outlined minimum criteria for acceptable respiratory protection. A respirator must be used that can be qualitatively or quantitatively fit tested in a reliable way, obtain a face seal leakage of no more than 10%, fit workers with different facial characteristics and sizes, and be checked for facepiece fit in accordance with OSHA standards. In addition, the respirator must be equipped with a filter that can capture 1-μm particles with an efficiency greater than or equal to 95% at a flow rate of 50 liters per minute in an unloaded state (28). All of the particulate respirators approved under 42 C.F.R. Part 84 NIOSH certification procedures and all atmosphere-supplying respirators meet these criteria (14). However, during procedures requiring a sterile field, the CDC recommends that respirators with exhalation valves and respirators that may be under positive pressure not be used. (Although NIOSH does not specify any type of respirator as positive pressure, respirators commonly referred to as such are pressure-demand SARs, continuous-flow SARs, PAPRs, and pressure-demand SCBAs.)

The 1994 CDC guidelines list a few circumstances in which the risk may be such that a level of respiratory protection exceeding the minimum criteria may be appropriate. These situations include, but are not limited to, bronchoscopy performed on patients with suspected or known tuberculosis and autopsy performed on deceased persons who were suspected of having or known to have tuberculosis.

In 1997, the CDC published proposed guidelines for working with *M. tuberculosis* in laboratories (28). The goal of this document was to present health and safety information, to be used in conjunction with BMBL, for those persons working with *M. tuberculosis* in laboratories. This document was a proposal, and public comments were collected. However, a revised or final draft was not published. With regard to respiratory protection, the CDC proposed that during collection of sputum specimens in an open laboratory the worker wear an air-purifying respirator with either N100 or HEPA filters. (However, if a patient is enclosed in a negative-pressure booth with HEPA-filtered exhaust, then respiratory protection is not necessary.) In addition, the guidelines recommended that all personnel working with *M. tuberculosis* in BSL3 laboratories wear an air-purifying respirator with N95 filters.

In February 1996, OSHA published enforcement procedures for occupational exposure to tuberculosis (29). This document indicated that inspections would be conducted in response to employee complaints, in response to fatalities and catastrophes, or in those workplaces where the incidence of tuberculosis infection is greater than in the general public. These workplaces include health care facilities, correctional institutions, long-term care facilities for the elderly, homeless shelters, and drug treatment centers. With regard to respiratory protection, a written program must be implemented according to 29 C.F.R. 1910.134 and meet the performance criteria for respiratory protection outlined in the 1994 CDC guidelines.

On 17 November 1997, OSHA published a proposed rule on occupational exposure to tuberculosis (30). In 1998, 29 C.F.R. 1910.134 was revised and became effective. The original standard was renamed 29 C.F.R. 1910.139 and was applied temporarily to occupational exposures to *M. tuberculosis*. On December 31, 2003, OSHA rescinded the temporary standard for respiratory protection for *M. tuberculosis*, 29 C.F.R. 1910.139, and withdrew the proposed standard for occupational exposure to *M. tuberculosis*. Currently respiratory protection for exposures to *M. tuberculosis*, and all airborne contaminants, is covered by 29 C.F.R. 1910.134.

Hantavirus

The CDC has issued guidelines specific to laboratories that handle agents associated with hantavirus pulmonary syndrome. LAIs have occurred among persons who handled infected wild or laboratory rodents (31). It is believed that these infections occurred through inhalation of virus-containing animal waste materials. The guidelines recommend that laboratory work that may result in propagation of hantaviruses be conducted in a BSL3 facility. For work not conducted in the laboratory, the CDC has recommended that those persons processing rodents (trapping or handling them or performing necropsies) wear a NIOSH-approved half-mask respirator or PAPR (32). These recommendations apply to workers cleaning homes of persons with confirmed hantavirus infection, workers cleaning buildings with heavy rodent infestation, and workers in affected areas who are regularly exposed to rodents (33).

Influenza Viruses

In the 5th edition of BMBL (1), there is a significant section on influenza viruses that should be reviewed by all laboratories that may be processing these strains. Influenza can be spread by multiple routes, including the airborne route, and historically has caused significant outbreaks; therefore, handling of novel and certain, highly pathogenic strains should be done very carefully and with consideration toward respiratory protection.

The BMBL recommends that any work with the 1818 strain of influenza or highly pathogenic avian influenza (HPAI) viruses include many control methods, including the rigorous adherence to the use of negative-pressure respirators with high-efficiency filters or PAPRs. However, it should be noted that high-efficiency or HEPA filters are no longer available for negative-pressure respirators. The closest equivalents for current NIOSH-approved respirators are N100 or P100 filters.

When new strains of influenza viruses emerge, as is the case with influenza A (H7N9) virus, the CDC may put forth special respiratory protection recommendations for laboratory workers and other potentially exposed workers as they did in "Interim Risk Assessment and Biosafety Level Recommendations for Working With Influenza A(H7N9) Viruses" (34). Respiratory protection is required for all activities at BSL3 and whenever the formation of aerosols is anticipated. Respirators should be selected on the basis of the type of activity being conducted. Suggested respirators include a PAPR with HE filter or a full face-piece respirator with an N95 or HEPA filter, "followed by demonstrating competence in wearing a HEPA-filtered powered air-purifying respirator (PAPR) or a properly fit tested full face respirator with HEPA or N95 particulate protection. The selection of an appropriate respirator depends upon the type of activity anticipated in the laboratory. Training in the effective use of respirators is mandatory and requires yearly certification by your institution's occupational health and safety program" (34).

They recommend workers wear an N95 respirator with eye protection or a faceshield or PAPR when processing samples and/or performing nonculture-based diagnostic testing on clinical specimens from patients and animals with suspected influenza A (H7N9) virus infection in a BSL2 laboratory.

Ebola Virus

In the "Guidance for U.S. Laboratories for Managing and Testing Routine Clinical Specimens When There is a Concern About Ebola Virus Disease," the CDC (35) sets forth considerations for selection and proper use of PPE. They emphasize that PPE should be selected on the basis of the risk assessment, work tasks, and user capabilities.

The CDC recommends that all laboratory staff be thoroughly trained in donning and doffing of PPE and that strict adherence to procedures be observed at all times. When performing laboratory testing, a certified Class I or Class II BSC should be used along with gloves, gowns, eye protection, and a surgical mask. If respiratory protection is selected instead of a surgical mask, employees must be medically cleared, fit tested, and trained, including training on donning and doffing. It is recommended that employees practice with all PPE before wearing them during work with Ebola specimens.

Employees collecting samples from patients should wear the same PPE recommended for health care workers (36). The CDC recommends that all health care workers entering the room of a patient with Ebola wear respiratory protection that would protect them during an aerosol-generating procedure. The suggestions range from an N95 filtering facepiece particulate respirator to a PAPR. The CDC recommendations around training, donning, and doffing, etc., are quite comprehensive. Those involved with sample collection should thoroughly review the recommendations and receive extensive training.

Middle East Respiratory Syndrome Coronavirus

In the CDC's "Interim Laboratory Biosafety Guidelines for Handling and Processing Specimens Associated with Middle East Respiratory Syndrome Coronavirus (MERS-CoV)—Version 2" (page last updated June 18, 2015) (37), laboratory workers that may be handling potentially infectious MERS-CoV specimens should wear PPE. Recommended PPE includes disposable gloves, laboratory coat/gown, eye protection, and a respirator. NIOSH-approved respirators are acceptable, including filtering facepiece respirators (N95 filters minimum) or a PAPR with HE filters.

RESPIRATOR SELECTION FOR EXPOSURES TO AEROSOLIZED MICROORGANISMS

The standard procedure for selecting respirators is based on a decision process that allows systematic elimination of available NIOSH-certified respirators until an appropriate respirator is found (12, 38). This procedure uses the ambient airborne concentration and OEL of a contaminant to determine the maximum use concentration and HR. A respirator class with an APF equal to or greater than the minimum APF is then selected. If an air-purifying respirator is chosen, the final step of the selection procedure is to select a filter or chemical cartridge that is appropriate for the contaminant of interest.

To use the NIOSH decision logic, information about the aerosol contaminant must be available. This includes the aerosol identity, probable airborne concentration, and OEL. At present, such information is generally not available for biological aerosols. For example, currently available sampling and analysis techniques have not been successful in evaluating airborne concentrations of many biological aerosols. OELs do not exist for such aerosols, and it is not always possible to identify the specific organism responsible for a set of health outcomes.

The first highly publicized case of selecting respiratory protection to control exposures to an aerosolized microorganism was to help reduce health care workers' exposure to tuberculosis. The CDC has recommended respiratory protection for the control of exposures to *M. tuberculosis*; however, due to the limitations mentioned above, the CDC could not follow the NIOSH decision logic. These recommendations did not thoroughly address exposure to various concentrations of *M. tuberculosis*. The guidelines state only the maximum allowable face seal leakage (10% = APF of 10) and minimum allowable filter performance (95% efficient for 1-μm particles) (27).

Historically, only a few methods for selecting the level of respiratory protection needed for controlling exposures to airborne microorganisms were proposed. One investigator, Nicas, developed a mathematical method for estimating exposure of a given worker by predicting the probability of tuberculosis infection (39). This method utilizes information regarding the number of tuberculosis patients in the room (I), the number of "quanta" (q) (disease-causing units) emitted per hour per tuberculosis patient, the worker's volumetric breathing rate (b), the worker's cumulative time (t) spent in the room, the fraction of inhaled quanta that deposit in the alveolar region (f), and the supply airflow rate into the room (Q_R): probability of infection $= 1 - \exp[-(Iqbtf/Q_R)]$. Whereas these variables are important in determining the probability of infection, some of them, such as the number of quanta emitted per hour and fraction of inhaled quanta that deposit in the alveolar region, are not available. Other variables, such as the worker's volumetric breathing rate, may be difficult to assess for each situation.

A qualitative method was also proposed for selecting respirators for protection in infectious aerosol environments (40), which is based on the NIOSH respirator decision logic. Modifications in the traditional respirator selection method are proposed for situations where information is absent regarding level of hazard and airborne concentrations. To estimate hazard, the authors suggest using risk-based rankings from several organizations, including the CDC, NIH, the Canadian Laboratory Centre for Disease Control, and the European Community and one proposed by the European Federation of Biotechnology. To estimate airborne concentration, assessment of the nature of the activity or procedure is combined with knowledge about room volume and airflow through the room to obtain a ranking of airborne concentration. All of this information is used to determine a minimum APF, which corresponds to a respirator class.

In recent years, two new methods have been developed in Canada that advance the science of respirator selection for biological aerosols. In 2011, the Canadian Standards Association (CSA) published Selection, Use and Care of Respirators, CAN/CSA Z94.4-11, which was updated in 2012, that included a method to select respirators for exposures to biological aerosols (41). To address the lack of OELs for biological aerosols, the standard uses control banding as a selection method. It is an expanded and more detailed process of the idea proposed by McCullough and Brosseau. As the standard is quite comprehensive, it should be reviewed thoroughly. The standard offers a methodology for selecting respirators in two types of workplaces: health care institutions and all other workplaces. To use the selection tool, the employer or biosafety officer must determine the risk group of the agent, generation of aerosol, and control methods available. The risk group (1–4) is assigned based on the transmissibility, infectivity, and health effects of the agent. The generation rate is based upon the patient activity (e.g., coughing, sneezing) or activity generating the aerosol. The control level is based upon the ventilation or level of wind. Once the appropriate level for each factor is identified, a matrixed wheel is consulted that identifies the minimum level of respiratory protection appropriate for the exposure. There are six levels of respiratory protection based upon APFs, ranging from 0 (no respiratory protection needed) to 6 (APF of 10,000, a self-contained breathing apparatus).

The most recent method to be developed was by Lavoie et al. (42), through the Institut de recherche Robert-Sauvé en santé et en sécurité du travail (IRSST), a scientific research organization in Quebec, Canada. This method was developed based upon the work of McCullough and Brosseau (40) and CAN/CSA Z94.4-11 (41). The IRSST method uses the same basic factors of agent risk group, generation, and control level, but with slightly different definitions. Five bands are used for the control levels and generation rates, each corresponding to a score. These scores are added together to provide an exposure level score. Using the microorganism risk group and the exposure level score, one can determine the associated minimum APF. This model was validated using multiple scenarios.

As with any PPE selection, once the minimum level of protection is determined, many other factors must be considered to select the appropriate piece of equipment for the particular situation. Several important considerations include the transmission modes, need for cleaning and decontamination, health status of the worker, and task being performed. For example, if a respirator with an APF of 10 is considered appropriate, a half-facepiece may be selected. If there is a risk of ocular transmission, then a full facepiece respirator may be selected to help protect the eyes. Cleaning considerations may result in either a disposable respirator being selected or one that can be easily cleaned. A worker with an underlying health condition may need a more protective system.

To help identify whether the equipment will be compatible with the tasks, it is always encouraged that the worker dons the full PPE ensemble in a clean area and conducts, or simulates, tasks. Practice donning and doffing the PPE without cross-contamination is also recommended.

While the Canadian Standard and IRSST method are best practices, in geographic areas outside of Canada, the individual biosafety officer will need to review the selection methods currently available and keep abreast of guidelines offered by relevant agencies. Respirator selection will be based on the best available practices, most current knowledge, and professional judgment and reemphasizes the need to employ engineering and administrative controls when possible.

References

1. **U.S. Department of Health and Human Services, Public Health Service, Centers for Disease Control and Prevention, National Institutes of Health.** 2009. *Biosafety in Microbiological and Biomedical Laboratories*, 5th ed. HHS Publication no. (CDC) 21-112. http://www.cdc.gov/biosafety/publications/bmbl5/BMBL.pdf.
2. **Birkner JS.** 1991. Supplied-air suits, p 65–66. *In* Colton CE, Birkner LR, Brosseau LM (ed), *Respiratory Protection: A Manual and Guideline*, 2nd ed. American Industrial Hygiene Association, Fairfax, VA.
3. **Moyer ES.** 1986. Respirator filtration efficiency testing, p 167–180. *In* Raber RR (ed), *Fluid Filtration: Gas*. ASTM, Philadelphia, Pa.
4. **Hinds WC.** 1982. Filtration, p 164–186. *In Aerosol Technology*. John Wiley and Sons, New York, N.Y.
5. **U.S. Department of Health and Human Services, National Institute for Occupational Safety and Health.** 1996. Approval of respiratory protective devices, p. 528–593. CFR Title 42, Part 84. U.S. Government Printing Office, Washington, DC.
6. **Pippin DJ, Verderame RA, Weber KK.** 1987. Efficacy of face masks in preventing inhalation of airborne contaminants. *J Oral Maxillofac Surg* **45:**319–323.
7. **Chen S-K, Vesley D, Brosseau LM, Vincent JH.** 1994. Evaluation of single-use masks and respirators for protection of health care workers against mycobacterial aerosols. *Am J Infect Control* **22:**65–74.
8. **Brosseau LM, McCullough NV, Vesley D.** 1997. Mycobacterial aerosol collection efficiency by respirator and surgical mask filters under varying conditions of flow and humidity. *Appl Occup Environ Hyg* **12:**435–445.

9. **Oberg T, Brosseau LM.** 2008. Surgical mask filter and fit performance. *Am J Infect Control* **36:**276–282.
10. **American Industrial Hygiene Association.** 1991. Glossary, p 123–125. *In* Colton CE, Birkner LR, Brosseau LM (ed), *Respiratory Protection: a Manual and Guideline,* 2nd ed. American Industrial Hygiene Association, Fairfax, VA.
11. **American National Standards Institute.** 1992. *American National Standard for Respirator Protection (ANSI Z88.2).* American National Standards Institute, New York, NY.
12. **National Institute for Occupational Safety and Health.** 1987. *NIOSH Respirator Decision Logic.* DHHS (NIOSH) publication no. 87-108. National Institute for Occupational Safety and Health, Cincinnati, OH.
13. **American Conference of Governmental Industrial Hygienists.** 2015. *2015 TLVs and BEIs.* American Conference of Governmental Industrial Hygienists, Cincinnati, OH.
14. **National Institute for Occupational Safety and Health.** 1996. *NIOSH Guide to the Selection and Use of Particulate Respirators Certified Under 42 CFR 84.* DHHS (NIOSH) publication no. 96-101. National Institute for Occupational Safety and Health, Cincinnati, OH.
15. **Occupational Safety and Health Administration.** 2006. Assigned protection factors. *Fed Regist* **71:**50122–50192. https://www.osha.gov/pls/oshaweb/owadisp.show_document?p_table=FEDERAL_REGISTER&p_id=18846.
16. **U.S. Department of Health and Human Services, National Institute for Occupational Safety and Health.** 1993. Respiratory protective devices; tests for permissibility; fees, p. 47–111. *In* CFR Title 30, Part 11. U.S. Government Printing Office, Washington, DC.
17. **Ivarsson R, Nilsson H, Santesson J (ed).** 1992. Protective equipment, p 58–64. *In* Ivarsson R, Nilsson H, Santesson J (ed), *A FOA Briefing Book on Chemical Weapons: Threat, Effects, and Protection.* Försvarets Forskningsanstalt, Sundbyberg, Sweden.
18. **California Occupational Safety and Health Administration.** 2009. California Code of Regulations, Title 8, §5199. Aerosol Transmissible Diseases. http://www.dir.ca.gov/title8/5199.HTML.
19. **Compressed Gas Association.** 1989. *Commodity Specification for Air (ANSI/CGA G-7.1).* Compressed Gas Association, Arlington, VA.
20. **Miller MJ, Astles R, Baszler T, Chapin K, Carey R, Garcia L, Gray L, Larone D, Pentella M, Shapiro DS, Weirich E.** 2012. Guidelines for safe work practices in human and animal medical diagnostic laboratories. *MMWR Surveill Summ.* **6:**1–102.
21. **Qian Y, Willeke K, Grinshpun SA, Donnelly J, Coffey CC.** 1998. Performance of N95 respirators: filtration efficiency for airborne microbial and inert particles. *Am Ind Hyg Assoc J* **59:** 128–132.
22. **Eninger RM, Honda T, Adhikari A, Heinonen-Tanski H, Reponen T, Grinshpun SA.** 2008. Filter performance of N99 and N95 facepiece respirators against viruses and ultrafine particles. *Ann Occup Hyg* **52:**385–396.
23. **Centers for Disease Control and Prevention.** 2004. Public health guidance for community-level preparedness and response to severe acute respiratory syndrome (SARS) version 2. Supplement F: laboratory guidance. Appendix F5—laboratory biosafety guidelines for handling and processing specimens associated with SARS-CoV. May 21, 2004. http://www.cdc.gov/ncidod/sars/guidance/f/pdf/f.pdf.
24. **Centers for Disease Control and Prevention.** 2004. Interim domestic guidance on the use of respirators to prevent transmission of SARS. May 6, 2003. http://www.cdc.gov/ncidod/sars/pdf/respirators-sars.pdf.
25. **Centers for Disease Control and Prevention.** 2001. Basic laboratory protocols for the presumptive identification of *Bacillus anthracis.* 4/18/01. http://www.bt.cdc.gov/Agent/Anthrax/Anthracis20010417.pdf.
26. **Centers for Disease Control and Prevention.** 2002. Comprehensive procedures for collecting environmental samples for culturing *Bacillus anthracis.* Revised April 2002. http://www.bt.cdc.gov/agent/anthrax/environmental-sampling-apr2002.asp.
27. **Centers for Disease Control and Prevention.** 1994. Guidelines for preventing the transmission of *Mycobacterium tuberculosis* in health-care facilities, 1994. *MMWR Recomm Rep* **43**(RR-13)**:** 1–132.
28. **Centers for Disease Control and Prevention and National Institutes of Health.** 1997. *Proposed Guidelines for Goals for Working Safely with M. tuberculosis in Clinical, Public Health, and Research Laboratories.* U.S. Department of Health and Human Services, Public Health Service, Atlanta, GA.
29. **Occupational Safety and Health Administration.** 1996. *CPL 2.106 Enforcement Procedures and Scheduling for Occupational Exposure to Tuberculosis.* Occupational Safety and Health Administration, Washington, DC.
30. **Occupational Safety and Health Administration.** 1997. Occupational exposure to tuberculosis; proposed rule. *Fed Regist* **62:**54159–54309.
31. **Centers for Disease Control and Prevention.** 1998. Hantavirus: laboratory information. http://www.cdc.gov/ncidod/diseases/hanta/labguide.htm#part2.
32. **Centers for Disease Control and Prevention.** 1995. Safety, p 7–13. *In* Mills JN, Childs JE, Ksiazek TG, Peters CJ, Velleca WM (ed), *Methods for Trapping and Sampling Small Mammals for Virologic Testing.* Centers for Disease Control and Prevention, Atlanta, GA.
33. **National Institute for Occupational Safety and Health.** 1993. *Martin County Courthouse and Constitutional Office Building, Stuart Florida. HETA 93-1110-2575.* NIOSH Publications Office, Cincinnati, OH.
34. **Centers for Disease Control and Prevention.** 2016. Interim Risk Assessment and Biosafety Level Recommendations for Working With Influenza A(H7N9) Viruses (page last updated January 26, 2016). http://www.cdc.gov/flu/avianflu/h7n9/risk-assessment.htm.
35. **Centers for Disease Control and Prevention.** 2015. Guidance for U.S. Laboratories for Managing and Testing Routine Clinical Specimens When There is a Concern About Ebola Virus Disease. Page last reviewed October 8, 2015. http://www.cdc.gov/vhf/ebola/healthcare-us/laboratories/safe-specimen-management.html.
36. **Centers for Disease Control and Prevention.** 2015. Guidance on Personal Protective Equipment (PPE) To Be Used By Healthcare Workers during Management of Patients with Confirmed Ebola or Persons under Investigation (PUIs) for Ebola who are Clinically Unstable or Have Bleeding, Vomiting, or Diarrhea in U.S. Hospitals, Including Procedures for Donning and Doffing PPE. Page last updated and reviewed November 17, 2015. http://www.cdc.gov/vhf/ebola/healthcare-us/ppe/guidance.html.
37. **Centers for Disease Control and Prevention.** 2015. Interim Laboratory Biosafety Guidelines for Handling and Processing Specimens Associated with Middle East Respiratory Syndrome Coronavirus (MERS-CoV)—Version 2, last updated May 15, 2015. http://www.cdc.gov/coronavirus/mers/guidelines-lab-biosafety.html.
38. **Johnston AR.** 1991. Introduction to selection and use, p 25–35. *In* Colton CE, Birkner LR, Brosseau LM (ed), *Respiratory Protection: a Manual and Guideline,* 2nd ed. American Industrial Hygiene Association, Fairfax, VA.
39. **Nicas M.** 1995. Respiratory protection and the risk of *Mycobacterium tuberculosis* infection. *Am J Ind Med* **27:**317–333.
40. **McCullough NV, Brosseau LM.** 1999. Selecting respirators for exposures to infectious aerosols. *Infect Control Hosp Epidemiol* **20:**136–144.

41. **Canadian Standards Association (CSA).** 2012. Selection, Use and Care of Respirators. CAN/CSA-Z94.4-11, A National Standard of Canada.
42. **Lavoie J, Neesham-Grenon E, Debia M, Cloutier Y, Marchand G.** 2013. *Development of a Control Banding Method for Selecting Respiratory Protection Against Bioaerosols, Report R-804.* ISSRT, Montréal, Québec. http://www.irsst.qc.ca/media/documents/PubIRSST/R-804.pdf.
43. **Centers for Disease Control and Prevention.** 1997. Compendium of psittacosis (chlamydiosis) control, 1997. *MMWR Recomm Rep* **46**(RR-13)**:**1–13.
44. **Centers for Disease Control and Prevention.** 2001. Interim recommendations for protecting workers from exposure to *Bacillus anthracis* in work sites where mail is handled or processed. 10/31/01. [http://www.bt.cdc.gov/documentsapp/anthrax/10312001/han51.asp.
45. **National Institute for Occupational Safety and Health.** 2001. Protecting investigators peforming environmental sampling for *Bacillus anthracis*: personal protective equipment. http://www.cdc.gov/niosh/unp-anthrax-ppe.html.
46. **Centers for Disease Control and Prevention (CDC).** 1999. Blastomycosis acquired occupationally during prairie dog relocation—Colorado, 1998. *MMWR Morb Mortal Wkly Rep* **48:** 98–100.
47. **Centers for Disease Control and Prevention.** 1997. Guidelines for prevention of nosocomial pneumonia. *MMWR Morb Mortal Wkly Rep* **46**(RR-1)**:**1–79. http://www.cdc.gov/mmwr/pdf/rr/rr4601.pdf.
48. **United States Army Environmental Hygiene Agency.** 1992. Managing health hazards associated with bird and bat excrement. USAEHA technical guideline no. 142. http://chppm-www.apgea.army.mil/ento/tg142.htm.
49. **Childs JE, Kaufmann AF, Peters CJ, Ehrenberg RL, Centers for Disease Control and Prevention.** 1993. Hantavirus infection—southwestern United States: interim recommendations for risk reduction. *MMWR Recomm Rep* **42**(RR-11)**:**1–13.
50. **Centers for Disease Control and Prevention.** 1993. Update: hantavirus pulmonary syndrome—United States, 1993. *MMWR Morb Mortal Wkly Rep* **42:**816–820.
51. **Lenhart SW, Schafer MP, Hajjeh RA.** 1997. Histoplasmosis: Protecting Workers at Risk. Centers for Disease Control and Prevention, National Institute for Occupational Safety and Health, National Center for Infectious Diseases. DHHS (NIOSH) Publication no. 97-146.
52. **Wisconsin Department of Health and Social Services.** 1987. *Control of Legionella in Cooling Towers: Summary Guidelines.* Wisconsin Department of Health, Madison, WI.
53. **New York City Department of Health.** 1993. *Guidelines on Assessment and Remediation of Stachybotrys atra in Indoor Environments.* New York City Human Resources Administration, New York.

Additional Resources

OSHA's Respiratory Protection Advisor: http://www.osha-slc.gov/SLTC/respiratory_advisor/mainpage.html
NIOSH's Respirator Information Website: http://www.cdc.gov/niosh/topics/respirators/
NIOSH Education and Research Centers: http://www.cdc.gov/niosh/oep/centers.html
OHSA training information: http://www.oshaslc.gov/Training/
American Industrial Hygiene Association: 703-849-8888; http://www.aiha.org/
American Conference of Governmental Industrial Hygienists: 513-742-2020; http://www.acgih.org/

22

Standard Precautions for Handling Human Fluids, Tissues, and Cells

DEBRA L. HUNT

A broad range of infectious agents can be found in the blood at different stages of infection in humans. Most agents are present at high levels during a brief amount of time (i.e., septicemic phase), rarely are transmitted by blood, and therefore, are not usually categorized as "blood-borne" pathogens. Some agents, particularly viruses that induce a latent-phase or long-term carrier state, can be transmitted to other humans through blood or body fluid contact. The three most common examples of viruses existing in long-term carrier states that frequently exist as an asymptomatic infection are human HIV-1, hepatitis B virus (HBV), and hepatitis C virus (HCV). Occupational infections with these blood-borne pathogens have been documented globally and can occur when blood or body fluids containing these agents are transferred directly to the worker, e.g., through needlestick exposures to contaminated needles or blood or body fluid contact with mucous membranes or nonintact skin. The study of how these infections occur provides insight into the risks associated with other blood-borne pathogens that may be present in a carrier state in the blood such as *Plasmodium* (malaria), West Nile virus, *Treponema pallidum* (syphilis), or viral hemorrhagic fever viruses (e.g., Ebola).

The risks for occupational transmission are dynamic, as evidenced by the trends in workplace infections over the past 2 decades.

Availability of vaccines, postexposure treatment options, and mandated precautionary measures have contributed to the reduced numbers of documented occupational infections with the most common blood-borne pathogens. Recently, health care workers (HCWs) around the world have experienced occupational infections with the Ebola virus, resulting in a focus on enhanced blood-borne pathogens precautions. This chapter attempts to review the risks associated with the blood-borne pathogens of major concern for workplaces handling human body fluids, tissues, or cells, and the evolution and efficacy of prevention methods developed to reduce exposures and transmission of infection.

OCCUPATIONAL RISK ASSESSMENT

Occupationally acquired infections from blood-borne pathogens have been recognized since 1949 when a laboratory worker was reported to have been infected with "serum hepatitis" in a blood bank (1). With the development of diagnostic tests for blood-borne agents (e.g., HBV, HIV-1 and HCV), studies documented that occupational infections with blood-borne pathogens were occurring. The risks for transmission of these pathogens are better defined in prospective employee exposure studies, seroprevalence studies, and in descriptive cases in literature. It is essential to understand how occupational infections occur in order to develop effective preventative measures.

Hepatitis Viruses

In the early 1970s, serological tests became available for the diagnosis of infection with hepatitis A and hepatitis B viruses. Seroprevalence studies were then able to document the distinct epidemiology of these two viruses and the extent of transmission to HCWs. For example, Skinhoj and Soeby (2) reported subsequent increases in occurrences of laboratory-acquired hepatitis B and found a 7-fold higher rate of hepatitis B in laboratory workers when compared with the general population.

Health care personnel are known to be at greater risk for HBV infection than the general population (3, 4). The incidence of clinical cases of hepatitis B in HCWs before the availability of the hepatitis B vaccine (i.e., before 1982) was reported to be between 50 and 120 per 100,000 (5, 6), much higher than that of the general population of <10 cases per 100,000 (7). The increased level of risk was related to several factors, including the frequency of exposure to blood, body fluids, or blood-contaminated sharps, the duration of employment in a high-risk occupation where blood exposure was common, and the underlying prevalence of HBV infection in the patient population. High prevalence of infection was found in occupations associated with the emergency department, laboratory, blood bank, intravenous team, and the surgical-house officers (8).

The probability of blood-borne pathogen transmission after occupational exposure is related to the concentration of virus in the material, the amount of the material transferred (i.e., the "dose"), and the route of exposure. HBV can be found in extremely high titers in blood (10^9 viruses per ml) (9), providing up to 100 infectious doses of HBV during a needlestick injury with a 22 gauge needle. Positive hepatitis Be antigen (HBeAg) status is indicative of high titers (10^7 to 10^9viruses/ml) (10). Without postexposure prophylaxis, a nonimmune, exposed HCW has an approximately 6% risk of infection with HBV after a needlestick exposure to an HBeAg-negative patient source and up to a 30% risk after a similar exposure to an HbeAg-positive source (11–13).

Since 1982, implementation of Standard (Universal) Precautions and the availability of the hepatitis B vaccine to at-risk workers have undoubtedly been responsible for the decline in the number of occupational hepatitis B infections from 12,000 in 1985 (14) to less than 100 in 2013 (15). The seroprevalence of hepatitis B in HCWs is at least 5-fold less than the U.S. population, possibly due to the Occupational Safety and Health Administration (OSHA) mandate to offer hepatitis B vaccination to HCWs (16, 17).

Although occupational HCV transmission occurs rarely, the consequences for the HCW are substantial. The seroprevalence rate in HCWs is only slightly higher than the corresponding general population. In general, most studies document a HCV seroprevalence of between 0.5 and 2% for HCWs compared with the rate of 0.3 to 1.5% in blood donors (the community rate) (18–20). A recent meta-analysis of 44 studies on hepatitis C in HCWs between 1989 and 2014 indicates an increased odds ratio (OR) for hepatitis C infection in HCWs relative to control populations.

Additionally, upon stratification by study region, the review found an increased OR (2.1) in HCWs in countries with a low prevalence of HCV infection (i.e., Europe and U.S.) and an OR of 2.7 for professionals at high risk of blood contact (21).

Prospective studies that record seroconversions after documented percutaneous exposures indicate the risk of HCV infection after a single injury to a HCW from a HCV seropositive source can range from 0.75% (22) to 10% when using PCR methodologies to determine HCV-RNA (23). The wide range of risk reflects the differences in study design, diagnostic tests used, number of cases followed, source-patient status, and community prevalence. In general, the average incidence of infection has been reported as 1.8% (range, 0–7%) after a percutaneous injury (PI) to a known anti-HCV-positive source (19). Information from 14 prospective studies indicates a lower transmission rate of 0.5% in more than 11,000 exposed HCWs (24). There have been at least two cases of HCV transmission in the occupational setting due to blood splash to the eyes (25, 26). Another case (confirmed by genetic analysis) indicates that a HCW was infected with HCV (in addition to HIV) via exposures of chapped and abraded hands to diarrheal stools, urine, and coffee ground emesis from a patient in a nursing home (27). Because an infectious dose concentration has not been established for HCV, infectivity markers have not been identified, and seroprevalence data and infection rates in HCWs are so varied, more studies are needed to further define risk rates for occupational transmission of HCV.

Human Immunodeficiency Virus (HIV)

The circumstances of transmission of HIV in the workplace are well defined, primarily due to the national efforts to collect information, such as prospective studies of HIV-exposed HCWs, seroprevalence surveys, and the U.S. surveillance data systems for reported AIDS and HIV. In addition, the national surveillance for occupationally acquired HIV infection analyzes all reports (published reports, information from physicians, etc.) indicating exposures resulting in HIV infection.

HIV prevalence studies conducted on cohorts of HCWs around the country provide indirect evidence that the risk of transmission of HIV-1 in the health care setting is small. Such studies examined 7,595 U.S. and European HCWs with reported HIV exposures and found nine seropositive individuals (0.12%) among workers with no identified community risk (18, 28–39). The prevalence of infection in HCWs does not appear to be any higher than that in the comparable population at large.

The lack of association of significant HIV transmission in the health care setting was demonstrated in a serosurvey of hospital-based surgeons in 21 hospitals in moderate to high AIDS incidence areas across the U.S., conducted by the CDC Serosurvey Study Group (40). This study also found a low prevalence of 0.14% (one seropositive in 740 surgeons with no community risk identified). This same low occupational risk has been demonstrated in prevalence studies from Kinshasa, Zaire (41, 42), where community prevalence of HIV is high (6–8%), infection control practices are limited, and needles and syringes are usually washed by hand, sterilized, then reused. No higher rates of seropositivity were found in the hospital staff, nor were there any significant differences among the medical, administrative, and manual workers (6.5%, 6.4%, and 6%, respectively). These findings reaffirm the apparent low risk for occupational transmission of HIV.

National surveillance data about reported AIDS cases also demonstrate there is not a high risk for working in the health care or laboratory setting. As of December 2001, 23,951 (5%) of reported AIDS cases (469,850), where work history was known, had related a history of working in a health care or laboratory setting since 1978 (43). Most of these workers (91%) had nonoccupational risks for HIV infection (i.e., intravenous drug abuse, sexual contact, transfusions). Some (2,050 or 8.6%) were lost to follow-up for a variety of reasons, such as death or declining participation. Only 199 (0.8%) were identified as having either documented or possible occupational infection after exposures or no societal risks for infection.

Data from the National Surveillance for Occupationally Acquired HIV Infection between 1985 and 2013 in the United States indicate a total of 58 documented cases of occupationally acquired HIV infections (i.e., documented seroconversion after positive-source exposures) with only one documented infection reported since 1999 (44). There are also 150 HCWs who have possible occupationally acquired HIV infections. Internationally, 344 occupational seroconversions have been documented or were considered a likely source of occupational infection (45). It is important to review the documented cases in detail in order to emphasize appropriate precautions for work with human specimens.

Most (49 or 84%) of the occupationally infected workers in the U.S. were exposed to blood, one was exposed to visibly bloody fluid, four were exposed to unspecified body fluids, and four had contact with concentrated HIV in a production or research laboratory. Percutaneous exposures accounted for 50 of the injuries, mucocutaneous exposures accounted for 5, two workers had both percutaneous and mucocutaneous exposures, and two workers had unknown exposures. The types of devices associated with the PIs are listed in Table 1, and indicate that 45 (87%) of the 52 PIs were associated with hollow-bore needles. More than half of these injuries occurred either during blood collection or during vascular access procedures (43, 44).

There were eight mucocutaneous exposures associated with the documented cases (including two workers

TABLE 1.

Sharp devices or objects causing 51 percutaneous injuries among 50 health care workers with documented occupationally acquired HIV infection[a]

Sharp device or object		**No. of injuries**
Hollow-bore needles (n = 45):		
For blood collection (n = 22):	Vacuum-tube device needle	9
	Hypodermic needle	6
	Arterial blood gas kit needle	3
	Winged steel (butterfly) needle	2
	Unspecified / unknown needle	2
For vascular access (n =11):	Intravenous needle	7
	Hypodermic needle	2
	Dialysis needle	1
	Trocar used for changing central line catheter	1
For vascular line connection (n = 1):	Heparin-lock connector needle	
For sampling tissue/ lesion aspirate (n = 2):	Biopsy needle	1
	Hypodermic needle	1
For other uses (n = 9):	Specimen sampling needle on lab machine	1
	Needle for cleaning debris in lab equipment	1
	Hypodermic needle (intra-muscular injection)	1
	Unknown use	6
Other sharps or objects (n = 6):	Broken glass from blood-collection tubes	2
	Scalpels	2
	Unknown sharp device	2

[a]Reported to the CDC as of December 2001 (adapted from reference 43).

who experienced percutaneous exposures at the same time). Five of these involved exposures of blood to chapped or abraded hands, face, or ear; three involved splashes to the eyes, nose, or mouth.

Of the 58 documented cases in the U.S., 16 (28%) were clinical laboratory technicians and four (6.9%) were nonclinical laboratory technicians. A list of the circumstances surrounding most of the conversions in laboratory workers worldwide is shown in Table 2. The reports of a laboratory worker infected with a laboratory strain of HIV (31, 46) considered the source of that exposure to be "contact of the individual's gloved hand with H9/HTLV-IIIB culture supernatant with unapparent and undetected exposure to skin." The subject worked with concentrated HIV and reported wearing gowns and gloves routinely. The subject admitted episodes when pinholes or tears in gloves required they be changed. The subject also related accounts of leakage of virus-positive culture fluid from equipment, and the subsequent decontamination efforts with a hand brush. The subject also recalled an episode of nonspecific dermatitis on the arm; however, a gown always covered the arm. A subgroup of 98 other laboratory workers who also worked with concentrated HIV was found to be seronegative. An incidence rate of 0.48 per 100 person-years exposure was calculated for prolonged laboratory exposure to concentrated virus, approximately the same magnitude of risk of infection as health care workers who experience needlestick HIV exposures (31).

TABLE 2.

Documented HIV seroconversions after occupational exposures to laboratory workers or phlebotomists, reported in literature[a]

Country	Year	Occupation	Source status	Description of exposure	PEP taken?	Reference
1. USA	1985	Research lab worker	Concentrated HIV	Inapparent exposure by unknown route; lab strain confirmed	No	46
2. USA	1986	Phlebotomist	HIV (+)	Vacuum tube accident during venipuncture, splash to face	No	81
3. USA	1986	Medical technologist	HIV (+)	Apheresis machine accident, blood on ungloved hands/arm	No	81
4. USA	1986	Clinical lab worker	AIDS	Cut with broken vial containing blood, through gloves	NR	47
5. USA	1987	HCW, unkn	AIDS	NS, 21ga needle during recapping	No	80
6. USA	1988	Research lab worker	Viral culture	Needle injury while cleaning elutriator used to concentrate the virus	No	31
7. USA	1988	HCW, unkn	AIDS	NS while filling vacuum tube after venipuncture	NR	82
8. Australia	1990	HCW, unkn	AIDS	Deep NS after taking blood	Yes	121
9. USA	1990	HCW, unkn	HIV (+)	Deep NS injury, performing phlebotomy, also contracted HCV	No	122
10. USA	1990	Lab worker	Concentrated HIV	Skin/mucous membrane exposure to concentrated virus; molecular match	NR	123
11. USA	1990	Phlebotomist	AIDS	Deep NS, phlebotomy	No	124
12. USA	1991	Phlebotomist	HIV (+)	NS from 22ga phlebotomy needle	Yes	125
13. UK	1992	Phlebotomist	HIV (+)	NS from 23ga needle, venipuncture	No	126
14. USA	1992	Clinical lab worker	AIDS	NS from 21ga needle, venipuncture	Yes	125
15. Australia	1992	HCW, unkn	HIV (+)	NS from 21ga "butterfly" needle after venipuncture	No	127
16. UK	1992	HCW, unkn	AIDS	NS from 21ga needle, venipuncture	No	126
17. USA	1993	HCW, unkn	HIV (+)	Cut with broken glass, vacuum tube	Yes	128
18. Australia	1994	HCW, unkn	HIV (+)	NS after venipuncture	No	127
19. Germany	2000	Clinical lab worker	AIDS	Splash of serum to eye	NR	129
20. USA	2003	Clinical lab worker	Multiple lab samples	Face exposure to blood splash from malfunctioning lab machine	Yes	130
21. Australia	2003	Phlebotomist	AIDS	NS from 21ga needle, venipuncture	Yes	131

[a]Adapted from references 44 and 64. Abbreviations: PEP, postexposure prophylaxis; HCW, health care worker; unk, unknown; NS, needlestick; 21ga, 21-gauge; NR, not reported.

Prevalence and epidemiological studies indicate that occupational HIV infection does not occur frequently; however, documented HIV seroconversions due to exposures demonstrate that an occupational risk of HIV transmission exists. Factors that may contribute to the magnitude of that risk include the type/extent of injury, the body fluid involved, the "dose" of inoculum, environmental factors, and recipient susceptibility. The interactions and additive effect of these factors on the individual laboratory worker are complex and unknown. However, some data are available that can help further define risks associated with several procedures or circumstances (Table 3).

Prospective cohort studies that document HIV exposure events with follow-up serological monitoring of the exposed HCWs provide the best direct measure of risk of HIV transmission. In 23 prospective studies of reported 6,202 percutaneous exposures in HCWs, 20 instances of seroconversion have been documented, for an overall risk of transmission per PI from an HIV infected patient of 0.32% (37, 47–60). The risk of 0.3% is the average of all the types of percutaneous exposures to blood from patients in various stages of HIV infection.

Certain factors contribute to a subset of exposures for which the risk can be higher than 0.3%. In 1995, a case-control study described risk factors associated with occupational HIV infection after percutaneous exposures in cases reported from national surveillance systems in the United States, France, Italy, and the United Kingdom (61). The findings indicated an increased risk for occupational infections following a percutaneous exposure if it involved a larger quantity of blood, such as a device visibly contaminated with the patient's blood, or a procedure that involved a large-gauge, hollow-bore needle, particularly if used for vascular access. This increased risk may be directly associated with the amount of blood exposure and is consistent with laboratory studies that have indicated that less blood is transferred by suture needles (solid bore) than by phlebotomy needles (hollow-bore) of similar diameter (62, 63). Other factors identified in the CDC case-control study as associated with increased risk included a source patient in the terminal stage of illness and lack of zidovudine prophylaxis of the HCW.

A summary of 21 prospective studies of mucous-membrane exposures reported only one seroconversion from 2,910 exposures (64). Therefore, the risk of transmission of HIV via mucous membrane is 0.03% (95% confidence interval, 0.006 to 0.19%), much lower than that of a PI, i.e., <0.3% per exposure.

There have been no documented cases of HIV transmission through the respiratory, ingestion, or vector route of exposure. Some have questioned the possibility of respiratory transmission of HIV (65), specifically with the research laboratory-acquired infection with no documented percutaneous exposure (31). It is well known that common laboratory procedures using blenders and centrifuges produce infectious aerosols. Prior to the CDC and NIH recommendations for biological containment in laboratories, agents such as rabies (66), that are not transmitted by aerosols in the community or clinical setting, were documented to cause infection under laboratory conditions when concentrated agents were aerosolized by blending or purification procedures. However, an expert safety review team, convened by the Director of NIH, addressed the "unknown exposure" in the HIV laboratory issue and agreed that the potential for direct-contact transmission was much greater than for aerosol transmission (46). Procedures that generated aerosols were carried out in biological safety cabinets. The team cited other instances involving overt aerosol exposure in laboratory and production facilities involving concentrated HIV that have not resulted in seroconversions in exposed workers (67). Nevertheless, the occurrence of infection due to an unknown exposure emphasizes the need for laboratory workers to strictly adhere to published safety guidelines.

TABLE 3.

Occupational HIV infection risks for health care workers

Exposure type	Documented (43, 44) (# through 12/2013)	Risk from single HIV (+) exposure (%) (64, 132)
Percutaneous	50	0.3
	Risk Factors Identified (61)	
	Large gauge, hollow-bore needle	
	Deep exposure	
	Visible blood on device	
	Vascular access	
	Patient source with terminal illness (i.e., AIDS)	
Mucocutaneous	6	0.03
Both percutaneous and mucocutaneous	2	
Intact skin	0	<0.03

Ebola Virus

The Ebola virus is another blood-borne pathogen spread through contact with infected blood or body fluids and through sharps exposures as with any other blood-borne pathogen. Although extremely rare, an outbreak in West Africa in 2014 was the largest in history, leading to 28,637 suspected and confirmed cases with 11,315 deaths (68).

A report from WHO summarizes the impact of the Ebola epidemic on the health care workforce of Guinea, Liberia, and Sierra Leone (69) between January 1, 2014, and March 31, 2015. HCWs accounted for 3.9% (815/20,955) of all confirmed and probable cases. Depending on their occupations in health care, the HCWs were between 21 and 32 times more likely to be infected than were people in the general adult population. More than half of the infected HCWs were nurses or nurses' aides. Laboratory workers accounted for 7% of the infected workers. Compared with non-HCWs, laboratory technicians were 29.3 times more likely to be infected (incidence rate of 40.4 per 1,000). It cannot be determined through this study if the HCW infections were occupationally acquired or community acquired. However, WHO found serious gaps in implementing infection prevention measures where HCWs were employed. These included inappropriate use, or lack, of personal protective equipment (PPE), lack of Standard Precautions, lack of triage for Ebola patients, inadequate isolation areas for infected patients, poor hygiene due to lack of hand-hygiene stations with soap and running water, HCWs touching mucous membranes while wearing PPE, staff shortages, and limited capacity or inadequate training of safe management and burial of the deceased.

Patients with viral hemorrhagic fever (VHF) have been cared for prior to diagnosis in the U.S. and western Europe during the last decade. Extensive follow-up of potentially exposed HCWs (including laboratory workers) indicate no infection transmission (70–72). Since March 10, 2014, four patients with Ebola were treated in the U.S. after being diagnosed in West Africa. Another patient was treated in the U.S. after developing symptoms upon arrival from Liberia. Two nurses with direct care of this fifth patient were infected and diagnosed with Ebola, even though they had used PPE during their care of the patient and did not report any overt exposures. During the investigation of these cases, the CDC developed new enhanced guidelines for direct care of Ebola patients (73).

The only reports in the literature of laboratory-acquired Ebola infections refer to those occurring prior to the implementation of Standard Precautions and the availability of safer sharps devices (74), or to infections acquired during the performance of animal necropsy and other animal experiments (75). No laboratory workers were infected from any of the imported Ebola cases during the 2014 outbreak. Laboratory personnel were deployed to West Africa in 2014 from Europe, Canada, the U.S. (CDC), and other international partner groups to assist with the epidemic. Although these laboratory workers processed 200 to 300 specimens per day, there were no documented laboratory-acquired Ebola infections reported from these workers (76). There are reports of laboratory-acquired infections in local West African laboratory workers who were handling human blood without gloves or appropriate PPE early in the outbreak. However, no infections were reported in those personnel wearing correct PPE and adhering to appropriate precautions (76).

The Ebola virus is blood borne and highly pathogenic with a low infectious dose. Clinical and research samples may contain high numbers of viral particles, and infection can result in severe disease (77). Risk assessments conducted by CDC and WHO indicate that the highest risk for occupational infection with Ebola are HCWs with direct patient care, who may lack or have inadequate PPE and who do not adhere to strict Standard Precautions. Workers in clinical or research laboratories, who process specimens of patients with Ebola and who wear appropriate PPE and follow biosafety precautions, are considered to be in the low (but not zero) risk category for infection (78).

STANDARD PRECAUTIONS

Until the recognition of AIDS and the concerns with management of infected patients and their specimens, the primary infectious occupational threat to laboratory workers from blood or body fluids was that of HBV. The prevailing philosophy was that patients who were possibly infected with hepatitis B could easily be identified and special precautions could be taken with them and their specimens. Health care facilities relied on the CDC for recommendations regarding these precautions. Within 1 year of the first-recognized cases of the newly defined disease, AIDS, the CDC issued guidelines for clinical and laboratory staff regarding appropriate precautions for handling specimens collected from AIDS patients (79). Later, the CDC re-emphasized previously recommended precautions for handling specimens from patients infected with hepatitis B, i.e., minimizing the risk for transmission by the percutaneous, mucous membrane, and cutaneous routes of infection (80).

Reports issued by CDC in May 1987 documented that laboratory workers and other clinical staff were occupationally infected with HIV via nonintact skin and mucous membrane exposures to blood (81). Because the HIV serostatus of the patient sources was unknown at the time of exposure and the exposures were nonparenteral, the CDC issued the "universal blood and body fluid precautions" recommendations in late 1987 (82). The major premise involved the careful handling of all human blood and certain body fluids as if *all* were contaminated with

HIV, HBV, or other blood-borne pathogens. This "Universal Precautions" concept formed the basis for all subsequent recommendations from CDC (14, 83).

Universal Precautions in a laboratory situation involves the *consistent* use of Biosafety Level 2 (BSL-2) facilities and practices as outlined in *Biosafety in Microbiological and Biomedical Laboratories* (BMBL) (84). The BSL-2 precautions are most appropriate for clinical settings or when exposure to human blood, primary human tissue, or cell cultures is anticipated. The BMBL, 5th edition, clarifies laboratory-specific materials, besides human blood, body fluids, or primary cells, to be handled under BSL-2 conditions. An Appendix H ("Working with Human, Non-Human Primate, and Other Mammalian Cells and Tissues") was added to this edition to address other human virus and tumorigenic human cell hazards. Standard microbiological practices form the basis for BSL-2 with additional protection available from PPE and biological safety cabinets (BSCs) when appropriate.

Universal Precautions have also been adopted throughout the world, in the United Kingdom (85), Canada (86, 87), Europe (88, 89), and the Far East (90). WHO issued guidelines reflecting this philosophy (91).

In 1996, the CDC and the Hospital Infection Control Practices Advisory Committee (HICPAC) expanded the Universal Precautions recommendations to include precautions to be taken not only with human blood and other "epidemiologically-significant" body fluids, but also with any moist body substance (92), and have referred to these recommendations as "Standard Precautions." While Universal Precautions shifted the emphasis on use of precautions with the "known" or "high-risk" patient to the philosophy of basic precautions taken with all blood and body fluids epidemiologically linked to blood-borne pathogen transmission, Standard Precautions expanded that philosophy to precautions with all human body fluids (including urine, feces, saliva, etc.) to reduce other nosocomial infections.

After anecdotal research laboratory-associated infections with HIV were reported, the CDC issued its first agent summary statement in 1986 for work with the virus (67). The statement included a summary of laboratory-associated infections with HTLV-III (HIV), the hazards encountered in the laboratory, and advice on the safety precautions that should be taken by laboratories.

Recommendations from the CDC and NIH (84) indicated that BSL-2 facilities (at a minimum) with BSL-3 practices and containment equipment must be used for laboratory activities involving research scale amounts of the virus. BSL-3 facility, practices, and equipment must be used for all work involving industrial-scale, large-volume concentrations of the virus (Table 4). In addition, laboratory workers must show proficiency in handling pathogenic organisms before working in these laboratories.

TABLE 4.

CDC/NIH Recommended Precautions for Laboratory Work with HIV-1 (46)

Facility	Practices and procedures	Activities involving:
BL2	BL2	Clinical specimens
		Body fluids
		Human/animal tissues infected with HIV
BL2	BL3	Growing HIV at research lab scale
		Growing HIV-producing cell lines
		Working with concentrated HIV preparations
		Droplet/aerosol production
BL3	BL3	HIV at industrial scale levels
		Large volume or high concentration
		Production and manipulation

In addition to the advisory nature of the CDC guidelines, OSHA issued a standard to regulate occupational exposure to blood-borne pathogens in 1991 (93). This Bloodborne Pathogen Standard builds on the implementation of Universal (or Standard) Precautions, specifying the need for control methods, training, compliance, and recordkeeping.

OSHA issued the Bloodborne Pathogen Standard as a "performance" standard. In other words, the employer has a mandate to develop an exposure control plan to provide a safe work environment but is allowed some flexibility in order to accomplish this task. OSHA embraces the basic philosophy of the CDC Universal Precautions and marries it with combinations of engineering controls, work practices, and PPE in order to accomplish the intent of the standard. Table 5 outlines the basic requirements of the OSHA Bloodborne Pathogens Standard.

Engineering Controls

Recognizing that human behavior is inherently less reliable than mechanical controls, OSHA advocates the use of available technology and devices to isolate or remove hazards from the worker. These controls are particularly relevant to the prevention of sharps-related injuries. Although Universal (Standard) Precautions is an important philosophy and reduces exposures to workers, the major recommended focus is on barriers (gloves, gowns, face protection) that protect skin and mucous membranes. These personal protective barriers cannot prevent sharps injuries. OSHA admits that 60%

TABLE 5.

Basic requirements of the OSHA Bloodborne Pathogens Standard (93)

I. Exposure control plan (ECP); the establishment's written or oral policy for implementation of procedures relating to control of infectious disease hazards

II. Components of the ECP include:

- A. Exposure risk determination for all employees
- B. Control methods:
    1. Universal Precautions; a method of infection control in which all human-derived blood and potentially infectious materials are treated as if known to be infectious for HIV or HBV
    2. Engineering controls; use of available technology and devices to isolate or remove hazards from the worker (safety sharps devices, puncture-resistant sharps containers, etc.)
    3. Work practice controls; alterations in the manner in which a task is performed to reduce the likelihood of exposure to the worker (standard microbiologic practices in laboratories, disposal of needles without recapping or breaking, etc.)
    4. Personal protective equipment (PPE); specialized clothing or equipment used by workers to protect themselves from exposures (gloves, gowns, laboratory coats, fluid-resistant aprons, face shields, masks, eye protection, and head and foot coverings)
    5. Additional requirements for HIV and HBV research laboratories and production facilities
- C. Housekeeping practices
- D. Laundry practices
- E. Regulated waste disposal
- F. Tags, labels, and bags
- G. Training and education programs
- H. Hepatitis B vaccination
- I. Postexposure evaluation and follow-up
- J. Record keeping; includes medical records, training records, and maintaining availability of records

III. Administrative controls; to develop the ECP; to provide support of the ECP and provide accessibility of control methods, monitor compliance, and survey for effectiveness, and to investigate exposures for prevention of future occurrences

of needle injuries would be unaffected by improved work practices and PPE (93). Prior to 1991, recommendations for prevention of needlesticks focused on appropriate design and placement of puncture-resistant sharps disposal containers, education of HCWs regarding risks, and avoidance of recapping, bending, or breaking needles (14, 79, 80, 83). It is now evident that better prevention strategies for sharps injuries are needed.

Sharps precautions

Precise national data are not available on the annual number of needlesticks or other PIs sustained by laboratory and other HCWs. However, Panlilio et al. (20) estimated the annual number of such exposures in hospitals as 384,325 (confidence interval: 311,091 to 463,922), based on data, adjusted for underreporting, from 15 National Surveillance System for Health Care Workers (NaSH) hospitals and 45 Exposure Prevention Information Network (EPINet) hospitals.

The risk of blood-borne pathogen transmission is highest for a PI, as indicated by the previously discussed prospective studies, case/control study, and documented cases of occupational transmission. Recognizing the risks inherent in needlestick and sharp exposures, Congress enacted the Needlestick Safety and Prevention Act of 2000 (Public Law No.106-430, 114 Stat 1901, 2000) that amended the OSHA Bloodborne Pathogens Standard to require the evaluation and implementation of safer sharps devices in the workplace. The revised Bloodborne Pathogens Standard (94) was issued on January 18, 2001, and added several new requirements (Table 6), clearly emphasizing the need to review, evaluate, and follow reported sharps injuries in the workplace, and implement appropriate safer sharps devices.

In keeping with these requirements, clinical and research laboratory safety plans should restrict the use of needles and other sharp instruments in the laboratory for use only when there is no alternative. For laboratory procedures, other means should be considered to achieve the job, such as the use of blunt cannulas, small-bore tubing, or plastic pipettes. If needles must be used, such as for phlebotomy, laboratory workers must prioritize, evaluate, and implement safety-engineered blood-drawing devices (i.e., higher risk, hollow-bore needles).

Addressing another type of sharp in the laboratory, a joint advisory notice was issued by OSHA, the Food and Drug Administration (FDA), and the National Institute for Occupational Safety and Health (NIOSH) in 1999 warning of the risks of injury and infection due to breakage of glass capillary tubes (95). These agencies now recommend use of devices that are not prone to breakage,

TABLE 6.

New requirements of the revised OSHA Bloodborne Pathogen Standard, 2001 (94)

- Expand the definition of "engineering controls" to include safer medical devices ("sharps with engineered sharps injury protections", or "SESIP")
- Solicit input from non-managerial health care workers to identify, evaluate, and select safety engineered sharp devices, and document this process in the exposure control plan
- Review and update the exposure control plan at least annually with the documentation of the evaluation and implementation of sharps safety devices
- Maintain a sharps injury log with specific information regarding percutaneous injuries reported by employees

such as nonbreakable capillary tubes or capillary tubes coated with plastic, such as Mylar.

The American Hospital Association (AHA) and NIOSH have issued guidelines to help health care or laboratory facilities develop sharps-injury prevention programs (96, 97). The CDC's Division of Healthcare Quality Promotion took on a challenge to eliminate preventable needlesticks sustained by HCWs by 2009 (20) and provided a "Workbook for Designing, Implementing, and Evaluating a Sharps Injury Prevention Program" on its website (www.cdc.gov/sharpssafety/index.html). These documents emphasize the need to reduce sharps injuries and provide guidance in selecting and evaluating safety devices. Important recommended program elements include:

- Analyze sharps-related injuries in the workplace to identify hazards and trends.
- Prioritize prevention strategies by reviewing local and national data about risk factors for sharps injuries and successful intervention efforts.
- Train employees in the safe use and disposal of sharps.
- Modify work practices that pose a sharps injury hazard to make them safer.
- Promote safety awareness in the work environment.
- Establish procedures for, and encourage reporting and timely follow-up of, all sharps-related injuries.
- Evaluate the effectiveness of prevention efforts and provide feedback on performance

More than 1,000 patents have been issued since 1984 for devices that incorporate safety features (98). Types of safety products designed to reduce risk of PIs due to blood collection or manipulation of blood or tissues in laboratories include the following:

- Shielded, self-blunting, or retracting needles for vacuum tube phlebotomy sets
- Plastic vacuum/specimen tubes resistant to breakage
- Shielded, self-blunting, or retracting winged-steel needles
- Blood-gas syringes with a hinged-needle recapping device
- Retracting finger/heelstick lancets
- Unbreakable plastic capillary tubes for hematocrit determination
- Mylar-wrapped plastic capillary tubes
- Rounded-tip, retracting, or shielded scalpel blades
- Disposable scalpels or quick-release scalpel blade handles
- Vacuum blood-tube devices for safe stopper removal

A listing of examples of these safety products and manufacturers is found on the website of the International Healthcare Worker Safety Center at the University of Virginia (http://www.medicalcenter.virginia.edu/epinet/new/safetydevice.html). This list is not meant to be exhaustive, because new devices are being developed, nor is it an endorsement of the products. Readers are urged to contact manufacturers for current information about safety products.

Effectiveness of safety-engineered sharps devices

Several studies have found that safety-engineered sharps devices reduce the number of injuries (20, 99–103). A multicenter study conducted by the CDC on effectiveness of phlebotomy devices reported a 76% reduction in injury rates for a vacuum tube phlebotomy device with a self-blunting needle, a 26% reduction with a winged-steel needle that incorporates a protective sliding shield compared to a conventional winged-steel needle, and a 66% reduction in needlestick injuries with the use of a vacuum-tube blood-collection needle with a hinged recapping sheath (99). In another study, Sohn et al. (102) compared pre-intervention PI data (1998–2000) with the PI data collected after implementation of several safety-engineered devices to allow for safe intravenous (i.v.) medication delivery, blood collection, i.v. insertion, and intramuscular and subcutaneous injection in February 2001. In this study, hollow-bore needles associated with PIs decreased 70.6%, from a rate of 26.33 PIs per 1,000 full time employees (FTEs) to 7.73 PIs per 1,000 FTEs. The "high-risk" injuries (those caused by hollow-bore needles used for vascular access and/or blood sampling) were reduced by 52.6%, targeting the injuries that place HCWs at greater risk of infection with blood-borne pathogens.

As noted, many devices can reduce the frequency of needle or sharps injuries but will not completely eliminate the risk. In most cases, safety features cannot be activated until after the needles or sharps are removed from the patient (i.e., an "active" device). Some devices can be activated in use without user activation (i.e., a "passive" device). In the study noted above (102), 27% of the PIs that occurred during the postintervention period were caused by safety-engineered devices. Most of these devices were those that required activation (active device), and the safety mechanisms were either not activated or improperly activated. Tosini et al. (103) found a greater sharps injury reduction with passive devices versus active devices. A sharps reduction program must include an intensive education program for proper handling and activation of the newly implemented devices. A detailed description of the PIs must be reviewed to target problem areas, whether it be educational needs or a search for a better safety-engineered device.

Several sources have described desirable characteristics of safety devices (Table 7). These characteristics should be evaluated for applicability in each laboratory and serve as a guideline for device design and selection.

TABLE 7.

Desired characteristics of safety features of sharps devices (96, 97, 99)

1. The device is needleless.
2. The safety feature is an integral part of the device.
3. The device works passively (requires no active steps by the user).
4. If the device requires activation by the user, the safety feature can be engaged with a single-handed technique.
5. The user's hand remains behind the sharp.
6. The user can easily tell whether the safety feature has been activated.
7. The safety feature cannot be deactivated and remains protective through disposal.
8. The device performs reliably.
9. The device is easy to use and practical.
10. The device is safe and effective for patient care.

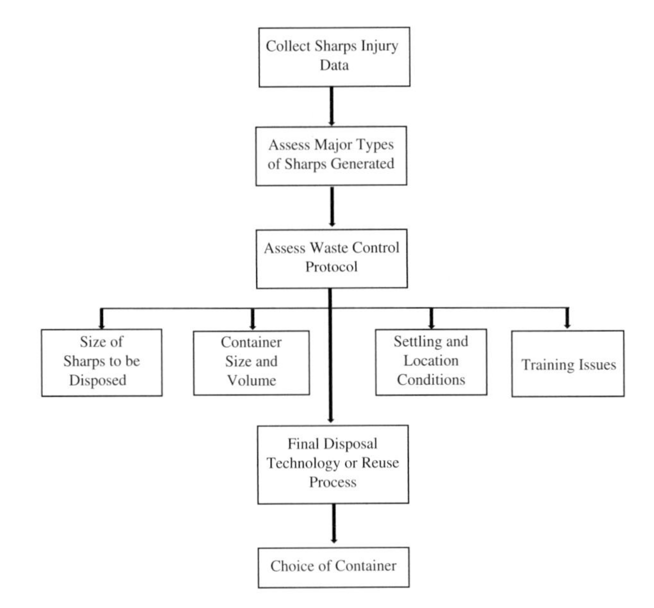

Figure 1: Decision logic for selecting sharps disposal containers. Adapted from NIOSH (107).

As facilities implement the use of safety devices, several methods of evaluation can be used (104–106). For example, evaluation forms developed by the Training for the Development of Innovative Control Technology Project (TDICT) provide written criteria developed by HCWs, product design engineers, and industrial hygienists for evaluation of specific safety devices (106). Each device evaluation form provides a quantitative score based on agreement of the evaluator with the desired criteria for the device. Forms are available on the TDICT website to evaluate a number of sharps devices, as well as protective equipment, such as eye protection, safety glasses, and safety gloves.

Sharps disposal containers are another important engineering control to consider in a sharps injury prevention program and are subject to performance standards to ensure their safe and effective use. The OSHA Bloodborne Pathogens Standard specifies that sharps disposal containers be (93):

1. closable
2. puncture resistant
3. leakproof on sides and bottom
4. labeled or color-coded according to specifics of the standard
5. easily accessible and located as close as is feasible to the area where sharps are used
6. maintained upright throughout use
7. replaced routinely and not be allowed to overfill

In addition to the OSHA requirements above, NIOSH issued guidelines for selection of sharps disposal containers based on site-specific hazard analysis (107). The decision logic developed by NIOSH is illustrated in Fig. 1 and includes components such as the assessment of the workplace hazards, size and types of sharps for disposal, volume of sharps at each point of use, frequency of emptying the containers, security requirements, etc. The NIOSH guidelines also establish criteria, such as the need to install container openings at a height between 52 and 56 inches to provide the ergonomically correct position for 95% of all adult female workers. Finally, the document provides evaluation tools to help select the most appropriate container for the facility.

Besides the disposal of needles, all disposable sharps encountered in the laboratory, including pipettes, microtome blades, micropipette tips, capillary tubes, and slides, should be carefully placed in conveniently located puncture-resistant, sharps disposal containers. Broken glassware should never be handled directly by hand but must be removed by mechanical means, such as a brush and dustpan, tongs, or forceps. Cotton swabs can be used to retrieve small slivers of glass. Nondisposable sharps should be placed in a hard-walled container for transport to a processing area and should never be manipulated by hand when retrieving for cleaning.

Another engineering control in the laboratory is the use of a properly maintained BSC to enclose work with a high potential for creating aerosols or droplets, i.e., blending, sonication, necropsy of infected animals, intranasal inoculation of animals, or opening lyophilized vials under pressure. All work with infectious material in an HIV research laboratory should be performed in a BSC or other physical containment device as per BSL-3 work practice requirements. In a laboratory handling human blood, body fluids, or human cells, the BSC provides excellent splash protection from droplets generated by procedures within the cabinet, such as removing rubber stoppers on vacuum tubes. Plastic shielding (i.e., salad-bar type shielding) can also be used to reduce the exposure to splatter or droplets from fluorescent-activated cell sorters or other automated laboratory equipment that might generate droplets of clinical material. Likewise, the Plexiglas radiation shield used in reverse transcriptase assays offers protection from splatter. However, if used in a BSC, the sloped top of the shield may divert airflow in the cabinet and must be removed to provide optimal protection by the BSC.

High-speed blenders and grinders that can contain aerosols of infectious material need to be opened in a BSC after processing. Enclosed electrical incinerators are preferable to open Bunsen burner flames for decontaminating bacteriological loops to prevent splatter and may be used within or outside of a BSC.

Work Practice Controls

The manner in which a task is performed can minimize the likelihood of exposures in the laboratory. For example, used needles should never be bent, broken, recapped, removed from disposable syringes, or otherwise manipulated by hand before disposal; rather, they should be carefully placed in conveniently located puncture-resistant containers (83).

Standard microbiological work practices have been recommended by CDC and NIH guidelines for all laboratory containment levels (84). Most of the practices are designed to prevent indirect transmission of infectious material from environmental surfaces to the hands and from hands to the mouth or mucous membranes. Such practices include prohibition of mouth pipetting, eating, drinking, smoking, applying cosmetics, or handling contact lenses in the laboratory, and attention to environmental decontamination.

One of the best work practices for any laboratory setting is that of frequent and adequate hand washing when hands are visibly contaminated, after completion of work, before leaving the lab, after removing gloves, and before eating, drinking, smoking, or changing contact lenses. Any standard hand washing product is adequate, but products that disrupt skin integrity should be avoided. When knee- or foot pedal-controlled faucets are not available, faucets should be turned off with the paper towels used for drying hands to prevent recontamination. Proper attention to hand washing will prevent inadvertent transfer of infectious material from hands to mucous membranes. (OSHA allows the use of antiseptic, waterless hand washing products only as a temporary measure during emergency situations until hands can be appropriately washed with soap and running water.)

In clinical settings, skin lesions may be covered by occlusive dressings and, if lesions are on the hands, gloves should be worn over the dressings to prevent contamination of nonintact skin. However, workers with skin lesions or dermatitis on hands or wrists should not perform procedures with potentially infectious material even if wearing gloves.

Other work practices can reduce the amount of splatter from laboratory procedures. Covering pressurized vials with plastic-backed or alcohol-soaked gauze when removing needles or when removing tops of pressurized vacuum tubes will minimize the exposure to splatter. To prevent popping stoppers on evacuated tubes or vials, blood should never be forced into the tube by exerting pressure on the syringe plunger; rather, tubes and vials should be filled by internal vacuum only. Extreme caution should be used when handling pressurized systems, such as continuous-flow centrifuges or apheresis or dialysis equipment. Use of imperviously-backed absorbent material ("lab diapers") can reduce the amount of splatter on laboratory work surfaces, when liquids accidentally leak or fall during lab procedures, and can aid in laboratory clean-up.

Safe transport of specimens or infectious material within the laboratory or to other areas can minimize the potential for accidental spills or injuries. Specimens should be contained in a closed, leakproof primary container, and placed in a secondary container (i.e., a plastic bag) to contain leaks during transport. The OSHA regulations do not mandate labeling or color-coding specimens if the specimens are handled only within the facility, a policy implementing "Universal Precautions" is in effect, and the containers are recognizable as human specimens. Bulk samples may be safely transported in a rack within a sealable plastic container, such as a modified "tackle box". The box may need to be labeled with a biohazard symbol or color-coded if the contents are not clearly visible as specimens. Luer caps should be used to transport syringes (needles removed with forceps or hemostats, or "unwinders" on the top of some sharps containers, and properly disposed of) or needles carefully recapped using a one-handed technique. Capillary tubes should be transported in a solid-walled secondary container, such as a screw-top test tube. Transport of cultures or hemocytometers from the BSC within the laboratory may be facilitated by placing them on a tray to limit the number of trips and opportunities for spillage.

Designation of "clean" versus "dirty" areas of the laboratory or within BSCs can help prevent inadvertent contamination. Work should be planned to move from clean areas to dirty areas.

Routine cleaning of work surfaces must be done after procedures are completed and at the end of each work shift, with additional decontamination as needed for spills. Routine cleaning can be accomplished using a variety of disinfectants including iodophors registered as hard surface disinfectants, phenolics, and 70% ethanol [with consideration given to the need for longer contact time when decontaminating dried viral cultures (108)]. Aldehydes, such as glutaraldehyde and formaldehyde, are not recommended for routine surface cleaning because of their potential toxicity.

Diluted bleach has been most widely used for routine disinfection [10% bleach (0.5% sodium hypochlorite) for porous surfaces and 1% bleach (0.05% sodium hypochlorite) for cleaned, hard, and smooth surfaces]. Weber et al. (109) demonstrated that the presence of blood in a spill interfered with the ability of the disinfectants to destroy herpes simplex virus type 2 (a prototype for lipophilic viruses, such as HIV). For large spills, in the presence of blood, a 1:10 bleach solution, 1:10 phenolic, and 1:10 or 1:128 quaternary ammonium salt were effective against the virus, and the authors recommend that large blood spills should be treated initially with hypochlorite to achieve at least a 1:10 dilution. This treatment may require the use of undiluted bleach to reach an effective concentration within a large spill.

Prompt decontamination is important following spills of infectious materials. Appropriate blood- or body-fluid spill clean-up in a clinical setting should involve the following steps:

1. Absorb the spill with towels or "lab diapers" to remove the extraneous organic material.
2. Clean with soap and water.
3. Decontaminate with an appropriate disinfectant. (CDC recommends an Environmental Protection Agency (EPA)-registered "hospital disinfectant" that is also "tuberculocidal," or a 1–10% bleach solution [83]. The EPA maintains a list of disinfectants approved for use in decontaminating blood spills on its website: www.epa.gov/oppad001/chemregindex.htm.)

Large spills of cultured or concentrated agents may be safely handled with an extra step:

1. Flood the spill with an appropriate disinfectant *or* absorb the spill with granular material impregnated with disinfectant.
2. Carefully soak up the liquid material with absorbent material (paper towels) or scrape up the granular absorbent material and dispose of it according to the waste disposal policy.
3. Clean the area with soap and water.
4. Decontaminate with fresh disinfectant.

Laboratory equipment (analyzers, centrifuges, pipettors) should be checked routinely for contamination and appropriately decontaminated. Any equipment sent for repair must also be decontaminated before leaving the laboratory or labeled as to the biohazard involved.

Because the intent of the OSHA Bloodborne Pathogens Standard is worker protection, the rules for appropriate waste disposal emphasize adequate packaging. Sharps disposal containers must be puncture- and leakproof, as well as easily accessible, as described previously. Other "Infectious" or "Medical" waste must be placed in leakproof containers or bags that are color-coded red or orange or labeled with the word "biohazard" or the universal biohazard symbol. All disposal containers should be replaced before they are full.

Blood or body fluids may be disposed of by carefully pouring down the sanitary sewer if local health codes permit but not by pouring into a sink where hand washing is performed. Liquid and solid culture materials, however, *must* be decontaminated before disposal, most commonly by steam sterilization (autoclaving). Tissues, body parts, and infected animal carcasses are generally incinerated. *All* laboratory waste from HIV, HBV, or HCV research-scale laboratories or production facilities and animal rooms must be decontaminated before disposal (BSL-3 practices). Additional "medical" or "infectious" waste definitions and requirements may exist locally and must be consulted for proper disposal policies.

Personal protective equipment (PPE)

Another strategy to minimize worker exposure to infectious material is the use of PPEs that are appropriate for the laboratory procedure and the type and extent of exposure anticipated. Examples include a variety of gloves, gowns, aprons, and face, shoe, and head protection. PPE may be used in combination with engineering controls and/or work practices for maximum worker protection.

OSHA requires gloves when hand contact with blood, other potentially infectious materials, mucous membranes, or nonintact skin is reasonably anticipated. The federal regulations also require gloves when handling or touching contaminated items or surfaces and for performing vascular access procedures. Gloves are appropriate in the laboratory when handling clinical specimens, infected animals, or soiled equipment, performing all laboratory procedures in research laboratories, cleaning spills, and handling waste.

For routine procedures, vinyl, nitrile, or latex gloves are effective for prevention of skin exposure to infectious materials, when used appropriately. Gloves are not intended to prevent puncture wounds from needles or sharps. However, there is evidence that a glove provides a "wiping" function that may reduce the amount of blood or infectious material exposure from the outside of the needle as it penetrates a glove or combination of gloves. Johnson et al. (110) found that two or three layers of latex gloves appeared to reduce the frequency of HIV-1-transfer by surgical needles to cell cultures. They also found that Kevlar gloves (untreated), Kevlar gloves (treated with the virucidal compound, nonoxynol-9), and nonoxynol-9-treated cotton gloves, used as intermediate layers between two layers of latex gloves, significantly reduced the amount of HIV-1 transfer when compared with a single latex-glove barrier. Gerberding et al. (111) reported that when surgeons wear double gloves, the rate of puncture of the inner glove is three times less than the rate of puncture of a single glove.

Other gloves that provide puncture "resistance" include stainless steel mesh (chain mail) gloves to protect against injury from large sharp edges, such as knife blades. Nitrile gloves (synthetic rubber) have some degree of puncture-resistance that may eliminate problems with rings or fingernails, yet retain the dexterity needed to perform laboratory procedures. Thin leather gloves, such as gardening gloves, can be worn under latex gloves for an additional barrier against cuts or animal bites. Even heavyweight utility gloves (dishwashing gloves) provide extra protection and should be worn when cleaning contaminated equipment or spills.

Gloves should frequently be inspected and changed to prevent contamination via undetected holes and leaks. The FDA has issued acceptable quality limits (AQL) for defects at 2.5% for surgeons' gloves and 4.0% for latex examination gloves (112), although the AQL varies widely among manufacturers. The reported defect percentage, due to holes for nonsterile latex gloves, ranges from 0 to 32% and, for nonsterile vinyl gloves, from 0 to 42%. Clearly, for high-risk situations, such as gross contamination of gloves with blood, bloody body fluid, or high concentrations of HIV-1, the use of double gloves will lower the risk of hand contamination from seepage through undetected glove defects. Although they are more puncture-resistant, nitrile gloves are designed to tear apart when any pressure is applied to a hole in the glove, so that any violation of the glove will be detected.

Gloves must never be washed or disinfected for reuse. Detergents may cause enhanced penetration of liquids through undetected holes, causing a "wicking" effect (14). Disinfectants, such as 70% ethanol, can also enhance the penetration of the glove barrier for polyethylene, polyvinyl chloride, and latex gloves, and facilitate deterioration (113).

Gloves must be changed when visibly contaminated, torn, defective, or when tasks are completed. Since hands may inadvertently be contaminated from laboratory surfaces, gloves should be removed before handling telephones, doorknobs, or "clean" equipment. Alternatively, "dirty" equipment may be designated and marked to be handled only with gloved hands. Laboratory workers should practice the aseptic technique for glove removal, i.e., gloves are removed inside out, so that the contaminated side remains on the inside, to protect the worker from skin contamination. Hands should always be washed after glove removal.

When soiling of clothing is anticipated, laboratory coats, gowns, or aprons are recommended. However, when a potential for splashing or spraying exists, solid-front, fluid-resistant gowns are appropriate. If the anticipated exposure involves soaking, solid-front fluid-proof gowns are required, as well as hoods/caps, facial protection, and shoe covers. Laboratory coats or gowns should not be worn outside of the laboratory.

Gowns with tightly fitting wrists or elasticized sleeves should be worn for work in BSCs. Alternatively, water-resistant "gauntlets" that provide a barrier between the glove and the laboratory coat are available to reduce skin exposure of the wrist and arm.

When splashing of blood or infectious material into the mucous membranes of the face is anticipated, a mask and goggles or face shield must be used. Most laboratory procedures involving this degree of exposure should be conducted within containment equipment, such as a BSC or behind a splash shield. Face protection might be needed for activities conducted outside a BSC, such as performing an arterial puncture, removing cryogenic samples from liquid nitrogen, or in some animal care areas. Masks and eye goggles or face shields also serve a passive function as a means of preventing accidental contact of contaminated gloved hands with the eyes, nose, and mouth during the course of work activities.

Whatever the PPE needs of any particular laboratory, OSHA requires that the employer provide an adequate supply of PPEs in the appropriate sizes. Hypoallergenic gloves must be available for employees who develop allergies to glove material or the powder inside gloves. Any defective PPE must be replaced, and reusable protective clothing must be laundered and maintained by the institution. Finally, all laboratory workers must be instructed in the proper use of PPEs and their location.

Employee Training and Monitoring

One of the most important components of an exposure control plan for the laboratory is a formal training program. "On-the-job" training is not acceptable as adequate safety training in the laboratory. The recommendations from CDC (83) that emphasized education of laboratory workers were incorporated into the OSHA Bloodborne Pathogens Standard (93).

Interactive training sessions must be conducted upon initial hire and with annual updates by a person knowledgeable about the Bloodborne Pathogens Standard. Employees must be educated regarding their risks and the institution's plan to control these risks. The training must be provided free of charge, during working hours, and be understood by all employees. Table 8 provides a list

TABLE 8.

Required elements of a training program for blood-borne pathogens (93)

- Accessibility of OSHA Bloodborne Pathogens Standard / Institutional Exposure Control Plan
- Blood-borne pathogen information (epidemiology, transmission, symptoms)
- Universal (or Standard) Precautions
- Selection, use and limitations of control methods (engineering, work practices, PPE)
- Emergency and post-exposure management
- HBV vaccination program
- Hazard communications

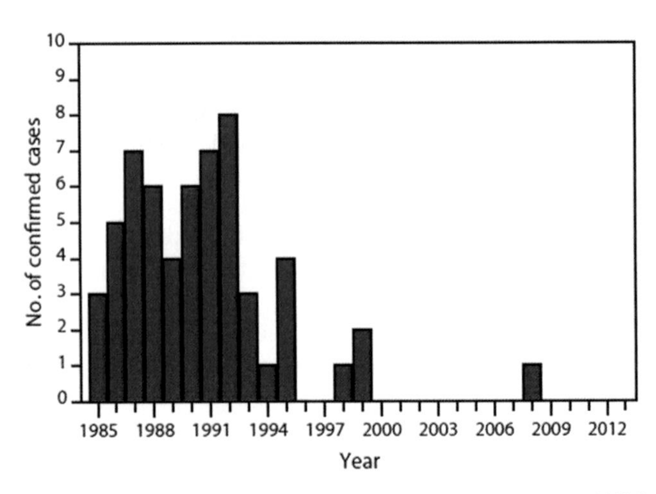

Figure 2: Number of documented occupationally acquired HIV infections by year of exposure or injury, as of December 2013. Adapted from reference 44.

of required elements of a training program for employees covered under the OSHA Bloodborne Pathogens Standard.

Employers must ensure compliance with the OSHA standard. The CDC (83) recommends that a biosafety expert monitor workplace practices at regular intervals through laboratory audits. The audit should also examine the adequacy of the laboratory facilities and equipment, the standard operating practices, and the written safety protocols. Corrective measures should be implemented, if needed. If breaches in protocol are detected, employees should be re-educated and, if necessary, disciplined.

Medical Care

The OSHA Bloodborne Pathogens Standard contains specific requirements for preventive and postexposure medical care evaluations, mostly taken directly from the U.S. Public Health Service recommendations. This standard includes requiring the employer to institute a hepatitis B vaccination program and providing adequate medical follow-up after an exposure incident, i.e., any testing, counseling, or appropriate prophylaxis, such as antiretrovirals, to reduce the risk of infection or transmission (14). Specific recommendations from the U.S. Public Health Service for postexposure treatment after HIV exposure are updated periodically, based on available drugs, HIV drugs used with patients, patient HIV status and receipt of HIV drugs, and potential drug toxicities (114).

Effectiveness of Standard (Universal) Precautions

The ultimate indication of compliance with Standard (Universal) Precautions is the reduction in workplace exposures and infection with blood-borne pathogens. As mentioned previously in this chapter, several studies have shown the effectiveness of safety devices used to prevent contaminated needlesticks. Data across the United States indicate that reported parenteral injuries in the workplace have declined substantially since the late 1980s since the implementation of Universal Precautions and the OSHA Bloodborne Pathogens Standard (115–117) that requires the implementation of safety-engineered devices.

Several studies have shown that implementation of Universal Precautions, coupled with training, have also resulted in the reduction of skin exposures to blood and body fluids when compared with previous practices (118, 119). Although these studies do not demonstrate a reduction in risky blood exposures, such data indicate that increase of barrier use prevented direct contact with blood and body fluids and probably prevented nonintact skin and mucous membrane exposures.

Most importantly, the incidence of occupationally acquired blood-borne pathogens has declined during the past 2 decades. Hepatitis B infection in HCWs in the United States has declined steadily since the vaccine became available in 1982 (15). As shown in Fig. 2, numbers of occupationally acquired HIV infection reported to the CDC have declined after implementation of the OSHA Bloodborne Pathogens Standard and the issuance of the guidelines for postexposure prophylaxis for exposures to HIV-positive sources (43, 44).

The reductions in PIs and blood-borne pathogen-occupational infections are encouraging. Several factors may influence the rarity of reported occupationally acquired blood-borne pathogen infections, such as underreporting, effectiveness of widespread treatment to reduce patient HIV viral loads, thus, reducing the "dose" of virus during exposures, possible effectiveness of postexposure treatment, improved safety-engineered devices, and worker training (44). However, the fact that needlesticks and other sharps injuries still occur at a rate of approximately 1,000 injuries per day in health care settings (20) reminds us to routinely evaluate employee blood and body fluid exposures in detail, and implement corrective measures as identified.

ENHANCED PRECAUTIONS FOR EBOLA VIRUS

All U.S. laboratories handling human blood or body fluids are required to comply with the OSHA Bloodborne Pathogens Standard. Strict adherence to these precautions should be the first step in providing protection to laboratory workers handling material from persons infected with any blood-borne pathogen, including Ebola virus. Because of the heightened concern in the U.S. about Ebola and the documented occupationally acquired infections during the 2014 Ebola outbreak, the CDC issued a new guidance document specifically for handling specimens from patients with Ebola (120). The Bloodborne Pathogens Standard covers most of these recommendations; however, the guidance document provides suggestions for other measures to further minimize risk to personnel. These include transferring specimens to designated public health laboratories for processing, if possible, developing controls based on a risk assessment regarding anticipated procedures, limiting the number of staff engaged in testing, evaluating and segregating equipment used for testing, and performing tests in a dedicated space.

OSHA also developed interim general guidance documents for occupations that are most likely to encounter people infected with Ebola or contaminated equipment or specimens from them (77). OSHA recognizes that employers will likely benefit from the provisions of both the OSHA standards and the CDC guidance documents to implement a comprehensive program to minimize worker exposure.

Engineering Controls

Previously discussed safer sharps precautions are the primary engineering controls to reduce the risk of laboratory-acquired infections with Ebola virus. In addition to those, both OSHA and CDC guidance documents recommend that BSCs be used when manipulating clinical material from Ebola patients to protect workers and prevent contamination of surfaces outside of the BSC (77, 120). These guidance documents define that the clinical or research labs handling specimens be designated as BSL-2 with BSL-3 precautions (77). The CDC also cautions that some laboratory equipment may not be suitable for testing specimens from Ebola-infected blood because the equipment may generate an aerosol, or recommended disinfectants to inactivate Ebola virus may affect the performance of the instrument or void the manufacturer's warranty for maintenance (120). Some considerations of equipment use may include:

- Use equipment with closed-tubes systems in which the specimen container stays capped during testing, such as automated hematology analyzers. Disinfect the outside of the sample before running the sample.
- If centrifugation is necessary, use sealed buckets or sealed rotors. Open the sealed bucket or rotor in a BSC.
- If point-of-care instruments are used, they should be placed inside a BSC or behind a barrier to prevent splashes or aerosols that might be generated.
- Ensure that all equipment used will withstand disinfection with disinfectants known to kill nonenveloped viruses.
- Do not use pneumatic tubes systems (automated or vacuum specimen delivery systems) for transporting specimens. They should be hand-carried to the laboratory.

Work-Practice Controls

Most of the work-practice controls routinely used in laboratories based on Standard Precautions are applicable for handling Ebola samples as well. The CDC notes some specific recommendations for Ebola (120), including:

- Use of recommended disinfectants that are known to kill nonenveloped viruses and are found on the EPA website in List L: http://www.epa.gov/oppad001/list-l-ebola-virus.html
- Any Ebola-contaminated waste material, such as patient specimens, sharps, linens, PPE, and cleaning material, should be disposed of as a Category A infectious substance according to the Department of Transportation (DOT); this waste material is regulated by the DOT's Hazardous Materials Regulations (HMR, 49 C.F.R., Parts 171–180). Regulations may require effective treatment of the waste, such as steam sterilization (autoclaving) before disposal as laboratory regulated medical waste. If no autoclave or other means are available for treatment, disposal arrangements may be made with a licensed waste contractor after obtaining proper permits. Since regulations are complex, and vary by state and local jurisdictions, the state's medical waste management program may need to be contacted for guidance.
- Work with Ebola samples must be done by properly trained laboratory workers, specifically on strict adherence to protocols and proper donning and doffing of PPE. Consideration should be given to the use of a buddy system to ensure that safe donning and doffing procedures are followed.

Personal Protective Equipment

The CDC recommendations in the latest guidance document (120) enhance the PPE required under Standard Precautions. CDC recommends a risk assessment of the procedures be done for potential exposures and a protocol be developed that combines the use of engineering controls, work practices, and PPE to protect the mouth, nose, eyes, and bare skin from coming into contact with Ebola materials. For example, the CDC recommends that, in addition to all work conducted in a BSC, the workers may use disposable gloves, solid-front wraparound gowns that are fluid resistant or fluid impermeable, a surgical mask to cover all of the nose and mouth, and eye protection, such as a full-face shield or goggles/safety glasses with side shields. Additional PPE, such as the use of an N-95 respirator, should be based on a risk assessment. Whatever specific PPE is determined to be required, the laboratory must provide sufficient training and practice on the use as well as donning and doffing protocols. The requirements should remain consistent to prevent inadvertent breaches in safe practices due to unfamiliarity or lack of training.

The OSHA guidance document (77) also contains some specific recommendations for laboratory PPE when handling Ebola specimens. OSHA recommends working in a BSC and using dedicated work clothing, such as surgical scrubs under PPE, dedicated washable footwear, double disposable gloves, face and eye protection, and a disposable fluid-resistant gown that extends to at least mid-calf. When higher-risk exposure is anticipated (i.e., a BSC cannot be used or bioaerosols are generated), the worker may need fluid-resistant coveralls instead of gowns, fluid-resistant head and neck covers, shoe/boot covers high enough to cover the lower leg, and respiratory protection (N-95 or better respirator). OSHA has developed a "PPE Selection Matrix for Occupational Exposures to Ebola Virus" that provides a task-based guidance to help employers select appropriate PPE for workers with potential exposures to Ebola virus (access through: http://www.osha.gov/ebola). Table 9 lists and compares the guidance for Ebola specimens from both the CDC and OSHA. The CDC and OSHA both agree that laboratory workers, who must retrieve specimens or perform point of care testing in patient rooms, must comply with the hospital requirements for PPE for direct patient care. Whatever combination of PPE is used, the CDC emphasizes that all workers must receive repeated training on and demonstrate competency in putting on and removing proper PPE before working with Ebola samples.

TABLE 9.

CDC (120) and OSHA (77) recommended PPE for Ebola in laboratories[a]

PPE (examples based on risk assessment)	CDC Guidance for Laboratories handling Ebola clinical samples	OSHA Guidance: Conducting Clinical Laboratory Work on samples from patients with suspected or confirmed Ebola	OSHA Guidance: Conducting Research Laboratory Work on research samples suspected of containing or known to contain Ebola virus
Typical precautions for normal work tasks	Standard Precautions	According to biosafety level; recommended BSL-2 with BSL-3 practices	According to biosafety level; recommended BSL-4
Dedicated clothing under PPE		✓	✓
Gloves, double (nitrile)	Double gloves not specified	✓	✓
Face and eye protection	✓	✓	✓
Head/neck cover	Not specified	✣ (fluid resistant)	✓ (impermeable)
Gown	✓ (fluid resistant or impermeable)	✓ (fluid resistant)	✓ (impermeable)
Coveralls	Not specified	✣ (fluid resistant)	✣ (impermeable)
Boot covers	Not specified	✓ (fluid resistant)	✓ (impermeable)
Disposable N95 respirator	✣ (based on risk assessment)	✣	✓
Powered Air-Purifying Respirator (PAPR)	✣ (based on risk assessment)	✣	✣
Full-body, air-supplied positive pressure suit	–	–	✓ (if not working in Class III BSC)

[a]Adapted from the OSHA PPE Selection Matrix for Laboratory Occupational Exposure to Ebola Virus (http://www.osha.gov/Publications/OSHA3761.pdf). Symbols: ✓, use at a minimum; ✣, use when higher risk exposure is present (exposure to bio-aerosols or BSC cannot be used).

SUMMARY

Current Standard Precautions have been developed and implemented based on risk assessments and proven mitigation strategies to minimize worker exposure to human blood and body fluids. Precautions for the most common blood-borne pathogens in the U.S., hepatitis viruses B and C and HIV, have been well characterized based on the risk for sharps precautions and mucous membrane and open skin contact. Standard Precautions have been incorporated into the OSHA Bloodborne Pathogens Standard since 1991 and have shown to be effective in reducing exposures. The same precautions should provide protection of workers from exposure to another blood-borne pathogen, Ebola virus. However, heightened concern after the 2014 Ebola epidemic in West Africa and subsequent care of Ebola patients in the U.S. brought attention to the need for strict attention to precautions and also the need to provide appropriate PPE and train workers thoroughly to enhance the Standard Precautions concept. Global accounts of laboratory-acquired infections with the Ebola virus demonstrate that lack of appropriate PPE and safety practices can present a risk of exposure to laboratory workers. The absence of laboratory-acquired Ebola infections in recent years in laboratories practicing Standard Precautions is a testimony to the effectiveness of these basic preventive measures.

References

1. **Leibowitz S, Greenwald L, Cohen I, Litwins J.** 1949. Serum hepatitis in a blood bank worker. *J Am Med Assoc* **140:**1331–1333.
2. **Skinhøj P, Søeby M.** 1981. Viral hepatitis in Danish health care personnel, 1974–78. *J Clin Pathol* **34:**408–411.
3. **Lewis TL, Alter HJ, Chalmers TC, Holland PV, Purcell RH, Alling DW, Young D, Frenkel LD, Lee SL, Lamson ME.** 1973. A comparison of the frequency of hepatitis-B antigen and antibody in hospital and nonhospital personnel. *N Engl J Med* **289:**647–651.
4. **Maynard JE.** 1978. Viral hepatitis as an occupational hazard in the health care professional, p 321. *In* Vyas GN, Cohen SN, Schmid R (ed), *Viral Hepatitis: A Contemporary Assessment of Etiology, Epidemiology, Pathogenesis and Prevention.* Franklin Institute Press, Philadelphia.
5. **Schneider WJ.** 1979. Hepatitis B: an occupational hazard of health care facilities. *J Occup Med* **21:**807–810.
6. **Hansen JP, Falconer JA, Hamilton JD, Herpok FJ.** 1981. Hepatitis B in a medical center. *J Occup Med* **23:**338–342.
7. **Centers for Disease Control.** 1991. Hepatitis B virus: a comprehensive strategy for eliminating transmission in the United States through universal childhood vaccination. Recommendations of the Immunization Practices Advisory Committee (ACIP). *MMWR Recomm Rep* **40**(RR-13):1–25.
8. **Dienstag JL, Ryan DM.** 1982. Occupational exposure to hepatitis B virus in hospital personnel: infection or immunization? *Am J Epidemiol* **115:**26–39.
9. **Hoofnagle JH.** 1995. Hepatitis B, p 2062–2063. *In* Haubrich WS, Schaffner F, Berk JE (ed), *Gastroenterology*, 5th ed. W. B. Saunders and Co, Philadelphia, PA.
10. **Shikata T, Karasawa T, Abe K, Uzawa T, Suzuki H, Oda T, Imai M, Mayumi M, Moritsugu Y.** 1977. Hepatitis B e antigen and infectivity of hepatitis B virus. *J Infect Dis* **136:**571–576.
11. **Grady GF.** 1976. Relation of e antigen to infectivity of hBsAg-positive inoculations among medical personnel. *Lancet* **1:** 492–494.
12. **Grady GF, Lee VA, Prince AM, Gitnick GL, Fawaz KA, Vyas GN, Levitt MD, Senior JR, Galambos JT, Bynum TE, Singleton JW, Clowdus BF, Akdamar K, Aach RD, Winkelman EI, Schiff GM, Hersh T.** 1978. Hepatitis B immune globulin for accidental exposures among medical personnel: final report of a multicenter controlled trial. *J Infect Dis* **138:**625–638.
13. **Werner BG, Grady GF.** 1982. Accidental hepatitis-B-surface-antigen-positive inoculations. Use of e antigen to estimate infectivity. *Ann Intern Med* **97:**367–369.
14. **Centers for Disease Control.** 1989. Guidelines for Prevention of transmission of human immunodeficiency virus and hepatitis B virus to health-care and public-safety workers. *Morb Mortal Wkly Rep* **38:**1–37.
15. **Centers for Disease Control and Prevention.** 2013. Surveillance for viral hepatitis-United States. www.cdc.gov/hepatitis/statistics/2013surveillance/index.htm. Accessed 12/21/2015.
16. **Beltrami EM, Williams IT, Shapiro CN, Chamberland ME.** 2000. Risk and management of blood-borne infections in health care workers. *Clin Microbiol Rev* **13:**385–407.
17. **Mahoney FJ, Stewart K, Hu H, Coleman P, Alter MJ.** 1997. Progress toward the elimination of hepatitis B virus transmission among health care workers in the United States. *Arch Intern Med* **157:**2601–2605.
18. **Gerberding JL.** 1994. Incidence and prevalence of human immunodeficiency virus, hepatitis B virus, hepatitis C virus, and cytomegalovirus among health care personnel at risk for blood exposure: final report from a longitudinal study. *J Infect Dis* **170:** 1410–1417.
19. **Centers for Disease Control and Prevention (CDC).** 1997. Notice to Readers Recommendations for follow-up of health-care workers after occupational exposure to hepatitis C virus. *MMWR Wkly Rep* **46:**603–606.
20. **Panlilio AL, Orelien JG, Srivastava PU, Jagger J, Cohn RD, Cardo DM; NaSH Surveillance Group; EPINet Data Sharing Network.** 2004. Estimate of the annual number of percutaneous injuries among hospital-based healthcare workers in the United States, 1997–1998. *Infect Control Hosp Epidemiol* **25:** 556–562.
21. **Westermann C, Peters C, Lisiak B, Lamberti M, Nienhaus A.** 2015. The prevalence of hepatitis C among healthcare workers: a systematic review and meta-analysis. *Occup Environ Med* **72:** 880–888.
22. **Puro V, Petrosillo N, Ippolito G, Jagger J, Lanphear BP, Linnemann CC Jr.** 1995. Hepatitis C virus infection in healthcare workers. *Infect Control Hosp Epidemiol* **16:**324–326.
23. **Mitsui T, Iwano K, Masuko K, Yamazaki C, Okamoto H, Tsuda F, Tanaka T, Mishiro S.** 1992. Hepatitis C virus infection in medical personnel after needlestick accident. *Hepatology* **16:**1109–1114.
24. **Jagger J, Puro V, De Carli G.** 2002. Occupational transmission of hepatitis C virus. *JAMA* **288:**1469, author reply 1469–1471.
25. **Sartori M, La Terra G, Aglietta M, Manzin A, Navino C, Verzetti G.** 1993. Transmission of hepatitis C via blood splash into conjunctiva. *Scand J Infect Dis* **25:**270–271.
26. **Ippolito G, Puro V, Petrosillo N, De Carli G, Micheloni G, Magliano E.** 1998. Simultaneous infection with HIV and hepatitis C virus following occupational conjunctival blood exposure. *JAMA* **280:**28. Letter
27. **Beltrami EM, Kozak A, Williams IT, Saekhou AM, Kalish ML, Nainan OV, Stramer SL, Fucci MC, Frederickson D, Cardo DM.** 2003. Transmission of HIV and hepatitis C virus from a

nursing home patient to a health care worker. *Am J Infect Control* **31:**168–175.
28. **Hirsch MS, Wormser GP, Schooley RT, Ho DD, Felsenstein D, Hopkins CC, Joline C, Duncanson F, Sarngadharan MG, Saxinger C, Gallo RC.** 1985. Risk of nosocomial infection with human T-cell lymphotropic virus III (HTLV-III). *N Engl J Med* **312:**1–4.
29. **Shanson DC, Evans R, Lai L.** 1985. Incidence and risk of transmission of HTLV III infections to staff at a London hospital, 1982–85. *J Hosp Infect* **6**(Suppl C)**:**15–22.
30. **Weiss SH, et al.** 1985. HTLV-III infection among health care workers. Association with needle-stick injuries. *JAMA* **254:** 2089–2093.
31. **Weiss SH, Goedert JJ, Gartner S, Popovic M, Waters D, Markham P, di Marzo Veronese F, Gail MH, Barkley WE, Gibbons J, Gill FA, Leuther M, Shaw GM, Gallo RC, Blattner WA.** 1988. Risk of human immunodeficiency virus (HIV-1) infection among laboratory workers. *Science* **239:**68–71.
32. **Boland M, Keresztes J, Evans P, Oleske J, Connor E.** 1986. HIV seroprevalence among nurses caring for children with AIDS/ARC. (Abstract THP.212) Presented at the 3rd International Conference on AIDS, Washington, D.C.
33. **Ebbesen P, Scheutz F, Bodner AJ, Biggar RJ.** 1986. Lack of antibodies to HTLV-III/LAV in Danish dentists. *JAMA* **256:**2199. letter
34. **Gilmore N, Ballachey ML, O'Shaughnessy M.** 1986. HTLV-III/LAV serologic survey of health care workers in a Canadian teaching hospital. (Abstract 200), Presented at the 2nd International Conference on AIDS, Paris, France.
35. **Harper S, Flynn N, VanHorne J, Jain S, Carlson J, Pollet S.** 1986. Absence of HIV antibody among dental professionals, surgeons, and household contacts exposed to persons with HIV infection. (Abstract THP215). Presented at the 3rd International Conference on AIDS, Washington, D.C.
36. **Lubick HA, Schaeffer LD, Kleinman SH.** 1986. Occupational risk of dental personnel survey. *J Am Dent Assoc* 113**:**10. letter.
37. **Gerberding JL, Bryant-LeBlanc CE, Nelson K, Moss AR, Osmond D, Chambers HF, Carlson JR, Drew WL, Levy JA, Sande MA.** 1987. Risk of transmitting the human immunodeficiency virus, cytomegalovirus, and hepatitis B virus to health care workers exposed to patients with AIDS and AIDS-related conditions. *J Infect Dis* **156:**1–8.
38. **Klein RS, Phelan JA, Freeman K, Schable C, Friedland GH, Trieger N, Steigbigel NH.** 1988. Low occupational risk of human immunodeficiency virus infection among dental professionals. *N Engl J Med* **318:**86–90.
39. **Marcus R.** 1988. The cooperative needlestick surveillance group: CDC's health care workers surveillance project: an update. Abstr. 9015. IV Int. Conf. AIDS. Stockholm, Sweden.
40. **Panlilio AL, et al,** Serosurvey Study Group**.** 1995. Serosurvey of human immunodeficiency virus, hepatitis B virus, and hepatitis C virus infection among hospital-based surgeons. *J Am Coll Surg* **180:**16–24.
41. **Mann JM, Francis H, Quinn TC, Bila K, Asila PK, Bosenge N, Nzilambi N, Jansegers L, Piot P, Ruti K, Curran JW.** 1986. HIV seroprevalence among hospital workers in Kinshasa, Zaire. Lack of association with occupational exposure. *JAMA* **256:**3099–3102.
42. **N'Galy B, Ryder RW, Bila K, Mwandagalirwa K, Colebunders RL, Francis H, Mann JM, Quinn TC.** 1988. Human immunodeficiency virus infection among employees in an African hospital. *N Engl J Med* **319:**1123–1127.
43. **Do AN, Ciesielski CA, Metler RP, Hammett TA, Li J, Fleming PL.** 2003. Occupationally acquired human immunodeficiency virus (HIV) infection: national case surveillance data during 20 years of the HIV epidemic in the United States. *Infect Control Hosp Epidemiol* **24:**86–96.
44. **Joyce MP, Kuhar D, Brooks JT, Centers for Disease Control and Prevention.** 2015. Notes from the field: occupationally acquired HIV infection among health care workers—United States, 1985–2013. *MMWR Morb Mortal Wkly Rep* **63:**1245–1246.
45. **Health Protection Agency Centre for Infections.** 2005. Occupational transmission of HIV. Summary of published papers. Health Protection Agency. London, UK. http://www.hpa.org.uk/infections/topics-az/bbv/pdf/intl_HIV-tables_2005.pdf. Accessed September 30, 2015.
46. **Centers for Disease Control.** 1988. Occupationally-acquired human immunodeficiency virus infections in laboratories producing virus concentrates in large quantities: Conclusions and recommendations of an expert team convened by the Director of the National Institutes of Health (NIH). *Morbid. Mortal. Weekly* 37:19–22.
47. **Henderson DK, Fahey BJ, Willy M, Schmitt JM, Carey K, Koziol DE, Lane HC, Fedio J, Saah AJ.** 1990. Risk for occupational transmission of human immunodeficiency virus type I (HIV-1) associated with clinical exposures. A perspective evaluation. *Ann Intern Med* **113:**740–746.
48. **Elmslie K, O'Shaughnessy JV.** 1987. National surveillance program on occupational exposure to HIV among health care workers in Canada. Can. Dis. Week. *Rep* **13:**163–166.
49. **Francavilla E, Cadrobbi P, Scaggiante P, DiSilvestro G, Bortolotti F, Bertaggia A.** 1989. Surveillance on occupational exposure to HIV among health care workers in Italy (Abstract D623) V International Conference on AIDS, Montreal, Quebec, Canada: 795.
50. **Heptonstall J, Porter KP, Gill ON.** 1995. *Occupational transmission of HIV: summary of published reports to July 1995.* Public Health Laboratory Service, London.
51. **Hernandez E, Gatell JM, Puyuelo T, Mariscal D, Barrera JM, Sanchez C.** 1988. Risk of transmitting the human immunodeficiency virus to health care workers exposed to HIV infected body fluids (Abstract 9003). Presented at the IV International Conference on AIDS, Stockholm, Sweden: 476.
52. **Ippolito G, Cadrobbi P, Carosi G.** 1989. Risk of occupational exposure to HIV- infected body fluids and transmission of HIV among health care workers: a multicenter study (Abstract MDP72). V International Conference on AIDS, Montreal, Quebec, Canada: 722.
53. **Jorbeck H, Marland M, Steinkeller E.** 1989. Accidental exposures to HIV- positive blood among health care workers in 2 Swedish hospitals. (Abstract A517). V International Conference on AIDS, Montreal, Quebec, Canada: 163.
54. **Kuhls TL, Viker S, Parris NB, Garakian A, Sullivan-Bolyai J, Cherry JD.** 1987. Occupational risk of HIV, HBV and HSV-2 infections in health care personnel caring for AIDS patients. *Am J Public Health* **77:**1306–1309.
55. **McEvoy M, Porter K, Mortimer P, Simmons N, Shanson D.** 1987. Prospective study of clinical, laboratory, and ancillary staff with accidental exposures to blood or body fluids from patients infected with HIV. *Br Med J (Clin Res Ed)* **294:**1595–1597.
56. **Ramsey KM, Smith EN, Reinarz JA.** 1988. Prospective evaluation of 44 health care workers exposed to human immunodeficiency virus-1, with one seroconversion. *Clin Res* **36:**1a. Abstract
57. **Rastrelli M, Ferrazzi D, Vigo B, Giannelli F.** 1989. Risk of HIV transmission to health care workers and comparison with the viral hepatitidies. (Abstract A503). Presented at the V International Conference on AIDS, Montreal, Quebec, Canada: 161.
58. **Tokars JI, Marcus R, The Cooperative Needlestick Surveillance Group.** 1990. Surveillance of health care workers exposed to blood from patients infected with the human immunodeficiency virus. (abstract 490). 30th Interscience Conference on Antimicrobial Agents and Chemotherapy, Atlanta, Ga.
59. **Wormser GP, Joline C, Sivak SL, Arlin ZA.** 1988. Human immunodeficiency virus infections: considerations for health care workers. *Bull N Y Acad Med* **64:**203–215.

60. **Pizzocolo G, Stellini G, Cadeo P, Casari S, Zampini PL.** 1988. Risk of HIV and HBV infection after accidental needlestick. Abstr. 9012. IV Int. Conf. AIDS. Stockholm, Sweden.
61. **Centers for Disease Control and Prevention (CDC).** 1995. Case-control study of HIV seroconversion in health-care workers after percutaneous exposure to HIV-infected blood—France, United Kingdom, and United States, January 1988-August 1994. *MMWR Morb Mortal Wkly Rep* **44:**929–933.
62. **Mast ST, Gerberding JL.** 1991. Factors predicting infectivity following needlestick exposure to HIV: an in vitro model. *Clin Res* **39:**58A.
63. **Bennett NT, Howard RJ.** 1994. Quantity of blood inoculated in a needlestick injury from suture needles. *J Am Coll Surg* **178:**107–110.
64. **Public Health Laboratory Service (PHLS) AIDS & STD Centre at the Communicable Disease Surveillance Centre.** 1999. Occupational Transmission of HIV, Summary of Published Reports. London, UK. (http://www.phls.co.uk)
65. **Johnson GK, Robinson WS.** 1991. Human immunodeficiency virus-1 (HIV-1) in the vapors of surgical power instruments. *J Med Virol* **33:**47–50.
66. **Winkler WG, Fashinell TR, Leffingwell L, Howard P, Conomy P.** 1973. Airborne rabies transmission in a laboratory worker. *JAMA* **226:**1219–1221.
67. **Centers for Disease Control.** 1986. HTLV III/LAV: agent summary statement. *MMWR Morb Mortal Wkly Rep* **35:**540–549.
68. **Centers for Disease Control and Prevention.** 2015. Ebola Outbreak in West Africa. Accessed on 12/29/2015. http://www.cdc.gov/vhf/ebola/outbreaks/2014-west-africa/index.html
69. **World Health Organization.** 2015. Health worker Ebola infections in Guinea, Liberia and Sierra Leone. A Preliminary Report: May 21, 2015. Accessed 12/30/2015. http://apps.who.int/csr/resources/publications/ebola/health-worker-infections/en/index.html.
70. **Centers for Disease Control and Prevention (CDC).** 2009. Imported case of Marburg hemorrhagic fever—Colorado, 2008. *MMWR Morb Mortal Wkly Rep* **58:**1377–1381.
71. **Timen A, Koopmans MPG, Vossen ACTM, van Doornum GJ, Günther S, van den Berkmortel F, Verduin KM, Dittrich S, Emmerich P, Osterhaus AD, van Dissel JT, Coutinho RA.** 2009. Response to imported case of Marburg hemorrhagic fever, the Netherland. *Emerg Infect Dis* **15:**1171–1175.
72. **Amorosa V, MacNeil A, McConnell R, Patel A, Dillon KE, Hamilton K, Erickson BR, Campbell S, Knust B, Cannon D, Miller D, Manning C, Rollin PE, Nichol ST.** 2010. Imported Lassa fever, Pennsylvania, USA, 2010. *Emerg Infect Dis* **16:**1598–1600.
73. **Centers for Disease Control and Prevention.** 2014. Ebola virus disease cluster in the United States-Dallas County, Texas, 2014. *Morbid. Mortal. Weekly Rep.* 63: 1087–1088. http://www.cdc.gov/mmwr/ebola_reports.html.
74. **Emond RTD, Evans B, Bowen ETW, Lloyd G.** 1977. A case of Ebola virus infection. *BMJ* **2:**541–544.
75. **Formenty P, Hatz C, Le Guenno B, Stoll A, Rogenmoser P, Widmer A.** 1999. Human infection due to Ebola virus, subtype Côte d'Ivoire: clinical and biologic presentation. *J Infect Dis* **179** (Suppl 1)**:**S48–S53.
76. **New York State Department of Health.** 2014. Revised NYS/NYC laboratory guidelines for handling specimens from patients with suspected or confirmed Ebola virus disease. Accessed 12/30/2015. http://www.health.ny.gov/diseases/communicable/ebola/docs/lab_guidelines.pdf.
77. **Occupational Safety and Health Administration.** 2015. Ebola Safety and Health Topics. Control and Prevention. Accessed 12/30/2015. https://www.osha.gov/SLTC/ebola/control_prevention.html.
78. **Centers for Disease Control and Prevention.** 2015. Interim US guidance for monitoring and movement of persons with potential ebola virus exposure. Updated 2015. Accessed 12/30/2015. http://www.cdc.gov/vhf/ebola/exposure/monitoring-and-movement-of-persons-with-exposure.html.
79. **Centers for Disease Control.** 1982. Current trends acquired immunodeficiency syndrome (AIDS): precautions for clinical and laboratory staffs. *MMWR Morb Mortal Wkly Rep* **31:**577–580.
80. **Centers for Disease Control.** 1987. Recommendations for prevention of HIV transmission in health care settings. *MMWR Morb Mortal Wkly Rep* **36**(suppl 2)**:**3S–18S.
81. **Centers for Disease Control (CDC).** 1987. Update: human immunodeficiency virus infections in health-care workers exposed to blood of infected patients. *MMWR Morb Mortal Wkly Rep* **36:**285–289.
82. **Centers for Disease Control (CDC).** 1988. Update: acquired immunodeficiency syndrome and human immunodeficiency virus infection among health-care workers. *MMWR Morb Mortal Wkly Rep* **37:**229–234, 239.
83. **Centers for Disease Control (CDC).** 1988. Update: universal precautions for prevention of transmission of human immunodeficiency virus, hepatitis B virus, and other blood-borne pathogens in health-care settings. *MMWR Morb Mortal Wkly Rep* **37:**377–382, 387–388.
84. **U.S. Department of Health and Human Services, Public Health Service, Centers for Disease Control and Prevention, National Institutes of Health.** 2009. Biosafety in Microbiological and Biomedical Laboratories. 5th ed. HHS Publication no. (CDC) 21-112. http://www.cdc.gov/biosafety/publications/bmbl5/bmbl.pdf .
85. **Advisory Committee on Dangerous Pathogens.** 1995. *Protection against blood-borne viruses in the workplace: hiv and hepatitis.* HMSO, London.
86. **Righter J.** 1991. Removal of warning labels from patient specimens. *Can J Infect Control* **6:**109.
87. **Osterman JW.** 1995. Beyond universal precautions. *CMAJ* **152:**1051–1055.
88. **Whitby M, McLaws ML, Slater K.** 2008. Needlestick injuries in a major teaching hospital: the worthwhile effect of hospital-wide replacement of conventional hollow-bore needles. *Am J Infect Control* **36:**180–186.
89. **Nelsing S, Nielsen TL, Nielsen JO.** 1993. Occupational blood exposure among health care workers: II. Exposure mechanisms and universal precautions. *Scand J Infect Dis* **25:**199–205.
90. **The Hospital Infection Control Group of Thailand.** 1995. Guidelines for implementation of universal precautions. *J Med Assoc Thai* **78**(Suppl 2)**:**S133–S134.
91. **Hu DJ, Kane MA, Heymann DL, World Health Organization.** 1991. Transmission of HIV, hepatitis B virus, and other blood-borne pathogens in health care settings: a review of risk factors and guidelines for prevention. *Bull World Health Organ* **69:**623–630.
92. **Garner JS, The Hospital Infection Control Advisory Committee.** 1996. Guideline for isolation precautions in hospitals. Part I. Evolution of isolation practices.. *Am J Infect Control* **24:**24–31.
93. **Occupational Safety and Health Administration (OSHA).** 1991. Occupational Exposure to Bloodborne Pathogens; Final Rule..*Fed Regist* **56:**64175–64182.
94. **Occupational Safety and Health Administration (OSHA).** 2001. 29 CFR Part 1910.1030. 2001. Occupational exposure to bloodborne pathogens; needlesticks and other sharp injuries; final rule. *Fed Regist* **66:**5317–5325.
95. **Occupational Safety and Health Administration (OSHA), Food and Drug Administration (FDA), and National Institute for Occupational Safety and Health (NIOSH).** 1999. Joint safety advisory about potential risk from glass capillary tubes. Feb. 22, 1999. (www.osha-slc.gov/SLTC/needlestick).
96. **Pugliese G, Salahuddin M.** 1999. Sharps Injury Prevention Program: A Step-by-Step Guide. Am. Hosp. Assoc. Cat. No. 196311. Chicago, IL.

97. **National Institute for Occupational Safety and Health (NIOSH).** 1999. NIOSH Alert: preventing needlestick injuries in health care settings. Department of Health and Human Services (NIOSH) Pub. No. 2000-108. November, 1999. http://www.cdc.gov/niosh/docs/2000-108/
98. **Kelly D.** 1996. Trends in US patents for needlestick prevention technology. *Adv Expos Prev* **2:**7–8.
99. **Centers for Disease Control and Prevention (CDC).** 1997. Evaluation of safety devices for preventing percutaneous injuries among health-care workers during phlebotomy procedures—Minneapolis-St. Paul, New York City, and San Francisco, 1993–1995. *MMWR Morb Mortal Wkly Rep* **46:**21–25.
100. **Tan L, Hawk JC III, Sterling ML.** 2001. Report of the Council on Scientific Affairs: preventing needlestick injuries in health care settings. *Arch Intern Med* **161:**929–936.
101. **Jagger J, Perry J.** 2003. Comparison of EpiNET data for 1993 and 2001 show marked decline in needlestick injury rates. *Adv Expos Prev* **6:**25–27.
102. **Sohn S, Eagan J, Sepkowitz KA, Zuccotti G.** 2004. Effect of implementing safety-engineered devices on percutaneous injury epidemiology. *Infect Control Hosp Epidemiol* **25:**536–542.
103. **Tosini W, Ciotti C, Goyer F, Lolom I, L'Hériteau F, Abiteboul D, Pellissier G, Bouvet E.** 2010. Needlestick injury rates according to different types of safety-engineered devices: results of a French multicenter study. *Infect Control Hosp Epidemiol* **31:** 402–407.
104. **Chiarello LA.** 1995. Selection of needlestick prevention devices: a conceptual framework for approaching product evaluation. *Am J Infect Control* **23:**386–395.
105. **Jagger J, Hunt EH, Brand-Elnaggar J, Pearson RD.** 1988. Rates of needle-stick injury caused by various devices in a university hospital. *N Engl J Med* **319:**284–288.
106. **Fisher J.** 1999. *Training for development of innovative control technology project (TDICT).* San Francisco General Hospital, San Francisco, Ca.
107. **National Institute for Occupational Safety and Health (NIOSH).** 1998. Selecting, Evaluating, and Using Sharps Disposal Containers. US Department of Health and Human Services (NIOSH) Pub. No. 97-111. January, 1998.
108. **Hanson PJV, Gor D, Jeffries DJ, Collins JV.** 1989. Chemical inactivation of HIV on surfaces. *BMJ* **298:**862–864.
109. **Weber DJ, Barbee SL, Sobsey MD, Rutala WA.** 1999. The effect of blood on the antiviral activity of sodium hypochlorite, a phenolic, and a quaternary ammonium compound. *Infect Control Hosp Epidemiol* **20:**821–827.
110. **Johnson GK, Nolan T, Wuh HC, Robinson WS.** 1991. Efficacy of glove combinations in reducing cell culture infection after glove puncture with needles contaminated with human immunodeficiency virus type 1. *Infect Control Hosp Epidemiol* **12:** 435–438.
111. **Gerberding JL, Littell C, Tarkington A, Brown A, Schecter WP.** 1990. Risk of exposure of surgical personnel to patients' blood during surgery at San Francisco General Hospital. *N Engl J Med* **322:**1788–1793.
112. **U.S. Food and Drug Administration (FDA).** 1990. Medical devices; patient examination and surgeons' gloves; adulteration—FDA. Final rule. *Fed Regist* **55:**51254–51258.
113. **Klein RC, Party E, Gershey EL.** 1989. Safety in the laboratory. *Nature* **341:**288.
114. **Kuhar DT, Henderson DK, Struble KA, Heneine W, Thomas V, Cheever LW, Gomaa A, Panlilio AL, US Public Health Service Working Group.** 2013. Updated US public health service guidelines for the management of occupational exposures to HIV and recommendations for postexposure prophylaxis. *Infect Control Hosp Epidemiol* **34:**875–892.
115. **Beekmann SE, Vlahov D, Koziol DE, McShalley ED, Schmitt JM, Henderson DK.** 1994. Temporal association between implementation of universal precautions and a sustained, progressive decrease in percutaneous exposures to blood. *Clin Infect Dis* **18:**562–569.
116. **Jagger J, Bentley M.** 1995. Substantial nationwide drop in percutaneous injury rates detected for 1995. *Adv Expos Prev* **2:**2.
117. **Dement JM, Epling C, Ostbye T, Pompeii LA, Hunt DL.** 2004. Blood and body fluid exposure risks among health care workers: results from the Duke Health and Safety Surveillance System. *Am J Ind Med* **46:**637–648.
118. **Fahey BJ, Koziol DE, Banks SM, Henderson DK.** 1991. Frequency of nonparenteral occupational exposures to blood and body fluids before and after universal precautions training. *Am J Med* **90:**145–153.
119. **Wong ES, Stotka JL, Chinchilli VM, Williams DS, Stuart CG, Markowitz SM.** 1991. Are universal precautions effective in reducing the number of occupational exposures among health care workers? A prospective study of physicians on a medical service. *JAMA* **265:**1123–1128.
120. **Centers for Disease Control and Prevention.** 2014. Guidance for U.S. laboratories for managing and testing routine clinical specimens when there is a concern about Ebola virus disease. Accessed 12/30/2015. http://www.cdc.gov/vhf/ebola/healthcare-us/laboratories/safe-specimen-management.html.
121. **Looke DFM, Grove DI.** 1990. Failed prophylactic zidovudine after needlestick injury. *Lancet* **335:**1280.
122. **Ridzon R, Gallagher K, Ciesielski C, Ginsberg MB, Robertson BJ, Luo CC, DeMaria A Jr.** 1997. Simultaneous transmission of human immunodeficiency virus and hepatitis C virus from a needle-stick injury. *N Engl J Med* **336:**919–922.
123. **Pincus SH, Messer KG, Nara PL, Blattner WA, Colclough G, Reitz M.** 1994. Temporal analysis of the antibody response to HIV envelope protein in HIV-infected laboratory workers. *J Clin Invest* **93:**2505–2513.
124. **Ridzon R, Kenyon T, Luskin-Hawk R, Schultz C, Valway S, Onorato IM.** 1997b. Nosocomial transmission of human immunodeficiency virus and subsequent transmission of multidrug-resistant tuberculosis in a healthcare worker. *Infect Control Hosp Epidemiol* **18:**422–423.
125. **Tokars JI, Marcus R, Culver DH, Schable CA, McKibben PS, Bandea CI, Bell DM, The CDC Cooperative Needlestick Surveillance Group.** 1993. Surveillance of HIV infection and zidovudine use among health care workers after occupational exposure to HIV-infected blood. *Ann Intern Med* **118:**913–919.
126. **Heptonstall J, Gill ON, Porter K, Black MB, Gilbart VL.** 1993. Health care workers and HIV: surveillance of occupationally acquired infection in the United Kingdom. *Commun Dis Rep CDR Rev* **3:**R147–R153.
127. **National Centre in HIV Epidemiology and Clinical Research (NCHECR).** 1995. *Australian HIV Surveillance Report.* 11:1, 3–7.
128. **Jochimsen EM.** 1997. Failures of zidovudine postexposure prophylaxis. *Am J Med* **102**(5B):52–55, discussion 56–57.
129. **Eberle J, Habermann J, Gürtler LG.** 2000. HIV-1 infection transmitted by serum droplets into the eye: a case report. *AIDS* **14:**206–207.
130. **Perry J, Jagger J.** 2005. Occupational co-infection with HIV and HCV in clinical lab via blood splash. *Adv. Expos. Prev.* **7:** 37–47.
131. **McDonald A.**2002. National Centre in HIV Epidemiology and Clinical Research, Australia. Cited in Health Protection Agency Centre for Infections, Occupational Transmission of HIV, March 2005 Edition, London, UK. http://www.hpa.org.uk/infections/topics-az/bbv/pdf/intl_HIV-tables_2005.pdf. Accessed September 30, 2015.
132. **Gerberding JL.** 1995. Management of occupational exposures to blood-borne viruses. *N Engl J Med* **332:**444–451.

23

Decontamination in the Microbiology Laboratory

MATTHEW J. ARDUINO

To protect laboratory workers, the general public, and the environment, as well as to avoid release of infectious agents into the environment, laboratories use a combination of work practices and engineering controls, including decontamination strategies for work surfaces, items, and spaces within the laboratory, to mitigate this risk. "Decontamination" is a general term that usually refers to a process that makes an item safe to handle, or a space safe to occupy, and can include processes ranging from simple cleaning with soap and water to sterilization. This chapter discusses the factors necessary for environmentally mediated transmission of infection to occur and methods for decontamination (which includes cleaning, disinfection, and sterilization). Emphasis is placed on the general approaches to decontamination practices and not on the detailed protocols and methods. The principles of sterilization and disinfection are discussed and compared in the context of the decontamination procedures used in laboratories.

ENVIRONMENTALLY MEDIATED INFECTION TRANSMISSION

When environmentally associated laboratory infections occur, they can be transmitted to laboratory staff directly or indirectly from environmental sources (e.g., air, contaminated fomites and laboratory instruments, and aerosols). Fortunately, laboratory-acquired infections are rare events (1, 2) due to the relatively controlled laboratory environment and routine safety practices. Also, for a laboratory-acquired infection to occur as a result of environmental transmission a number of requirements must be met (3). These requirements, commonly referred to as the "chain of infection," include the presence of a pathogen of sufficient virulence, a relatively high concentration of the pathogen (i.e., infectious dose), a mechanism of transmission of the pathogen from the environment to the host, a correct portal of entry on the host, and a susceptible host. To accomplish successful transmission from an environmental source, all of these requirements for the chain of infection must be present. Because the absence of any of these elements will minimize environmental transmission from occurring, the requirements for the chain of infection comprise the targets of various approaches of environmental control to prevent transmission. Additionally, the pathogen in question must be able to cope with environmental stresses to retain its viability, virulence, and the capability to initiate infection in the host. In the laboratory setting, high concentrations of pathogens can be common. Overkill methods are often employed to remove any potential for infection transmission, even though the reduction of environmental microbial contamination by conventional cleaning procedures (with either soap and water or a detergent disinfectant) is often enough to interrupt the potential for environmentally mediated transmission. Overkill methods usually employ longer treatment times than are required to kill the bioburden present on or in an item being sterilized.

PRINCIPLES OF STERILIZATION AND DISINFECTION

Understanding the principles of decontamination, cleaning, disinfection, and sterilization is important for implementing a laboratory biosafety program. The definitions of antisepsis (disinfection of skin and tissues), decontamination, disinfection, and sterilization are sometimes confused and misused. The definitions and implied capabilities of each inactivation procedure are discussed with an emphasis on achieving and, in some cases, monitoring each stage of microbial reduction. Antisepsis is not covered in this chapter. To understand the application of these concepts to laboratory biosafety, one needs first to understand them in the context of their application to patient safety in health care.

Sterilization

Any item, device, or solution is considered to be sterile when it is completely free of all living microorganisms and viruses, or because it is impossible to prove a negative, that its probability of being contaminated is below a stringent standard. The definition is categorical and absolute (i.e., an item either is sterile or is not). Sterilization can be accomplished by heat; such gaseous materials as ethylene oxide, chlorine dioxide (CD), hydrogen peroxide gas plasma, ozone, or irradiation (e.g, gamma, e-beam, etc.) (4–16). Many of these methods are used in industry or health care settings. From an operational standpoint, a sterilization procedure cannot be defined categorically. Rather, the procedure is defined as a process, after which the probability of a microorganism surviving on an item subjected to the sterilization procedures is less than one in one million (10^{-6}). This is referred to as the "sterility assurance level" (17, 18).

Disinfection

In general, disinfection is a less lethal process than sterilization. It is important to note that the margin of overkill achieved by a disinfection practice is less than sterilization and lacks the margin of overkill achieved by sterilization. Disinfection eliminates nearly all recognized pathogenic microorganisms but not necessarily all microbial forms (e.g., bacterial spores, prions) on inanimate objects. The effectiveness of disinfection is impacted by a number of factors, each one of which may have a pronounced effect on the end result. Among these are (i) the nature and number of contaminating microorganisms (especially the presence of bacterial spores); (ii) the amount of organic matter present (e.g., soil, feces, blood, and even culture media in the laboratory); (iii) the type and condition of the instruments, devices, and materials to be disinfected; and (iv) the temperature.

Disinfection is a procedure that reduces the level of microbial contamination, but there is a broad range of activity that extends from sterility at one extreme to a minimal reduction in the number of microbial contaminants at the other. By definition, high-level chemical disinfection differs from chemical sterilization by its lack of sporicidal activity. Actually, there are some chemical germicides used as disinfectants that do, in fact, kill large numbers of spores, although high concentrations and several hours of exposure time may be required. Nonsporicidal disinfectants may differ in their capacity to accomplish disinfection or decontamination. Some germicides rapidly kill only the ordinary vegetative forms of bacteria, such as staphylococci and streptococci, some forms of fungi, and lipid-containing viruses (e.g., enveloped viruses), whereas others are effective against such relatively resistant organisms as *Mycobacterium tuberculosis* subsp. *bovis*, nonlipid viruses (nonenveloped viruses), and most forms of fungi.

Spaulding Classification

In 1972, Earl Spaulding (19) proposed a system for classifying liquid chemical germicides and inanimate surfaces that has been used subsequently by the Centers for Disease Control and Prevention (CDC), the Food and Drug Administration (FDA), and experts in the United States. This system, as it applies to device surfaces, can be divided into three general categories on the basis of the theoretical risk of infection if the surfaces are contaminated at the time of use. From the health care perspective, these categories are as follows.

- Critical—instruments or devices that are exposed to normally sterile areas of the body and which require sterilization;
- Semicritical—instruments or devices that touch mucous membranes and may be either sterilized or disinfected;

- Noncritical—instruments or devices that touch skin or come into contact with persons only indirectly and can be either cleaned and then disinfected with an intermediate-level disinfectant, sanitized with a low-level disinfectant, or simply cleaned with soap and water (this is most likely the major category as it relates to laboratory safety)

In 1991, microbiologists at the CDC proposed an additional category—environmental surfaces (e.g., floors, walls, and other "housekeeping surfaces") that do not make direct contact with a person's skin (20, 21). These four categories provide a rational basis for decontamination that can help to prevent unnecessary overkill and expense.

Spaulding also classified chemical germicides by activity level in the following process categories.

(i) High-level disinfection uses disinfectants that kill vegetative microorganisms and inactivate viruses, but not necessarily high numbers of bacterial spores. They are capable of sterilization after prolonged contact time (e.g., 6 to 10 hours). As high-level disinfectants, they are used for relatively short periods of time (e.g., 10 to 30 minutes). These chemical germicides are very potent sporicides and, in the United States, are classified by the FDA as sterilants/disinfectants and as sterilizers by the Environmental Protection Agency (EPA). They are formulated for use on medical devices, but not on environmental surfaces such as laboratory benches or floors (20).

(ii) Intermediate-level disinfection kills vegetative microorganisms, including *M. tuberculosis* and all fungi, and inactivates most viruses. Chemical germicides used in this procedure often correspond to EPA-registered "hospital disinfectants" that are also "tuberculocidal." They are used commonly in laboratories for disinfection of laboratory benches and are the detergent germicides used for housekeeping purposes.

(iii) Low-level disinfection kills most vegetative bacteria except *M. tuberculosis* and some fungi, and it inactivates some viruses. Chemical germicides used in this procedure are approved in the United States by the EPA as "hospital disinfectants" or "sanitizers."

Disinfection Hierarchy

Microbes also demonstrate different degrees of resistance to both chemical and physical methods of inactivation due to their physical state and their biochemical-biophysical structure, bacterial spores being the most resistant (requiring sterilizing agents). The makeup of the outermost structure, a waxy cell wall that is presented to the germicidal agent, provides for increased resistance in mycobacteria. The lack of lipids in the outer capsid of a virus renders it hydrophilic and more resistant to some germicides than the lipophilic viruses. These structural differences among microorganisms lead to what is known as the disinfection hierarchy (Fig. 1) (6, 22, 23). There are a few exceptions to the hierarchy and these include members of the *Rickettsiae*, *Chlamydiae*, and mycoplasma. These are typically not placed on the hierarchy because information about the efficacy of disinfectants is limited, but because they contain lipids and are similar in structure to other bacteria, one can expect disinfectants to be effective that destroy both viruses and bacteria (23). The exception here is the known resistance of *Coxiella burnetii* (24).

DECONTAMINATION IN THE MICROBIOLOGY LABORATORY

Decontamination in the microbiology laboratory requires great care. It may entail disinfection of work surfaces or decontamination of equipment so that it is safe to handle. On the other hand, it may require sterilization, as is done prior to removing infectious waste from the biosafety level (BSL) 3 or BSL4 facility. Regardless of the method, the purpose of decontamination in the microbiology laboratory is to protect the laboratory worker as well as those who enter the laboratory or handle laboratory products away from the laboratory, and to protect the environment. Reduction of cross-contamination in the laboratory is an added benefit.

Decontamination and Cleaning

Decontamination renders an area, device, item, or material safe to handle, that is, reasonably free from a risk of disease transmission. The primary objective of a decontamination procedure is to reduce the level of microbial contamination such that the risk of the transmission of infection is minimized. The decontamination process may be as simple as cleaning an instrument, device, or area with ordinary soap and water. In laboratory settings, decontamination of items, spent laboratory materials, and regulated laboratory wastes is often accomplished by a sterilization procedure such as steam autoclaving, which may be the most cost-effective method. However, sterilization methods are almost always ultraconservative and constitute overkill. This is especially true with emerging infectious diseases and etiologic agents that cause serious and fatal diseases. In these instances, it is often assumed, incorrectly, that more serious or emerging infectious agents are resistant to standard microbial inactivation methods. In addition to these incorrect assumptions regarding these serious/emerging infectious agents, there is an additional incorrect assumption, that all environmental contaminants that are etiologic agents of human disease will be transmitted to susceptible hosts.

Bacterial spores
Bacillus subtilis
Clostridium sporogenes
⇩
Coccidia
(e.g. *Cryptosporidium*)
⇩
Mycobacteria
Mycobacterium tuberculosis subsp. *bovis*
Nontuberculous mycobacteria
⇩
⇩
Nonlipid or small viruses
Poliovirus
Coxsackievirus
Rhinovirus
⇩
⇩
Fungi
Trichophyton, *Cryptococcus*, *Aspergillus*, and *Candida* spp.
⇩
⇩
Vegetative bacteria
Pseudomonas aeruginosa
Staphylococcus aureus
Salmonella enterica serovar Choleraesuis
Enterococci
⇩
⇩
Lipid or medium-size viruses
Herpes simplex virus
Cytomegalovirus
Respiratory syncytial virus
Hepatitis B virus
Hepatitis C virus
HIV
Hantavirus
Ebola virus
Influenza viruses

Figure 1: The hierarchy of resistance to disinfection (20, 23). There are some exceptions to this list. *Pseudomonas* spp. are sensitive to high-level disinfectants, but if they grow in water and form biofilms on surfaces, the protected cells can approach the resistance of bacterial spores to the same disinfectant. The same is true for the resistance to glutaraldehyde by some nontuberculous mycobacteria, some fungal ascospores of *Microascus cinereus* and *Chaetomium globosum*, and the pink-pigmented *Methylobacterium*. Prions are also resistant to most liquid chemical germicides and are discussed in the last part of this chapter.

The presence of any organic matter can interfere with the disinfection and necessitates longer contact time if the item or area is not precleaned. Sterilization cycles are, unlike disinfection, dependent on a number of factors including type and size of packaging, volume, load size and placement, amount of organic soil (whether items are precleaned or contaminated materials), etc. For example, a steam sterilization cycle used to sterilize precleaned items is 20 minutes at 121°C. When steam sterilization is used to decontaminate items that have a high bioburden or large packaging and there is no precleaning (i.e., infectious waste), the cycle time is considerably longer. Decontamination in laboratory settings often requires longer cycles and exposure times because pathogenic microorganisms may be protected from contact with the decontaminating agents by organic material, airspaces, or packaging. Validation of the sterilization cycles involves placing a biological indicator (BI) in the worst-case location in the autoclave and obtaining negative results.

Chemical germicides used for decontamination range in activity from high-level disinfectants (i.e., high concentrations of sodium hypochlorite, or household bleach), which might be used to decontaminate spills of cultured or concentrated infectious agents in research or clinical laboratories, to low-level disinfectants or sanitizers (i.e., a dilute bleach solution) for general housekeeping purposes or spot decontamination of environmental surfaces in health care settings. Examples are shown in Table 1. If dangerous and highly infectious agents are contained in a laboratory, the methods for decontamination of spills, laboratory equipment, biological safety cabinets, or infectious waste are very significant and may include prolonged autoclave waste cycles or incineration prior to disposal to assure extra margins of safety. Once treated, these materials are no longer considered to be infectious and waste items may enter the wastestream (solid waste, sanitary sewer, etc.) as per state and local regulations.

DECONTAMINATION OF LARGE SPACES

Another category of decontamination is area or space decontamination (e.g., biological safety cabinets, large bioaerosol or environmental chambers used for aerosol studies, laboratories, animal facilities, hospital rooms, etc.). This activity should be performed by trained specialists using appropriate protective equipment and process measures (25, 26). The most common approaches employ formaldehyde, chlorine dioxide, or hydrogen peroxide (27, 28). Decontamination requirements for BSL3 and BSL4 laboratory spaces have an impact on the design of these facilities (29). The interior surfaces of BSL3 laboratories must be water resistant for them to be easily cleaned and

TABLE 1.

Categories of chemical disinfectants based on their activity (20, 23)

Level	Bacterial spores	Nonenveloped and small viruses	Mycobacteria (tubercle bacilli)	Fungi[a]	Vegetative bacteria	Enveloped viruses
High level	+[b]	+	+	+	+	+
Intermediate level	±[c]	±[d]	+	+	+	+
Low level	–	–	–	±	+	+

[a]Includes asexual spores; may not include chlamydospores or sexual spores.
[b]Plus sign indicates that kill can be expected when product label instructions are followed; minus sign indicates no activity. With extended contact time, some of the high-level disinfectants (HLD) are capable of sterilization. These are cleared by FDA as chemical sterilants and registered by EPA as chemical sterilizers. Most HLDs are sterilizers used for shorter contact times.
[c]Some intermediate-level disinfectant (ILD) products may have activity against some bacterial spores.
[d]Some ILD products have activity against nonenveloped viruses.

decontaminated. Penetrations in these surfaces should be sealed or capable of being sealed for decontamination purposes.

Thus, in the BSL3 laboratory, liquid decontamination is assumed to be the norm (work and frequently touched surfaces), and fumigation is not considered the primary means of decontaminating the space. Fumigation in the laboratory is typically performed following either an intentional or accidental release of an infectious agent resulting in widespread contamination of surfaces and equipment. Fumigation of biological safety cabinets is typically performed prior to changing the HEPA filters or when internal repairs are needed (29). Care should be taken that penetrations (e.g., electrical, plumbing, network ports, phone/fax jacks, etc.) in the walls, floors, and ceilings are kept to a minimum and "sight sealed." Verification of the seals is usually not required for most BSL3 laboratories; an exception is the agricultural BSL3 (BSL3Ag) laboratory described in chapter 35. In some instances, air monitoring for the chemical agent being used is conducted both inside and outside of the area being treated. This is to ensure that the appropriate concentration has been reached and for safety of personnel during treatment and following neutralization or aeration at the end of the treatment. The BSL-4 laboratory design requires that the interior surfaces be water resistant and sealed to facilitate fumigation. These seals require testing and verification to ensure containment. The BSL4 laboratory is designed to allow both liquid disinfection and fumigation. Periodic fumigation is required in the BSL4 suit laboratory to allow routine maintenance and certification of equipment. Procedures for decontamination of large spaces, such as rooms or walk-in incubators, are varied and influenced significantly by the type of etiologic agent involved, the characteristics of the structure of the space, and the materials present in the space. The primary methods for space decontamination are listed below.

Formaldehyde and Paraformaldehyde

Historically, aqueous formaldehyde has been used to flood areas that required decontamination (30–32). The procedure, although effective in killing microbial pathogens, is difficult to control and produces fumes that are toxic to humans and animals. In addition, formaldehyde has been determined to be a group I carcinogen by the International Agency for Research on Cancer (33). Before one can use formaldehyde or paraformaldehyde, a state or federal agency must apply for a section 18 emergency exemption (Federal Insecticide, Fungicide, and Rodenticide Act, 40 C.F.R. Part 166; 7 USC §136 *et seq.*).

For decontamination purposes, 4% to 8% formaldehyde for a contact time of 30 minutes is recommended to achieve a minimum of high-level disinfection. Formaldehyde gas at a concentration of 0.3 g/ft^3 for 4 hours is often used for space decontamination. Gaseous formaldehyde can be generated by heating flake paraformaldehyde (0.3 g/ft^3) in a frying pan, thereby converting it to formaldehyde gas. The humidity must be controlled to prevent an explosion; the system works optimally at 80% relative humidity (RH). As with liquid formaldehyde applications, the gaseous method is effective in killing microorganisms, but toxicity issues are of concern (1, 34) as well as the extensive cleaning required to clean up the residuals of formaldehyde neutralization (28, 35).

Hydrogen Peroxide Vapor

Hydrogen peroxide can be vaporized and used for the decontamination of glove boxes as well as walk-in incubators and small rooms or areas. Vapor-phase hydrogen peroxide (VHP) was also used during the anthrax cleanups in 2002 and was used successfully in the remediation of the U.S. Department of Justice mail facility (36). VHP has been shown to be an effective sporicide at concentrations ranging from 0.5 to <10 mg/liter. The optimal concentration of this agent is about 2.4 mg/liter with a contact time of at least 1 hour. An advantage of this system is that the end products (i.e., water and oxygen) are not toxic. Low RH (not greater than 30% RH) must be used to prevent condensation and ensure that the air contains an effective concentration of the VHP (37–42).

Chlorine Dioxide Gas

CD gas sterilization can be used for decontamination of laboratory rooms, equipment, glove boxes, and incubators (27, 28, 36, 43). The concentration of CD at the site of decontamination is approximately 10 mg/liter, and there is a contact time of 1 to 2 hours. CD possesses the bactericidal, virucidal, and sporicidal properties of chlorine, but unlike chlorine, it does not lead to the formation of trihalomethanes or combine with ammonia to form chlorinated organic products (chloramines). CD gas cannot be compressed and stored in high-pressure cylinders, but it is generated upon demand using a column-based solid-phase generation system. The gas is diluted to the use concentration, usually between 10 and 30 mg/liter. Within reasonable limits, a CD gas generation system is unaffected by the size or location of the ultimate destination for the gas. RH does need to be controlled, however; the optimum is at 50% or higher RH. The CD gas exits the gas generator at a modest positive pressure and flow rate, so, in actuality, the destination enclosure (e.g., isolator, glove box, sealed biosafety cabinet, or room) needs to be sealed to prevent gas egress (44).

Biological Indicators

When performing fumigation, process measures should be used to ensure effectiveness; this is most often accomplished with the use of BIs placed in selected areas within the space being treated. The chemical being used determines which BI is used as a process measure. Traditionally, *Bacillus atrophaeus* spores have been used with the use of paraformaldehyde (45). In some instances both *B. atrophaeus* and *Geobacillus stearothermophilus* indicators have been used for CD gas (46). Studies of the comparative effectiveness of VHP against spores have shown that *G. stearothermophilus* is generally more resistant than either *Bacillus anthracis* or *B. atrophaeus* (47, 48). The manufacturers of VHP systems recommend the use of *G. stearothermophilus* indicators when employing their technology.

Placement of the BIs is also important. The National Sanitation Foundation International (NSF) has general guidance for the decontamination of biological safety cabinets and the placement of BIs therein (49, 50). The NSF recommends using six pairs of indicators: one pair of BIs is placed between the pleats on the downstream (clean) side of the exhaust HEPA filter near the center; two more pairs of BIs are at the opposite corners of the filter; a pair of BIs is placed within a potentially contaminated positive-pressure plenum; another pair of BIs is placed beneath the work surface in the plenum below the cabinet work area; and the last pair of BIs is placed between the pleats near the center of the upstream (dirty) side of the downflow HEPA filter (50). When decontaminating larger spaces, multiple BIs may be used. The BIs would be placed in a variety of locations, including corners of the room, various locations on wall faces, underneath horizontal surfaces, and also placed at various heights (51). Decontamination is considered successful when growth BIs after decontamination are negative.

DECONTAMINATION OF SURFACES

Liquid chemical germicides formulated as disinfectants are used for the decontamination of spills of infectious material as well as for the decontamination of large areas. The usual procedure is to flood the area with a disinfectant for periods up to several hours (some of these agents may be applied as foams/gels). This approach is messy and, with some disinfectants, can be a toxic hazard to laboratory staff and is rarely if ever performed in the laboratory. For example, most of the high-level disinfectants on the U.S. market are formulated to be used on instruments and medical devices and not on environmental surfaces (see Table 1). Intermediate- and low-level disinfectants are formulated to be used on fomites and environmental surfaces but lack the potency of high-level disinfectants. For the most part, intermediate- and low-level disinfectants can be safely used as long as the manufacturer's use instructions are closely followed (21, 52). Disinfectants that have historically been used for decontamination include sodium hypochlorite solutions at concentrations of 500 to 6,000 ppm; oxidative disinfectants, such as hydrogen peroxide and peracetic acid; phenolic disinfectants; and iodophor disinfectants. Concentrations and exposure times vary depending on the formulation and the manufacturer's instructions for use (17, 18, 20, 21, 52). Table 2 lists chemical germicides and their activity levels.

A spill control plan must be available and should describe how a spill of any agent in the laboratory should be handled. This plan should include the rationale for selecting the disinfecting agent, the approach to applying the disinfectant, the contact time, and other parameters. Spills of BSL3 and BSL4 agents pose a high risk to workers and must be dealt with by a well-informed professional

TABLE 2.

Activity levels of selected liquid germicides[a]

Procedure/product	Aqueous concentration	Activity level
Sterilization		
Glutaraldehyde	Variable	
Hydrogen peroxide	6–30%	
Formaldehyde	6–8%[b]	
Chlorine dioxide	Variable	
Peracetic acid	Variable	
Disinfection		
Glutaraldehyde	Variable	High to intermediate
ortho-Phthalaldehyde	0.5%	High
Hydrogen peroxide	3–6%	High to intermediate
Formaldehyde	1–8%	High to low
Chlorine dioxide	Variable	High
Peracetic acid	Variable	High
Chlorine compounds[c]	500–5,000 mg of free/available chlorine per liter	Intermediate
Alcohols (ethyl, isopropyl)[d]	70%	Intermediate
Phenolic compounds	0.5–3%	Intermediate to low
Iodophor compounds[e]	30–50 mg of free iodine per liter; up to 10,000 mg of available iodine per liter	Intermediate to low
Quaternary ammonium compounds	0.1–0.2%	Low

[a]This list of chemical germicides centers on generic formulations. A large number of commercial products based on these generic components can be considered for use. Users should ensure that commercial formulations are registered with the EPA or cleared by the FDA.

[b]Because of the ongoing controversy of the role of formaldehyde as a potential occupational carcinogen, the use of formaldehyde is limited to certain specific circumstances under carefully controlled conditions, e.g., for the disinfection of certain hemodialysis equipment. There are no FDA-cleared liquid chemical sterilants/disinfectants that contain formaldehyde.

[c]Generic disinfectants containing chlorine are available in liquid or solid form, e.g., sodium or calcium hypochlorite. Although the indicated concentrations are rapid acting and broad spectrum (tuberculocidal, bactericidal, fungicidal, and virucidal), no proprietary hypochlorite formulations are formally registered with the EPA or cleared by the FDA. Common household bleach is an excellent and inexpensive source of sodium hypochlorite. Concentrations between 500 and 1,000 mg of chlorine per liter are appropriate for the vast majority of uses requiring an intermediate level of germicidal activity. Higher concentrations are extremely corrosive as well as irritating to personnel, and their use should be limited to situations where there is an excessive amount of organic material or unusually high concentrations of microorganisms (e.g., spills of cultured material in the laboratory).

[d]The effectiveness of alcohols as intermediate-level germicides is limited because they evaporate rapidly, resulting in very short contact times, and also lack the ability to penetrate residual organic material. They are rapidly tuberculocidal, bactericidal, and fungicidal, but may vary in spectrum of virucidal activity (see text). Items to be disinfected with alcohols should be carefully precleaned and then totally submerged for an appropriate exposure time (e.g., 10 minutes).

[e]Only those iodophors registered with the EPA as hard-surface disinfectants should be used, and the instructions of the manufacturer regarding proper dilution and product stability should be closely followed. Antiseptic iodophors are not suitable for disinfecting medical instruments or devices or environmental surfaces.

staff trained and equipped to work with spills of concentrated biohazardous material.

SPECIAL INFECTIOUS AGENT ISSUES AND BIOTERRORISM

The subject of bioterrorism, although discussed for many years, was not of direct concern to the general public until the fall of 2001. *B. anthracis* is unique among the agents of bioterrorism because the etiologic agent is a bacterial spore that is more resistant than other pathogens (see Table 1). All other agents that are considered potential weapons of biological warfare are vegetative bacteria or viruses, all of which are susceptible to the common array of chemical germicides used in hospitals or in the home. For example, smallpox virus infections are highly transmissible, but the virus can be killed by any low- to high-level disinfectant.

However, in the fall of 2001, anthrax spores were deliberately sent through the U.S. mail to various government officials and news organizations, resulting in 22 confirmed cases of infections with *B. anthracis*, 11 cutaneous and 11 inhalation anthrax cases (5 deaths) (53). The subject is included here because the press and the general public seem to misunderstand the application of the principles of sterilization and decontamination in such situations. Conventional disinfection and sterilization procedures are more than adequate to kill these emerging threats and potential agents of bioterrorism.

The problem with anthrax in the United States is a good example. Recommendations for the decontamination of items and areas contaminated with *B. anthracis* (anthrax) spores are based on two historical sources. The first is the industrial setting, where animal hides and hairs are processed. The second is the laboratory setting, where biological safety protocols have been developed to address decontamination of high concentrations of anthrax spores after accidental release in the laboratory. In both settings, procedures known to be sporicidal are recommended. They include incineration, steam autoclaving, and exposure to formaldehyde, paraformaldehyde, peracetic acid, pH adjusted sodium hypochlorite, β-propiolactone, ethylene oxide, and, more recently, CD.

Historically, most of the inactivation strategies as well as environmental microbiology principles (i.e., airborne studies) for anthrax are based on *Bacillus subtilis* subsp. *niger* (*B. atrophaeus*) spores that are used as a surrogate for *B. anthracis* spores. The standard BIs for dry heat, ethylene oxide, and VHP sterilization systems use *B. atrophaeus* spores. Consequently, there are data to demonstrate kill. Spores of *B. atrophaeus* and *B. anthracis* are not unusually resistant to physical and chemical agents (54). Any sterilization procedure will quickly kill them. Sterilization systems (e.g., autoclaves, ethylene oxide [EtO] sterilizers, irradiation systems, low temperature plasma, etc.) include any system (e.g., washer disinfectors) approved for marketing by the FDA in the United States. The chemical agents (e.g., disinfectants) are registered with EPA.

There is no scientific reason to extend the conventional cycles of sterilizers. For example, the standard steam autoclave sterilization cycle of 15 minutes at 121°C is used for previously cleaned material. Waste, on the other hand, is typically treated with longer sterilization cycles and, sometimes, higher temperatures. However, cycle times can be influenced by the materials being treated (55). Sterilization systems that will kill anthrax spores include the following:

- Steam sterilization
- EtO gas sterilization
- Hydrogen peroxide gas plasma sterilization
- Dry heat sterilization
- Radiation sterilization—cobalt (gamma) and e-beam
- CD gas sterilization

Liquid chemical germicides formulated as sterilants/high-level disinfectants will also kill anthrax spores when used according to the manufacturer's instructions. In clinical settings, such as endoscopy units, the normal infection control precautions are adequate to care for patients who have clinical anthrax. These patients do not have spores of *B. anthracis* in blood, tissues, or feces but, rather, vegetative (nonspore) forms of *B. anthracis*, which are very susceptible to conventional disinfection procedures. For example, endoscope cleaning and high-level disinfection protocols with the conventional high-level disinfectants do not need to be changed for instruments used for treating patients diagnosed with any of the three forms of anthrax.

Methods that would not be expected to be effective against anthrax spores or have no supporting data include the following:

- Ultraviolet (UV) radiation
- Boiling
- Exposure to alcohols, and low-level disinfectants (i.e., phenolics and quaternary ammonium compounds) and liquid chemical germicides formulated as antiseptics (chlorhexidine, iodophors, etc.)
- Superoxidized water
- Microwave ovens
- Ironing

ANTIBIOTIC-RESISTANT ORGANISMS AND EMERGING PATHOGENS

Outbreaks of disease caused by newly discovered microorganisms or microorganisms that have acquired resistance to antimicrobial agents are usually accompanied by disease control strategies that erroneously assign to these agents extraordinary resistance to commonly used sterilization and disinfection procedures. Examples include severe acute respiratory syndrome-associated coronavirus, the AIDS virus (HIV), hepatitis B virus, Ebola virus, hantavirus, multidrug-resistant *M. tuberculosis*, vancomycin-resistant enterococci, and methicillin-resistant *Staphylococcus aureus*. There is no relationship between the resistance of an organism to antimicrobial agents used for therapy and its innate resistance to chemical germicides or sterilization. Current protocols used in laboratories are conservative and do not need to be altered when they are used on devices or environmental surfaces that are exposed to new or antimicrobial-resistant microorganisms.

TRANSMISSIBLE SPONGIFORM ENCEPHALOPATHY AGENTS (PRIONS)

The major exception to the rule in the previous discussion of microbial inactivation and decontamination is the causative agent of Creutzfeldt-Jakob disease (CJD) or other related prions responsible for certain fatal degenerative diseases of the central nervous system (CNS) in humans and animals (e.g., prions) (29, 56, 57). One must remember that occupational transmission of prions has not been reported (29). The recommendations for the sterilization of instruments and medical devices exposed to patients with prion disease (CJD) are based on studies that show that prions are resistant to heat and chemical germicides. However, these studies use prion challenges that are enormous and unrealistic in that materials subjected to the inactivation processes are slurries and bits of tissue. Also, in virtually every reported study, the posttreatment recovery of active prions is done by injecting the treated material directly into the brains of susceptible animal models. The experimental portal of entry in these studies may not accurately reflect downstream transmission risks for any surface exclusive of neurosurgical instruments. Furthermore, because the empirical principles of instrument cleaning are not taken into consideration in these studies, most recommendations are extraordinarily conservative. Clearly the invariably fatal outcome of CJD infection influences the ultraconservative nature of many current recommendations (23, 56–59).

The transmissibility of the CJD agent has been demonstrated by the induction of disease in laboratory animals through the intracerebral inoculation of infective material (i.e., brain tissue or cerebrospinal fluid) but not by simple direct contact. Transmission of CJD has not been associated with environmental contamination or fomites (with the exception of neurosurgical instruments and electroencephalogram [EEG] depth electrodes). Person-to-person transmission via skin contact has not been documented. Routine environmental surfaces would not be expected to be associated with transmission of CJD to laboratory workers. Floors, walls, countertops, or other housekeeping surfaces in medical wards, autopsy rooms, and laboratories in the absence of contamination with high-risk tissues, such as brain and CNS tissue, should be cleaned with a suitable detergent-disinfectant in the conventional fashion. For a spill containing a high-risk tissue (e.g., cerebrospinal fluid, brain, and CNS tissues) onto a housekeeping surface, the bulk of the tissue residue should be removed carefully using disposable absorbent material and discarded into laboratory waste for incineration. Apply 1 N solution of NaOH or dilution of sodium hypochlorite containing 10,000 to 20,000 ppm to spot decontaminate the spill site/contaminated surfaces before cleaning (23).

Decontamination protocols for laboratories processing prion-infected tissues would not differ from the standard laboratory recommendations), and these protocols are adequate to prevent transmission to staff and the environment. The exception is if there is contamination following spill procedures for surfaces contaminated with high-risk tissues (from an individual with prion disease), in which case the spill procedures above should be followed. Items or laboratory instruments that are exposed to high-risk tissue and are impossible or difficult to clean should be decontaminated by autoclaving at 132 to 134°C for 18 minutes in a prevacuum sterilizer, or at 121°C for 1 hour in a gravity displacement sterilizer, or soaked in 1 N NaOH for 1 hour before sterilization. Under no circumstances should items be placed in containers of 1 N NaOH and then autoclaved. This procedure is dangerous to laboratory staff and ruins autoclaves (23).

In the United States there are guidelines that are evidenced based, and these include the following:

- *Biosafety in Microbiological and Biomedical Laboratories*, 5th edition (29)
- *Comprehensive Guide to Steam Sterilization and Sterility Assurance in Health Care Facilities* (60)
- Guideline for disinfection and sterilization in health care facilities, 2008 (23)
- *Protection of Laboratory Workers from Occupationally Acquired Infections*, approved guideline, 3rd ed. (61)

Laboratory investigations involving prion agents such as those causing scrapie or CJD can be conducted in BSL2 laboratory settings. Laboratory work with the agent responsible for bovine spongiform encephalopathy and variant CJD can be conducted in BSL2 labs with BSL3 practices or in the BSL3 facility (29). Nevertheless, decontamination protocols for laboratories processing prion-infected tissues would not differ significantly from the recommendations for handling these materials in the clinical setting (23, 29). These protocols are adequate to prevent transmission to staff and the environment, provided that prion-containing tissues and homogenates are handled carefully to avoid spills and unnecessary contamination of working surfaces. There have been several recent studies that show that alkaline cleaners, alone or in combination with certain biocides or hydrogen peroxide gas plasma sterilization, can significantly reduce or eliminate a high prion challenge (62–67).

Appropriate materials and methods for decontamination, disinfection, and sterilization are required for the safe conduct of work with biohazardous agents. Further information may be found in Vesley et al. (1), Favero and Bond (20), Hawley and Kozlovac (40), and Rutula (68).

I would like to thank my former supervisor and coauthor Dr. Martin S. Favero for his mentorship and dedication in disinfection and sterilization and for his work in environmental infection control in the health care setting.

Disclaimer

The conclusions, findings, and opinions expressed by authors in this chapter do not necessarily reflect the official position of the U.S. Department of Health and Human Services, the Public Health Service, the Centers for Disease Control and Prevention, or the authors' affiliated institutions

References

1. **Vesley D, Lauer JL, Hawley RJ.** 2000. Decontamination, sterilization, disinfection, and antisepsis, p 383–402. *In* Fleming DO, Hunt DL (ed), *Biological Safety: Principles and Practices*, 3rd ed. ASM Press, Washington, DC.
2. **Harding AL, Byers KB.** Epidemiology of laboratory-associated infections, p 53–77. *In* Fleming DO, Hunt DL (ed), *Biological Safety: Principles and Practices*, 4th ed. ASM Press, Washington, DC.
3. **Rhame FS.** 1998. The inanimate environment, p 299–324. *In* Bennett JV, Brachmann PS (ed), *Hospital Infections*, 4th ed. Lippincott-Raven, Philadelphia, PA.
4. **Halls N.** 1992. The microbiology of irradiation sterilization. *Med Device Technol* **3:**37–45.
5. **Griffiths N.** 1993. Low-temperature sterilization using gas plasmas. *Med Device Technol* **4:**37–40.
6. **Favero MS.** 1994. Forum: disinfection & sterilization procedures used in hospitals in the U.S. *Asepsis* **16:**16–19.
7. **Crow S, Smith JH III.** 1995. Gas plasma sterilization—application of space-age technology. *Infect Control Hosp Epidemiol* **16:** 483–487.
8. **Rutala WA, Weber DJ.** 2001. New disinfection and sterilization methods. *Emerg Infect Dis* **7:**348–353.
9. **Mendes GC, Brandão TR, Silva CL.** 2007. Ethylene oxide sterilization of medical devices: a review. *Am J Infect Control* **35:** 574–581.
10. **Murphy L.** 2006. Ozone—the latest advance in sterilization of medical devices. Can Oper Room Nurs 24:28, 30–32, 37–38.
11. **Hasanain F, Guenther K, Mullett WM, Craven E.** 2014. Gamma sterilization of pharmaceuticals—a review of the irradiation of excipients, active pharmaceutical ingredients, and final drug product formulations. *PDA J Pharm Sci Technol* **68:**113–137.
12. **Wallace CA.** 2016. New developments in disinfection and sterilization. *Am J Infect Control* **44**(Suppl):e23–e27.
13. **Aydogan A, Gurol MD.** 2006. Application of gaseous ozone for inactivation of Bacillus subtilis spores. *J Air Waste Manag Assoc* **56:**179–185.
14. **Iwamura T, Nagano K, Nogami T, Matsuki N, Kosaka N, Shintani H, Katoh M.** 2013. Confirmation of the sterilization effect using a high concentration of ozone gas for the bio-clean room. *Biocontrol Sci* **18:**9–20.
15. **Bertoldi S, Farè S, Haugen HJ, Tanzi MC.** 2015. Exploiting novel sterilization techniques for porous polyurethane scaffolds. *J Mater Sci Mater Med* **26:**182.
16. **Rediguieri CF, Pinto TdeJ, Bou-Chacra NA, Galante R, de Araújo GL, Pedrosa TdoN, Maria-Engler SS, De Bank PA.** 2016. Ozone gas as a benign sterilization treatment for PLGA nanofiber scaffolds. *Tissue Eng Part C Methods* **22:**338–347.
17. **Favero M.** 1998. Developing indicators for sterilization, p 119–132. *In* Rutala WA (ed), *Disinfection, Sterilization and Antisepsis in Health Care.* Association for Professionals in Infection Control and Epidemiology, Inc, Champlain, NY.
18. **Favero M.** 2001. Sterility assurance: concepts for patient safety, p 110–119. *In* Rutala WA (ed), *Disinfection, Sterilization and Antisepsis: Principles and Practices in Healthcare Facilities.* Association for Professionals in Infection Control and Epidemiology, Inc, Washington, DC.
19. **Spaulding EH.** 1972. Chemical disinfection and antisepsis in the hospital. *J Hous Res* **9:**5–31.
20. **Favero M, Bond W.** 2001. Chemical disinfection of medical surgical material, p 881–917. *In* Block SS (ed), *Disinfection, Sterilization, and Preservation*, 5th ed. Lippincott, Williams and Wilkins, Philadelphia, PA.
21. **Centers for Disease Control and Prevention.** 2003. Guidelines for environmental infection control in health-care facilities. Recommendations of CDC and the Healthcare Infection Control Practices Advisory Committee (HICPAC). *Morb Morta. Wkly Rep* **52**(RR-10):1–48. http://www.cdc.gov/ncidod/dh8p/pdf/guidelines/Enviro_guide_03.pdf.
22. **Spalding EH.** 1968. Chemical disinfection of medical and surgical materials, p 517–531. *In* Lawrence C, Block SS (ed), *Disinfection Sterilization and Preservation.* Lea & Febiger, Philadelphia, PA.
23. **Rutala WA, Weber DJ, Healthcare Infection Control Practices Advisory Committee.** 2008. *Guideline for disinfection and sterilization in healthcare facilities, 2008.* Centers for Disease Control and Prevention, Atlanta, GA. http://www.cdc.gov/hicpac/pdf/guidelines/Disinfection_Nov_2008.pdf.
24. **Scott GH, Williams JC.** 1990. Susceptibility of *Coxiella burnetii* to chemical disinfectants. *Ann N Y Acad Sci* **590**(1 Rickettsiolog)**:** 291–296.
25. **Tearle P.** 2003. Decontamination by fumigation. *Commun Dis Public Health* **6:**166–168.
26. **Environmental Protection Agency.** 2005. *Compilation of available data on building decontamination alternatives. EPA/600/R-05/036 EPA.* National Homeland Security Research Center, Washington, DC.
27. **Beswick AJ, Farrant J, Makison C, Gawn C, Frost G, Crook B, Pride J.** 2011. Comparison of multiple systems for laboratory whole room fumigation. *Appl Biosaf* **16:**139–157.
28. **Gordon D, Carruthers B-A, Theriault S.** 2012. Gaseous decontamination methods in high-containment laboratories. *Appl Biosaf* **17:**31–39.
29. **Centers for Disease Control and Prevention and National Institutes of Health.** 2006. Section VIII-H: Prion Diseases, p 282–289. *In* Chosewood LC, Wilson DE (ed), *Biosafety in Microbiological and Biomedical Laboratories*, 5th ed. U.S. Department of Health and Human Services, Washington, DC.
30. **Trujillo R, David TJ.** 1972. Sporostatic and sporocidal properties of aqueous formaldehyde. *Appl Microbiol* **23:**618–622.
31. **Meyer HH, Gottlieb R, Halsey JT.** 1914. General antiseptics, pp 506–509. In *Pharmacology, Clinical and Experimental, A Groundwork of Medical Treatment, Being a Text-book for Students and Physicians.* JB Lippincott, Philadelphia, PA.
32. **Rayburn SR.** 1990. Principles of decontamination and sterilization, p 44–65. *In The Foundations of Laboratory Safety: A Guide for the Biomedical Laboratory.* Springer-Verlag, NY.
33. **International Agency for Research on Cancer.** 2006. *Formaldehyde, IARC Monographs—100F.* IARC, Lyon, France. http://monographs.iarc.fr/ENG/Monographs/vol100F.
34. **Fink R, Liberman DF, Murphy K, Lupo D, Israeli E.** 1988. Biological safety cabinets, decontamination or sterilization with paraformaldehyde. *Am Ind Hyg Assoc J* **49:**277–279.
35. **Luftman HS.** 2005. Neutralization of formaldehyde gas by ammonium bicarbonate and ammonium carbonate. *Appl Biosaf* **10:** 101–106.
36. **Canter DA, Gunning D, Rodgers P, O'connor L, Traunero C, Kempter CJ.** 2005. Remediation of *Bacillus anthracis* contamination in the U.S. Department of Justice mail facility. *Biosecur Bioterror* **3:**119–127.
37. **Klapes NA, Vesley D.** 1990. Vapor-phase hydrogen peroxide as a surface decontaminant and sterilant. *Appl Environ Microbiol* **56:** 503–506.

38. **Graham GS, Rickloff JR.** 1992. Development of VHP sterilization technology. *J Healthc Mater Manage* **10:**54, 56–58.
39. **Johnson JW, Arnold JF, Nail SL, Renzi E.** 1992. Vaporized hydrogen peroxide sterilization of freeze dryers. *J Parenter Sci Technol* **46:**215–225.
40. **Hawley RJ, Kozlovac JP.** 2004. Decontamination, p 333–348. *In* Lindler L, Lebeda F, Korch G (ed), *Biological Weapons Defense: Infectious Diseases and Counterbioterrorism.* Humana Press, Totowa, NJ.
41. **Heckert RA, Best M, Jordan LT, Dulac GC, Eddington DL, Sterritt WG.** 1997. Efficacy of vaporized hydrogen peroxide against exotic animal viruses. *Appl Environ Microbiol* **63:**3916–3918.
42. **Krause J, McDonnell G, Riedesel H.** 2001. Biodecontamination of animal rooms and heat-sensitive equipment with vaporized hydrogen peroxide. *Contemp Top Lab Anim Sci* **40:**18–21.
43. **Czarneski MA, Lorcheim K.** 2011. A discussion of biological safety cabinet decontamination methods: formaldehyde, chlorine dioxide, and vapor phase peroxide. *Appl Biosaf* **16:**26–33.
44. **Knapp JE, Battisti DL.** 2001. Chlorine dioxide, p 215–227. *In* Block SS (ed), *Disinfection, Sterilization, and Preservation,* 5th ed. Lippincott, Williams and Wilkins, Philadelphia, PA.
45. **Taylor LA, Barbeito MS, Gremillion GG.** 1969. Paraformaldehyde for surface sterilization and detoxification. *Appl Microbiol* **17:**614–618.
46. **Luftman HS, Regits MA.** 2008. *B. atrophaeus and G, stearothermophilus* biological indicators for chlorine dioxide gas decontamination. *Appl Biosaf* **13:**143–157.
47. **Rogers JV, Choi YW, Richter WR, Rudnicki DC, Joseph DW, Sabourin CL, Taylor ML, Chang JC.** 2007. Formaldehyde gas inactivation of *Bacillus anthracis, Bacillus subtilis,* and *Geobacillus stearothermophilus* spores on indoor surface materials. *J Appl Microbiol* **103:**1104–1112.
48. **Krause J, McDonnell G, Riedesel H.** 2001. Biodecontamination of animal rooms and heat-sensitive equipment with vaporized hydrogen peroxide. *Contemp Top Lab Anim Sci* **40:**18–21.
49. **National Sanitation Foundation International.** 2014. Biosafety cabinetry: design, construction, performance, and field certification. NSF/ANSI 49-2014, Ann Arbor, MI.
50. **National Sanitation Foundation International.** 2008. Protocol for the validation of a gas decontamination process for biological safety cabinets. NSF International, Ann Arbor, MI. http://standards.nsf.org/apps/group_public/download.php/2726/NSF%20General%20Decon%20revision%203-24-08.pdf.
51. **Environmental Protection Agency.** 2015. Protocol for room sterilization by fogger application. Environmental Protection Agency, Washington, DC. https://www.epa.gov/sites/production/files/2015-09/documents/room-sterilization.pdf.
52. **Weber DJ, Rutala WA.** 1998. Occupational risks associated with the use of selected disinfectants and sterilants, p 211–226. *In* Rutala WA (ed), *Disinfection, Sterilization and Antisepsis in Health Care.* Polyscience Publications, Champlain, NY.
53. **Jernigan DB, Raghunathan PL, Bell BP, Brechner R, Bresnitz EA, Butler JC, Cetron M, Cohen M, Doyle T, Fischer M, Greene C, Griffith KS, Guarner J, Hadler JL, Hayslett JA, Meyer R, Petersen LR, Phillips M, Pinner R, Popovic T, Quinn CP, Reefhuis J, Reissman D, Rosenstein N, Schuchat A, Shieh WJ, Siegal L, Swerdlow DL, Tenover FC, Traeger M, Ward JW, Weisfuse I, Wiersma S, Yeskey K, Zaki S, Ashford DA, Perkins BA, Ostroff S, Hughes J, Fleming D, Koplan JP, Gerberding JL, National Anthrax Epidemiologic Investigation Team.** 2002. Investigation of bioterrorism-related anthrax, United States, 2001: epidemiologic findings. *Emerg Infect Dis* **8:**1019–1028.
54. **Whitney EAS, Beatty ME, Taylor TH Jr, Weyant R, Sobel J, Arduino MJ, Ashford DA.** 2003. Inactivation of *Bacillus anthracis* spores. *Emerg Infect Dis* **9:**623–627.
55. **Lemieux P, Sieber R, Osborne A, Woodard A.** 2006. Destruction of spores on building decontamination residue in a commercial autoclave. *Appl Environ Microbiol* **72:**7687–7693.
56. **World Health Organization.** 2000. WHO infection control guidelines for transmissible spongiform encephalopathies. Report of a WHO consultation, Geneva, Switzerland, 23–26 March 1999. http://www.who.int/csr/resources/publications/bse/WHO_CDS_C SR_APH_2000_3/en/.
57. **Baron H, Prusiner SB.** 2006. Biosafety of prion diseases, p 461–485. *In* Fleming DO, Hunt DL (ed), *Biological Safety: Principles and Practices,* 4th ed. ASM Press, Washington, DC.
58. **Weinstein RA, Rutala WA, Weber DJ.** 2001. Creutzfeldt-Jakob disease: recommendations for disinfection and sterilization. *Clin Infect Dis* **32:**1348–1356.
59. **Taylor DM.** 2003. Preventing accidental transmission of human transmissible spongifom encephalopathies. *Br Med Bull* **66:** 293–303.
60. **Association for Advancement of Medical Instrumentation.** *Comprehensive Guide to Steam Sterilization and Sterility Assurance in Health Care Facilities,* in press. Association for Advancement of Medical Instrumentation, Arlington, VA.
61. **Clinical and Laboratory Standards Institute.** 2014. *Protection of Laboratory Workers from Occupationally Acquired Infections. Approved Guideline,* 4th ed. CLSI M29-A4. Clinical and Laboratory Standards Institute, Wayne, PA.
62. **Baier M, Schwarz A, Mielke M.** 2004. Activity of an alkaline 'cleaner' in the inactivation of the scrapie agent. *J Hosp Infect* **57:** 80–84.
63. **Fichet G, Comoy E, Duval C, Antloga K, Dehen C, Charbonnier A, McDonnell G, Brown P, Lasmézas CI, Deslys J-P.** 2004. Novel methods for disinfection of prion-contaminated medical devices. *Lancet* **364:**521–526.
64. **Jackson GS, McKintosh E, Flechsig E, Prodromidou K, Hirsch P, Linehan J, Brandner S, Clarke AR, Weissmann C, Collinge J.** 2005. An enzyme-detergent method for effective prion decontamination of surgical steel. *J Gen Virol* **86:**869–878.
65. **Lemmer K, Mielke M, Pauli G, Beekes M.** 2004. Decontamination of surgical instruments from prion proteins: *in vitro* studies on the detachment, destabilization and degradation of PrP^{Sc} bound to steel surfaces. *J Gen Virol* **85:**3805–3816.
66. **Race RE, Raymond GJ.** 2004. Inactivation of transmissible spongiform encephalopathy (prion) agents by environ LpH. *J Virol* **78:**2164–2165.
67. **Yan ZX, Stitz L, Heeg P, Pfaff E, Roth K.** 2004. Infectivity of prion protein bound to stainless steel wires: a model for testing decontamination procedures for transmissible spongiform encephalopathies. *Infect Control Hosp Epidemiol* **25:**280–283.
68. **Rutala WA (ed).** 2004. *Disinfection, Sterilization, and Antisepsis: Principles, Practices, Challenges, and New Research.* Association for Professionals in Infection Control and Epidemiology, Washington, DC.

24

Packing and Shipping Biological Materials

RYAN F. RELICH AND JAMES W. SNYDER

GOVERNING AUTHORITIES AND REGULATIONS

Laboratory workers who ship or transport dangerous goods, in general, and diagnostic specimens and infectious substances, in particular, by a commercial land or air carrier are required to follow a complex and often confusing set of national and international regulations and requirements. The purpose of these regulations and requirements is to protect the public, emergency responders, laboratory workers, and personnel involved in the transportation industry from accidental exposure to the contents of the packages (1–3).

Statistical data show that these regulations are effective in protecting both the contents of packages and the persons who handle the packages. To date, there are no reported cases of illness due to the release of a diagnostic specimen or infectious substance during transport. In addition, only 106 (0.002%) of the 4,920,000 primary containers shipped in 2003 to worldwide laboratories and other destinations were reported broken during transit. In each of the 106 reported breakages, absorbent material in appropriately prepared packages contained the leaking material, and none of the secondary or outer containers were damaged (4).

An important non-safety-related benefit of adherence to these regulations and requirements is to minimize the potential for damage to the contents of the package during transport and to reduce the exposure of the shipper to the risks of criminal and civil liability associated with the improper shipment of dangerous goods. Shipping regulations and requirements are developed and published by many authorities, the most notable of which are shown in Table 1 (1–3).

Most regulations for the air transport of dangerous goods throughout the world originate as decisions (called Model Regulations) made by the United Nations Committee of Experts. The ICAO uses these decisions to develop formal, legally binding, and standardized regulations for use in international aviation (4). These specific ICAO regulations (*Technical Instructions for the Safe Transportation of Dangerous Goods by Air*) are the standards for the international shipment of dangerous goods by air. IATA uses these IACO *Technical Instructions* to develop *Dangerous Good Regulations*, which are used by essentially all commercial airlines involved in the transport of dangerous goods (1, 2, 5). IATA requirements have become the most widely recognized, copied, and used

TABLE 1.

Agencies governing transportation of dangerous goods

Governing authority	Agency	Regulations (reference)
United Nations	ICAO[a]	*Technical Instructions for the Safe Transport of Dangerous Goods by Air*
Commercial airline industry	IATA[b]	*Dangerous Goods Regulations*
United States	DOT[c]	*United States Hazardous Materials Regulations*
United States	USPS[d]	*Domestic Mail Manual. Publication 52: Hazardous, Restricted and Perishable Mail*
Canada	Transport Canada	*Transportation of Dangerous Goods Regulations*
Other nations		Individual national regulations

[a]International Civil Aviation Organization.
[b]International Air Transport Association.
[c]Department of Transportation.
[d]United States Postal Service.

packing and shipping guidelines in the world. Most national and international regulations (except those issued by the DOT) are based on, or are at least in substantial agreement or harmonization with, IATA requirements. Individual nations of the world often issue additional (usually more restrictive) national regulations for the shipment of dangerous goods into these individual nations.

In the United States, the DOT regulates the commercial transportation of dangerous goods by both air and ground carriers. Just as IATA derives its requirements from ICAO, the DOT also derives its regulations from ICAO (6–8). In 2002 and 2004, the DOT revised its regulations for the transportation of diagnostic specimens and infectious substances to be in substantial agreement with IATA regulations (1, 6, 7). In May, 2005, the DOT published another Notice of Proposed Rulemaking to continue the harmonization of federal regulations with those of IATA (4). For practical purposes, shippers of diagnostic specimens and infectious substances can consider compliance with IATA requirements to be compliance with DOT regulations. It is important to note that the transportation of diagnostic specimens is exempt from DOT regulations if the specimens are transported by a private or contract carrier in a motor vehicle used exclusively to transport diagnostic specimens or biological products (6). An example of this exemption is the transport of specimens in a private courier van to, from, and between local hospitals and laboratories or a local core laboratory.

The United States Postal Service (USPS) publishes its own regulations in the USPS *Domestic Mail Manual* (9). The USPS regulations for mailing hazardous materials generally adhere to DOT regulations. In some cases (e.g., lower allowable volume limits), USPS regulations are more restrictive than those of DOT or IATA, and USPS does not transport Category A infectious substances. USPS regulations are not addressed in this chapter but can be found in USPS Publication 52—Hazardous, Restricted, and Perishable Mail (10).

IATA requirements and DOT regulations mandate the minimum requirements for packaging and shipping diagnostic specimens and infectious substances that can pose a threat to humans, animals, or the environment. The safe and legal transport of these substances is based on the following mandated activities:

- classification and naming of the material to be shipped
- selection of packaging that will contain the contents if the package is damaged and, thus, will protect carrier personnel if the package is damaged
- correct packing of the shipment
- application of appropriate markings and labels onto the outer package to alert carrier personnel to the hazardous contents of the package and to identify contacts if an accident occurs. Be sure to check the regulations for the required size and site of application of these labels.
- documentation of relevant aspects of each package and its contents
- training of individuals on the requirements for appropriate packaging and shipping of diagnostic specimens and infectious substances and subsequent certification to document all required training, including OSHA and security training

Each of the aforementioned activities is presented in detail in this chapter.

CLASSIFICATION OF A SUBSTANCE

Classification is a mandatory four-step process to define dangerous goods that are shipped by commercial carriers (1–3). Classification serves two purposes: (1) it allows the shipper to select the proper packing instructions (PI) to use, and (2) if the substance is a Category A infectious substance, it provides important information necessary to complete the Shipper's Declaration.

First, the material must be classified into one of the nine U.S. federally specified hazard classes (Class 1 through Class 9) of dangerous goods (Table 2). Infectious and toxic substances are Class 6 dangerous goods; dry ice is a Class 9 dangerous good. Class 6 and Class 9 substances usually are the only dangerous goods shipped by clinical microbiologists; however, some specimen preservatives may be classified in a different hazard class.

Second, Class 6 substances must be divided into either Division 6.1 (toxic substances) or Division 6.2 (infectious substances).

TABLE 2.

IATA-defined classes of dangerous goods

Class	Substance
1	Explosives
2	Gases
3	Flammable liquids
4	Flammable solids
5	Oxidizing substances and organic peroxides
6	Toxic and infectious substances Division 6.1: toxic substances Division 6.2: infectious substances[a]
7	Radioactive materials
8	Corrosive substances
9	Miscellaneous dangerous goods (e.g., dry ice)[a]

[a]Addressed in detail in this chapter; DOT refers to this as "hazardous material."

Third, Division 6.2 infectious substances must be divided into one of seven IATA-specified groups (Table 3) (2):

- Category A infectious substances
- Category B infectious substances
- biological products
- genetically modified microorganisms and organisms
- medical or clinical wastes (not addressed in depth in this chapter)
- infected animals (not addressed in depth in this chapter)
- patient specimens

Fourth, if the substance is determined to be in one of the above groups, other than Category A or Category B, the shipper must determine if the substance contains Category A or B infectious substances. If a Category A or B substance is present within the material, the sender is required to follow the appropriate regulations regarding transport of that material. For example, if a shipment of medical waste is known to contain Ebola virus (a Category A infectious substance), the medical waste then meets the criteria for

TABLE 3.

Types, proper shipping names, UN numbers, and packaging instructions (or directions) for IATA Division 6.2 infectious substances (2)

Infectious substance	Proper shipping name	UN number[a]	Packing instructions
Category A	Infectious Substance, Affecting Humans Infectious Substance, Affecting Animals	UN2814[b] UN2900[c]	620
Category B	Biological Substance, Category B	UN3373	650
Biological Products		UN2814, UN2900, or UN3373[d]	620 or 650
Genetically Modified Microorganisms and Organisms			
Meets Category A criteria	Infectious Substance, Affecting Humans Infectious Substance, Affecting Animals	UN2814 UN2900	620 620
Meets Category B criteria	Biological Substance, Category B	UN3373	650
Does not meet either Category A or B criteria	Genetically Modified Organism	UN3245	959
Medical or Clinical Wastes	Biomedical Waste, n.o.s.; Clinical Waste, Unspecified, n.o.s.; Medical Waste, n.o.s., or Regulated Medical Waste, n.o.s.	UN3291[e]	622
Infected Animals		UN2814 or UN2900	620
Patient Specimens			
Meets Category A criteria	Infectious Substance, Affecting Humans Infectious Substance, Affecting Animals	UN2814 UN2900	620 620
Meets Category B criteria	Biological Substance, Category B	UN3373	650

[a]If packages of infectious substance contain dry ice, they must also be assigned to UN1845 and labeled with a Class 9 label. Packing Instructions 954 apply to shipments containing dry ice.
[b]Infectious substances assigned to UN2814 are those affecting humans or both humans and animals (e.g., Ebola virus).
[c]Infectious substances assigned to UN2900 are those affecting animals only (e.g., foot-and-mouth disease virus).
[d]Biological products that are known or reasonably believed to harbor an infectious substance that meet criteria for inclusion in Category A or Category B must have the appropriate UN number assignment.
[e]Medical or Clinical Wastes must be assigned to UN2814 or UN2900 if they contain Category A infectious substances and to UN3373 if they contain Category B infectious substances.

Category A packing and shipping. Decisions made in the fourth step can be subjective and difficult; however, these decisions will determine exactly how the substance must be packed and shipped. Although the decisions in this classification process can be difficult, the shipper must not arbitrarily classify all shipments as either an infectious or a diagnostic (or clinical) specimen to avoid having to make important discriminatory shipping decisions to make packing easier or less expensive. Such cavalier classification is illegal and can be overly expensive (1, 3).

Category A Infectious Substances

A Category A infectious substance (pathogen or agent) is defined by the IATA as "an infectious substance which is transported in a form that, when exposure to it occurs, is capable of causing permanent disability, or life-threatening or fatal disease to otherwise healthy humans or animals" (2). Category A substances are specifically designated and listed (by IATA) as pathogens, which can be dangerous to both individual and public health (Table 4). This list is not all-inclusive, and a thorough risk assessment must be performed before assigning a substance to Category A. These pathogens are essentially the same as those previously known as "forbidden substances." Category A pathogens, and substances likely to contain Category A pathogens, must be assigned the UN number UN2814 (Infectious Substance, Affecting Humans) or UN2900 (Infectious Substance, Affecting Animals). Agents assigned to UN2814 are those that are capable of causing disease in humans and animals, and those assigned to UN2900 are known to cause disease only in animals.

IATA requirements allow shippers to use their discretion and professional judgment when deciding if a substance meets Category A criteria. IATA *Dangerous Goods Regulations* state the following:

- regarding judgment—"Assignment to UN2814 or UN2900 must be based on the known medical history and symptoms of the source human or animal, endemic local conditions, or professional judgement concerning individual circumstances of the source human or animal."
- regarding assigning infectious agents which, in the shipper's opinion, meet category A criteria, but which are not specifically listed as a Category A agent—". . . infectious substances . . . which do not appear in the table but which meet the same criteria must be assigned to Category A." An example of such an agent would be an isolate of a novel hemorrhagic fever virus that is known to be easily transmitted from person to person and causes significant morbidity and mortality.
- regarding uncertainty of Category A criteria—". . . if there is doubt as to whether or not a substance meets the criteria [of Category A] it must be included in Category A" (2).

Some Category A pathogens have been designated as agents of bioterrorism and are known as select agents (Appendix B). United States federal regulations require shippers to have special permits to possess, transfer, and receive these agents (11, 12). As of July 17, 2015, Federal Express no longer transports select agents; however, the company continues to transport Category A, nonselect agent infectious substances. A list of couriers who currently (February, 2016) offer biomedical specimen transport, many of whom also transport select agents, can be found by joining the Medical Courier Connection (https://integrity delivers.webconnex.com/MedCourierCnxMemberList).

Category B Infectious Substances

A Category B substance is defined by IATA as "an infectious substance which does not meet the criteria for inclusion in Category A" (2). In the authors' opinions, examples of Category B substances are the following:

- typical clinical or patient specimens (e.g., blood, biopsies, swab specimens, excreta, secreta, body fluids, tissues, etc.) not in a form that should be classified as Category A or patient specimens that are not "harmless" enough to fall into the "exempt" or "not subject" classification. Consult IATA regulations for a list of pathogens that are considered Category A in any form.
- cultures (usually on solid media) of nonCategory A microorganisms (e.g., *Staphylococcus aureus*). Consult IATA regulations for a list of pathogens that are considered Category A in culture form.

Category B substances are assigned UN number UN 3733 and the proper shipping name is Biological Substance, Category B (Table 3).

A number of exceptions have been identified by the IATA that are not subject to the provisions of the Dangerous Goods Regulations unless they fall into another class or division. Substances not known to contain infectious material include:

- substances that contain nonpathogenic microorganisms
- substances that contain inactivated or neutralized pathogens
- environmental samples, including food and water, that do not pose a significant risk of infection
- dried blood spots and fecal occult blood screening samples
- blood, blood components, and tissues that are intended for transfusion or transplantation

Other substances may be subject to "Exempt human (or animal) specimen" provisions, and these include

TABLE 4.

Examples of infectious substances included in Category A in any form unless otherwise indicated [a]

UN number and proper shipping name	Organism
UN 2814	*Bacillus anthracis* (cultures only)
Infectious Substance, Affecting Humans	*Brucella abortus* (cultures only)
	Brucella melitensis (cultures only)
	Brucella suis (cultures only)
	Burkholderia mallei (cultures only)
	Burkholderia pseudomallei (cultures only)
	Chlamydophila psittaci (avian; cultures only)
	Clostridium botulinum (cultures only)
	Coccidioides immitis (cultures only)
	Coxiella burnetii (cultures only)
	Crimean-Congo hemorrhagic fever virus
	Dengue virus (cultures only)
	Eastern equine encephalitis virus (cultures only)
	Escherichia coli, verotoxigenic (cultures only)
	Ebola virus
	Flexal virus
	Francisella tularensis (cultures only)
	Guanarito virus
	Hantaan virus
	Hantaviruses causing hemorrhagic fever with renal syndrome
	Hendra virus
	Hepatitis B virus (cultures only)
	Herpesvirus type B (cultures only)
	Human immunodeficiency virus (cultures only)
	Highly pathogenic avian influenza virus (cultures only)
	Japanese encephalitis virus (cultures only)
	Junin virus
	Kyasanur Forest disease virus
	Lassa virus
	Machupo virus
	Marburg virus
	Monkeypox virus
	Mycobacterium tuberculosis (cultures only)
	Nipah virus
	Omsk hemorrhagic fever virus
	Poliovirus (cultures only)
	Rabies virus (cultures only)
	Rickettsia prowazekii (cultures only)
	Rickettsia rickettsii (cultures only)
	Rift Valley fever virus (cultures only)
	Russian spring-summer encephalitis virus (cultures only)
	Sabiá virus
	Shigella dysenteriae type 1 (cultures only)
	Tick-borne encephalitis virus (cultures only)
	Variola virus

(continued)

TABLE 4.

Examples of infectious substances included in Category A in any form unless otherwise indicated [a] (*Continued*)

UN number and proper shipping name	Organism
	Venezuelan equine encephalitis virus (cultures only)
	West Nile virus (cultures only)
	Yellow fever virus (cultures only)
	Yersinia pestis (cultures only)
UN2900	African swine fever virus (cultures only)
Infectious Substance, Affecting Animals	Avian paramyxovirus type 1—velogenic Newcastle disease virus (cultures only)
	Classical swine fever virus (cultures only)
	Foot-and-mouth disease virus (cultures only)
	Goatpox virus (cultures only)
	Lumpy skin disease virus (cultures only)
	Mycoplasma mycoides—contagious bovine pleuropneumonia (cultures only)
	Peste des petits ruminants virus (cultures only)
	Rinderpest virus (cultures only)
	Sheep-pox virus (cultures only)
	Swine vesicular disease virus (cultures only)
	Vesicular stomatitis virus (cultures only)

[a]Adapted from reference 2. Please note that this list is not all-inclusive and is subject to change; consult the Federal Select Agent Program website (http://www.selectagents.gov/) for up-to-date details.

patient specimens that pose a minimal likelihood of containing pathogens. An example of "exempt human specimen" would be a screening sample drawn from a healthy individual.

For packaging of these substances, (i) substances must be placed inside of a leakproof primary container, (ii) the primary container must be sealed inside of a leakproof secondary container with sufficient absorbent material to contain the entire substance carried within the primary container if the primary container is breached, and (iii) the secondary container must be packed using a container of adequate strength to sustain the weight of the substance being transported and must have at least one surface that measures, at a minimum, 10 cm × 10 cm (2). Be aware that the above classifications have progressively more restrictive packaging requirements. Refer to the most up-to-date IATA *Dangerous Goods Regulations* for more information.

Biological Products

Virtually all commercially available biological products, as defined by IATA (see definitions in Appendix B), are exempt from the packing and shipping regulations presented in this chapter. However, if a biological product is determined to meet the criteria of one of the aforementioned infectious substances (Category A, Category B, Exempt Human or Animal Specimen, etc.), it must be packed and shipped as such (Table 3) (2).

Genetically Modified Microorganisms and Organisms

Genetically modified microorganisms and organisms usually meet the criteria of one of the aforementioned infectious substances (Category A [UN2814 or UN 2900] or Category B [UN3373]). If this is not the case, the substance or organism is assigned to UN3245 and packed and shipped as such (Table 3) (2).

Medical or Clinical Wastes

Medical waste that contains Category A or Category B infectious substances must be packed and shipped as such and assigned UN2814, UN2900, or UN3373 (Table 3) (2). Medical waste, which is reasonably believed to have a low probability of containing infectious substances, must be packed and shipped as Medical Waste, n.o.s. (UN 3291) (2). Other proper shipping names for medical or clinical wastes are "Biomedical Waste, n.o.s.," "Clinical Waste, Unspecified, n.o.s.," and "Regulated Medical Waste, n.o.s."

Infected Animals

A live intentionally infected animal that is known to contain or reasonably expected to contain an infectious substance cannot be transported by air unless the substance cannot be transported by any other means (2). Consultation with individual commercial carriers is advised if either live or dead infected animals need to be shipped.

Patient Specimens

IATA has defined a "patient specimen" as material collected directly from humans or animals for diagnostic, treatment, prevention, investigational, or research purposes (2). All specimens (and cultures) should be classified according to the IATA classification flowchart to determine how they should be packaged.

Patient specimens that meet Category A or Category B criteria should be classified, packed, and shipped as Category A (UN2814 or UN2900) or Category B (UN3373) infectious substances (Table 3). Patient specimens that do not meet Category A nor Category B criteria should be packed and shipped as Exempt Human (or Animal) Specimens, or they may not be subject to provisions of the Dangerous Goods Regulations.

NAMING A SUBSTANCE

After classifying the substance, the shipper must identify (officially name) a Category A and Category B substance by assigning the substance a proper shipping name. Proper shipping names and their associated UN numbers are specifically listed and published internationally by IATA so that most carriers around the world will recognize the general group or kind of infectious agent or dangerous good they are handling. This list provides 11 informational items for each of the proper shipping names (Table 5). The 11 items correspond conveniently to the information needed to complete the Shipper's Declaration. Fortunately, only five of the 3,000 proper shipping names are used by most clinical laboratories: one name for Category A infectious substances which affect humans (liquids or solids), one name for Category A infectious substances which affect animals (liquids or solids), one name for Category B diagnostic (or clinical) substances, one name for genetically modified microorganisms or organisms, and two names for dry ice (Table 6).

TABLE 5.

Information provided in the IATA alphabetical List of Dangerous Goods and applicable to completing a Shipper's Declaration for Dangerous Goods

Column[a]	Information
A	United Nations ID number of the proper shipping name/description
B	Proper shipping name/description
C	Class or division of dangerous good
D	N/A[b]
E	The hazardous label required on the outer package
F	N/A
G	N/A
H	N/A
I	Packing instructions to use for *passenger and cargo aircraft*
J	Maximum allowable amounts to be shipped in *passenger and cargo aircraft*
K	Packing instructions to use for *cargo aircraft only*
L	Maximum allowable amounts to be shipped in *cargo aircraft only*
M	Applicable special provisions and exceptions
N	Emergency response code

[a]Refers to the 14 columns in the IATA alphabetical List of Dangerous Goods.
[b]Not applicable to infectious substances.

PACKING INSTRUCTIONS AND PACKING SUBSTANCES

DOT regulations, USPS regulations, IATA requirements, and IATA Packing Instructions (PI) describe the minimum standards for the safe transport of various biological materials. Shippers are legally responsible for complying with these regulations, for following prescribed PI, and for packing substances correctly to ensure the safety of all personnel who handle the package before, during, and after shipment to the point of acceptance of the package by the consignee. After determining the exact nature and category of the substance to be shipped, the shipper must select the most appropriate PI and packing directions to use (Fig. 1 and Table 3). Generally, the PI used in clinical laboratories are those that relate to shipping Category A infectious substances (PI 620); Category B infectious substances (PI 650); diagnostic, clinical, or biological substances, category B substances (PI 650); and dry ice (PI 954). There are no specifically numbered PI for specimens classified as Exempt Human or Animal Specimens; however, IATA provides specific directions which must be followed. See Table 7 for a comparison of the details of packing instructions and directions.

PI 620 and PI 650 provide several similar instructions, the most notable of which are those that mandate the use of triple packaging for both diagnostic specimens and infectious substances (Table 7). The major differences are those associated with documentation and with marking and labeling outer containers. The similarities of, and differences between, PI 620 and PI 650 are shown in Table 7. Triple packaging required by both PI 620 and PI 650 consists of a primary container, a secondary container, absorbent material, list of contents, and an outer shipping package (Table 7).

- A leakproof primary container made of glass, metal, or plastic and, if it contains a Category A infectious substance, sealed by a positive method (e.g., heat seal,

TABLE 6.

Selected examples of dangerous goods from the IATA alphabetical list of Dangerous Goods[a]

UN						Passenger or cargo aircraft				Cargo aircraft only			
						Ltd quantity							
ID #	Proper shipping name/description	Class	SR	Hazard labels	Pk gp	Pk inst	Max net qt/pk	Pack inst	Max net qt/pk	Pack inst	Max net qt/pk	Spec prov	ERC
A	*B*	*C*	*D*	*E*	*F*	*G*	*H*	*I*	*J*	*K*	*L*	*M*	*N*
2814	Infectious substance, affecting humans	6.2	—	IS	—	—	—	620	50 ml/ 50 g	620	4 l/4kg	A81 A140	11Y
2900	Infectious substance, affecting animals only	6.2	—	IS	—	—	—	620	50 ml/ 50 g	620	4 l/4 kg	A81 A140	11Y
3373	Biological substance, category B	6.2	—	3373	—	—	—	650	4 l/4 kg	650	4 l/4 kg	—	6L
1845	Dry ice or Carbon dioxide, solid	9	—	misc	—	—	—	954	200 kg	954	200 kg	A48 A151 A805	9L

[a]SR, subsidiary class; Ltd, limited; Pk gp, packing group; Max net qt/pk, maximum net quantity per package; Pack inst, packing instructions; Spec prov, special provisions; ERC, emergency response drill code (formerly known as emergency response guide).

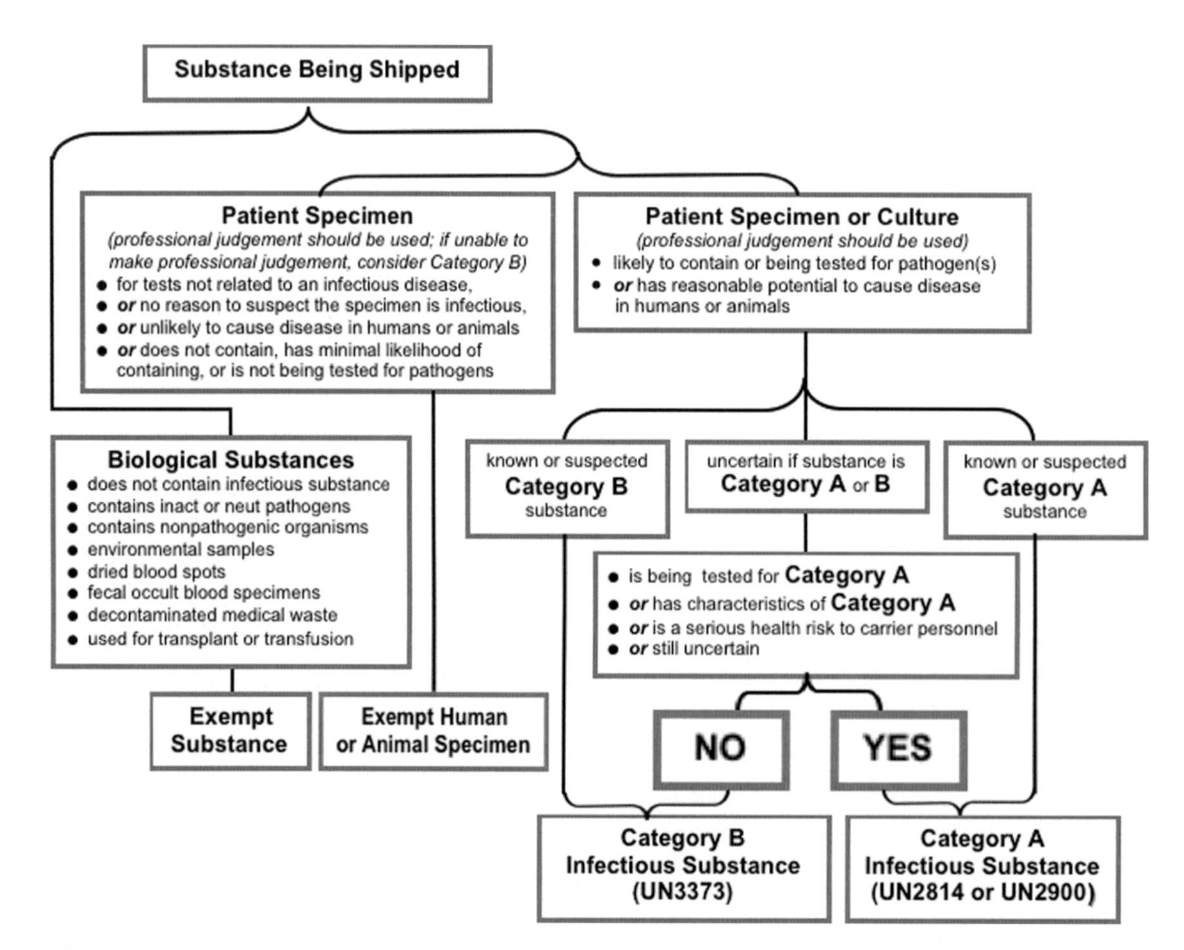

Figure 1: Algorithm to help shippers select appropriate packing instructions. The algorithm represents the authors' interpretations of IATA and DOT regulations.

TABLE 7.

Comparison of IATA packing instructions 650 and 620, and packing directions for exempt human specimens

Requirement	Exempt human specimens[a]	650[b]	620[c]
Leakproof primary (1°) and secondary (2°) containers	Yes	Yes	Yes
Pressure-resistant 1° or 2° container	—[d]	Yes	Yes
Absorbent between 1° and 2° containers[e]	Yes	Yes	Yes
List of contents between 2° and outer package	—	Yes	Yes
Rigid outer packaging	—	Yes	Yes
Positively sealed 1° container	—	No	Yes
Name, address, and number of responsible person on outer package	—	Yes	Yes
Shipper's Declaration for Dangerous Goods	—	No	Yes
Outer packaging			
Marking labels	—	Less	More
Strict manufacturing specifications	—	No	Yes
Quantity limits for passenger and cargo aircraft			
Maximum for each 1° container	—	1 liter (4 kg)	50 ml (50 g)
Total maximum for outer package	—	4 liters (4 kg)	50 ml (50 g)[f]
Cost of labor and materials to pack substance	Least	More	Most

[a]Includes substances with minimal likelihood of causing disease in humans and animals, and substances not likely to contain pathogens (see text).
[b]Packing instructions for Biological Substances, Category B.
[c]Packing instructions for Category A infectious substances.
[d]Requirement not specified by IATA.
[e]Not required for solid substances, such as tissue and solid agar media cultures or slants.
[f]Applies to passenger aircraft.

metal crimp, or taped screw-cap lid). For diagnostic specimens and infectious substances being shipped in either passenger or cargo aircraft, the maximum volume per primary container must not exceed 50 ml (50 g) and 1 liter (4 kg) for Category A and for Category B infectious substances, respectively.

- Absorbent material sufficient to absorb all liquid contained within the primary container(s) in case of breakage; placed between the primary and secondary containers. Absorbent material is not required if the material being shipped is a solid. Absorbent material should be used with frozen liquids shipped in a frozen state.
- A leakproof secondary container that contains the primary container(s).
- When shipping liquids or solids classified as Category A, or liquids classified as Category B, either the primary or secondary container must be able to withstand an internal pressure of at least 95 kPa (13.8 lbs/in^2) because shipments are likely to be placed into unpressurized cargo sections of aircraft, which fly at high altitudes.
- A list of the contents and quantities of the primary container(s) must be attached to the outside of the secondary container.
- Rigid and durable outer containers of adequate strength for their intended use and constructed of cardboard, wood, or material of equivalent strength, and which measure at least 10 cm × 10 cm on at least one surface. For shipping Category A infectious substances, these outer containers must meet United Nations manufacturing and testing specifications (see below). In contrast, packages containing Category B infectious substances must be able to withstand the IATA drop test.

Requirements for packaging used with Exempt Human or Animal Specimens are less strict than the aforementioned requirements in packing instructions 650 and 620. However, packaging must be composed of four elements: (a) a leakproof primary container, (b) a leakproof secondary container, (c) for liquid substances, absorbent material of sufficient quantity to absorb the entire liquid, placed between the primary and secondary containers, and (d) outer packaging of sufficient strength for its capacity, mass, and intended use (Table 7) (2).

MARKING AND LABELING PACKAGES

Marking is the act of writing or typing information onto the surface of an outer package. Labeling is the act of placing informational labels or stickers onto the surface of an outer package. The shipper is responsible for the proper marking and labeling of the outer shipping container as described in the DOT and IATA regulations

(1–3, 6). The marking and labeling on the outer container communicates essential information regarding the shipper and consignee of the package, nature and weight of the contents of the package, the potential hazard of the substance, how the substance is packed, and information to be used in case of an emergency. The outer package must display markings and labels appropriate for the particular shipment. Some of these markings and labels can be seen in the IATA *Dangerous Goods Regulations* (2). These labels and markings include the following:

- The shipper's and consignee's name and address
- If the substance is a Category A infectious substance, the name and telephone number of a "person responsible" (IATA quotation), who is knowledgeable about the contents of the shipment and can provide emergency information in case the package is damaged and the contents escape their containment, must be on the outer container. This emergency contact must be available 24 hours a day. If the substance being shipped is a Category B infectious substance, this information may be provided on either the air waybill or the outer package, and the person responsible must be available during normal duty hours.
- If the substance is a Category A infectious substance: (a) the Class 6 diamond-shaped "Infectious Substance. In Case of Leakage . . ." label, complete with the international biohazard symbol, and (b) a label which shows the proper shipping name, UN number, and quantity of the substance (Fig. 2). Please note that these labels must no longer carry the phrase "In U.S.A. Notify Director-CDC, Atlanta, GA 1-800-232-0124."
- If the substance is a Category B substance: (a) "Biological Specimen, Category B" and (b) the marking or label "UN3373" (Fig. 3)
- If dry ice is used: a Class 9 "Miscellaneous Dangerous Goods" label and the weight of dry ice (Fig. 4)
- Package orientation label (Fig. 5). Orientation labels (arrows) must be placed on opposite sides of all packages which contain >50 ml of an infectious substance to indicate the correct orientation of the package.
- "Cargo Aircraft Only" label if the substance (because of its quantity) can be transported only by cargo aircraft (Fig. 6). This label is used if infectious substance amounts to more than 50 ml (5g) but less than 4 liters (4 kg) per outer package are shipped.
- "Overpack" markings if overpacks are used (Fig. 7)
- Patient specimens classified as "Subject to 'exempt human (or animal) specimen' provisions" must be labeled clearly as "Exempt Human Specimen" or "Exempt Animal Specimen" (Fig. 8).
- All outer packaging used to ship Category A infectious substances and substances considered by the shipper

Figure 2: Labels that indicate an infectious substance (Class 6), proper shipping name, UN identification number, and quantity of substance. In accordance with DOT regulations, the Class 6 Infectious Substance labels with the text "In U.S.A. Notify Director-CDC, Atlanta, GA 1-800-232-0124" are no longer permissible for labeling containers of infectious substances.

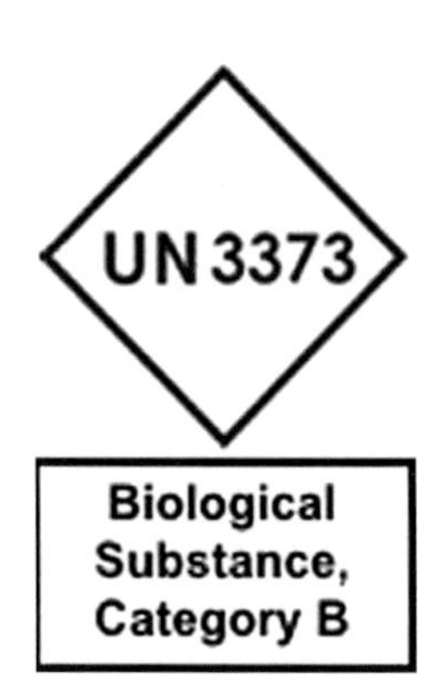

Figure 3: Markings that indicate a Biological Substance, Category B, and appropriate UN number.

Figure 4: Label that indicates a miscellaneous (Class 9) dangerous good (2 kg of dry ice).

Figure 5: Label that indicates the correct orientation of package during shipping.

to be an infectious risk to the health of carrier personnel must meet specifications established by the United Nations and must be marked as such by the manufacturer (Table 3). Packaging that meets the UN specifications are marked by a "UN" inside of a circle and a series of letters and numbers, which indicate the type of package, class of goods the package is designed to carry, manufacturing date, authorizing agency, and the manufacturer (Fig. 9). The designation "Class 6.2" in the marked code indicates that the container is approved for shipping infectious substances. These containers are commercially available and are preprinted with the appropriate UN marking. The strict UN specifications for outer packaging are not required when shipping Category B substances. Outer boxes used to ship diagnostic or clinical specimens need only to be rigid and strong enough for their intended purpose.

- All corresponding labels should be placed alongside one another, and all labels should not overlap or be masked or defaced in any way.
- Any unnecessary or inapplicable labels should be removed or covered.

Figure 6: Label that indicates a substance must be transported only in a cargo (not passenger) aircraft.

OVERPACK

Figure 7: Marking that indicates an overpack is used and inner packages comply with regulations.

Figures 10, 11, and 12 show completely labeled and marked outer shipping containers that contain an Exempt Human Specimen; a Biological Substance, Category B; and a Category A infectious substance, respectively. Packages in Fig. 11 and 12 also contain dry ice. For convenience and lower costs, one or more triple packages, packed in full compliance with IATA regulations, may be shipped within a single overpack that does not have to meet UN specifications. However, the overpack must be labeled "Overpack" and must be completely labeled according to applicable IATA regulations (Fig. 7).

DOCUMENTATION

Shipper's Declarations are required for all Category A infectious materials, regardless of transport method (i.e., air or ground). A Shipper's Declaration is a legal contract between the shipper and carrier, is required to document the shipment of all dangerous goods, must be accurate, and must be legible, or the carrier may reject the package for transport. Some carriers require the Shipper's Declaration to be typed; some require multiple copies, some in color. The original Shipper's Declaration given to the carrier must have red candy stripes along the left and right edges of the document. Shippers must retain a copy for their records for a minimum of 2 years. All corrections must be neatly "lined out," and all changes must be signed by the same person who signed the document. A carrier may reject a shipment if each field on the Shipper's Declaration is not completed exactly to the carrier's satisfaction. Commercial carriers and the Federal Aviation Administration often exercise their authority at airports to examine Shipper's Declarations for compliance with applicable regulations and to open and inspect any package (whether or not the package is leaking) which contains, or is suspected of containing, an infectious substance. In addition, these agencies can

**Exempt
Human
Specimen**

Figure 8: Label that indicates an exempt human specimen (e.g., blood) being transported for routine diagnostic testing (e.g., cholesterol testing).

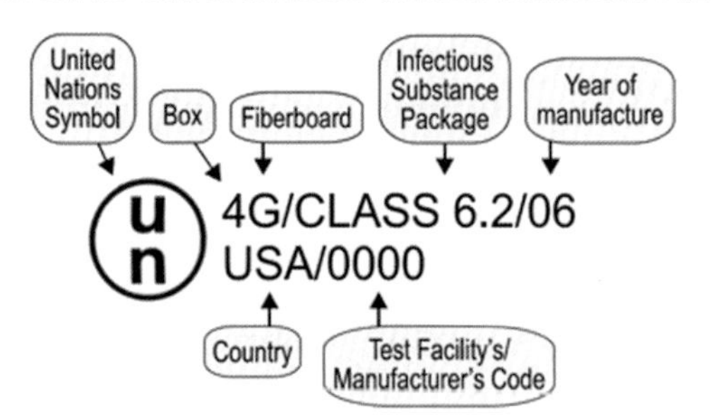

Figure 9: Label that indicates the outer container has met IATA-specified manufacturing standards.

and do examine documentation of perfectly packaged shipments, go to the facilities from which the packages originated, and request a copy of the Shipper's Declaration and documentation of adequate training of employees. Figure 13 shows a blank Shipper's Declaration and the 13 sections that shippers must complete. Essentially all of the IATA-specified technical information required to complete the seven subsections of section nine (Nature and Quantity of Dangerous Goods) of the document can be found in Table 6 and reference 4 (2). Figure 14 shows a completed and acceptable Shipper's Declaration. If a Shipper's Declaration is not correct to the carrier's satisfaction, the shipment may be rejected by the carrier (1, 3).

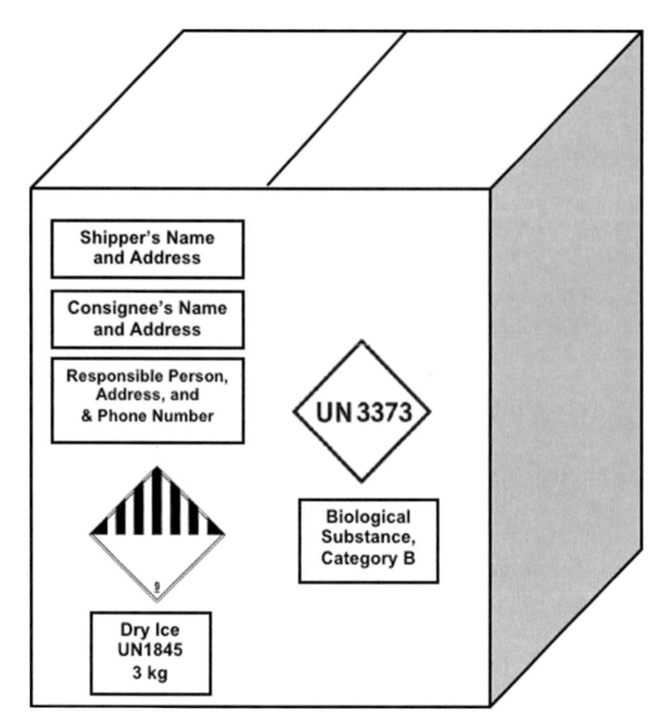

Figure 11: A completely labeled outer package. The primary container inside the package contains a Category B infectious substance (diagnostic or clinical specimen) and is packed according to PI 650.

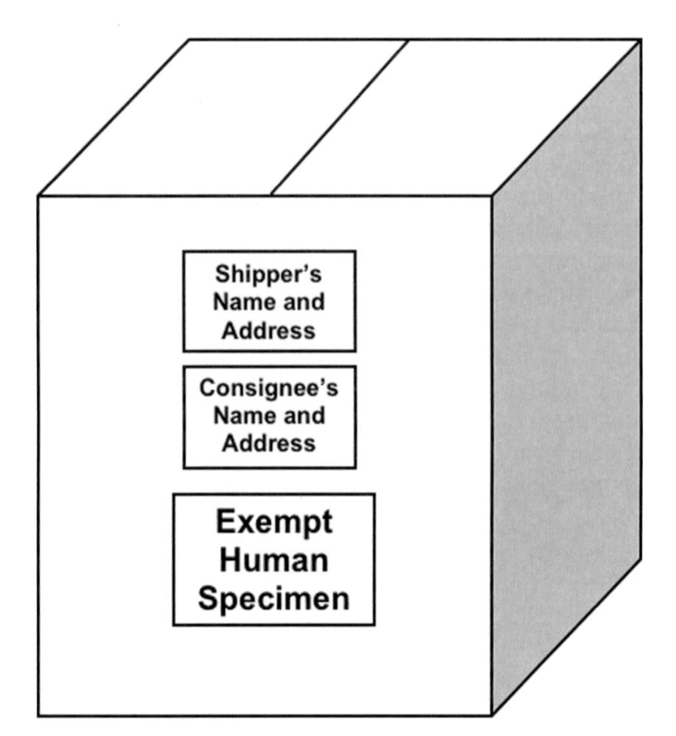

Figure 10: A completely labeled outer package. The primary container inside the package contains an Exempt Human Specimen and is packed according to IATA directions.

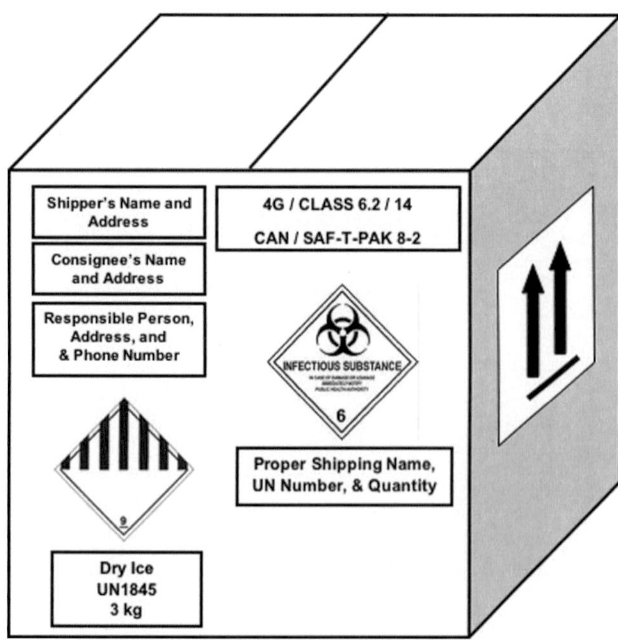

Figure 12: A completely labeled outer package. The primary container inside the package contains a Category A infectious substance and is packed according to PI 620.

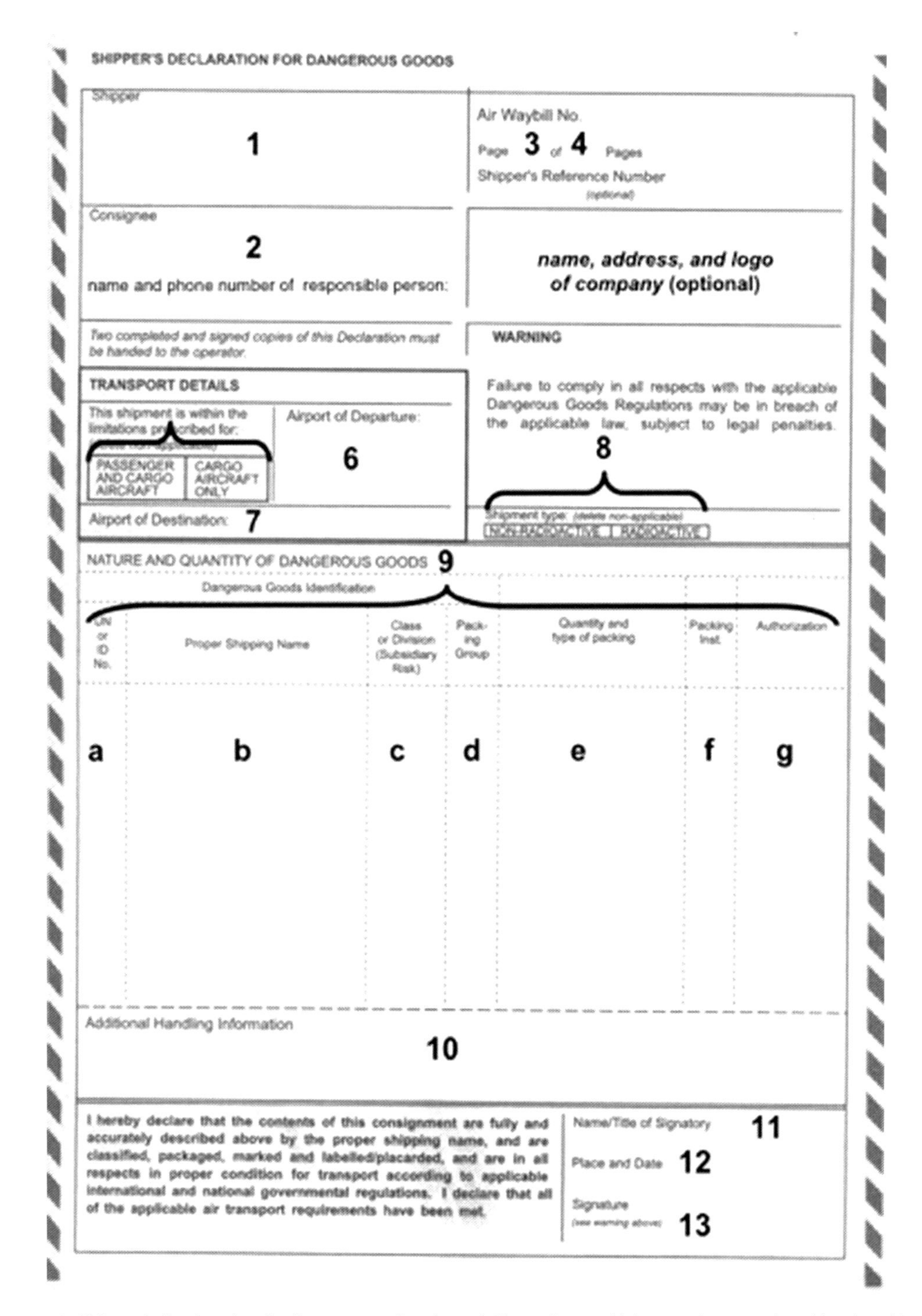
SHIPPER'S DECLARATION FOR DANGEROUS GOODS

Shipper
1

Air Waybill No.
Page 3 of 4 Pages
Shipper's Reference Number
(optional)

Consignee
2
name and phone number of responsible person:

name, address, and logo of company (optional)

Two completed and signed copies of this Declaration must be handed to the operator.

WARNING

Failure to comply in all respects with the applicable Dangerous Goods Regulations may be in breach of the applicable law, subject to legal penalties.
8

TRANSPORT DETAILS

This shipment is within the limitations prescribed for:

PASSENGER AND CARGO AIRCRAFT | CARGO AIRCRAFT ONLY

Airport of Departure:
6

Airport of Destination: 7

Shipment type: (delete non-applicable)
NON-RADIOACTIVE | RADIOACTIVE

NATURE AND QUANTITY OF DANGEROUS GOODS 9

Dangerous Goods Identification

UN or ID No.	Proper Shipping Name	Class or Division (Subsidiary Risk)	Packing Group	Quantity and type of packing	Packing Inst.	Authorization
a	b	c	d	e	f	g

Additional Handling Information
10

I hereby declare that the contents of this consignment are fully and accurately described above by the proper shipping name, and are classified, packaged, marked and labelled/placarded, and are in all respects in proper condition for transport according to applicable international and national governmental regulations. I declare that all of the applicable air transport requirements have been met.

Name/Title of Signatory 11

Place and Date 12

Signature
(see warning above) 13

Figure 13: Shipper's Declaration for Dangerous Goods and 13 sections which must be completed by the shipper.

The DOT requires that an "emergency response telephone number" be provided on Shipper's Declarations which accompany shipments of Category A infectious substances (Fig. 14) (2). The telephone number must be monitored at all times by a person who has knowledge of the following: (i) the hazards of the material being shipped, (ii) emergency response and accident mitigation information, in case a handler contacts the released contents of the package, or (iii) appropriate first aid information. Alternatively, the telephone number can be that of a person who has immediate access to a person who has such knowledge and information. The name and number

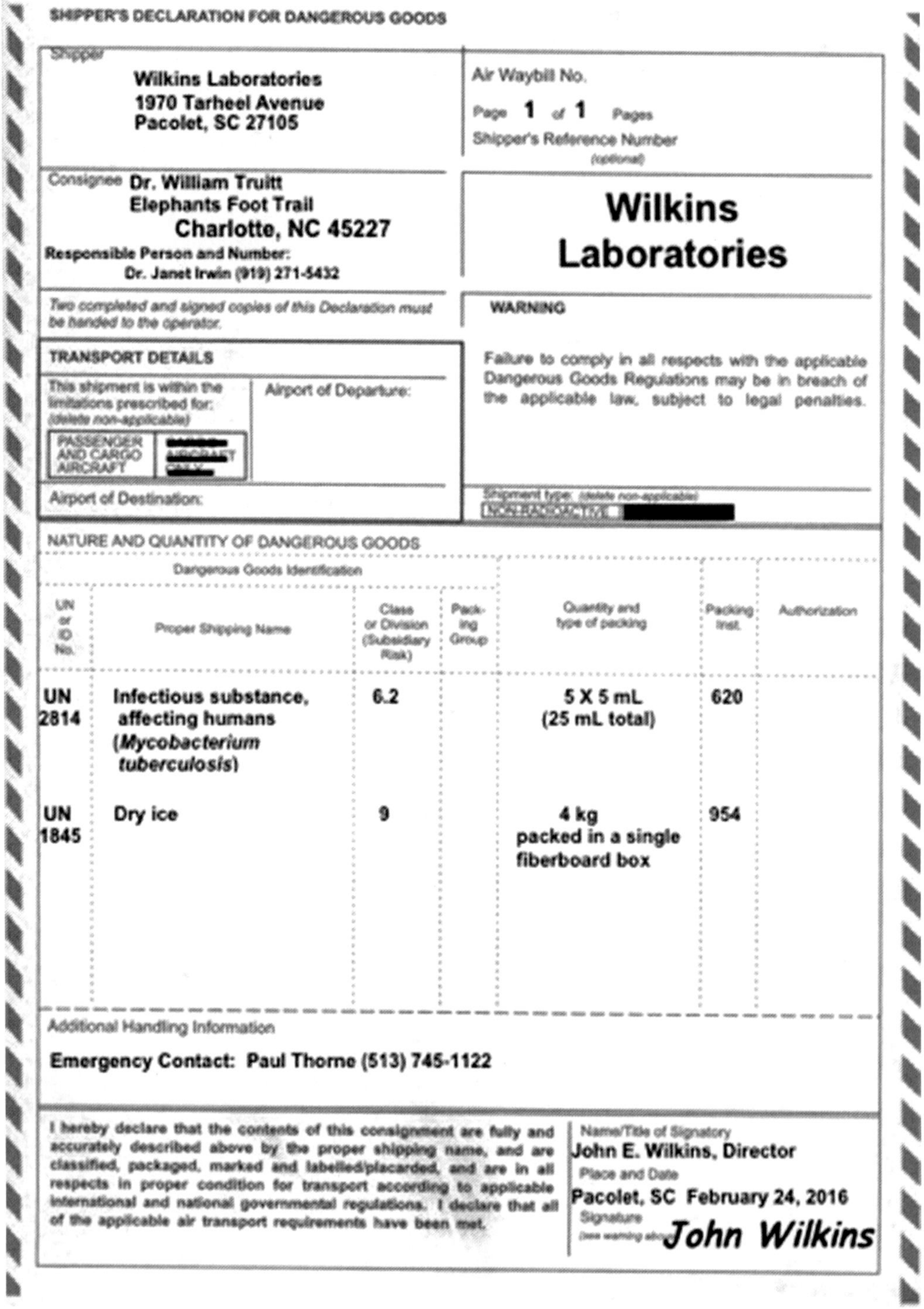

SHIPPER'S DECLARATION FOR DANGEROUS GOODS

Shipper
Wilkins Laboratories
1970 Tarheel Avenue
Pacolet, SC 27105

Air Waybill No.
Page **1** of **1** Pages
Shipper's Reference Number
(optional)

Consignee **Dr. William Truitt**
Elephants Foot Trail
Charlotte, NC 45227
Responsible Person and Number:
Dr. Janet Irwin (919) 271-5432

Wilkins Laboratories

Two completed and signed copies of this Declaration must be handed to the operator.

WARNING

Failure to comply in all respects with the applicable Dangerous Goods Regulations may be in breach of the applicable law, subject to legal penalties.

TRANSPORT DETAILS

This shipment is within the limitations prescribed for: *(delete non-applicable)*
PASSENGER AND CARGO AIRCRAFT | ~~AIRCRAFT ONLY~~

Airport of Departure:

Airport of Destination:

Shipment type: *(delete non-applicable)*
NON-RADIOACTIVE

NATURE AND QUANTITY OF DANGEROUS GOODS

Dangerous Goods Identification						
UN or ID No.	Proper Shipping Name	Class or Division (Subsidiary Risk)	Packing Group	Quantity and type of packing	Packing Inst.	Authorization
UN 2814	**Infectious substance, affecting humans (*Mycobacterium tuberculosis*)**	**6.2**		**5 X 5 mL (25 mL total)**	**620**	
UN 1845	**Dry ice**	**9**		**4 kg packed in a single fiberboard box**	**954**	

Additional Handling Information

Emergency Contact: Paul Thorne (513) 745-1122

I hereby declare that the contents of this consignment are fully and accurately described above by the proper shipping name, and are classified, packaged, marked and labelled/placarded, and are in all respects in proper condition for transport according to applicable international and national governmental regulations. I declare that all of the applicable air transport requirements have been met.

Name/Title of Signatory
John E. Wilkins, Director
Place and Date
Pacolet, SC February 24, 2016
Signature
(see warning above) *John Wilkins*

Figure 14: Completed Shipper's Declaration for Dangerous Goods.

of an agency, organization, or commercial company may be used instead of the aforementioned persons if the shipper can ensure the agency, organization, or company can supply the required aforementioned emergency information in a timely manner.

REFRIGERANTS

Two common refrigerants used to ship diagnostic specimens and infectious substances are wet and dry ice. The DOT requires that the packaging must be leakproof when wet ice is used. Dry ice is a Class 9 dangerous good, must be packaged according to PI 954, and, when it is the only dangerous substance being shipped, does not require a Shipper's Declaration to accompany the parcel. The secondary container must be secured such that it doesn't become loose as the dry ice sublimates. Outer packages must be labeled with a miscellaneous "9" diamond, UN number, and proper shipping name on the outside of the outer package and be recorded on the Shipper's Declaration (Fig. 4, 11, and 12). The maximum permitted net weight of dry ice per outer package is 200 kg. NOTE: Dry ice must never be placed into a tightly sealed container (explosion hazard!). Dry ice must be placed outside the secondary container, and the outer packaging must permit the release of CO_2!

TRAINING AND CERTIFICATION

Anyone involved in the shipping or transportation of dangerous goods, including diagnostic specimens and infectious substances, must be trained in the shipment of dangerous goods (1–3, 6). Acceptable training materials and methods include manuals, training courses, and workshops, all of which are commercially available from professional organizations and commercial suppliers of packaging materials for dangerous goods. Alternatively, a training program or workshop that includes hands-on training and demonstrations can be developed by any hospital, laboratory, school, institution, or other facility through the direction of a certified trainer. All training programs should be designed to provide initial and regular follow-up training to each employee responsible for shipping and packing infectious substances. The essential components of a training program must include the following: (a) general awareness and familiarity with the many aspects of shipping dangerous goods; (b) importance, nature, and contents of IATA and DOT regulations; (c) function-specific training (hands-on and/or demonstrations) on packaging, marking, labeling, and documentation of shipments of dangerous goods; (d) safety training, including blood-borne pathogens training; (e) security awareness training, and in-depth security awareness training for entities possessing select agents; (f) testing; and (g) issuance of a certificate after successful completion of the training (1, 2). IATA requires all aspects of training to be documented. The most important document used to prove appropriate and timely training is a certificate, which is issued after training is complete. Employers should keep a record for each employee who is trained and keep the most current training record throughout employment plus 90 days following termination of employment. The record should include employee's name, location and date of training, name of the certified trainer, course content, documentation of testing, and a copy of the certificate of training. IATA and DOT certification is valid for 2 and 3 years, respectively.

WHO 2015–2016 guidelines state that persons who pack and ship Category A infectious substances must receive the aforementioned formal training and certification (4). Persons who pack and ship Category B infectious substances and exempt human and animal specimens need to receive only general and practical training, such as "clear instructions on the use of packaging" and "training and awareness" of the importance of packing substances appropriately certification (4). Such persons should receive clear instructions, guidance, and training appropriate for packing and shipping Category B infectious substances and diagnostic specimens, addressing spills, and protecting themselves certification (4).

The DOT and the Federal Aviation Administration have authority to perform unannounced inspections of facilities (e.g., clinical laboratories) that ship dangerous goods, to inspect these facilities for compliance with the training regulations, and to inspect training records at these facilities. Facilities that do not comply with prescribed regulations are subject to substantial fines.

Training and training material for the transportation of dangerous goods and infectious substances is available at the following sources:

- American Society for Microbiology (www.asm.org)
- International Air Transport Association (training manuals) (www.iata.com)
- Regional and national clinical microbiology meetings (workshops and presentations)
- Most major universities and medical centers
- Many state departments of health and public health
- SafTPak (training courses and CDs) (www.saftpak.com)
- CARGOpak (one-day seminars) (www.cargopak.com)
- Dangerous Goods International (www.dgitraining.com)
- ICC The Compliance Center (www.thecompliancecenter.com)
- World Courier Training Course (www.worldcourier.com)
- Casing Scientific (www.casingcorp.com)

CONCLUSION

There are several simple but extremely important points that must be regularly and strongly emphasized to persons who pack and ship diagnostic specimens and infectious substances. (a) The purpose of correct packing and shipping is to protect carrier personnel who are not scientists. These personnel are helping shippers; however, they usually do not know the danger or lack of danger involved in the substances being shipped. (b) Shippers must be trained, and maintain current and accurate training documentation. (c) Retraining is mandated every 2 or 3 years, depending on regulatory or accreditation agency, and when regulations change. (d) Shippers should regularly search for new and revised regulations. Internet searches, and monitoring IATA and packaging materials manufacturers' websites, can provide access to changes in the transport regulations. IATA regulations change annually, and the updates are available every January. As this chapter goes to press, WHO has released a guidance document on regulations for transport of infectious substances (4) available online at http://apps.who.int/iris/bitstream/10665/149288/1/WHO_HSE_GCR_2015.2_eng.pdf. (e) Shippers must prepare each package correctly and with great care. When in doubt, ask someone with expertise. (f) The shipper is totally responsible for the integrity of the package from the shipper's door to the recipient's door. All aspects of packing and shipping are the responsibility of the shipper.

APPENDIX A

Definitions of Terms Related to Packing and Shipping

Biological Product—a substance which originated from living organisms (including humans and other mammals) and has been manufactured and distributed in accordance with compliance and licensing requirements set forth by the federal government; can be classified as an infectious substance or a diagnostic (or clinical) specimen, if such is appropriate. Biological products can be finished or unfinished, are intended for use in the prevention, treatment, or diagnosis of disease in humans or animals, and are be used for investigational, experimental, or development purposes. Biological products include common items such as clinical microbiology reagents and kits, serological reagents, diagnostic reagents, and vaccines. In certain parts of the world, some licensed biological products are regarded as biohazardous and are subject either to compliance criteria specified for infectious substances or must adhere to other restrictions imposed by the government of that country.

biological substance, Category B—any infectious substance which does not meet the criteria of a category A substance. See diagnostic (or clinical) specimen.

carrier (operator)—individual or organization engaged in the commercial transportation of goods (e.g., DHL, Federal Express, United Parcel Service, and Northwest Airlines).

Category A substance—an infectious substance or microorganism which is transported *in a form* that, when exposure to it occurs, is capable of causing permanent disability or life-threatening or fatal disease in an otherwise healthy human or animal. Category A substances are individually designated and specifically listed by IATA.

Category B substance—an infectious substance which does not meet Category A criteria. In the authors' opinion, Category B substances generally are considered to be the following: (a) patient or clinical specimens reasonably expected to contain, or being cultured or otherwise tested for, a pathogen, and (b) microorganisms not specifically listed in Category A.

***Code of Federal Regulations* (CFR)**—United States laws published in the *Federal Register* and available electronically at http://www.ecfr.gov/cgi-bin/text-idx?tpl=%2Findex.tpl.

consignee—the receiver of the shipment (e.g., a reference laboratory).

culture—the result of a process by which pathogens are intentionally propagated. This definition refers to typical laboratory cultures of microorganisms grown in broth or on solid media and does refer or apply to specimens as defined in this chapter. Typical clinical cultures may be classified either as Category A or Category B, depending on the organism concerned and the professional judgment of the shipper (4).

dangerous goods—material which, when not properly handled and contained, can pose a risk to the health, safety, property, or environment and which are shown on the list of dangerous goods in IATA *Dangerous Goods Regulations*.

***Dangerous Goods Regulations* (DGR)**—a commercially available book of IATA requirements; published by IATA; based on and incorporates ICAO regulations; provides packaging and shipping regulations for dangerous goods (e.g., diagnostic specimens and infectious substances); recognized and generally accepted worldwide.

diagnostic (or clinical) specimen—a Category B infectious substance; an infectious substance that does not meet the criteria for Category A; generally considered to be clinical specimens, such as swabs, tissue, and body fluids, commonly encountered in a clinical laboratory and being cultured or otherwise tested for a pathogen.

genetically modified microorganism (GMO)—microorganisms that have had their genetic material purposely modified or altered through genetic engineering in a manner that does not occur naturally; must be classified in the same manner and to the same extent as any potentially infectious substance or diagnostic (or clinical) specimen (Table 3).

Hazard Materials Regulations—49 Code of Federal Regulations (CFR) parts 171–177.

International Air Transport Association (IATA)—a trade organization of the commercial airline industry; governs international aviation; publishes *Dangerous Goods Regulations* for use by anyone who packs, ships, transports, or handles dangerous goods.

International Civil Aviation Organization (ICAO)—a specialized agency of the United Nations; governs international aviation; regulates the transportation of dangerous goods for all international civil air carriers; the source of IATA requirements and DOT regulations.

infectious substance—a substance which is known, or reasonably expected, to contain pathogens (microorganisms which can cause disease in humans and animals); material known to contain, or reasonably suspected of containing, a Category A or B pathogen or substance; can be a class (Class 6), a division (Division 6.2), or a category (Category A or B) of dangerous goods as defined by IATA.

overpack—the outmost packaging used to enclose more than one complete package, each of which contains dangerous goods; usually used for convenience and to reduce shipping costs.

package—end product of the packing process.

packaging—all of the numerous materials used to contain a shipped substance and to prepare the substance for shipping; the container (receptacle) and its associated components (e.g., tubes, containers, absorbent material, boxes, and labels) used to contain and pack a substance and to ensure compliance with packing requirements.

packing—the physical action and method by which packaging is used to secure articles or substances for shipment.

packing instructions—IATA-defined directions shippers must follow to select, assemble, mark, label, and document the packing process for shipping dangerous goods, including diagnostic specimens and infectious substances; includes manufacturing, testing, and performance specifications for packaging materials.

pathogen—a microorganism (bacterium, fungus, protozoan, proteinaceous infectious particle [prion], virus, worm, etc.) that is known to cause, or is reasonably expected to be able to cause, disease in humans or animals.

patient specimen—material collected from humans or animals including, but not limited to, excreta, secreta, blood and its components, tissue, body fluids, body organs and parts, and swabs of human material being transported for purposes, such as research, diagnosis, investigational activities, and disease treatment and prevention.

primary specimen container—the innermost packaging containing a diagnostic specimen or infectious substance; composed of glass, metal, or plastic; must be leakproof; must be positively sealed if it contains an infectious substance.

proper shipping name—any of more than 3,000 IATA-listed and internationally recognized names of dangerous goods.

secondary specimen container—the container that contains the primary specimen container.

shipper—anyone who ships goods by a commercial carrier (usually an employee of a company or healthcare facility [e.g., laboratory staff member]); anyone who offers goods for transport to a member of IATA; anyone who completes and signs the Shipper's Declaration. The person who signs the Shipper's Declaration is the person who accepts responsibility for the accuracy of the information on the document.

Shipper's Declaration for Dangerous Goods (Shipper's Declaration)—an IATA-defined and mandated form which must accompany each shipment of dangerous goods; contains information which describes the dangerous goods; is helpful to persons who handle the shipment; must be completed by the shipper.

UN certified container—packaging material (usually a cardboard box) that has passed UN certification tests (and IATA's construction and testing criteria) and is labeled by the manufacturer as such for the transport of certain dangerous goods.

United States Department of Transportation (DOT)—the federal agency which regulates domestic transportation of all dangerous goods into and within the United States through regulations published in the *Federal Register*, publishes regulations that are based on, and that are in, substantial agreement with ICAO regulations.

APPENDIX B

Select Agents

Select agents are microorganisms, biological agents, or biological toxins that have been deemed by the United States Government to be major threats to public health and safety because they could be used as agents of bioterrorism. Examples of select agents include most hemorrhagic fever viruses (e.g., Crimean-Congo hemorrhagic fever virus, Ebola virus, Marburg virus, etc.), *Bacillus anthracis*, *Yersinia pestis*, *Brucella abortis*, *Francisella tularensis*, variola virus, *Clostridium botulinum* neurotoxin, and all agents of bioterrorism, including zoonotic agents and agents of significant animal diseases (11, 12). If a select agent or a specimen or item suspected of containing a select agent must be shipped or otherwise transported from one facility to another, the shipper must contact the appropriate state and federal authorities for guidance and instructions before shipping. In addition, the shipper must confirm that the recipient is approved for receiving select agents. Up-to-date select-agent regulations and a list of select agents can be found at http://www.selectagents.gov.

References

1. **Denys GA, Gray LD, Snyder JW.** 2004. *Cumitech 40, Packing and shipping diagnostic specimens and infectious substances.* Sewell DL (ed). ASM Press, Washington, DC.
2. **International Air Transport Association.** 2015. *Dangerous Goods Regulations*, 56th ed. International Air Transport Association, Montreal, Canada.
3. **Snyder JW.** 2002. Packaging and shipping of infectious substances. *Clin Microbiol Newsl* **24:**89–93.
4. **World Health Organization.** 2015. *Guidance on regulations for the transport of infectious substances 2015 – 2016.* World Health Organization, Geneva (http://apps.who.int/iris/bitstream/10665/149288/1/WHO_HSE_GCR_2015.2_eng.pdf)
5. **McKay J, Fleming DO.** 2000. Packaging and shipping biological materials, p 411–423. *In* Fleming DO, Hunt DL (ed), *Biological Safety: Principles and Practices*, 3rd ed. American Society for Microbiology, Washington, D.C.
6. **United States Department of Transportation, Research and Special Programs Administration.** 2002. Hazardous materials: revision to standards for infectious substances and genetically modified organisms; final rule. *Fed Regist* **67:**53118–53144.
7. **United States Department of Transportation, Research and Special Programs Administration.** 2004. Harmonization with the United Nations Recommendations, International Maritime Dangerous Goods Code, and International Civil Aviation Organization's Technical Instructions; final rule. *Fed Regist* **69:**76043–76187.
8. **United States Department of Transportation, Pipeline and Hazardous Materials Safety Administration.** 2005. Hazardous materials: infectious substances; harmonization with the United Nations recommendations; proposed rule. *Fed Regist* **96:**29170–29187.
9. **United States Postal Service.** 2016. Mailing standards of the United States postal service, domestic mail manual. (http://about.usps.com/manuals/welcome.htm)
10. **United States Postal Service.** 2015. Publication 52: Hazardous, restricted, and perishable mail. (http://pe.usps.com/text/pub52/welcome.htm)
11. **Animal and Plant Health Inspection Service, United States Department of Agriculture.** 2016. Agricultural bioterrorism protection act of 2002: Possession, use, and transfer of biological agents and toxins; final rule (7 CFR Part 331; 9 CFR Part 121). *Fed Regist* **70:**13242–13292 http://www.selectagents.gov/.
12. **Centers of Disease Control and Prevention and the Office of the Inspector General, United States Department of Health and Human Services.** 2016. Possession, use, and transfer of select agents and toxins; final rule (42 CFR Part 73). *Fed. Regist.* 70:13316–13325. http://www.selectagents.gov/

SECTION

IV Administrative Control

Developing a Biorisk Management Program to Support Biorisk Management Culture

25

LOUANN C. BURNETT

The past few decades have seen a consistent evolution of approaches to safety and security risk management across a diversity of industries. In their review of this evolution, a U.S. National Academies of Sciences panel reviewing safety culture in academic chemistry laboratories (1) summarized, from safety science literature, three "epochs" that arose in response to accidents: (i) technology, (ii) systems, and (iii) culture.

- Technology is the application of engineering or other technical measures to control hazards and prevent injuries. This approach presumes that a linear chain of events causes accidents.
- Systems uses understanding of interactions between people, tasks, technology, and the environment to pursue risk reduction goals. Accidents result from complex interactions.
- Culture targets organizational shared values, assumptions, and beliefs toward and relative importance of workplace safety; it is highlighted by management commitment and involvement. This approach recognizes that it is rare that a single individual bears the entire responsibility for an undesirable outcome.

All of these approaches are relevant and critical to reducing risk. However, this evolution acknowledges that for technical and systems approaches to succeed, they must be supported by an organizational culture that holds safety as an integral and imperative part of its mission.

Approaches to biosafety and biosecurity also follow this evolution. The life sciences have seen an accelerated pace of technological change, new types of hazards, accessibility to materials by those who are not life scientists, and increased complexity. These additional pressures have made it even more important that biosafety and biosecurity approaches evolve. This chapter proposes some considerations for facilitating the transition of a technology-based "biosafety program" to support a broader and more encompassing "biorisk management culture."

The benefits of an effective biorisk management program that moderates the top-down approach of biorisk management culture with the bottom-up technology-driven approach include not only biorisk reduction or elimination, but also efficiency and cost reduction, increased integration of biorisk management into the organization, better responsiveness to unanticipated events, and enhanced goodwill with the community and oversight bodies.

EVOLVING APPROACHES FOR MANAGING BIORISKS

Technology

The concept of a "biosafety program" has traditionally been focused specifically on the efforts of an organization, primarily laboratorians and biosafety professionals, toward containment of potentially hazardous biological agents to the lab or work area where they are stored, manipulated, or disposed and the prevention of unintentional release ("biosafety"). The WHO defines biosafety as "the containment principles, technologies and practices that are implemented to prevent the unintentional exposure to biological agents and toxins, or their accidental release" (37).

Since 2001, the prevention of intentional misuse ("laboratory biosecurity") has also been included in many "biosafety programs" which are defined by the WHO as "the protection, control, and accountability for biological agents and toxins within laboratories, in order to prevent their loss, theft, misuse, diversion of, unauthorized access, or intentional unauthorized release" (37). Guidelines such as the publication *Biosafety in Microbiological and Biomedical Laboratories* (2) jointly published by the U.S. CDC and NIH, the NIH *Guidelines for Research Involving Recombinant or Synthetic Nucleic Acid Molecules* (3), and regulations such as the Occupational Safety and Health Administration (OSHA) Bloodborne Pathogens Standard (4) and the Federal Select Agents and Toxins Program (5) are examples in the United States of the technical, laboratory-centered focus on environments, equipment, and operations.

This approach has relied on the use of risk groups and biosafety levels to define the control measures for potentially hazardous biological agents. In the use of these categories, there is implicit acknowledgment that "the risk assessment had already been done by the experts, who had categorized the agents into risk groups and established specific controls for each of the biosafety levels" (6). Top management generally delegates responsibility to "biosafety officers" and "institutional biosafety committees," which are generally relied upon by their hiring organizations to assure that the measures prescribed under these categorizations are in place. Management tends to become involved only when something goes wrong.

Another hallmark of the technology-based approach is the singular emphasis on scientific expertise (and self-governance) over inclusion of expertise in governance, risk management, and organizational behavior. Palmer et al. argue that "[l]eadership biased towards those that conduct the work in question can promote a culture dismissive of outside criticism and embolden a culture of invincibility" (7).

Management Systems

The 2006 version of this chapter in the 4th edition of this book (8) introduced the integration of biosafety programs into a management system approach. This paralleled and, to some extent, sprang from the requirement by the U.S. Environmental Protection Agency (EPA) of the adoption of environmental management systems (ISO 14001) for managing hazardous wastes at U.S. universities—many of which took the opportunity to also utilize management systems to define and operate other environmental health and safety programs, such as biosafety. Concurrently, an international effort to develop a biorisk management system approach was under way, resulting in the European Committee for Standardization (CEN) Workshop Agreement (CWA) 15793:2008 (Laboratory Biorisk Management) (11) and 16393:2012 (Laboratory Biorisk Management—Guidelines for the Implementation of CWA 15793:2008) (38) documents in 2008 and 2012, respectively. (Note that a CWA is an agreement developed and approved in a CEN Workshop; the latter is open to the direct participation of anyone with an interest in the development of the agreement. See http://www.cen.eu/work/products/CWA/Pages/default.aspx.)

These efforts trailed and were informed by those of other industries, such as the chemical industry, nuclear power plants, off-shore drilling, aviation, and health care, among others. Such efforts were prompted by significant incidents (deemed preventable) and were meant to develop and implement management systems to improve safety and security (1, 6, 9, 10).

Resulting safety and security management systems tend to be structured similarly, using similar key steps (8):

1. Set a policy.
2. Plan actions to support the policy.
3. Implement the plan.
4. Monitor and measure the performance of the implementation.
5. Establish corrective and preventive action to address areas where performance requires improvement.
6. Conduct regular program and management review of the entire system.

Biorisk Management Systems

A biorisk management system approach enables an organization to "effectively identify, monitor and control the laboratory biosafety and biosecurity aspects of its activities" according to CWA 15793 (11). Biorisk management, then, comprises both biosafety and biosecurity and is an integrated approach to reducing risks arising from unintentional and intentional incidents involving biological materials. A management system approach is highlighted by the responsibility of many roles in an organization to contribute to the success of the system, including the direct involvement of top management. The introduction to CWA 15793 states that key factors in establishing and implementing a successful biorisk management system include:

"Commitment by top management to:

- providing adequate resources, prioritization and communication of biosafety and biosecurity policy;
- integrating of biorisk management throughout the organization;
- identifying opportunities for improvement and prevention, determining root causes and
- preventing recurrence."

As noted by Brodsky and Müeller-Doblies (12), the adoption of a systems approach and, thus, CWA 15793, has been slow, particularly in the United States where laboratory-centered guidelines such as the BMBL and *NIH Guidelines* have traditionally provided the foundation for biosafety program implementation and oversight. Interest, however, is growing in a more systems-based perspective, driven, perhaps, by the persistent incidents seen at even the most sophisticated and well-resourced laboratories. The decision to move the CWA documents into an ISO technical document, in progress, signaled the intent by and interest of the international biorisk management community to retain and, even expand, the biorisk management system approach.

Safety and Security Culture

More recently, investigations of incidents such as the 2011 Fukushima disaster and fatal accidents in academic chemistry laboratories (1), among others, have specifically indicted the lack of a "safety culture" as a significant cause for failure in these situations. (The Fukushima Daiichi nuclear disaster at the Fukushima I Nuclear Power Plant in Japan was initiated primarily by the tsunami that was triggered by the Tōhoku earthquake on March 11, 2011. The damage caused by the tsunami produced equipment failures, and without this equipment a loss-of-coolant accident followed, resulting in three nuclear meltdowns and the release of radioactive material beginning on March 12, 2011. With respect to accidents in academic laboratories, in 1997, a chemistry professor from Dartmouth College died from acute mercury toxicity upon exposure to dimethyl mercury through her gloved hand. In 2008, a staff research assistant at UCLA died of injuries sustained in a fire caused by pyrophoric reagents. In 2010, a Texas Tech student was seriously injured during an experiment involving nickel hydrazine perchlorate.)

The term "safety culture" is not new. The first reported appearance of the term in 1985 was in reference to the 1984 Bhopal disaster (10), and then it was commonly used in the follow-up to the Chernobyl incident. (The Bhopal disaster, also referred to as the Bhopal gas tragedy, was a methyl isocyanate gas leak incident in India in 1984 and was considered the world's worst industrial disaster. The Chernobyl disaster was a catastrophic nuclear accident that occurred on April 26, 1986, at the Chernobyl nuclear power plant in the town of Pripyat, in Ukraine (then USSR). An explosion and fire released large quantities of radioactive particles into the atmosphere, which spread over much of the western USSR and Europe.)

Metrics and indicators can be subverted but culture cannot (10). At Fukushima, investigations noted that at the operational level, procedures were generally followed and were appropriate, but the organizational level failed to think beyond "business as usual," leaving workers unprepared and unequipped for the emerging incident. In an organization with a strong safety culture, the mission of the organization is imperative, but if things go wrong, safety takes precedence.

The life sciences community has experienced fewer catastrophic incidents than those listed above, but the consequences of an unintentional or intentional release of even a small amount of certain pathogens could conceivably have a more devastating effect than those described from the chemical and nuclear industry above. Unintentional exposures to or releases from laboratory biocontainment have been documented involving severe acute respiratory syndrome coronavirus (SARS-CoV) (13), *Bacillus anthracis* (14), influenza H5N1 (15), *Yersinia pestis* (16), and *Francisella tularensis* (17), among others (18, 19).

The "failures" identified with these incidents and those across other industries tend to be similar:

- Prioritization of production over safety
- Unclear lines of responsibility and accountability
- Adherence to only minimal standards with no question as to whether additional measures might be necessary to address risk
- Lack of recognition of hazards and risks
- Safety not considered in decision-making
- Corrective action addresses symptoms not causes
- Failure to integrate and apply experience and lessons learned
- Lack of effective training programs
- Lack of nonpunitive incident or near-miss reporting
- Disregard for safety and security expertise

While there is still not a consistent definition for safety (or security) culture across the many industries that are currently discussing it, key features that define a safety culture are (1, 10):

- Strong leadership and management for safety—"walk the talk"
- Lines of authority and accountability are clear
- Two-way communication and free exchange
- Continuous learning about safety with incentivized information flow and self-criticism
- Strong safety attitudes, awareness, and ethics
- Learning from incidents, with greater importance assigned to problem-solving than finding blame
- Collaborative efforts to build safety culture; responsibility for success shared by all
- Promoting and communicating safety
- Institutional support for funding safety
- Prioritize expertise, not seniority, in decision-making

In an editorial reflecting on recent lapses in the life sciences, Trevan (20) calls for the adoption of the best practices of high-reliability organizations (HROs) to improve biosafety. An HRO is an organization that has succeeded in avoiding catastrophes in an environment where normal accidents can be expected due to risk factors and complexity—preventing failure rather than maximizing output. The essential components of HROs comprise a list quite similar to those of safety culture listed above. Palmer et al. comment that U.S. government recommendations designed to address lapses involving infectious agents, "identif[y] important gaps and constructive steps, but graft recommendations onto inadequate institutional structures and fail[s] to address underlying systemic needs (7)."

This chapter does not intend to abandon the core and essential principles of biosafety and biosecurity described by the foundational, technical documents of the disciplines nor the broader implementation of a management system such as CWA 15793 and its successors. Rather it seeks to integrate these concepts with pragmatic and implementable organizational practices that together contribute to an effective and sustainable biorisk management culture. Please note that many of the references cited in this chapter refer to "safety" or "safety culture." The literature on security culture is not as developed, but the information presented here is intended to be applicable to both safety and security and for the biorisks resulting from either unintentional or intentional actions.

Figure 1 illustrates the interreliance of the core principles of containment on (i) the competence and critical thinking skills of those who impact or influence this containment, (ii) the physical, financial, and human capacity of the organization to apply sufficient resources where they will be effective to reduce or eliminate biorisk, and (iii) the commitment and intent of the organization to take action reduce or eliminate unacceptable biorisks, irrespective of other priorities, and the integration of that commitment into the function and mission of the organization. In addition, communication must be flowing effectively and unimpeded across all of these elements. Of course, there are other elements that could be called out; this model comprises the author's categorization of the features required for safety culture, as summarized from the literature and listed above.

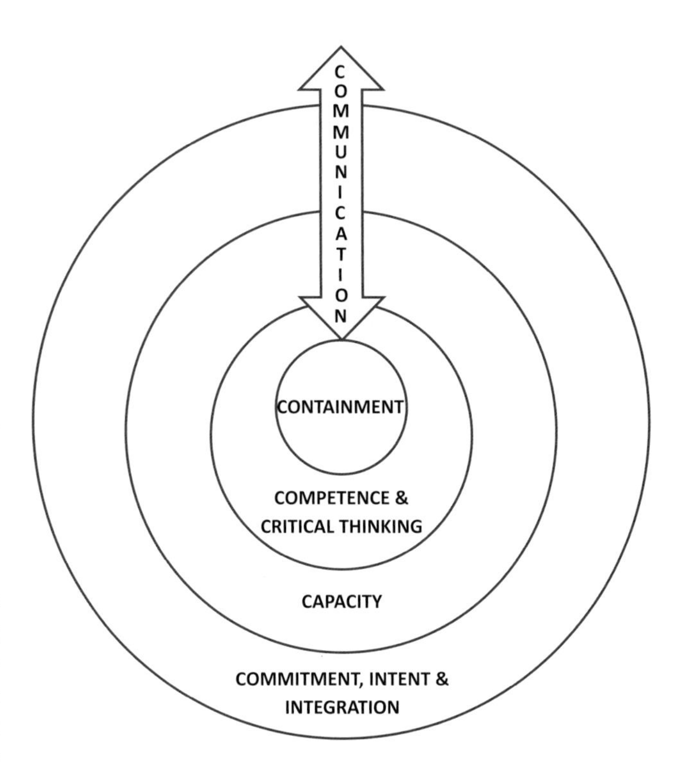

Figure 1: Influencing factors on biorisk management culture.

This concept of a biorisk management culture is scalable—it can be applied in a single laboratory, department, institution, ministry or agency, country, or region. Likewise, it is applicable in organizations with different missions—patient care (animal or human), diagnostics, research, and field (animal or human disease investigations or surveillance, agriculture). These elements are essential regardless of size or mission. A robust biorisk management system requires continuous collaboration as new and refined inputs and responses are encountered.

COMPLIANCE VERSUS CULTURE

Biosafety programs have traditionally been established within organizations on the basis of a need to comply with various guidelines, contract requirements, and regulations. Personnel responsible for biosafety programs are often called "biosafety officers" and are tasked with assuring compliance. The primary risk that a biosafety program addresses under this configuration is the risk of noncompliance and the associated consequences (fines, citations, loss of funding and credibility, etc.). Although these are not trivial risks and consequences, a biosafety program focused and supported only for this goal will easily miss other biorisks not anticipated in the reactive regulatory environment. Biosafety professionals charged with addressing compliance are often caught between these minimum requirements and those that might actually address risk, for example, in the case of evolving technology or approaches that outpace regulatory oversight. One of the oft-cited failures in the incidents described above was the attitude that there was no need to assess risks or mitigate them beyond that stated (or implied) in the legal requirement.

Regulations cannot ensure safe practices; organizations must support norms that cause people to act properly even when no one is looking or if the action is above the minimum required (9). Building a biorisk management culture expands the role of a biosafety program to provide key expertise in biorisk assessment, mitigation, and performance. However, the discussions below will soon show that assuring the organizational expectations encompassed in a biorisk management system and culture is well beyond the capacity and reasonable expectation of a "biosafety officer" or even an environmental, health, and safety office. Assuring a functional biorisk management system and culture requires ongoing engagement and collaboration from top management downward. However, biorisk management decisions must not rest solely with any one role, even the leadership. All roles must carry some expectation to "do the right thing" to reduce biorisk, even in the face of failed systems or unanticipated situations. This requires equipping all workers, leadership downward and including those staffing a biorisk management program, with enough knowledge and context sufficient to guide and inform their actions in the absence of clear direction or normal operations.

EVALUATING BIORISK MANAGEMENT SYSTEM PERFORMANCE

One of the hallmarks of a systems approach is the measurement of performance of the various components of the system. An honest systems approach acknowledges that the entire system will naturally degrade subtly or drift toward failure if not maintained. In fact, without designed continuous evaluation and improvement, a systems approach is not markedly different from the technology epoch described above.

Too often, the performance indicators for biosafety or biosecurity are based on failure data, such as incidents or actions of nonconformance (21). In these cases, the system has already failed. Establishing performance indicators and metrics that can provide an early warning against failures is imperative in the systems approach. Although biosafety programs have traditionally focused on implementing control measures for biorisk, this chapter posits that evaluating performance should be targeted and strengthened in biorisk management programs. Measuring and evaluating the actual function and outcomes of biorisk management actions will feed and inform the system and culture and assure its continuity.

Many models for establishing performance indicators exist. A helpful reference to begin a focus on performance is the Organisation for Economic Co-operation and Development (OECD) Environment Directorate publication on safety performance indicators for the chemical industry (22). Although this publication targets chemical manufacturers, many of the functions measured are similar to biorisk management (21). As described in this document, performance indicators (for any target) start with setting a desired outcome (outcome indicator) and then determining the actions that must be in place to influence the completion of the outcome (activity indicator). Indicators in this context are "observable measures that provide insights into a concept" (22).

Metrics are used to measure the extent to which the activity or outcome is in place (occurring). Here metrics are the way in which data will be compiled and reported. Examples include descriptive metrics (sums, percentages, composite), threshold metrics or tolerances (does the measurement exceed a given value?), trended metrics (change of descriptive metric over time), etc. A metric can also be binary, such as a "yes" or "no" answer to a question (22) (see Fig. 2). This pairing of indicators

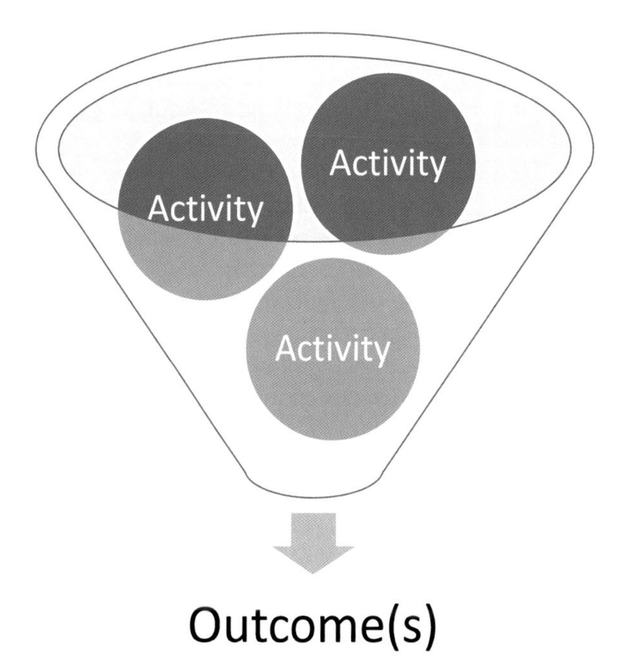

Figure 2: Model demonstrating how performance indicators (actions) converge to achieve a desired outcome(s).

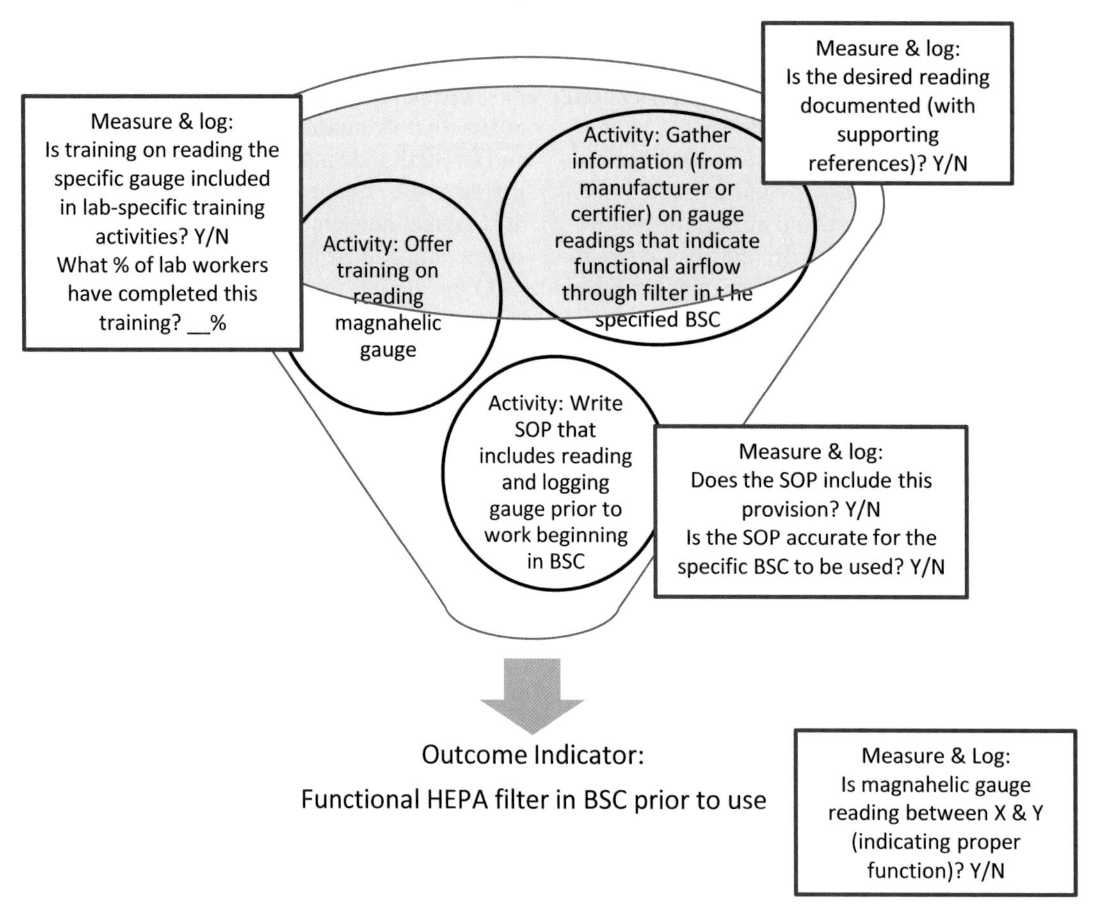

Figure 3: Laboratory example of indicators and metrics: checking airflow through the HEPA filter for safe and proper biosafety cabinet function.

addresses one of the difficulties of measuring system performance. In the most simple, technology-driven approach, failures are assumed to be related to a linear causality ("*x*" action causes "*y*" outcome). However, this is too simplistic for a systems approach. As a start to recognizing that performance is impacted by a variety of factors and not linearly, activity indicators measure the system components that must be in place to assure a positive outcome. This is borne out by investigations of incidents where errors or mistakes made by frontline workers (active failures) can be directly influenced by actions or decisions taking place (or not taking place) at higher levels of the organization (latent failures) (1). In a simple example, an inspection observation that a worker is not wearing "appropriate" personal protective equipment (PPE) ignores whether that PPE was actually provided or accessible, if a risk assessment has been used to define what "appropriate" PPE is, or if workers have received training on how, when, and why to use PPE. Corrective action requiring a lab to assure that workers wear appropriate PPE might actually require action at an organizational level where an individual lab has very little influence and no authority. Using paired indicators and metrics allows concurrent review of the lab level and the organizational system.

Two examples of paired indicators are provided in Figs. 3 and 4. Figure 3 is an illustration of indicators and metrics for a familiar prudent practice when using a biological safety cabinet (BSC)—checking that airflow through the HEPA filter is within margins that indicate that the filter and BSC are working properly. This set of performance indicators and metrics revolves around an outcome of a specific individual desired behavior by a worker in a lab. However, by also using activities indicators, several additional activities must be put into place by the organization for the worker to complete the desired behavior. If there is a failure of the worker to check the function of the BSC, these other activities might also be absent. Note the use of "yes/no" metrics—a metric does not have to be a number, but rather a way of collecting data that can be analyzed over time or compared to other data.

Figure 4 focuses on a more organizational activity frequently missing—assigning clear responsibilities to

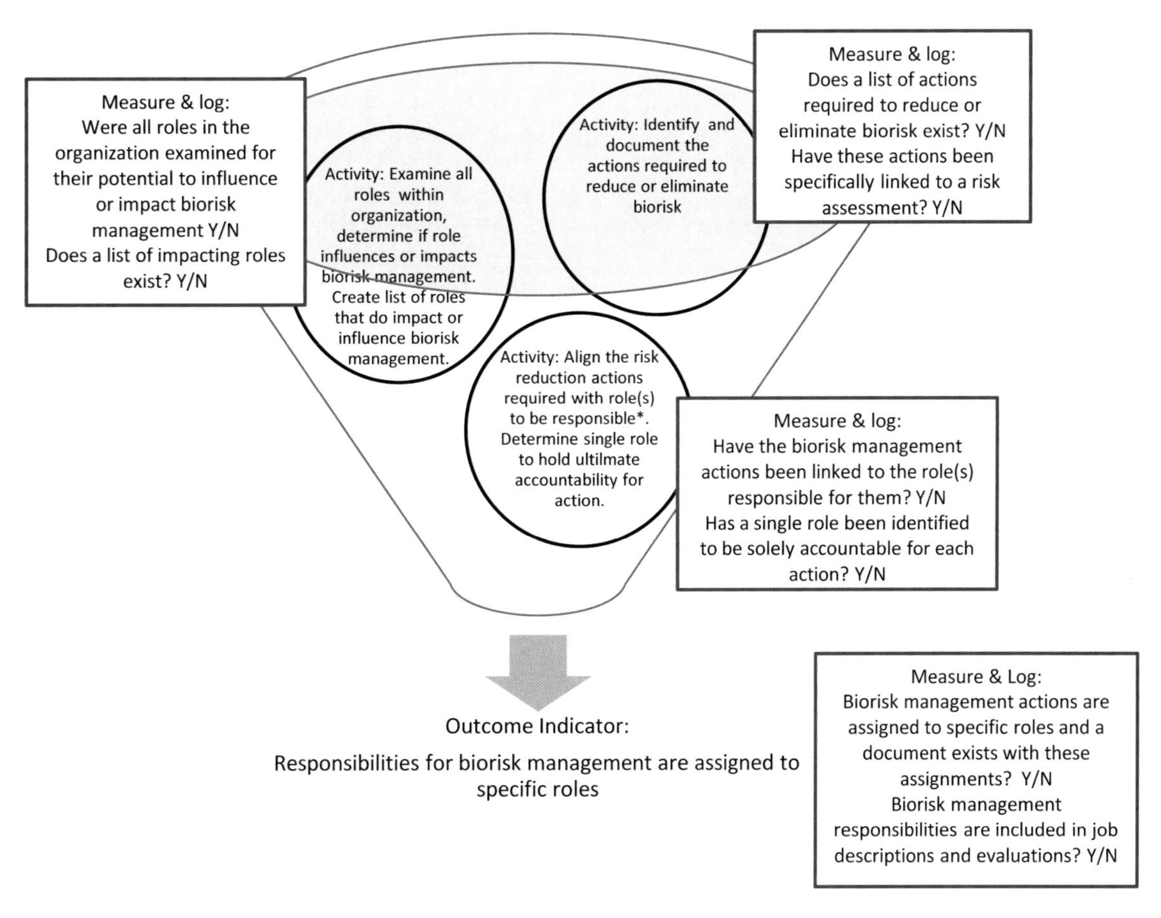

Figure 4: Administrative example of indicators and metrics: developing, evaluating, and assigning clear responsibilities in biorisk management to specific individual roles.

specific roles. The desired outcome is the documentation, assignment, and communication of biorisk management responsibilities to specific organizational roles. Activities that must be in place to influence this outcome involve identifying roles that impact or influence biorisk management, actions required to reduce or eliminate biorisk, and an analysis that links the role with the action. The absence or partial implementation of these activities will directly impact the outcome.

Mapping outcomes and activities and their accompanying indicators and metrics as part of planning along with establishing policy, goals, and objectives increases the likelihood that successful measures and those that are not as successful can be identified and supported or remedied before an incident occurs. A program that is based on robust risk assessment and on integrated performance measurements and evaluation is more likely to be focusing on improvements that make a demonstrable difference for the individual organization. A finding in investigations of organizations with robust safety cultures is that well-integrated risk assessments and performance evaluations will guide resource allocation. Proactive risk-based resource allocation can minimize, substantially, the impact of an unexpected event.

BIORISK MANAGEMENT—THE ASSESSMENT, MITIGATION, AND PERFORMANCE (AMP) MODEL

At its core containment mission and as illustrated in Fig. 5, a biorisk management program functions to define and assess risks (assessment), assign controls to reduce or eliminate identified unacceptable risks (mitigation), and assure that assessments are accurate and mitigation strategies are working as planned (performance). Table 1 provides examples of containment and objectives within the functions of assessment, mitigation, and performance.

Assessment

First, biorisks must be identified and characterized. If biorisk does not exist or is determined to be acceptably low, then the program does not need to address that risk.

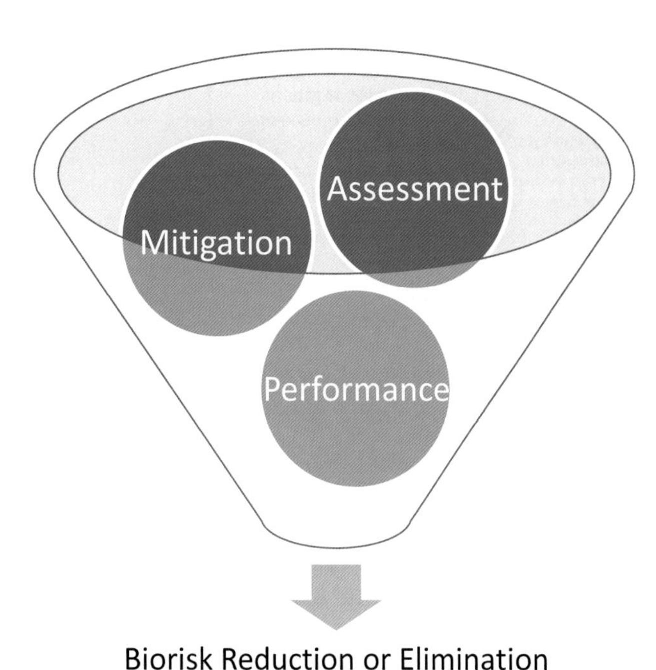

Figure 5: The assessment, mitigation, and performance (AMP) model of biorisk management.

This identification and characterization are commonly referred to as biorisk assessment (23). Risk assessment (characterization) involves asking a series of questions, in a formal, documented manner, of each activity involving biological materials (as well as those activities that support work with biological materials):

1. What can go wrong?
   a. Define potential biorisks. Examples of biorisks include infection of lab workers; release of uncontained or active biological materials beyond the laboratory into the community and environment, resulting in infection of humans, animals, or plants in the community; theft by an outsider; and theft by an insider.
2. How likely is it that something will go wrong and how likely are we to see it coming?
   a. Which activities involving the biological material might cause exposure or release (unintentionally or intentionally)?
   b. How likely is that activity to cause exposure or release?
   c. What are the vulnerabilities and threats that might lead to intentional theft, misuse, and/or release?
   d. What existing control measures might prevent or limit exposure or release?
3. What are the consequences?
   a. How hazardous is the biological material in the concentration and form utilized?
   b. What would be the result of exposure to or release of the biological material?
      i. To workers?
      ii. To community outside laboratory?
      iii. To environment (animals, plants, etc.)?
   c. What existing control measures might prevent or limit the consequences of an exposure or release?

The outcomes of biorisk assessment guide the selection of relevant biosafety and biosecurity measures and allow prioritization of resources toward the activities with greatest risk. Owners and operators of nuclear power plants discovered that conducting risk assessments helped the plants run more cost-effectively because the assessments highlighted what was most important to protect. The same owners and operators reported that employing a risk assessment allowed productive and constructive rollback of controls and oversight that were unnecessary on the basis of risk profile of the activity or facility (10). Risk assessment is also an example of

TABLE 1.

Containment (technical measures in place to reduce or eliminate biorisk)

Assessment	Mitigation	Performance
Identify hazards, threats, manipulations, and existing control measures. Define biorisk to be assessed. Characterize risk (e.g,, high, medium, low) based on probability of occurrence and consequence after occurrence. Evaluate risk acceptance. Identify additional mitigation measures needed to reduce existing risk to acceptable risk.	Implement biorisk-reducing controls: Elimination/substitution Engineering controls Administrative controls Work practices Personal protective equipment	Outcome indicators: Extent hazards are identified. Extent risk assessment is used to identify biorisk-reducing control measures. Extent incidents are reported Activities indicators: Are there systematic procedures for hazard identification and for risk assessment? Are implemented control measures linked to documented risk assessment results?

advance planning, which will be obvious and beneficial if a crisis evolves.

Mitigation

Second, once a risk is identified, characterized, and determined to require reduction or elimination, control measures suitable to reduce or eliminate that specific risk must be chosen and put into place. This is referred to as biorisk mitigation. Table 2 lists examples of common biorisk mitigation strategies used for laboratory containment. Specific references describing these strategies are plentiful and, thus, are not discussed in depth here. The success of biorisk mitigation strategies for reducing or eliminating biorisk depends on the targeted risk and an accurate risk assessment. For example, the use of aerosol prevention devices may have limited success in preventing exposure to blood-borne pathogens, which are rarely transmitted by aerosol. The use of risk mitigation measures linked to the risk assessment is crucial to assuring that limited resources are used wisely and toward effective biorisk reduction or elimination. This is the limitation to wholesale and indiscriminate use of the entirety of mitigation measures defined by biosafety level. Not all risks will be mitigated effectively by all of the measures defined in a given biosafety level. Likewise, some controls in biosafety levels may unnecessarily over-mitigate certain risks.

Performance

Last, the mere existence of risk assessment and risk mitigation strategies does not assure that risk is reduced or eliminated or that the initial assessment or mitigation assignments are accurate or still valid. Checks are required to assure that these earlier actions are indeed implemented, are functional, and are working toward the goal of risk reduction (21). These checks can be called biorisk management performance, which was emphasized in more detail above.

TABLE 2.

Common biorisk mitigation strategies

Category of mitigation strategy	Examples[a] of mitigation measures
Elimination	Choosing not to work with a given pathogen
Substitution	Using a less hazardous pathogen
Engineering controls	Physical lab space, including doors, walls, etc.
	Locks, access control systems
	Ventilation systems
	Biological safety cabinets
	Sharps and waste disposal containers
	Safer sharps
	Sealed centrifuge rotors or cups
Administrative controls	Signs and labels (hazard communication)
	Training and mentoring programs
	Assigning authorized access
	Assigning materials control and accountability
	Hazard inventory
	Job descriptions and evaluations
	Policy
	Written roles and responsibilities
	Program documents
	Standard operating procedures
	Strategic and operational plans
Work practices	Following standard operating procedures
	Hand washing
	Sterile technique
	Waste disposal
Personal protective equipment	Gloves Lab-specific clothing (e.g., lab coats) Masks or respirators Protective eyewear Face shields

[a]Please note that these examples are far from exhaustive.

THE AMP MODEL

Collectively, these three steps comprise the AMP model for biorisk management—Assessment, Mitigation, Performance. The desired outcome of a biorisk management program is to reduce or eliminate biorisks, and the activities necessary to accomplish that outcome require the implementation and maintenance of these three functions (see Fig. 5). Although each activity may be executed in sequence as a cycle (assessment, then mitigation, then performance), the reality is that the functions necessarily overlap and each is required for the other to be successful and for the overarching goal of biorisk management to be supported. This concept might be visualized as a three-legged stool where the fitness of each leg is required to keep the stool upright. If any component is overlooked or not addressed effectively, the system is, at the least, unstable and, at worst, fails completely (24).

APPLYING THE AMP MODEL BEYOND BIORISK CONTAINMENT TO ELEMENTS INFLUENCING BIORISK MANAGEMENT CULTURE

The above paragraphs describe the application of the AMP model to the more familiar technical aspects of laboratory containment. The sections below expand the concept of AMP to include the earlier model of elements that influence biorisk management culture (see Fig. 1). Each goal—(i) commitment, (ii) capacity, (iii) competence and critical thinking, and (iv) communication—has unique objectives relating to assessment, mitigation (addressing the gaps between existing and desired states), and performance. This discussion is presented as a step for those maintaining a biorisk management program to examine opportunities to review organizational focus beyond containment and compliance.

BIORISK MANAGEMENT COMPETENCE AND CRITICAL THINKING

Competence infers that a worker has skills and knowledge to complete an assigned task successfully. Experience further dictates that, to be truly competent, a worker must also have the will (and permission) to deploy those skills and knowledge.

Several efforts to define biosafety and biosecurity competence have been recently published. In 2011, the CDC and Association of Public Health Laboratories (APHL) published guidelines for biosafety laboratory competency (25). They defined competencies as "measurable and include not only knowledge, skills, and abilities but also judgment and self-criticism." More recently, the CDC and APHL published competency guidelines for public health laboratory professionals (25). While not specifically targeting biosafety or biosecurity, these more recent guidelines include a section on safety competency. Another CWA (16335:2011) was developed with the European Biosafety Association (EBSA) to define biosafety professional competencies (26). Of note, all of these documents were developed with consultation and review of a large group of subject matter experts. None of these guidelines may be directly applicable to every setting, but they provide a solid starting point for meaningful development of role-based competence.

Despite the efforts of the authors of the publications described above, biorisk management competence in the laboratory has often been equated only with the ability to execute standard operating procedures (SOPs) correctly. Although this is an important part of maintaining containment, an aggressive effort to prescribe correct behavior within narrow limits has often led to system failures (10) because workers were not given the capacity or flexibility to address an unexpected issue. Factors defining an HRO include assuring the workers are not just mindlessly following SOPs, but they also have sufficient knowledge of the abstract principles guiding the procedures to transfer that learning effectively to a novel situation, should it arise (20). Simply learning to perform procedures and in only a single context (e.g., a specific laboratory setting) does not promote this flexibility. Effective transfer, and thus critical thinking, comes from a balance of specific examples and general principles, not from either one alone. The ability to transfer learning increases or creates the capability to make complex decisions under stress (27).

When workers are aware of what they are supposed to know and do and they know the principles supporting that knowledge and behavior, they tend to participate in self-assessment over time. This, in turn, prompts a desire for and action toward additional improvement, learning, or practice (10, 27). Organizations that support this type of learning and the resulting desire for improvements find that their workers are not only better prepared for novel situations, but also tend to be more satisfied and more easily retained. Studies have also shown that in environments with more uncertainty and specialization workers given higher autonomy for their own behavior demonstrate more positive safety behavior (28).

Despite the importance of biorisk management skills and knowledge to reducing or eliminating biorisks and being able to apply that knowledge in unanticipated situations, most job descriptions for workers handling potentially hazardous biological agents do not include any requirements for biorisk management competence. Thus, recruiting, hiring, and job performance evaluations do not include ongoing monitoring and expectations for biorisk management competence, actions, and behaviors. Table 3 outlines objectives toward using the AMP approach to address competence and critical thinking, including the addition of biorisk management expectations in the hiring and performance review process.

Improving biorisk management competence does not mean only establishing or improving a biorisk management training program (although effective training is essential), but also exploring and developing options like mentoring, peer review and observations, external professional development, etc.

BIORISK MANAGEMENT CAPACITY

Any organization must prioritize resources to execute its mission most effectively. Importantly, conducting biorisk assessments guides the placement of resources—people, physical infrastructure (facilities, equipment, etc.), and finances—toward activities where these resources will be most effective at reducing or eliminating biorisk. In cases where the risk is greater than the resources to be leveraged, the organization leadership may have to make the difficult decision to postpone or eliminate the activities involved or to down-size or eliminate other activities to allow for resource allocation to areas of higher risk. Table 4 presents some options to consider for developing and improving biorisk management capacity.

TABLE 3.

Competence and critical thinking (knowledge, skills, and abilities in place to conduct normal operations but also to take informed action in unanticipated situations)

Assessment	Mitigation	Performance
Identify education, experience, skills of each role in existing workforce. Determine desired biorisk management (BRM) actions and behaviors for each role. Determine desired education, experience, skills required for each role. Analyze gaps between existing and desired competence for each role.	Include desired BRM competence, actions, and behaviors in job descriptions for recruiting, hiring and evaluating workers. Recruit and hire new workers for each role who meet BRM competences defined. Execute professional development plan to increase BRM competence and critical thinking among existing workers.	<u>Outcome indicators:</u> Extent to which recruiting, hiring, and evaluation documents and processes include biorisk management competence Extent to which employees are offered and enrolled in programs to increase biorisk management competence and critical thinking Extent to which workers demonstrate desired biorisk management behavior(s) Extent to which workers demonstrate critical thinking to reduce biorisks when faced with an unanticipated situation <u>Activity indicators:</u> Are there mechanisms to assure that the scope, content, and quality of training programs are adequate for the specific role the worker plays in biorisk management? Are there mechanisms used to demonstrate that training or other competence-building activities result in desired operational behaviors?

Determining the human, physical, and financial resources that are necessary to reduce or eliminate biorisk in given activities requires a close collaboration between many roles in the organization. The biorisk management program leadership and the managers directly responsible for activities with biorisks must engage with top management, as well as business, financial, human resource, and facilities management, at the least, to assure that all involved are aware of the biorisks involved and the resources needed to reduce or eliminate the biorisk. In addition, legal counsel may be required to determine if there are legal requirements that impact resource allocation. Ultimately, top management must be responsible and accountable for the extent to which resources will be allocated or if activities will be postponed or cancelled, if biorisk cannot be managed. In a culture where biorisk management is prioritized with, or even elevated over, other priorities, the only options are to either allocate

TABLE 4.

Capacity (resource allocation to reduce biorisk)

Assessment	Mitigation	Performance
Identify existing human, physical, and financial resources directed toward BRM. Determine desired human, physical, and financial resources directed toward BRM, appropriate for the biorisks identified or anticipated. Analyze gaps between existing and desired resources.	Assure that existing human, physical, and financial resources are prioritized for current activities with the highest biorisk. Implement plans for improving infrastructure in settings where those improvements will reduce biorisk. Explore nontraditional options for controlling biorisk that may be more compatible with existing resources. Hire additional workers or redistribute workers for areas with highest biorisk. If resources are inadequate to reduce risk, stop or limit operations that present biorisk.	<u>Outcome indicators:</u> Extent to which risk assessment is linked to assignment of human, physical, and financial resources Extent to which resources are directed toward higher risk operations Extent to which operations are limited or eliminated if sufficient resources for risk reduction do not exist <u>Activities indicators</u> Is there a mechanism to determine relative biorisk between different organizational operations? Is there a mechanism for workers to identify situations where adequate resources for biorisk reduction are not applied or available? Is there a mechanism for stopping or limiting operations where resources adequate to reduce biorisk to acceptable levels are not applied or available?

resources that will effectively reduce or eliminate biorisk or to change, postpone, or cancel the activity.

That said, the biorisk management program leadership and collaborators should explore nontraditional options for reducing or eliminating biorisk that may be more compatible with existing resources. This approach is increasingly used in countries with fewer resources, but equal or greater biorisks. Biorisk mitigation strategies are layered and redundant to prevent an exposure or release if one strategy fails (29). A different layering of strategies may effectively reduce risk (30). The thinking behind alternative mitigation strategies is apparent, for example, when considering the layering of strategies for sample collection and manipulation in the field as opposed to the same activity in the laboratory. The protection offered from locks, doors, walls, and ventilation in the laboratory is not available in the field. Controls to reduce risk in the field are more reliant on portable or administrative measures, such as PPE, chain of custody for materials, accountability for personnel, and sealed containment such as triple-packaging similar to that used for shipping. The use of alternate mitigation strategies must be accompanied by a documented biorisk assessment that can be used to communicate the equal or greater reduction of anticipated biorisk.

BIORISK MANAGEMENT COMMITMENT, INTENT, AND INTEGRATION

As stated earlier, a tangible commitment to biorisk management and a culture of biorisk management must be initiated and maintained by the top leadership. This commitment, however, must be more than a written policy stating that the organization is committed to reducing or eliminating biorisks. This has been referred to as "satisficing"—where an organization develops documents merely to address regulatory or business expectations, rather than as a true organizational strategy (10).

The organization must also demonstrate visibly the intent to manage biorisks to assure that these statements extend beyond pieces of paper and are actually integrated into the actions of the organization, especially the leadership. This visibility can be demonstrated by leadership involvement and by active participation in day-to-day activities supporting biorisk management. Safety science literature provides many studies demonstrating that where safety is a central goal for managers in an organization and when a good social relationship exists between managers and workers, members of that organization are less likely to be involved in incidents or unsafe behaviors (31).

While a commitment statement can seem frustratingly high-level and vague, a visible, supported statement sets the tone and expectation for organizational behavior, regardless of what may occur. An example of a commitment statement that leaves no question about expectations regarding safety can be found in the policy from Electronuclear (a Brazilian nuclear power company): "Nuclear safety is a priority, precedes productivity and economic aspects, and should never be impaired for any reason" (10). A strong commitment statement that is embodied by the organization can help avoid the paralysis and indecision that can happen when things go wrong or when events occur that were not or could not be anticipated. Every worker knows that the right decision to be made is the one that prioritizes safety over productivity or economics and that they are encouraged and supported to step beyond their functional responsibilities, if needed. If the organization truly supports this commitment, that worker's decision, regardless of impact, will be supported by top management downward.

There is case after case where the "right" commitment and intent is not actually integrated into the work of the organization. Workers can read a policy and sign that they understand it, but do not really know or embrace the actions required of them. As outlined earlier, job descriptions and evaluations that contain expectations for actions to reduce or eliminate biorisk can support, at each level of an organization, the overarching commitment to biorisk management. Essentially, these actions comprise "responsibilities" and it is the bedrock of a management system that all or nearly all roles in an organization have direct responsibility for the actions and outcomes of the system.

Written responsibilities must be specific and actionable. A common approach to assuring this is to use the acronym SMART—actions that are specific (S), measurable (M), achievable (A), reasonable (R), and timely (T). Responsibilities must also be targeted toward a role. A role is not a title. A role can be defined as "a person who ____." For example, a biorisk management advisor or biosafety officer could be defined as "a person who provides advice and expertise on biorisk management to the organization." Some workers in an organization will have several roles and thus several responsibilities toward biorisk management. Part of the work around competence, critical thinking, and capacity must assure that workers are in a position to be successful in all the roles they are assigned. A system that is not designed to be successful will inevitably fail. Even the best system degrades without attention and maintenance.

Bridging the often-ignored gap between a policy and actual actions is the perception of the workers toward the system. In general, the concept of "perception" is viewed negatively—perception often implies a biased opinion that may or may not have a basis in reality. However, the primary tool to measure "safety climate" is a survey of employee perception. Even though it may seem counterintuitive, studies in safety science have shown the worker perception, despite its subjective nature, is the closest indicator of actual system performance (32). Safety policies and programs, open communication, and organization support have been shown to be the top contributors to positive safety perception by workers (33). In addition, workers who can tangibly recognize the risks involved in their work are more likely to practice safe behaviors (34).

Table 5 lists examples of actions supporting strong commitment, intent, and integration of biorisk management within an organization.

BIORISK MANAGEMENT COMMUNICATION

As shown in Fig. 1 (above), communication is a thread that runs through all components that influence and impact biorisk management culture. An organization can work diligently on commitment, capacity, competence, and containment initiatives, but if these cannot be transmitted to the workforce and translated into behaviors, the effect will be limited. Likewise, if there is not an effective feedback loop so that the workforce at any level can provide input freely into the system—identifying unsafe situations, providing suggestions for improvements, and collaborating on problem solving—the top-down initiatives will never be fully integrated into the organization. Open communication has been cited frequently as a top contributing factor to positive worker perception and safety culture (1, 10, 28, 33).

Table 6 outlines examples of actions toward improving biorisk management communication. Critical factors to increasing effective communication include assuring that representatives of the entire organization are involved in shaping biorisk management communication; this cannot be just a top-down activity. If biorisk management is truly integrated into the organizational mission, no communication, on any subject, will conflict with the intent to reduce or eliminate biorisk. Intentionally excluded from this discussion is the critical role of communication regarding biorisk management to the public (anyone outside of the direct employment of the organization). Although this exclusion is not intended to imply that this focus is not imperative, this section is targeted toward internal communication.

Because communication for biorisk management is, by default, communication about risk and because one of the

TABLE 5.

Commitment, intent, and integration (priority given biorisk reduction and integration into organizational function and mission)

Assessment	Mitigation	Performance
Identify existing top and local management visibility in and awareness of BRM commitment and intent. Identify existing documented roles and responsibilities for BRM. Identify existing means of communicating expectations for BRM to entire workforce and to stakeholders outside the organization. Identify workforce perceptions on status of BRM. Analyze gaps between existing and desired expectations toward BRM.	Establish an organization-wide, documented commitment, intent, roles, and responsibilities toward BRM. Hold top and local management responsible for demonstrating commitment and intent. Assure visibility and communication of BRM commitment and intent. Monitor adoption of BRM commitment and intent. Assure that demonstrations of commitment and intent by any worker are supported (e.g., give workers the ability to stop work if biorisk is not addressed). Determine perception of workforce toward biorisk management commitment and intent on an ongoing basis.	Outcome indicators: Extent of workforce familiarity with commitment and intent for biorisk management Extent of management and worker expectations and support toward biorisk management roles and responsibilities Extent of top and local management demonstration of commitment and intent toward biorisk management Extent of collection of worker perception of organization's commitment and intent toward biorisk management Activity indicators: Is there a process in place to develop and/or review biorisk management policy that states organizational commitment and intent? Is there a mechanism in place to develop, review, and communicate organizational roles and responsibilities with regard to biorisk management? Is there a mechanism to integrate responsibilities for biorisk management into job descriptions and evaluation processes? Is there an ongoing mechanism to survey worker perception of organizational commitment and intent toward biorisk management?

TABLE 6.

Communication

Assessment	Mitigation	Performance
Identify the key biorisk management messages to be delivered. Identify the mechanisms by which the various targeted roles receive communication (media). Identify any gaps in communication through review of safety climate (worker perception) surveys and worker suggestions, feedback, ombudspersons, etc.	Involve representatives of those who will be receiving key messages in development and delivery. Assure and test access to all key biorisk management communication by targeted roles (if not entire workforce). Utilize multiple methods of delivery to emphasize importance and to help assure delivery. Review communication tools periodically and revise as needed to assure relevance.	Outcome indicators: Extent to which roles are aware of and familiar with the biorisk management messages targeted toward them Extent to which workers perceive that messages are accurate and relevant to their work Extent to which workers participate in planning development and delivery of key biorisk management messages Extent to which workers participate in surveys, forums, and other feedback mechanisms Extent to which workers act in accordance with biorisk management expectations, as communicated. Activity indicators: Are communication tools written clearly and concisely in language accessible to targeted role(s)? Are communication tools relevant to the work of the targeted role? Are the requested actions communicated linked to relevant biorisk reduction?

factors required for a culture that effectively addresses biorisks is that those risks must be perceived and acted upon by workers as credible, special considerations for communication are needed. Conveniently, many of the tools discussed in previous sections for containment, competence, capacity, and commitment are tools that can also facilitate communication. Primary among these are the biorisk assessments that catalog and characterize, at a minimum, the potential risks, the nature of the hazards, the activities that influence the likelihood of incidents, the actors wishing to acquire hazards for malicious purposes, and the mitigation strategies that are currently in place. To be trusted, these assessments must be developed in an unbiased, robust, and repeatable manner, using expertise from a variety of roles throughout the organization (35).

Tools that can be used to aid in two-way biorisk management communication include:

- Organizational commitment and intent (policy statement)
- Biorisk management roles and responsibilities
- Biorisk assessments
- Biorisk-reducing mitigation strategies
- Standard operating procedures
- Warning signs
- Training and other competence and critical thinking development methods
- Perception surveys
- Audits and audit results
- Suggestion boxes
- Committee meeting agenda and minutes

An imperative communication tool is the reporting and investigation of incidents in an open, nonpunitive manner. All the investigations of failures used in research for this chapter have identified the need for meaningful reporting and investigation. and the debate for the best model is still unresolved for incidents involving biological materials. This lack of clarity in the discipline does not remove the obligation of an organization to seek to learn from incidents and near misses.

SUMMARY AND CONCLUSIONS

Relying on compliance with technical requirements for laboratories is an arguably easier path than developing a biorisk management culture that permeates an entire organization, but experience and incidents across many industries, including the life sciences, have clearly demonstrated that a solely technical focus is not enough to reduce risk. Today's biorisk management programs must be more fully integrated into the organization rather than viewed as a standalone component.

Applying a methodology of AMP not only to containment but also to biorisk management competence and critical thinking, capacity, commitment and intent, and two-way communication should enhance this organizational integration, serving to elevate biorisk management as an overarching priority. This focus and attention may be uncomfortable and unwieldy at first, but with continued emphasis and persistence, biorisk management culture will, over time, become second nature. An improved culture will be apparent as leadership and staff are aware of and articulate risks, acknowledge that things can go wrong, that mistakes can occur at all levels, but that through the efforts and commitment of the entire workforce, the organization is committed to be able to learn from those mistakes and take organization-wide action to prevent recurrence (36).

No longer can this responsibility be delegated solely to a biorisk management professional or office. Biorisk management professionals must be highly valued for their expertise and experience and relied on as a critical partner in improving biorisk management. However, top management must follow the lead of other industries and recognize and shoulder their responsibilities to set the tone, expectation, and example that effective biorisk management is of paramount importance to the success of the organization.

References

1. **Committee on Establishing and Promoting a Culture of Safety in Academic Laboratory Research; Board on Chemical Sciences and Technology; Division on Earth and Life Studies; Board on Human-Systems Integration; Division of Behavioral and Social Sciences and Education; National Research Council.** 2014. *Committee on Establishing and Promoting a Culture of Safety in Academic Laboratory Research. Safe Science: Promoting a Culture of Safety in Academic Chemical Research.* National Academies Press, Washington, D.C.
2. **U.S. Department of Health and Human Services, Public Health Service, Centers for Disease Control and Prevention, National Institutes of Health.** 2009. *Biosafety in Microbiological and Biomedical Laboratories*, 5th ed. HHS Publication No. (CDC) 21-1112. http://www.cdc.gov/biosafety/publications/bmbl5/BMBL.pdf.
3. **National Institutes of Health.** 2013. *Guidelines for Research Involving Recombinant or Synthetic Nucleic Acid Molecules.* U.S. Department of Health and Human Services, Washington, DC.
4. **U.S. Code of Federal Regulations.** 29 CFR 1910. 1030 - Bloodborne Pathogens Standard.
5. **U.S. Code of Federal Regulations.** 7 CFR 331, 9 CFR 121, 42 CFR 73 - Federal Select Agent Program.
6. **Salerno RM, Gaudioso J.** 2015. Introduction: the case for biorisk management, p 1–29. *In* Salerno RM, Gaudioso J (ed), *Laboratory Biorisk Management: Biosafety and Biosecurity.* CRC Press, Boca Raton, FL.
7. **Palmer MJ, Fukuyama F, Relman DA.** 2015. A more systematic approach to biological risk. *Science* **350:**1471–1473.
8. **Burnett LC.** 2006. Biological safety program management, p 405–415. *In* Fleming DO, Hunt DL (ed), *Biological Safety: Principles and Practices*, 4th ed. ASM Press, Washington, DC.
9. **Committee on the Effectiveness of Safety and Environmental Management.** 2012. *Evaluating the Effectiveness of Offshore Safety and Environmental Management Systems.* Transportation Research Board Special Report 309. National Academies Press, Washington, DC.
10. **Rusek B, Lowenthal M.** 2015. *Brazil-U.S. Workshop on Strengthening the Culture of Nuclear Safety and Security: Summary of a Workshop.* National Academies Press, Washington, DC.
11. **European Committee on Standardization (CEN).** 2011. *CEN Workshop Agreement (CWA) 15793:2011 - Laboratory Biorisk Management.* European Committee on Standardization. http://www.eubarnet.eu/?post_type=library&p=493.
12. **Brodsky B, Müeller-Doblies U.** 2015. Future development of biorisk management: challenges and opportunities, p 205–228. *In* Salerno RM, Gaudioso J (ed), *Laboratory Biorisk Management: Biosafety and Biosecurity.* CRC Press, Boca Raton, FL.
13. **Lim PL, Kurup A, Gopalakrishna G, Chan KP, Wong CW, Ng LC, Se-Thoe SY, Oon L, Bai X, Stanton LW, Ruan Y, Miller LD, Vega VB, James L, Ooi PL, Kai CS, Olsen SJ, Ang B, Leo YS.** 2004. Laboratory-acquired severe acute respiratory syndrome. *N Engl J Med* **350:**1740–1745.
14. **Centers for Disease Control and Prevention.** 2014. *Report on the Potential Exposure to Anthrax.* Centers for Disease Control and Prevention, Atlanta, GA.
15. **Centers for Disease Control and Prevention.** 2014. Report on the Inadvertent Cross-Contamination and Shipment of a Laboratory Specimen with Influenza Virus H5N1. *CDC Lab Safety-Related Reports and Findings.* http://www.cdc.gov/about/lab-safety/reports-updates.html.
16. **Centers for Disease Control and Prevention.** 2011. Fatal laboratory-acquired infection with an attentuated yersinia pestis strain-Chicago, Illinois, 2009. *MMWR* **60:**201–205.
17. **Barry MA.** 2005. *Report of Pneumonic Tularemia in Three Boston University Researchers.* Boston Public Health Commission, Boston, MA.
18. **Gaudioso J, Caskey SA, Burnett L, Heegaard E, Owens J, Stroot P.** 2009. *Strengthening Risk Governance in Bioscience Laboratories.* Sandia Report: SAND2009-8070. Sandia National Laboratories, Albuquerque, NM.
19. **Salerno RM.** 2015. Three recent case studies: the role of biorisk management, p 191–204. *In* Salerno RM, Gaudioso J (ed), *Laboratory Biorisk Management: Biosafety and Biosecurity.* CRC Press, Boca Raton, FL.
20. **Trevan T.** 2015. Rethink biosafety. *Nature* **527:**155–158.
21. **Burnett L, Olinger P.** 2015. Evaluating biorisk management performance, p 145–168. *In* Salerno RM, Gaudioso J (ed), *Laboratory Biorisk Management: Biosafety and Biosecurity.* CRC Press, Boca Raton, FL.
22. **Organisation for Economic Co-operation and Development (OECD) Environment Directorate.** 2008. *Guidance on Developing Safety Performance Standards related to Chemical Accident Prevention, Preparedness and Response*, vol 19, p 137. OECD Environment, Health, and Safety Publications, Paris.
23. **Caskey S, Sevilla-Reyes EE.** 2015. Risk assessment, p 45–64. *In* Salerno RM, Gaudioso J (ed), *Laboratory Biorisk Management: Biosafety and Biosecurity.* CRC Press, Boca Raton, FL.
24. **Gribble LA, Tria ES, Wallis L.** 2015. The AMP model, p 31–44. *In* Salerno RM, Gaudioso J (ed), *Laboratory Biorisk Management: Biosafety and Biosecurity.* CRC Press, Boca Raton, FL.
25. **Ned-Sykes R, Johnson C, Ridderhof JC, Perlman E, Pollock A, DeBoy JM, Centers for Disease Control and Prevention.** 2015. Competency guidelines for public health laboratory professionals. *MMWR Surveill Summ* **64:**1–81.
26. **European Committee for Standardization (CEN).** 2011. Biosafety Professional Competence, CWA 16335:2011.
27. **Committee on Developments in the Science of Learning.** 2000. *How People Learn: Brain, Mind, Experience, and School: Expanded Edition.* National Academies Press, Washington, D.C.

28. **Martínez-Córcoles M, Gracia F, Tomás I, Peiró JM.** 2011. Leadership and employees' perceived safety behaviours in a nuclear power plant: a structural equation model. *Safety Sci* **49:**1118–1129.
29. **Gaudioso J, Boggs S, Griffith NK, Haddad H, Jones L, Khaemba E, Miguel S, Williams CV.** 2015. Rethinking mitigation measures, p 87–99. *In* Salerno RM, Gaudioso J (ed), *Laboratory Biorisk Management: Biosafety and Biosecurity.* CRC Press, Boca Raton, FL.
30. **Richmond JY, Jackman J.** 2015. *Anthology of Biosafety XIV: Sustainability.* American Biological Safety Association, Mundelein, IL.
31. **Luria G.** 2010. The social aspects of safety management: Trust and safety climate. *Accid Anal Prev* **42:**1288–1295.
32. **Choudry R, Fang D, Mohamed S.** 2007. Developing a model of construction safety culture. *J Manag Eng* 23:207–212.
33. **DeJoy DM, Schaffer BS, Wilson MG, Vandenberg RJ, Butts MM.** 2004. Creating safer workplaces: assessing the determinants and role of safety climate. *J Safety Res* **35:**81–90.
34. **Arezes PM, Miguel AS.** 2008. Risk perception and safety behaviour: A study in an occupational environment. *Safety Sci* **46:**900–907.
35. **Makvandi M, Shigematsu M.** 2015. Communication for biorisk management, p 169–190. *In* Salerno RM, Gaudioso J (ed), *Laboratory Biorisk Management: Biosafety and Biosecurity.* CRC Press, Boca Raton, FL.
36. **Allen S, Chiarella M, Homer CSE.** 2010. Lessons learned from measuring safety culture: an Australian case study. *Midwifery* **26:**497–503.
37. **World Health Organization.** 2006. *Biorisk Management: Laboratory Biosecurity Guidance,* WHO/CDS/EPR/2006.6. World Health Organization, Geneva.
38. **European Committee on Standardization (CEN).** 2012. CEN Workshop Agreement (CWA) 16393:2012—Guidelines for the Implementation of CWA 15793:2008. European Committee on Standardization, CWA 16393:2012.

26 Occupational Medicine in a Biomedical Research Setting

JAMES M. SCHMITT

The purpose of an occupational medicine program is to promote a safe and healthy workplace through the provision of work-related medical services. In a biomedical research setting that involves biohazardous materials, those services should include a preplacement medical evaluation, job-specific counseling and immunizations, and a practical plan for responding to suspected exposures to workplace health hazards and caring for work-related injuries. Before discussing these core elements of an occupational medical program, a review of the prerequisites for these services is in order.

It is essential that the occupational medicine health care provider be an integral part of the institution's occupational health and safety team and be familiar with applicable guidance (1, 2). At many institutions, the occupational medical function is provided in concert with employee health and wellness services and/or student health services. Although this approach may prove satisfactory, care must be taken to avoid compromising the focus of the occupational medical responsibilities or weakening the requisite bond with the institution's safety specialists.

As an active member of the institution's health and safety team, the health care provider should be aware of the potential health hazards present in the workplace. The services offered by the institution's health care provider should be tailored to meet the organization's needs and based upon a detailed, protocol-driven risk assessment. The assessment must address the unique aspects of the project, such as the nature of the infectious agent and the animal model used, as well as how the infectious agent is handled (see chapter 5 on risk assessment). The occupational medical services should be designed by the health care provider in consultation with representatives from the safety program and principal investigators. This approach increases the likelihood that the services properly address workplace health hazards and maximize eventual participation in the program. The resulting medical services should be provided for all personnel regardless of employment status. Work sites with multiple employers may find it challenging to accomplish this goal. Contracted workers, students, guest workers, etc., should be provided with access to medical care equivalent to that provided by the host institution for its own employees. Institutional policies, agreements, and contract language should reflect this need.

PREPLACEMENT MEDICAL EVALUATIONS

A preplacement medical evaluation is recommended for individuals who are at risk for exposure to human pathogens, including zoonotic agents, at work. The evaluation may include a questionnaire or a medical interview that identifies current and past medical conditions, current medications and allergies, and prior immunizations. Individuals who will need to wear a respirator must complete an initial questionnaire for respirator use (3). Regardless of the mechanism used to gather the worker's medical history, the health care provider should assess the individual's general state of health and previous immunizations with the worker's proposed work responsibilities in mind. In general, the preplacement medical evaluation should not involve a physical examination. A hands-on physical examination typically does not offer any practical value to a preplacement medical evaluation and usually is not warranted. It may be prudent perform a behavioral health screen for individuals who will work in Biological Safety Level (BSL) -3 and -4 laboratories. The National Institutes of Health has implemented this approach in hopes of building a safety culture that promotes trust, respect, and reliability (4).

Laboratory testing usually is not a necessary ingredient of the preplacement medical evaluation. Exceptions to this general rule include: testing workers, who will have contact with nonhuman primates, for prior infection with *Mycobacterium tuberculosis* (MTB) and for immunologic protection to rubeola, and testing women of child-bearing capacity, who will work with cats, for prior infection with *Toxoplasmosis gondii* (5). Testing for antibodies to hepatitis B-surface antigen is warranted only within two months of completing the vaccine series (6). Pulmonary function testing is not required as part of the medical clearance for respirator use, and rarely is there a clinical indication for testing as part of the medical assessment. Similarly, there is no sound occupational medical basis for routinely checking a position-applicant's chemistry values, blood counts, or serologic evidence of protection to infectious agents that are not found in the workplace. Laboratory screening for evidence of allergies to laboratory animal proteins is not clinically useful (7, 8). Similarly, routine storage of serum during the preplacement medical evaluation is not recommended (9).

When available, and clinically indicated, immunizations for hazards in the workplace should be provided, including hepatitis B immunization for those who may be exposed to human body fluids. In addition to immunizing workers for infectious agents that may be encountered in the workplace, it may be advisable to offer immunization for influenza for workers conducting research with highly pathogenic strains of influenza virus. In some situations, a noncommercial investigational new drug vaccine may be offered; however, receipt of an investigational new drug vaccine cannot be required (10).

Counseling provided during the preplacement medical evaluation is the most valuable part of the medical visit. Ideally, the counseling is tailored to address the potential health hazards in the work area. These hazards may include infectious agents studied in the laboratory, zoonotic illnesses associated with animals used in the research, allergies to animal proteins, and specific chemical hazards. The health care provider should describe the earliest signs and symptoms suggestive of a laboratory-acquired infection and allergies to animal proteins. The counseling should stress that workers must promptly report all work-related injuries and suspected illnesses and should include a detailed description of the steps the worker should take in the event of a recognized exposure, including appropriate first aid measures and the process for accessing medical attention promptly, regardless of when the accident occurs. Ideally, the health care provider should supplement the counseling with printed handouts or electronic documents, for future reference.

Given the nature of the work and access restrictions to BSL-3 and -4 laboratories, individuals medically cleared to work in the laboratories should not have any medical conditions that could result in an altered state of consciousness, impaired judgment or concentration, or inability to utilize personal protective equipment (PPE) or to perform the physical requirements of the position. When a question regarding a worker's ability to meet this requirement, the health care provider should withhold medical clearance until the worker can provide medical records that establish that the medical condition is sufficiently controlled to permit him or her to enter the work area.

ROUTINE PERIODIC MEDICAL EVALUATIONS

In general, workers do not need to return to the occupational medical provider on a regular basis. For example, the interval medical questionnaire for respirator use may be administered by the person providing the training and respirator fit testing. Only workers who indicate that they have a positive response to one of the seven questions on the questionnaire need to be evaluated by a health care provider. There are a couple of exceptions to the assertion that routine medical evaluations are not needed. Workers who have contact with nonhuman primates and have not been previously infected with MTB should return annually for periodic skin or serologic testing to confirm that they have not been infected in the interim. This practice is intended to protect the nonhuman primates from inadvertent infection with MTB (11). Annual recall is also recommended for workers that have access to infectious agents and toxins handled in BSL-3 and -4 laboratories. These visits permit the health care provider to update his or her understanding of the worker's role in the research and the worker's medical history. The visit also provides an opportunity to review the counseling provided during the preplacement medical evaluation and to ask about injuries and illnesses that occurred in the prior year (1, 12, 13).

The health care provider should have an effective mechanism for recalling workers to complete and boost immunizations that are justified by workplace health hazards. In addition, the health care provider should periodically remind workers of potential health hazards in their work environments (for example allergenic animal proteins), first aid measures, and the steps needed to access emergency medical assistance for occupational injuries.

MEDICAL CARE FOR OCCUPATIONAL INJURIES AND ILLNESSES

Employers should insist that all occupational injuries and any suspected exposures or occupational illnesses be reported to the institution's designated occupational medical provider. The health care provider must be aware not only of the potential health hazards in the workplace but also must have a detailed understanding of the various clinical manifestations of infection with the biohazards and remain alert for subtle evidence of a work-acquired infection. Because the mechanism of exposure to biohazards in a biomedical research environment may vary from mechanisms of natural environmental exposures, the clinical signs and symptoms of infection may be unusual. To properly handle such challenging cases, the clinician must have immediate access to appropriate subject matter experts and infectious disease consultants.

Workers should be able to access the designated health care provider, without delay, following a suspected exposure, so that evaluation and treatment can be provided in a timely fashion. The institution should review the provision of medical support services to identify and minimize barriers to prompt patient evaluation and treatment. Such barriers may include limited clinic hours, staffing limitations, transportation concerns, access to pharmaceutical supplies, and lack of an integrated safety and health approach.

First aid measures should be defined in advance, and related information should be widely disseminated to employees who may need to respond to a potential exposure. First aid materials, as appropriate, should be made immediately available at the work site. Suspected exposures to some health hazards encountered in biomedical research facilities warrant immediate chemoprophylaxis. Examples of such hazards include the neurotoxin MPTP (1-methyl-4-phenyl-1,2,3,6-tetrahydropyridine), hydrofluoric acid, phenol, and *Macacine herpesvirus 1* (formerly Cercopithecine herpesvirus 1 [CHV-1], B virus) (14, 15). The strategies for responding to such exposures must be designed in advance of the incident, shared with the workgroups, and practiced. Following the provision of work-site first aid, the worker should proceed directly to the health care provider for further evaluation. The treating facility, which may be the occupational medical clinic or designated local health care facility, should be prepared to provide definitive medical care appropriate to the exposure incident or illness.

In assessing the significance of an incident involving an infectious agent, the health care provider should consider separately the risk of exposure (RoE) and the subsequent risk of disease (RoD). This approach helps organize the assessment and clarify related communications.

The RoE is an estimate of the likelihood that a worker sustained an exposure to an infectious agent. Two conditions must be met for a worker to be at risk for an exposure to an infectious agent. First, a biologically active infectious agent must be present. Second, the worker's PPE or innate protection (for example intact skin) must be breached. The worker, safety specialist, principal investigator (PI), and health care worker work cooperatively to determine the RoE. Table 1 stratifies RoE and provides additional details.

Occasionally the circumstances of an accident warrant an estimate of the RoE for more than one infectious agent. For example, a percutaneous injury that occurs while dissecting neurologic tissue from a macaque known to be infected with B virus and experimentally infected with an infectious agent would warrant separate estimates of RoE to B virus and to the infectious agent. Similarly, a bite injury involving an animal infected experimentally with an infectious agent would warrant an assessment for the RoE to animal's oral biota and a second estimate for the RoE to the infectious agent.

The RoD is an estimate of the probability that an exposure will result in an illness or a measurable immunologic response to the infectious agent in the expected timeframe following the incident. This estimate is made by the health care provider in consultation with infectious disease specialists and subject matter experts. The estimate is shared with the injured worker but not with safety specialists or other institutional officials. This risk is stratified in a similar fashion to the RoE (i.e., no, negligible, minimal, moderate, and high risk). The RoD frequently is the same as the RoE; however, the estimate

TABLE 1.

Conditions that determine the RoE[a]

Risk of exposure level	Description
No Risk RoE = 0	1. Either the infectious agent was not present; or 2. There was no plausible route of exposure (e.g., the worker's PPE or intact skin was not breached, or it is not reasonable to suspect that the worker could have sustained an exposure to the contaminated fluid/material).
Negligible Risk RoE = 1	1. The infectious agent may have been present; and 2. The worker's PPE was breached. 3. The risk is too small to quantify, but cannot be excluded.
Minimal Risk RoE = 2	1. The infectious agent may have been present; and 2. The worker's PPE was breached and the worker may have sustained an exposure. 3. The risk is theoretically possible, but is unlikely.
Moderate or High Risk RoE = 3 or 4	1. The infectious agent likely was present; and 2. The worker's PPE was breached and the worker possibly or likely sustained an exposure to the infectious agent.

[a]The estimated RoE is based on the likelihood that a biologically active infectious agent was present and the worker's personal protective equipment or innate protection was compromised in the event.

may be higher or lower based on the following factors (see Table 2 for additional details):

- the virulence of the infectious agent
- the volume (size of inoculum) and concentration of the infectious agent in fluid
- the route of exposure
- the adequacy of the first aid (e.g., timeliness, technique, duration, agent used)
- the worker's protection to the infectious agent from prior immunization or infection
- the worker's medical conditions and treatments
- the adequacy of postexposure chemoprophylaxis (e.g., timeliness, effectiveness)

TABLE 2.

Factors that affect the RoD

Virulence	Is the bioagent a known human pathogen (e.g., simian immunodeficiency virus [SIV], a close relative of HIV-2, is known to infect humans without causing illness)? How likely is it to cause disease in humans (e.g., only 10% of people with latent TB infection develop tuberculosis)? How likely is it to cause severe health consequences (e.g., case-fatality rates of influenza strains differ significantly)? Has the infectious agent been modified to enhance or diminish the parent strain's pathogenic potential (e.g., nonvirulent Sterne strain of *B. anthracis*)?
Volume and concentration	What was the estimated dose of exposure (e.g., decreased size of inoculum if the needle passes through gloves first, the entire bevel of the needle does not penetrate the worker, or the syringe plunger is not depressed during injury)? How much viable pathogen may have been introduced (e.g., was source material chemically or physically inactivated prior to the incident)? What is the agent's minimum infectious dose in a healthy host?
Route of exposure	How likely is it that the incident (e.g., spill or splash) led to a plausible exposure? Was there inhalational or mucosal contact or non-intact skin contact with agent?
First aid	How much time passed from exposure to starting cleansing the affected body part (immediate cleansing of agent may reduce disease risk)? Was first aid delivered appropriately (e.g., duration, technique, and use of an appropriate disinfectant)?
Pre-exposure protection	Was the individual vaccinated against the infectious agent? Does the individual have protective antibody titers? How effective is the vaccine (prior vaccination may lower the disease risk)?
Worker's medical conditions and treatment	Does the worker have an illness or take medications that increase the risk for disease (immune suppressing medical conditions and medications) or complications (valvular heart disease and *C. burnetii*)?
Postexposure medical countermeasures	Are pharmacologic or immunologic agents available that are known to be effective against the bioagent (e.g., antibiotics, antivirals or specific IgG therapy)? Are there any investigational medical treatments available (e.g., a vaccine, serum from survivors of a disease)?

Decisions on whether to offer treatment can be linked to the RoD, presuming treatment is available and there are no significant clinical contraindications to administering it. Treatment is not offered when there is "no" or "negligible" (too low to calculate) risk. If the risk is considered to be "minimal," treatment decisions are made on a case-by-case basis. Treatment, when available, is encouraged when the risk it thought to be either "moderate" or "high." Treatment, when indicated, should be provided promptly, and there should be an agreed-upon plan to monitor the individual's clinical course.

Ongoing clinical assessment may involve additional laboratory testing. In some circumstances, it may be appropriate to utilize tests that are not commercially available. If noncommercial or unlicensed tests are going to be utilized in evaluating a potential exposure, the health care provider should submit the patient's specimens and appropriate internal negative controls to the testing laboratory in a blinded fashion. The laboratory providing the testing service should also run positive and negative control samples and report results for all samples to the requesting health care provider. The health care provider must be circumspect in the interpretation of noncommercial, unlicensed laboratory test results because, in many cases, he or she may not have information regarding the predictive value of the test used or the sensitivity and/or specificity of the assay employed.

If the injury involves an infectious agent in a BSL-3 or -4 laboratory and advanced medical care is warranted, the health care provider must have a strategy for safely transporting the worker to an appropriate facility and transferring clinical responsibility for the injured or ill worker to an infectious disease specialist (16, 17). The transportation team and receiving facility will require notification of the need and briefing on appropriate PPE for those involved. The health care provider will need to keep the worker fully informed of the steps that are being taken to ensure safe transport and care and respond to the worker's needs (e.g., alerting family members, arranging for children to be picked up from school, etc.).

OTHER ASPECTS OF AN OCCUPATIONAL MEDICAL SUPPORT SERVICE

Training and Drills

The health care provider and his or her staff should regularly review the infectious agents used in the protocols the providers support. Emergency response plans should be reviewed and revised regularly. The procedures for infectious agents handled in BSL-3 and -4 laboratories should be updated annually. In addition, the emergency medical response plan should be reviewed and drills conducted annually. The training and drills ideally will involve all parties likely to be involved in an incident response. Exposure drills and mock incident response activities are useful in ensuring that appropriate and timely medical services are provided in the event of a worker exposure (2). The most frequent finding from such drills is that communications could be improved. Finally, each institution should collect accident, injury, and exposure data; review the data at least annually; and include a review of the appropriateness of the medical support services provided.

Communications

The practice of occupational medicine is unique in that the practitioner is responsible to both the employee-patient and the worker's employer. The practice can reasonably be considered as the provision of public health services, including patient care responsibilities, in a work setting. Although the health care provider's first responsibility is to safeguard the health and safety of the worker, he or she is also responsible for minimizing the chance of significant injury to others at the workplace. Communications from the health care provider to the employer are constrained by these often-competing needs. Typically, the clinician strikes a balance by restricting communications to a description of a worker's functional restrictions and their anticipated duration. This approach provides the employer with administrative actionable information without compromising the worker's right to maintain the confidentiality of his or her medical condition. This approach is utilized in the preplacement medical evaluation, as well as in subsequent visits for work-related medical concerns and for return to work following a personal medical concern.

The health care provider must have a clear plan for communicating with responsible officials when a worker is not medically cleared to work in a BSL-3 or -4 laboratory or sustains a potential exposure to an infectious agent or develops symptoms suggestive of a laboratory-acquired infection. The health care provider must define the RoE level that will precipitate a call, who will receive notice, and what information will be shared. Typically, an injury with a negligible risk of exposure to an infectious agent (RoE = 1) will only be shared with the safety specialist and principal investigator. Incidents with a RoE greater than 1 typically trigger prompt notification of the lead safety specialist, the director for the laboratory, the office responsible for public communications, and the county public health officer. The following information is shared with those officials: the circumstances of the incident, the route of exposure, PPE used at the time of the accident, the timeliness and nature of the first aid provided, the infectious-disease consultants' consensus-estimate of the risk that the potential exposure will result in a risk to public health, and the plan for mitigating the risk to others, such as isolation and monitoring or transportation to an appropriate health care facility.

Medical Record Keeping and Workers' Compensation

The employer is responsible for maintaining employees' work-related medical records just as the records would be maintained in a conventional medical practice. The records should adequately reflect all of the medical care provided, including the medical history obtained by the practitioner, the findings on physical examination and diagnostic testing, the clinician's assessment of that information, and the treatment recommended and provided. The Occupational Safety and Health Administration (OSHA) requires that medical records containing information regarding exposures be retained for the length of employment plus 30 years. The OSHA standards on blood-borne pathogens (18) and access to employee exposure and medical records (19) should be consulted for a full discussion of record-keeping requirements and the circumstances in which they apply. In addition, all OSHA 300 logs, recording reportable injuries and illnesses, must be retained for 5 years and, during this storage period, must be updated to include newly discovered information (20).

Serum Storage

The health care provider may wish to store a worker's serum following a recognized or suspected exposure to a significant human pathogen in the workplace. The purpose of storing workers' sera is to permit the health care provider to perform serological testing of paired acute and convalescent serum specimens to determine whether or not the worker has been infected. If the health care provider offers this service, the sera should be stored at −20°C or lower in a non-self-defrosting freezer. Access to the sera should be strictly limited to maintain the workers' right to privacy. Any testing of the sera, when the worker could be identified, should be permitted only with the worker's informed, written consent. The testing methodology and findings should be recorded in the worker's medical record.

RESEARCH INVOLVING ANIMALS

Individuals working with animals in a laboratory setting should be aware of the potential health risks posed by animal proteins, as well as zoonotic hazards associated with animals used in the research (21).

It is estimated that 20 to 30% of individuals who work with laboratory animals will develop allergic symptoms. One in twenty workers with allergies to animal proteins will develop asthma as a result of their contact with laboratory animals. The proteins associated with these allergic reactions are found in the animals' urine, saliva, and dander (13). Most animals used in research have been identified as the source of workers' allergy symptoms. Because mice and rats are the animals most frequently used in research studies, there are more reports of allergies to rodents than to other laboratory animals. A personal history of allergies to other animals (typically cats and dogs) is the best predictor for who will develop an allergy to animals found in research laboratories. Other factors associated with allergic reactions to laboratory animals include the intensity, frequency, and route of exposure to animal proteins. See also animal allergens in chapter 15 of this book. Activities, such as handling animals and cleaning their cages, may be associated with an increased risk of exposure to the animal proteins, and thereby place the worker at greater risk of developing an allergic reaction. Inhalation exposures are particularly hazardous in this regard.

The earliest symptoms of an allergic reaction include nasal stuffiness; a "runny" nose; sneezing; red, irritated eyes; and hives. Symptoms suggestive of asthma include coughing, wheezing, and shortness of breath. Rarely, a worker with allergic symptoms will have an anaphylactic reaction following an animal bite. Most workers who develop allergic reactions to laboratory animals will do so within the first 12 months of working with them. Infrequently, reactions occur only after working with the animals for several years. Initially, the symptoms are present within minutes of the worker's exposure to the animals. Approximately half of allergic workers will have their initial symptoms subside and then recur 3 to 4 hours following the exposure.

The best approach for reducing the likelihood that a worker will develop an allergic reaction, or to control it once it has occurred, is to eliminate or minimize the exposure to animal proteins (22). Desensitization injections are not particularly effective. Medications may control allergy symptoms; however, the goal should be to eliminate opportunities for workers to inhale or have skin contact with those proteins. In addition to using well-designed air handling and waste management systems (e.g., chemical fume hood, biosafety cabinet, or downdraft table), workers should reduce their risk of exposure by routinely using appropriate personal protective equipment, such as dust/mist masks, gloves, and gowns.

Animals used in research may harbor a variety of zoonotic biohazards, including, among others, viruses, rickettsiae, and bacteria. In addition to naturally occurring infections, these animals may be intentionally exposed to other infectious agents as part of the research or testing protocol. Although transmission of these zoonotic agents in research laboratories is uncommon, some laboratory-acquired infections, such as B-virus (cercopithecine herpesvirus 1), lymphocytic choriomeningitis virus, and *Coxiella burnetii*, may have devastating consequences for the laboratory worker. As a result, it is imperative that the institution's health care provider be familiar with any zoonotic hazards that may be present in the workplace and have an understanding of subtle manifestations of a laboratory-acquired infection. A thorough review of these health hazards is beyond the scope of this chapter and can be found in other reference texts.

References

1. **Centers for Disease Control and Prevention and National Institutes of Health.** 2009. *Biosafety in Microbiological and Biomedical Laboratories*, 5th ed. U.S. Government Printing Office, Washington, D.C.
2. **Centers for Disease Control and Prevention (CDC) Division of Select Agents and Toxins and Animal and Plant Health Inspection Service (APHIS) Agriculture Select Agent Program.** 2013. Occupational Health Program Guidance Document for Working with Tier 1 Select Agents and Toxins. 7 CFR Part 331, 9 CFR Part 121, 42 CFR Part 73.
3. **Department of Labor. Occupational Safety and Health Administration (OSHA).** Respiratory Protection Standard 1998. 29 CFR 1910.134 https://www.osha.gov/pls/oshaweb/owadisp.show_document?p_table=STANDARDS&p_id=12716
4. **Skvorc C, Wilson DE.** 2011. Developing a behavioral health screening program for BSL-4 laboratory workers at the National Institutes of Health. *Biosecur Bioterror* **9:**23–29.
5. **Institute for Laboratory Animal Research.** 2003. *Occupational Safety in the Care and Use of Nonhuman Primates*. National Research Council, National Academy Press, Washington, D.C.
6. **Schillie S, Murphy TV, Sawyer M, Ly K, Hughes E, Jiles R, de Perio MA, Reilly M, Byrd K, Ward JK. Centers for Disease Control and Prevention.** 2013. Guidance for Evaluating Health-Care Personnel for Hepatitis B Virus Protection and for Administering Postexposure Management—Recommendations and Report, December 20. *MMWR Morb Mortal Wkly Rep* **62** (RR-10):1–19.
7. **Nicholson PJ, Mayho GV, Roomes D, Swann AB, Blackburn BS.** 2010. Health surveillance of workers exposed to laboratory animal allergens. *Occup Med* **60:**591–597.
8. **Ferraz E, Arruda LK, Bagatin E, Martinez EZ, Cetlin AA, Simoneti CS, Freitas AS, Martinez JAB, Borges MC, Vianna EO.** 2013. Laboratory animals and respiratory allergies: the prevalence of allergies among laboratory animal workers and the need for prophylaxis. *Clinics (Sao Paulo)* **68:**750–759.
9. **Lehner NDM, Huerkamp MJ, Dillehay DL.** 1994. Reference serum revisited. *Contemp Top Lab Anim Sci* **33:**61–63.
10. **Rusnak JM, Kortepeter MG, Hawley RJ, Boudreau E, Aldis J, Pittman PR.** 2004. Management guidelines for laboratory exposures to agents of bioterrorism. *J Occup Environ Med* **46:** 791–800.
11. **Centers for Disease Control and Prevention (CDC).** 1993. Tuberculosis in imported nonhuman primates—United States, June 1990-May 1993. *MMWR Morb Mortal Wkly Rep* **42:**572–576.
12. **Stave GM, Darcey DJ.** 2012. Prevention of laboratory animal allergy in the United States: a national survey. *J Occup Environ Med* **54:**558–563.
13. **Bush RK, Stave GM.** 2003. Laboratory animal allergy: an update. *ILAR J* **44:**28–51.
14. **Gupta A, Dhir A, Kumar A, Kulkarni SK.** 2009. Protective effect of cyclooxygenase (COX)-inhibitors against drug-induced catatonia and MPTP-induced striatal lesions in rats. *Pharmacol Biochem Behav* **94:**219–226.
15. **Cohen JI, Davenport DS, Stewart JA, Deitchman S, Hilliard JK, Chapman LE, B Virus Working Group.** 2002. Recommendations for prevention of and therapy for exposure to B virus (cercopithecine herpesvirus 1). *Clin Infect Dis* **35:**1191–1203.
16. **Rusnak JM, Kortepeter MG, Aldis J, Boudreau E.** 2004. Experience in the medical management of potential laboratory exposures to agents of bioterrorism on the basis of risk assessment at the United States Army Medical Research Institute of Infectious Diseases (USAMRIID). *J Occup Environ Med* **46:**801–811.
17. **Henkel RD, Miller T, Weyant RS.** 2012. Monitoring select agent theft, loss and release reports in the United States—2004–2010. *Appl Biosaf* **17:**171–180.
18. **Code of Federal Regulations.** 1992. Title 29. Labor. Chapter XVII. Occupational Safety and Health Administration. Subpart Z. Toxic and hazardous substances. Bloodborne pathogens. *29 CFR* 1910.1030. http://www.ecfr.gov/cgi-bin/text-idx?SID=b9621763e037d7871af053893e122319&mc=true&node=se29.6.1910_11030&rgn=div8
19. **Code of Federal Regulations. Title 29. Labor. Chapter XVII. Occupational Safety and Health Administration.** 1996. Subpart Z. Toxic and hazardous substances. Part 1910. Access to employee exposure and medical records. *29 CFR* 1910.1020. http://www.ecfr.gov/cgi-bin/text-idx?SID=b9621763e037d7871af053893e122319&mc=true&node=se29.6.1910_11020&rgn=div8
20. **Code of Federal Regulations. Title 29. Labor. Chapter XVII. Occupational Safety and Health Administration.** 2001. Recording and reporting occupational injuries and illnesses. *29 CFR* 1904. http://www.ecfr.gov/cgi-bin/retrieveECFR?gp=&SID=2ada6b8d00ee27d013de97cfb51d4ac0&r=PART&n=29y5.1.1.1.4
21. **Institute for Laboratory Animal Research.** 1997. *Occupational Safety in the Care and Use of Research Animals*. National Research Council, National Academy Press, Washington, D.C.
22. **Gordon S, Preece R.** 2003. Prevention of laboratory animal allergy. *Occup Med (Lond)* **53:**371–377.

Measuring Biosafety Program Effectiveness

27

JANET S. PETERSON AND MELISSA A. MORLAND

BIOSAFETY PROGRAM COMPONENTS

A biosafety program consists of many components, which are determined by the type of research being conducted at the institution as well as by current regulations and guidelines. A biosafety program may include oversight of blood-borne pathogens, research involving infectious materials or recombinant or synthetic nucleic acid molecules (rDNA), biosafety cabinet (BSC) certifications, and/or high-containment laboratories and select agents. Many institutions will not have all elements, but a means is needed to evaluate the success of each component present.

Blood-Borne Pathogens

Laboratories working with human materials, such as blood and unfixed tissue, are required to comply with the Occupational Safety and Health Administration's (OSHA) Bloodborne Pathogen Standard (1). Effectiveness of biosafety programs covering such labs can be assessed by evaluating compliance with requirements of the OSHA standard. This can be accomplished by annual site visits to evaluate adherence to biosafety level 2 (BSL2) containment criteria. A checklist that may be used to evaluate BSL2 laboratories can be found in Appendix A. The biosafety manager should also conduct a records review to ensure that administrative requirements of the regulations are being met. This should include review of training records to document that annual training on preventing occupational transmission of blood-borne pathogens has been provided to individuals in all job classifications identified in the exposure determination table, as well as documentation that these individuals have been offered the hepatitis B vaccination series. In addition, an annual review and update of the facility's written Exposure Control Plan should be conducted.

Wild-Type Infectious Agents

Evaluation of the effectiveness of biosafety programs for laboratories working with infectious agents that are not genetically modified and therefore not subject to the *NIH Guidelines for Research Involving Recombinant or Synthetic Nucleic Acid Molecules* (2) can be measured by site visits conducted when work is initiated and annually thereafter. Site visits can assess the lab's compliance with the containment level at which the research has been approved. An example of a checklist that may be used to evaluate BSL2 laboratories may be found in Appendix A. As in the case of blood-borne pathogens, the biosafety professional must conduct a program audit as well as a site visit. The program audit should include review of training records to confirm that all personnel have been trained in biosafety concepts. Perhaps one of the most important aspects of the program audit is confirmation that the research has met the institution's requirements for Institutional Biosafety Committee (IBC) or biosafety manager oversight.

Institutional Biosafety Committees

An in-depth review of IBC effectiveness can be conducted using the self-assessment tool provided by the NIH Office of Biotechnology Activities (OBA) (http://osp.od.nih.gov/office-biotechnology-activities). Although NIH OBA also conducts site visits to review IBCs, they are unable to visit all registered IBCs on a regular basis; therefore, facilities need to make use of the online assessment tool. One criterion for evaluating the IBC is to confirm that adequate periodic training has been provided to IBC members, especially those who represent the community. An annual review of the IBC membership roster should be conducted to ensure it reflects the current membership of the committee and provides adequate expertise. Additionally, the IBC must be registered with the NIH OBA using the online IBC Registration Management System, and registrations must be updated at least annually. However, rosters should be updated continuously as members are added or removed from the committee. Last, the IBC charter should be reviewed and updated annually to check that it adequately addresses (i) the charge of the committee, (ii) processes for approval, denial, or suspension of work, and (iii) potential member issues such as conflict of interest.

Laboratories that work with rDNA should be evaluated for adherence to the requirements of the *NIH Guidelines for Research Involving Recombinant or Synthetic Nucleic Acid Molecules* (2). The NIH states in the *Guidelines* that the institution is responsible for periodically reviewing recombinant or synthetic nucleic acid research conducted at the institution to ensure compliance with the *NIH Guidelines* [Section IV-B-2-b-(5)]. Site visits to the lab, in the form of postapproval monitoring, play an important role in assessing the effectiveness of the IBC program. Site visits allow the biosafety team to verify that appropriate containment criteria are being used. The site visitor must be familiar with the approved protocol so they can confirm that the approved containment facility level, equipment, and practices are being followed. This visit also allows the biosafety staff a chance to talk with the research team about their project to make sure all vectors, genes of interest, and biological material are approved. It also allows the biosafety team to make sure that everyone working on the project is aware of the hazards of the project, knows what to do in the event of an emergency, and is aware of the need to report exposures or spills outside of containment. Last, the site visit can be used to follow up on past facility audits and confirm that corrective actions are in place. An example of a postapproving monitoring checklist can be found as Appendix B.

In addition, compliance with the *NIH Guidelines* can be reviewed administratively (see section below) to assess whether all recombinant DNA research conducted by the institution is being reviewed by the IBC. Training records should be reviewed to assess whether training on the requirements of the *NIH Guidelines* has been provided to all researchers. Laboratory containment criteria may be found in Appendix G of the *NIH Guidelines*.

Biosafety Cabinet Certification

An examination of the certification stickers on a BSC during a site visit will confirm if the cabinet has been certified within the past 12 months. Therefore, the site visit checklist should include a line for the date of the last certification. This information can also be obtained by a review of BSC certification records, which are often maintained in the biosafety office. Program effectiveness can be measured by comparing the number of BSCs that have undergone annual certification with the total number of units in the facility.

Occupational Health

Coordination between the occupational health physician and the biosafety team is a key component to a successful biosafety program. It is often the responsibility of the biosafety manager to alert the occupational health physician of the infectious agents being used in the institution. This allows the occupational health provider to be prepared to handle potential exposures, especially for diseases that would not be routinely diagnosed or treated in the area (not endemic). In addition, all researchers working with biological materials should be trained on emergency response and reporting procedures, because timely medical follow-up is crucial to the management of exposure to infectious agents. When work is conducted with high-risk pathogens, a medical surveillance program should be instituted. The elements of this program will be dependent on agents in use but may include training on signs and symptoms of disease, monitoring for symptoms (e.g., temperature logs), physicals, and/or infectious agent alert cards.

Evaluating the effectiveness of the occupational health portion of the biosafety program may include a review of the number of workplace accidents or near misses. Evaluating these incidents to identify a root cause will shape the biosafety program through application of improved practices and procedures. In addition, program effectiveness can be measured by periodic coordination with the occupational health clinic to confirm that it is up to date with the list of infectious agents in use at the institution.

High-Containment Laboratories

Facilities that include high-containment laboratories (BSL3 containment) should include these laboratories in an evaluation of their program's effectiveness. New BSL3 laboratories require verification of BSL3 facility design and operational parameters prior to operation. These parameters must be reverified annually to ensure that the facility containment features continue to operate as originally designed. Annual reverification is discussed later in this chapter (see chapter 14 for initial certification requirements). Requirements for maximum-containment laboratories (BSL4) are not discussed in this chapter.

Select Agent Program

The Select Agent Program (SAP) regulations (3, 4) require several types of program evaluations that can be used to assess the effectiveness of the program. The responsible official (RO) is responsible for ensuring that the facility conducts annual site visits to the select agent laboratories. Annual drills or exercises are also required to evaluate the effectiveness of security, biosafety, and incident response plans. In addition to conducting annual training and site visits, administrative responsibilities include maintaining records of inventory, access, and training, as well as suitability and occupational health clearance, if required. In addition to the program evaluations delegated to the facility, the SAP conducts site visits prior to issuance of a registration certificate and at least every 3 years thereafter. The results of these agency site visits are a good measure of the effectiveness of the SAP. The agency inspection is described in more detail elsewhere in this chapter.

EVALUATION MECHANISMS

Mechanisms used to evaluate biosafety program effectiveness can be grouped into two broad categories—direct observation (site visits) and program review (review of records). Both appear in each of the program components discussed above. Because site visits and records review play such an important role in measuring program effectiveness, each will be described in more detail below. See Table 1 for an overview of mechanisms that can be used to evaluate biosafety program components.

TABLE 1.

Evaluation mechanisms to measure effectiveness of biosafety program components

Program components	Evaluation mechanism			
Blood-borne pathogens	Training conducted annually	Training encompasses all job classifications identified in exposure control plan	Number of exposures is reviewed annually	Annual audit conducted of labs using human material
Wild-type infectious agents	Use of infectious agents is reviewed by IBC or other mechanism	All use of infectious agents (IA) is being captured	Research involving IA is conducted at appropriate containment level	Site visits are conducted annually
IBC/rDNA	OBA's IBC Self-Assessment Tool is used	Recommendations of OBA site visit, if any, are reviewed and implemented	Annual site visits are conducted by staff	IBC meetings are conducted in person or by teleconference
BSC management	All BSCs are certified annually	If not certified annually, BSCs are labeled as unsafe for use with infectious agents	BSCs are repaired promptly if they fail certification	Certification of BSC is checked at annual site visit
High-containment labs	Are audited annually	Are reverified annually	Have site-specific lab manual	Personnel are trained annually
Select Agent Program	Results of internal audits (e.g., biosafety and security) as well as regulatory inspection reports are reviewed and implemented	Labs are audited annually by biosafety team	Annual training is provided for personnel	Annual tabletop drill is conducted

BBPs, blood-borne pathogens; IBC, Institutional Biosafety Committee; rDNA, recombinant DNA or synthetic nucleic acid molecules; OBA, Office of Biotechnology Activities; BSC, biosafety cabinet.

Laboratory Site Visits

Site visits are the most common method used to evaluate the effectiveness of a biosafety program. During a site visit, lab practices, procedures, and the laboratory facility are reviewed. On-site evaluations may also provide information on the location of new equipment or addition of new personnel or uncover a new line of research needing IBC review. Site visits may be conducted by the biosafety manager, by regulatory agencies, by outside consultants, or as a self-audit. Site visits may also be conducted to verify that a new laboratory meets the criteria to which it was designed.

A standard checklist should be used for site visits and should be provided to the principal investigator (PI) or lab manager in advance of the visit. This allows laboratory personnel to become familiar with the criteria to be evaluated and provides the opportunity to make needed changes before the visit. In addition, some facilities find it useful to provide a brief point-by-point document that explains in more detail the information that is being sought, although this is a matter of personal preference. The checklist may contain criteria based both on regulations and best practices.

It is becoming more widespread for biosafety professionals to emphasize safe laboratory practices and to help the researcher achieve them rather than holding to strict compliance with regulations. The biosafety manager's job is to make the distinction of when to be flexible and when the potential risk level is too high for flexibility. Records for site-specific training provided by the laboratory director, vaccination records (such as the mandatory offering of the hepatitis B vaccination for workers who handle human blood), and equipment records (such as annual certification of Class II BSCs) should also be checked during the site visit. The auditor should comment on both the strengths and areas needing improvement; a negatively oriented site visit often will not lead to constructive remediation. It is useful to try to address as many of the findings during the site visit as possible, as many of the negative observations are merely oversights of the lab, not a deliberate intent to condone unsafe conditions. Most importantly, if deficiencies are identified during a site visit, it is important to conduct follow-up surveys to confirm that deficiencies have been corrected. A lab site visit program that simply identifies gaps without timely follow-through is not an effective program. Therefore, it is critical to confirm and document that identified problems have been remedied. The following year's audit can be used to confirm that the changes are still in effect.

Types of Auditors

Regardless of the biological research conducted in the laboratory, the site visit can be conducted in several ways depending on who performs it. See Table 2 for a comparison of strengths and weaknesses of various types of auditors. Each program manager determines the type of auditor (biosafety manager, self-audit, or generalist; see section on laboratory audits below) that will be most effective for the program. If resources permit, the type of auditor may be varied from year to year.

TABLE 2.

Factors to consider when selecting auditors

Inspector	Strengths	Weaknesses
Biosafety manager	Technical knowledge Familiarity with agents, procedures	Time-consuming Multiple labs may need simultaneous audit
Self-inspection	Timely information Deficiencies easily corrected	Lack of knowledge to correct deficiencies Record keeping may be lax Inspections may be missed
Safety generalist	Timely dissemination of information to other safety professionals	Lack of in-depth knowledge May miss deficiencies
Outside consultant	Objectivity Technical knowledge	Cost Not familiar with institutional procedures

Biosafety Manager

The biosafety manager has access to IBC/research registration records, is familiar with regulations and guidelines as well as the location and use of biological agents in the facility, and is therefore the usual choice to conduct internal site visits to assess effectiveness of the biosafety program.

Self-Audits

The laboratory manager's periodic self-audit is useful because immediate corrections and familiarity with the facility will minimize the potential for a laboratory incident. In this type of audit, the checklist is usually provided by the biosafety manager and may contain other areas of lab safety, such as chemical and radiation safety. Following the self-audit, the lab manager returns the completed form to the biosafety manager, who may provide follow-up if any areas are found to be unsafe.

Safety Generalist

Regularly scheduled visits can be carried out by representatives from the safety office on behalf of the biosafety manager. The team concept in which various members of the safety staff provide a general safety audit, a portion of which is devoted to biosafety, is a useful alternative to frequent laboratory visits by the safety program managers, who may be tied up with more technical and administrative matters. Conducting audits simultaneously is a more efficient use of the researchers' time, and is often seen by them as preferable to multiple individual site visits.

Outside Consultant
An outside consultant may be retained to review and evaluate the effectiveness of the biosafety program. This consultant may be hired to conduct the site visits within the facility on behalf of the IBC and biosafety manager. An outside vendor/consultant/contractor is the one who is often chosen to provide a "validation" type of inspection, such as commissioning of a new BSL3 facility that may require engineers and other specialists in addition to a biosafety professional as part of the outside inspection team.

Peer Review
Biosafety program managers may occasionally ask a biosafety professional from a similar institution to conduct an in-depth peer review of their program. This individual will be provided in advance with a description of their program and during a site visit will talk to researchers, check program records, and assess adequacy of staffing and funding. Although this type of review is time-consuming, it can provide a fresh and objective view of the program.

Regulatory Agency
A different type of program evaluation is provided when a regulatory agency visits the institution to review its program. With the implementation of the Select Agent Program regulations (3, 4), this type of evaluation by the Centers for Disease Control and Prevention (CDC) or Animal Plant Health Inspection Service (APHIS) is becoming more common. A site visit by a regulatory agency can provide an opportunity to evaluate the program in the most compliance-based light. The biosafety manager can use this opportunity to evaluate and improve the program. An in-depth discussion of how to prepare for and benefit from an inspection by a regulatory agency is provided later in this chapter.

Types of Facilities

There will be variations in an evaluation program based on the type of facility in which the laboratory is located.

Academia
An academic biosafety program must evaluate teaching laboratories as well as research laboratories. Most undergraduate teaching laboratories perform experiments with well-characterized agents to illustrate key techniques or important procedures (e.g., a teaching laboratory may work with *Bacillus subtilis* to demonstrate a spore stain). During a site visit, the evaluation of teaching laboratories should focus primarily on the storage, proposed use, and disposal of biological agents. Instructors and teaching assistants must complete some formal biosafety training and notify the biosafety manager if new agents will be considered for classroom instruction. (For more information on teaching laboratories, see chapter 29.) Research laboratories often work with more complex biological systems, which pose a higher risk than teaching labs. The evaluation of research labs requires the auditor to be familiar with the research conducted in the lab, and with the risks it presents. Academic scientists are used to working more independently, with freedom to pursue their research wherever it leads. Therefore, it takes a tactful auditor to work with the researcher effectively to support the research, while helping the investigator work safely.

Industry
Industrial biosafety programs may have different issues, especially with regard to risks associated with working with large volumes of culture. Larger volumes of culture may require large-scale containment levels and additional safeguards to decrease the risk of an inadvertent release of biological material to the environment, as noted in Appendix K of the *NIH Guidelines* (2). Large-scale biosafety considerations are covered in more detail in chapters 30 and 31. Another factor in industrial biosafety programs is the existence of multiple sites within one company, and therefore the evaluation process may be multiplied in terms of time and resources. Other issues that need to be considered involve quality assurance and good work practices (e.g., ISO 14000) and increased government oversight, such as the FDA for pharmaceutical companies. However, there are many laboratories in industrial settings that do not culture large volumes or high concentrations of agents, and these laboratories may be audited like academic facilities.

Government
Federal government facilities are not subject to the same level of surveillance as private industry, as these facilities are not under the jurisdiction of state or local regulatory agencies. Therefore, the internal evaluation process may need to be more frequent to ensure a safe and healthy environment. Some laboratory facilities on military installations may be subject to additional military regulations, especially with regard to access and security. State laboratory facilities (as well as private industry) must comply with local guidelines, and state safety standards will often be enforced by regulators at the city or county level.

The emphasis on the internal evaluation program should focus on those standards that would be enforced at the local level, since a local inspector can easily travel to a facility to conduct unannounced inspections.

Hospital and Health Care Facilities

Hospital and health care facilities operate clinical labs that require biosafety site visits. In these facilities, biosafety managers can reach out to the infection control office, which may provide a collaboration that will benefit both staff and patients. Good biosafety work practices are essential in clinical laboratories and in the handling of medical waste in the hospital setting. Disinfectant selection may also be an issue since harsh disinfectants may adversely affect patients and workers. These biosafety issues should be examined throughout the hospital on a regular basis. See chapter 37 of this volume for additional information on clinical labs.

Frequency of Audits

The frequency of the audit process will vary depending on the circumstances and the person conducting the audit. A formal site visit by the biosafety manager can be scheduled on an annual basis or more frequently if there are indications of a deficiency in safe work practices. In addition, as a prerequisite for IBC approval, the biosafety manager should visit the facilities of any new applicant to ensure that they meet the standards of the IBC and the organization's biosafety policy.

An urgent time for an internal evaluation of biosafety standard operating procedures may be before a major outside organization conducts a site visit. Examples of external organizations visiting laboratory facilities include the Joint Commission for Accreditation of Healthcare Organizations, the CDC, or the APHIS of the U.S. Department of Agriculture (USDA) conducting a select agent inspection.

An expedited evaluation may also be appropriate after a major health and safety accident or incident, such as personnel exposure or a spill of biological agents outside of containment. The public is wary of the credibility of any facility that has an accidental release of agents, and the adverse publicity will take time to overcome. The offices of environmental safety, public safety, or other groups that may have responded to the accident should take immediate action in investigating the incident and recommending actions to prevent a recurrence.

Administrative Review

One logical way to evaluate the components of the biosafety program is through review of records and documentation rather than a site visit. This is often the case for review of an institution's rDNA program and management of a BSC certification program, where records of rDNA registrations and BSC certifications may be reviewed.

Recombinant and Synthetic DNA

In accordance with the *NIH Guidelines for Research Involving Recombinant or Synthetic Nucleic Acid Molecules* (2), the IBC is responsible for the review and approval of most experiments involving the construction and use of genetically modified organisms. To evaluate whether the IBC or biosafety team is reviewing all rDNA research, the biosafety manager can review research registration files to confirm that all registration forms have been approved. Additionally, records can be reviewed to confirm that the institution's procedures for research registration renewal, updates, and/or amendments are being followed. Many institutions require that a research registration be renewed within a specified period of time, e.g., registration valid for 3 years with an annual review to determine whether the experiment or personnel have changed. Appendix C provides an example of an amendment form that may be used to notify the biosafety office/IBC that the experiment has changed. Another parameter that lends itself to administrative review is the biosafety incident log to confirm that all incidents involving rDNA were promptly reported to the NIH OBA. Incidents involving recombinant DNA must be reported to the OBA within 30 days or immediately if the accident occurs in a BSL2 or BSL3 facility. It may be helpful to review all of the accident reports sent to the NIH on an annual basis to determine if changes in administrative policy or procedures are warranted.

BSC Certification Program

The *NIH Guidelines* (along with the National Sanitation Foundation 49 standard and the *Biosafety in Microbiological and Biomedical Laboratories* [BMBL]) requires annual certification of all Class II BSCs. In many institutions, the biosafety program manages the BSC certification program. Periodic review of the certification records will highlight whether the program is effective, i.e., all cabinets have been certified within the past year. Other aspects of the certification records that should be considered are whether all cabinets not meeting certification standards are either repaired or posted with a prominent sign indicating that they are unsafe for use. Once established, these records are simple to maintain, and they provide validation that the program is working as planned.

SELECT AGENT INSPECTION

The Final Rules for the Possession, Use and Transfer of Select Agents were published March 18, 2005, and became effective on April 17, 2005 (3, 4). Facilities that possess select agents (see chapter 29) can expect to be inspected by their lead agency (either the CDC or APHIS) prior to receiving their certificate of registration, and at least every 3 years thereafter as part of the 3-year renewal cycle of their facility registration. In addition, most facilities will experience both scheduled and surprise inspections on a somewhat irregular basis. Depending on the size of the program and changes requested by the institution, some facilities are averaging approximately one regulatory inspection per year. The regulations also require the RO or a representative to conduct annual audits of laboratories where select agents are used (3, 4).

Regulatory/External Audits

Although select agent inspections are usually scheduled in advance, the agencies have recently begun conducting unannounced inspections. Therefore, it is important to have all records ready for review at any time. Those inspections that are announced usually provide several dates from which to choose so the inspection can be scheduled at the convenience of both parties. The RO is the point of contact for the agency, not the scientists working with the select agents and toxins.

The number of inspectors sent to one site varies according to the agency doing the inspection and the size and complexity of the laboratory. In the past, APHIS has traditionally sent one inspector, usually the field veterinarian from the geographical area where the laboratory is located. The CDC, on the other hand, often sends two, three, or more individuals to the site.

Preparation for Inspection

Because an inspection could happen at any time, facilities should always be prepared. It is imperative that records are kept up to date and organized. In addition, any deficiencies found during internal audits should be quickly remediated so that facilities, security, and inventory are audit ready at all times. Records reviews will include:

- Biosafety, security, and incident response plans
- Annual plan review and drill/exercise information
- Training curriculum and training records for Security Risk Assessment (SRA)-approved individuals as well as laboratory visitors (as required under Section 15)
- Refresher training records after significant plan updates
- Electronic and/or manual access records for all points of access to registered space
- Inventory records
- Intraentity transfer records and chain-of-custody records (if intraentity transfers are conducted)
- Form 2s (Transfer); Form 3s (Theft, Loss & Release); Form 4s (Notification of Identification of Select Agent/Toxin); Form 5s (Request for Exemption) (if applicable). All forms are available on the select agent webpage at http://www.selectagents.gov.
- Records of due diligence used when transferring toxin [as required under 42 CFR 73(16)(l)(1)] (if applicable)
- Annual internal inspection records ensuring compliance with all parts of the Select Agent Program regulations
- Annual BSC certification records
- Annual HEPA filter certification records for laboratory exhaust air, ventilated caging systems, Biobubble, etc., as applicable
- Annual BSL3 design and operational reverification records
- Occupational Health Program (medical surveillance plan)
- OSHA 300 occupational exposure records (if applicable)
- IBC applications/minutes/approvals (if applicable)
- IACUC applications/minutes/approvals (if applicable)

For scheduled audits, the RO may be able to communicate with the inspector to determine the agenda for the inspection. A usual scenario is to have an opening briefing where the laboratory directors give a short overview of their select agent research, followed by a walk-through of the facility, records reviews, and an out-briefing. In addition, the inspector will normally request interviews with support staff (e.g., HVAC technicians, emergency managers). Because this is an inspection with significant consequences, it is useful to get help from the institution's experts. However, it is also important to limit attendance to those experts who are able to answer accurately questions that may arise during the inspection. In addition to the scientists, RO, and alternate RO(s), a reasonable list of attendees might include representatives from the following:

- Information technology to discuss cybersecurity
- Public safety to discuss incident response
- Building security to discuss physical security
- Human resources to discuss personnel reliability
- Laboratory personnel to discuss inventory control

These individuals should be selected according to their specialized knowledge of the facility being reviewed. For scheduled audits, it is a good idea to meet with individuals who may be called for interviews to review the requirements and their responsibilities as they pertain to the Select Agent Program.

Another important aspect of preparation is to walk through the facility several times before the inspection date. It is essential for the scientists to be present on these occasions, to provide immediate answers to questions that arise. Use the walk-through to:

- Check that recommendations in the security plan have been implemented
- Test the security system to reduce the likelihood of malfunctions during the inspection (note that self-closing, self-locking doors have a tendency for failure)
- Look for problems and provide remedies on the spot if possible (e.g., look for cracks in paint due to building settling)

In preparation for the inspection, it is useful to prepare a "briefing book" to give to the inspector(s) upon arrival at the facility. This book should contain the pre-inspection checklists, if any; the biosafety, incident response, and security plans; and inventory, training, transfer, and access records. Having all documents assembled in one place will save time during the inspection.

Inspection Checklists

Checklists used by the Select Agent Program are available on their website (www.selectagents.gov). These provide a valuable tool for preparation before the inspection.

Inspection

At the opening meeting, the usual order of business is for a dean or other individual in upper management to give the welcome and introductions. Each project leader then gives a short (15-minute) presentation of the research planned or ongoing in the facility. Then the inspector takes over. Each inspection will differ depending on the agency and the inspectors. At some point, a walk-through of the facility will be conducted. It is usually best to limit the number of individuals involved in the walk-through to have a more manageable process. One scenario would be to include only the scientists, RO, and inspector.

During a closing meeting, the inspector states his/her immediate concerns. Action should be taken to correct any major items noted during the closing briefing. Following the audit, audit findings will be sent to the RO and alternate ROs. Once the inspection report is received, a response documenting corrective actions will be required within 15 to 30 days.

Self-Audits

As noted above, the Select Agent Program regulations require the RO or representative to conduct annual audits of each laboratory to ensure compliance with all parts of the regulations. This requires laboratory audits, security audits, inventory audits, and records/plans reviews. Using the same checklists that the regulatory agency will use (provided on www.selectagents.gov) will meet requirements and document program review. These audits must be documented and must include any findings and follow-up/corrective actions taken. In 2011, the Select Agent Program provided guidance on their requirements for initial and annual verifications. This information can be found at http://www.selectagents.gov/resources/BSL3%20and%20ABSL3%20Verification%20Guidance.pdf.

ANNUAL REVERIFICATION OF BSL3 LABORATORIES

In addition to testing facility operations of newly constructed or renovated BSL3 laboratories before they are put into use, BMBL (5) requires that they be reverified on an annual basis.

> The BSL-3 facility design, operational parameters, and procedures must be verified and documented prior to operation. Facilities must be re-verified and documented at least annually (5).

The facility reverification consists of (i) a review of the laboratory facility by the biosafety manager, (ii) a review and testing of the HVAC system operation by facility engineers, and (iii) testing of the HVAC safety features designed to prevent positive pressurization, such as redundant exhaust fans and emergency power. Alternatively, a consultant or an outside entity that did not participate in the original construction and design of the facility may be retained to assess the facility independently with the assistance of safety and facility personnel.

Preparation

The biosafety manager should meet with the project leaders of the BSL3 facility to review all of the existing standard operating procedures, biosafety manual, and any other written safety procedures for the facility. These documents should be updated as necessary to include changes in the following:

- Location of registered work
- IBC approvals
- Agent inventory
- Institutional animal care and use committee approvals
- Laboratory standard operating procedures
- Autoclave validation records

Accidents or injuries that may have occurred in the BSL3 facility in the past year should be evaluated closely to determine whether processes or procedures that led to the accident have been modified, e.g., use of plasticware instead of glassware in the event of an injury involving broken glass. Any changes should be reflected in documentation of SOPs so that they accurately reflect the activity within the facility.

Training records should also be evaluated to ensure that all workers have received safety training and annual update training (e.g., blood-borne pathogen training) as needed. The auditor should verify the existence of a competency training protocol for new laboratory workers and review those records. If nonlaboratory personnel, such as facility workers or repair contractors, are performing the maintenance procedures within the BSL3 facility, the biosafety manager should ensure that they have received safety training and respirator fit testing (if necessary) prior to entering the area.

Facility Reverification

The annual revalidation should be scheduled well in advance to allow for shutdown of laboratory experiments. Some BSL3 facilities choose to have the laboratory suite space decontaminated before reverification. Alternatively, the lab's established protocol for maintenance procedures will be followed, which may include:

- Biological waste autoclaved
- Cultures stored in incubators or freezers
- Surfaces disinfected
- Sufficient time for adequate air changes

This is also a convenient time for routine maintenance that can be consolidated and conducted during this pause in experimental work.

The biosafety manager's site visit (see Appendix D for checklist) should include an examination of the following:

- Cracks or openings in the walls, ceilings, or floors
- Evidence of water leakage, such as discoloration or observable moisture
- Cracking or flaking of epoxy sealant paint on walls or ceiling panels
- Frayed rubber gaskets around pipes, wiring, or other sealed openings
- Damaged wall covings, especially around compressed gas cylinder storage areas
- Batteries for alarms and directional airflow meters

A work order should be placed for immediate repair if any of these breaches of containment have been identified.

Other items that should be checked and verified by the biosafety manager include the following:

- Eyewash functioning and regularly tested
- Fire extinguishers tested according to schedule
- Verification of directional airflow (e.g., use of smoke stick to determine if laboratory is under negative pressure)
- Proper operation of automatic door closure
- Continued functioning of security features, such as key card readers, motion detectors, and cameras
- Certification of Class II biohazard cabinetry and testing of chemical fume hoods
- Posting and accuracy of biohazard warning signs and labeling; verification of accurate point-of-contact information needed in the event of an emergency
- Adequacy of spill kit supplies
- Emergency shower (if applicable) tested
- Posting of doffing procedures
- Decontamination materials and spill kit available

HVAC Reverification

Facilities and maintenance personnel will check the following within the facility:

- Balancing of ventilation system
- Continued function and maintenance of redundant fans
- Calibration of magnahelic gauges, meters, monitors, and other alarm systems
- Function of emergency power outlets and generators
- Function of supply and exhaust air interlock
- Function of alarms during emergency situations, including loss of power and failure of exhaust fan

Inspection of Utilities Outside BSL3 Facility

Key areas outside the BSL3 facility that will be checked include:

- Vacuum line systems if there is a HEPA filter and disinfectant trap at the point of entry
- Emergency power generators
- Fire suppression sprinkler systems and fire alarms
- Exhaust HEPA filter annual certification (if applicable)

Each facility is different; therefore it is essential to review all of the main utilities provided to the laboratory area.

In conclusion, a BSL3 facility should be reverified on an annual basis to ensure that it continues to meet the BSL3 criteria stated in BMBL. As part of this assessment, administrative procedures and controls should also be checked and updated, and facility workers should check utility systems, such as the HVAC system and electrical systems, to ensure that they are functional and maintained properly.

SUMMARY

Biosafety programs must have a mechanism to evaluate whether they are working effectively. This evaluation is critical because it provides accurate information about an organization's work using biohazardous agents and allows the organization to respond appropriately to changing regulations and guidelines. Knowledge of agents in use or proposed for future use can identify the need for new facilities or additional training as the use of new and exotic agents occurs.

Evaluations may be informal, such as a self-inspection checklist completed by the laboratory director and forwarded to the biosafety manager, or more formal, such as a scheduled audit by the biosafety manager that includes a walk-through of a laboratory facility as well as a check of records or other administrative files. Other safety and administrative personnel may assist as part of the process. Before any inspection or audit is conducted, the procedures for conducting biohazardous work must be clearly stated in the biosafety manual, in institutional policy statements, or on the inspection form itself to clearly delineate what is expected of laboratory personnel.

The success of any evaluation program depends on communication among the laboratory director, biosafety manager, IBC, and senior management. Monitoring research is made much easier when the biosafety manager has a partnership with researchers and an understanding of the goals of the research community and its work.

The authors would like to recognize the valuable contributions of Robert J. Hashimoto, who passed away in 2014.

References

1. **Occupational Safety and Health Administration.** 1991. Occupational exposure to bloodborne pathogens, final rule (29 CFR 1910.1030). *Fed Regist* **56:**64175–64182.
2. **National Institutes of Health.** 2013. *NIH Guidelines for Research Involving Recombinant or Synthetic Nucleic Acid Molecules (NIH Guidelines), as amended.* http://osp.od.nih.gov/office-biotechnology-activities/biosafety/nih-guidelines.
3. **Animal and Plant Health Inspection Service.** 2012. Agricultural Bioterrorism Protection Act of 2002; Possession, Use and Transfer of Biological Agents and Toxins; final rule (7 CFR Part 331 and 9 CFR Part 121). *Fed Regist* **70:**13278–13292.
4. **Centers for Disease Control and Prevention.** 2012. Possession, Use, and Transfer of Select Agents and Toxins; final rule. (42 CFR Part 73). *Fed Regist* **70:**13316–13325.
5. **U.S. Department of Health and Human Services, Public Health Service, Centers for Disease Control and Prevention, National Institutes of Health.** 2009. *Biosafety in Microbiological and Biomedical Laboratories,* 5th ed. HHS Publication No. (CDC) 21-1112. U.S. Government Printing Office, Washington, DC. http://www.cdc.gov/biosafety/publications/bmbl5/bmbl.pdf.

APPENDIX A: Sample BSL2 checklist

Principal Investigator:	**Date Approved**:
IBC #:	**Surveyed by**:
Protocol Title:	**Survey Date**:
Labs/Office:	**Lab Personnel Present**:

Section 1: Facilities:				
Question	**Yes**	**No**	**N/A**	**Comments**
1.1) Laboratory contains a sink for hand washing.	❑	❑	❑	
1.2) An eyewash station is readily available.	❑	❑	❑	
1.3) A Class II biological safety cabinet is available in room.	❑	❑	❑	
Section 2: Access policies:				
Questions	**Yes**	**No**	**N/A**	**Comments**
2.1) Access to the laboratory is limited when work with infectious agents is in progress.	❑	❑	❑	
2.2) The biohazard sign is posted on the entrance to the laboratory.	❑	❑	❑	
2.3) All employees are advised of potential hazards before entering the laboratory.	❑	❑	❑	
Section 3: Material-handling procedures:				
Questions	**Yes**	**No**	**N/A**	**Comments**
3.1) Precaution is taken with any contaminated sharp item.	❑	❑	❑	
3.2) Cultures, tissues, specimens of body fluids, or potentially infectious wastes are placed in a container with a cover that prevents leakage during collection, handling, processing, storage, transport, or shipping.	❑	❑	❑	
3.3) Laboratory equipment and work surfaces are decontaminated after spills, splashes, or contamination and after work with infectious materials is finished.	❑	❑	❑	
3.4) A Class II biological safety cabinet is used for procedures with a potential for creating infectious aerosols or for procedures that involve high concentration or large volumes of infectious agents.	❑	❑	❑	

Questions	Yes	No	N/A	Comments
3.5) Personal protective equipment is used based on the hazard associated with the agent and the procedure. This could include gloves, face protection, and/or protective clothing.	❑	❑	❑	

Section 4: Training practices:

Questions	Yes	No	N/A	Comments
4.1) Employees receive annual laboratory safety training. Records of this training are available.	❑	❑	❑	
4.2) Employees are advised of the potential hazards associated with the work involved, the necessary precautions to prevent exposures, and the exposure evaluation procedures.	❑	❑	❑	

The PI certifies that he/she will: (i) ensure that only qualified scientists work with these materials in proper facilities; (ii) comply with all applicable federal, state, or local laws and regulations pertaining to these materials or their handling, storage, use, transportation; and (iii) destroy all materials according to accepted practices for destruction of microbiological cultures upon completion of work.

Yes	No
❑	❑

Additional Comments

APPENDIX B: Sample postapproval monitoring checklist

Principal Investigator:	**Date Approved**:
IBC #:	**Surveyed by**:
Protocol Title:	**Survey Date**:
Labs/Office:	**Lab Personnel Present**:

Section 1: General responsibilities of the principal investigator				
Question	**Yes**	**No**	**N/A**	**Comments**
1.1) Have there been any significant problems, violations of the *NIH Guidelines*, or any significant research-related accidents?				
1.1a) Events involving a personal injury or loss of containment?	❑	❑	❑	
1.1b) Accidental needlesticks?	❑	❑	❑	
1.1c) Escape or improper disposal of animals used in research?	❑	❑	❑	
1.1d) Spills of high-risk recombinant materials outside of the biosafety cabinet?	❑	❑	❑	
1.2) Does the principal investigator (PI) have the most recent version of the protocol?	❑	❑	❑	
1.3) Are all appropriate personnel listed in the protocol?	❑	❑	❑	
1.4) Have all persons working in the lab received the necessary training?				
1.4a) Lab safety?	❑	❑	❑	
1.4b) Bloodborne pathogens?	❑	❑	❑	
1.4c) DOT infectious and Biological Shipping?	❑	❑	❑	
1.5) Have there been any modifications or changes in research conducted in the lab since you have received approval?	❑	❑	❑	
1.6) Does the PI understand his/her responsibilities?	❑	❑	❑	
Section 2: Responsibilities of PI to lab staff				
Questions	**Yes**	**No**	**N/A**	**Comments**
2.1) Has the PI made the protocols available to all laboratory staff that describes the potential biohazards and the precautions to be taken?	❑	❑	❑	
2.2) Has the PI informed the laboratory staff of the reasons and provisions for any precautionary medical practices advised or requested (e.g., vaccinations or serum collection)?	❑	❑	❑	

Questions	Yes	No	N/A	Comments
2.3) Has the PI provided instruction and training for laboratory staff in the practices and techniques required to ensure safety, and the procedures for dealing with accidents? Do you have a signed risk communication form?	❑	❑	❑	

Section 3: Research

Questions	Yes	No	N/A	Comments
3.1) Are the procedures in use consistent with the approved protocol?	❑	❑	❑	
3.2) Are there any corrective actions needed to reduce the risk of release of recombinant DNA materials?	❑	❑	❑	
3.3) Have there been any deviations from the approved protocol?	❑	❑	❑	
3.4) Is the work being conducted at the approved biosafety level using the approved facility?	❑	❑	❑	
3.5) Is required personal protective equipment available and being used?	❑	❑	❑	
3.6) Are there any binding conditions of this Institutional Biosafety Committee approval?	❑	❑	❑	
3.6a) If so, are those measures being followed?	❑	❑	❑	
3.7) Are the Institutional Biosafety Committee-approved emergency plans for handling accidental spills and personnel contamination being adhered to?	❑	❑	❑	
3.8) Do the biosafety cabinets (BSCs) in use have current certification dates?	❑	❑	❑	

Section 4: Compliance

Questions	Yes	No	N/A	Comments
4.1) Date of last audit: ____________				
4.2) Are there any pending issues from the previous audit?	❑	❑	❑	
4.3) Are there any items of concern?	❑	❑	❑	See below.

Additional Comments

Previous Audit Findings:

Date	Building	Room	Auditor	Observation	Notes	PI

APPENDIX C: Sample amendment form Request to Amend Previously Approved Research Registration Involving the Use of Biohazardous Agents or Recombinant or Synthetic Nucleic Acid Molecules

Principal Investigator:	Dept.:
Building/Room:	Phone:
Email address:	Date:
Title of original research project:	
Approval number of IBC project:	
Biosafety Level of approved work:	
Section of the NIH Guidelines (e.g., Section III-D-1):	
Original biohazardous agents used:	

Reason for requested amendment:

❑ Add/delete recombinant DNA (if recombinant DNA molecules are added, please complete the template below):

Host:
Vector:
Nature of inserted sequences:
Source of inserted sequences:
Types of manipulation:
Attempt to express foreign gene:
Protein produced:
Containment:
Section of Guidelines:

❑ Add/delete biological agents

❑ Amend scope of work or procedure involving biohazardous agent use

❑ Add/delete laboratory rooms, work areas location (complete back of form)

❑ Add/delete lab personnel (complete back of form)

❑ Add/delete the use of human blood or other human materials

❑ Add/delete use of select agents

❑ Add/delete use of animals and rDNA and/or biohazardous agents

❑ Project is complete; IBC approval no longer necessary

Summary of changes/scope of Work (include animal procedures if necessary):

____

____

____

____

Signature: ______________________________ Date: ______________

APPENDIX D: Sample BSL3 annual validation checklist

BSL3 Annual Validation Checklist	
Building/Room:	**Department:**
Facility Contact:	**Lab Director:**
Name of Inspector:	**Date:**

Biosafety Level 3	Yes	No	N/A	Comments
Laboratory equipment				
Class II or III biosafety cabinets (BSCs) are certified annually, located away from air supplies and doors, and pass visible smoke test.				
A double-door autoclave is convenient to the animal rooms. State certification as a pressure vessel is up-to-date. Door interlock functional.				
An eyewash station and shower are readily available. These items have been tested within the past 12 months.				
Chemical fume hood has been tested within the past 12 months and passes smoke test.				
Laboratory facilities				
The facility is separated from areas that are open to unrestricted personnel traffic within the building by a series of two self-closing and self-locking doors.				
Security features function properly. Doors cannot be opened by pulling or pushing.				
The facility is designed, constructed, and maintained to facilitate cleaning and housekeeping. The interior surfaces (walls, floors, and ceilings) are water resistant. Penetrations in floors, walls, and ceiling surfaces are sealed and openings around ducts and the spaces between doors and frames are capable of being sealed to facilitate decontamination.				
A hands-free washing sink is provided in the anteroom as well as near the exit door in each laboratory within the suite. The hot and cold water valves operate as designed.				
Internal facility light fixtures, air ducts, and utility pipes are arranged to minimize horizontal surface areas.				
All windows closed and sealed.				
Illumination is adequate for all activities, avoiding reflections and glare that could impede vision.				
Biosafety vacuum lines protected by liquid disinfectant traps and HEPA filters				
Central vacuum line is protected by HEPA filtration				

Biosafety Level 3	Yes	No	N/A	Comments
HVAC system				
A dedicated, ducted exhaust system provides directional airflow which draws air into the laboratory from "clean" areas and toward "contaminated" areas.				
If laboratory exhaust air is HEPA filtered, the HEPA filter has been certified within the past 12 months.				
When one fan fails, the redundant fan maintains the directional airflow.				
During complete exhaust failure, supply dampers close and the suite becomes neutral in respect to the corridor.				
Visible alarms activate when laboratory exhaust is lost.				
Differential pressure indicators function properly.				
Visible directional airflow indicators function properly.				
Electrical system				
Circuit breaker panels and controls are labeled.				
Electrical outlets are properly grounded.				
Access systems and emergency door release are functional.				
Back-up generator functioned properly.				
Fire Alarm system functional.				

ADDITIONAL COMMENTS:

____

____

____

____

____

28

A "One-Safe" Approach: Continuous Safety Training Initiatives

SEAN G. KAUFMAN

A "ONE-SAFE" APPROACH

Training prepares people to behave, and it is behavior that connects plans to desired outcomes. The four phases of biological risk mitigation are (i) risk identification, (ii) risk assessment, (iii) risk management, and (iv) risk communication. During the risk identification process, both the agent and processes of working with the agent are reviewed to determine risk, which is assessed and managed primarily through the development of standard operating procedures (SOPs). However, the greatest risk that is often overlooked is the people interacting with agents on a daily basis. How these individuals perceive laboratory risks—their experiences, educational levels, comfort, skills—and the culture of the organization in which they work influence safety attitudes and behaviors. SOPs provide the process to achieve the desired outcome and training ensures consistency of behavior among many individuals with vast differences in education and experience.

Biosafety programs depend on the integration of four primary controls: (i) engineering, (ii) personal protective equipment, (iii) SOPs, and (iv) training. Millions of dollars are spent on engineering controls to ensure that biological agents remain within laboratory environments using mechanical devices. Thousands of dollars are spent on personal protective equipment (PPE) to ensure biological agents cannot enter the needed human portals of entry to make laboratory staff sick. Hundreds of hours are spent writing SOPs to ensure consistency of behavior among different people doing the same thing, thereby ensuring consistency of safety outcomes. However, one poorly trained staff member can negate these controls in an instant, leading to the potential loss of life, illness, damage to institutional reputation, increased regulation, and decreases in scientific research funding.

Adult learning theory, measured learning objectives, and training evaluations must be incorporated in any training initiative, and these issues will not be discussed in this chapter. Rather, this chapter will focus on why we train, who we train, when we train, and how we train to prepare and protect laboratory staff. Human risk factors, the blending of multiple existing cultures, opportunities for learning, and strategies for workforce preparedness will be discussed. Training is a tool that can be used to form a "One-Safe" culture with all laboratory staff.

WHY DO WE TRAIN?

There are three goals of training: (i) to increase awareness of laboratory risks, (ii) to increase skills and abilities of laboratory staff, and (iii) to increase the application of biosafety practices and procedures. Simply identifying goals of training is not enough. Laboratory staff working with biological agents should know the risks they are working with, have the skills to protect themselves, and adequately respond to issues which are unexpected.

Laboratories today are filled with staff coming from different generations and with different educational backgrounds, and their training needs differ on the basis of the level of laboratory experience, defined here as novice, practitioner, or expert. Generally speaking, training for novices should emphasize an understanding of risk. The training for practitioners should focus on skills and abilities. The training for experts requires the ability to articulate, share, and motivate their colleagues using "lessons learned." We should not train novices as though they are experts and experts as though they are novices. The reasons for this are simple. It is very difficult to change the behavior of experts. Experts have experience enough to know success can be gained by using a variety of techniques. Experts have been successful, so they view differences in processes as a challenge to their preferences and may not see adherence to a SOP as a professional responsibility. However, most practitioners have tasted minimal success and believe their processes are the only way to achieve success. Experts have developed a reputation; however, practitioners are in process of developing their professional reputation and may not be open to changing processes that have demonstrated success. This lack of experience among practitioners leads to a great amount of resistance to change, even if the change is a safety practice that will not adversely affect the conduct of the research. In short, one approach does not fit all training needs or audiences within an organization. Laboratory staff will have different perceptions of risk, based on their previous experiences and education. These perceptions of risk directly influence safety attitudes and behaviors. For example, someone who enters the laboratory for the first time views risks far differently than someone who has worked in the laboratory for many years. These differences produce different perceptions that will ultimately lead to different behaviors. This is human nature and can be controlled with SOPs that are validated cognitively and behaviorally. Additionally, training controls for many other human risk factors, as being human is a risk factor in itself.

Until human beings are replaced by robots, knowing human risk factors is critical in understanding why training is needed. Although there are many human risk factors, this section will focus on three that call for the greatest need for training.

Human Risk Factors

Being human is not a fault. However, expecting humans to be free of error is unrealistic. When something goes wrong, the cause overwhelmingly points to human error—airplane crashes (70%), car wrecks (90%), and workplace accidents (90%) (1). When the blame falls on human error, a more intensive review of systemic issues (root causes) within an organization is avoided. The question leaders should be asking today is, "Did the organization do enough to control for human risk factors?"

Some of the most dangerous human risk factors are perceived mastery, complacency, and apathy. Failure to control these three human risk factors in an organization working with biological agents could result in serious problems.

Perceived mastery

The overestimation of skill and ability is the most serious human risk factor. Research finds that with difficult tasks, people overestimate their actual performances (2). Laboratory staff believe they are better at containment strategies than they actually are. This overestimation of skills and abilities places them and those they work with at an increased risk.

Perceived mastery occurs as habit begins to form. Usually with a novice and typically within the first year of learning a new behavior, laboratory staff believe in their skill, and the attention to detail (which was a critical component of mastering the behavior) decreases as steps become habitual processes. It is the lack of attention to detail that leads to an incident, accident, or near miss, thereby calling workers back to a state of increased awareness where they are once again paying greater attention to details that keep them safe.

Complacency

After long periods of time, habits are shaped, and practices are formed. A high level of confidence develops within oneself. This happens as a result of extended periods where no accidents, incidents, or near misses occur. Laboratory staff begin to believe falsely that they are safe and secure, even when those around them are far less experienced and trained. While it is true that laboratory staff sometimes fail to observe potential hazards, studies monitoring behavior show "that not even optimal staff can detect all abnormal events. Such an analysis emphasizes the importance of alarm systems rather than monitoring for the efficient supervisory control of highly reliable systems" (3). With new technologies, the future may include a process for monitoring body temperature and respiratory and heart rates. Monitoring these variables would allow for fever screening prior to working in a biological laboratory while identifying unexpected events with increased heart and breathing rates among laboratory staff.

Complacency produces false beliefs about individual safety and security. Risk is no longer viewed as a communal perspective; rather it is seen as an individual issue. For example, if a laboratory staff member fails to comply with a SOP, a leader should ask the laboratory staff a very simple question. What is the risk of not following the SOP? Most laboratory staff will answer defining risk specific to themselves as individuals. However, the answer is far greater than just the risk to themselves. Complacency threatens the health and safety of others, the reputation of the organization, and even the general public or loved ones of the complacent laboratory staff member.

Apathy

Apathy is best characterized in behavioral terms as an absence of responsiveness to stimuli as demonstrated by a lack of self-initiated action (4), in other words, acceptance and tolerance of risks as a result of the inability to change them. Apathy occurs when laboratory staff become concerned about the behavior of others or laboratory practices, report them, and are ignored and instructed to continue working with their concerns unaddressed.

Apathy starts slowly but grows quickly within a bubble (as most organizations are). Laboratory staff are aware of incompetence, low compliance to SOPs, and serious containment breaches and yet remain quiet as though they accept the unacceptable. Although perceived mastery and complacency produce the majority of human error incidents and accidents, apathy serves as the greatest risk to health, safety, and security. Apathy fosters an environment where the unacceptable to many becomes the norm for all within the organization.

Training is not a "one and done" approach, to satisfy a regulatory requirement. Training is a strategic approach aimed at minimizing human risk factors by increasing awareness to details, teaching the skills to behave safely, and fostering the ability to apply calculated strategies when unexpected events occur.

If we know new laboratory staff experience perceived mastery, would it not be a good idea to provide a training program that challenges their newly formed habits with an unexpected event? Would a training program that recreates incidents, accidents, and near misses not decrease complacency among the most experienced laboratory staff? Training cannot replace good leadership. Training prepares the workforce and leadership protects the workforce. Leaders listen and address identified risks rather than ignoring them. Good leadership prevents organizational apathy.

WHO DO WE TRAIN?

Just as science and safety must be blended, the leadership and workforce of an organization must also be blended for an effective culture of safety to exist. Leadership sets the tone for the culture of the organization. Like a parent in a family, leaders influence the acceptable attitudes and behaviors within an organization. Culture is the set of beliefs, values, and behaviors, both explicit and implicit, that underpin an organization and provide the basis of action and decision making and is neatly summarized as "the way we do things around here" (5). It is when the leadership and workforce blend that a culture of safety develops.

Like a parent, the leadership expects the workforce to report problems. Like a child, the workforce expects the leadership to protect, prepare, support, and assist them. If the leadership punishes the workforce for reporting problems, then no additional problems will be reported. The blending of the leadership and workforce means the two are trained together. It is during this training process that trust is established and roles are clearly defined. Both the leadership and workforce must have clear roles and expectations within the organization. Training clarifies these roles and connects the leadership and workforce, while strengthening the relationship needed for outstanding safety attitudes and behaviors (safety culture).

Leadership Training

Leadership has the right to expect specific behaviors of the workforce; however, the workforce has the right to expect leadership to prepare, protect, and promote them (6). Leaders prepare the workforce by providing resources and training. They protect the workforce by defending laboratory staff who are complying with SOPs when they make mistakes and addressing staff who insubordinately choose to break SOPs. Leaders promote the workforce when members of the organization are reminded success is owned by them, not just promoting the success of one, but the success as a whole.

Preparing the workforce

The first question in preparing the workforce deals with time. How much time is being spent to train the workforce? For example, if a group of employees work 46 weeks a year, on average they work 8 hours a day for 5 days a week. This means they work (8 hours * 5 days a week * 46 weeks) = 1,840 hours per year. Let's say the organization trains the group 3 full days each year. That is 24 hours of safety training per year. Is that enough?

Now imagine a jury is listening. An employee who worked at the organization was accidentally infected with hepatitis C virus. The employee is suing your organization for substantial damages. An expert witness is called. It is during this testimony you discover the employee had only averaged 1% of on-the-job safety training over the 5 years they worked at your organization. That's right—3 full days of training (24 hours) is equivalent to only 1% of on-the-job training. In fact, the expert witness says the industry standard for high-risk jobs falls between 3 and 5%, requiring a minimum of at least 60 hours per year to be at industry standard. Working with agents that can make healthy people sick, including their family members, would be perceived as a high-risk job. A 1% investment in training may be considered negligent and most juries would agree with this observation.

Laboratory staff working with dangerous biological agents require high-performance skills. A high-performance skill is defined as one for which more than 100 hours of training are required, substantial numbers of individuals fail to develop proficiency, and the performance of the expert is qualitatively different from that of the novice (7). In using the analogy above, 100 hours of training for someone who works 1,840 hours per year equals 5% of on-the-job training.

Leaders must identify how much time they are willing to dedicate to training and ensure that the workforce receives enough training to be prepared adequately. Preparedness not only means providing resources, but plans for how to use those resources safely, and provides the time required to learn, practice, and master the behaviors needed to successfully follow the plans. This requires time and leaders must support the time it takes to prepare the workforce.

Even when humans are replaced by robots, maintenance will be required. Humans require maintenance as well. Leaders should believe the workforce requires constant training (human maintenance). In short, leaders recognize the workforce needs frequent training and is never prepared but remains in a state constant preparation.

Protecting the workforce

Leadership training will separate the process of leading from the process of managing. The difference between a leader and manager is simple to distinguish. A manager assigns responsibility when failure occurs and assumes responsibility for success. A leader practices with a different philosophy, promoting a culture of responsibility rather than a culture of blame.

A culture of blame fosters the feeling that all is well as long as someone is in charge: if something goes wrong we have someone to blame, and we will be well again if we get rid of the person who caused the problem (8). The greatest contribution a leader can make is to encourage a transition from a culture of blame to one of responsibility and accountability. Rather than looking at *who* was wrong, focus on *what* went wrong and apply that lesson to the entire organization. Leaders own failure, and the responsibility of success is assigned to others by a leader. Unfortunately, many scientists are at great risk—not from the agents they work with but from the lack of leadership within organizations suffering from systemic shortcomings and cultures of blame.

Leaders protect the workforce in two ways. First, leaders understand the workforce is human and mistakes are going to happen. Leaders cannot prevent mistakes from happening but they can foster an environment that encourages mistakes to be reported. This environment is created when leaders are trained to support those who make mistakes and protect them from scrutiny or even dismissal. If the workforce is doing their best to follow procedures, it may be an issue of organizational culture. Leaders will be the first to see this because they look at themselves first when failure occurs. Again, the protection of the workforce means leaders focus on what happened rather than who did it.

Second, leaders must address insubordinate members of the workforce. The difference between the workforce members mentioned above and those who intentionally violate safety rules and procedures is vast. Safety is only as good as its most unsafe workforce member. Any toleration or acceptance of insubordinate behavior has a profound impact on current and future safety levels in the laboratory environment.

Tolerating nonadherence to SOPs puts the laboratory staff at risk and demonstrates to team members that insubordination is an accepted practice. A leader must take this lesson into the laboratory and realize their organization is only as safe as the most unsafe accepted behavior. Leadership is not a title or a position; it is a philosophy and practice that requires training.

Promoting the workforce

Leaders must be trained to recognize the team effort and promote communal success. Scientists rarely succeed on their own. Animals must be cared for, mechanical devices must be maintained, grants must be written, and laboratory staff must be protected. A successful project is a team effort and leaders must ensure the promotion of the entire workforce rather than the success of one.

Training leaders to promote the success of the entire workforce not only increases retention levels among staff, but begins to build a sense of respect and appreciation for what the entire workforce contributes to the endeavor.

Safety is achieved with the blending of the leadership and workforce. For this blending to occur, the workforce must trust the leadership. Leadership must be trained to behave in manners that increase workforce respect, trust, and appreciation. If the workforce does not respect, trust, or appreciate leadership, silos and gaps become numerous, breeding an organization full of systemic safety risks. (A "silo" is a term for working within an individual context rather than an organizational one, focusing on personal implications rather than the implications of one's behavior for the organization as a whole.) Training must start with leaders and include the workforce.

Workforce Training

A neighborhood is similar to an organization with multiple laboratories. Neighborhoods have rules and each family house within the neighborhood has rules as well. For the benefit of all people living in the neighborhood and family members residing in each house, both sets of rules must be followed.

Safety cultures exist when a group of people shares common attitudes and behaviors for a common cause. This solidifies the need to train the workforce on two different levels. First, the workforce comes together and is trained on a set of organizational expectations. These expectations are the glue of the safety culture. The goal of this training is to raise awareness and ensure all staff understand what is expected of them while working at the organization.

Families are different from one another and may have different house rules as a result of many differences (like the differences between each laboratory). Second, the workforce must be trained on laboratory-specific expectations. These expectations protect them from immediate risks found in the laboratory environment. The goal of this training is to increase skill, ability, and application of safety plans and practices. This training also solidifies a culture of safety within the laboratory (among staff) by ensuring safety equality regardless of the levels of laboratory staff education or experience.

Organizational expectations connect all laboratories, providing a consistency of expectations among all laboratory staff within the organization. Expectations for each laboratory must exist, but they must not outweigh or override the organization as a whole. The workforce must be trained on both levels.

Organizational expectations

The following are a set of expectations that have been gathered as a result of past safety lapses. Lessons will continue to present themselves until they are learned. A death or illness as a result of a repeated safety lapse within the laboratory is unacceptable. Organizations must communicate expectations and provide the training needed to adhere to them. Consider the following list of organizational expectations, developed to minimize incidents that have led to past death or injuries in biological laboratories (6).

1. *I will follow all SOPs to the best of my ability.* SOPs are standardized behaviors, ensuring consistency of practice and outcome among individuals who may vary drastically in education and experience levels. Plans alone do not produce skills and confidence needed to adhere successfully to a SOP. Laboratory staff must be provided training and time to practice the skill before demonstrating mastery and being allowed to work independently in the laboratory environment.
2. *I will ensure others follow all SOPs to the best of their ability.* Laboratories are communal environments where the behaviors of one affect the well-being and safety of others. This highlights the importance of staff maintaining a heightened awareness of other laboratory staff behavior. Laboratory staff must be trained to speak up and speak out when someone does not follow a SOP. Addressing failed SOP compliance will establish a social norm of expectation among peers within the laboratory while ensuring a higher level of compliance with laboratory SOPs.
3. *I will report all incidents, accidents, and near misses.* Although this seems like an easy request, laboratory staff must be trained to identify incidents, accidents, and near misses. Organizations may define these differently and have unique processes for reporting and responding to each situation. The distinction between an incident, accident, and near miss must be clarified along with the processes for reporting each.
4. *I will report the clinical presentation of symptoms matching agents in the laboratory.* Laboratories that house radiological and chemical agents have alarms and indicators that notify laboratory staff should an environment become contaminated. This contamination serves as a risk to laboratory staff, and the ability to know when contamination occurs is a critical factor of laboratory staff protection. Unfortunately, there are no alarms that sound in biological laboratories. There are no indicators deployed until after a contamination of the laboratory environment is suspected. Therefore, it is imperative that laboratory staff receive agent-specific training and each staff member can cognitively describe clinical symptoms of the agents they work with along with the SOP for reporting illness should the symptoms present themselves.
5. *I will report any new medical conditions.* Laboratory staff age, and as they age new medical conditions appear. Pregnancy, diabetes, new medications, and undiagnosed conditions can produce dangerous risks to the health and safety of laboratory staff. Staff should be trained to understand that risk is not static; as new conditions appear, risks change, so changes in health need to be reported. These changes may require additional risk assessment and management strategies. Medical conditions are not a privacy issue, they are a risk issue and should be handled as such for the protection of organizational reputation and laboratory staff.

Developing a "One-Safe" Culture

Culture can be defined as a group of people who share a common vision and have similar attitudes and behaviors. Today, within one organization, there are four subcultures: the leadership, workforce, science, and safety. Development of a "One-Safe" culture requires the blending of all four subcultures within an organization. The struggles seen today are not a matter of the lack of culture within an organization, but rather the inability to blend the variety of subcultures that exist within the organization. The European Committee for Standardization (CEN) Workshop Agreement, International Organization for Standardization (ISO) Standards, and National Institutes of Health (NIH)/Centers for Disease Control and Prevention (CDC) *Biosafety in Microbiological and Biomedical Laboratories* (BMBL), along with many other biosafety guidelines, attempt to connect all subcultures. In the absence of a concerted effort, the subcultures will not be blended.

Imagine a family of four eating dinner around the table. The mother is the leader, the father is the workforce, the daughter represents safety, and the son represents science. All of them are eating together, but at the same time they are all on their cell phones. They are BEING together but they are not DOING together. A "One-Safe" culture requires the leadership and workforce within both the professions of science and safety to blend and DO safety together (Fig. 1).

The workforce must trust leadership. Trust between the two increases workforce empowerment (9). Trust is gained when protection and preparedness are offered. Safety cannot practice in a manner that increases dependency on the safety officer; rather, safety must be implemented in a way that increases the capacity and empowerment of the laboratory staff to do safety for themselves. Blending safety and science means a safety officer does biosafety with and by the scientists instead of to and for them. A safety officer must lead rather than manage; serve rather than direct.

Blending subcultures to form a "One-Safe" culture means training together. Providing safety assignments where the leadership and workforce within the professions of safety and science must work together to achieve success provides a blended training experience. After identifying why and who we need to train, the next challenge is determining when training is needed.

WHEN DO WE TRAIN?

The simple answer to the question of "when do we train?" is, before any expected behavior occurs. In short, training

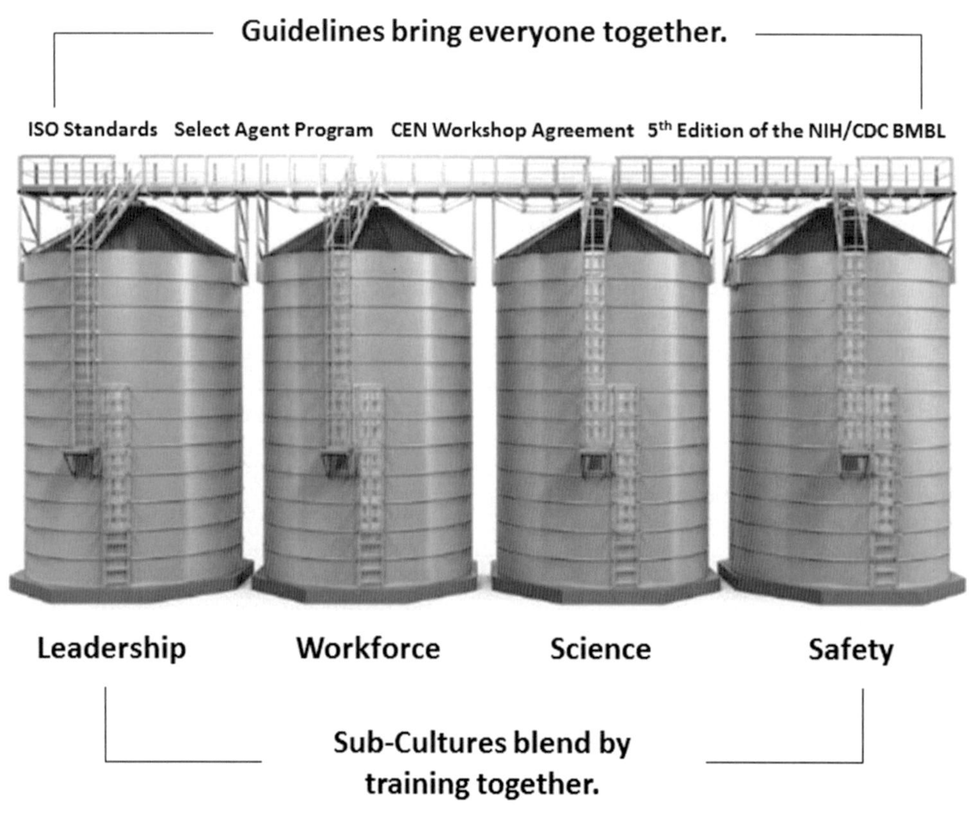

Figure 1: Blending subcultures.

must occur with verification by demonstration of the expected behavior before laboratory staff can be called "trained." Is someone ever trained or does he or she remain in a constant state of training? Organizations who provide a "one and done" training approach may fail to address human risk factors (mentioned earlier in this chapter) that periodically and unexpectedly occur throughout any given year or career. The concept of "behavioral evolution" provides a diagram demonstrating how behaviors are developed and timing when training may be optimal.

The Concept of Behavioral Evolution

Our heads have shrunk as a result of an evolutionary process where our brains have become smaller and more efficient. Wrapped up inside your brain are the basal ganglia responsible for the development of habits. Anytime a new behavior occurs, the brain is stimulated and aware of details surrounding the behavior (10). The concept of behavioral evolution diagrams how individuals learn and master specific behaviors. This concept can be applied to staff working in laboratories.

Behavioral evolution starts when a new behavior is learned and full attention and awareness are given to that behavior. Unfortunately, in a short period of time, the individual begins to form habits and believes he or she has mastered the behavior. This decrease in awareness and attention to detail leads to an incident, accident, or near miss. If an accident occurs and the individual evolves, a change in behavior takes place. This change in behavior leads to new behaviors, which brings the person back to an increased state of awareness. However, because the person has gained valuable experience (prior to the incident), he or she is at a new level of mastery, typically better than before. Unfortunately, as time passes, individuals are challenged with complacency or a state of decreased attention and awareness to detail. This again leads to an incident, accident, or near miss, leading to change and an infinite process of evolution throughout life. Behavioral evolution provides evidence that someone is never "trained" and demonstrates a great need for individuals to remain in a constant state of training throughout their career in the laboratory (11) (Fig. 2). For example, when someone first learns to drive a car, the radio may serve as a huge distraction. In time, the basal ganglia will allow human beings to develop habits and do multiple things at one time without much thought.

When someone has developed a skill behaving and achieves success repeatedly, a belief and mindset begins to form. "I have driven for months now and no accidents have occurred. I've got this." Behaviors begin to change as a result of the basal ganglia taking over, a phone call here and there, singing along with the radio, followed by a text message, and then WHAM—an accident occurs. When working with laboratory novices (someone new to the expected behaviors), strong attention to detail occurs as they begin their work experience. Depending on the frequency of behavior, habits can form within weeks. The formation of these habits along with limited success forms a sense of perceived mastery.

During the stage of perceived mastery, attention to detail slips and laboratory staff become more reliant on

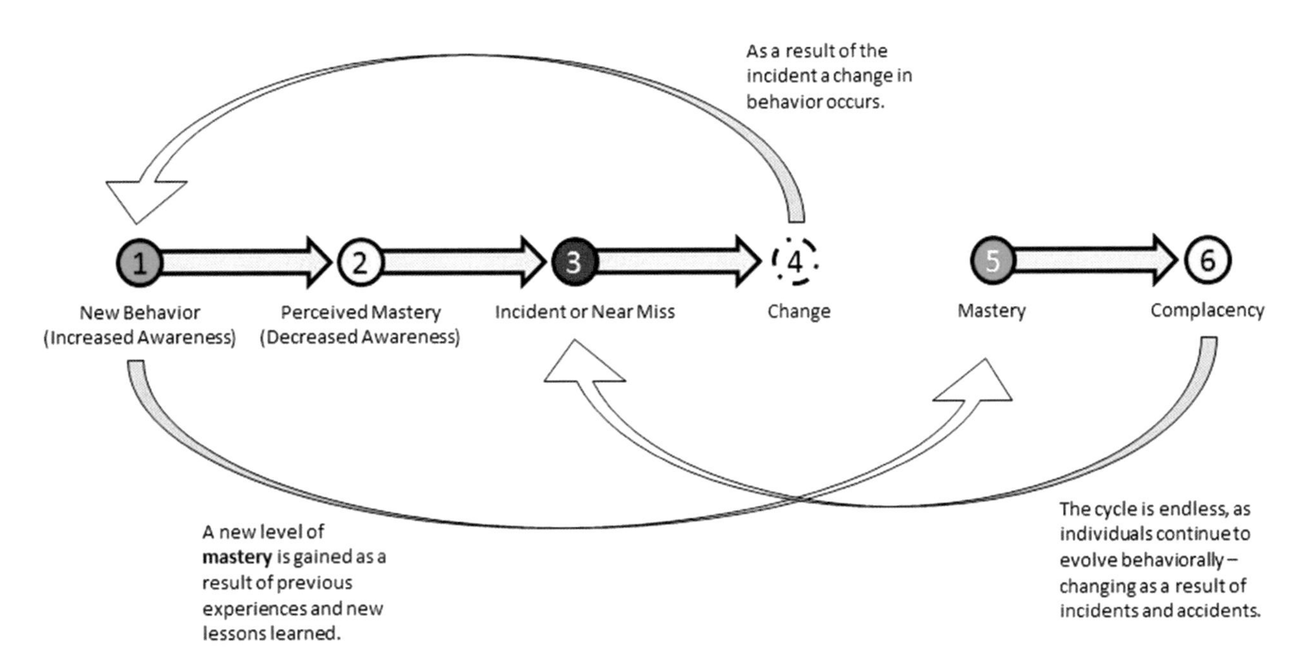

Figure 2: Behavioral evolution.

formed habits. This decrease in attention to detail leads to an increased risk for either an accident or incident to occur. The German philosopher Friedrich Nietzsche said, "That which does not kill us makes us stronger." When an accident occurs, behaviors change, and those changes lead to new behavioral habits that make laboratory staff safer.

Laboratory staff who learn from their mistakes do not return to the same level they were at when the behaviors were just learned. With their previous experiences, developed habits, and lessons learned as a result of the incident, attention levels increase as they enter a new level of behavioral mastery. This increased attention and fine-tuning of habits minimizes risk. As the laboratory staff continue to work accident free, new habits begin to form and a false sense of security begins to set in (complacency).

Complacency paired with lowered levels of attention to detail produces higher risks of incidents and accidents. As seen in the behavioral evolution diagram (Fig. 2), the cycle is infinite and continuous throughout the career of laboratory staff. Every incident and accident serves as an opportunity to become better, to learn, improve, and develop modified behaviors leading to improved habits. Understanding this cycle can provide great insights as to when laboratory staff need training. It also demonstrates the need for a continuous training approach rather than a "one and done" approach.

Continuous training should be a strategic and scheduled event. To ensure all laboratory staff have been adequately prepared to do their jobs safely, laboratory staff should receive: (i) initial training, (ii) on-the-job training, and (iii) postincident training.

Initial training

Regardless of the level of education or on-the-job experience, laboratory staff who are new to the organization or laboratory environment must not only receive initial training, but must also verify mastery of safety expectations before being allowed to work. Each organization may operate in different regions, states, or localities and may follow different guidelines, polices, or laws. Agent-specific donning and doffing of PPE, laboratory emergency response, waste management, and laboratory equipment training occur during an initial training.

Verification of skills and abilities must occur at the initial stage of training. Laboratory staff must be asked to demonstrate a specific set of skills prior to working with and around risks inside a laboratory. There is a difference between knowing and doing. A human risk factor is perceived mastery, or the belief they know how to do something and yet cannot do it. Initial training teaches about the laboratory risks and skills needed to mitigate those risks. Initial training also ensures laboratory staff can demonstrate (verify) the skills needed to remain safe.

On-the-job training

Effective training shakes up existing habits, increases attention to detail, and ensures that laboratory staff return to a heightened state of awareness. We know once a new behavior is learned, habits form within weeks and a false sense of mastery occurs. We also know after months and years of repetitive behavior with little to no unexpected incidents or accidents, a false sense of security develops within laboratory staff (complacency). Developing high levels of staff competency is one of the most challenging issues for organizations (12). On-the-job training is not classroom PowerPoints or computer-driven modules. It can be as simple as walking into the laboratory environment, blindfolding a staff member, telling him he has just splashed an agent in his eyes and cannot see, and asking him to find an eye wash station. This 5-minute simulation exercise not only has a profound impact on the individual who is blindfolded but also on those witnessing it. The increased attention to the laboratory environment has been accomplished, returning the laboratory staff to a heightened state of awareness.

Another example is simply asking laboratory staff to demonstrate safety behaviors at random times. Spills and other laboratory emergencies are not scheduled events. Training, if done frequently enough, will lead to habits as well. Asking a staff member to respond to a spill and demonstrate her response at unexpected times leads to a variety of lessons learned.

On-the-job training monitors high-frequency behaviors to ensure compliance to existing SOPs and that risky habits are not developed. However, low-frequency behaviors must go beyond monitoring and should be exercised. Low-frequency behaviors are situations that require responses one to two times per year, if that. Although these situations may not occur often, chances are the level of consequence may be high, requiring specific attention to detail and behavioral mastery. Deliberate practice with both high- and low-frequency behaviors results in a strong improvement of the learner's early procedural skill acquisition (13). The more feedback provided to learners, however, results in even better and smoother performance. On-the-job training provides laboratory staff with opportunities to improve and fine-tune safety behaviors.

On-the-job training is different from initial training. Initial training focuses on teaching laboratory staff the risks, benefits of following SOPs, and skills needed to follow the SOPs. On-the-job training ensures laboratory staff can actually do the expected behaviors within the laboratory environment when they need to be done. On-the-job training focuses on the direct interaction between staff, their skill, and the laboratory environment. An intact synergy between staff, their skill, and the laboratory environment is essential in minimizing laboratory incidents and accidents.

Postincident training

Organizations should consider adopting a philosophy of identifying "what" went wrong rather than "who" went wrong when incidents occur. Adopting this philosophy would provide opportunities for organizational growth in a monumental manner. When an organization focuses on "who" rather than "what," it becomes an issue of competency. Additionally, the incident is being managed by passing blame onto an individual rather than leading by owning the incident as an organization.

Individuals may develop bad habits and fail to comply with SOPs. However, this is an issue of leadership accepting poor practices and failing to protect the workforce by addressing insubordinate behaviors. This has very little to do with "who" and is a "what" issue. An organization may learn it does not lead and in fact manages by blaming individuals for organizational failures.

Postincident training should focus on having laboratory staff articulate why they took the actions they did during the accident. This allows the leaders to gauge whether the staff performed similarly to the expectations of the industry. The open-ended questioning method of articulation training is described as a common and easy-to-employ technique. Postincident scenarios used in articulation training should be challenging enough to require the application of learned skills and should describe incidents that could actually occur.

Postincident training uses real incidents to provide training to laboratory staff within the organization. Incidents can be recreated and staff can be asked to participate and demonstrate what they would do following the incident. Organizations would explore best practices, develop situational expectations, and shape the behavioral responses of laboratory staff to ensure the same incident does not happen again.

By shifting away from the "who" to the "what" went wrong philosophy, organizations gain trust and capitalize on learning opportunities. When incidents happen, laboratory staff are not punished or scrutinized. Instead, there is a sense of contribution to the betterment of the organization by offering a lesson learned.

Training occurs from the moment staff arrive (initial), while staff are working (on-the-job), and following an accident, incident, or near miss (postincident). Training should be a continuous, strategic, and scheduled component of workforce preparedness. Are laboratory staff adequately prepared to work with biological agents that can make them, their families, or the general public sick?

HOW DO WE TRAIN?

Training is one of the most misunderstood opportunities today. How a training program is facilitated and implemented within an organization will determine its success or failure. Within each organization training generations exist. These different generations have unique training needs, preventing organizations from delivering a "one-size fits all" training approach. Training must be delivered through modules that increase cognitive awareness, behavioral skills, and situational responses. Training must utilize adult learning principles and be evaluated on multiple levels.

Training Generations

Organizations are very similar to families. Just like families, organizations have policies, procedures, and philosophies that lead to the development of a specific culture. Families also have generations—children needing to know why they do something, parents needing to know how to do something, and grandparents needing to know how to engage and guide both the parents and children with the lessons they have gained from their experiences. Organizations have generations who have specific training needs (Table 1).

TABLE 1.

Training generations

Novice	Practitioner	Expert
I'm just starting to learn and cannot do it on my own just yet.	I can work in the laboratory on my own—I don't need any help.	I can do it on my own and can explain how to do things successfully in multiple ways.
Open to training	Resistant to training	Uninterested in training
Open to change	Initially resistant to change	Resistant to change
I absorb as much as and as quickly as I can.	I use my experience to mentor and teach others.	I use my experience to solve problems and issues at my organization.
Expected to follow SOPs and report those who don't	Expected to escort those who do not follow SOPs—preparing the workforce	Expected to increase SOP compliance—protecting the workforce

Novices: Why is this important?
Novices are new to the laboratory environment. Their training begins with explanations as to why specific behaviors are expected of them. Although the explanation has very little to do with short-term compliance (as regulatory compliance programs and culture have greater impacts), laying the foundation and providing a justification for the requested behavior gives reasoning to the specific expectation. This reasoning increases the likelihood of behavioral compliance and development of safe habits. Novices have to learn why and most of their training can occur in a classroom or online setting. Novices tend to be the easiest group to train and change.

Practitioners: How do I do it safely?
Practitioners are either new or have been working in the laboratory for a long period of time. Their training consists of skill development, verification of capabilities, and SOP compliance. Practitioners develop habits and experience limited amounts of success. In contrast to novices, practitioners are the hardest group to train because they fear changing any behavior that has led to their success. This fear makes practitioners extremely resistant to any change. In short, novices "learn" to make a name for themselves while practitioners "sustain" to make a reputation for themselves.

Experts: How do I share what I know?
Experts have not only worked in the laboratory environment for more than 15 years, but they have been successful in most of their scientific endeavors. Novices learn, practitioners sustain, and experts retain critical lessons to make a reputation for themselves. They understand success in science can be achieved in multiple ways, making them the hardest to change (rather than train). Experts understand preferences and can be very open to training. However, changing habits that have led to successful outcomes over many years is something much different. Experts training includes solving safety problems within organizations and mentoring both practitioners and novices. It is the process of sharing lessons learned from years of experience that provides experts an opportunity to fine-tune their problem-solving skills and leadership capabilities, while learning new behaviors and developing safer habits.

When talking about how we train, we must understand that within an organization there are generations who may have vastly different training needs. Understanding the differences between generations of learners (novices, practitioners, and experts) means that different training approaches may be needed. One cannot train an expert like a novice or a novice like an expert without causing significant learner frustration and disengagement. Both generations require unique training experiences. Although there may be some overlap in training between generations, the benefits of training each generation specifically are numerous.

Cognitive Awareness

Laboratory risks are identified and assessed. For laboratory staff, risks are typically managed with the adherence to strategic plans when working with and around laboratory hazards. These plans call for standardized behaviors, and sustained behavior requires several things. Laboratory staff must understand the risks they face, as defined by the organization. It is important to note here that this is risk as defined by the organization and not the laboratory staff. Risk cannot be defined by individual laboratory staff because individual perceptions of the risk will produce individual behavior surrounding the risk. The differences in perceptions will lead to differences in behaviors, and those differences will produce variances in safety outcomes. Risk must be understood by laboratory staff through the perceptions of the organization and not the individual.

After discussing risk, laboratory staff should have an understanding of the benefits they receive by following the SOP. Benefits are not only inclusive of the individual, but of the organization as well. An understanding as to why the organization is making the request and what will be gained as a result of the request is critical when asking for sustained behavior to occur. Most laboratory staff will be working independently and with little to no supervision. This fact alone reinforces the need for all laboratory staff to understand the risks they face and benefits gained by following all SOPs.

Cognitive awareness can be increased in classroom and online settings. The main goal of building cognitive awareness is to ensure consistency of perception and attitude within the organization (the beginning of a safety culture). Classroom and online instruction can assist in developing consistent perceptions of risk across all three learning generations. Once there is an understanding of the organization's perception of risk, laboratory staff must continue to the next step of cognitive awareness training—attitude. Attitude connects how staff view risks (perceptions) and what they do about risk (behaviors) in the laboratory environment.

Attitude is developed by understanding the benefits received by following the SOPs. Some organizations use punishment to influence attitude—focusing on stopping rather than starting behavior. Punishment builds resentment, hides true behavior, and teaches no new behavior. Rather than using punishment, negative reinforcement may be a stronger and more effective attitude influencer. An example of negative reinforcement is an escort system. If someone in the laboratory is not following a SOP, banning him from the laboratory and subjecting him to more training can be wasteful. Leaders may wish to consider developing an escort system, where someone who is not following SOPs is escorted by someone who can ensure that she does; this is a strong way to influence overall attitude surrounding compliance to safety protocols.

Although perception training would allow generations to train together, separating the generations for attitude training is imperative. Novices must be trained to understand they are expected to not only follow SOPs but also to report any laboratory staff not following SOPs. Practitioners must be trained to be escorts for those who are not following SOPs, because it is unnecessary for them to receive training about adherence to SOPs. Experts must be trained to facilitate a discussion with those not following SOPs, leading to increases in SOP compliance throughout the organization.

Consistent perceptions and attitudes are the first two ingredients of a safety culture. Cognitive awareness is verified when laboratory staff (regardless of generation) provide consistent lists of organizational risks and benefits for following SOPs. Cognitive awareness training sets the tone for ensuring consistent views of risk and desire to follow all SOPs connecting risk perceptions, staff attitudes, and behaviors (Table 2).

Behavioral Preparation

Once laboratory staff perceptions and attitudes are consistent, behavioral preparation training should occur. Behavioral preparation training includes four stages: (i) listing the resources needed, (ii) listing the steps of the plan, (iii) demonstrating the skill, and (iv) verifying mastery of skill.

1. *Listing resources needed.* Laboratory staff cannot complete specific SOPs without the resources needed to do so. Therefore, it is important that all laboratory staff be able not only to list resources needed to comply with SOPs, but also to discuss what they would do if a resource is missing and where they would go to acquire the resources needed for them to do their job. Prior to laboratory staff learning the SOPs, they should be able to list all resources needed to follow the SOPs.
2. *Listing the steps of the plan.* Once laboratory staff can list all resources needed to follow a SOP, staff should be asked to list the steps in chronological order. When laboratory staff are capable of properly listing the steps of the SOP in order, they have the cognitive "know-how" of following the SOP. However, knowing and doing are very different things, and failing to separate the two can lead to serious lapses in SOP compliance.
3. *Demonstrating the skill.* As mentioned above, listing the steps of the SOP is very different from demonstrating the SOP. Connecting the cognitive "know-how" with the behavioral skills is absolutely critical in preparing laboratory staff. Trainers should start by demonstrating the skill. Next, trainers should ask laboratory staff to guide them through the process. Once the laboratory staff guide the trainer perfectly, the trainer asks laboratory staff to demonstrate the behavior three times. During this demonstration, the trainer is watching and fine-tuning laboratory staff behavior. Once the laboratory staff have demonstrated the behavior three times, the trainer observes them individually until they demonstrate the behavior perfectly.
4. *Verifying mastery of skill.* Laboratory staff who can list the steps of a SOP and demonstrate each step of the SOP as expected are ready to begin work. To ensure proper habits are formed, trainers inform the laboratory staff that an observational examination will occur 3 months from their start date to verify mastery of specific skills. After 3 months of practicing the SOP, trainers ask laboratory staff to demonstrate the SOP. This verification process ensures the development of good safety habits and catches variations of behavior before they become solidified habits.

TABLE 2.

Laboratory training

Cognitive awareness	Behavioral preparation	Situational response
Consistency of laboratory staff risk perceptions and compliance attitudes	Consistency of laboratory staff skills and abilities	Consistency of laboratory staff application of safety practices
Risk is communal, not individual	Behavioral consistency is needed for consistent safe outcomes	Capability of responding when and where a situation occurs
Understand laboratory risks	Listing of resources needed and steps of SOP	Awareness of location and resources readily available
Understand need to follow laboratory SOPs	Demonstration and verification of skill	Ability to adapt when environmental conditions prevent planned response
Delivered in classroom and/or online settings	Delivered in a simulation or training laboratory	Delivered in laboratory environment

Training all generations together during the behavioral preparation phase fosters a sense of unity and respect among experts, practitioners, and novices. Safety is not achieved by one, but a collective group working as one. Behavioral preparation allows for laboratory staff to learn, work, and succeed together. Up to this point, laboratory members have received cognitive awareness training, which ensures perceptions and attitudes are consistent with one another. Completing behavioral preparation adds a consistency of behavior, which is the inclusion of all ingredients within a "One-Safe" culture—a group of laboratory staff with consistent risk perceptions, attitudes, and behaviors within an organization.

Situational Response

The environment is one of the most underestimated influencers on behavioral success. Laboratory staff may know the SOP and have the skills needed to follow the SOP, but something in the laboratory prevents the SOP from being followed. Situational response training occurs in the laboratory environment, at random and unexpected times, for the sake of learning how the environment impacts the abilities of laboratory staff to follow SOPs. For example, how do laboratory staff behave if an item of PPE is missing? Demonstrate cleaning a spill when the spill kit is missing. Wash your eyes out when you cannot see. There are several situations that may occur, requiring additional skills for laboratory staff to make decisions at critical moments which could lead to high-risk situations.

Situational response training is the continuous component of the training cycle. Training people to drive cars in parking lots makes people better parking lot drivers. Is that the goal though? Training laboratory staff in classrooms may make them better in classroom settings, but the real hazards are found in laboratory environments, which is where situational response training must take place. Preparing laboratory staff cognitively, behaviorally, and situationally is a comprehensive training approach that ensures consistent perceptions of risk, attitudes about SOP compliance, and levels of behavioral skills among laboratory staff needing to work safely.

CONCLUSIONS

Training is an administrative control of biosafety. It takes the written process of risk management and transfers it into human behavior. Human risk factors remain the greatest challenge to safety in the laboratory environment, and addressing these challenges calls for a process of continuous training. Training programs provide opportunity to minimize risks associated with perceived mastery, complacency, and apathy.

Unfortunately, within many organizations, multiple silos have formed, preventing the development of a "One-Safe" culture. Today, one culture does not exist within an organization. In fact, there are many, including the cultures of leadership, workforce, science, and safety. Training is the strategy organizations can use to blend existing cultures forming a consistent approach to science and safety throughout the workforce and leadership. Existing cultures must move from being together to doing together, and training is the tool for cultural blending.

Risk is not static and is always changing. Mitigating risk is a continuous process calling for a continuous training approach. Laboratory staff need initial, on-the-job, and postincident training programs. The provision of SOPs is not enough to mitigate laboratory risks. Continuous training that is scheduled, strategic, and provided in multiple stages ensures that laboratory staff are prepared to mitigate all laboratory risks.

There are multiple training generations that exist within organizations. Differences between novices, practitioners, and experts require unique training approaches. All training generations are in need of cognitive awareness, behavioral preparation, and situational response training. The preparedness and protection of the workforce should be the most important goal for any organization. Leadership must support continuous training initiatives for ample workforce protection.

References

1. **Hallinan JT.** 2009. *Why We Make Mistakes.* Broadway Books, New York.
2. **Moore DA, Healy PJ.** 2008. The trouble with overconfidence. *Psychol Rev* **115:**502–517.
3. **Moray N.** 2003. Monitoring, complacency, scepticism and eutactic behaviour. *Int J Ind Ergon* **31:**175–178.
4. **Stuss DT, Van Reekum RJMK, Murphy KJ.** 2000. Differentiation of states and causes of apathy, p 340–363. *In* Borod JC (ed), *The neuropsychology of emotion.* Oxford University Press, New York, NY.
5. **Lumby J, Foskett N.** 2008. Leadership and Culture, p 43–60. *In* Crow J, Petros G, Lumby P (ed), *International handbook on the preparation and development of school leaders.* Routledge, New York, NY.
6. **Kaufman SG.** 2014. Bioerror and safety culture: the leadership commitment to the preparedness, protection, and promotion of scientists. *Cultures* **1:**38–45.
7. **Schneider W.** 1985. Training high-performance skills: fallacies and guidelines. *Hum Factors* **27:**285–300.
8. **Bogue EG.** 2010. *The Leadership Choice: Designing Climates of Blame Or Responsibility.* WestBow Press, New York, NY.

9. **Gómez C, Rosen B.** 2001. The leader-member exchange as a link between managerial trust and employee empowerment. *Group Organ Manage* **26:**53–69.
10. **Duhigg C.** 2002. *The power of habit: Why we do what we do in life and business.* Random House, New York, NY.
11. **Kaufman SG.** 2013. BSL4 workforce preparedness in hemorrhagic fever outbreaks, p 149–158. *In* Singh SK, Ruzek D (ed), *Viral Hemorrhagic Fevers.* CRC Press, Boca Raton, FL.
12. **Jacob RL.** 2003. *Structured on-the-job training: Unleashing employee expertise in the workplace,* 2nd ed. Berrett-Koehler Publishers, San Francisco, CA.
13. **Bosse HM, Mohr J, Buss B, Krautter M, Weyrich P, Herzog W, Jünger J, Nikendei C.** 2015. The benefit of repetitive skills training and frequency of expert feedback in the early acquisition of procedural skills. *BMC Med Educ* **15:**22.

Biosafety and Biosecurity: Regulatory Impact

29

ROBERT J. HAWLEY AND THERESA D. BELL TOMS

Biological agents have been documented as instruments of warfare and terror (bioterrorism) to produce fear and harm in vulnerable and susceptible populations for thousands of years. The ultimate goal for those using these agents was to inflict harm upon selected individuals or the general human population as well as upon animals and plants (1, 2). The Federal Bureau of Investigation (FBI) defines terrorism as the "unlawful use of force against persons or property to intimidate or coerce a government, the civilian population, or any segment thereof, in the furtherance of political or social objectives" (3). Basically, bioterrorism is a form of biological warfare. Biological warfare is the intentional use of etiologic agents, such as viruses, bacteria, fungi, or toxins derived from living organisms, to produce death or disease in humans, animals, or plants (4). An etiologic agent is "a viable microorganism or its toxin that causes or may cause human disease, and includes those agents listed in 42 C.F.R. 72.3 of the U.S. Department of Health and Human Services (DHHS) regulations and any material of biologic algorithm that poses a degree of hazard similar to those organisms" (5). A toxin, also included as an etiologic agent, is defined as "toxic material of biologic origin that has been isolated from the parent organism. The toxic material of plants, animals, or microorganisms" (6). Potential agents that could be used in a bioterrorist event include those causing the diseases anthrax (*Bacillus anthracis*), plague (*Yersinia pestis*), tularemia (*Francisella tularensis*), the equine encephalitides (Venezuelan equine encephalitis and eastern equine encephalitis), hemorrhagic fever viruses (arenaviruses, filoviruses, flaviviruses, and bunyaviruses), and variola virus (smallpox). Some of the toxins that could be used in a bioterrorism event include botulinum toxin from *Clostridium botulinum*; ricin toxin from the castor bean *Ricinus communis*; the trichothecene mycotoxins from *Fusarium*, *Myrothecium*, *Trichoderma*, *Stachybotrys*, and other filamentous fungi; staphylococcal enterotoxins from *Staphylococcus aureus*; and the toxins from marine organisms such as dinoflagellates, shellfish, and blue-green algae. The list of potential etiologic agents is quite extensive (7). However, the list of agents that could cause mass casualties by the aerosol route of exposure is considerably smaller (8–17).

HISTORICAL ACCOUNTS OF BIOLOGICAL WARFARE AND BIOTERRORISM

There have been reports of biowarfare ever since antiquity (see http://www.penn.museum/documents/publications/expedition/pdfs/47-1/fleming.pdf). Attacking Tatar forces hurled cadavers of their diseased soldiers into the city to initiate an epidemic of plague during the 14th-century siege of Caffa by the use of trebuchets (18). Ancient aboriginal people, the forefathers of South American tribesmen, used poison darts to subdue their enemy. Some poisons were obtained from plants; others came from animals such as the poison arrow frog (also known as the poison dart frog). The British deliberately used smallpox virus against Native American tribes during the French and Indian War (1754–1763). On June 24, 1763 during the Pontiac Rebellion, two blankets and a handkerchief contaminated with smallpox virus from an outbreak at Fort Pitt, PA, were given to the immunologically naive Delaware Indians (15, 19). A brief chronology of some additional accounts of biological warfare is provided in Table 1 (2, 20).

There have been several accounts of the use or attempted use of additional agents for the purpose of bioterrorism. An example of the use of a toxin as a bioterrorism agent was the failed attempt to assassinate the Bulgarian exile Vladimir Kostov in Paris, France, in 1978. A ricin-containing pellet was discharged from an umbrella gun into his back (see https://coldwarradios.blogspot.com

TABLE 1.

Chronology of additional accounts of biological warfare and terrorism

Time period and location (reference)	Intent or action	Agent(s) or action
World War I; Germany (9)	Contaminate animal feed and infect livestock for export to Allied forces, and infect Romanian sheep for export to Russia	*Bacillus anthracis* (anthrax), *Burkholderia mallei* (glanders)
World War I; Germany (9)	Operatives in Mesopotamia planned to inoculate mules and infect horses of the French cavalry in France	*B. mallei*
World War I; Germany (9)	Livestock in Argentina for export to Allied forces were infected, resulting in the death of more than 200 mules	*B. anthracis* and *B. mallei*
1932–1945; Japan, Unit 731; Ping Fan, Manchuria (19, 87, 88, 89)	Biological warfare experimentation on prisoners	*B. anthracis*, *Neisseria meningitidis*, *Shigella* spp., *B. mallei*, *Salmonella typhosa*, *Vibrio cholerae*, *Yersinia pestis*, and smallpox virus
	Chinese cities attacked (cultures tossed into homes and sprayed from aircraft; potentially infected fleas harvested in the laboratory were released from aircraft)	*Y. pestis* and other biological warfare agents
	Water supplies and food items contaminated	*B. anthracis*, *Shigella* spp., *Salmonella* spp., *V. cholerae*, and *Y. pestis*
World War II; Germany (82)	Prisoners in Nazi concentration camps forcibly infected; treated with investigational vaccines and drugs	*Rickettsia prowazekii*, *Rickettsia mooseri*, hepatitis A virus, and *Plasmodium* spp.
World War II (May 1945); Germany (90)	Polluted a large reservoir in northwestern Bohemia	Sewage
1941–1942; Great Britain; Gruinard Island (91)	Conducted bomb experiments with weaponized agent	*B. anthracis*
1941; Koch Foundation Laboratories, Paris, France (92)	With German experts, containerized toxin for delivery by airplane	*Clostridium botulinum* toxin
1943; United States (Camp Detrick, MD; Horn Island, MS; Granite Peak, UT) (93)	Offensive and defensive biological warfare program	*B. anthracis* and *Brucella suis*
1950–1953; United States (Pine Bluff, AR) (21, 94)	Production facility constructed	
1955; United States (Camp Detrick, MD) (95)	Aerosol studies (1-million-liter spherical aerosolization chamber) and efficacy of vaccines, prophylaxis, and therapies under development	*Francisella tularensis* and *Coxiella burnetii*
	Production and storage procedures and techniques, aerosolization methods, behavior of aerosols over large geographic areas, and effects of solar irradiation and climatic conditions	Simulants used were *Aspergillus fumigatus*, *Bacillus subtilis* (formerly called *Bacillus globigii*), and *Serratia marcescens*

/2012/06/murder-attempt-on-vladimir-kostov.html). Ten days later, the Bulgarian exile Georgi Markov was assassinated in London, England, with a ricin-filled polycarbonate ball (21). An umbrella gun, developed by the KGB and deployed by the Bulgarian Secret Service, discharged the tiny pellet (the size of the head of a pin) into the subcutaneous tissue of the exile's leg while he was waiting at a bus stop. He died 3 days later from multiple organ failure despite care administered during his hospitalization (9).

Shoko Asahara, the leader of the Aum Shinrikyo (Supreme Truth) cult, sought to establish a theocratic state in Japan (22). The cult disseminated the chemical agent sarin (agent GB, which inhibits acetylcholinesterase, resulting in a disruption of nerve impulse transmission) in the Tokyo subway system in 1995. This incident resulted in 12 deaths and over 5,000 injuries—both physical and psychological—but could have resulted in hundreds of thousands had the cult been more effective. Hospitals and doctors' offices were overwhelmed with casualties. What is not well known is that on 10 prior occasions between 1993 and 1995, the cult had been unsuccessful in its attempts to disperse various quantities of either anthrax or botulinum toxin in downtown Tokyo (19). The cult attempted to obtain Ebola virus from Zaire in 1993, and in 1994, it discussed the possibility of using Ebola virus as a biological weapon. It also cultured and experimented with the agents of anthrax, Q fever, and cholera and botulinum toxin (23, 24).

Similar acts of bioterrorism in the United States were carried out by followers of the Indian guru Bagwan Shree Rajneesh. The Rajneeshee cult, an Indian religious group, contaminated restaurant salad bars with *Salmonella enterica* serovar Typhimurium in Oregon in 1984. Over 750 cases of enteritis resulted from this attack; 45 individuals required hospitalization. The cult's motivation was to incapacitate voters so that they could win a local election and seize political control of Wasco County (in fact, they imported over 2,000 homeless people to vote in the election). It is important to note that the biological attack was not discovered until 1985, when a cult member confessed the contamination incident (19).

Larry Wayne Harris, with links to the Christian Identity and the Aryan Nation (a white supremacist group), wanted to alert Americans to the Iraqi biological warfare threat and sought a separate homeland for whites in the United States. Harris made vague threats against U.S. federal officials on behalf of right-wing "patriot" groups. He had obtained the *B. anthracis* vaccine strain as well as *Y. pestis* through the mail by ordering from a national depository and discussed the dissemination of biological warfare agents with crop duster aircraft and other methods. In 1995, he was arrested and detained in Ohio for possessing plague bacteria (*Y. pestis*) but could only be convicted of mail fraud. At the time, there was no law that prohibited the possession of these types of organisms. Harris was again arrested in 1998 after making threatening remarks to U.S. officials and openly talking about biological warfare terrorism (24). He was arrested in Las Vegas, NV, for possession of anthrax bacilli, but the organism was found to be the vaccine strain (25).

A terrorism issue related to anthrax disease surfaced in October of 2001 in Boca Raton, FL. Robert Stevens, a newspaper photo editor employed at American Media, Inc., died on October 5 of pulmonary anthrax, 1 day after being admitted to a local hospital (see http://www.npr.org/2011/02/15/93170200/timeline-how-the-anthrax-terror-unfolded). His illness was subsequently determined to be the result of his exposure to *B. anthracis* spores sent through the company mail system (26). An office of a U.S. Senator received a letter that tested positive for anthrax (*B. anthacis*) on October 15, 2001. On October 20, traces of *B. anthracis* were found in a mail bundling machine in a House of Representatives office building near the Capitol in Washington, DC. More than 2,000 postal employees at two mail facilities were tested and treated for possible exposure to *B. anthracis* on October 21. A postal worker who worked at the Brentwood mail facility in Fairfax, VA, was diagnosed on October 21 with inhalation anthrax. The Brentwood facility processes almost all mail to the District of Columbia. On October 22, two of four postal workers employed at the Brentwood mail facility that had been diagnosed with inhalational anthrax died as a result of *B. anthracis* infection (27). Six individuals in New Jersey and New York were subsequently diagnosed with symptoms of cutaneous anthrax, allegedly from primary or associated exposure to mail contaminated with *B. anthracis* spores (28). These series of letters targeting the media and Congress resulted in 22 individuals becoming infected and the death of five (29).

Subsequent to the initial mailing of *B. anthracis* spores in October of 2001, the 2001 anthrax attacks investigation was given the name Amerithrax from its FBI case name. According to the FBI, the ensuing investigation became "one of the largest and most complex in the history of law enforcement" (30). In August 2002, it became known that Dr. Steven J. Hatfill was a person of interest. While working as a researcher at the U.S. Army Medical Research Institute of Infectious Diseases (USAMRIID) from 1997 to 1999, Dr. Hatfill had unrestricted access to the Ames strain of *B. anthracis*, the same strain used in the 2001 mailings. During the investigation, scientific breakthroughs (genetic sequencing) led investigators to conclude that the RMR-1029 flask was the parent material to the anthrax powder used in the mailings. It was then determined that Dr. Hatfill could not have been the mailer because he never had access to the particular biocontainment suites at USAMRIID that held the RMR-1029 flask.

The FBI's genetic analysis of the organism used in the attacks eventually led investigators to exclude him conclusively as a suspect (30). Dr. Hatfill denied that he had anything to do with the anthrax letters and said irresponsible news media coverage based on government leaks had destroyed his reputation (31). Dr. Hatfill settled with the Justice Department for $5.8 million (consisting of a $2.825 million one-time payment in cash and an annuity paying $150,000 a year for 20 years) (32). During the 8 years of inquiries, the Department of Justice (DOJ) conducted a far-ranging, technically complex, and expensive investigation and eventually identified the mailer as Dr. Bruce E. Ivins, a microbiologist at USAMRIID. On the basis of their evidence, the investigation into the anthrax letters attacks of 2001 was concluded on February 19, 2010 (30). However, to this day, there are many questions and facts that challenge the involvement of Dr. Ivins in the anthrax letter attacks (32–40). Dr. Ivins was never charged because he apparently committed suicide in July of 2008.

The federal government has spent over $60 billion on biodefense efforts between 2001 and 2011 (41). Questions to consider are: Is the United States safer from a bioterrorism incident? Has the spending been worth it? For the first question, the answer is yes, but gaps still exist. For the second question, research results cannot be certain in advance, and it is not credible to expect every dollar invested to produce the anticipated promising result. However, several investments in biosecurity have clearly provided benefits (42). Despite this huge investment, a few bioterrorism incidents have occurred subsequent to the Amerithrax incident. For instance, at least eight incidents in the United States involving ricin have been documented (43). As of June 1, 2016, we are not aware of any other publicly documented incidents unrelated to the Amerithrax that involved biological agents.

It is interesting to note that during the period from 2010 to 2011, 55% of local health departments reduced or eliminated programs to protect Americans and keep them safe, including emergency preparedness and immunizations. This reveals a major challenge in this economic crisis and its effect on health departments. Jobs that were lost or reduced in state and local health departments included health workers such as public health physicians and nurses, laboratory specialists, and epidemiologists (44, 45). Although the reduction or elimination of jobs is a sad occurrence, this may be perhaps considered fortuitous as services from these health care providers were not needed. A handbook was recently developed by the FBI and Centers for Disease Control and Prevention (CDC) to make possible the use of resources and maximize communication and interaction between law enforcement and public health officials. This combined effort will serve to minimize potential barriers prior to and during the response to a biological threat. The handbook enhances the appreciation and understanding of the expertise of the FBI and the CDC, discusses the procedures and methods involved in a response to a biological threat, provides potential solutions that may be adapted to meet the needs of various agencies and jurisdictions, and demonstrates effective collaboration between law enforcement and public health officials (46).

U.S. PROGRAMS TO DETER BIOTERRORISM

Select Agent Program

The Antiterrorism and Effective Death Penalty Act of 1996 was passed by Congress after the Larry Wayne Harris incident to control the transport and receipt of hazardous biological agents. The law directed the DHHS and CDC to establish new regulatory requirements governing the receipt and transfer of those etiologic agents classified as Select Agents (47). Select Agents are biological agents that have been declared by the U.S. Department of Health and Human Services or by the U.S. Department of Agriculture (USDA) to have the "potential to pose a severe threat to public health and safety" (48).

The initial rule, Title 42 Code of Federal Regulations (C.F.R.) Part 72, was issued on April 15, 1997, and was updated in 2002 as an interim rule with Title 42 C.F.R. Part 73 (49) to include possession, use, and transfer of Select Agents under the regulatory requirements. Likewise, the USDA issued similar interim rules to control the use, possession, receipt, and transfer of agents posing a serious threat to animals or plants (50). The DHHS and the USDA published final rules on March 18, 2005. The final rules addressed public responses to the interim rule and harmonized the DHHS and USDA regulations (51, 52). The Select Agents and Toxins regulated by the final rule are listed in Table 2 (53). Shiga-like ribosome-inactivating proteins and certain genetically modified organisms are also regulated under the Select Agent rule.

Since the inception of the Select Agent Program and regulations in 1997, costs at the laboratory level have increased due, in part, to preparation for inspections and costly facility modifications and safety procedural changes. The initial thought is that these costs would tend to have a negative impact on the final product(s) of a research facility and its programs—as far as the cost to produce the product(s) and the timeline involved to provide the product(s) to the public. It has been suggested that entities registered with the Select Agent Program join together to coordinate training activities in an effort to defray costs while sharing lessons learned to make overall improvements to facilities and operations (54). These constraints provide for an in-depth analysis, and the reader is encouraged to read a further discussion on the Select Agent Program (55).

TABLE 2.

Select Agents and Toxins[a]

Agency	Virus(es)	Bacterium	Fungus	Toxin(s)
HHS	Crimean-Congo hemorrhagic fever (CCHF)	*Coxiella burnetii*** (causative agent of Q fever)	Diacetoxyscirpenol	Abrin
	Eastern equine encephalitis	*Francisella tularensis**		
	Ebola*	*Rickettsia prowazekii***		Conotoxins
	Lassa fever	Staphylococcal enterotoxins (A, B, C, D, E subtypes)		Ricin
	Lujo	*Yersinia pestis**		Saxitoxin
	Marburg*	Botulinum neurotoxins*		T-2
	Monkeypox	Botulinum neurotoxin-producing species of *Clostridium**		Tetrodotoxin
	Reconstructed 1918 pandemic influenza virus	*Bacillus cereus* biovar *anthracis****		
	Severe acute respiratory syndrome coronavirus (SARS-CoV)			
	South American hemorrhagic fever viruses (Chapare, Guanarito, Junin, Machupo, Sabia)			
	Tick-borne encephalitis complex (flavi-) virus: (Far Eastern and Siberian subtypes)			
	Kyasanur Forest disease virus			
	Omsk hemorrhagic fever virus			
	Variola major virus* (smallpox)			
	Variola minor virus* (alastrim)			
DHHS and USDA	Hendra virus	*Bacillus anthracis**		
	Nipah virus	*Bacillus anthracis* Pasteur strain**		
	Rift valley fever virus	*Brucella abortus***		
	Venezuelan equine encephalitis virus	*Brucella melitensis***		
		*Brucella suis***		
		*Burkholderia mallei**		
		*Burkholderia pseudomallei**		
USDA	African horse sickness virus	*Mycoplasma capricolum*		
	African swine fever virus	*Mycoplasma mycoides*		
	Avian influenza virus			
	Classical swine fever virus			
	Foot-and-mouth disease virus			
	Goat pox virus			
	Lumpy skin disease virus			
	Newcastle disease virus			
	Peste des petits ruminants virus			
	Rinderpest virus			
	Sheep pox virus			
	Swine vesicular disease virus			

[a]From 9 C.F.R. Part 121 and 42 C.F.R. Part 73 (52) and the *Federal Register* Proposed Rule, January 19, 2016 (57). *Denotes Tier 1 agent; **denotes proposed for removal from the DHHS list of Select Agents; ***, Interim Final Rule addition 14 September 2016 (96).

In addition to the above-mentioned microorganisms and toxins, there are also additional "overlap Select Agents and Toxins." Overlap agents are agents that are covered by both the USDA and DHHS regulatory requirements and pose risks to both humans and animals (50). Overlap agents are also listed in Table 2. Again, certain genetically modified organisms and nucleic acids are also covered in this category of Select Agents.

However, after receiving requests for exclusions and careful review of research material, the DHHS (CDC) and USDA have determined that some attenuated strains are not subject to the Select Agent requirements of Titles 42 C.F.R. Part 73 and 9 C.F.R. Part 121. The final rule published on October 5, 2012, identified criteria for exclusion of low pathogenic strains, Select Agents or Toxins in their naturally occurring environment, and animals inoculated with or exposed to a DHHS Select Toxin (56). Agents identified as Select Agents and Toxins are subject to Select Agent regulations until testing can document exclusion for the particular agent by "strain, subtype, or pathogenicity level." The excluded strains are listed in Table 3.

In January 2016, the DHHS issued a proposed rule in the *Federal Register* to amend the current list of DHHS Select Agents. The proposal included the removal of six biological agents, while including "provisions to address the inactivation of Select Agents." The agents proposed for exclusion are denoted in Table 2 (57, 58).

The CDC, included within the DHHS, is the agency responsible for implementing the Select Agent Program. The Select Agent regulation is comprehensive in scope. It applies to all entities that possess, use, transfer, or store Select Agents and/or Toxins. The Federal Select Agent Program is in close collaboration with the FBI Criminal Justice Information Services (CJIS) implementing Section 7A of the C.F.R., under which each entity must be issued a certificate of registration by the federal Select Agent Program (CDC and/or the USDA) (59). The application process involves the facility designating a responsible official and possibly an alternate responsible official, and declaration of all personnel who will have access to Select Agents covered by the regulation. A security risk assessment on each person with access must be performed by the FBI in cooperation with the DOJ, which includes background security checks, assignment of a unique DOJ identifying number, and submission of fingerprints. An inspection of the registered facility by the CDC and/or USDA is mandatory during the time period covered by the facility's registration. In addition to biological safety (biosafety) requirements, the regulation specifies compliance parameters for site threat assessments, inspections, access, training, registration, transfer, record keeping, physical security provisions, and incident reporting, to include notification of loss, theft, or release of agent. Inspections are conducted to ensure that all requirements of the final regulations are met, and that facilities are constructed and managed according to the 5th edition of *Biosafety in Microbiological and Biomedical Laboratories* (BMBL) (60). The BMBL contains the biosafety guidelines and recommendations for work with microbial agents, as determined by the CDC and NIH.

Security Requirements

Existing guidelines were changed to develop and upgrade security plans with implementation procedures for

TABLE 3.

Attenuated strains of DHHS Select Agents and Toxins excluded[a]

Agency	Virus(es)	Bacterium	Fungi	Toxin(s)
DHHS	Eastern equine encephalitis	Botulinum neurotoxins		Conotoxins
	Ebola	*Coxiella burnetii*		Tetrodotoxin
	Junin virus	*Francisella tularensis*		
	Lassa fever virus	Staphylococcal enterotoxins (SE)		
	Monkeypox virus	*Yersinia pestis*		
	Severe acute respiratory syndrome coronavirus (SARS-CoV)			
DHHS and USDA	Rift Valley fever virus	*Bacillus anthracis*		
	Venezuelan equine encephalitis virus	*Brucella abortus*		
		Brucella melitensis		
		Burkholderia mallei		
		Burkholderia pseudomallei		
USDA	Avian influenza virus (highly pathogenic)			

[a]See http://www.selectagents.gov/SelectAgentsandToxinsExclusions.html.

the development and implementation of risk assessments, security plans, and incident response plans (61, 62). To meet the new final federal rules, those facilities using, storing, and receiving Select Agents must now perform a vulnerability assessment to identify both internal and external threats, provide a plan to mitigate those threats, and develop an implementation program to address security concerns and Select Agent pathogen protection (51). Both internal threats (disgruntled employees, financial motivation, or personal threats) and external threats (intruders such as terrorists or others with the means to manipulate agents) must be identified. The biosecurity risk assessment must include an asset assessment to include the motivations to target the assets (63). An incident response action plan must then be developed and implemented. Furthermore, primary assets (the agents themselves) and secondary assets (anything that may assist the adversary in acquiring the primary assets) must be identified to prioritize security risks. For those individuals working with Tier 1 agents (those agents that have the potential to pose a severe threat to public health and safety, to animal health, or to animal products) (64), a suitability assessment is required before granting authorized access to these higher-level agents. The highest levels of security are assigned to those assets involving the most risk.

Security plans provide procedures to prevent unauthorized access, theft, loss, or release of Select Agents and Toxins. The plans must also include procedures for the removal of suspicious persons. Emergency response plans address reactionary processes for responding to theft, loss, release, and security breaches as a result of weather-related events, inventory discrepancies, workplace violence, bombs, suspicious packages, emergencies involving utility outages, and other natural, man-made events. Cameras with 24-hour real-time monitoring and intrusion detection systems with monitoring systems may also be included, depending on the agents in possession.

Issues of physical security and time to respond to an unauthorized intrusion can be addressed in the incident response plan. Other points to be addressed include potential emergency situations involving Select Agents, such as accidental release or theft, and the identification of those individuals responsible for the agents, those with access, and those maintaining tracking and inventory logs for accountability of Select Agent volumes and storage locations. Individuals with appropriate emergency response training should be identified and listed on an emergency contact list within the containment areas to facilitate prompt access to those knowledgeable in responding to a biological incident (spill, contamination, or employee medical emergency). Annual drills to include carrying out worst-case scenarios in which security of assets is compromised are prudent to ensure procedures are comprehensive and should involve local emergency responders to obtain input and guidance (65). Training ensures that proper personal protective equipment (PPE) is donned and functioning properly prior to entry into any containment suite; such equipment may include gloves, Tyvek coveralls, autoclavable shoes or boots, protective eyewear, face masks, and negative- or positive-pressure respirators on the basis of appropriate biosafety level determination.

In addition to the security risk assessment required by the Select Agent Program, a personnel reliability program (PRP) or biosurety program is another method that may be used to ensure that Select Agents are secure in a facility (66, 67). The PRP is designed to determine an employee's individual integrity and involves a background security investigation and personnel screening that must be executed by the employer. A PRP limits access to those employees who are cleared to enter the areas where Select Agents are stored and used. These employees can be identified with badges or biometrics specific to that individual that may be used to grant containment access. Procedures for visitor access must also be implemented, including issuance of visitor badges, escorting of visitors at all times, and employees reporting any suspicious activity immediately to those in charge of securing the site. Biosecurity includes physical security, Select Agent accountability, and personnel reliability in an effort to prevent unauthorized access to biological Select Agents and Toxins (68).

A site-specific risk assessment must also be conducted to determine procedures for information systems security control (69). Procedures must be established to determine requirements for general systems access to include log-ins, server security, and software security.

Inventory of Select Agents and Toxins

Requirements for the inventory of Select Agents has evolved imposing stricter requirements to include long-term storage. Over the years since inception of the Select Agent regulations, the struggle to define a "working" quantity of viable Select Agents in the laboratory has been a continuing challenge (70). The federal Select Agent Program has issued the Guidance on the Inventory Requirements for Select Agents and Toxins (71) to delineate requirements for facility inventory and reporting procedures.

Transport of Select Agents and Toxins

Provisions within the Select Agent rule allow for both importation and interstate transport without the need to obtain an importation permit. Titles 42 C.F.R. 73.16 (CDC) and 9 C.F.R. 121.16b (USDA) allow for the transfer of Select Agents and Toxins without an importation permit.

Entities must establish procedures to address outgoing and inbound shipments of biological Select Agents and Toxins as well as how to address unexpected packages. There must be protocols to address step-by-step how to inspect packages and how to handle intraentity transfers, to include packaging, labeling, marking, transportation documents to accompany the parcel, and delivery by authorized and cleared individuals (72).

Employee Medical Surveillance

The occupational health hazards associated with working with Select Agents make a medical surveillance program as recommended in BMBL for the specific organisms a necessity. Although not required by the Select Agent regulation, most employees working with Select Agents may have a drug screen upon hire, an annual physical, initial blood sample obtained before working in containment, serum banked for reference, and enrollment in an immunization program (as applicable). A reporting mechanism must be in place to guarantee notification of appropriate personnel should a work-related injury or illness occur. Standard operating procedures (SOPs) must be implemented and training conducted for medical incident emergency response and emergency contacts. Interim medical treatment measures with necessary medical supplies may need to be provided in a convenient location within the containment suite. Capabilities to administer postexposure prophylaxis protocols are imperative to address after-hours exposures and to include diagnostic tests and expert medical evaluation (73). A designated medical care facility should be identified, and all risks associated with Select Agent work should be communicated to the medical care staff, as well as to local health departments. If an exposure victim should not be removed from containment because of the agent involved, alternative medical treatment methods need to be established and coordinated. The final Select Agent rules added the requirement that drills or exercises of security, biosafety, and incident response plans be conducted at least annually (51, 52).

Additional Regulatory Requirements

In addition to the Select Agent Program, there are several other regulatory requirements, recommendations, and guidelines that complement and reinforce DHHS mandates. Title 42 C.F.R. Part 1003 (51), a DHHS regulation from the Office of the Inspector General, issues monetary penalties for noncompliance based on assessments. Title 42 C.F.R. Part 71.54 (74) provides importation requirements applicable to Select Agents. Title 40 C.F.R. Part 300, National Oil and Hazardous Substances Pollution Contingency Plan, provides regulatory requirements issued by the U.S. Environmental Protection Agency (EPA) (75) to be used in the case of an accidental release. For work with recombinant organisms, the risk assessment and biological containment requirements are specified in the *NIH Guidelines* (76). The U.S. guidelines and regulations (60, 77, 78) apply to a broad scope of pathogens and bacterial toxins, including the Select Agents.

Army Regulation 385-10 contains Department of Defense regulations for the handling of toxins (79), and Department of the Army Pamphlet 385-69 *Biological Defense Safety Program* (80) addresses technical safety requirements for Select Agent work in biocontainment laboratories. The U.S. Department of Transportation (DOT) Title 49 C.F.R. Parts 171–180, specifically Part 173.134 Division 6.2 (Infectious Substances) (81), applies to the domestic shipment of infectious agents, including Select Agents, via ground transportation. The International Air Transport Association (IATA) regulates both domestic and international shipment of Select Agents by air (82). These regulations describe proper packaging, labeling, markings, and shippers' declaration documentation requirements for transport and shipment of all Select Agents. The USDA departmental manual 9610-1 (83) has also defined parameters for security policy and procedures for biosafety with respect to "pathogens of consequence to agriculture." The USDA manual further defines a new biosafety level 3 agriculture (BSL3-Ag) containment suite for Select Agent work specific to agricultural research with larger animals. For example, the USDA requires a BSL3-Ag environment to accommodate additional provisions to enhance containment, including filtration of supply and exhaust air, sewage decontamination, an exit shower, and facility integrity testing (83). Finally, the White House administration issued a joint memo in an effort to enhance biosafety and biosecurity at infectious disease laboratories to include reviews and recommendations with specific plans and timelines to address implementation of each recommendation, with semiannual reviews on progress with program implementation (84).

As the environmental parameters and other conditions within the regulatory requirements, guidelines, and recommendations change, facility review committees (e.g., institutional biosafety committees) and timely updates are necessary to reflect those changes. The DOT and IATA shipping and transport regulations, as well as the BMBL, are continuously being updated to accommodate any significant changes in the federal rules. Additions include a revised list of dangerous goods in accordance with the United Nations recommendations, new training requirements regarding dangerous goods security, revised shipper's declarations, and reclassification criteria for infectious substances and diagnostic specimens. The

BMBL contains additional provisions for strengthening biosafety and biosecurity, decontamination methods, a risk management approach to security, occupational medicine recommendations, and new Select Agent summary statements (60).

CONCLUSION

The potential for the use of biological weapons by many countries (85) and the documented incidents of biological terrorism since antiquity have resulted in an increased awareness of the threat posed by many agents of microbial origin. This increased level of awareness has resulted in numerous changes that have affected the daily lives of millions worldwide. These changes include an increase in physical security as seen in areas of high personnel traffic and gathering and the security of biological agents at research and development institutions, hospitals, and pharmaceutical facilities. Perhaps the greatest impact of biological security and accountability is presently being experienced by individuals involved in basic medical research. Many regulations, guidelines, and policies have been adopted to safeguard both the American public, the environment, and our collaborating nations. Although intrusive at times and inconvenient, most of these requirements are necessary to reduce the possibility of Select Agent use as a biothreat. The approach to protection must be scientifically and risk based (86). This requires the continued involvement of knowledgeable persons and organizations in the regulatory process to ensure that the regulations reflect a reasonable approach to this difficult and complex issue.

We gratefully acknowledge Evelyn M. Hawley for her critical review of the manuscript.

References

1. **Hawley RJ, Kozlovac JP.** 2013. A perspective of biosecurity: past to present, p 27–34. *In* Burnette R (ed), *Biosecurity: Understanding, Assessing, and Preventing the Threat.* John Wiley & Sons, Hoboken, NJ.
2. **Carus WS.** 2015. The history of biological weapons use: what we know and what we don't. *Health Secur* **13:**219–255. http://www.swissbiosafety.ch/wp-content/uploads/2015/09/The-History-of-Biological-Weapons.pdf.
3. **National Institute of Justice.** 2011. Terrorism. http://www.nij.gov/topics/crime/terrorism/pages/welcome.aspx.
4. **Stern J.** 1999. Definitions, p 20. *In* Stern J (ed), *The Ultimate Terrorists.* Harvard University Press, Cambridge, MA.
5. **Headquarters, Department of the Army.** 2013. Army Regulation 385–10, Safety, The Army Safety Program. Section II, Terms, p. 129. Washington, DC. The current version can be accessed at http://www.apd.army.mil/pdffiles/r385_10.pdf.
6. **Headquarters, Department of the Army.** 2013. Army Regulation 385–10, Safety, The Army Safety Program. Section II, Terms, p. 145. Washington, DC. The current version can be accessed at http://www.apd.army.mil/pdffiles/r385_10.pdf.
7. **Stern J.** 1999. Antipersonnel biological warfare agents, p 164–165. *In* Stern J (ed), *The Ultimate Terrorists.* Harvard University Press, Cambridge, MA.
8. **Franz DR.** 1997. Defense against toxin weapons, p 603–619. *In* Sidell FR, Takafuji ET, Franz DR (ed), *Medical Aspects of Chemical and Biological Warfare. (TMM series. Part I; warfare, weaponry, and the casualty.).* Border Institute, Walter Reed Army Medical Center, Washington, DC.
9. **Eitzen EM, Takafuji ET.** 1997. Historical overview of biological warfare, p 415–423. *In* Sidell FR, Takafuji ET, Franz DR (ed), *Medical Aspects of Chemical and Biological Warfare. (TMM series. Part I, warfare, weaponry, and the casualty.).* Borden Institute, Walter Reed Army Medical Center, Washington, DC.
10. **Eitzen E, Pavlin J, Cieslak T, Christopher G, Culpepper R.** 1998. *Medical Management of Biological Casualties Handbook.* U.S. Army Medical Research Institute of Infectious Diseases, Fort Detrick, MD.
11. *NATO Handbook on the Medical Aspects of NBC Defensive Operations AMedP-6(B), Part II—Biological.* 1996. Medical classification of potential biological warfare agents, p. A-3–A-4. Departments of the Army, the Navy, and the Air Force, Annex A. See https://fas.org/irp/doddir/army/fm8-9.pdf.
12. **Anonymous.** 1998. What are aerosols and sprays and how are they used to deliver biological weapons? p. 39. *In Chem-Bio: Frequently Asked Questions—Guide to Better Understanding Chem-Bio,* version 1.0. Tempest Publishing, Alexandria, VA.
13. **Burrows WD, Renner SE.** 1998. Biological warfare agents as potable water threats, p 1–5. *In* Burrows WD, Renner SE (ed), *Medical Issues Information Paper No. IP-31–017.* U.S. Army Center for Health Promotion and Preventive Medicine, Aberdeen Proving Ground, MD.
14. **Peters CJ, Dalrymple JM.** 1990. Alphaviruses, p 713–761. *In* Fields BN, Knipe DM (ed), *Virology,* 2nd ed. Raven Press, Ltd, New York, NY.
15. **Christopher GW, Cieslak TJ, Pavlin JA, Eitzen EM Jr.** 1997. Biological warfare. A historical perspective. *JAMA* **278:**412–417.
16. **Christopher GW, Eitzen EM Jr.** 1999. Air evacuation under high-level biosafety containment: the aeromedical isolation team. *Emerg Infect Dis* **5:**241–246.
17. **Heymann DL.** 2015. Arboviral Hemorrhagic Fevers, p 43–54. *In* Heymann DL (ed), *Control of Communicable Diseases Manual,* 20th ed. American Public Health Association, Washington, DC.
18. **Wheelis M.** 2002. Biological warfare at the 1346 siege of Caffa. *Emerg Infect Dis* **8:**971–975.
19. **Langford RE.** 2004. Brief history of biological weapons, p 139–150. *In* Langford RE (ed), *Introduction to Weapons of Mass Destruction.* John Wiley & Sons, Inc, Hoboken, NJ.
20. **Thalassinou E, Tsiamis C, Poulakou-Rebelakou E, Hatzakis A.** 2015. Biological warfare plan in the 17th century—the siege of Candia, 1648–1669. *Emerg Infect Dis* **21:**2148–2153.http://wwwnc.cdc.gov/eid/article/21/12/13-0822_article
21. **Mangold T, Goldberg J.** 1999. Truth and reconciliation, p 261–262. *In* Mangold T, Goldberg J (ed), *Plague Wars: a True Story of Biological Warfare.* Macmillan Publishers Ltd, London.
22. **Stern J.** 1999. Getting and using the weapons, p 60–65. *In* Stern J (ed), *The Ultimate Terrorists.* Harvard University Press, Cambridge, MA.
23. **Olson KB.** 1999. Aum Shinrikyo: once and future threat? *Emerg Infect Dis* **5:**513–516.
24. **Tucker JB.** 1999. Historical trends related to bioterrorism: an empirical analysis. *Emerg Infect Dis* **5:**498–504.
25. **Snyder JW, Check W.** 2001. *Bioterrorism Threats to Our Future. The Role of the Clinical Microbiology Laboratory in Detection, Identification, and Confirmation of Biological Agents.* American Academy of Microbiology, Washington, DC.

26. **Traeger MS, Wiersma ST, Rosenstein NE, Malecki JM, Shepard CW, Raghunathan PL, Pillai SP, Popovic T, Quinn CP, Meyer RF, Zaki SR, Kumar S, Bruce SM, Sejvar JJ, Dull PM, Tierney BC, Jones JD, Perkins BA, Florida Investigation Team.** 2002. First case of bioterrorism-related inhalational anthrax in the United States, Palm Beach County, Florida, 2001. *Emerg Infect Dis* **8:**1029–1034.
27. **Dewan PK, Fry AM, Laserson K, Tierney BC, Quinn CP, Hayslett JA, Broyles LN, Shane A, Winthrop KL, Walks I, Siegel L, Hales T, Semenova VA, Romero-Steiner S, Elie C, Khabbaz R, Khan AS, Hajjeh RA, Schuchat A, and members of the Washington, D.C., Anthrax Response Team.** 2002. Inhalational anthrax outbreak among postal workers, Washington, D.C., 2001. *Emerg Infect Dis* **8:**1066–1072.
28. **Morse SA.** 2002. Bioterrorism: the role of the clinical laboratory. *BD Newsletter* **13:**1–10.
29. **Saathoff G, DeFrancisco G.** 2011. Report of the Expert Behavioral Analysis Panel. Research Strategies Network, Vienna, VA. http://www.dcd.uscourts.gov/dcd/sites/dcd/files/unsealed Doc031011.pdf.
30. **United States Department of Justice (USDOJ).** 2010. Amerithrax Investigative Summary. http://www.justice.gov/archive/amerithrax/docs/amx-investigative-summary.pdf.
31. **Lichtblau E.** 2008. Scientist officially exonerated in anthrax attacks, *The New York Times*, August 8, 2008. http://www.nytimes.com/2008/08/09/washington/09anthrax.html.
32. **Johnson C.** 2008. U.S. settles with scientist named in anthrax cases. *The Washington Post*, June 28, 2008. http://www.washingtonpost.com/wp-dyn/content/story/2008/06/27/ST2008062702767.html.
33. **Shachtman N.** 2011. Anthrax redux did the feds nab the wrong guy? Wired Magazine, March 24, 2011. http://www.wired.com/2011/03/ff_anthrax_fbi/.
34. **Garrett L.** 2011. The Anthrax Letters. Council on Foreign Relations. http://www.cfr.org/terrorist-attacks/anthrax-letters/p26175.
35. **National Academy of Sciences.** 2011. *Review of the Scientific Approaches Used During the FBI's Investigation of the 2001 Anthrax Letters. National Academy of Sciences,* The National Academies Press, Washington, DC. http://www.nap.edu/download.php?record_id=13098.
36. **Engelberg S, McClatchy G, Gilmore J, Wiser M.** 2011. Probublica. New evidence adds doubt to FBI's case against anthrax suspect. http://www.propublica.org/article/new-evidence-disputes-case-against-bruce-e-ivins.
37. **Gordon G, Wiser M.** 2014. Frontline. New report casts doubt on fbi anthrax investigation. http://www.pbs.org/wgbh/pages/frontline/criminal-justice/anthrax-files/new-report-casts-doubt-on-fbi-anthrax-investigation/.
38. **WashingtonsBlog.** 2015. HEAD of the FBI's anthrax investigation says the whole thing was a SHAM. http://www.washingtonsblog.com/2015/04/head-fbis-anthrax-investigation-calls-b-s.html.
39. **Dillon KJ.** 2015. *FBI v. Bruce Ivins: The Missing Pieces.* Scientia Press. http://www.scientiapress.com/fbi-ivins.
40. **Willman D.** 2011. Epilogue, p 323–337. *In* Willman D (ed), *The Mirage Man: Bruce Ivins, the anthrax attacks, and America's rush to war.* Bantam Books, New York.
41. **Check Hayden EC.** 2011. Biodefence since 9/11: the price of protection. *Nature* **477:**150–152.http://www.nature.com/news/2011/110907/pdf/477150a.pdf.
42. **Cole LA.** 2012. Bioterrorism: still a threat to the United States. *CTC Sentinel* 5:8–12. https://www.ctc.usma.edu/posts/bioterrorism-still-a-threat-to-the-united-states.
43. **Roxas-Duncan VI, Smith LA.** 2012. Ricin Perspective in Bioterrorism, p 133–158. *In* Morse SA (ed), *Bioterrorism.* InTech, Rijeka, Croatia. http://library.umac.mo/ebooks/b28113433.pdf.
44. **National Association of County and City Health Officials.** 2011. More than half of local health departments cut services in first half of 2011. http://archived.naccho.org/press/releases/100411.cfm.
45. **Khan AS.** 2011. Public health preparedness and response in the USA since 9/11: a national health security imperative. *Lancet* **378:**953–956. and http://www.cdc.gov/phpr/documents/Lancet_Article_Sept2011.pdf.
46. **Centers for Disease Control and Prevention and the Federal Bureau of Investigation.** 2015. Joint Criminal and Epidemiological Investigations Handbook. FBI (WMD Directorate, Biological Countermeasures Unit) and the CDC (National Center for Emerging and Zoonotic Infectious Diseases, Division of Preparedness and Emerging Infections). https://www.fbi.gov/about-us/investigate/terrorism/wmd/criminal-and-epidemiological-investigation-handbook.
47. **Ferguson JR.** 1997. Biological weapons and US law. *JAMA* **278:**357–360.
48. **U.S. Department of Health and Human Services (DHHS), and Centers for Disease Control and Prevention.** 1997. Additional Requirements for Facilities Transferring or Receiving Select Agents, Title 42 CFR Part 72 and Appendix A; 15 April 1997. https://grants.nih.gov/grants/policy/select_agent/42CFR_Additional_Requirements.pdf.
49. **U.S. Department of Health and Human Services (DHHS).** 2002. Part IV Title 42 CFR Part 73, Office of the Inspector General, Title 42 CFR Part 1003. Possession, Use and Transfer of Select Agents and Toxins; interim final rule. *Fed Regist* **67:** 76886–76905. http://www.gpo.gov/fdsys/pkg/FR-2002-12-13/html/02-31370.htm.
50. **U.S. Department of Agriculture, Animal and Plant Health Inspection Service USDA-APHIS).** 2002. Title 7 CFR Part 331 and Title 9 CFR Part 121. Agricultural Bioterrorism Protection Act of 2002; Possession, Use, and Transfer of Biological Agents and Toxins; interim final rule. Fed Regist **67:**76908–76938. http://www.gpo.gov/fdsys/pkg/FR-2002-12-13/pdf/02-31373.pdf.
51. **U.S. Department of Health and Human Services.** 2005. Part III. Title 42 CFR 72 and 73, Office of the Inspector General 42 CFR Part 1003. Possession, Use, and Transfer of Select Agents and Toxins; final rule. *Fed Regist* **70:**13294–13325. http://www.gpo.gov/fdsys/pkg/FR-2005-03-18/html/05-5216.htm.
52. **U.S. Department of Agriculture-Animal and Plant Health Inspection Service (USDA-APHIS).** 2005. Part II. Title 7 CFR Part 331 and Title 9 CFR Part 121. Agricultural Bioterrorism Protection Act of 2002; Possession, Use, and Transfer of Biological Agents and Toxins; final rule. *Fed Regist* **70:**13242–13292. http://www.gpo.gov/fdsys/pkg/FR-2005-03-18/pdf/FR-2005-03-18.pdf.
53. **Federal Select Agent Program.** 2014. Select Agents and Toxins List. http://www.selectagents.gov/SelectAgentsandToxinsList.html.
54. **Centers for Disease Control and Prevention, Animal and Plant Health Inspection Service.** 2013. Training required, p 5. In Guidance for Meeting the Training Requirements of the Select Agent Regulations. http://www.selectagents.gov/resources/Guidance_for_Training_Requirements_v3-English.pdf.
55. **Franz DR.** 2015. Implementing the Select Agent Legislation: Perfect Record or Wrong Metric? *Health Secur* **13:**290–294. http://online.liebertpub.com/doi/abs/10.1089/hs.2015.0029?journalCode=hs.
56. **Federal Select Agent Program.** 2012. Select Agents and Toxins Exclusions. http://www.selectagents.gov/SelectAgentsand ToxinsExclusions.html.
57. **U.S. Department of Health and Human Services (DHHS).** 2016. Proposed Rule: Possession, Use, and Transfer of Select Agents and Toxins; Biennial Review of the List of Select Agents and Toxins and Enhanced Biosafety Requirements. *Fed Regist* **81:**2805–2818. http://www.gpo.gov/fdsys/pkg/FR-2016-01-19/html/2016-00758.htm.
58. **U.S. Department of Agriculture (USDA)/Animal and Plant Health Inspection Service (APHIS) and U.S. Department of Health and Human Services (HHS)/Centers for Disease Con-**

trol and Prevention (CDC). 2015. Exclusion Guidance Document. http://www.selectagents.gov/resources/Guidance_for_Exclusion_version_1.pdf.

59. **Federal Select Agent Program.** 2014. FSAP Policy Statement: Registration of Select Agents and Toxins. http://www.selectagents.gov/regpolicystatement.html.
60. **Wilson DE, Chosewood LC.** 2009. *Biosafety in Microbiological and Biomedical Laboratories*, 5th ed., HHS Publication no. (CDC) 21-1112. http://www.cdc.gov/biosafety/publications/bmbl5/bmbl.pdf.
61. **Richmond JY, Nesby-O'Dell SL, Centers for Disease Control and Prevention.** 2002. Laboratory security and emergency response guidance for laboratories working with select agents. *MMWR Recomm Rep* **51**(RR-19):1–6.
62. **Blaine J.** 2014. Incident response plan, p. 46. *In* Blaine J (ed), *Responsible Official Resource Manual.* Centers for Disease Control and Prevention Atlanta, Georgia and Animal and Plant Health and Inspection Service. http://www.selectagents.gov/resources/RO_Manual_2014.pdf.
63. **Salerno R, Gaudioso J.**2015. *Laboratory Biorisk Management Biosafety and Biosecurity.* CRC Press, Taylor & Francis Group, Boca Raton, FL.
64. **Executive Order 13546.** 2010. Optimizing the Security of Biological Select Agents and Toxins in the United States. https://www.gpo.gov/fdsys/pkg/FR-2010-07-08/pdf/2010-16864.pdf.
65. **Wilson DE, Chosewood LC.**2009. Elements of a biosecurity program, p 112. *In Biosafety in Microbiological and Biomedical Laboratories*, 5th ed, HHS Publication no. (CDC) 21-1112. http://www.cdc.gov/biosafety/publications/bmbl5/bmbl.pdf.
66. **Crow R.** 2004. *Personnel Reliability Programs.* Project Performance Corporation, McLean, Virginia. http://www.ppc.com/assets/pdf/white-papers/Personnel-Reliability-Programs.pdf.
67. **Defense Science Board.** 2009. Department of Defense Biological Safety and Security Program. Office of the Undersecretary of Defense for Acquisition, Technology and Logistics, Washington, DC. http://www.acq.osd.mil/dsb/reports/ADA499977.pdf.
68. **Burnette RN, Hess JE, Kozlovac JP, Richmond JY.** 2013. Defining biosecurity and related concepts, p 1–16. *In* Burnette R (ed), *Biosecurity: Understanding, Assessing, and Preventing the Threat.* John Wiley & Sons, Hoboken, NJ.
69. **Animal and Plant Health Inspection Service (APHIS) Agricultural Select Agent Services and Centers for Disease Control and Prevention (CDC) Division of Select Agents. 2014.** Risk management and computer security incident management, p 20. In Information Systems Security Control Guidance Document. http://www.selectagents.gov/resources/Information_Systems_Security_Control_Guidance_version_3_English.pdf.
70. **Wilson DE, Chosewood LC.** 2009. Elements of a biosecurity program, p 110. *In Biosafety in Microbiological and Biomedical Laboratories*, 5th ed, HHS Publication no. (CDC) 21-1112. http://www.cdc.gov/biosafety/publications/bmbl5/bmbl.pdf.
71. **U.S. Department of Agriculture (USDA)/Animal and Plant Health Inspection Service (APHIS) and U.S. Department of Health and Human Services (HHS)/Centers for Disease Control and Prevention (CDC).** 2015. Guidance on the Inventory of Select Agents and Toxins. http://www.selectagents.gov/resources/Long_Term_Storage_version_5.pdf.
72. **Federal Select Agent Program.** 2015. FSAP Policy Statement: When APHIS and CDC Import Permits Not Required for the Importation or Interstate Transportation of Select Agents. http://www.selectagents.gov/regpermits.html.
73. **Wilson DE, Chosewood LC.** 2009. Occupational health support service elements, p. 117. *In Biosafety in Microbiological and Biomedical Laboratories*, 5th ed, HHS Publication no. (CDC) 21-1112. http://www.cdc.gov/biosafety/publications/bmbl5/bmbl.pdf.
74. **U. S. Public Health Service (USPHS).** 2007. Title 42 CFR Part 71. Foreign Quarantine. Subpart F: Importations, Section 71.54, Etiological agents, hosts, and vectors. http://www.gpo.gov/fdsys/pkg/CFR-2007-title42-vol1/pdf/CFR-2007-title42-vol1-sec71-54.pdf.
75. **U. S. Environmental Protection Agency (USEPA).** 2011. Title 40 CFR Part 300. National oil and hazardous substances pollution contingency plan. http://www.gpo.gov/fdsys/pkg/CFR-2011-title40-vol28/pdf/CFR-2011-title40-vol28-part300.pdf.
76. **National Institutes of Health.** 2013. *The NIH Guidelines for Research Involving Recombinant or Synthetic Nucleic Acid Molecules (NIH Guidelines).* The current amended version of the *NIH Guidelines* can be accessed at http://osp.od.nih.gov/sites/default/files/NIH_Guidelines_0.pdf.
77. **U. S. Department of Labor (USDOL).** 2011. Title 29 CFR Part 1910. Occupational Safety and Health Standards. Subpart Z: Toxic and Hazardous Substances. Section 1910.1450 Occupational exposure to hazardous chemicals in laboratories, p. 484–499. http://www.gpo.gov/fdsys/pkg/CFR-2011-title29-vol6/pdf/CFR-2011-title29-vol6-sec1910-1450.pdf.
78. **U. S. Department of Labor.** 2016. Title 29 CFR Part 1910. Occupational Safety and Health Standards. Subpart Z: Toxic and Hazardous Substances. Section 1910.1030. http://www.ecfr.gov/cgi-bin/text-idx?SID=4f5383c3668a88224102236d19d5f0c0&mc=true&node=se29.6.1910_11030&rgn=div8.
79. **Headquarters, Department of the Army.** 2013. Army Regulation 385–10, Safety, The Army Safety Program. Chapter 20. Infectious Agents and Toxins, p 87. Washington, DC. The current version can be accessed at http://www.apd.army.mil/pdffiles/r385_10.pdf.
80. **Headquarters, Department of the Army.** 2013. Department of the Army Pamphlet 385–69, Safety. Safety Standards for Microbiological and Biomedical Laboratories. Washington, DC. http://www.apd.army.mil/pdffiles/p385_69.pdf.
81. **U.S. Department of Transportation (USDOT).** 2011. Title 49 C.F.R. Part 173.134, Class 6, Division 6.2-Definitions and exceptions. p 539–543. http://www.gpo.gov/fdsys/pkg/CFR-2011-title49-vol2/pdf/CFR-2011-title49-vol2-sec173-134.pdf.
82. **International Air Transport Association.** 2016. *Dangerous Goods Regulations*, 57th ed, International Air Transport Association, Montreal-Geneva. http://www.iata.org/publications/dgr/Pages/manuals.aspx.
83. **U. S. Department of Agriculture-Agricultural Research Service.** 2002. USDA Security Policies and Procedures for Biosafety Level 3 Facilities. USDA Department Manual 9610-1. http://www.ocio.usda.gov/sites/default/files/docs/2012/DM9610-001_0.pdf.
84. **Monaco LO, Holdren JP.** 2015. A National Biosafety and Biosecurity System in the United States. https://www.whitehouse.gov/blog/2015/10/29/national-biosafety-and-biosecurity-system-united-states.
85. **Biological Weapons Convention.** 1925. Convention on the Prohibition of the Development, Production and Stockpiling of Bacteriological (Biological) and Toxin Weapons and on their Destruction. http://www.state.gov/www/global/arms/treaties/bwc1.html
86. **Inglesby TV, Relman DA.** 2015. How likely is it that biological agents will be used deliberately to cause widespread harm? *EMBO Rep*, published online December 18, 2015. http://onlinelibrary.wiley.com/doi/10.15252/embr.201541674/abstract.
87. **Williams P, Wallace D.** 1989. The secret of secrets, p 31–50. *In* Williams P, Wallace D (ed), *Unit 731: Japan's Secret Biological Warfare in World War II.* MacMillan, Inc, New York
88. **Williams P, Wallace D.** 1989. Waging germ warfare, p 68–70. *In* Williams P, Wallace D (ed), *Unit 731: Japan's Secret Biological Warfare in World War II.* MacMillan, Inc, New York.
89. **Harris SH.** 1995. Human experiments: "secrets of secrets," p 59–66. *In* Harris SH (ed), *Factories of Death, Japanese Biological Warfare 1932–45 and the American Cover-up.* Routledge, New York.
90. **Mitscherlich A, Mielke F.** 1983. *Medizin ohne Menschlichkeit: Documente des Nurnberger Arztepoozesses.* Fischer Taschenbuchverlag, Frankfurt am Main, Germany.

91. **Manchee RJ, Stewart WDP.** 1988. The decontamination of Gruinard Island. *Chem Br* **24:**690–691.
92. **Harris SH.** 1995. The United States BW Program, p 149–159. *In* Harris SH (ed), *Factories of Death, Japanese Biological Warfare 1932–45 and the American Cover-up.* Routledge, New York.
93. **Endicott S, Hagerman E.** 1998. World War II origins, p 31–35. *In* Endicott S, Hagerman E (ed), *The United States and Biological Warfare: Secrets from the Early Cold War and Korea.* Indiana University Press, Bloomington.
94. **Mangold T, Goldberg J.** 1999. Arms race, p 34–36. *In* Mangold T, Goldberg J (ed), *Plague Wars: a True Story of Biological Warfare.* Macmillan Publishers Ltd, London.
95. **Franz DR, Parrott CD, Takafuji ET.** 1997. The U.S. Biological Warfare and Biological Defense Programs, p 428. *In* Sidell FR, Takafuji ET, Franz DR (ed), *Medical Aspects of Chemical and Biological Warfare. (TMM series. Part I, warfare, weaponry, and the casualty.).* Borden Institute, Walter Reed Army Medical Center, Washington, DC. http://www.au.af.mil/au/awc/awcgate/medaspec/Ch-19electrv699.pdf.
96. **Health and Human Services Department.** 2016. Possession, use, and transfer of Select Agents and Toxins—addition of Bacillus cereus biovar anthracis to the HHS List of Select Agents and Toxins. *Fed Regist* https://www.federalregister.gov/documents/2016/09/14/2016-22049/possession-use-and-transfer-of-select-agents-and-toxins-addition-of-bacillus-cereus-biovar-anthracis

SECTION

V

Special Environments

30

Biological Safety and Security in Teaching Laboratories

CHRISTOPHER J. WOOLVERTON AND ABBEY K. WOOLVERTON

INTRODUCTION

Any discussion of biological safety within undergraduate basic science and clinical teaching laboratories should be prefaced by several important caveats. First, the foundation of any science laboratory experience should include a solid understanding of safety. Practicing safety teaches responsibility and respect for life and property. Regardless of unique institutional conditions and oversight agencies, adherence to common safety and biosafety practices demonstrates a good-faith effort to students, parents, faculty, administrators, and even accreditors, that academic laboratory users are valued. Additionally, most safety practices were developed in response to documented need and thus serve to mitigate hazards. At its core, the teaching laboratory should be an environment where students are challenged by the science of microbiology, not by fear of infection. Furthermore, as the science of microbiology evolves, time-tested biosafety practices continue as a microbiology legacy, passed down to subsequent generations of microbiologists as résumé staples.

Second, while overarching federal regulations provide sound advice to establish best practices and common standards, a "one-size-fits-all" biosafety directive is not feasible, especially as state and local regulations create unique requirements that preclude a single, national guidance document. While many laboratory practices promoting biosafety have been codified for research, hospital, and industry environments, the academic teaching laboratories have been historically self-policing. However, concern over the "escape" of risk group 2 microorganisms from undergraduate laboratories into communities and continued efforts to prevent crime and terrorism resulting from misuse of microorganisms have increased scrutiny of microbiology teaching laboratories. Whereas some of the legislated biosafety regulations for nonteaching laboratories readily extend to teaching labs, they may interfere with traditional teaching and learning of microbiology. Figure 1 demonstrates the hierarchy of typical safety oversight within U.S. academic environments; not all apply to every type of academic institution. Although larger institutions of higher education employ safety officers (who are not always biosafety officers), and accept most of the liability associated with regulation compliance, insight from microbiology faculty is often requested in

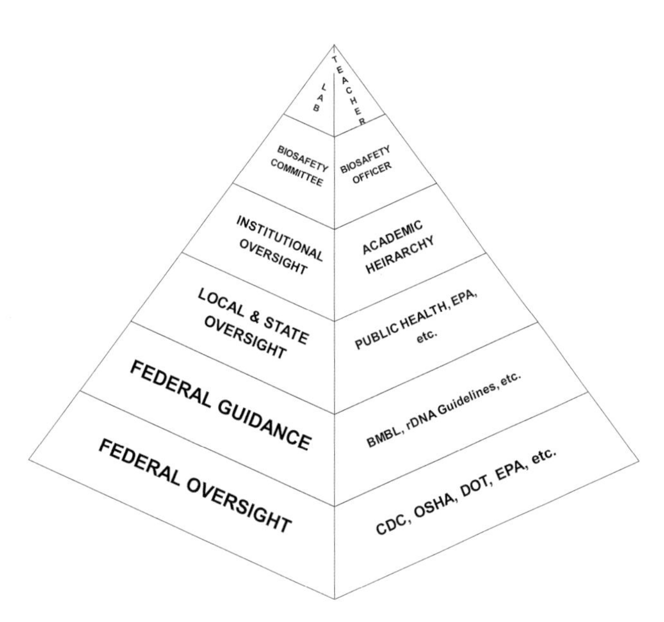

Figure 1: Typical model of biosafety oversight in academia. Note that the instructor, being the closest to daily laboratory activities, has the highest level of responsibility to ensure a safe environment (adapted from reference 32).

the establishment of institutional practice. This can be a greater burden on microbiologists of smaller institutions who have no formally trained safety officers.

Third, an institution of higher education is indeed a microcosm of the world and, as such, reflects the spectrum of human immune competency. Some users of the microbiology laboratory (students, staff, and faculty) may, in fact, be immunosuppressed due to medications, pregnancy, or HIV/AIDS. Immunosuppression increases susceptibility to infectious agents. Thus, educators today have an additional consideration in planning laboratory exercises; they should be judicious in their choice of microorganisms and procedures so as to meet pedagogical goals, while also protecting this (often unidentified) immunosuppressed population. Finally, a laboratory-acquired infection should be treated as a rare and preventable occupational hazard. Sound safety practices would certainly keep it this way.

EVIDENCE FOR IMPROVED BIOSAFETY NEED IN TEACHING LABORATORIES

Is biological safety specifically taught as a principle? How is biosafety taught? How can the culture of biosafety be measured? Perhaps it is easier to instead ask about the incidence of laboratory-acquired infections as an indication of biological safety lapses. This, too, is a difficult question to answer because there is no requirement for reporting all laboratory-acquired infections to public health authorities, and subclinical infections are rarely identified. Furthermore, unless the afflicted specifically associate the infection with their laboratory activities, the infection may be determined to be community-acquired (as often occurs with *Streptococcus pneumoniae* or *Salmonella enterica* serovar Typhimurium, for example). Nonetheless, various reports by Sulkin and Pike from 1949 through 1978 describe the types of laboratory-acquired infections and their impact (1–6). These studies identified 4,079 (reported) laboratory-acquired infections between 1930 and 1978. The five most frequently reported laboratory-acquired infections in that time period were caused by *Brucella* (507 cases), *Coxiella burnetii* (456 cases), *Mycobacterium tuberculosis* (417 cases), hepatitis viruses (380 cases), and *Salmonella* (324 cases). Of these, 168 fatalities occurred. Some may argue that these infections did not all occur in teaching laboratories. The better point to make is that the infections occurred in microbiologists who presumably were trained (as undergraduates) in biosafety.

A subsequent study reviewed 270 published reports between 1997 and 2004 that identified 1,448 (symptomatic) laboratory-acquired infections occurring in clinical, public health, research, and teaching laboratories. Of these, 36 fatalities and 17 secondary infections were noted. When the distribution of infections was correlated to the primary purpose of work, it was found that 14 of the 1,448 laboratory-acquired infections occurred in teaching laboratories (7).

Certainly, lapses in biosafety practice are noted when laboratory incidents are reported in the local news media. However, not widely reported were the 2010–2011 and the 2013–2014 academic year outbreaks of *Salmonella* Typhimurium among microbiology educators and their students. A total of 150 *Salmonella* cases (109 between August 2010 and April 2011 in 38 states, and 41 between November 2013 and May 2014 in 13 states) associated with undergraduate teaching laboratories were reported. There were 23 hospitalizations, with one fatality. The infections were caused by a commonly used "laboratory strain" of *Salmonella* (8). One wonders how these many infections occurred lest lapses in biosafety practices be the fault.

REGULATORY OVERSIGHT PERTINENT TO TEACHING LABORATORIES

Diverse issues, such as outdated laboratory design, available renovation funds, pedagogical goals, multipurpose space, the definition of "employee," private versus public institution, and clinical versus nonclinical environments, to name a few, stir countless debates in favor of and against compliance with various biosafety standards and guidelines. However, it seems necessary that the microbiology community agree, at its grassroots, that the most important

Figure 2: Biosafety regulations, standards, and guidance pertinent to various academic laboratories (adapted from reference 32).

issue is the safety of those who use the laboratory in which there are microorganisms. Various U.S. regulations and guidance documents, which may not be perceived as having governance over teaching laboratories, nonetheless contain common standards that can be applied as a general means of instilling a culture of safety at universities and colleges (Fig. 1 and 2). For example, the Occupational Safety and Health Act of 1970 stipulates that certain precautions be observed to protect the safety and health of employees on the job (9). The employee designation includes all faculty employed by private and public school systems in states that have occupational safety and health plans accepted by the Occupational Safety and Health Administration (OSHA). The purview of oversight documents is to direct institutional biosafety and to promote a culture of safety.

The Bloodborne Pathogens Act of 1991 takes the protection of workers further as it requires engineering controls, work practices, and personal protective equipment to protect workers from potentially infectious material (10). Additionally, the Clinical and Laboratory Standards Institute (CLSI) published *Protection of Laboratory Workers From Occupationally Acquired Infections* to promote good laboratory practice and provide guidance on the risk of transmission of infectious agents in laboratory settings and to identify specific precautions for preventing the transmission of infectious agents by aerosols, droplets, blood, and body substances (11, 12). Furthermore, the International Organization for Standardization (ISO) established two standards in 2003, and revised in 2007, to ensure a high-quality and competent laboratory workforce (13) and safe practices in medical laboratories (14). Included in the standards are (i) general safety responsibilities for all personnel of the laboratory, (ii) laboratory inspection mandates to reduce hazards, (iii) production of an up-to-date safety manual, and (iv) initial and annual safety training. Together, these standards establish sound practices to protect against laboratory-acquired infection (Fig. 2). As such, they provide a solid theoretical framework for every microbiology laboratory. However, application of the aforementioned standards is not always possible and/or appropriate for all teaching environments.

Guidance from the Public Health Sector

Sound advice for preventing laboratory infections and accidents is offered through a safety guidance document authored by the U.S. Department of Health and Human Services. It has its origin in teaching best practices to prevent laboratory-acquired infection. In 1974, the Centers for Disease Control and Prevention (CDC) published *Classification of Etiologic Agents on the Basis of Hazard* (15), as a means to disseminate hazard characteristics of microorganisms requiring increasing levels of physical containment. Included in the document were the pioneering biosafety techniques established by Arnold G. Wedum, M.D., Ph.D. (former Director of Industrial Health and Safety, Army Biological Research Laboratory, Fort Detrick, MD). Years of hazard recognition and risk reduction led Wedum to develop work practices, equipment, engineering controls, and trainings necessary to contain some of the world's most deadly infectious agents studied at Fort Detrick, then housing the U.S. biodefense program (16). Subsequent publication of safety standards to study oncogenic viruses (17) and guidelines for use of recombinant DNA (18) further cemented the criteria for establishing ascending levels of physical containment and specific work practices to prevent human exposure to infectious agents. These early documents quickly evolved into a codified "manual of biosafety guidance," and were published together in 1984 as *Biosafety in Microbiological and Biomedical Laboratories*, or BMBL. The BMBL was subsequently published as a second edition in 1988, a third in 1993, a fourth in 1999, and the current fifth edition in 2007 with revisions in 2009 (19). Alas, the BMBL is often misperceived as guidance for research laboratories alone. In fact, "the intent was and is to establish a voluntary code of practice, one that all members of a laboratory community will together embrace to safeguard themselves and their colleagues, and to protect the public health and environment" (19). Although the BMBL does not specifically reference teaching laboratory precautions and practices, it nonetheless serves as a well-vetted manual of best microbiology practices. As such it can be a substantial resource for microbiology and molecular biology teaching laboratories.

RISK ASSESSMENT

The principles of laboratory biosafety integrate the aforementioned regulations and guidance documents with best practices gleaned from recommendations vetted through organizations such as the American Society for Microbiology (ASM), American Biological Safety Association (ABSA), CDC, and Association of Public Health Laboratories (APHL). Importantly, the recommendations place biosafety equipment use and work practices in the context of risk assessment, rather than in a static, one-size-fits-all approach. This focuses all aspects of biological safety through the lens of evidence-based science. It is through this lens that the microbiology laboratory instructor can not only reduce the likelihood of laboratory-acquired infection, but also discuss the culture of safety, basic microbiology, pathogenicity and parasitism, dual-use agents, and bioethics.

Realistically, oversight regulations and standards do not expect, nor require, the elimination of all risk in the laboratory. However, they establish a reasonable expectation that people working within the laboratory environment are protected from most hazards. In general, the risk assessment process assists in reducing hazards to laboratory users by focusing on five discrete components of the process. First, laboratory hazards should be identified by a thorough analysis of the facility, equipment, microbiological agents, and procedures to be used. The examination should involve all users of the laboratory; insight from users often identifies hazards, or at least perceived hazards. This is not an expensive process, nor one that requires certified safety officers. Faculty at all types of educational institutions should be able to identify and mitigate the most obvious safety violations. Additionally, the monitoring of experimental work occurring throughout a typical day can be used to pinpoint potential practice hazards that may not be recognized, or are ignored in the attempt to rapidly complete work.

Second, individuals at risk should be identified so as to minimize their risk. Risk due to use of human pathogens, or various pieces of equipment, can be mitigated often with little impact on the pedagogical activity. For example, reasonable accommodations can be made for laboratory users who are immunocompromised. Most institutions of higher education have offices for assisting students with special needs. Reasonable accommodations can be made without violation of the Health Insurance Portability and Accountability Act of 1996 (20). Alternative activities can often achieve the pedagogical goals through simulation or use of class data outside the laboratory.

Third, methods to attenuate or eliminate a hazard, such as lab-specific training or use of specific personal protective equipment (PPE), should be identified. This too often presents academics with fodder for debate. OSHA guidance to industry is to provide PPE, accommodating the employee. Academic institutions are not mandated to purchase PPE for students, but reasonable methods are available to provide it. PPE can be purchased from course fees, and PPE suppliers may be able to donate products for tax benefits. It may also be possible for instructors to modify the laboratory experiment to reduce the risk.

Fourth, the methods used and their effectiveness should be documented to demonstrate risk-reduction attempts. One common process is to use a published experimental manual or guide that directs laboratory activities. The techniques, reagents, and safety recommendations are typically evaluated prior to publication or distribution. Thus, the manual serves as a collection of standard operating procedures of pedagogical techniques that have been evaluated for risk by other users. Importantly, many of the published laboratory manuals have a safety code or even agreement, to which students can be asked to follow. Fifth, periodic reviews of the risk assessment process should be scheduled so that laboratory activities can be modified as needed (21).

Risk assessment within the microbiology laboratory is essentially a problem-solving process comparing the probability of hazard exposure with the severity of its consequences. Therefore, one can evaluate the likelihood of a laboratory-acquired infection in context with the microorganism's infectivity (risk group and host susceptibility), transmissibility within the laboratory, specific procedures and equipment that might increase exposure to the agent, as well as the severity of disease that might result from infection. Inherent in this assessment is not just the potential to cause harm to the user, but also the potential for environmental contamination. The CDC has produced a valuable risk assessment resource to assist in determining the appropriate biosafety level (BSL) necessary to conduct a particular procedure using a specific microorganism. The form is appended to this chapter (Appendix A) and can be obtained from the CDC website (http://www.cdc.gov/biosafety/publications/BiologicalRiskAssessmentWorksheet.pdf). Risk assessment, then, is a means to determine one's tolerance of negative outcomes based on their potential frequency of occurring. However, the risk assessment process in the microbiology teaching laboratory can serve as a learning exercise that reveals specific strategies to reduce risk of infection or other safety outcomes.

Appropriately applied, risk assessment can reveal potential problems, establish response plans to hazards, target necessary human and physical resources, and provide credible evidence for specific work practices. For example, interpreting the results of a laboratory risk assessment can determine if appropriate safeguards exist to minimize potential negative outcomes associated with the use of a specific agent, in a specific space, using specific equipment and work practices. In other words, the risk assessment should produce awareness of all information necessary to anticipate a laboratory-acquired infection and the severity of its consequences. Nonetheless, prior knowledge necessary to anticipate a laboratory-acquired infection may not be sufficient to prevent one from occurring. This is the "predictable surprise" concept defined by Bazerman and Watkins (22). The risk assessment process should prepare the laboratory user(s) to respond quickly to potential laboratory hazards, as well as to manage or mitigate them. Risk assessment may not always eliminate a hazard, but it can prepare those using the laboratory to expect it, even if the actual time of occurrence is unknown.

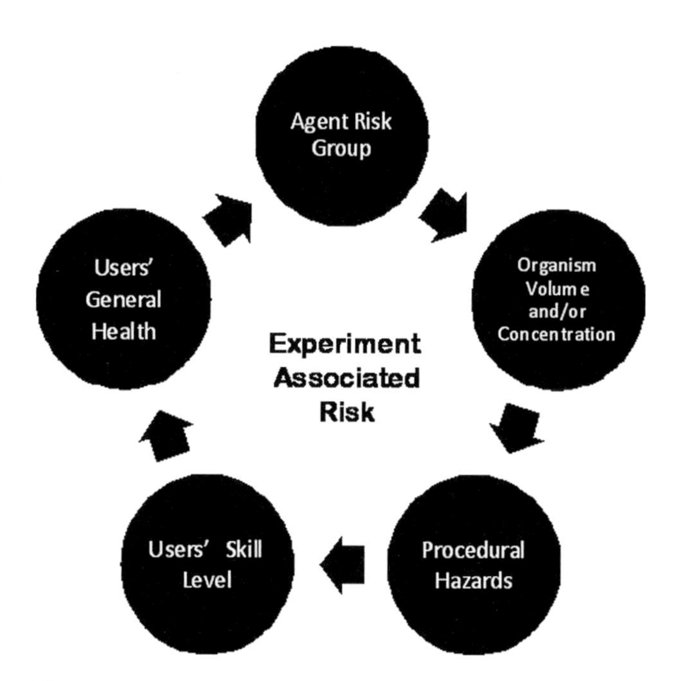

Figure 3: Example of the iterative process used to assess the risk associated with each laboratory experiment.

Application of Risk Assessment in the Teaching Laboratory

The assessment of risk in the microbiology teaching laboratory can begin before the first student interaction. In general, the environment is surveyed to determine the laboratory design (airflow, sink access, decontamination processes), biosafety cabinet (BSC) access (if used), and any obvious physical hazards (such as poorly functioning equipment, trip hazards, exposed electrical wires, etc.), so as to remedy the hazard or communicate its presence to incoming users. Second, the instructor can prepare an assessment of potential risks associated with all experiments in advance, including follow-up data collection (i.e., manipulation of concentrated organisms). This part of the risk assessment includes (i) a correct determination of the agent risk group relying only on risk group information from a credible source; if there is no written documentation of an organism's pedigree, it should be destroyed; organisms passed down may be genetically altered resulting in a change of risk group designation; (ii) a thorough evaluation of the procedure(s) used with each agent to identify hazard-generating techniques (any process that generates aerosols or involves "sharps," for example); and (iii) an appreciation for the skill level and general health (potential for immunocompromise, for example) status of all laboratory users (Fig. 3).

It is important to note that there is no internationally accepted list of microorganism risk groups. Each country assigns a risk group to a microorganism based on the amount of risk acceptable to that country. For example, *Lactobacillus* spp. are assigned as risk group 1 in the United States but risk group 2 in the United Kingdom. In general, the risk group determination of a microorganism is determined by its pathogenicity or infectivity and virulence [the 50% lethal dose (LD_{50}) is one measure], its host range, its method(s) of transmission, availability of vaccines, and sensitivity to antimicrobial agents. Most countries recognize four risk groups of microorganisms. The risk group definitions used by the United States and the World Health Organization are presented in Table 1. It should be emphasized, however, that the risk group of any microorganism can change with the insertion of genetic material (native and/or recombinant DNA). For example, organisms that are inherently risk group 1 can become risk group 2 by the addition of genes from risk group 2 and/or risk group 3 organisms (so-called "gain of function"). This is especially problematic when the transferred genes encode antimicrobial resistance.

To readily obtain information pertinent for a risk assessment, the BMBL contains agent summary sheets for many organisms (bacterial, fungal, parasitic, rickettsial and viral) that define the general (i) characteristics and pathogenicity, (ii) occupational infection, (iii) natural modes of infection, (iv) laboratory safety, (v) containment recommendations, and (vi) special issues (vaccines and agent transfer, for example). ABSA hosts a risk group database (http://www.absa.org/riskgroups/index.html) that summarizes agent risk groups in various countries and serves as a portal to the "pathogen safety data sheets" (analogous to material safety data sheets) produced by the Public Health Agency of Canada (PHAC) (http://www.phac-aspc.gc.ca/lab-bio/res/psds-ftss/index-eng.php).

The PHAC reports substantially more microorganisms, covering the spectrum of bacterial, viral, fungal, and parasitic agents. These "pathogen data sheets" serve as a quick reference to important information to assist in the risk assessment process. The information is distributed through nine short sections: (i) synonyms and general characteristics, (ii) health hazard (pathogenicity, epidemiology, host range, transmission, etc.), (iii) dissemination (vectors and

TABLE 1.

Risk group classifications defined by the United States and the World Health Organization[a]

Risk group classification	U.S. National Institutes of Health	World Health Organization
Risk group 1	Not associated with disease in healthy adult humans	No or low individual and community risk
Risk group 2	Associated with human disease, rarely serious, prevention methods and treatments *often* available	Moderate individual risk, low community risk
Risk group 3	Associated with serious or lethal human disease, prevention methods and treatments *may* be available	High individual risk, low community risk
Risk group 4	Likely to cause serious or lethal human disease, prevention methods and treatments *not usually* available	High individual and community risk

[a]Adapted from U.S. Department of Health and Human Services, Public Health Service. 2009: Section II, Table 1.

reservoirs, etc.), (iv) viability (drug and disinfectant susceptibility, drug resistance, survival out of a host, etc.), (v) medical (first aid, surveillance, immunization, etc.), (vi) laboratory hazards (laboratory-acquired infections, sources and specimens, etc.), (vii) recommended precautions (containment and protective clothing recommendations, etc.), (viii) handling information (spills, disposal and storage), and (ix) miscellaneous information. Neither the BMBL nor the PHAC has lists that contain all of the risk group 1 and 2 agents that might be used in teaching laboratories. However, information about a specific microorganism can often be additionally obtained from a type collection (American Type Culture Collection, for example), scientific supplier, or colleagues who may share microorganisms.

Risk Groups and Biosafety Levels

As mentioned previously, risk groups are assigned to microorganisms on the basis of pathogenicity, mode of transmission, availability of protective vaccines and chemotherapeutic agents, alteration of the native genotype, and others. The four risk groups span the hazard spectrum (Table 1). However, undergraduate microbiology and biology teaching laboratories often limit student access to risk group 2 organisms, replacing them with risk group 1 organisms that teach the same morphological, biochemical, and/or genetic principle, when possible. Risk group 1 agents are often sufficient for demonstrating various phenotypic traits, including Gram reactions, culture characteristics, and biochemical reactions. Risk group 2 agents are sometimes necessary for teaching specific microbiological principles not expressed by risk group 1 surrogates, and often require additional biosafety precautions. Table 2 conveys biosafety precautions as one shifts between BSL1 and BSL2 environments when working with respective organisms. Because teaching laboratories are not likely to study organisms of higher risk category, the remainder of this chapter will only focus on risk groups 1 and 2. Table 3 lists some example organisms that are often used in basic science and clinical teaching laboratories (personal communications to authors).

Microorganisms to be used for teaching should be chosen judiciously. Once the pedagogical goal is established, the lowest risk group level possible should be selected so

TABLE 2.

Biosafety recommendations for BSL1 and BSL2 microbiology laboratories[a]

Biosafety level	Agents	Work practices	Safety equipment	Facilities
1	Not known to cause diseases consistently in healthy adults Low community risk	Standard microbiology practices	No primary barriers required PPE: laboratory coats and gloves and eye and face protection, as needed	Laboratory bench and sink required
2	Agents associated with human disease Routes of transmission include percutaneous injury, ingestion, mucous membrane exposure	BSL1 practice plus limited lab access Biohazard warning signs "Sharps" precautions Biosafety manual defining any needed waste decontamination or medical surveillance policies	BSCs or other physical containment devices used for all manipulations of agents that cause splashes or aerosols of infectious materials PPE: laboratory coats and gloves and face and eye protection, as needed	BSL1 facilities plus autoclave available

[a]Adapted from U.S. Department of Health and Human Services, Public Health Service, 2009: Section IV, Table 1).BSL, biosafety level; PPE, personal protective equipment; BSC, biosafety cabinet.

TABLE 3.

Examples of microorganisms often used in teaching laboratories

Risk group 1	Risk group 2
Aquaspirillum serpens	*Alcaligenes faecalis*
Caulobacter vibrioides	*Citrobacter freundii*
Bacillus stearothermophilus	*Escherichia coli* (non K-12)
Bacillus subtilis	*Klebsiella pneumoniae*
Clostridium butyricum	*Mycobacterium smegmatis*
Corynebacterium xerosis	*Moraxella catarrhalis*
Enterobacter cloacae	*Proteus vulgaris*
Escherichia coli K-12	*Pseudomonas aeruginosa*
Lactobacillus spp.	*Salmonella enterica* serovar Typhimurium
Micrococcus luteus	*Serratia marcescens*
Staphylococcus saprophyticus	*Shigella flexneri*
Streptococcus lactis	*Shigella sonnei*
Streptococcus griseus	*Staphylococcus aureus*
Vibrio fischeri	*Streptococcus pneumoniae*
T-even bacteriophage	Adenovirus
Baculoviruses	*Aspergillus niger*[a]
Penicillium notatum	*Cryptococcus neoformans*[a]
Saccharomyces cerevisiae	*Trichophyton* spp.

[a]Mold cultures should be sealed and slides fixed to further reduce exposure.

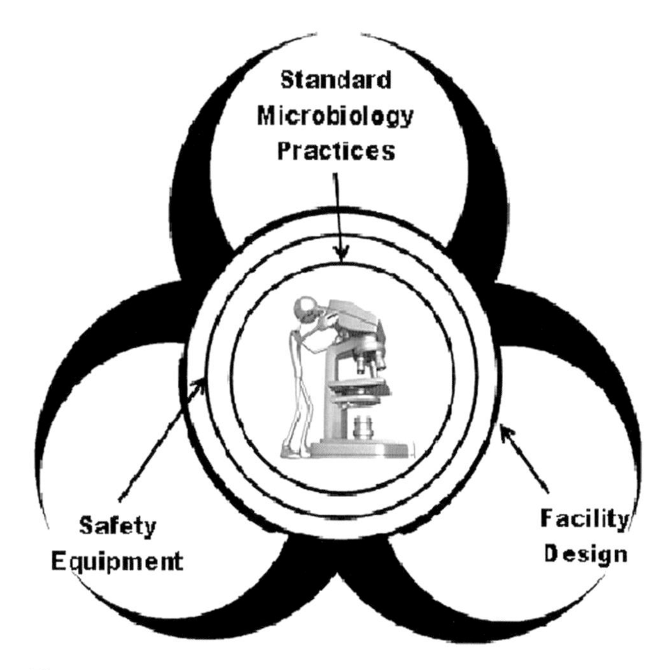

Figure 4: The biosafety level (BSL) is determined from the risk assessment (including the organism risk group), the work practices and techniques to be used, the safety equipment, and the facility design.

as to facilitate the desired learning outcome. As a caution, recall that risk groups can change when organisms are transfected with genetic material from other organisms, for example, when antibiotic resistance genes are used to demonstrate bacterial transformation. In general, microorganisms should be selected to meet the following criteria (ranked from least pathogenic to most): (i) avirulent, presenting the least possible potential to cause human disease (typically risk group 1 organisms); (ii) attenuated, having weakened ability to cause human disease; (iii) opportunistic, requiring a deficient host immune response to cause disease; and (iv) virulent, capable of causing human disease under specific circumstances. Attenuated, opportunistic, and virulent organisms are further ranked into the risk groups 2–4. Risk group 3 (and 4) organisms should not be used to train introductory microbiology students.

To this point, the discussion has only focused on the risk group designation and not the BSL. The two are often confused or thought to be the same. The risk group is a comment on the organism, whereas the BSL considers a combination of: (i) the ability of the organism to cause disease; (ii) severity of disease, ability to prevent or treat illness; (iii) the containment facility; (iv) procedures; and (v) safety equipment. In other words, the risk assessment will determine the BSL, and the BSL will dictate the minimum hazard protection available through the type of facility, safety equipment, and the specialized work practices necessary to minimize exposure to the organism (Fig. 4). The BSL designation indicates that a specific set of engineering controls, safety equipment, and work practices is expected in the designated laboratory to achieve hazard reduction. BSL1 and BSL2 laboratory designs are discussed in greater detail in the Supplemental Material section below.

The risk assessment will govern all other requirements for entry and work in the microbiology laboratory. Remember that protection from infectious hazards should mirror other risk reduction practices defined by the hierarchy of controls; this is three-layered. Layer one is the standard microbiological practices that are controlled by the laboratory user and offer the most immediate method for hazard reduction. Layer two is the safety equipment (mechanical devices and PPE) that acts as the first physical barrier between the infectious hazard and the laboratory user. Layer three is the facility design reflecting engineering controls (Fig. 5).

Biosecurity

Any discussion of laboratory safety and risk assessment would be remiss in today's terrorism-aware climate if it did not comment on biosecurity. Although there is little evidence to suspect that risk group 1 and risk group 2 organisms would support widespread bioterrorism, there is sufficient history documenting the misuse of risk group 2 microorganisms in biocrimes. For example, microbiology historians often point to the Rajneeshee Cult of The Dalles, Oregon, who, in 1984, added *Salmonella* to restaurant salad

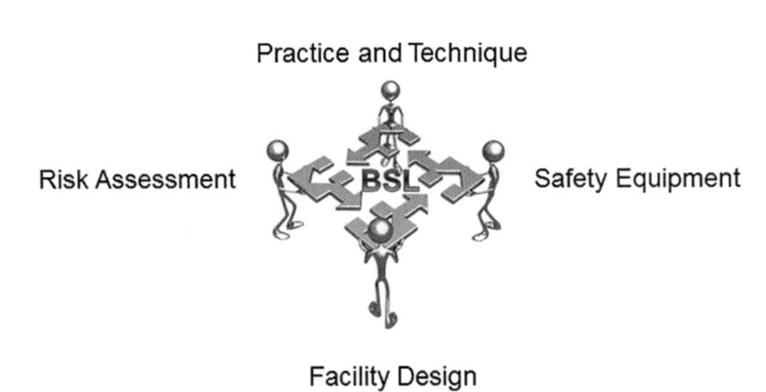

Figure 5: Schematic depicting three layers of biohazard reduction whose specific components are determined by risk assessment.

bars in hopes of incapacitating Wasco County voters on Election Day so that their own candidates would be voted into offices. The contaminated salads resulted in 751 cases of *Salmonella* poisoning, with 45 hospitalizations (23). Certainly, history will also remember the infamous 1996 conflict in a Dallas, Texas, hospital laboratory that resulted in the intentional contamination of break room food with *Shigella dysenteriae* type 2. Twelve of the 45 laboratorians fell ill with shigellosis and four of the 12 had to be hospitalized (23). These types of crimes and ultimately the U.S. anthrax attacks in 2001 led to restrictive legislation regulating the purchase, use, ownership, and transport of risk group 3 and 4 microorganisms and select toxins (24, 25). There is no widespread oversight regulation of risk group 1 and 2 organisms, but several countries restrict access to some risk group 2 organisms. Of note is the enhanced scrutiny of risk group 2 organism purchase from scientific suppliers in the United States. In some cases, a signed acknowledgment is required of how the laboratory facility, access policy, procedures, and training fulfill BMBL criteria, so that risk group 2 organisms will be used safely.

Biosecurity is a relatively new and somewhat contradictory idea on academic campuses. Academic institutions typically operate as bastions of free access, disseminating information and materials for the good of learning. Biosecurity implies restricted access and control. Faculty, in general, had not seen the need for a security-focused campus until the 2007 Virginia Tech and the 2012 Newtown, Connecticut, massacres demanded it. Yet, ongoing internal debate revolves around the balance of knowledge and open access versus prevention of misuse. Procedures to safeguard laboratory cultures should certainly be undertaken. A traceable audit trail should be established to account for all organisms removed from or introduced to the laboratory. Stock cultures should be stored away from the general laboratory supplies and preferably with "need-to-use" access. Waste should be decontaminated as soon as possible. A laboratory and/or institutional policy should be implemented and strictly enforced to prevent persons from knowingly removing microorganisms from the institution without expressed permission from someone responsible to the institution.

TEACHING AND LEARNING IN SAFETY

The microbiology teaching laboratory should be a place where teaching and learning occur in safety. The microbiology laboratory instructor should therefore be trained specifically in microbiology. Adjunct professors, graduate students, or faculty who do not routinely teach microbiology courses may not wish to accept the responsibility for teaching safety and/or may not necessarily believe that they have the authority to enforce biosafety regulations and policies.

General safety instructions should be given by the laboratory instructor for the specific work to be completed in a specific laboratory space. These should reflect the results of a risk assessment identifying the microbiological, chemical, and physical hazards within the laboratory. Additional instructions often include a brief review of emergency procedures and the location of the laboratory safety manual (if available). An initial opportunity should be provided for practice of techniques in the absence of hazardous materials, so that if a student is going to fail, the student can "fail in safety." In other words, once the student becomes proficient practicing a procedure or technique absent a hazard, or with nonhazardous surrogates, higher-risk group microorganisms, noxious chemicals, and/or more difficult techniques can be introduced. This allows the student the freedom to gain proficiency prior to the introduction of the potential hazard. Proper decontamination and disposal of contaminated waste can be taught in a similar manner.

Curricular Design

Laboratory instruction offers pedagogical opportunities for focused problem-solving, skill acquisition, and self-assessment. Having defined the parameters for working within a microbiology laboratory, the risk assessment process will reveal the specific combinations of standard and specialized practices, safety equipment, and facility design to minimize laboratory-acquired infection. Uniting the knowledge and skills of the microbiology curriculum with those of the biosafety competency domains, students can move away from "cookbook" exercises and toward the critical thinking and self-criticism recommendations designed to enhance educational outcomes (26, 27).

It is important that pedagogical goals and learning outcomes of the laboratory exercises be tied to safety. This is readily accomplished by designing student-centered laboratory experiences in light of the entry-level, biosafety laboratory competencies (28). For example, the integration of an initial exercise directing students to answer the question, "Are you safe in this laboratory?" can focus their attention on basic risk assessment and risk reduction. By combining risk assessment as part of the laboratory exercise, the instructor can create a problem-based learning experience that inculcates a culture of safety. Such exercises can be completed prior to laboratory entry and used to suggest newer/better solutions to previously explored methods.

For example, one could ask leading questions, such as, "How might you (re)design the proposed experiment to prevent accidental contamination of (i) the lab, (ii) your lab partner, and (iii) yourself?" Or, "What techniques/methods could be used to help prevent aerosolization of bacteria during the Ziehl-Neelsen staining technique?" Both questions imply an increased risk with negative consequences. Exploration of the potential outcomes, with and without the experimental redesign, can lead to fruitful discussion about why microbiologists practice standard techniques, wear personal protective equipment, and work on water-resistant benchtops, for example. Thus, the original pedagogy behind the laboratory exercise can be augmented with the addition of open-ended, safety problems to solve prior to, or while engaging in, the laboratory experience. The drawing of a "concept map" linking terms associated with biosafety, risk reduction, and common laboratory hazards can similarly demonstrate concept connection and organizational skills. Furthermore, repeating the "Are you safe in this laboratory?" question to elicit responses that stimulate higher-ordered thinking—such as analysis ("Differentiate the typical specimens you will manipulate."), evaluation ("Identify the hazards within a 5-foot radius of your work space."), and creation ("Generate a strategy to decrease the risk of aerosols when using the vortex.")—can assess deeper levels of student understanding (29).

A Hypothetical Example: Preparing to Use *Salmonella* Typhimurium in the Laboratory

An example case study that integrates biosafety with a laboratory exercise is provided below. Students are provided with an exercise and the Pathogen Safety Data Sheet (Fig. 6) and asked to address the following:

1. Assemble information on the (a) organism, (b) procedure(s), and (c) laboratory facility.

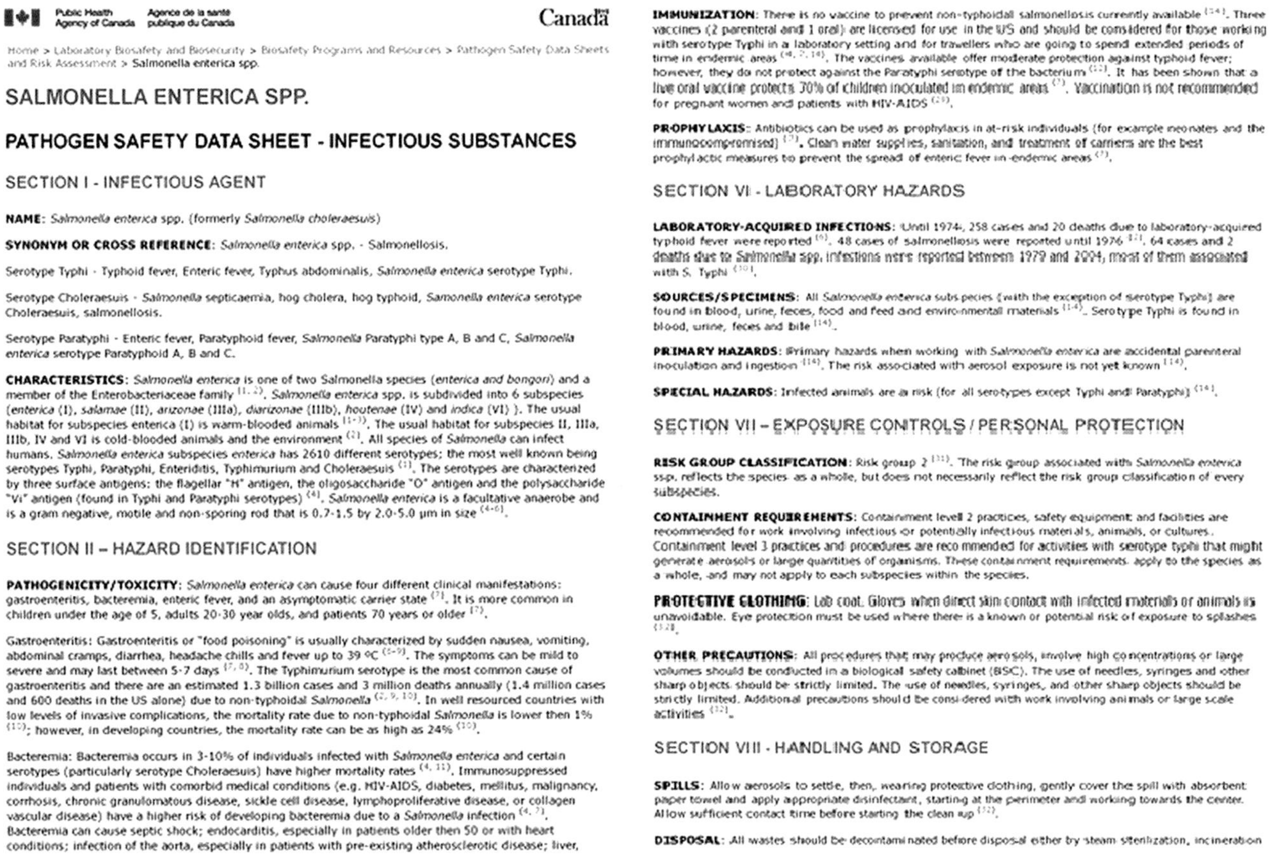

Public Health Agency of Canada | Agence de la santé publique du Canada | Canada

Home > Laboratory Biosafety and Biosecurity > Biosafety Programs and Resources > Pathogen Safety Data Sheets and Risk Assessment > Salmonella enterica spp.

SALMONELLA ENTERICA SPP.

PATHOGEN SAFETY DATA SHEET - INFECTIOUS SUBSTANCES

SECTION I - INFECTIOUS AGENT

NAME: *Salmonella enterica* spp. (formerly *Salmonella choleraesuis*)

SYNONYM OR CROSS REFERENCE: *Salmonella enterica* spp. - Salmonellosis.

Serotype Typhi - Typhoid fever, Enteric fever, Typhus abdominalis, *Salmonella enterica* serotype Typhi.

Serotype Choleraesuis - *Salmonella* septicaemia, hog cholera, hog typhoid, *Samonella enterica* serotype Choleraesuis, salmonellosis.

Serotype Paratyphi - Enteric fever, Paratyphoid fever, *Salmonella* Paratyphi type A, B and C, *Salmonella enterica* serotype Paratyphoid A, B and C.

CHARACTERISTICS: *Salmonella enterica* is one of two Salmonella species (*enterica and bongori*) and a member of the Enterobacteriaceae family [1, 2]. *Salmonella enterica* spp. is subdivided into 6 subspecies (*enterica* (I), *salamae* (II), *arizonae* (IIIa), *diarizonae* (IIIb), *houtenae* (IV) and *indica* (VI)). The usual habitat for subspecies enterica (I) is warm-blooded animals [1-3]. The usual habitat for subspecies II, IIIa, IIIb, IV and VI is cold-blooded animals and the environment [2]. All species of *Salmonella* can infect humans. *Salmonella enterica* subspecies *enterica* has 2610 different serotypes; the most well known being serotypes Typhi, Paratyphi, Enteriditis, Typhimurium and Choleraesuis [3]. The serotypes are characterized by three surface antigens: the flagellar "H" antigen, the oligosaccharide "O" antigen and the polysaccharide "Vi" antigen (found in Typhi and Paratyphi serotypes) [4]. *Salmonella enterica* is a facultative anaerobe and is a gram negative, motile and non-sporing rod that is 0.7-1.5 by 2.0-5.0 μm in size [4-6].

SECTION II – HAZARD IDENTIFICATION

PATHOGENICITY/TOXICITY: *Salmonella enterica* can cause four different clinical manifestations: gastroenteritis, bacteremia, enteric fever, and an asymptomatic carrier state [7]. It is more common in children under the age of 5, adults 20-30 year olds, and patients 70 years or older [7].

Gastroenteritis: Gastroenteritis or "food poisoning" is usually characterized by sudden nausea, vomiting, abdominal cramps, diarrhea, headache chills and fever up to 39 °C [6-9]. The symptoms can be mild to severe and may last between 5-7 days [7, 8]. The Typhimurium serotype is the most common cause of gastroenteritis and there are an estimated 1.3 billion cases and 3 million deaths annually (1.4 million cases and 600 deaths in the US alone) due to non-typhoidal *Salmonella* [2, 9, 10]. In well resourced countries with low levels of invasive complications, the mortality rate due to non-typhoidal *Salmonella* is lower then 1% [10]; however, in developing countries, the mortality rate can be as high as 24% [10].

Bacteremia: Bacteremia occurs in 3-10% of individuals infected with *Salmonella enterica* and certain serotypes (particularly serotype Choleraesuis) have higher mortality rates [4, 11]. Immunosuppressed individuals and patients with comorbid medical conditions (e.g. HIV-AIDS, diabetes, mellitus, malignancy, corrhosis, chronic granulomatous disease, sickle cell disease, lymphoproliferative disease, or collagen vascular disease) have a higher risk of developing bacteremia due to a *Salmonella* infection [4, 7]. Bacteremia can cause septic shock; endocarditis, especially in patients older then 50 or with heart conditions; infection of the aorta, especially in patients with pre-existing atherosclerotic disease; liver,

IMMUNIZATION: There is no vaccine to prevent non-typhoidal salmonellosis currently available [14]. Three vaccines (2 parenteral and 1 oral) are licensed for use in the US and should be considered for those working with serotype Typhi in a laboratory setting and for travellers who are going to spend extended periods of time in endemic areas [4, 7, 14]. The vaccines available offer moderate protection against typhoid fever; however, they do not protect against the Paratyphi serotype of the bacterium [12]. It has been shown that a live oral vaccine protects 70% of children inoculated in endemic areas [7]. Vaccination is not recommended for pregnant women and patients with HIV-AIDS [23].

PROPHYLAXIS: Antibiotics can be used as prophylaxis in at-risk individuals (for example neonates and the immunocompromised) [7]. Clean water supplies, sanitation, and treatment of carriers are the best prophylactic measures to prevent the spread of enteric fever in endemic areas [7].

SECTION VI - LABORATORY HAZARDS

LABORATORY-ACQUIRED INFECTIONS: Until 1974, 258 cases and 20 deaths due to laboratory-acquired typhoid fever were reported [6]. 48 cases of salmonellosis were reported until 1976 [12]. 64 cases and 2 deaths due to *Salmonella* spp. infections were reported between 1979 and 2004, most of them associated with S. Typhi [30].

SOURCES/SPECIMENS: All *Salmonella enterica* subspecies (with the exception of serotype Typhi) are found in blood, urine, feces, food and feed and environmental materials [14]. Serotype Typhi is found in blood, urine, feces and bile [14].

PRIMARY HAZARDS: Primary hazards when working with *Salmonella enterica* are accidental parenteral inoculation and ingestion [14]. The risk associated with aerosol exposure is not yet known [14].

SPECIAL HAZARDS: Infected animals are a risk (for all serotypes except Typhi and Paratyphi) [14].

SECTION VII – EXPOSURE CONTROLS / PERSONAL PROTECTION

RISK GROUP CLASSIFICATION: Risk group 2 [31]. The risk group associated with *Salmonella enterica* ssp. reflects the species as a whole, but does not necessarily reflect the risk group classification of every subspecies.

CONTAINMENT REQUIREMENTS: Containment level 2 practices, safety equipment and facilities are recommended for work involving infectious or potentially infectious materials, animals, or cultures. Containment level 3 practices and procedures are recommended for activities with serotype typhi that might generate aerosols or large quantities of organisms. These containment requirements apply to the species as a whole, and may not apply to each subspecies within the species.

PROTECTIVE CLOTHING: Lab coat. Gloves when direct skin contact with infected materials or animals is unavoidable. Eye protection must be used where there is a known or potential risk of exposure to splashes [32].

OTHER PRECAUTIONS: All procedures that may produce aerosols, involve high concentrations or large volumes should be conducted in a biological safety cabinet (BSC). The use of needles, syringes and other sharp objects should be strictly limited. The use of needles, syringes, and other sharp objects should be strictly limited. Additional precautions should be considered with work involving animals or large scale activities [32].

SECTION VIII - HANDLING AND STORAGE

SPILLS: Allow aerosols to settle, then, wearing protective clothing, gently cover the spill with absorbent paper towel and apply appropriate disinfectant, starting at the perimeter and working towards the center. Allow sufficient contact time before starting the clean up [32].

DISPOSAL: All wastes should be decontaminated before disposal either by steam sterilization, incineration

Figure 6: Partial screen captures of the Pathogen Safety Data Sheet for *Salmonella enterica* spp. obtained from the Public Health Agency of Canada (http://www.phac-aspc.gc.ca/lab-bio/res/psds-ftss/index-eng.php).

   a. Determine *Salmonella enterica* serovar Typhimurium risk group; http://www.phac-aspc.gc.ca/lab-bio/res/psds-ftss/index-eng.php (see Fig. 6).
   b. Evaluate procedures for Gram stain, culture amplification, and biochemical testing for species confirmation.
   c. Facility is a dedicated microbiology laboratory built in 1975, two sinks, autoclave, BSC (certified within the last year), tabletop centrifuge, glass and plasticware, and manual pipetting devices.
2. Perform risk assessment (see Appendix A).
   a. Organism is risk group 2, with transmission through ingestion and parenteral inoculation.
   b. Some reagents are flammable, some with low-level toxicity, standard procedures of low risk and low consequence.
   c. Standard microbiological procedures apply; special procedures to avoid ingestion of organism; PPE includes lab coat, goggles, face shield for open bench work; BSC for all work with amplified cultures and any procedures that generate aerosols, including initial slide preparations (until fixation complete).
3. Determine BSL (as BSL2) and implement process for approvals (if needed) and laboratory setup, add signage to door.
4. Write and/or review all procedures, including waste processing.
5. Train lab staff and students using demonstrations and exercises prior to using BSL2 *Salmonella* activity; assess competency.
6. Assemble all materials and supplies, PPE options, and waste containers (confirm proper equipment function and validation).
7. Disinfect BSC, place all materials for initial culture transfer and staining procedures into the BSC and allow it to equilibrate (see Table 4).
8. Follow procedures to complete culture transfers and slides for staining.
9. Discard plasticware into bags to be autoclaved.
10. Choose appropriate chemical decontamination agent; 70% ethanol or 5% hydrogen peroxide sufficient for intermediate killing of risk group 2 pathogen, decontaminate all materials exiting the BSC and the BSC surfaces.
11. Incubate cultures and stain slides.
12. Autoclave waste.
13. Record and analyze data.
14. Review successful, near-miss, and failed safety procedures.

Biosafety Guidelines for the Teaching Laboratory

The teaching laboratory challenge, however, is to create and maintain a laboratory culture that rewards safety and prevents the relaxing of rules and safeguards. While higher-ordered, critical thinking is a welcomed learning outcome indicating content familiarity and skill mastery, care must be taken to prevent these positive outcomes from fostering overconfidence when considering biosafety practices.

Regardless of BSL, users of a (basic science or clinical) microbiology teaching laboratory should always follow standard microbiology practices (19) and a few common sense guidelines:

1. Work with hair that is above the shoulders, or is tied back.
2. Wear closed-toed shoes (no canvas shoes or sandals).
3. Protect your eyes from spills and splashes.
4. Wash hands, especially after removing gloves and before exiting the laboratory.
5. Decontaminate and clean spills of microorganisms (decontamination strategies should be validated; in general, it is easier to inactivate lipid viruses, vegetative bacteria, fungi, naked viruses, mycobacteria, and spores, in this order).
6. Prevent the obstruction of exits, aisles, and emergency equipment.

TABLE 4.

Guidance for the use of the biological safety cabinet

Do	Do not
Let cabinet equilibrate for 4 minutes prior to use.	Block the grills.
Set the window sash to armpit height.	Use open flames inside.
Place ALL needed items inside, 4 inches from the front grill, BEFORE the equilibration step.	Place grossly contaminated items inside the cabinet.
Establish "clean-to-dirty" gradient.	Place anything on top of the cabinet.
Use the cabinet one person at a time.	Use sweeping or flailing movements.
Use slow perpendicular movements.	Leave or store things in the cabinet when finished.
Disinfect cabinet and grills when finished.	Leave bleach on the metal; rinse with sterile water.
Leave cabinet running when work is complete.	Use ultraviolet light when people will be in the same room.

7. Seek advice if you believe you have been exposed to an infectious agent, especially if you may be immunocompromised.
8. Alert another if you will be alone in the laboratory, especially at odd hours.
9. Ask if you don't understand a procedure, instruction, or written information.
10. Keep hands and other potentially contaminated surfaces away from your face.

Biosafety Cautions for the Teaching Laboratory

One of the most popular, if not traditional, laboratory-based, critical-thinking exercises is the solving of "the unknown." Potential learning outcomes certainly justify this time-honored capstone experience. However, the absence of specific BSL2 training, personal protective equipment, and biological safety cabinets in most U.S. basic science microbiology laboratories makes one thing quite clear—the potential for laboratory-acquired infection from human pathogens increases when the human body is used as a source of the "unknown." While a search of the literature reveals only one peer-reviewed publication of a laboratory-acquired infection from an "unknown" (30), the data of Harding and Byers (7), as well as anecdotal evidence shared with the authors, suggest that (while rare) infections from "unknowns" may occur more frequently than reported. The presentation of an abstract identifying the laboratory-acquired infection during an unknown exercise, resulting from the mislabeling of a *Salmonella* culture (as *Citrobacter*), further substantiates anecdotal stories (31). The risk assessment for unknown activities should factor in the increased susceptibility to infection of the unidentified compromised student, and the increasing human carriage of methicillin-resistant *Staphylococcus aureus* (MRSA), vancomycin-resistant enterococci, carbapenem-resistant *Enterobacteriaceae*, and *Clostridium difficile*. The culture and amplification of bacteria arising from human bodies should be monitored by those well trained in BSL2 practices. Yet even then, infections from human body sites occur (1–7). Furthermore, the teaching laboratory-associated *Salmonella* outbreaks of 2010 and 2014 (8) can serve as a lesson that even attenuated "laboratory strains" of risk group 2 microorganisms can be potent human pathogens.

In contrast to the use of "human body unknowns," the isolation of "environmental unknowns (from streams, bathrooms, keyboards, pens, etc.)" is often pursued as a less risky alternative. While it may be that environmental specimens mostly contain risk group 1 organisms that can be studied, it is prudent to treat environmental specimens as if they contained higher-level risk group organisms, until proven otherwise. If the pedagogical goal of environmental sampling is to demonstrate the ubiquity of microorganisms, any culture system (plates, tubes, etc.) can be sealed to prevent human access without impairing observation. [Cellulose bands are commercially available to shrink around the closed petri dish (or test tube) and seal it to prevent dehydration and reduce exposure.] If the goal is to amplify and identify the environmental isolate, risk group 1 organisms may be specifically recruited using well-defined selection media and techniques. It is worth noting, however, that an organism only has to be "unknown" to the student; specimens can be "contrived" to meet the pedagogical goals and still be intellectually exciting. Thus, the lesson for all microbiology educators is to weigh carefully the pedagogical goals of any student exercise against potential risks. Whenever possible, risk group 1 organisms should be used when it can achieve the same (or similar) learning objective as the use of a higher risk group organism. It should certainly be stated that when necessary, risk group 2 organisms should be used to train students on the specifics of that organism. The risk assessment should be used, though, to lower the risk of exposure.

Additional Resources for Teaching Laboratories

The May 2011 Report from the Task Force on Curriculum Guidelines for Undergraduate Microbiology Education, commissioned by the Education Board of ASM in 2010, reaffirmed its 1997 Laboratory Core Curriculum recommendations on the importance of teaching safety skills and competencies, adding "where students' development of competency would have enduring and lasting value beyond the classroom and laboratories" (27). Additionally, ASM has created a number of pedagogical strategies to teach laboratory safety skills and competencies. As previously mentioned, most microbiology laboratory manuals have preface material that summarizes the standard microbiological techniques. Microbiology textbooks integrate commentaries on risk groups, BSLs, and safety procedures when referencing specific laboratory activities. There are also a growing number of safety manuals and safety video clips available through the Internet. Importantly, the ASM recommendation is to teach safety skills and competencies that have lifelong value.

In response to the reports of the Trans-Federal Task Force (32) and the CDC Blue Ribbon Panel (33), the CDC and the APHL created a set of biosafety laboratory competency and skill domains around which educational goals, training standards, safety assessments, professional development, and certifications could be developed. The competency and skill domains reflect the goals of the BMBL in that they promote hazard reduction through risk assessment, but they are not tasks to accomplish or check off. The domains tie job responsibility to risk and user experience (i.e., entry level [some training in the life sciences], midlevel [experienced laboratory user], and senior level [supervisor or manager]). The domains focus on (i) potential hazards, (ii) hazard controls, (iii) administrative controls, and (iv) emergency preparedness and response. They can be used to develop curricular materials that measure the essential skills, knowledge, abilities, judgment, and self-criticism required for working in laboratories containing potentially infectious material (28). Similarly, the June 2012 Report from the Task Committee on Guidelines for Biosafety in Teaching Laboratories, commissioned by the Education Board of ASM in 2011, offers microbiology instructors and students a guidance document which highlights best teaching practices and thus facilitates laboratory safety (34). The CDC, additionally, produces educational materials that foster safety in teaching laboratories (for example, see http://www.cdc.gov/salmonella/pdf/CDC_LAI_Prevention_Poster_012313_508.pdf and Appendix B).

SUPPLEMENTAL MATERIALS

To aid in developing a culture of safety in the teaching laboratory, the topics of standard microbiology practices, safety equipment, BSL1 and BSL2 laboratory design, and biosafety manuals are discussed in the following sections. This section additionally contains a list of "frequently asked questions" to assist the microbiology educator.

Standard Microbiological Practices or Techniques

Regardless of other conditions to meet BSL requirements and recommendations, microbiologists have universally practiced "standard microbiological techniques" [sometimes confused with the similar "universal precautions," defined by blood-borne pathogen legislation (10)], as the first means of hazard reduction. These have been honored as one means to minimize laboratory-acquired infection and are the minimal conditions required for working in any laboratory where microorganisms are studied. It is recommended that these standard techniques, and any specialized work practices, be included in the laboratory protocol as a reminder to the student (and faculty) of their importance in reducing risks associated with working in the laboratory. The standard microbiological techniques (19) include:

1. The supervisor must enforce the institutional policies that control access to the laboratory.
2. Persons must wash their hands after working with potentially hazardous materials and before leaving the laboratory.
3. Eating, drinking, smoking, handling contact lenses, applying cosmetics, and storing food for human consumption must not be permitted in the laboratory.
4. Mouth pipetting is prohibited; mechanical pipetting devices must be used.
5. Policies for the safe handling of sharps (such as needles, scalpels, pipettes, and broken glassware) must be developed and implemented.
6. Perform all procedures to minimize the creation of splashes and/or aerosols.
7. Decontaminate work surfaces after completion of work and after any spill or splash of potentially infectious material with appropriate disinfectant.
8. Decontaminate all cultures, stocks, and other potentially infectious materials before disposal using an effective method.
9. A sign incorporating the universal biohazard symbol must be posted at the entrance to the laboratory when infectious agents are present. The sign may include the BSL, name of the agent(s) in use, and the name and phone number of the laboratory supervisor or other responsible personnel. Information should be posted according to institutional policy.
10. An effective integrated pest management program is required.
11. The laboratory supervisor must ensure that laboratory personnel receive appropriate training regarding their duties, the necessary precautions to prevent exposures, and exposure evaluation procedures. Personnel must receive annual updates or additional training when procedural or policy changes occur.

Safety Equipment (Primary Barrier)

In addition to specific procedures or techniques that the laboratory user can immediately control to effect safety, the next level of hazard reduction is called the primary barrier. The primary barrier is envisioned as the first layer of physical protection separating the laboratory user from hazards and is composed of specific "equipment" that serves to protect the laboratory user from being contaminated or infected by the organisms in the laboratory. Appropriate safety equipment is determined by the risk assessment, not arbitrarily assigned or used. The equipment must be user friendly, functional and in good order, acceptable to the user, and readily available in or near the laboratory. Importantly, safety equipment must be specific for the task.

Safety equipment typically refers to specialized containment devices (such as clean benches, biological safety cabinets, mechanical pipetting devices, removable centrifuge rotors, and sealable rotor cups) and PPE worn on the body (laboratory coats or smocks, eye or face protection, and gloves, for example). Clean benches, mechanical pipetting devices, and safety cups and rotors are self-explanatory. More in-depth discussion of BSCs and PPE is found below.

Biological safety cabinets

The BSC is a leak-tight container designed to control airflow and contain infectious particles that might be generated within the cabinet during specimen manipulation. Particles that circulate within the cabinet will be collected on the HEPA filter prior to exiting the BSC. In general, the BSC should be positioned away from external sources of air movement (laboratory doors, refrigerator doors, ventilation gratings, etc.) (Table 4). There are three "classes" of BSC designated by the CDC, having increasingly greater protection for both the user and the specimen or culture. Each category of BSC is required to undergo annual certification to ensure adherence to standards established in NSF/ANSI 49 (35). A comprehensive review of the different classes of BSCs is presented elsewhere in this book and will not be included in this section. Typically, however, most BSL1 and BSL2 laboratories use a Class I or Class II BSC, respectively.

Personal protective equipment

As its name implies, PPE is a unique combination of safety equipment designed to meet the specific safety needs of the wearer. PPE is included as part of the primary barrier because it is the first line of defense against laboratory hazards. The risk assessment will determine the amount and type of body protection, evaluating the specific need depending on the agent(s) and procedures used. The risk assessment, by design, permits the justification for PPE required for the job and helps prevent excessive PPE waste. PPE selection and use are the responsibility of the laboratory instructor, with input from the user (who comments on allergies, fit, etc.). Correctly used, PPE reduces the exposure potential. The effectiveness of PPE, however, is predicated on proper training and use. Recall that OSHA regulations require employers to provide PPE and to have employees trained on its care and use (10).

PPE includes (i) protective coats, gowns, or uniforms; (ii) protective eyewear (typically covering the side, as well as the front access to the eyes); and (iii) gloves (with latex alternatives available). Additional PPE (for example, face shield, respiratory protection, etc.) may be required depending of the risk assessment results. All PPE should be closely inspected before donning to ensure its integrity and protection of the user.

Protective outerwear is worn over street clothes to prevent potentially infectious materials from (unknowingly) leaving the laboratory. There are a number of options when one seeks to protect street clothes from exposure to microorganisms (Fig. 7). The typical choice is the long-sleeved, laboratory coat that buttons in front (Fig. 7A). Newer versions of the laboratory coat include those with plastic linings, elastic cuffs, and high, buttoning collars (Fig. 7B). Laboratory "coveralls" made from a water-repellent material (e.g., Tyvek) are also useful in protecting street clothes, but may retain heat (Fig. 7C). This latter PPE option is rarely used in teaching laboratories that operate at a BSL1/BSL2 level. It is not uncommon to also see laboratory users wearing "surgical scrubs" over street clothes, or even in the absence of street clothes. It is important to remember that the role of this outer layer

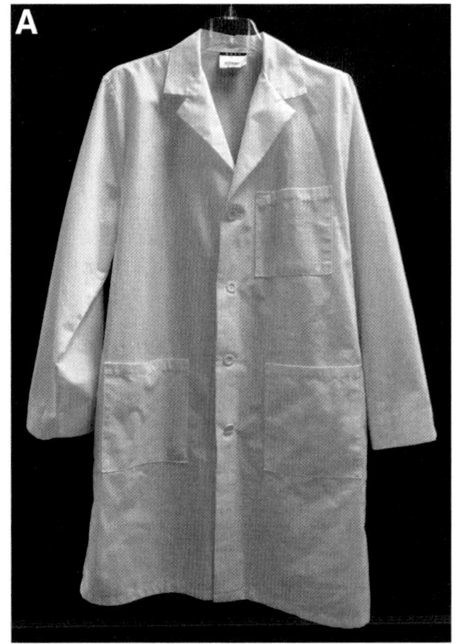

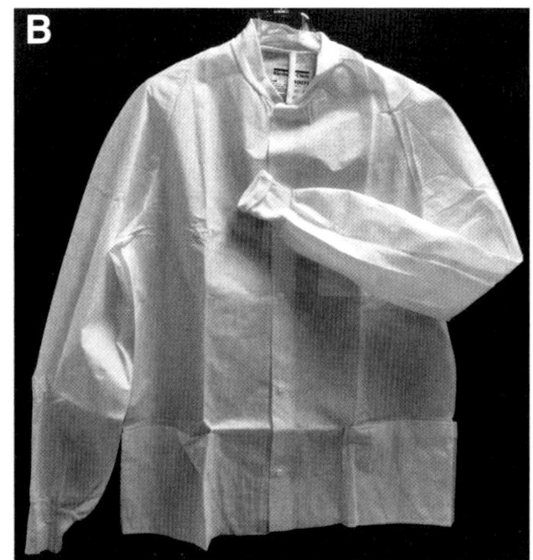

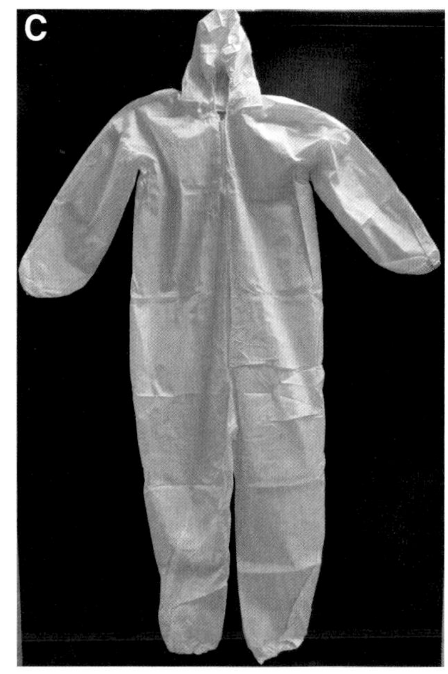

Figure 7: Examples of materials and designs to wear over street clothes while working in the laboratory. (A) Standard cotton laboratory coat; (B) high collar and cuffed sleeves incorporated into a water-resistant jacket; and (C) waterproof coveralls.

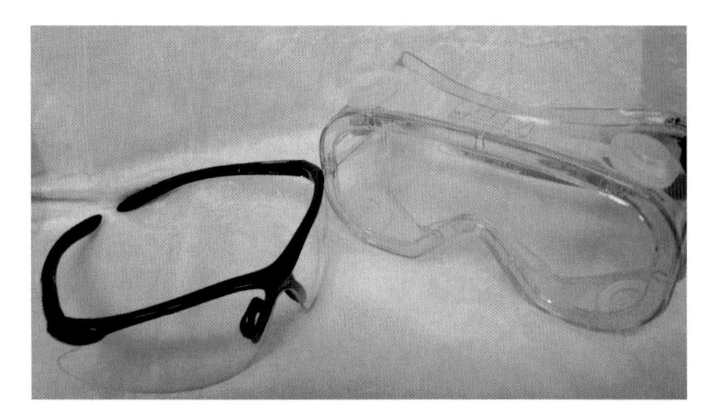

Figure 8: Examples of safety glasses (left) and safety goggles (right).

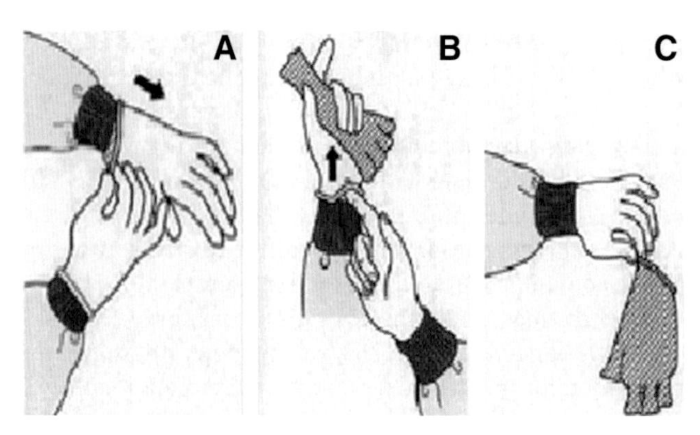

Figure 9: Technique for the aseptic removal (doffing) of contaminated gloves (19).

is to be a barrier preventing microorganisms from being transported out of the laboratory. Therefore, this covering should NOT be worn outside of the laboratory.

Eye protection in the laboratory is essential. Protective eyewear is worn when conducting procedures with the potential to generate spills or splashes, preventing transmission of potentially infectious agents and hazardous materials through the eyes of the user. Safety goggles that wrap around the temples provide appropriate front and side protection during splashes and spills (Fig. 8). Safety glasses, even with side shields, are often insufficient in the protection from splashes and spills (Fig. 8); however, they are better than no eye coverage at all. Impact resistant (shatterproof) plastic materials are preferred over materials that shatter. Face shields or visors made of shatterproof plastic may be used, especially by those with prescription eyewear to protect. Similar to laboratory coats and coveralls, protective eyewear should not be worn outside of the laboratory. For further information, see ANSI Z87.1-2003, American National Standard for Occupational and Educational Eye and Face Protection Devices (36).

Gloves are worn to protect the hands from contact with hazardous materials, especially when a break in the skin exists; they are not a replacement for hand washing and should never be washed for continued use, unless specifically purposed for that process. A number of different materials (latex, vinyl, nitrile, butyl, neoprene, etc.) are used to manufacture gloves. The more common material is disposable latex. However, there appears to be an increase in the number of people allergic to latex, and persons that continuously use latex gloves are more likely to develop latex allergies. Oil-based lotions should not be used with latex gloves because it increases the rate of deterioration (http://www.grainger.com/Grainger/static/latex-allergies-reactions-126.html). Therefore latex glove alternatives such as vinyl and nitrile are necessary in the laboratory.

Nitrile surgical gloves are an appropriate alternative to latex because they do not appear to elicit allergic responses and can be used with harsh chemicals in addition to microorganisms. Recall, though, that the risk assessment will determine the need for gloves and the conditions under which they are used. Most laboratory work involves both biologicals and chemicals, thus consideration should be given to the type of glove that will protect against both hazards. Gloves are rated according to their permeation, breakthrough time, and degradation (for more information, see ANSI/ISEA 105-2005, American National Standard for Hand Protection Selection Criteria) (37). Several glove manufacturers provide chemical-resistance guides for gloves, and this information should be taken into consideration when choosing gloves.

Gloves should not be worn outside after being used inside of the laboratory. In all cases, hands should be washed as soon as gloves are removed. Removal (doffing) of gloves to avoid hand contamination is presented in Fig. 9. Aseptic doffing of gloves starts by carefully pealing the glove of one hand, from the wrist toward the fingers, using the other gloved hand (Fig. 9A). The removed glove is held in the gloved hand, as a finger from the bare hand carefully slides under the cuff of the remaining glove (Fig. 9B), sliding the glove forward and inverting the glove over the fingers, capturing the other glove (Fig. 9C). The gloves are then properly disposed as contaminated dry waste.

Laboratory Design

The BSL designation is meant to provide awareness for the laboratory user, in that it indicates that a specific set of engineering controls, safety equipment, and work practices is expected in the designated laboratory to achieve hazard reduction. Table 2 provides a summary of the BMBL-recommended BSL1 and BSL2 safety equipment and facility design specifications, as a quick reference. Table 3 lists some of the agents that might be used in the BSL1 or

BSL2 teaching laboratory. Specific recommendations follow.

BSL1 environment

The BSL1 environment is defined as one in which use of "well-characterized agents not known to cause disease consistently in immunocompetent adult humans and present minimal potential hazard to laboratory personnel and the environment." These spaces may be shared with other laboratory-based science teaching and not necessarily separated from the normal building traffic. By the nature of the BSL1 risk assessment results, work in this space does not require BSCs, because most work is appropriate for open benchtop processing, using standard microbiological techniques. The risk assessment for the specific facility and procedures will determine any special equipment or facility design necessary for unique applications. Importantly, laboratory users must be supervised by an instructor with training in microbiology or a related science (19).

One model of a BSL1 laboratory design is found in Fig. 10 (38). Note that the facility design includes a door and sinks, the required elements. The floor appears to be covered with a material that would facilitate spill cleanup. The lab is uncluttered. Laboratory users wear lab coats to protect street clothes and safety goggles (especially contact lens wearers). Gloves might be worn to protect hand wounds. Most important for this BSL1 laboratory is the diligent practice of standard microbiological techniques.

BSL2 environment

The BSL2 environment is defined for use of agents that are associated with human disease and whose routes of transmission include ingestion, mucous membrane exposure, and percutaneous injury. These agents pose moderate threat to laboratory personnel and the environment. The BSL2 space, therefore, builds upon the BSL1 design (Fig. 11), but has several notable enhancements (38). It is intentionally isolated from the normal building traffic. The laboratory should have self-closing doors that restrict access when work with risk group 2 agents is being conducted. Doors should display a sign with the universal biosafety symbol, the biological safety level, entry and exit requirements, and the emergency contact information, as well as lock according to institutional policy. A sink should be near the exit door to facilitate hand washing just prior to exit. An emergency eyewash station and an autoclave must be readily available. The biosafety manual should be adopted as policy (19).

BSL2 safety equipment is also enhanced and integrated as a result of the risk assessment. Procedures in which aerosols may be generated are performed inside a BSC or using other appropriate containment devices and PPE. The BSC must be installed away from normal foot traffic in the laboratory and away from windows, doors, and vents. The BSC must be certified annually. A validated method must be used for the decontamination of laboratory waste. Large volumes or high concentrations of agents to be centrifuged are done so using sealed rotor heads or

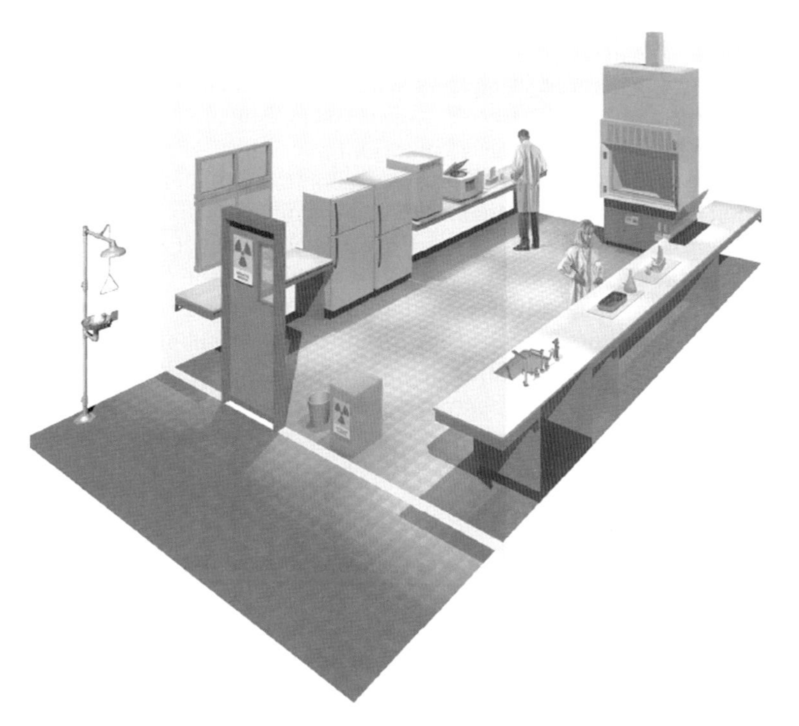

Figure 10: Diagram of a shared teaching laboratory space that might be used as a BSL1 laboratory (38).

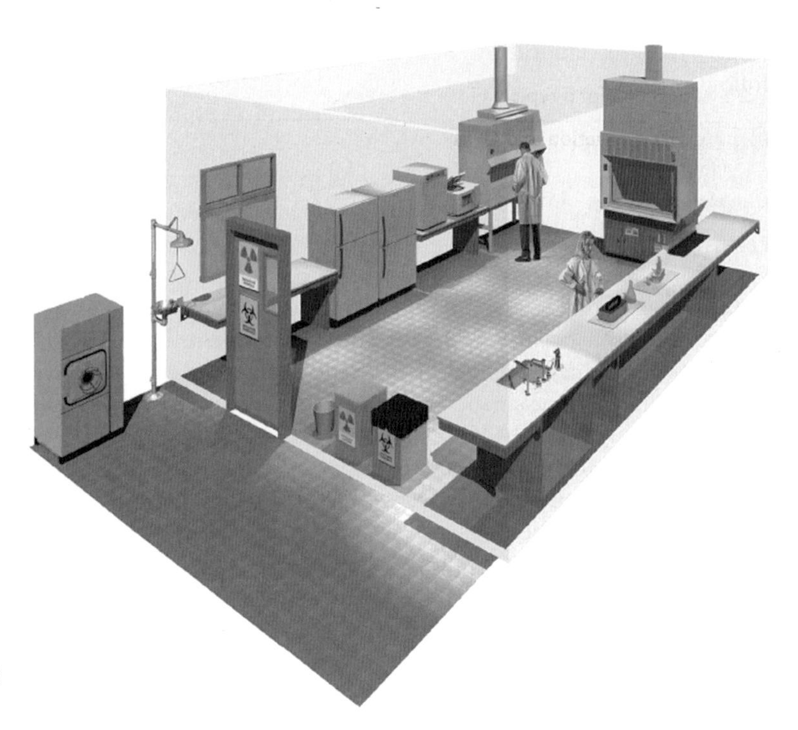

Figure 11: Diagram of a BSL2 laboratory design (adapted from reference 38).

centrifuge safety cups. Protective clothing is removed before entering nonlaboratory spaces and is destroyed or laundered by the institution (not taken home). In the absence of laundry facilities, protective clothing should be kept in the laboratory for future use or until decontaminated. (Laboratory coats/aprons/smocks/etc. can be folded dirty sides to each other and then further folded to reduce size. These can then be placed in zip-lock plastic bags and/or in laboratory drawers, or on shelves.) Non-cloth (i.e., paper/plastic)-based protective clothing is another alternative and can be added to the appropriate waste stream when contaminated or soiled. Eye/face protection must be used when manipulation of agents outside of the BSC may result in spills or splashes. Eye, face, and respiratory protection must be worn in areas with infected animals. All laboratory users are to have specific training in the handling of pathogenic agents and decontamination procedures, in addition to being supervised by scientists competent in the handling of potentially infectious material. Waste removal should be approved by institutional policy and municipal regulation. Again, the most important function within the BSL2 laboratory is the diligent practice of standard microbiological techniques (19).

Biosafety Manual

In addition to the required standard microbiology practices, safety equipment, and facility design determined from the risk assessment, a very strong recommendation for all laboratories is the development and maintenance of an up-to-date biosafety manual. (A biosafety manual may be required for clinical laboratory accreditation.) The biosafety manual can be a snapshot of the organisms, procedures, equipment, and responsible person(s) of the laboratory. It can also indicate (i) the specific microbiological practices, (ii) personal protective equipment, including instructions for use, (iii) information regarding care and use of BSCs and autoclaves, and (iv) safety competencies and trainings required and allowable in the laboratory space.

The general contents of the biosafety manual do not need to be exhaustive, but rather a "ready reference" that can refer the reader to other and more specific information as needed (such as a separate binder of all chemical Material Data Safety Sheets [MSDS] or standard operating procedures for the laboratory). Table 5 suggests topics (not all being pertinent for every laboratory) that could serve as the table of contents for a laboratory biosafety manual. The manual could then serve as a local guidance document to direct laboratory activities rapidly on the basis of BSL. Importantly, it can and should be used to educate new laboratory users and emergency responders about laboratory risks. Its location should therefore be somewhere near the laboratory entrance for easy access, especially to emergency responders entering from outside the laboratory.

Frequently Asked Questions

1. Can stains, microorganisms, etc., be flushed down the sink?
   Depends on local and state regulations. Contact your state health department or Environmental Protection Agency office for guidance.

TABLE 5.

Table of contents of a typical biosafety manual[a]

Chapter 1 Introduction and Overview of Manual
Definitions
Chapter 2 General Information and Structure
Reporting structure and responsibility
Research agents listing
Laboratory access and security
Personnel training
Facilities and engineering controls
Posting and labeling requirements
Storage of biohazardous materials
Transport of hazardous materials to storage areas
Chapter 3 BSL1 Laboratory Procedures
General safety and biosafety procedures
Personal protective equipment care and use
Standard microbiological practices
Use of pipetting aids
Use and disposal of sharps
Minimization of aerosol production
Decontamination methods
Chapter 4 BSL2 Laboratory Procedures
General safety and biosafety procedures
Personal protective equipment care and use
Standard and specialized microbiological practices
Use of pipetting aids
Use and disposal of sharps
Minimization of aerosol production
Biological safety cabinet use
Decontamination methods
Chapter 5 Equipment and Facility Management
Equipment care and use (microscope, autoclave, centrifuge, etc.)
Other physical and chemical hazards
Cleaning and decontamination
Chapter 6 Emergency Management
Emergency procedures
Accidents and accidental exposure to an infectious agent
Spills inside and spills outside the biological safety cabinet
First aid
Reporting and record keeping
Appendices
Emergency call list
Laboratory schematic
PSDS for infectious agents located in laboratory
Accident report forms
Recombinant DNA registration form
Standard operating procedure template

[a]Chapter 4 may be deleted for laboratories that do not use BSL2 conditions.
BSL, biosafety level; PSDS, Pathogen Safety Data Sheets.

2. Are spores inactivated by the chemicals used in the stains?
   Spore inactivation is the result of an appropriate disinfectant used for an effective time period. EPA-registered, sporicidal disinfectants should be used according to the label specifications for concentration and contact time. Typical spore staining procedures therefore do not inactivate spores.
3. Does "fixing" a microorganism to a slide inactivate it?
   Inactivation of microorganisms results when metabolic and/or reproductive functions are prevented. Heat fixing typically inactivates microorganisms. Inactivation of microorganisms by chemical fixation techniques is a function of the agent and exposure time.
3. Are goggles necessary or just safety glasses?
   Safety glasses should always be worn when working in a science laboratory. The risk assessment will determine when goggles are required, for example, when diluting bleach. In general, goggles prevent side access to eye; safety glasses may not.
4. Gloves or no gloves?
   Gloves should be used to cover open wounds and to prevent exposure of the skin to potentially infectious materials. Other uses are determined from the risk assessment.
5. What items are permitted in the lab? What items can leave the lab?
   This is one of the most debated questions (after glove use). Technically, anything that could transport microorganisms out of the lab should not be expected to leave the lab, unless decontaminated. This includes pens and markers (designated for laboratory use), cell phones, iPads, textbooks, lab notebooks, etc. However, this may be an impractical expectation. Novel solutions to this problem have arisen, and the problem itself is used as a critical thinking exercise for laboratory students. Data from experiments certainly can be emailed from a lab computer, captured by a laptop camera, collected in waterproof notebooks that are decontaminated prior to exiting the lab, etc.
6. What methods are approved for decontamination of waste?
   In addition to autoclaving to decontaminate microbiological waste, there are a number of chemical decontamination methods. Licensed medical waste vendors may also remove properly packaged infectious waste for decontamination and disposal. Most local and state jurisdictions have regulations for waste decontamination and disposal. Contact your institutional and state environmental protection authorities for guidance.
7. How is decontamination validated?
   The most reliable method is a bioassay. There are variations of the phenol-inactivation assay to validate chemical decontamination methods and spore-inactivation assays to validate heat (autoclave) decontamination methods. Guidance should be sought from local and state waste regulators to determine the recommended methods.

The authors thank Lyssa Rickard, M.Ed., Neil Baker, Ph.D., Richard Green, M.Sc., C.T.M., and Rebecca Buxton, M.S., M.T.(ASCP) for insights and critical review of the manuscript.

References

1. **Sulkin SE, Pike RM.** 1949. Viral infections contracted in the laboratory. *N Engl J Med* **241:**205–213.
2. **Sulkin SE, Pike RM.** 1951. Survey of laboratory-acquired infections. *Am J Public Health Nations Health* **41:**769–781.
3. **Pike RM, Sulkin SE, Schulze ML.** 1965. Continuing importance of laboratory-acquired infections. *Am J Public Health Nations Health* **55:**190–199.
4. **Pike RM.** 1976. Laboratory-associated infections: summary and analysis of 3921 cases. *Health Lab Sci* **13:**105–114.
5. **Pike RM.** 1978. Past and present hazards of working with infectious agents. *Arch Pathol Lab Med* **102:**333–336.
6. **Pike RM.** 1979. Laboratory-associated infections: incidence, fatalities, causes, and prevention. *Annu Rev Microbiol* **33:**41–66.
7. **Harding AL, Byers KB.** 2006. Epidemiology of laboratory-associated infections, p 53–77. *In* Fleming DO, Hunt DL (ed), *Biological Safety*, 4th ed. ASM Press, Washington, DC.
8. **Centers for Disease Control and Prevention.** 2011. Investigation announcement: Multistate outbreak of human Salmonella Typhimurium infections associated with exposure to clinical and teaching microbiology laboratories. April 28, 2011. Available at http://www.cdc.gov/salmonella/typhimurium-laboratory/042711, accessed May 19, 2011.
9. **Occupational and Safety Health Administration.** 1970. Occupational Health and Safety Act, 29 CFR Section 654. U. S. Department of Labor. Public Law 91-596, Dec 29, 1970.
10. **Occupational and Safety Health Administration.** 1991. Occupational exposure to bloodborne pathogens; final rule. 29 CFR 1910. U.S. Department of Labor. *Fed Regist* **56:**64003–64182.
11. **Clinical and Laboratory Standards Institute (CLSI).** 2005. *Protection of Laboratory Workers from Occupationally Acquired Infections; Approved Guideline,* 3rd ed. Clinical and Laboratory Standards Institute. CLSI document M29-A3.
12. **Clinical and Laboratory Standards Institute (CLSI).** 2010. *Protection of Laboratory Workers from Occupationally Acquired Infections; Approved Guideline,* 4th ed. Clinical and Laboratory Standards Institute. CLSI document M29-A4.
13. **International Organization for Standardization (ISO).** 2007. Medical laboratories—Particular requirements for quality and competence. Geneva. ISO 15189, 2007.
14. **International Organization for Standardization (ISO).** 2007. Medical laboratories—Particular requirements for quality and competence. Geneva. ISO 15190, 2007.
15. **Centers for Disease Control.** 1974. *Classification of Etiological Agents on the Basis of Hazard,* 4th ed. Centers for Disease Control, Atlanta, GA.
16. **Wedum AG.** 1964. Laboratory safety in research with infectious aerosols. *Public Health Rep* **79:**619–633.
17. **National Cancer Institute.** 1974. *Safety Standards for Research involving Oncogenic Viruses.* (DHEW pub; 75-790). The National Cancer Institute, Bethesda, MD.
18. **U.S. National Institutes of Health.** 1976. Recombinant DNA Research Guidelines. *Fed Regist* **41:**27902–27943.
19. **U.S. Department of Health and Human Services.** Public Health Service, Centers for Disease Control and Prevention, National Institutes of Health. 2009. *Biosafety in Microbiological and Biomedical Laboratories*, 5th ed. Wilson DE, Chosewood LC (ed). HHS Publication no. (CDC) 21-1112. U.S. Government Printing Office, Washington, DC.
20. **Civic Impulse.** 2016. H.R. 3103—104th Congress: Health Insurance Portability and Accountability Act of 1996. https://www.govtrack.us/congress/bills/104/hr3103, accessed Jan. 7, 2016.
21. **Health and Safety Executive.** 2006. *Five Steps to Risk Assessment.* HSE Books, Suffolk, United Kingdom.
22. **Bazerman M, Watkins M.** 2008. *Predictable Surprises: the Disasters You Should Have Seen Coming and How To Prevent Them.* Harvard Business Press, Boston, MA.
23. **Carus S.** 2002. *Bioterrorism and Biocrimes: The Illicit Use of Biological Agents since 1900.* Fredonia Books, Amsterdam, The Netherlands.
24. **Centers for Disease Control and Prevention.** 2002. Public Health Security and Bioterrorism Response Act. 42 CFR 73 (42 USC 262 and PL107-188). U.S. Department of Health and Human Services. U.S. Government Printing Office, Boulder, CO.
25. **Centers for Disease Control and Prevention.** 2005. Possession, use, and transfer of select agents and toxins. Final rule. 42 CFR Parts 72 and 73. U.S. Department of Health and Human Services. *Fed Regist* **70:**13293–13325.
26. **Association for the Advancement of Science.** 2011. *Vision and Change in Undergraduate Biology Education: A Call to Action.* Association for the Advancement of Science, Washington, DC.
27. **American Society for Microbiology.** May 2011. Report from the Task Force on Curriculum Guidelines for Undergraduate Microbiology Education. Washington, DC. http://www.asm.org/images/Education/website%20may%202011%20final.pdf, accessed May 30, 2011.
28. **Delany JR, Pentella MA, Rodriguez JA, Shah KV, Baxley KP, Holmes DE, Centers for Disease Control and Prevention.** 2011. Guidelines for biosafety laboratory competency. *MMWR Suppl* 60:1–23. http://www.cdc.gov/mmwr/preview/mmwrhtml/su6002a1.htm?s_cid=su6002a1_w, accessed 19 May 2011.
29. **Forehand M.** 2005. Bloom's taxonomy: original and revised. *In* Orey M (ed), *Emerging Perspectives on Learning, Teaching, and Technology.* http://projects.coe.uga.edu/epltt/, accessed June 7, 2011.
30. **Boyer B, DeBenedictis KJ, Master R, Jones RS.** 1998. The microbiology "unknown" misadventure. *Am J Infect Control* **26:** 355–358.
31. **Said MA, Smyth S, Wright-Andoh J, Myers R, Razeq J, Blythe D.** 2013. *Salmonella enterica* serotype Typhimurium gastrointestinal illness associated with a university microbiology course—Maryland, 2011. *2013 EIS Conference Abstracts*, p. 103.
32. **US Department of Health and Human Services, US Department of Agriculture.** 2009. Report of the Trans-Federal Task Force on Optimizing Biosafety and Biocontainment Oversight. Agriculture Research Service, Washington, DC. http://www.ars.usda.gov/is/br/bbotaskforce/biosafety-FINAL-REPORT-092009.pdf, accessed May 19, 2011.
33. **Miller JM.** 2011. Guidelines for Safe Work Practices in Human and Animal Clinical Diagnostic Laboratories. Report of the CDC Blue Ribbon Panel. Available at http://www.asm.org/images/pdf/CDCCompleteSafetyDocument.pdf, accessed May 19, 2011.
34. **Emmert EAB, ASM Task Committee on Laboratory Biosafety.** 2013. Biosafety guidelines for handling microorganisms in the teaching laboratory: development and rationale. *J Microbiol Biol Educ* **14:**78–83.
35. **International NSF.** 2010. *NSF/ANSI 2010, NSF Standard 49, Class II (laminar flow).* Biohazard Cabinetry. NSF International, Michigan.
36. **American National Standards Institute.** 2003. Industrial Eyewear Impact Standard. Occupational and Educational Eye and Face Protection Devices. ANSI Z87.1-2003.
37. **American National Standards Institute/International Safety Equipment Association.** 2005. Hand Protection Standard. American National Standard for Hand Protection Selection Criteria. ANSI/ISEA 105-2005.
38. **World Health Organization.** 2004. *Laboratory Biosafety Manual,* 3rd ed. World Health Organization, Geneva, Switzerland.

APPENDIX A

Biological Risk Assessment Worksheet

Tracking # ______________ **Building/Lab Room #** __________ **PI Name** ____________________

Laboratory protocols consist of one or more procedures. Each procedure in the protocol needs an agent-specific Biological Risk Assessment. Once an agent-specific Biological Risk Assessment has been completed for the procedure, it can be used for multiple protocols by referencing its tracking number. The procedure may be performed with additional precautions, if desired, but must be no less stringent than what is calculated below at Section II.

Keep a completed copy of this worksheet in your Biosafety Manual. The ***Biosafety in Microbiological and Biological Laboratories*** (BMBL) 5th Edition has additional guidance on facilities, work practices, PPE, and medical surveillance.

Section I: Complete All Data Entry in this Section

1. **Agent Used** ______________________________
2. **Is a vaccine available?** **Yes** O **No** O
3. **Risk Group of Agent** (check www.absa.org) 1 O 2 O 3 O 4 O {Inactivated agents = Risk Group 1}
4. **Procedure** ______________________________
5. **For Risk Group 2-3, is there a splash potential?** **Yes** O **No** O
6. **For Risk Group 2-3, does the procedure generate aerosol or large concentration?** **Yes** O **No** O
   (e.g., cell culture, vortex, centrifuge, aerosol chamber, sonicate)

Section II: Data will be calculated in this Section according to the answers entered above in Section I

1. **Facility and Work Practices Biological Safety Levels (BSLs)**

   Facility BSL 1 ☐ 2 ☐ 3 ☐ 4 ☐ Work Practices BSL 1 ☐ 2 ☐ 3 ☐ 4 ☐
2. **Biological Safety Cabinet** Class I/II ☐ Class III ☐
3. **Personal Protective Equipment Needed for Procedure:** (left to right = increased protection)
   a. **Gloves** latex/nitrile required
   b. **Eye** safety glasses ☐ goggles + face shield ☐
   c. **Lab coat** white ☐ blue smock/coveralls ☐ space suit ☐
   d. **Respirator*** N-95/PAPR ☐ space suit ☐
4. **Medical Protection and Surveillance**
   a. **Medical Monitoring required** ☐ **b. Hearing Conservation Program** ☐
   c. **Vaccine recommended*** ☐ **d. Respiratory Protection Program** ☐
5. **Comments** ______________________________

   Note: *Vaccines and respirators require separate risk assessments.

Biosafety Officer's Signature

[ ]

Adapted from CDC:
http://www.cdc.gov/biosafety/publications/BiologicalRiskAssessmentWorksheet.pdf

APPENDIX B

What You Work With Can Make You Sick

Follow safe lab practices—and don't bring germs home with you.

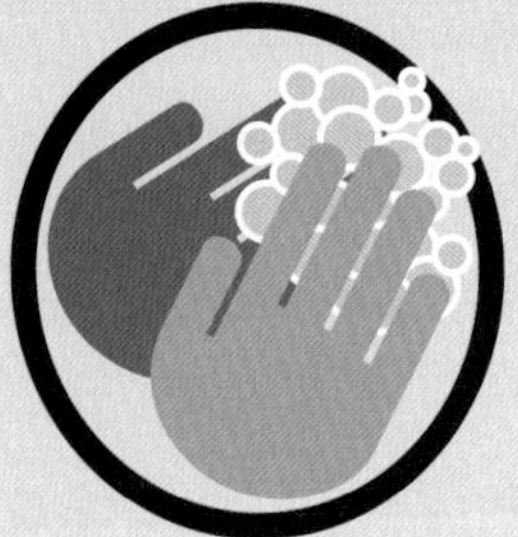

Always wash your hands with soap and water...

- Right after working in the lab
- Just before you leave the lab

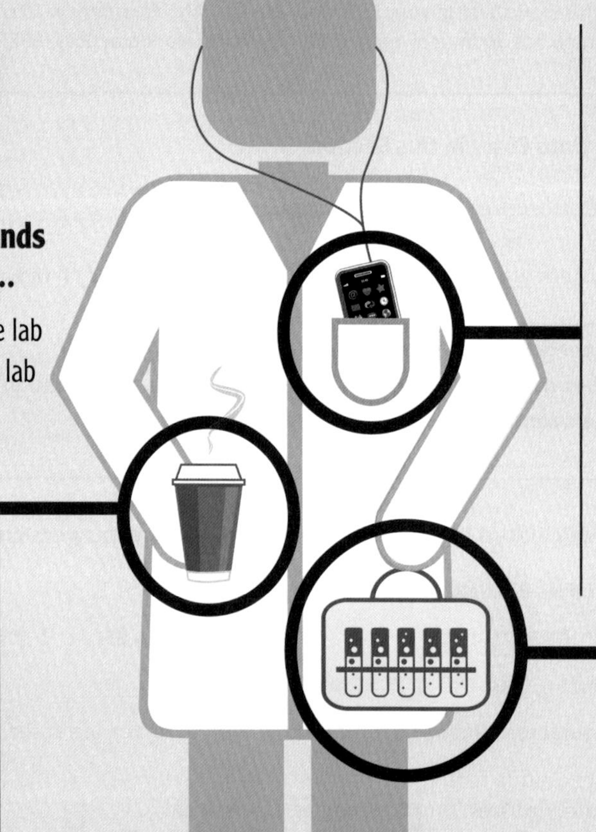

Don't carry dangerous germs from the laboratory home with you.

Leave personal items outside of the lab so you don't contaminate them: cell phone, car keys, tablet or laptop, MP3 player

Keep work items off of bench areas where you do experiments: backpacks, notebooks, pencils, pens

Avoid contamination while in the lab.

Don't eat, drink, or put things in your mouth (such as gum)

Don't touch your mouth or eyes

Don't put on cosmetics (like lip balm) or handle your contact lenses

Leave lab supplies inside the lab.

If you must take supplies out of the lab, keep them in a separate bag so you don't contaminate anything else

Leave your experiment inside the lab so you can stay healthy outside the lab.

Centers for Disease Control and Prevention
National Center for Emerging and Zoonotic Infectious Diseases

CS237165

31 Biosafety in the Pharmaceutical Industry

BRIAN R. PETUCH

INTRODUCTION

Pharmaceutical companies that employ pathogenic microorganisms to produce vaccines and sometimes pharmaceuticals must establish a broad range of biosafety practices to ensure the safety of their employees as well as their products. At the drug discovery stage, especially during the search for candidates from natural sources, these safety practices must allow the research laboratories to cultivate myriad microorganisms, many of which are initially unknown. During scale-up, the biosafety practices employed should be in harmony with international guidelines to ensure that the manufacturing process and product may be implemented and sold, respectively, in other countries. Because the biosafety concerns experienced in pharmaceutical research laboratories are quite similar to those discussed in earlier chapters, they will not be repeated here. Therefore, this chapter briefly addresses the biosafety challenges commonly experienced in cultivating recombinant and pathogenic microbes. The use of mammalian cells for the production of therapeutic proteins and viruses will also be addressed.

MICROORGANISMS

The microorganisms used in pharmaceutical companies run the gamut from viruses to bacteria to fungi. Their place in pharmaceutical research and development is summarized in the following sections.

Viruses

Both attenuated and recombinant viruses are in current or projected use as vaccine materials. The classic "Jennerian" approach to vaccine generation is through the use of antigenically related viruses that are not permissive in human hosts (e.g., the use of cowpox, later vaccinia virus, to immunize against smallpox). For a number of human viruses, deliberate attenuation has been accomplished by passage in other vertebrates or vertebrate embryos until they are unable to replicate successfully in human cells (e.g., oral polio vaccine) or replicate poorly in human cells (e.g., varicella vaccine strains).

Recombinant viral agents have become a significant starting material for both gene therapy and vaccine production. The list of viruses under consideration is extensive and includes retroviruses and lentiviruses (e.g., equine infectious anemia virus and HIV), adenoviruses and adeno-associated viruses (AAV), and poxviruses (e.g., vaccinia). Process scale-up employing these viruses presents novel problems for biocontainment, even if the agent is either replication defective or nonpathogenic. Although retroviruses are capable of infecting and producing their gene product in nondividing cells, stable transfection of a retrovirus requires integration of the virus into the host genome. Insertional mutagenesis could result in the activation of proto-oncogenes or inactivation of tumor suppressor genes in the trial participant or the manufacturing employees (1). Therefore, careful consideration of what is the proper containment for a defective retrovirus containing a therapeutic gene is required.

Adenoviruses and AAV have the advantage of nonnuclear replication without integration into the genome of the host, but are also only temporary for the same reason. Repeated administration of the virus generates an immune response to the adenoviral vector, with a reduced effectiveness. Moreover, the early generations of adenovirus were deleted in the E1 or E1 and E3 genes. Growth in adeno-transformed human embryonic kidney cells (HEK-293) was required to complement the mutation, but it also resulted in replication-competent adenoviruses, contaminating the resultant stock (2). With certain additional knowledge, such as the fact that adenovirus can infect via the aerosol route, the available concentration will be approximately 10^{10} to 10^{12}/ml, and the infectious dose is unknown, a definition of the appropriate containment level for the vector can be problematic. Where vaccine candidate genes are also used in clinical assays to determine the presence of the disease state (e.g., the gp 120/160 gene from HIV is used in ongoing clinical trials and is screened for in commercial HIV enzyme-linked immunosorbent assay [ELISA] tests), containment is not a simple issue that can be determined by reading the appropriate regulations or guidelines. The biosafety aspects of a program must be an outgrowth of consultation of existing guidelines as well as careful consideration of the biology of the agent in question.

The use of baculovirus has not, to date, been significant within the pharmaceutical industry. However, the ability of insect cells to grow at room temperature and to process intron-containing RNA, and nascent proteins to add carbohydrates, fold, and excrete into the extracellular environment, makes the baculovirus/lepidopteran cell system attractive to pharmaceutical researchers. The supposition that baculovirus could not infect human cells has led to the extensive use at biosafety level 1 (BSL-1) containment. However, recent papers have suggested that the baculovirus is capable of transfecting human cells, both primary and transformed lines (3). The ramifications of that research on large-scale growth of baculovirus-infected insect cells have not yet been observed at the federal level in the United States.

Bacteria

Bacteria are used for both the production of metabolites as drugs or drug precursors as well as vaccines. Pathogenic microbes grown for their use as vaccines include *Streptococcus pneumoniae, Neisseria meningitidis, Haemophilus influenzae, Clostridium tetani, Corynebacterium diphtheriae, Bordetella pertussis*, and *Bacillus anthracis*. Either inactivated organisms or "liberated" cellular material is used for the vaccines. As expected, the microbes must be inactivated prior to processing and in a manner that does not affect the immunogenicity of the immunogen. Failure in either task exposes workers to highly pathogenic agents or yields vaccines with unacceptable levels of impurities. Containment for large-scale growth of these organisms has been established at BSL-2 Large Scale (BSL2-LS) (4). Informal benchmarking by those involved in the cultivation of these microbes suggests that applicable guidelines are met or exceeded in the production areas.

Bacteria are also used to produce pharmacologically active metabolites that are frequently modified chemically to yield the final drug candidate. For example, *Streptomyces avermitilis* produces a potent series of antihelminthitic agents, the ivermectins (the chemically modified avermectin for human use is Mectizan) (5). Although this microorganism does not create a biosafety problem for humans, it does produce a compound that is toxic to aquatic life, requiring BSL-2 containment for the fermentation broth.

Recombinant bacteria are frequently used to produce pharmacological proteins, enzymes and plasmid DNA. The two bacteria frequently employed are *Escherichia coli* and *Bacillus subtilis*. Currently, of the 79 "biotechnology drugs" listed by the Biotechnology Industry Organization on its website (www.bio.org), 18 approved drugs can be found to have been produced in *E. coli*. The ease of use, low containment (Good Large Scale Practice [GLSP] or BSL-1, depending on the gene expressed), and high concentration of organisms (10^8/ml) are offset by the inability to have most proteins excreted or cleaved by the bacterium (6).

The recombinant guidelines, however, can be problematic when dealing with strains that do not fit neatly into the published risk groups. An example is the commonly used *E. coli* strain BL21. The *National Institutes of Health Guidelines for Research Involving Recombinant DNA Molecules* (*NIH Guidelines*) do provide that *E. coli* strains that are ". . . enteropathogenic, enterotoxigenic, enteroinvasive and strains bearing the K1 antigen . . ." are classified as risk group 2 agents, whereas K-12-derived strains lacking transducing phages are not (in the absence of plasmids bearing toxins or other genetic inserts that would increase risk). Because most group B and C strains have not been tested specifically for pathogenicity *in vivo* or by molecular means, but have been in large-scale use for years, this has left the risk group placement on a site-by-site or country-by-country basis. Recently, a paper was published that looked for the presence or absence of "pathogenicity islands" in K-12 and a series of group B and C strains, including BL21, a group B strain (7). International harmonization and recognition of molecular approaches to pathogenesis are needed to resolve risk assessment issues in a timely and cost-effective manner.

Fungi

The yeast *Saccharomyces cerevisiae* has assumed a significant role in the pharmaceutical industry as a producer of recombinant DNA-based vaccines (e.g., the cloned hepatitis B surface antigen in Recombivax [8] and Engerix-B [9]) and recombinant drugs (e.g., insulin, granulocyte/macrophage-colony stimulating factor [10], and platelet-derived growth factor [11]). The technology to grow the organism in large quantities and high densities and the low biological risk have made work with this organism attractive. The disadvantages of working with *S. cerevisiae* include the fact that many heterologous proteins are not secreted, requiring disruption of the cells to obtain the product. Also important is the consideration that individuals allergic to yeast may not be able to tolerate proteins made in yeast if they are not highly purified.

The yeast genus *Pichia* is used more frequently than *E. coli* because it is capable of producing disulfide bonds and glycosylations. *Pichia* is a methylotroph, meaning it can use methanol as the only energy source. It can also grow to very high cell densities, and under ideal conditions, it multiplies to the point that the cell suspension is nearly a paste (12). In the case of Albrec (Bipha Co., Chitose City, Hokkaido, Japan), a human serum albumin (HSA) expression plasmid was integrated into the chromosome of *Pichia pastoris*. The alcohol oxidase 2 (AOX2) promoter region was modified for efficient recombinant (r) HSA expression, where the expression of the rHSA gene and the subsequent secretion of rHSA into the culture medium will take place upon induction by methanol (13). In 2006, a research group managed to create a strain that produces erythropoietin in its normal glycosylation form (14). This was achieved by exchanging the enzymes responsible for the fungal type glycosylation, with the mammalian homologs. Table 1 lists some of the products currently produced using *Pichia*.

Other fungi (many filamentous) are frequently screened for their ability to produce interesting pharmacologically active compounds. The statins, multibillion-dollar drugs significant for their ability to lower serum cholesterol levels, are isolated from *Aspergillus* species (15, 16). Likewise, Sandimmune (cyclosporin A), a standard drug in preventing rejection of transplanted organs, is derived from the fungus *Beauveria nivea* (17). During a screening program, unidentified microbes are frequently cultivated at BSL-2. This containment level is usually maintained until the microbe is fully characterized.

Mammalian Cells

Mammalian cells may be used for the direct generation of pharmacological proteins, usually created by recombinant DNA techniques. According to the Biotechnology Industry Organization's website (www.bio.org), 12 of 79 "Biotechnology Drugs" approved by the U.S. FDA by the spring of 1999 were recombinant DNA products grown in Chinese hamster ovary (CHO) cells and seven additional products were generated in other mammalian cell lines. The generation of monoclonal antibodies, to date, is only commercially feasible in cultured mammalian cells.

Mammalian cells are also frequently employed as hosts for producing large numbers of viruses for human vaccine production. Rabies, rubella, varicella, hepatitis A, and polio vaccine strains are current virus vaccines cultured in mammalian cells. The bulk of the viruses are grown in normal human fibroblasts, such as MRC5 and WI38 cells that require only BSL-1 containment, according to the American Type Culture Collection website (www.atcc.org). In reality, the cell lines are at greater risk of contamination from environmental agents carried by humans than the cell lines or the viruses pose to the staff. Use of mammalian cell lines to produce virus traditionally suffers from low density requirements for successful infection, use of complex media containing fetal bovine serum, and a requirement for disruption of the cells to release the vaccine virus. The "comfort level" reached by most production staffs in handling these cells and national drug regulations regarding "significant process changes" make it unlikely that significant increases in efficiencies will occur with existing vaccines. Further information regarding the culturing of mammalian cells may be found elsewhere in the book, and the reader is also directed to reference 18 for more specific information on the growth of mammalian cells.

Early batches of polio vaccine were contaminated with SV40 from the monkey kidney cells used for growth. This, and the finding of endogenous retroviral elements in a variety of organisms, including avian and porcine species, have led for a call to determine the risk to humans from vaccines and potential xenotransplants from these organisms (19).

TABLE 1.

Some of the products currently produced using *Pichia*[a]

Product	Company	Use
Kalbitor (DX-88 ecallantide, a recombinant kallikrein inhibitor protein)	Dyax (Cambridge, MA)	Hereditary angioedema treatment
Insugen (recombinant human insulin)	Biocon (India)	Diabetes therapy
Medway (recombinant human serum albumin)	Mitsubishi Tanabe Pharma (Japan)	Blood volume expansion
Shanvac (recombinant hepatitis B vaccine)	Shantha/Sanofi (India)	Hepatitis B prevention
Shanferon (recombinant interferon-alpha 2b)	Shantha/Sanofi (India)	Hepatitis C and cancer treatment
Ocriplasmin (recombinant microplasmin)	ThromboGenics (Belgium)	Vitreomacular adhesion (VMA) treatment
Nanobody ALX-0061 (recombinant anti-interleukin-6 receptor single-domain antibody fragment)	Ablynx (Belgium)	Rheumatoid arthritis treatment
Nanobody ALX00171 (recombinant anti-respiratory syncytial virus (RSV) single-domain antibody fragment)	Ablynx (Belgium)	RSV infection treatment
Heparin-binding epidermal growth factor (EGF)-like growth factor (HB-EGF)	Trillium (Canada)	Treatment of interstitial cystitis/bladder pain syndrome (IC/BPS) treatment
Purifine (recombinant phospholipase C)	Verenium/DSM (San Diego, CA/Netherlands)	Degumming of high-phosphorus oils
Recombinant trypsin	Roche Applied Science (Germany)	Digestion of proteins
Recombinant collagen	Fibrogen (San Francisco, CA)	Medical research reagents/dermal filler
Aquavac IPN (recombinant infectious pancreatic necrosis virus capsid proteins)	Merck/Schering Plough Animal Health (Summit, NJ)	Vaccines for infectious pancreatic necrosis in salmon
Recombinant phytase	Phytex, LLC (Sheridan, IN)	Animal feed additive
Superior Stock recombinant nitrate reductase	The Nitrate Elimination Co. (Lake Linden, MI)	Enzyme-based products for water testing and water treatment
Recombinant human cystatin C	Scipac (United Kingdom)	Research reagent

[a]From http://www.pichia.com/science-center/commercialized-products/.

Insect Cells

There is only one approved biopharmaceutical product containing recombinant proteins from the infected insect cell line Hi Five. This product is Cervarix, which consists of recombinant papillomavirus carboxy-terminal truncated major capsid protein L1 types 16 and 18 (20). Nonetheless, this expression system has been used extensively in structural studies, as correctly folded eukaryotic proteins can be obtained in a secreted form in serum-free media, which enormously simplifies protein capture in purification protocols (21).

Select Agents are handled at large scale for production of pharmaceuticals and animal vaccines. *Clostridium botulinum* is grown at large scale to produce botulinum toxin types A and B used in medicine for, among others, upper motor neuron syndrome, focal hyperhidrosis, blepharospasm, strabismus, chronic migraine, and bruxism. It is also widely used in cosmetic treatments as Botox (22). Animal vaccines to prevent Eastern equine encephalitis and Venezuelan equine encephalitis (Encevac T, Encevac T+VEE) are also produced at large scale (23). The Select Agent Program has adopted the NIH rDNA Guidelines, Appendix K Large Scale, as a compliance checklist for BSL2-LS and BSL3-LS operations (24).

SCALE-UP TO MANUFACTURING

Once a product candidate becomes of "high" interest at the research level, efforts begin at making this candidate in larger quantities to support process research and development activities. Frequently, the laboratory method of preparing the candidate is initially scaled up in 10-liter to 70-liter bioreactors. Subsequently, this process is usually scaled up rapidly in the pilot plant to meet material requirements for safety assessment and clinical studies. Concurrent with making material, the process is subjected to further research and development, paving the way for factory introduction. Since the *NIH Guidelines* Appendix K refers to cultivation of viable organisms in volumes of 10 liters or greater as large-scale, the scale-up efforts just described are performed under the Large Scale Biosafety Guidelines. These guidelines address only the biological hazards associated with the microorganism, not the other hazards that may accompany large-scale cultivation (such as chemical, physical, and mechanical-related process hazards).

When scale-up involves biohazardous microbes, current Good Manufacturing Practices (cGMP) or national equivalent features must be considered along with good biosafety practices (25). The focus on good biosafety practices assures protection of both personnel and the environment from uncontrolled release of the biohazardous microbe. Conversely, adherence to cGMP principles assures product protection from the environment. In actual practice, the requirements for aseptic operations overlap and complement the features of cGMP. Throughout scale-up activities, a biosafety committee that consists of members from research and production can provide guidance for consistent adherence to good biosafety practices. This section is organized to discuss a number of biosafety issues that are routinely addressed during the transition of a process from laboratory-scale investigations to pilot-scale production of clinical material to manufacturing.

The production of bioactive molecules, such as monoclonal antibodies, interferons, and toxins/toxoids, creates unique problems related to safe handling during isolation and purification. Many times, the producing organisms are handled at GLSP or BSL1-LS, with the bioactive molecule requiring a high level of control as it is purified and concentrated. Control banding is a common method of assigning employee protection (26). It is a process of assigning a biomolecule to a hazard category that corresponds to a range of airborne concentrations—and the engineering controls, administrative controls, and personal protective equipment—needed to ensure safe handling. In many cases, the control band assignment is low as the processing system is closed and the final product remains a liquid. The control band may be adjusted upward if the process involves lyophilization. Another concern to be aware of is process upsets that can generate aerosols.

Technology Transfer

Once a product candidate has been selected for manufacturing, the process developed in the pilot plant to make this candidate is now implemented in the factory. All of the information learned during research and development studies, including the safe handling of the process microorganism, must be communicated in an understandable and concise format. The institutional memory needs to be maintained, whether by codifying on paper and handing it out widely to the staff, or by securing it on an internal Internet page. Biosafety information in this environment needs to be shared widely and securely to protect proprietary information at the same time. Each pharmaceutical corporation must deal with these issues, including information and permit requirements in different countries and sharing of information with joint ventures or other groups not formally part of the organization. The type of information to be shared also needs to be determined: training, manufacturing standard operating procedures (SOPs), pathogen risk assessments, and facility design are some of the materials that are generated routinely and may be able to be shared. The method of communication has been developed independently in each pharmaceutical company, as no broad consensus exists on the best practices in this area.

The construction and use of recombinant and pathogenic organisms for pharmaceutical use poses another unique situation for biosafety. In the generation of a vaccine from an attenuated agent, there may be a point where it is reasonable to reduce containment. This may occur upon transfer to scale-up or after phase I clinical trials. Conversely, biosafety information generated may suggest an increased stringency of containment. In either case, these situations should include review by members of the corporate biosafety committee before implementing the change. One of the key roles of this committee is to help interpret the guidelines and regulations, both national and international. Once properly reviewed, the biosafety approach selected becomes part of the technology transfer package.

Included in this technology transfer is information about the pathogenicity of the organism, disinfection and inactivation procedures, and links to medical surveillance programs or treatment protocols obtained from medical professionals, as well as a summation of "lessons learned" from small-scale experiments. In the event the agent is identified as a pathogen for which a vaccine exists, or is a vaccine strain, the biosafety guidelines and legal issues become more difficult. The *NIH Guidelines* and the Centers for Disease Control and Prevention/NIH *Biosafety in Microbiological and Biomedical Laboratories* (BMBL) both recommend vaccination for workers handling human pathogens (27). Moreover, the World Health Organization (WHO) guidelines (28), incorporated into a number of national regulations, require vaccinations for workers in vaccine seed areas (agent and host, if used), to prevent wild-type strains from being cocultivated. However, the U.S. Supreme Court, in the *Automobile Workers v. Johnson Control* decision in 1991 [499 US 187(1991)], precluded a U.S. employer from prohibiting a worker from an area due to a potential health risk. In this specific case, it prohibited an employer from gender-based prohibitions that were established to protect a female worker's fetus. Many U.S. pharmaceutical corporations have viewed that ruling as limiting their abilities to mandate vaccination without a specific legislative mandate. They, therefore, may recommend (or strongly recommend) vaccines or titers where appropriate in research and production areas. The exceptions are where blood-borne pathogens are in use and the employer is required to offer hepatitis B vaccination, and in production areas where required by the FDA or WHO requirements for licensing (29).

Culture Identification

Prior to initial scale-up in laboratory fermenters, the microorganism of interest should be properly identified and characterized, focusing especially on its level of pathogenicity, if any. This information will dictate the containment needed or BLS to ensure personnel safety. It is prudent to document the transfer of knowledge officially about the organism in question and biohazard status via the use of formal reports.

Establishing an antibiotic sensitivity spectrum can prove to be a valuable tool for corporate medical services in treating infections acquired by unexpected exposure to bacterial agents. If the microbe is difficult to identify and to establish its pathogenicity, scale-up to laboratory fermenters may proceed at an elevated containment level while proper characterization is obtained. Other health surveillance may be required if the agent may generate allergies or may not be preventable via vaccines (e.g., *Mycobacterium tuberculosis*).

Disinfection and Inactivation Procedures

A key component in handling a microbe safely is understanding how to inactivate for both unexpected and controlled-release situations, especially if it is a pathogen. Frequently, the unexpected release is addressed with the use of an effective disinfectant. The controlled release, which includes removing pathogenic microbes from a contained bioreactor for downstream processing, is safely performed by inactivation of the culture prior to release. This inactivation is commonly achieved by exposing the microbe to an inactivation agent under controlled conditions for a specified time period (dictated by preestablished inactivation kinetics). As a safety practice, the exposure period is usually extended well beyond the time required for inactivation. Nonpathogens are usually pasteurized (65°C, 5–10 minutes) before they are released into the environment (30, 31), depending on local environmental regulations.

Changeover Procedures

Multiuse facilities are used initially in drug development both at laboratory and pilot plant scales (32), allowing development of different processes in the same equipment. From a biosafety standpoint, successful operation of these facilities requires that SOPs be in place to ensure against the release and cross-contamination of the microbes used in different processes. In the laboratory, these SOPs follow good laboratory procedures.

Accidental Release

Regardless of how well process equipment and the surrounding building are designed, an accidental release to the environment could occur. Again, the *NIH Guidelines* provide some useful definitions in Appendix K-VII to define and characterize the release.

As opposed to certain environmental regulations (e.g., the U.S. Nuclear Regulatory Commission [USNRC] regulations on environmental release), the NIH does define a *de minimis* release as one that does not result in the establishment of disease in humans, plants, or animals. The volume is not specified; rather it is the result of the release that is the key in determining whether it is *de minimis* or significant. The use of a performance standard makes the risk analysis more critical for determining the requirements for containment. The appropriateness of the requirements for containment will strongly depend on the experience and professional training of the person(s) generating the analysis.

Secondary containment is enhanced to preclude release to the environment of even *de minimis* amounts of material (e.g., the use of HEPA filtration on the exhaust of BSL2-LS room air and sealed dikes capable of containing more than the entire fermenter load and disinfectant). Biosafety training for the participating staff and establishment of emergency SOPs are required to allow the staff to respond safely to biosafety-related emergencies. Equally important is the prudent practice of keeping health services and the responding emergency group(s) appropriately informed as to the biosafety issues that exist in the facility. Regular training exercises for the staff and yearly drills for all response groups allow all involved to become familiar with their responsibilities in the event of a release.

Security of Stock Cultures

Stock cultures are frequently stored in more than one location to minimize the risk of catastrophic loss. At each location, the stock cultures are maintained in secured containment with complete records, including data on culture deposition, removal, and transfer. The BSL assigned to the microbe dictates the facility and method(s) that should be employed in handling it.

Startup/qualification

Training staff and verification of training

Each member of the startup team must have a clear understanding and appreciation of the relative importance of the potential hazards that are associated with the process. This training must be a thorough review of the operations with presentations from both the biosafety and GMP compliance groups. The review should include such topics as characterization of the microbe used in the process; proper handling of the microbe, including possible sources of exposure; required work practices; engineering controls; emergency procedures to follow in the event of an accident or unexpected release; proper disposal of biowaste; and decontamination of surfaces. Verification of this training may be achieved by practical examinations, giving the staff member the opportunity to demonstrate a thorough understanding of the biosafety issues and the recommended approaches for addressing them. Thorough training does not eliminate biosafety risks, but it greatly reduces them. New members to an existing team must be trained thoroughly prior to their initial handling of pathogens. Use of a mentoring system that allows oversight of new staff by experienced personnel is essential in minimizing the hazard new, inexperienced personnel present to both themselves and others.

Identification of deficiencies in existing equipment

Once trained, team members will want to examine the facility and equipment for conformity to the biosafety guidelines. All deficiencies identified must be appropriately addressed before start-up activities commence. Documentation is a key component of ensuring equipment readiness. Ongoing preventive maintenance of equipment must be recognized as an essential component of routine operations.

Verification of inactivation process at scale

Although effective disinfectants and culture inactivation kinetics are initially developed at laboratory scale, verification of this information is required at scale, both in the target facility and equipment. For controlled release, the liquid biowaste is inactivated in the primary containment equipment, usually at the end of the cell cultivation stage, employing culture inactivation kinetics established at scale. Solid biowaste is placed in appropriate biosafety containers, sealed, labeled, and eventually rendered noninfectious by a validated process (e.g., incineration, autoclave, or irradiation). Process equipment must be designed to allow effective decontamination of the primary containment devices, including fermentation or cell culture reactors, recovery equipment, process piping; condensate lines, piping to decontamination vessel, and the inactivation vessel. Commonly, Sterilize in Place (SIP) measures are employed in upstream process equipment. Special attention must be given to recovery unit operations used in biohazardous processing. Aseptic processing is not normally a primary design feature of downstream processing equipment (33).

A critical feature for attention during process transfer is the treatment of spills of biologically contaminated materials during routine operations. Chemical disinfectant treatment procedures and personnel training should be addressed prior to the initiation of any large-scale processing. Importantly, the treatments used to disinfect the organism must be tested under actual conditions. Thus, factors such as the nature of the agent, surface, temperature, and environmental effects can influence the rate and extent of inactivation.

Closed processing

The key to closed processing is containment of the microbe or biohazardous agent at all times. All culture transfers for inoculum development, production tank inoculation, culture purity testing, etc., are performed in appropriate biosafety cabinets or by direct vessel-to-vessel transfer. It must be noted that by containment we mean physical containment. In the pharmaceutical industry, as elsewhere, the term "biological containment" is also used, although mostly in limited areas. The definition of biological containment is provided in Appendices E and I of the *NIH Guidelines*, which are found elsewhere in this book. Physical containment for pharmaceutical companies, as is done elsewhere, defines both primary and secondary containment. An excellent summary of the minimum containment requirements for the four different risk-level agents can be found in the *NIH Guidelines*, Appendix K, Table 1.

Liquid transfers during processing may occur via SIP piping or through sterile welded tubing attached to sterile containers. Only contained sampling is employed to follow cultivation performance in the bioreactor. Filtration or incineration is employed to prevent release of the microbe in the exhaust gas.

Routine Operations

A variety of standard practices are employed to maintain environmental control within pilot scale or manufacturing areas that process biological agents. These practices include facility design and control, facility changeover, cleaning procedures, personnel controls, standardized manufacturing processes, specific standard operating procedures, preventive maintenance schedules, and quality control procedures.

SIP and Clean in Place (CIP) procedures, if verified experimentally and validated, provide adequate biological decontamination measures for process equipment. These practices must be written as SOPs. Validation of filter sterilization procedures must be established during facility startup. System integrity checks and verification of filter integrity after use are important features of normal aseptic processing operations and serve to verify the containment features of a system. Annual revalidation activities must be executed to assure that containment features have not been compromised (34).

International Regulations

Internationally, the biosafety issues related to the production of bioprocess-based pharmaceuticals are governed by a large variety of regulations. Those regulations in developed countries (United States, Canada, and European Union) are very specific, addressing both the protection of workers and the environment.

The number and scope of biosafety regulations governing the laboratory and industrial scale use of microorganisms vary widely around the world. Many countries do not have specific regulations, but the use of microorganisms is covered under general workplace regulations. Many other countries are in the midst of developing a biosafety system. Several of them seem to be patterning their system on existing U.S., Canadian, and European regulations. The ultimate goal would be to achieve international harmonization of the biosafety guideline, and encouraging progress is being made.

New Technologies

With the rise of biotherapeutic molecules, especially antibodies, the demand for new bioreactor technology has grown. Past efforts to produce proteins expressed by bacteria, yeast, and mammalian cells in commercial quantities employed costly, reusable hardware, requiring extensive cleaning and resterilization. This historical approach relies heavily on SOPs and inevitably entails high costs and extensive personnel training.

Disposable components and systems are increasingly favored, both for improved process reliability and economic advantages. Currently, many biotech producers of protein molecules are evaluating disposable modules, which come sterilized and qualified according to GMP requirements. Disposable components are readily accessible, and their availability allows switching cell line production platforms, as well as target proteins, quickly and inexpensively.

Integrated single-use bioreactor systems

The bioreactor has progressed from manually operated models into computer-controlled descendants, which can be controlled from the office or home. Today's sophisticated technology allows microorganisms to be cultivated under controlled conditions so that their cells or metabolic products may be harvested for the manufacture of pharmaceuticals. Many manufacturers have developed bioreactors as disposable systems meeting the regulatory requirements of the biopharmaceutical industry.

For example, one manufacturer has developed disposable bioreactor and mixing vessel sensors, the beginning steps of a totally disposable process platform. Included in this process platform are disposable oxygen, pH, and carbon dioxide sensors. The devices use immobilized pH-, CO_2-, and O_2-responsive dyes; fermentation parameters are measured noninvasively with commercially available LEDs and photodetectors.

The sensors are inexpensive; thus, they can be incorporated into disposable components of the bioreactors. They come precalibrated and do not require further calibration. Low-cost sensors are essential for the standardization process when new products are in development, as many production batches will be required. Without disposable sensor technology, the choices are cleaning in-place, discarding reusable sensors, or avoiding sensors, which make process monitoring and control impossible.

A number of new fermentation systems employ single-use plastic bags, which mix by rocking, or incorporate disposable mixing technology, including gas sparger systems, filters, impellers, and bearing assemblies. Some designs use reusable shafts and impellers that can be cleaned and sanitized using standard chemicals and procedures. With various size options, up to 1,000 liters can be accommodated.

To date, disposable bioreactors have been used primarily in cell cultures. The shift in their usage in microbial or yeast systems will require significant technical advances. Improvements are required in mixing, sparging, and heating and cooling of the current models. These are essential to meet the demands of the high-titer processes, especially at larger volumes. Disposable bioreactors will become the dominant technology in biomanufacturing operations. Their adoption is being driven by two trends—reduced manufacturing costs and the development of high-titer processes.

While disposable, these technologies bring up issues during the risk assessment process, such as disposal of bags used with infectious organisms, sparge and overlay gas pressure control to prevent bag rupture, and developing spill control plans should a rupture occur.

The Dutch Office Genetically Modified Organisms (GMO) Office commissioned a study in 2010 on the risks of single-use bioreactors (35). Although the use of single-use bioreactors reduces some of the risks, new potential risks are introduced compared to conventional bioreactors. The GMO Office document is an excellent reference for a risk assessment.

Miniature bioreactor systems

Miniature systems have been designed to mimic conventional bioreactors, yet permit high-throughput cell culturing and process optimization. One system utilizes a microplate 24-chamber bioreactor; the supporting robotic equipment automatically loads, samples, and feeds cells growing under 24 independent reaction conditions. A second system uses specially designed, 50-ml tubes to be used as minibioreactors. The manufacturer states that by using tubes with semipermeable caps for CO_2 exchange, they were able to achieve high throughput and mimic both batch and fed-batch processes. While this technology uses multiple miniature bioreactors, the overall footprint is still large. If it is used with BSL-2 organisms, one needs to confirm if the unit will fit in a standard biological safety cabinet (BSC) or if a custom-designed BSC is required.

Disposable tubing and sterile sampling

On the one hand, sampling may seem to be a simple procedure—just open a manual sample valve of the bioreactor vessel, collect as much fermentation broth as required for the sample, and close the tap! With this type of sampling, we can easily guarantee batch contamination will occur and employees will be exposed via spray or aerosol. Sampling systems must be designed with both concerns in mind. Several sampling designs exist. Autoclavable reusable bioreactors employ a metal sampling tube installed in the fermenter head plate. The other end has a fixed metal hood with internal vial screws. Flexible tubing connects the metal tubing and a pinch clamp used to prevent liquid flow (Fig. 1). The metal hood has a flexible bulb attached to create a vacuum and draw sample into the vial. The entire assembly is sterilized together with the bioreactor vessel, and it remains sterile until

Figure 1: Simple fermenter sampling device.

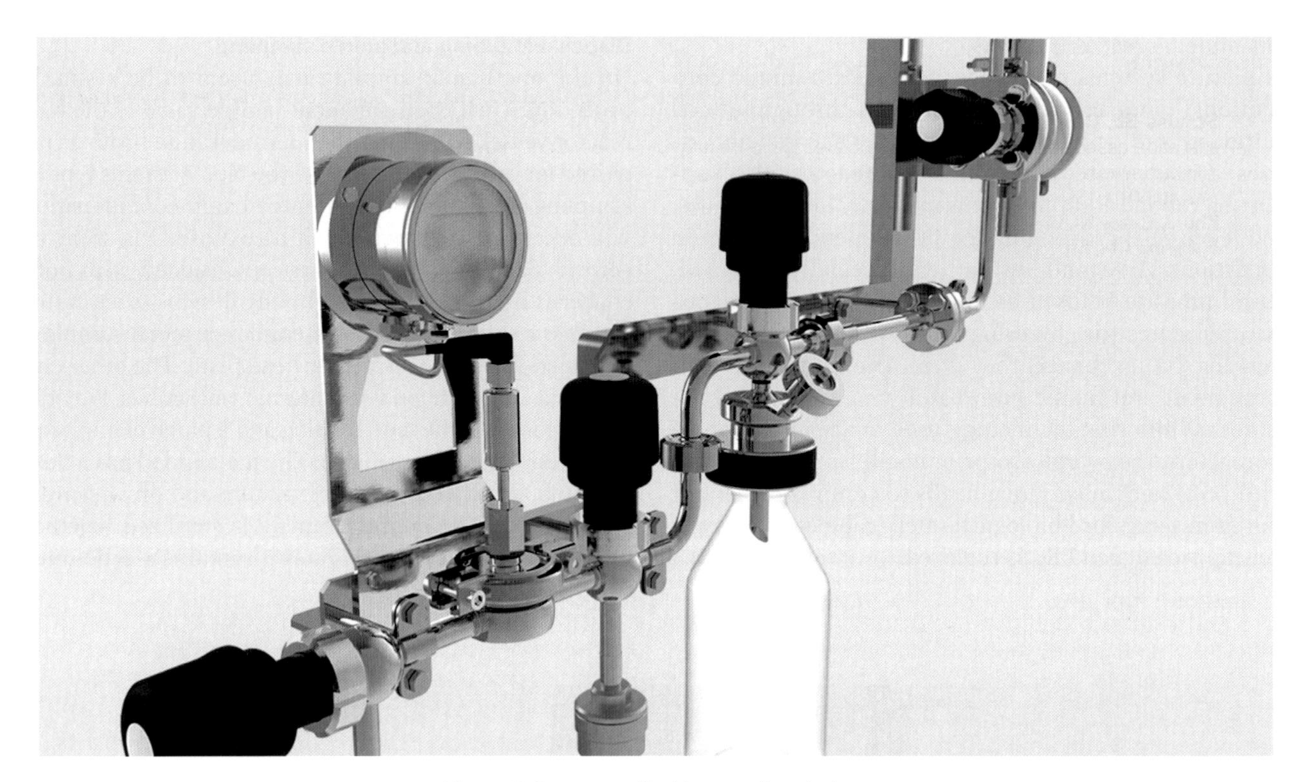

Figure 2: Steam sterilizable sampling device.

sampling. To start sampling, the pinch clamp remains closed and the bulb is squeezed while the vial is loosened. The vial is twisted to seal shut, and then the pinch clamp is loosened to pull medium into the vial. The vial is removed and replaced aseptically. Drawbacks of this method are limited sample size, ease of batch contamination, and leakage of infectious material resulting in employee exposure.

Many *in situ*-sterilized bioreactors employ a three-valve assembly that permits sample collection, and then sample path decontamination using steam (Fig. 2). The valve assembly is sterilized while installed in the fermenter. While the medium is being sterilized, steam flows through the valve components. During sampling, a filter-vented sterile glass or plastic bottle is attached to one port of the valve. The internal fermenter valve is opened and sample flows into the bottle. When filled, the valve is closed and the bottle replaced. The valve is now flushed with steam, purging the remaining liquid to drain. Leakage of sample to generate splash or aerosols is one drawback. In addition, if the process involves an infectious agent, the remaining sample must be properly inactivated before disposal.

Disposable process equipment demands reliable and fast sterile connecting devices. Two of the important issues when using disposable process technology are ensuring quick, safe, and sterile transfer of fluids and not forgetting every transfer needs to be disconnected after use. Early use of disposable tubing employed materials made from silicone. Silicone hose cannot be used with thermal welders, because it lacks thermal bonding characteristics. Today's thermal tube fusing technology is one of the safest and most flexible ways to meet those requirements, and it uses well-known heat sterilization technology. Because the existing thermoplastic tubing can be used for the connection of presterilized assemblies, it also avoids costly additional sterile connectors and clamping-off procedures. Thermoplastic hose materials meet these requirements, have been successfully used for such applications, and have a long proven history of biocompatibility. The important weld parameters are always temperature, compression, and temperature contact time.

One of the major issues for employee contamination prevention is ensuring the tubing is completely compressed where the hot blade cuts through. The hot sterile blade must melt through the compressed section, and then seal it immediately on both sides. This also prevents remaining medium in the hose lumen from contacting the hot blade and causing burned medium or gases (particles) inside the sterile fused hose. It is well known that dry heat is effective for sterilization. Depyrogenization procedures typically require an exposure of 30 seconds at 250°C or 3 seconds at 320°C. Using an exposure over 350°C for 1 second exceeds this requirement.

References

1. **Strauss BE, Costanzi-Strauss E.** 2007. Combating oncogene activation associated with retrovirus-mediated gene therapy of X-linked severe combined immunodeficiency. *Braz J Med Biol Res* **40:**601–613.
2. **Zhu J, Grace M, Casale J, Bordens R, Greenberg R, Schaefer E, Chang AT-I, Musco ML, Indelicato SR.** 1999. Characterization of replication-competent adenovirus isolates from large-scale production of a recombinant adenoviral vector. *Hum Gene Ther* **10:**113–121.
3. **Condreay JP, Witherspoon SM, Clay WC, Kost TA.** 1999. Transient and stable gene expression in mammalian cells transduced with a recombinant baculovirus vector. *Proc Natl Acad Sci USA* **96:**127–132.
4. **National Institutes of Health.** 2013. Guidelines for research involving recombinant DNA molecules (NIH Guidelines). http://osp.od.nih.gov/sites/default/files/NIH_Guidelines_0.pdf, accessed 3/29/2016.
5. **Campbell WC (ed).** 1989. *Ivermectin and Abamectin.* Springer-Verlag, New York.
6. **Bent R.** 1997. Protein expression, p 16.0.1–16.0.3. *In* Ausubel FM, Brent R, Kingston RE, Moore DD, Seidman JG, Smith JA, Struhl K (ed). *Current Protocols in Molecular Biology,* vol. 3. John Wiley & Sons, Inc., New York, NY.
7. **Kuhnert P, Hacker J, Mühldorfer I, Burnens AP, Nicolet J, Frey J.** 1997. Detection system for *Escherichia coli*-specific virulence genes: absence of virulence determinants in B and C strains. *Appl Environ Microbiol* **63:**703–709.
8. **Merck.** 2014. Recombivax HB® prescribing circular. Merck & Co. Inc., Whitehouse Station, NJ. https://www.merck.com/product/usa/pi_circulars/r/recombivax_hb/recombivax_pi.pdf, accessed 3/29/2016.
9. **GlaxoSmithKline.** 2015. Engerix-B® prescribing circular. GlaxoSmithKline Research Triangle Park, NC. https://www.gsksource.com/pharma/content/dam/GlaxoSmithKline/US/en/Prescribing_Information/Engerix-B/pdf/ENGERIX-B.PDF, accessed 3/29/2015
10. **Sanofi-Aventis.** 2015. Leukine® prescribing circular. sanofi-aventis U.S. LLC Bridgewater, NJ. http://products.sanofi.us/Leukine/Leukine.html, accessed 3/29/2016.
11. **Smith & Nephew.** 2014. Regranex® product insert. Smith & Nephew, Inc. Fort Worth, Tx. http://www.regranex.com/pdf/PI_Full_Version.pdf, accessed 3/29/2016.
12. **Cregg JM, Tolstorukov I, Kusari A, Sunga J, Madden K, Chappell T.** 2009. Expression in the yeast Pichia pastoris. *Methods Enzymol* **463:**169–189.
13. **Kobayashi K.** 2006. Summary of recombinant human serum albumin development. *Biologicals* **34:**55–59.
14. **Hamilton SR, Davidson RC, Sethuraman N, Nett JH, Jiang Y, Rios S, Bobrowicz P, Stadheim TA, Li H, Choi BK, Hopkins D, Wischnewski H, Roser J, Mitchell T, Strawbridge RR, Hoopes J, Wildt S, Gerngross TU.** 2006. Humanization of yeast to produce complex terminally sialylated glycoproteins. Science 313**:**1441–1443.
15. **Merck.** 2015. Zocor® prescribing circular. Merck & Co. Inc., Whitehouse Station, NJ. https://www.merck.com/product/usa/pi_circulars/z/zocor/zocor_pi.pdf, accessed 3/29/2016.
16. **Parke-Davis.** 2015. Lipitor® prescribing circular. Parke-Davis, Division of Pfizer, NY, NY. http://labeling.pfizer.com/ShowLabeling.aspx?id=587, accessed 3/29/2016.
17. **Novartis.** 1998. Sandimmune® prescribing circular. Novartis Pharmaceuticals Corp., East Hanover, NJ. https://www.pharma.us.novartis.com/product/pi/pdf/sandimmune.pdf, accessed 3/29/2016.
18. **Freshney RI.** 2016. *Culture of Animal Cells: A Manual of Basic Technique and Specialized Applications,* 10th ed. John Wiley and Sons, New York, NY.
19. **Food and Drug Administration.** 1999. Evolving scientific and regulatory perspectives on cell substrates for vaccine development. US Food and Drug Administration Workshop Report, Rockville, MD.
20. **GlaxoSmithKlein.** 2015. Cervarix® Highlights of Prescribing Information GSK, Research Triangle Park, NC. http://us.gsk.com/products/assets/us_cervarix.pdf, accessed 3/29/2016.
21. **Ferrer-Miralles N, Domingo-Espín J, Corchero JL, Vázquez E, Villaverde A.** 2009. Microbial factories for recombinant pharmaceuticals. Microb Cell Fact **8:**17–25.
22. **Allergan.** 2015. BOTOX® prescribing information. Allergan, Inc., Irvine, CA. https://www.botoxchronicmigraine.com/?cid=sem_goo_43700007526995097, accessed 3/29/2016.
23. **Merck.** 2015. ENCEVAC® TC-4 with HAVLOGEN® technical information. Merck Animal Health, Madison, NJ. http://www.merck-animal-health-usa.com/products/130_120671/productdetails_130_121133.aspx, accessed 3/29/2016.
24. **US Federal Select Agent Program.** 2014. Inspection checklists. http://www.selectagents.gov/resources/Checklist-NIH-BL2-LS.pdf, http://www.selectagents.gov/resources/Checklist-NIH-BL3-LS.pdf, accessed 3/29/2016.
25. **Food and Drug Administration.** 2015. 21CFR Part 211. Current Good Manufacturing Practices for finished pharmaceuticals. US Code of Federal Regulations. https://www.accessdata.fda.gov/scripts/cdrh/cfdocs/cfCFR/CFRSearch.cfm?CFRPart=211, accessed 3/29/2015.
26. **Naumann BD, Sargent EV, Starkman BS, Fraser WJ, Becker GT, Kirk GD.** 1996. Performance-based exposure control limits for pharmaceutical active ingredients. *Am Ind Hyg Assoc J* **57:** 33–42.
27. **Centers for Disease Control and Prevention.** 2009. *Biosafety in Microbiological and Biomedical Laboratories.* Government Printing Office, Washington, DC.
28. **World Health Organization.** 2014. *Expert Committee on Specifications for Pharmaceutical Preparations, 49th Report.* WHO, New York, NY.
29. **Occupational Safety and Health Administration.** 1999. *(29 CFR 1910.1030). Bloodborne Pathogen Standard.* US Code of Federal Regulations.
30. **Liberman DL.** 1993. Biowaste management in bioprocessing, pp. 769–787. *In* Stephanopolos G (ed). *Biotechnology,* 2nd ed., vol. 3. VCH Verlagsgesellschaft mbH, Wienheim, Germany.
31. **Lieberman DL, Fink R, Schaefer F.** 1999. Biosafety and biotechnology, p 300–308. *In* Demain AL, Davies JE (ed). *Manual of Industrial Microbiology and Biotechnology,* 2nd ed. American Society for Microbiology, Washington, D.C.
32. **Barta J, Blum A, Inloes D, Lindsay J, Nash A, Olson M, Staub L, Walcroft J.** 1998. Environmental control and monitoring in bulk manufacturing facilities for biological products. *Pharm Technol* **22:**40–46.
33. **Sinclair A, Ashley MHJ.** 1995. Sterilization and Containment, p 553–588. *In* Asenjo JA, Merchuk JC (ed). *Bioreactor System Design.* Marcel Dekker, Inc., New York, New York.
34. **Vesley D.** 1986. Decontamination, sterilization, disinfection, and antisepsis in the microbiology laboratory, p 182–198. *In* Miller BM (ed in chief). *Laboratory Safety: Principles and Practices.* American Society for Microbiology, Washington, D.C.
35. **Neeleman R.** 2010. GMO containment risks evaluation of single-use bioreactors. *Xendo Process* **2010:**1–28.

32

Biosafety Considerations for Large-Scale Processes

MARY L. CIPRIANO, MARIAN DOWNING, AND BRIAN PETUCH

DEFINING LARGE SCALE

The notion of scale-up or large-scale processing of microorganisms is currently associated with recombinant DNA (rDNA) technology, but in fact it has been common practice for many years. Microorganisms have been scaled up for the manufacture of foods and beverages for centuries. In the past hundred years, the large-scale production of antibiotics, vaccines, and biological products has become commonplace. The relative numbers of laboratory-acquired infections from the production environment are extremely low, approximately 3.4% of the total numbers documented (1). Part of the reason for these low numbers may be the reduction in virulence of the cultured organism, but they are most likely attributable to the extensive use of primary and secondary containment barriers, i.e., containment equipment and facilities, which are generally required to maintain the integrity of the product.

There are a number of guidance documents addressing biosafety requirements: Advisory Committee on Dangerous Pathogens, 1995 (2) and 1998 (3); Public Health Agency of Canada, 2015 (4); NIH, 2013 (5); Organisation for Economic Co-operation and Development (OECD), 2000 (6) and 2009 (7); CDC/NIH *Biosafety in Biological and Biomedical Laboratories* (BMBL), 2009 (8); Prime Minister, 1991 (9); and World Health Organization (WHO), 2003 (10) and 2004 (11). However, none of them go into extensive detail for large-scale operations. They require that a risk assessment be carried out on the basis of the organism, procedures, equipment, and facilities to be used. Some suggest that a biological safety professional be consulted. Many biosafety professionals understand the agent and are capable of performing a risk assessment; however, they may have limited experience with the large-scale processes and be unfamiliar with the equipment, processes, and facilities to be used.

The purpose of this chapter is to provide points to consider in the selection and design of equipment and facilities to achieve a safe work environment. There are few absolutes that can be applied across the board, because the decisions to be made are dependent on the risk assessment of the organism and the processes used. Fortunately, similar equipment and facility design criteria are used, so there are common biosafety principles that can be employed, which are discussed herein.

In discussing large-scale processes, the term "large-scale" must be defined. According to the NIH rDNA guidelines, greater than 10 liters constitutes "large scale." The Canadian Biosafety Standards and Guideline recommends that determination of cutoff values between laboratory and large-scale volumes be made in consultation with the Public Health Agency of Canada. In Japan, the Ministry of Health has designated more than 20 liters as large scale. In the United Kingdom, the Advisory Committee on Dangerous Pathogens states that it is not the volume but the intent of the work that determines the scale.

Earlier editions of the CDC/NIH publication, e.g., the 3rd edition, defined "production quantities" as a volume or concentration of infectious organisms "considerably in excess of those used for identification and typing." It stated that there is no finite volume or concentration that can be universally cited, and that the laboratory director must make the assessment on the basis of the organism, process, equipment, and facilities used. Certainly, in an ideal world that is the best solution. Unfortunately, not all laboratory directors have the depth of knowledge to make the assessments of the appropriate biological containment required and/or access to a biosafety professional for input. It is hoped that the ideas and suggestions discussed here will be of assistance to individuals making those decisions.

AGENT CONSIDERATIONS

An agent-based risk assessment is the starting point for any large-scale work. The scale of the work can influence the risk assessment. For example, a nonpathogenic organism that produces an extracellular toxin may not pose a problem at 40 ml but can create significant concerns when one is dealing with 10,000 liters. Even if there is not a "real" health risk, the potential for negative publicity from an accidental release or exposure incident is a major concern. The general public has a heightened sensitivity regarding infectious organisms because of media attention paid to recent outbreaks involving Ebola, severe acute respiratory syndrome (SARS), Norwalk virus, mad cow disease, anthrax-tainted letters, renewed threats of bioterrorism, and so forth. There are many people who are deeply concerned about the potential negative impact of recombinant organisms on people and the environment. Although there may be some scientific data and a solid risk assessment indicating limited adverse effects, these points may be unable to assuage the concerns raised. Thus, institutions should consider additional containment features for a large-scale facility to minimize the occurrence of such an event.

Three categories that need to be taken into consideration are agent, process, and external environment. These considerations may not come into play for every assessment, but should be reviewed to determine their relevance to the specific situation. In doing the risk assessment, a number of questions need to be answered about the organism(s) that will be used in the facility. These include, but are not limited to, the following:

- What is the highest biosafety level needed for containment of the agent(s) that will be used in the facility?
- What is the mode of transmission?
- What is the infectious dose?
- How communicable is the agent?
- Is the agent an opportunistic pathogen that could infect immunocompromised individuals?
- Is the organism a Select Agent, or does it have characteristics that warrant increased security and oversight?
- Has the disease that the organism causes been eradicated such that release could cause a serious public health threat by reintroducing it into the community, e.g., polio, smallpox, etc.?
- Does the agent produce any toxic, biologically active, or allergenic compounds?
- Are vaccines, prophylaxis, or therapeutic measures available to prevent or mediate an infection?
- Is the agent endemic in the area?
- How well does the agent survive outside of the culture system?
- Can the organism transfer genetic traits to other organisms in the environment?
- Is the agent disseminated through vectors, e.g., insects, fomites, etc.?

PROCESS CONSIDERATIONS

Once the above information is gathered, some specific information must be put together about the process. Table 1 provides some standard process flows and typical procedures used.

- Will the facility be dedicated to one agent, or will a number of agents be used?
- What volume of active agents will be present in the facility?
- Will the process be continuous or batch?
- Will the equipment be stationary, movable, or disposable?
- What type of equipment will be used?
- What types of manipulations need to be carried out?
- Does any of the equipment or do any of the manipulations generate aerosols?
- Will the facility be required to comply with competent authority drug or device regulations, NIH rDNA guidelines, or other governmental regulations?

TABLE 1.

Examples of some standard process flows

Unit process	Procedure
Bacterial and yeast processes (inclusion product)	
Cell production	Fermentation
Cell separation	Filtration/centrifugation
Cell lysis and removal	Homogenization
	Centrifugation
Protein concentration and refolding	Ultrafiltration
	Precipitation
Buffer exchange/adjustment	Ultrafiltration
Purification	Chromatography
	Two-phase extraction
Product formulation	Buffer formulation
	Lyophilization
	Packaging
Mammalian and *Pichia* (secreted product)	
Cell production	Fermentation
Cell removal	Filtration
	Centrifugation
Supernatant concentration	Ultrafiltration
Buffer exchange/adjustment	Ultrafiltration
	Precipitation
Purification	Chromatography
	Two-phase extraction
Product formulation	Buffer formulation
	Lyophilization
	Packaging

- What type of cleaning, disinfection, decontamination equipment, etc., is needed?

ENVIRONMENTAL CONSIDERATIONS

The last category can roughly be termed environmental considerations. These include questions about the local environment external to the facility:

- What are the climatic conditions in the area, e.g., temperature, humidity, etc.?
- What is the geography of the site?
- What are the native flora and fauna?
- Are the air supply intake and exhaust close to other facilities?
- How near is the facility to private property? What is that property, e.g., industrial, school, housing, etc.?
- Is the site physical security adequate for the types of organisms handled?

GENERAL BIOSAFETY RECOMMENDATIONS FOR LARGE-SCALE WORK

Although it is not the intent to review the exhaustive procedural requirements for work at each designated biosafety level here, it is important to review the basic concepts to understand the criteria for selection of the equipment and facility options. A copy of large-scale biosafety guidelines is appended to this chapter.

- Good large scale practice (GLSP) is designated for well-characterized organisms that are not pathogenic and do not produce compounds that are toxic, allergenic, or biologically active (12). These are Risk Group 1 agents that meet the above criteria and will have been used safely over a period of time, or have been designated as safe, and will not survive or cause adverse effects in the environment per NIH and OECD. There are no specific biosafety containment requirements for GLSP facilities. Procedures should be done in a way that does not adversely affect the health and safety of the employee; e.g., splashing, spraying, and generation of aerosols are minimized.
- Biosafety level 1–large scale (BSL1-LS) is for nonpathogenic organisms, but can include organisms that can cause sensitization or are opportunistic pathogens. These are Risk Group 1 organisms that do not meet the criteria for work at GLSP. The goal at this level is to minimize the release of viable organisms.
- Biosafety level 2–large scale (BSL2-LS) is used with moderate-risk pathogens that occur naturally in an area, i.e., Risk Group 2 agents. The operations should be designed to prevent release and employee exposure to splashing and spraying.
- Biosafety level 3–large scale (BSL3-LS) is for Risk Group 3 agents that may be aerosol transmissible or able to spread by insect vectors. These agents cause serious, potentially lethal diseases in humans or animals. Equipment and facilities used for this level must be designed to prevent employee exposure and aerosol release of the agent within the facility, and release of the agent outside the facility.

The requirements for large-scale production of Risk Group 4 agents are not addressed in this chapter because they are highly specialized and are in limited use.

PRIMARY CONTAINMENT

Special Practices

Primary containment is provided by the equipment used and appropriate biosafety practices, administrative practices, and personal protective equipment. In general, all of the standard and special practices and safety equipment identified in BMBL, along with the recommendations in the NIH rDNA guidelines, are applicable to large-scale processes. However, additional requirements should be considered depending on the agent used and the processes involved. These issues are addressed in the large-scale biosafety guidelines appended to this chapter. They include the following:

- Respirators capable of protecting from the organism in use
- Physicals, health screening, and immunizations, if appropriate
- Written procedures for the process, e.g., standard operating procedures
- Emergency response plans
- Additional gowning as necessary to provide adequate safety and to maintain product integrity, e.g., shoe covers, hairnets, coveralls, etc.

The most critical components of primary containment are the actions taken by the workers. It is imperative that the workers receive adequate training and have an understanding of the risks associated with the work that they perform. Workers who handle infectious organisms must receive specific training on how to safely carry out the specific techniques and procedures used in the facility.

Written procedures, and compliance with those requirements, are key elements to establishing a safe, controlled work environment. The procedures need to cover safety requirements, operating requirements, emergency response, and, where applicable, security issues.

Equipment Selection and Usage

The large-scale equipment employed provides most of the containment required, because the materials need to be protected from external contamination. Where the containment is not adequate for the risks associated with the agents used, additional barriers may need to be utilized.

Most large-scale industrial operations have the benefits of having process safety specialists who can assist with the safety evaluation of the equipment and procedures. (13)

Fermenters and cell culture vessels

Bioreactors used for growth of microorganisms, referred to here as fermenters, and those used for cell culture can share many attributes. To maintain the integrity of the culture, the culture vessel must provide an appropriate level of containment. The vessel must be constructed (14) to be able to withstand rigorous cleaning and decontamination procedures. See Fig. 1 for an example of a sterilizable bioreactor. It must be insulated and have heating and cooling capabilities to maintain the proper growth temperature and be capable of protecting the contents from contamination. Although glass and plastic systems are used for smaller volumes, most large-scale units are constructed of metal, generally food-quality stainless steel. Stainless steel minimizes corrosion, obviates the adverse effects of metallic ions on cultures, and is accepted by most competent authorities as suitable for direct contact with food and drugs, for which most of these processes are used. To facilitate cleaning and decontamination, the interior of the tank should be designed to be smooth, without dead legs, ledges, or inaccessible areas. The vessel may need to meet the applicable boiler/pressure vessel requirements, as it may be operated at a slight positive pressure or be pressurized during a sterilization cycle.

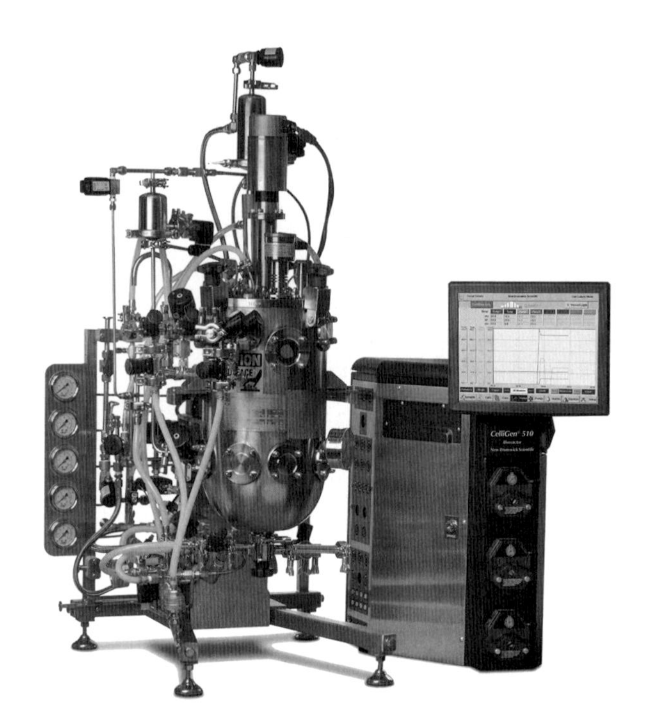

Figure 1: Sterilizable bioreactor. (Courtesy of New Brunswick Scientific Co.)

With increasing demands for biotherapeutic molecules, especially monoclonal antibodies, innovative bioreactor technology has become widespread. Standard fermentation technology used to produce proteins expressed by bacteria, yeast, and mammalian cells in commercial quantities employed costly, reusable hardware, requiring extensive cleaning and resterilization. This longstanding approach places heavy demands on standard operating procedures (SOPs) and inevitably entails high costs and extensive personnel training.

Disposable components and systems are increasingly favored, both for improved process reliability and for the economic advantages. Examples of single-use bioreactors can be seen in Fig. 2. Many biotech producers of protein molecules are moving to disposable modules, presterilized and qualified according to regulatory

Figure 2: Single-use, stirred-tank bioreactor. (Courtesy of New Brunswick Scientific Co.)

requirements. Disposable components are accessible at a moment's notice, and their ready availability makes it easy to change cell lines and target proteins in a production process quickly and inexpensively.

Treatment or filtration of the exhaust from the culture system is not generally warranted for GLSP systems. Beyond that level, the exhaust gases must be filtered or treated. The filters used need to be capable of removing the organism, allergen, toxin, or biologically active compounds present. It may be desirable to pretreat the exhaust air before filtration by passing it through a condenser, separator, or preheating system, particularly if HEPA-rated filters must be used. Note that some guidelines specify the use of a HEPA-rated filter for all such applications, regardless of the size of the organism being cultured. Where HEPA filters are not specified, the general practice is to use a single 0.2-μm sterilizing-grade air vent filter to reduce the potential for the escape of viable organisms (the requirement for BSL1-LS), and to use two of these filters in sequence to prevent the escape of viable organisms (the requirement for BSL2-LS and BSL3-LS). Most fermenters use an agitator system that is connected to the tank via a rotating seal. For higher levels of containment, i.e., above BSL2-LS, a double mechanical seal is generally called for. There is some question as to the increased reliability of a double versus a single seal (15, 16). For infectious agents, the seal must be designed to prevent leakage from the bioreactor. Where processes involve toxic or biologically active materials, or require additional containment measures because of the agents involved, liquids or steam can be used as the lubricant between the seals. The lubricant flow can be sent to a biowaste kill system. The location of the drive is another issue, i.e., bottom versus top mounted. The bottom-mounted systems facilitate maintenance of the units, and where a magnetically coupled agitator system can be used, actually provide better containment. However, where the mixing requirements are demanding, either because of the nature of the cells or the higher viscosity of the material, top-driven units may need to be considered, particularly for BSL2-LS and BSL3-LS operations.

The type of bioreactor used for cell culture depends on whether or not the cells are anchorage dependent. Those that are not anchorage dependent can be grown in vessels very similar to fermenters. These vessels might use impellers to ensure proper mixing of the cells and nutrients, although they must be carefully designed to prevent them from shearing the cells. Other cell culture systems use a bubble column or bubble column with a draft tube, also called an airlift reactor, to achieve proper mixing and aeration of the culture. Using an air perfusion system or a magnetic coupling for the agitator facilitates the containment of the unit; however, their application may be limited by the size and/or viscosity of the material.

If the cells are anchorage dependent, they must be grown in roller bottles, cell factories, microcarriers, or hollow fiber systems. Most of these systems are designed with adequate integral containment to maintain a sterile growth environment, although containment of accidental leakage from these systems should be considered based on the volumes and organisms involved.

It is not generally feasible to operate a fermenter or a cell culture vessel under negative pressure due to the obvious problems of foaming and product contamination. For processes where escape from the system must be prevented, the unit should be equipped with devices that monitor the pressure in the chamber and sound an alarm if the set level is exceeded.

Pressure vessels must be equipped with a pressure relief device (PRD), which consists of a rupture disk and/or spring-loaded pressure relief valve. When dealing with Risk Group 2 agents, it is desirable to have the PRDs located so that they release away from the work area. Depending on the agent in use, some type of shrouding should be considered for BSL2-LS. For BSL3-LS, the PRD should be vented to a waste decontamination tank or some other contained system. Another option would be to use a pressure sensor that shuts off the air supply when the unit exceeds the normal operating pressure.

Sampling devices used should maintain the integrity of the culture, as well as meet the containment requirements. For GLSP and BSL1-LS, a steamable sampling valve should be used because the aim is to minimize release. The use of a needle needs to be evaluated for each agent in use to ensure that it does not create potential employee exposure problems. When working with pathogens,

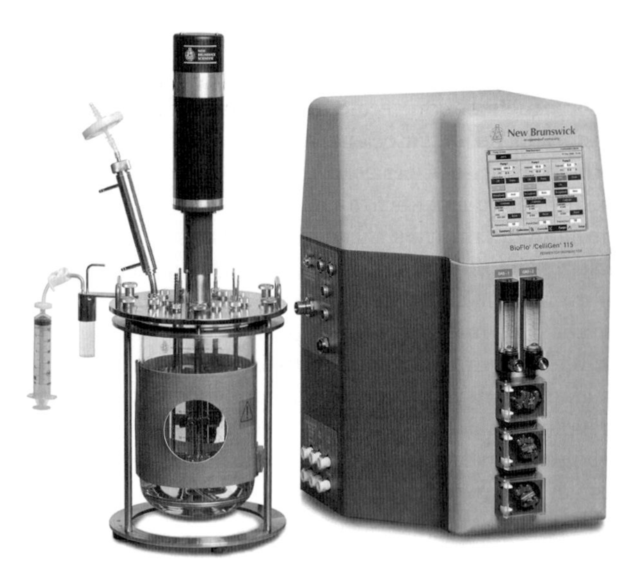

Figure 3: Sampling assembly on a bioreactor. (Courtesy of New Brunswick Scientific Co.)

where the goal is to prevent release, a sampling device that provides containment should be used. In some cases, secondary containment of the sampling device may be used. See Figs. 3 and 4 for examples of sampling devices. A variety of tubing welders/sealers that can be used to connect and disconnect aseptically are also available. All connections to the vessels are to be secured to prevent leakage or release. Depending upon the agent and/or vessel size, hard piping may be indicated. All connections must be designed to facilitate cleaning and decontamination, e.g., flush mounted, steamable, etc. If the fermenter does not have a sufficient degree of containment built into the unit for certain higher-risk materials, the entire unit may need to be placed within a containment device.

Despite its advantages, disposable technology poses new concerns regarding the ability to control or prevent aerosols, especially when producing viral vectors. Overpressurization during gas sparging or liquid level control can result in bag rupture. Exhaust filters are not the same robust design as in sterilize-in-place fermenters. The bags cannot be sterilized in place; waste liquid must be transferred to other vessels for inactivation by chemical or thermal means. Bag sampling does not employ steam-sterilized valves; instead, aseptic thermal welding technology is used to attach and remove sample bags.

Recovery and purification

In the downstream processing of the culture material, containment of the material is necessary to protect the product. If the organisms are inactivated in the culture system and there are no toxic, allergenic, or biologically active products (as is the case with GLSP processes),

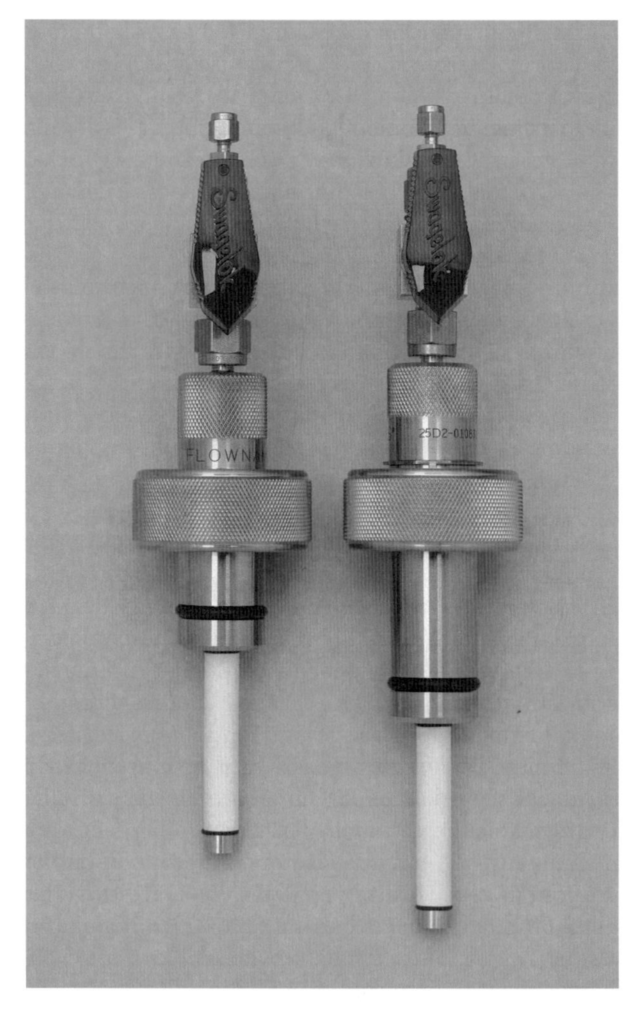

Figure 4: Sterilizable *in situ* sampling probes. (Courtesy of Merck & Co.)

additional containment measures should not be needed. If the organisms are not killed prior to processing, BSL1-LS requires that the equipment being used for processing viable organisms be designed to reduce the potential for the escape of viable organisms. At the higher containment levels, the equipment needs to be designed to prevent release. The risk assessment of the organism, which includes an analysis of any harmful characteristics of the organism, is the most important consideration in the choice of the containment features of the equipment.

Downstream processing can be divided into three basic categories, although some of the equipment can be used for more than one purpose: cell separation, cell disruption, and purification. Some of the equipment includes the following:

- Cell separation equipment
  - Depth filtration
  - Centrifuges
  - Filter presses

- Cell disruption equipment
  - Homogenizers
  - Sonicators
- Purification equipment
  - Ultrafiltration-buffer exchange
  - Chromatography columns
  - Lyophilization
  - Depth filtration

Rather than review each type of equipment separately, general containment design approaches to achieve the different levels of containment are discussed. These approaches can be applied to a wide variety of equipment based on the required level.

At BSL1-LS, the containment objective is to reduce the potential for release of viable organisms to minimize release. In more practical terms, this means that the equipment used should be designed to prevent spraying, splashing, or significant release of material. Where not designed into the equipment, shielding or placing barriers around the equipment or points where this release can occur may achieve this level of containment. If the equipment generates aerosols, the use of shielding with an exhaust vent placed near the point of aerosol generation could be adequate to minimize release to the work area.

At the higher biosafety levels, the requirement is to prevent release of aerosols. However, the rationale for doing that is different for agents transmissible by the respiratory route (typically Risk Group 3) than for agents transmitted through contaminated fomites, e.g., surfaces, equipment, etc. (typically Risk Group 2). In the latter case, if the employees can be vaccinated and the agent is not transmissible by the aerosol route, the use of venting/shielding/barriers may provide adequate containment of the agents.

Where prevention of aerosol escape is required, because of either the agent or the nature of the product, more rigorous containment measures must be utilized. One method for achieving this is to place the entire piece of equipment in a containment device or a room. For example, flowthrough centrifuges are known to generate aerosols, and where containment is needed, the unit can be placed in a separate room or containment device. Alternatively, negative-pressure isolators may be considered. These can be either flexible or rigid plastic or metal (usually stainless steel). A biological safety cabinet (BSC) can be used for smaller equipment that does not generate much turbulence. In some instances, BSC manufacturers, or other specialty equipment fabricators, can make specialized containment devices for specific equipment. A containment device can consist of plastic shrouding or a type of cabinet with HEPA supply and exhaust to dissipate the heat load. The device needs to provide accessibility to the equipment for operation, routine maintenance and servicing, material loading and removal, and cleaning. This may necessitate the use of access panels, portholes, and gloves. Depending on the size, it may need to be moved into the facility prior to the installation of walls or doors. Obviously, all of these issues need to be considered in the design of the device so that it can be used in the manner intended.

The topic of BSCs and the choice of the appropriate unit is discussed in another chapter of this book and in BMBL. For drug and device applications, all air coming into contact with open product must be class 100, which usually requires the unit to have a HEPA filter directly over the work surface.

The use of horizontal, laminar-flow, clean air stations should be limited to media preparation. For application where a BSC cannot be used but class 100 conditions are required, a vertical, laminar-flow unit may be utilized provided that additional shielding, curtains, low-level returns, etc., are used to reduce employee exposure. This may be an option for Risk Group 2 agents or below, and perhaps some Risk Group 3 organisms that are not transmissible by the respiratory route, but will depend on the specific agent involved and the processes being performed.

Mechanisms for Cleaning and Disinfection of Equipment

All reusable equipment is required to be cleaned and sterilized to prevent product contamination between runs, where the sterilization cycle has been validated to kill the organism being manipulated. A more appropriate term would be "decontaminate," because one does not always need to eliminate all viable organisms (17).

For equipment that is used for cell culture, where bovine-derived raw materials may be used, the decontamination cycle must be able to achieve 134°C to 138°C for 18 minutes, or as prescribed by the competent authority.

Some facilities may use a clean-in-place (CIP) system that employs detergent and an extreme pH. In some cases, that treatment may be adequate to kill the organism in use, but the process would have to be validated. Where the CIP process does not kill the agent, all effluents must be piped to a tank for treatment and disposal. Many facilities also use a steam/sterilize-in-place system that injects steam into pipes and processing equipment to a specific temperature, in some cases 121°C. The temperature achieved, the length of time that the steam is held in the system, and the testing and monitoring of the process dictate whether the process can be referred to as a sterilization process.

For some equipment, treatment with chlorine, other disinfectants, or an acidic or caustic solution may be used for decontamination. For processes above the GLSP level, the organisms must be inactivated before the equipment is opened.

SECONDARY CONTAINMENT

The facility design and construction provide the secondary containment that protects people outside of the immediate work area, both in other parts of the facility and in the community at large. The process, agent, and environmental issue determinations, as outlined above, will dictate many of the design parameters. It is important to keep the basic facility design criteria in mind throughout the process: GLSP, no special facility requirements; BSL1-LS, facility designed to contain large spills/releases of organisms; BSL2-LS, facility designed to contain all spills/releases of organisms; BSL3-LS, facility designed to contain all spills/releases of organisms, including aerosols.

One of the most significant issues is whether the facility will be used for the manufacture of a drug or device and must meet good manufacturing practice (GMP) requirements. Some additional considerations of this issue are addressed in chapter 31, "Biosafety in the Pharmaceutical Industry."

There are a number of ways to achieve containment, and various approaches to design issues are provided. These containment concepts are generally not applicable to GLSP facilities, but certain features may be usable, e.g., waste treatment.

Construction and Finishes

All surfaces within the facility should be designed to withstand regular cleaning and decontamination. Decontamination usually takes the form of application of a disinfectant solution; however, BSL3-LS facilities generally require that the facility be capable of withstanding fumigation.

The floors should provide a durable, slip-resistant, sanitary surface. If large equipment will be moved around in the facility, concrete floors with polyacrylate topping or an architectural epoxy can be used. Sheet vinyl flooring with welded seams, or another monolithic system that is coved, should be considered for pathogenic agents where environmental release is a major concern, or where required to meet GMP cleanability requirements. The facility must be capable of containing releases from fermenters or bioreactors in case of a large spill. If the equipment is not movable, it can be surrounded by a diked area. If the propagation vessels are movable, they can be placed in a diked area with ramps, or in an area with sunken or sloped floors that can collect any leaking material. The dike or depressed area must be of sufficient volume to allow for the contents of the tank(s) and sufficient disinfectant to decontaminate the material if there is no liquid biowaste treatment task. Similarly, drains in the containment area should be capped or raised unless connected to a biowaste treatment system.

Walls and ceilings should be smooth, nonporous, and capable of withstanding cleaning and disinfection. Walls may need additional shielding, such as rub rails or wainscots, to prevent damage if large equipment is moved around the area. Ceilings can be epoxy-coated hard plaster, welded vinyl sheet systems, or other ceiling systems that are cleanable and sealable, although sealability is not an issue at BSL1-LS. Rigid, walkable (weight-bearing) ceiling panels can provide enhanced safety for facility support personnel for BSL3 facilities, if they allow servicing of lights, HVAC, and other utilities from outside of the containment area.

Penetrations into the floor, walls, and ceilings should be minimized to facilitate cleaning and prevent leakage through the floor. All penetrations into a BSL3 facility must be sealed to prevent the escape of aerosols and to allow for fumigation of the facility. Penetrations for conduit and cable trays should be sealed internally.

Work surfaces must be impervious to water and resistant to chemicals, particularly those used for decontamination. Work surfaces should be finished with smooth edges to minimize injuries to the employees. Furniture used in the facility must be sturdy, capable of being cleaned and decontaminated, and positioned to facilitate cleaning of the area.

Doors should be of flush design and smooth, nonporous material that can withstand repeated cleaning and treatment with disinfectants. Doors should be self-closing and swing into the more hazardous room. Windows should be sealed to the frame. Sloping sills help to promote cleanability.

Heating, Ventilation, and Air Conditioning (HVAC)

Directional airflow created by negative pressure differentials is used to create an air barrier between production and adjoining areas. While that is sufficient for BSL1-LS, work with pathogens requires additional containment. This can be achieved in a number of ways. There are two basic designs that are most commonly used (18). One is an envelope system, where the internal production areas are maintained at positive pressure and are completely surrounded by a negative-pressure corridor to prevent the migration of the agent/product from the facility. This design may be preferable for operations that are more vulnerable to contamination, for which product cross-contamination must be prevented and/or for which stringent GMP requirements must be met. In most cases, these same criteria can be met by using negative-pressure gradients in the production area and a pressure bubble air lock; i.e., air is pressurized to provide containment of hazardous material in the production area. For BSL3-LS areas where more stringent containment requirements must be met and/or where there is need to prevent product cross-contamination from a GMP perspective, two adjacent air locks can be used. The first should be a pressure bubble air lock off of the corridor, adjacent to the second air lock, that is, a cascading negative-pressure flow air lock connected to the work area. In some cases, the corridor can serve as the first air lock. For facilities with multiple rooms, the room pressure should be most negative in the area of greatest hazard, which is usually the fermenter/bioreactor. Depending on the techniques or processes involved, culture starter areas may require a similar level of containment. Generally, work with open product in these culture starter areas will be contained in a BSC to prevent contamination of the culture.

The number of air changes per hour (ACH) significantly impacts the quality of the air. In facilities that must meet class 100,000 conditions, an ACH of 20 is not uncommon. For facilities above BSL1-LS, an ACH of 10 to 15 should be targeted. The ventilation in the rooms should be designed to maximize the air exchange in the room, generally with ceiling supply and low-level returns.

The HVAC system should be sized to dissipate the heat load generated by the equipment and provide a comfortable atmosphere for employees wearing personal protective equipment.

BSL1-LS facilities do not require any specialized supply or exhaust features. Most of the following features are critical for BSL3-LS facilities and may be considered for BSL2-LS:

- A dedicated air supply is desirable to facilitate system control and balancing.
- If the supply is shared with other areas, HEPA filters at the room supply vent or airtight dampers should be used to prevent contamination of the supply system.
- Supply air should be HEPA filtered if recirculated in the facility.
- If HEPA filters are used, suitable prefilters should be used to extend the life of the HEPA filter.
- Ports should be provided in the filter housing to allow for periodic testing of the filters.
- HEPA filtration of exhaust air should be considered depending on the agent and environmental concerns.
- Provision for testing and decontaminating HEPA filters, or use of bag-in/bag-out assemblies, should be made.
- Where exhaust air is not filtered at the point of discharge from the room, the exhaust ductwork should be welded/sealed up to the HEPA filter.
- Exhaust air from BSCs and other containment devices should be HEPA filtered prior to discharge, preferably to the outside of the facility. Exhausting these devices through the room exhaust system can create problems in air balance (19) and the containment function of the devices. Class II/A2 biosafety cabinets should not be directly connected to an exhaust; a canopy connection with an airflow alarm is required for ANSI/NSF certification. Placing Class II Type B cabinets on a separate exhaust system allows them to be used to maintain negative airflow in the facility should the room exhaust system fail.
- The exhaust and supply systems should be interlocked to prevent the facility from sustained positive pressurization.

Utilities and Maintenance Issues

All critical equipment and systems supporting the facility should be placed on a preventative maintenance program. Control panels and items that require regular maintenance should be positioned to allow repair and adjustments to be performed outside of the facility where possible. Provisions for supply of critical replacement parts should be available, particularly for long-lead-time items, to prevent extended facility shutdown.

HVAC, autoclave, and other equipment and utility support systems should be designed so that maintenance personnel should not have to enter the facility for repairs and scheduled maintenance, especially in facilities where infectious materials are handled.

Sufficient lighting should be provided for all activities, with efforts made to minimize reflections and glare. Lights should be covered with a cleanable surface and sealed for BSL3-LS facilities.

Liquid and gas utility services, if not dedicated to the facility, should be protected with backflow preventers or other devices to prevent contamination, e.g., a bump tank for steamable distilled water systems, a liquid disinfectant trap and HEPA or equivalent filter at point of use for vacuum systems, etc. A separate vacuum system should be considered for any BSL2-LS or BSL3-LS facilities. If a dedicated system is not available, vacuum lines must be protected with a liquid disinfection trap and appropriate filtration at the point of use.

Facility Layout and Support Systems

Ideally, the facility should be designed so that there is unidirectional flow of materials and personnel. In many cases, that is not possible because it requires a separate way in and out, or "clean" versus "dirty" corridors. This is confusing terminology and highlights an area where biosafety and GMP guidance generally diverge. From a biosafety perspective, "dirty" signifies the area of the highest concentration of organisms, whereas in the GMP sense, "dirty" signifies the crudest form of the product, i.e., raw materials. When the material and personnel flow is unidirectional, most of the biosafety and GMP criteria can be met. Facilities that are used to manufacture multiple agents at the same time will need additional features to prevent potential cross contamination. One way to accomplish this is to use an entry-exit corridor system with each of the various rooms/suites having entrance and exit air locks, or to use the double-air lock system described above. When there is no "clean/dirty" corridor system, operational procedures need to be adopted to prevent contamination of the adjacent production areas and potential cross-contamination.

Adequate space must be provided for change rooms, storage areas for raw materials, equipment supplies, a janitor's closet for housekeeping supplies, equipment for cleanup/decontamination, toilets and showers, freezer/refrigerator space, gas supplies/servicing, etc. In general, large-scale facilities utilize special garbing, so the entry air lock is typically designed as a change room. Office areas should be located outside of the large-scale facility. It is understood that paperwork areas and computer terminals are necessary in a large-scale production area; however, office areas should be separated from production areas by full-height walls and doors.

Large-scale facilities should be separated from high-traffic areas to assist in access restriction and to promote cleanliness. A controlled access system should be considered for all facilities above the GSLP level to protect the product. At higher levels, controlling access helps to protect outside personnel from inadvertently being exposed by wandering into the area, and it also helps to prevent unauthorized access to infectious materials. This system can range from an electronic card entry system to a combination lock or a key system. More rigorous systems, e.g., biometrics, may be required if the facility physical security requires this level of protection.

Each facility needs to contain all of the required safety equipment. Hand washing facilities, eyewash stations, and emergency showers must be provided. Sinks should generally be automatic or capable of being operated by foot, knee, or elbow. In some facilities, GMP requirements may not allow sinks in the production area. In those cases, hand sanitizers should be provided in the production areas, and a sink should be available in the change room. The sinks can discharge to the sanitary sewer, provided that they are not used for disposal of viable materials.

Provisions should be made to equip the work area with telephones, computer terminals, fax machines, etc., to facilitate information and data transfer outside of the facility and to minimize the need for personnel and paperwork to leave the area. Where required, data packets can be autoclaved with a quick-dry vacuum cycle for removal from areas where pathogens are used.

An integrated pest management program should be developed for the facility. Fortunately, the design criteria focusing on cleanability help in that effort.

All critical systems and equipment should be alarmed. This includes loss of supply air system, loss of exhaust system, and failure of containment device exhaust. These alarms should be audible and visible both inside and outside of the facility so that people will not enter the facility unprepared if the containment has been breached. Critical process equipment must be alarmed when breaches of containment are noted at BSL3-LS.

After completion, the facility must be commissioned. This process documents that the facility, as built, meets the design criteria established. Other terms that have similar meanings include validation, containment verification, and qualification. Regardless of the term used, the items that must be included in the commissioning package include a set of drawings, the defined use and purpose of the facility, the equipment requirements, and test results. The testing must cover the HVAC system, including controls; BSCs, fume hoods, and other containment devices; the alarms and failure mode testing; and liquid waste treatment systems and autoclaves. Other controls that may be critical for containment, such as those that monitor pressure on fermenters, must also be tested.

The term "certification" has a different connotation from a biological safety standpoint. According to WHO, it is the systematic examination of all safety features and processes within the laboratory (engineering controls, personal protective equipment, and administrative controls). Biosafety practices and procedures are also examined. Laboratory certification is an ongoing quality and safety assurance activity that should take place on a regular basis. This is considered the final step before starting up a facility. Most guidance documents require an annual reverification of critical containment parameters.

Waste Treatment

Guidance documents for GLSP facilities have not required that the discharge of viable GLSP organisms be decontaminated; however, some local authorities may require that. Many state and local regulations and biosafety guidelines may require that stock cultures be decontaminated. There may be additional local regulations governing specific waste parameters, e.g., biological oxygen demand and level of solids, which may require further processing prior to disposal. Because these organisms are considered "safe," waste from these operations is generally exempt from treatment and may actually be considered for further use, e.g., fertilizer.

All discharges of viable organisms and waste from BSL1-LS to BSL3-LS recombinant processes require decontamination prior to disposal per OECD and U.S. NIH rDNA guidelines. If the organism contains mobile/transferable genes that code for undesirable traits, e.g., confer antibiotic resistance, the potential for transfer of the trait in the environment needs to be assessed. For most systems currently used for large-scale purposes, this should not be a concern. However, should it be identified as a potential risk, the inactivation process should be capable of degrading the DNA (20).

For large-scale work with infectious organisms, an autoclave, or other method for decontamination, should be available to process contaminated materials. It should be located within the facility for BSL3-LS. The material generally poses a higher risk due to the concentration and volumes involved. A double-door autoclave with access from inside and outside the containment facility is preferred.

A decontamination tank or liquid waste treatment system, also called biowaste tanks or biokill systems, may be needed to inactivate any viable organisms from the process, if that cannot be carried out in the fermenter or bioprocesser. Waste tanks are often placed inside diked areas, pits, or other physical containment devices of sufficient capacity to hold the contents of the tank(s) until the waste can be treated or recycled to another tank for treatment.

The decontamination method generally involves the application of heat or chemicals to the material. Generally, some type of stirring or spraying system needs to be used in the tank to provide better heat or chemical distribution and reduce treatment time. These waste tanks may be used to decontaminate equipment cleaning solutions, process rinses, material from spills, etc. All of these factors should be considered in sizing these tanks.

It is important to have procedural safeguards built into the processing to prevent the release of viable organisms into the sanitary sewers when inactivation prior to disposal is required. For infectious agents, procedural and operational measures must be implemented.

Prior to release, the treated material must be assessed for pH, harmful chemicals involved in the process, whether the material can be construed as pharmaceutical wastes, biological oxygen demand, solids, etc., to assure conformance to local/federal regulations.

Biosecurity/Biorisk Management

Appropriate security measures are important to protect product and prevent inadvertent release. While bioterrorism may come to mind in this category, it is probably a minimal risk. Incidents such as unauthorized access, loss, theft, deliberate release, adulteration of seed stocks/master cell banks, etc., pose a more likely risk. However, the highest risk is probably due to natural disasters, i.e., floods, hurricanes, fires, tsunamis, and earthquakes. Depending on the location, the facility may need to be designed to take the natural risks into account. (21)

The biosecurity measures needed should be based on a site-specific risk assessment that takes into account the agents processed and the type of operation, e.g., a facility making a costly biological product or vaccine.

When Select Agents or other high-risk agents are involved, there are specific security requirements established by the registering body. These requirements are extensive and are not covered in this chapter. Additional information can be found on the CDC Select Agent website which provides a Security Information Document and a Security Plan Template to assist in the development of a written security plan.

However, the implementation of basic biosecurity measures makes sense to minimize the chance of product tampering, whether intentional or not. Some measures that should be considered are limiting access to the area to employees who have been appropriately trained in large-scale biosafety operations and requirements and locking cell bank or stock culture freezers and further limitation of access to those materials. Maintenance of cell/stock bank inventories in multiple locations should also be considered.

CONCLUSION

Large-scale processes have the potential for increased risk of exposure because of the volume and concentration of the agents used. However, appropriate practices, equipment, and facility design can reduce the risks significantly.

There is not one "right" way to achieve an acceptable level of containment. Depending on the agent and the process, a number of techniques can be used. It is hoped that the concepts presented here will help in those decisions.

I thank Jon Ryan, Jay Jackson, Danielle Caucheteux, and Robert Hawley for their assistance in this work.

APPENDIX: Large-Scale Guidelines

Biosafety guidelines for work with small volumes of infectious agents, i.e., those amounts typically used for diagnosis, characterization, or basic research, have been established by the CDC, NIH, and WHO. Additionally, guidelines for working with recombinant DNA (rDNA) molecules in both small and large volumes exist in the *NIH Guidelines for Research Involving Recombinant DNA Molecules*. However, no specific biosafety guidelines had been established for large-scale work with organisms that do not contain rDNA molecules. This document has been developed by including additional equipment and practices for safe large-scale work to existing guidelines. It serves as an effort to collect best practices for maximizing the safety for large-scale work, and can be used by an institutional biosafety committee and/or a biological safety officer to develop biosafety procedures for the work to be done, in conjunction with the risk assessment.

It is understood that the organism, quantity, and process have a significant impact on the choice of an appropriate biosafety level for the work to be conducted. There is no specific volume that constitutes "large scale" for microbial agents. Certain CDC/NIH guidance documents have referred to "large scale" as volumes typically in excess of those used for identification, typing, assay performance, or testing. The risk analysis must include an assessment of the infectivity of the agent, the routes of transmission, the severity of infection, the availability of prophylaxis, the level of containment afforded by the process and equipment used, etc., and not just the volume of material being handled. Similarly, there is little scientific evidence to support the premise that only volumes greater than 10 liters merit large-scale requirements. Certainly, that is not true for biosafety level 2 (BSL2) and BSL3 organisms. The CDC/NIH guideline recommends raising the biosafety level for culturing and purification of many BSL2 organisms; however, that was only done in an effort to provide considerations for the biosafety officer and scientists in the establishment of the appropriate level of protection. The NIH rDNA guidelines provide guidance for the large-scale use of recombinant organisms to protect the environment, but do not adequately address the level of containment necessary to protect the personnel working with infectious agents.

These guidelines will cover four different levels for large-scale work: good large scale practices (GLSP), biosafety level 1–large scale (BSL1-LS), biosafety level 2–large scale (BSL2-LS), and biosafety level 3–large scale (BSL3-LS). The containment conditions for biosafety level 4–large scale are not defined here but should be determined on a case-by-case basis.

Only the biological hazard of the organism is addressed here. Other hazards, such as the toxicity or biological activity of the products produced, should be considered separately. These guidelines do not specifically address animal or plant pathogens; however, the containment principles and practices may be useful for some of those agents.

All institutions that engage in large-scale research or production with microorganisms should appoint a biological safety officer (BSO) to oversee the procedures, facilities, and equipment used. The services of a BSO are critical at BSL2-LS and above, where knowledge and experience with handling pathogenic organisms, biosafety practices, containment equipment, and design criteria are required.

I. GLSP

The GLSP level is recommended for certain Risk Group 1 organisms that are not known to cause disease in healthy adults, are nontoxigenic, are well characterized, and/or have an extended history of safe large-scale work. These organisms should not be able to transfer antibiotic resistance to other organisms. Examples of these organisms include *Saccharomyces cerevisiae* and *Escherichia coli* K-12. These organisms should have limited survival and/or no known adverse consequences if released into the environment.

A. Standard microbiological practices
   1. Individuals wash their hands after handling viable material. Hand disinfectants may be used as an interim measure if a sink is not readily accessible.
   2. Eating, drinking, smoking, handling contact lenses, and applying cosmetics are not allowed in the work area.
   3. Mouth pipetting is prohibited. Only mechanical pipetting devices are used.
   4. Work surfaces are capable of being cleaned and disinfected.
   5. An effective integrated pest management program is required.

B. Special practices
   1. Institutions that engage in large-scale work should have a health and safety program for their employees.
   2. Written instructions and training are provided for personnel who work at GLSP conditions.
   3. Processing, sampling, transfer, and handling of viable organisms are done in a manner that minimizes employee exposure and the generation of aerosols.
   4. Discharges containing viable organisms are disposed of in accordance with applicable local, state, and federal requirements.
   5. The facility should have an emergency response plan that includes the handling of spills and accident handling.

C. Safety equipment
   1. Protective clothing, e.g., uniforms, laboratory coats, etc., is provided to minimize the soiling of personal clothing.
   2. Safety glasses are worn in the facility.

D. Facilities
   1. Each facility contains a sink for hand washing or has hand disinfectant available. If present, the sink should be located near the exit doorway. An eyewash station and emergency shower are provided in the work area or easily accessible from it.

II. BSL1-LS

BSL1-LS is recommended for the large-scale growth of Risk Group 1 organisms that are not known to cause disease in healthy adult humans and pose minimal hazard to personnel and the environment, but otherwise do not qualify for the GLSP level.

A. Standard microbiological practices
   1. Access to the work area may be restricted at the discretion of the project manager when work is ongoing. A warning sign should be placed on the door that lists the agent(s) being used, the names and telephone numbers of persons knowledgeable about and responsible for the facility, and entry requirements, if any.
   2. Persons wash their hands after they handle viable organisms, after removing gloves, and on leaving the work area. Hand disinfectant may be used as an interim measure if a sink is not readily accessible.
   3. Eating, drinking, smoking, handling contact lenses, and applying cosmetics are not permitted in the work area.
   4. Food is stored outside of the work area in cabinets or refrigerators designated and used for this purpose only.
   5. Mouth pipetting is prohibited. Only mechanical pipetting devices are used.
   6. Work surfaces are decontaminated on a routine basis and after any spill of viable organisms.
   7. Procedures are performed carefully in a manner that minimizes aerosol generation.
   8. The use of sharps should be minimized, and procedures for their safe handling should be adopted.
   9. All discharges of the viable organisms should be inactivated prior to disposal, unless allowed by applicable local, state, or federal regulations.
   10. All stocks and cultures of microorganisms should be decontaminated prior to disposal.
   11. An effective integrated pest management program is required.

B. Special practices
   1. Institutions that engage in large-scale work have a health and safety program for their employees.
   2. Written procedures and training in basic microbiological practices are provided and documented. Personnel must receive updated training when procedural or policy changes occur
   3. Medical evaluation, surveillance, and treatment are provided where indicated; e.g., determine functional status or competency of employees' immune system when working with opportunistic pathogens, etc.
   4. Spills and accidents that result in overt exposure to viable organisms are reported to the facility supervisor/manager. Medical evaluation, surveillance, and treatment are provided as appropriate, and written records are maintained.
   5. Emergency response plans shall include methods and procedures for handling spills and employee exposures.
   6. Cultures of viable organisms are handled in a closed system or other primary containment equipment, e.g., biological safety cabinet (BSC), which is designed to reduce the potential for the release of viable organisms.
   7. Sample collection, material addition to a closed system, and transfer of culture materials from one closed system to another are conducted in a manner that minimizes employee exposure, the release of viable material, and the generation of aerosols.
   8. Culture fluids may be removed from a closed system or other primary containment system in a manner that minimizes employee exposure, the release of viable material, and the generation of aerosols.
   9. Removal of exhaust gases from a closed system or other primary containment system minimizes the release of viable organisms to the environment by the use of appropriate filters or procedures. Note: Some country-specific guidance documents require the use of HEPA or equivalent filtration. In the absence of that requirement, the filter must be capable of minimizing the release of the organisms.
   10. A closed system or other primary containment equipment that has contained viable organisms shall not be opened for maintenance or other purposes until it has been decontaminated.

C. Safety equipment
   1. Protective clothing, e.g., uniforms, laboratory coats, etc., is provided to prevent the contamination or soiling of personal clothing.
   2. Safety glasses must be worn.
   3. Gloves are recommended for any contact with process materials. Gloves must be worn if the skin on the hands is broken, irritated, or otherwise not intact. Alternatives to latex gloves must be provided. The gloves used must also provide adequate protection from any of the chemicals used.

D. Facilities
   1. Each facility contains a sink for hand washing or has hand disinfectant available. If present, the sink should be located near the exit doorway. An eyewash station and emergency shower should be provided in the work area or easily accessible from it.
   2. The work area has a door that can be closed when large-scale work is ongoing.
   3. The work area is designed to be easily cleaned.

4. Floors are able to be cleaned and disinfected in case of spills of viable organisms. Rugs are not allowed.
5. Work surfaces are impervious to water and resistant to acids, alkali, organic solvents, and moderate heat.
6. Furniture in the work area is sturdy and placed so that all areas are accessible for cleaning, and should be able to undergo frequent cleaning and decontamination.
7. If the work area has windows that open, they are fitted with fly screens.
8. Facilities are designed to contain large spills of viable materials within the facility until appropriately decontaminated. This can be accomplished by utilizing a dike, or sloping or lowering the floor where the process vessels are located to allow adequate containment of the vessel and the required amount of disinfectant. The design should minimize the release of viable organisms directly to the sanitary sewer system.
9. All discharges of viable organisms are inactivated by a validated process, i.e., one that has been demonstrated to be effective using the organism in question, or with an indicator organism that is known to be more resistant to the physical or chemical methods used.
10. An effective integrated pest management program is required.

B. Special practices
1. Institutions that engage in large-scale work have a health and safety program for their employees.
2. Doors to the work area are kept closed when work is ongoing.
3. Access to the work area is restricted to personnel whose presence is required and who meet entry requirements, i.e., immunization, if any. Individuals who cannot take or do not respond to the vaccine, who cannot take the recommended prophylaxis in the event of an exposure incident, who are at increased risk of infection, or for whom infection may prove unusually hazardous are not allowed in the work area until their situation has been reviewed by appropriate medical personnel. The individuals are informed of the potential risks and may be asked to review the issue with their personal physician to ensure that they understand and accept the potential risk.
4. Personnel are able to demonstrate proficiency in standard microbiological practices and procedures and handling of human pathogens at BSL2. This can consist of previous experience and/or training. Training in the hazards associated with the organisms involved and in the practices and operations specific to the large-scale work area is provided and documented.
5. Appropriate immunizations, medical evaluation surveillance, and treatment are provided where indicated, i.e., immunization, survey of immune status, etc.
6. A hazard warning sign, incorporating the universal biohazard symbol, identifying the infectious agents, and listing the names and telephone numbers of the persons knowledgeable about and responsible for the work area, along with any special entry requirements for entering the work area, is posted at the entry to the work area.
7. Baseline serum samples are generally not warranted, but may be considered depending on the agents handled and/or the product made.
8. Written facility and process specific biosafety policies/procedures that detail required safety practices for handling infectious materials, spill cleanup, handling of accidents, emergency procedures, and other appropriate safety information are readily available.

III. BSL2-LS

BSL2-LS is recommended for the propagation and cultivation of Risk Group 2 infectious organisms that would be handled at BSL2 in laboratory scale. The following guidelines have been developed for facilities that handle large volumes of these materials.

A. Standard microbiological practices
1. Access to the work area is restricted to personnel who meet the entry requirements. Doors are locked in accordance with the institutional policies.
2. Persons wash their hands after they handle viable organisms, after removing gloves, and before leaving the work area.
3. Eating, drinking, smoking, handling contact lenses, and applying cosmetics are not permitted in the facility.
4. Food is stored outside of the facility in cabinets or refrigerators designated and used for this purpose only.
5. Mouth pipetting is prohibited. Only mechanical pipetting devices are used.
6. Work surfaces are decontaminated on a routine basis and after any spill of viable organisms. Absorbent towels/coverings that are used on work surfaces are discarded and decontaminated after use.
7. Procedures are performed carefully in a manner that prevents aerosol and/or splash generation.
8. All contaminated materials are decontaminated by an approved method prior to removal from the facility, reuse, or disposal. Wastes must be handled, packaged, and transported in accordance with the applicable regulations.

9. The use of sharps is avoided. Policies for the safe handling of sharps, such as needles, scalpels, pipettes, and broken glassware, must be developed and implemented. If required, additional safer engineered devices or work practice controls and/or personal protective equipment is used to prevent accidental exposure. Plastic laboratory ware is substituted for glassware whenever possible. If glassware is used, it is coated or shielded to minimize the potential for breakage.
10. Viable organisms are placed in a container that prevents leakage during collection, handling, processing, and transport.
11. Viable organisms are handled in a closed system or other primary containment equipment that prevents their release into the environment.
12. Sample collection, material addition to a closed system, and transfers of culture materials from one closed system to another are conducted in a manner that prevents employee exposure and the release of viable material from the closed system.
13. Culture fluids shall not be removed from a closed system (except as allowed in item 12) unless the viable organisms have been inactivated by a validated procedure or the organism/viral vector is the desired product.
14. Exhaust gases removed from a closed system or other primary containment systems are filtered or otherwise treated to prevent the release of viable organisms into the environment. Note: Some country-specific guidance documents require the use of HEPA or equivalent filtration. In the absence of that requirement, the filter must be capable of preventing the release of the organisms.
15. A closed system that has contained viable organisms will not be opened for maintenance or other purposes unless it has been decontaminated.
16. Rotating seals and other mechanical devices directly associated with a closed system used for the propagation of viable organisms are designed to prevent leakage or are fully enclosed in ventilated housings that are exhausted through filters or otherwise treated to prevent the release of viable organisms into the environment.
17. Closed systems used for the propagation of viable organisms and other primary containment equipment are tested for the integrity of the containment features prior to use and following any changes/modifications to the system that could affect the containment characteristics of the equipment. These systems are equipped with a sensing device that monitors the integrity of the containment while in use. Containment equipment for which the integrity cannot be verified or monitored during use is enclosed in ventilated housings that are exhausted through filters or otherwise treated to prevent the release of viable organisms.
18. Closed systems that are used for propagation of viable organisms or other primary containment equipment are permanently identified. This identifier is used on all records regarding validation, testing, operation, and maintenance.
19. Contaminated equipment and work surfaces are decontaminated with a suitable disinfectant on a routine basis, after spill cleanup, etc. Contaminated equipment is decontaminated prior to servicing or transport. Absorbent toweling/coverings used on work surfaces to collect droplets and minimize aerosols are discarded and decontaminated after use.
20. Individuals seek medical attention immediately after an exposure incident. Spills and accidents that result in overt exposure to infectious materials are immediately reported to the facility supervisor/manager and the BSO. Appropriate medical treatment, medical evaluation, and surveillance are provided, and written records are maintained.
21. Emergency response plans shall include provisions for handling employee exposures and decontamination and cleanup of all spills/releases of viable material, including proper use of personal protective equipment.
22. Animals and plants not involved in the work being performed are not permitted in the work area.

C. Safety equipment
   1. Protective clothing, e.g., laboratory coats, protective coveralls, etc., is worn to prevent contamination of personal clothing. If skin contact with the organism poses an exposure risk, consideration should be given to the use of waterproof solid-front, wraparound, or back- or side-tie coats. Protective clothing is removed when leaving the work area.
   2. Protective eyewear is worn at all times in the work area. Protective face protection, i.e., face shield or goggles and face mask/respirator, is worn for any procedures that may involve splashing or spraying. If reusable, the protective eye/face protection should be decontaminated before reuse.
   3. Impervious gloves are worn at all times in the work area when work is ongoing. Alternatives to latex gloves must be provided. Double gloving is considered if personnel are working over extended periods of time or with processes that may require direct contact with the infectious material. Gloves must not be washed, treated with disinfectant, or reused. Gloves must be removed and hands washed if the integrity of the gloves has been compromised and upon leaving the work area.

The gloves used must also provide adequate protection from any of the chemicals used.

4. The selection of a respirator/face mask is made based on the agent. If the agent is capable of respiratory transmission, a respirator with a filtration efficiency capable of protecting the individual from the organism is used, e.g., HEPA for viruses, N95 for *Mycobacterium tuberculosis*, etc. If the agent is transmitted through mucous membrane contact, a face mask that prevents droplet penetration, e.g., fluid resistant, can be used. Personnel are trained in the use of respirators and face masks for procedures that may involve aerosol generation and for emergency situations that involve the release of viable organisms in the work area.
5. BSCs or other ventilated containment devices can be used to contain aerosol-generating processes or to prevent contamination of viable organisms when removed from a closed system.
   a. BSCs must be installed so that fluctuations of the room air supply and exhaust do not interfere with proper operations.
   b. BSCs should be located away from doors, heavily traveled work areas, and other possible airflow disruptions.
   c. HEPA-filtered exhaust air from a Class II BSC can be safely recirculated back into the work area environment if the cabinet is tested and certified at least annually and operated according to manufacturer's recommendations.
   d. BSCs can also be connected to the facility contaminated exhaust system by either a thimble (canopy) connection or directly exhausted to the outside through a hard connection.
   e. Provisions to assure proper safety cabinet performance and air system operation must be verified.
6. Only centrifuge units with sealed rotor heads or safety cups that can be opened in a BSC are used, or the centrifuge is placed in a containment device.

D. Facilities
1. Each facility contains a sink for hand washing, an eyewash station, and an emergency shower. The sink is foot-, elbow-, or knee-operated, automatic, or otherwise not hand-operated, and located near the door of each room in the work area.
2. The work area has self-closing doors that are closed when large-scale work is ongoing.
3. The work area is designed to be easily cleaned and disinfected. Work surfaces and stationary equipment are sealed to the floor or raised to allow for cleaning and disinfection of the facility. Chairs should be able to undergo frequent cleaning and decontamination.
4. Floors, walls, and ceilings are made of materials that are smooth, impermeable to liquids, and resistant to the chemicals and disinfectants normally used. Light fixtures are covered with a cleanable surface.
5. Work surfaces are impervious to water and resistant to acids, alkali, organic solvents, and moderate heat.
6. Windows to the facility are kept closed and sealed while work is ongoing.
7. Vacuum lines are protected with liquid disinfectant traps and HEPA or equivalent filters.
8. General laboratory-type work areas are designed to have a minimum of six air changes per hour. For large-scale facilities, the number of air changes per hour will depend on the size of the area, the chemicals and agents handled, the procedures and equipment utilized, and the microbial/particulate requirements for the area.
9. The ventilation in the work area is designed to maximize the air exchange in the area, i.e., the supply and exhaust are placed at opposite ends of the room, ceiling supply with low-level exhaust, etc.
10. The work areas in the facility where the infectious organisms are handled are at negative pressure relative to the surrounding areas.
11. Provisions are made to contain large spills of viable organisms within the facility until appropriately decontaminated. This can be accomplished by placing the equipment in a diked area, or sloping or lowering the floors in those areas to allow for sufficient capacity to contain the viable material and disinfectant.
12. Drainage from the facility is designed to prevent the release of large volumes of viable material directly to the sanitary sewer; e.g., floor drains are capped, raised, or fitted with liquid-tight gaskets.
13. The facility and equipment should be tested to verify that they meet the design criteria established. This testing should include the HVAC system and controls, BSCs and other containment devices, closed culture or processing equipment, alarms, autoclaves, and waste treatment systems. These should be retested periodically and after any changes to the equipment or facilities that could affect their operation. Depending on the risk assessment, a facility security plan may need to be developed.
14. At a minimum, access to the facility should be limited to trained employees. Visitors should only be allowed when accompanied by a trained employee. Consideration should be given to additional access controls to freezers where stock cultures are stored.

IV. BSL3-LS

BSL3-LS is recommended for the propagation and cultivation of infectious organisms designated as Risk Group 3. The following guidelines have been developed for facilities that handle large volumes of these materials.

A. Standard microbiological practices

1. Access to the facility is restricted to personnel who meet the entry requirements. Individuals who have not been trained in the operating and emergency procedures of the facility are accompanied by trained personnel at all times while in the facility.
2. Persons wash their hands after they handle viable materials, after removing gloves, and before leaving the work area.
3. Eating, drinking, smoking, handling contact lenses, and applying cosmetics are not permitted in the work area.
4. Food is stored outside of the work area in cabinets or refrigerators designated and used for this purpose only.
5. Mouth pipetting is prohibited. Only mechanical pipetting devices are used.
6. Work surfaces are decontaminated on a routine basis and after any spill of viable material.
7. All procedures involving the manipulation of infectious material must be carried out within a BSC or other containment device.
8. All contaminated materials are decontaminated by an approved method prior to removal from the facility, reuse, or disposal. An autoclave and gas decontamination area should be located within the facility for this purpose. Provisions should be made for emergency decontamination support in the event of equipment/system failures.
9. All discharges of the viable materials are inactivated by a validated process, i.e., one that has been demonstrated to be effective using the organism in question, or with an indicator organism that is known to be more resistant to the physical or chemical methods used.
10. An effective integrated pest management program is required.

B. Special practices

1. Institutions that engage in large-scale work must have a health and safety program for their employees.
2. Doors to the facility are kept closed and locked.
3. Access to the facility is restricted to personnel whose presence is required and who meet entry requirements, i.e., training, immunization, if any, and comply with all entry and exit procedures. Individuals who cannot take or do not respond to the vaccine available, who are at increased risk of infection or for whom infection may prove unusually hazardous, or individuals who cannot take the recommended prophylaxis in the event of an exposure incident are not allowed in the work area until their situation has been reviewed by appropriate medical personnel.
4. All personnel working at BSL3-LS must demonstrate proficiency in standard microbiological practices and techniques and in handling human pathogens at BSL3. This can consist of previous experience and/or an on-the-job training program that covers the hazards associated with the materials involved and the practices and operations specific to the facility. The training should be documented.
5. Appropriate immunizations, medical evaluation surveillance, and treatment are provided where indicated, e.g., immunization, survey of immune status, etc.
6. A hazard warning sign, incorporating the universal biohazard symbol, identifying the infectious agents, and listing the names and telephone numbers of the persons knowledgeable about and responsible for the facility, along with any special entry requirements for entering the work area, is posted at the entry to the facility.
7. Baseline serum samples or other appropriate specimens are generally not warranted, but may be considered depending on the agents handled and/or the product made.
8. Written facility- and process-specific biosafety policies/procedures that detail required safety practices for handling infectious materials, spill cleanup, handling of accidents, emergency procedures and other appropriate safety information are readily available.
9. The use of sharps is avoided. Policies for the safe handling of sharps, such as needles, scalpels, pipettes, and broken glassware, must be developed and implemented. If required, additional safer engineered devices, work practice controls, and/or personal protective equipment is used to prevent accidental exposure. Plastic laboratory ware is substituted for glassware whenever possible. If glassware is used, it is coated or shielded to minimize the potential for breakage.
10. Viable organisms are placed in a container that prevents leakage during collection, handling, processing, and transport.
11. Viable organisms are handled in a closed system or other primary containment equipment that prevents their release into the environment.
12. Sample collection, material addition to a closed system, and transfer of culture materials from one

closed system to another are conducted in a manner that prevents employee exposure and the release of viable material from the closed system.

13. Culture fluids shall not be removed from a closed system (except as allowed in item 12) unless the viable organisms have been inactivated by a validated procedure or the organism/viral vector is the desired product.
14. Exhaust gases removed from a closed system or other primary containment systems are filtered or otherwise treated to prevent the release of viable organisms into the environment. Note: Some country-specific guidance documents require the use of HEPA or equivalent filtration. In the absence of that requirement, the filter must be capable of preventing the release of the organisms being grown.
15. A closed system that has contained viable organisms will not be opened for maintenance or other purposes unless it has been decontaminated.
16. Rotating seals and other mechanical devices directly associated with a closed system used for the propagation of viable organisms are designed to prevent leakage or are fully enclosed in ventilated housings that are exhausted through filters or otherwise treated to prevent the release of viable organisms.
17. Closed systems used for the propagation of viable organisms and other primary containment equipment are tested for the integrity of the containment features prior to use and following any modifications to the system that could affect the containment characteristics of the equipment. These systems are equipped with a sensing device that monitors the integrity of the containment while in use. Containment equipment for which the integrity cannot be verified or monitored during use is enclosed in ventilated housings that are exhausted through filters or otherwise treated to prevent the release of viable organisms.
18. Closed systems that are used for propagation of viable organisms or other primary containment equipment are permanently identified. This identifier is used on all records regarding validation, testing, operation, and maintenance.
19. Contaminated equipment and work surfaces are decontaminated with a suitable disinfectant on a routine basis, after spill cleanup, etc. Contaminated equipment is decontaminated prior to servicing or transport. Absorbent toweling/coverings used on work surfaces to collect droplets and minimize aerosols should be decontaminated and discarded after use.
20. Individuals seek medical attention immediately after an exposure incident. Spills and accidents that result in overt exposure to infectious materials are immediately reported to the facility supervisor/manager and the BSO. Appropriate medical treatment, medical evaluation, and surveillance are provided, and written records are maintained.
21. Emergency response plans include provisions for handling employee exposures and decontamination and cleanup of all spills/releases of viable material, including proper use of personal protective equipment.
22. Animals or plants not involved in the work being performed are not permitted in the work area.

C. Safety equipment

1. Persons entering the facility will exchange or completely cover their clothing with garments, such as solid-front or wraparound gowns, coveralls, etc. If skin contact with the organism poses an exposure risk, the protective clothing must be waterproof. Head and shoe covers or captive shoes are provided. Protective clothing is to be removed when leaving the facility and decontaminated before disposal or laundering.
2. Protective eyewear is worn at all times in the work area. Protective face protection, i.e., face shield or goggles and face mask/respirator, is worn for any procedures that may involve splashing or spraying. If reusable, the protective eye/face protection should be decontaminated before reuse.
3. Impervious gloves are worn at all times in the work area when work is ongoing. Double gloving is considered if personnel are working over extended periods of time or with processes that may require direct contact with the infectious material. Gloves are discarded upon leaving the work area. Alternatives to latex gloves must be provided. The gloves used must also provide adequate protection from any of the chemicals used.
4. The selection of a respirator/face mask is made on the basis of the transmissibility of the agent. If the agent is transmitted through the respiratory route, a respirator with a filtration efficiency capable of protecting the individual from the organism is used, e.g., HEPA for viruses, N95 for *M. tuberculosis*, etc. If the agent is transmitted through mucous membrane contact, a face mask that prevents droplet penetration, e.g., liquid resistant, is preferred. Personnel are trained in the use of respirators/face masks for procedures that may involve aerosol generation and for emergency situations that involve the release of viable organisms in the work area.

5. Class II or III BSCs or other ventilated containment devices are used to contain processes of viable materials if removed from a closed system.
   a. BSCs must be installed so that fluctuations of the room air supply and exhaust do not interfere with proper operations.
   b. BSCs should be located away from doors, heavily traveled work areas, and other possible airflow disruptions.
   c. HEPA-filtered exhaust air from a Class II BSC can be safely recirculated back into the work area environment if the cabinet is tested and certified at least annually and operated according to manufacturer's recommendations.
   d. BSCs can also be connected to the facility contaminated exhaust system by either a thimble (canopy) connection or directly exhausted to the outside through a hard connection.
   e. Provisions to assure proper safety cabinet performance and air system operation must be verified.
6. Only centrifuge units with sealed rotor heads or safety cups that can be opened in a BSC are used, or the centrifuge is placed in a containment device that can provide adequate containment of the unit.
7. Continuous-flow centrifuges or other aerosol-generating equipment are contained in devices that are exhausted through filters or otherwise treated to prevent the release of viable organisms.
8. Vacuum lines are protected with liquid disinfection traps and HEPA filters or the equivalent, which are routinely maintained and replaced as needed.

D. Facilities

1. The facility is separated from areas that are open to unrestricted traffic flow within the building. The facility has a double-doored entry area, such as an air lock or pass-through. The change room may be part of the entry area.
2. Each major work area contains a sink for hand washing which is not hand operated, e.g., is automatic or foot or elbow operated.
3. An eyewash station and emergency shower are available in the facility.
4. The facility is designed to be easily cleaned and disinfected. Furniture and stationary equipment are sealed to the floor, raised, or placed on wheels to allow for cleaning and disinfection of the facility.
5. Work surfaces are impervious to water and resistant to acids, alkali, organic solvents, and moderate heat.
6. Floors, walls, and ceilings are made of materials that are smooth, impermeable to liquids, and resistant to the chemicals and disinfectants normally used. Floors should be monolithic, slip resistant, and coved. Light fixtures are sealed, or recessed and covered with a cleanable surface.
7. Penetrations into the containment facility are kept to a minimum and sealed to maintain the integrity of the facility.
8. Windows to the facility are kept closed and sealed.
9. Liquid and gas services are dedicated to the facility, or they are protected from backflow. Fire protection sprinkler systems do not require backflow prevention devices.
10. The ventilation system for the facility is designed to control air movement.
    a. The position of the supply and exhaust vents is designed to maximize the air exchange in the area, i.e., the supply and exhaust are placed at opposite ends of the room, ceiling supply with low-level exhaust, etc.
    b. General laboratory-type work areas are designed to have a minimum of six air changes per hour. For large-scale facilities, the number of air changes per hour will depend on the size of the area, the chemicals and agents handled, the procedures, equipment utilized, the anticipated heat load, and the microbial/particulate requirements for the area.
    c. The facility is at negative air pressure relative to the surrounding areas or corridors. The system shall create directional airflow that draws air from the "clean" areas of the facility into the "contaminated" areas. If there are multiple contaminated areas, the area of highest potential contamination is the most negative.
    d. The exhaust air from the facility is not recirculated to any other area in the facility and is discharged to the outside through HEPA filters or other treatments that prevent the release of viable microorganisms.
    e. The facility has a dedicated air supply system. If the supply system is not dedicated to the facility, it contains HEPA filter gas-tight isolation dampers, which can protect the system from potential backflow in the event of a system failure.
    f. HEPA filter housings should have gas-tight isolation dampers, decontamination ports, and/or bag-in/bag-out capability. The housing should allow for leak testing of each filter and assembly. The filters and housings should be certified at least annually.
    g. The supply and exhaust systems for the facility are designed to prevent the room pressure from going positive in the event of power or equipment failure.

h. The system is alarmed to indicate system failures or changes in desired airflow. A visual monitoring device that indicates and confirms directional airflow is provided at the entry to the facility to facilitate verification of proper airflow by personnel. Visual and audible alarms should be available to notify personnel of any HVAC system failure.

11. A method for decontaminating all wastes is available in the facility, i.e., autoclave, chemical disinfection, incineration, or other validated method.
12. Provisions are made to contain large spills of viable organisms within the facility until appropriately decontaminated. This can be accomplished by placing the equipment in a diked area or sloping or lowering the floors in those areas to allow for sufficient capacity to contain the viable organisms and disinfectant.
13. Drainage from the facility is designed to prevent the release of viable organisms directly to the sanitary sewer; e.g., floor drains are capped, raised, or fitted with liquid-tight gaskets to prevent release of untreated organisms. A liquid effluent decontamination system may be considered depending on the agent, process, and volumes handled.
14. The facility and equipment should be tested to verify that they meet the design criteria established. This testing should include the HEPA filters and filter housings, HVAC system and controls, BSCs and other containment devices, closed culture or processing equipment, alarms, autoclaves, and waste treatment systems. These should be verified at least annually and after any changes to the equipment or facilities that could affect their operation.
15. Additional consideration should be given to the need for:
    a. computer networks, phones, etc., and other communication and data transfer systems within the facility
    b. an area decontamination system to decontaminate large equipment and/or rooms
    c. an anteroom for clean storage of equipment
    d. a change room with dress-in, shower-out capability
    e. advanced access control devices, e.g., biometrics
    f. spaces around doors and ventilation openings that can to be sealed to allow for gas decontamination of the facility
16. Depending on the risk assessment, a facility security plan may need to be developed. At a minimum, access to the facility should be limited to trained employees. Visitors should only be allowed when accompanied by a trained employee. Consideration should be given to additional access controls to freezers where stock cultures are stored.

This guideline was prepared by the former ASM Subcommittee on Laboratory Safety Task Force on Biosafety: M. Cipriano, D. Fleming, R. Hawley, J. Richmond, J. Coggins, B. Fontes, C. Thompson, and S. Wagener. Special thanks to M. Downing, B. Petuch, R. Hawley, P. Meecham, J. Gyuris, R. Rebar, D. Caucheteux, M.E. Kennedy, H. Sheely, S. Gendel, C. Carlson, and R. Fink for their comments and suggestions.

References

1. **Harding AL, Byers KB.** 2006. Epidemiology of laboratory-associated infections, p 53–77. *In* Fleming DO, Hunt DL (ed), *Laboratory Safety: Principles and Practices*, 4th ed. ASM Press, Washington, D.C.
2. **Advisory Committee on Dangerous Pathogens.** 1995. *Categorization of Biological Agents According to Hazards and Categories of Containment*, 4th ed. Her Majesty's Stationery Office, London, United Kingdom.
3. **Advisory Committee on Dangerous Pathogens.** 1998. *The Large Scale Contained Use of Biological Agents.* Her Majesty's Stationery Office, London, England.
4. **Public Health Agency of Canada.** 2015. *Canadian Biosafety Standards and Guidelines*, 2nd ed. Health and Welfare Canada, Ottawa, Canada.
5. **National Institutes of Health.** 2013. *NIH Guidelines for Research Involving Recombinant DNA Molecules (NIH Guidelines)*, as amended.
6. **Organisation for Economic Co-operation and Development.** 2000. Directive 2000/54/EC of the European Parliament and of the Council of 18 September 2000 on the protection of workers from risks related to exposure to biological agents at work. OECD Publications, Paris, France.
7. **Organisation for Economic Co-operation and Development.** 2009. Directive 2009/41/EC of the European Parliament and of the Council of 6 May 2009 on the contained use of genetically modified micro-organisms. *Off J Eur Union L* **125:**75–97.
8. **U.S. Department of Health and Human Services, Public Health Service, Centers for Disease Control and Prevention, National Institutes of Health.** 2009. *Biosafety in Microbiological and Biomedical Laboratories*, 5th ed. HHS Publication No. (CDC) 21-1112. U.S. Government Printing Office, Washington, DC.
9. **Prime Minister.** 1991. *Guidelines for Recombinant DNA Experiments.* Ministry of Health, Tokyo, Japan.
10. **World Health Organization.** 2004. *Laboratory Biosafety Manual*, 3rd ed. World Health Organization, Geneva, Switzerland.
11. **World Health Organization.** 2003. *Guidelines for the Safe Production and Quality Control of IPV Manufactured from Wild Polio Virus.* World Health Organization, Geneva, Switzerland.
12. **McGarrity GJ, Hoerner CL.** 1995. Biological safety in the biotechnology industry, p 119–129. *In* Fleming DO, Richardson JH, Tulis JJ, Vesley D (ed), *Laboratory Safety: Principles and Practices*, 2nd ed. ASM Press, Washington, D.C.
13. **American Institute of Chemical Engineers (AICE).** 2011. *Guidelines for Process Safety in Bioprocess Manufacturing Facilities.* John Wiley & Sons, Inc, Hoboken, New Jersey.
14. **Bailey JE, Ollis DF.** 1977. *Biochemical Engineering Fundamentals.* McGraw Hill, New York, N.Y.

15. **Hambleton P, Melling J, Salusbury TT (ed).** 1994. *Biosafety in Industrial Biotechnology*. Blackie, Glasgow, Scotland.
16. **Liberman DF, Fink R, Schaefer F.** 1986. Biosafety in biotechnology, p 402–408. *In* Solomon AL, Demain NA (ed), *Industrial Microbiology and Biotechnology*. American Society for Microbiology, Washington, D.C.
17. **National Research Council.** 1989. *Biosafety in the Laboratory*. National Academy Press, Washington, D.C.
18. **Odum J.** 1995. Fundamental guidelines for biotech multiuse facilities. *Pharm. Eng.* **15:**8–20.
19. **Ghidoni DA.** 1999. HVAC issues in secondary containment, p 63–72. *In* Richmond JY (ed), *Anthology of Biosafety*. American Biological Safety Association, Mundelein, Ill.
20. **Safety in Biotechnology Working Party of the European Federation of Biotechnology.** 2000. Safe biotechnology 10: DNA content of biotechnological process waste. Trends Biotechnol 18:141–146.
21. **World Health Organization.** 2006. *Biorisk Management. Laboratory Biosecurity Guidance*. World Health Organization, Geneva, Switzerland.

33

Veterinary Diagnostic Laboratories and Necropsy

TIMOTHY BASZLER AND TANYA GRAHAM

INTRODUCTION

Similar to human clinical microbiology laboratories, the work performed in veterinary diagnostic laboratories has inherent risk to laboratory workers. According to the 2008 American Veterinary Medical Association "One Health Initiative" task force report (www.avma.org/onehealth), 60% of infectious diseases in humans are due to multihost pathogens that move across species lines (1). Over the last 30 years, 75% of the emerging human pathogens (e.g., West Nile fever, avian influenza, Lyme disease) have been zoonotic (transmitted between animals and humans) (2). Thus, veterinary diagnostic laboratorians are at risk for laboratory-acquired infections (LAIs) from multiple host species pathogens.

Potentially infectious agents in human diagnostic specimens are by definition human pathogens. Not all infectious agents in host species submitted to a veterinary diagnostic laboratory are human pathogens (e.g., are zoonotic). The key to managing biosafety risk in veterinary diagnostic laboratories involves good general biosafety practices, secure facilities, and a practical risk analysis of diagnostic samples that may or may not include zoonotic organisms. Within the veterinary laboratory and animal facility environment, there will be the inevitable presence and handling of biological materials that can pose biological risks for both animal and human populations. The primary reasons to contain potential pathogens during diagnostic laboratory testing are prevention of LAI in humans and prevention of release of pathogens into the environment, potentially infecting animals and humans. Therefore, it is of critical importance that laboratory facility managers ensure that biological risks in their facility are clearly identified, understood, controlled, and communicated to the appropriate stakeholders.

Many of the biosafety practice guidelines listed in other sections of this text and the *Biosafety in Microbiological and Biomedical Laboratories* (BMBL), 5th edition (3), will suffice for use in veterinary diagnostic laboratories. Important changes and clarifications to the practices associated with biosafety level 2 (BSL2) laboratories are in the 5th (latest) edition of the BMBL released in 2007.

Another excellent reference is a CDC publication, "Guidelines for Safe Work Practices in Clinical Diagnostic Laboratories," which is directed specifically toward clinical laboratories handling diagnostic specimens of unknown infectious agent risk (diagnostic "unknowns") (4). Comprehensive biosafety guidelines, such as the BMBL, the Animal Research Services (ARS) Facilities Design Standards (5), and the NIH Design Policy Guidelines (6), emphasize the use of (i) engineering controls such as directional airflow, appropriately designed laboratory facilities, and biosafety cabinets (BSCs); (ii) administrative controls, such as signage, standard operating procedures (SOPs) that outline safe work practices, and frequent hand washing requirements; and (iii) guidelines for the use of personal protective equipment (PPE).

Because the BMBL is written primarily for research laboratories and gives "agent-specific" biosafety guidelines when working with known infectious agents, the purpose of this chapter is to provide practical guidelines and work practices that minimize biosafety hazards associated with veterinary diagnostic specimens—clinical material that has unknown contents but may be infected with pathogens that could be hazardous to humans and pose a threat to animal populations. In general, veterinary diagnostic laboratories should use BSL2 practices and facilities for routine clinical/diagnostic work. Diagnosticians should perform a practical risk assessment with each incoming accession to determine whether or not it is warranted to handle the specimens using decreased (BSL1) or increased (BSL3) biosafety facilities, procedures, and practices.

BIOLOGICAL RISK CLASSIFICATION AND ASSESSMENT

Risk Classifications

Table 1 lists the risk groups (G1–G4) and biosafety level classifications (BSL1–BSL4) from the World Organisation for Animal Health (WOAH; formerly known as the Office International des Epizooties [OIE]) (7) and the CDC/NIH (3), respectively. The recommended laboratory practices, equipment, and facilities for handling microbes within each risk group are listed in Table 2. In both the OIE and CDC/NIH classification systems, increasing risk levels (e.g., increasing numbers) imply greater biosafety and biocontainment risks associated with occupational exposure to an agent. With each increase in the risk level, there is a requirement for additional, more stringent containment during procedures designed to identify, isolate, or propagate the agent(s) present in a specimen.

Tables 1 and 2 facilitate the assessment of risk from different microbes and recommend appropriate safety practices for the handling of those microbes. The risk group classifications in Table 1 from the OIE are based upon level of risk with relation to biocontainment of animal pathogens in the laboratory, whereas the risk group classifications from CDC/NIH are in relation to biosafety, namely availability of effective treatment and prevention for human infection. The CDC/NIH guidelines in Table 2

TABLE 1.

Classifications: WOAH/OIE Risk Groups (G1–G4) and CDC/NIH Biosafety Levels (RG1–RG4)[a]

	Characterization
OIE Risk Group animal pathogen	
1	Disease-producing organisms that are enzootic but not subject to official control
2	Disease-producing organisms that are either exotic or enzootic but subject to official control and have a low risk of spread from the laboratory
3	Disease-producing organisms that are either exotic or enzootic but subject to official control and have a moderate risk of spread from the laboratory
4	Disease-producing organisms that are either exotic or enzootic but subject to official control and have a high risk of spread from the laboratory and the national animal population
CDC/NIH Risk Group	
1	Agents not associated with disease in healthy adult humans
2	Agents associated with human disease that is rarely serious and for which preventive or therapeutic interventions are *often* available
3	Agents associated with serious or lethal human disease for which preventive or therapeutic interventions may be available (high individual risk but low community risk)
4	Agents likely to cause serious or lethal human disease for which preventive or therapeutic interventions are not usually available (high individual risk and high community risk)

[a]Data from references 3 and 7.

TABLE 2.

CDC/NIH BSL1–BSL4: practices and equipment[a]

BSL	Criteria	Practices	Safety equipment and facilities
1	Not known to cause disease in normal healthy adults	Standard precautions[b]	Laboratory bench and sink PPE: lab coat and gloves; eye and face protection as needed
2	Agents associated with human disease Routes of transmission: percutaneous injury, ingestions, mucous membrane exposure	BSL1 practices plus limited access Display biohazard signs Sharps precautions Staff trained with pathogens Safety manual available	BSL1 plus: BSC used for specimen processing and work producing aerosols or splashes Autoclave available
3	Known to cause serious or lethal disease in humans Known potential for aerosol transmission Agent may be indigenous or exotic OR Agent is a "high-containment agricultural pathogen"	BSL2 practices plus controlled access Decontamination of all waste Decontamination of laboratory clothing before laundering	BSL2 equipment/facilities BSC used for work with all specimens and cultures PPE (gowns and masks) as needed Negative-pressure airflow Self-closing double doors Exhaust air not recirculated Entry through airlock or anteroom Physical separation from access corridors Hand washing sink near laboratory exit
4	Known to cause life-threatening disease in humans Aerosol transmission Agent may be indigenous or exotic No vaccine or therapy available	BSL3 practices plus: Clothing change before entering Shower on exit Decontaminate all waste on exit	BSL3 equipment/facilities plus: Class III BSC or Class I or II BSC with full body, air-supplied positive-pressure suite for all procedures Separate building or isolated zone Specialized ventilation and decontamination systems

[a]PPE, personal protective equipment; BSL, biosafety level; BSC, biosafety cabinet. Data from references 3 and 43.

[b]Standard precautions are based on the principle that all blood, body fluids (except sweat), excreta, nonintact skin, and mucous membranes may contain transmissible infectious agents (20).

propose four biosafety levels along with recommendations for appropriate containment practices for agents known to cause LAIs. Risk group classifications (RG1–RG4) and laboratory biosafety classifications (BSL1–BSL4) should never be used interchangeably. Risk group classifications (RG1–RG4 or G1–G4) are generally derived based on the severity of the consequences of a breach in biosafety and/or biosecurity. Thus, risk groups may not match biosafety levels (BSL1–BSL4) because they may not take the same mitigation factors (e.g., engineering controls, administrative controls, and PPE) or procedures into account. Neither classification makes allowance for persons with increased susceptibility to infections due to preexisting conditions, such as a compromised immune system or pregnancy. For further information and reference sources, see Sauri's guest editorial about medical surveillance in biomedical research facilities (8).

Routine work in veterinary diagnostic laboratories should assume that clinical specimens contain RG2 agents and handling of the specimens should proceed with BSL2 practices and procedures, unless a risk assessment would indicate otherwise. Although the infectious nature of the specimens may be unknown, diagnostic specimens may contain a variety of unknown agents, some of which could be extremely hazardous to human health or pose a significant threat to animal populations. Until the specimen is characterized as noninfectious or risk assessment indicates relevant biocontainment and risk controls, veterinary laboratories should take precautions to prevent exposure using standard BSL2 practices and barriers (Table 2). Once the laboratory has identified a specific agent or toxin, further work is carried out using relevant biocontainment and risk controls and indicated for the infectious agent. On occasion, veterinary diagnostic laboratories may use BSL3 practices and procedures. Only under extraordinary circumstances would veterinary diagnostic specimens contain RG4 agents, which will not be included in this chapter. Some RG2 and RG3 agents commonly encountered in veterinary diagnostic laboratories are listed in Table 3; the list is not meant to be all-inclusive. There are online resources with more extensive

TABLE 3.

Examples of common risk group 2 and risk group 3 zoonotic pathogens that may be present in specimens submitted to the veterinary diagnostic laboratory[a]

Risk group 2

Viruses: Influenza virus types A, B, C; Newcastle disease virus; parapoxvirus (Orf); West Nile virus

Bacteria: *Alcaligenes* spp., *Arizona* spp., *Campylobacter* spp., *Chlamydophila psittaci* (nonavian), *Clostridium tetani, Clostridium botulinum, Corynebacterium* spp., *Erysipelothrix rhusiopathiae, Escherichia coli, Haemophilus* spp., *Leptospira* spp., *Listeria monocytogenes, Moraxella* spp., *Mycobacterium avium, Pasteurella* spp., *Proteus* spp., *Pseudomonas* spp., *Salmonella* spp., *Staphylococcus* spp., *Yersinia enterocolitica, Yersinia pseudotuberculosis*

Fungi: *Aspergillus fumigatus, Microsporum* spp., *Trichophyton* spp., *Blastomyces dermatitidis, Coccidioides immitis, Cryptococcus neoformans, Histoplasma capsulatum, Sporothrix schenckii*

Risk group 3

Viruses: Rabies virus, equine encephalomyelitis virus (Eastern, Western, Venezuelan), Japanese encephalitis virus, louping ill virus

Bacteria: *Bacillus anthracis, Burkholderia mallei, Brucella* spp., *Chlamydia psittaci* (avian strains), *Coxiella burnetii, Mycobacterium bovis.*

Fungi: *Blastomyces dermatitidis* spores (cultures only), *Coccidioides immitis* spores (cultures only), *Histoplasma capsulatum* spores (cultures only)

[a]Data from reference 44.

listing of pathogens important to animals such as "OIE Listed Diseases and Other Diseases of Importance to International Trade" (9), as well as searchable databases of microbes, their risk groups, and suggested biosafety best practices based upon pathogenicity, mode of transmission and host range, available preventative measures, available effective treatments, and other factors (American Biological Safety Association [ABSA] at http://my.absa.org/tiki-index.php) (10).

Risk Assessment

For the purposes of this chapter, risk assessment is defined generically as the evaluation or estimation of biological risk and does not propose a specific method. When performing a risk assessment, veterinary diagnostic laboratories must consider the possible presence of "high-consequence livestock pathogens." The Federal Select Agent Program administered jointly by the U.S. Department of Agriculture (USDA) Animal and Plant Health Inspection Service (APHIS) and Department of Health and Human Services CDC defines high-consequence livestock pathogens. The Select Agent and Toxin regulations are published in the U.S. Federal Register by the U.S. Department of Health and Human Services (42 C.F.R. part 73) and by the USDA (9 C.F.R. part 121 and 7 C.F.R. part 331). Livestock pathogens that fall into the "high-consequence" classification are identified as such because of the potentially severe detrimental impact on agricultural animal health or animal products, the virulence and transmissibility of the agent, the availability of an effective treatment method(s), and economic considerations. Although not necessarily zoonotic, "high-consequence livestock pathogens" require handling using BSL3 facilities and/or practices to prevent accidental or intentional release/dispersement of the pathogen into the environment. Table 4 shows the Select Agents and Toxins List from the Federal Select Agent Registry as of 2016. The list combines currently classified Select Agents that pose a severe threat to human or animal health and includes USDA pathogens (animal health threats), Health and Human Services pathogens (human health threats), and "Overlap" pathogens (both animal and human health threats) (11).

A biological risk assessment should include evaluation of both biosafety and biosecurity. Traditionally the term "biosecurity" referred to agricultural biosecurity—the objective being to prevent diseases from entering a herd or flock. In 2008, during the Biological Weapons Convention it was stated that ". . . biosafety protects people from dangerous germs, while biosecurity protects germs from dangerous people." More specifically, the WHO defines biosafety as the "containment principles, technologies and practices that are implemented to prevent *unintentional* exposure . . . or accidental release." Biosecurity is defined as the "institutional and personal security measures designed to prevent the loss, theft, misuse, diversion or *intentional* release of pathogens and toxins" (12).

Risk assessments can be formulated in a variety of ways. There is no one official or standardized approach to conducting a risk assessment, but there are several strategies available. These include using a risk prioritization matrix, conducting a job hazard analysis, or simply listing the potential scenarios of what could go wrong while conducting a procedure/task/activity (13). The WHO suggests that a common-sense approach be taken when dealing with clinical specimens because laboratory personnel do not know which agent(s) may or may not be present. Handling and transporting unknowns should follow, at a minimum, BSL2 policies and procedures. Specimens should always be transported within the laboratory (internally) in approved leakproof containers. When transporting or shipping a specimen to another facility, all shipping and transport regulations must be complied with (14).

Risk can be defined as a function of the likelihood of an adverse event occurring and the resulting undesirable consequences. The likelihood of an adverse event happening is based on (i) the current science known about an agent, (ii) the properties of the laboratory facilities, (iii) the procedure(s) to be performed, and (iv) the miti-

TABLE 4.

HHS and USDA Select Agents and Toxins[a]

HHS Select Agents and Toxins
Abrin
Botulinum neurotoxins*
Botulinum neurotoxin-producing species of *Clostridium**
Conotoxins (short, paralytic alpha conotoxins containing the amino acid sequence $X_1CCX_2PACGX_3X_4X_5X_6CX_7$)
Coxiella burnetii
Crimean-Congo hemorrhagic fever virus
Diacetoxyscirpenol
Eastern equine encephalitis virus
Ebola virus*
*Francisella tularensis**
Lassa fever virus
Lujo virus
Marburg virus*
Monkeypox virus
Reconstructed replication-competent forms of the 1918 pandemic influenza virus containing any portion of the coding regions of all eight gene segments (reconstructed 1918 influenza virus)
Ricin
Rickettsia prowazekii
Severe acute respiratory syndrome-associated coronavirus (SARS-CoV)
Saxitoxin
South American hemorrhagic fever viruses:
Chapare
Guanarito
Junin
Machupo
Sabia
Staphylococcal enterotoxins A, B, C, D, E subtypes
T-2 toxin
Tetrodotoxin
Tick-borne encephalitis complex (flavi-) viruses:
Far Eastern subtype
Siberian subtype
Kyasanur Forest disease virus
Omsk hemorrhagic fever virus
Variola major virus (smallpox virus)*
HHS Select Agents and Toxins *(continued)*
Variola minor virus (Alastrim)*
*Yersinia pestis**
Overlap Select Agents and Toxins
*Bacillus anthracis**
Bacillus anthracis Pasteur strain
Brucella abortus
Brucella melitensis
Brucella suis
*Burkholderia mallei**
*Burkholderia pseudomallei**
Hendra virus
Nipah virus
Rift Valley fever virus
Venezuelan equine encephalitis virus
USDA Select Agents and Toxins
African horse sickness virus
African swine fever virus
Avian influenza virus
Classical swine fever virus
Foot-and-mouth disease virus*
Goat poxvirus
Lumpy skin disease virus
Mycoplasma capricolum
Mycoplasma mycoides
Newcastle disease virus
Peste des petits ruminants virus
Rinderpest virus*
Sheep pox virus
Swine vesicular disease virus
USDA Plant Protection and Quarantine (PPQ) Select Agents and Toxins
Peronosclerospora philippinensis (*Peronosclerospora sacchari*)
Phoma glycinicola (formerly *Pyrenochaeta glycines*)
Ralstonia solanacearum
Rathayibacter toxicus
Sclerophthora rayssiae
Synchytrium endobioticum
Xanthomonas oryzae

[a] *Denotes Tier 1 Agent. Data from reference 45.

gation factors in place. Each mitigation factor is designed to lessen the likelihood that an adverse event will occur. Examples of effective mitigation factors, beginning with those that are most effective, include engineering and administrative controls, laboratory safety policies and procedures, and the use of PPE.

Mitigation factors can be used to lessen the likelihood of an LAI. Once an adverse event occurs, the consequences—i.e., the severity of the LAI—are not easily mitigated. Preexposure immunization of laboratorians is one of the few ways in which the severity of an adverse event (in this case an LAI) can be altered (15). Figure 1 provides an

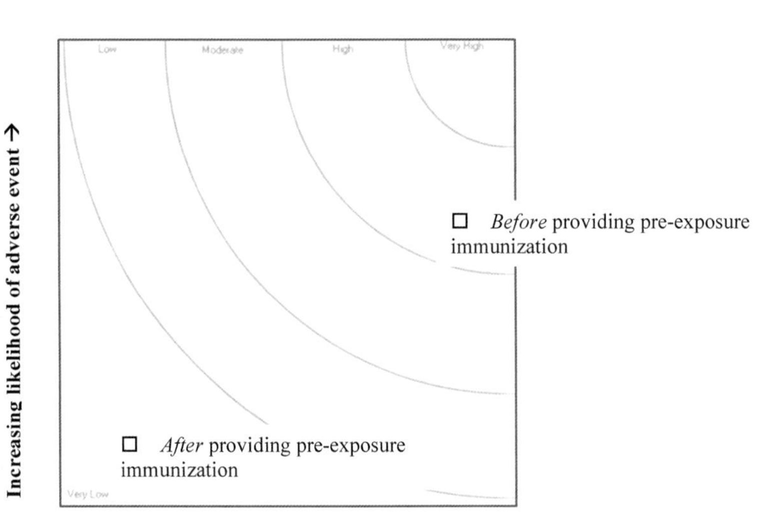

Figure 1: Using biorisk assessment models (BioRAM). The effect of reducing the risk of laboratory-acquired rabies virus infection by providing a mitigation factor—in this case, preexposure rabies virus immunization for laboratory employees handling suspect brain tissue. Demonstration of how a BioRAM user can simulate changing various mitigation factors in their laboratory (in this example, providing preexposure rabies virus immunization for laboratory personnel handling brain tissue) to evaluate which changes will be the most effective in reducing the risk. In this particular example, the reduction in risk is striking; this is not unexpected given that the presence of prophylactic immunization can mean the difference between life and death in the case of a rabies virus exposure. If the mitigation factor did not also affect the consequences, the change would not be as striking. Adapted from reference 15.

example of how the risk associated with an LAI can be lessened when employees are given preexposure prophylactic vaccinations.

The purpose of the risk assessment is to determine which mitigation strategies are necessary, given a particular hazard and/or threat (15). Table 5 lists the information for preparing and conducting a risk assessment. A new risk assessment should be performed whenever a new agent is introduced and/or facility infrastructure change(s) occur. For the veterinary diagnostic laboratory, which deals with unknown agents, a series of risk assessments could be completed in anticipation of the agents that are typically isolated from specimens submitted to the laboratory. Following completion of the risk assessment(s), SOPs or other documents can be used to educate laboratorians about mitigation factors that should be in place before performing a particular procedure. For example, the laboratory SOP(s) on how to handle rabies-suspect cases should include information about what PPE is required when handling the animal/specimen.

Sandia National Laboratories' International Biological Threat Reduction Department has developed a performance-based biorisk assessment methodology program (BioRAM) that allows users to evaluate the biosafety and biosecurity risks associated with handling/manipulating, storing, and disposing of pathogens and toxins in the laboratory setting. By answering a set of straightforward multiple choice or true/false questions, human bias is removed because the software's numerical engine converts the answers into a specific score for each question. This risk assessment tool is particularly useful when evaluating emerging diseases, implementing biosafety protocols for new procedures, or whenever changes in the facility have occurred. (See Fig. 1 as an example of the BioRAM results.) This biorisk assessment program is also extremely useful for updating biorisk assessments whenever new information about a pathogen or toxin is made available (13).

The BioRAM methodology is designed to "... [determine] the likelihood of infection by the agent and the likelihood of exposure through an infectious route based on the procedures and work practices [in the laboratory] and the consequences of disease assuming infection" (15). The development of this biorisk assessment tool required the assistance of a global partnership of individuals and organizations, including the American Biological Safety Association, the Public Health Agency of Canada, Colorado State University, the Asia Pacific Biosafety Association, the European Biosafety Professionals, Laboratory Speiz;WHO, the Belgium Biosafety Association, fellows from NIH, individuals from the CDC and U.S. Department of Homeland Security, the Universities of Chicago, Iowa, Kentucky, and Pennsylvania, Vanderbilt University, Wageningen University and Research Center, Emory University, and many more. A global creation, the BioRAM program is freely available for use by all laboratories and takes only minutes to download onto any computer (15).

An added benefit of using BioRAM is the ability of a user to manipulate specific mitigation criteria and observe the resulting changes in the likelihood of an adverse event occurring. BioRAM allows users to consider changes to laboratory policies and procedures—before they are implemented—to determine which change(s) will best reduce the risk. For example, is it best to (i) require the use of a BSC to perform certain procedures, (ii) offer immunization to employees, or (iii) change the waste management policies, etc.? In the example shown in Fig. 1, the model indicates that the user (i.e., the laboratory manager) has reduced both the likelihood and the consequences of an LAI by offering rabies vaccinations for laboratory personnel.

TABLE 5.

Information needed for a laboratory risk assessment

General

The properties of the agent and its environment
The activities to be performed in the laboratory
The details of the laboratory's infrastructure
The training and abilities of the laboratory personnel

Specific

Agent properties: What is the pathogenicity, concentration, and infectious dose of the agent? What are the route(s) of natural infection? What is the time range for incubation period? What are the primary symptoms of disease? Is the agent stable in the environment? Is/are suitable host(s) present in the environment? Are prophylactic/therapeutic interventions available for the host(s)? What quantity and concentration of the agent (i.e., *in vivo* versus *in vitro*) are involved? What is the biological decay rate (e.g., how long will the agent survive in the environment outside of a susceptible host or culture medium)?

Laboratory activities: What is the manner in which an agent will be manipulated (centrifugation, aerosolization, other)? Will laboratory activities make the agent more or less likely to be infectious to the normal or potentially expanded host range? What mitigation measures will the laboratory use to lessen the likelihood that an adverse event will occur? What is the volume of specimen and/or agent being manipulated? (Larger volumes generally have a greater associated risk.)

Laboratory infrastructure: What environmental controls are currently in place at the facility?

Human inputs: What is the level of training, experience, and technique (e.g., good laboratory practice [GLP]) of the laboratorian? What is the immune status of the workforce? Do predisposing conditions (i.e., heart valve replacement, etc.) or ethnic factors among the workforce make them more susceptible to infection?

Mitigation measures: What are the engineering and administrative controls in place? What policies, procedures, and personal protective equipment will be used?

Environmental/community factors: Is the agent normally present in the environment? Are there suitable hosts or vectors in the environment? Does the life cycle of the agent require multiple hosts in succession? Will suitable hosts or vectors be present in the environment if the agent is genetically modified? What is the immune status of the community? What is the availability of effective prophylactics? Is therapeutic treatment available?

Risk assessment in the veterinary diagnostic laboratory setting must take into account the likelihood of various unknown microorganisms being present within clinical specimens and the likelihood that routine processing of the specimen could expose laboratory workers to infectious agents. The typical risk assessment should take into consideration the host species, the known medical condition and clinical history of the patient, previous vaccinations and/or treatments, clinical signs in the rest of the herd/flock, and the endemic diseases prevalent in the geographic area where the animal resides or recently resided. Examples of locally endemic zoonotic diseases include tularemia in western gray squirrels in the western United States and anthrax in cattle in the north central region of the United States.

The risk assessment may indicate a reduction of biosafety practices from routine BSL2 practices. An example would be animal blood samples submitted to the serology laboratory. Unlike human blood samples that can harbor blood-borne human pathogens such as human immunodeficiency virus (HIV) or hepatitis virus, animal serum does not (in general) contain zoonotic blood-borne pathogens. (Nonhuman primates are an obvious exception.) On the basis of a risk assessment, the blood samples submitted for serological analysis may be handled using BSL1 practices.

The assessment of clinical history and signalment provided on a laboratory submission form depends upon professional judgment and should be conducted or overseen by a qualified veterinarian familiar with the zoonotic and Select Agents, such as those listed in Tables 3 and 4 and online sources (9, 10). Knowledge regarding typical clinical signs, host range, basic epidemiology, and geographical distribution of diseases caused by these agents is essential. If questions arise during case accessioning and login that cannot be clarified from the submission form, the submitting veterinarian should be contacted by telephone for further information.

For most procedures performed in veterinary diagnostic laboratories, Universal Precautions should be followed. Universal Precautions in a laboratory setting involve the consistent use of BSL2 practices and procedures as outlined in the BMBL (3). In 1996, Standard Precautions were established by the CDC by combining the major features of Universal Precautions and Body Substance Isolation. Standard Precautions are based on the principle that human blood, all nonblood body fluids (except sweat), excreta, nonintact skin, and mucous membranes may contain transmissible infectious agents. In clinical/diagnostic laboratories human patients/specimens are handled using Standard Precautions. These precautions are followed continuously and are considered the foundation for preventing infectious pathogen transmission.

In veterinary diagnostic laboratories, not all submissions contain zoonotic agents, so risk group/biosafety level assessments must be performed with each new submission. In most cases, diagnosticians in veterinary diagnostic laboratories are already performing these risk assessments with each case as the differential diagnosis is being formulated, and Universal Precautions (the use of BSL2 practices and procedures) are routinely being followed. For example, when a horse with central nervous system signs is submitted for necropsy, the differential diagnosis may include rabies virus, West Nile virus, equine herpesvirus, and equine protozoal encephalomyelitis. The diagnostician may order a rabies virus test before performing the necropsy in order to mitigate the risk of an

LAI should the horse be infected with rabies virus. Similarly, a bison may be tested for possible anthrax infection before the carcass is even moved/manipulated.

It is also important to remember that, regardless of the risk group and biosafety level practices assigned to a procedure, there is no way to remove all risks completely from clinical diagnostic laboratories. Therefore, anytime an LAI does occur, a thorough root-cause analysis and corrective action should be performed. It is not enough to assume that the mitigation measures must have failed or that human error was the only factor involved. The postincident review must follow the animal and/or specimen(s) from delivery at the unloading dock all the way through to the final treatment(s) of waste management stream(s). During this review, all members of the risk assessment team should be involved in establishing why an adverse event occurred (or almost occurred) and how it can be prevented in the future.

GENERAL BIOSAFETY GUIDELINES IN THE VETERINARY DIAGNOSTIC LABORATORY

Biosafety can refer to reducing the risks associated with handling biological materials, manipulating genomes, creating synthetic organisms, or bioterrorism. In the laboratory setting, all employees must be committed to a culture of safety at all times. In addition to the obvious components of safety, such as engineering controls, administrative policies/procedures, and PPE, the laboratorian must be knowledgeable about the agent(s) they may handle. Developing a set of required competencies for employees (appropriate to their individual education, training, and experience) will, in turn, allow management to use their biosafety guidelines as a resource tool when developing in-house training program(s). An excellent set of recommended competencies (for each biosafety level) is listed in the 2011 *Morbidity and Mortality Weekly Report,* "Guidelines for Biosafety Laboratory Competency" (16).

The facility must also provide an effective method of communicating the potential risks to all employees via the use of appropriate training, signage, policies and/or SOPs, and employee feedback. At no time should an employee perceive that the organization is acting in a punitive manner or is unwilling to address their safety concerns fully. The goal of communicating what is and is not known about all laboratory hazards must be part of an ongoing, continuous training program designed to promote a meaningful culture of safety within the laboratory.

During initial case accessioning ("login"), all specimen containers should be considered contaminated and potentially infectious at the time of receipt. These specimens should be handled with BSL2/Standard Precautions using the appropriate PPE—usually a laboratory coat and gloves—until the container is properly cleaned and disinfected (4).

All receiving areas in the laboratory should be considered "dirty" or contaminated. The exterior of all containers being received should be examined for leaks. If the container is broken, cracked, or is received with a loose cap, this observation should be documented in the record. Leaking primary containers should be placed inside another clean container before being moved out of the receiving area and into other areas of the laboratory. If the contents of the container are unusable, the submitter should be contacted and asked to submit a new specimen. If receipt of damaged specimens is a frequent occurrence, the laboratory may need to revisit its submission and specimen receipt policies. Alternatively, the laboratory may need to discuss improved specimen collection and transport procedures with the offending submitter(s) (4).

All contaminated specimen containers should be examined again for leaks before placing the specimen in/on automated equipment, rockers, or centrifuges as part of testing procedures. Laboratory operations should, whenever possible, provide for unidirectional flow of specimens, employees, and procedures to maximize the effectiveness of the engineering controls (i.e., always moving from a "clean" environment to a "dirty" environment, to avoid contaminating "clean" testing areas) (4). Documentation, such as equipment manuals, instrument logs, and SOPs, should be kept in an orderly fashion at the workbench. If a computer keyboard is present at the workstation, a protective cover that is easily cleaned and disinfected should be used at all times. Medical waste containers and sharps containers should be kept within easy reach at each workstation. Clean and tidy workbenches should be the expected "norm" within the laboratory. Instructions for cleaning and disinfecting laboratory workbenches should be included in the appropriate policy or procedure documents and should be posted at the laboratory bench for easy reference. Employees should be reminded regularly not to bring briefcases, backpacks, purses, books, e-readers, ipods, or other personal items into the laboratory because these items are difficult, and in some cases impossible, to disinfect (4). A storage area for personal materials should be provided for employees away from the laboratory.

Prior to performing laboratory procedures, employees must determine what PPE is required for the procedure(s) to be performed. This determination should be based on the risk assessment and the appropriate laboratory SOP(s) and policy document(s). If an organism requiring increased respiratory protection must be manipulated outside the BSC, appropriate respiratory protective equipment (i.e., user seal-checked/fit-tested N95 respirator or powered air-purifying respirator [PAPR]) should be worn as a part of the required PPE (4).

Laboratory procedures that commonly cause aerosolization of an infectious agent include, but are not limited to, pipetting, opening screw-capped containers, inoculations with a wire loop, touching the medium with a hot wire probe/loop containing a bacterial colony, inoculating eggs or mice (injections), lyophilization, blending, vortex mixing, centrifuging, use of sonicators (ultrasonic devices), withdrawing material from a vacuum bottle, and, in some cases, malfunctioning equipment (4, 17, 18). These procedures can generate inhalable particles that may remain airborne for protracted periods of time (4). An effective rule of thumb is the following: Low-energy input procedures produce aerosols containing large particles and high-energy procedures generate small particles.

Because any energy input has the potential to cause aerosolization, most laboratory procedures should be performed inside a BSC. An excellent, detailed source of information on how to prevent aerosol generation when using laboratory equipment is provided in the 2012 *Morbidity and Mortality Weekly Report* "Guidelines for Safe Work Practices in Clinical Diagnostic Laboratories" (4). All laboratorians should be particularly interested in the information provided about generation of aerosols during the use of routine laboratory equipment.

In the laboratory, potential routes of exposure are limited. These include inhalation of respirable droplets containing infectious agent(s), ingestion of large-droplet infectious material, and percutaneous transmission by sharps or dermal absorption through skin or mucous membranes. In 1989, the National Research Council (NRC) Committee on Hazardous Biological Substances in the Laboratory recommended seven basic prudent biosafety practices designed to minimize the likelihood of exposure to infectious agents via these routes. These seven basic practices, which form the foundation of general laboratory biosafety, are outlined in Table 6. The NRC recommendations were identified as essential biosafety work practices and should be supplemented by additional engineering and administrative controls, SOPs, and safety equipment (including PPE) as needed, depending on the level of risk associated with a particular biohazard. These biosafety work practices should be applied to quality control and proficiency testing samples in the same manner they are applied to diagnostic specimens (19, 20).

Common explanations for LAIs include failure to receive proper training, assuming the organism is no longer viable, working too fast, being disorganized while working, and using defective equipment (4). LAIs may be caused by needlesticks/sharps, touching contaminated hands to mucous membranes (eyes, nose, or mouth), or the generation of uncontained aerosols. Tables 6 and 7 show how continuous biosafety training and adherence to appropriate policies and procedures can effectively minimize the likelihood of an LAI by sharps or contaminated hands (17). For example, laboratory policies and procedural documents such as SOPs should limit the use of needles to operations where no alternative exists. Because needlestick injuries occur most often when needles are resheathed, this practice should be prohibited. Instead, needles and syringes should be placed directly into an appropriate sharps container for decontamination and disposal (4). The 2012 *Morbidity and Mortality Weekly Report* "Guidelines for Safe Work Practices in Clinical Diagnostic Laboratories" provides excellent information about how to reduce the incidence of aerosol generation during routine laboratory procedures (4).

TABLE 6.

Prudent biosafety practices for facilities that handle biohazardous agents[a]

Prudent practices and/or barrier protections used	Exposure route(s) blocked by biosafety practices
Do not mouth pipette.	A, I, C
Manipulate infectious fluids carefully to prevent aerosol production.	A, C
Restrict use of sharps (needles, syringes).	P, A
Use appropriate personnel protective equipment (PPE).	C, A, I
Wash hands frequently.	C, I
Decontaminate work surfaces.	C, I
Do not put contaminated droplets into mouth, eyes, or nose.	C, I
No eating, drinking, storing foods, smoking, or applying cosmetics or contact lenses in the laboratory.	C, I

[a]Includes diagnostic specimens and quality control/proficiency testing samples. Data from reference 19. A, airborne; C, contact (skin, mucous membrane); I, ingestion; P, percutaneous.

Hand Washing

Aerosols are generated from most laboratory tasks, and the contamination is often greatest on the hands of the laboratorian and any surfaces (or individuals) in the immediate work area. Hand contamination also occurs when handling

TABLE 7.

When should laboratory employees wash their hands?[a]

After obvious contamination or anytime gloves are torn, punctured, or otherwise degraded
Immediately after removing gloves, to minimize the risk of contaminating nonworkspace surfaces or equipment in the lab
Before contact with nonintact skin, eyes, or mucous membranes
Before leaving the laboratory
Before and after using lavatory facilities

[a]Data from reference 20.

office equipment such as telephones, computer keyboards, and writing utensils (4). Table 7 lists the situations in which laboratory personnel should wash their hands. Hand washing is the most important way to reduce the duration of exposure to an infectious agent and prevent dissemination of the infectious agent. Laboratory employees should wash their hands by rubbing wet hands vigorously for 20 seconds, "... covering all surfaces of the hands and fingers. Rinse hands with water and dry thoroughly with a disposable towel. Use towel to turn off the faucet" (21, 22).

The intercellular lipids of the stratum corneum create a natural skin barrier (provided skin is intact). Detergents used in hand washing remove the intercellular lipids for the next 6 hours to several days. Hand washing also removes the transient bacterial flora on hands. In the case of hand washing with non-antimicrobial soaps, the skin irritation and dryness created with frequent washing may result in paradoxical increases in the numbers of transient bacteria on the skin. The use of hand lotion or cream containing humectants, fats, and oils at least twice a day has been proven to replace depleted lipids and thus help restore the normal barrier function of the skin (4).

Because hand washing with soap and water can result in an increased number of transient bacterial on the skin, alcohol-based hand-rub products (rinses, gels, or foams) may be an acceptable alternative when hands are not visibly soiled. CDC states that these alcohol-based hand-rub products are "... more effective for standard hand washing or hand antisepsis by [health care workers] than soap or antimicrobial soaps" (21). However, recent studies have demonstrated reduced efficacy of alcohol-based hand-rub products with repeated use throughout the day, and Triclosan, a bacteriostatic component of many antimicrobial soaps, had been found to weaken cardiac and skeletal muscle contractility (23). Therefore, traditional hand washing with an antiseptic agent or non-antimicrobial soap should be performed periodically throughout the day, with alcohol-based hand-rub products used between soap and water hand washing (24).

When using an alcohol-based product, hands should be rubbed together until all of the alcohol has evaporated. If hands are dry after 10 to 15 seconds of rubbing, an insufficient amount of the product was used; 3 ml or more is usually needed to be effective. Hand-rub products with 60% to 95% alcohol have demonstrated excellent *in vitro* germicidal activity against vegetative Gram-positive and Gram-negative bacteria (including methicillin-resistant *Staphylococcus aureus* [MRSA] and vancomycin-resistant enterococcus) and *Mycobacterium tuberculosis*, as well as several fungi and enveloped viruses. Alcohol-based products may not be effective against nonenveloped (nonlipophilic) viruses (21, 24). Because none of the agents (alcohols, chlorhexidine, hexachlorophene, iodophors, etc.) used in antiseptic hand wash or antiseptic hand-rub preparations are reliably sporicidal against *Clostridium* spp. or *Bacillus* spp., it is important to wash hands with soap and water to remove these organisms (21).

Personal Protective Equipment

The Occupational Safety and Health Administration (OSHA) defines PPE as appropriate for the task at hand if the PPE "does not permit blood or other potentially infectious materials to pass through or reach the employee's street clothes, undergarments, skin, eyes, mouth or other mucous membranes under normal conditions of use" (4).

For routine work in veterinary diagnostic and clinical laboratories, PPE appropriate for the task (generally BSL2 practices and procedures) must be provided, used, and maintained in the laboratory workspace. Employers are encouraged (or in some states required) by OSHA to ensure that employees know when to use PPE and which PPE item(s) are appropriate for a given hazard. Employees must also be trained on how to (i) properly don (i.e., put on) and doff (i.e., take off) PPE; (ii) properly care for, use, and dispose of PPE; and (iii) know the limitations of the PPE, including the useful life span for a specific item (4). PPE use in veterinary diagnostic laboratory work should be sufficient for the level of biosafety practices necessary for the identified risk and at a minimum should include gloves, protective clothing, and protective shoes or shoe covers. Use of eye and face protection is also encouraged, whenever splattering of blood or body fluids can be reasonably expected, as a mechanism to prevent mucous membrane contamination (see Table 8).

Disposable, medical-grade, latex/vinyl/nitrile gloves protect the wearer from exposure to potentially infectious material. These gloves should be changed frequently. Disposable medical-grade gloves should not be washed and reused; washing degrades the protective function of disposal gloves. Nondisposable utility/household cleaning gloves, when used for housekeeping chores, may be disinfected and reused. Anytime cracks, tears, or discoloration are present, the gloves should be discarded, regardless of their intended use (20). Laboratory personnel should practice aseptic techniques when donning and doffing gloves and be diligent about not contaminating "clean areas" of their workstation and/or the laboratory. Personnel should wear gloves when handling or manipulating samples and during waste disposal procedures (20).

The laboratory biosafety officer or a member of the safety team should require that glove manufacturers provide documentation that the gloves meet the standards necessary for the anticipated hazards, including information about how long the glove can be worn prior to product degradation (4). Depending on the chemicals used,

TABLE 8.

Personal protective equipment recommended for use during necropsy performed outside a biosafety cabinet[a]

BSL2-level necropsies
Scrub suit covered with a long-sleeved gown or a long-sleeved coverall suit
Fluid-impervious apron (if gown or coverall is not fluidproof)
Fluid-impervious shoe covers.
Fluid impervious gloves (preferably double gloves)
Facial protection (when there is a risk of generating fine aerosols): face shield, OR goggles[b] AND a fluid-resistant face mask
BSL3-level necropsies (requires active respiratory protection program)
Scrub suit covered with a long-sleeved gown or a long-sleeved coverall suit
Fluid-impervious apron (if gown or coverall is not fluidproof)
Fluid-impervious shoe covers
Double gloves (preferably including one pair of heavy-duty gloves)[c]
Respirator (N95 face mask AND goggles that form a protective seal against wearer's face; OR powered air purifying respirator (PAPR)

[a]Data from references 28 and 32.
[b]If goggles are worn, the user must also wear a fluid-resistant face mask (to reduce the risk of droplet contamination to mucous membranes).
[c]The outer glove should cover the cuff of the sleeve. Tape may be necessary to hold the outer glove in place over the cuff of the sleeve. Stainless steel mesh glove should be worn on the noncutting hand when cutting blindly, or whenever saws, chisels, or bone cutters are used.

personnel in different sections of the laboratory may need to change gloves at different intervals to prevent product degradation.

Gauntlet/surgical-length gloves provide better protection against exposure to skin between the user's hand and the cuff of the sleeve than wrist/exam-length gloves. During autopsy, approximately 8% of gloves are punctured, and one out of three of these punctures (biosafety failures) is not detected until after doffing the gloves (4). Double gloving should be considered whenever gross contamination of blood or body substances is anticipated, in part because less skin contamination is observed when using a double-glove technique (20). If using a double-glove technique, use two different colors of gloves so that one can more easily see if a puncture/leak has occurred.

Protective clothing appropriate to the task should be worn at all times inside the laboratory. Open-toed shoes should not be worn in the laboratory to prevent accidental spillage on bare skin. Protective clothing should include fully closable long-sleeved coats or gowns that extend below the level of the workbench. Front-opening laboratory coats must be closed when working in the laboratory. Solid-front gowns with snug cuffs offer better protection against sprays and splatter than front-closing coats; in some situations a solid-front lab coat may be required (e.g., during manipulation of Select Agents) (4). If there is a potential for splashing or spraying, solid-front, fluid-resistant gowns or laboratory coats should be worn. If there is a potential for soaking the clothing, fluid-proof material such as a plastic apron or plastic-lined surgeon's gown should be worn. If visibly contaminated with blood or body substances, the protective clothing should be changed immediately to prevent blood from seeping through the fabric and onto the inner garments or skin. Protective clothing should not be worn outside the laboratory or taken home for cleaning or laundering.

To protect the mucous membranes of the eyes, nose, and mouth, a face shield should be worn whenever splattering or splashing of blood or bodily fluids can be reasonably expected during laboratory procedures or cleanup. If a face shield is not available, a fluid-resistant facemask and goggles (or splash guard) should be used.

Respiratory devices should be used as part of biosafety practices to prevent inhalation of potentially infectious aerosols. The decision to use respiratory protection may come from a risk assessment that identifies the inherent risk in a particular clinical specimen (e.g., suspicion that RG3 infectious agents are present) or from laboratory manipulations that generate aerosols and may necessitate the use of respiratory protection, such as a properly fitted N95 respirator or a PAPR. Clinical and diagnostic laboratories should establish a written respiratory protection program, particularly as it pertains to aerosolized infectious agents. In the United States, respiratory protection standards are published in the *Federal Register* (OSHA 29 C.F.R. 1910.134) and administered by OSHA. The basic elements would include (i) fit testing procedures for tight-fitting respirators; (ii) medical evaluations of employees required to use respirators, (iii) procedures for proper use of respirators (e.g., N95), (iv) procedures for maintaining respirators (e.g., cleaning, disinfection, storage of PAPR), (v) training on use of respirators (annual or biannual retraining recommended), and (vi) recordkeeping, particularly of fit testing and medical evaluations (25). The CDC National Personal Protective Technology Laboratory (NPPTL)–National Institute of Occupational Safety and Health (NIOSH) has an online key resource for respirator users (http://www.cdc.gov/niosh/npptl/respusers.html).

The N95 disposable particulate respirator is the most commonly used respirator in the diagnostic laboratory setting. This respirator, when properly fitted, is certified by the NIOSH to protect the wearer against 95% of an aerosolized test substance, typically within <1 to >100 μm in diameter, and over a range of approximately 10 to 100 liters of air/minute (airflow). When worn incorrectly (e.g., with facial hair/beard) or improperly fitted, the majority of the particles that enter the mask do so through the inadequate face seal. An improper fit can be due to recent weight loss/gain, the development of skin wrinkles during aging, scars, or simply the wrong style of respirator for a person's facial structure. For this reason, Grinshpun et al. recommend that improving the fit of a respirator is more important to the wearer (from a biosafety perspective) than using the most efficient filter. In other words, the N95 respirator that fits correctly provides better protection to the wearer than a poorly fitting N100 (26). Employers should provide employees with multiple styles of N95 respirator because a single style will not fit all faces the same.

Biological Safety Cabinets

Every diagnostic microbiology or pathology laboratory should have BSCs to serve as a primary means of containment for working safely with infectious organisms. There are three basic types of BSCs that are designated as Class I, Class II, and Class III, and their functions and components are thoroughly reviewed in a recent CDC biosafety guideline document for clinical diagnostic laboratories (4). The Class II-A1 or Class II-A2 BSC is best suited and recommended for the clinical diagnostic laboratory. Trained professionals must certify all BSCs at least annually and any time the unit is moved. Moving the cabinet can damage the filter at the glue joint and result in dangerous leaks, so filter integrity must be tested after each move.

Class II BSCs are designed to be fully contained work areas that prevent laboratorian exposure to infectious aerosols and protect materials inside the work area of the cabinet from environmental contamination. BSCs are based on the principles of laminar airflow, relative negative air pressure, and high-efficiency particulate air (HEPA) filtration of both the supply and exhaust air. Laminar airflow employs sheets of moving air to carry particulate matter away from the user and into an area of decontamination. Negative air pressure ensures that air will be drawn into the cabinet, thus precluding the outflow of infectious agents. HEPA filtration is used to remove environmental particles from the airflow in the work area of the cabinet and is the usual method of containment for infectious particles generated in the cabinet.

Proper loading of the BSC and proper access by the laboratorian is described in the BMBL, and some basic rules of BSC use are listed in Table 9. For example, when using a 6-foot-wide BSC, two employees may safely work side by side only after a careful risk assessment. Such a

TABLE 9.

Working safely inside a biosafety cabinet[a]

The DOs	The DON'Ts
Do install biosafety cabinet (BSC) away from walking traffic, doors, and fans.	Do not turn off BSC during periods of nonuse if this will affect room air balance.
Do allow the BSC to warm up for at least 4 minutes after turning on. *Allows BSC to purge suspended particles.*	Do not begin manipulations during the first minute after putting arms inside a BSC. *Allows air curtain to be reestablished.*
Do disinfect the BSC before and after use. *Disinfect while the unit is running.*	
Do limit the supplies and equipment inside the BSC.	Do not stack items inside the BSC or use as storage space.
Do locate supplies toward the rear edge of the work surface as practical and away from the front grille.	Do not work with large volumes of volatile chemicals without a careful risk assessment of staff safety and fire hazards. HEPA filters provide protection from particles, not volatile chemicals, and biosafety cabinet motors are not explosion proof.
Do use absorbent pads that can be incinerated after use or autoclaved prior to disposal.	Do not block the grates at the front or back of the BSC with absorbent material, papers, discarded packaging, equipment, or the user's arms.
Do move in and out of the BSC slowly, perpendicularly to the window opening.	Do not lean on the front of the BSC with arms or body.
Do work at least 4 inches "inside" front grate/opening.	Do not make rapid movements or sweeping motions with your arms.
Do remove outer gloves before pulling hands out of the BSC.	Do not bring items out of the BSC until they have been disinfected.
Do work from clean to dirty areas inside BSC.	Do not use an open flame in the biosafety cabinet. The convection currents from heat disrupt airflow.
Do open sealed rotors or safety cups on high-speed and ultracentrifuges inside a BSC.	
Do have the biosafety cabinet recertified annually. Adjustments may be required to maintain personnel safety and protect materials from environmental contamination.	
Do decontaminate the BSC using formaldehyde gas, hydrogen peroxide vapor, or chlorine dioxide gas prior to maintenance.	

[a]Based upon MMWR "Guidelines for Safe Work Practices in Human and Animal Clinical Diagnostic Laboratories" (4) and *Biosafety in Microbiological and Biomedical Laboratories*, 5th edition (3).

risk assessment would evaluate whether or not cross-contamination affecting the results of work performed is prevented (e.g., laboratorians working on same activity) and penetration of the "air curtain" is minimized to ensure the protection provided by the biosafety downflow cubic feet per minute (cfm) is not exceeded. When using a 4-foot-wide BSC, only one employee should be working at the BSC (4). Other employees should avoid working behind or in the area of the BSC because this can interfere with the creation of a uniform air curtain inside the BSC. For more information, the reader is referred to *Laboratory Biosafety Guidelines*, which has an excellent chapter on the various classes of BSCs and details about the direction(s) of airflow within the BSC (27).

Staff Training

Essential to creating a safe work environment is regular, ongoing biosafety training for laboratory employees. Initial training should take place before a new individual enters the laboratory, regardless of the position they are hired for (i.e., laboratorian, graduate student, visiting scientist, maintenance personnel, etc.). Training should also take place prior to an employee rotating into a new section or performing new tasks. All training must occur during regular working hours and be provided free of charge. The size of the biosafety training program will vary with the needs of the individual laboratory sections and the overall laboratory safety program (28). The items listed in Table 10 may be appropriate for inclusion in the laboratory training program.

Biosafety training should be provided at least annually and should address new information on pathogens (if available) and also why and how new biosafety equipment or procedures are being implemented (if applicable). Annual biosafety training should provide a refresher of existing biosafety procedures and policies within the laboratory. All training should be appropriate to the employees' educational background, literacy, and language (20). Documentation of the training (date and content of training) should be placed in the employee's training record and maintained for a minimum of 3 years (28). Biosafety training for laboratorians should include the use of BSCs.

The effectiveness of laboratory safety training should be evaluated periodically. Safety assessments could include safety audits, inspections or audits by outside agencies, review of accident or incident reports, and observations and suggestions made by employees (28). Suggested information to include in a safety audit is listed in Table 11.

TABLE 10.

Topics to be included in the biosafety training program[a]

Basic understanding of standard precautions
Selection, use, and limitations of personal protective equipment
Management of biohazardous waste
Postexposure accident reporting and incident investigation
Blood-borne pathogen information
Basic understanding of risk groups/risk assessment
How to perform a risk assessment
Procedures for biohazardous spill cleanup
Respiratory protection training (required for employees who use a respirator)

[a]Data from references 3 and 46.

TABLE 11.

Information to include in an audit of the laboratory biosafety program[a]

The use and effectiveness of the training program
The competence of the trainer
The adequacy of the facilities and equipment
The adequacy of the documentation and record-keeping system(s) used for safety-related activities
Adherence to and understanding of standard operating procedures (SOPs) and safety policies
Adherence to the visitor policy requirements

[a]Data from reference 28.

Biological Spill Management

The management of biological spills in clinical laboratories must account for the specific infectious agent(s) (if known), the volume of infectious material spilled, any chemicals spilled, and the presence of potential aerosols. Aerosols may be created during and/or following the spill and during the cleanup process. For accidents involving RG3 agents, occupant(s) should immediately stop breathing and evacuate the area. Do not inhale in preparation for holding your breath. As the occupant(s) leave the area, close all doors. Do not reenter the area for 30 to 60 minutes. For a typical BSL2 or BSL3 laboratory room, with 10 or more air changes/hour, 99% of the aerosols will be removed in 28 minutes or less (28). Similarly, when breakage occurs in a centrifuge (which inherently would produce aerosols), the equipment (sealed rotor or centrifuge bucket with sealed safety covers) should be placed inside a BSC and remain closed for 30 minutes before decontamination of individual canisters commences.

PPE for biological spills should include puncture-resistant gloves, fluid-impenetrable shoe covers, coats, or gowns, and facial protection. Selection of PPE should be based on the typical route(s) of exposure for the agent spilled, if known. The laboratory employees also may wish to look up the Material Safety Data Sheet (MSDS) information for details on the appropriate disinfection method and contact time prior to beginning the cleanup procedure. For RG3 agents, a PAPR or HEPA-filtered

TABLE 12.

Typical biological spill cleanup procedure involving possible aerosol agent[a]

Stop breathing (do not inhale in preparation to holding breath).
Alert personnel in area and evacuate.
Close doors and do not reenter area for 30–60 minutes. Post sign forbidding entry to the area.
Alert laboratory supervisor (who in turn will notify laboratory director, biosafety officer, and others as required by institutional policies).
Don personal protective equipment (PPE) appropriate for type of spill; e.g., gown, gloves, eye/face protection, and respirator.
Check to make sure that the HEPA filters on powered air-purifying respirators (PAPRs) are designed for both biological and chemical filtration in case the spill contains both components.
Slowly cover spill with towels soaked in disinfectant. Allow contact time as indicated for the disinfectant.
Remove and discard broken glass or other objects. Do not touch objects directly with hands; use broom or forceps. Discard in sharps container.
Discard contaminated material in a biohazardous waste container.
Decontaminate area again with appropriate disinfectant (see Material Data Safety Sheet [MSDS]). Leave disinfectant on area for recommended contact time.
Rinse spill site with water and allow site to dry.
Copy contaminated laboratory forms and discard into the biohazard waste container.
Place all disposable contaminated cleanup material in the biohazard bag and treat as infectious waste.
Remove gloves and wash hands.
Prepare a spill/incident report, identify cause of spill, and determine remedial action.
Replace/restock spill kit.

[a]Data from reference 28.

respirator (i.e., N95 respirator) and goggles must be used. (Note: If a spill may contain both biological and chemical components, a PAPR must be used.) A typical biological spill clean-up procedure involving a possible RG3/aerosol agent is listed in Table 12.

For biological spills outside a BSC, three people should work as a team when performing cleanup. There are three containment zones: the central hot zone; the middle decon zone; and the outer clean zone (see Fig. 2). The employee cleaning up the spill is considered to be in the hot zone and as such is required to wear the appropriate PPE. This hot zone employee first removes broken glass, equipment, or other solid material using a broom or forceps from the spill kit. The hot zone employee then covers the spill with paper towels and pours disinfectant slowly on the paper towels, beginning at the outer edges and moving inward toward the center of the spill; e.g., pour "low and slow." The disinfectant must make contact with all material involved in the spill and must remain in place for the appropriate contact time.

A second employee (also in appropriate PPE) operates in the decon zone. The decon zone employee provides support for the individual in the hot zone but does not participate in the actual cleanup. The decon zone (second) employee may be tasked with relaying information and/or with asking for additional supplies or equipment from a third employee who operates only in the clean zone and does not wear PPE. This clean zone employee acts as an errand/supply runner when necessary and is tasked with documenting the procedures used to clean up the spill. After the spill has been successfully cleaned up, the hot-zone employee moves to the decon zone. Both the hot- and decon-zone employees decontaminate themselves and doff and discard their PPE in the decon zone before moving into the clean zone. Last, the involved individuals should complete an accident/incident report and restock the laboratory spill kit. Because PPE (and some disinfectants) can degrade over time, the contents of spill kits contents should be inspected annually and items should be replaced as needed.

If a spill occurs inside a BSC, do not turn off the cabinet fan. Minor spills in a BSC can be absorbed with the absorbent paper already in place in the BSC. If the adsorbent pad is insufficient to contain the spill, cover the spill with paper towels. Slowly pour the appropriate disinfectant around the outer edges of the spill and move toward the center; again, pour "low and slow." (To minimize splashing and/or aerosolizing the spill, avoid pouring the disinfectant directly onto the spill.) Thoroughly soak the towels with disinfectant. Spray the cabinet walls, work surfaces, and inside the BSC sash with disinfectant. All items in the cabinet should be sprayed or wiped with disinfectant. If infectious material flows into the grill, the drain valve should be closed and disinfectant poured onto the surface and through the grill into the drain pan. Allow 30 minutes of contact time with the disinfectant. Disassemble the grill and top work surface of the BSC (with assistance from another person) and clean both sides of the work surface. Remove the top work surface from the BSC and use a Swiffer to clean the plenum under the work surface. Properly dispose of waste. If PPE is contaminated, spray or wipe with disinfectant for the appropriate contact time. Remove PPE (after the appropriate amount of contact time with the disinfectant) and wash hands. Allow the BSC to run for at least 10 minutes before resuming activities.

If a specimen tube breaks within the plastic screw-capped canister in a centrifuge, follow the procedure outlined in Table 13.

Immunizations

According to the Healthcare Personnel Vaccination Recommendation, laboratory personnel working with human

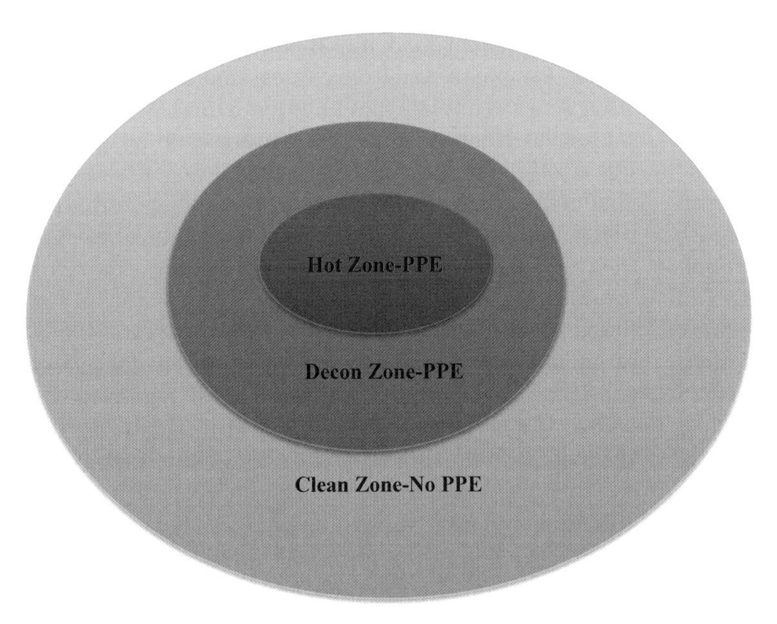

Figure 2: Zones of containment for a biological spill using a three-person team. Each individual stays in their assigned zone until cleanup is complete. The two individuals in the hot and decon zones wear personal protective equipment (PPE). The individual in the clean zone does not wear PPE. Once cleanup is complete, the individual in the hot zone moves to the decon zone and both individuals perform decontamination followed by removal of PPE.

pathogens should receive vaccinations for hepatitis B, influenza, measles, mumps, rubella, varicella (chicken pox), meningococcus, tetanus, diphtheria, and pertussis (4). None of these human pathogens is present routinely in animal diagnostic specimens, with the exception of nonhuman primates. Laboratorians in veterinary diagnostic laboratories should consider immunization against rabies virus, especially if personnel are routinely processing central nervous system tissues or cerebrospinal fluid from animals with neurological diseases compatible with rabies. The CDC publishes rabies preexposure prophylaxis guides (4, 29). Laboratory directors should also consider providing animal handlers with immunizations for tetanus, and, if working with nonhuman primates, measles, mumps, and rubella. Because of the frequent zoonotic nature of both avian and swine strains of influenza virus, employees should be offered annual flu immunizations.

TABLE 13.

Procedure for cleanup when a specimen tube breaks within the plastic screw-capped canister in a centrifuge[a]

Turn the motor off.
Remove the canister immediately and place under a biological safety cabinet (BSC).
Notify senior person in charge and other colleagues working in the area.
While wearing protective clothing, open the canister under the BSC.
Pour a 1:10 dilution of bleach or a noncorrosive disinfectant into the canister to decontaminate all surfaces; let the canister soak in bleach or disinfectant solution for 10 minutes or amount of time appropriate for the disinfectant.
Remove broken glass from canister. Do not pick up broken glass with gloved hands. Use forceps or cotton held in forceps or tongs or hemostats and dispose into a biosafety sharps container.
Discard all nonsharp contaminated materials from canister into a biohazard bag for biohazard waste disposal.
Unbroken capped tubes should be swabbed with the same disinfectant, then swabbed again, washed with water, and dried.
Clean canister thoroughly.
All materials used during the cleanup must be treated as infectious waste and disinfected (in the case of nondisposable materials) or disposed of as biohazardous waste (for disposable materials).

[a]Note: If the specimen tube breaks in a centrifuge that does not have individual canisters, but does have a biohazard cover and sealed rotor, follow the manufacturer's instructions for cleaning and decontamination. Data from reference 4.

The 5th edition of the BMBL states: "If the potential consequences of infections are substantial and the protective benefit from immunization is proven, acceptance of such immunization may be a condition for employment. Current, applicable vaccine information statements must be provided whenever a vaccine is administered. Each worker's immunization history should be evaluated for completeness and currency at the time of employment and re-evaluated when the individual is assigned job responsibilities with a new biohazard" (3). If an employee elects not to receive an offered nonmandatory vaccine, documentation to that effect (e.g., signed statement) should be kept with the employee's personnel file. A copy of the signed documentation should be given to the employee. Employees should be made aware that they can change their mind and receive an offered vaccine at any time in the future. Laboratory directors should consider reviewing the vaccination policy with an employee who has previously refused an offered vaccine any time the individual is assigned new and/or different job responsibilities, regardless of whether there is a change in the biohazard(s) involved.

Serum banking is the practice of collecting and storing frozen serum obtained from laboratorians who may be at risk of developing an LAI. The goal of serum banking is to compare serum obtained during a recent exposure to "banked" serum (e.g., serum obtained before the exposure). Comparing pre- and postexposure titers may aid in determining the cause of an LAI. Serum banking should be conducted only when there is a clear reason for obtaining specimens and a plan to analyze data as part of a risk assessment strategy. In general, serum banking is recommended only in research settings when (i) the risk of exposure is known, (ii) possible exposure involves a finite number of infectious agents, and (iii) serological tests to those specific infectious agents are available. Serum banking is not generally recommended for employees in veterinary diagnostic laboratories, where exposure risk from clinical specimens is to unknown agents for which validated human serological tests may or may not be available.

The decision process to begin a serum banking program in a veterinary diagnostic laboratory should include (i) assessment of the risk of the biohazardous material or biologic agent exposure; (ii) evaluation of the potential usefulness of diagnostic tests that could be performed on banked serum; and (iii) consideration of the privacy risks and benefits. If banked serum is available for review of a potential LAI, serum samples from the current exposure should be collected only after the host (in this case the laboratory employee) has had time to develop an antibody response to the pathogen(s) involved in the potential LAI. Alternatively, a "baseline" serum sample could be obtained at the time of exposure in the event the exposure leads to disease. Serological markers for disease, if available, could develop "anew" after a primary exposure, or rise to a higher concentration after a secondary exposure depending upon the immunological "history" of the individual.

NECROPSY AND SURGICAL PATHOLOGY

Facility requirements for performing necropsies in BSL1–BSL4 laboratories are covered in depth elsewhere and will not be covered in this chapter (3). The following information refers to the recommended policies and procedures for BSL2 or BSL3 veterinary diagnostic laboratories.

In the typical veterinary diagnostic laboratory, specimens collected in the field or at distant locations are shipped overnight to the diagnostic laboratory following appropriate procedures for packaging and shipping potential infectious substances. The American Society for Microbiology and WHO have published comprehensive guidelines for packing and shipping infectious substances (30, 31). They are based on guidelines issued by the International Air Transport Association (IATA) and the U.S. Department of Transportation (DOT). Miller et al. provide an excellent guideline for packaging and shipping clinical samples to and from clinical laboratories abridged from the IATA, WHO, and ASM guidelines (4). It is helpful for veterinary diagnostic laboratories to provide online packaging and shipping instructions for laboratory clients to ensure handling and receipt of diagnostic samples following appropriate government regulations. Submission forms and other documents submitted with the samples may arrive at the laboratory grossly contaminated with blood, bodily fluids, and/or excreta. If contaminated, submission forms and/or other documents should be placed inside a biohazard bag. If possible, these contaminated documents should be reproduced (digitally scanned or photocopied) and the original contaminated document(s) discarded as biohazardous waste.

Necropsy Loading Dock

One area of the veterinary diagnostic laboratory that differs markedly from that of laboratories dealing with human specimens is the loading dock area for receipt of animal cadavers or live animals for necropsy (particularly agricultural animals originating directly from owner premises). A well-demarcated, preferably physical separation (such as an outside wall/doorway/walkway) should be provided to separate the necropsy area from the unloading area or dock. Individuals not employed by the laboratory should be cautioned to remain in the "public" section of the unloading area and should not be allowed to enter beyond this separation area. If physical separation of the area is not possible, then adequate signage should be provided to ensure that all parties recognize where the public area begins and ends. Diagnostic laboratory employees should be required to disinfect their footwear and any visibly soiled clothing or PPE and wash their hands each time they move between the laboratory and public areas of the unloading dock. To facilitate this transition, foot traffic between the laboratory and public areas should require laboratory employee(s) to walk through a foot bath or other disinfection system prior to walking to or from the public area. The foot bath should be cleaned regularly, and the disinfectant used must be effective against the microorganisms commonly seen at the diagnostic laboratory. The disinfectant should be changed in accordance with the active "shelf-life" of the product (32).

Sources of Injury on the Necropsy Floor

Whether working with a naturally or experimentally infected animal (or an uninfected animal), there are inherent risks associated with necropsy procedures. Most laboratories also have live animal holding and restraint areas to accommodate humane euthanasia prior to necropsy (in particular agricultural animals). Accidental self-inoculations can occur when working with live animals during these procedures. Animal handlers should be knowledgeable about the specific behaviors and traits of the species being handled. Animals can bite, kick, scratch, or crush their human handlers. Even a cadaver, if large, can cause crush injuries, as can restraining equipment such as squeeze chutes. Careful animal handling by employees, trained in the proper method of restraint for the species involved, will help to minimize injuries to both the employee(s) and the animal (32).

Injuries can also be caused by electrical equipment, chemicals (i.e., fixatives and disinfectants), instruments used in cutting or sawing, the sharp ends of cracked bones, electrical equipment, and slippery floors. Accidents and injuries can be minimized by allowing only properly trained employees access to the necropsy floor. These individuals must be trained on the use of all equipment (both hand-held and stationary items such as the necropsy table or the chain hoist). Individuals working on the necropsy floor must also know how to clean and disinfect the necropsy room properly. Protocols for entry and exit of the necropsy area, protocols for movement of specimens and carcasses, protocols for maintenance and use of PPE, specific assignments for each individual, and protocols for emergency situations should be incorporated into the laboratory's policies and procedures (32).

When working with certain species (e.g., live swine), laboratory personnel should consider wearing hearing protection. NIOSH recommends employers require hearing protection if an employee is exposed to an average daily noise level >85 dB (33).

Dermal allergies and, less frequently, respiratory allergies are common and occur in as many as one out of five laboratory personnel that routinely have contact with rodent saliva, fur, and bedding or wastes (34). Laboratorians could conceivably develop a similar allergy risk with long-term routine exposure to sentinel laboratory rodents submitted for a diagnostic necropsy. Thus a respirator is recommended anytime a rodent necropsy cannot be performed inside a BSC (32).

Procedures Performed Inside a BSC

If, on the basis of the risk assessment, the procedures such as sample processing, tissue grossing, or necropsy are to be performed in a BSC, the specimen should be safely transported in the original container (if applicable) to the BSC. After the specimen has been placed in the BSC, sufficient time should elapse to allow the air curtain to be reestablished; recommendations range from 4 to 10 minutes (4, 27).

It is important to follow the correct procedures for working inside a BSC (see Table 9). When using a 6-foot-wide BSC, two employees working side by side should occur only after a careful risk assessment. Such a risk assessment would evaluate whether or not cross-contamination affecting the results of work performed is prevented (e.g., laboratorians working on same activity), and penetration of the "air curtain" is minimized to ensure the protection provided by the biosafety downflow cfm is not exceeded. When using a 4-foot-wide BSC, only one employee should be working at the BSC (4). Other employees should avoid working behind or in the area of the BSC because this can interfere with the creation of a uniform air curtain inside the BSC. For more information, the reader is referred to *Laboratory Biosafety Guidelines*, which has an excellent chapter on the various classes of BSCs and details about the direction(s) of airflow within the BSC (27).

Routine disinfection of the BSC should be performed before and after procedures are completed inside the BSC. Alcohols should not be used because they have minimal effect on nonenveloped viruses and because the alcohol evaporates too quickly (4). Dilute bleach (1:100) or iodophors (0.5% concentration) can be used to disinfect the work surface, interior walls, and inside of the window/sash. Residual chlorine can corrode stainless steel; if dilute bleach is used, the chlorine must be removed following disinfection by wiping with sterile water followed by wiping with 70% ethanol. Iodophor dilutions are not corrosive but may leave a brown film on surfaces. This can be removed by wiping with sterile water after disinfection. In addition to daily cleaning, the floor plate and front grate should be removed monthly or bimonthly for routine cleaning. The plenum below the floor plate should also be disinfected.

Necropsy Floor PPE

Most necropsies performed in veterinary diagnostic laboratories can be done under BSL2 conditions. The prosector(s) (i.e., diagnosticians, technicians, students, or others assisting in the necropsy) and observers should wear the appropriate PPE listed in Table 8. If risk assessment determines that a necropsy requires BSL3 conditions and cannot be performed inside a BSC (e.g., the carcass is too large), the prosector(s) must use a respirator (i.e., N95 face mask and goggles that form a protective seal against the wearer's face or a PAPR) in addition to wearing the PPE described for BSL2 conditions. BSL3-level practices are used in a BSL2 facility whenever there is a risk of aerosolization occurring before, during, and/or after the necropsy (i.e., during cleanup). If the BSL2 necropsy suite consists of one space with a single air heating, ventilation, and air conditioning (HVAC) system, it is important to remember that if multiple necropsies are being performed in that necropsy space, then the PPE level to be used by all individuals in the room should be protective against the suspect/confirmed organism(s) with the greatest biorisk. For example, if one case involves a suspected RG2 organism and another case involves a suspected RG3 organism, then all persons in that space should be wearing PPE for the RG3 organism throughout the necropsy and during cleanup. Although studies of aerobiology and transmission of infectious agents indicate that aerosol transmission of infectious agents is multifaceted and "case-by-case" dependent (dependent upon droplet size, droplet velocity, environmental conditions, HVAC systems, pathogen "environmental" resistance, and others), studies of LAI indicated aerosol transmission of infectious agents is a significant source of LAI in clinical laboratories (4). In 2007, Weiss et al. pointed out that the purported "safe distance" for droplet spray of ≥3 feet was carried down incorrectly through the literature from a quote in 1910 and that there was no scientific evidence to support the "3-foot" claim (35). Rather, literature reports from as early as 1900 found that people talking could spray viable bacteria up to 20 feet and that coughing or sneezing could spray bacteria up to 40 feet or more. Thus, it is prudent to take appropriate best practice precautions against aerosol transmission of infectious agents in the necropsy laboratory settings.

Potentially infectious aerosols can be generated from carcasses or their tissues and fluids. By definition, aerosolization is "... the generation of liquid droplets or particles, 5 microns [μm] in diameter or less, that can be inhaled and retained in the lungs" (4). Typically, bacterial cells and spores range from 0.3 to 10 μm, fungal spores range from 2.0 to 5.0 μm, and viruses range from 0.02 to 0.30 μm. Aerosolization of infectious material in the necropsy and surgical pathology suites can occur during spraying of the carcass and/or adjacent bloody floor, sawing, aspirating fluids, scraping dried blood or body fluids off of surfaces during cleanup, and by using high-pressure hoses during the wash-down phase of cleanup (4).

Performing a Necropsy Outside a BSC

If an employee is present during the necropsy to act as a circulator (e.g,. a "clean person"), this individual should wear protective clothing, including a gown or lab coat, respiratory PPE if appropriate, waterproof and slip-resistant footwear, and (in some laboratory situations) gloves as a part of the PPE required for all individuals involved in the necropsy (Table 8). The circulator assists the prosector(s) with the collection of specimens and cultures, photography, making telephone calls, recording written communications during the necropsy, ordering tests, and performing other tasks with the goal of minimizing contamination of equipment not on the necropsy table or in the hands of the prosector(s). If the protective clothing or hands of the circulator become soiled, they should immediately be changed or washed, respectively, prior to donning new protective clothing.

Wetting the feathers or fur of birds and small laboratory animals, respectively, is recommended even if the necropsy will be performed inside a BSC, because feathers and fur tend to detach from the skin easily and become a ready source of aerosolized, infectious material (32). Bone surfaces should be wetted with water prior to cutting so that bone dust dispersal is minimized (28).

Dissection should be limited to one individual (prosector) using a knife at the time, no matter how large the carcass. Additional individuals may assist in lifting or holding specific parts of the body (i.e., ribcage, limb, or pluck), but should not be cutting with a knife. The prosector should avoiding cutting blindly (i.e., by feel) and should never cut blindly if another person is assisting or holding part of the carcass. Stainless steel mesh gloves should be worn on the noncutting hand during periods of "blind" tissue/organ removal. A latex glove may cover the mesh glove to provide protection from fluids and slip resistance (28).

If specimens are collected by needle and syringe (i.e., cardiac puncture to collect blood) during the necropsy, do not leave the needle (and attached syringe) on the necropsy table; discard it immediately after use in an appropriate sharps container. Collection tubes and vials should never be held in the hand of the prosector or the circulator while the specimen is being introduced into the tube/vial. (Collection tubes and vials should be placed in a tube rack or secondary container.) As soon as the collection is made and the tube or vial is sealed, it should be disinfected and removed from the necropsy table. All other contaminated nonsharps devices, such as fixative containers, Whirl-Pak bags, etc., should remain on the necropsy table along with the tissues/organs that are not being relocated for further rendering or incineration. When the necropsy is finished, these containers and/or bags should be disinfected prior to removing them from the necropsy table.

All unfixed tissues (i.e., tissues that have not undergone formalin fixation), specimen tubes or vials, swabs, and similar items should be handled as contaminated material that may contain one or more infectious agents. All unfixed, retained material should be properly identified as biohazardous, and the remaining portions of the carcass and the outside of containers used during necropsy should be considered infectious (32). Incineration, digestion, rendering, tissue autoclaving, radiation, or other disposal method(s) used for this biohazardous waste should be done in accordance with local and state regulations. The technology used must be proven to be effective for the type of pathogens that may reasonably be anticipated to be present in the specimen(s) (32).

Carcass Disposal

Following necropsy, the carcass should be disposed of in accordance with laboratory policies, the risk assessment, and local legal requirements. Common methods of disposal include rendering, incineration, alkaline hydrolysis, autoclaving, composting, and disposal at approved municipal landfills. Laboratories must become familiar with and adhere to federal, state, and local requirements for disposal of particular animal species (e.g., no small ruminant carcasses or tissues in rendering containers).

Carcasses containing radioactive isotopes or antineoplastic drugs require special procedures during necropsy and subsequent disposal and will not be discussed in this chapter. For more detailed information about waste regulation and disposing of radioactive waste see the National Academy of Science's *Prudent Practices in the Laboratory: Handling and Management of Chemical Hazards, Updated Version (2011)* (http://www.nap.edu/catalog/12654/prudent-practices-in-the-laboratory-handling-and-management-of-chemical). Note that the Environmental Protection Agency (EPA) places the burden on the diagnostic laboratory to determine if a waste is hazardous. If the waste is hazardous, the diagnostic laboratory must also determine what waste classification is most appropriate. Both the DOT and EPA regulations for transporting and disposing of hazardous waste, respectively, must be followed. Veterinary diagnostic laboratories are encouraged to contact their Environmental Health Services staff (or chemical hygiene officer) before disposing of biomedical waste.

Necropsy Cleanup

Following necropsy or sectioning of surgical pathology specimens (or whenever there is a spill), all instruments, equipment, and surfaces used should be cleaned and disinfected, using a proven disinfectant that meets EPA guidelines. More information about EPA-registered disinfectants is available at http://www.epa.gov/oppad001/chemregindex.htm. Necropsy and/or surgical pathology SOPs should include instructions for disinfecting laboratory work areas, what PPE should be worn during cleanup, how to clean the surfaces involved, what disinfectant(s) to use, and how to dispose of the cleaning materials (4, 32).

The choice of disinfectant(s) to use for cleaning depends on a number of factors such as the degree of microbial killing required, the composition of the item being cleaned and disinfected, and the costs, safety, and ease of application (32). Alcohols and alcohol-based solutions should not be used to disinfect surfaces because they evaporate too quickly (4). Disinfectants should be used according to specific manufacturer's recommendations with regard to concentration and contact time with organic material (e.g., wet or dried blood, tissues, body fluids, feces, etc.). Preliminary washing using a low-pressure water source and a general purpose disinfectant or detergent can be performed, followed by decontamination with an appropriate disinfectant (32). A 1:10 or 1:100 dilution of household bleach (using 10 minutes of contact time) or other chemical germicide for stainless steel and aluminum (using the manufacturer's recommended contact time) is commonly used in autopsy suites. Regardless of the disinfectant used, it important not to scrape dried blood or bodily fluids excessively from surfaces, because scraping can generate aerosols (4).

The laboratory should have written policies or procedures for necropsy effluent decontamination. The release of decontaminated liquids must meet all applicable municipal and/or regional and state regulations. These regulations typically are concerned with the chemical and/or metal content, the volume of suspected solids present, and the biochemical oxygen demand (32).

During the cleanup procedures, all individuals on the necropsy floor should wear the appropriate PPE, as previously determined during the risk assessment. Generation of aerosolized pathogens may be greatest during the cleanup period, particularly if high-pressure washing is employed. Nonwashable items such as cameras, computer keyboards, and phones should be considered contaminated and handled with gloves at all times. During the cleanup procedures, these items should be wiped with a 10% bleach solution after each use (4).

Once the necropsy suite cleanup procedures are completed, soiled disposable PPE items should be placed in a biohazard bag for future incineration. Nondisposable clothing, such as scrubs and coveralls, should be removed and placed in the appropriate container for laundry or autoclaving. To prevent blood or fecal stains from being "locked in" by the autoclave, heavily soiled areas should be washed/cleaned inside the containment area prior to being placed in a container for autoclaving (32).

Nondisposable, nonlaunderable items, such as plastic aprons, should be cleaned and disinfected by first removing the item and then submerging it in the appropriate disinfectant. Hand wash these items (while still wearing gloves) until grossly visible soiled areas are clean. The item(s) should remain in contact with the disinfectant for the appropriate contact time, on the basis of the manufacturer's recommendations.

If the necropsy was performed on a carcass suspected of harboring an RG3 pathogen or Select Agent, the employee(s) performing the necropsy should shower (e.g., entire body, including washing hair and beards) upon exiting the necropsy floor. Eyeglasses must be disinfected at the containment barrier (32).

Handling of Diagnostic Samples in the Veterinary Diagnostic Laboratory from Animals with Suspect Prion Diseases

Scrapie, chronic wasting disease (CWD), transmissible mink encephalopathy (TME), and bovine spongiform encephalopathy (BSE) belong to the transmissible spongiform encephalopathy (TSE) family of diseases. Scrapie is a nearly worldwide disease of domestic sheep, goats, and mouflon (*Ovis musimon*). CWD naturally affects mule deer (*Odocoileus hemionus*), white-tailed deer (*Odocoileus virginianus*), and Rocky Mountain elk (*Cervus elaphus nelsoni*) in the United States and Canada. TME is a rare disease of ranched mink (*Mustela vison*). BSE naturally affects domestic cattle, captive nondomestic bovids, domestic and nondomestic felids, and laboratory primates. The agents of scrapie, CWD, and TME require BSL2 practices and procedures (3). The BMBL states that "The most prudent approach is to study BSE prions at a minimum in a BSL2 facility utilizing BSL3 practices." Biosafety guidelines for BSE in veterinary diagnostic laboratories are previously published and readily available online (32).

Veterinary laboratories are a critical element to ensure that infected carcasses are identified so that the location of the disease can be mapped accurately for management purposes by state wildlife agencies. At the same time, it is in the interest of animal health laboratories to demonstrate to regulatory agencies and the public that they take effective, responsible measures to ensure that they do not concentrate and recirculate TSE agents into the environment. Although animal TSEs are RG2 agents, the unique biology of prions warrants special recommendations to assist veterinary diagnostic laboratories in processes for reducing exposure of TSE agents to laboratory personnel and disposing and inactivating TSE-contaminated wastes.

The veterinary diagnostic laboratory areas most commonly working with prion-infected specimens are the necropsy floor where specimens from infected carcasses are collected, as well as the histology and/or immunohistochemistry laboratory where formalin-fixed tissue is processed and analyzed, and the laboratory section where a rapid enzyme-linked immunosorbent assay (ELISA) test is used to identify prion-infected fresh tissues. Each of these laboratory areas should use standard BSL2 safety precautions for scrapie, CWD, and TME, such as restricted access to laboratories, protective clothing, facial protection, special care with sharp instruments, and avoidance of aerosols (3). All technicians working in the laboratory areas dealing with prions should wear BSL2 level PPE, including two pairs of water-impervious surgical gloves any time there is the opportunity for skin contact with infectious tissues and fluids from a prion-suspect specimen. If accidental contamination of skin does occur, the area should be swabbed with 1:10 dilute bleach or 1 N sodium hydroxide for 1 to 5 minutes and then washed with copious amounts of water (3). Some laboratories treat formalin-fixed tissues "en-block" to inactivate prions from cases of suspected prion disease by immersing for 30 to 60 minutes in 96% to 100% formic acid before histology processing (3, 36). Table 14 lists a variety of disinfection/inactivation methods when dealing with laboratory surfaces and wastes that are potentially contaminated with prions. Disposable supplies and dedicated equipment/dedicated laboratory areas should be used when possible (37).

TABLE 14.

Inactivation and disposal of prion-contaminated materials, equipment, and laboratories

Material	Method
Tissues, carcasses, and wastes (including microtomy and disposal wastes)	Incineration @ 1,000°C (1,832°F).
	Digestion via alkaline hydrolysis for 6 hours @ 300°F.
	Landfill in accordance with municipal and state regulations.
Equipment	Soak in 1:1 bleach solution[a] for 1 hour, rinse with water, autoclave @ 121°C (250°F) for 1 hour.
	Soak in 1 N NaOH for 1 hour, rinse with water, autoclave @ 121°C (250°F) for 1 hour.
	Autoclave (gravity displacement steam autoclaving) for 1 hour @ 134°C (274°F).
	Autoclave (porous load steam autoclaving) for one 18-minute cycle @ 30 lbs psi.
	Autoclave (porous load steam autoclaving) for six 3-minute cycles @ 30 lbs psi.
	Soak in 1% aqueous acid phenol[b] for 16 hours.
	Soak in 10% aqueous acid phenol[b] for 30–60 minutes.
Laboratory surfaces and necropsy floor	Soak with 10% aqueous acid phenol[b] for 30–60 minutes. Surfaces must remain wet.
	Soak with 1:1 bleach solution (2% or 20,000 ppm free chlorine)[a] for 30–60 minutes. Surfaces must remain wet.
Formalin and other reagents	Incinerate with tissues and carcasses.
	Dispose of remaining tissues/carcasses via alkaline hydrolysis.
	Collect in a 4-liter waste bottle containing 600 ml of 6 N sodium hydroxide, followed by chemically neutralizing the pH and disposing of liquid in accordance with municipal, state, and federal regulations.
Biosafety cabinets	Vapor-phase hydrogen peroxide (VHP) followed by incineration of HEPA filter whenever the BSC must be serviced, certified, or moved. This is a relatively new method of inactivation. See current literature for details.

[a]Free chlorine: 2% or 20,000 ppm. Household bleach: 2% free chlorine = 2 parts bleach to 3 parts water. Commercial bleach: 2% free chlorine = 1 part bleach to 4 parts water.
[b]STERIS Environ LpH.

Tissue Fixation

Formaldehyde (3.7–4.0%) is the most commonly used fixative in veterinary diagnostic laboratories. Used most frequently as 10% neutral buffered formalin (NBF) (which equates to 3.7–4.0% formaldehyde), this chemical stops (i.e., "fixes") protein denaturation and thus prevents continued postmortem autolysis when used at 10 times the volume of tissue. The chemical is volatile; fumes can cause irritation to the mucous membranes, eyes, and skin, and its use is associated with increased risk for various cancers. If formaldehyde can be detected by smell, exposure may be occurring at a concentration beyond the OSHA-acceptable limits of 0.75 ppm (as an 8-hour time-weighted average) and 2.0 ppm for short-term (15 minute) exposures (4). To minimize exposure to formaldehyde/formalin, the employee(s) handling the tissues should attempt to keep the tops on all containers of 10% NBF and trim tissues ("gross in") inside a fume hood or downdraft table. Gloves and appropriate PPE should be worn to avoid skin exposure. Contact lenses should not be worn when working with formalin. Laboratorians may wish to monitor formaldehyde exposure periodically in areas of the diagnostic laboratory where formalin is commonly used, such as the necropsy floor and histology laboratory. Several commercial companies can, for a fee, provide the end user with a chemical monitoring badge to wear during the day. After a specified period of use, the wearer seals the badge in a foil envelope and returns the badge to the monitoring company. The results provided by the company may serve as documentation for both the individual employee and for the laboratory's biosafety assurance program.

Most laboratory grossing stations are placed inside fume hoods and not BSCs. Thus, if there is the potential to aerosolize pathogens at the time of sectioning (for example, during any sawing or high-energy procedures), PPE appropriate to the risk assessment should be worn. If there is a concern for potential splashing or spilling of formalin (or other chemicals), a disposable face shield (or goggles and a fluid-resistant face mask) should be worn at the grossing bench (4). Tubes and vials containing body fluids, aspirates, or exudates should be aliquoted by disposable pipette (rather than pouring) to avoid splashing or spilling (4).

Incineration is the preferred method of disposal for formaldehyde/formalin-fixed specimens. When 10% NBF is flushed down the drain, the laboratory must comply with all municipal regulations concerning the amount of formaldehyde that may be dumped over a given time period (4). If tissues and other items collected during necropsy/surgical pathology will not be fixed (i.e., foreign bodies, calculi, teeth, unused serum or blood tubes, etc.), they should be double bagged, identified as biohazardous, and refrigerated or frozen under the appropriate laboratory numbering system (4).

In the Histology Laboratory

Surgical pathology risks are associated with manipulating large amounts of fresh tissues from unknown and/or potentially infectious sources. These procedures may result in punctures, cuts, and splashes of blood and body fluids when fresh organs must be viewed at a grossing table and "breadloaved" or cut into sections thin enough for fixation. Other risks include the use of cryostat cutting equipment or freezing spray (generates infectious aerosols when sectioning frozen tissue) and exposure to large volumes of formaldehyde. When handling/processing surgical specimens, these specimens should be considered potentially infectious and should be handled with the appropriate PPE until completely fixed with a germicidal fixative or stained and covered with a coverslip (in the case of frozen section slides).

Cytology receives large- and small-volume body fluids, bone marrow samples, or needle aspirate specimens, most of which are received in fixative. Aliquoting or pouring off large volumes of bodily fluids or fixatives can result in splashing and spills and/or the potential for aerosolization; thus, these procedures should be performed inside a BSC. Procedures such as centrifugation and cytospin processing can also produce aerosols and should be performed inside a BSC if the containers are not sealed. Cytology specimen PPE is dependent on the specimen and the preparatory procedures to be performed. The laboratorian should wear fluid-resistant laboratory clothing, an apron, and two pairs of gloves until the slides are fixed and stained. Opening of small-volume body fluids or aspirates submitted in tubes should occur inside a BSC or with the use of a splash guard or a face shield and should be aliquoted by disposable pipette rather than pouring to avoid splashing and spill. Air-dried, unstained slides should also be considered infectious, as should the stains used for cytology specimens (4).

Frozen sectioning, performed on fresh, unfixed tissue, is a high-risk procedure for infectious agent exposure. Freezing tissue does not kill organisms, and the use of the cryostat cutting blade creates potentially dangerous aerosols. The true clinical necessity for frozen-sectioning should be discussed with the surgical team. Cryostats do not use negative airflow, and only some models have a downdraft feature. Thus, whenever the cryostat is in use, the ultrathin tissue sections (and their pathogens) have the potential to be aerosolized upward and outward toward the operator's face and into the room. For this reason, freezing propellants should not be used in the cryostat (4, 28).

The CLSI's *Protection of Laboratory Workers from Occupationally Acquired Infections: Approved Guidelines*, 4th edition, of 2014 (28) requires that cryostat operators wear gloves and an N95 respirator, even when the current case does not involve RG3 pathogens. The cryostat cabinet contains potential pathogens from all specimens sectioned since the last time the cryostat was properly defrosted, cleaned, and disinfected (at the appropriate temperature for the disinfectant used). Thus, the cryostat should be defrosted and decontaminated regularly. Daily defrosting and disinfection with 70% ethanol and weekly disinfection with a germicidal disinfectant for *Mycobacterium tuberculosis* is recommended in facilities that deal with human tissues (4) as well as animal tissues because accidental zoonotic infection by *M. tuberculosis* from animal tissues has been reported (38). Stainless steel mesh gloves should be worn when changing the knife blades.

PARASITOLOGY

Individuals working directly with animals should consider using methods to minimize their exposure to ecto- or endoparasites that may be present (32). Protective clothing, tick and flea repellents, and self-examination to identify any attached parasites are each recommended. Herwaldt provides an excellent review of LAIs involving parasites (39). Over 44% of the laboratory accidents (for which a likely route of transmission was determined) are believed to have occurred following percutaneous exposure via a needlestick or other sharp object (39). Note that detailed information about specific diagnostic tests, availability of vaccines for occupational exposure, and treatment modalities will not be included in this chapter. Individuals are encouraged to contact their health care provider for specific information about testing and/or treatment of both zoonotic and nonzoonotic parasitic diseases.

Primary containers should be cleaned and disinfected prior to handling. Whenever formalin or xylene is present in the specimen or is used as a part of the procedure, the specimen and reagents should be manipulated inside a fume hood to reduce exposure to formaldehyde or xylene (4). Procedures or manipulations involving cultures, tissue, blood and other bodily fluids, gastrointestinal tract specimens, urine, and cerebrospinal fluid that contain parasitic organisms as well as the parasites themselves may require manipulation in a BSC in conjunction with BSL2-level PPE (lab coat, gloves, and face shield to protect against splashes) (3, 4). A BSC should be used when processing fresh specimens or culturing for parasite isolation (4). For more information about reagents (i.e., stains) used in the parasitology laboratory, the reader is referred to "Guidelines for Safe Work Practices in Human and Animal Medical Diagnostic Laboratories. Recommendations of a CDC-Convened, Biosafety Blue Ribbon Panel" (4).

MYCOLOGY

In the veterinary diagnostic laboratory, mycology activities should be performed in a separate, closable room with negative airflow moving from the main microbiology laboratory into the mycology room/area (4).

All plates and slants should be observed prior to opening. Shrink seals should be used with petri plates, especially once growth of the organism(s) begins, to prevent accidentally opening and spreading fungal elements. Cultures growing yeast-like colonies (except *Cryptococcus neoformans* suspects) can be read on the open bench in a BSL2 laboratory. "Reading" and manipulation of *Cryptococcus* suspects should always be conducted inside a BSC (4).

Fungal cultures should never be sniffed in an attempt to determine the presence/absence of a potential odor. Accidental release of mold spores is the most common cause of a laboratory-acquired mycotic infection. This includes both primary pathogens infecting immunocompetent laboratory employees and opportunistic fungal pathogens infecting individuals with a compromised immune system (3). The BMBL recommends BSL2-level practices with strict use of a BSC when handling any unidentified mold, processing any clinical soil or animal tissue samples, or handling any culture containing dimorphic fungi. BSL3-level practices, containment equipment, and facilities are required when handling, propagating, or manipulating cultures known to contain deep or systemic fungi (e.g., *Blastomyces dermatitidis*, *Coccidioides* spp., and *Histoplasma capsulatum* var. *capsulatum*), as well as any sample known or suspected to contain infectious (arthro)conidia (3).

Detailed information about specific diagnostic tests, availability of vaccines for occupational exposure, and treatment modalities will not be included in this chapter. Individuals are encouraged to contact their health care provider for specific information about testing and/or treatment of both primary and opportunistic mycotic diseases.

Cutaneous Mycosis

Cutaneous mycotic infections involve the dermis, hair, or nails. In the veterinary laboratory, the most common lab-acquired dermal mycotic infection involves a dermatophyte ("ringworm"). The dermatophytes *Microsporum* and *Trichophyton* are most commonly seen in canine and feline hair samples submitted to veterinary diagnostic laboratories.

Subcutaneous Mycosis

Subcutaneous infections require a breach in the integrity of the skin (usually caused by trauma such as a splinter or thorn). Fungi that cause phaeohyphomycosis and *Sporothrix schenckii* are the most common causes of subcutaneous infections following a "break" in the dermis; however, percutaneous exposure via needlestick could conceivably result in an LAI. Because cat tissues infected with *S. schenckii* contain numerous organisms, they are believed to be more infectious than tissues from infected dogs. Most LAIs involving *S. schenckii* have been associated with splashes (into the eye), scratches, bites, or injection of infectious material into the skin; therefore BSL2-level safety precautions are recommended with handling cats or any potentially infected specimens (3).

Pulmonary or Systemic Mycosis

Pulmonary or deep/systemic mycosis (which begins in the lungs and may disseminate to other organs) involves *C. neoformans*, *B. dermatitidis*, *Coccidioides* spp., *H. capsulatum* var. *capsulatum*, and *Aspergillus* spp. If a filamentous culture or *Coccidioides immitis* is suspected, or the sample is being transported to another laboratory, screw-cap tubes containing "slants" should be used (4). Before setting up a slide culture, a wet preparation of cultured molds should be made to detect physical characteristics indicative of highly pathogenic, systemic, or "deep" fungi. If a culture isolate is suspected of being *H. capsulatum*, *B. dermatitidis*, *C. immitis*, *Coccidioides posadasii*, *Paracoccidioides brasiliensis*, *Penicillium marneffei*, or *Cladophialophora bantiana*, do not set up a slide culture. Mold-form cultures and environmental specimens that could contain infectious conidia should be handled using BSL3 practices, containment equipment, and facilities (4). In general, primary systemic mycoses are caused by dimorphic fungi, including *C. neoformans*, *B. dermatitidis*, *Coccidioides* spp., and *H. capsulatum* var. *capsulatum*, which have a transmissible mold form at room temperature and nontransmissible yeast form at body temperature. Yeast forms are not infectious, thus the danger of LAI from these systemic mycoses from necropsy prosection is extremely low.

For more information about reagents (i.e., stains) used in the mycology laboratory the reader is referred to "Guidelines for Safe Work Practices in Human and Animal Medical Diagnostic Laboratories. Recommendations of a CDC-Convened, Biosafety Blue Ribbon Panel" (4).

VIROLOGY

Potentially zoonotic viruses, including latent and adventitious agents, may be present in animal tissues and cells. Thus, appropriate biosafety practices are necessary when manipulating specimens and cell cultures. Furthermore, the BMBL recommends that nonhuman primate cell cultures be manipulated using BSL2 practices, containment equipment, and facilities. Individuals working with nonhuman primate cells and tissues should receive bloodborne pathogen training and be offered hepatitis B immunizations. Baseline serum sampling may be warranted in certain situations (3).

In virology laboratories, the preparation areas for preparing cell cultures and reagents are often referred to as the "clean" areas. No specimens or control materials are allowed in the "clean" area. This is done in an effort to prevent cell culture contamination that could result in inaccurate/false results during testing. From a biosafety perspective, these areas are no different than the "dirty" areas of the virology laboratory and appropriate biosafety precautions should still be applied (4). Because laboratory manipulation of cell cultures can contaminate the outside of the culture vessels and/or other containers in the area, all containers should be handled as if they are contaminated (4) (see Table 15). Cell lines with known or potential viral contaminants should be handled at the containment level appropriate for the contaminating agent with the highest risk (27).

All cell cultures, whether inoculated with a specimen or not, are potentially infectious. These cells may contain unanticipated viral agents from latently infected primary tissue, the animal products (e.g., fetal calf serum), or agents such as retroviruses used to transform cell lines. Alternatively, cell lines may also be infected with bacteria (including *Mycoplasma* spp.), fungi, and prions (27). It is important to remember that many of these agents do not produce cytopathic effects.

A risk assessment should be performed each time a new cell line is introduced. During cell passage unanticipated pathogens are passed repeatedly (4). Primary cell lines have the greatest risk, as do cell lines that are the closest phylogenetically to humans. For example, the greatest risk is associated with human autologous cell lines followed by human heterologous, nonhuman

TABLE 15.

Laboratory procedures when working with viral cultures[a]

Wear gloves and a lab coat.

Wear a face shield or goggles and fluid-resistant face mask (i.e., N95 respirator) when splashing can be reasonably anticipated.

Perform all manipulations inside a biosafety cabinet.

- Includes culture inoculation/feeding/passage; hemadsorption and hemagglutination testing; virus dilutions and titrations; cell fixation; immunofluorescent staining; preparing controls/control slides; and opening cytocentrifuge/sealed centrifuge heads containing specimens or cultures.

Decontaminate all culture tubes and materials that come into contact with cell cultures by using appropriate chemical disinfection or autoclaving before disposal.

[a]Data from references 3 and 4.

primate, mammalian, avian, and invertebrate cell lines. In the case of hybridomas, the characteristics of each separate component cell line should be evaluated as a part of the risk assessment (27). Veterinary diagnostic laboratories should not test any unusual or unapproved specimens and should not accept environmental specimens unless the laboratory is qualified/certified to do this type of testing (4).

For detailed information about biosafety-related issues involving the use of liquid nitrogen, stains, chemicals, and antibiotics in the virology laboratory, the reader is referred to "Guidelines for Safe Work Practices in Human and Animal Medical Diagnostic Laboratories. Recommendations of a CDC-Convened, Biosafety Blue Ribbon Panel" (4). This document also is an excellent source of information concerning electron microscopy-associated procedures in the laboratory.

MOLECULAR DIAGNOSTICS AND RAPID TESTS

After complete inactivation or extraction, nucleic acids can be handled in a BSL2 laboratory. Extracted nucleic acids may or may not, however, be sterile. For example, the genome of positive-strand RNA viruses (e.g., poliovirus) is infectious and can replicate inside the cell without the typical proteins necessary for virus replication. For this reason, it is recommended that nucleic acids be handled as if they are infectious (4). Biosafety procedures described elsewhere in this text for operations performed in BSL2 facilities are applicable to the molecular diagnostics section of the veterinary diagnostic laboratory.

For detailed information about the chemical hazards associated with the molecular diagnostic/PCR laboratory, see "Guidelines for Safe Work Practices in Human and Animal Medical Diagnostic Laboratories. Recommendations of a CDC-Convened, Biosafety Blue Ribbon Panel" (4).

MICROBIOLOGY

The reader is directed to other areas of this text and "Guidelines for Safe Work Practices in Human and Animal Medical Diagnostic Laboratories. Recommendations of a CDC-Convened, Biosafety Blue Ribbon Panel" (4) for information concerning biosafety in the microbiology laboratory suite.

STORAGE, PACKAGING, AND SHIPPING

Laboratorians have the responsibility to ensure that materials leaving their facilities protect public safety. The regulations associated with shipping requirements are very detailed, and laboratory personnel who ship infectious substances must be trained and certified annually (for air transport) or every other year (for other forms of transport) (28, 30). Shippers may reuse packages, provided they are appropriately disinfected and all markings and labels are removed or covered. If the package is being returned to the shipper, it must be disinfected to nullify any hazard. Labels or other markings that indicate the package contained an infectious substance must be removed or covered (7).

It is the responsibility of the employer (and not the trainer) to certify that the laboratory individuals responsible for shipping are knowledgeable about the packaging requirements. Penalties for failing to meet the packaging and shipping requirements of the national and international regulatory agencies (i.e., DOT in the United States and the IATA) can be levied against both the individual and the institution. Because the regulations vary from country to country and changes are made frequently, the reader is referred to a listing of training organizations for more information. The Office of Hazardous Materials Safety provides a list of training organizations at http://phmsa.dot.gov/hazmat, as does the Transportation Safety Institute at http://www.tsi.dot.gov/.

EMPLOYER AND EMPLOYEE RESPONSIBILITIES

It is the responsibility of the laboratory director/employer to ensure that each new employee receives safety training. Such training should include safe handling and transportation practices, the use of PPE, the proper use of BSCs, spill cleanup and decontamination procedures, the proper use of disinfectants and disinfecting procedures (i.e., autoclave operation) where applicable to the employee's job, information about proper waste disposal, self-monitoring practices for potential LAIs, and information about reporting possible exposure and/or illness to the laboratory's management and, where necessary, appropriate state and federal animal health and public health authorities (4). Furthermore, it is the responsibility of the employer to identify and instruct employees about laboratory hazards and the specific practices and procedures that will minimize or eliminate these hazards (3, 4). According to the 5th edition of BMBL, "The laboratory director is specifically and primarily responsible for assessing the risks and applying the appropriate biosafety levels" (3).

The laboratory director is also responsible for ensuring that a laboratory-specific biosafety manual is developed and utilized by all employees. The employer is also required to ensure that all personnel are proficient in the necessary standard practices and techniques required for the tasks they perform.

With regard to PPE, employers are encouraged or (in some states) required by OSHA to ensure that employees know when to use PPE and which PPE item(s) are appropriate for a given hazard. Employees must also be trained on how to don and doff PPE properly, how to care for, use, and dispose of PPE properly, and the limitations of the PPE, including the useful life span for each item of PPE (4).

The facility must also provide an effective method of communicating the potential risks to all employees via the use of appropriate training, signage, policies/SOPs, and employee feedback. The efforts of laboratorians working at the bench provide the foundation of a safe work environment; as such, their input and participation is critical to the effectiveness of the laboratory's biosafety program. At no time should an employee perceive that the organization is punitive or unwilling to address their safety concerns.

It is the responsibility of both the employer and the employees to maintain a culture of safety. This is particularly important in diagnostic laboratories where a definitive diagnosis may be lacking. A culture of safety "... embraces the idea that all specimens may contain a dangerous pathogen and that the specimen container may be as dangerous as its contents" (4). As such, "... it is the responsibility of all laboratory workers to ensure the effective use of products for decontamination of materials, equipment, and samples from containment zones; of surfaces and rooms; and of spills of infectious materials" (27). Failure to decontaminate equipment and facilities adequately can result in occupational exposures to infectious agents and/or the unintentional release of infectious agents into the environment/community (27). Reviewing the literature and current science is important; it is not sufficient to rely on the manufacturer's claims, because these tests may not have been performed under conditions similar to that of the veterinary diagnostic laboratory. In-house testing of disinfectant efficacy may be necessary in some situations to confirm that the disinfectant being used is both appropriate and adequate to the task, since the manufacturer's original testing may not have included a soil load (27).

BIOSAFETY EDUCATION AND TRAINING

Requiring that biosafety and biosecurity be addressed, as a part of accreditation requirements, may be the best way to ensure that biosafety education and training are part of an on-going process within the laboratory. In the 2012 American Association of Veterinary Laboratory Diagnosticians' (AAVLD's) *Requirements for an Accredited Veterinary Medical Diagnostic Laboratory*, biosafety (and biosecurity) is a physical requirement for the facility. An accredited laboratory must also "...ensure the establishment and maintenance of safety, biosafety, biocontainment and biosecurity programs relevant to present and anticipated needs. The programs will provide staff training and address all necessary elements to ensure a safe work environment" (40).

The laboratory director is responsible for ensuring that a laboratory-specific biosafety manual is developed and utilized. The biosafety manual should be reviewed annually by all employees and updated as needed. All employees should have easy access to the biosafety manual, and refresher training for all employees with access to the facility should be held at least yearly. Topics that should be considered during annual biosafety training and/or biosafety inspections are previously published (4). The biosafety manual requirements may vary from institution to institution or agency to agency but most contain common information as outlined in Table 16. In

TABLE 16.

Topics to be considered during annual biosafety manual review and employee biosafety training and/or biosafety inspections[a]

Topics
Physical environment, including cleanliness presence of safety equipment, chemical storage
Institutional and laboratory safety policies and standard operating procedures
Personnel responsibilities for management, supervisors, and technicians
Regulations and guidelines, including regional, national, and, where applicable, international requirements
Routes of exposure in the laboratory
The need for continuous risk assessments
Biosafety practices and principles
Standard precautions when handling potentially infectious materials
Hazard communication and biohazard signage
Utilizing engineering controls, including HEPA-filtered air, routes of airflow, showers, and room design
Administrative and work practice controls, including separation of clean and dirty areas, standard operating procedures
Facility access controls, who can go where within the laboratory
Use of personal protective equipment, including annual fit testing, donning, and doffing
When the use of a biosafety cabinet is required
How to work safely inside a biosafety cabinet
Shipping and/or transporting biohazardous and hazardous materials
Biohazardous waste decontamination and disposal
Safety, biosafety, and biosecurity/biocontainment training programs and documentation
Emergency response procedures
Medical surveillance program/immunization program (if used)
Laboratory exposure, evaluation procedures
Illness associated with possible laboratory exposure, what to do after hours
Communications with regulatory agencies

[a]Data from references 4 and 27.

addition to regular biosafety training, it is important for the organizational leadership to make biosafety and infection control an integral part of the laboratory's culture of safety. Single-source intervention strategies, such as putting up posters to remind everyone to wash their hands regularly, are ineffective at permanently changing employee behaviors (20).

It is also important to document the biosafety and biosecurity training and competency of equipment-repair personnel, facilities maintenance personnel (if these individuals are not routinely employed on site), janitors, and other individuals in similar positions before they enter the laboratory. All internal and external audit requirements for reporting who enters and exits the laboratory should be strictly followed (4).

Biosafety in the veterinary laboratory is not limited to infectious agents. A significant variety of chemicals may be used in veterinary laboratories. Many of these chemicals may be carcinogenic, mutagenic, or toxic. Some chemicals can be absorbed by intact skin. Alternatively, chemical vapors may be hazardous. The WHO's *Laboratory Biosafety Manual,* 3rd edition, contains information about recommended safety precautions for many of the chemicals used in veterinary diagnostic laboratories (12). The WHO/OIE requires that a list of hazardous chemicals be maintained, and a file record kept of chemicals to which individual staff members could be exposed. In the United States, OSHA requires employers to keep a record of the identity of substances to which employees are exposed for 30 years. If the MSDS itself is not kept, the location of use and the time period used for each substance must also be recorded. Because the MSDS is considered an acceptable alternative record and does not require the "when and where" to be kept by the employer, laboratories are encouraged to keep the MSDS records themselves (41).

As a part of their biosafety training, all employees should be encouraged to seek medical evaluation for symptoms they believe may be associated with their work involving infectious agents, without fear of condemnation or reprisal. Because the modes of transmission and the clinical presentation of an occupationally acquired infection may be different from a naturally acquired infection, employees must remain alert for evidence of potential exposures (3). Employees should be given information listing potential zoonotic pathogens to which they may routinely be exposed. This resource should contain information that would be useful to a medical practitioner in the event of illness and employees must be able to readily share this information with their health care provider. Any special vaccinations that have been given to the employee should also be recorded and accessible by the employee, along with the date administered (7). Because not all hazards in the laboratory are infectious, laboratory directors may also wish to include a record of hazardous chemicals to which the employee is routinely exposed.

MANAGEMENT OF LAIs

Frequently the source of a LAI is not identified; only that the person was "working with the agent [or] being in or around the laboratory, or being exposed to infected animals." In many cases, the route of infection (inhalation, direct contact, ingestion, or percutaneous) is not the standard (i.e., expected) route of infection. For example, 20% of arbovirus-related LAIs are believed to be the result of inhalation exposure, rather than the traditional mosquito vector/injection by needle stick.

The employees' understanding of how laboratory accidents will be managed should be documented (e.g., by written exam or signed statement). This documentation should become a part of each individual's personnel record. All occupational injuries should be reported, and the laboratory biosafety plan should include procedures on how an employee should handle exposures and/or the development of symptoms—both during normal working hours and after normal working hours (3). In addition to providing the employee(s) with information on how to report an accident and who to contact, it is also important to ensure that local medical facilities have information about the potential zoonotic pathogens that may be present at the veterinary diagnostic laboratory.

Because it may be difficult to determine the significance of an exposure in some cases, physicians may need to base their decisions upon knowledge of similar agents, the circumstances surrounding an exposure, and information obtained from subject matter experts. Postexposure prophylactic treatment is pathogen dependent and exposure dependent. It may also be host-factor-dependent or even contraindicated in some cases (due to past medical history such as an allergic reaction). The clinical risk and treatment decision process should be explained and all of the worker's questions addressed with relevant information and educational materials (3). Table 17 provides a list of items that should be documented following a laboratory exposure, regardless of whether this exposure ultimately results in an infection or not.

Following any skin puncture involving a possible infectious agent, the individual should immediately wash the site with soap and water while permitting bleeding to continue; then, if appropriate, the site should be bandaged. Potentially contaminated mucosal surfaces should be washed with a large volume of water (28). Once first aid has been rendered, the employee should contact his

TABLE 17.

Documenting a laboratory exposure: information to include[a]

Date, time, and location of the exposure
Employee's name and any identifying numbers
Names of other individuals in the laboratory at the time of the exposure, including witnesses to the event
Details of the incident or exposure
Brand names and serial numbers of any devices or instruments involved
Relevant health information/status of the exposed employee at the time of the exposure
Immediate or remedial actions which were taken following the exposure such as first aid; posting "do not use" signs on relevant equipment, etc.
Listing of all recommendations given to the employee at the time of the exposure, including visits to the emergency department; administration of chemoprophylaxis; consultations with physicians; etc.
Results of discussions with employee physicians
Monitoring/follow-up plan
Results of the monitoring and follow-up plans (to be completed at a later date)
Appropriate signatures, including, at a minimum, the employee and the employee's immediate supervisor

[a]Data from reference 4.

or her immediate supervisor and follow the procedures outlined in the laboratory's biosafety manual.

If the exposure involves an employee who was working in a BSC, all other individuals in the laboratory who also work at this same BSC should be tested and/or monitored for evidence of exposure to the pathogen in question (4). Following confirmation of a laboratory exposure, the equipment evaluation must include BSC inspection and recertification. The employer should also evaluate the policies and procedures for techniques performed in the BSC and, if necessary, provide retraining for the employees working in this section/suite (4).

As a part of the management of LAIs, a system of nonretaliatory reporting and shared learning and analysis should be used to improve the laboratory's ability to "fix what needs to be fixed." The ultimate goal of any reporting system should be to aid in the prevention of future LAIs or release of pathogens into the environment. Collecting information on hazards, accidents, and incidents (including "near misses") will help identify the issues that led to the situation being reported. Furthermore, nationwide voluntary sharing of this information (in an anonymous fashion) would serve as a source of prevention for all laboratorians. All veterinary diagnostic laboratories could use such a system to improve biosafety and biosecurity continuously in their facilities (42).

References

1. **Torrey EF, Yolken RH.** 2005. *Beasts of the Earth: Animals, Humans, and Disease.* Rutgers University Press, New Brunswick, NJ.
2. **Taylor LH, Latham SM, Woolhouse ME.** 2001. Risk factors for human disease emergence. *Philos Trans R Soc Lond B Biol Sci* **356:**983–989.
3. **U.S. Department of Health and Human Services PHS, Centers for Disease Control and Prevention, National Institutes of Health.** 2009. *Biosafety in Microbiological and Biomedical Laboratories,* 5th ed. HHS Publication No. (CDC) 21-112. http://www.cdc.gov/biosafety/publications/bmbl5/BMBL.pdf.
4. **Miller JM, Astles R, Baszler T, Chapin K, Carey R, Garcia L, Gray L, Larone D, Pentella M, Pollock A, Shapiro DS, Weirich E, Wiedbrauk D, Biosafety Blue Ribbon Panel.** 2012. Guidelines for safe work practices in human and animal medical diagnostic laboratories. Recommendations of a CDC-convened, Biosafety Blue Ribbon Panel. *MMWR Suppl* **61**(Suppl):1–102.
5. **U.S. Department of Agriculture Agricultural Research Service.** 2012. ARS Facilities Design Standards. ARS-242.1, Facilities Division, Facilities Engineering Branch AFM/ARS
6. **U.S. Department of Health and Human Services NIoH, Office of Research Facilities.** 2003. *NIH Design Policy and Guidelines.* U.S. Department of Health and Human Services, Washington, DC.
7. **World Organisation for Animal Health.** 2012. *Manual of Diagnostic Tests and Vaccines for Terrestrial Animals, Biosafety and Biosecurity in the Veterinary Diagnostic Laboratory and Animal Facilities,* 7th ed, vol 1. World Organization for Animal Health, Paris, France.
8. **Sauri M.** 2007. Medical surveillance in biomedical research. *Appl Biosaf* **12:**214–216.
9. **World Organisation for Animal Health.** 2015. *Manual of Diagnostic Tests and Vaccines for Terrestrial Animals 2015.* World Organization for Animal Health, Paris, France. http://www.oie.int/en/international-standard-setting/terrestrial-manual/access-online/.
10. **American Biological Safety Assocation.** 2016. Risk Group Database. http://my.absa.org/tiki-index.php?page=Riskgroups.
11. **Centers for Disease Control and Prevention, Animal and Plant Health Inspection Service.** 2016. Federal Select Agent Program. Select Agents and Toxins List. http://www.selectagents.gov/SelectAgentsandToxinsList.html.
12. **World Health Organization.** 2004. *Laboratory Biosafety Manual,* 3rd ed. World Health Organization, Geneva, Switzerland.
13. **Gaudioso J, Caskey SA, Burnett L, Heegaard E, Owens J, Stroot P.** 2009. *Strengthening Risk Governance in Bioscience Laboratories.* Sandia National Laboratories, Albuquerque, NM.
14. **World Health Organization.** 2012. *Guidance on Regulations for the Transport of Infectious Substances. 2013–2014.* World Health Organization, Geneva, Switzerland.
15. **Caskey SA, Gaudioso J, Salerno R, Wagener S, Shigematsu M, Risi G, Kozlovac J, Halkjaer-Knudsen V, Prat E.** 2010. *Biosafety Risk Assessment Methodology.* Sandia National Laboratories, Albuquerque, NM.
16. **Delany JR, Pentella MA, Rodriguez JA, Shah KV, Baxley KP, Holmes DE.** 2011. Guidelines for biosafety laboratory competency: CDC and the Association of Public Health Laboratories. *MMWR Suppl* **60:**1–23.
17. **Bennett A, Parks S.** 2006. Microbial aerosol generation during laboratory accidents and subsequent risk assessment. *J Appl Microbiol* **100:**658–663.
18. **Pike RM.** 1979. Laboratory-associated infections: incidence, fatalities, causes, and prevention. *Annu Rev Microbiol* **33:**41–66.
19. **Adelberg EA, Austrian R, Bachrach HL, Barkley WE, Burnet JP, Fleming DO, Fuchs RL, Ginsburg H, Goldman R, Hughes J, Mikell WG, Richardson JH, Schmidt JP, Smith JW, Walton TE.**

1989. *Biosafety in the Laboratory: Prudent Practices for the Handling and Disposal of Infectious Materials.* National Academy of Sciences Press, Washington, DC.
20. **Siegel JD, Rhinehar E, Jackson M, Chiarello L.** 2007. 2007 Guideline for Isolation Precautions: Preventing Transmission of Infectious Agents in Healthcare Settings. http://www.cdc.gov/ncidod/dhqp/pdf/isolation2007.pdf.
21. **Boyce JM, Pittet D.** 2002. Guideline for Hand Hygiene in Health-Care Settings. Recommendations of the Healthcare Infection Control Practices Advisory Committee and the HICPAC/SHEA/APIC/IDSA Hand Hygiene Task Force. *MMWR Recomm Rep* **51**(RR-16):1–45, quiz CE1–CE4.
22. **National Association of State Public Health Veterinarians.** 2007. Compendium of Measures to Prevent Disease Associated with Animals in Public Settings, 2007; Appendix C—HandWashing Recommendations to Reduce Disease Transmission From Animals in Public Settings. *MMWR* **56:**16–17. https://www.cdc.gov/mmwr/preview/mmwrhtml/rr5605a1.htm.
23. **Cherednichenko G, Zhang R, Bannister RA, Timofeyev V, Li N, Fritsch EB, Feng W, Barrientos GC, Schebb NH, Hammock BD, Beam KG, Chiamvimonvat N, Pessah IN.** 2012. Triclosan impairs excitation-contraction coupling and Ca2+ dynamics in striated muscle. *Proc Natl Acad Sci USA* **109:**14158–14163.
24. **Sickbert-Bennett EE, Weber DJ, Gergen-Teague MF, Sobsey MD, Samsa GP, Rutala WA.** 2005. Comparative efficacy of hand hygiene agents in the reduction of bacteria and viruses. *Am J Infect Control* **33:**67–77.
25. **Occupational Safety and Health Administration, U.S. Department of Labor.** 2006. Major Requirements of OSHA'S Respiratory Protection Standard 29 C.F.R. 1910.134.
26. **Grinshpun SA, Haruta H, Eninger RM, Reponen T, McKay RT, Lee SA.** 2009. Performance of an N95 filtering facepiece particulate respirator and a surgical mask during human breathing: two pathways for particle penetration. *J Occup Environ Hyg* **6:**593–603.
27. **Public Health Agency of Canada.** 2004. *Laboratory Biosafety Guidelines*, 3rd ed., p 1–113. Minister of Health, Population and Public Health Branch, Centre of Emergency Preparedness and Response, Ottowa, Ontario, CA. http://www.phac-aspc.gc.ca/publicat/lbg-ldmbl-04/index-eng.php.
28. **Clinical and Laboratory Standards Institute.** 2014. *Protection of Laboratory Workers From Occupationally Acquired Infections: Approved Guideline*, 4th ed. Clinical and Laboratory Standards Institute, Wayne, PA.
29. **Manning SE, Rupprecht CE, Fishbein D, Hanlon CA, Lumlertdacha B, Guerra M, Meltzer MI, Dhankhar P, Vaidya SA, Jenkins SR, Sun B, Hull HF.** 2008. Human rabies prevention—United States, 2008: recommendations of the Advisory Committee on Immunization Practices. *MMWR Recomm Rep* **57**(RR-3):1–28.
30. **World Health Organization.** 2012. *Guidance on regulations for transport of infectious substances.* World Health Organization, Geneva, Switzerland.
31. **American Society for Microbiology.** 2011. *Sentinel Laboratory Guidelines for Suspected Agents of Bioterrorism and Emerging Infectious Disease.* American Society for Microbiology, Washington, DC.
32. **Public Health Agency of Canada.** 2015. *Canadian Biosafety Standards (CBS) Second Edition for Facilities Handling and Storing Human and Terrestrial Animal Pathogens and Toxins.* Public Health Agency of Canada, Ottawa, Ontario, Canada.
33. **National Institute for Occupational Safety and Health, U.S. Department of Health and Human Services.** 1996. *Preventing Occupational Hearing Loss-A Practical Guide.* U.S. Department of Health and Human Services, Washington, DC.
34. **Seward JP.** 2001. Medical surveillance of allergy in laboratory animal handlers. *ILAR J* **42:**47–54.
35. **Weiss MM, Weiss PD, Weiss DE, Weiss JB.** 2007. Disrupting the transmission of influenza a: face masks and ultraviolet light as control measures. *Am J Public Health* **97**(Suppl 1):S32–S37.
36. **Taylor DM, Brown JM, Fernie K, McConnell I.** 1997. The effect of formic acid on BSE and scrapie infectivity in fixed and unfixed brain-tissue. *Vet Microbiol* **58:**167–174.
37. **Saunders SE, Bartelt-Hunt SL, Bartz JC.** 2008. Prions in the environment: occurrence, fate and mitigation. *Prion* **2:**162–169.
38. **Posthaus H, Bodmer T, Alves L, Oevermann A, Schiller I, Rhodes SG, Zimmerli S.** 2011. Accidental infection of veterinary personnel with Mycobacterium tuberculosis at necropsy: a case study. *Vet Microbiol* **149:**374–380.
39. **Herwaldt BL.** 2001. Laboratory-acquired parasitic infections from accidental exposures. *Clin Microbiol Rev* **14:**659–688.
40. **American Association of Veterinary Laboratory Diagnosticians (AAVLD).** 2010. *Requirements for an Accredited Veterinary Medical Diagnostic Laboratory, Version 6.2.* American Association of Veterinary Laboratory Diagnosticians, Visalia, CA.
41. **National Institute for Occupational Safety and Health, U.S. Department of Health and Human Services.** 1985. *Memorandum: Retention of Exposure Records Under 29 C.F.R. 1910.1020.*
42. **Gronvall GK, Fitzgerald J, Chamberlain A, Inglesby TV, O'Toole T.** 2007. High-containment biodefense research laboratories: meeting report and center recommendations. *Biosecur Bioterror* **5:**75–85.
43. **Sewell DL.** 2004. Laboratory Safety, p 446–472. *In* Garcia LS (ed), *Clinical Laboratory Management.* ASM Press, Washington, DC.
44. **World Organisation for Animal Health.** 2008. Biosafety and Biosecurity in the Veterinary Microbiology Laboratory and Animal Facilities, p 15–26. *In* Commission OBS (ed), *Manual of Diagnostic Tests and Vaccines for Terrestrial Animals*, 6th ed. Office International des Epizooties, Paris, France.
45. **Centers for Disease Control and Prevention, U.S. Department of Agriculture.** 2014. Select Agents and Toxins List. http://www.selectagents.gov/SelectAgentsandToxinsList.html. Accessed June 2, 2016.
46. **Fleming DO, Hunt DL (ed).** 2006. *Biological Safety: Principles and Practices*, 4th ed. ASM Press, Washington, DC.

Special Considerations for Animal Agriculture Pathogen Biosafety

34

ROBERT A. HECKERT, JOSEPH P. KOZLOVAC, AND JOHN T. BALOG

The food and agriculture industry in some countries is often a concentrated, highly accessible, vertically integrated, global, and complex system that relies on a sophisticated agricultural infrastructure. These characteristics make some agricultural systems very productive and efficient; however, these same qualities make this industry inherently vulnerable to foreign/transboundary animal, emerging, and zoonotic disease outbreaks that could threaten the stability of the economy, food security, and the nation's public health. Thus, there is a continuing need to ensure that basic and applied research in agricultural biosafety and biosecurity be adequately supported to ensure that the agricultural system remains productive, economical, and, most of all, safe (food security).

Veterinary diagnostic and research activities are critical and fundamental components of every nation's veterinary and public health systems, and institutions conducting these vital operations need to have a robust biorisk management program in place to support the safe and successful day-to-day operation of these institutions. Toward that goal, this chapter will discuss the unique risks and operational and infrastructure challenges related to veterinary research and diagnostic activities related to endemic and transboundary animal diseases of agricultural significance.

RISK ASSESSMENT

Many institutions and organizations throughout the world must consider the risks presented by proposed research with agricultural pathogens and make decisions regarding the placement of these pathogens into proper biocontainment and/or biosafety categories (1). There are a number of risk assessment and hazard assessment modalities that have been employed to assess hazards that can impact the worker, community, animal health, and environment. These methodologies vary and may either be quantitative or qualitative in nature. Risk assessment and management methodologies that have been used in animal agriculture and the food production industries, as related to food safety and food defense, include operational risk management, hazard analysis and critical control points (HACCP), CARVER + Shock (a six-step approach to conducting security vulnerability assessments on critical infrastructure), and consequence modeling (2). It has been the authors' experience, as biosafety professionals, that the operational risk management approach is the one that most closely emulates the process used at various life science research facilities to manage research involving biological hazards. Operational risk management is a qualitative process that can be broken down into six key steps:

1. identifying hazards,
2. assigning risks,
3. analyzing risk controls,
4. making control decisions,
5. implementing selected controls against the risks, and
6. supervising and reviewing the process.

Most biosafety professionals are familiar with the risk management approach described above as it is applied to biomedical research and clinical diagnostic operations conducted in support of human health. Indeed, the vast majority of currently available industry standards and guidance documents on biosafety describe facility design, work practices, and engineering controls to protect laboratory workers and the population in the surrounding community by containing biological hazards and sterilizing laboratory waste prior to disposal. Whereas these strategies are successful in human biomedical and clinical operations, they do not adequately address the additional risks of pathogen escape to the environment and the potential consequences to the agricultural economy (e.g., loss of trade in agricultural commodities, morbidity and mortality to livestock, long-term environmental contamination, etc.).

The U.S. Public Health Service publication *Biosafety in Microbiological and Biomedical Laboratories* (BMBL) biosafety guidance is designed mainly to protect people—those working directly with or around the biohazardous agent and also those in the community (3). It was not until its 5th edition that the BMBL included a description of agricultural pathogen safety in Appendix D (3, 4). However, this new Appendix in the BMBL did not offer any guidance on risk management of agricultural pathogens used in studies in biomedical environments. Therefore, this chapter identifies the different risks and approaches to risk assessment that need to be considered in agricultural research and diagnostic operations that support animal health.

Biosafety professionals and others engaged in the conduct of risk assessment for research and diagnostic activities involving livestock pathogens must recognize that the rationale for agricultural standards will differ from those for human public health standards and those for worker protection. Because the pathogens of concern for humans are either zoonotic or human disease agents, there are always susceptible hosts within (worker risk) and outside (community risk) the local laboratory environment, as compared to most agricultural pathogens (except those that are zoonotic) for which humans are not susceptible hosts. A risk assessor or manager has additional risk management options available for controlling an agricultural hazard that does not represent a direct risk to worker or public health. In an agricultural setting, considerations for seasonal separation between the agent of concern and the susceptible agriculture commodity include climatic and geographic factors. Host and/or vector availability outside the research environment can influence the risk assessment and be a critical factor when determining the appropriate biocontainment level.

It is important that biosafety professionals understand that the primary risks in agriculture research are (i) the potential economic impact on animal and plant morbidity and mortality and (ii) the international trade implications of disease presence in the country (1). There are many veterinary pathogens (Table 1) that infect agricultural species (including bees, mollusks, amphibians, and aquatic species) that could cause economic problems to a region, state, or nation. In 2015, the list provided by the World Organization for Animal Health (OIE) included 117 diseases (5). The United States and other developed countries have invested significant resources, both human and financial, to eliminate many of the most economically damaging disease agents from their national herds. However, many of these pathogens are still widely distributed around the world and continue to cause agriculture production losses and could be reintroduced into the United States (6). Additionally, newly emerging pathogens that impact livestock species, such as Hendra virus (equine), Nipah virus (swine), and Ebola Reston virus (swine), may cause significant agricultural losses (7, 8). For this reason, it has been difficult to devise a single global risk ranking for agricultural pathogens due to the variable factors of economic and trade impacts should an outbreak occur as well as the differences in disease status between countries and regions within countries (9). Several nations have developed biocontainment guidance specific to animal agriculture; however, the risk ranking of transboundary animal diseases and biocontainment recommendations are not uniform.

The lack of U.S. specific agricultural biocontainment guidance was identified in the 2009 Report of Transfederal Taskforce on Biosafety and Biocontainment Oversight (10). This taskforce recommended the development of comprehensive biocontainment guidelines comparable to the BMBL to cover research, including high- and maximum-containment research, on plant, livestock, and other agriculturally significant pests and pathogens (an agriculture-specific BMBL). Since the original recommendation in 2009, the USDA has made progress in developing this document. A USDA steering committee comprising

TABLE 1.

Agents of greatest concern to the health and productivity of agricultural animals in the world[a]

Multiple species diseases, infections, and infestations

Anthrax
Bluetongue
Brucellosis (*Brucella abortus*)
Brucellosis (*Brucella melitensis*)
Brucellosis (*Brucella suis*)
Crimean-Congo hemorrhagic fever
Epizootic hemorrhagic disease
Equine encephalomyelitis (eastern)
Foot-and-mouth disease
Heartwater
Infection with Aujeszky's disease virus
Infection with *Echinococcus granulosus*
Infection with *Echinococcus multilocularis*
Infection with rabies virus
Infection with Rift Valley fever virus
Infection with rinderpest virus
Infection with *Trichinella* spp.
Japanese encephalitis
New World screwworm (*Cochliomyia hominivorax*)
Old World screwworm (*Chrysomya bezziana*)
Paratuberculosis
Q fever
Surra (*Trypanosoma evansi*)
Tularemia
West Nile fever

Sheep and goat diseases and infections

Caprine arthritis/encephalitis
Contagious agalactia
Contagious caprine pleuropneumonia
Infection with *Chlamydophila abortus* (enzootic abortion of ewes, ovine chlamydiosis)
Infection with peste des petits ruminants virus
Maedi-visna
Nairobi sheep disease
Ovine epididymitis (*Brucella ovis*)
Salmonellosis (*S. abortusovis*)
Scrapie
Sheep pox and goat pox

Swine diseases and infections

African swine fever
Infection with classical swine fever virus
Nipah virus encephalitis
Porcine cysticercosis
Porcine reproductive and respiratory syndrome
Transmissible gastroenteritis

Lagomorph diseases and infections

Myxomatosis
Rabbit hemorrhagic disease

Other diseases and infections

Camelpox
Leishmaniasis

Fish diseases

Epizootic hematopoietic necrosis
Infection with *Aphanomyces invadans* (epizootic ulcerative syndrome)
Infection with *Gyrodactylus salaris*
Infection with highly polymorphic region (HPR)-deleted or HPR0 infectious salmon anemia virus
Infection with salmonid alphavirus
Infectious hematopoietic necrosis
Koi herpesvirus disease
Red sea bream iridoviral disease
Spring viremia of carp
Viral hemorrhagic septicemia

Cattle diseases and infections

Bovine anaplasmosis
Bovine babesiosis
Bovine genital campylobacteriosis
Bovine spongiform encephalopathy
Bovine tuberculosis
Bovine viral diarrhea
Enzootic bovine leukosis
Hemorrhagic septicemia
Infectious bovine rhinotracheitis/infectious pustular vulvovaginitis
Infection with *Mycoplasma mycoides* subsp. *mycoides* SC (contagious bovine pleuropneumonia)
Lumpy skin disease
Theileriosis
Trichomonosis
Trypanosomosis (tsetse-transmitted)

Equine diseases and infections

Contagious equine metritis
Dourine
Equine encephalomyelitis (western)
Equine infectious anemia
Equine influenza
Equine piroplasmosis
Glanders
Infection with African horse sickness virus
Infection with equid herpesvirus-1 (EHV-1)
Infection with equine arteritis virus
Venezuelan equine encephalomyelitis

Avian diseases and infections

Avian chlamydiosis
Avian infectious bronchitis
Avian infectious laryngotracheitis
Avian mycoplasmosis (*Mycoplasma gallisepticum*)
Avian mycoplasmosis (*Mycoplasma synoviae*)
Duck virus hepatitis
Fowl typhoid
Infection with avian influenza viruses
Infection with influenza A viruses of high pathogenicity in birds other than poultry, including wild birds
Infection with Newcastle disease virus
Infectious bursal disease (Gumboro disease)
Pullorum disease
Turkey rhinotracheitis

Bee diseases, infections, and infestations

Infection of honey bees with *Melissococcus plutonius* (European foulbrood)
Infection of honey bees with *Paenibacillus larvae* (American foulbrood)
Infestation of honey bees with *Acarapis woodi*
Infestation of honey bees with *Tropilaelaps* spp.
Infestation of honey bees with *Varroa* spp. (varroosis)
Infestation with *Aethina tumida* (small hive beetle).

Mollusc diseases

Infection with abalone herpesvirus
Infection with *Bonamia exitiosa*
Infection with *Bonamia ostreae*
Infection with *Marteilia refringens*
Infection with *Perkinsus marinus*
Infection with *Perkinsus olseni*
Infection with *Xenohaliotis californiensis*

Crustacean diseases

Crayfish plague (*Aphanomyces astaci*)
Infection with Yellowhead virus
Infectious hypodermal and hematopoietic necrosis
Infectious myonecrosis
Necrotizing hepatopancreatitis
Taura syndrome
White spot disease
White tail disease

[a]From reference 5.

Agricultural Research Service (ARS) and Animal and Plant Health Inspection Service (APHIS) staff was formed to move this project forward. The USDA steering committee developed a draft table of contents that was put forward for public comment, and revisions to the table of contents were made. Over the next several years, the entire document will be developed by groups of subject matter experts. Appendix D, which first appeared in the 5th edition of the BMBL, will be updated for the 6th edition by the USDA. It is anticipated that the USDA biocontainment guidance document will be released prior to the 7th edition of the BMBL and Appendix D will then not be incorporated into that edition.

Risk assessment with biohazardous material should always start with the proposed agent and its characteristics. With regard to pathogens of veterinary significance, some issues to consider would be:

1. Is the agent endemic or exotic to the country or region?
2. What is known regarding the morbidity and mortality caused by the agent in all susceptible species, including humans?
3. What is known regarding the infectious dose, shedding patterns, and epidemiology of the disease agent in relevant host species?
4. Are there effective prophylaxes, treatments, or vaccines available in animals and humans?
5. Are there active control, eradication, or outbreak response programs for the disease, locally, nationally, or internationally?
6. What is known about the environmental stability, quantity, and concentration of the agent?
7. How will the agents be used in animals (large or small) and/or in the laboratory?
8. What is the host range of the agent, and is there ongoing surveillance testing in the region or country?

Chapter 1.1.3 of the OIE *Manual of Diagnostic Tests and Vaccines for Terrestrial Animals* discusses the need for each country to adopt a risk analysis approach to the management of biological risks for biosafety and biosecurity in veterinary laboratories and animal facilities. This would provide each country with a means of tailoring their relevant national animal health policies and procedures regarding their laboratories to their particular circumstances and priorities (5). This chapter states,

> The risk analysis approach moves towards a comprehensive biological risk management framework that is science-based and specific to an individual country and laboratory's circumstances. The process could accommodate the assigning of pathogens to risk groups relevant to the country and the subsequent restriction of the associated work to laboratory facilities defined by containment levels tailored to the types of risk identified if this suits an individual country's requirements as identified by its biological risk analysis.

The factors the OIE considers to be important in making a risk assessment for work with animal diseases include:

1. Epidemiology of the biological agent; routes of transmission, including aerosol, direct contact, fomites, vectors; infectious dose, susceptible species, and likely extent of transmission.
2. Origin of the agent outside the host.
3. May cause human or animal disease (severity of harm for laboratory workers, public health, and animal health).
4. Impacts associated with animal population morbidity and mortality and associated economic consequences (e.g., trade, food security, costs of disease control and movement controls, destocking, or vaccination) dependent on whether the agent is exotic or endemic to the country or region.
5. Nature of the laboratory procedures to be conducted in a facility (e.g., small- versus large-scale amplification, use and storage of the agent).
6. Use of animals in association with the biological agent or toxin.

More details on each factor can be found in Appendix 1.1.3.2, Considerations Used in Evaluating and Implementing Biological Risk Control Measures (5).

Considering the need for each facility and country to conduct its own localized risk assessment, the OIE does not describe specific containment levels for laboratory and animal facilities appropriate for each risk group (5). Instead, risk management guidance is provided. Two examples of how this might be done are described in Guideline 3.5 of the OIE *Manual of Diagnostic Tests and Vaccines for Terrestrial Animals* (5). In this guideline, risk management falls under four broad categories: (i) administrative controls, (ii) operational controls, (iii) engineering controls, and (iv) personal protective equipment (PPE). More details of each control can be found in the guideline.

One must also take into account the differences in facilities that conduct veterinary research versus facilities that conduct veterinary diagnostic activities. In research facilities, the agent and activity risks are known and these can be addressed in the risk assessment of the proposed work, whereas the nature of diagnostic activities requires individuals to work with samples of unknown status. The Biosafety Principles and Practices for the Veterinary Diagnostic Laboratory chapter (11) and U.S. CDC Guidelines for Safe Work Practices in Human and Animal Medical Diagnostic Laboratories (12) are two recommended resources that can assist veterinary diagnostic facilities with risk assessment and mitigation measures specific to their activities.

DEFINITIONS OF BIOSAFETY AND BIOSECURITY IN AGRICULTURE

Terminology such as "biosafety," "biosecurity," "biocontainment," and "biosurety" can mean different things in different contexts to different individuals. In the context of the international animal health or public health arenas (or where they overlap), it is important to have a clear understanding of the concepts behind the words. However, regardless of the terminology used, both the animal health and public health research communities have a responsibility to ensure that research is conducted responsibly and that effective biosafety and biocontainment practices and oversight of research activities at life science facilities are recognized as critical components of the research enterprise. It has been observed for many years that infectious agents do not respect human-derived borders and that the emergence of disease is based on the interactions among animals, humans, and pathogens in complex ecosystems. It has been recognized that many of the emerging diseases that have impacted public health were and continue to be zoonotic in nature and had emerged from human interaction with livestock or wildlife. Therefore, those involved in animal or agricultural research have a duty to ensure that appropriate biosafety and biocontainment practices are in place for the manipulations of these agents to protect the researcher and community from accidental exposure and the environment from accidental release. There is also a duty shared by all life science institutes to ensure that appropriate laboratory security measures are in place to assure that working stocks and archival collections/inventories of infectious agents are protected from theft or misuse.

Biosafety and Biocontainment

Biosafety, or biological safety, is a concept that promotes safe laboratory practices and procedures and the proper use of containment equipment and facilities by workers in the life science environment to prevent occupationally acquired infections or release of organisms to the environment. The development of the tenets of biosafety has paralleled the development of the science of microbiology and its extension into new and related areas (tissue culture, recombinant DNA, animal studies, biotechnology, and synthetic biology). A number of international and national organizations have developed guidance documents on this specific topic of biosafety; however, very few regulatory or guidance documents address the issues related to animal pathogens in agricultural research and diagnostic environments. Therefore, this chapter focuses on biosafety and biocontainment as they relate to protecting the environment and the susceptible animals in that environment.

The term "biocontainment," which was first coined in 1985 (although the underlying concept has been recognized since the 1940s), is defined as the containment of extremely pathogenic organisms (such as viruses), usually by isolation in secure facilities to prevent their accidental release to the environment—a definition that is very applicable to the agricultural situation (13).

Biosecurity

Biosecurity, specifically in the agricultural setting, has been defined in many different ways. Depending upon the situation, background, or organization, different terms have been used over the years and are now often confused. The original use of biosecurity in agriculture started with farmers preventing disease transmission within and between farms. This is still used today and often called "on-farm biosecurity." A common and simple definition is "those measures taken to keep diseases out of populations, herds, or groups of animals where they do not currently exist or to limit the spread of disease within the herd." This term is also used commonly in "specific-pathogen-free" stocks of laboratory animals in research environments where extraneous diseases are being kept out of the colony. Another definition of agricultural biosecurity introduces the term bioexclusion. Mee et al. defined biosecurity as having two components. Bioexclusion relates to preventive measures (risk reduction strategies) designed to avoid the introduction of pathogenic infections (hazards), whereas biocontainment relates to measures to limit within-farm transmission of infectious hazards and onward spread to other farms (14).

The Food and Agriculture Organization of the United Nations defines biosecurity in a much broader way as "a strategic and integrated approach that encompasses the policy and regulatory frameworks (including instruments and activities) that analyze and manage risks in the sectors of food safety, animal life and health, and plant life and health, including associated environmental risk. Biosecurity covers the introduction of plant pests, animal pests and diseases, and zoonoses, the introduction and release of genetically modified organisms (GMOs) and their products, and the introduction and management of invasive alien species and genotypes. Biosecurity is a holistic concept of direct relevance to the sustainability of agriculture, food safety, and the protection of the environment, including biodiversity" (15).

In the research laboratory or vivarium settings where highly infectious animal pathogens are being used, biosecurity is often used and defined as "the protection of microbial agents from loss, theft, diversion or intentional misuse." This definition is consistent with the American Biological Safety Association (ABSA) usage of this term, the current World Health Organization (WHO) (16) usage, and the U.S. government's usage (3) and is also very relevant to the agricultural research and diagnostic setting as well.

FIELD BIOSAFETY

Working with large animals in a field setting is a common and necessary task for those who handle livestock. Large agricultural animals are often kept in open pens or on the open range because of their size and difficulties housing them. People often need to capture, control, and work closely with these species. Such activities generate extra hazards. The first is the physical challenge of dealing with large animals that are often fractious and potentially life threatening.

Physical Injury

When working with large animals, controlling the animals through physical restraints, such as penning and gating, is most important to ensure the animals are secured and contained so that they cannot cause harm to the handler or to themselves. Once the animal is corralled, the handler can further restrain the animal via physical means, such as ropes, halters, snares, squeeze chutes, etc., or via chemical means, such as injectable or inhalant anesthetics. However, these restraint items in themselves can be hazards (chemical or physical) and need to be handled with caution and experience. It is important to know the specifics of the species that you are working with, which dictate the type of physical or chemical restraint needed and the comfort zone of an animal (their flight zone and reaction to fear). Each species has different comfort zones and degrees of domestication (flight zone), which need to be understood and respected when interacting with them. An animal handler who has a good understanding of the species flight zone and the concept of point of balance can move animals safely without the need for prodding while reducing the risk to themselves. The size of the flight zone varies depending on the tameness or wildness of the livestock, which can vary greatly within the same species (Fig. 1) (17).

Some animals require very tall fences; some require narrow spaces between bars in the penning, whereas others require "hides" (places where they can go that are not in a direct line of sight with a human) so as not to hurt themselves, etc. Any personnel attending large animals should always have an escape route in mind at all times; this can include manways (places where people can go but animals cannot) and going behind or over a fence. The situations and the species are too varied to provide any

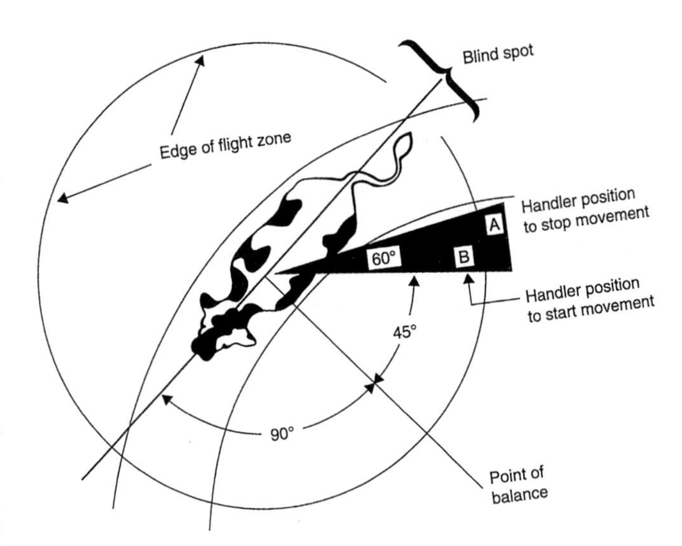

Figure 1: Livestock behavioral response characteristics. (Reprinted from the *Journal of Professional Animal Science* [17] with permission of the publisher.)

single guidance in this chapter. However, there are many good textbooks and monographs written about handling domestic livestock and wildlife that the reader should seek out, including the *Guide for the Care and Use of Agricultural Animals in Research and Teaching* (18). A summary of hazards and concerns when working in the field with animals has been published (19). The institution via the staff veterinarian should make sure that all staff are adequately trained to work with the particular animal species being handled. However, ultimately it is the responsibility of the animal care workers and their supervisor to understand the species being dealt with and take the necessary precautions as part of doing a thorough risk assessment. If animal care staff are uncomfortable or uncertain, they should stop work and request further training or expert guidance.

Preventing Zoonotic Disease Exposure

In addition to the potential for physical injuries to occur when working with large animal species, there is also the risk of exposure to zoonotic diseases. Reports cataloging human disease agents found that of 1,415 agents causing disease in humans, 868 (61%) are zoonotic (20, 21). Because of their frequent contact with a wide variety of species in clinical and field settings, veterinary personnel, researchers, diagnosticians, and animal care workers are at particularly high risk of contracting zoonotic diseases (22, 23). In addition to existing and known zoonotic diseases, new zoonoses are continually emerging, making it even more difficult to protect oneself (24, 25). Despite these facts, infection control practices in the veterinary field and clinic are variable and often are not sufficient to prevent zoonotic disease transmission (26). Therefore, it would be prudent for veterinarians and animal care workers to apply standard infection control procedures when working in the field similar to those used for hospital infection control (27, 28). Although eliminating all risk from zoonotic pathogens during fieldwork is not achievable, this section will provide some guidance for minimizing disease and injury.

In 2015, the National Association of State Public Health Veterinarians (NASPHV) Veterinary Infection Control Committee published a compendium of veterinary standard precautions addressing zoonotic disease prevention in veterinary personnel (23). This document does an excellent job of reviewing the standard precautions one should take while working with animals in the field or veterinary clinic and describing an infection control plan. All entities working with animals infected with zoonotic diseases should have a written infection control plan, which should be specific to the facility and disease being studied. A model infection control plan is available in electronic format from the NASPHV website (23).

Below are some points from that compendium with additional recommendations for work, specifically with large agricultural species in the field.

Hand hygiene

Consistent, thorough hand hygiene is the single most important measure veterinary personnel can take to reduce the risk of disease transmission. Hands should be washed between animal contacts and after contact with blood, body fluids, secretions, excretions, and equipment or articles contaminated by them. When running water is not available, antimicrobial-impregnated wipes (i.e., towelettes), followed by alcohol-based gels, may be used. Used alone, wipes are not as effective as alcohol-based hand gels or washing hands with soap and running water. Therefore, when working in the field in remote locations, it would be prudent to take along a sufficient supply of water for effective hand washing.

Use of protective gloves and sleeves

Use of gloves and sleeves reduces the risk of pathogen transmission by providing barrier protection. They should be worn when touching blood, body fluids, secretions, excretions, mucous membranes, and nonintact skin. However, wearing gloves (including sleeves) does not replace hand washing. The gloves should be removed promptly after use, avoiding contact between skin and the outer glove surface. Disposable gloves should not be washed and reused. Hands should be washed immediately after glove removal. Gloves come in a variety of materials and should be strong enough to prevent tearing while working in the field with large animal species. This may require thicker gloves than the normal latex exam gloves or double gloving (the outer pair being more robust).

Facial protection

Facial protection prevents exposure of mucous membranes of the eyes, nose, and mouth to infectious materials. Facial protection should be used whenever exposures to splashes or sprays are likely to occur and should include a mask worn with either goggles or a face shield. A surgical mask may provide adequate protection from noninfectious particulates generated during most procedures that do not involve potentially infectious aerosols. It should be noted, however, that wearing this extra protection can increase breathing difficulty. Glasses may fog easily and shields can be knocked off. Therefore, a careful determination should be made as to when to use the extra protection to reduce risk and when the extra protection may actually increase risk.

Respiratory protection

Respiratory protection is designed to protect the respiratory tract from zoonotic organisms transmitted via inhalation of infectious aerosols. It may be necessary in certain situations, such as when investigating abortion storms in small ruminants (Q fever), significant poultry mortality (avian influenza), ill psittacines (avian chlamydiosis), potential anthrax-spore-containing soil, or other circumstances where there is concern about aerosol transmission. The N95-rated disposable particulate respirator is a mask that is inexpensive, readily available, and easy to use. However, there may be challenges in getting staff to wear the required respirator under field settings, which are either very hot or cold, or where there is a lot of exertion required. Work with large animal species often requires excessive physical exertion and therefore may make wearing a respirator very difficult for certain work activity. Administrative controls, such as limiting the time an individual works while wearing PPE and having an area upwind of animals designated as the area for staff to take a break from the work, may be a natural and simple way of reducing potential exposure to biohazardous aerosols. When respirators are used in the United States, compliance with the Occupational Safety and Health Administration (OSHA) Respiratory Protection Standard (29 C.F.R. 1910. 134) 117 is required. This includes a medical evaluation of the user, annual fit testing of the respirator on the user, and training in proper use (29).

Protective outerwear

Lab coats, smocks, and coveralls

This apparel is designed to protect street clothes or scrubs from contamination but is generally not fluid resistant. Thus, such apparel should not be used in situations where splashing or soaking with potentially infectious liquids is anticipated. In the field, coveralls are often the most practical and effective outerwear that protects the entire body while not overly impeding the worker's range of motion.

Garments should be changed promptly whenever visibly soiled or contaminated with body fluids or feces. These garments should not be worn outside of the work area, especially not in a vehicle after use or back home. After use in the field, they should be placed in an autoclave bag and taken to an autoclave for sterilization.

Footwear

Footwear should be suitable for the specific working conditions (e.g., rubber boots for farm work) and should protect veterinary personnel from exposure to infectious material as well as trauma. Recommendations include shoes or boots with thick soles and closed toe construction (including steel toe inserts), which are impermeable to liquid and easily cleaned. Footwear should be cleaned to prevent transfer of infectious material from one environment to another, such as between farm visits and before returning from a field visit to a veterinary facility or home. Disposable plastic shoe covers or booties add an extra level of protection when small quantities of infectious materials are present or expected.

Head covers

Disposable head covers provide a barrier when gross contamination of the hair and scalp is expected. Disposable head covers should not be reused.

LARGE ANIMAL NECROPSY

Doing a full necropsy or collecting samples in a research setting is a common task in the research vivarium or field. However, this activity generates additional risk to the workers doing the procedure because of increased potential for cuts and needle sticks during the process and the potential for exposure to pathogens (commensal or experimental) that the animal may contain. When possible, it is recommended that a primary containment device (ventilated enclosure) be used for the smaller species (e.g., chickens, mice, rats, rabbits, fish, etc.). However, this is not possible with larger species, and the animals are manipulated in the open on a table or on the floor. In these cases, extra precautions need to be taken by the personnel to prevent aerosol, percutaneous, mucous membrane, or oral exposure to the infectious agents. The areas in which to take precautions and some suggestions to reduce exposure include:

1. Aerosol exposure:
   a. Minimize procedures that create aerosols during necropsy such as opening gas-distended pouches, excessive bone sawing, spraying water, etc.
   b. Use a downdraft table or other similar ventilated engineering control if appropriate for the species and procedure to draw infectious aerosols away from the face and breathing zone.
   c. Use a respirator (e.g., N95 mask, powered-air-purifying respirator, half- or full-face respirator with HEPA filters, etc.) during certain aerosol-generating procedures of the necropsy.
   d. Use a ventilated enclosure (e.g., flexible film isolator, hanging curtains with directional airflow, etc.) for procedures known to create infectious aerosols (e.g., cutting bone).
2. Percutaneous exposure via cuts or needle stick:
   a. Use extreme caution when using knives and scissors when doing the necropsy to avoid injury.
   b. Use protective cut-resistant clothing, in particular gloves, that help minimize minor cuts and punctures to the skin.
   c. Use engineered safe sharps whenever possible, such as retractable needles, sheathed scalpels, blunt-tipped scissors, etc.
   d. Do not recap needles unless absolutely required by the protocol or process. If recapping is required, then use a one-handed technique.
3. Mucous membrane exposure:
   a. Use face shields, goggles, face masks, or other personal protective equipment to cover as much of the mucous membranes (mouth, nose, and eyes) as possible during procedures that can create splashes.
4. Oral exposure:
   a. Use a face mask to cover the mouth and nose to ensure nothing enters the oral cavity. This will also protect from accidentally touching the mouth and nose with gloved hands that may carry infectious organisms.
   b. Never attempt to recap needles by holding the cap between your teeth while inserting the needle into the cap.

Because of the difficulty in physically manipulating a large animal species, wearing a lot of personal protective equipment is challenging. Workers become overheated and may be tempted to remove the extra layers of protective equipment, or protective equipment may get dislodged due to contact with other heavy equipment in the necropsy area (e.g., hoists, hanging chains, cart handles, etc.). Due to the size and weight of the animal carcasses, heavy lift equipment and carts are often needed to elevate and move the carcass. In many cases, the whole carcass is too big to fit into a cart, destruction equipment charging head, freezer, or doorway and needs to be cut into smaller parts, adding further risk to the operation.

Field Necropsy

Doing a necropsy is one of the most common procedures when working in the field. To determine the cause of morbidity or mortalities affecting livestock, necropsies and appropriate samples need to be collected for diagnostic testing. However, when conducting a field necropsy it is important to keep biosafety and biosecurity in mind (the same as for a necropsy in the vivarium). All carcasses should be handled as if they were harboring potentially dangerous zoonotic diseases, and precautions for personal safety should be exercised. If possible, minimize the degree and amount the carcass is opened and the amount of liquid material (blood, serum, cerebrospinal fluid, amniotic fluid, etc.) that is released. Taking a biopsy sample of the tissue or fluid required is the least invasive and safest procedure if the carcass does not need to be opened and fully examined for the work being conducted.

Minimal protective clothing includes coveralls, gloves, a mask that covers the nose and mouth, and rubber boots. If infectious aerosols are possible, then a respirator should also be used. Diseased animals should be always be handled in a way to minimize exposure of other wild and domestic animals.

In the field, the best place to necropsy an animal is:

- Away from other animals, food storage areas, and workers on the property.
- Accessible for easy and safe disposal of the carcass.
- An area that can be easily and thoroughly disinfected, preferably a concrete pad, which can be cleaned fairly easily with a good disinfectant.
- If no concrete pad is available, a dirt area would be the next likely option. Unlike concrete, the dirt area cannot be easily disinfected. For this reason it is best to have the area in direct sunlight because the heat and light will help kill many pathogens (30).

Carcass Disposal

In field or the vivarium, there will be a need to dispose of animal carcasses safely and securely. The following discussion will focus on worker biosafety, inactivation/sterilization of pathogens, and environmental safety in carcass disposal. Conducting a risk assessment as outlined at the beginning of this chapter for the respective carcass disposal methods will inform the biosafety professional on the appropriate worker PPE. Similarly, the recommendations of the NASPHV Veterinary Infection Control Committee compendium of veterinary standard precautions along with local safety policy will assist in generating a safety plan for the carcass disposal operation (23).

There are a number of methods commonly used to dispose of animal carcasses, including burial, composting, open-pit burning, air-curtain incineration, incineration in a pathological waste or medical waste incinerator, rendering, and tissue digestion via alkaline hydrolysis (31). Each of these methods requires the use of basic personal protective equipment (gloves, coveralls, and eye, face, and mucous membrane barriers) along with additional measures like procedural controls that minimize the potential for pinch-point injuries in operating chain hoists or heat-resistant garments used for controlled burn methods. A well-designed and robust worker training program is essential to the effective implementation of the biosafety measures intended to support carcass disposal in both routine and disease outbreak response operations. Open-air disposal methods and those conducted in an enclosed facility require an operating permit from state and/or federal environmental protection agencies and in some instances a state department of agriculture permit. Selecting the method for disposing animal carcasses is influenced by a number of factors, including the number and size of animals to be disposed of, whether the animals perished due to disease (either experimental or naturally occurring infection), and the availability of sufficient resources to perform the selected method safely, i.e., sufficient land for burial, equipment, device capacity, appropriate fuel to burn or hydrolyze the biomass (carcasses), and appropriate vehicles should carcasses need to be transported off site. Also of critical importance to the success of any carcass disposal operation is the availability of personnel trained on the nuances of the disposal method. The disposal of a large number of carcasses conducted as the result of a disease outbreak managed by government officials according to a state-specific plan is beyond the scope of this chapter, but can be found elsewhere (32, 33). We will focus the balance of this discussion on the disposal of experimentally infected animal carcasses in a research or diagnostic environment.

Carcass disposal in the field

In the field, the most common and practical option is burial. If possible, high ground should be found where the water table is low; however, in most cases, the carcass is buried close to the location where the animal died. If a highly infectious and transmissible disease (e.g., anthrax) is suspected, the carcasses should be covered with a disinfectant and buried at least 4 to 6 feet deep (if possible) to prevent scavenging (34).

If there is sufficient fuel and the carcasses are small, burning may also be an option, but careful attention needs to be taken to ensure a complete burn of the carcasses and that no unintentional fires are created. Additionally, composting may be an option if sufficient materials are available to make a proper compost pile and people will be available to turn the pile, as needed. As mentioned above, it is essential that site selection and method of disposal be evaluated in parallel to develop, communicate, and implement an effective plan that results in the safe and environmentally acceptable operation.

Carcass disposal in the vivarium

Disposal of carcasses in the vivarium can be simple or complex depending upon local regulations, available infrastructure, the species and number of animals being disposed of, as well as the nature of the infecting organism. In the case of small numbers of agricultural animals (e.g., chickens, quail, rabbits, fish, young stock, etc.), the biomass is not considerable and disposal is manageable. In most instances, autoclaving is sufficient to sterilize the infected carcass of small species prior to final disposal in a municipal landfill or by burning in a licensed medical waste or pathological incinerator. However, larger species or large numbers of carcasses can easily exceed the capacity of a standard-sized laboratory autoclave, resulting in the need to store carcasses temporarily. Additional options include burying, incineration, rendering, digesting, or using a combination of steam and maceration. If working in a high-containment vivarium [animal biosafety level 3 (ABSL3) and above], the charging head for the destruction equipment (renderer or alkaline hydrolysis) is usually inside of biocontainment and the operational side is outside of biocontainment to allow for easy servicing. The method of sterilization used must be validated to demonstrate that sterility of the carcass and the pathogens they might contain is achieved. Additionally, the discharge (air or effluent) must be acceptable to the local authorities (e.g., publicly owned treatment works) for release to the environment (e.g., temperature, biological or chemical oxygen demand, heavy metal concentrations, etc.) or further treated on site to ensure agreed-upon discharge parameters.

Rendering

Rendering is an environmentally friendly, multistep process that cooks butchered carcasses and animal parts, resulting in the recovery of meat and bone meal used to make animal feed and fat used for lubrication. The economic benefits of rendering have a basis in locating these facilities in close proximity to slaughterhouses, thus ensuring a steady supply of carcasses and animal parts. Some high-containment animal research facilities have onsite rendering equipment (also known as tissue autoclaves), while those working with noninfected or low-risk agricultural agents may use the services of mobile rendering units. Rendering is acceptable at all biocontainment levels as long as there is validation of the complete destruction of the pathogens under study.

Alkaline hydrolysis

Tissue digestion by alkaline hydrolysis utilizes a strong alkali in solution (sodium hydroxide or potassium hydroxide) at high temperature in a sealed system to solubilize and hydrolyze animal tissues. The process results in an effluent solution of amino acids and sugars that can be disposed of through the municipal sewer system when the temperature cools and the pH is adjusted to meet system requirements. Several different commercial companies make this equipment, and it comes in a variety of sizes.

Incineration

The Environmental Protection Agency (EPA) defines two types of waste eligible for incineration:

1. Pathological waste is "waste material consisting of only human or animal remains, anatomical parts, and/or tissue, the bags/containers used to collect and transport the waste material, and animal bedding."
2. Medical waste is cultures and stocks of infectious agents, human pathological wastes (e.g., tissues, body parts), human blood and blood products, used sharps (e.g., hypodermic needles and syringes used in animal or human patient care), certain animal wastes, certain isolation wastes (e.g., wastes from patients with highly communicable diseases), and unused sharps (e.g., suture needles, scalpel blades, hypodermic needles).

Regardless of the disposal method, animal carcasses are regulated waste under state environmental protection agencies and sometimes state departments of agriculture. In addition to federal regulations, the biosafety professional should research and become familiar with their local and state regulations to determine applicability.

BIOCONTAINMENT RESEARCH FACILITIES

In the agricultural research and diagnostic setting, worker protection is always a consideration; however, the emphasis is placed on reducing the risk of agents under study escaping to the environment (5, 35, 36). Some countries have described their own standards and practices for how they define and then contain high-risk agricultural pathogens in research facilities (36, 37). Here we describe the facility requirements and work practices of biosafety level 3 agriculture (BSL3-Ag) as used in the United States. This BSL, unique to agriculture, was developed to protect the environment from economy-threatening, high-risk pathogens (9) in situations where studies employ large animals or other situations where facility barriers, which normally serve as secondary barriers, must serve as primary containment. The USDA ARS was the first agency to define BSL3-Ag as a containment level providing detailed information on the design and construction of these specialized high-containment livestock facilities (38). More recently the BMBL 5th edition clarifies the effect that agricultural concerns have on BSL facility design (3). The new Appendix D, Agriculture Pathogen Biosafety, provides U.S.-specific guidance on what laboratory and vivarium features are necessary when work is done with veterinary agents of concern. Later in this chapter, we will also describe some of the enhancements beyond BSL3 that may be required by the USDA APHIS when working in the laboratory or vivarium with agricultural agents of concern. In the United States, special conditions apply for the possession and use of certain high-risk agricultural pathogens called select agents. Containment conditions required for work with these specific agricultural agents are provided by the USDA APHIS at the same time that USDA APHIS approves and permits a location to work with an agent (see Appendix and Table 2).

BSL3-Ag for Work with Loose-Housed Animals

In the United States, special biocontainment features are required for certain types of research involving high-consequence livestock pathogens (listed in Table 2) in animal species or other research where the room provides primary containment. To support such research, a special standard for facility design, construction, and operation called BSL3-Ag was developed. This type of facility was first defined in USDA ARS manual 242.1M-ARS, _Facilities Design Standards_, using the containment features of the standard ABSL3 facility as a starting point and including many of the features ordinarily used for ABSL4 facilities as enhancements (38).

All BSL3-Ag containment spaces must be designed, constructed, commissioned, and annually reverified as primary containment barriers (Fig. 2). The BSL3-Ag facility can be a separate building, but more often, it is an isolated zone contained within a facility operating at a lower BSL, usually BSL3. This isolated zone has strictly controlled access and special physical security measures, and it functions on the "box within a box" principle. Ideally the BSL3-Ag facility would be separate from other parts of the building with its own access from the outside to aid in separation of spaces. All BSL3-Ag facilities employing animals that cannot be readily housed in primary containment devices require the features typical of an ABSL3 facility, with the enhancements (typical of BSL4 facilities), as shown in Appendix D of the BMBL (3).

TABLE 2.

Pathogens of veterinary significance in the United States

Pathogen or disease	Pathogen or disease	Pathogen or disease
African horse sickness[a,b]	*Foot-and-mouth disease virus[a,b,c]	Porcine reproductive and respiratory syndrome
African swine fever virus[a,b,c]	*_Francisella tularensis_[a,b]	Pseudorabies virus[b]
Akabane virus	Goatpox[a,b]	Rift Valley fever virus[a,b,c]
Avian influenza virus (highly pathogenic)[a,b,c]	Hendra virus[a,b,d]	*Rinderpest virus[a,b,c]
*_Bacillus anthracis_[a,b]	_Histoplasma_ (_Zymonema_) _farciminosum_	Sheeppox[a,b]
Besnoitia besnoiti	Infectious salmon anemia virus	Spring viremia of carp virus
Bluetongue virus (exotic)[b]	Japanese encephalitis virus[b]	Swine vesicular disease virus[a,b]
Borna disease virus	Louping ill virus[b]	Teschen disease virus[b]
Bovine infectious petechial fever agent	Lumpy skin disease virus[a,b,c]	_Theileria annulata_
Bovine spongiform encephalopathy	Lyssa virus (rabies)[b]	_Theileria bovis_
Brucella abortus[a,b]	Malignant catarrhal fever virus (exotic strains or alcelaphine herpesvirus type 1)	_Theileria hirci_
Brucella melitensis[a,b]	Menangle virus	_Theileria lawrencei_
Brucella suis[a,b]	_Mycobacterium bovis_	_Trypanosoma brucei_
*_Burkholderia mallei_ (_Pseudomonas mallei_—glanders)[a,b]	_Mycoplasma agalactiae_	_Trypanosoma congolense_
*_Burkholderia pseudomallei_[a,b]	_Mycoplasma mycoides_ subsp. _Mycoides_ (small colony type)[a,b,c]	_Trypanosoma equiperdum_ (dourine)
Camelpox virus	_Mycoplasma capricolum_ subsp. _capripneumoniae_[a,b,c]	_Trypanosoma evansi_
Classical swine fever[a,b,c]	Nairobi sheep disease virus (Ganjam virus)	_Trypanosoma vivax_
Coccidiodes immitis	Newcastle disease virus (velogenic strains)[a,b,c]	Venezuelan equine encephalomyelitis[a,b]
Cochliomyia hominivorax (screwworm)	Nipah virus[a,b,d]	Vesicular exanthema virus
Cowdria ruminantium (heartwater)	Peste des petits ruminants (plague of small ruminants)[a,b,c]	Vesicular stomatitis (exotic)[b]
Coxiella burnetii (Q fever)[a,b]		Viral hemorrhagic disease of rabbits
Eastern equine encephalitis[a,b]		Wesselsbron disease virus
Ephemeral fever virus agent		Western equine encephalitis virus[b]

*Denotes Tier 1 Select Agent (see http://www.selectagents.gov for further information about U.S. select agents).

[a]Agents regulated as select agents under the Bioterrorism Act of 2002 (Public Health Security and Bioterrorism Preparedness and Response Act of 2002, Public Law 107-188). Possession of these agents requires registration with either the CDC or APHIS and a permit issued for interstate movement or importation by APHIS Veterinary Services. Most require BSL3/ABSL3 or higher containment (enhancements or on a case-by-case basis as determined by APHIS-Veterinary Services.

[b]Export license required from the U.S. Department of Commerce under authority EAR/CCL—Listed Biological Agents and Toxins, 15 C.F.R. 774, Suppl. 1 (IC 351, IC 352, IC 353, IC 354).

[c]May require BSL3-Ag containment for all work with the agent in loose-housed animals, based upon a USDA APHIS risk assessment.

[d]Requires BSL4 containment for all work with the agent.

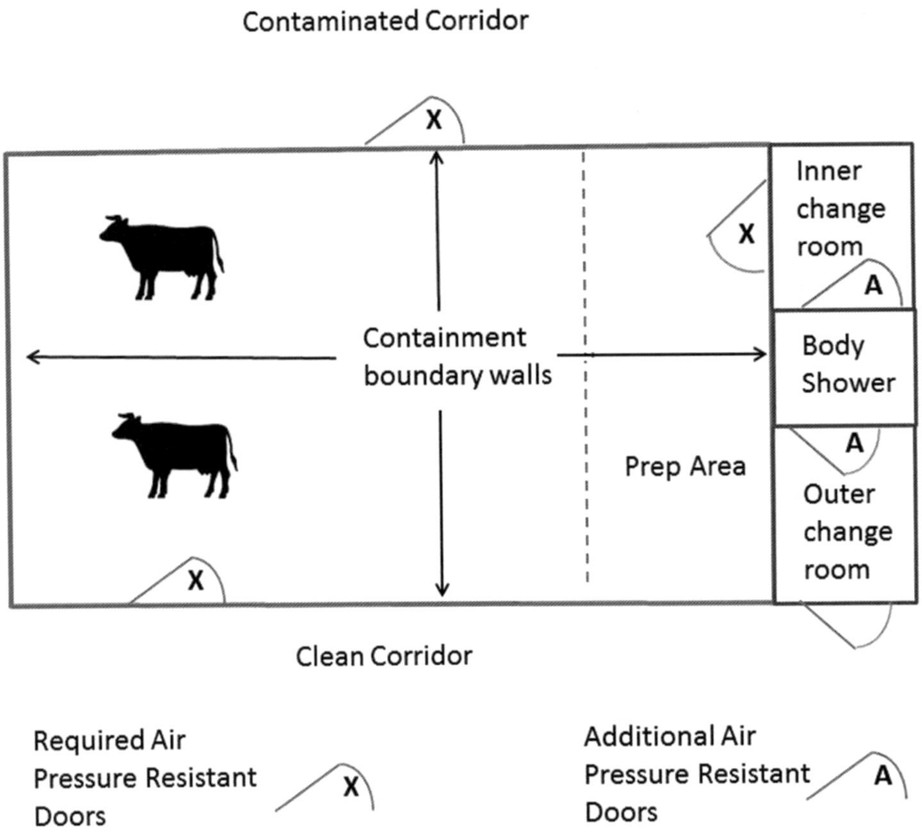

Figure 2: Large-animal BSL3-Ag room containment features.

Laboratory BSL3 and contained ABSL3 plus enhancements

The descriptions and requirements listed above for BSL3-Ag studies are based on the use of high-risk organisms in animal systems or other types of agriculture research where the facility barriers, usually considered secondary barriers, now act as primary barriers. There are circumstances where certain high-consequence livestock pathogens may be studied in animals within primary-containment devices in an enhanced BSL3 laboratory or enhanced ABSL3. In these situations, where the facility is not needed as the primary barrier, the design and testing requirements for laboratory space can reflect the difference between the two situations without compromising the required environmental protection. When working with high-consequence livestock pathogens in the laboratory or small animal facility, all manipulations with pathogenic material are conducted in an appropriately designed facility while using appropriate engineering controls and work procedures. These must meet the requirements of BSL3 or ABSL3 with the additional enhancements unique to agriculture. Therefore, in addition to meeting the basic ABSL3 requirements, the facility may be required to have the following features based on a risk assessment and local permit requirements:

- A personnel entry and exit through a clothing change and shower room
- A double-door autoclave and/or fumigation chamber
- HEPA filter supply and exhaust air, and all ductwork serving BSL3-enhanced spaces airtight and certified by pressure testing
- A liquid effluent decontamination system (preferably central heat sterilization)
- Sealed penetrations and the capability of sealing the area for gas- or vapor-phase decontamination (construction materials should be appropriate for the intended end use and compatible with the decontamination method of choice; however, because all work with infectious material is conducted within primary containment, it is not necessary to certify the room using pressure decay testing).

ABSL4 with Large Livestock and Wildlife Species

Many of the facility features that are used in a BSL3-Ag facility are also employed in an ABSL4 large-animal facility. The major difference is the provision of a life support system for the protective suits and the need for a chemical decontamination system. The suit and the associated air line that are common with standard BSL4 design present a potential hazard in the large-animal maximum-containment facility. The potential for entanglement of the line in conjunction with penning and gating and other restraint devices or with the animal is a significant concern that needs to be addressed. Ceiling-mounted tension tethers for the air lines and ceiling-mounted self-coiling air lines have been used at the few existing ABSL4 large-animal facilities worldwide. Both approaches have advantages and disadvantages. An appropriately tensioned tether suspends the line out of the reach of the animal's mouth (especially important for work with swine). This method does have limitations because the distance it is possible to travel in the room is directly related to the length of the tether. Another consideration is that the further out from the attachment the suited individual moves, the more tension the suited individual is under. Suspended coiled lines work best to keep the lines out of short animals' mouths. These systems can work with well-trained employees who know how to react and come to the assistance of fellow staff in case of line entanglement, which can occur when moving penning and gating (J. Copps, 2010, personal communication). In ABSL4 large-animal facilities, use of a buddy system or two-person rule for safety reasons is highly recommended by the authors.

Penning and gating or other animal restraints must be designed to ensure that these devices minimize pinch points and eliminate sharp edges to protect both the animal and the individuals in the laboratory. The design of this equipment needs to ensure that the equipment is sealed and uses finishes or coatings that are resistant to common disinfectants used during routine cleaning of the animal facilities.

BSL3-Ag facilities require individuals exiting the animal room to remove all protective equipment and clothing, take a body shower, and don clean clothing prior to accessing other areas within the animal facility. An ABSL4 facility needs to have a similar process in place to decontaminate the exterior of the protective suit to reduce the spread of the agents that are generated in this high-risk animal space and minimize the potential for cross-contamination if other agents are utilized within the maximum-containment suite. For this reason, it is advisable to consider including a shower (water only or chemical disinfectant) to remove gross contamination experienced in the large-animal room if returning to a BSL4 laboratory. If returning to lower containment and removing the protective suit, a chemical disinfectant shower of the suit is required. The shower room (regardless of shower type) should have interlocked air pressure-resistant doors on the shower entry and exit.

Because ABSL4 large-animal facilities are facilities in which the secondary or facility barriers are now serving as primary containment barriers, it is recommended by the authors that these facilities must undergo pressure decay testing similar to a BSL3-Ag facility. As with any other ABSL4 facility, the design and operational parameters must be met and verified prior to operation and reverified annually. Large-animal maximum-containment facilities are extremely expensive to construct and operate; thus, any entity considering such an endeavor must carefully examine the need and potential benefits against the risks and costs associated with these facilities.

FACILITY COMMISSIONING

This section provides the requirements for testing and commissioning that should be conducted at the factory and/or in the field to verify the containment integrity of the critical components of the biological containment systems. It is beneficial to have a set of as-built architectural drawings to ensure the construction conforms to the design requirements and to develop and implement a comprehensive commissioning plan.

Testing and Certification of Biosafety Cabinets

Biosafety cabinets (BSCs) shall be tested prior to initial use and annually thereafter in accordance with the latest version of NSF/ANSI Standard 49, Biosafety Cabinetry: Design, Construction, Performance, and Field Certification (39).

Testing and Certification of HEPA Filter Assemblies

Factory testing

The filter housing pressure boundary shall undergo factory testing per American Society of Mechanical Engineers (ASME) N510-1989 to a 10-inch water gauge (w.g.) with a maximum permissible leak rate of 0.2% of the housing volume per hour. The filter element sealing surface shall be factory tested by the pressure decay method as specified in ASME N510-1989.

In-place HEPA filter particulate challenge

Field test and provide written certification of all HEPA filter units with dispersed oil particulate as a particulate challenge after filter installation to verify that there are no leaks in the filter media, in the bond between the filter media in the filter frame, and in the filter frame gasket to filter housing.

Filter testing is intended to be completed in a manner consistent with industry standards for certification of HEPA filters in BSCs. An alternate procedure may be used as outlined in the USDA ARS *Facilities Design Standards* (38).

Testing and Commissioning of a Containment Room

General

The purpose of testing the containment room or envelope is to determine if the walls, floors, ceilings, penetrations, and other containment barrier features have adequate integrity to prevent leakage of air from the containment space. Testing is typically completed by subjecting the containment area to negative or positive air pressure in excess of the anticipated operating conditions and monitoring the containment air pressure over a test period.

Testing and certification will typically consist of three progressive steps: (i) pretesting for gross leaks by raising or lowering the containment space air pressure to about a 0.5-inch water column (125 Pa) and then looking and listening for major leaks, (ii) soap bubble pretesting, and (iii) pressure decay testing for final certification per USDA ARS *Facilities Design Standards* (38).

Pretesting

The integrity of the containment space to prevent leakage will largely be the result of the care used by the contractor and subcontractors to install products in accordance with the plans and specifications. The project quality assurance and quality control measures should include documentation of pretesting steps prior to acceptance testing, even if the contractor preparing the area or conducting pretesting is not responsible for final acceptance testing and certification.

Prior to testing, supply and exhaust ventilation openings shall be sealed closed, and all doors and other openings through the containment perimeter shall be placed in their normal closed positions. If the doors in the containment perimeter are not gasket sealed, they will need to be caulked temporarily or otherwise sealed to complete the testing. The testing plan should address how the openings are to be sealed.

A calibrated digital or inclined manometer shall be installed across the containment perimeter in a manner to minimize interference with wind or ventilation turbulence and to accurately represent the interior and exterior differential air pressure. The manometer shall have a display with capabilities to be easily read to an accuracy of a 0.05-inch water column (10 Pa) and to accurately read pressures to a 3-inch water column (750 Pa).

When pretesting for large leaks, the containment space may be pressurized or depressurized by installing a variable-speed "blower door" or other approved means to generate a nominal 0.5-inch water column (125 Pa) differential pressure across the containment perimeter. The building surfaces, joints, penetrations, etc., are then inspected for air leakage and sealed in accordance with the plans and specifications.

Following sealing of all leaks identified at a 0.5-inch water column (125 Pa), pretesting may proceed to soap bubble testing. Depending on the location of the containment barrier and construction, soap bubble testing may be completed under positive or negative differential pressure. Typically, testing is completed under negative pressure when the soap bubbles are readily visible on the inside surface of the containment barrier (see Appendix 9B of USDA ARS *Facilities Design Standards*) (38).

Final pressure decay testing

Prepare for testing by closing openings at the perimeter of the containment envelope and setting up testing equipment as described for pretesting. The fan/blower unit shall be capable of creating a 2-inch water column (500 Pa) pressure differential in the containment zone, and shall have a ball valve in the piping to the containment zone to allow the room or zone to be sealed once the testing pressure differential has been reached.

Testing shall be completed under generally stable conditions of outside wind, temperature, barometric pressure, and humidity. Testing shall be under negative differential pressure with respect to the surrounding environment. Air pressure testing ports/openings for the digital or inclined manometer instruments shall be located where the readings will not be affected by wind, air disturbances, or traffic.

Pressure decay testing procedure

Operate the fan/blower unit to slowly bring the differential pressure to a 2-inch water column (500 Pa), over a period of 5 to 10 minutes. Close the valve between the fan/blower and the test zone to seal the containment zone at a 2-inch water column negative pressure with respect to the adjacent areas. Record the differential pressure each minute for 20 minutes. Slowly open the seal valve to allow the room or containment zone to return to normal pressure. Decay testing may be repeated after a 20-minute wait period. Visually inspect the containment surfaces between testing and make repairs as necessary. If the acceptance criterion is not met, repeat the soap bubble testing and make repairs before retesting.

Acceptance criteria

Two consecutive pressure decay tests demonstrating a minimum of a 1-inch water column (250 Pa) negative differential pressure remaining after 20 min from an initial negative pressure differential of a 2-inch water column (500 Pa) is acceptable.

Testing and Certification of Gas-Tight Ductwork and Isolation Valves

Testing shall include all portions of the gas-tight ductwork and filter systems that may be exposed to contamination from the rooms to the respective isolation dampers on the upstream side of the supply HEPA filters and on the downstream side of the exhaust HEPA filters. Perform in-place positive-pressure testing and written certification. All welds and/or duct joints shall remain fully exposed and accessible for inspection and repair until testing is completed and certified.

Preliminary testing

Preliminary testing shall be completed using soap bubble leak detection and/or helium gas to detect leaks for repair prior to final testing and certification. Use of Freon or other chlorofluorocarbon gas is not acceptable.

Certification testing

Certification testing shall be completed using helium gas and a leak detector. The detector shall be an industrial type, adjusted for detection of leaks of 10^{-7} cc/s. Pressurize duct or assemblies to a 4-inch water column (1,000 Pa) with a helium concentration adequate to ensure that leaks will be detected. Scan the exterior surfaces of all ducts, seams, joints, gaskets, and other areas of possible leakage at a distance of ¼ to ½ inch from the surface and at an approximate rate of 1 inch/second. Currently, the acceptance is no detected leaks in excess of 10^{-5} cc/s. In many instances, the alternate pressure testing described next is the more common method used.

Alternative pressure testing

An alternative pressure testing may also be used if temperature and other environmental conditions will not affect the test. Pressure testing shall be completed by pressurizing the gas-tight assembly or ductwork to the specified pressure criteria, closing all valves and monitoring for pressure drop. Acceptance shall be zero pressure drop in 1 hour. In addition, recent recommendations and clarifications to the above from the U.S. regulations on constructing and maintaining ABSL3 facilities include the following.

ABSL3 Initial HVAC Verification

Initial HVAC design verification must be performed and documented by someone with experience and expertise with the HVAC system prior to operation. This initial HVAC design verification ensures that secondary containment is maintained under failure conditions to prevent possible exposure of personnel outside the containment boundary. After HVAC verification is documented initially, the testing need not be repeated, providing no major changes have been made to, or major problems noted with, the HVAC system.

Documentation must be provided of verification of HVAC design functionality under failure conditions. The failure conditions for verification include:

1. Mechanical failure of exhaust fan or fan component(s):
   a. If redundant fans are present, the ability to transition to the alternate fan without reversal of airflow from potentially contaminated laboratory space into "clean" areas surrounding the laboratory must be verified.
   b. If no redundancy is present in the laboratory HVAC system, the capacity to transition from sustained inward airflow into the laboratory to a "static" condition, i.e., no airflow out of the laboratory, must be verified.
2. Simultaneous power failure supporting supply and exhaust fan components:
   a. If emergency power supply is available for the laboratory HVAC system, the ability to transition from "normal" power to the backup system without a reversal of airflow from the laboratory should be verified.
   b. If no backup power supply is available, the ability of the HVAC system to transition to a "static" condition, i.e., no outward airflow, should be verified.
   c. Return from power failure to "normal" operating conditions.
   d. If emergency power supply is available, it should be verified that the ability exists to transition from backup power to normal power without a reversal of airflow from the laboratory.
   e. If no backup power supply is available, the ability of the HVAC system to return to normal operating conditions without a reversal of airflow from laboratory spaces to clean areas surrounding the laboratory should be verified.

ABSL3 Repeat HVAC verification

Once the ABSL3 HVAC verification has been completed and approved, HVAC failure conditions testing need not be repeated, providing there have been no major changes made to the HVAC system and no major problems noted with HVAC performance. Examples of major changes to the HVAC system that may require reverification of HVAC design functionality under failure conditions by someone with expertise with the system include replacement of exhaust or supply fans that serve the ABSL3 containment areas, replacement of ductwork valves or dampers that serve these areas, replacement or repair of HVAC system control wiring, building automation system logic programming changes, structural changes to the ABSL3 rooms, or addition or removal of hard-ducted BSCs or fume hoods. Examples of major problems with HVAC performance that may require reverification of HVAC design functionality under failure conditions include frequent failures of the HVAC system, supply-exhaust interlocking system failure, observation that directional airflow is reversed under normal conditions, observation that HVAC alarms are not working, and any BSCs with an HVAC connection are not working properly.

Acceptance Criteria for HVAC Verification

The documentation provided must demonstrate that under exhaust fan or normal power failure conditions, or during normal power start-up, there is no reversal of air that originates within the ABSL3 vivarium room and travels all of the way outside the containment boundary. A facility may be considered to pass the HVAC verification tests as long as vivarium air does not exit the containment barrier of the facility. The ABSL3 anteroom is considered to be within the containment envelope. A positive-pressure excursion is not necessarily an airflow reversal; if a brief, weak, positive-pressure excursion is noted, a repeat test may be performed with airflow observation using an airflow indicator, such as a smoke stick or dry ice in a container of water, at the base of the closed laboratory door to confirm whether airflow reversal is occurring.

ABSL3 Initial Facility Verification and Annual Reverification

In addition to initial HVAC verification and reverification as described above, the following are the minimum facility verification requirements that are expected to be performed and documented initially for an ABSL3 facility and again at least annually. Some entities may choose to perform additional facility verification beyond what is listed below:

1. The means of detecting airflow ("tell-tale," magnehelic or digital gauge, or Baulin tube) has been confirmed to reflect observed airflow accurately. It is recommended, but not required, that digital or magnehelic gauges be calibrated annually.
2. Inward directional airflow has been confirmed by observation for the laboratory.
3. Decontamination systems (autoclave, room decontamination systems, digesters, liquid effluent systems, etc.) have been validated to be operating correctly.
4. If a building automation system (BAS) has the capacity to monitor and record performance measurements, e.g., differential pressures, the entity is encouraged to capture and store data from potential failure events, drills, etc. This information may provide verification of system performance. In addition, any programmed BAS alarms should be verified for proper functioning.
5. All alarms (fire, airflow, security, etc.) have been checked and are functioning according to established specifications.
6. Laboratory HVAC HEPA filters, if present, have been certified annually.
7. Exhaust fan motors have been checked and routine monitoring and preventive maintenance conducted.
7. The laboratory has been checked for unsealed penetrations, cracks, breaks, etc., and these have been repaired if present.
8. All BSCs have been certified within the past 12 months.
9. Seals on centrifuges, Class III cabinets, gloves on Class III cabinets, etc., have been checked and replaced if required.
10. Drench showers, eye wash stations, and hands-free sinks have been confirmed to be operating properly.

In addition to the above guidance on high-containment laboratory and vivarium HVAC validation and testing, the reader is also referred to the ANSI/ASSE Z9.14-2014, Testing and Performance-Verification Methodologies for Ventilation Systems for Biosafety Level 3 (BSL-3) and Animal Biosafety Level 3 (ABSL-3) Facilities (40).

The Agricultural Subcommittee for the CDC/NIH Biosafety in Microbiological and Biomedical Laboratories, *5th edition, established some of the original guidance for this chapter.*

APPENDIX

Some pathogens of livestock, poultry, and fish may require special laboratory design, operation, and containment features. This may be BSL3, BSL3 plus enhancements, or BSL4 and for animals ABSL2, ABSL3, ABSL3 plus enhancements, ABSL4, or BSL3-Ag. The importation, exportation, possession, or use of some of the agents shown in Table 1 are prohibited or restricted by law or by USDA regulations or administrative policies. Manipulation of diagnostic samples is not covered by this table; however, if a foreign animal disease agent is suspected, samples should be immediately forwarded to a USDA diagnostic laboratory (the National Veterinary Services Laboratories, Ames, IA, or the Foreign Animal Disease Diagnostic Laboratory, Plum Island, NY).

A USDA APHIS import or interstate movement permit is required to obtain any infectious agent of animals or plants that is regulated by USDA APHIS. An import permit is also required to import any livestock or poultry product such as blood, serum, or other tissues.

Live Animals

If you have any questions or require further information related to imports or export of live animals, birds, or germplasm, please contact the National Center for Import and Export at 301-851-3300 or email at VS.Live.Animal.Import.Export@aphis.usda.gov

Animal Products

If you have any questions or require further information related to imports of animal products or by-products, please contact the National Center for Import and Export at 301-851-3300 or email at AskNIES.Products@aphis.usda.gov.

Plants and Plants Products

Phone, 301-851-204; toll free, 1-877-770-5990; or email at plantproducts.permits@aphis.usda.gov.

Pest Permits and Noxious Weeds

Phone, 301-851-2046; toll free, 866-524-5421; or email at Pest.Permits@aphis.usda.gov.

Genetically Engineered Organisms

Phone, 301-851-3877; or email at biotechquery@aphis.usda.gov.

References

1. **Heckert RA, Kozlovac JP.** 2007. Biosafety levels for animal agriculture pathogens. *Appl Biosaf* **12:**168–174.
2. **Buchanan RL, Appel B.** 2010. Combining analysis tools and mathematical modeling to enhance and harmonize food safety and food defense regulatory requirements. *Int J Food Microbiol* **139**(Suppl 1)**:**S48–S56.
3. **U.S. Department of Health and Human Services, Public Health Service, Centers for Disease Control and Prevention, National Institutes of Health.** 2009. *Biosafety in Microbiological and Biomedical Laboratories*, 5th ed. HHS Publication No. (CDC) 21-1112. U.S. Government Printing Office, Washington, D.C. http://www.cdc.gov/biosafety/publications/bmbl5/bmbl.pdf.
4. **Kray R.** 2010. The BMBL 5th edition: a model of continuity and change. Animal Laboratory News Magazine, March 2010. http://www.alnmag.com/article/bmbl-5th-edition-model-continuity-and-change, accessed June 7, 2016.
5. **OIE Manual of Diagnostic Tests and Vaccines for Terrestrial Animals.** 2015. http://www.oie.int/international-standard-setting/terrestrial-manual/access-online, accessed June 7, 2016.
6. **Knight-Jones TJD, Rushton J.** 2013. The economic impacts of foot and mouth disease—what are they, how big are they and where do they occur? *Prev Vet Med* **112:**161–173.
7. **Cutler SJ, Fooks AR, van der Poel WHM.** 2010. Public health threat of new, reemerging, and neglected zoonoses in the industrialized world. *Emerg Infect Dis* **16:**1–7.
8. **Arzt J, White WR, Thomsen BV, Brown CC.** 2010. Agricultural diseases on the move early in the third millennium. *Vet Pathol* **47:**15–27.
9. **Rusk JS.** 2000. Biosafety classification of livestock and poultry animal pathogens, p 13–22. *In* Brown C, Bolin C (ed), *Emerging Diseases of Animals.* ASM Press, Washington, DC.
10. **Transfederal Taskforce on Biosafety and Biocontainment Oversight.** 2009. HHS/USDA Transfederal Taskforce Report. http://www.ars.usda.gov/is/br/bbotaskforce/biosafety-FINAL-REPORT-092009.pdf, accessed June 7, 2016.
11. **Kozlovac JP, Schmitt B.** 2015. Biosafety principles and practices for the veterinary diagnostic laboratory, p 31–41. *In* Cunha MV, Inácio J (eds), *Veterinary Infection Biology: Molecular Diagnostics and High-Throughput Strategies, Methods in Molecular Biology*, vol. 1247, Springer Science+Business Media, New York, NY.
12. **Miller JM, Astles R, Baszler T, Chapin K, Carey R, Garcia L, Gray L, Larone D, Pentella M, Pollock A, Shapiro DS, Weirich E, Wiedbrauk D.** 2012. Guidelines for safe work practices in human and animal medical diagnostic laboratories. *MMWR Surveill Summ* 61**:**1–103. http://www.cdc.gov/mmwr/preview/mmwrhtml/su6101a1.htm.
13. **Kozlovac JP, Thacker EL.** 2012. Introduction to biocontainment and biosafety concepts as they relate to research with large livestock and wildlife species, p 9–19. *In* Richmond J (ed), *ABSA anthology XIII.* American Biological Safety Association, Mundelein, IL.
14. **Mee JF, Geraghty T, O'Neill R, More SJ.** 2012. Bioexclusion of diseases from dairy and beef farms: risks of introducing infectious agents and risk reduction strategies. *Vet J* **194:**143–150.
15. **Food and Agricultural Organization.** Biosecurity. http://www.fao.org/biosecurity, accessed June 7, 2016.
16. **World Health Organization.** 2006. Biorisk management, Laboratory biosecurity guidance. http://www.who.int/csr/resources/publications/biosafety/WHO_CDS_EPR_2006_6.pdf, accessed June 7, 2016.
17. **Grandin T.** 1989. Behavioral Principles of Livestock Handling. *Prof Anim Sci* **5:**1–11.
18. **Federation of Animal Science Societies.** 2010. Guide for the Care and Use of Agricultural Animals in Research and Teaching, 3rd ed. http://www.fass.org/docs/agguide3rd/Ag_Guide_3rd_ed.pdf, accessed June 7, 2016.
19. **Laber K, Kennedy BW, Young L.** 2007. Field studies and the IACUC: protocol review, oversight, and occupational health and safety considerations. *Lab Anim (NY)* **36:**27–33.
20. **Taylor LH, Latham SM, Woolhouse MEJ.** 2001. Risk factors for human disease emergence. *Philos Trans R Soc Lond B Biol Sci* **356:**983–989.
21. **Jones KE, Patel NG, Levy MA, Storeygard A, Balk D, Gittleman JL, Daszak P.** 2008. Global trends in emerging infectious diseases. *Nature* **451:**990–993.
22. **Langley RL, Pryor WH Jr, O'Brien KF.** 1995. Health hazards among veterinarians: a survey and review of the literature. *J Agromed* **2:**23–52.
23. **Williams CJ, Scheftel JM, Elchos BL, Hopkins SG, Levine JF.** 2015. Compendium of veterinary standard precautions for zoonotic disease prevention in veterinary personnel: National Association of State Public Health Veterinarians: Veterinary Infection Control Committee 2015. *J Am Vet Med Assoc* **247:**1252–1277.
24. **Chastel C.** 2014. [Middle East respiratory syndrome (MERS): bats or dromedary, which of them is responsible?]. *Bull Soc Pathol Exot* **107:**69–73.
25. **Marano N, Pappaioanou M.** 2004. Historical, new, and reemerging links between human and animal health. *Emerg Infect Dis* **10:**2065–2066.
26. **Dowd K, Taylor M, Toribio JA, Hooker C, Dhand NK.** 2013. Zoonotic disease risk perceptions and infection control practices

of Australian veterinarians: call for change in work culture. *Prev Vet Med* **111:**17–24.

27. **Garner JS, The Hospital Infection Control Practices Advisory Committee.** 1996. Guideline for isolation precautions in hospitals. *Infect Control Hosp Epidemiol* **17:**53–80.
28. **Wright JG, Jung S, Holman RC, Marano NN, McQuiston JH.** 2008. Infection control practices and zoonotic disease risks among veterinarians in the United States. *J Am Vet Med Assoc* **232:**1863–1872.
29. **Occupational Safety and Health Administration.** Respiratory Protection Standard 29 CRF 1910.134. https://www.osha.gov/dte/library/respirators/major_requirements.html, accessed June 7, 2016.
30. **Severidt JA, Madden DJ, Mason G, Garry F, Gould D.** 2002. Dairy cattle necropsy on the farm. Integrated Livestock Management, Colorado State University. http://www.cvmbs.colostate.edu/ilm/proinfo/cdn/2002/CDNnov02insert.pdf, accessed June 7, 2016.
31. **National Agricultural Biosecurity Center Consortium, Carcass Disposal Working Group for USDA APHIS.** 2004. Carcass disposal: a comprehensive review. http://krex.k-state.edu/dspace/bitstream/handle/2097/662/Chapter17.pdf?sequence=1, accessed June 7, 2016.
32. **OIE Terrestrial Animal Health Code.** 2015. Disposal of dead animals: Chapter 4.12. http://www.oie.int/index.php?id=169&L=0&htmfile=chapitre_disposal.htm, accessed June 7, 2016.
33. **US Department of Agriculture.** 2016. Animal and Plant Health Inspection Service. Carcass Management During a Mass Animal Health Emergency, 81 FR 15678 Pages 15678 -15679, Docket No. APHIS-2013-0044, Document Number: 2016-0665. https://federalregister.gov/a/2016-06657Publication
34. **Munson L.** Necropsy of wild animals. Wildlife Health Center School of Veterinary Medicine University of California, Davis. https://www.yumpu.com/en/document/view/11713447/munson-necropsy-uc-davis-school-of-veterinary-medicine-, accessed June 7, 2016.
35. **Barbeito MS, Abraham G, Best M, Cairns P, Langevin P, Sterritt WG, Barr D, Meulepas W, Sanchez-Vizcaíno JM, Saraza M, Requena E, Collado M, Mani P, Breeze R, Brunner H, Mebus CA, Morgan RL, Rusk S, Siegfried LM, Thompson LH.** 1995. Recommended biocontainment features for research and diagnostic facilities where animal pathogens are used. First International Veterinary Biosafety Workshop. *Rev Sci Tech* **14:**873–887.
36. **Public Health Agency of Canada.** 2015. Canadian Biosafety Standards and Guidelines, 2nd ed. http://canadianbiosafetystandards.collaboration.gc.ca/cbsg-nldcb/assets/pdf/cbsg-nldcb-eng.pdf, accessed June 7, 2016.
37. **AS/NZS 2243.3.** 2010. Safety in laboratories—Microbiological safety and containment. Standards Australia. https://law.resource.org/pub/nz/ibr/as-nzs.2243.3.2010.pdf, accessed June 7, 2016.
38. **U.S. Department of Agriculture, Agricultural Research Service.** 2002. ARS Facilities Design Standards. Manual 242.1M-ARS. http://www.afm.ars.usda.gov/ppweb/pdf/242-01m.pdf, accessed June 7, 2016.
39. **International NSF.** 2014. *NSF/ANSI Standard No. 49-2014: Biosafety Cabinetry: Design, Construction, Performance, and Field Certification.* NSF International Standard/American National Standard. NSF International, Ann Arbor, MI. http://www.techstreet.com/nsf/products/1893278, accessed June 7, 2016.
40. **American Society for Safety Engineers.** 2014. *ANSI/ASSE Z9.14-2014. Testing and Performance-Verification Methodologies for Ventilation Systems for Biosafety Level 3 (BSL-3) and Animal Biosafety Level 3 (ABSL-3) Facilities.* American National Standards Institute, Washington, D.C. http://webstore.ansi.org/RecordDetail.aspx?sku=ANSI%2fASSE+Z9.14-2014, accessed June 7, 2016.

Biosafety of Plant Research in Greenhouses and Other Specialized Containment Facilities

35

DANN ADAIR, SUE TOLIN, ANNE K. VIDAVER, AND RUTH IRWIN

INTRODUCTION

Biosafety evaluations of plant research require the assessment and analysis of the plants alone as well as of the biological organisms associated with the plants, either naturally or introduced in planned experiments. Hence, the term "plant" refers to both the plant and its associated biological organisms. Plant research discussed in this chapter is conducted in specialized facilities that allow for plant growth and manipulation, collectively referred to here as containment facilities. Such facilities may be greenhouses, growth chambers, and modified laboratories that serve as places for growing plants under controlled conditions. Some of these facilities are specialized to isolate plants from biotic risks in the environment or to control fluctuations in abiotic or environmental factors. Although a particular plant can exist in an environment with wide variability in ambient temperature, light, nutrition, and other essential growth components, environmental conditions must be controlled to ensure scientific reproducibility. It is generally accepted that reducing variability, in this case by controlling environmental conditions, results in better scientific predictability. Furthermore, these actions enable other researchers to reproduce the experiments.

Other types of facilities are designed and operated specifically to prevent escape of plants or associated organisms that cause or have the potential to cause harm to agricultural enterprises or to the environment. In cases in which a plant could become an invasive weed or has been designated a noxious weed, containment is generally designed to quarantine the plant and its propagules, e.g., seeds, spores, pollens, etc. Current regulations require that research with noxious weeds be conducted in appropriate containment facilities. Quarantines are also imposed on imported plants because of the potential for contamination with associated pests, i.e., exotic insects and disease-causing pathogenic agents that may not exist in the importing country. Once plants are found free of associated pests, propagative stocks are grown in a containment facility (see above) to assure that they remain free of diseases or pests. Plant pathologists, entomologists, and greenhouse curators have developed standards of practice as well as diagnostic methods for healthy plant production and maintenance. These practices include biosafety considerations during pathogen culture, sometimes in plants, and pathogenicity testing on plants. Containment facilities are also recommended for use when generating and conducting research with plants and associated microorganisms that have been genetically modified using recombinant DNA technology (1).

Plant research biosafety is thus viewed from two perspectives: (i) experimental reproducibility and (ii) mitigation of the risk of an accidental release of organisms potentially harmful to agriculture or natural resources—not deleterious to humans directly. The risk of actual harm or the relative impact posed by an accidental release depends upon the specific agents involved, i.e., the pests, pathogens, and biocontrol organisms that might survive outside the containment area. Kahn (2) illustrated the potential relative impact on the environment of a disease or infestation by modifying the classic "disease triangle" (Fig. 1). The total area within the triangle is reduced when different sides (B, C, D) are shortened, which represents the reduced impact when one of the factors (environment, pathogen, or host) is limited. A disease or infestation can only occur if these three factors are present—susceptible host, sufficient and viable inocula, and compatible environmental conditions. In contained facilities, human intervention could be added as a fourth factor.

Plant research programs that may require containment include, but are not limited to, the following situations. Most of these are recognized by governmental agencies and addressed by regulations or guidelines. For details, the reader is referred to agency websites, which are updated when changed.

- Plants and associated organisms that have been genetically modified by recombinant DNA (1)
- Plants and plant pests that are subject to regulation under the Plant Protection and Quarantine (PPQ) program within the Animal and Plant Health Inspection Service (APHIS) of the U.S. Department of Agriculture (USDA) (3). PPQ safeguards agriculture and natural resources from risks associated with the entry, establishment, or spread of plant pests and noxious weeds to ensure an abundant, high-quality, and varied food supply.

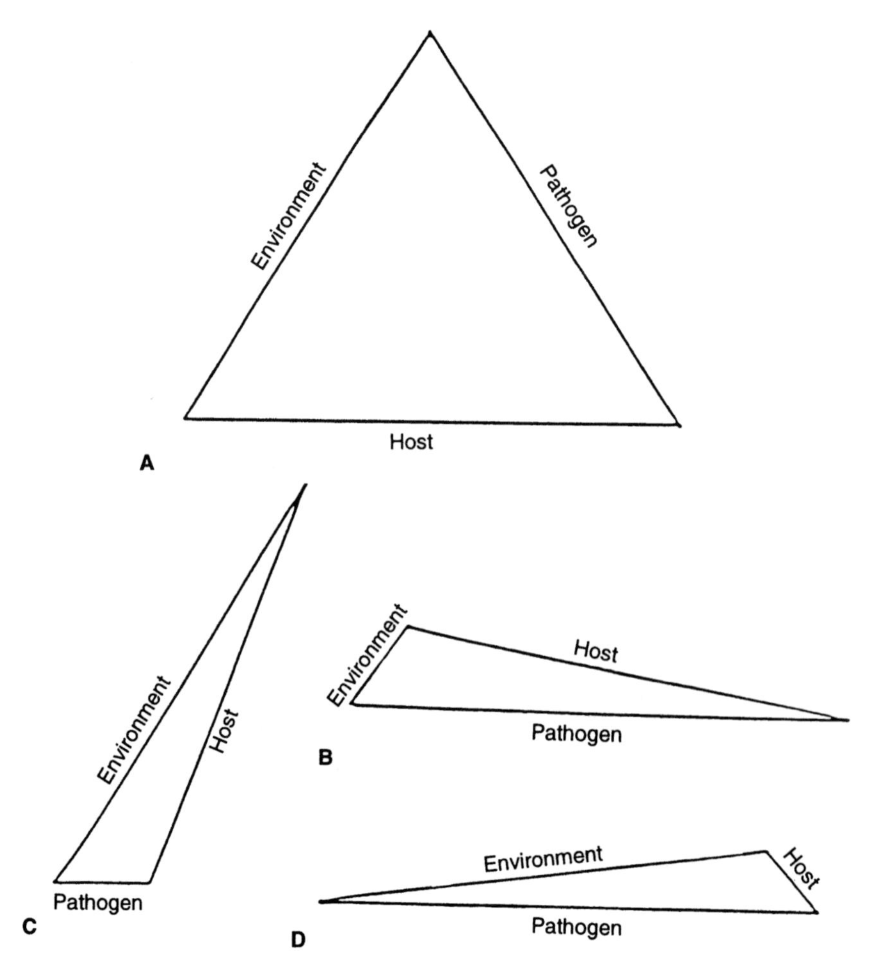

Figure 1: Weighted disease triangle. The shortened sides represent the limiting effects of the respective factors on disease, represented by triangle area. Reprinted from *Containment Facilities and Safeguards for Exotic Plant Pathogens and Pests* (2) with permission of the publisher.

- Certain plants and plant pathogens and pests that have been imported across national boundaries, i.e., exotic organisms (3)
- Plant pathogens and pests that have been moved across state boundaries under a required permit (3)
- Pathogenicity studies with fungal, bacterial, viral, or other domestic microbes
- Arthropods and nematodes that are phytophagous, pathogen-vectoring
- Parasitoids and predators of such invertebrates
- Microbes, nematodes, and arthropods used for biological control of pest species
- Plants intended for planting (nursery stock, seeds, etc.) that are subject to USDA APHIS quarantine (3)
- Microbial pathogens that are considered of especially high consequence to plants that may be purposefully introduced as biological weapons, i.e., Select Agents (3)
- Testing of potentially useful or toxic chemicals applicable to plants or associated microorganisms (4)
- Plants requiring a protected facility because they are difficult to grow in the environment where the research is to be conducted
- Rare and endangered plant species

Sequence-specific nuclease (SSN) technologies offer new methods of genome engineering that often do not involve transgenesis. Currently there are four major classes of SSNs: meganucleases, zinc finger nucleases (ZFNs), transcription activator-like effector nucleases (TALENs), and clustered regularly interspersed palindromic repeats (CRISPR)/Cas9 reagents (5). Plant breeders exploit these technologies to create mutants in plant populations where they are uncommon, among other uses. Regulators are beginning to address the use of these technologies, ruling in some cases to exempt them from regulation. However, some research institutions treat these engineered plants as if they were created by transgenesis.

Well-designed and -operated containment facilities are highly effective. This success can be attributed to researchers, facility managers, and a design based on knowledge of the biological systems involved. In only one documented incident has a contained plant pathogen escaped and been reported to cause some harm (6). However, there are many examples of environmental or economic harm caused by introduced or exotic plants, plant pathogens, or insects that have become invasive weeds and/or pests (7). Containment facilities must be highly effective, yet practical, to protect the surrounding environment from a containment breach.

Biosafety measures must also protect the occasional sensitive researcher (e.g., an allergic or immunocompromised individual). It is beyond the scope of this chapter to illustrate human biosafety hazards associated with workers in a plant containment facility or plant diagnostic laboratory. Yet there may be exposure to a variety of skin and respiratory irritants as well as dust, ergonomic challenges, pesticide use, mechanical and electrical hazards, or plant pathogens that affect humans.

The information that follows describes biosafety regulations and containment guidelines and is intended to provide guidance on designing, maintaining, and operating effective containment facilities. Many types of the experiments requiring containment also require an APHIS permit. Containment details that are specified in the permit must, by law, be followed. For many major commodities, foreign germplasm is initially quarantined in U.S. government facilities. If regulators deem the material to pose little or no environmental risk, it may be distributed for research or commercial purposes. Otherwise, the material is destroyed.

GUIDELINES AND REGULATIONS

Guidelines, rules, and regulations applicable to plant research in containment facilities are limited in number and scope, in comparison to those regarding the field release of genetically modified plants and transboundary movement of plant pests. Most countries subscribe to the International Plant Protection Convention (IPPC), a treaty developed by the Food and Agriculture Organization of the United Nations. IPPC has issued the *International Standards for Phytosanitary Measures* (8), used to control the introduction of pests or putative pests into new ecosystems. Ten regional phytosanitary organizations cooperatively implement the IPPC. The United States, Canada, and Mexico are member participants in the North American Plant Protection Organization (9). In this chapter, most of the references apply to U.S. agencies, but the principles and their application are applicable globally.

It is critical to distinguish between guidance and regulation. In the United States, primary guidance for safe handling of transgenic plant material in laboratories, growth chambers, and greenhouses can be found in Appendices G and P of the National Institutes of Health's *Guidelines for Research Involving Recombinant or Synthetic Nucleic Acid Molecules* (1). Appendix G in the *Guidelines* describes physical facilities and microbiological safety practices for research with human pathogens in four risk groups, but can also be applied to plants if they are grown on laboratory benches or in growth rooms. Appendix P specifically describes physical and biological biosafety principles for greenhouse-grown plants or for other instances where the experiment expands and resembles the scale common to a greenhouse. Other U.S. Federal research funding agencies formally comply with the *NIH Guidelines* for all recombinant DNA research, including plant containment. The Coordinated Framework for Regulation of Biotechnology (10) clarified Federal Agency responsibilities for regulating products of biotechnology, including transgenic plants. Noncompliance can result in loss of U.S. Federal research funding. The *NIH Guidelines*, first promulgated over 30 years ago, became a primary guidance document for biotechnology worldwide and have been adopted by the World Health Organization (WHO). Other guidance documents for contained research with recombinant DNA have now been adopted in Canada, the United Kingdom (11), and many other countries.

Institutional oversight of recombinant DNA research is the responsibility of local institutional biosafety committees (IBCs), which are described in the *NIH Guidelines* and authorized by the NIH Office of Biotechnology Activities. Research institutions are required to maintain IBCs if U.S. Federal research funds have been accepted. Many other organizations worldwide utilize an IBC for recombinant DNA and other research on plants. The *NIH Guidelines* state that an IBC is to appoint at least one plant expert to the committee if the institution conducts recombinant DNA plant research (1). An institutional biosafety officer (BSO) is appointed to the IBC to serve as a technical contact when a high level of containment is required or when permitted materials are involved. USDA-inspected high-security containment facilities must have a designated containment or quarantine officer. This individual could be the BSO, principal investigator (PI), or greenhouse manager. As is often the case, the PI holds ultimate responsibility for adhering to all necessary guidelines and regulations. Under the *NIH Guidelines*, the PI must file a notification with the IBC when his or her laboratory is conducting work involving recombinant DNA research and must receive prior approval from the IBC for certain experiments.

The mission of USDA APHIS is to protect and promote U.S. agricultural safety and includes regulating agricultural applications of genetically engineered organisms (12). APHIS' Biotechnology Regulatory Services (BRS) regulates the importation and field release or "introduction" of transgenic plants that may pose a risk to plant health (12). APHIS maintains PPQ, a group devoted to plants and related organisms. PPQ provides safeguards and issues permits regarding "the risks associated with the entry, establishment, or spread of animal and plant pests and noxious weeds to ensure an abundant, high-quality, and varied food supply" (3). Specific containment facility guidelines for several types of PPQ-permitted work can be found on their website (13). Nearly all imported plants are subject to regulation under the Plant Pest Act. PPQ coordinates a number of crop biosecurity and emergency response programs and manages various accreditation and certification services (3). Canada's Food Inspection Agency has also established an excellent set of containment standards for handling plant pests (14).

CONTAINMENT OBJECTIVES

The standard reference *Biosafety in Microbiological and Biomedical Laboratories* (BMBL) describes containment as methodology to ". . . reduce or eliminate exposure of . . . workers, other persons, and the outside environment to potentially hazardous agents," and addresses only human and zoonotic pathogens (15). The *NIH Guidelines* succinctly states the goals to "avoid unintentional transmission of recombinant DNA-containing plant genome . . . or release of recombinant DNA-derived organisms . . ." and to ". . . minimize the possibility of an unanticipated deleterious effect on organisms and ecosystems outside of the experimental facility . . ." (1). We therefore re-emphasize here that the protection of agricultural crops as well as the natural environment is crucial. Transgenic plants do not generally pose a risk to human or animal health; the biosafety risks are primarily ecological and economic. A potential threat to humans or animals would be the deliberate engineering of plants to produce compounds affecting these species, such as pharmaceuticals or toxins. However, because plant pathogenic fungi and bacteria have been associated with human diseases, precautions for worker safety should be taken in the laboratory and plant growth areas if these organisms are used, even if they do not appear on risk groupings of human pathogens. All risks and subsequent containment measures can only be assessed with a working knowledge of the biological systems involved. The containment goals of the *NIH Guidelines* can also apply to nonengineered plants, as Appendix P Biosafety Levels (BLs) 1-P and 2-P are essentially the same as those used to grow healthy plants, by preventing the invasion of facilities by arthropod or microbial pests. In principle, higher BSLs are used when the goal is to provide confinement of higher risk plant-associated organisms.

Although containment goals and approaches vary, the primary elements of containment for plants parallel those of basic laboratory biosafety:

- Administrative or management-based controls
- Engineering or design-based controls for physical containment
- Adherence to standard or performance-based procedures to minimize escape

Management tools are employed to achieve administrative controls and good laboratory practices. Personal protective equipment (PPE) is standard for the general biosafety of laboratory workers. For plant work, PPE can help to avoid the inadvertent transfer of organisms outside of a containment facility. Common white lab coats are recommended for most research activities. Blue or yellow lab coats or street clothing may attract certain insects, whether they be pests or research organisms. Gloves, eye protection, and shoe covers are occasionally used as well; however, face protection and inhalation filters, common in biomedical or animal research labs, are uncommon in plant research. Engineering controls are achieved with good design and construction methods, but may need to be adjusted because of the greater diversity and dissemination mechanisms of plants and associated organisms in comparison to human and animal pathogens. Formal standard operating procedures (SOPs) are critical for understanding containment needs and serve as a mechanism for adhering to procedures required for maintaining containment. As we use these controls, we remain cognizant of the disease triangle (Fig. 1), understanding that a disease, infestation, or, even worse, an epidemic, cannot occur without the contribution of all factors.

Breaches of containment occur most often by movement associated with people, pests, water, and air currents. Air currents can carry spores of bacterial and fungal pathogens, insects, and pollen, but attachment to a person's skin or clothing is the most likely means by which organisms escape from a facility (2). It is advantageous that pollen is generally short-lived and that both pollen and airborne propagules must find a suitable environment and/or host for survival. Hardware that serves to mitigate these risks is described below, yet basic management practices that include changing lab coats, hand washing, and visual inspections upon exiting are highly effective and recommended.

Because of the small size of the organisms involved, engineering controls cannot be relied on to prevent the breach of containment completely. Good design of facilities and proper choice of materials can certainly minimize the probability of loss. Figure 2 illustrates the relative size of particles under consideration. A 100-micrometer grain of pollen dwarfs a virus particle, but the tiny Western flower thrips (*Frankliniella occidentalis*), a major greenhouse pest and vector of specific viruses, is 100 times larger than the grain of pollen. As described below, using screens to restrict the movement of even small insect pests presents engineering challenges.

The advent of synthetic biology and nanotechnology may alter containment procedures and provide challenges in taxonomy of affected microbes (16).

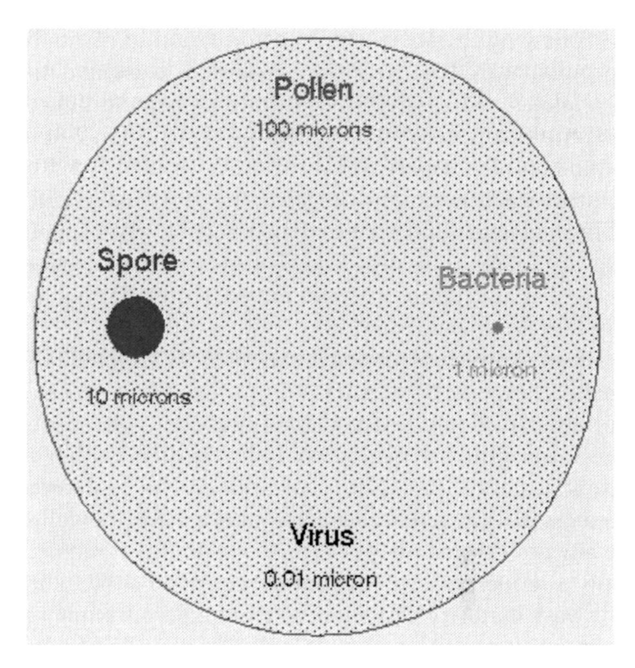

Figure 2: Relative sizes of a pollen grain, fungal spore, bacterium, and virus particle.

BIOSAFETY LEVELS BL1-PLANTS (P), BL2-P, BL3-P, AND BL4-P

Since the original publication of BMBL in 1984, the concept of biosafety levels has become a standard of guidance for research and clinical diagnostic laboratories handling human and zoonotic pathogens (15). Biosafety levels BSL1 through BSL4 describe general practices, facility features, and equipment that enable safe working conditions in medical laboratories for activities with increasing risk. This publication was preceded by *Classification of Etiologic Agents on the Basis of Hazard* (17), first assembled in 1969. Early versions of the BMBL and its predecessor made no mention of plant materials. The BMBL now assigns infectious agents to biosafety levels based on activities performed and the risk to personnel, to the environment, and to the community, and includes descriptions of the most commonly encountered human and zoonotic pathogens.

The Center for Disease Control's (CDC) guidance for the risk assessment of genetically engineered organisms was used for recombinant DNA applications in the 1978 *NIH Guidelines*, on the basis of the premise that potential risk is related to risk level of the non-modified organism. The *Guidelines* include an Appendix B entitled "Classification of human etiologic agents on the basis of hazard," and uses the term "Risk Group" for bacterial, fungal, viral, and parasitic agents. Assignment to a BSL level parallels the risk group (RG) classification, but can be modified by Biological Containment (Appendix I). Appendix G is very design specific in its descriptions of laboratory facilities, equipment, and microbiological practices for BSL1 through BSL4. Research with certain organisms and host-vector systems is considered such a low risk that they are exempt from the requirement for a BSL containment level (Appendix C).

Following the successful transformation of plants by recombinant DNA, NIH was requested to expand the *Guidelines* and their BSL descriptions beyond microorganisms to accommodate plant research. Over a 5-year period and in cooperation with USDA, Appendix P was approved as a part of the *Guidelines* in 1990. Understandably, the format of Appendix P paralleled that of Appendix G, with four biosafety levels based on potential risk to managed and natural ecosystems. Greater details are given in Adair and Irwin (18).

If a risk assessment concludes there is no likelihood that a transgenic organism would survive and reproduce in nature, or if released would not cause any type of serious environmental risk, it would be classified either as exempt or as BL1-P. BL1-P research can be conducted in most modern research greenhouses and labs as well as in any growth chamber or growth room. BL1-P requires only a moderate level of containment and suffices for the vast majority of genetically engineered plant research. The addition of a biological containment measure in a BL1-P facility permits experiments with slightly higher risk to be conducted. It can be designated as BL1-P + biological containment (BC). An example of BC is the introduction of "gene silencing" into the subject organism, which can greatly reduce or even eliminate functions, e.g., reproduction.

BL2-P is a higher level of containment, but it can still be operational in standard facilities when the consequence of containment loss is predictably minimal. An important consideration for BL2-P facilities is the locale. For instance, special attention is required to work with disease resistance in wheat in an area adjacent to wheat production, e.g., by stipulating either higher containment or conducting the experiments when wheat is not actively growing in fields. Documentation, signage, and access to an autoclave may be needed. A growth chamber would generally satisfy BL2-P requirements. As per BL1-P+BC, a BL2-P facility may occasionally house higher-risk experiments, with the addition of BC measures (BL2-P+BC). No special apparel or PPE is suggested at either BL1-P or BL2-P, although lab coats are not uncommon.

Highly specialized facilities are required to attain BL3-P and higher containment. Experiments done in a BL3-P facility would have a recognized potential for serious detrimental environmental impact. Although the plant itself would likely pose no threat, an associated pathogen or pest, or a genetically expressed protein, could require this containment level. Besides several additional management requirements, facility design would also include the decontamination of any items that leave the facility, HEPA filtration of air, complete clothing changes and showering when exiting, negative air pressurization, specialized air conditioning systems, thorough sealing of penetrations, and other features discussed below (18). Material leaving a BL3-P or higher facility is autoclaved or otherwise treated to ensure that high-risk organisms do not survive. Even apparel worn in these facilities must be autoclaved prior to laundering or disposing.

BL4-P and BSL-3Ag facilities are similar and require a few more features added to the BL3-P design. BSL-3Ag is a designation created by the USDA Agricultural Research Service (ARS), but the research may resemble that being done at BL3-P or BL4-P for other organizations. Regardless, all three are used for conducting research posing a potentially serious environmental risk or, in the case of BL4-P or BSL3-Ag, potential risks to the researchers themselves. APHIS refers to such facilities as High Security Containment Facilities (HCSFs) and is authorized to issue permits for research that must be conducted on serious exotic pathogens and/or their insect vectors. There are very few facilities built to meet these standards, as they are expensive to build, operate, and maintain. A greenhouse used as the primary containment barrier requires very specialized construction techniques (see below). Growth chambers and growth rooms within a secured facility would seldom be a primary containment barrier and may be preferred to greenhouses for high-containment facilities, especially for BSL3-Ag housing Select Agent Programs.

The ARS has published design standards for their containment facilities as well (19). The ARS BSL-3Ag designation is for both plant and animal high-security containment where the facility walls, as in a greenhouse, are the primary containment barriers to the outdoors (19). The U.S. Select Agent Program, which is administered by both APHIS and CDC, also uses the BSL-3Ag designation.

APHIS does not assign plant pest organisms to risk groups and corresponding biosafety levels that relate to containment measures. Instead, APHIS may prescribe containment measures when a permit is issued or suggest containment in their guidelines. APHIS may inspect a facility prior to issuing a permit and has the right to inspect at any time while the permit is active. Containment facility guidelines include construction standards with an accompanying list of the suggested measures needed to meet a standard (13). The guidelines are specific to certain program areas but not as specific as an issued permit.

PHYSICAL CONTAINMENT: FACILITY CHOICE

A greenhouse can be defined as a "means of overcoming climatic adversity, using a free energy source, the sun" (20). A growth chamber, on the other hand, is a controlled environment that must artificially supply the required light energy for plant growth. Two recommended publications on growth chambers are *Plant Growth Chamber*

Handbook (21) and the earlier text, *A Growth Chamber Manual* (22). An assessment must be made to determine when the natural light of a greenhouse is preferred over a growth chamber. Although light energy may be considered "free," the associated heat that gets trapped within a greenhouse must be addressed. The decision of whether to use a greenhouse or a growth chamber is often based on growth area desired and performance needs, but energy considerations are increasingly prominent. A well-maintained growth chamber can provide consistent environmental conditions that cannot be found in most greenhouses or, for that matter, in nature. Initial cost, maintenance, and energy inputs can be higher for a growth chamber than a greenhouse, per unit growth area. Yet, as stated above, using a greenhouse as the primary containment barrier for BL3-P poses an engineering challenge. The vast majority of APHIS-permitted containment facilities are laboratories in which organisms are generally manipulated and cultured. A laboratory may also house plants grown in tissue culture or whole plants grown in relatively small quantities or in containers, but seldom is the laboratory considered as the primary site for plant growth. Growth chamber facilities are commonly placed in locations with convenient access to laboratories.

Design

A successful plant growth facility design team includes architects and/or greenhouse designers, engineers, commissioning staff, researchers, and support staff. Regulators should also be kept informed, even though they may not be able to provide significant input. The team remains in place from the point of early facility design through completion and testing. The experience and quality of all contractors becomes increasingly important as containment levels increase.

It is recommended that SOPs be formulated before the design of a containment facility. Research activities will define the traffic patterns, equipment required, needs for decontamination, and many other important factors. One can use SOPs for known and future experiments. BL1-P and BL2-P facilities are less likely to need detailed SOPs, but high-containment facilities would benefit immensely.

Greenhouse Glazing and Structure

There are a large variety of translucent materials available to cover a greenhouse. The National Greenhouse Manufacturers Association provides relevant performance data on glazing (23). Unlike commercial greenhouses that generally utilize inexpensive plastic coverings, a research greenhouse is normally covered with rigid panels or glass. Structured panels made of polycarbonate or acrylic plastic are acceptable for containment structures at BL1-P and BL2-P as well as for many APHIS-permitted experiments. BL3-P and higher facilities and APHIS HCSFs are required to use sealed, laminated glass or insulated glass units that are strong enough to withstand hail damage. The clearest possible glazing is always recommended, because this maximizes the amount of natural light, which is the primary reason for using a greenhouse. Glass is also the most durable material, but because it is heavier than plastic glazing, it requires a more robust and thus a more expensive structure. High-containment facilities must also have alarms that alert managers of broken glazing and have a documented procedure for replacing the glazing.

A steel or aluminum structure is preferred for all research greenhouses. Interior surfaces that are smooth and easy to clean and can withstand disinfectants are always preferred and are required for high containment. High-containment greenhouses may employ integrated structure and glazing systems, i.e., "curtainwall" or "structural silicone" systems. These are specialized construction methods and are rarely used for greenhouse applications. Therefore, they require careful design and installation to be successful.

Floors and Drains

Regardless of whether the material is grown on a lab bench, in a growth chamber, or a greenhouse, the growing area is routinely cleaned with disinfectants. Liquid runoff collection and decontamination are generally not required for BL1- and BL2-P, although this requirement is decided on a case-by-case basis, depending upon the organisms present. BL3-P and higher facilities must have floors that are impervious to organisms, can tolerate repeated disinfectant treatment, and can collect waste for decontamination. BL4-P and BSL-3Ag require HEPA filtration of the sewage vent. A tightly sealed greenhouse, especially in cold climates, can be subject to great amounts of condensation. For BL3-P and higher facilities, this condensate must remain in the greenhouse and be treated along with all other liquid waste before disposal. Careful greenhouse design is required to minimize condensation for maximal sunlight capture; in addition, a capture system is needed to route this to the wastewater stream. For other research greenhouses, condensate or any rainwater that migrates inside is normally allowed to move to the outside.

Caulking and Sealing

It is often assumed that a containment greenhouse or specialized growth chamber can be built "airtight" or sealed after construction to become impermeable. In fact, greenhouses and growth chambers are not airtight, but

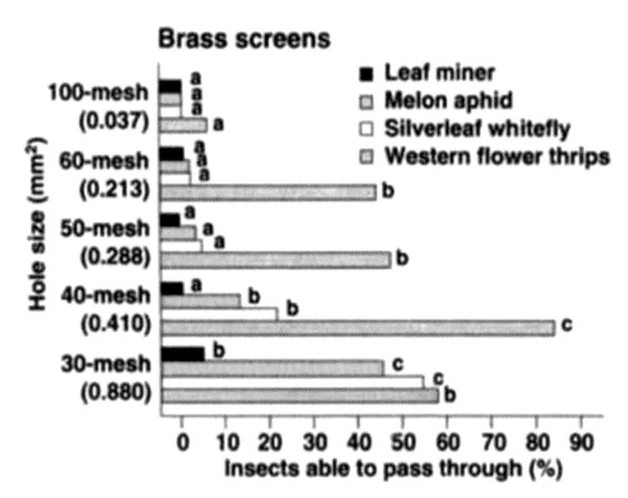

Figure 3: Effectiveness of brass screens in preventing the movement of leaf miner, melon aphid, silverleaf whitefly, and western flower thrips. For a particular insect, screens with the same letter are not significantly different from one another in their ability to exclude that insect ($P<0.05$ analysis of variance [ANOVA] and Ryan's multiple range Q-test). All of the brass screens tested prevented passage of green peach aphids. Reprinted from *California Agriculture* with permission of the publisher (25).

Figure 4: Drain sock for catching seed. Release of seeds from a facility can be controlled by the use of a fabric "sock" attached to the drain opening. (Photo courtesy of Darrin Rose, University of Sheffield.)

good design and tight-fitting construction that is sealed with proper sealant and caulk can enhance containment and environmental control significantly. High-containment facilities have all penetrations properly sealed between containment and noncontainment areas. Silicone and firestop products are commonly used, although greenhouse glazing often uses butyl gaskets and sealants.

Screens and Filters

Containment guidelines at BL1-P and BL2-P levels recommend that openings should be screened or otherwise restricted to exclude "small flying animals (e.g., arthropods and birds)" but do not require any special barrier to exclude pollen or microorganisms (1). Screens and filters are commonly used to include or exclude motile organisms, but they may restrict air movement, which can contribute to poor environmental control. For most greenhouses or growth chambers, securing insect screen and loose filter material on supply air vents does a reasonably good job of preventing the ingress of most pests. Organisms held under containment cannot be allowed to leave the growth area, and therefore egress through the exhaust air needs to be restricted. The western flower thrips are able to pass through very fine mesh screen, so a suitable restrictive material has been manufactured (24). APHIS suggests using an 80-mesh metallic screen for some arthropod containment, which is over twice as fine as common home window screens. The exhaust surface area must be carefully calculated to assure adequate ventilation when using such a fine screen. Figure 3 (above) describes some greenhouse screen sizes used to restrict pest entry.

Screened cages for insect work are common, and NIH specifies that these and "other motile macroorganisms" be housed in them (1). Cages are commercially available but are commonly constructed on site.

Pollen and spores are more readily trapped with HEPA or other filtration systems that are incorporated into the supply and/or exhaust air connected to a growth chamber or air conditioned greenhouse.

Figure 4 shows a sock attached to a specially manufactured drain basket to prevent small seeds from leaving the facility through a drain.

Cooling, Heating, and Ventilation

Precision environmental control is at minimum important and at most critical to plant research. Light, either natural or artificial, is often accompanied by the addition of normally unwanted heat. Therefore, the predominant engineering task is to develop a way to cool the facility. The additional challenge of designing a good containment system is to be able to move adequate amounts of treated air to maintain a proper environment, while at the same time controlling the egress of pathogens, pests, or plant propagules.

The least costly approach to greenhouse cooling uses shade materials and natural ventilation. This approach is inadequate for high BSL containment, because the facility must be sealed to the outside environment and air must be filtered on exit. Mechanical cooling, i.e., air conditioning, must be employed to achieve the needed temperature setpoint and containment.

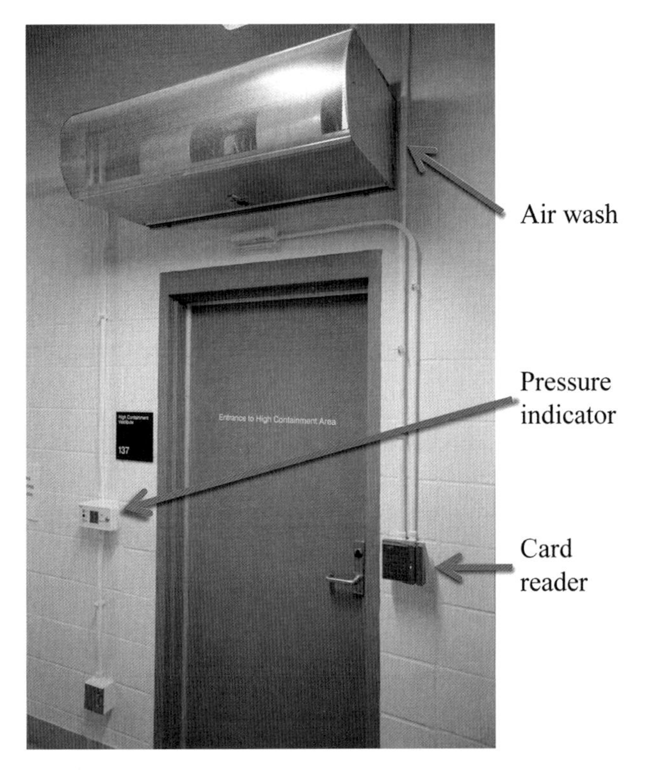

Figure 5: Precautions at the barrier door in a BL3-P facility. Arrows indicate the location of the manometer/pressure indicator for visualization of directional air flow, an air wash that blows air directly onto personnel during entry or exit, thereby reducing the likelihood of introducing insects, and a card reader for restricting/recording access. (Photo courtesy of David Hansen, University of Minnesota.)

Figure 6: Plant Containment Symbol. A new symbol is available to designate when plant material requires containment. This is an alternate to the widespread and often inappropriate use of the universal biohazard symbol for transgenic plant material. Creation of Spoon Creative Inc. and Adair Consulting.

Negative air pressure needs to be maintained within the "hottest"—highest risk—areas for high-level containment and may be employed for lower-containment needs as well. Maintaining negative air pressure is a significant engineering challenge. A manometer or differential pressure readout can be mounted adjacent to the door when one needs to enter a space that requires a maintained (generally negative) air pressure (Fig. 5).

An emergency backup generator should be included for especially sensitive experiments and high-level containment. Dedicated circuits are assigned to the backup generator, which would exclude most if not all energy expensive supplemental plant lighting.

Entry and Exit

Containment breaches occur most frequently at entry and exit points. Self-closing and locking doors are recommended. Personnel safety is always the priority; therefore, doors that allow reasonably quick exit in an emergency are required, regardless of the programmatic needs. In addition to well-fitting doors, special construction at entry and exit points may be required. APHIS commonly suggests erecting and using vestibules, anterooms, or corridors. Vestibules allow one to don and doff PPE and may include a shoe wash or sticky floor mats. They also provide a convenient location for hand-washing sinks, as simple hand washing with ordinary soap and water eliminates plant viruses and many other organisms. Insect quarantine facilities benefit from a dark vestibule with a light trap. Note the air wash in Fig. 5. This is useful for removing small items from clothing, which can then be captured in the building's filtered air system.

Card readers have become commonplace and provide both personnel logging and security services at reasonable cost. Signage is also commonly employed to indicate the presence of either transgenic or permitted material. Unfortunately, the universal biohazard symbol (26) is often chosen. This universal symbol indicates the presence of biological hazards to humans and research subjects, a situation that is extremely rare in plant research. There are situations where the universal biohazard symbol is entirely appropriate, i.e., for human pathogens cultured on plants. As a response to the improper use of the symbol, a new plant containment symbol was created (Fig. 6). Just as the universal biohazard symbol quickly became the standard for biohazards worldwide, the authors of this chapter anticipate widespread use of this new symbol. Postings can be made on the doors of greenhouses, growth chambers, or labs. Augmenting the symbol with text below stating "CONTAINMENT," an NIH biosafety level (e.g BL2-P) for transgenic material, USDA APHIS PPQ-permitted material, or other special contaiment is encouraged. The new symbol is freely available to the public.

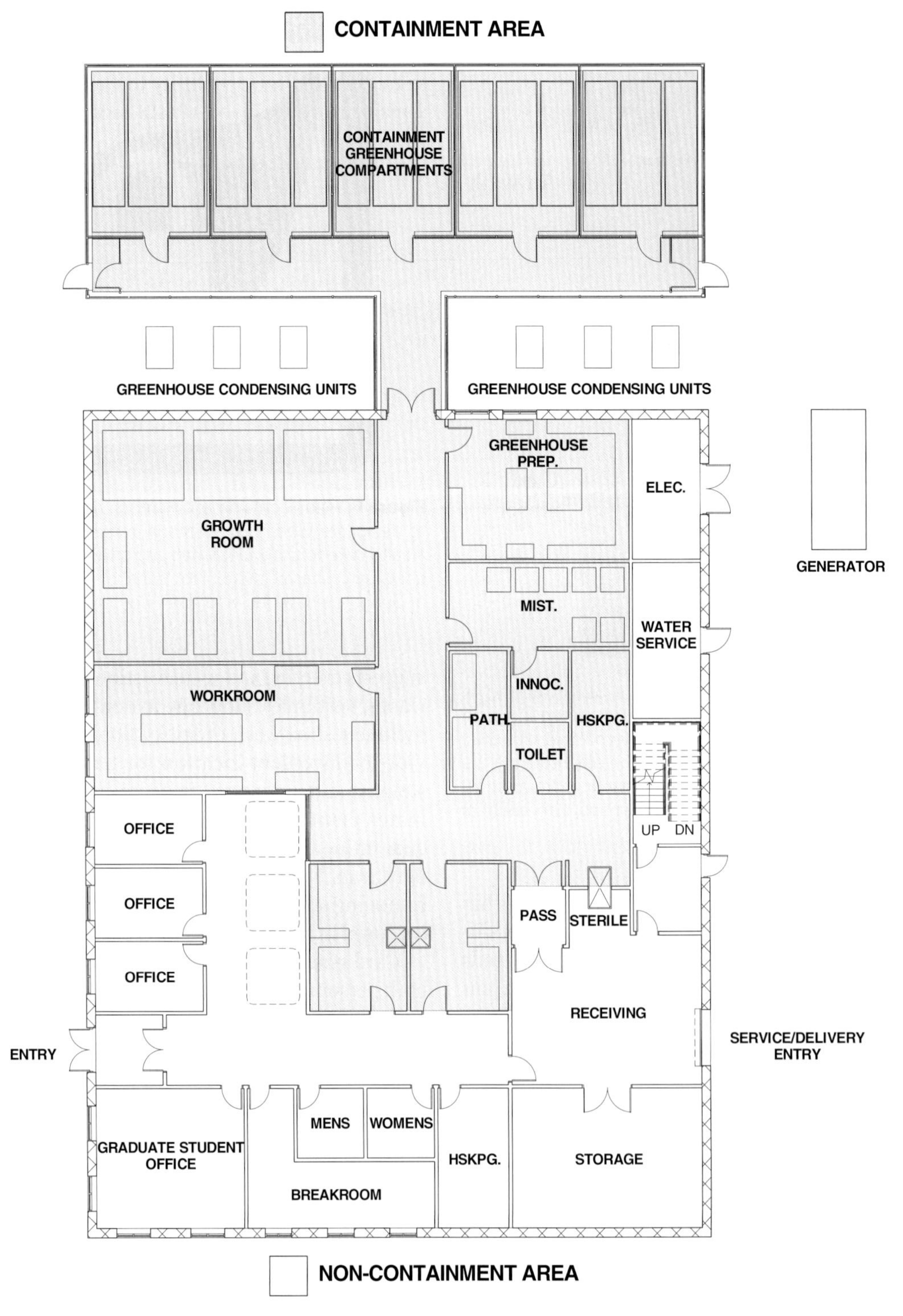

Figure 7: Design of high-containment facility. (Reprinted with permission from RSP Architects, Minneapolis, MN.)

Layout

It is essential to design that facility layout is both efficient and secure. Figure 7 illustrates such a design for a high-containment facility, with contained and noncontained areas indicated. Additional designs can be found in *Containment Facilities and Safeguards for Exotic Plant Pathogens and Pests* (27).

BL1- and BL2-P greenhouses would also benefit from the design (above). A layout of individual rooms connected by a common corridor creates separately conditioned, isolated growing areas, which allow smaller experiments

Figure 8: Double-loaded greenhouse corridor. (Reprinted with permission from Samuel Roberts Noble Foundation, Ardmore, OK.)

to be both contained and conducted under differing environmental conditions. The "double-loaded" corridor design (Fig. 8) is another efficient layout, because it places greenhouse rooms on either side of the corridor.

Good exposure to natural light is highly important when siting a greenhouse facility. Ease of access by researchers and the proximity to their laboratories is also a major consideration. APHIS recommends and may require that greenhouses be built in areas that minimize risk to the local environment. They also discourage building containment greenhouses on rooftops due to exposure to higher winds. Water drainage can also be a problem. Placing a facility in nonagricultural or natural environments within an urban setting generates a unique set of issues regarding security, light pollution, or access to natural light. A facility should not be sited next to adjacent farms or fields. Growth chambers are less affected by these factors, but need security and ease of access.

Decontamination and Pest Control

The *NIH Guidelines* instruct that plant materials shall be made biologically inactive, thus leaving a broad interpretation on how to keep propagules from spreading, both within and outside the facility. The intent is to acknowledge that plants can be killed simply by withholding water or by treating with herbicides or other chemicals. BL2-P regulations state that an autoclave be available for treating greenhouse materials. Higher-containment facilities must essentially have a double-door pass-through autoclave and other sterilization equipment. Although technically there could be alternatives to an autoclave, it is the most practical. Surfaces can be decontaminated with vaporized hydrogen peroxide or other similar chemical treatments. Table 1 provides a summary of surface disinfectants used against many plant pathogens.

Greenhouse research projects commonly fail because of pest infestation. Control of pests within the facility is

TABLE 1.

Surface disinfectants for plant pathogens[a, b]

Disinfectant	Trade names	Comments	Contact time
Alcohols (ethyl and isopropyl) 60–85%	Lysol	Evaporates quickly so that adequate contact time may not be achieved. High concentrations of organic matter diminish effectiveness. Is flammable.	10–15 minutes
Phenolics (0.4–5%)	**Pheno-cen**	Phenol penetrates latex gloves. Is an eye/skin irritant. Remains active upon contact with organic soil; may leave residue.	10–15 minutes
Quaternary ammonium (0.5–1.5%)	**Consan Triple Action 20** **Physan 20** Green-Shield 20	Effective for nonporous surface sanitation (floors, walls, benches, pots). Low odor, irritation. Use according to labels.	10–15 minutes
Chlorine (100–1,000 ppm)	**10% Clorox** **10% Bleach**	Is inactivated by some organic matter. Fresh solutions of hypochlorite (Clorox) should be prepared every 8 hours or more frequently if exposed to sunlight. Is corrosive and irritating to eyes and skin. **Exposure to sunlight further reduces hypochlorite efficacy. Keep solution opaque container.**	10–15 minutes

[a]Items in boldface are effective against a current serious pathogen, *Phytophthora ramorum*, which causes Sudden Oak Death.
[b]Reprinted with permission from USDA-APHIS-PPQ (28).

not only an aid to research but is an NIH requirement at all biosafety levels and for APHIS permits. Good sanitation practices can both decontaminate a facility of test organisms and also reduce or eliminate pest levels.

MANAGEMENT

In biosafety parlance, administrative controls and good laboratory practice are the "soft" elements of containment facility management. They pertain to discretionary access and who grants it, the type of apparel required, personnel logging, record keeping, and the creation and updating of manuals. Initial and periodic follow-up training is also needed for users and maintenance staff. Updating and following SOPs not only comprise good laboratory practices but may also be a requirement under permit.

Contingency plans, including procedures for notifying IBCs or other local authorities, APHIS, or NIH as appropriate, must be formulated and periodically updated. Inspections are conducted for those facilities regulated by APHIS before work begins under a permit. APHIS has the authority to reinspect without notice. It is recommended that knowledgeable local committees such as an IBC or safety committee, or individuals such as a BSO or quarantine officer, periodically inspect the facility to prevent any larger issues from developing.

Security is prominent when working with plant pathogens designated as Select Agents (3, 15). Select Agent designation for plant pathogens that could be used as bioweapons requires a high level of management in addition to operating under very secure BL3-P or higher containment. The USDA Agricultural Research Service offers security guidelines when operating at this biosafety level (29).

CONCLUDING REMARKS

Plant research biosafety consists of paying attention to details involving plants, pests, pathogens, and nondesirable organisms in plant containment facilities. The details presented here are efficacious and have a very high safety record for people and the environment. The creation of containment facilities for plant research requires careful planning, quality construction, and dedicated maintenance. Most of these facilities involve research conducted at low-to-moderate biosafety containment levels. Except for high-containment needs, regulations and guidelines are not particularly challenging to follow, nor are the facilities more expensive to build. The key to avoiding excessive cost while maintaining good containment is to have a thorough understanding of the biological systems involved and to match them to existing regulations and guidelines. Creating SOPs prior to facility design and keeping them updated is a straightforward way to assess needs and manage expectations. Flexibility is key to garnering the skills required to address future needs. For example, climate change is shifting the ecological niches of many species, which invariably includes their challenge organisms.

References

1. **National Institutes of Health.** 2013. Guidelines for Research Involving Recombinant or Synthetic Nucleic Acid Molecules. http://osp.od.nih.gov/sites/default/files/NIH_Guidelines.html, accessed January 2016.
2. **Kahn RP.** 1999. Biological concepts, p 8–16. *In* Kahn RP, Mathur SB (ed), *Containment Facilities and Safeguards for Exotic Plant Pathogens and Pests.* American Phytopathology Society Press, St. Paul, MN.
3. **United States Department of Agriculture (USDA) Animal and Plant Health Inspection Service (APHIS).** 2011. Plant Protection and Quarantine (PPQ). https://www.aphis.usda.gov/wps/portal/aphis/ourfocus/planthealth, accessed January 2016.
4. Federal Insecticide, Fungicide, and Rodenticide Act, 7 U.S.C. §136 et seq, 1996.
5. **Voytas DF, Gao C.** 2014. Precision genome engineering and agriculture: opportunities and regulatory challenges. PLoS Biol 12: e1001877.
6. **McKeen WE.** 1989. *Blue Mold of Tobacco.* American Phytopathology Society Press, St. Paul, MN.
7. **Kahn RP.** 1989. *Plant Protection and Quarantine: Biological Concepts.* CRC Press, Boca Raton, FL.
8. International Standards for Phytosanitary Measures 1-24. 2005. Secretariat of the International Plant Protection Convention, FAO, Rome, 2006.
9. **North American Plant Protection Organization.** 2011. http://www.nappo.org/files/2514/3781/8218/NAPPO_IAS_Discussion_Doc_03_12-07-2012-e.pdf, accessed January 2016.
10. **Office of Science and Technology Policy.** 1986. *Coordinated Framework for Regulation of Biotechnology.* https://www.aphis.usda.gov/brs/fedregister/coordinated_framework.pdf, accessed January 2016.
11. **Health and Safety Executive.** 2007. *SACGM Compendium of Guidance.* Revised 2014. http://www.hse.gov.uk/biosafety/gmo/acgm/acgmcomp, accessed January 2016
12. **United States Department of Agriculture (USDA) Animal and Plant Health Inspection Service (APHIS).** Biotechnology Regulatory Services (BRS). https://www.aphis.usda.gov/wps/portal/aphis/ourfocus/biotechnology, accessed January 2016
13. **USDA-APHIS-PPQ.** 2012. *Containment facility guidelines for noxious weeds and parasitic seed plants.* https://www.aphis.usda.gov/plant_health/permits/downloads/noxiousweeds_containment_guidelines.pdf, accessed January 2016.
14. **Canadian Food Inspection Agency.** 2007. *Containment Standards for Facilities Handling Plant Pests.* Updated 2014. http://www.inspection.gc.ca/plants/plant-pests-invasive-species/biocontainment/containment-standards/eng/1412353866032/1412354048442, accessed January 2016.
15. **U.S. Department of Health and Human Services. Public Health Service, Centers for Disease Control and Prevention, National Institutes of Health.** 2009. *Biosafety in Microbiological and Biomedical Laboratories,* 5th ed. HHS Publication No. (CDC) 21-1112. http://www.cdc.gov/biosafety/publications/bmbl5/bmbl.pdf.

16. **Organization for Economic Cooperation and Development.** 2010. *Symposium on Opportunities and Challenges in the Emerging Field of Synthetic Biology*, U.S. National Academies, the Organization for Economic Cooperation and Development, and The Royal Society. http://www.oecd.org/sti/biotech/45144066.pdf, accessed January 2016.
17. **Centers for Disease Control. Office of Biosafety.** 1974. *Classification of Etiologic Agents on the Basis of Hazard*, 4th ed. U.S. Department of Health, Education and Welfare, Public Health Service, Centers for Disease Control, Washington, DC.
18. **Adair D, Irwin R.** 2008. *A Practical Guide to Containment: Plant Biosafety in Research Greenhouses*, 2nd ed. Information Systems for Biotechnology, Blacksburg, VA.
19. **USDA-ARS.** 2012. *ARS Facility Design Standards 242.01*. http://www.afm.ars.usda.gov/ppweb/pdf/242-01m.pdf, accessed January 2016.
20. **Hanan JJ.** 1989. *Greenhouses: Advanced Technology for Protected Horticulture*. CRC Press, Boca Raton, FL.
21. **Langhans RW, Tibbitts T.** 1997. *Plant Growth Chamber Handbook*. Iowa Agriculture and Home Economics Experiment Station, Ames, IA.
22. **Langhans RW.** 1978. *A Growth Chamber Manual: Environmental Control for Plants*. Cornell University Press, Ithaca, NY.
23. **National Greenhouse Manufacturers Association.** 2012. https://www.ngma.com, accessed January 2016.
24. **Bell ML, Baker JR.** 2000. Comparison of greenhouse screening materials for excluding whitefly (Homoptera: Aleyrodidae) and thrips (Thysanoptera: Thripidae). *J Econ Entomol* **93:**800–804.
25. **Bethke JA, Redak RA, Paine TD.** 1994. Screens deny specific pests entry to greenhouses. *Calif Agric* **48:**37–40.
26. **Baldwin CL, Runkle RS.** 1967. Biohazards symbol: development of a biological hazards warning signal. *Science* **158:** 264–265.
27. **Kahn RP, Mathur SB.** 1999. *Containment Facilities and Safeguards for Exotic Plant Pathogens and Pests*. American Phytopathology Society Press, St. Paul, MN.
28. **USDA-APHIS-PPQ.** 2010. *Official Regulatory Protocol for Wholesale and Production Nurseries Containing Plants Infected with Phytophthora ramorum*, p 56. Revised 2014. https://www.aphis.usda.gov/plant_health/plant_pest_info/pram/downloads/pdf_files/ConfirmedNurseryProtocol.pdf, accessed January 2016.
29. **USDA.Agricultural Research Service (ARS).** 2002. *Security Policy and Procedures for Biosafety Level-3 Facilities*. http://www.ocio.usda.gov/sites/default/files/docs/2012/DM9610-001_0.htm, accessed January 2016.

Biosafety Guidelines for Working with Small Mammals in a Field Environment

36

DARIN S. CARROLL, DANIELLE TACK, AND CHARLES H. CALISHER

INTRODUCTION

Discussions of biosafety in the laboratory setting can be found in numerous texts, including other chapters in this book. Several published guidelines exist, with the *Biosafety in Microbiological and Biomedical Laboratories* (BMBL) (1) text serving as the general authority. There are a wealth of guidelines available for biosafety in the laboratory, but only a handful of references exist regarding biosafety considerations for individuals performing fieldwork, i.e., work done outside a laboratory with materials containing or potentially containing infectious agents, and there is no formal text describing structured risk assessment strategies for fieldwork that either incidentally or intentionally involves contact with zoonotic agents that have pathogenic potential. This text is meant to serve as a reference for existing guidelines, as well as a tool that can be used to help determine what risk levels exist for a planned field activity and how those risks may be mitigated. Protection of fieldworkers should be the prime focus of both supervisor and the workers themselves.

The last 2 decades have seen the discovery or rediscovery of several infectious diseases referred to as "emerging." In 2001, 175 diseases were classified as emerging, with 75% identified as zoonotic (2). As a result, field studies targeting the ecology of emerging zoonotic diseases increased. In 1993, an outbreak of a previously unrecognized disease occurred in the southwestern United States with an initial case-fatality rate reaching 76% (3, 4). This hitherto unrecognized disease was eventually described as hantavirus pulmonary syndrome and linked to Sin Nombre virus (family *Bunyaviridae*, genus *Hantavirus*) (5). In the years that followed, multiagency efforts involving state and local health departments, universities, the Centers for Disease Control and Prevention (CDC), and others investigated the etiology of the disease and the risks associated with human infection. While conducting these studies, multiple local public health departments simultaneously were trained to safely trap and sample rodents associated with the disease safely. In an effort to ensure safety and uniformity among scientists targeting hantaviruses, a set of guidelines for working with rodents was published in 1995 by CDC personnel (6) and methods were suggested for working with small mammals to be tested for virus (7). In the absence of other guidelines, many institutions have applied these recommendations to research involving any small mammal, regardless of its capacity to harbor a hantavirus or any other zoonotic agent. As one of the few such texts available for biosafety considerations for conducting fieldwork, these guidelines have often been applied to projects outside their original intended scope. To distinguish disease ecology studies from other types of field biology and ecological research not specifically targeting zoonotic diseases, many organizations have developed guidelines for use in investigations of small mammal ecology.

These guidelines are appropriate based on the type of research being conducted, but do not address the topic of biosafety in a more general format that takes into account the species, geographical location, and risk of zoonotic disease as it pertains to field activities in general. Some institutions or agencies, such as the U.S. National Park Service, have recognized a need for such guidance and have developed risk assessment tools for employees to use in identifying potential hazards, including biohazards that may be encountered while conducting activities in the field.

SUGGESTED GUIDELINES

Current guidelines addressing zoonotic disease biosafety have been consolidated into a framework based on risk management principles established by the International Organization for Standardization (ISO 31000; http://www.iso.org/iso/home/standards/iso31000.htm). Risk management consists of the following steps:

1. Establishment of context
2. Identification of risk
3. Assessment of risk
4. Avoidance and mitigation of the risk
5. Creation of a risk management plan
6. Implementation of the plan

The guidelines in this chapter provide the framework for conducting a risk assessment as it pertains to the risk of contact with zoonotic diseases. Unlike laboratory research, field research is conducted in a minimally controlled environment. It is impossible to contain all risks or control the natural environment. Therefore, deliberate planning as well as preparation and execution of specific safety measures are required to reduce the probability of infection or injury. Planning includes conducting a risk assessment to identify hazards that may exist in the environment or as a result of the activity that is undertaken. Once these hazards have been identified, the researcher identifies the items that are needed to ensure the health and well-being of those conducting the research. Then, in the field, these safety precautions are executed to ensure mitigation of the risks identified. This entails fully informing the research team and associated volunteers of the identified risks, the mitigation strategies to be employed, the training necessary for proper use of specialized gear that may be used, and specific activities to be performed. Approaches used to prevent human injury or illness when handling small mammals vary with numerous factors, such as species, age and sex of the animal, reason for handling, handler's level of experience, and presence of enzootic zoonotic disease(s). Handling devices and anesthesia can usually decrease the risk of physical injury but, depending on the circumstances, may also increase the risk. Therefore, proper training and familiarity with the procedures as well as the risks to be taken are required for appropriate and safe use of these tools.

In general, the reduction of zoonotic disease risk requires using the appropriate barrier between the animal, animal sample or animal biologic excretions and secretions, and the handler.

Additional methods include:

- Awareness of potential animal handling risks
- Developing animal handling protocols
- Discussing with one's personal physician or prophylactic vaccination or medications as indicated for the level of risk
- Daily or less periodic project safety briefings and post-handling debriefing
- Training about appropriate methods of animal restraint
- Training on proper use of personal protection equipment (PPE)
- Contingency plans, i.e., emergency contact information and contact information for obtaining medical advice or assistance
- Hand hygiene; washing hands after handling animals and prior to eating or touching one's body parts (alcohol-based sanitizers can be used, especially when running water is not available, but are not a substitute for thorough washing of hands with soap and water)
- Proper disinfection and disposal of equipment and samples
- Having an appropriate first-aid kit available
- Designating a "clean" area for eating, drinking, and smoking

Establishment of Context

Numerous hazards with varying risks occur with any situation; therefore, the initial step in conducting a risk assessment is to establish the context of the research being conducted. The goal of this assessment is to prevent the acquisition of a zoonotic disease. There are four basic components involved in zoonotic disease transmission: (i) biological agent, (ii) host, (iii) route of infection, and

(iv) environment. For field research, we must consider each of these components to develop our context. Because a zoonotic disease is specifically being addressed here, we can assume that the animal host, researcher, and research assistants are all potentially susceptible to disease; we also assume that the researcher and assistants are healthy, nonimmune adults, and are not immunocompromised. Modifications may need to be made to the assessment in the event that this assumption is not correct. Biologic agents are often geographically or taxonomically confined to particular location(s) and particular hosts. For field research, the context is established by identifying the risks inherent to the geographic location(s) in which research will occur, the risks inherent in handling both the animals of interest and any others that will be incidentally contacted, the specific activities being conducted, and the conditions under which those activities are performed. Once the context is established, each component can be evaluated to identify the risk of zoonotic disease transmission associated with that component.

Identification of Risk

Each biologic agent poses inherent levels of risk based on the level of exposure. To determine the risk of zoonotic disease transmission to a researcher or research assistant, each biologic agent should first be identified, if possible. Following this, the route of transmission as well as the type of exposure (dictated by the activities being conducted) and the conditions under which activities will occur also must be established. The following sections describe some of the variables that might need to be considered in determining risk.

Biologic agent

Many zoonotic organisms are naturally confined to a relatively specific geographic location depending on their reservoir hosts, biological vectors, and climate. The first step in establishing the risks is to identify the known zoonotic diseases that may be encountered by a field researcher in the region where the research is being conducted. General environmental conditions may be applied in the event that information is not available for the location where research is being conducted (Fig. 1).

The potential animal host must also be well considered to identify possible biologic agents to guard against. It is impractical to list all hosts associated with one or even some of the 800 zoonotic diseases (2), or even the scores of "emerging" zoonoses, especially given the rapid rate of discovery of new agents and diseases (8). However, Table 1 provides the names of diseases of most concern and the broad categories of mammals affected; the geographic range of the possible host is also provided. It is the responsibility of the investigator to extract the most recent relevant information from the available literature and consult with other experts in the field. Once an actual or potential pathogenic agent has been identified, the researcher must give consideration to the means of transmission by which the agent of concern enters the human host. There are four main routes for zoonotic disease transmission: (i) direct contact, (ii) indirect contact, (iii) aerosols, and (iv) vector-borne. Direct contact requires that the source animal or its excretions and secretions be infected with the organism and enter the human host through mucous membranes, breaks in the skin, or ingestion. Indirect contact transmission occurs in the human host the same way but involves an object (biological or otherwise) that is contaminated with the pathogen. Aerosol transmission occurs when the pathogen is inhaled. Vector-borne transmission occurs when a feeding arthropod transfers the pathogen from an infected animal to a human host. Not all animals that are encountered are infected or infectious, but it is safest to assume that all potential hosts are infected and that they should be handled appropriately (9, 10).

Activity and conditions

Every research project is unique and may entail multiple activities, each with different levels of animal contact depending on the protocol-specific conditions in which the activity is being conducted. Conditions include where the activity is being conducted, i.e., confined space versus in the open; disease status of the potential zoonotic host, i.e., healthy versus ill; or disease status of the surrounding population, i.e., whether there is a current epizootic or outbreak. It is important to note that not all infected hosts are ill or show clinical signs of being ill; in fact, oftentimes reservoir hosts will not be ill. These and other activities and conditions will determine the level of risk of transmission. Table 2 lists common research activities and the conditions in which they may occur.

Assessment of Risk

After identifying the pathogens that may be encountered (and taking into consideration the possible presence of unrecognized agents of disease), the activity being conducted, and the conditions under which the activity is taking place, a risk level can be assigned for each agent based on a review of the four risk groups the World Health Organization (WHO), the U.S. National Institutes of Health (NIH), and the BMBL (1) have established for known and potentially infectious microorganisms. The groups vary slightly among these entities, but the general properties can be combined from the three sources into four outcomes of infection: risk group 1 (mild outcome) includes agents not currently associated with human disease in healthy humans with a low individual and

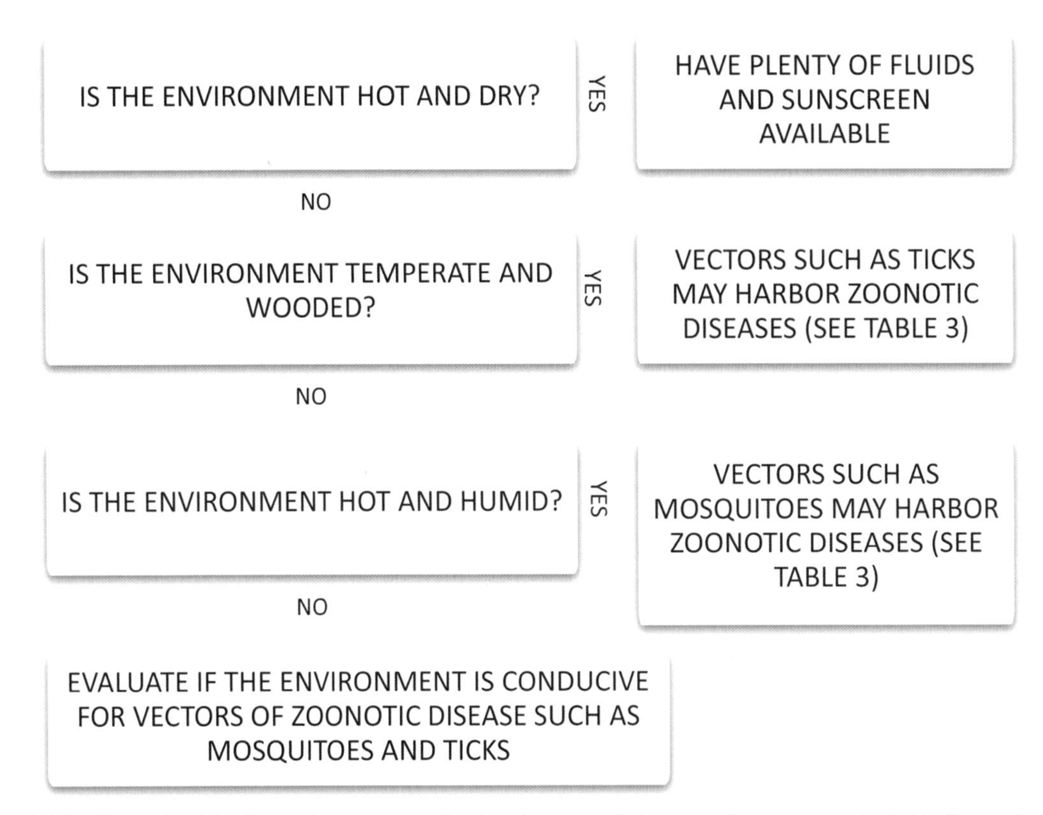

Figure 1: Identifying the risks. Example of a generalized decision matrix for zoonotic disease potential in the environment.

community risk; risk group 2 consists of agents associated with mild to moderate human disease for which there are widely available effective treatments for an illness with moderate risk to individuals but low risk to communities; risk group 3 comprises agents known to cause severe to lethal disease that does not readily spread between humans and for which treatments have limited availability, so there is a high risk to individuals but a low community risk; risk group 4 contains the pathogenic agents likely to cause serious or fatal disease that is readily transmitted between humans and for which prophylaxis and therapeutics are rare or nonexistent, so there is both a high (catastrophic) individual and community risk (1).

A risk level is assigned to every combination of activity, condition, and route of transmission for each biologic agent. The matrix in Fig. 2 is an example that can be used to estimate the risk of acquiring an infectious agent.

Mitigation of Risk

Proper planning and an awareness and understanding of the general surroundings in which the work is to be done are essential to any successful risk management protocol. The first layer of protection involves having on hand the supplies necessary for maintaining basic hygienic practices, such as hand washing. The wide availability of hand sanitizers is helpful, but proper washing with soap and clean water (which can then be followed by the use of a hand sanitizer) is preferred. Additional considerations include the availability of a purification system for drinking water, sunscreen, and insect repellent, as well as having appropriate vaccinations or prophylactic medications (antimalarials, etc.) as determined by the location of the task being undertaken.

Further mitigation strategies for the risk of zoonotic disease are dependent on the specific field protocol, the possible zoonotic agents involved, and how the infectious agent(s) is/are transmitted. Therefore, the risk level can be used to ascertain the PPE that should be worn for the specified activity and conditions. Much like in the laboratory setting, PPE requirements in the field can range from wearing appropriate clothing to requiring the use of powered air-purifying respirators (PAPR). Regardless of the situation, the use of appropriate PPE can significantly reduce the risk of zoonotic disease transmission for any pathogenic agent. Basic barrier precautions (appropriate clothing and latex or nitrile gloves) are generally recommended as the minimum PPE level, which can be expanded as needed. The U.S. Occupational Safety and Health Administration (OSHA) has prepared guidelines for the use of PPE in various hazardous working conditions, articulated in 29 Code of Federal Regula-

TABLE 1.

Some important zoonotic diseases that may pose a risk for wildlife biologists working with small mammals in the field, their natural hosts, and their general geographic distributions[a]

Disease	Host species	Distribution
Anthrax	Many mammals	Worldwide
Argentine hemorrhagic fever	Drylands vesper mouse (*Calomys musculinus*)	Central Argentina
Bolivian hemorrhagic fever	Large vesper mouse (*Calomys callosus*)	Northern Bolivia
Brazilian hemorrhagic fever	Unknown	Southern Brazil
Cowpox	Voles, domestic cats, livestock	Europe, Asia
Ebola hemorrhagic fever	Unknown (bats suspected)	Africa, Philippines
Hemorrhagic fever with renal syndrome	Muroid rodents	Europe, Asia
Hantavirus pulmonary syndrome	Cricetid rodents	North, South, and Central Americas
Leptospirosis	Rodents, marsupials, other mammals	Worldwide
Listeriosis	Various mammals and birds	Worldwide
Lymphocytic choriomeningitis	*Mus musculus* (rodent)	Africa, Asia, Americas
Marburg hemorrhagic fever	Unknown (bats likely)	Central, Southern Africa
Monkeypox	Unknown mammal (rodents likely)	Africa
Murine typhus	*Rattus* spp. (rodent)	Worldwide
Nipah virus encephalitis	*Pteropus* spp. (bats)	Southeast Asia, Indian subcontinent
Pasteurellosis	Birds, rodents, carnivores	Worldwide
Plague	Various rodents	Asia, Africa, South America, western United States
Q fever	Various mammals and birds	Worldwide
Rabies	Various mammals	Worldwide
Rat-bite fever	Various rodents	Worldwide
Salmonellosis	Many reptiles, birds, mammals	Worldwide
Tularemia	Various rodents and lagomorphs	North America, Europe, Asia
Venezuelan hemorrhagic fever	*Zygodontomys brevicauda* (rodent)	Venezuela
Zika disease	Unknown (probably vertebrates)	Essentially worldwide

[a]Modified from reference 11.

tions (C.F.R.) 1910.120 Appendix B, which divides PPE into categories A, B, C, and D (https://www.osha.gov/pls/oshaweb/owadisp.show_document?p_table=STANDARDS&p_id=9767; http://www.ecfr.gov/cgi-bin/text-idx?node=29:5.1.1.1.8.8.33.14). In general, Category A consists of the highest levels of skin, respiratory system, and eye protection, involving complete encapsulation with a self-contained air supply or PAPR and protective boots. Category B also involves the greatest levels of respiratory system protection [full-face air supply or positive-pressure high-efficiency particulate air (HEPA)-filtered], but instead of complete encapsulation, the use of double

TABLE 2.

Potential activities associated with fieldwork and the conditions under which they may occur

Activity	Conditions
Movement to study site	Open area
Observations	Confined or open areas
Tag and release	Healthy or ill animals
Collection of carcasses	Healthy or ill animals collected for laboratory studies
Collection of biologic samples	Current epizootic or human outbreak; excretions, secretions
Collection of biologic samples	Invasive, i.e., tissues, blood; ectoparasites

Probability of Exposure	Outcome Severity if Exposed (Effect)			
	MILD	MODERATE	SEVERE	CATASTROPHIC
UNLIKELY				
SPORADIC				
LIKELY				
FREQUENT				

Probability of Exposure

UNLIKELY	Never reported, but possible or unknown
SPORADIC	Reported transmission but rare
LIKELY	Reported transmission or conditions favor transmission
FREQUENT	Reported transmission based on current conditions

Outcome Severity

MILD	Illness is short-term and usually resolves without care
MODERATE	Illness is treatable, but requires care from a medical professional
SEVERE	Illness results in hospitalization and long-term (>3mo) effects
CATASTROPHIC	Death

Risk Level

	MILD
	MODERATE
	HIGH

Figure 2: Estimating the risk of acquiring an infectious agent.

gloves and protective clothing, such as a disposable hooded Tyvek suit or other relevant impermeable barrier (determined by the nature of the concern), with protective boots is sufficient. Category C is applicable to situations where exposure risks (agents and concentrations) are known and that may or may not require full-face respiratory protection; thus, half-face masks can be used along with double gloves, disposable impermeable suit, and protective boots. A face shield also may be appropriate. Category D involves basic precautions for "nuisance contamination" and involves appropriate clothing (i.e., coveralls) and footwear, single gloves, and optional face shield, as needed.

Creation of a Risk Management Plan

The risk assessment and plan can be conducted by anyone who will be involved with the field activities; however, it is the primary researcher's responsibility to review the assessment and sign off on the plan. The plan should be a written document that includes all of the biologic agents considered, how the risk level was ascertained, and what PPE will be used to mitigate risk. In some cases, recommended PPE for one disease may result in an increased risk for another disease. In this situation, recommended PPE may need to be adjusted. The plan should reflect this and provide an explanation. The PPE chosen should be adequate for the agent with the highest level of risk. If this is not possible, an alternative activity should be considered. A risk matrix template is provided in Table 3 to aid in the development of a written plan.

Implementation of the Plan

No plan is complete until it is actually put into use. Once a plan is shown to be suitable, the investigator should distribute it amongst the field team and ensure that all questions that the team may have are resolved. It is the responsibility of the investigator to provide the appropriate PPE as determined in the risk analysis and to ensure that all participants are educated regarding the proper use of all PPE.

SUMMARY AND CONCLUSIONS

No matter the risk management approach that is taken, the idea behind all available approaches is to err on the side of caution. Those who do not participate enthusiastically in safety programs should not be allowed to be involved in what may be dangerous work. Also, it should be recognized that "low probability" is not the same as low risk (9).

TABLE 3.

Risk matrix template

Location of study			**Research vertebrate or incidental pathogenic exposure**				
Risk factor (biologic agent)	Activity	Conditions	Route of transmission	Probability of exposure	Severity of disease	Risk level	Mitigation (PPE needed)
Assessment created by:			Research lead:				
Date:							

References

1. **U.S. Department of Health and Human Services, Public Health Service, Centers for Disease Control and Prevention, National Institutes of Health.** 2009. *Biosafety in Microbiological and Biomedical Laboratories*, 5th ed. HHS Publication No. (CDC) 21-1112. http://www.cdc.gov/biosafety/publications/bmbl5/BMBL.pdf.
2. **Taylor LH, Latham SM, Woolhouse ME.** 2001. Risk factors for human disease emergence. *Philos Trans R Soc Lond B Biol Sci* **356:**983–989.
3. **Hughes JM, Peters CJ, Cohen ML, Mahy BW.** 1993. Hantavirus pulmonary syndrome: an emerging infectious disease. *Science* **262:**850–851.
4. **Duchin JS, Koster FT, Peters CJ, Simpson GL, Tempest B, Zaki SR, Ksiazek TG, Rollin PE, Nichol S, Umland ET, Moolenaar RL, Reef SE, Nolte KB, Gallaher MM, Butler JC, Breiman RF, The Hantavirus Study Group.** 1994. Hantavirus pulmonary syndrome: a clinical description of 17 patients with a newly recognized disease. *N Engl J Med* **330:** 949–955.
5. **Nichol ST, Spiropoulou CF, Morzunov S, Rollin PE, Ksiazek TG, Feldmann H, Sanchez A, Childs J, Zaki S, Peters CJ.** 1993. Genetic identification of a hantavirus associated with an outbreak of acute respiratory illness. *Science* **262:**914–917.
6. **Mills JN, Yates TL, Childs JE, Parmenter PR, Ksiazek TG, Rollin PE, Peters CJ.** 1995. Guidelines for working with rodents potentially infected with hantavirus. *J Mammal* **76:** 716–722.
7. **Mills JN, Childs JE, Ksiazek TG, Peters CJ, Velleca WM.** 1995. *Methods for Trapping and Sampling Small Mammals for Virologic Testing.* U.S. Department of Health and Human Services, Atlanta, GA.
8. **Jones KE, Patel NG, Levy MA, Storeygard A, Balk D, Gittleman JL, Daszak P.** 2008. Global trends in emerging infectious diseases. *Nature* **451:**990–993.
9. **Calisher CH.** 2015. Rabies: low probability, not low risk. *Bat Res. News.* **56:**15–17.
10. **Johnson B.** 2001. Understanding, assessing, and communicating topics related to risk in biomedical research facilities, p 149–166. *In ABSA Anthology of Biosafety IV—Issues in Public Health.* American Biological Safety Association, Mundelein, IL. http://www.absa.org/0100johnson.html.
11. **Mills JN, Carroll DS, Revelez MA, Amman BR, Gage KL, Henry S, Regnery RL.** 2007. Minimizing the infectious disease risks in the field. *Wildl Prof* **1:**30–35.

Components of a Biosafety Program for a Clinical Laboratory

37

MICHAEL A. PENTELLA

The risk of exposure to infectious agents exists in every clinical laboratory; however, the goal of every clinical laboratory should be to minimize that risk and conduct its activities as safely as possible. To achieve this, a strong culture of biosafety must be in place. The biosafety culture depends on the opinions, beliefs, views, and feelings of all the laboratory staff. When there is a strong culture of biosafety, then every employee accepts responsibility and is accountable to maintain biosafety practices that protect both the employee and coworkers. The major question is how to achieve a strong culture of biosafety. Management has the responsibility to build and sustain that culture. This chapter will explain the required components.

In the United States, biosafety is defined as the development and implementation of administrative policies, work practices, facility design, and safety equipment to prevent transmission of biohazardous agents to workers, other persons, and the environment (1). Several organizations, such as the Occupational Safety and Health Administration (OSHA), the Centers for Disease Control and Prevention (CDC), College of American Pathologists, Clinical and Laboratory Standards Institute (CLSI), and the Joint Commission on Accreditation of Healthcare Organizations (formerly JCAHO, now The Joint Commission), provide guidelines for biosafety practices. It is the responsibility of each laboratory to have in place a written biosafety plan that describes the biosafety practices. To establish a strong culture of biosafety in the laboratory requires a systematic approach to build a biosafety program that supports the quality of the laboratory. An essential element in building a strong biosafety culture is that staff must be trained in safe practices, and their adherence to these practices must be monitored.

LABORATORY-ACQUIRED INFECTIONS

Fundamentally, biosafety is an infection control problem for the clinical lab. As in all infection control issues, the critical prevention step is breaking the chain of infection. For example, the extent of laboratory-acquired tuberculosis occurring in laboratory workers is unknown because laboratory-acquired infections are not reportable conditions. The lack of reporting deprives us the opportunity to learn from those experiences and determine best practices. Recent events (2) regarding biosafety-related issues indicate that not every facility is prepared to handle emerging infectious agents or to protect the employees from acquiring an infection. The chain of infection requires an infectious agent, susceptible host, entry site, means of spread, portal of exit, and reservoir. There are several points along the chain where an intervention is necessary to break the chain.

Of great assistance in building a culture of safety in the laboratory is to study instances of laboratory-acquired infections in the clinical lab. For example, the incident of laboratory-acquired brucellosis (3) can serve as an excellent learning tool for preventing laboratory-acquired infections in the future.

How do laboratorians get infected? Laboratory-acquired infections have been reported from needlestick and sharps injuries, ingestion, ocular or mucosal splash, inhalation of aerosols (aerosols may affect persons in adjacent workspaces), or unknown routes (4). When researchers looked at the distribution of injury types among public health laboratories and Minnesota hospital laboratories in 1986 (5), they found that 63% of hospital lab injuries were needlesticks, followed by cuts/scrapes at 21%. Inhalation, ingestion, or skin injury accounted for only 1%. Later, Sewell (4) looked at the agents most frequently reported in laboratory-acquired infections in the United States and Great Britain. Topping the list was typhoid fever, followed by Q fever, brucellosis, and tuberculosis.

STEPS TO A SUCCESSFUL BIOSAFETY PROGRAM

The foundation of building a culture of biosafety in a laboratory requires a successful biosafety program. Deciding where to begin the process of building a biosafety program in a busy laboratory can be daunting because critical laboratory procedures cannot be interrupted. However, looking at building the program as a project that has multiple steps can make this achievable. The steps to building a biosafety program are:

1. Perform risk assessments.
2. Select mitigation tools based on risk assessment.
3. Establish biosafety competencies.
4. Provide safety orientation and ongoing training.
5. Establish a safety committee, perform regular audits, and monitor compliance.
6. Work with the occupational health program.
7. Foster the culture of biosafety.

Perform Risk Assessments

To establish facility-specific policies and practices, each laboratory must perform a risk assessment that relies on factual information to define the health effects of exposure to workers from exposure to hazardous materials and situations (6). Data to conduct risk assessment for the biological laboratory may be limited due to underreporting or even a unique laboratory setting, resulting in a subjective approach of risk management that incorporates hazard probability and severity to provide recommended actions to ensure a safe environment. Integral to the risk management process is knowledge of the biological agent, facility, safety equipment, and work processes (7). Qualitative risk assessment is encompassed in the biosafety levels (BSL) established in *Biosafety in Microbiological and Biomedical Laboratories* (BMBL) (2). Most clinical laboratories are at the BSL2 level. In the BSL2 environment, work is performed using organisms that are transmitted primarily by percutaneous exposure, mucous membrane exposure, or ingestion. BSL2 practices call for containment in a biosafety cabinet (BSC), especially when there is a risk of splash, splatter, and the production of aerosols. An example is the subculture of blood culture bottles. Some clinical labs have a BSL3 laboratory for work with organisms that are efficient in producing infections at low concentrations by the aerosol route of transmission.

The risk assessment is the process of gathering all available information on a hazardous substance and evaluating it to determine the possible risks associated with exposure (8). This is followed by determining the mitigation strategies necessary to provide protection. The risk assessment should be completed before a new test is implemented in the lab (9), bringing on a new instrument, or beginning facility alterations that involve construction and/or renovation. It should also be performed whenever there is a significant turnover of staff in the lab. Periodic review of risk assessments is important so that practices and processes are renewed and refreshed. There is no single standard approach to the risk assessment process. There are several tools available to assist the laboratory in completing the tasks. The risk will never be zero, but it can be mitigated. The goal of performing the risk assessment is to predict, identify, and mitigate risk. Through the risk assessment process, the laboratory is able to determine the mitigation tools needed to reduce the risk.

The risk assessment is the first step in the process of building the biosafety program because information and issues that are identified in the risk assessment process can then inform other steps in the biosafety program. The risk assessment should cover preanalytical activities from the time the specimen is collected, transported, unpackaged, centrifuged, aliquoted, and moved through the lab. Analytical activities are whether the test is performed manually or on an instrument, and postanalytical activities include cleanup of the lab and destruction of the specimen and lab-generated waste.

The person assigned to perform the risk assessment should be an experienced laboratorian who is familiar with the methods used in the clinical laboratory, but not necessarily someone who does the testing every day. By bringing an outsider's perspective, the laboratory is more likely to identify issues that may not be obvious to someone who is extremely familiar with the present processes. That is not to say that those doing the testing every day should not participate in the risk assessment process; they definitely bring invaluable experience to the risk assessment process. It is important that staff be fully engaged so that any changes in procedures will be readily implemented. Another benefit of including staff in the risk assessment process is that it allows for the identification of training needs. Besides the identification of training needs, the risk assessment allows for evaluation of procedural changes, ensures compliance with regulatory agencies, justifies space and equipment needs, and evaluates emergency plans.

An important first step in completing the risk assessment is identifying microbial agent hazards that are reasonably expected to be encountered in the laboratory. This can vary by region of the country where the laboratory is located or the region from which specimens are submitted. For example, if the laboratory is located in the southwestern area of the United States, it is reasonable to expect that sputum specimens containing *Coccidioides immitis* may be cultured in the laboratory. This may also be expected in the laboratory whose state residents travel to that region of the country and return home to be diagnosed with pneumonia due to *C. immitis*. A good starting point for determining what organisms may be encountered in the laboratory is the reviewing the state's Department of Public Health for recent reportable diseases. On the basis of the list of commonly encountered organisms, the laboratory can identify potential microorganisms that may be encountered and perform an initial risk assessment.

Once a list of microbial agent hazards is established, the laboratory must consider the specimens that are received by the laboratory and the testing methods performed on those specimens. Through the risk assessment process, protocols are reviewed to determine the agent concentration in specimens and suspension volumes. If the specimens are cultured and microbial agent concentration further amplified, then the agent concentration significantly changes, which greatly impacts the risk assessment process. A good example of this is cerebrospinal fluid (CSF) containing *Neisseria meningitidis*. CSF with *N. meningitidis* is not a significant risk of infection to the laboratorian performing Gram stains and cell counts. However, once *N. meningitidis* is cultured on plates and broth media, these cultures represent a significant risk to the laboratorian, and testing should not be performed on the benchtop but should be performed in a BSC.

During the risk assessment process, every procedure should be reviewed to determine the potential for the generation of aerosols, droplets or droplet nuclei, protocol complexity, and the use of sharps. The risk assessor should determine the risk of laboratory-acquired infection from each step performed on the basis of the microbial agent and the task performed.

Risk assessment findings should be "immortalized" in writing and reviewed by the laboratory leadership as part of the quality assurance program. Having a written document also helps with future reviews of the risk assessment. The risk assessment should be reviewed with staff and management.

A risk assessment should be performed before work begins, whenever there is a move or renovation, when there are new employees, when the lab is going to encounter a new infectious agent or reagent, and when new equipment is purchased. The risk assessment should be repeated when changes are made in agents, practice, employees or facilities.

The steps to complete a risk assessment are:

1. Create a risk assessment matrix that includes the procedure, agent, a potential hazardous procedure, staff susceptibility to disease, and the control measures or protection that are necessary.
2. Identify activities that could cause exposure to agent or material.
3. Determine risk mitigation strategies.
4. Evaluate staff competency and utilization of safety equipment.
5. Review assessment with staff and management.

An example of a risk assessment tool can be found in Fig. 1. It comes down to the dual responsibilities of risk and work performance. Reasonable safety practices must be developed to mitigate risk, while allowing the laboratory to provide accurate data efficiently for patient diagnosis and treatment. Fundamentally, the primary benefit of the risk assessment is keeping the laboratorians, their families, and the community safe.

SELECTION OF MITIGATION TOOLS

From the risk assessment, the mitigation tools are selected to prevent laboratory-acquired infections. For example, building a barrier between the hazard and the employees is an engineering control; changing the way in which employees perform their work is a work practice control. The mitigation tools include the determination of the biosafety level of laboratory in which to perform the testing, engineering controls, personal

Procedure	Potential Hazard(s)	Control/Protection	Additional Information
Package receipt at specimen receiving	Leaking Package	• Place leaking package in plastic bag and transfer to a Biological Safety Cabinet (BSC). • PPE: nitrile/latex gloves, lab coat, safety glasses	• Disinfect exterior of sealed plastic bag prior to transfer to testing area.

Figure 1: Example of risk assessment tool.

protective equipment (PPE), laboratory practices, and the flow of medical waste (10). The mitigation tools to be used in the performance of testing should be written into the procedure manual.

For the selection of the BSL in which the work is to be performed, the risk assessment dictates the process. BSLs represent the conditions under which an agent can be handled safely. There are four BSLs that consider the agent, practices, equipment, and facilities. BSL1 is basic containment practices for organisms not known to cause disease in the healthy adult. Work with organisms such as *Staphylococcus epidermidis* would occur in the BSL1 lab environment. This is the basic level of containment where standard microbiological practices are followed. There is no need for special barriers. Most clinical laboratories are BSL2, which is for moderate risk agents, such as *Staphylococcus aureus*. BSL2 laboratory space has a BSC available, aerosol containment practices are used, and PPE is worn. The PPE includes lab coats, gloves, goggles, and face mask when indicated. BSL2 labs have restricted access to the laboratory. BSL3 laboratory space is required for organisms that cause serious or lethal diseases or exotic agents. For example, work with *Mycobacterium tuberculosis* cultures requires BSL3 laboratories. The typical BSL3 laboratory has an antichamber where staff members don appropriate PPE. Employees working within BSL3 laboratories wear N95 respirators or powered air-purifying respirators (PAPRs) and are part of the respirator program. In the BSL2 laboratory, there is directional airflow into the room and out through dedicated high-efficiency particulate air (HEPA) exhaust fans. There is typically a dedicated autoclave with a pass-through design. The BSL3 lab operations are verified annually. The floors of the BSL3 lab are a single sheet for easy decontamination. BSL4 laboratory space, the highest level of containment, is required for work with dangerous and exotic agents that cause life-threatening disease, such as Ebola virus culture. There are two main types of BSL4, either the cabinet lab, where all agents are manipulated in a class III BSC, or a suit laboratory, where personnel wear a positive-pressure-supplied air-protective suits.

The risk assessment determines the selection of engineering controls. Examples of engineering controls used for primary containment include BSC, sharps containers, sharps safety devices, sealed rotor centrifuge, and aerosol-resistant pipette tips. Examples of engineering controls used for secondary containment include the floor plan of the laboratory space, sinks for hand washing, and self-closing and interlocking doors.

On the basis of the risk assessment, the PPE practices can be determined for the laboratory and tests performed. The selection process includes gloves (latex and nitrile), laboratory coats or gowns and, whether reusable or single-use disposable, eye protection, face masks, shoe covers, sleeve covers, and respiratory protection. When selecting PPE, standards provided by various agencies must be considered. ASTM provides standards on selection of laboratory coats in WK38455 "New Specification for Healthcare Worker Protective Uniforms" and for selection of gloves in D6319 "Standard Specification for Nitrile Examination Gloves for Medical Application." The American National Standards Institute (ANSI) provides Z87.1 on "Occupational and Educational Personal Eye and Face Protection Devices." The OSHA 1910.1030 Bloodborne Pathogens Standard covers selection of PPE in 29 C.F.R. 1910.1030(d)(3)(i) and hand sanitation in 29 C.F.R.1910.1030(d)(2)(v).

The risk assessment informs the adoption of laboratory safety practices. The safety practices are dictated by any number of factors that are often unique to the laboratory design and processes. These practices can include the use of PPE in specific situations, such as wearing gloves to read culture plates; disinfectant practices, such as daily disinfectant of the BSC, counters, and centrifuge; the use of capped centrifuge tubes, splash-proof containers, UV lights, and disposable loops; allowing slides to dry in a BSC; and the spill cleanup procedure.

An important part of the mitigation tools is the biosafety manual. This should include the safety policies, roles and responsibilities, regulations, routes of exposure, risk assessment process, reporting incidents, and biosafety practices. The biosafety manual should be reviewed and updated at least annually and after significant events or incidents. It should be read by all staff working in the laboratory at the beginning of their work tenure, annually, and whenever it is updated.

ESTABLISH BIOSAFETY COMPETENCIES

An important step to building the culture of biosafety is connecting biosafety competencies to required skills. Competencies are defined as action-oriented statements that delineate the essential knowledge, skills, and abilities required for the performance of work responsibilities. The CDC (11) competencies are designed for working with biologic agents in BSL2–4. The competencies are tiered to a worker's experience and position in the facility whether entry level, experienced, or management positions (12). In 2015, the CDC and Association of Public Health Laboratories (APHL) published a comprehensive set of competencies (13) for laboratory professionals. The 2011 and 2015 publications are companion documents for biosafety. Whereas the 2011 guideline has task-level details, the 2015 guidelines contain the same content, which has been revised and restructured.

Competencies can be useful for assessing current skills, creating career development plans, and planning specific training to meet educational needs. The intent of developing biosafety competencies is to define the essential competencies needed by laboratory personnel to work safely with biologic materials and other hazards commonly found in biologic laboratory, reduce the risk of exposures at all levels, and provide essential baseline information in a format that can be used to develop facility-specific competencies.

Similar to competencies developed to perform laboratory testing, biosafety competencies are necessary to build the skills and improve the performance of staff. The 2011 *Morbidity and Mortality Weekly Report* (MMWR) guidelines define essential competencies needed by laboratory personnel to work safely with biologic materials and other hazards commonly found in the biologic laboratory. By incorporating competencies into the biosafety program, the laboratory is expected to reduce the risk of exposures at all levels and provide essential baseline information for a format to develop facility-specific competencies. The MMWR guidelines provide competencies that are tiered to three professional levels of practitioners. The titles for these professionals vary by institution. The three levels are (i) the entry-level laboratory scientist; (ii) the mid-level scientist, which is the chief or head scientist or medical technologist, laboratory specialist, or laboratory manager; and (iii) the senior-level scientist, which is the laboratory manager, chief technologist, or hospital or clinic director.

In the 2011 guidelines, the biosafety competencies are separated into four skill sets: Skill Domain I, potential hazards; Skill Domain II, hazard controls; Skill Domain III, administrative controls; and Skill Domain IV, emergency preparedness and response. In Skill Domain I, potential hazards, the focus is on competencies involved with understanding the hazards. There are four subdomains to Skill Domain I: biologic materials, research animals, chemical materials, and radiologic materials. In Skill Domain II, hazard controls, the focus is on use of primary and secondary barriers to prevent exposure. The subdomains of Skill Domain II are PPE, engineering controls–equipment (primary barriers), engineering controls–facility (secondary barriers), and decontamination and waste control management. In Skill Domain III, administrative controls, the focus is on administrative controls to reduce the duration, frequency, and severity of exposure to hazardous materials or situations. The subdomains for Skill Domain III are: hazard communication and signage, guidelines and regulatory compliance, safety program management, occupational health–medical surveillance, and risk management. In Skill Domain IV, emergency preparedness and response, the focus is on management of emergencies. The subdomains for Skill Domain IV are emergencies and incident response, exposure prevention and hazard mitigation, and emergency response–exercises and drills.

To build biosafety competencies into a competency assessment program, first review the competencies and then select the competencies from each domain that are applicable to the laboratory on the basis of the risk assessment. The laboratory biosafety competency assessment form in Fig. 2 serves as an example.

PROVIDE SAFETY ORIENTATION AND ONGOING TRAINING

Training and education are the anchors to the biosafety program in building the culture of safety in the clinical laboratory. The training needs are guided by regulatory requirements, the risk assessments, and competencies. Each facility needs to determine the method of training

Laboratory Biosafety Competency Assessment Form – Entry Level Name: ____________ Date: ________

Skill Domain*	Biosafety Competency – abbreviated from the Guidelines for Biosafety Laboratory Competency	Competency Level Ranking	Importance	Frequency	Comment
I Bio 3a	Describe PPE used when handling biologic materials				
II PPE 1	List PPE required for general laboratory entry				
II PPE 2	Describe specific PPE to be used for each procedure				
II PPE 4a	Demonstrate proper donning and doffing of gloves and gown				
II PPE 4b	Describe the limitations of PPE				
II Decon 3e	Describe routine surface decontamination procedures				
II Decon 1	Describe waste segregation procedures				
II Decon 2a	Describe proper disposal of different types of biological waste				
III Occ Health 4	Describe signs and symptoms following exposure				
III Risk Mgmt 3	Describe the risk assessment process				
IV Emer Resp 2	Describe reporting requirements for emergencies				
IV Drills	Participate in drills and exercises				

Reviewed by: ____________ Date: ________

Legend:

Competency Level: Entry Level: Laboratory Scientist or Medical Technologist; **Midlevel**: Chief/Lead Scientist or Medical Technologist, Laboratory Specialist or Laboratory Manager; **Senior Level**: Laboratory Manager, Chief Technologist, or Hospital or Clinical Director.

Competency Level Ranking:

1 = Awareness: You have no training or experience.

2 = Basic: You have received basic training.

3 = Intermediate: You have repeated successful experiences.

4 = Advanced: You can perform the actions associated with this skill without assistance.

5 = Expert: You can train others in this competency

Importance to the Position:

1 = An important competency for position

2 = Neutral

Frequency Competency Performed:

D = Daily W = Weekly M = Monthly R = Rarely A = As Needed

Figure 2: Example of competency assessment form.

that best meets the needs of the employees. In some cases, it is best to determine what outside training is available and what site-specific training is needed. The best format for the training will vary depending on the employees and their availability to participate in training. It is best to have written materials and exams. Any hazards identified in the risk assessment serve as an excellent starting point for the development of trainings. It is always important to train staff on use of safety practices; i.e, engineering controls, PPE, and laboratory practices. If the risk assessment process results in changes to the procedures, then staff must be required to review changes to the procedures. Routine observations of staff with regard to biosafety practices serve as an excellent means to determine staff level of knowledge. It is important to develop a training plan for the laboratory and include biosafety topics in your overall plan.

A broad net should be cast regarding who should receive the training in biosafety topics. Besides the laboratory and transport staff working in the facility, it may be important to include maintenance and cleaning staff, external first responders, security staff, fellows, students, and other visitors. Training should be commensurate with the roles and responsibilities and authorities of staff.

ESTABLISH A SAFETY COMMITTEE, PERFORM REGULAR AUDITS, AND MONITOR COMPLIANCE

Part of the plan includes exercise and training on the safety procedures. The biosafety plan should have the following components: audits of the program by an internal auditing process and potentially external audits; management must monitor staff and safety equipment performance; the biosafety plan must be reviewed and revised at regular intervals; and the plan must include mandated reporting of all incidents, accidents, and near misses. The plan should define the process for follow-up on accidents, incidents, and near misses, and these events should serve to inform revision to the plan. Currently the only national requirement for reporting of safety incidents is for Possession, Use, and Transfer of Select Agents and Toxins (42 C.F.R. Part 73). Because there is not a requirement, incidents of laboratory-acquired infection are underreported, and this contributes to insufficient data in addressing any changes needed in practices or equipment.

SAFETY AUDITS

Perform regular safety audits at least annually for each laboratory section. This is often performed by members of the safety committee who do not work in the laboratory section to offer a fresh perspective and question practices. To perform the audit, using a biosafety checklist provides standardization and structure to the process. The auditors can observe for unsafe practices and processes. Involving colleagues from other laboratory sections serves to remind individuals of how important following the safety practices are.

Examples of safety audit questions are:

1. Has the person performing the risk assessment received training and are they experienced in risk assessments?
2. Is there a written procedure for appropriate donning and doffing PPE, including laboratory coats, gloves, protective eyewear, face shields, N95, and/or PAPRs?
3. Are the Biosafety Laboratory Competencies used for annual staff reviews?
4. Do all new personnel receive safety training before they begin working in their assigned laboratory?
5. Are internal safety audits performed at least annually and after significant safety breaches?
6. Are biohazard signs posted by the entrance of laboratories where infectious agents are processed and tested and in other areas where indicated?

SAFETY COMMITTEE

An important component of the laboratory's biosafety plan is the safety committee, which has the responsibility to perform audits of lab sections, to review deidentified information from all incidents for the purpose of determining if there are trends or patterns of incidents, make recommendations to improve safety practices to prevent incidents, and monitor that issues that cause incidents have been resolved. Membership on the safety committee is a useful training ground in biosafety principles and procedures.

Collaborate with the Occupational Health Program

The provision of occupational health services in a laboratory setting is critical to promoting a safe and healthy workplace. This is accomplished by limiting opportunities for exposure, promptly detecting and treating exposures, and using information gained from work injuries to enhance safety precautions further. Immunization services are an important part of the occupational health plan and should be decided upon based on the risk assessment and regulatory requirement, for example, the hepatitis B vaccine. Other vaccines (14) to consider the basis of the risk assessment of the hazards faced in the laboratory include tetanus-dipththeria-acellular pertussis vaccine (Tdap); measles, mumps, rubella (MMR) vaccine;

meningococcal vaccine; typhoid vaccine; anthrax vaccine; smallpox vaccine; and rabies vaccine. Occupational health is a shared responsibility of occupational health care providers, laboratory directors, laboratory managers, laboratory supervisors, and staff. It is important the biosafety program include medical surveillance procedures, prophylactic measure, and response procedures for laboratory occupational illnesses and exposures.

The laboratory should review with the occupational health care provider the risk assessment and mitigation tools determined for the laboratory.

- Meet with occupational health services to review the risk assessment.
- Implement preexposure prevention activities, including vaccinations.
- Create a postexposure management plan.
- Review procedure for staff access to occupational health services.
- Review reports from occupational health.
- Train staff on when to connect with occupational health.

FOSTERING A CULTURE OF BIOSAFETY

Building the culture of safety requires a commitment from administration and lab leadership. A biosafety program is only effective if it is embraced by staff. Everyone needs to be engaged to improve biosafety practices. It is critical that hazards be identified ahead of time to minimize exposures. Each lab creates a culture of safety that is open and nonpunitive, encourages questions, and is willing to be self-critical or introspective. No regulation or guideline can ensure safe practices; it is the individuals and organizational attitudes that determine safe practices. Everyone must be committed to safety, be aware of risks, act to enhance safety, and be adaptable. To build the culture of biosafety, discuss biosafety at the regularly held meetings of the laboratory. Address concerns from staff in labs not impacted by the testing for an emerging pathogen. For example, hold a meeting for all staff and discuss safety-related issues to the emerging pathogen. Be sure to take every safety question/concern seriously. The APHL has developed a biosafety checklist (15) to assist laboratories in assessing the biosafety measures that they have in place.

References

1. **Centers for Disease Control and Prevention and National Institutes of Health.** 2009. *Biosafety in Microbiological and Biomedical Laboratories*, 5th ed. HHS Publication no. (CDC) 21-112. http://www.cdc.gov/biosafety/publications/bmbl5/BMBL.pdf.
2. **Association of Public Health Laboratories. 2016.** Lab matters: amp up your biosafety system. http://digital.aphl.org/publication/?i=290925&p=20, accessed March 1, 2016.
3. **Centers for Disease Control and Prevention.** 2008. Laboratory-acquired brucellosis—Indiana and Minnesota, 2006. *MMWR* 57:39–42. http://www.cdc.gov/mmwr/preview/mmwrhtml/mm5702a3.htm.
4. **Sewell DL.** 1995. Laboratory-associated infections and biosafety. *Clin Microbiol Rev* **8:**389–405.
5. **Vesley D, Hartmann HM.** 1988. Laboratory-acquired infections and injuries in clinical laboratories: a 1986 survey. *Am J Public Health* **78:**1213–1215.
6. **Boa E, Lynch J, Lilliquist DR.** 2000. *Risk Assessment Resources.* American Industrial Hygiene Association, Fairfax, Virginia.
7. **Ryan TJ.** 2003. Biohazards in the work environment, p 363–393. *In* DiNardi SR (ed), *The Occupational Environment: Its Evaluation, Control, and Management*, 2nd ed. American Industrial Hygiene Association, Fairfax, VA.
8. **Dunn JJ, Sewell DL.** 2014. Laboratory safety, p 515–544. *In* Garcia L (ed), *Clinical Laboratory Management*, 2nd ed. ASM Press, Washington, DC.
9. **Pentella MA.** 2008. Overview of the biosafety risk assessment process, p 145–156. *In* Richmond JY (ed), *Anthropology of Biosafety XI: Worker Safety Issues.* American Biological Safety Association, Mundelein, IL.
10. **Miller JM, Astles R, Baszler T, Chapin K, Carey R, Garcia L, Gray L, Larone D, Pentella M, Pollock A, Shapiro DS, Weirich E, Wiedbrauk D, Biosafety Blue Ribbon Panel, Centers for Disease Control and Prevention (CDC).** 2012. Guidelines for safe work practices in human and animal medical diagnostic laboratories. Recommendations of a CDC-convened, Biosafety Blue Ribbon Panel. *MMWR Suppl* **61:**1–101.
11. **Delany J, Pentella MA, Rodriguez J, Shah KV, Baxley KP, Holmes DE.** 2011. Guidelines for biosafety laboratory competency: CDC and the Association of Public Health Laboratories. *MMWR Suppl* **60(2):**1–23. PMID:21490563.
12. **Shah K, Pentella MA.** 2010. Laboratory biosafety competency development for the BSL-2, 3, and 4, p 67–74. *In* Richmond JY (ed), *Anthropology of Biosafety XII: Worker Safety Issues.* American Biological Safety Association, Mundelein, IL.
13. **Ned-Sykes R, Johnson C, Ridderhof JC, Perlman E, Pollock A, DeBoy JM, Centers for Disease Control and Prevention (CDC).** 2015. Competency guidelines for public health laboratory professionals: CDC and the Association of Public Health Laboratories. *MMWR Suppl* **64:**1–81. http://www.cdc.gov/mmwr/preview/ind2015_su.html.
14. **Centers for Disease Control and Prevention.** Recommended adult immunization, United States. CDC. http://www.cdc.gov/vaccines/schedules/hcp, accessed March 1, 2016.
15. **Association of Public Health Laboratories.** 2015. A biosafety checklist: developing a culture of biosafety. http://www.aphl.org/AboutAPHL/publications/Documents/ID_BiosafetyChecklist_42015.pdf, accessed March 1, 2016.

Safety Considerations in the Biosafety Level 4 Maximum-Containment Laboratory

38

DAVID S. BRESSLER AND ROBERT J. HAWLEY

The number of biosafety level 4 (BSL4) maximum-containment laboratories (facilities) worldwide has increased significantly. In the early 1980s, only two such laboratories existed in North America, one at the Centers for Disease Control and Prevention (CDC) in Atlanta, GA, and the other at the U.S. Army Medical Research Institute of Infectious Diseases (USAMRIID), Fort Detrick, MD. By early 2005, there were at least six operational BSL4-capable laboratories in the United States and over a dozen worldwide (1). As of September 2011, there are at least 13 operational or planned BSL4 facilities within the United States (see also http://fas.org/programs/bio/research.html#USBSL4). Canada also has operational BSL4 laboratories in Winnipeg, Manitoba, for the study of both human and animal disease agents. Worldwide there are at least 27 operational BSL4 facilities (2).

Some diseases associated with high morbidity and mortality, such as Nipah and Hendra viruses, are recent discoveries, whereas still others, such as Ebola virus, continue to reemerge (see http://www.cdc.gov/about/ebola/index.html). The recent Ebola outbreak (2014–2015) is the largest ever reported and demonstrates the potential for spread across international borders, as well as widespread and severe economic and health care implications of an unrestrained hemorrhagic fever epidemic. News reports during the outbreak point out how little we still know about high consequence pathogens such as Ebola (3–5; http://www.virology.net/ATVemerinf.html). Concerns that bioterrorists may use weaponized versions of an eradicated disease, smallpox, and the increased funding for studies on the Select Agents identified by the CDC and U.S. Department of Agriculture (USDA), have contributed to the burgeoning growth of these specialized laboratories. Maximum-containment laboratories are considered necessary for research on naturally evolving diseases, as well as those agents that can be used as potential biothreats.

An increasing number of outbreaks caused by organisms requiring maximum containment (Risk Group [RG] 4 or "BSL4" viruses) for study, along with increased workloads for government laboratories that normally handle the response to these events, have necessitated redistribution of resources and allowed more private and commercial laboratories to become involved in research on BSL4 viruses. Resources have been made available for bioterrorism preparedness and response, and what was once the domain of government laboratories has become almost routine research in the public and private sectors. Within the last decade, a growing number of universities and private organizations have begun maximum-containment research programs (see http://www.fas.org/programs/bio/research.html#USBSL; 4, 6).

With more resources directed to constructing new maximum-containment (BSL4) laboratories, a limiting factor has become the number of personnel and biosafety specialists trained in proper safety and risk assessment procedures. There are few personnel and safety managers with the knowledge of safety operations and "real-world" hands-on (skills-based) experience in a BSL4 laboratory. This has led to a demand for personnel with commensurate BSL4 safety and hazard assessment training and experience. Little has been published about the specific hazards and risks associated with BSL4 laboratory activities, other than the obvious issues of preventing and dealing with autoinoculation or other exposures to RG4 agents (7, 8). It is important for maximum-containment laboratory workers to recognize specific hazards and to take appropriate action because of the potentially serious consequences associated with an accident occurring inside the BSL4 laboratory.

The purpose of this chapter is to address some of the basic safety issues and risk assessment considerations for those individuals who will be affiliated with BSL4 laboratory operations. Training is a major component of BSL4 laboratory safety, and specific training aspects are discussed here. The minimization of hazards early in the design and construction of BSL4 laboratories is just as important as recognizing an immediate potentially life-threatening situation in a working laboratory. This chapter also discusses basic laboratory design and engineering considerations for reducing daily operational risks in these unique laboratories.

BSL4 CONTAINMENT OVERVIEW

The concept of BSL4 containment is, in principle, simple: separate the infectious agent from personnel working with it by either (i) enclosing the agent in a containment enclosure or box (Class III cabinet), manipulating it from outside the enclosure via a flexible, nonpermeable interface (usually gloves), or (ii) encapsulating personnel in self-contained, positive-pressure protective suits with supplied air. Accepted guidelines allow combinations of these approaches, but all serve to prevent exposure to the organism and allow for manipulation, without contaminating the laboratory worker or the environment (9). If everything is operating properly, the worker will have no contact with the infectious agents being studied. In addition, BSL4 laboratories with properly trained workers who correctly execute BSL4 operations should be the safest of all microbiological laboratories.

The Class III cabinet (Figs. 1 and 2) offers the most economical protection available for working with RG4 agents. Factors such as specific research objectives (e.g., handling multiple agents) and personnel issues (e.g., number of research personnel, physical fatigue, vulnerability of the cabinet gloves, operational space, the need to avoid cross-contamination, etc.) limit the operational flexibility of the Class III cabinet.

Alternatively, positive-pressure protective suit laboratories allow almost unconstrained movement within the laboratory. Depending on the design of the laboratory, protective suits allow multiple personnel to work with more than one agent at the same time with less discomfort and fatigue. The BSL4 laboratory, in effect, takes on the characteristics of a large Class III cabinet (10), with the sealed, internal shell of the laboratory acting as the demarcation of the containment zone or envelope. Significant disadvantages of protective suit laboratories are

Figure 1: Class III biological safety cabinet (CDC).

Figure 2: Class III biological safety cabinet (USAMRIID).

the construction costs and the subsequent maintenance, sustainability, and support. Both types of laboratories may also attract community attention, usually as a result of insufficient risk communication about the work being conducted, the safety features of the laboratory, or reports of laboratory-acquired infections.

Each approach to BSL4 containment makes these types of laboratories unique and deserving of special safety considerations. The ultimate goal in the BSL4 laboratory is to provide the safest environment possible for personnel who work with agents that cause diseases for which there are no known cures or available treatments. The most serious incident in a BSL4 laboratory would likely involve direct autoinoculation of a viable agent or breach of the protective suit or gloves leading to a potential aerosol or mucous membrane exposure of personnel. However, many common hazards that exist in the BSL4 laboratory could inadvertently lead to an exposure incident if not addressed. Personnel within the BSL4 environment must contend with risks that laboratory workers familiar with BSL1, BSL2, and BSL3 laboratories would dismiss as routine, and either preventable or easily treated with antibiotics should an accident occur. Routine hazards in the BSL4 laboratory can be life-threatening and have far-reaching consequences (for the individual, coworkers, and potentially the community), because vaccines or treatments are not available for most of the viruses handled under BSL4 conditions. Although there are no examples of RG4 infectious agents having been spread to the surrounding community, infections and mortality of laboratory workers have been reported. There were two ProMED-mail reports of BSL4 laboratory accidents in 2004 involving work with Ebola virus; one which resulted in the death of a laboratory worker was from a needlestick (11; http://www.promedmail.org/?p=2400:1202:4740070175186804; http://www.cidrap.umn.edu/news-perspective/2004/05/russian-scientist-dies-ebola-after-lab-accident). Recently, a scientist working in a laboratory in Germany pricked herself with a needle that had just been used to infect a mouse with the Ebola virus (12). While accidents result from obvious hazards (contaminated sharps and bites from infected animals), there are less obvious hazards from seemingly simple operations (such as cutting paper with a pair of scissors or removing tape from a dispenser) to be recognized and avoided. The physical isolation of the worker in a BSL4 laboratory not only makes it difficult to get help when needed but also can have a psychological impact on some individuals (13).

RG4 Biological Agents

Currently, viruses to be handled under BSL4 maximum containment are found in the following seven taxonomic families: *Arenaviridae* (Junin virus, Lassa virus, Machupo virus, Sabia virus, etc.), *Bunyaviridae* (hantaviruses and nairoviruses, such as Crimean-Congo hemorrhagic fever virus), *Filoviridae* (Ebola virus and Marburg virus), *Flaviviridae* (Central European tick-borne encephalitis virus complex, Omsk hemorrhagic fever virus, Kyasanur Forest disease virus, and Russian spring-summer encephalitis virus), *Herpesviridae* (*Cercopithecine herpesvirus 1*, previously known as herpes B virus), *Paramyxoviridae* (Nipah virus and Hendra virus), and *Poxviridae* (variola major, i.e., smallpox virus). They are considered to be highly virulent for humans, potentially infectious by the aerosol route, and generally exotic to the United States. They are capable of direct transmission from person to person (they are highly infectious but NOT highly contagious) and produce diseases for which no accepted treatment or prevention is available, with the following exceptions. The antiviral drug ribavirin is an effective treatment for Lassa fever if given early on in the

TABLE 1.

Maximum-containment conditions for work in the laboratory (BSL4) or with animals (ABSL4)[a]

Agents	Practices	Safety equipment (primary barriers)	Facilities (secondary barriers)
Agents include those that are (i) dangerous or exotic and pose a high risk of life-threatening disease, (ii) responsible for aerosol-transmitted laboratory infections, and (iii) related to known BSL4 organisms but have an unknown risk of transmission.	BSL4 practices are required, including a complete change of clothing before entering the laboratory and a wet personal shower upon exit. All laboratory waste material must be decontaminated before removal from the primary barriers and again when exiting the facility.	Personnel within a BSL4 cabinet laboratory contain all procedures and activities involving an agent in Class III BSC primary barriers. Personnel within a BSL4 suit laboratory wear a full-body, air-supplied, positive-pressure protective suit and contain all procedures and activities involving an agent in Class I and II BSCs or other equivalent primary barriers.	All BSL4 secondary barriers as described in BMBL, 4th ed. (CDC/NIH 1999), for the BSL4 cabinet laboratory Additional facility safeguards such as chemical exit showers, redundant breathing air systems, and serial HEPA filtration of the room exhaust air are required for the BSL4 suit laboratory. See BMBL, 4th ed., for other recommendations.

[a]BSL, biological level; BSC, biosafety cabinet.

course of clinical illness (14); the outcome can be enhanced by combining this treatment with immune plasma (15). Passive therapy with immune plasma was shown to be efficacious in cynomolgus monkeys (16). A vaccine for Lassa fever is in development (17). Since the early 1990s, several new viruses have been discovered. Studies of Hendra and Nipah viruses, as well as production-level work with hantavirus, are carried out under maximum containment (18). Other RG4 viruses, such as Ebola Reston, have been identified in unexpected locations of the world due to animal importation and human travel (19). At this time, research using viable smallpox virus is conducted only at two BSL4 laboratories—at the CDC in Atlanta, GA, and at the State Research Center of Virology and Biotechnology in Novosibirsk, Russia (20). Although no bacteria are classified as belonging to RG4, some risk assessments for multidrug-resistant bacteria call for what are essentially maximum-containment conditions (1).

BSL4 Biosafety Guidelines from the BMBL

The CDC and NIH have published recommended laboratory biosafety guidelines in *Biosafety in Microbiological and Biomedical Laboratories* (BMBL) (9). The four biosafety levels for work involving infectious microorganisms, biological toxins, and laboratory animals are described in ascending order to provide a stepwise increase in protection to personnel, the environment, and the community on the basis of a risk assessment. Implementation of these guidelines does not guarantee all aspects of maximum-containment safety.

The BMBL recommends that BSL4 principles, practices, and containment be used for any procedures involving known or potentially infectious materials from arthropods, animals, or humans suspected of harboring viral agents that are normally handled under BSL4 conditions. Clinical specimens from persons suspected or known to be infected with these highly pathogenic viral agents are also to be submitted to BSL4-capable laboratories for workup. If evidence of aerosol transmission is known or suspected, preparations of new or uncharacterized viral agents with high mortality should be handled at BSL4 containment as well. A risk assessment based on all known characteristics must be conducted prior to working with any new or novel agents in the BSL4 laboratory (21). The recommended conditions for working with infectious microorganisms and laboratory animals at maximum containment are summarized in Table 1. Readers are encouraged to review the specific criteria for each biosafety level in the BMBL (9).

BSL4 Risk Assessment and Risk Management

An important aspect to consider when planning work with biological materials (e.g., microorganisms, toxins, diagnostic samples, and environmental samples) is a worker's familiarity with the items being handled and the procedures to be followed. Investigators and support personnel (e.g., animal handlers and facility engineers) will need detailed information from current literature and colleagues prior to work. Biological agent summary statements (including risk group classification), research publications, and other documents may contain information needed for a risk assessment and the subsequent management of the identified risk(s). Investigators should conduct a job hazard analysis, taking into consideration all of the written step-by-step procedures that laboratory and support personnel will follow while working with the material(s) or supporting the work. Each step is analyzed with a job hazard analysis using the risk management process (see Table 2 for sample form). Information on a risk management process is also provided in *Risk Management* (22) as described in the chapter on toxins. Individuals must review and sign this

TABLE 2.

Example of BSL4 job hazard analysis[a]

NAME OF PROCEDURE: **BSL4 Suckling Mouse Brain Inoculation for Antigen Production**

DEPARTMENT: Virology

LOCATION: BSL4 Laboratory, Building 123, Room 456

All operations with live virus and/or infected animals will be conducted under BSL4 containment (Protective Suit Laboratory). All individuals involved in inoculation and harvest procedures will be appropriately trained in BSL4 operations and associated concepts. All individuals handling BSL4 infected mice will have prior training to handle mice under BSL4 containment conditions. All individuals involved in these procedures will be trained to handle sharps under BSL4 containment prior to involvement in these procedures.

Principal step(s)	Potential safety or health hazard(s)	Recommended control(s)	Equipment to be used	Inspection requirement(s)	Training requirement(s)
Virus dilution	Spill, aerosol generation	Perform all dilutions in Class II BSCs; use mechanical/automatic pipettors; disinfectant	BSC, virus stock vial, dilution tubes, diluent, automatic pipettors, tube holders, appropriate disinfectant, wet ice	Laboratory exhaust filter and BSC filter testing; BSC annual certification; training records; agent accountability records	Basic BSL4 operations; BSC usage; pipette handling technique
Preparation for virus inoculation	Aerosol generation, autoinoculation	Fill syringes in Class II BSC or on downdraft table; fill syringe barrel; using forceps or a hemostat, attach needle just prior to inoculation; use sterile 15-ml centrifuge tube to keep syringe sterile before adding needle	BSC, dilution tubes, wet ice, tube holders, 15-ml sterile centrifuge tubes, 0.5-ml syringes, 28-gauge single-use Luer Lock needle and syringe combination (or similarly engineered sharps safety inoculation device), sharps disposal containers, appropriate disinfectant	Training records	Basic BSL4 operations; BSC usage; sharps handling under BSL4 containment; aerosol containment; pipette handling technique
Virus inoculation into suckling	Aerosol generation, autoinoculation, bite from adult mouse	Use only single-use retractable Luer Lock needle and syringe combination or similarly engineered safety devices; use animal restraint device; use cut-resistant gloves; restrain suckling mouse with forceps or other appropriate method	Suckling mouse litter (~10–12 suckling mice + adult female), 0.5-ml syringes, 28-gauge single-use Luer Lock needle and syringe combination (or similarly engineered sharps safety inoculation device), forceps, alcohol pads, clean mouse cage to transfer inoculated mice, litter husbandry documentation, appropriate disinfectant	Training records	Sharps handling under BSL4 containment; mouse handling technique; animal restraint procedures; laboratory animal husbandry procedures; waste handling and disposal procedures

Analyzed and prepared by: ____________________ Date: ________________

Reviewed by: ______________________________ Date: ________________

[a]BSL, biosafety level; BSC, biosafety cabinet.

document, attesting that they have read and understood its contents. This workplace job hazard analysis serves to identify the hazards (biological agents and environmental, psychological, physical, and medical) that may be associated with working in a maximum-containment laboratory or with animals.

Biological agent operations at this level of containment involve working with highly pathogenic microorganisms characterized as dangerous or exotic and posing a high risk of life-threatening disease as a result of exposure. Currently, with the exception of smallpox and some of the Central European tick-borne encephalitis viruses, vaccines are not available to protect against RG4 agents (23). The management of the risks associated with working with biological agents, including information on primary and secondary protection, is addressed elsewhere in this book.

Environmental hazards include physical hazards (slips, trips, and falls) and those risks associated with working with animals, sharps, and high- and low-temperature materials. Other hazards that must also be addressed for working in a maximum-containment laboratory include the following:

- Crush, pinch, and puncture (airlock doors, quick-disconnects, movement of equipment, gas cylinders, and suit puncture and tear)
- Water and other liquids (spills, slips, trips, and falls)
- Physical effects (heat and cold injury, dehydration from physical exertion and dry air supply, hypothermia, and hot surfaces associated with autoclaves and other equipment)
- Psychological effects (physical and sensory isolation and claustrophobia)
- Hypoxia (disruption of supply or distribution of breathing air within the protective suit)
- Electrical (water baths and grounded equipment)
- Sharps (needles, scalpels, scissors, sharp laboratory surfaces, and glass items [reagent bottles and vials, blood tubes, capillary tubes, and microscope slides])
- Chemical (formaldehyde vapor decontamination and phenol hypersensitivity), fumes (such as non-human primate odors), and outside air contamination (such as diesel and gas fumes)
- Cryogenic materials (liquid nitrogen and dry ice)
- Compressed gas cylinders
- Radiation (radioactive material and mixed-waste handling and gamma radiation hazards)
- Animal handling (animal temperament, anesthesia and chemical restraint, bites, scratches, needlesticks, escaped animals, and necropsy safety [scalpels, needles, and bone saws])

THE CLASS III CABINET LABORATORY

Class III Cabinet

At the First Biological Safety Conference (precursor to the American Biological Safety Association) in 1955, Orin Miller described the Class III modular biosafety cabinet that was designed by Fort Detrick's Engineering Division and Safety Division, and fabricated and installed by the Blickman Company in 1954 in Building 550, Fort Detrick (24). The Class III biological safety cabinet was designed for work at BSL4 with microbiological agents, and it offers the highest degree of personnel and environmental protection from infectious aerosols, as well as protection of research materials from microbiological contaminants (25). The Class III cabinet is a specialized type of ventilated biosafety cabinet that is totally enclosed, gastight, and self-contained. It is often erroneously called a glove box, but a glove box may be neither ventilated nor characterized (see below description), and it may not have many of the other attributes of the Class III biological safety cabinet.

Work within the cabinet is performed through attached, arm-length rubber gloves (nonpermeable, flexible, and 18 to 20 mils thick) as shown in Fig. 1. Alternatively, work can be performed using a half-suit attached to a sealed front panel (nonopening view window) ports in the cabinet, allowing manipulation of materials in the cabinet while providing a physical barrier between the agent and worker.

The Class III cabinet is operated under negative pressure (0.5-inch water gauge) and is ventilated with supply air that is HEPA filtered. The exhaust air is filtered through two HEPA filters in series before being discharged to the outside environment. Any equipment required by the activity in the cabinet, such as aerosol devices, incubators, refrigerators, and centrifuges, must be contained within the cabinet system. An integral component of the Class III cabinet is a double-door autoclave to sterilize all materials exiting the cabinet and to allow supplies to enter the cabinet. A chemical dunk tank is incorporated to sterilize or disinfect materials exiting the cabinet (25).

A Class III cabinet may also be referred to as a cabinet line, when several cabinets are set up as interconnected systems. There are no nationally recognized standards governing construction and performance for a Class III cabinet. Representative Class III cabinets are seen in Figs. 1 and 2. A Class III cabinet laboratory schematic is depicted in Fig. 3. The advantages and disadvantages of using a Class III biological safety cabinet are summarized in Table 3.

Class III Cabinet Laboratory Requirements

The Class III cabinet laboratory should be located in a separate building or isolated laboratory within a building housing other operations. Personnel must pass through a minimum of two doors to get to a Class III cabinet. A BSL4 cabinet laboratory contains a personal shower that separates change rooms. The showers and change rooms may be separate for male and female workers or configured as a single unisex facility. An interlocked, double-door autoclave, dunk tank, fumigation chamber, or ventilated anteroom is provided for movement of material into or out of the containment

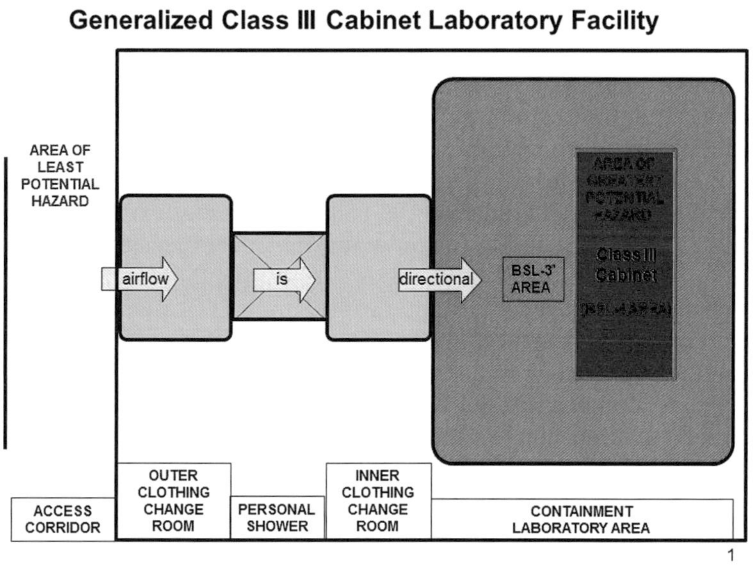

Generalized Class III Cabinet Laboratory Facility

Figure 3: Generalized Class III biological safety cabinet laboratory facility.

zone. Additional criteria of a Class III cabinet laboratory include the following:

- A daily inspection of containment parameters and life-critical functions is conducted prior to entry and commencement of work.
- Walls, floors, and ceilings are specially constructed to form a sealed internal shell for decontamination.
- All penetrations are sealed. Openings around doors are capable of being sealed for decontamination.
- Drains are connected directly to a waste decontamination system.
- Sewer vents and lines are HEPA filtered and protected against pests.
- Bench tops are sealed and impervious to liquids.
- Furniture is constructed of nonporous materials that can be easily decontaminated.
- A hands-free sink is located near the entryway of the cabinet system and change rooms.
- Any central vacuum is HEPA filtered and does not serve areas outside the cabinet room.
- Water fountains (sinks) are automatic or foot operated, isolated from supply to the laboratory, and located outside the cabinet room. All water supply pipes are equipped with a backflow preventer.
- Entry doors are self-closing and lockable.
- Windows are break-resistant and sealed.
- Double-door autoclaves are provided for materials passing out of the Class III cabinet or room.
- Autoclaves opening outside the containment area are sealed to the wall of the containment barrier with a bioseal.
- Autoclave doors are interlocked so they can only be opened after a sterilization cycle is complete.
- Dunk tanks or other devices are provided for nonautoclavable materials. A dunk tank is also used to transfer material to another area operating at a similar containment level.
- Liquid effluents from contaminated or possibly contaminated areas, or other sources within the cabinet and cabinet room, are decontaminated by heat or other effective methods. A validated method

TABLE 3.

Advantages and disadvantages of using a Class III biological safety cabinet

Advantages	Disadvantages
All infectious/toxic materials manipulated within hermetically sealed chamber	Personnel manipulate materials through heavy-duty, impervious rubber gloves or half-suit (reduced dexterity)
No direct contact with agent(s)	Difficult working conditions
Maximizes personal safety	Susceptibility to sharps
Particulate-free air within cabinet	Turbulent airflow within cabinet
Self-sufficient: contains refrigerators, incubators, special equipment (aerosol apparatus), animal cages, centrifuge, autoclave, and dunk tank	Minimum working area(s) and decreased operational efficiency
Less costly construction and maintenance	Less flexibility for use of multiple agents

of decontamination is established prior to discharge of any effluent to the sewer system.

- Effluents from clean areas may be discharged without treatment if allowed by state and local regulations.
- There is a dedicated, nonrecirculating, directional laboratory ventilation system.
- The alarm system is monitored prior to entry. The pressure differential must be visually monitored prior to entry.
- The cabinet exhaust system is designed and connected to prevent positive pressurization.
- The supply and exhaust air is HEPA filtered; the HEPA filters are tested and certified annually.
- Design and operational procedures are documented.
- The cabinet system is verified annually or according to operational experience.
- Appropriate communication systems between the cabinet area and noncontainment area must be provided.

Class III Cabinet Safety Considerations

When the main hazard is confined to the cabinet, the biological safety considerations involve prevention of exposure due to cuts, tears, or holes in the gloves or through some other means associated with the cabinet containment system.

Other aspects to consider for safety awareness in a BSL4 Class III cabinet laboratory are as follows:

- Cabinet operating parameters
- Cabinet maintenance and certification
- Sharps handling inside the cabinet
- Incident response
- Glove replacement
- Biological agent handling
- Animal handling

Since it is extremely tiring to work in the Class III cabinet for long periods of time, safety is directly correlated to the detailed planning and organization of experiments, the proficiency of the workers, and their responses to normal operations and any incidents that might occur.

THE PERSONAL PROTECTIVE SUIT LABORATORY

Protective Suit

Work in the BSL4 environment requires separation from the virulent microorganisms that are being handled. The protective suit is a specially designed garment made of chemically resistant plastic material containing one-way valves and a connection to accommodate externally supplied air. A few designs of protective suits are in use by BSL4 programs in the United States and abroad (see http://www.absaconference.org/pdf54/SessionXI-Harbourt.pdf). One type is called the Chemturion reusable level A suit, which is manufactured by ILC Dover, Inc., in Frederica, DE, and has been used extensively for many years (http://www.ilcdover.com/sites/default/files/MKT-0065_RevA_Chemturion_2.pdf, http://www.ilcdover.com/Chemturion-Suit/). The Chemturion suit is constructed of chlorinated polyethylene and weighs between 10 and 18 lb. Another type of protective suit is manufactured by Honeywell and is being used more frequently in the United States (http://www.honeywellsafety.com/Products/Protective_Clothing/BSL_4.aspx?site=/americas). This type of protective suit was previously manufactured by Delta Protection (France) and is now produced by Sperian Protection (ex-Bacou Dalloz). Sperian Protection later became part of Honeywell. This suit is constructed of a lighter polyamide material coated with polyvinyl chloride and weighs approximately 8 to 10 pounds. For the Chemturion protective suit, breathing air is delivered through an attached HEPA-filtered regulator hose that can be quickly connected and disconnected from air lines distributed throughout the containment area (a BSL4 protective suit laboratory is depicted in Fig. 4). Air is distributed to the appendages and head area of the body by an internal five-prong manifold. Sixty percent of the air is distributed to the head region, and each appendage receives 10% of the air volume. Outside air is prevented from entering the suit by four one-way exhaust valves located on the back and legs of the suit. Air-line connections, as well as the HEPA filter canister and hose system, should be checked for leakage and tightness prior to entry into the BSL4 laboratory (the external HEPA filter canister is replaced on the Honeywell suit by an internal, integral HEPA filter). Personnel wearing the suit are instructed to verify continually that the suit is connected to an air line, except when changing connections within the laboratory, to prevent possible asphyxiation. It is also necessary to maintain positive air pressure within the protective suit to prevent intake of potentially contaminated laboratory air into the suit through unseen holes or punctures. Depending on the manufacturer or laboratory specifications, boots may be incorporated as an integral part of the suit design. If not, a pair of waterproof boots or overshoes with thick, incorporated soles must be worn to protect the feet of the wearer. Another protective suit, developed by India and Germany, is a gastight suit with an adjustable external air supply, is available in various sizes, and is made from a double-sided coated special fabric. The suit has seams that are sealed on both sides, a gastight zipper, and removable flange-mounted gloves (see http://www.oshodefence.com/BSL4S.php for additional details).

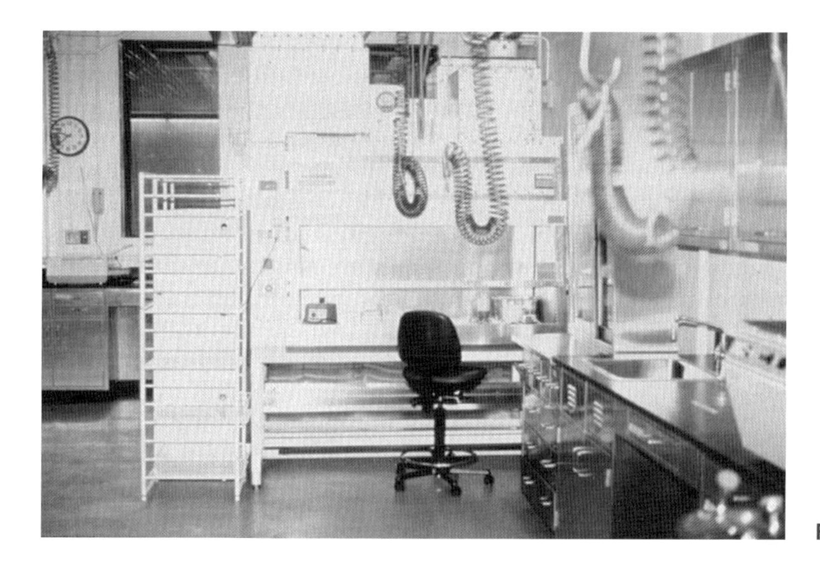

Figure 4: Personal protective suit laboratory (CDC).

Personnel and trainees must be thoroughly familiar with the care, operation, and maintenance of their protective suit. New personnel should learn how to choose the best suit for their body type and how to properly maintain the suit. The major parts of a typical protective suit are as follows:

- Hood, bodice, and foot portions
- Zipper or suit closure
- Visor
- Integral air distribution manifold and lines
- Breathing air lines, HEPA filter canister, or integral HEPA filter and air line couplings
- One-way air exhaust valves
- Molded wrist cuffs

Gloves

An outer pair of thick (approximately 18–20 mils) protective latex gloves must be worn over an inner pair of surgical gloves. The outer pair of gloves is attached directly to the cuff of the protective suit using adhesive polypropylene tape, duct tape, or elastic rubber rings. The inner surgical gloves are taped (using masking tape) to the scrub garments worn under the protective suit. Although the extra layer of gloving adds further protection to the hands, it contributes to a loss of tactile dexterity and sensitivity. The two layers of gloves can also cause muscle cramping of the hands and fingers unless the proper size is provided. Talcum powder may be used as a dry lubricant between the glove layers to help reduce friction and cramping. Personnel with sensitivity to latex or known latex allergies should be provided nonlatex options and monitored closely through a medical surveillance program.

As with the Class III cabinet, the gloves are generally considered to be the weakest component of the protective suit due to constant exposure, abrasion, and handling of materials. Rings and other jewelry that may snag or tear the suit material or gloves must be removed in the outer change room prior to donning the protective suit.

Disinfectant chemicals can also damage glove materials. Personnel should be trained to recognize deterioration, dry rot, and material incompatibility effects and to repair rips and tears of the gloves within the maximum-containment area. The extent of the glove damage and whether or not skin integrity has been breached will determine the course of action. Close inspection of the site, rinsing the area with an appropriate disinfectant, and exiting the maximum-containment laboratory as soon as possible are recommended. Site-specific standard operating procedures (SOPs) regarding glove and suit repair must be developed and included in the BSL4 laboratory safety manual.

Hearing Protection

Air entering the ILC Dover Chemturion protective suit has a noise level of approximately 85 to 90 dB and requires the use of hearing protection (26). A hearing conservation program is required (27). BSL4 laboratory workers enrolled in the facility hearing conservation program should be taught how to use hearing protection while working in a suit (28). Normally, one of two types of hearing protection is used—foam or latex rubber inserts or earmuffs. Comfort and unobtrusiveness are important considerations for deciding which type of protection works best for the individual. The insertion types are preferred because bulky earmuff protectors are easily dislodged and often slip from the ears while in the suit, resulting in discomfort and dangerous distraction during work.

Protective Suit Laboratory Requirements

The containment requirements for the protective suit laboratory are comparable to those of the cabinet laboratory in that similar measures are in place to prevent unintentional exposure to or transmission of biological agents. Because workers are essentially wearing a Ziploc bag within a space equal in containment to the Class III cabinet (only larger), the laboratory differences are mainly in infrastructure and life support components.

In the protective suit laboratory, all work with live agents is confined to the specially designed suit area and accessed by suited personnel through airlocks. Personnel are required to exit the laboratory through a chemical disinfectant shower before leaving the containment area and then proceed to the personal shower. As with the Class III cabinet laboratory, all penetrations into the internal shell of the suit area are sealed. Critical components of the life support system are redundant breathing air compressors, alarms, and backup emergency breathing air tanks.

The air pressure within the suit area is maintained lower than that of the adjacent laboratory areas, and emergency lighting and communication systems are provided. The laboratory air exhaust system, life support systems, alarms, lighting, entry and exit controls, and other critical containment systems are connected to a redundant backup emergency power source. For a full comparison between the two types of BSL4 laboratories, see "Laboratory Biosafety Level Criteria" in BMBL (29, 30). The advantages and disadvantages of using a personal protective suit are summarized in Table 4.

Personnel Suitability

The BSL4 scientific (laboratory supervisor or designee) and safety staff select qualified candidates to work in the maximum-containment environment. The laboratory supervisor is responsible for ensuring that appropriate safety orientation and training are conducted for every employee whose duties may require entry into the maximum-containment environment, including maintenance staff and response personnel. The laboratory supervisor, or a designee, must ensure that all details of operations, maintenance, and response to emergencies are thoroughly ascribed to, and understood by, every laboratory worker. The supervisor makes the final decision regarding suitability of a maximum-containment laboratory worker candidate. A safety official knowledgeable in maximum-containment operations may also be involved in this decision. Operations involving any entry into the laboratory while under containment ("hot") conditions should be approved by the laboratory supervisor and a safety office official, such as the biological safety officer (BSO). An access list includes assigned laboratory technicians and microbiologists, facility engineers, repair technicians, and any other support personnel.

The maximum-containment laboratory is not a suitable environment for a novice or an impatient microbiologist. Candidate BSL4 laboratory workers are evaluated for their psychological suitability, containment practices, familiarity with the risks involved, and microbiological proficiency at BSL2 and BSL3 operations before being considered for work in the maximum-containment laboratory. This assessment of experience should be done by both the laboratory supervisor and the BSO before initiating BSL4 training and before ending the period of apprenticeship.

Although the safety record for maximum-containment laboratories is good, all personnel need to be aware of the potential personal and community risks associated with carelessness or poor technique in handling of infectious materials. An accident occurring in a BSL2 laboratory may not have the same consequences as one in a maximum-containment laboratory. Laboratory personnel should be afforded the opportunity to work and learn at lower levels of containment before being considered

TABLE 4.

Advantages and disadvantages of using a personal protective suit

Advantages	Disadvantages
Relative ease of movement	Personnel enclosed in protective suit have diminished sensory awareness, dehydration concerns
No direct contact with agent(s)	Susceptibility to sharps
Maximizes personal safety	Laboratory is isolated from general access and emergency help
Work activities are comparable to those in lower-containment laboratories	Time and resource intensive
Self-sufficient: contains refrigerators, incubators, special equipment (aerosol apparatus), animal cages, and centrifuge	Multiple levels of redundancy required to system infrastructure and safety
Allows compartmentalization and multiple-agent experimentation	High construction and subsequent maintenance costs

for work in maximum-containment conditions, where even the slightest miscalculation can have grave consequences. All candidates, including experienced personnel from other maximum-containment laboratories, should be referred to the laboratory safety office for initial orientation. Some candidates may have established experience in support disciplines such as animal husbandry or laboratory equipment maintenance and repair. Equipment and system repair technicians may be required to enter the laboratory under containment conditions and may also need to be trained to wear the protective suit.

Medical Evaluation and Surveillance

All qualified maximum-containment laboratory personnel must be evaluated medically prior to being admitted to work under containment conditions and informed of the hazards of the work they will be conducting. The employee's overall physical health, emotional stability, medical condition, and any prescription medications the employee may be taking should be considered during this evaluation. A thorough preplacement examination should be conducted to determine the worker's capabilities and physical fitness (including respiratory function) for the job. Workers should also have periodic evaluations to confirm their physical fitness to safely complete assigned tasks. Fatigue, claustrophobia, mental alertness, and other factors can directly affect the safety of individuals and their coworkers, as can those who suffer seizures and others who cannot work comfortably and safely in the protective suit.

Prescription and over-the-counter medications may have side effects or cause short-term physiological or mental changes that may contraindicate work within the maximum-containment laboratory. Persons who have had recent surgical procedures, are physically disabled, or are pregnant should be evaluated on a case-by-case basis, according to the policy for BSL4 work. Strenuous activity and the dehydrating effects of air supplied to the protective suit require that all suit workers be attentive to proper water intake, and this should be explained to all prospective workers. An on-call physician knowledgeable in maximum-containment activities, disease symptoms, and treatment options is a valuable resource for maximum-containment facilities.

Although there are few protective vaccines against viruses worked with in maximum containment, laboratory personnel should be offered them as well as other vaccines such as hepatitis B vaccine (31), on the basis of the risk assessment and laboratory requirements.

Procedures must be established for reporting and investigating any unexplained employee illness or fever; at-risk employees should be trained on the importance of following these procedures. Baseline serum collection and storage are recommended by BMBL, but should be based on institutional policies. Specific recommendations for handling suspected viral hemorrhagic fever exposures have been published (32–35; see also http://www.cdc.gov/HAI/pdfs/bbp/VHFinterimGuidance05_19_05.pdf, http://emergency.cdc.gov/han/han00364.asp, and http://www.cdc.gov/vhf/ebola/healthcare-us/ppe/guidance.html).

TRAINING

As the number of operational maximum-containment laboratories increases, the proper training of personnel to work in these laboratories as well as those associated with facility operations, facility management, and biosafety is critical. The maximum-containment laboratory is a specialized containment area with unique work practices and demanding requirements for access, egress, and agent and waste handling. The worker must completely understand facility design and secondary containment, how to recognize normal and abnormal parameters, and how to deal with emergencies while simultaneously avoiding exposure to potentially lethal pathogens (29). Because of the unique features of each maximum-containment laboratory, the facility BSO must tailor training to meet local conditions and regulatory requirements. The laboratory supervisor is responsible for ensuring the proficiency of each laboratory worker. Training, including proficiency testing, must be documented and records maintained for all persons granted access to the laboratory. A procedure to validate this training should also be available to the facility security office or the office controlling access to the facility.

Maximum-Containment Laboratory Safety Manual

The laboratory-specific maximum-containment safety and operations manual and its comprehensive protocols are required reading for trainees and may be used for refresher training of experienced personnel. Because individuals inside the laboratory must be self-sustaining during normal operations, as well as during the initial stages of an emergency, the biosafety manual should describe the duties and responsibilities of the various personnel that operate and support the laboratory and the procedures for working safely, as well as how to deal with emergencies and how to move materials into and out of the maximum-containment laboratory. SOPs for handling suit malfunctions, medical emergencies, potential exposures to agents, spills, fire, etc., must be described in detail. Other items that should be addressed include animal handling, bite management, and necropsy procedures; storage and transport requirements of infectious agents; waste management that includes handling and disposal of radioactive and/or highly toxic materials; equipment operation essentials; specialized room functions; and the special design and containment aspects of the laboratory.

The laboratory operations section provides information regarding the physical characteristics of the facility, especially the ventilation and decontamination systems, and how to recognize and correct any malfunctions. Floor plans and suggested placarding of room doors of the laboratory should be included in the safety manual to help familiarize personnel with the location of equipment and emergency alarm sites, autoclaves, entry and exit points, and other components. An emergency contact list should be provided in the manual and reviewed at least semiannually, or when there are key personnel changes. Any changes to safety protocols must be reviewed and approved by senior safety staff and the maximum-containment laboratory supervisor and be fully explained to all staff before being implemented. The biosafety manual is a living document, reviewed and endorsed at least annually, and must reflect experience, training, and current information. Examples of training items for maximum-containment work and proficiency are found in Table 5. The required training forms, design parameters of the biological containment areas, checklists, equipment operation manuals, injury claim forms, and airlock and laboratory decontamination procedures can be included in an appendix to the manual.

BSL4 Laboratory Training

Candidates for entry into the maximum-containment laboratory receive information about the laboratory and associated support spaces and how they operate (see the BSL4 protective suit laboratory schematic depicted in Fig. 5) as well as procedures for dealing with any anomalies (36). This includes the normal operational parameters for access and security; air handling and heating, ventilation, and air-conditioning issues; breathing air; alarm systems; containment integrity; fire control; and waste management systems, including safety, life support, and other containment parameters that must be verified daily.

Although a video can be used to illustrate BSL4 practices, containment equipment, and facility operations, such as laboratory entry, exit, and emergency procedures, it is not a substitute for apprenticeship with an experienced maximum-containment laboratory worker. Inexperienced personnel should always be accompanied in the laboratory to ensure that the trainee clearly understands the differences that separate the maximum-containment environment from other environments. A review of the most current edition of BMBL is required reading.

Apprenticeship

A supervised apprenticeship is completed when the trainer is satisfied that the trainee fully understands all of the principles of BSL4 practices, containment equipment, and facility operations, and has satisfactorily demonstrated the requisite skills and temperament to work in the maximum-containment environment. The trainer is responsible for the welfare of the trainee, whose life may depend on what he or she observes and learns from the trainer. The trainee must understand that experience outweighs seniority in this environment. The trainer must also address issues of anxiety or fitness for work inside the laboratory. Because workers learn at different rates and have different skills, the apprenticeship phase of training lasts as long as is necessary. The length of this apprenticeship may be determined on a case-by-case basis, but also must meet a minimum time or work entry period.

A checklist documenting the individual tasks and topics that each laboratory worker has been trained to do is a necessary component of maximum-containment laboratory safety program management. The checklist should be one of the final items signed by the laboratory supervisor and the BSO upon completion of the supervised apprenticeship and should be maintained as part of the trainee's official file. When the trainer decides that the trainee is ready to work independently, the checklist can be signed to indicate satisfactory completion of training.

Outside the Maximum-Containment Envelope

Laboratory entry and exit procedures and critical systems checklist

The SOP for entry into the maximum-containment laboratory is used for training and to develop a critical laboratory function checklist. This checklist is reviewed daily, prior to entry of personnel into the laboratory to monitor the status of the life support systems and equipment controls for the laboratory. The facility systems documented should include breathing air and laboratory air handling, containment status, liquid waste, physical security, and fire and emergency notification systems, as well as other indicators located outside of the containment envelope.

The clothing change room

For almost all maximum-containment laboratories, normal entry into the laboratory is through a single door leading to the clothing change room, or locker room, where street clothes, including underwear, are removed. A supply of surgical scrubs, undergarments, or other appropriate

TABLE 5.

List of training and proficiency items for the BSL4 environment[a]

Training item	Laboratory personnel	Facility support personnel	Animal care personnel	Guest researchers/ visitors
BMBL BSL4 stand and special practices, safety equipment, and facility criteria	•	•	•	As applicable
BSL4 laboratory-specific biosafety manual	•	•	•	•
BSL4 BSC specific operation	•	•	•	As applicable
BSL4 agent familiarization (potential reservoir, transmission, symptoms, treatment, decontamination strategies)	•	As applicable	•	As applicable
Class III cabinet orientation and familiarization	•	As applicable	As applicable	As applicable
Protective suit orientation and familiarization	•	•	•	As applicable
BSL4 communication	•	•	•	As applicable
General emergency and incident communication and response	•	•	•	As applicable
Medical emergency response	•	•	•	As applicable
BSL4 facility security and agent accountability	•	•	•	As applicable
BSL4 sharps handling and remediation	•	As applicable	•	As applicable
Equipment familiarization (BSCs, autoclaves, centrifuges, gamma source sterilizers, hematology and blood chemistry instrumentation, incubators and roller bottle apparatus, LN_2 tanks, and ultra-low-temperature freezers)	•	•	•	As applicable
Laboratory and equipment decontamination	•	•	As applicable	As applicable
BSL4 animal handling techniques, husbandry, monitoring, and restraint	As applicable	As applicable	•	As applicable
Animal equipment familiarization (caging, protective equipment, surgical devices)	As applicable	As applicable	•	As applicable
Autoclave and waste stream operations	•	As applicable	•	As applicable
Sample decontamination and removal, dunk tank use	•	As applicable	•	As applicable
Chemical decontamination shower operation and emergency exit	•	•	•	As applicable
Air lock operations	•	•	•	As applicable
Risk assessment, communication, and management	•	•	•	As applicable
Critical facility systems familiarization and checklists	•	•	•	As applicable
Protective suit and Class III cabinet malfunction and response	•	As applicable	As applicable	As applicable
Necropsy, euthanasia, and animal restraint procedures	As applicable	As applicable	•	As applicable
BSL4 personnel responsibilities	•	•	•	As applicable
BSL4 entry and exit procedures	•	•	•	As applicable
Guest researcher and visitor policies	•	As applicable	•	•
Radioactive material usage and management (if applicable)	As applicable	As applicable	As applicable	As applicable
Spill remediation	•	•	•	As applicable

[a]BSL, biosafety level; BSC, biosafety cabinet; BMBL, *Biosafety in Microbiological and Biomedical Laboratories* (9).

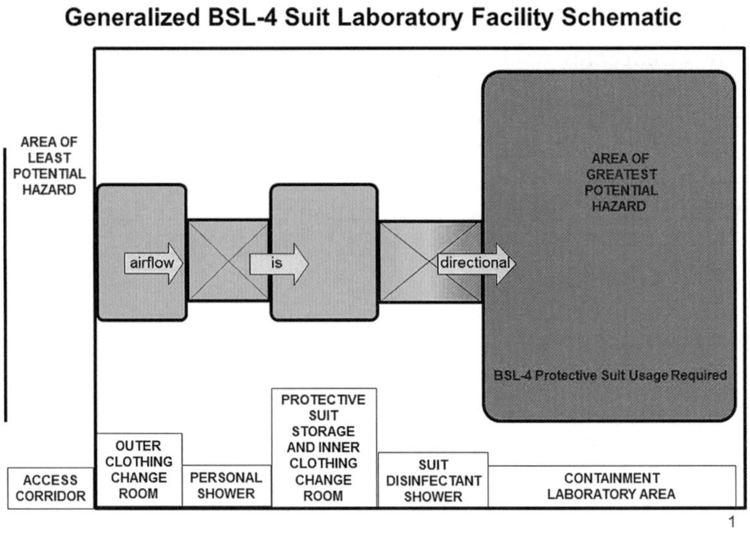

Generalized BSL-4 Suit Laboratory Facility Schematic

Figure 5: Generalized BSL4 protective suit laboratory facility schematic.

laboratory clothing is provided, and workers follow an established dress protocol for entry. The change room represents the crossover from a clean environment to the containment side of the laboratory. Directional airflow and other systems status indicators may be located here.

The personal shower

After donning scrubs, the laboratory worker walks through a personal shower, or a one-way door that bypasses the shower, and enters the protective suit storage and change room. At this point, there is no return to the outer change room without taking a complete soap-and-water shower as is required upon exiting the containment laboratory. The shower is to be provided with enough soap and towels for the number of workers and their exit frequency.

The protective suit storage and change room

The protective suit storage and change room is generally used as the staging area or "gray area" prior to entering the actual maximum-containment laboratory. Protective suits are normally stored in the hanging position during periods of nonuse. Supplies kept inside the protective suit room include various sizes of heavy latex or neoprene outer protective gloves, waterproof adhesive-backed tape for attaching the outer gloves to the protective suit, spare surgical gloves, talcum powder, towels, masking tape for taping surgical gloves to the scrub garments, and emergency repair items (such as blunt-end scissors and duct tape). Air delivery line HEPA filters and air regulators that are not integral to the suit may also be stored in the gray area.

Protective suit entry preparation

A suit integrity test, which can identify unseen or minute holes by using detergent solution, may be performed by sealing the exit valves (e.g., with adhesive polypropylene tape) and attaching the suit to an air hose. This test should be done as necessary, or as determined by a risk assessment or SOP. Trainers should demonstrate how to detect hidden leaks by looking for wet areas on their laboratory clothing after exiting from the disinfectant shower. If a suit breach is detected, SOPs should be followed and the laboratory supervisor should be notified. A repair kit for the ILC Dover Chemturion suit is available from the manufacturer. After periods of use, the visor of the suit may become abraded and scratched. Personnel should periodically clean both the inside and outside of the visor with 70% ethyl alcohol, which easily removes facial oils and dirt without damaging the visor. The zipper closure may need to be cleaned and lubricated prior to entry. A detailed account of protective suit preparation and use may be found in reference 37.

Personnel must also be trained to handle leaks or tears that occur while inside the laboratory. Punctures or tears to the suit material may be temporarily managed by patching with duct tape or similar pressure-sensitive tape (adhesive polypropylene tape) after ensuring that there has been no breach of skin integrity or mucous membrane exposure. In such situations, personnel should maintain the positive air pressure to the suit by not disconnecting from the air supply and exit the laboratory immediately.

Durability studies of protective suits used at the Australian Animal Health Laboratory suggest that activity levels and usage requirements of the laboratorians have a direct influence on suit failures. Suits used by personnel in the research laboratory were compared to those suits used by individuals working in animal areas for component failure. Failure rates of suit components were highest where individuals were repeatedly engaged in more demanding physical activity and exposed to caging and animal support equipment (38).

The chemical disinfectant shower

The disinfectant, or chemical, shower is a critical barrier for maintaining the maximum-containment envelope. It functions as an airlock and is the threshold between the noncontainment ("cold") and containment ("hot") sides of the laboratory. The shower contains a minimum of two breathing air connections for personnel. The chemicals used most often for the disinfectant shower are quaternary ammonium detergent compounds, such as Micro-Chem Plus, or phenolic compounds, such as Lysol. Most labs no longer use Lysol because of the chemical sensitization and exposure issues and its tendency to clog the disinfectant shower nozzles.

The chemical shower area should be equipped with an indicator panel for the operational status of the shower system (e.g., sufficient amount of disinfectant). Personnel are familiarized with the door interlocks, automatic shower cycle, entry and exit sequence through the shower, mixing of the decontamination solution, operation of the manual shower mechanism during an emergency situation, and all warning indicators. If the chemical shower mechanism is automatic, a backup manual decontamination system is maintained as well. Workers entering the laboratory should ensure that the shower activation mechanism starts the decontamination cycle to remove any potential contamination of the airlock after securing the door. They should be trained on the proper procedure in the event of a malfunction and immediately report the condition to the laboratory supervisor and maintenance personnel. Depending on the type of protective suit used, laboratory workers may need to don additional waterproof and slip-resistant protective foot covering upon entry to the laboratory. Overshoes or boots protect the feet of wearers and the protective suit. Over time, the use of this additional protective footwear may cause stress on the seams in the foot portion of the suit, which should always be checked for leaks or tears before use. The protective footwear is removed before exiting the laboratory through the disinfectant shower.

Inside the Containment Envelope

Laboratory and microbiological security

Due to international threats of bioterrorism and incidents of domestic terrorism, the physical security of the laboratory structure must now be considered early in the design process. The laboratory is located either in a separate building or in a "controlled area of the building" and "access is limited by means of secure, locked doors; accessibility is managed by the laboratory director, BSO, or other person responsible for the physical security of the laboratory" (NIH). Access should be allowed only to those individuals designated by the laboratory supervisor as being properly trained and required for the program mission or its support needs. Even experienced personnel from other maximum-containment laboratories must complete local laboratory-specific training. If necessary, access for temporary personnel may be restricted to regular work hours, when such personnel are accompanied by resident staff members. Security clearances may also be required. All workers must be alert to security system failures, unexplained changes in protocols, or other issues and report any incidents promptly.

At present, most pathogens handled in the maximum-containment laboratory are designated as Select Agents and are viruses. These agents are recognized as public health or environmental threats that could be used as bioterrorism agents. Title 42 Code of Federal Regulations (CFR) Part 73 (39; Public Health Service, Department of Health and Human Services, 2010), and Titles 7 and 9 CFR (Department of Agriculture, Animal and Plant Health Inspection Service, 2008; 40), the Select Agent Rule, deal in depth with the specifics of the security and handling of this group of biological agents. The security and biological containment measures in and around maximum-containment laboratories must be understood and followed by those who work in or near the laboratory, especially those in training. Security awareness must involve laboratory management as well as personnel involved in the security decision-making processes. Most threats to assets and facility security are from authorized insiders (the "insider threat"). Therefore, all employees must be taught to recognize the signs of a potential security risk in coworkers and associated personnel, such as unusual behavior (41). Workers new to the maximum-containment laboratory must become familiar with the roles and responsibilities of other approved personnel to be able to recognize unauthorized individuals in the laboratory area.

Any facility registered under Title 42 CFR Part 73 must maintain and routinely verify records giving a complete account of all activities related to the Select Agents used there (42; Public Health Service, Department of Health and Human Services, 2010). Personnel must be instructed on methods for tracking agent usage, keeping an inventory, and recognizing indicators of unusual usage. They must keep freezers, refrigerators, cabinets, and other equipment that contains stocks of biological agents secured when not under direct supervision. For example, they should try to anticipate circumstances when materials could be out of their direct control and maintain control of such biological materials during manipulations inside or outside of the laboratory facility. The laboratory-specific maximum-containment biosafety manual contains information that should be kept secure from unauthorized use.

Communication training

Effective communication is crucial for safety and coordination of activities. New workers must be instructed about the proper operation of all communication equipment used within and around the containment envelope. Communication between workers in the protective suit laboratory environment can be a challenge because of the noise level of breathing air entering the protective suit (43), but it can be handled in a number of ways: (i) voice-actuated communicators tuned to facility-specific frequencies, (ii) hand signals, (iii) talking loudly enough to overcome the background noise of the suit, and, as a last resort, (iv) physically disconnecting or crimping the breathing air supply momentarily while communicating with another worker(s). All personal communication equipment should be durable, lightweight, and comfortable to prevent dangerous distractions. New trainees should be allowed to explore their options during preliminary training, when an experienced trainer can be alert to signs of discomfort. For safety reasons, redundant systems of communication are installed and used as a means of contact with the outside. Proficiency in the use of telephonic, intercom, facsimile (fax), and electronic mail (email) systems should be part of the documented entry and access requirement training checklist. Emergency numbers, contact lists, and emergency response procedures may be posted near telephone stations within and outside the laboratory. Instructions for responding to other emergency situations, such as an exposure to an infectious agent, medical emergency, equipment malfunction, spill, or fire, must be trained under realistic conditions. The response to these situations should be practiced regularly to maintain proficiency. Such instructions may also be posted near the telephone or intercom area.

Microbiological practices within the maximum-containment laboratory

The risks associated with work conducted on highly pathogenic viruses require modification of "standard" microbiological techniques used in the maximum-containment laboratory to complete the laboratory work as safely and efficiently as possible. Good work practices protect the integrity of samples or cultures and the safety of the researchers working with them, as well as the outside community.

The skill sets acquired by personnel at lower biosafety levels are prerequisites for work using BSL4 principles, practices, and equipment. The ability to recognize unique situations that may arise under maximum containment is a valuable trait. Experienced personnel can often share time- and labor-saving procedures that do not compromise the integrity of the work or the safety of personnel with new workers. The proper planning of experiments and careful arrangement of experimental and support materials, instruments, and accessories in the maximum-containment laboratory are critical for successful, efficient, and safe operations. The simplest BSL4 procedures can become extremely time-consuming, tedious, and unsafe without careful planning. Even with proper planning, it can take considerably longer to perform procedures while wearing the protective suit than for similar operations conducted at BSL2 or BSL3. The value of planning and equipment checklists becomes immediately apparent when one forgets to bring a needed item into the laboratory and the worker has to leave the containment laboratory and return later. The worker must exit the laboratory, take a personal shower, retrieve the item, and then return to the suit or cabinet area to begin work. Workers should be taught the value of rehearsal outside of containment and setting up staging areas where items can be gathered prior to laboratory entry. The consolidated items can then be passed to workers inside the maximum-containment laboratory or into the work area of the Class III cabinet through a sterile autoclave or dunk tank.

General procedures

The loss of dexterity and control due to the layers of gloves is a significant issue. Working in a protective suit is similar to putting a jigsaw puzzle together while wearing oven mitts (43). Delicate procedures, such as sample preparations, handling potential sharps, and changing the cages of small (or large) animals, are made more difficult. Indeed, many of the senses, such as hearing, sight, tactile sensitivity, and motor coordination, are limited due to the constraints of the protective equipment necessary at this containment level. Repetitive pipetting along with wearing multiple pairs of gloves can often lead to hand cramps, which may be more frequent or intense when working in the Class III cabinet. Using automated pipetting devices, as well as using talcum powder as a dry lubricant between the surgical gloves and the gloves attached to the cabinet, can help reduce their occurrence. Appropriate ergonomic design of laboratory components, such as cabinet and glove aperture heights and spacing along with appropriate furniture, can help prevent fatigue when working in the Class III cabinet.

When primary containers of infectious agents must be moved to various locations within the laboratory, they should be transported in unbreakable plastic trays or secondary containers, ideally, sealed with liquid-tight closures to prevent spattering of materials if dropped. Small polypropylene picnic-type coolers with lockable covers are ideal. Items with pointed or sharp edges, such as scissors or tape dispensers, are usually blunted, filed down, or shielded to prevent damage to the protective suits. Pens and pencils generally do not represent a hazard to suits, however, indelible ink markers are preferred because they can allow easier reading of notations and markings. Because the ink is waterproof, the markers can be used to label the outside of double-bagged and heat-sealed items that are passed through dunk tanks containing an appropriate disinfectant.

Spills in the maximum-containment laboratory involving infectious agents, radioisotopes, and chemicals that may cause damage to the protective suit must be dealt with promptly to eliminate the hazard. Spills of potentially infectious material should be treated with an appropriate disinfectant to inactivate any infectious agent, and then handled according to the laboratory SOP.

The use of flames (Bunsen burners) or other sources of open high heat is to be avoided. The protective suits (constructed of polyester fabric with PVC coating) can easily melt or burn if exposed to ignition or heat sources. Flammable chemicals and gases should be handled with extreme care in this environment.

Biological safety cabinets

All work with infectious agents is conducted within the confines of a properly operating Class II biological safety cabinet in the protective suit laboratory. New workers must be taught the practices specific to operation of the cabinet in the maximum-containment laboratory, including attention to differential pressure gauges and what to do in case of malfunction, i.e., immediately stop work and post a sign or placard to alert other workers and repair personnel to the status of the cabinet.

Autoclaves

Autoclaves should only be operated by personnel who are trained to operate them and to recognize autoclave or interlock malfunctions as a potential breach of containment. Personnel must be trained to load and unload materials properly in large or small autoclaves and to recognize that autoclave doors present pinch and crush hazards during opening and closing. Metal discard pans are often used for autoclaving waste because of their efficient heat conduction, but sharp or rough edges must be minimized or avoided. The heat generated by autoclaves can cause the plastic of the protective suit to melt on contact, causing a hole and possible exposure to unfiltered, potentially contaminated air.

Chemicals and cryogenic liquids

Chemical solvents, such as acetone, for procedures such as fixing tissue culture slides do not generally present an inhalation hazard within the maximum-containment laboratory due to the protection provided by the suit or Class III cabinet. However, if spilled, they can potentially cause damage to the protective suit or glove material.

Liquid nitrogen containers (Dewar flasks) and ultra-low-temperature freezers used for long-term storage of biological agents and samples are hazardous to personnel, requiring training in the correct methods of removing freezer boxes and vials using cryogenic safety gloves. Extreme cold can cause the outer gloves of the protective suit to become brittle and crack, allowing potential exposure, while contact with a cryogenic liquid can cause direct frostbite injury to immersed hands and arms even while wearing the protective suit and gloves. Condensation around cryogenic piping may also create slip and fall hazards inside the laboratory. All infectious or toxic materials stored in refrigerators or freezers should be properly labeled and stored in containers capable of withstanding the thermal shock of freezing and thawing; glass is not an appropriate storage container.

Sharps handling

Personnel must be trained in the handling and disposal of sharps and on the importance of substituting blunt instruments for sharp items and plastic items for glass whenever possible. The use of sharps should be strictly controlled and limited to areas of the laboratory where their use is required by personnel. Glass bottles, blood tubes, and vials, needles, scalpels, and any other items with the potential to break into sharp fragments need to be treated with care during use and disposal (see reference 29 for additional sharps precautions).

Sharps disposal containers should be placed for easy access within the laboratory and properly disposed of before being completely filled. For tissue culture work in the maximum-containment laboratory, cell culture media preparations may be formulated outside of the laboratory whenever possible, filter sterilized, and then transported to the laboratory in plastic containers. When there is no other choice, small glass reagent bottles may be made safer by wrapping the container with adhesive-backed tape, such as duct tape, to prevent shattering if dropped. Plastic-coated glass bottles are commercially available but should be tested prior to use. Alternatively, glass reagent bottles can be secured and transported in secondary plastic containers. Glass should only be used with the specific approval of the laboratory supervisor and in accordance with laboratory protocols. Should an incident with glass occur in the maximum-containment laboratory, all sharps and broken glassware are removed to a sharps container by mechanical means (i.e., with brush and dustpan, or forceps), never by hand. If laboratory operations continue, subsequent personnel entering the laboratory must be alerted to the potential of residual sharp materials in the area of the incident.

Centrifuges

The hazards presented by centrifuges are the same at BSL4 as at lower BSLs and can be prevented by proper operation and maintenance and an awareness of failure indicators, including metal fatigue, cracked buckets or rotors, or broken or cracked O-rings. Trainees should be observed in the use of each tabletop centrifuge, high-speed centrifuge, and ultracentrifuge before being allowed to use them without supervision. Trainers should demonstrate the proper operation of all potential aerosol-generating devices used in the laboratory, such as sonicators, tissue grinders, and pipetting devices.

Dunk tanks

Laboratory personnel are trained in the proper removal of materials from the maximum-containment laboratory, including the operation and maintenance of dunk tanks or other passthrough devices. A chemical dunk tank is a disinfectant-filled container that allows passthrough and removal (without loss of containment) and surface decontamination of materials that cannot otherwise be autoclaved from the containment area of the Class III cabinet or protective suit laboratory. The disinfectant solution in the dunk tank reservoir must be shown to be efficacious against the viral agent(s) used in the laboratory, maintained at the proper level and in the required concentration to be effective.

Animals in the BSL4 environment

Careful risk analysis and management are essential when evaluating the safety of a maximum-containment protocol that includes animal studies. Written protocols for the routine care of animals within the containment environment may be included as appendices in the maximum-containment safety manual or animal care SOPs. Only personnel knowledgeable and trained in handling infected animals should be permitted to handle animals within the maximum-containment laboratory. Even experienced technicians benefit from practicing a new technique on pristine animals outside of the containment area where the constraints do not interrupt the learning process. Those responsible for training new personnel must be competent in animal-handling techniques and husbandry procedures.

Risk management criteria can also be used to determine the essential training requirements for handling animals under BSL4 principles and practices. These criteria include the following:

- Characteristics of the pathogen
- Animal model to be used
- Age and size of the animal(s)
- Pathogen delivery mechanism(s)
- Caging requirements and maintenance
- Tissue and sampling methods
- Postmortem, necropsy, or recovery requirements
- Anesthesia and euthanasia requirements

The care of animals within the BSL4 laboratory must conform to standards described in the *Guide for the Care and Use of Laboratory Animals* (44) and can be accomplished by using animal handlers experienced with these procedures. Non-human primates are of special concern in the maximum-containment laboratory, and only the most competent animal handlers should be tasked with their husbandry requirements. When manual handling of these animals is indicated, at least two trained personnel should always be present.

Skills necessary for safely handling and maintaining research animals under maximum-containment conditions include, but are not limited to, the following:

- Specific animal-handling techniques
- Proper inoculation and sharps handling techniques
- Feeding, bedding, cage, and animal housing requirements
- Tissue and sample harvesting methods (e.g., venipuncture and necropsy)
- Maintenance and troubleshooting of isolators and other animal housing units
- Documentation requirements
- Chemical restraint procedures

Considerable one-on-one training must be devoted to proper sharps handling techniques used when inoculating animals with infectious materials, as well as anesthesia and necropsy procedures performed at maximum containment (45). An individual that inoculates suckling mice on a daily basis under lower-containment conditions may not be competent to inoculate a New Zealand White rabbit on the first attempt using BSL4 principles and practices. Animals should always be restrained chemically or by other methods for inoculation of an infectious agent. With more experienced laboratory workers and animal caretakers, some routine manipulations on laboratory animals may be performed without anesthesia and with a minimum level of restraint. Animal inoculation, necropsy, specimen retrieval, and other operations with infectious materials that can be aerosolized should be carried out in biosafety cabinets or on downdraft tables. Personnel should be taught caging types, specific operation and maintenance requirements, proper feeding and handling procedures, and methods of dealing with loose or escaped animals. Special procedures required for working with large or unique animal species can be found in references 46 and 45.

Necropsy techniques must be practiced at lower-containment levels, using pristine animals under the supervision of a maximum-containment-qualified investigator or veterinarian, prior to performing the procedure. Conditions should be duplicated as closely as possible for effective training. Additional precautions, such as the use of an extra pair of gloves (cut resistant [e.g., Kevlar] or chain mail) and the sequence of procedures to be performed, can be practiced and assessed for satisfactory performance (see references 45 and 47).

Equipment maintenance activities

If equipment cannot be decontaminated and removed from the containment envelope, a decision may be made to repair the item inside the laboratory. To prevent operational delays, it may be necessary to train maintenance personnel to do repairs under maximum-containment conditions. They should be trained to the same standards as other workers, especially regarding emergency response procedures, recognition of normal system operational parameters, and risk assessment techniques. Repairs that involve sharp objects and tools must be discussed prior to beginning work, and all maintenance and support personnel, even if qualified to work under maximum-containment conditions, should be escorted while performing repairs in the laboratory. Regular laboratory workers should be trained in basic maintenance and repair procedures.

Waste management

All personnel must be trained on the maximum-containment laboratory waste stream, including the handling, segregation process, and discard of general laboratory waste, sharps, mixed waste, radioactive waste, and animal materials. All infectious or toxic materials and other contaminated waste must be placed directly into the autoclave or held in covered containers until autoclaved or otherwise rendered sterile (e.g., by disinfectant treatment) prior to final disposal by appropriate personnel.

BSL4 Facility Emergency Response Issues

Class III cabinet laboratory

Entry and egress from the Class III cabinet laboratory are easier for personnel but they still must be familiar with specific emergency procedures, including basic lifesaving techniques and cardiopulmonary resuscitation (CPR). Workers should be trained to recognize signs of fatigue or illness to prevent any compromise in their own or their coworkers' performance. Response to critical situations in the cabinet laboratory, such as agent exposures, other medical emergencies, equipment malfunctions, spills, and fires, must be practiced under realistic conditions until proficiency is achieved.

The protective suit laboratory

Rapid response to a life-threatening medical emergency is crucial, and immediate access to the protective suit laboratory would be difficult for rescue personnel; therefore, all regular laboratory personnel should be trained in basic first aid, lifesaving, and CPR techniques. Training and rehearsal for getting medically affected laboratorians outside of the containment envelope to where first responders can assist the situation should also be practiced. Prior familiarization of local first responders with the laboratory may significantly reduce access and medical treatment delays during an actual medical emergency (48).

For either type of laboratory, management of a potentially exposed employee requires advance planning to identify hospitals with appropriate facilities and treatment resources before operations within the laboratory begin. Arrangements for transport to medical facilities should also be made in advance (49, 50).

The definition of an exposure is a part of the risk assessment process that helps to identify the proper course of action and prevent unnecessary medical treatment and concern. Personnel must report any potential exposure(s) to their laboratory supervisor, who should arrange for medical evaluation, treatment, and subsequent medical surveillance if needed. Workers must also inform their supervisor of any prolonged febrile illness or symptoms consistent with possible exposure. Medical management decisions should be based on the events leading up to the exposure, the actual exposure scenario, agents in use in the laboratory at the time, the social situation of the exposed person, and medical measures available for the particular agent. For potential exposures, the response may range from no action to isolation and treatment with prophylactic drugs or other supportive measures.

Hypoxia is an immediate threat unique to laboratory workers in protective suits. Effective treatment to prevent suffocation must take precedence over the unquantified threat of the microbial environment. Specific responses depend on actual laboratory conditions at the time of the occurrence, but all reasonable measures should be rehearsed in advance. In the event of significant symptoms (e.g., chest pain, shortness of breath, and severe pain in other parts of the body), the worker should immediately summon help, notify coworkers within the laboratory, telephone outside the laboratory to request help, or sound the summon-aid alarm and proceed to the decontamination shower and exit. Consideration may also be given to installation of automatic external defibrillation (AED) devices in the maximum-containment laboratory for use in particular medical situations.

Dehydration is a concern for those working for long periods in the protective suit laboratory. Normal exertion combined with the dehumidified air delivered to the protective suit can allow significant water loss from the body over a period of a few hours. The effect is further exacerbated by strenuous physical activity and the separation from drinking water while working in the suit. Workers need to be aware of the signs of dehydration, such as increased thirst, dry lips, and difficulty concentrating. To reduce the potential for dehydration, they should be instructed to hydrate themselves properly before entering the laboratory. A maximum length of time for working in the protective suit may be considered as an administrative control.

Just as for the cabinet laboratory, training in the protective suit laboratory regarding hazard communication and emergency situations, potential exposures, medical emergencies, equipment malfunctions, spills, and fires must occur regularly and under realistic conditions to maintain proficiency.

Emergency exit

In an emergency situation, all persons working in the maximum-containment laboratory should be notified immediately by use of the alarm system or by any other means possible. If an emergency requires an immediate exit, more than two persons can shower out of the protective suit laboratory because of the availability of at least two breathing air connections or by sharing breathing air lines in the disinfectant shower. The SOP or biosafety manual should provide specific procedures to be taken for immediate evacuation.

RISK MITIGATION THROUGH BSL4 DESIGN

Laboratory design and construction should consider the basic maximum-containment concerns of limited access, limited communication, and aspects of disease agents that will be manipulated. On-site visits to other operating maximum-containment laboratories by personnel involved in design and operations of the laboratory should be incorporated as part of the design process (51, 52).

Large, single-room, protective suit laboratories facilitate movement of equipment and personnel and make housekeeping and decontamination easier, but operational flexibility may be reduced unless separate independent laboratories are planned. Multiple rooms allow segregation of operations and reduce the likelihood of cross-contamination, but they may restrict movement of personnel and equipment and hinder communication and visual accessibility. As a general rule, the containment area should be open and obstruction free. Corridors and aisles must be wide enough for large laboratory equipment. Narrow doors and thresholds are obstacles that may contribute to injury (crush, pinch, and puncture) when moving large or unwieldy equipment within the maximum-containment laboratory. The amount of space dedicated to support areas of the laboratory is as important as that for the working laboratory. For example, the protective suit change and storage room of a protective suit laboratory should be designed to accommodate supplies and suit storage. Space allowing at least two individuals to don and remove protective suits, as well as a table for suit repairs and integrity testing, should be considered at minimum. The addition of windows and view ports around and within the laboratory can increase safety by allowing visual communication between workers in the laboratory and safety personnel outside. Visual access also serves as a training tool whereby prospective workers and visitors can observe individuals, procedures, and operations taking place in the maximum-containment laboratory. Glass incorporated into design elements of the laboratory (including windows and cabinetry doors) must be shatterproof. Windows must be sealed and capable of withstanding pressure differentials expected during normal laboratory operations (including pressure decay testing, if necessary). The compatibility of window materials with decontamination chemicals should also be determined prior to installation. Incorporating closed-circuit television monitoring can enhance safety by compensating for a lack of windows in otherwise inaccessible areas of the maximum-containment laboratory.

Laboratory casework and other permanent equipment should be constructed of durable materials, such as plastic composite or stainless steel with smooth surfaces and rounded corners, to reduce the potential of snagging gloves and protective suits. Soft, porous construction materials should be avoided because they are difficult to decontaminate and are potential sources of absorbed area gaseous decontaminants. Casework and other furniture may be sealed to walls and floors to limit the spread of spills and potential contamination, or may be mounted on locking casters for easy movement. Walls and ceilings should be smooth, and coatings should be durable and impervious to chemicals.

Flooring should be resilient and monolithic, with integral coving for spill management and decontamination. A lightly textured, chemical-resistant architectural epoxy may be considered to minimize the potential for slips and falls. Removable floor coverings should be considered for areas of special use, such as around liquid nitrogen tanks, to prevent damage from spillage. Foot-operated or automatic sinks should be checked for proper operation and leakage during scheduled laboratory maintenance.

Because of the storage limitations and processing times for large amounts of potentially contaminated liquid waste, floor drains in protective suit laboratories may not be recommended. However, if the substantial waste-handling drawbacks can be overcome; floor drains, especially when placed in animal areas, can alleviate much of the arduous work in animal area housekeeping. If floor drains are included, provision must be made to keep the traps filled with an appropriate disinfectant at all times.

Lighting must be adequate for all areas of the laboratory. Light fixtures and bulbs located outside the containment envelope allow repair and replacement without necessitating entry into the laboratory itself (53). Emergency lighting operation must be verified regularly. Pressurized gas cylinders for carbon dioxide and other specialty gases should be maintained outside the containment envelope. All penetrations for communication conduits, liquid supply, and decontamination systems should be sealed and fitted with backflow prevention devices. Supply conduits for cryogenic liquids entering the containment envelope not only require backflow prevention but also may require additional insulation and compatible sealants to reduce condensation and ice formation inside the laboratory and interstitial spaces.

Alarms to notify workers of a system malfunction should employ at least a visual signal, as well as an audible component, during activation to counteract the sensory isolation of workers in protective suits. A centralized station within the laboratory allows individuals to identify the exact nature of the alarms. "Summon for assistance" buttons may be installed at strategic locations throughout the protective suit laboratory and should be especially considered for multiple-room laboratory plans.

BSL4 LABORATORY AND EQUIPMENT DECONTAMINATION CONSIDERATIONS

Maximum-containment laboratories may be decontaminated prior to renovation, periodic servicing, changes in research protocols, or for emergency reasons. Routine decontaminations may occur on a rotating monthly or annual schedule. The decontamination of a single piece of equipment, isolated from the rest of the laboratory in a specially designed room (e.g., an airlock) and fumigated with a sterilant, can be done more frequently. Emergency decontamination showers for rapid exit of personnel may also be located in fumigation rooms.

A process hazard analysis of the safety procedures for all decontamination processes should include autoclaves, fumigation rooms, gas and vapor decontamination systems, and irradiation sources. The safety checks and services to be included in the laboratory shutdown and decontamination protocols include the following:

- Establish and review personnel and laboratory responsibilities prior to decontamination.
- Review decontamination protocols (including laboratory equipment shutdown items).
- Review emergency response procedures.
- Determine maintenance and repair issues for critical life support systems.
- Determine maintenance and repair issues for laboratory equipment and infrastructure.
- Schedule laboratory housekeeping and waste (biological, chemical, and mixed) removal.
- Schedule routine equipment certification (biological safety cabinets, HEPA filters, and alarm systems).
- Schedule required biosafety and biosecurity audits and inspections.

Equipment and systems that often require special attention during the laboratory shutdown are as follows:

- Autoclaves (bioseals and gasket and door function)
- Pneumatic door gaskets and seals
- Disinfectant shower supply mechanisms, piping, and nozzles
- Air-supply hoses and connections
- Alarm and control panel lights and switches
- Air-handling units, air ducts and plenums, dampers, and HEPA filter housings
- Containment envelope integrity
- Dunk tanks

After decontamination, and prior to returning the laboratory to a containment ("hot" or operational) status, all critical life safety, infrastructure, emergency, and backup systems must be validated for operational parameters. The laboratory may be operated under noncontainment ("cold") conditions for a few days prior to introduction of infectious agents to allow easier troubleshooting and repair of any system malfunctions. Because training time under realistic BSL4 principles and practices conditions is limited, the downtime of scheduled laboratory decontaminations may provide an opportunity to introduce prospective workers to the laboratory environment. Individuals may be afforded the experience of protective suit conditions in an environment free of the pathogens normally handled in the laboratory. This opportunity may also be used to train and familiarize community health care and local emergency response personnel (risk communication) who may be called on to respond to a maximum-containment laboratory emergency.

CONCLUSION

The increase in the number of newly constructed maximum-containment laboratories is the result of the need to study new and emerging diseases associated with high morbidity and mortality, as well as the concern that bioterrorists may use weaponized versions of exotic disease agents. The challenge now is to increase the number of experienced and trained personnel and biosafety specialists with knowledge of BSL4 principles, practices, equipment, and facilities. This chapter is meant to provide a practical introduction to the training issues and risk assessments for individuals affiliated with BSL4 facilities. The laboratory design and engineering considerations are also important if daily operational risks are to be mitigated or minimized in these unique facilities. Staffed with competent, well-trained researchers and support personnel, BSL4 facilities can be the safest of all microbiological laboratories.

We gratefully acknowledge Dr. Gene Olinger for his critical review of the manuscript.

References

1. **McSweegan E.** 1999. Hot times for hot labs. *ASM News* **65:** 743–746.
2. **Defense Science Board.** 2009. Department of Defense Biological Safety and Security Program. Office of the Under Secretary of Defense for Acquisition. Technology, and Logistics, Washington, D.C.
3. **Fischer R, Judson S, Miazgowicz K, Bushmaker T, Prescott J, Munster VJ.** 2015. Ebola virus stability on surfaces and in fluids in simulated outbreak environments. *Emerg Infect Dis* 21:1243.
4. **Varkey JB, Shantha JG, Crozier I, Kraft CS, Lyon GM, Mehta AK, Kumar G, Smith JR, Kainulainen MH, Whitmer S, Ströher U, Uyeki TM, Ribner BS, Yeh S.** 2015. Persistence of Ebola virus in ocular fluid during convalescence. *N Engl J Med* **372:**2423–2427.
5. **Mate SE, Kugelman JR, Nyenswah TG, Ladner JT, Wiley MR, Cordier-Lassalle T, Christie A, Schroth GP, Gross SM, Davies-Wayne GJ, Shinde SA, Murugan R, Sieh SB, Badio M, Fakoli L, Taweh F, de Wit E, van Doremalen N, Munster VJ, Pettitt J, Prieto K, Humrighouse BW, Ströher U, DiClaro JW, Hensley LE, Schoepp RJ, Safronetz D, Fair J, Kuhn JH, Blackley DJ, Laney AS, Williams DE, Lo T, Gasasira A, Nichol ST, Formenty P, Kateh FN, De Cock KM, Bolay F, Sanchez-Lockhart M, Palacios G.** 2015. Molecular evidence of sexual transmission of Ebola virus. *N Engl J Med* **373:**2448–2454.
6. **Tradeline Publications.** 2005. Operating a BSL-4 laboratory in a university setting: Georgia State University lab studies deadly alpha herpes virus. *Appl Biosaf* **10:**253–257.
7. **Wilhelmsen CL, Hawley RJ.** 2007. Biosafety, p 515–541. *In* Dembek Z (ed), *Medical Aspects of Biological Warfare.* Defense Department of the Army, Office of the Surgeon General, Borden Institute, Walter Reed Medical Center, Washington, DC.
8. **Le Duc JW, Anderson K, Bloom ME, Estep JE, Feldmann H, Geisbert JB, Geisbert TW, Hensley L, Holbrook M, Jahrling PB, Ksiazek TG, Korch G, Patterson J, Skvorak JP, Weingartl H.** 2008. Framework for leadership and training of Biosafety Level 4 laboratory workers. *Emerg Infect Dis* **14:**1685–1688.
9. **U.S. Department of Health and Human Services, Public Health Service, Centers for Disease Control and Prevention, National Institutes of Health.** 2009. *Biosafety in Microbiological and Biomedical Laboratories*, 5th ed. HHS Publication No. (CDC) 21-1112. U.S. Government Printing Office, Washington, DC. http://www.cdc.gov/biosafety/publications/bmbl5/bmbl.pdf.
10. **Stuart D, Hilliard J, Henkel R, Kelley J, Richmond J.** 1999. Role of the Class III biological safety cabinet in achieving biological safety level 4 containment, p 149–160. *In* Richmond JY (ed), *Anthology of Biosafety I. Perspectives on Laboratory Design.* American Biological Safety Association, Mundelein, Ill.
11. **International Society for Infectious Diseases.** 2004. Ebola, lab accident death—Russia (Siberia), May 22, 2004. Archive no. 20040522.1377. http://www.promedmail.org/
12. **Feldmann H.** 2010. Are we any closer to combating Ebola infections? *Lancet* **375:**1850–1852.
13. **Jahrling P, Rodak C, Bray M, Davey RT.** 2009. Triage and management of accidental laboratory exposures to biosafety level-3 and -4 agents. *Biosecur Bioterror* **7:**135–143.
14. **McCormick JB, King IJ, Webb PA, Scribner CL, Craven RB, Johnson KM, Elliott LH, Belmont-Williams R.** 1986. Lassa fever. Effective therapy with ribavirin. *N Engl J Med* **314:** 20–26.
15. **Jahrling PB, Peters CJ, Stephen EL.** 1984. Enhanced treatment of Lassa fever by immune plasma combined with ribavirin in cynomolgus monkeys. *J Infect Dis* **149:**420–427.
16. **Jahrling PB, Peters CJ.** 1984. Passive antibody therapy of Lassa fever in cynomolgus monkeys: importance of neutralizing antibody and Lassa virus strain. *Infect Immun* **44:**528–533.
17. **Fisher-Hoch SP, Hutwagner L, Brown B, McCormick JB.** 2000. Effective vaccine for lassa fever. *J Virol* **74:**6777–6783.
18. **Centers for Disease Control and Prevention.** 1994. Laboratory management of agents associated with hantavirus pulmonary syndrome: interim biosafety guidelines. *MMWR Recomm Rep* **43**(RR-7)**:**1–7.
19. **Centers for Disease Control (CDC).** 1989. Ebola virus infection in imported primates—Virginia, 1989. *MMWR Morb Mortal Wkly Rep* **38:**831–832, 837–838.
20. **LeDuc JW, Jahrling PB.** 2001. Strengthening national preparedness for smallpox: an update. *Emerg Infect Dis* **7:**155–157.
21. **Wilson DE, Chosewood LC (ed).** 2009. Biosafety Level 4, p 45. *In Biosafety in Microbiological and Biomedical Laboratories.* 5th ed. HHS Publication No. (CDC) 21-1112. U.S. Government Printing Office, Washington, DC.
22. **Headquarters, Department of the Army.** 1998. Risk Management. Field Manual 100–14, p. 2-0–2-24.
23. **Heymann DL.** 2015. Arboviral Hemorrhagic Fevers, p 43–54. *In* Heymann DL (ed), *Control of Communicable Diseases Manual*, 20th ed. American Public Health Association, Washington, D.C.
24. **Barbeito MS, Kruse RH.** 1997. A history of the American Biological Safety Association. Part I. The first ten biological safety conferences 1955–1965. *J. Am. Biol. Saf Assoc.* **2:**7–19.
25. **Wilson DE, Chosewood LC (ed).** 2009. Appendix A—Primary containment for biohazards: selection, installation and use of biological safety cabinets. *In Biosafety in Microbiological and Biomedical Laboratories*, 5th ed. HHS Publication No. (CDC) 21-1112. U.S. Government Printing Office, Washington, DC.
26. **Hawley RJ, Pittman PR, Nerges JA.** 2000. Maximum containment for researchers exposed to biosafety level 4 agents, p 35–53. *In* Richmond JY (ed), *Anthology of Biosafety II. Facility Design Considerations.* American Biological Safety Association, Mundelein, Ill.
27. **U.S. Department of Labor.** 2010. *Occupational noise exposure. Title 29 Code of Federal Regulations Part 1910.95.* Occupational Safety and Health Administration, Washington, D.C.
28. **Hawley RJ, Pittman PR, Nerges JA.** 2000. Maximum containment for researchers exposed to biosafety level 4 agents, p 35–53. *In* Richmond JY (ed), *Anthology of Biosafety II. Facility Design Considerations.* American Biological Safety Association, Mundelein, Ill.
29. **Wilson DE, Chosewood LC (ed).** 2009. Biosafety level criteria. p 30–59. *In Biosafety in Microbiological and Biomedical Laboratories*, 5th ed., HHS Publication No. (CDC) 21-1112. U.S. Government Printing Office, Washington, DC.
30. **Crane JT.** 2002. BSL-4 laboratory guidelines, p 253–271. *In* Richmond JY (ed), *Anthology of Biosafety V. BSL-4 Laboratories.* American Biological Safety Association, Mundelein, Ill.
31. **Centers for Disease Control and Prevention.** 1997. Immunization of health-care workers: recommendations of the Advisory Committee on Immunization Practices (ACIP) and the Hospital Infection Control Practices Advisory Committee (HICPAC). *MMWR Recomm Rep* **46**(RR-18)**:**1–42.
32. **Centers for Disease Control (CDC).** 1988. Management of patients with suspected viral hemorrhagic fever. *MMWR Suppl* **37**(S-3)**:**1–16.
33. **Centers for Disease Control and Prevention (CDC).** 1995. Update: management of patients with suspected viral hemor-

rhagic fever—United States. *MMWR Morb Mortal Wkly Rep* **44:** 475–479.

34. **Centers for Disease Control and Prevention and World Health Organization.** 1998. *Infection Control for Viral Haemorrhagic Fevers in the African Health Care Setting.* Centers for Disease Control and Prevention, Atlanta, GA.
35. **Kortepeter MG, Martin JW, Rusnak JM, Cieslak TJ, Warfield KL, Anderson EL, Ranadive MV.** 2008. Managing potential laboratory exposure to Ebola virus by using a patient biocontainment care unit. *Emerg Infect Dis* **14:**881–887.
36. **Alderman L.** 2000. Construction and commissioning guidelines for biosafety level 4 (BSL-4) facilities, p 82–87. *In* Richmond JY (ed), *Anthology of Biosafety II. Facility Design Considerations.* American Biological Safety Association, Mundelein, Ill.
37. **Wilhelmsen CL, Jaax NK, Davis K III.** 2002. Animal necropsy in maximum containment, p 361–408. *In* Richmond JY (ed), *Anthology of Biosafety V. BSL-4 Laboratories.* American Biological Safety Association, Mundelein, Ill.
38. **Le Blanc Smith PM, Edwards SF.** 2002. Working at biosafety level 4—contain the operator or contain the bug, p 209–236. *In* Richmond JY (ed), *Anthology of Biosafety V. BSL-4 Laboratories.* American Biological Safety Association, Mundelein, Ill.
39. **Centers for Disease Control and Prevention and Office of the Inspector General, U.S. Department of Health and Human Services.** 2005. Possession, Use and Transfer of Select Agents and Toxins; final rule (42 CFR Part 73). *Fed Regist* **70:** 13316–13325.
40. **Animal and Plant Health Inspection Service, U.S. Department of Agriculture.** 2005. Agricultural Bioterrorism Protection Act of 2002: Possession, Use and Transfer of Biological Agents and Toxins; final rule (7 CFR Part 331; 9 CFR Part 121). *Fed Regist* **70:**13278–13292.
41. **Royse C, Johnson B.** 2002. Security considerations for microbiological and biomedical facilities, p 131–148. *In* Richmond JY (ed), *Anthology of Biosafety V. BSL-4 Laboratories.* American Biological Safety Association, Mundelein, Ill.
42. **Centers for Disease Control and Prevention and Office of the Inspector General, U.S. Department of Health and Human Services.** 2005. Possession, Use and Transfer of Select Agents and Toxins; final rule (42 CFR Part 73). *Fed Regist* **70:**13316–13325.
43. **Hawley RJ, Pittman PR, Nerges JA.** 2000. Maximum containment for researchers exposed to biosafety level 4 agents, p 35–53. *In* Richmond JY (ed), *Anthology of Biosafety II. Facility Design Considerations.* American Biological Safety Association, Mundelein, Ill.
44. **National Research Council.** 2010. *Guide for the Care and Use of Laboratory Animals,* 8th ed. National Academy Press, Washington, D.C.
45. **Wilhelmsen CL, Jaax NK, Davis K III.** 2002. Animal necropsy in maximum containment, p 361–408. *In* Richmond JY (ed), *Anthology of Biosafety V. BSL-4 Laboratories.* American Biological Safety Association, Mundelein, Ill.
46. **Abraham G, Muschialli J, Middleton D.** 2002. Animal experimentation in level 4 facilities, p 343–359. *In* Richmond JY (ed), *Anthology of Biosafety V. BSL-4 Laboratories.* American Biological Safety Association, Mundelein, Ill.
47. **Copps J.** 2005. Issues related to the use of animals in biocontainment research facilities. *ILAR J* **46:**34–43.
48. **Kaufman SG, Alderman LM, Mathews HM, Augustine JJ, Berkelman RL.** 2009. Review of the Emory University Applied Laboratory Emergency Response Training (ALERT) Program. *J Am Biol Saf Assoc* **14:**22–32.
49. **Best M.** 2002. Medical emergency planning for BSL-4 containment facilities, p 295–299. *In* Richmond JY (ed), *Anthology of Biosafety V. BSL-4 Laboratories.* American Biological Safety Association, Mundelein, Ill.
50. **Jahrling P, Rodak C, Bray M, Davey RT.** 2009. Triage and management of accidental laboratory exposures to biosafety level-3 and -4 agents. *Biosecur Bioterror* **7:**135–143.
51. **Kelley JA.** 1999. Building a maximum containment laboratory, p 121–133. *In* Richmond JY (ed), *Anthology of Biosafety I. Perspectives on Laboratory Design.* American Biological Safety Association, Mundelein, Ill.
52. **Trans-Federal Task Force on Optimizing Biosafety and Biocontainment Oversight.** 2009. Report of the Trans-Federal Task Force on Optimizing Biosafety and Biocontainment Oversight. http://www.ars.usda.gov/is/br/bbotaskforce/biosafety-FINAL-REPORT-092009.pdf.
53. **Crane JT, Bullock FC, Richmond JY.** 1999. Designing the BSL4 laboratory, p 135–147. *In* Richmond JY (ed), *Anthology of Biosafety I. Perspectives on Laboratory Design.* American Biological Safety Association, Mundelein, Ill.

Index